CRC Handbook of Solubility Parameters and Other Cohesion Parameters

Second Edition

Author

Allan F. M. Barton, Ph.D.
Associate-Professor of Chemistry
Murdoch University
Perth
Western Australia

CRC Press
Boca Raton Ann Arbor Boston London

Library of Congress Cataloging-in-Publication Data

Barton, Allan F. M.
CRC handbook of solubility parameters and other cohesion parameters / Allan F. M. Barton. -- 2nd ed.
p. cm.
Includes bibliographical references and index.
ISBN 0-8493-0176-9
1. Solubility. 2. Cohesion. I. Title
QD543.B22 1991
541.3′42--dc20 91-18630
CIP

Direct all inquiries to CRC Press, Inc., 2000 Corporate Blvd., N.W., Boca Raton, Florida 33431.

International Standard Book Number 0-8493-0176-9

Library of Congress Card Number 91-18630

Printed in the United States of America 1 2 3 4 5 6 7 8 9 0

PREFACE TO SECOND EDITION

I based my original comprehensive overview of cohesion parameters (published in 1983) on about 1700 reports published up to mid-1980. Activity in the application of cohesion parameters to a wide range of materials has continued at a high level, and this Second Edition includes reference to the work of an additional 1200 reports, most having been published during the past decade. The increasing use of solubility parameters in patent specifications is a particularly interesting illustration of the general acceptance of the concept.

I have taken the opportunity of completely revising the Handbook, while retaining the same general chapter structure. Some sections have been rearranged and rewritten to provide a more logical development of the subject, but a concordance of section numbers, table numbers and figure numbers is included to facilitate use of the Second Edition by those familiar with the First Edition. Additional material is provided in 17 new sections and 40 new data tables. Areas expanded in this way include liquid-liquid and supercritical systems, as well as polymer permeation, cosolvency and polymerization reactions.

The single alphabetical reference list at the end of the First Edition has been replaced by separate lists in citation order for each chapter, but the 400 general literature and review references in Chapter 1 are arranged alphabetically by first author as a bibliographic resource.

Another CRC Handbook (*Polymer-Liquid Interaction Parameters and Solubility Parameters,* published in 1990) includes complementary numerical information on polymer cohesion parameter values, but does not include the solubility maps and diagrams which have been retained in Chapter 15 of this Handbook.

I have incorporated numerous suggestions made to me by readers, and I would appreciate receiving new information and comments to assist me in future compilations.

Allan Barton
Perth, Australia

THE AUTHOR

Associate Professor Allan Barton is a university teacher, research scientist, author and consultant in the areas of physical science, data compilation, resource utilization, and product innovation.

He is the author of four major books — *The Dynamic Liquid State* (Longman, London), *Resource Recovery and Recycling* (Wiley-Interscience, New York), *Polymer-Liquid Interaction Parameters and Solubility Parameters* (CRC Press, Florida), and *Solubility Parameters and Other Cohesion Parameters* (CRC Press, Florida) — the last classified by the publisher as a technical ''best-seller''.

Associate Professor Barton's research achievements in chemistry gained him a Ph.D. degree (University of Auckland, 1965), election as a Fellow of the New Zealand Institute of Chemistry (1972), award of the New Zealand Association of Scientists' Research Medal (1975), and election as Fellow of the Royal Australian Chemical Institute (1978).

His publications include 60 research papers on topics as diverse as reaction rates, liquid miscibilities, resources recycling, and heritability of oil yield from eucalypts. He has addressed research conferences in many countries, and as a widely cited author and international authority on solubility parameters was invited recently to participate in a workshop on pharmaceutical formulation in Switzerland.

Associate Professor Barton joined Murdoch University during its final planning and building stage, and he has participated in decision-making at most levels of the University as well as contributing to research and teaching in several areas.

He was a foundation member of the Commission on Solubility Data of the International Union of Pure and Applied Chemistry, and he contributed to the Commission for 10 years, including taking responsibility for compiling the guidelines for contributors and chairing the liquid-liquid systems committee.

In addition to his present position, Associate Professor Barton has worked at the University of Southampton (U.K.), Victoria University of Wellington (N.Z.), Auckland University (N.Z.), Imperial College of Science and Technology (London, U.K.) and Pergamon Press Ltd. (Oxford, U.K.). Currently, he is a consultant in product innovation and intellectual property to Technology and Innovation Management Pty. Ltd. (Perth).

His interests extend to non-technical subjects, including a publication in *Patterns* (Fremantle Arts Centre) and recent attendance at television script-writing and World Competitive Manufacturing management courses. Other interests include local government and community issues: his wife, June Barton, is currently mayor of the City of Melville. They live with their two daughters in Bicton, Western Australia.

ACKNOWLEDGEMENTS

I wish to express my gratitude to

- the Radcliffe Science Library and Pergamon Press in Oxford, where I began preparing the first edition during study leave from Murdoch University in 1979 while assisting with the Solubility Data Series of the International Union of Pure and Applied Chemistry
- the numerous academic and industrial scientists and technologists who responded generously to my requests for reprints and further information during my research for both editions
- the library staff of Murdoch University who assisted patiently with on-line bibliographic data base searches and inter-library loans
- the Murdoch University graphics staff and the secretarial staff of the School of Mathematical and Physical Sciences, Mrs. Lian Chan in particular, for help in the preparation of this second edition

Dedicated to

Rowena Elizabeth Jane Barton
Charmian Victoria Mary Barton

LIST OF TABLES

LIST OF TABLES (continued)

LIST OF TABLES (continued)

LIST OF TABLES (continued)

LIST OF TABLES (continued)

LIST OF TABLES (continued)

LIST OF TABLES (continued)

LIST OF FIGURES

LIST OF FIGURES (continued)

LIST OF FIGURES (continued)

LIST OF FIGURES (continued)

LIST OF FIGURES (continued)

LIST OF FIGURES (continued)

LIST OF FIGURES (continued)

LIST OF FIGURES (continued)

LIST OF FIGURES (continued)

LIST OF FIGURES (continued)

LIST OF FIGURES (continued)

LIST OF FIGURES (continued)

TABLE OF CONTENTS

Chapter 1

INTRODUCTION

The use of solubility parameters and other cohesion parameters has just one aim: to provide a simple method of correlating and predicting the cohesive and adhesive properties of materials from a knowledge of the properties of the components only. There are more sophisticated theories which can be employed if a precise description of a system is required, but there is not one that can be used more easily to estimate the properties of a mixture from those of its constituents.

A very useful analogy was used by Darby, Touchette, and Sears[71] in discussing the use of relative permittivities for the prediction of plasticizer-poly(vinyl chloride) behavior, and it can be applied to cohesion parameters. A *predictor* of mixing ability (solubility, compatibility) differs from a *measure* of mixing ability just as a falling barometer differs from a rain gauge: two components of a mixture do not need to be brought together before predictions can be made on their miscibility. The analogy can be taken further: neither the falling barometer as a predictor of rain nor the solubility parameter as a predictor of mixing ability is fully adequate on its own, but each predictor provides useful advance information. A barometer is even more valuable when combined with additional meteorological information, and cohesion parameters are of greater value when considered along with other chemical and physical data.

There have been numerous reviews on the subject of Hildebrand solubility parameters and other cohesion parameters in various languages, concentrating on various aspects, as indicated in Table 1. In addition, many manufacturers of polymers, solvents, and other products include cohesion parameters in their specifications and supply information sheets on their use (Table 2). Cohesion parameters also are being used in patent specifications (Table 3).

Despite this extensive activity, prior to the First Edition of this *Handbook* in 1983 there was no comprehensive overview of the diverse applications of cohesion parameters, and it is clear that some groups of scientists and technologists are still unaware of valuable developments being made by other groups. The 1984 Symposium on Solubility Parameters organized by the American Chemical Society at the 188th National Meeting in Philadelphia was helpful in this respect, and an exhaustive compilation of polymer cohesion parameters (as well as polymer-liquid interaction parameters) has been published as a companion *CRC Handbook.*[21]

Major problems confront any attempt to bring together all the information on cohesion parameters, as the tremendous range of applications make a complete literature search extremely difficult. As far as possible, publications cited in *Chemical Abstracts* and *Science Citation Index* through 1989 have been included in this Second Edition.

My intention is to bring together as much as possible of the published information on cohesion parameters in one volume, and to provide the bibliographic data necessary to find the remainder. At the same time I have converted all numeric data into the preferred SI unit and developed a uniform terminology and notation.

One of the methods of classifying the diverse applications of cohesion parameters is based on the states of matter involved. Solubility parameters were developed originally to explain liquid state properties, and these have proven convenient to deal with in the first nine chapters of this Handbook. This is followed by gases (Chapter 11), solids (Chapter 12), and polymers (Chapter 13 to 16). Surfaces and interfaces are introduced in Chapter 17, and this subject is extended to chromatography in the subsequent chapter. The last chapter concentrates on biological systems. In this arrangement, each chapter draws primarily on the concepts and data of previous chapters.

TABLE 1
A Selection of General Reviews on Cohesion Parameters and Related Topics

Author	Topic	Year	Language	Ref.
Acree	General	1984	English	1
Ahmad, Yaseen, Jayasri	Group contributions	1974—80	English	2,3,5,156, 157
Ahmad, Yaseen	Polymer group contributions	1977	English	4,5
Alessi et al.	Drying of coatings	1989	English	6a
Ashton, McDermott, Brench	Solvent	1983	English	11
Askadskii	Polymer group contributions	1976—7	English, Russian	12,13
ASTM	Polymers, resins	1972, 1976	English	14
Barton	General	1975, 1987	English	15,19
Barton	Wetting, adhesion	1982	English	16
Barton	Polymers	1983—90	English	17,18,20,21
Beerbower, Kaye, Pattison	Elastomer-liquid interactions	1967	English	24
Beerbower, Dickey	Elastomer-liquid interactions	1969	English	22
Beerbower, Jensen	General	1983	English	23
Beezer	Pharmaceutical	1989	English	25
Bello	Polymer solubility	1985	Spanish	26
Bennett	Polymers	1960	English	27
Bernardo, Burrell	General; plasticization	1972	English	28
Blanks	Polymer solution thermodynamics	1976—7	English	29
Blanks, Prausnitz	Polymer solution thermodynamics	1964	English	30
Brandts-Buys, Lhoest	Polymer solubility	1971	French	33
Brydson	Plastics compounding	1961—2	English	35,36
Burrell	Polymers, general	1962—75	English	37—45
Bustamante, Parera, Selles	Pharmaceutical	1983	Spanish	46,47
Cammarata, Yau, Rogers	Biological	1971—3	English	48,49
Chen	Alkyds	1985	Chinese	50
Christensen	Protective gloves	1982	English	53
Christian, Lane	Molecular	1976	English	54
Conder, Young	Gas chromatography	1979	English	55
Corish	Elastomers	1967, 1978	English	58,59
Cosaert	General	1971	French	60
Dack	Solvents	1970—6	English	62—67
Dangel	Solvents	1981	French	69
Daniel	Editorial	1984	English	70
Dayantis	Polymer solubility	1977	French	72
Doolittle	General	1954	English	73
Faidutti	Polymers	1982	Italian	77
Feng, Zhu, Hu	Chromatography	1987	Chinese	78
Frank	Polymer blends	1981	English	79
Frank, Gashgari, Semerak	Polymer blends	1986	English	80
Fuchs, Suhr	Polymer solvents	1975	English	81
Fukunaga, Shirai, Ide, Kimura	Phase transfer catalysis	1980	Japanese	85
Gardon	General	1964	English	87,88
Gardon	Surfaces, adhesion	1966, 1977	English	88,89
Gardon, Teas	General	1976	English	91
Gaylord	Polymer blend compatibilization	1989	English	93,94
Geczy	General	1962	Hungarian	95
Gilles	General	1968	French	96
Grossman	Lubricants for elastomers	1989	English	97
Guzman	General	1964	Spanish	98
Hadert	Polymer solvents	1969	German	99
Hampe	Solvent extraction	1985—6	English, German	101

TABLE 1 (continued)
A Selection of General Reviews on Cohesion Parameters and Related Topics

Author	Topic	Year	Language	Ref.
Hansen	Hansen parameters	1967—8	English	102—105
Hansen	General	1968—9, 1981	English	105,106,110
Hansen	Coatings	1971—82	English	107—109, 111
Hansen, Beerbower	General	1971	English	112
Hansen, Skaarup	Hansen parameters	1967	English	113,114
Hansen, Wallström	Surfaces	1983	English	115
Harazaki	General	1970	Japanese	117
Harazaki	Polymers	1987	Japanese	118
Harris	Polymers	1981	English	119
Harrison	Adhesives	1978—9	English	120,121
Henriksen	Protective gloves	1982	Danish, English	122
Hildebrand	General	1936—77	English	124—127
Hildebrand, Prausnitz, Scott	General	1950—70	English	128—130
Hoy	Liquids	1969—85	English	134,135
Iguchi	Cryogenics	1971	Japanese	136
Immergut, Mark	Polymer plasticization	1965	English	137
Imoto	General	1969, 1972	Japanese	138,139
Irving	Solvent extraction	1974	English	141
Jackson	Cleaning, adhesion	1974—9	English	145—152
Jayasri, Yaseen	Vaporization enthalpy	1979, 1980	English	155—157
Jobert	General	1981	French	158
Johannsen	General	1963	German	159
Johnston	Supercritical fluids	1989	English	159a
Kaelble	General; adhesion	1971	English	160
Kakabadse	Solvents	1977	English	161
Kambour et al.	Polymer crazing	1972—4	English	163—169
Karger, Snyder, et al.	Chromatography	1973—6	English	171—173
Karickhoff	Aquatic pollutants	1984	English	174
Keller, Karger, Snyder	Chromatography	1971, 1973	English	179,180
Kendall	Solids	1980	English	181
King	Supercritical fluid chromatography	1989	English	183
Klein, Smith	Membranes	1972	English	186,187
Ko	Adhesives	1978	Japanese	188
Koningsveld, et al.	Polymer thermodynamics	1987	English	189a
Korol	Chromatography	1982	Russian, English	190
Krause	Polymer blends	1972, 1978	English	191
Kumar, Prausnitz	Solvents	1975	English	192
Lee	Polymer solvents	1988	Korean	196
Lee, Sewell	Polymer glass transition	1968	English	197
Lewis	Polymers	1968	English	198
Li, Long	Plasticization by permeates	1970	English	199
Liao	Dyeing	1984	Chinese	200
Lonsdale	Membrane technology	1982	English	201
Lucas	General	1971	French	202
Maeda	Oil-rubber interactions	1978, 1985	Japanese	203,204
Mandik et al.	General	1967, 1972	Czech	208,210
Mandik et al.	Coatings	1965, 1972	Czech	209,211
Marron	Solvent formulation	1981	French	212
Marti	General	1967	German	213
Martin et al.	General pharmaceutical	1983, 1988	English	214,215
Mellan	General	1968	English	222

TABLE 1 (continued)
A Selection of General Reviews on Cohesion Parameters and Related Topics

Author	Topic	Year	Language	Ref.
Meyer zu Bexten	General	1972	German	223
Mikuchi	Paints	1982	English	224
Mikos, Peppas	Biomedical copolymer	1988	English	223a
Mitomo, Teshirogi	General	1979—80	Japanese	225
Moiseev, Pilyagin	Spraying solvents	1971	Russian	226
Moore	Fiber polymers	1957	English	227
Mori	Conductive coatings	1988	Japanese	228
Nakanishi	Solutions	1980, 1982	Japanese	233,234
Nakanishi	Chemical engineering	1987	Japanese	235
Neau	Organic solids	1988	English	237
Nelson et al.	General	1970	English	239—241
Novák, Matouš, Pick	Liquid-liquid	1987	English	244
Nunn	General	1971	English	246
Ohm	Rubber curative diffusion	1985	English	248
OCCAA	Surface coatings	1983	English	249
Olabisi	Polymer blends	1982	English	251
Orwoll	Polymer-solvent	1977	English	254
Paul	Polymer transport	1985	English	257
Perepechkin	Membrane preparation	1988	Russian	260
Poole	Gas chromatography	1989	English	262
Prausnitz et al.	General	1969, 1986	English	263,264
Price	Inverse gas chromatography	1989	English	265
Pytela	Solution processes	1988	English	266
Ramsbotham	Coatings formulation	1978—80	English	268—271
Rebenfeld et al.	Polymers	1976	English	273
Reichardt	General	1978, 1979	English	274—276
Reid, Prausnitz, Sherwood	General	1977	English	277
Rey-Mermet	Pharmaceutical solids	1989	English	278
Reynolds, Larson	General	1962	English	279
Rider	Solutions	1985	English	280
Riedel	Dyes	1976	German	281
Robu	General	1969	Romanian	282
Ross	Polymer swelling	1983	English	284
Sakamoto	Fabric adhesives	1971	Japanese	287
Sandholm	Polymers	1970	English	288
Sato	Solvent extraction	1986	English	290
Sattelmeyer	Plastics	1976	German	291
Sausaman	General	1973	English	292
Schoenmakers	Liquid chromatography	1981	English, Dutch	293
Schoenmakers	General	1986	English	295
Schoenmakers, Billiet, De Galan	Chromatography	1982	English, German	296
Schreiber, Lloyd	Inverse gas chromatography	1989	English	297
Sears, Touchette, Darby	Plasticizers	1982, 1985	English	299—301
Senich	Polymer	1981	English	302
Sevestre	General	1964	English, French	303,304
Seymour	Polymers	1968—84	English, Spanish	307—313
Seymour, Sosa	General	1974	English	314
Seymour, Sosa	Polymers	1977	English	315
Sheehan, Bisio	Polymer solvents	1966	English	317
Shell	Solvents	1967—79	English	318—321
Shen	General	1980	Chinese	322
Shinoda	General	1966—78	English, Japanese	323

TABLE 1 (continued)
A Selection of General Reviews on Cohesion Parameters and Related Topics

Author	Topic	Year	Language	Ref.
Shvarts	Rubber swelling	1965	Russian, English	324
Siddiqui	Dyes	1981, 1983	English	326—328
Siddiqui, Needles	Textiles	1982	English	329
Skelly	Textiles	1975	English	331
Slinn	Refrigerants	1971	English	332
Sliwka	Micro-encapsulation	1975	English	333
Small	Group contributions	1953	English	334
Snyder	Chromatography	1968, 1971	English	335, 336
Snyder	General	1974—80	English	337—340
Snyder, Kirkland	Chromatography	1974	English	341
Sörensen	Pigments, coatings	1967—8	English, Danish, German	342—349
Somerville, Lopez	Solvent formulation	1973	English	350
Sonnich Thomsen	General	1965—69	English, Danish	351—354
Stanton	Paste rheology	1983	English	355
Staverman	Binary liquid systems	1941	English	357
Stawiszynski, Zawadzki	General	1979	Polish	358
Sugihara	Resin solvents	1971	Japanese	359
Suzuki, Hagiwara	Metals	1977	English	361
Svoboda	Liquids	1982	Czech	363
Sweeney et al.	Gas solvents	1988	English	364
Tager, Kolmakova	General	1980	Russian, English	365
Takada	General	1971	Japanese	366
Tanaka	Solvent extraction	1971, 1976	English, Japanese	369,370
Taylor	General	1957	English	372
Teas	Resins	1968, 1971	English	373—375
Tess	Solvent formulation	1975, 1985	English	376,377
Tlusta, Zelinger	General	1971	Czech	378
Tortorello, Kinsella, Gearon	Copolymers	1982—3	English	379,380
Unishi	Polymer solubility	1981	Japanese	385
Van Dyk et al.	Polymer solubility	1985	English	386
van Krevelen	Polymers	1976	English	387
Vaughan	Cosmetics	1985	English	388
Vavruch	Solvents	1984	German	389
Verneret	Solvents	1970	Spanish	390
Vernon	Gas-liquid chromatography	1978	English	391
Vilcu, Leca	Inverse gas chromatography	1989—90	English	392
Vinther	Pigments	1971	English	394
Wágner, Majer, Svoboda	Liquids	1983	English	395
Wallström et al.	General	1982	Swedish	396
Walz, Emrich	Resins	1974, 1975	English	397
Watanabe	General	1975	Japanese	399,400
Weast	General	1988	English	401
Williamson	Nonelectrolyte solutions	1964	English	404
Wilson	Group contributions	1985	English	405
Wolf	Polymers	1985	English	407
Yoshida	General	1971	Japanese	414
Yoshioka	Polymer tackifiers	1983	Japanese	418

TABLE 2
A Selection of Manufacturers' Bulletins and Publications Which Include Cohesion Parameter Information

Company	Author	Ref.
Albert	Walz	397,398
American Cyanamid	Nelson	239
Ashland	Teas	373—375
BASF	Riedel	281
	Schröder	298
Bendix	Jackson	145—149
Bostik	Harrison	120,121
CIBA-Geigy	Skelly	331
Commercial Solvents	Sausaman	292
Diamond Shamrock	Dale, Millar	68
Dow Corning	Wessling	402
	Yerrick, Beck	413
Dunlop	Corish	58,59
du Pont	—	74,75
	Cosaert	60
	Hagman	100
	Starkweather	275
	Zunker	419
Eastman	Crowley	61
Esso	Beerbower, Dickey	22
	Beerbower, Kaye, Pattison	24
	Sheehan, Bisio	317
Exxon	Funk	86
General Electric	Kambour	163—169
General Motors	Wyzgowski	408
Goodrich	Oetzel	247
Harcross Chemical	Meusberger	221
Hercules	—	123
	Orr	253
Hoechst	Bohn	32
	Sattelmeyer	291
Humble	Kirkland	184
ICI	Small	334
	Vincent, Raha	393
Imperial Smelting	Slinn	332
Inmont	Burrell	37—45
Isotron	—	142
Köge (KVK)	Sörensen	342—348
M & T Chemicals	Vinther	394
	Gardon	89
Mobay Corp. (Bayer)	Rains, Bailey	267
Monsanto	Anagnostopoulos, Coran, Gamrath	9,10
	Darby, Touchette, Sears	71
	Wilt	406
Neville Chemical	—	242
Pennsylvania Industrial Chemical	—	259
Pennwalt Corp.	Kamath, Sargent	162
Philips	Schoenmakers	294—296
Rohm & Haas	—	403
	Brendley	34
	Gardon	87,88
Shell	—	318—321
	Nelson	240,241
	Nunn	246

TABLE 2 (continued)
A Selection of Manufacturers' Bulletins and Publications Which Include Cohesion Parameter Information

Company	Author	Ref.
	Ramsbotham	268—271
	Reynolds, Larson	279
Technicon Instruments	Snyder	335—341
Union Carbide	Hoy	134,135
	Shaw	316
Upjohn	Yalkowsky et al.	410

TABLE 3
Some Patents Which Incorporate Cohesion Parameter Specifications

Author	Topic	Ref.
Aikawa, Ikeda	Zinc stearate removal	6
Alfrey	Microporous polymers	7
Chervin	Plastisols	51
Chisso	Liquid cleaning of propylene copolymer	52
Corcoran	Coal desulfurization	57
Fujita, Kuroda	Optical fibers	82
Fujiwara, Ishitobi	Aramid Fibers	83
Fuji-Xerox	Toners and photoreceptors	84
Gaylord	Compatibilization of polymers	92
Hara	Polymerization	116
Hirose et al.	Optical recording medium	132
Hirose et al.	Thermoplastic polymers	131
Howard, Young	Polyolefin fiber plasticizers	133
Irani	Drive fluids	140
Iwamoto	Polyurethane leather substitutes	143
Iwatomo	Toners for electrostatic copiers	144
Japan Synthetic Rubber Co.	Polymerization	153
Japan Styrene Paper Corp.	Cellular polystyrene	154
Kanebo	Cosmetics	170
Kashihara et al.	Polymerization	175
Kashihara et al.	Silicon- and fluorine-containing polymers	176
Kato	Polyamides	177,178
Kikuchi et al.	Resin coatings	182
Kito	Epoxy adhesives	185
Komabashiri et al.	Scale-resistant coatings	189
Kurita et al.	Polyester fibers	193
Kuwajima et al.	Metallic pigment coatings	194
Larsen, Kindt-Larsen	Shaped hydrogel articles	195
Maeda	Polymers resisting environmental stress cracking	205
Maeda	Resins	206
Manabe	Propylene copolymers	207
Martin et al.	Pressure-sensitive adhesives	216
Mason, Smith	Polyamides	217
Masutani et al.	Laminates	218
Matson	Gas-separation membranes	219
Matsuhisa	Pressure-transfer inks	220
Mori	Conductive coatings	228
Mori et al.	Poly(vinyl chloride) surface modification	229
Morris	Sizing hydrophobic yarns	230
Nakagawa	Rubber-wire adhesion	231
Nakane	Thermoplastic polymers	232
Nasune, Sano	Cosmetics	236

TABLE 3 (continued)
Some Patents Which Incorporate Cohesion Parameter Specifications

Author	Topic	Ref.
Nehashi et al.	Dyeing agents	238
Nippon Paint Co.	Metallic coating materials	243
Nozaki	Coatings	245
Okamoto et al.	Nonwoven fabric	250
Oota	Heat-resistant polyimide	252
Osajima	Ethylene—vinyl acetate copolymer films	255
Overfield	Nickel and vanadium extraction from oils	256
Peck	Recovery of carbonaceous liquid	258
Persson	Polymer blending	261
Randen	Mosquito repellent	272
Saida, Narahara	Unsaturated polyesters	285
Sakamoto	Resin coatings	286
Sasaki et al.	Methacrylimidepolymers	289
Schober, Schroeder	Photosensitive compositions	293
Shimada, Susumu	Maleic adhesives	325
Suzuki, Harada	Powdered polyimides	360
Suzuki	Photosensitive compositions	362
Takenaka	Photosensitive compositions	367
Tanabe	Automobile coatings	368
Tayama	Acrylics	371
Toyota	Coatings for plastic moldings	381
Tsukuda et al.	Gelation of liquid oils	382
Uchida	Toners for electrostatic copiers	383
Uhara	Porous polymer films	384
Williams	Sheet for impact printer	403
Yabuta	Polymer dispersions	409
Yamamoto et al.	Photoreceptors	411
Yazaki, Iida	Laminates	412
Yoshida et al.	Poly(vinyl chloride) compositions	415
Yoshifuji	Phenolic resins	416
Yoshihara et al.	Vinyl polymer in hair treatment preparation	417

General conclusions for all systems may be summarized:

1. Hildebrand parameters or total cohesion parameters (δ) provide a broad, qualitative indication of behavior for most systems, and good results for a very small number of hydrocarbon systems.
2. Hansen parameters which separate dispersion (δ_d), polar (δ_p) and hydrogen-bonding (δ_h) interactions and similar sets of two- or three-component cohesion parameters give an approximate quantitative measure of the properties of most systems (excluding some aqueous and alcoholic mixtures) and good results for a limited number of systems without significant chemical interactions.
3. A set of component cohesion parameters representing dispersion (δ_d), orientation (δ_o), induction (δ_i), Lewis acid (δ_a), and Lewis base (δ_b) properties is necessary for a reliable quantitative evaluation of the behavior of multicomponent systems. This is not just because ''more adjustable parameters are better'', but because induction and acid-base interaction are ''unsymmetrical'', involving participating molecules with complementary rather than similar roles. It is also necessary to allow for differences in the molar volumes of the components, and there is scope for extension to ion-dipole and ion-induced dipole electrostatic cohesion parameters.

In a few applications (particularly chromatography) the subject has developed to stage 3; in others (such as polymer-liquid interactions and membrane performance) it has reached stage 2; but in many other areas only single-component Hildebrand parameters have been used.

The literature citations for this chapter are arranged alphabetically by author, as a bibliographic resource.

REFERENCES

1. **Acree, W. E., Jr.,** *Thermodynamic Properties of Nonelectrolyte Solutions,* Academic Press, Orlando, FL, 1984.
2. **Ahmad, H. and Yaseen, M.,** Solubility parameter. Its concept and determination, *J. Colour Soc.,* 13(2), 7, 1974.
3. **Ahmad, H. and Yaseen, M.,** Estimation of the solubility parameters of low molecular weight compounds by a chemical group contribution technique, *J. Oil Colour Chem. Assoc.,* 60, 99, 1977.
4. **Ahmad, H. and Yaseen, M.,** Determination of solubility parameters of polymers by a group contribution technique, *J. Oil Colour Chem. Assoc.,* 60, 488, 1977.
5. **Ahmad, H. and Yaseen, M.,** Determination of solubility parameters of low and high molecular weight compounds, *J. Colour Soc.,* 16(2), 19, 1977.
6. **Aikawa, J. and Ikeda, M.,** Agents for steam cleaning of zinc stearate, *Jpn. Pat.,* 01101400, 1989.

6a. **Alessi, P., Orlandini-Visalberghi, M., and Torriano, G.,** New applications of gas chromatography in the field of paints and varnishes, *J. Oil Colour Chem. Assoc.,* 72, 15, 1989.

7. **Alfrey, T., Jr. and Lloyd, W. G.,** Microporous polymers, *U.S. Pat. Re-Issue,* 27026, 1971 (Reissue of 3322695).
8. **Anagnostopoulos, C. E. and Coran, A. Y.,** Polymer-diluent interactions. II. Polyvinyl chloride-diluent interactions, *J. Polym. Sci.,* 57, 1, 1962.
9. **Anagnostopoulos, C. E., Coran, A. Y., and Gamrath, H. R.,** Polymer-diluent interactions. I. A new micromethod for determining poly(vinyl chloride)-diluent interactions, *J. Appl. Polym. Sci.,* 4, 181, 1960.
10. **Anagnostopoulos, C. E., Coran, A. Y., and Gamrath, H. R.,** Microtest for PVC-plasticizer compatibility, *Mod. Plast.,* 43, 141, 1965.
11. **Ashton, N. F., McDermott, C., and Brench, A.,** Chemistry of extraction of nonreacting solutes, in *Handbook of Solvent Extraction,* Lo, T. C., Baird, M. H. I., and Hanson, C., Eds., John Wiley & Sons-Interscience, New York, 1983, 3.
12. **Askadskii, A. A.,** Influence of chemical structure on the properties of polymers, *Pure Appl. Chem.,* 46, 19, 1976.
13. **Askadskii, A. A.,** Computational procedures for the determination of the physical characteristics of polymers, *Usp. Khim.,* 46, 1122, 1977; *Russ. Chem. Rev.,* 46, 589, 1977.
14. **ASTM,** *Standard Test Method for Solubility Range of Resins and Polymers,* ANSI/ASTM D3132-72, American Society for Testing and Materials/American National Standards Institute, New York, reapproved 1976.
15. **Barton, A. F. M.,** Solubility parameters, *Chem. Rev.,* 75, 731, 1975.
16. **Barton, A. F. M.,** The application of cohesion parameters to wetting and adhesion — a review, *J. Adhesion,* 14, 33, 1982.
17. **Barton, A. F. M.,** Cohesion parameters, in *Polymer Yearbook,* Elias, H.-G. and Pethrick, R. A., Eds., Harwood, Chur, Switzerland, 1983, 149.
18. **Barton, A. F. M.,** Applications of solubility parameters and other cohesion parameters in polymer science and technology, *Pure Appl. Chem.,* 57, 905, 1985.
19. **Barton, A. F. M.,** Cohesion parameters, in *Encyclopedia of Physical Science and Technology,* Vol. 3, Academic Press, Orlando, FL, 1987, 140.
20. **Barton, A. F. M.,** Polymer interaction parameters and cohesion parameters, *Polymer Yearbook 6,* Pethrick, R. A., Ed., Harwood, Chur, Switzerland, 1990, 87.
21. **Barton, A. F. M.,** *Handbook of Polymer-Liquid Interaction Parameters and Solubility Parameters,* CRC Press, Boca Raton, FL, 1990.
22. **Beerbower, A. and Dickey, J. R.,** Advanced methods for predicting elastomer/fluids interactions, *Am. Soc. Lubric. Eng. Trans.,* 12, 1, 1969.
23. **Beerbower, A. and Jensen, W. B.,** The HSAB principle and extended solubility theory, *Inorg. Chim. Acta,* 75, 193, 1983.

24. **Beerbower, A., Kaye, L. A., and Pattison, D. A.,** Picking the right elastomer to fit your fluids, *Chem. Eng. (New York),* 74(26), 118, 1967.
25. **Beezer, A. E.,** Solubility and pharmacological importance, *Pure Appl. Chem.,* 61, 153, 1989.
26. **Bello, A.,** Science and technology of plastics, X. Solubility, *Rev. Plast. Mod.,* 50, 577, 1985.
27. **Bennett, R. R.,** Nitroparaffins in coatings, 3. Measuring their solvent power, *Can. Paint Varn.,* 34(2), 30, 1960.
28. **Bernardo, J. J. and Burrell, H.,** Plasticization, in *Polymer Science,* Jenkins, A. D., Ed., North-Holland, Amsterdam, 1972, chap. 8.
29. **Blanks, R. F.,** Engineering applications of polymer solution thermodynamics, *Chemtech,* 6, 396, 1976; Thermodynamics of polymer solutions, *Polym. Plast. Technol. Eng.,* 8, 13, 1977.
30. **Blanks, R. F. and Prausnitz, J. M.,** Thermodynamics of polymer solubility in polar and nonpolar systems, *Ind. Eng. Chem. Fundam.,* 3, 1, 1964.
31. **Bohn, L.,** Incompatibility and phase formation in polymer mixtures and graft and block copolymers in the solid state, *Kolloid-Z. Z. Polym.,* 213, 55, 1966; *Rubber Chem. Technol.,* 41, 495, 1968.
32. **Bohn, L.,** Compatible polymers, in *Polymer Handbook,* 2nd ed., Brandrup, J. and Immergut, E. H., Eds., John Wiley & Sons, New York, 1975, III-211.
33. **Brandts-Buys, L. and Lhoest, G.,** Solubility parameters and graphic representation of contours of the solubility of polymers, *Chim. Peintures,* 34, 261, 1971.
34. **Brendley, W. H.,** Fundamentals of acrylic polymers, *Paint Varnish Prod.,* 63(7), 19, 1973.
35. **Brydson, J. A.,** Selection of compounding ingredients, *Plastics (London),* 26, 107, 1961.
36. **Brydson, J. A.,** *Plastics Materials,* Van Nostrand, Princeton, NJ, 1966; 4th ed., Butterworths, London, 1982.
37. **Burrell, H.,** The use of the solubility parameter concept in the United States, *FATIPEC (Fed. Assoc. Tech. Ind. Peint. Vernis, Emaux Encres Impr. Eur. Cont.) Congr.,* 6, 21, 1962.
38. **Burrell, H.,** Solubility parameter values, in *Polymer Handbook,* Brandrup, J. and Immergut, E. H., Eds., John Wiley & Sons-Interscience, New York, 1966, IV-341.
39. **Burrell, H.,** The challenge of the solubility parameter concept, *J. Paint Technol.,* 40, 197, 1968.
40. **Burrell, H.,** Entropy: the hidden ingredient, *J. Paint Technol.,* 42(540), 3, 1970.
41. **Burrell, H.,** Solubility of polymers, in *Kirk-Othmer Encyclopedia of Science and Technology,* Vol. 12, 2nd ed., John Wiley & Sons, New York, 1970, 618.
42. **Burrell, H.,** Trends in solvent science and technology, in *Solvents Theory and Practice (Adv. Chem. Ser.* 124), Tess, R. W., Ed., American Chemical Society, Washington, D.C., 1973, chap. 1.
43. **Burrell, H.,** Solubility parameter values, in *Polymer Handbook,* 2nd ed., Brandrup, J. and Immergut, E. H., Eds., John Wiley & Sons-Interscience, New York, 1975, IV-337.
44. **Burrell, H.,** Pigment wetting parameter, *Am. Chem. Soc. Div. Org. Coat. Plast. Chem. Pap.,* 35(2), 18, 1975.
45. **Burrell, H.,** Pigment dispersion parameter, *Am. Chem. Soc. Div. Org. Coat. Plast. Pap.,* 35(2), 25, 1975.
46. **Bustamante, P., Parera, A., and Selles, E.,** Solubility parameter and its applications in pharmacy, *Ars. Pharm.,* 24, 303, 1983.
47. **Bustamante, P., Parera, A., and Sellés, E.,** Applications of the multicomponent parameter in pharmaceutical liquid formulation, *An. R. Acad. Farm.,* 49, 221, 1983.
48. **Cammarata, A., Yau, S. J., and Rogers, K. S.,** Structure-activity relationships having a basis in regular solution theory, *J. Med. Chem.,* 14, 211, 1971.
49. **Cammarata, A., Yau, S. J., and Rogers, K. S.,** Some biological applications of regular solution theory, *Pure Appl. Chem.,* 35, 495, 1973.
50. **Chen, J.,** Calculation and application of solubility parameters of alkyd resins, *Tuliao Gangye,* 84, 21, 1985.
51. **Chervin, V. G., Zubenko, L. N., Bychkov, R. A., and Krasheninnkov, A. I.,** Production of plastisol from methyl methacrylate — methacrylic acid copolymer, *USSR Pat.,* 794031, 1981.
52. **Chisso Corp.,** Purification of propylene random copolymers for films, *Jpn. Pat.,* 85/47010, 1985.
53. **Christensen, U. L. and Banke, O.,** Gloves and solubility parameter, *Dan. Kem.,* 63(3), 70, 1982.
54. **Christian, S. D. and Lane, E. H.,** Solvent effects on molecular complex equilibria, in *Solutions and Solubilities,* Part 1, Dack, M. R. J., Ed., John Wiley & Sons-Interscience, New York, 1975, chap. 6.
55. **Conder, J. R. and Young, C. L.,** *Physicochemical Measurement by Gas Chromatography,* John Wiley & Sons-Interscience, New York, 1979, chap. 5.
56. **Coran, A. Y. and Anagnostopoulos, C. E.,** Polymer-diluent interactions. III. Polyethylene-diluent interactions, *J. Polym. Sci.,* 57, 13, 1962.
57. **Corcoran, W. H., Vasilakos, N. P., and Lawson, D. D.,** Coal desulfurization, *U.S. Pat.,* 4334888A, 1982.
58. **Corish, P. J.,** Fundamental studies of rubber blends, *Rubber Chem. Technol.,* 40, 324, 1967.
59. **Corish, P. J.,** Elastomer blends, in *Science and Technology of Rubber,* Eirich, F. R., Ed., Academic Press, New York, 1978, chap. 12.

60. **Cosaert, E.,** Solubility parameters allowing the prediction of the solubility of a solid in a solvent system, *Chim. Peintures,* 34, 169, 1971.
61. **Crowley, J. D., Teague, G. S., Jr., and Lowe, J. W.,** A three-dimensional approach to solubility, *J. Paint Technol.,* 38, 269, 1966; 39, 19, 1967.
62. **Dack, M. R. J.,** Organic solvents — their effect on chemical reactivity, *Chem. Br.,* 6, 347, 1970.
63. **Dack, M. R. J.,** Solvent effects and reactivities, *Chemtech,* 1, 108, 1971.
64. **Dack, M. R. J.,** The influence of solvent on chemical reactivity. An alternative approach, *J. Chem. Educ.,* 51, 231, 1974.
65. **Dack, M. R. J.,** The importance of solvent internal pressure and cohesion to solution phenomena, *Chem. Soc. Rev.,* 4, 211, 1975.
66. **Dack, M. R. J.,** Solvent structure. The use of internal pressure and cohesive energy density to examine contributions to solvent-solvent interactions, *Aust. J. Chem.,* 28, 1643, 1975.
67. **Dack, M. R. J.,** The influence of solvent on chemical reactivity, in *Solutions and Solubilities,* Dack, M. R. J., Ed., Part II, John Wiley & Sons-Interscience, New York, 1975, chap. 11.
68. **Dale, J. A. and Millar, J. R.,** Cross-linker effectiveness in styrene copolymerization, *Macromolecules,* 14, 1515, 1981.
69. **Dangel, B.,** Round table on solubility parameters. Solvent formulation by a graphical method, *Double-Liaison — Chim. Peintures,* 28(306), 75, 1981.
70. **Daniel, F. K.,** The obstacle course from mill base to finished coating, *J. Coatings Technol.,* 56(708), 38, 1984.
71. **Darby, J. R., Touchette, N. W., and Sears, K.,** Dielectric constants of plasticizers as predictors of compatibility with polyvinyl chloride, *Polym. Eng. Sci.,* 7, 295, 1967.
72. **Dayantis, J.,** The Hildebrand parameter. Solubility of polymers, *Plast. Mod. Elastomers,* 29(2), 58, 1977.
73. **Doolittle, A. K.,** *The Technology of Solvents and Plasticizers,* John Wiley & Sons, New York, 1954, 807.
74. **du Pont de Nemours, E. I. and Co.,** *A New Dimension in Solvent Formulation — Elvacite Acrylic Resins,* A44369, du Pont Electrochemicals Dept., Wilmington, DE, 1965.
75. **du Pont de Nemours, E. I. and Co.,** *Solvent Formulating Maps for Elvacite Acrylic Resins,* Bulletin PA-12-770, A70562, Wilmington, DE, 1970; PA-12-174, 1974.
76. **Eastman Kodak,** *Three Dimensional Approach to Solubility,* Rochester, NY (undated).
77. **Faidutti, M.,** Solubility parameters of solvents and polymers, *Ind. Vernice,* 36(2), 12, 1982.
78. **Feng, Y. Q., Zhu, P. L., and Hu, Z.,** Applications of solubility parameter in chromatography, *Fenzi Huaxue,* 15, 469, 1987.
79. **Frank, C. W.,** Excimer fluorescence: a new tool for studying polymer alloys, *Plast. Compounding,* Jan/Feb, 67, 1981.
80. **Frank, C. W., Gashgari, M. A., and Semerak, S. N.,** Polymer blend thermodynamics: Flory-Huggins theory and its application to excimer fluorescence studies, *NATO ASI Ser., Ser. C,* 182 (Photophys. Photochem. Tools Polym. Sci.: Conform., Dyn., Morphol.), 523, 1986.
81. **Fuchs, O. and Suhr, H. H.,** Solvents and non-solvents for polymers, in *Polymer Handbook,* Brandrup, J. and Immergut, E. H., Eds., John Wiley & Sons-Interscience, New York, 2nd ed., 1975, IV-241.
82. **Fujita, H. and Kuroda, T.,** Clad optical fibers with polycarbonate cores, *Jpn. Pat.,* 01061706, 1989.
83. **Fujiwara, T. and Ishitobi, T.,** High tenacity poly(*p*-phenyleneterephthalamide) fibers with low cutting strength and their manufacture, *Jpn. Pat.,* 62/21814, 1987.
84. **Fuji-Xerox Co. Ltd.,** Electrophotographic toners and toner image receptors, *Jpn. Pat.,* 57/30848, 1982.
85. **Fukunaga, K., Shirai, H., Ide, S., and Kimura, M.,** Solubility parameters of phase-transfer catalysts and their applications, *Nippon Kagaku Kaishi,* 1148, 1980.
86. **Funk, E. W.,** Study of heavy hydrocarbons by inverse-phase chromatography, *Ind. Eng. Chem. Prod. Res. Dev.,* 16, 115, 1977.
87. **Gardon, J. L.,** Cohesive-energy density, in *Encyclopedia of Polymer Science and Technology,* Vol. 3, John Wiley & Sons, New York, 1964, 833.
88. **Gardon, J. L.,** The influence of polarity upon the solubility parameter concept, *J. Paint Technol.,* 38(492), 43, 1966.
89. **Gardon, J. L.,** Cohesive energy density and adhesion, *Proc. Chem. Inst. Can. Symp. Polym. Interfaces: Focus Adhes.,* 33, 1976.
90. **Gardon, J. L.,** Critical review of concepts common to cohesive energy density, surface tension, tensile strength, heat of mixing, interfacial tension and butt joint strength, *J. Colloid Interface Sci.,* 59, 582, 1977; *Prog. Org. Coat.,* 5, 1, 1977.
91. **Gardon, J. L. and Teas, J. P.,** Solubility parameters, in *Treatise on Coatings,* Vol. 2, *Characterization of Coatings: Physical Techniques,* Part II, Myers, R. R. and Long, J. S., Eds., Marcel Dekker, New York, 1976, chap. 8, 413.
92. **Gaylord, N. G.,** Compatibilization of polymers by adding graft copolymers, *U.S. Pat.,* 3485777, 1969.
93. **Gaylord, N. G.,** Compatibilizing agents: structure and function in polyblends, *J. Macromol. Sci. Chem.,* A26, 1211, 1989.

94. **Gaylord, N. G.,** Use surfactants to blend polymers, *Chemtech,* 19, 435, 1989.
95. **Geczy, I.,** The use of solubility parameters as technological characteristics in selecting industrial chemicals for the manufacture of plastics, *Kolor. Ert.,* 4, 99, 1962.
96. **Gilles, P.,** Solubility parameter, *Double-Liaison,* 152, 457, 1968.
97. **Grossman, R. F.,** Internal and external lubricants enhance processing of elastomers, *Elastomerics,* 121(4), 18, 1989.
97a. **Grulke, E. A.,** Solubility parameter values, in *Polymer Handbook,* Brandrup, J. and Immergut, E. H., Eds., John Wiley & Sons — Interscience, New York, 3rd ed., 1989, VII-519.
98. **Guzman, G. M.,** Solutions and solvents for polymers. I., *Rev. Plast. Mod.,* 15, 489, 1964.
99. **Hadert, H.,** Polymer solvents, *Coating,* 2(2), 39; (3), 83; (4), 108, 1969.
100. **Hagman, J. F.,** *Solvent Systems for Neoprene — Predicting Solvent Strength,* A35117, du Pont Elastomer Chemicals Dept., Wilmington, DE, March, 1964.
101. **Hampe, M. J.,** Selection of solvents in liquid/liquid extraction according to physico-chemical aspects, *Chem. Ing. Tech.,* 57, 669, 1985; *Ger. Chem. Eng.,* 9, 251, 1986.
102. **Hansen, C. M.,** The three dimensional solubility parameter — key to paint component affinities. I. Solvents, plasticizers, polymers and resins, *J. Paint Technol.,* 39, 104, 1967.
103. **Hansen, C. M.,** The three dimensional solubility parameter — key to paint component affinities. II. Dyes, emulsifiers, mutual solubility and compatibility, and pigments, *J. Paint Technol.,* 39, 505, 1967.
104. **Hansen, C. M.,** On the application of the three dimensional solubility parameter concept to the prediction of mutual solubility and compatibility, *Skand. Tidskr. Färg Lack,* 13, 132, 1967.
105. **Hansen, C. M.,** The three dimensional solubility parameter and its use. I. Introduction and non-pigmented systems, *Skand. Tidskr. Färg Lack,* 14(2), 28; 14(3), 45, 1968.
106. **Hansen, C. M.,** The universality of the solubility parameter, *Ind. Eng. Chem. Prod. Res. Dev.,* 8, 2, 1969.
107. **Hansen, C. M.,** Solubility in the coatings industry, *Skand. Tidskr. Färg Lack,* 17, 69, 1971.
108. **Hansen, C. M.,** Surface dewetting and coatings performance, *J. Paint Technol.,* 44(570), 57, 1972.
109. **Hansen, C. M.,** Solvents for coatings, *Chemtech,* 2, 547, 1972.
110. **Hansen, C. M.,** Some answered and unanswered questions about the solubility parameter, in *Macromolecular Solutions: Solvent-Polarity Relationships in Polymers* (Symp. Proc., Am. Chem. Soc., New York, 1981), Seymour, R. B. and Stahl, G. A., Eds., Pergamon, New York, 1982, 1; *Org. Coat. Plast. Chem.,* 45, 227, 1981.
111. **Hansen, C. M.,** Organic solvents in high solids and water-reducible coatings, *Prog. Org. Coat.,* 10, 331, 1982.
112. **Hansen, C. M. and Beerbower, A.,** Solubility parameters, in *Kirk-Othmer Encyclopedia of Chemical Technology,* Suppl. Vol., 2nd ed., Standen, A., Ed., Interscience, New York, 1971, 889.
113. **Hansen, C. M. and Skaarup, K.,** The three dimensional solubility parameter — key to paint component affinities. III. Independent calculation of the parameter components, *J. Paint Technol.,* 39, 511, 1967.
114. **Hansen, C. M. and Skaarup, K.,** Aspects of the three-dimensional solubility parameter, *Dan. Kemi,* 48(6), 81, 1967.
115. **Hansen, C. M. and Wallström, E.,** On the use of cohesion parameters to characterise surfaces, *J. Adhesion,* 15, 275, 1983.
116. **Hara, T., Horino, H., and Kita, R.,** Curable, trans-rich polypentenamers, *Jpn. Pat.,* 49/2900, 1974.
117. **Harazaki, Y.,** Solubility parameters, *Kami Pa Gikyoshi,* 24, 231, 1970.
118. **Harazaki, Y.,** Prediction of the solubility of polymers, 2, *Toso Gijutsu,* 26, 119, 1987.
119. **Harris, F. W.,** Introduction to polymer chemistry, *J. Chem. Educ.,* 58, 837, 1981.
120. **Harrison, K. W.,** Solvents in polymer-based adhesives, *Polym. Paint. Col.,* 168, 602, and 885, 1978.
121. **Harrison, K. W.,** Solvents in polymer-based adhesives, in *Adhesion,* Vol. 3, Allen, K. W., Ed., Applied Science Publishers, Barking, England, 1979, chap. 9.
122. **Henriksen, H. R.,** *Selection of Materials for Protective Gloves. Polymer Membranes to Protect Against Contact with Epoxy Products,* translated from *Materials for Protective Gloves,* Danish Directorate of Labor Inspection Services, Copenhagen, 1982.
123. **Hercules, Inc.,** *Solubility Parameters — Synthetic Resins, Related Materials and Solvents,* Tech. Bull., Hercules Inc., Wilmington, DE (undated).
124. **Hildebrand, J. H.,** *Solubility of Non-Electrolytes,* 2nd ed., Reinhold, New York, 1936; GONTI, Moscow-Leningrad, 1938.
125. **Hildebrand, J. H.,** A critique of the theory of solubility of non-electrolytes, *Chem. Rev.,* 44, 37, 1979.
126. **Hildebrand, J. H.,** Order from chaos, *Science,* 150, 441, 1965.
127. **Hildebrand, J. H.,** Operation on swollen theories with Occam's razor, in *Structure-Solubility Relationships in Polymers* (Proc. Symp., 1976), Harrison, F. W. and Seymour, R. B., Eds., Academic Press, New York, 1977, 1.
128. **Hildebrand, J. H., Prausnitz, J. M., and Scott, R. L.,** *Regular and Related Solutions,* van Nostrand-Reinhold, Princeton, NJ, 1970.

129. **Hildebrand, J. H. and Scott, R. L.,** *Solubility of Non-Electrolytes,* 3rd ed., Reinhold, New York, 1950; Dover, New York, 1964.
130. **Hildebrand, J. H. and Scott, R. L.,** *Regular Solutions,* Prentice-Hall, Englewood Cliffs, NJ, 1962.
131. **Hirose, T., Ishihara, K., and Kinoshita, M.,** Thermoplastic polyurethane resin compositions, *Jpn. Pat.,* 01031863, 1989.
132. **Hirose, S., Ozawa, H., Abe, K., and Hosono, Y.,** Manufacturing optical recording medium with phthalo/naphthalocyanine dye and organic solvent therefrom, *Jpn. Pat.,* 63/158294, 1988.
133. **Howard, R. E. and Young, J.,** Microporous filaments and fibers, process of making same, and articles made therefrom, *Eur. Pat.,* 322169, 1989.
134. **Hoy, K. L.,** *Tables of Solubility Parameters,* Union Carbide Corp., Chemicals and Plastics Research and Development Dept., South Charleston, WV, 1969, 1975, and 1985.
135. **Hoy, K. L.,** New values of the solubility parameters from vapor pressure data, *J. Paint Technol.,* 42(541), 76 and 115, 1970.
136. **Iguchi, A.,** Predicting cryogenic vapor-liquid equilibrium, *Kagaku Sochi,* 13(11), 85, 1971.
137. **Immergut, E. H. and Mark, H. F.,** Principles of plasticization, *Adv. Chem. Ser.,* 48, 1, 1965.
138. **Imoto, M.,** Introduction to theory of adhesion. III. Cohesion and solubility parameter values, *Nippon Secchaka Kyokai Shi,* 5, 156, 1969.
139. **Imoto, M.,** Senriyama evening talks, *Secchaka,* 16, 436, 1972.
140. **Irani, C. A.,** Solubilizing surfactants in miscible drive solvents, *U.S. Pat.,* 4828029, 1989.
141. **Irving, H. M. N. H.,** Application of the solubility concept in liquid-liquid extraction, in *Ion Exchange and Solvent Extraction,* Vol. 6, Marinsky, J. A. and Marcus, Y., Eds., Marcel Dekker, New York, 1974, chap. 3, 139.
142. **Isotron Chemicals Department,** *Propellent Solvency Values,* Isotron Tech. Inf. File, 3(2), Pennwalt Chemical Corp., Philadelphia, PA, 1964.
143. **Iwamoto, T.,** Polyurethane leather substitutes having smooth surfaces, *Jpn. Pat.,* 75/39363, 1975.
144. **Iwatomo, K., Kadowski, S., Moribe, I., Kumagai, Y., and Higashida, O.,** Toners for electrostatic image development, *Jpn. Pat.,* 60/194461, 1985.
145. **Jackson, L. C.,** *Solubility Parameter Technology and Evaporative Rate Analysis Used to Evaluate Contaminants,* Bendix Report BDX-613-859 (Rev.), Bendix Corp., Kansas City, MO, 1973.
146. **Jackson, L. C.,** How to select a substrate cleaning solvent — contaminant removal using solubility parameter technology, *Adhes. Age,* 17(12), 22, 1974.
147. **Jackson, L. C.,** *Solubility Parameters and Evaporation Rate Analysis in Organic Residue Characterization,* Bendix Report BDX-613-1099 (Rev.), Bendix Corp., Kansas City, MO, 1974.
148. **Jackson, L. C.,** A scientific technique for adhesive development, *Adhes. Age,* 18(3), 30, 1975.
149. **Jackson, L. C.,** Surface characterization based on solubility parameters, *Adhes. Age,* 19(10), 17, 1976.
150. **Jackson, L. C.,** Solvent cleaning process efficiency — contaminant removal in the electronics industry, *Adhes. Age,* 19(7), 31, 1976.
151. **Jackson, L. C.,** Removal of silicone grease and oil contaminants, *Adhes. Age,* 20(4), 29, 1977.
152. **Jackson, L. C.,** Contaminant detection, characterization and removal based on solubility parameters, in *Proc. Int. Symp. Contam. Control,* 4th, Inst. Environ. Sci., Mt. Prospect, IL, 1978, 226; *Surface Contamination: Its Genesis, Detection and Control,* Vol. 2, Mittal, K. L., Ed., New York, 1979, 527.
153. **Japan Synthetic Rubber Co. Ltd,** High-cis 1,4-polybutadienes, *Jpn. Pat.,* 59/117512, 1984.
154. **Japan Styrene Paper Corp.,** Cellular polystyrene sheets, *Jpn. Pat.,* 55/71533, 1980.
155. **Jayasri, A. and Yaseen, M.,** Heat of vaporization: its determination and application in evaluation of the solubility parameter: a review, *Prog. Org. Coat.,* 7, 167, 1979.
156. **Jayasri, A. and Yaseen, M.,** Solubility parameter values suggested using the reported and calculated values for organic compounds, *J. Oil Colour Chem.,* 63(2), 61, 1980.
157. **Jayasri, A. and Yaseen, M.,** Nomograms for solubility parameter at 25°C and at different temperatures, *J. Coatings. Technol.,* 52(667), 41, 1980.
158. **Jobert, R.,** Solubility parameters, *Double Liaison — Chim. Peintures,* 28(304-5), 29, 1981.
159. **Johannsen, R.,** Importance and application of the solubility parameter, *Dtsch. Farben Z.,* 17, 264, 1963.
159a. **Johnston, K. P.,** New directions in supercritical fluid science and technology, *Am. Chem. Soc. Symp. Ser.,* 406, 1, 1989.
160. **Kaelble, D. H.,** *Physical Chemistry of Adhesion,* John Wiley & Sons-Interscience, New York, 1971.
161. **Kakabadse, G., Ed.,** *Solvents — The Neglected Parameter, 2nd Solvents Symp.,* University of Manchester Institute of Science and Technology, Manchester, U.K., 1977.
162. **Kamath, V. R. and Sargent, J. D., Jr.,** Production of high solids acrylic coating resins with t-amyl peroxides: a new way to meet VOC requirements, *J. Coatings Technol.,* 59(746), 51, 1987.
163. **Kambour, R. P.,** Environmental stress cracking of thermoplastics, in *Corrosion Fatigue: Chemistry, Mechanics and Microstructure,* Devereux, O. F., McEvily, A. J., and Staehle, R. W., Eds., National Association of Corrosion Engineers, Houston, TX, 1972, 681.

164. **Kambour, R. P.,** A review of crazing and fracture in thermoplastics, *Macromol. Rev.*, 7, 1, 1973.
165. **Kambour, R. P. and Gruner, C. L.,** Effects of polar group incorporation on crazing of glassy polymers: styrene-acrylonitrile copolymer and a dicyano bisphenol polycarbonate, *J. Polym. Sci. Polym. Phys.*, 16, 703, 1978 (General Electric Reprint 8346).
166. **Kambour, R. P., Gruner, C. L., and Romagosa, E. E.,** Solvent crazing of 'dry' polystyrene and 'dry' crazing of plasticized polystyrene, *J. Polym. Sci. Polym. Phys.*, 11, 1879, 1973 (General Electric Reprint 7435).
167. **Kambour, R. P., Gruner, C. L., and Romagosa, E. E.,** Bisphenol-A polycarbonate immersed in organic media. Swelling and response to stress, *Macromolecules*, 7, 248, 1974 (General Electric Reprint 7514).
168. **Kambour, R. P. and Robertson, R. E.,** The mechanical properties of plastics, in *Polymer Science*, Vol. 1, Jenkins, A. D., Ed., North-Holland, Amsterdam, 1972, chap. 11.
169. **Kambour, R. P., Romagosa, E. E., and Gruner, C. L.,** Swelling, crazing and cracking of an aromatic copolyether-sulfone in organic media, *Macromolecules*, 5, 335, 1972 (General Electric Reprint 7171).
170. **Kanebo Ltd,** Improved oil foundation compositions, *Jpn. Pat.*, 80/57510, 1980.
171. **Karger, B. L., Snyder, L. R., and Eon, C.,** An expanded solubility parameter treatment for classification and use of chromatographic solvents and adsorbents. Parameters for dispersion, dipole and hydrogen bonding interactions, *J. Chromatogr.*, 125, 71, 1976.
172. **Karger, B. L., Snyder, L. R., and Eon, C.,** Expanded solubility parameter treatment for classification and use of chromatographic solvents and adsorbents, *Analyt. Chem.*, 50, 2126, 1978.
173. **Karger, B. L., Snyder, L. R., and Horvath, C.,** *An Introduction to Separation Science*, John Wiley & Sons, New York, 1973.
174. **Karickhoff, S. W.,** Organic pollutant sorption in aquatic systems, *J. Hydraul. Eng.*, 10, 707, 1984.
175. **Kashihara, A., Otsuki, C., and Ishikawa, K.,** Process for preparing particles having monodisperse particle size, *Eur. Pat.*, 275899, 1988.
176. **Kashihara, A., Otsuka, C., and Yabuuchi, N.,** Silicon- and/or fluorine-containing polymer spherical particles, *Eur. Pat.*, 314158, 1989.
177. **Kato, M., Kamisaka, T., and Fukai, T.,** Preparation of polyamide compositions with good mechanical strength, *Jpn. Pat.*, 01069639, 1989.
178. **Kato, M., Kamisaka, T., and Fukai, T.,** *Jpn. Pat.*, 01031867, 1989.
179. **Keller, R. A., Karger, B. L., and Snyder, L. R.,** Use of the solubility parameter in predicting chromatographic retention and eluotropic strength, *Gas Chromatogr.* (Proc. Int. Symp., Eur., 1970), 8, 125, 1971.
180. **Keller, R. A. and Snyder, L. R.,** Relation between the solubility parameter and the liquid-solid solvent strength parameter, *J. Chromatogr. Sci.*, 9, 346, 1971.
181. **Kendall, K.,** Sticky solids, *Contemp. Phys.*, 21, 277, 1980.
182. **Kikuchi, M., Kubo, S., Yamaguchi, M., and Yoshikawa, M.,** Resin compositions for coating, *Jpn. Pat.*, 60/179472, 1985.
183. **King, J. W.,** Fundamentals and applications of supercritical fluid extraction in chromatographic science, *J. Chromatogr. Sci.*, 27, 355, 1989.
184. **Kirkland, D. O.,** *Parameter of Solubility*, Humble Oil and Refining Bulletin, New Orleans, 1968.
185. **Kito, O., Suzuki, S., and Higashikubo, S.,** Epoxy resin adhesive tapes for spot welding of metals, *Jpn. Pat.*, 01104681, 1989.
186. **Klein, E. and Smith, J. K.,** The use of solubilty parameters for solvent selection in asymmetric membrane formation, in *Reverse Osmosis Membrane Research*, Lonsdale, H. K. and Podall, H. E., Eds., Plenum Press, New York, 1972, 61.
187. **Klein, E. and Smith, J. K.,** Asymmetric membrane formation. Solubility parameters for solvent selection, *Ind. Eng. Chem. Prod. Res. Dev.*, 11, 207, 1972.
188. **Ko, K.-U.,** Adaptability of adhesives to substrates, surface energy, cohesive energy density, and material structure, *Nenchakuzai, Setchakuzai Sogo Semina* [Ronbunshu], 3rd *(Setchaku Riron to Jissai e no Tekiyo)*, 38, 1978.
189. **Komabahiri, T., Tsujinaka, M., and Mitani, T.,** Prevention of scale deposition in polymerization reactors by coatings containing diamine-quinone polymers and silica sols, *Eur. Pat.*, 317306, 1989.
189a. **Koningsveld, R., Kleintjens, L. A., and Nies, E.,** Polymers and thermodynamics, *Croat. Chem. Acta*, 60, 53, 1987.
190. **Korol, A. N.,** The possibility of predicting the retention parameters in gas-liquid chromatography, *Usp. Khim.*, 51, 1225, 1982; *Russ. Chem. Rev.*, 51, 698, 1982.
191. **Krause, S.,** Polymer compatibility, *J. Macromol. Sci. Rev. Macromol. Chem.*, C7, 251, 1972; Polymer-polymer compatibility, in *Polymer Blends*, Vol. 2, Paul, D. R. and Newman, S., Eds., Academic Press, New York, 1978, chaps. 2, 19.
192. **Kumar, R. and Prausnitz, J. M.,** Solvents in chemical technology, in *Solutions and Solubilities*, Part 1, Dack, M. R. J., Ed., John Wiley & Sons-Interscience, New York, 1975, chap. 5.

193. **Kurita, K., Tsuruno, T., and Teramoto, Y.,** Manufacture of ethylene terephthalate polyester fibers with high orientation and low specific gravity, *Jpn. Pat.,* 01201519, 1989.
194. **Kuwajima, T., Uenoyama, K., Takae, M., and Muramoto, H.,** Water-thinned, weather-resistant coating compositions containing metallic pigments, *Jpn. Pat.,* 63/175079, 1988.
195. **Larsen, H. O. and Kindt-Larsen, T.,** Shaped hydrogel articles, *Eur. Pat.,* 182659, 1986.
196. **Lee, J. O.,** Use of the solubility parameter concept in the estimation of polymer-solvent interactions, *Pollimo,* 12(1), 1, 1988.
197. **Lee, W. A. and Sewell, J. H.,** Influence of cohesive forces on the glass transition temperatures of polymers, *J. Appl. Polym. Sci.,* 12, 1397, 1968.
198. **Lewis, O. G.,** *Physical Constants of Linear Homopolymers* (Monographs on the Chemistry, Physics and Technology of Plastics, Vol. 2), Springer-Verlag, New York, 1968.
199. **Li, N. N. and Long, R. B.,** Plasticizing effect of permeates on membrane permeation and separation, in *Progress in Separation and Purification,* Vol. 3, Perry, E. S. and Van Oss, C. J., Eds., John Wiley & Sons-Interscience, New York, 1970, 153.
200. **Liao, S. K.,** Solubility parameters, *Fang Chih K'o Hsueh,* 48, 91, 1984.
201. **Lonsdale, H. K.,** The growth of membrane technology, *J. Membr. Sci.,* 10, 81, 1982.
202. **Lucas, M. T.,** Solubility parameters and problems in their application, *Chim. Peintures,* 34, 125, 1971.
203. **Maeda, M.,** Oil resistance of rubber. Treatment of solubility parameters, *Gosei Gomu,* 19, 21, 1978.
204. **Maeda, M.,** Oil swelling of rubber and the recent approach to solubility parameters, *Nippon Gomu Kyokaishi,* 58(3), 122, 1985.
205. **Maeda, T.,** Thermoplastic compositions resistant to environmental stress cracking, *Jpn. Pat.,* 63/191850, 1988.
206. **Maeda, T.,** Compositions of acrylic resins and rubber-modified styrene resins and their manufacture, *Jpn. Pat.,* 63/199749, 1988.
207. **Manabe, K., Sato, Y., and Muroi, R.,** Plasma treatment of propylene polymer moldings for coating, *Jpn. Pat.,* 01085172, 1989.
208. **Mandik, L.,** Quantitative characterization of cohesion forces. I. Low molecular weight organic substances, *Chem. Prum.,* 22, 243, 1972.
209. **Mandik, L. and Kaspar, F.,** *The Selection of Solvents for Film-Forming Materials,* Research Institute of Synthetic Resins and Lacquers, Pardubice, Czechoslovakia, 1972.
210. **Mandik, L. and Korinsky, J.,** Cohesion energy of nonpolar liquids, *Chem. Listy,* 61, 727, 1967.
211. **Mandik, L. and Stanek, J.,** Selection of solvents for coating systems with the aid of the solubility parameter, *Chem. Prum.,* 15, 223, 1965.
212. **Marron, M.,** Solubility parameters, *Double-Liaison — Chim. Peintures,* 28(306), 77, 1981.
213. **Marti, B.,** Solubility parameter, *Schweiz. Arch. Angew. Wiss. Tech.,* 33, 297, 1967.
214. **Martin, A., Swarbrick, J., and Cammarata, A.,** Solubility and distribution phenomena, *Physical Pharmacy,* 3rd ed., Lea & Febiger, Philadelphia, PA, 1983, chap. 12.
215. **Martin, A. and Mauger, J.,** The curious solubility of phenobarbital: how to use solubility parameters, *Am. J. Pharm. Ed.,* 52, 68, 1988.
216. **Martin, M., Moon, J. D., and Stark, F. M., Jr.,** Pressure-sensitive alkyl acrylate copolymer adhesives, *Eur. Pat.,* 303430, 1989.
217. **Mason, C. D. and Smith, W. F.,** Polyamide compositions resistant to fluorocarbon and hydrocarbon permeation, *World Pat.,* 8904348, 1989.
218. **Masutani, N., Fujimura, Y., Saka, I., and Matsumoto, T.,** Manufacture of uniform laminates from thin polyimide films, *Jpn. Pat.,* 01063071, 1989.
219. **Matson, S. L., Lee, E. K. L., Friesen, D. T., and Kelly, D. J.,** *U.S. Pat.,* 4824443, 1989.
220. **Matsuhisa, H.,** Inks for pressure-sensitive transfer sheets, *Eur. Pat.,* 306946, 1989.
221. **Meusberger, K. E.,** Computerized optimization of emulsifiers for pesticide emulsifiable concentrates, in *Advances in Pesticide Formulation Technology,* Scher, H., Ed., American Chemical Society, 1984, 121.
222. **Mellan, I.,** *Compatibility and Solubility,* Noyes, Park Ridge, NJ, 1968.
223. **Meyer zu Bexten, J. H.,** Solubility parameters and their practical application, *Farbe Lack,* 78, 813, 1972.
223a. **Mikos, A. G. and Peppas, N. A.,** Flory interaction parameter χ for hydrophilic copolymers with water, *Biomaterials,* 9, 419, 1988.
224. **Mikucki, V.,** Evaluation of modern paint systems as tank linings for chemical carriers. The significance of solubility parameters, *Polym. Paint. Colour J.,* 172, 706, 1982.
225. **Mitomo, T. and Teshirogi, T.,** A simple method for the calculation of solvent solubility parameters. I and II, *Kagaku Kyoiku,* 27, 449, 1979; 28, 70, 1980.
226. **Moiseev, E. V. and Pilyagin, M. V.,** Selection of solvents based on the solubility parameter for materials applied by spraying in an electrical field, *Lakokrasoch. Mater. Ikh. Primen.,* No. 4, 45, 1971.
227. **Moore, W. R.,** The swelling and solution of synthetic fibre-forming polar polymers in liquids, *J. Soc. Dyers Colour.,* 73, 500, 1957.
228. **Mori, K.,** Conductive coatings, *Kobunshi Kako,* 37, 373, 1988.

229. **Mori, K., Koshiishi, K., Tomosue, T., and Masuhara, K.,** Process for modifying the surface of poly(vinyl chloride) compositions, *World Pat.*, 8900172, 1989.
230. **Morris, H. C.,** Sizing hydrophobic yarns, and sizing composition, *Eur. Pat.*, 52949, 1982; *U.S. Pat.*, 4448839, 1984.
231. **Nakagawa, M.,** Adhesions between rubber and wire, *Jpn. Pat.*, 74/60382, 1974.
232. **Nakane, T., Kageyama, Y., and Hijikata, K.,** Thermoplastic resin compositions containing modifier and polysiloxane-polyacetal, *Eur. Pat.*, 319290, 1989.
233. **Nakanishi, K.,** Chemistry of liquids and solutions. VIII. Theory on regular solutions and solubility parameters, *Kagaku (Kyoto)*, 35, 647, 1980.
234. **Nakanishi, K.,** Solubility parameters, *Netsu Sokutei*, 9(2), 61, 1982.
235. **Nakanishi, K.,** Regular solution theory of Hildebrand and its application to chemical engineering thermodynamics, *Kagaku Kogaku*, 51, 768, 1987.
236. **Nasuno, T. and Sano, I.,** Oily powder make-up compositions, *Jpn. Pat.*, 80/51011, 1980.
237. **Neau, S. H.,** *On the Application of Regular Solution Theory to the Estimation of Solubility Parameters of Crystalline Organic Compounds from Solubility Data*, Thesis, University of Michigan, Ann Arbor, MI; *Diss. Abstr. Int. B*, 49, 5319, 1989.
238. **Nehashi, T., Iguchi, K., and Ishikawa, Y.,** Nonaqueous deep color dyeing agents, *Jpn. Pat.*, 63/165588, 1988.
239. **Nelson, P. D. and Ohm, R. F.,** Fluid resistance of Norsorex® polynorbornene (PNR), *Elastomerics*, 111(1), 28, 1979; Paper No. 34, Polymer and Chemicals Dept., American Cyanamid Co., Bound Brook, NJ, 1979.
240. **Nelson, R. C., Figurelli, V. F., Walsham, J. G., and Edwards, G. D.,** Solution theory and the computer. Effective tools for the coatings chemist, *J. Paint Technol.*, 42, 644, 1970.
241. **Nelson, R. C., Hemwall, R. W., and Edwards, G. D.,** Treatment of hydrogen bonding in predicting miscibility, *J. Paint Technol.*, 42, 636, 1970.
242. **Neville Chemical Co.,** *Solubility Parameters of Selected Neville Resins*, Pittsburgh, 1966.
243. **Nippon Paint Co. Ltd.,** Metallic coating materials, *Jpn. Pat.*, 56/24074, 1981.
244. **Novák, J. P., Matouš, J., and Pick, J.,** *Liquid-liquid Equilibria*, Elsevier, Amsterdam, 1987, 239.
245. **Nozaki, Z.,** Inorganic decorative panels with multilayer coatings, *Jpn. Pat.*, 01164477, 1989.
246. **Nunn, C. J.,** Solubility parameters: some practical considerations, *Chim. Peintures*, 34, 215, 1971.
247. **Oetzel, J. T.,** *Predicting Plasticizer Compatibility for Hydrin 200 by Solubility Parameter*, B. F. Goodrich Chemical Co., Avon Lakes, OH, 1967.
248. **Ohm, R. F.,** Practical aspects of curative diffusion, *Rubber World*, 192(3), 34, 1985.
249. **Oil and Colour Chemists' Association Australia,** *Surface Coatings*, Vol. 1 (Raw Materials and Their Usage), TAFE Educational Books, Randwick, Australia, 2nd ed., 1983, chap. 24.
250. **Okamoto, M., Watanabe, K., and Kubo, M.,** Method for producing nonwoven fabric with improved properties, *Jpn. Pat.*, 48/41787, 1973.
251. **Olabisi, O.,** Polyblends, in *Kirk-Othmer Encyclopedia of Chemical Technology*, Vol. 18, 3rd ed., John Wiley & Sons, New York, 1982, 455.
252. **Oota, T.,** Heat-resistant polyimide coating compositions, *Jpn. Pat.*, 01095155, 1989.
253. **Orr, A. A.,** Improved infrared technique for measurement of hydrogen bonding parameter of solvents, *J. Paint Technol.*, 47(607), 45, 1975.
254. **Orwoll, R. A.,** The polymer-solvent interaction parameter χ, *Rubber Chem. Technol.*, 50, 451, 1977.
255. **Osajima, Y.,** Ethylene-vinyl acetate films for inner linings on food containers, *Jpn. Pat.*, 01056739, 1989.
256. **Overfield, R. E.,** Extracting nickel and vanadium compounds from oil, *U.S. Pat.*, 4618413, 1986.
257. **Paul, D. R.,** Transport properties of polymers, in *Applied Polymer Science*, 2nd ed. (Am. Chem. Soc. Symp. Ser. 285), Tess, R. W. and Pohlein, G. W., Eds., American Chemical Society, Washington, D.C., 1985, 253.
258. **Peck, L. B.,** Recovery of a carbonaceous liquid with a low fines content, *U.S. Pat.*, 4572777, 1986.
259. **Pennsylvania Industrial Chemical Corp.,** *Solubility Contours of Hydrocarbon Resins*, Clairton, PA (undated).
260. **Perepechkin, L. P.,** Methods for obtaining polymeric membranes, *Usp. Khim.*, 57, 951, 1988; *Russ. Chem. Rev.*, 57, 539, 1988.
261. **Persson, S.,** Blending of mixtures of nonpolar polyolefins and polar polymers, especially rubber, *Swed. Pat.*, 458686, 1989.
262. **Poole, C. F. and Poole, S. K.,** Characterization of solvent properties of gas chromatographic liquid phases, *Chem. Rev.*, 89, 377, 1989.
263. **Prausnitz, J. M.,** *Molecular Thermodynamics of Fluid Phase Equilibria*, Prentice-Hall, Englewood Cliffs, NJ, 1969.
264. **Prausnitz, J. M., Lichtenthaler, R. N., and Gomez de Azevedo, E.,** *Molecular Thermodynamics of Fluid Phase Equilibria*, 2nd ed., Prentice-Hall, Englewood Cliffs, NJ, 1986.

265. **Price, G. J.,** Calculation of solubility parameters by inverse gas chromatography, in *Inverse Gas Chromatography Characterization of Polymers and Other Materials (Am. Chem. Soc. Symp. Ser.* 391), Lloyd, D. R., Schreiber, H. P., and Ward, T. C., Eds., American Chemical Society, Washington, D.C., 1989, 48.
266. **Pytela, O.,** Empirical approach to description of solvent effect of processes in solution: a review, *Collect. Czech. Chem. Commun.*, 53, 1333, 1988.
267. **Rains, R. C. and Bailey, R. J.,** Solvent aging studies on polyester urethane elastomers, *J. Elastomers Plast.*, 15, 124, 1983; *Proc. SPI Annu. Tech./Mark. Conf.*, 27th (Meet. Minds: Mark. Polyurethane Technol.), 1982, 181.
268. **Ramsbotham, J.,** Quantitative polymer solubility maps, *FATIPEC (Fed. Assoc. Tech. Ind. Peint. Vernis, Emaux Encres Impr. Eur. Cont.) Congr.*, 14, 567, 1978.
269. **Ramsbotham, J.,** Solvent system design for epoxy resin, *Paint Manuf. Resin News,* 49 (March), 26; (April) 18, 1979.
270. **Ramsbotham, J.,** Solvent loss from paint films, *J. Oil Colour Chem. Assoc.*, 62, 359, 1979.
271. **Ramsbotham, J.,** Solvent formation for surface coatings, *Prog. Org. Coatings,* 8, 113, 1980.
272. **Randen, N. A.,** Mosquito repellant compositions comprising an acrylic polymer, *Eur. Pat.*, 265087, 1988.
273. **Rebenfeld, L., Makarewicz, P. J., Weigmann, H.-D., and Wilkes, G. L.,** Interactions between solvents and polymers in the solid state, *J. Macromol. Sci. Rev. Macromol. Chem.*, C15, 297, 1976.
274. **Reichardt, C.,** Solvent scales and chemical reactivity, in *Organic Liquids: Structure, Dynamics, Chemical Properties,* Buckingham, A. D., Lippert, E., and Bratos, S., Eds., John Wiley & Sons, Chichester, 1978, chap. 16.
275. **Reichardt, C.,** Empirical parameters of solvent polarity as linear free-energy relationships, *Angew. Chem. Int. Ed.*, 18, 98, 1979.
276. **Reichardt, C.,** *Solvent Effects in Organic Chemistry* (Monographs in Modern Chemistry, Vol. 3), Verlag Chemie, Weinheim, 1979.
277. **Reid, R. C., Prausnitz, J. M., and Sherwood, T. K.,** The *Properties of Gases and Liquids,* McGraw-Hill, New York, 1977.
278. **Rey-Mermet, C.,** *Determination of Partial Cohesion Parameters from Binary Thermodynamic Quantities,* Doctoral thesis, School of Pharmacy, University of Lausanne, Switzerland, 1989.
279. **Reynolds, W. W. and Larson, E. C.,** The fundamentals of solvent selection, *Off. Dig. Fed. Soc. Paint Technol.*, 34, 311, 1962.
280. **Rider, P. E.,** Theories of solvency and solution, in *Applied Polymer Science,* 2nd ed. (*Am. Chem. Soc. Symp. Ser.*, Vol. 285), Tess, R. W. and Poehlein, G. W., Eds., 1985, chap. 27, 643.
281. **Riedel, G.,** Solubility of dyes in organic solvents, *Farbe Lack,* 82, 281, 1976.
282. **Robu, C.,** Solubility parameters, *Bul. Teh.-Inf. Lab. Cent. Cercet. Lacuri Cerneluri Bucuresti,* No. 2, 57, 1969.
283. **Rohm & Haas,** *Acryloid Thermoplastic Acrylic Ester Resins,* 1983.
284. **Ross, J. H.,** Demonstration of solvent differences by visible polymer swelling, *J. Chem. Educ.*, 60, 169, 1983.
285. **Saida, K. and Narahara, T.,** Unsaturated polyesters, *Jpn. Pat.*, 74/41485, 1974.
286. **Sakamoto, M. and Maruyama, T.,** Radiation-curable resin coating compositions, *Jpn. Pat.*, 52/27437, 1977.
287. **Sakamoto, Y.,** Saving of sewing labor with a low melting point polyamide copolymer, *Kobunshi Kako,* 20(11), 674, 1971.
288. **Sandholm, B.,** Evaluation of solubility parameters of polymers, *Suomen Kem. Tied.*, 79, 14, 1970.
289. **Sasaki, I., Anzai, H., Nishida, K., Makino, H., and Morimoto, M.,** Transparent and heat-resistant methacrylamide polymers, *Jpn. Pat.*, 62/41203, 1987.
290. **Sato, T.,** Solvent extraction of some metals from aqueous solutions, *J. Radioanalyt. Nucl. Chem.*, 101, 77, 1986.
291. **Sattelmeyer, R.,** The solubility parameters of synthetic resins and elastomers for the plastics sector, *Adhäsion,* 20, 278, 1976.
292. **Sausaman, D. K.,** A practical approach to solvent applications in coatings and inks, in *Solvents Theory and Practice (Adv. Chem. Ser.* 124), Tess, R. W., Ed., American Chemical Society, Washington, D.C., 1973, chap. 8.
293. **Schober, M. and Schroeder, H. L.,** Photosensitive compositions for preparing flexographic plates, *Ger. Pat.*, 374243, 1989.
294. **Schoenmakers, P. J.,** *A Systematic Approach to Mobile Phase Effects in Reversed Phase Liquid Chromatography,* Thesis, Delft Technical College, Netherlands, 1981.
295. **Schoenmakers, P. J.,** *Optimization of Chromatographic Selectivity: A Guide to Method Development* (Journal of Chromatography Library, Vol. 35), Elsevier, Amsterdam, 1986.
296. **Schoenmakers, P. J., Billiet, H. A. H., and de Galan, L.,** Solubility parameter as a tool in understanding liquid chromatography, *Chromatographia,* 15, 205, 1982 (English); 15, 397, 1982 (German).

297. **Schreiber, H. P. and Lloyd, D. R.,** Overview of inverse gas chromatography, *Am. Chem. Soc. Symp. Ser.,* 391, 1, 1989.
298. **Schröder, J.,** Considerations on the energetics of pigment dispersions containing solvents and binders, *Progr. Org. Coatings,* 12, 339, 1984.
299. **Sears, J. K. and Darby, J. R.,** *The Technology of Plasticizers,* John Wiley & Sons-Interscience, New York, 1982.
300. **Sears, J. K. and Touchette, N. W.,** Plasticizers, in *Kirk-Othmer Encyclopedia of Chemical Technology,* 3rd ed., Vol. 18, John Wiley & Sons, New York, 1982.
301. **Sears, J. K., Touchette, N. W., and Darby, J. R.,** Plasticizers, in *Applied Polymer Science,* 2nd ed., Tess, R. W. and Poehlein, G. W., Eds. (*Am. Chem. Soc. Symp. Ser.,* Vol. 285), 1985, 611.
302. **Senich, G. A.,** Migration to and from plastics, *Chemtech,* 11, 360, 1981.
303. **Sevestre, J.,** Solubility of polymers, *Double-Liaison,* No. 105, 81, 1964.
304. **Sevestre, J.,** On some correlations involving the solubility parameter "δ", *Double-Liaison,* No. 109, 67, 1964.
305. **Seymour, R. B.,** Solubility parameters for polymers and hydrocarbon solvents, *Austin Paint J. Austin Finish. Rev.,* 13(10), 18, 1968.
306. **Seymour, R. B.,** Plastics vs. solvents, *Mod. Plast.,* 48, 150, 1971.
307. **Seymour, R. B.,** *Introduction to Polymer Chemistry,* McGraw-Hill, New York, 1971, section 2.2.
308. **Seymour, R. B.,** Structure/solubility relation in polymers, *Rev. Plast. Mod.,* 34, 851, 1977.
309. **Seymour, R. B.,** Characterization of polymers by turbidimetry, *Rev. Plast. Mod.,* 33, 228, 1977.
310. **Seymour, R. B.,** Structure-solubility relationships in polymers. The importance of solubility parameters in research, *Pop. Plast.,* 23(5), 32, 1978.
311. **Seymour, R. B.,** Applications of the solubility parameter concepts in polymer science, in *Macromolecular Solutions: Solvent-Property Relationships in Polymers* (Symp. Proc., Am. Chem. Soc., New York, 1981), Seymour, R. B. and Stahl, G. A., Eds., Pergamon, New York, 1982, 10.
312. **Seymour, R. B.,** Solubility parameters: yesterday and today, *Polym. Mater. Sci. Eng.,* 51, 512, 1984.
313. **Seymour, R. B.,** Progress in predicting the solubility of polymers, *Pop. Plast.,* 29(6), 25, 1984.
314. **Seymour, R. B. and Sosa, J. M.,** Estimation of solubility parameters of nonelectrolytes, *Nature (London),* 249, 759, 1974.
315. **Seymour, R. B. and Sosa, J. M.,** Plastics from plastics, *Chemtech,* 7, 507, 1977.
316. **Shaw, M. T.,** Studies of polymer-polymer solubility using a two-dimensional solubility parameter approach, *J. Appl. Polym. Sci.,* 18, 449, 1974.
317. **Sheehan, C. J. and Bisio, A. L.,** Polymer/solvent interaction parameters, *Rubber Chem. Technol.,* 39, 149, 1966.
318. **Shell Chemical Co.,** *Solvent Notes,* Bulletin IC67-64-SN, 1967.
319. **Shell Chemicals,** *Solubility Parameters,* 2nd ed., Tech. Bull. ICS(X)/78/1, 1978.
320. **Shell Chemicals,** *Solvent System Design,* Tech. Bull. ICS(X)/78/1, 1978.
321. **Shell Chemicals,** *Solvent Power,* Tech. Bull. ICS(X)/79/2, 1979.
322. **Shen, B.-E. and Jin, S.-S.,** Theory of non-electrolyte solutions. (II), *Hua Hsueh Tung Pao,* No. 2, 119, 1980.
323. **Shinoda, K.,** in collaboration with translator Becher, P., *Principles of Solution and Solubility,* Marcel Dekker, New York, 1978; translated from *Yoeki Toyokaido,* 1966, revised 1974.
324. **Shvarts, A. G.,** Use of the swelling method for the study of properties of various rubbers, *Kauch. Rezina,* 24(4), 39, 1965; *Soviet Rubber Technol.,* 24(4), 38, 1965.
325. **Shimada, H. and Miyoshi, S.,** Preparation of low molecular weight maleic copolymers for adhesives, coatings, and dispersing agents, *Jpn. Pat.,* 01051415, 1989.
326. **Siddiqui, S. A.,** Studies of solvent dyeing. I. Preparation of disperse dyes and determination of their solubility parameters, *Text. Res. J.,* 51, 527, 1981.
327. **Siddiqui, S. A.,** Determining the solubility parameter of dyes, *Am. Dyestuff Rep.,* 72(3), 30, 1983.
328. **Siddiqui, S. A.,** Dye solubility by solubility method, *Am. Dyestuff Rep.,* 72(4), 36, 1983.
329. **Siddiqui, S. A. and Needles, H. L.,** Solubility parameters, *Text. Res. J.,* 52, 570, 1982.
330. **Skaarup, K.,** The three-dimensional solubility parameter and its use. II. Pigmented systems, *Skand. Tidskr. Färg Lack,* 14(2), 28; 14(3), 45, 1968.
331. **Skelly, J. K.,** Alternatives to water in textile processing, *J. Soc. Dye Colour.,* 91, 177, 1975.
332. **Slinn, D. S. L.,** Factors in the choice of refrigerants for air conditioning, *Refrig. Air Cond.,* 74(880), 49, 1971.
333. **Sliwka, W.,** Microencapsulation, *Angew. Chem. Int. Ed.,* 14, 538, 1975.
334. **Small, P. A.,** Some factors affecting the solubility of polymers, *J. Appl. Chem.,* 3, 71, 1953.
335. **Snyder, L. R.,** *Principles of Adsorption Chromatography: The Separation of Nonionic Organic Compounds,* Marcel Dekker, New York, 1968, chap. 8.
336. **Snyder, L. R.,** The role of the mobile phase in liquid chromatography, in *Modern Practice of Liquid Chromatography,* Kirkland, J. J., Ed., John Wiley & Sons, New York, 1971, chap. 4.

337. **Snyder, L. R.,** Classification of the solvent properties of common liquids, *J. Chromatogr.*, 92, 223, 1974.

338. **Snyder, L. R.,** Solvent selection for separation processes, in *Separation and Purification,* 3rd ed., Perry, E. S. and Weissberger, A., Eds., John Wiley & Sons-Interscience, New York, 1978, 25.

339. **Snyder, L. R.,** Classification of the solvent properties of common liquids, *J. Chromatogr. Sci.*, 16, 223, 1978.

340. **Snyder, L. R.,** Solutions to solution problems, *Chemtech,* 9, 750, 1979; 10, 188, 1980.

341. **Snyder, L. R. and Kirkland, J. J.,** *Introduction to Modern Liquid Chromatography,* John Wiley & Sons, New York, 1974.

342. **Sörensen, P.,** The solubility parameter concept in the formulation of liquid inks, *J. Oil Colour Chem. Assoc.*, 50, 226, 1967.

343. **Sörensen, P.,** The solubility parameter concept in pigment-resin-solvent systems, *Deut. Farben-Z.*, 24, 473, 1970; *PIRA Translation* No. 876(PR), Print Abstr. 963/1971, Research Association for the Paper Board, Printing and Packaging Industries, Leatherhead, Surrey, U.K., 1971.

344. **Sörensen, P.,** Pigment charges and the solubilty parameter concept, *Am. Ink Maker,* 49 (November), 30, 1971.

345. **Sörensen, P.,** Electrical charge and solubility parameter concept used for pigmented system, *Skand. Tidskr. Färg. Lack,* 17, 200, 1971.

346. **Sörensen, P.,** New aspects on the application of the solubility parameters, *Skand. Tidskr. Färg. Lack,* 20(2), 9, 1974.

347. **Sörensen, P.,** Application of the acid/base concept desribing the interaction between pigments, binders and solvents, *J. Paint Technol.*, 47(602), 31, 1975.

348. **Sörensen, P.,** New formulation technology can reduce high costs, *Pigm. Resin Technol.*, 4(7), 5, 1975.

349. **Sörensen, P.,** Cohesion parameters in formulation of coatings for optimum properties, *Färg Lack Scand.*, 34(4), 81, 1988.

350. **Somerville, G. R. and Lopez, J. A.,** Technique for reformulating solvent mixtures in epoxy resin coatings, in *Solvents Theory and Practice* (*Adv. Chem. Ser.* 124), Tess, R. W., Ed., American Chemical Society, Washington, D.C., 1973, chap. 12, 175.

351. **Sonnich Thomsen, E.,** *The Energy of Mixing of Non-Polar Liquids. Reflections upon the Hildebrand Theory,* Thesis, Royal Danish School of Pharmacy, Copenhagen, 1965.

352. **Sonnich Thomsen, E.,** The energy of mixing of non-polar liquids. Reflections upon the theories of Prigogine and Hildebrand, *Dansk Kemi,* 47(3), 35, 1966.

353. **Sonnich Thomsen, E.,** Solubility parameters, *Farmaceuten,* 29(3), 44, 1966.

354. **Sonnich Thomsen, E.,** Present state of the solubility parameter theory, *Mitt. Chem. Gesell. DDR,* 15, 150, 1969.

355. **Stanton, R. M.,** Rheological aspects of thick film technology: an investigation of the flow properties of ethyl cellulose vehicle systems, *Int. J. Hybrid Microelectronics,* 6, 419, 1983.

356. **Starkweather, H. W., Jr.,** The sorption of chemicals by perfluorocarbon polymers, *Macromolecules,* 10, 1161, 1977.

357. **Staverman, A. J.,** Cohesion forces and phase theory of binary liquid systems, *Rec. Trav. Chim. Pays-Bas,* 60, 827, 1941.

358. **Stawiszynski, A. and Zawadzki, J.,** Use of the solubility parameter δ for evaluation and selection of organic solvents, *Powloki Ochr.*, 7, 26, 1979.

359. **Sugihara, A.,** Selection of solvents for paint resins, *Kagaku To Kogyo (Osaka),* 45, 501, 1971.

360. **Suzuki, A. and Harada, K.,** Manufacture of powdered polyimides, *Jpn. Pat.*, 64/000122, 1988.

361. **Suzuki, T. and Hagiwara, M.,** A survey of regular solution parameters for the systems among the first transition metals, *Metall. Trans. A,* 8A, 371, 1977.

362. **Suzuki, Y., Takeuchi, S., Sakai, E., and Fujimaki, Y.,** Preparation of electrophotographic photoreceptors, *Jpn. Pat.*, 63/55552, 1988.

363. **Svoboda, V.,** Cohesional energy of liquids, *Chem. Listy,* 76, 389, 1982.

364. **Sweeney, C. W., Ritter, T. J., and McGinley, E. B.,** A strategy for screening physical solvents, *Chem. Eng.*, 95(9), 119, 1988.

365. **Tager, A. A. and Kolmakova, L. K.,** Solubility parameter. Methods of its estimation. A relation with a polymer solubility, *Vysokomol. Soedin. Ser. A,* 22, 483, 1980; *Polym. Sci. U.S.S.R.*, 22, 533, 1981.

366. **Takada, M.,** Three dimensional solubility parameters and their applications, *Hyomen,* 9, 177, 1971.

367. **Takenaka, S. and Imahashi, S.,** Photosensitive image-forming materials, *Jpn. Pat.*, 63/249840, 1988.

368. **Tanabe, H., Umemoto, H., and Takagawa, R.,** Oil-free polyesters for two-coat-one-bake coatings, *Jpn. Pat.*, 63/258672, 1988.

369. **Tanaka, M.,** Partition coefficient and extraction constant of non-charged species as considered from the regular solution theory, *Proc. Intl. Solvent Extraction Conf., The Hague, 1971,* Society of Chemical Industry, London, 1971, 18.

370. **Tanaka, M.,** Separation by solvent extraction. A study by regular solution theory, *Bunseki Kiki,* 14, 327, 1976.

371. **Tayama, S., Tamura, M., Nishida, K., and Morimoto, M.,** Treatment of imido ring-containing acrylic polymer moldings with solvent mixtures and acrylic compositions, *Jpn. Pat.*, 01038174, 1989.
372. **Taylor, N. W.,** in *Modern Chemistry for the Engineer and Scientist,* Robertson, G. R., Ed., McGraw-Hill, New York, 1957, 183.
373. **Teas, J. P.,** Graphic analysis of resins solubilities, *J. Paint Technol.*, 40(516), 19, 1968.
374. **Teas, J. P.,** Re solubility parameters, *J. Paint Technol.*, 40, 104, 1968.
375. **Teas, J. P.,** *Predicting Resin Solubilities,* Ashland Chemical Co., Tech. Bull. 1206, Ashland Chemical Co., Ashland, KY, 1971.
376. **Tess, R. W.,** Chemistry and technology of solvents, in *Applied Polymer Science,* Craver, J. K. and Tess, R. W., Eds., American Chemical Society, Washington, D.C., 1975, chap. 44.
377. **Tess, R. W.,** Solvents, in *Applied Polymer Science,* 2nd ed., (*Am. Chem. Soc. Symp. Ser.*, 285), Tess, R. W. and Poehlein, G. W., Eds., 1985, 661.
378. **Tlusta, H. and Zelinger, J.,** Miscibility of polymers, *Chem. Listy,* 65, 1143, 1971.
379. **Tortorello, A. J. and Kinsella, M. A.,** Solubility parameter concept in the design of polymers for high performance coatings. I. Acrylic copolymers. II. Urethane copolymers, *J. Coatings Technol.*, 55(696), 99, 1983; 55(697), 29, 1983.
380. **Tortorello, A. J., Kinsella, M. A., and Gearon, R. T.,** The solubility parameter concept in the design of polymers for high performance coatings, in *Macromolecular Solutions: Solvent-Property Relationships in Polymers* (Symp. Proc., Am. Chem. Soc., New York, 1981), Seymour, R. B. and Stahl, G. A., Eds., Pergamon, New York, 1982, 26.
381. **Toyota Motor Co. Ltd,** Crazing materials for plastic moldings, *Jpn. Pat.*, 58/91726, 1983.
382. **Tsukuda, K., Kimura, A., Amo, A., Yamazaki, S., and Koyabu, E.,** Method for transparent gelation of liquid oils, *Jpn. Pat.*, 63/223083, 1988.
383. **Uchida, M., Takahashi, J., Takagiwa, H., Shirase, A., and Akimoto, K.,** Electrostatographic toners, *Jpn. Pat.*, 61/62045, 1986.
384. **Uhara, K., Tanaka, S., and Yabe, K.,** Manufacture of porous polypropylene films, *Jpn. Pat.*, 63/221140, 1988.
385. **Unishi, T.,** Numerical evaluation of the solubility for polymers, *Nippon Kagaku Kaishi,* 8, 1342, 1981.
386. **Van Dyk, J. W., Frisch, H. L., and Wu, D. T.,** Solubility, solvency, and solubility parameters, *Ind. Eng. Chem. Prod. Res. Dev.*, 24, 473, 1985.
387. **van Krevelen, D. W.,** *Properties of Polymers: Their Estimation and Correlation with Chemical Structure,* 2nd ed., Elsevier, Amsterdam, 1976.
388. **Vaughan, C. D.,** Using solubility parameters in cosmetics formulation, *J. Soc. Cosmet. Chem.*, 36, 319, 1985.
389. **Vavruch, I.,** Miscibility of solvents; Solvent effects and solvent selection, *Chem. Labor Betr.*, 35, 385 and 536, 1984.
390. **Verneret, H.,** Solvents to be used in the paints, varnishes, adhesives, and printing inks industries, *Afinidad,* 27(274), 87, 1970.
391. **Vernon, F.,** The characterization of solute-solvent interactions in gas-liquid chromatography, *Dev. Chromatogr.*, 1, 1, 1978.
392. **Vilcu, R. and Leca, M.,** Thermodynamic characterization of polymers by chromatography with gas mobile phase, *Rev. Roum. Chim.*, 34, 387, 1989; *Polymer Thermodynamics by Gas Chromatography,* Elsevier, Amsterdam, 1990.
393. **Vincent, P. I. and Raha, S.,** Influence of hydrogen bonding on crazing and cracking of amorphous thermoplastics, *Polymer,* 13, 283, 1972.
394. **Vinther, A.,** Application of the concepts solubility parameter and pigment charge, *Chim. Peintures, Engl.*, 34, 363, 1971.
395. **Wágner, Z., Majer, V., and Svoboda, V.,** Calculation of solubility parameters and their temperature dependences for a selected group of substances, *Chem. Listy,* 77, 1150, 1983.
396. **Wallström, E., Svenngsen, I., and Hansen, C. M.,** Research Report T5-82, Scandinavian Paint and Printing Ink Research Institute, 1982.
397. **Walz, G.,** The choice of solvents for 'high solid' systems with the aid of the solubility parameter model, *Kunstharz Nachr. (Hoechst),* 33(6), 30, 1974.
398. **Walz, G. and Emrich, G.,** Solubility parameter diagrams of selected alkyd and melamine resins, *Kunstharz Nachr. (Hoechst),* 34(8), 19, 1975.
399. **Watanabe, A.,** Aspects of solubility theory, *Kagaku No Ryoiki,* 29, 15, 1975.
400. **Watanabe, A.,** Solubility theory, *Kagaku No Ryoiki,* 29, 167, 1975.
401. **Weast, R. C., Ed.,** Solubility parameters of organic compounds, in *CRC Handbook of Chemistry and Physics,* 69th ed., Boca Raton, FL, 1988.
402. **Wessling, R. A.,** The solubility of poly(vinylidene chloride), *J. Appl. Polym. Sci.*, 14, 1531, 1970.
403. **Williams, D. J.,** Receptor sheet for impact printers, *U.S. Pat.*, 4713280, 1987.
404. **Williamson, A. G.,** Solutions of nonelectrolytes, *Ann. Rev. Phys. Chem.*, 15, 63, 1964.

405. **Wilson, G. M.,** From Redlich-Kister to local composition models, group solutions, and electrolytes, *Fluid Phase Equil.*, 24, 77, 1985.

406. **Wilt, M. H.,** *Use of Cohesive Energy Density Measurements in Correlating the Swell of Elastomers in Fluids,* Monsanto Chemical Co., Tech. Bull., St. Louis, MO, April 1, 1954; July 15, 1955.

407. **Wolf, B. A.,** Solubility of polymers, *Pure Appl. Chem.*, 57, 323, 1985.

408. **Wyzgoski, M. G.,** The role of solubility in stress cracking of nylon 6,6, in *Macromolecular Solutions: Solvent-Property Relationships in Polymers* (Symp. Proc., Am. Chem. Soc., New York, 1981), Seymour, R. B. and Stahl, G. A., Eds., Pergamon, New York, 1982, 41; General Motors Research Publication 3779.

409. **Yabuta, M.,** Microparticle polymer dispersions with high refractive index, *Jpn. Pat.*, 62/250001, 1987.

410. **Yalkowsky, S. H., Flynn, G. L., and Slunick, T. G.,** Importance of chain length on physicochemical and crystalline properties of organic homologs, *J. Pharm. Sci.*, 61, 852, 1972.

411. **Yamamoto, T., Goto, S., Suzuki, N., Tomiyasu, H., and Kobayashi, Y.,** Manufacture of photosensitive lithographic plates with good coating uniformity, *Jpn. Pat.*, 01050040, 1989.

412. **Yazaki, K. and Iida, S.,** Preparation of plastic containers, *Jpn. Pat.*, 63/143943, 1988.

413. **Yerrick, K. B. and Beck, H. N.,** Solvent resistance of silicone elastomers: solvent-polymer interactions, *Rubber Chem. Technol.*, 37, 261, 1964.

414. **Yoshida, T.,** Solubility parameters, meaning and application, *Shikizai Kyokaishi,* 44, 186, 1971.

415. **Yoshida, C., Nabekura, H., and Shimizu, T.,** Chlorinated poly(vinyl chloride) compositions with good processability, *Jpn. Pat.*, 01033148, 1989.

416. **Yoshifuji, N.,** Phenolic resin molding compositions, *Jpn. Pat.*, 62/209158, 1987.

417. **Yoshihara, T., Kawase, J., and Fukuyama, Y.,** Vinyl polymer-containing hair preparation for reducing greasiness of scalp and hair, *Eur. Pat.*, 295903, 1988.

418. **Yoshioka, A.,** Compatibility of polymers with tackifiers from the viewpoint of solubility parameter and hydrogen bonding index, *Nippon Setchaku Kyokaishi,* 19, 504, 1983.

419. **Zunker, D. W.,** Unique aqueous nitrocellulose/acrylic air-dry systems for industrial finishing, *J. Coatings Technol.*, 48(616), 37, 1976.

Chapter 2

THERMODYNAMICS

Molecular materials exist as liquids and solids over certain ranges of temperature and pressure because in some circumstances the liquid and solid states are more stable than the gaseous state; that is, there are energetic advantages in having the molecules arranged in either random or ordered close-packed configurations. In these condensed phases, the strong attractive or cohesive forces existing between the molecules give rise to considerable negative potential energies relative to vapor phase molecules which have negligible potential energy arising in this way. (Ionic liquids and crystals have even stronger attractive forces arising from coulombic interactions.)

2.1 COHESIVE ENERGY

If U is defined as the molar internal energy (the molar potential energy of a material relative to the ideal vapor at the same temperature), then U has a numerically negative value for a condensed material. It follows, therefore, that the molar *cohesive energy* (the energy associated with the net attractive interactions of the material and defined as $-U$) has a positive value.

For a liquid, if it is assumed that the intramolecular properties (those associated with individual molecules) are identical in gaseous and liquid states (which is true except in the case of complex molecules: see Section 14.9) it can be seen that the molar cohesive energy can be divided into two parts:

1. The molar vaporization energy, ${}_{l}\Delta_{g}U$, required to vaporize one mole of the liquid to its saturated vapor
2. The energy, ${}_{g}\Delta_{\infty}U$, required to expand the saturated vapor to infinite volume at constant temperature; that is, the energy necessary to completely separate the molecules

As presented by Polak,[1] this can be expressed

$$-U = {}_{l}\Delta_{g}U + {}_{g}\Delta_{\infty}\mathrm{U} = {}_{l}\Delta_{g}\mathrm{u} + \int_{V={}^{g}V}^{V=\infty} (\partial U/\partial V)_{\mathrm{T}}\, \mathrm{d}V \tag{1}$$

where V is the molar volume. The molar cohesive energy $-U$ can be subdivided also according to the relationship

$$-U = {}_{l}\Delta_{g} H + {}_{g}\Delta_{\infty}H - RT + p_{s}{}^{l}V \tag{2}$$

where ${}_{l}\Delta_{g}H$ is the molar vaporization enthalpy; ${}_{g}\Delta_{\infty}H$ is the enthalpy change (increase) on isothermally expanding 1 mol of saturated vapor to zero pressure; p_s is the saturation vapor pressure at temperature T; ${}^{l}V$ is the molar volume of the liquid (the superscript l is frequently omitted if there is no chance of ambiguity); and R gas constant (8.31441 J K^{-1} mol^{-1}). At pressures below atmospheric pressure (that is, at temperatures below the normal boiling point) ${}_{g}\Delta_{\infty}H$ and $p_{s}{}^{l}V$ are usually negligible compared with ${}_{l}\Delta_{g}H$ and RT:

$$-U = {}_{l}\Delta_{g} U = {}_{l}\Delta_{g}H - RT \tag{3}$$

However, at higher pressures the other terms cannot be neglected, and in fact at the critical point ${}_{l}\Delta_{g}H$ is zero, so Equation 3 erroneously leads to a negative value for the cohesive energy while the full Equation 2 correctly predicts a small positive cohesive energy at the critical point. Values of ${}_l\Delta_g H$, ${}_l\Delta_g{}^g U$, and $-U$ for various liquids at their normal boiling points in Table 1 illustrates typical variations in these quantities.

Svoboda and co-workers[2-7] have recently considered the cohesive energies of liquids in some detail, and Table 2 summarizes their recommended 25°C values.

2.2 COHESIVE PRESSURE AND THE HILDEBRAND PARAMETER

The stabilizing or cohesive effect in condensed phases can be expressed in terms of the *cohesive pressure* which is dimensionally identical with the *cohesive energy density* (cohesive energy per unit volume),

$$c = -U/V \tag{4}$$

Cohesive energy density was the basis of the original definition by Hildebrand and Scott[8-10] of what is now generally called the Hildebrand solubility parameter or *Hildebrand parameter*,

$$\delta = c^{1/2} = (-U/V)^{1/2} \approx ({}_l\Delta_g\, U/V)^{1/2} \tag{5}$$

This parameter was intended for nonpolar, nonassociating systems, but the concept has been extended to all types of systems.

The term "solubility" parameter, which has been used widely, is really too restrictive for a quantity that may be used to correlate such a wide range of physical and chemical properties.[11] The name "cohesion parameter" is preferred by the author for the group of parameters with dimensions of (pressure)$^{1/2}$ that includes the Hildebrand parameter as defined in Equation 5. Use of the proposed alternative title "interaction parameter"[12] would result in confusion with the polymer-liquid interaction parameter χ (Chapter 13) and several other binary interaction parameters characterizing *pairs* of substances.

The title "solubility parameter" and the form of Equation 5 suggest a close link between the phenomena of "solubility" or "miscibility" and those of "cohesion" and "vaporization." This similarity can be appreciated by considering what happens in a mixing process: the "like" molecules of each component in a mixture become separated from one another by what approximates to an infinite distance, comparable in some respects to what happens in the vaporization process. The Hildebrand parameter is sometimes called the "total" cohesion parameter, δ_t, because there are various "component" cohesion parameters, but the subscript "t" is usually omitted if this can be done without ambiguity.

From Equations 3 and 5 it is clear that the Hildebrand parameter of a liquid may be readily evaluated if the molar volume and molar vaporization enthalpy have been determined at the required temperature, and if that temperature is well below the normal boiling point of the liquid:

$$\delta = ({}_l\Delta_g\, H - RT)^{1/2}/V^{1/2} \tag{6}$$

This density and enthalpy information is readily available for some liquids, but for many other liquids and for all polymers, solids, and surfaces it is necessary to use indirect evaluation methods, described in subsequent chapters, for the estimation of cohesion parameters.

Table 3 lists selected values of Hildebrand parameters, molar volumes and molar vaporization enthalpies at 25°C, reported by Hildebrand, Prausnitz, and Scott[10] and presented here in SI units (see Section 2.5). The vaporization enthalpies are corrected for expansion

TABLE 1
Calculated Values of Molar Vaporization Enthalpies, Molar Vaporization Energies and Molar Cohesive Energies for Liquids at their Normal Boiling Points

Liquid	t_b/°C	${}_l\Delta_g H$/kJ mol^{-1}	${}_l\Delta_g U$/kJ mol^{-1}	$-U$/kJ mol^{-1}
Acetone, 2-propanone	56.1	29.7	27.0	27.2
Ammonia	−33.3	23.5	21.5	21.7
Aniline	184.2	43.9	40.2	40.5
Benzene	80.1	30.8	27.9	28.2
Bromine	58.7	29.3	26.6	26.8
Bromobenzene	155.9	37.6	34.2	34.5
Bromoethane, ethyl bromide	38.4	26.8	24.3	24.5
Butane	−0.5	22.4	20.3	20.5
1-Butanol	117.5	43.6	40.5	40.8
2-Butanol	108.0	42.9	39.9	40.2
tert-Butanol	82.9	39.9	37.1	37.4
1-Butene	−6.3	22.1	20.0	20.2
cis-2-Butene	3.7	23.3	21.1	21.3
trans-2-Butene	0.9	22.8	20.6	20.8
Carbon disulfide	46.2	26.7	24.1	24.3
Chlorobenzene	131.7	35.4	32.2	32.5
Chloroethane, ethyl chloride	12.2	24.6	22.3	22.5
1-Chloropropane, propyl chloride	46.6	27.5	25.0	25.2
o-Cresol, 2-methylphenol	190.8	45.5	41.8	42.2
m-Cresol, 3-methylphenol	202.2	48.1	44.2	44.6
p-Cresol, 4-methylphenol	201.8	48.0	44.2	44.5
Cyclohexane	80.7	30.0	27.2	27.4
Cyclopentane	49.3	27.3	24.7	25.0
1,2-Dibromoethane, ethylene dibromide	131.5	36.1	32.8	33.1
Dichlorodifluoromethane	−29.8	20.1	18.1	18.3
1,1-Dichloroethane	57.3	29.0	26.3	26.6
1,2-Dichloroethane, ethylene dichloride	83.7	32.2	29.3	29.5
Dichloromethane	39.8	28.1	25.6	25.8
Diethyl ether	34.6	26.5	24.1	24.3
Dimethylamine	6.9	24.7	22.5	22.7
2,2-Dimethylbutane	49.7	26.3	23.7	24.0
2,3-Dimethylbutane	58.0	27.3	24.7	25.0
Dimethyl ether	−24.8	21.5	19.5	19.7
2,2-Dimethylpropane, neopentane	9.5	22.7	20.5	20.8
Ethanol	78.3	39.4	36.6	36.9
Ethyl acetate	77.1	32.1	29.4	29.7
Ethylbenzene	136.2	35.8	32.5	32.9
Ethyl formate	54.0	29.9	27.3	27.5
Ethyl propionate	98.9	33.9	31.0	31.3
Heptane	98.4	31.9	28.9	29.3
Hexane	68.7	28.9	26.2	26.6
1-Hexene	63.3	28.3	25.7	26.0
cis-2-Hexene	68.8	29.0	26.3	26.6
trans-2-Hexene	68.0	28.9	26.2	26.5
cis-3-Hexene	66.8	28.8	26.1	26.4
trans-3-Hexene	67.3	29.0	26.3	26.6
Hydrazine	113.1	41.5	38.4	38.5
2-Propanol	82.2	40.6	37.8	38.1
Methanol	64.5	36.4	33.7	33.9
Methyl acetate	56.9	30.4	27.7	28.0
Methylamine	−6.4	26.2	24.0	24.2
2-Methylbutane, isopentane	27.9	24.7	22.3	22.6
Methyl formate	31.8	28.5	26.0	26.2

TABLE 1 (continued)
Calculated Values of Molar Vaporization Enthalpies, Molar Vaporization Energies and Molar Cohesive Energies for Liquids at their Normal Boiling Points

Liquid	t_b/°C	${}_l\Delta_g H$/kJ mol^{-1}	${}_l\Delta_g U$/kJ mol^{-1}	$-U$/kJ mol^{-1}
2-Methylpentane	60.3	27.9	25.2	25.5
2-Methylpropane, isobutane	−11.4	21.2	19.1	19.3
2-Methylpropene, isobutylene	−6.9	22.1	20.0	20.2
Methyl propionate	79.4	32.1	29.3	29.6
Octane	125.7	34.6	31.4	31.9
Pentane	36.1	25.8	23.4	23.7
1-Pentene	29.5	25.3	22.9	23.1
cis-2-Pentene	36.5	26.4	23.9	24.1
trans-2-Pentene	36.3	26.1	23.7	24.0
Phenol	181.8	46.6	43.0	43.4
Propane	−42.1	18.8	17.0	17.1
1-Propanol	97.2	41.1	38.4	38.7
Propionic acid	141.0	42.3	39.0	39.3
Propyl acetate	101.5	34.1	31.2	31.5
Propylene, propene	−47.7	18.5	16.7	16.8
Tetrachloromethane	76.6	30.4	27.6	27.9
1,2,3,5-Tetramethylbenzene	198.0	42.5	38.8	39.2
1,2,4,5-Tetramethylbenzene	196.8	42.4	38.7	39.1
Toluene, methylbenzene	110.6	33.4	30.3	30.6
Trichloromethane, chloroform	61.7	29.1	26.4	26.7
Trimethylamine	2.9	24.1	21.9	22.1
1,2,3-Trimethylbenzene	176.1	39.9	36.3	36.7
1,2,4-Trimethylbenzene	169.4	39.3	35.8	36.2
Water	100.0	40.8	37.8	37.9
o-Xylene,1,2-dimethylbenzene	144.4	36.7	33.4	33.7
m-Xylene,1,3-dimethylbenzene	139.1	36.2	32.9	33.3
p-Xylene,1,4-dimethylbenzene	138.3	35.9	32.7	33.0

Adapted from Polak, J., *Collect. Czech. Chem. Commun.*, 31, 1483, 1966

to the ideal gas state where possible. Substances that are solids at 25°C have been treated as subcooled liquids (see Chapter 12). Similar information on cohesive energy densities was provided by Varushchenko, Loseva, and Druzhina[41] for 1,1,- 1,1- and α,ω-dichloro-*n*-alkanes, from which the Hildebrand parameters in Table 3a were evaluated.

The basis of the cohesion parameter approach to interactions may be stated as follows. A material with a high δ value requires more energy for dispersal than is gained by mixing it with a material of low cohesion parameter, so immiscibility results. On the other hand, two materials with similar δ values gain sufficient energy on mutual dispersion to permit mixing. This concept is attractive for practical applications because it aims to predict the properties of a *system* using only the properties of its individual *components*: in principle no information on the properties of the mixed system is required.

It is necessary to emphasize that the Hildebrand parameter is fundamentally a liquid state property. When gases are considered (Chapter 11) they are treated as hypothetical "liquid" solutes at atmospheric pressure, and substances that are solids at normal temperatures are treated as subcooled liquids (Chapter 12). As defined here, Hildebrand parameters cannot be calculated directly from vaporization enthalpies or sublimation enthalpies without taking into account their liquid-state basis. Lawson's list of "solubility parameters" of the elements[13,14] (Table 4) is a useful set of *cohesion* parameters, (although they are not *Hildebrand* parameters if they are calculated directly from enthalpies of vaporization or sub-

TABLE 2
Rounded Values of Cohesive Energies at 25°C

Formula	Compound	$-U$/kJ mol^{-1}
C_2H_6	Ethane	7.7
C_3H_8	Propane	13.9
C_4H_{10}	Butane	19.2
	Isobutane	17.6
C_5H_{12}	Pentane	24.3
	Isopentane	22.8
	Neopentane	19.9
C_6H_{14}	Hexane	29.3
	Branched hexanes	26—28
C_7H_{16}	Heptane	34.2
	Branched heptanes	30—33
C_8H_{18}	Octane	39.1
	Branched octanes	33—41
C_9H_{20}	Nonane	44.0
	Branched nonanes	36—40
$C_{10}H_{22}$	Decane	48.9
	Branched decanes	45—47
$C_{11}H_{24}$	Undecane	54.0
	Branched undecanes	48—52
$C_{12}H_{26}$	Dodecane	59.0
	2,2,4,6,6-Pentamethylheptane	46.5
$C_{13}H_{28}$	Tridecane	64.0
$C_{14}H_{30}$	Tetradecane	68.8
$C_{15}H_{32}$	Pentadecane	73.6
$C_{16}H_{34}$	Hexadecane	78.9
$C_{17}H_{36}$	Heptadecane	83.5
C_3H_6	Cyclopropane	15.7
C_4H_8	Cyclobutane	21.5
C_5H_8	Spiropentane	25.2
C_5H_{10}	Cyclopentane	26.3
C_6H_{12}	Cycloalkanes	29—31
C_7H_{12}	1-Methylbicyclo[3.1.0]hexane	32.4
C_7H_{14}	Cycloalkanes	32—34
C_8H_{16}	Cycloalkanes	36—39
C_9H_{16}	1,4-Dimethylbicyclo[2.2.1]heptane	36.4
C_9H_{18}	Cycloalkanes	42—43
$C_{10}H_{20}$	Cycloalkanes	45
$C_{11}H_{22}$	Pentylcyclohexane	51.4
$C_{12}H_{22}$	Cyclohexylcyclohexane	55.5
C_3H_6	Propylene	13.7
C_4H_6	1,2-Butadiene	21.2
	1,3-Butadiene	19.0
	1-Butyne	21.3
C_4H_8	1-Butene	18.4
	2-Butenes	20
C_5H_{10}	Alkenes	23—25
C_6H_{12}	2,3-Dimethyl-2-butene	30.2
C_8H_{14}	Alkynes	40—42
C_8H_{16}	1-Octene	38.0
$C_{10}H_{20}$	1-Decene	48.0
$C_{12}H_{24}$	1-Dodecene	58.3
$C_{16}H_{32}$	1-Hexadecene	77.8
C_6H_6	Benzene	31.5
C_6H_{10}	Cyclohexene	31.1
C_7H_8	Toluene	35.6
C_8H_{10}	Alkylbenzenes	40—41

TABLE 2 (continued)
Rounded Values of Cohesive Energies at 25°C

Formula	Compound	$-U$/kJ mol^{-1}
C_8H_{12}	4-Vinyl-4-cyclohexene	37.1
	cis,cis-1,5-Cyclooctadiene	40.9
C_9H_{10}	Alkylbenzenes	43—48
	cis-5-Ethylidenebicyclo[2.2.1]-2-heptene	41.7
C_9H_{14}	2,3-Dimethylbicyclo[2.2.1]-2-heptene	39.7
$C_{10}H_{14}$	Alkylbenzenes	45—49
$C_{12}H_{16}$	Cyclohexylbenzene	57.5
C_6F_6	Hexafluorobenzene	33.3
C_6HF_5	Pentafluorobenzene	33.9
$C_6H_4F_2$	Difluorobenzenes	32—34
C_6H_5F	Fluorobenzene	32.2
$C_7H_3F_5$	2,3,4,5,6-Pentafluorotoluene	38.7
$C_7H_5F_3$	(Trifluoromethyl)benzene	35.2
C_7H_7F	4-Fluorotoluene	37.0
C_8F_{18}	Octadecafluorooctane	38.7
$C_8H_{17}F$	1-Fluorooctane	47.2
C_9F_{18}	Octadecafluoropropylcyclohexane	40.6
$C_{10}F_{18}$	Octadecafluorodecahydronaphthalenes	43
$C_{10}F_{20}$	Perfluoro-2-methylpropylcyclohexane	44.3
CCl_4	Tetrachloromethane	30.1
$CHCl_3$	Trichloromethane	28.9
CH_2Cl_2	Dichloromethane	26.5
C_2Cl_4	Tetrachloroethylene	37.2
C_2HCl_3	Trichloroethylene	32.1
$C_2H_2Cl_2$	1,1-Dichloroethylene	24.3
$C_2H_2Cl_4$	1,1,2,2-Tetrachloroethane	43.3
$C_2H_3Cl_3$	1,1,1-Trichloroethane	30.1
	1,1,2-Trichloroethane	37.8
$C_2H_4Cl_2$	1,1-Dichloroethane	28.3
	1,2-Dichloroethane	32.7
$C_3H_6Cl_2$	1,3-Dichloropropane	38.3
C_3H_7Cl	1-Chloropropane	26.1
$C_4H_8Cl_2$	1,2-Dichlorobutane	37.1
	1,4-Dichlorobutane	43.9
C_4H_9Cl	1-Chlorobutane	31.2
	Isobutyl chloride	29.3
	sec-Butyl chloride, 2-chlorobutane	29.2
	tert-Butyl chloride	26.7
$C_5H_{10}Cl_2$	1,2-Dichloropentane	41.4
	1,5-Dichloropentane	48.2
$C_5H_{11}Cl$	Chloropentanes	34—36
C_6H_5Cl	Chlorobenzene	38.5
$C_6H_{12}Cl_2$	1,2-Dichlorohexane	45.7
$C_6H_{13}Cl$	1-Chlorohexane	40.4
$C_7H_{15}Cl$	1-Chloroheptane	45.2
$C_8H_{17}Cl$	1-Chlorooctane	49.9
$C_{12}H_{25}Cl$	1-Chlorododecane	68.7
$C_{16}H_{33}Cl$	1-Chlorohexadecane	89.3
$CHBr_3$	Tribromomethane	43.6
CH_2Br_2	Dibromomethane	34.6
CH_3Br	Bromomethane	20.8
$C_2H_4Br_2$	1,2-Dibromoethane	39.3
C_2H_5Br	Bromoethane	25.8
C_3H_5Br	Allyl bromide	30.4
$C_3H_6Br_2$	1,2-Dibromopropane	39.2
	1,3-Dibromopropane	45.0

TABLE 2 (continued)
Rounded Values of Cohesive Energies at 25°C

Formula	Compound	$-U$/kJ mol^{-1}
C_3H_7Br	Bromopropanes	28—30
$C_4H_8Br_2$	1,4-Dibromobutane	50.6
	1,2-Dibromo-2-methylpropane	40.9
C_4H_9Br	Alkyl bromides	30—34
$C_5H_{11}Br$	1-Bromopentane	38.8
C_6H_5Br	Bromobenzene	42.1
$C_6H_{13}r$	1-Bromohexane	43.4
$C_7H_{15}Br$	1-Bromoheptane	48.1
$C_8H_{17}Br$	1-Bromooctane	53.3
$C_{12}H_{25}Br$	1-Bromododecane	72.3
$C_{16}H_{33}Br$	1-Bromohexadecane	91.9
C_2H_5I	Iodoethane	29.6
C_3H_7I	Iodopropanes	32—34
C_4H_9I	Iodobutanes	33—38
$C_5H_{11}I$	1-Iodopentane	42.8
$C_6H_{13}I$	1-Iodohexane	47.3
$C_2Br_2ClF_3$	1,2-Dibromochlorotrifluoroethane	32.6
$C_2Br_2F_4$	1,2-Dibromotetrafluoroethane	26.1
$C_2Cl_3F_3$	Trichlorotrifluoroethanes	26
$C_2HBrClF_3$	Bromochlorotrifluoroethanes	27—28
C_2H_4BrCl	1-Bromo-2-chloroethane	35.7
$C_3Cl_2F_6$	1,2-Dichlorohexafluoropropane	24.8
$C_3H_2Cl_3F_3$	1,1,1-Trichloro-3,3,3-trifluoropropane	34.3
$C_3H_3Cl_2F_3$	1,1-Dichloro-3,3,3-trifluoropropane	31.7
C_3H_6BrCl	1-Bromo-3-chloropropane	41.6
C_6ClF_5	Chloropentafluorobenzene	38.6
CH_5N	Methylamine	21.4
C_2H_7N	Dimethylamine	23.0
$C_2H_8N_2$	1,2-Ethanediamine, ethylene diamine	42.5
C_3H_9N	Propylamine	29.0
	Isopropylamine	26.2
	Trimethylamine	19.7
$C_3H_{10}N_2$	1,3-Propanediamine	47.7
	N-Methyl-1,2-ethanediamine	42.7
$C_4H_{11}N$	Butylamines	29—33
$C_5H_{13}N$	Propylamines	31—38
C_6H_7N	Aniline	53.4
$C_6H_{13}N$	Cyclohexylamine	41.2
$C_6H_{15}N$	Hexylamines	32—43
C_7H_9N	Benzylamine	57.7
$C_7H_{17}N$	Heptylamines	40—48
$C_8H_{11}N$	*N,N*-Dimethylaniline	50.4
$C_8H_{19}N$	Octylamines	47—50
$C_9H_{21}N$	Tripropylamine	43.7
C_2N_2	Ethanedinitrile	18.4
C_2H_3N	Ethanenitrile, acetonitrile	30.9
C_3H_5N	Propanenitrile, propionotrile	33.7
C_4H_5N	Butenenitriles, allyl cyanide	37—38
	Cyclopropanecarbonitrile	39.5
C_4H_7N	Butanenitriles	35—37
C_5H_7N	Pentenenitriles	41—43
	Cyclobutanecarbonitrile	41.9
C_5H_9N	Pentanenitriles	35—41
C_6H_7N	1-Cyclopentenecarbonitrile	42.5
C_6H_9N	Cyclopentanecarbonitrile	41.0
$C_6H_{11}N$	Hexanenitrile, capronitrile	45.4

TABLE 2 (continued)
Rounded Values of Cohesive Energies at 25°C

Formula	Compound	$-U$/kJ mol^{-1}
C_7H_9N	1-Cyclohexenecarbonitrile	51.1
$C_7H_{11}N$	Cyclohexanecarbonitrile	49.4
$C_8H_{15}N$	Octanenitrile	54.3
$C_{10}H_{19}N$	Decanenitrile	64.4
$C_{11}H_{21}N$	Undecanenitrile	68.7
$C_{12}H_{23}N$	Dodecanenitrile	73.6
$C_{14}H_{27}N$	Tetradecanenitrile	82.8
$C_4H_4N_2$	Pyridazine	51.0
	Pyrimidine	47.3
	Pyrrole	42.9
C_4H_9N	Pyrrolidine	35.0
C_5H_5N	Pyridine	37.7
C_6H_7N	Methylpyridines, picoline	40—42
C_7H_9N	Dimethylpyridines	43—48
$C_8H_{11}N$	Trimethylpyridines	48
$C_8H_{16}N_2$	3,3,6,6-Tetramethyl-3,4,5,6-tetrahydropyridazine	47.6
CH_6N_2	Methylhydrazine	38.0
$C_2H_8N_2$	Dimethylhydrazines	33—37
$C_5H_{12}N_2$	1-(Methylazo)butane	33.9
$C_6H_{14}N_2$	Azopropane	37.4
	2-(Isopropylazo)propane	33.4
$C_8H_{18}N_2$	Azobutane	46.8
	2-(*tert*-Butylazo)-2-methylpropane	36.6
$C_{12}H_{26}N_2$	2-(*tert*-Butylazo)-2,4,4-trimethylpentane	51.1
$C_{16}H_{34}N_2$	2,2′-Azo-(2,4,4-trimethylpentane)	64.1
C_2H_6O	Dimethyl ether	16.8
$C_4H_{10}O$	Ethers	25
$C_4H_{10}O_2$	1,2-Dimethoxyethane	34.0
$C_5H_{12}O$	Butyl methyl ethers	28—30
	Ethyl propyl ethers	28—29
$C_5H_{12}O_2$	1-Ethoxy-2-methoxyethane	37.4
	Diethoxymethane	33.3
$C_6H_{12}O$	Butyl vinyl ethers	32—34
$C_6H_{14}O$	Methyl pentyl ethers	33—34
	Butyl ethyl ethers	31—34
	Dipropyl ethers	30—33
$C_6H_{14}O_2$	2-Methoxy-1-propoxyethane	41.2
	Diethoxyethane	40.8
$C_6H_{14}O_3$	Bis(ethoxymethyl)ether	42.2
C_7H_8O	Methyl phenyl ether	44.4
$C_7H_{16}O$	Ethers	32—40
$C_7H_{16}O_2$	Diethers	43—45
$C_7H_{16}O_4$	3,5,7,9-Tetraoxoundecane	51.2
$C_8H_{10}O$	Ethyl phenyl ether	48.6
$C_8H_{18}O$	Ethers	35—44
$C_8H_{18}O_2$	Diethers	48—49
$C_8H_{18}O_3$	Diethylene glycol diethyl ether	55.9
$C_9H_{20}O$	Ethers	42—50
$C_9H_{20}O_2$	1-Butoxy-2-propoxyethane	52.2
$C_{10}H_{22}O$	Ethers	48—55
$C_{10}H_{22}O_2$	1,2-Dibutoxyethane	56.3
$C_{11}H_{24}O$	Ethers	58—60
$C_{12}H_{26}O$	Ethers	62—63
CH_4O	Methanol	35.4
C_2H_6O	Ethanol	40.0
C_3H_8O	1-Propanol	45.0
	2-Propanol	43.0

TABLE 2 (continued)
Rounded Values of Cohesive Energies at 25°C

Formula	Compound	$-U$/kJ mol^{-1}
$C_4H_{10}O$	1-Butanol	49.9
	Other butanols	44—48
$C_5H_{10}O$	Cyclopentanol	55.2
$C_5H_{12}O$	1-Pentanol	54.6
	Other pentanols	48—53
$C_6H_{12}O$	Cyclohexanol	59.5
$C_6H_{14}O$	1-Hexanol	59.1
	Other hexanols	52—58
$C_7H_{16}O$	1-Heptanol	64.3
$C_8H_{10}O$	Dimethylphenols	62—80
$C_8H_{18}O$	1-Octanol	68.5
$C_9H_{20}O$	1-Nonanol	74.4
$C_{10}H_{22}O$	1-Decanol	79.0
$C_{11}H_{24}O$	1-Dodecanol	89.5
$C_{14}H_{30}O$	1-Tetradecanol	99.7
C_2H_4O	Ethanal, acetaldehyde	23.7
C_3H_6O	Propanal	27.5
	2-Propanone, acetone	28.8
C_4H_8O	2-Butanone, methyl ethyl ketone	32.4
C_5H_8O	Cyclopentanone	40.3
	Cyclopropyl methyl ketone	37.0
$C_5H_8O_2$	2,4-Pentanedione	40
$C_5H_{10}O$	Pentanones	34—36
$C_6H_{10}O$	Cyclohexanone	43.4
$C_6H_{12}O$	Hexanones	36—41
$C_7H_{10}O$	Dicyclopropyl ketone	51.2
$C_7H_{14}O$	Heptanones	39—45
$C_8H_{16}O$	2,2,4-Trimethyl-3-pentanone	40.8
$C_9H_{14}O$	2-Hexahydroindanone	54—55
$C_9H_{16}O$	Dimethyl-3,5-heptanediones	53.6
$C_9H_{18}O$	Nonanones	43—54
$C_{10}H_{16}O$	*trans*-8-Methyl-2-hexahydroindanone	55.8
$C_{10}H_{18}O_2$	2,2,6-Trimethyl-3,5-heptanedione	55.3
$C_{10}H_{20}O$	2,2,5,5-Tetramethyl-3-hexanone	46.3
$C_{11}H_{20}O_2$	2,2,6,6-Tetramethyl-3,5-heptanedione	57.1
$C_{11}H_{22}O$	Undecanones	50—65
$C_{12}H_{24}O$	2-Dodecanone	69.4
CH_2O_2	Formic acid	43.8
$C_2H_4O_2$	Acetic acid	49.1
$C_3H_6O_2$	Propionic acid	52.5
$C_4H_8O_2$	Butyric acids	51—56
$C_2H_4O_2$	Methyl formate	26
$C_3H_6O_2$	Esters	30
$C_4H_8O_2$	Esters	33—35
$C_5H_8O_2$	Methyl cyclopropanecarboxylate	38.8
$C_5H_{10}O_2$	Esters	35—37
$C_6H_{10}O_2$	Methyl cyclobutanecarboxylate	42.3
	Ethylene glycol diacetate	59.0
$C_6H_{12}O_2$	Esters	36—41
$C_7H_{14}O_2$	Esters	39—46
$C_8H_8O_2$	Methyl benzoate	53.1
$C_8H_{14}O_4$	Ethylene glycol dipropanoate	65.1
$C_8H_{16}O_2$	Esters	49
$C_9H_{14}O_6$	Triacetin	83.3
$C_9H_{18}O_2$	Methyl octanoate	53.9
$C_{10}H_{18}O_4$	Ethylene glycol dibutanoate	70.7

TABLE 2 (continued)
Rounded Values of Cohesive Energies at 25°C

Formula	Compound	$-U$/kJ mol^{-1}
$C_{10}H_{20}O_2$	Methyl nonanoate	59.5
$C_{11}H_{22}O_2$	Methyl decanoate	64.3
$C_{12}H_{20}O_6$	Tripropionin	88.9
$C_{12}H_{24}O_2$	Methyl undecanoate	69.0
$C_{13}H_{26}O_2$	Methyl dodecanoate	74.7
$C_{14}H_{28}O_2$	Methyl tridecanoate	80.2
$C_{15}H_{26}O_6$	Glcyerol tributyrate	104.6
$C_{15}H_{30}O_2$	Methyl tetradecanoate	84.5
$C_{16}H_{32}O_2$	Methyl pentadecanoate	91.0
C_2H_4O	Oxirane, ethylene oxide	23.0
$C_3H_4O_2$	β-Propiolactone	44.6
C_3H_6O	Methyloxirane, propylene oxide	25.8
	Oxetane, trimethylene oxide	27.7
C_4H_4O	Furan	25.2
$C_4H_4O_2$	Diketene	40.4
C_4H_8O	Tetrahydropyran	29.7
$C_4H_8O_2$	1,3-Dioxane	36.7
	1,4-Dioxane	36.2
$C_5H_{10}O$	3,3-Dimethyloxetane	31.6
	Tetrahydropyran	32.2
$C_5H_{10}O_3$	1,3,6-Trioxacyclooctane	46.3
$C_8H_{16}O_4$	1,4,7,10-Tetraoxacyclododecane	63.2
$C_{10}H_{20}O_5$	1,4,7,10,13-Pentaoxacyclopentadecane	77.1
$C_3H_8O_2$	2-Methoxyethanol, methyl cellosolve®	42.7
$C_4H_{10}O_2$	2-Ethoxyethanol, cellosolve	45.8
$C_5H_{10}O_3$	Diethyl carbonate	41.1
	2-Methoxyethyl acetate methyl cellosolve acetate	47.8
$C_5H_{12}O_2$	Propoxyethanols	48—50
$C_6H_{12}O_3$	2-Ethoxyethyl acetate cellosolve acetate	50.2
	Ethoxymethyl propanoate	47.4
$C_6H_{14}O_2$	2-Butoxyethanol, butyl cellosolve	54.1
$C_7H_{14}O_3$	2-Propoxyethyl acetate	53.1
$C_8H_{16}O_3$	2-Butoxyethyl acetate, butyl cellosolve acetate	57.1
CS_2	Carbon disulfide	25.2
C_2H_6S	Dimethyl sulfide	26.5
$C_2H_6S_2$	Dimethyl disulfide	35.4
C_3H_8S	Ethyl methyl sulfide	29.5
$C_4H_{10}S$	Methyl propyl sulfide	33.8
	Isopropyl methyl sulfide	31.8
	Diethyl sulfide	33.4
$C_4H_{10}S_2$	Diethyl disufide	42.7
$C_5H_{12}S$	Dialkyl sulfides	33—38
$C_5H_{12}S_2$	Bis(ethylthio)methane	48.3
$C_6H_{14}S$	Dialkyl sulfides	37—43
$C_6H_{14}S_2$	1,2-Bis(ethylthio)ethane	57.0
	Dipropyl disulfide	56.7
$C_8H_{18}S$	Dialkyl sulfides	41—50
$C_8H_{18}S_2$	Dialkyl disulfides	52—60
C_2H_6S	Ethanethiol, ethyl mercaptan	25.1
$C_2H_6S_2$	1,2-Ethanedithiol	42.2
C_3H_8S	Propanethiols	27—30
C_3H_8S	1,3-Propanedithiol	47.2
$C_4H_{10}S$	1,4-Butanedithiol	29—34
$C_4H_{10}S_2$	Cyclopentanethiol	39.0
$C_5H_{12}S$	Pentanethiols	33—39
$C_5H_{12}S_2$	1,5-Pentanedithiol	56.8

TABLE 2 (continued)
Rounded Values of Cohesive Energies at 25°C

Formula	Compound	$-U$/kJ mol^{-1}
C_6H_6S	Benzenethiol	45.1
$C_6H_{12}S$	Cyclohexanethiol	42.1
$C_{10}H_{22}S$	1-Decanethiol	63.0
C_3H_6S	Thiacyclobutane	33.5
C_4H_4S	Thiophene	32.3
C_4H_6S	Dihydrothiophenes	35—38
C_4H_8S	Thiacyclopentane	37.0
C_5H_6S	Methylthiophenes	36—37
$C_5H_{10}S$	Thiacyclohexane	40.1
CN_4O_8	Tetranitromethane	47.5
CH_3NO	Formamide	57.7
$CH_3\ NO_2$	Nitromethane	35.9
C_2H_5NO	*N*-Methylformamide	53.7
C_3H_7NO	*N*-Ethylformamide	56.0
	N,N-Dimethylformamide	44.4
C_4H_9NO	*N*-Ethylacetamide	62.4
	N,N-Dimethylacetamide	47.8
	N-Methylpropionamide	62.4
	N,N-Diethylformamide	47.8
$C_5H_{11}NO$	*N*-Propylacetamide	67.3
	N-Isopropylacetamide	63.9
	N-Methylisobutyramide	64.6
$C_6H_5NO_2$	Nitrobenzene	52.5
$C_6H_9NO_3$	Triacetamide	57.9
$C_6H_{13}NO$	*N*-Butylacetamide	72.5
	N,N-Diethylacetamide	51.6
$C_6H_{14}\ N_2O$	*N*-Nitrosodipropylamine	49.2
$C_8H_8N_2O$	*N*-Nitroso-di-*tert*-buylamine	43.5
$C_{10}H_{11}NO_2$	*N,N*-Diacetylaniline	68.1
C_4H_8OS	Ethyl thiolethanoate	37.5
$C_5H_{10}OS$	Propyl thiolethanoate	41.6
	1-Methylethyl thiolethanoate	39.8
$C_6H_{10}OS$	Butyl thiolethanoate	45.6
	1,1-Dimethylethyl thiolethanoate	40.5
C_4H_5NS	4-Methylthiazole	41.4
—	Halogen-substituted ethers	30—52
—	Halogen-substituted esters	44—49
—	Halogen-substituted diones	28—35

Selected and adapted from Majer, V. and Svoboda, V., *Enthalpy of Vaporization of Organic Compounds*, IUPAC Chemical Data Series No. 32, Blackwell, Oxford, 1985.

limation of solids) and thermodynamics of liquid metal solutions can be discussed in terms of cohesion parameters.[40,42] Alternative definitions of cohesion parameters for compressed gases are described in Section 11.3.

2.3 THERMODYNAMIC EQUATION OF STATE

Any expression linking the state properties of a material is known as an "equation of state". The most fundamental equation of state is the *thermodynamic equation of state*, which follows from basic thermodynamic relationships and involves pressure p, molar volume v, absolute temperature T, and the molar internal energy U:

TABLE 3
Selected Values of Hildebrand Parameters at 25°C

Formula	Substance	$V/cm^3\ mol^{-1}$	${}_l\Delta_g H$ or ${}_l\Delta_\infty H$ kJ mol^{-1}	$\delta/MPa^{1/2}$
	Elements			
Br_2	Bromine	51	30.7	23.5
I_2	Iodine	59	—	28.8
S_8	Sulfur	135	—	25.4
P_4	Phosphorus	70	52.7	26.8
	Tetrahalides			
CCl_4	Tetrachloromethane, carbon tetrachloride	97	32.8	17.6
$SiCl_4$	Silicon tetrachloride, tetrachlorosilane	115	30.1	15.5
$SiBr_4$	Silicon tetrabromide, tetrabromosilane	127	43.4	18.0
$GeCl_4$	Germanium tetrachloride	115	33.8	16.6
$SnCl_4$	Stannic chloride	118	40.0	17.8
SnI_4	Stannic iodide	151	82.1	23.9
	Other Inorganic compounds			
OsO_4	Osmium tetroxide	58	41.0	25.8
MoF_6	Molybdenum hexafluoride	84	26.6	17.0
WF_6	Tungsten hexafluoride	88	26.2	16.4
UF_6	Uranium hexafluoride	96	30	18.2
$Si(CH_3)$	Tetramethylsilane, silicon tetramethyl	136	24.3	12.7
	Aliphatic Hydrocarbons			
C_5H_{12}	Pentane	116	26.8	14.5
	Isopentane, 2-methylbutane	117	25.2	13.9
	Neopentane, 2,2-dimethylpropane	122	22.4	12.7
C_6H_{14}	Hexane	132	31.7	14.9
C_7H_{16}	Heptane	148	36.6	15.1
C_8H_{18}	Octane	164	41.5	15.3
	Isooctane, 2,2,4-trimethylpentane	166	35.1	14.1
$C_{16}H_{34}$	Hexadecane	294	81.1	16.4
C_5H_{10}	Cyclopentane	95	28.7	16.6
C_6H_{12}	Cyclohexane	109	33.1	16.8
C_7H_{14}	Methylcyclohexane	128	35.4	16.0
C_6H_{12}	1-Hexene	126	30.7	14.9
C_8H_{16}	1-Octene	158	40.6	15.5
C_6H_{10}	1,5-Hexadiene	118	31.8	15.8
	Aromatic Hydrocarbons			
C_6H_6	Benzene	89	33.9	18.8
C_7H_8	Toluene, methylbenzene	107	38.0	18.2
C_8H_{10}	Ethylbenzene	123	42.3	18.0
	o-Xylene, 1,2-dimethylbenzene	121	43.4	18.4
	m-Xylene, 1,3-dimethylbenzene	123	42.7	18.0
	p-Xylene, 1,4-dimethylbenzene	124	42.4	18.0
C_9H_{12}	Propylbenzene	140	46.2	17.6
	Mesitylene, 1,3,5-trimethylbenzene	140	47.5	18.0
C_8H_8	Styrene, ethenylbenzene	116	43.9	19.0
$C_{10}H_8$	Naphthalene	123	—	20.3
$C_{14}H_{10}$	Anthracene	(150)	—	20.3
	Phenanthrene	158	—	20.0

TABLE 3 (continued)
Selected Values of Hildebrand Parameters at 25°C

Formula	Substance	$V/cm^3\ mol^{-1}$	${}_l\Delta_g H$ or ${}_l\Delta_\infty H$ kJ mol^{-1}	$\delta/MPa^{1/2}$
	Fluorocarbons			
C_6F_{14}	Perfluorohexane	205	32.4	12.1
C_7F_{16}	Perfluoroheptane	226	36.4	12.3
C_6F_{12}	Perfluorocyclohexane	170	28.9	12.5
C_7F_{14}	Perfluoro(methylcyclohexane)	196	33.1	12.5
	Other Fluorochemicals			
$(C_4F_9)_3N$	Perfluorotributylamine	360	54.4	12.7
$C_4Cl_2F_6$	Dichlorohexafluorocyclobutane	142	—	14.5
$C_4Cl_3F_7$	2,2,3-Trichloroheptafluorobutane	165	35.6	14.1
$C_2Cl_3F_3$	1,1,2-Trichloro-1,2,2,-trifluoroethane	120	27.5	14.5
$C_7F_{15}H$	Pentadecafluoroheptane	215	37.7	12.9
	Other Aliphatic Halogen Compounds			
CH_2Cl_2	Dichloromethane, methylene dichloride	64	28.6	20.0
$CHCl_3$	Trichloromethane, chloroform	81	31.0	18.8
CCl_4	Tetrachloromethane, carbon tetrachlorie	97	32.8	17.6
$CHBr_3$	Tribromomethane, bromoform	88	43.1	21.5
CH_3I	Iodomethane, methyl iodide	63	28.0	20.3
CH_2I_2	Diiodomethane, methylene diiodide	81	—	24.1
C_2H_5Cl	Chloroethane, ethyl chloride	74	23.8	17.0
C_2H_5Br	Bromoethane, ethyl bromide	75	27.2	18.2
C_2H_5I	Iodethane, ethyl iodide	81	32.2	19.2
$C_2H_4Cl_2$	1,2-Dichloroethane, ethylene dichloride	79	34.7	20.3
	1,1-Dichloroethane, ethylidene dichloride	85	32.2	18.6
$C_2H_4Br_2$	1,2-Dibromoethane, ethylene dibromide	90	41.4	20.9
$C_2H_3Cl_3$	1,1,1-Trichloroethane	100	32.6	17.4
$C_2H_2Cl_2$	*cis*-1,2-Dichloroethylene	76	28.9	18.6
	trans-1,2-Dichloroethylene	78	28.5	18.4
C_2Cl_4	Tetrachloroethylene	103	39.7	19.0
	Other Aliphatic Compounds			
CS_2	Carbon disulfide	61	28.0	20.5
$C_3H_8O_2$	Dimethoxymethane, methylal	89	27.6	16.8
$C_4H_{10}O$	Diethyl ether	105	26.6	15.1

Adapted from Hildebrand, J. H., Prausnitz, J. M., and Scott, R. L., *Regular and Related Solutions*, van Nostrand-Reinhold, Princeton, NJ, 1970.

$$(\partial U/\partial V)_T = T(\partial p/\partial T)_V - p$$

Many liquids have values of $(\partial p/\partial T)V$ and $(\partial U/\partial V)T$ which are functions only of the molar volume within experimental precision over reasonably wide temperature and pressure ranges.[15,16] Because they show this simple behavior, these functions have been given special names and symbols. The *internal pressure* is

$$\pi \equiv (\partial U/\partial V)_T \qquad (8a)$$

and the isochoric (constant volume) *thermal pressure coefficient* is

TABLE 3a
Molar Volumes and Hildebrand Parameters of Dichloro-Substituted *n*-Alkanes at 25°C

	$V/\text{cm}^3\ \text{mol}^{-1}$	$\delta/\text{MPa}^{1/2}$
1,1-Dichloroethane	84.8	18.3
1,1-Dichloropropane	100.5	18.0
1,1-Dichlorobutane	116.9	17.8
1,1-Dichloropentane	133.8	17.9
1,1-Dichlorohexane	150.7	17.8
1,1-Dichloroheptane	167.3	17.8
1,1-Dichlorooctane	183.6	17.8
1,1-Dichlorononane	199.2	17.7
1,1-Dichloroundecane	233.2	17.6
1,2-Dichloroethane	79.4	20.3
1,2-Dichloropropane	98.2	18.5
1,2-Dichlorobutane	114.3	18.0
1,2-Dichloropentane	131.2	17.8
1,2-Dichlorohexane	148.1	17.6
1,2-Dichloroheptane	164.4	17.8
1,2-Dichlorooctane	181.2	17.8
1,2-Dichlorononane	197.9	17.7
1,2-Dichloroundecane	232.0	17.6
1,3-Dichloropropane	95.8	20.0
1,4-Dichlorobutane	112.1	19.8
1,5-Dichloropentane	128.7	19.4
1,6-Dichlorohexane	145.9	19.7
1,7-Dichloroheptane	162.6	19.6
1,8-Dichlorooctane	179.5	19.6
1,10-Dichlorodecane	213.2	19.4
2,2-Dichloropropane	104.2	16.9

$$\beta \equiv (\partial p/\partial T)_V \tag{9a}$$

Thus, Equation 7 can be written

$$\pi = t\beta - p \tag{8b}$$

The thermal pressure coefficient β is related to the isothermal compressibility κ and the thermal expansion coefficient α by

$$\beta = \alpha/\kappa \tag{9b}$$

and from Equation 7, neglecting p, which is usually much smaller than the other term,

$$\pi = T\beta = T\alpha/\kappa \tag{10}$$

The internal pressure results from the forces of attraction between molecules in condensed phases exceeding the forces of repulsion, and although there is obviously a close connection between internal pressure and cohesive pressure, they are not equivalent: the cohesive pressure c is a measure of the total molecular cohesion per unit volume (an integral quantity), while the internal pressure π is the instantaneous isothermal volume derivative of the internal energy (a differential quantity). The expression

$$\pi = nc \tag{11}$$

TABLE 4
Cohesion Parameters and Atomic Volumes of the Elements

Element	Atomic number	Atomic volume $cm^3\ mol^{-1}$	$[(\Delta H - RT)/V]^{1/2}$ $MPa^{1/2}$	T/K
Actinium	39	22.6	139	298
Aluminum	13	10	180	298
Antimony	51	18.2	119	298
Argon	18	23.9	16	28
Arsenic	33	13.1	149	28
Astatine	85	[a]	[a]	—
Barium	56	39.2	57	298
Beryllium	4	4.4	258	298
Bismuth	83	21.3	99	298
Boron	5	4.7	346	298
Bromine	35	(25.5)	66	298
Cadmium	48	13.0	93	298
Calcium	20	26.0	83	298
Carbon	6	5.3	366	298
Cerium	58	20.7	141	298
Cesium	55	70.1	34	298
Chlorine	17	(19.3)	79	298
Chromium	24	7.2	235	298
Cobalt	27	6.7	253	298
Copper	29	7.1	219	298
Dysprosium	66	19.0	121	298
Erbium	68	13.3	127	298
Europium	63	29.0	78	298
Fluorine	9	(10.3)	88	298
Francium	87	73.0	32	298
Gadolinium	64	20.0	131	298
Gallium	31	11.8	152	298
Germanium	32	13.6	166	298
Gold	79	10.2	190	298
Hafnium	72	13.6	227	298
Helium	2	19.5	2.2	1
Holmium	67	18.8	125	298
Hydrogen	1	(6.7)	254	298
Indium	49	15.7	124	298
Iodine	53	(25.7)	64	298
Iridium	77	8.5	279	298
Iron	26	7.1	243	298
Krypton	36	32.0	17	121
Lanthanum	57	22.4	138	298
Lead	82	18.3	104	298
Lithium	3	13.0	112	298
Lutetium	71	17.8	153	298
Magnesium	12	14.0	103	298
Manganese	25	7.4	196	298
Mercury	80	14.8	64	298
Molybdenum	42	9.4	265	298
Neodymium	60	20.6	124	298
Neon	10	16.8	11	24
Nickel	28	6.6	255	298
Niobium	41	10.8	260	298
Nitrogen	7	(11.4)	203	298
Osmium	76	8.4	305	298
Oxygen	8	(8.5)	171	298
Palladium	46	8.9	206	298
Phosphorus	15	16.9	139	298

TABLE 4 (continued)
Cohesion Parameters and Atomic Volumes of the Elements

Element	Atomic number	Atomic volume $cm^3\ mol^{-1}$	$[(\Delta H - RT)/V]^{1/2}$ $MPa^{1/2}$	T/K
Platinum	78	9.1	249	298
Polonium	84	22.6	80	298
Potassium	19	45.5	45	298
Praseodymium	59	20.8	131	298
Promethium	61	20.3	115	298
Protoactinium	91	15.0	192	298
Radium	88	38.8	67	298
Radon	86	50.5	60	208
Rhenium	75	8.9	297	298
Rhodium	45	8.3	259	298
Rubidium	37	55.9	38	298
Ruthenium	44	8.3	279	298
Samarium	62	20.1	102	298
Scandium	21	15.1	150	298
Selenium	34	16.5	112	298
Silicon	14	12.1	194	298
Silver	47	10.3	167	298
Sodium	11	23.7	68	298
Strontium	38	33.7	70	298
Sulfur	16	15.5	134	298
Tantalum	73	10.9	267	298
Technetium	43	8.6	273	298
Tellurium	52	20.5	98	298
Terbium	64	19.3	140	298
Thallium	81	17.3	102	298
Thorium	90	19.9	170	298
Thulium	69	18.2	116	298
Tin	50	16.3	136	298
Titanium	22	10.6	211	298
Tungsten	74	9.5	296	298
Uranium	92	12.5	205	298
Vanadium	23	8.4	248	298
Xenon	54	36.8	19	166
Ytterbium	70	24.9	82	298
Yttrium	39	19.9	144	298
Zinc	30	9.2	120	298
Zirconium	40	14.1	208	298

[a] No data available.

Adapted from Lawson, D. D., *Proc., DOE Chemical/Hydrogen Energy Contractor Review Systems,* CONF-771131, National Technical Information Service, Springfield, VA, 1978, 109. See also Tables 24 and 25, Chapters 5, for liquid metals.

with n an empirical parameter can be used (see next section). Although n approaches a value of unity for nonpolar liquids, it can be considerably less than or greater than unity for other liquids, as shown in Table 12, Chapter 7.

The internal pressure π is in some ways a more satisfactory quantity than the cohesive pressure c or the Hildebrand parameter ($\delta = c^{1/2}$) to describe the macroscopic resultant of molecular interactions. This is because π is defined thermodynamically and because it may be evaluated directly and unambiguously in some situations. The effects on reaction kinetics and other properties of change in solvent internal pressure parallel those resulting from external pressure variation, and are experimentally simpler to achieve.[17] The relationship between π and c is considered in more detail in Section 7.5, where the internal pressure is shown to provide one method of estimating the Hildebrand parameter values for simple liquids, and component cohesion parameters for more complex liquids. Sharma and Das[18-21] and Wilson[22,23] are among those who have considered various aspects of equations of state, cohesion paramcters, and internal pressure.

2.4 EMPIRICAL EQUATIONS OF STATE

In the course of numerous attempts to analyze and rationalize the nature of intermolecular forces, many empirical equations of state have been proposed. The most used equation, of course, is the ideal gas law,

$$pV = RT \tag{12}$$

but the next best known is the van der Waals equation of state[24-26]

$$(p + a/V^2)\,(V - b) = RT \tag{13}$$

The thermodynamic equation of state (Equation 7) and the van der Waals equation can be rewritten as follows to yield a direct comparison:

$$p + (\partial U/\partial V)_T = T(\partial p/\partial T)_V \tag{14}$$

$$p + a/V^2 = TR/\,(V - b) \tag{15}$$

It is then possible to define a *van der Waals liquid* as one that obeys the equation

$$(\partial U/\partial V)_T = a/V^2 \tag{16}$$

Integrating,

$$U = -\,a/V^2 \tag{17}$$

If $U = -\,{}_l\Delta_g{}^gU$ (Equation 3), then

$$a = V_l\Delta_g\,U = V^2\,c = V^2\,\delta^2 \tag{18}$$

and

$$(\partial U/\partial V)_T = {}_l\Delta_g U/V = c = \delta^2 \tag{19}$$

This very simple relationship between the Hildebrand parameter and van der Waals a parameter has been shown approximately true for some liquids.[8-10,27-29] More generally,[30-32]

$$U = -\,a/V^n \tag{20}$$

which, with the assumption that the vapor is ideal, leads to

$$(\partial U/\partial V)_T = n({}_l\Delta_g U/V) \tag{21}$$

so the deviation of n from unity (Table 12, Chapter 7) can be used as a measure of the extend of deviation of a real liquid from a van der Waals liquid.

McGowan[33] classified three main types of property (or combination of properties) for liquids:

- Those which are constant for all unassociated liquids, such as the characteristic pressure, p_x (4455 MPa)
- Those (most properties) which vary with temperature
- Those which vary between compounds but which do not depend on temperature, such as the parachor (Section 6.3) and also the van der Waals a and b parameters

The estimation of both the parameters a and b and cohesion parameters from characteristic atomic volumes and characteristic pressure was discussed.

Equations of state are considered further in Chapter 7.

2.5 UNITS AND CONVERSION FACTORS

The internal pressure is expressed in pressure units (preferably in the SI unit, the pascal, 1 Pa ≡ 1 N m^{-2}), but in the past the cohesive pressure or cohesive energy density has been given units of energy per unit volume, often cal cm^{-3}. As internal pressure and cohesive pressure are dimensionally identical, it is logical to use a common unit. Also, although the units cal cm^{-3} for cohesive pressure and $cal^{1/2}$ $cm^{-3/2}$ for solubility parameters are still used widely, eventual conversion to the SI units is inevitable, and these units are used throughout this book. It is not now appropriate to honor the founder of the solubility parameter concept by adopting the "hildebrand" as the title of the non-SI unit $cal^{1/2}$ cm,$^{-3/2}$ as originally suggested by Taylor,[34] and a more permanent form of recognition is desirable. This is achieved by calling the original thermodynamic or "total" cohesion parameter, as defined in Equation 5 or 6, the *Hildebrand parameter*.

From many points of view, the most appropriate and convenient unit for cohesion parameters is $MPa^{1/2}$, which is numerically identical with $J^{1/2}$ $cm^{-3/2}$ and with $MJ^{1/2}$ $m^{-3/2}$. This conforms to the SI conventions, is of convenient numerical size (1 $cal^{1/2}$ $cm^{-3/2}$ being approximately 2 $MPa^{1/2}$), and can be written in compact form. Tables 5 and 6 list some conversion factors for pressure units and for cohesion parameter units.

In Chapter 6 the quantity $(-UV)^{1/2}$ is introduced, with dimensions

$$(\text{energy})^{1/2}\,(\text{volume})^{1/2}\,(\text{amount})^{-1} \equiv (\text{pressure})^{1/2}\,(\text{volume})\,(\text{amount})^{-1}$$

and the conversion to SI units is

$$\begin{array}{c} (\text{energy})^{1/2}\,(\text{volume})^{1/2}\,(\text{amount})^{-1} \equiv (\text{pressure})^{1/2}\,(\text{volume})\,(\text{amount})^{-1} \\ 1\ \text{cal}^{1/2}\,\text{cm}^{3/2}\,\text{mol}^{-1} = 2.0455\ \text{U}^{1/2}\,\text{cm}^{3/2}\,\text{mol}^{-1}\ (\text{MPa}^{1/2}\,\text{cm}^{3}\,\text{mol}^{-1}) \end{array}$$

In Chapter 7, dimensionless coefficients have been converted from $cal^{1/2}$ $cm^{-3/2}$ $atm^{-1/2}$ using

$$1\ \text{cal}^{1/2}\,\text{cm}^{-3/2}\,\text{atm}^{-1/2} = 6.4260$$

and the parameter $A = \delta^2 V^{1/3}\,\gamma^{-1}$ in Chapter 17 has been converted to $mol^{-1/3}$ from cal erg^{-1} $mol^{-1/3}$ by using 4.184 × 10^7 as a multiplying factor. In Chapter 6, for the normal lyoparachor and true lyoparachor,

$$1\ \text{cal}^{4/5}\,\text{g}^{-4/5}\,\text{cm}^{3}\,\text{mol}^{-1} = 3.143\ \text{J}^{4/5}\,\text{g}^{-4/5}\,\text{cm}^{3}\,\text{mol}^{-1}$$

$$1\ \text{cal}^{3/4}\,\text{g}^{-3/4}\,\text{cm}^{3}\,\text{mol}^{-1} = 2.925\ \text{J}^{3/4}\,\text{g}^{-3/4}\,\text{cm}^{3}\,\text{mol}^{-1}$$

The dipole moment is frequently quoted in the non-SI debye unit,

TABLE 5
Pressure Units and Conversion Factors

	cal cm^{-3}	MPa ($J\ cm^{-3}$)	atm	bar	kg cm^{-2}	mmHg (0°C)
1 cal cm^{-3}	—	4.184 0	41.292 9	41.840	42.665	31 383
1 MPa ($J\ cm^{-3}$)	0.239 01	—	9.869 2	10.000	10.197 2	7 500.6
1 atm	0.024 217	0.101 325	—	1.013 25	1.033 23	760.00
1 bar	0.023 901	0.100 000	0.986 92	—	1.019 72	750.06
1 kg cm^{-2}	0.023 439	0.098 066	0.967 84	0.9806	—	735.56
1 mmHg (0°C)	0.000 031 865	0.000 133 322	0.001 315 78	0.001 332 22	0.001 359 5	—

TABLE 6
Cohesion Parameter Units and Conversion Factors

	$cal^{1/2}\ cm^{-3/2}$	$MPa^{1/2}$ ($J^{1/2}\ cm^{-3/2}$)	$atm^{1/2}$
1 $cal^{1/2}\ cm^{-3/2}$	—	2.045 5	6.4260
1 $MPa^{1/2}$ ($J^{1/2}\ cm^{-3/2}$)	0.488 88	—	3.1415
1 $atm^{1/2}$	0.155 62	0.318 32	—

$$1\ D = 3.336 \times 10^{-30}\ C\ m$$

Density conversions are made using

$$1\ lb\ (US\ gal)^{-1} = 0.11983\ g\ cm^{-3}$$

2.6 MIXTURES

For mixing to be possible, the Gibbs free energy of mixing at constant pressure must be negative:

$$\Delta_m G_p = \Delta_m H_p - T\Delta_m S_p < 0 \tag{22}$$

The entropy change $\Delta_m S_p$, of a mixing process is usually positive, but in order to predict if mixing will take place, it is necessary to evaluate the enthalpy term, $\Delta_m H_p$. When this term is negative, or positive and less than $T\Delta_m S_p$, mixing can occur. Because of the temperature dependence of the entropy term, if the temperature of a mixture is decreased spontaneous "unmixing" (phase separation) may occur, although it is also possible for metastable homogeneous systems to exist.

Through thermodynamic relationships, cohesion parameters can also provide information on the properties of the components within mixtures. Differentiation of the Gibbs free energy of mixing with respect to the amount of substance i provides the chemical potential ${}^i\mu^\circ$ in the pure liquid. The chemical potential is also known as the relative partial molar Gibbs free energy, or the Gibbs free energy of dilution, and can be subdivided into enthalpy of dilution and entropy of dilution terms. The *activity*, ${}^ia$, of component i follows from

$$RT \ln {}^ia = {}^i\mu - {}^i\mu^\circ \tag{23}$$

The usual approach to the study of thermodynamic properties of mixtures or solutions is to determine the changes in the values of certain characteristic properties when the components are mixed. In doing this, the concept of an *ideal mixture* or *ideal solution* is valuable in describing the idealized behavior of mixtures, as in the same way the ideal gas law describes the idealized, limiting behavior of expanded gases. It is demonstrated in most general physical chemistry textbooks that the thermodynamic definition of an ideal mixture (a mixture in which the activity equals the mole fraction composition over the entire composition range and over a nonzero range of temperature and pressure) leads to the following properties:

1. There is *no volume change* during the formation of an ideal mixture from its components, $\Delta_m V = 0$. (The volume change on mixing can be determined experimentally by dilatometry.)
2. There is *no enthalpy change* in the system when the components are mixed at a fixed total pressure; that is, there is a zero heat of mixing, $\Delta_m H = 0$. (Experimentally, there would be no temperature change observed in a thermally isolated system during an ideal mixing process.)
3. There is an *entropy change* during mixing equal to that occurring during the formation of an ideal gas mixture due to the extra degrees of freedom created by the mixing process. This is sometimes called the combinatorial entropy, and for equal-sized, low molecular mass components the molar entropy of mixing is

$$\Delta_m S = -R \Sigma_i {}^i x \ln {}^i x \tag{24}$$

where ${}_i x$ is the mole fraction of component i. (Each mole fraction ${}^i x$ is less than unity, so $\ln {}^i x$ is negative and the overall $\Delta_m S$ term is positive.) Different relative molecular sizes of the components reduce the number of possible combinations, and $\Delta_m S$ is less than the ideal value (Section 13.2).
4. The resulting molar *Gibbs free energy change* during the formation of an ideal mixture is therefore completely provided by the entropy gained by each component:

$$\Delta_m G = \Delta_m H - T\Delta_m S = RT \Sigma_i {}^i x \ln {}^i x \tag{25}$$

5. The components forming an ideal mixture are always *completely miscible* in all proportions.

Another way of considering an ideal mixture is on the molecular level: an ideal mixture is one in which the different types of molecules, *i* and *j* for example, behave exactly as if they were surrounded by molecules of their own kind; that is, all intermolecular interactions are equivalent. This is discussed further below.

In *nonideal mixtures*, the Gibbs free energy change for the mixing process is not equal to the ideal value, and the "excess" Gibbs free energy change on mixing is[35]

$$^E\Delta_m G = \Delta_m G - RT \Sigma_i {}^i x \ln {}^i x \tag{26}$$

This function may be considered either from the point of view of the mixed *system* as the excess G^E of the Gibbs free energy of the nonideal mixture relative to that of the ideal mixture, or from the point of view of the mixing *process* as the excess $^E\Delta_m G$ of the nonideal Gibbs free energy of mixing relative to the ideal Gibbs free energy of mixing.[10] Similarly, the excess entropy of mixing is defined

$$^E\Delta_m S = \Delta_m S + R\,({}^i x \ln {}^j x\ {}^i x \ln {}^j x) \tag{27}$$

An ideal mixture follows *Raoult's law*, which states that the partial pressure, ${}^i p$, of any component i is given by

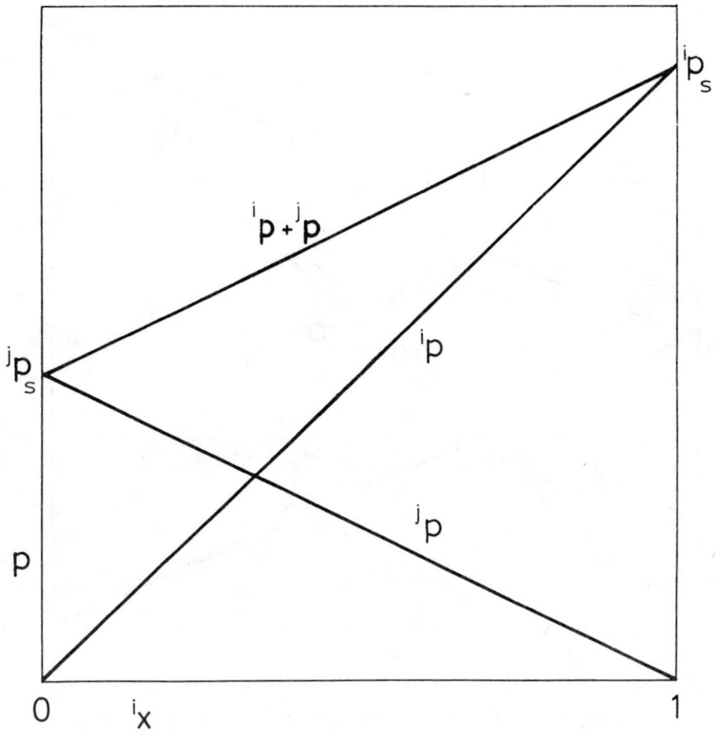

FIGURE 1. Total and partial vapor pressures in an ideal binary mixture at constant temperature.

$$^ip = {}^ix^ip_s \tag{28}$$

where ^ix is the mole fraction of component i in the mixture and ^ip_s is the saturation pressure of component i. This ideal behavior is illustrated in Figure 1. A real system with behavior close to ideal is tetrachloromethane-cyclohexane.

Negative deviations from Raoult's law (Figure 2) occur when interactions between unlike molecules ($i - j$ interactions) are markedly stronger than like-pair interactions. *Moderate positive deviations* (Figure 3) are usual, and occur when there is little or no specific interaction between any of the molecules. *Strong positive deviations* result from situations where molecules of one or more of the components undergo strong self-interaction, but only normal interactions with other components, as in alcohol-hydrocarbon mixtures. *Very strong positive deviations* from Raoult's law lead to liquid-liquid immiscibility (Section 2.7).

Nonideality is also frequently described in terms of activity coefficients. Except for the special case of an ideal mixture, the activity ^ia of component i is not equal to its mole fraction value^1x, and it is therefore convenient to define the activity coefficient ^if_x such that

$$^if_x = {}^ia^ix \tag{29}$$

The Gibbs free energy change on mixing is then expressed

$$\Delta_m G = RT\ \Sigma_i{}^i\mathrm{x}\ \ln {}^ix + RT\ \Sigma_i{}^i\ a = \ln {}^ix + RT\ \Sigma_i{}^ix \ln {}^if_x \tag{30}$$

and from Equation 26,

$$^E\Delta_m G = RT\ \Sigma_i\ {}^ix \ln {}^if_x \tag{31}$$

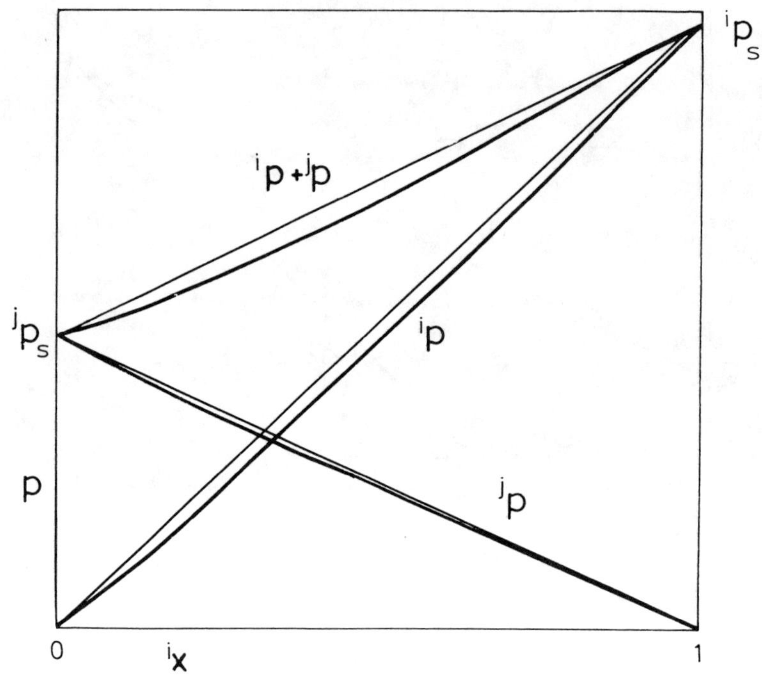

FIGURE 2. Total and partial vapor pressures at constant temperature for a system with negative deviations from Raoult's law.

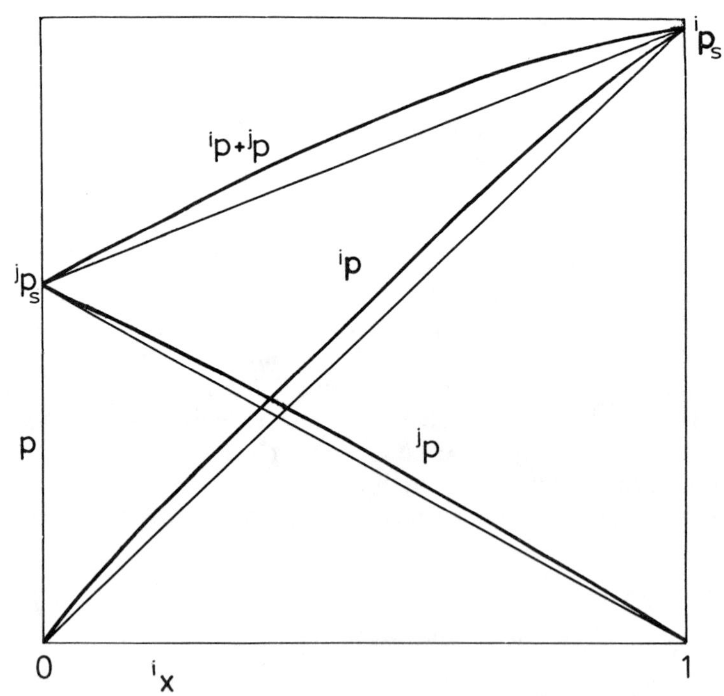

FIGURE 3. Total and partial vapor pressures at constant temperature for a system with moderate positive deviations from Raoult's law.

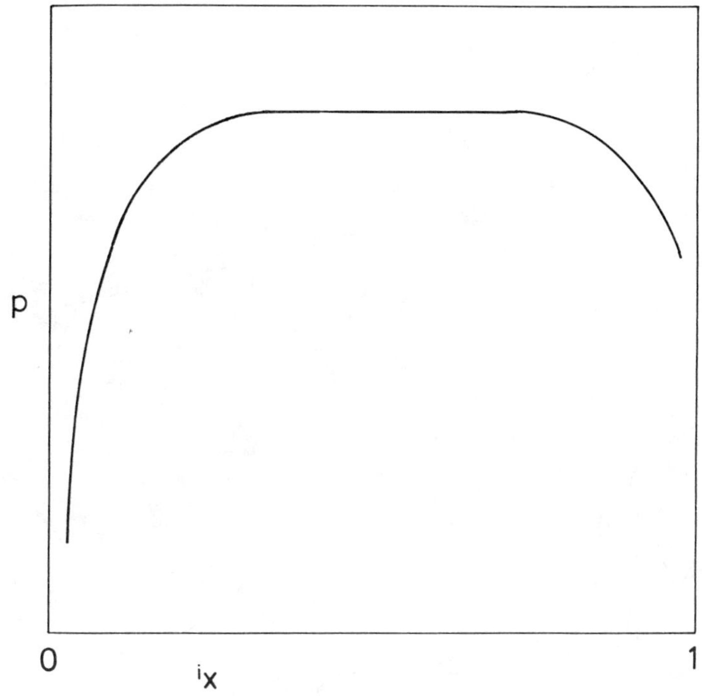

FIGURE 4. Total vapor pressure in equilibrium with a partially miscible pair of liquids, at constant temperature.

Hildebrand's definition[36,37] of a *regular solution* arises from the separation of the Gibbs free energy of mixing at constant pressure into entropy and enthalpy components (Equation 22): "A regular solution is one involving no entropy change when a small amount of one of its components is transferred to it from an ideal solution of the same composition, the total volume remaining unchanged." In other words, a solution of mixture described as "regular" is one which, despite a nonideal (nonzero, either positive or negative) enthalpy of formation, has an ideal entropy of formation. This can occur only if the random distribution of molecules persists even in the presence of $i-j$ interactions which differ from the original $i-i$ and $j-j$ interactions. The concept has proven valuable in the development of an understanding of miscibility criteria and of deviations from ideality. Unfortunately, although the term "regular solution" has come into general use, the usage does not always conform with its original definition.[38] The definition in terms of an ideal entropy of mixing forms the most useful reference state and should be retained. When molecules of different sizes are mixed, an alternative entropy of mixing term is required, as described subsequently (Section 13.2).

2.7 PHASE SEPARATION

It was stated in the previous section that when there are strong self-interactions in the components of a mixture and weaker interactions between components, there occur strong positive deviations from Raoult's law which can lead to liquid-liquid immiscibility. In Figure 4, which can be compared with Figures 1 to 3, the flat portion of the pressure-composition curve represents the miscibility gap.

The possibility of this type of behavior can be investigated in an alternative way by plotting the excess Gibbs free energy of mixing (Equation 26) against the mole fraction.[39] If this curve has no point of inflexion and is concave upward (Figure 5a), the binary mixture

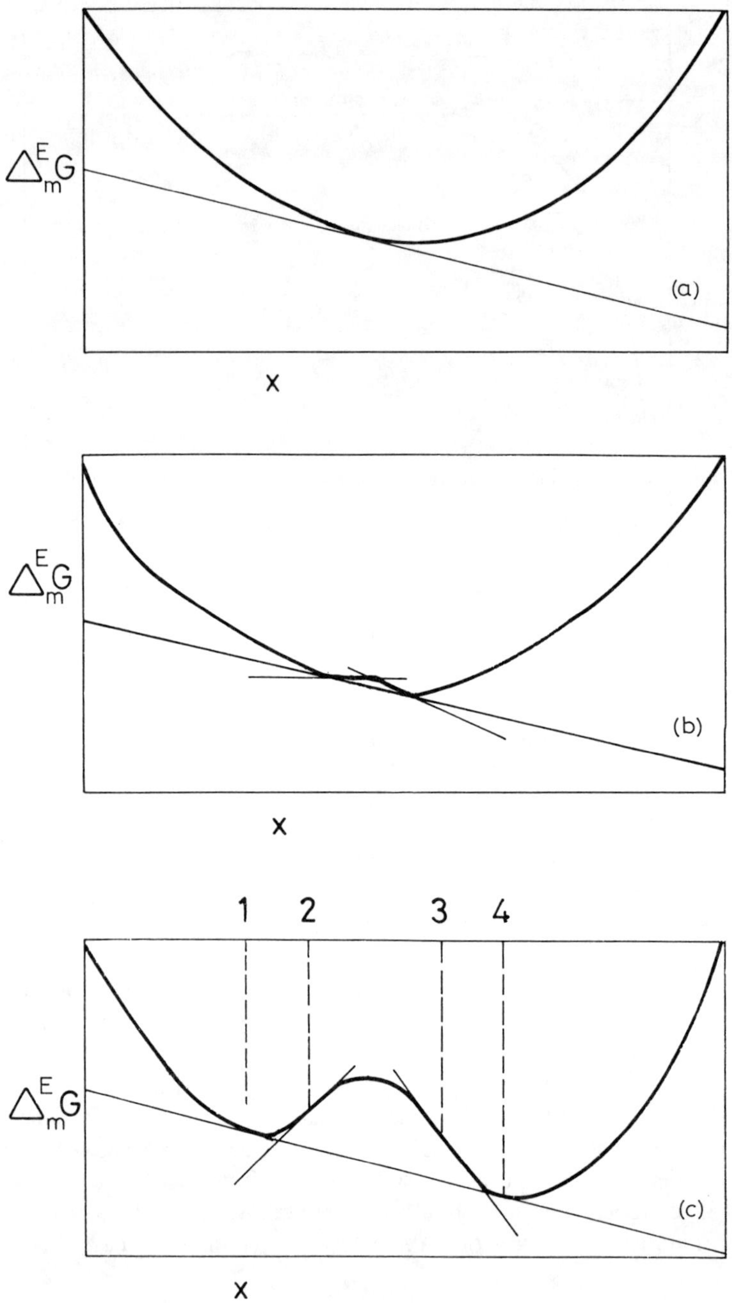

FIGURE 5. Excess Gibbs free energy of a binary mixture as a function of mole fraction: (a) miscibility in all proportions; (b) onset of demixing; (c) existence of two regions of stability, two regions of metastability, and one region of instability. (Adapted from Dayantis, J., *Plast. Mod. Elastomeres*, 29(2), 58, 1977.)

is stable at all compositions and no phase separation occurs. If the curve has two upward-facing concavities separated by a convex section and two points of inflexion (Figure 5c), there is a region of total instability, two metastable regions and, at either end of the composition range, two regions of binary mixture stability. An attempt to obtain a mixture with a composition between points 2 and 3 leads to two phases with compositions represented by points 2 and 4 defined by the double tangent to the curve. Curve 5b is the limiting case, defining a critical miscibility situation where the two points of inflexion coincide. As the temperature of some liquid-liquid systems is decreased, the behavior may be represented by each of Figures 5a, b, and c in turn as complete miscibility at higher temperatures changes to phase separation at lower temperatures. The temperature corresponding to diagram (b) is then called the (upper) critical solution temperature, UCST. Polymer-liquid, polymer-polymer systems (Section 13.3) and a very few liquid-liquid systems, like aqueous amine solutions, exhibit a second critical solution temperature as the temperature of the mixture is *raised.*

It should also be noted that useful polymer-polymer and polymer-liquid dispersions and metastable solutions can be produced from thermodynamically incompatible systems if the ingredients can be mixed in the appropriate thermodynamic conditions and then prevented from demixing when the conditions are changed (Chapter 16).

REFERENCES

1. **Polak, J.,** Heat of vaporization and cohesion energy of liquids, *Collect. Czech. Chem. Comm.,* 31, 1483, 1966.
2. **Majer, V., Svoboda, V., Pošta, A., and Pick, J.,** Enthalpy data of liquids. XXII. Determination of heats of vaporization and some other thermodynamic quantities for three fluorinated halogen ethanes, *Collect. Czech. Chem. Commun.,* 46, 817, 1981.
3. **Svoboda, V., Charvatova, V., Majer, V., and Pick, J.,** Enthalpy data on liquids. XXIV. Determination of heats of vaporization and other thermodynamic quantities for four alklycycloparaffins, *Collect. Czech. Chem. Commun.,* 46, 2983, 1981.
4. **Svoboda, V.,** Cohesion energy of liquids, *Chem. Listy,* 76, 389, 1982.
5. **Uchytilová, V., Majer, V., and Svoboda, V.,** Enthalpies of vaporization and cohesive energies of 2-chloro-1,1,2-trifluoroethyl ethyl ether, 2-chloro-1,1,2-trifluoroethyl propyl ether, 2-chloro,1,2-trifluoroethyl chloromethyl ether and 2-chloro-1,1,2-trifluororthyl difluoromethyl ether, *J. Chem. Thermodyn.,* 16, 475, 1984.
6. **Majer, V., Svoboda, V., and Lencka, M.,** Enthalpies of vaporization and cohesive energies of pyridine and isomeric methylpyridines, *J. Chem. Thermodyn.,* 16, 1019, 1984.
7. **Majer, V. and Svoboda, V.,** *Enthalpies of Vaporization of Organic Compounds,* IUPAC Chemical Data Series No. 32, Blackwell, Oxford, 1985.
8. **Hildebrand, J. H. and Scott, R. L.,** *Solubility of Non-Electrolytes,* 3rd ed., Reinhold, New York, 1950; Dover, New York, 1964.
9. **Hildebrand, J. H. and Scott, R. L.,** *Regular Solutions,* Prentice-Hall, Englewood Cliffs, NJ, 1962.
10. **Hildebrand, J. H., Prausnitz, J. M., and Scott, R. L.,** *Regular and Related Solutions,* Van Nostrand-Reinhold, Princeton, NJ, 1970.
11. **Teas, J. P.,** Re solubility parameters, *J. Paint Technol.,* 40, 104, 1968.
12. **Gardon, J. L. and Teas, J. P.,** Solubility parameters, in *Treatise on Coatings,* Vol. 2, Part II, Myers, R. R. and Long, J. S., Eds., Marcel Dekker, New York, 1976, chap. 8, 413.
13. **Lawson, D. D.,** Applications of solubility parameters. Part I, *Proc. DOE Chemical/Hydrogen Energy Contractor Review Systems,* (Pasadena, CA, 1977), CONF-771131, National Technical Information Service, Springfield, VA, 1978, 109.
14. **England, C., Lawson, D. D., and Hrubes, J. D.,** Superconductivity, cohesive energy density, and electron-atom ratio in metals, *J. Appl. Phys.* 52, 2923, 1981.
15. **Barton, A. F. M.,** Internal pressure. A fundamental liquid property, *J. Chem. Educ.,* 48, 156, 1971.
16. **Barton, A. F. M.,** *The Dynamic Liquid State,* Longman, London, 1974.
17. **Ouellette, R. J. and Williams, S. H.,** Internal solvent pressure. I. A demonstration of the effect of internal solvent pressure on conformational equilibria, *J. Am. Chem, Soc.,* 93, 466, 1971.

18. **Sharma, B. K.,** A new method of estimation of solubility parameter of liquids, *Indian J. Pure Appl. Phys.*, 124, 939, 1976; Solubility parameter and evaluation of some thermodynamic properties of simple liquids, *Indian J. Pure Appl. Phys.*, 15, 633, 1977.
19. **Sharma, B. K.,** Internal pressure-temperature behaviour and lattice Gruneisen constant of polymers, *Acustica,* 48, 121, 1981.
20. **Sharma, B. K.,** Statistical thermodynamics of liquids. I. An equation of state and solubility parameter, *Acustica,* 49, 164, 1981.
21. **Sharma, B. K. and Das, N.,** Internal pressure-temperature behavior and solubility parameter of liquids, *Indian J. Pure Appl. Phys.*, 19, 668, 1981.
22. **Wilson, G. M.,** Interpretation of Trouton's law, solubility parameters, molecular size and polarity in relation to equation of state properties, *Abs. Pap. Am. Chem. Soc.*, 189, (Apr.), 110, 1985.
23. **Wilson, G. M.,** Interpretation of Trouton's law in relation to equation of state properties, *Am. Chem. Soc. Symp. Ser.*, 300, (Equations of State), 520, 1986.
24. **Del Rio, F., Guzman, F., and Mier y Teran, L.,** THe van der Waals model and the theory of solutions, *Chem. Phys. Lett.*, 35, 274, 1975.
25. **Haward, R. N.,** Modified van der Waals equation for liquids, *Trans. Faraday Soc.*, 62, 828, 1966.
26. **Rowlinson, J. S.,** Legacy of van der Waals, *Nature (London),* 224, 414, 1973.
27. **Hildebrand, J. H.,** Compressibilities and thermal pressure coefficients of certain liquids, *Phys. Rev.*, 34, 649, 1929.
28. **Hildebrand, J. H.,** *Solubility of Non-Electrolytes,* 2nd ed., Reinhold, New York, 1936; (Russian translation), Gonti, Moscow-Leningrad, 1938.
29. **Hildebrand, J. H. and Carter, J. M.,** A study of van der Waals forces between tetrahalide molecules, *J. Am. Chem. Soc.*, 54, 3592, 1932.
30. **Frank, H. S.,** Free volume and entropy in condensed systems. II. Liquids, *J. Chem. Phys.* 13, 493, 1945.
31. **Fried, V. and Schneier, G. B.,** Some comments on cohesion energies of liquids, *J. Phys. Chem.*, 72, 4688, 1968.
32. **Amoras, J., Solana, J. R., and Villar, E.,** Behaviour of the internal pressure of liquids in accordance with variations in temperature, volume and cohesive energy density, *Mater. Chem. Phys.*, 10, 557, 1984.
33. **McGowan, J. C.,** The estimation of solubility parameters and related properties of liquids, *J. Chem. Tech. Biotechnol.*, *34A*, 38, 1984.
34. **Taylor, N. W.,** in *Modern Chemistry for the Engineer and Scientist,* Robertson, G. R., Ed., McGraw-Hill, New York, 1957, p 183.
35. **Eckert, C. A.,** Introduction to theories of solutions and application to phase equilibria, in *Solutions and Solubilities,* Part II, Dack, M. R. J., Ed., John Wiley & Sons-Interscience, New York, 1976, chap. 9.
36. **Hildebrand, J. H.,** Solubility. XII. Regular solutions, *J. Am. Chem. Soc.*, 51, 66, 1929.
37. **Hildebrand, J. H.,** Order from chaos, *Science,* 150, 441, 1965.
38. **Hildebrand, J. H.,** The term 'regular solution,' *Nature (London),* 168, 868, 1951.
39. **Dayantis, J.,** The Hildebrand parameter. Solubility of polymers, *Plast. Mod. Elastomers,* 29(2), 58, 1977.
40. **Bowersox, D. F.,** Thermodynamic properties of selected solutes in liquid metal solutions, *U.S.A.E.C.*, CONF-690801, 565-75, 1969.
41. **Varushchenko, R. M., Loseva, O. L., and Druzhinina, A. I.,** Energies of evaporation of dichloroalkanes, *Zh. Fiz. Khim.*, 62, 2329, 1988; *Russ. J. Phys. Chem.*, 62, 1204, 1988.
42. **Yatsenko, S. P. and Dieva, G. N.,** Solubility of high-melting metals in molten indium, *Zh. Fiz. Khim.*, 47, 2948, 1973; *Russ. J. Phys. Chem.*, 47, 1658, 1973.

Chapter 3

MOLECULAR INTERACTIONS

As an introduction to the behavior of bulk or macroscopic materials it is useful to briefly consider some properties at the molecular level, and in particular the electrical properties of molecules.

The net energy of interaction or internal energy, U between nonbonded atoms or molecules is the result of both attractive and repulsive effects. There is one major repulsive interaction, the tendency of two molecules to avoid occupying the same space. This energy term rises very steeply to high positive values when the intermolecular separation falls below a certain distance, but otherwise has little effect on the internal energy. There are several possible attractive interactions, collectively called van der Waals forces, all having weaker dependences on intermolecular separation than the repulsive interactions and, therefore, dominating at larger distances. Consequently,

1. The equilibrium *intermolecular separation* is largely determined by the repulsive interactions.
2. The net *energy of interaction* at equilibrium is mainly a function of the attractive interaction.

Both these facts are of great assistance in describing molecular interactions by means of simple models, but it is the latter which is significant in the development of cohesive energy and cohesion parameter relations.

3.1 ELECTRICAL PROPERTIES

When two charges q and $-q$ are separated by a fixed distance r they form an electric *dipole moment*, μ, of magnitude

$$\mu = qr \tag{1}$$

The permanent molecular dipole moment is one of the important quantities determining the extent of interaction and, therefore, controlling miscibilities.

When two charges are separated by a distance r in a vacuum, their potential energy of interaction is inversely proportional to $\varepsilon_o r$ where ε_o is a constant (the permittivity of vacuum). But, when they are separated by a solvent medium the potential energy is reduced proportionally according to $\varepsilon r = \varepsilon_o \varepsilon_r r$, where $\varepsilon_r = \varepsilon/_o$ is the *relative permittivity* or dielectric constant of medium. The value of ε_r depends on the nature of the solvent molecules, particularly their polarities. Values of both μ and ε are included in Table 5, Chapter 8.

Nonpolar molecules can acquire a temporary dipole moment while exposed to an electric field, because their electron distribution becomes distorted. The magnitude of the induced dipole moment, μ_i of a molecule is proportional to the electric field strength E, provided the field strength is not too great:

$$\mu_i = \alpha \varepsilon_o E \tag{2}$$

The proportionality coefficient, α, is called the electric *polarizability* of the molecule. The polarizability is related to the refractive index, n, by the Lorentz-Lorenz equation, and to the molar refraction R,

$$R = (M/\rho)(n^2 - 1)/(n^2 + 2) = V(n^2 - 1)/(n^2 + 2) \tag{3}$$

R is an additive property (Section 6.3), related to cohesion properties in such a way that the Hildebrand parameter (or at least its dispersion component) can be obtained from the refractive index function $(n^2 - 1)/(n^2 + 2)$ as discussed by various authors,[1,2] or from equations such as that of Böttcher.[3,4] Djordjevic and Porter[5] reported an empirical correlation between cohesive energy density and the material parameter g^2, reflecting the intensity of the Onsager reaction field, where $g = (n^2 - 1)/(2n^2 + 1)$. For nonpolar n-alkanes, and for both nonpolar and polar 1-substituted n-alkanes, the cohesive energy density, c, changed linearly with g^2. Substituents modify both c and g^2; change of g^2 reflects the effect of the substituent on the purely dispersive interactions.

A more complete list of reports on the empirical and semiempirical relationships between molecular electrical properties and cohesion parameters appears in Section 8.2.

The attractive, cohesive van der Waals interactions between molecules can be discussed in terms of these electrostatic properties, but to do so it is convenient to divide them into three types.

3.2 DIPOLE-DIPOLE INTERACTIONS

Dipole-dipole or *Keesom*[6] interactions have an orientating effect on molecules, one dipole tending to align the other into energetically favorable arrangements. To a good approximation, the average orientation energy, $-U_o$, of the two molecules with permanent dipole moment ${}^i\mu$ and ${}^j\mu$ is found to be

$$-U_o = 2\,{}^i\mu^{2\,j}\mu^2/(3kTr^6) \quad (4)$$

where r is the intermolecular separation. The factor kT appears in the expression because the randomizing, thermal motion of the molecule must be overcome if orientation is to occur. *Quadrupole* interactions must also be considered in some cases, for example, carbon dioxide (Section 5.7).

3.3 DIPOLE-INDUCED DIPOLE INTERACTIONS

Dipole-induced dipole or *Debye*[7,8] interactions are caused by the polarization of a molecule (polar or nonpolar) in the vicinity of a polar molecule. The induced and permanent dipoles are mutually attracted, and the thermal molecular motion does not disorientate the direction of the induced moment from that of the inducing moment because the electric moment adjusts simultaneously and is independent of molecular rotation. The effect persists even when the polar molecule is rotating freely in the vicinity of the polarizable molecule, so the effect is independent of temperature:

$$-{}^iU_i = ({}^i\alpha^j\mu^2 + {}^j\alpha^i\mu^2)/r^6 \quad (5)$$

3.4 INDUCED DIPOLE-INDUCED DIPOLE INTERACTIONS

Induced dipole-induced dipole, temporary dipole, dispersion or *London*[9,10] forces exist between all adjacent pairs of atoms and molecules. Their origin is the instantaneous electrical dissymmetry of electrons in one molecule polarizing the electron clouds in adjacent molecules and inducing instantaneous dipoles of opposite polarity, resulting in intermolecular attraction. Although the molecules continue to rotate and so change the dipole direction, one molecule tends to follow the other, and because of this correlation the attractive effect does not average to zero. The calculation of the magnitude of the dispersion cohesive energy, $-U_d$, is an involved process, but a reasonable approximation (the London formula) predicts

$$-U_d = 3\,{}^iI\,{}^jI\,{}^i\alpha\,{}^j\alpha/[2({}^iI + {}^jI)\,r^6] \tag{6}$$

so the magnitude of the dispersion energy depends on both the polarizability, α, and the first ionization potential, I, of each molecule.

3.5 SPECTROSCOPIC STUDIES

Spectroscopic techniques can be classified as "absorption spectroscopy" when quanta of radiation are completely absorbed by a material, and as "scattering spectroscopy" when they are not completely absorbed. When energy is imparted to molecules or given off from molecules by radiation, the scattering and absorption processes are either associated with quantized transitions and discrete energy "steps", or with nonquantized translational and rotational-diffusion motion, which provides information on the distribution of particles as functions of time and distance.

In the optical absorption spectroscopy techniques, light quanta (photons) of the appropriate energy or frequency are completely absorbed. Absorption of visible and ultraviolet light causes electronic transitions, infrared absorption corresponds to vibrational processes, and far infrared and microwave spectroscopy provides information on particle rotation. The precise frequencies at which energy is absorbed or emitted by a solute depend on the nature of the solvent because of solute-solvent interactions; for example, the use of infrared absorption spectroscopy to provide an empirical hydrogen bonding parameter is described in Section 5.12.

Cohesion parameter relationships may be used for this purpose also. The Hildebrand parameter has been used by Parker and Gould[11,12] to describe the solvent dependence of the infrared-active carbonyl stretching frequencies of dimanganese decacarbonyl, $Mn_2(Co)_{10}$, and some of its transdisubstituted derivatives, and of $Re_2(CO)_{10}$. It is possible to construct a straight line graph of the solvent Hildebrand parameter against frequency for all CO bands in nonpolar solvents, but polar solvents deviate from this relationship. The solvent frequency shift of the emission of a fluorescent probe (2-toluidinonaphthalene-6-sulfonate) was found by Reeves, Maggio, and Costa[13] to correlate well in a linear fashion with the Hildebrand parameters of a wide range of solvents. Zachariasse, Duveneck, and Busse[14] discussed the solvent effect on fluorescence decay of 1,3-di(1-pyrenyl)propane in terms of Hildebrand parameters. Photostimulated charge transfer across organic photoconductor/insulating liquid interfaces has been discussed by Ahuja and Hauffe[15] in terms of several properties, including Hildebrand parameters.

A molecular energy transition may take only a portion of the energy from incident radiation from a *scattering* process: when monoenergetic radiation (neutrons, X-rays, or light) interacts with molecules during the scattering process, energy is exchanged so that a spectrum of energies is produced. In particular, neutron scattering[16-18] provides information on coil dimensions and structure of polymers that can be correlated by means of cohesion parameters and Flory-Huggins theory (Chapter 13). Nuclear magnetic resonance chemical shift observations reflect the solvent environments of the active nuclei and are related to solvent cohesion parameters. Solvent effects on fluorine-19 nuclear magnetic resonance chemical shifts were correlated with cohesive energy density by Muller[19], and by Koehler and co-workers.[20,21] Chang and Krimm[22] correlated the Raman longitudinal acoustic mode frequencies of freeze-dried polyethylene single crystals with Hildebrand parameter differences between polyethylene and decalin, xylene, and silicone oil.

Studies on the interaction of radiation with polymers (discussed further in Section 16.5) include a fluorescence method for the determination of Hildebrand parameters of soluble polymers by Li, Winnick, and Guillet[23], and a semiquantitative relation between polymer sidechain dynamics and solvent Hildebrand parameters by Levy and Wang[24], who reported

^{13}C NMR spin-lattice relaxation times and nuclear Overhauser effects for poly(butyl methacrylate) as a function of solvent Hildebrand parameter. Law[25] used a *p-N,N-d*ialkylaminobenzylidenemalonitrile derivative as a fluorescence probe, discussing the results of vapor swelling in terms of cohesion parameters; the effect of vapors on the swelling properties of 83/17 vinyl chloride-vinyl acetate copolymer showed no correlation between fluorescence quantum yields and liquid Hildebrand parameters. Polymer Hildebrand parameters have been determined by microwave absorption in wave guides,[26] and interferometry has been used for studying polymer dissolution rates.[27] Frank and co-workers[28-30] have developed excimer fluorescence as a molecular probe of polymer blend compatibility (Section 16.9).

3.6 POSITRON ANNIHILATION

The positron, which can be thought of as a positively charged electron, is produced as a decay product of radioactive nuclei where the proton/neutron ratio is greater than unity. Positrons interact with electrons, resulting in the annihilation of both particles and the generation of γ-rays, the number and angular distribution of the γ quanta depending on whether the spins of positron and electron are parallel or antiparallel.

If the energy of positron has been reduced by interaction with the medium, it can also combine with an electron to form a positronium (Ps) atom, which acts in some respects as a form of hydrogen in which the nucleus has been replaced by a positron. The spins of positron and electron may be either antiparallel (in parapositronium) or parallel (in orthospositronium). Positronium atoms eventually annihilate with electrons, and the spectrum can be interpreted mathematically as a multiexponential function with lifetime components of about 0.125 ns for parapositronium and about 140 ns for orthopositronium.

Studies of orthopositronium lifetimes provide information on a variety of structural properties[31-35] such as the molecular weight, degree of crystallinity and glass transition temperature of polymers, and the microstructure of polymers, amino acids and proteins. Correlations have been made with properties such as refractive index, viscosity, and surface tension. Comparison of orthopositronium lifetimes of members of homologous series with cohesion parameters shows that in *n*-alkanes the lifetime tends to decrease as the Hildebrand parameter increases, while in primary alcohols this trend is reversed.[32] Jensen[36] proposed a correlation of the ortho-Ps annihilation rate with the dispersion component of the cohesion parameter, and this has been explored further by Mogensen and co-workers.[37-40] Mogensen[41] subsequently demonstrated that a good correlation with various components of the cohesion parameters was found only if the hydrogen bonding component was neglected. It was further suggested that the positron method could be used for evaluation of the components of the cohesion parameter (Section 5.12).

REFERENCES

1. **Bonn, R. and van Aartsen, J. J.,** Solubility of polymers in relation to surface tension and index of refraction, *Eur. Polym. J.*, 8, 1055, 1972.
2. **Sewell, J. H.,** *A Study of Cohesive Forces, Refractive Indexes and Glass Transition Temperature Data of Polymers,* Royal Aircraft Establishment Technical Report No. 66185, Farnborough, Hampshire, 1966, AD 641308.
3. **Böttcher, C. J. F.,** *Theory of Electric Polarisation,* Elsevier, New York, 1952.
4. **Liszi, J.** Cohesion energy of a liquid due to permanent dipoles, *Acta Chim. Acad. Sci. Hung.*, 98, 93, 1978.

5. **Djordjevic, M. B. and Porter, R. S.,** New empirical relation between cohesion energy density and Onsager reaction field for several classes of organic compounds, in *Macromolecular Solutions: Solvent-Polarity Relationships in Polymers* (Symp. Proc., Am. Chem. Soc., New York, 1981), Seymour, R. B. and Stahl, G. A., Eds., Pergamon, New York, 1982, 198.
6. **Keesom, W. H.,** Various papers, *Leiden Comm.,* 1912—1915; *Proc. Akad. Wetenschappen Amsterdam,* 1913—1916.
7. **Debye, P.,** van der Waals' cohesion forces, *Phys. Z.,* 21, 178, 1920.
8. **Debye, P.,** Molecular forces and their electrical interpretation, *Phys. Z.,* 22, 302, 1921.
9. **London, F.,** Properties and applications of molecular forces, *Z. Phys. Chem. Abt. B,* 11, 222, 1930.
10. **London, F.,** The general theory of molecular forces, *Trans. Faraday Soc.,* 33, 8, 1937.
11. **Gould, N. J. and Parker, D. J.,** Use of the solvent parameter δ_s, to describe the solvent dependence of the i.r. active carbonyl stretching vibrations of dimanganese decarbonyl and some of its trans-disubstituted derivatives, *Spectrochim. Acta,* 31A, 1785, 1975.
12. **Parker, D. J.,** Solvent effects on the infrared active CO stretching frequencies of some metal carbonyl complexes — I. Dimanganese decacarbonyl and dirhenium decacarbonyl, *Spectrochim. Acta,* 39A, 463, 1983.
13. **Reeves, R. L., Maggio, M. S., and Costa, L. F.,** Importance of solvent cohesion and structure in solvent effect on binding site probes, *J. Am. Chem. Soc.,* 96, 5917, 1974.
14. **Zachariasse, K. A., Duveneck, G., and Busse, R.,** Intramolecular excimer formation with 1,3-di(1-pyrenyl)propane. Decay parameters and influence of viscosity, *J. Am. Chem. Soc.,* 106, 1045, 1984.
15. **Ahuja, R. C. and Hauffe, K.,** Charge transfer across organic photoconductor/insulating liquid interfaces. II. Metal free phthalocyanine/dodecane containing electron donor/acceptor molecules, *Ber. Bunsenges. Phys. Chem.,* 84, 129, 1980.
16. **Jelenic, J., Kirste, R. G., Schmitt, B. J., and Schmitt-Strecker, S.,** Structure and thermodynamics of polymer blends from neutron scattering experiments, *Makromol. Chem.,* 180, 2057, 1979; Neutron scattering experiments on polymer mixtures, International Union of Pure and Applied Chemistry Symposium, Mainz, 1979.
17. **Schmitt, B. J. and Kirste, R. G.,** Molecular dispersion and thermodynamics of amorphous polymer blends, International Union of Pure and Applied Chemistry Meeting, Tashkent, 1978.
18. **Schmitt, B. J., Kirste, R. G., and Jelenic, J.,** Investigation of the thermodynamics and the conformation of macromolecules in polymer blends in the vicinity of phase separation by neutron scattering, *Makromol. Chem.,* 181, 1655, 1980.
19. **Muller, N.,** Volume of mixing effect on solvent shifts in fluorine magnetic resonance spectra, *J. Magn. Resonance,* 22, 479, 1976; Solvent effects on fluorine chemical shifts in chloroalkanes and alcohols, *J. Magn. Resonance,* 23, 327, 1976; Fluorine chemical shift of 6,6,6-trifluoro-1-heptanol in mixtures of water with organic cosolvents, *J. Magn. Resonance,* 28, 203, 1977.
20. **Koehler, K. A., Stone, E. E., Shelton, R. A., Jarnagin, F., Koehler, L. S., and Fossel, E. T.,** Interaction of fluorinated ether anesthetics with solvents, *J. Magn. Resonance,* 30, 75, 1978.
21. **Koehler, L. S., Curley, W., and Koehler, K. A.,** Solvent effects on halothane: ^{19}F nuclear magnetic resonance in solvent and artificial membranes, *Mol. Pharmacol.,* 13, 113, 1977.
22. **Chang, C. and Krimm, S.,** Longitudinal acoustic mode in polymers. VII. Surface structure from swelling studies of freeze-dried polyethylene single crystals, *J. Polym. Sci. Polym. Phys. Ed.,* 22, 1871, 1984.
23. **Li, X.-B., Winnick, M. A., and Guillet, J. E.,** A fluorescence method to determine the solubility parameters δ_H of soluble polymers at infinite dilution. Cyclization dynamics of polymers. I, *Macromolecules,* 16, 992, 1983.
24. **Levy, G. C. and Wang, D.,** Semiquantitative relationship between polymer side-chain dynamics and solvent solubility parameters. Variable-frequency ^{13}C NMR relaxation study of poly(*n*-butyl methacrylate), *Macromolecules,* 19, 1013, 1986.
25. **Law, K.-Y.,** Fluorescence probe for microenvironments: effect of solvent vapour on the properties of vapour-swollen polymers, *Polymer,* 25, 399, 1984.
26. **Tatsumi, M., Yamaguchi, Y., Moriwaki, S., Usami, S., and Yamamoto, S.,** Determination of solubility parameter of polymer by microwave absorption in waveguide, *Kobunshi Ronbunshu,* 45, 161, 1988.
27. **Rodriguez, F., Krasicky, P. D., and Groele, R. J.,** Dissolution rate measurements, *Solid State Technol.,* 28, 125, 1985; **Rodriguez, F., Groele, R. J., and Krasicky, P. D.,** Dissolution rates of thin polymer films using laser interferometry, *Proc. Soc. Photo-Opt. Instrum. Eng.,* 539, 14, 1985.
28. **Frank, C. W.,** Excimer fluorescence: a new tool for studying polymer alloys, *Plast. Compounds,* No. 1/2, 67, 1981.
29. **Frank C. W. and Gashgari, M. A.,** Excimer fluorescence as a molecular probe of polymer-polymer interactions in the amorphous solid state, *Ann. N.Y. Acad. Sci.,* 366, 387, 1981.
30. **Frank, C. W., Gashgari, M. A., and Semerak, S. N.,** Polymer blend thermodynamics: Flory-Huggins theory and its application to excimer fluorescence studies, *NATO Adv. Sc. Ser., Ser. C,* 182 (Photophys. Photochem. Tools Polym. Sci.: Conform., Dyn. Morphol.), 523, 1986.

31. **Chuang, P. L. and Tao, S. J.,** Positron annihilation in amino acids and proteins, *J. Phys. Chem.,* 78, 1261, 1974.
32. **Jilek, J. H.,** Ortho-positronium annihilation — a new tool for coatings research?, *Prog. Org. Coatings,* 5, 97, 1977.
33. **Khan, M. N. G. A., Carswell, D. J., and Bell, J.,** Measurement of positron lifetimes in solid polymers at room temperature, *Phys. Lett.,* 29A, 237, 1969.
34. **Levay, B.,** Chemical structure studies with positrons and mesons, *At. Energy Rev.,* 17, 413, 1979.
35. **Tao, S. J.,** Positronium annihilation in molecular substances, *J. Chem. Phys.,* 56, 5499, 1972; Positronium in molecular substances, *Appl. Phys.,* 3, 1, 1974.
36. **Jensen, B. S.,** Positronium pickoff annihilation in pure liquids, *Fourth Int. Conf. Positron Annihilation* (Helsinger, Denmark, August 1976), Paper G16.
37. **Mogensen, O. E. and Jacobsen, F. M.,** Positronium yields in liquids determined by lifetime and angular correlation measurements, *Chem. Phys.,* 73, 223, 1982.
38. **Wikander, G. and Mogensen, O. E.,** A study of picosecond dehalogenation of chlorobenzene anions in liquids by positronium inhibition measurements, *Chem. Phys.,* 72, 407, 1982.
39. **Mogensen, O. E., Pedersen, N. J., Malhorta, B. D., and Pethrick, R. A.,** The temperature dependence of the positronium in bubble state. An angular correlation and Doppler broadening study of a phenyl ether, *Chem. Phys.,* 87, 139, 1984.
40. **Mogensen, O. E.,** Personal communication, 1984.
41. **Mogensen, O. E.,** Information on the structure of liquids and solutions obtained by positron annihilation, *Electrochim. Acta,* 33, 1203, 1988.

Chapter 4

REGULAR SOLUTIONS AND THE HILDEBRAND PARAMETER

The Hildebrand parameter developed from regular solution theory, but whereas a regular solution is an idealized concept (a mixture in which the partial excess entropy is zero), the Hildebrand parameter, the geometric mean approximation, and the component cohesion parameters to be introduced in the next chapter are general concepts. These originated in regular solutions, but also can be applied to varying extents of other types of solution. This distinction is important, but is not always made.

4.1 GEOMETRIC MEAN APPROXIMATION

As introduced in Section 2.6, a regular solution has an ideal entropy of formation, arising from a completely random molecular distribution, despite the existence of interactions which lead to a nonideal (non-zero) enthalpy of formation. This effectively restricts regular mixtures to those systems in which only dispersion forces are important, because the orientation effects of polar molecules cause nonrandom molecular distributions. The extent of the dispersion effect depends on the ionization potential and the polarizability of the molecules concerned (Equation 6, Chapter 3). Ionization potentials, I, and intermolecular distances, r, do not usually vary greatly for different pairs of adjacent molecules, and to a good approximation

$$^{i}I + {}^{j}I \approx 2({}^{i}I\,{}^{j}I)^{1/2} \tag{1}$$

and

$$r \approx 2({}^{i}r\,{}^{j}r)^{1/2} \tag{2}$$

This forms the basis of the geometric mean relation for dispersion interactions,

$$^{ij}U_{\mathrm{d}} = {}^{i}U_{\mathrm{d}}^{1/2}\,{}^{j}U_{\mathrm{d}}^{1/2} \tag{3}$$

which is a major approximation in cohesion parameter theory. Early in the development of the cohesion parameter concept, the geometric mean approximation was tested experimentally on eight mixtures of tetrahalomethanes. Hildebrand and Carter[1] verified to within 1% the relationship

$$^{ij}a = ({}^{i}x\,{}^{i}a^{1/2} + {}^{j}x\,{}^{j}a^{1/2})^{2} \tag{4}$$

where $a = TV^2\beta$ (Equations 10 and 16, Chapter 2) for the mixture $i-j$ and the components i and j, which would follow from

$$^{ij}a = {}^{i}x^{2}\,{}^{i}a + {}^{j}x^{2}\,{}^{j}a + 2\,{}^{i}x\,{}^{j}x\,{}^{ij}a \tag{4a}$$

if $^{ij}a = ({}^{i}a\,{}^{j}a)^{1/2}$. Further tests were made by Scatchard et al.[2] and Staveley et al.[3], and the geometric mean approximation can be considered to be justified, at least in favorable circumstances, although discussion on its validity continues[4-9]. Empirical correction methods for deviations are described in Section 5.7.

4.2 HILDEBRAND-SCATCHARD EQUATION

The regular solution equation for the internal energy of mixing at constant volume, based on the pioneering work of van der Waals[10] and van Laar,[11,12] was derived on semitheoretical grounds by Scatchard[13,14] and Hildebrand[15-19] and popularized by J. H. Hildebrand, R. L. Scott, J. M. Prausnitz, and others.[20-35]

The energy of mixing for 1 mol of solution is

$$\Delta_m U_v = ({}^ix\,{}^iV + {}^jx\,{}^jV)\,{}^{ij}A\,{}^i\phi\,{}^j\phi \tag{5}$$

$$= {}^iV\,({}^ix + m\,{}^jx)\,{}^{ij}A\,{}^i\phi\,{}^j\phi \tag{6}$$

In these equations, x is the mole fraction, V is the molar volume, m is the ratio of molar volumes, and ϕ is the volume fraction:

$$m = {}^jV/{}^iV = ({}^jM/{}^iM)({}^i\rho/{}^j\rho) \tag{7}$$

$${}^i\phi = {}^iV\,{}^ix/({}^jV\,{}^jx + {}^iV\,{}^ix) = {}^ix/(m\,{}^jx + {}^ix) \tag{8}$$

$${}^j\phi = {}^jV\,{}^jx/({}^jV\,{}^jx + {}^iV\,{}^ix) = m\,{}^jx/(m\,{}^jx + {}^ix) \tag{9}$$

The superscripts i and j identify the components of the binary mixture, i being the solvent and j the solute if this distinction is made. ${}^{ij}A$ is a measure of the change of the cohesion pressure or energy density associated with the $i-j$ mixing process, called the *exchange energy density* or *exchange cohesive pressure*, and given by

$${}^{ij}A = {}^ic + {}^jc - 2\,{}^{ij}c \tag{10}$$

where the ${}^{ij}c$ term is the cohesive pressure characteristic of the intermolecular forces acting between molecules of type i and j. This equation can be appreciated in a simple fashion by considering what happens when unit volumes of components i and j are mixed: two $i-j$ interactions are formed for each pair of $i-i$ and $j-j$ interactions broken. It should be noted that ${}^{ij}A$ is not simply the difference between the cohesive pressure of components i and j; this quantity (${}^ic - {}^jc$) has been termed [36] the "mutual cohesion factor", but has not proved of great value in correlating the properties of materials.

The interchange cohesive pressure ${}^{ij}A$ can be used simply as an empirical parameter for a particular $i-j$ pair of series of mixtures, like the physical interaction parameter of Harris and Prausnitz,[37] but this does not make full use of the opportunity to develop a method to provide information on *mixtures* from data on individual *components*. This is where the relationship between ${}^{ij}A$ and the individual Hildebrand parameters ${}^i\delta$ and ${}^j\delta$ is important.

The Hildebrand parameters for the substances are defined by Equation 5, Chapter 2, so

$${}^i\delta^2 = {}^ic = \Delta^iU/{}^iV \text{ and } {}^j\delta^2 = {}^jc = \Delta^jU/{}^jV \tag{11}$$

Using the geometric mean approximation, Equation 3,

$${}^{ij}c = ({}^ic\,{}^jc)^{1/2} \tag{12}$$

so

$${}^{ij}A = ({}^ic^{1/2} - {}^jc^{1/2})^2 = ({}^i\delta - {}^j\delta)^2$$

$$= {}^i\delta^2 + {}^j\delta^2 - 2\,{}^{ij}c \tag{13}$$

and the Hildebrand-Scatchard equation can be written

$$\Delta_m U_v = ({}^ix\,{}^iV + {}^jx\,{}^jV)\,{}^{ij}A\,{}^i\phi\,{}^j\phi$$
$$= ({}^ix\,{}^iV + {}^jx\,{}^jV)\,({}^i\delta - {}^i\delta)^2\,{}^i\phi\,{}^j\phi \tag{14}$$

This provides the basis of the cohesion parameter approach to liquid miscibility and other properties. The Hildebrand-Scatchard equation, although a very simple predictive relationship, is still proving useful.[88-91]

One of the most important derived properties is the activity coefficient. The partial molar energy of transfer of component j is obtained by differentiating Equation 5 with respect to the amount of j, and assuming an ideal entropy of transfer, the resulting expression for the activity coefficient^j is

$$RT \ln {}^jf_x = RT \ln ({}^ja/{}^jx) = {}^jV\,{}^i\phi^2\,{}^{ij}A = {}^jV\,{}^i\phi^2\,({}^i\delta - {}^j\delta)^2 \tag{15}$$

Similarly,

$$RT \ln {}^if_x = RT \ln ({}^ia/{}^ix) = {}^iV\,{}^j\phi^2\,{}^{ij}A = {}^jV\,{}^j\phi^2\,({}^i\delta - {}^j\delta)^2 \tag{15a}$$

Expanding Equation 15,

$$RT \ln {}^jf_x = {}^jV/[({}^iV\,{}^ix)\,/\,({}^jV\,{}^j\propto + {}^iV\,{}^ix)]^2\,({}^i\delta - {}^j\delta)^2 \tag{15b}$$

In dilute solutions, ${}^jV{}^jX \ll {}^iV{}^iX$, and in infinitely dilute solutions ${}^i\phi = 1$, so

$$RT \ln {}^jf_x^\infty = {}^jV\,{}^{ij}A = {}^iV\,({}^i\delta - {}^j\delta)^2 \tag{16}$$

A more general expression for the limiting infinite dilution activity coefficient of component j with molar volume ${}^jV$ in a multicomponent mixture is

$$RT \ln {}^jf_x^\infty = {}^jV\,({}^i\delta - \bar{\delta})^2 \tag{17}$$

where $\bar{\delta}$ is the volume fraction average Hildebrand parameter (Equation 41). If the i and j molecules differ appreciably in size, inclusion of the Flory-Huggins size effect term is necessary (Section 13.2). This involves replacing $-R \ln {}^ix$ in the entropy of mixing with

$$-R\,[\ln {}^i\phi + {}^j\phi(1 - {}^jV/{}^jV)]$$

and leads to

$$RT \ln {}^if_x = {}^iV\,{}^j\phi^2\,{}^{ij}A + RT[\ln {}^i\phi/x + {}^j\phi(1 - {}^iV/{}^jV)] \tag{18}$$

or

$$\ln {}^ia = {}^iV\,{}^j\phi^2\,{}^{ij}A/(RT) + \ln {}^i\phi + {}^j\phi\,(1 - {}^iV/{}^jV) \tag{19}$$

For solute j,

$$RT \ln {}^jf_x = {}^jV\,{}^i\phi^2\,{}^{ij}A + RT[\ln {}^j\phi/{}^jx + {}^i\phi(1 - {}^jV/{}^iV)] \tag{20}$$

and for infinitely dilute solutions of j in i where ${}^i\phi = 1$ and ${}^j\phi/{}^jx = {}^jV/{}^iV$,

$$RT \ln {}^j f_x^\infty = {}^jV\, {}^{ij}A + RT[\ln({}^jV/{}^iV) + 1 - {}^jV/{}^iV] \tag{21}$$

or

$$\ln {}^j f_x^\infty = {}^jV/(RT)\, {}^{ij}A + {}^{ij}d \tag{21a}$$

where the Flory-Huggins combinatorial size effect term is

$${}^{ij}d = \ln({}^jV/{}^iV) + 1 - {}^jV/{}^iV \tag{21b}$$

The upper critical solution temperature, UCST, (Section 2.7) for an $i-j$ mixture can be found from the Gibbs free energy of mixing (excess term plus ideal term: see Section 2.6),

$$\Delta_m G = {}^E\Delta_m G + {}^I\Delta_m G = RT\,({}^ix \ln {}^ix\, {}^if_x + {}^jx \ln {}^jx\, {}^jf_x) \tag{22}$$

Insertion of the Hildebrand-Scatchard expression for a binary mixture gives

$$\Delta_m G/(RT) = ({}^ix\, {}^iV + {}^jx\, {}^jV)\, {}^{ij}A\, {}^i\phi\, {}^j\phi + {}^ix \ln {}^ix + {}^jx \ln {}^jx \tag{23}$$

$$\Delta_m G/(RT) = [{}^ix\, {}^jx\, {}^iV\, {}^jV\, ({}^i\delta - {}^i\delta)^2]/({}^ix\, {}^iV + {}^jx\, {}^jV) + {}^ix \ln {}^ix + {}^jx \ln {}^jx \tag{23a}$$

The second and third derivatives of the Gibbs free energy of mixing with respect to mole fraction are set equal to zero, corresponding to the situation illustrated in Figure 5b, Chapter 2, and the simultaneous solution yields

$${}^ix = [{}^iV - ({}^iV^2 + {}^jV^2 - {}^iV\, {}^jV)^{1/2}]/({}^iV - {}^jV) \tag{24}$$

$$T_{cs} = 2\, {}^ix\, {}^jx\, {}^iV^2\, {}^jV^2\, {}^{ij}A/[R({}^ix\, {}^iV + {}^jx\, {}^jV)^3] \tag{25}$$

or

$$T_{cs} = 2\, {}^ix\, {}^jx\, {}^iV^2\, {}^jV^2\, ({}^i\delta - {}^i\delta)^2/[R({}^ix\, {}^iV + {}^jx\, {}^jV)^3] \tag{25a}$$

If the molar volumes are close in value, ${}^iV = {}^jV$, and

$$T_{cs} = V({}^i\delta - {}^j\delta)^2/(2R) \tag{26}$$

The relationships between T_{cs} and δ for hydrocarbons and film-forming materials have been explored by Mandik[38]. The rest of the liquid (1)-liquid (2) phase boundary can also be determined, with the equations

$$RT \ln({}^ix_1/{}^ix_2) = {}^jV\, ({}^j\phi_2{}^2 - {}^j\phi_1^2)({}^i\delta - {}^j\delta)^2 \tag{27}$$

$$RT \ln({}^jx_1/{}^jx_2) = {}^iV\, ({}^i\phi_2{}^2 - {}^i\phi_1^2)({}^i\delta - {}^j\delta)^2 \tag{27a}$$

or by means of the corresponding equations with ${}^{ij}A$ replacing $({}^i\delta - {}^i\delta)^2$. The experimental UCST may be used to evaluate ${}^{ij}A$, which is then used to generate the whole phase boundary curve. In partially miscible systems, dispersion forces usually dominate over polar and specific effects, and the regular solution model is usually a good approximation.

If Flory-Huggins entropy is substituted for ideal mixing entropy, the corresponding expression for the critical solution temperature is

$$T_{cs} = 2\, {}^{ij}A\, {}^{i}V\, {}^{j}V/({}^{i}V^{1/2} + {}^{j}V^{1/2})^2$$

or

$$T_{cs} = 2\, ({}^{i}\delta - {}^{j}\delta)^2\, {}^{i}V\, {}^{j}V/({}^{i}V^{1/2} + {}^{j}V^{1/2})^2 \tag{28}$$

For the situation of the limiting miscibility of component j (such as water) in component i (such as a low permittivity organic liquid), where the high-j (organic) solution can be considered to be in equilibrium with pure i (water), it is possible to derive an expression for the solubility of j in i (water in the organic phase). If ${}^{j}f_x = {}^{j}f^{\infty}$, Equation 21 can be written in terms of the mole fraction solubility of j in i, ${}^{j}x_s$ (solubility of water in the organic phase):

$$-\, RT \ln {}^{j}x_s = {}^{j}V\, {}^{ij}A + RT(\ln {}^{j}V \mid {}^{i}V + 1 - {}^{j}V \mid {}^{i}V) \tag{29}$$

If ${}^{j}V \approx {}^{i}V$,

$$-RT \ln {}^{j}x_s = {}^{j}V\, {}^{ij}A \tag{29a}$$

The interchange cohesive pressure ${}^{ij}A$ may be expanded, (Equation 13),

$${}^{ij}A = {}^{j}\delta^2 + {}^{i}\delta^2 - 2\,{}^{ij}c$$

and the value of ${}^{ij}c$ can be used as a measure of the degree of interaction or liquid-liquid (water-organic, for example) complexing.[39] Equation 29a further simplifies with the geometric mean approximation to

$$-RT \ln {}^{j}x_s = {}^{j}V({}^{j}\delta - {}^{i}\delta)^2 \tag{30}$$

$$-RT \ln {}^{i}x_s = {}^{i}V\, ({}^{i}\delta - {}^{j}\delta)^2 \tag{30a}$$

as used in early studies.[40]

4.3 LIMITATIONS OF THE HILDEBRAND-SCATCHARD EQUATION

It should be noted that the Hildebrand-Scatchard equation provides an expression in terms of the constant volume internal energy of mixing, not the constant pressure enthalpy of mixing which is the quantity usually measured in experimental determinations. Although these are identical if there is no volume change on mixing, the effect of volume changes,[41] which are particularly significant at high temperatures, is to produce a disparity between $\Delta_m U_v$ and $\Delta_m H_p$ which is typically of the same order of magnitude as $\Delta_m U_v$.[42]

In the opinion of the author, the Hildebrand-Scatchard expression and the equations derived from it are best considered as relatively simple semi-empirical equations that have sufficient theoretical foundation on a molecular level to work reasonably well, so no formal derivation is provided here. Accounts emphasizing physical significance and making comparisons with other liquid theories have been published.[43-48] Several modifications incoporate polar and association effects (Chapter 5).

It is instructive to summarize the assumptions made in the original derivation as they highlight the strengths and weaknesses of the resulting equation:

1. It is assumed that the interaction forces act between the *centers* of the molecules.
2. It is assumed that the interactions are *additive*: the interaction between a pair of molecules is not influenced by the presence of other molecules.

3. It is assumed that the mixing is *random*, with neither $i-j$, $j-j$, nor $i-j$ nearest neighbor situations being favored, and the distribution is temperature independent.
4. The constant pressure change of volume on mixing is assumed *zero* with the numbers of nearest neighbors of a molecule in the mixture and in the pure state being considered to be the same.
5. The *geometric mean approximation*, Equation 12, already has been discussed.

These assumptions are not, of course, generally valid, but they produce an equation which has proven valuable both in its own right and as a starting point for other empirical expressions.

Entropy factors are not taken into account in regular solution theory, but when cohesion parameters are used to define the limits of solubility (which corresponds to $\Delta_m G_p$ in Equation 22, Chapter 2, being equal to zero), it follows that $\Delta_m H_p = T\Delta_m S_p$, and entropy is an inherent part of the resulting prediction.[49] In order to establish phase conditions, Gibbs free energies or chemical potentials must be evaluated. Lee[45] has developed "irregular solution theory" by incorporating "hard convex body" entropic effects, and has carried out numerical calculations for binary mixtures (Section 7.6).

As well as the theoretical shortcomings, there is a computational problem with cohesion parameters due to the lack of accurate data. Bagley and Scigliano[42] have demonstrated this in the case of a mixture of two typical organic liquids, both with $V = 100$ cm^3 mol^{-1}, but with ΔU values of 3.3 and 3.8 mol^{-1}. If the ΔU value used for the first component is only 5% too high, the correct $\Delta_m U_v$ would be 100% higher. Variations of several percent in reported values of vaporization energies are not unusual, even for common solvents. Another assumption implicit in the Hildebrand-Scatchard equation is that the quantity $(-UV)^{1/2}$ is additive. Not only does this quantity appear to be additive in many systems on a solute-solvent basis, but also to a considerable extent for atoms or groups of atoms within molecules. These additive constants, named group molar attraction constants (Section 6.4), are of considerable value in the estimation of cohesion parameters.

An interesting extension[50-52] of the Hildebrand-Scatchard equations assumes that the intermolecular forces between polyatomic molecules are not of a central type but peripheral in nature, so it is possible to refer the intermolecular potential energy or cohesion energy to the surface of the molecules and to derive a "surface" modification of the Hildebrand-Scatchard equation. Although it has been claimed that this method is more widely applicable than the original, correlations have not been studied extensively so far.

4.4 MIXED LIQUIDS AND MULTICOMPONENT SYSTEMS

Practical good solvents may be blends of poor solvents or even of nonsolvents,[92] and it is therefore important to be able to evaluate the effective cohesion parameters of liquid mixtures. Although other factors such as viscosity, volatility, and cost must also be considered in solvent formulation, the effective cohesion parameters are particularly valuable.

For ternary systems, an example of the simplest Hildebrand parameter approach is the prediction that for three phases to coexist

$$|{}^j\delta - {}^i\delta|/\mathrm{MPa}^{1/2} \leq 5 \leq |{}^k\delta - {}^j\delta|/\mathrm{MPa}^{1/2}$$

as observed for diamyl ether-acetonitrile-glycerol.[53] In a few cases, vaporization enthalpies (1,4-dioxane-water,[54] hexane-cyclohexanone[55]) or condensation enthalpies (trichloromethane-acetone, trichloromethane-ether[56]) of binary mixtures have been determined experimentally, but usually it is necessary to evaluate cohesion parameters from the properties of the pure components. Mixed solvents are commonly encountered in polymer systems(Section

16.6), and in this connection Table 9 of Chapter 8 contains Hildebrand parameters of various liquid mixtures. Binary liquids are also of pharmaceutical importance, and cohesion parameters have been used extensively for this purpose by Martin and co-workers.[57-62] (See also Sections 5.7 and 12.2.)

On the basis of the assumptions made in the derivation of cohesion parameter expressions,[29] the effective Hildebrand parameter $\bar{\delta}$ of a binary liquid mixture is

$$\bar{\delta} = ({}^i\phi\,{}^i\delta + {}^j\phi\,{}^j\delta)/({}^i\phi + {}^j\phi) \tag{31}$$

where ϕ is the volume fraction (Equation 8), predicting that the effective Hildebrand parameter of a mixture is volume-wise proportional to the Hildebrand parameters of its components. When the value of the Hildebrand parameter ${}^k\delta$ of solute k in the mixed solvent lies between the Hildebrand parameters ${}^i\delta$ and ${}^j\delta$ of the two component liquids, the solute should be completely miscible when the ratio ${}^i\phi/{}^j\phi$ is adjusted so that ${}^k\delta = \bar{\delta}$, even if liquids i and j individually are nonsolvents for k.

By definition, the volume fractions must total unity,

$${}^i\phi + {}^j\phi + {}^k\phi = 1$$

so if ${}^k\phi$ is small,

and

$$\bar{\delta} = {}^i\phi\,{}^i\delta + {}^j\phi\,{}^j\delta \tag{32}$$

As a special case, if ${}^iV \approx {}^jV$,

$$\bar{\delta} \approx {}^ix\,{}^i\delta + {}^jx\,{}^j\delta \tag{32a}$$

The more general expression for the effective Hildebrand parameter is

$$\bar{\delta} = \Sigma_i\,{}^i\phi\,{}^i\delta/\Sigma_i\,{}^i\phi \tag{32b}$$

It is of interest that the critical solution temperatures have been shown (for the hydrocarbon-aniline systems) also to be additive on a volume fraction basis:[62]

$$\bar{T}_{cs} = \Sigma_i\,{}^iT_{cs}\,{}^i\phi$$

Equation 32 has been found to give excellent results in predicting the properties of C_4 and C_5 hydrocarbon vapor-liquid systems (Section 11.5) and the solubilities of gases in mixtures of benzene and 1,1,2-trichloro-1,2,2-trifluoroethane (Freon 113) (Section 11.2), but this equation is not always quantitatively accurate.[63] It is certainly true in general terms that the addition of a nonsolvent may improve the solubility of a solute in a solvent, and even that a solute may be soluble in a mixture of two or more nonsolvents. An example of this "cosolvency"[64] is diethyl ether (δ = 15.1 MPa$^{1/2}$) and ethanol (δ = 26.0 MPa$^{1/2}$) as a solvent mixture for cellulose nitrate (δ = 23 MPa$^{1/2}$). However, frequently when the solute solubility (${}^kx_s$) is plotted against $\bar{\delta}$ (a curve which by regular solution theory should be a parabola, as indicated in Section 12.1), asymmetrical curves are observed, with the maximum value of ${}^kx_s$ not appearing at ${}^k\delta$ and the maximum being lower than ideal. For example, the solubility of 2-nitro-5-methylphenol (${}^k\delta$ = 21.9 MPa$^{1/2}$) is shown in Figure 1 for two binary liquid systems, cyclohexane-diiodomethane and ethanol-hexane. Also, it is possible for the "solvent power" of a mixture to be less than that of its components, a phenomenon which has been called "cononsolvency".[65]

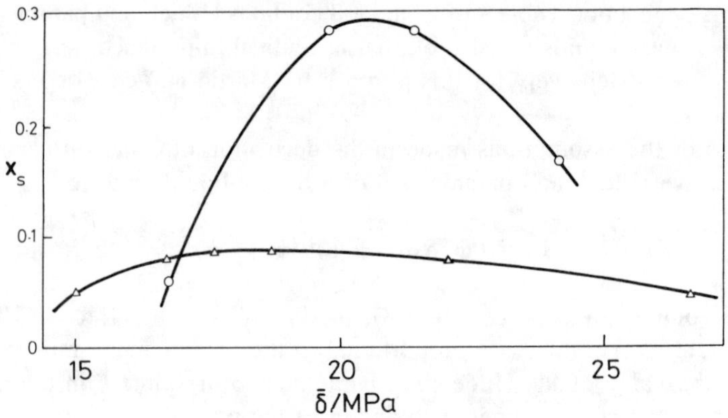

FIGURE 1. Variation in the solubility of 2-nitro-5-methylphenol as a function of the effective Hildebrand parameter of the binary solvent mixtures cyclohexane-diiodomethane (○) and ethanol-hexane (△) at 25°C. (Adapted from Buchowski, H., Domanska, U., and Ksiazczak, A., *Pol. J. Chem.*, 53, 1127, 1979.)

The relations between Hamaker constants, originally developed in a microscopic model for adhesion[66,67] for bodies of materials i and j embedded in medium k are of interest in connection with this, showing on the basis of the geometric mean rule that the interaction of two different materials with a third medium in which they are immersed can be stronger than the interaction between the materials themselves, so spontaneous separation can occur as a result of dispersion forces only.

Purkayastha and Walkley[68] attempted to improve on Equation 32 by defining an effective volume fraction, ϕ^*, which reflects the preferential "solvation" by one of the liquid components:

$$\bar{\delta} = {}^i\delta\,{}^i\phi^* + {}^j\delta\,{}^j\phi^* \tag{33}$$

In those cases where kϕ is small, then

$${}^i\phi^* + {}^j\phi^* = 1 \tag{33a}$$

and the effective volume fractions are related to the true volume fractions ϕ by

$${}^i\phi^*/{}^j\phi^* = [{}^i\phi({}^k\delta - {}^j\delta)^2]/[{}^j\phi({}^k\delta - {}^i\delta)^2] \tag{34}$$

In other words, the effective volume fractions are themselves related to the bulk volume fractions in terms of the Hildebrand parameters, in such a way that they are inversely proportional to the ratios of the enthalpies of mixing of the solute k in each of the pure solvents i and j. An alternative equation based on Christian's[69] idea of a Boltzmann factor in the enthalpy of complex formation was proposed by Nakanishi and Asakura[70] for iodine in mixed liquids. Other approaches to the problem of complex formation in solution are discussed in section 7.9.

More generally, for a multicomponent mixture[29] the excess Gibbs free energy of mixing is

$${}^E\Delta_m G = (\Sigma_i\,{}^ix\,{}^iV)({}^1\!/_2\,\Sigma_i\,\Sigma_j\,{}^{ij}A\,{}^i\phi) + RT\,\Sigma_i\,{}^ix\ln({}^i\phi/{}^ix) \tag{35}$$

where ${}^{ij}A = {}^{ji}A$ and ${}^{ii}A = {}^{jj}A = 0$. The second term is the Flory-Huggins correction for size effect (Section 13.2), but if ${}^ix = {}^i\phi$ for every component, that is if

$$^{i}V = {}^{j}V = {}^{k}V \ldots$$

then

$$^{E}\Delta_{m}G = (\Sigma_{i}\, {}^{i}x\, {}^{i}V)(^{1}/_{2}\, \Sigma_{i}\, \Sigma_{j}\, {}^{ij}A\, {}^{i}\phi\, {}^{j}\phi) \tag{36}$$

The activity coefficient of component n is obtained from the partial derivative (constant T, p, and ${}^{i}n$ for $i \neq k$)

$$RT\ \ln({}^{k}a/{}^{k}x) = RT\ \ln\, {}^{k}f_{x} = [(\partial\ {}^{t}n\ {}^{E}\Delta_{m}G)/\partial\ {}^{k}n] \tag{37}$$

where ${}^{i}n$ is the amount of substance i and ${}^{t}n$ is the total amount of substance. This can be shown to be

$$RT\ \ln\, {}^{k}f_{x} = {}^{k}V\, \Sigma_{i}\Sigma_{j}\, ({}^{ik}A - {}^{1}/_{2}\, {}^{ij}A)\, {}^{i}\phi\, {}^{j}\phi + RT\, [\ln\, ({}^{k}\phi/{}^{k}x) + 1 - {}^{k}\phi/{}^{k}x] \tag{38}$$

Only if ${}^{i}x = {}^{i}\phi$ for all i does this equation reduce to

$$RT\ ln\, {}^{k}f_{x} = {}^{k}V\, \Sigma_{i}\, \Sigma_{j}\, ({}^{ik}A - {}^{1}/_{2}\, {}^{ij}A)\, {}^{i}\phi\, {}^{j}\phi \tag{39}$$

and with Equation 13,

$$RT\ \ln\, {}^{k}f_{x} = {}^{k}V\, ({}^{k}\delta - \bar{\delta})^{2} \tag{40}$$

where the effective Hildebrand parameter of the mixture is

$$\bar{\delta} = \Sigma_{i}\, {}^{i}\phi\, {}^{i}\delta \tag{41}$$

the summation extending over all components, including the solute k when the volume fraction ${}^{k}\phi$ is appreciable.

There has been considerable discussion on the prediction of optimum solvent mixtures for polymers (Section 16.6), and Hansen parameters have proven very convenient for representation as three-dimensional vectors (Section 14.5). The vector representation was introduced by Froehling, Koenhen, Bantjes, and Smolders;[71] this was improved by Rigbi[72] and by Froehling and Hillegers.[73] Behnken[74] extended the analysis to the problem of finding the best solvent mixtures with any number of components by means of readily available multiple regression programs. The only requirement is that the program should permit fitting the general equation

$$y = b_{1}\, x_{1} + b_{2}\, x_{2} + \ldots = b_{k}\, x_{k} \tag{42}$$

without a leading constant term, and the input variables needed for the computer can be calculated by hand from the Hansen parameters of the polymer and liquid components.

Equations incorporating experimentally determined composition dependence of vaporization enthalpy may be used[75] for calculating the Hildebrand parameter of binary liquid mixtures in an extension of the methods described in Section 7.2.

4.5 SOLVENT SPECTRA

A list of liquids may be compiled with gradually increasing Hildebrand parameter values

TABLE 1
Liquids for Hildebrand Parameter Spectra, Divided According to Hydrogen-Bonding Capability

Liquids with Poor Hydrogen-Bonding Capability	
Pentane	14.3
Heptane	15.1
Methylcyclohexane	16.0
Solvesso® 150	17.4
Toluene	18.2
Tetrahydronaphthalene, tetralin	19.4
o-Dichlorobenzene	20.5
1-Bromonaphthalene	21.7
Nitroethane	22.7
Acetonitrile	24.1
Nitromethane	26.0
Liquids with Moderate Hydrogen-Bonding Capability	
Diethyl ether	15.1
Diisobutyl ketone	16.0
Butyl acetate	17.4
Methyl propionate	18.2
Dibutyl phthalate	19.0
1,4-Dioxane	20.3
Dimethyl phthalate	21.9
2,3-Butylene carbonate	24.8
Propylene carbonate	27.2
Ethylene carbonate	30.1
Liquids with Strong Hydrogen-Bonding Capability	
2-Ethylhexanol	19.4
Methyl isobutyl carbinol, 1,2-dimethyl-propanol	20.5
2-Ethylbutanol	21.5
1-Pentanol	22.3
1-Butanol	23.3
1-Propanol	24.3
Ethanol	26.0
Methanol	29.7

Adapted from Burrell, H., *Polymer Handbook*, Brandrup, J. and Immergut, E. H., Eds., John Wiley & Sons, New York, 1966, IV—341 and 1975, IV—337.

to form a ''solvent spectrum''[64,76-83] In its most common form (Table 1), it includes subdivision into categories of hydrogen-bonding ability. The Hildebrand parameter of a solute is taken as the midpoint of the range of liquid Hildebrand parameters which provides complete miscibility. ''Fine control'' over liquid Hildebrand parameter values can be provided using mixtures of liquids, the effective Hildebrand parameter of a mixture being calculated by Equation 31, or to some extent by changing the temperature. The ASTM D3121 test method for polymer solubility ranges (Table 9, Chapter 8) uses liquid mixtures to provide a spectrum of closely spaced Hildebrand parameters. For polymers (section 14.3), either the dissolution behavior of a polymer of the swelling of a slightly cross-linked analog of the polymer of

TABLE 2
Hildebrand Parameters and Molar Volumes for Cholesterol and Triglycerides

		$\delta/MPa^{1/2}$	$V/cm^3\ mol^{-1}$
Cholesterol		20.7	362
Triglyceride	C24:0	20.4	476
	C30:0	20.0	572
	C36.0	19.4	669
	C42:0	18.9	765
	C48:0	18.6	862
	C54:0	17.9	959
	C54:3	18.5	943
	C54:6	19.1	927

Adapted from Arul, J., Boudreau, A., Makhlouf, J., Tardif, R., and Gressier, B., *J. Dairy Res.*, 55, 361, 1988.

interest in a series of liquids may be studied, the polymer being assigned the δ value of the liquid providing the greatest solubility or the maximum swelling coefficient. This method may also be applied to high-molecular-weight mineral oils, dyes and similar compounds, from maxima in solubilities determined in liquids with various Hildebrand parameters. Thus, for the study of the distribution of cholesterol between the liquid fractions of milk fat,[87] Hildebrand parameters were determined by solvent spectrum and molar volumes by group contribution (Table 2). Other physical properties that can be studied include viscosity and related characteristic quantities such as grease dropping point (Section 14.7). In the case of solids (Section 12.1), the solute mole fraction solubility can be plotted against the solvent Hildebrand parameter to provide the solute δ value at the peak of the curve.

4.6 WILSON EQUATION

Both the Hildebrand-Scatchard model for molecules of similar size and the Flory-Huggins model for very dissimilar molecules (Section 13.2) assume zero enthalpy of mixing: athermal solutions.

Wilson[9,46,84] consider the situation where components differ in intermolecular forces as well as molecular size; that is, where there are non-zero enthalpies of mixing. It was assumed that the Gibbs free energy change on mixing was given by a relation similar to the athermal Flory-Huggins equation, expressed in terms of "local" volume fractions ξ, and that probabilities of molecules occurring in each others' vicinities were given by Boltzmann factors, including interaction energies $({}^{ij}g - {}^{ii}g)$ and $({}^{ij}g - {}^{jj}g)$.

For polymers, a similar equation using local volume fractions and interaction energy differences has been developed by Heil and Prausnitz[85,86] (Section 13.8).

REFERENCES

1. **Hildebrand, J. H. and Carter, J. M.,** A study of van der Waals forces between tetrahalide molecules, *J. Am. Chem. Soc.*, 54, 3592, 1932.
2. **Scatchard, G., Wood, S. E., and Mochel, J. M.,** Vapor-liquid equilibrium, V. Carbon tetrachloride-benzene mixtures, *J. Am. Chem. Soc.*, 62, 712, 1940.
3. **Staveley, L. A. K., Tupman, W. I., and Hart, K. R.,** Some thermodynamic properties of the systems benzene + ethylene chloride, benzene + carbon tetrachloride, acetone + chloroform and acetone + carbon disulfide, *Trans. Faraday Soc.*, 51, 323, 1955.

4. **Reed, T. M., III,** The theoretical energies of mixing for fluorocarbon-hydrocarbon mixtures, *J. Phys. Chem.*, 59, 425, 1955; The polarizabilities of molecules in liquid mixtures, *J. Phys. Chem.*, 63, 1798, 1959.
5. **Sonnich Thomsen, E.,** *The Energy of Mixing of Non-Polar Liquids. Reflections upon the Hildebrand Theory,* Thesis, Royal Danish School of Pharmacy, Copenhagen, 1965.
6. **Sonnich Thomsen, E.,** The energy of mixing of non-polar liquids. Reflections upon the theories of Prigogine and Hildebrand, *Dansk Kemi,* 47(3), 35, 1966.
7. **Sonnich Thomsen, E.,** Solubility Parameters, *Farmaceuten,* 29(3), 44, 1966; Present state of the solubility parameter theory, *Mitt. Chem. Gesell, DDR,* 16, 150, 1969.
8. **Sonich Thomsen, E.,** Interaction between unlike nonpolar molecules: correction of the geometric mean rule. 1. Basic derivations. 2. Dilute systems. 3. Energies of mixing, *Acta Chem. Scand.*, 25, 260, 1971; 25, 265, 1971; 26, 2100, 1972.
9. **Acree, W. E., Jr.,** *Thermodynamic Properties of Nonelectrolyte Solutions,* Academic Press, Orlando, FL, 1984.
10. **van der Waals, J. D.,** *Z. Phys. Chem.*, 5, 133, 1890.
11. **van Laar, J. J.,** *Sechs Vortrage uber das Thermodynamische Potential,* F. Vieweg, Braunschweig, 1906.
12. **van Laar, J. J.,** The vapor pressure of binary mixtures, *Z. Phys. Chem.*, 72, 723, 1910; Theory of vapor pressure of binary mixtures, *Z. Phys. Chem.*, 83, 599, 1913; Relation between the deviation of the vapor pressure of binary mixtures of normal compounds from the straight line, and the heat of mixing, *Z. Phys. Chem.*, A. 137, 421, 1928.
13. **Scatchard, G.,** Equilibria in non-electrolyte solutions in relation to the vapor pressures and densities of the components, *Chem. Rev.*, 8, 321, 1931.
14. **Scatchard, G.,** Equilibrium in nonelectrolyte mixtures, *Chem. Rev.*, 44, 7, 1949.
15. **Hildebrand, J. H.,** Solubility, *J. Am. Chem. Soc.*, 38, 1452, 1916.
16. **Hildebrand, J. H.,** Solubility. III. Relative values of internal pressures and their practical application, *J. Am. Chem. Soc.*, 41, 1067, 1919.
17. **Hildebrand, J. H.,** Compressibilities and thermal pressure coefficients of certain liquids, *Phys. Rev.*, 34, 649, 1929.
18. **Hildebrand, J. H.,** Solubility. XII. Regular solutions, *J. Am. Chem. Soc.*, 51, 66, 1929.
19. **Hildebrand, J. H.,** *Solubility of Non-Electrolytes,* 2nd ed., Reinhold, New York, 1936; Russian translation, GONTI, Moscow-Leningrad, 1938.
20. **Hildebrand, J. H.,** A critique of the theory of solubility of non-electrolytes, *Chem. Rev.*, 44, 37, 1949.
21. **Hildebrand, J. H.,** The term 'regular solution', *Nature (London),* 168, 868, 1951.
22. **Hildebrand, J. H.,** A simple correlation of gas solubilities, *J. Phys. Chem.*, 58, 671, 1954.
23. **Hildebrand, J. H.,** Order from chaos, *Science,* 150, 441, 1965.
24. **Hildebrand, J. H.,** Operation on swollen theories with Occam's razor, in *Structure-Solubility Relationships in Polymers* (Proc. Symp., 1976), Harris, F. W. and Seymour, R. W., Eds., Academic Press, New York, 1977, 1.
25. **Hildebrand, J. H.,** Absence of outer non-bonding electrons in methyl groups affects solubility parameters, *Ind. Eng. Chem. Fundam.*, 17, 365, 1978.
26. **Hildebrand, J. H.,** An improvement in the theory of regular solutions, *Proc. Natl. Acad. Sci. U.S.A.*, 76, 6040, 1979.
27. **Hildebrand, J. H. and Dymond, J. H.,** Effect of methyl groups upon the solvent power of aliphatic liquids, *Proc. Natl. Acad. Sci. U.S.A.*, 54, 1001, 1965.
28. **Hildebrand, J. H. and Lamoreaux, R. H.,** Solubility of gases in liquids; fact and theory, *Ind. Eng. Chem. Fundam.*, 13, 110, 1974.
29. **Hildebrand, J. H., Prausnitz, J. M., and Scott, R. L.,** *Regular and Related Solutions,* van Nostrand-Reinhold, Princeton, NJ, 1970.
30. **Hildebrand, J. H. and Scott, R. L.,** *Solubility of Non-Electrolytes,* 3rd ed., Reinhold, New York, 1950; Dover, New York, 1964.
31. **Hildebrand, J. H. and Scott, R. L.,** *Regular Solutions,* Prentice-Hall, Englewood Cliffs, NJ, 1962.
32. **Scott, R. L.,** *Thesis,* Princeton, 1945.
33. **Scott, R. L.,** The solubility of fluorocarbons, *J. Am. Chem. Soc.*, 70, 4090, 1948.
34. **Scott, R. L.,** Theoretical models and real solutions, *Disc. Faraday Soc.*, 15, 44, 1953.
35. **Staverman, A. J.,** Cohesion forces and phase theory of binary liquid systems, *Rec. Trav. Chim. Pays-Bas,* 60, 827, 1941.
36. **Edelmann, E.,** The cohesion factor concept and the solubility of macromolecular substances, *FATIPEC (Fed. Assoc. Tech. Ind. Peint. Vernis, Emaux Encres Impr. Eur. Cont.) Congr.*, 7, 160, 1964.
37. **Harris, H. G. and Prausnitz, J. M.,** Thermodynamics of solutions with physical and chemical interactions, *Ind. Eng. Chem. Fundam.*, 8, 180, 1969.
38. **Mandik, L.,** Dissolving capacity of hydrocarbon solvents for film forming substances. Critical solution point and solubility parameter, *Chem. Prum.*, 16, 545, 1966.

39. **Kirchnerova, J. and Cave, G. C. B.,** The solubility of water in low-dielectric systems. *Can. J. Chem.*, 54, 3909, 1976.
40. **Black, C., Joris, G. G., and Taylor, H. S.,** The solubility of water in hydrocarbons, *J. Chem. Phys.*, 16, 537, 1948.
41. **Battino, R.,** Volume changes on mixing for binary mixtures of liquids, *Chem. Rev.*, 71, 5, 1971.
42. **Bagley, E. B. and Scigliano, J. M.,** Polymer solutions, in *Solutions and Solubilities,* Part II, Dack, M. R. J., Ed., John Wiley & Sons-Interscience, New York, 1976, Chap. 16.
43. **Ashton, N. F., McDermott, C., and Brench, A.,** Chemistry of extraction of nonreacting solutes, in *Handbook of Solvent Extraction,* Lo., T. C., Baird, M. H. I., and Hanson, C., Eds., John Wiley & Sons-Interscience, New York, 1983, chap. 1, 3.
44. **Kaelble, D. H.,** *Physical Chemistry of Adhesion,* John Wiley & Sons-Interscience, New York, 1971.
45. **Lee, L. L.,** A molecular theory of solubility parameters: generalization to polar fluid mixtures, *Fluid Phase Equil.*, 35, 77, 1987.
46. **Prausnitz, J. M.,** *Molecular Thermodynamics of Fluid Phase Equilibria,* Prentice-Hall, Englewood Cliffs, NJ, 1969.
47. **Prausnitz, J. M., Lichtenthaler, R. N., and Gomez de Azevedo, E.,** *Molecular Thermodynamics of Fluid Phase Equilibria,* 2nd ed., Prentice-Hall, Englewood Cliffs, NJ, 1986.
48. **Brandts-Buys, L. and Huyskens, P.,** Limitations of the concept of 'solubility parameter', *FATIPEC (Fed. Assoc. Tech Ind. Peint. Vernis, Emauz Encres Impr. Eur. Cont.) Congr.*, 16 (Liege), 191, 1982.
49. **Hansen, C. M.,** Characterization of surfaces by spreading liquids, *J. Paint Technol.*, 42, 660, 1970.
50. **Drahos, J.,** Liquid-vapor equilibrium. LXXVII. Comparison of the volume and surface modifications of the regular solution model, *Collect. Czech. Chem. Commun.*, 43, 805, 1978.
51. **Drahos, J., Wichterle, I., and Hala, E.,** A generalized method for calculation and prediction of vapor-liquid equilibria at high pressures, *Fluid Phase Equil.*, 1, 1973, 1977/78.
52. **Erdös, E.,** Liquid-vapor equilibriums. XV. Calculation of the constants in the van Laar equation from the properties of the pure components, *Collect. Czech. Chem. Commun.*, 21, 1528, 1956.
53. **Sazonov, V. P. and Gudkina, L. V.,** Liquid-phase equilibria between the components of the ternary system diamyl ether-acetonitrile-glycerol, *Zh. Prik. Khim.*, 47, 2515, 1974; *J. Appl. Chem. U.S.S.R.*, 47, 2587, 1974.
54. **Stallard, R. D. and Amis, E. S.,** Heat of vaporization and other properties of dioxane, water and their mixtures, *J. Am. Chem. Soc.*, 74, 1781, 1952.
55. **Iscan, M.,** The solubility of aniline in the mixture of *n*-hexane + cyclohexane, *Istanbul Univ. Fen. Fak. Mecm. Ser.C.*, 39, 195, 1974—1976 (pub. 1977).
56. **Lui, P. C. K. and Donnelly, H. G.,** Determination of liquid-mixture solubility parameters and interaction energy densities, *Am. Inst. Chem. Eng. J.*, 19, 574, 1973.
57. **Martin, A., Paruta, A. N., and Adjei, A.,** Extended Hildebrand solubility approach: methylxanthines in mixed solvents, *J. Pharm. Sci.*, 70, 1115, 1981.
58. **Martin, A. and Miralles, M. J.,** Extended Hildebrand solubility approach: solubility of tolbutamide, acetohexamide, and sulfisomidine in binary solvent mixtures, *J. Pharm. Sci.*, 71, 439, 1982.
59. **Martin, A., Wu, P. L., Adjei, A., Lindstrom, R. E., and Elworthy, P. H.,** Extended Hildebrand solubility approach and the log linear solubility equation, *J. Pharm. Sci.*, 71, 849, 1982.
60. **Martin, A., Wu, P. L., Adjei, A., Mehdlzadeh, M., James, K. C., and Metzler, C.,** Extended Hildebrand solubility approach: testosterone and testosterone propionate in binary solvents, *J. Pharm. Sci.*, 71, 1334, 1982.
61. **Wu, P. L. and Martin, A.,** Extended Hildebrand solubility approach: *p*-hydroxybenzoic acid in mixtures of dioxane and water, *J. Pharm. Sci.*, 71, 849, 1982.
62. **Mandik, L.,** Additivity of the critical solution temperatures of hydrocarbons with aniline, *Collect. Czech. Chem. Commun.*, 31, 3869, 1966.
63. **Purkayastha, A. and Walkley, J.,** Studies in solubility parameter theory for mixed solvent systems, *Can. J. Chem.*, 50, 834, 1972.
64. **Burrell, H.,** Solubility parameters, *Interchem. Rev.*, 14, 3—16 and 31—46, 1955.
65. **Wolf, B. A. and Willms, M. M.,** Measured and calculated solubility of polymers in mixed solvents: cononsolvency, *Makromol. Chem.*, 179, 2265, 1978.
66. **Hamaker, H. C.,** The London-van der Waals attraction between spherical particles, *Physica (Utrecht),* 4, 1058, 1937.
67. **Visser, J.,** Colloid and other forces in particle adhesion and particle renewal (a review), in *Deposition and Filtration of Particles from Gases and Liquids,* Society of Chemical Industry, London, 1978.
68. **Purkayastha, A.,** *Studies in Regular Solution Theory,* Thesis, Simon Fraser University, British Columbia, 1971; *Diss. Abstr. Int. B,* 32, 1406, 1971.
69. **Christian, S. D.,** Solvent effects on molecular complex formation. Pyridine-iodine system, *J. Am. Chem. Soc.*, 91, 6514, 1969.

70. **Nakanishi, K. and Asakura, S.,** Solubility of iodine in mixed solvents. A case study of preferential solvation in non-polar and associated solutions, *J. Phys. Chem.*, 81, 1745, 1977.
71. **Froehling, P. E., Koenhen, D. M., Bantjes, A., and Smolders, C. A.,** Swelling of linear polymers in mixed swelling agents; predictability by means of solubility parameters, *Polymer,* 17, 835, 1976.
72. **Rigbi, Z.,** Prediction of swelling of polymers in 2 and 3 component mixtures, *Polymer,* 19, 1229, 1978.
73. **Froehling, P. E. and Hillegers, L. T.,** Solubility parameters of ternary solvent mixtures. Calculation of the solvent composition with maximum polymer interaction, *Polymer,* 22, 261, 1981.
74. **Behnken, D. W.,** Calculation of optimum multicomponent solvent compositions for polymers using a multiple regression program, *Polymer,* 24, 193, 1983.
75. **Mizerovskii, L. N. and Belova, O. V.,** Calculation of the solubility parameter of binary mixtures of liquids and copolymers, *Vysokomol. Soedin. Ser. A.*, 29, 1236, 1987; *Polym. Sci. U.S.S.R.*, 29, 1987.
76. **Burrell, H.,** Solubility parameters for film formers, *Off. Dig. Fed. Paint Varn. Prod. Clubs,* 27, 726, 1955.
77. **Lazurin, E. A., Samorodov, V. T., and Kosmodemyanskii, L. V.,** Determination of solubility parameters of mineral oils, *Promst. Sint. Kauch.* No.8, 10, 1980; *Int. Polym. Sci. Tech.*, 8(12), T12, 1981.
78. **Piskarev, V. A. and Radushkevich, R. L.,** Method for determining the compatibility of bitumens with elastomers, *Polim. Stroit. Mater.*, 42, 163, 1975.
79. **Vasilakos, N. P. and Lawson, D. D.,** Solubility parameter spectroscopy in coal extraction, *Liq. Fuels Technol.*, 3, 305, 1985.
80. **Siddiqui, S. A.,** Studies of solvent dyeing. Part I: Preparation of disperse dyes and determination of their solubility parameters, *Text. Res. J.*, 51, 527, 1981.
81. **Siddiqui, S. A. and Needles, H. L.,** Solubility parameters, *Text. Res. J.*, 52, 570, 1982.
82. **Siddiqui, S. A.,** Determining the solubility parameter of dyes; Dye solubility by solubility method, *Am. Dyest. Rep.*, 72(3), 30, 1983; 2(4), 36, 1983.
83. **Burrell, H.,** Solubility parameter values, in *Polymer Handbook,* Brandrup, J. and Immergut, E. H., Eds., John Wiley & Sons-Interscience, New York, 1966, IV—341; 2nd ed., 1975, IV—337.
84. **Wilson, G. M.,** Vapor-liquid equilibrium. XI. A new expression for the excess free energy of mixing, *J. Am. Chem. Soc.*, 86, 127, 1964.
85. **Heil, J. F.,** *Phase Equilibria in Multi-Component Polymer Solutions,* Doctoral Dissertation, University of California, Berkeley, 1965; *Diss. Abstr.*, 26, 7196, 1966.
86. **Heil, J. F. and Prausnitz, J. M.** Phase equilibria in polymer solutions, *Am. Inst. Chem. Eng. J.*, 12, 678, 1966.
87. **Arul, J., Boudreau, A., Makhlouf, J., Tardif, R., and Gressier, B.,** Distribution of cholesterol in milk fat fractions, *J. Dairy Res.*, 55, 361, 1988.
88. **Tucker, S. A., Murral, D. J., Oswalt, B. M., Halmic, J. L., and Acree, W. E., Jr.,** Solubility of anthracene in binary toluene and alkane solvent mixtures, *Phys. Chem. Liq.*, 18, 279, 1988.
89. **Bissell, M., Chittick, C. E., and Acree, W. E., Jr.,** Solubility of anthracene in binary solvent mixtures containing tetrahydrofuran, *Fluid Phase Equil.*, 41, 187, 1988.
90. **McCargar, J. W. and Acree, W. E., Jr.,** Thermochemical investigations of associated solutions, VI. Mole fraction versus volume fraction based equilibrium constants; Solubility of carbazole in binary chloroalkane and dibutyl ether solvent mixtures, *J. Solution Chem.*, 17, 1081, 1988 and 18, 151, 1989.
91. **Wallach, J. R., Tucker, S. A., Oswalt, B. M., Murral, D. J., and Acree, W. G., Jr.,** Solubility of pyrene in binary solvent mixtures containing dibutyl ether, *J. Chem. Eng. Data,* 34, 70, 1989.
92. **Durbut, P. and Broze, G.,** Application of the solubility parameter theory to fatty soil removal, *Commun. Jorn. Cons. Esp. Deterg.*, 20, 383, 1989.

Chapter 5

EXPANDED COHESION PARAMETERS

So far in this development of the cohesion parameter approach to miscibility and other interaction properties, the existence of polar interactions and of specific effects such as hydrogen bonding has been neglected while the traditional Hildebrand description of regular or near-regular systems was explored.

Accommodation of interactions which do not conform to geometric mean behavior (Section 4.1) has been approached in two main ways. On the one hand empirical corrections have been made for geometric mean deviations (Section 5.7), and on the other the Hildebrand parameter has been subdivided into various component cohesion parameters. The most general formalism was introduced by Karger, Keller, Snyder, and Eon,[1-8] who developed it for the optimization of chromatographic selectivity, although it has proved too cumbersome for widespread use so far. The most widely used method has been a three-component parameter proposed by Hansen[9-30] for the empirical description of polymer-liquid systems (Section 5.9). There have been several other variations developed for practical applications, the one with most promise being perhaps that of Beerbower, Martin, and Wu,[31,32] (Section 5.12) which provides an acceptable compromise between rigor and simplicity.

To be generally useful, theories or models attempting to systematize the behavior of matter must deal with molecular interactions by providing information about their origins and natures as well as about their strengths. The cohesive properties characteristic of the condensed states of matter are produced by the various intermolecular forces described in Chapter 3. The cohesive pressures ${}^ic$, ${}^jc$, and ${}^{ij}c$ represent the resultant effect of these forces acting between molecules of types i and j, and from Equation 10, Chapter 4, the cohesive pressure of the mixture relative to the components is

$$ {}^{ij}A = {}^ic + {}^jc - 2\,{}^{ij}c $$

where

$$ {}^ic = {}^i\delta^2 \text{ and } {}^jc = {}^j\delta^2 $$

5.1 DISPERSION

Dispersion or London forces, which can be considered as arising from the fluctuating dipoles which result from a positive nucleus and a negative electron ''cloud'' in each atom, occur in all molecules, whether polar or not. The dispersion cohesive pressure of a pure material i is here denoted ${}^ic_d$, and the corresponding cohesion parameter, δ_d, is defined by

$$ -\,{}^iU_d/{}^iV = {}^ic_d = {}^i\delta_d^2 \qquad (1) $$

It can be shown on the basis of London theory that the nonpolar, dispersive interactions between unlike molecules of types i and j provide a contribution to the cohesive pressure which is approximated by the geometric mean of the individual values and is given by

$$ {}^{ij}c_d = ({}^ic_d\,{}^jc_d)^{1/2} = {}^i\delta_d\,{}^j\delta_d \qquad (2) $$

A simple interpretation of this ''geometric mean'' behavior is that the interaction is of a ''symmetrical'' nature: each member of a pair of molecules interacts by virtue of the same property, the polarizability. (The ionization potentials, which also appear in Equation 6, Chapter 3, have similar values for most organic compounds, i.e., ${}^iI \approx {}^jI$.) It follows that

$$^{ij}A_d = {}^i\delta_d^2 + {}^j\delta_d^2 - 2\,{}^i\delta_d\,{}^j\delta_d = ({}^i\delta_d - {}^j\delta_d)^2 \tag{3}$$

For nonpolar molecules, dispersion forces should make the only contributions to the cohesive pressure, so from Equation 5, Chapter 2,

$$^i\delta_d = (\Delta\, U \,/\, {}^iV)^{1/2} \tag{4}$$

However, it is of interest that hydrocarbons possessing methyl groups show iodine solution behaviors consistent with Hildebrand parameter values significantly greater than those given by Equation 4, even though the solutions are violet in color and therefore "non-associated" (unlike the straw-colored aqueous solutions, for example). In fact, the extent of deviation from Equation 4, $[\delta - (\Delta\, U/{}^iV)^{1/2}]$, is proportional to the number of methyl groups in these molecules,[33,34] a fact attributed to a variation in the extent of the dispersion forces in molecules with a high proportion of saturated C–H bonding.

The dispersion cohesion parameter can be calculated by the homomorph method (Sections 6.7 and 6.8) or from the refractive index (Section 8.2). Beerbower and Jensen[35] have reviewed the correlation between dispersion properties and the "softness" in the Pearson "hard or soft" acid-base concept (Section 5.12).

5.2 ORIENTATION

Orientation effects that result from dipole-dipole or Keesom interactions occur between molecules which have permanent dipole moments. The orientation cohesive pressure of a pure material i is denoted ${}^ic_o$, and the corresponding orientation cohesion parameter, ${}^i\delta_o$, is defined by

$$-\,{}^iU_o \,/\, {}^iV = {}^ic_o = {}^i\delta_o^2 \tag{5}$$

Like dispersion forces, these are "symmetrical" interactions depending on the same property of each molecule, which in this case is the dipole moment as shown in Equation 4, Chapter 3. It follows that the geometric mean approximation is well obeyed for orientation interactions even between unlike molecules.[36] For polar molecules that may be represented by spherical force fields with small ideal dipoles at their centers, this contribution to the cohesive pressure in mixtures of i and j molecules is

$$^{ij}c_o = ({}^ic_o\,{}^jc_o)^{1/2} = {}^i\delta_o\,{}^j\delta_o \tag{6}$$

The interchange cohesive pressure due to orientation is

$$^{ij}A_o = ({}^i\delta_o - {}^j\delta_o)^2 \tag{7}$$

Keller, Karger, and Snyder[5] and Munafo, Buchmann, Hô Nam-Tran and Kesselring[37] are among those to have discussed the evaluation of orientation and induction components from dipole moments and relative permittivities: see also Section 8.2.

5.3 INDUCTION

Dipole induction effects arise from dipole-induced dipole or Debye interactions occurring between molecules with permanent dipole moments and any other neighboring molecules, whether polar or not, and result in an induced nonuniform charge distribution. In contrast to dispersion and orientation interactions, dipole induction interactions are "unsymmetrical", involving the dipole moment of one molecule and the polarizability of the other (Chapter

3). Consequently, the cohesive pressure term for induction in a pure material i involves the product ${}^i\delta_i\,{}^i\delta_d$, where ${}^i\delta_i$ is the induction cohesion parameter. Similarly, in a mixture of i and j

$$^{ij}c_i = {}^i\delta_i\,{}^j\delta_d + {}^j\delta_i\,{}^i\delta_d \tag{8}$$

Therefore it can be shown that

$$^{ij}A_i = 2\,{}^i\delta_i\,{}^i\delta_d + 2\,{}^j\delta_i\,{}^j\delta_d - 2\,{}^j\delta_i\,{}^i\delta_d - 2\,{}^i\delta_i\,{}^j\delta_d \tag{9}$$

$$= 2({}^i\delta_d - {}^j\delta_d)({}^i\delta_i - {}^j\delta_i) \tag{10}$$

5.4 LEWIS ACID-BASE

Lewis acid-base or electron donor-acceptor interactions, which have been reviewed frequently,* can be denoted by the equation

A	+	D	$\rightleftharpoons$	$\overset{\delta-}{\mathrm{A}}$........$\overset{\delta+}{\mathrm{D}}$
Lewis acid		Lewis base		Lewis acid-base complex
Electron-pair acceptor (EPA)		Electron-pair donor (EPD)		Addition compound
Electrophile		Nucleophile		Adduct

The Lewis acid-base complex is formed by an overlap between a filled electron orbital of sufficiently high energy in the donor molecule and a vacant orbital of sufficiently low energy (high electron affinity) in the acceptor molecule. This type of interaction differs from a "normal" chemical bond in that only one type of molecule (the donor) supplies the pair of electrons, rather than each type of molecule supplying one electron. More than one electron must be involved, and coordination of the Lewis acid to the Lewis base must occur. Electron-pair donors are of three types:

1. n-EPD molecules, where the electrons donated are the lone pair of n (nonbonding) electrons of hetero atoms in compounds such as R_2O, R_3N, R_2SO, where R is an alkyl group
2. σ-EPD molecules, which donate the electron-pair of a σ-bond, as in alkyl halides and cyclopropane
3. π-EPD molecules, using the pair of π-electrons of unsaturated and aromatic compounds such as alkenes, alkylbenzenes and polycyclic aromatics

Electron-pair acceptor molecules also may be divided into three groups:

1. v-EPA, in which the lowest orbital is a vacant (v) valence orbital of a metal atom, such as Ag^+ and some organometallics
2. σ-EPA, which use a nonbonding σ orbital, as in halogens or interhalogens
3. π-EPA, in the case of molecules with a π-bond system (aromatic and unsaturated compounds) with electron-withdrawing substituents, for example, aromatic polynitro compounds, halogenated benzoquinones, and tetracyanoethylene

* For example: Arnett;[38] Bent;[39,40] Coetzee and Ritchie;[41] Drag, Lim, and Matwiyoff;[42,43,47] Jensen;[44-46] Reichardt;[48] and Wrona.[49]

All nine types of EPD-EPA complexes resulting from combinations of these donors and acceptors exist, with bond strengths ranging from high values (n-EPD/v-EPA) to very weak interactions (π-EPD/π-EPA effects between neutral molecules). Ion solvation (Section 12.3) also may be treated in terms of Lewis acid-Lewis base interactions.

Many authors have pointed out that electron donating and accepting properties can be discussed in terms of acid-base cohesion parameters.[1-6,50-55] Lewis acid-base interactions are "unsymmetrical", involving a donor and an acceptor with different roles (rather than two equivalent participants as in dispersion interactions, which are "symmetrical"). It is apparent, therefore, that it is necessary to use two separate cohesion parameters for each partner to characterize these interactions, and this may be done in terms of a Lewis acid cohesion parameter (δ_a) and a Lewis base cohesion parameter (δ_b), in a manner analogous to that for induction interactions:

$$ {}^{ij}A_{ab} = 2\,({}^{i}\delta_a - {}^{j}\delta_a)({}^{i}\delta_b - {}^{j}\delta_b) \qquad (12) $$

Clearly from this equation the maximum interaction (${}^{ij}A_{ab}$ large and negative) occurs when

$$ {}^{i}\delta_a = {}^{j}\delta_b = 0,\; {}^{ij}A_{ab} = -\,2\,{}^{j}\delta_a\,{}^{i}\delta_b \qquad (13) $$

and when

$$ {}^{j}\delta_a = {}^{i}\delta_b = 0,\; {}^{ij}A_{ab} = -\,2\,{}^{i}\delta_a\,{}^{j}\delta_b \qquad (14) $$

When ${}^{ij}A_{ab}$ is large and negative, exothermic mixing is possible, in contrast to mixing being restricted to athermic or endothermic processes, which is the case when only dispersion and polar forces exist. These acid-base cohesion parameters have something in common with the four-parameter acid-base equation of Drago (Section 7.12).

In those situations where donor-acceptor complexes are formed to such an extent that there is an appreciable proportion of identifiable new compound in the mixture, it may be necessary to evaluate the cohesion parameters of this new species so that it can be used in the estimation of thermodynamic properties such as activity coefficients. Approaches to this problem have included using the arithmetic mean of the donor and acceptor parameters[56] and volume-weighted averages (Equation 31, Chapter 4).[57] The topic is considered further in Section 7.9.

Acid-base interactions are not restricted to solutions, but occur in all types of systems, including pigment dispersions[54,55] (as illustrated in later chapters).

5.5 HYDROGEN BONDING ASSOCIATION

Hydrogen-bonding interaction is a particular type of Lewis acid-base reaction in which the electron acceptor is a Brönsted acid. A convenient definition is that a hydrogen bond is a second bond formed to another atom by a covalently bound hydrogen atom. In the following scheme, atoms X and Y have electronegativities (relative tendencies of the bonded atom to attract electrons) higher than that of H, for example, C, N, P, O, S, F, Cl, Br, or I:

R—X—H +	:Y—R ⇆	R—X—H......Y=R	
Electron-pair acceptor	Electron-pair donor	Hydrogen bond	
(EPA)	(EPD)		(15)
Proton donor	Proton acceptor		
Hydrogen-bond donor	Hydrogen-bond acceptor		
Brönsted acid	Base		

The classification by Pimentel and McClellan[58] of liquids according to their hydrogen-bonding characteristics has been used widely:[53,59]

- *Proton donor*, such as trichloromethane
- *Proton acceptor*, such as ketones, aldehydes, esters, ethers, tertiary amines, aromatic hydrocarbons, alkenes
- *Proton donor/acceptor*, such as alcohols, carboxylic acids, water, primary and secondary amines
- *Proton non-donor/non-acceptor* such as alkanes, carbon disulfide, tetrachloromethane

On this basis, it is possible to predict immediately in a qualitative way the extent of contribution that hydrogen bonding is likely to make to the interchange cohesive pressure, ${}^{ij}A_{ab}$. Any net increase in the extent of hydrogen bonding as a result of mixing decreases ${}^{ij}A_{ab}$ and so favors miscibility. Thus, miscibility is enhanced if hydrogen bonds are formed when two liquids without hydrogen bonding (such as trichloromethane and acetone) are mixed, while it is reduced if hydrogen bonding in the component liquids is destroyed (as in water being mixed with an alkane).

It is apparent that two pairs of parameters to characterize these interactions are necessary, and this may be done using the Lewis acid-base formalism, Equation 12.

$$ {}^{ij}A_{ab} = 2\,({}^{i}\delta_a - {}^{j}\delta_a)({}^{i}\delta_b - {}^{j}\delta_b) \qquad (17) $$

In this situation the parameters δ_a and δ_b are then equivalent to the parameters σ and τ proposed by Small.[60] Rider[61,62] developed an alternative two-parameter donor-acceptor model for predicting polymer-liquid interactions (Section 7.12); the parameters in this model are determined from experimental information including frequency shifts, solubilities, and acid-base properties.

Variations in properties resulting from hydrogen-bonding interactions can be looked at from other points of view. Chastrette and co-workers[63] have observed from multivariate statistical analysis of solubility information that a liquid may have more than one point in hyperspace associated with it: for different chemical applications one point is more relevant than the other.

Another approach to hydrogen bonding is to consider it as a chemical reaction (as suggested in the previous section), producing new molecules which then interact with the other components. Renon and Prausnitz,[64] Wiehe and Bagley,[65] and Bagley and Chen[66] have studied the relatively simple situation of an alcohol mixed with an alkane. This can be considered as a self-associated component (the alcohol) interacting with a nonassociated component (the alkane), and suggests a clear division into "chemical" (subscript C) and "physical" (subscript P) interactions. It is assumed that the alcohol exists in solution in the form of linear, hydrogen-bonded polymers which interact with each other by physical processes only. In terms of excess Gibbs free energies of mixing,

$$ {}^{E}\Delta_m G = {}^{E}\Delta_m G_C + {}^{E}\Delta_m G_P \qquad (18) $$

where ${}^{E}\Delta_m G_P$ is given by an expression of the form of Equation 35, Chapter 4, and ${}^{E}\Delta_m G_C$ is an expression involving an equilibrium constant describing the hydrogen bonding. A similar procedure was used for acetylene in various hydrogen-bonding liquids.[67] Kirchnerova and Cavé[68] assumed water-organic complexes when discussing the solubility of water in low permittivity liquids in terms of cohesion parameters, and defined $-U_h$ as the cohesive energy corresponding to the energy change associated with the notional process of real water being converted into a hypothetical liquid of dipolar monomers.

In situations where only hydrogen bond dimerization is significant, correction can be made in the cohesion parameter values in the simpler manner[69] illustrated in Equation 20, Chapter 6. Examples are carboxylic acids,

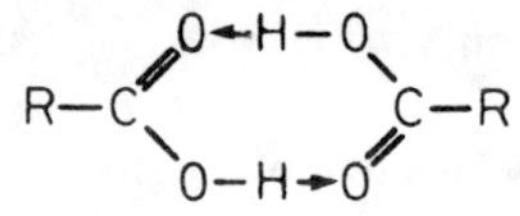

and glycol ethers such as Cellosolves® and Carbitols®

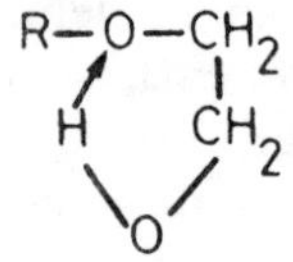

In such cases the result is that the dimer has a cohesion parameter significantly different from that of the monomer, and so is expected to have different solubility characteristics. This is in agreement with the observation that a material like acetic acid is soluble in such diverse liquids as water and heptane. Thus, in polar liquids these compounds are capable of interacting as if they are polar, while in nonpolar liquids the polar interactions are ''self-contained'' or ''internal'' and the dimeric molecules tend to behave in a nonpolar manner. Hoy[69] proposed that this ability to assume the character of the environment be termed *chameleonic*, after the reptile which is able to adopt the color of its background. The dimerization of several carboxylic acids has been assessed in terms of cohesion parameters.[70-72]

Huyskens, Haulait-Pirson, Siegel and Kapuku[153,154,162] preferred to consider hydrogen bonds in liquids as the cohesive forces with the longest lifetime, and introduced the concept of ''mobile disorder''. Accommodation of such variations in cohesion properties by means of corrections to the geometric mean approximation is discussed in Section 5.7. Nisbet[75] took intramolecular hydrogen bonding into account by considering the formation of all possible pseudo ring structures in alcohol molecules, with interaction energies evaluated for dispersion, polar, and hydrogen-bonding components. Three- and four-membered rings are illustrated by:

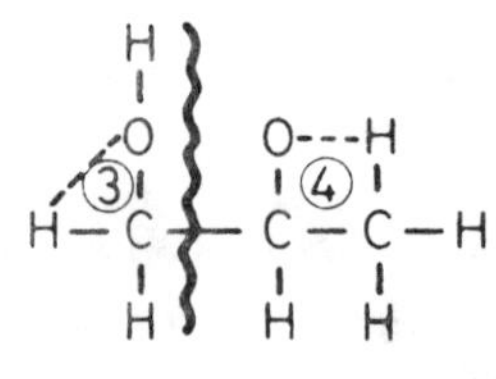

Martin, Wu, Liron, and Cohen[73] proposed an approach to the chameleonic behavior problem in drug solutions which assumed a virtual solute cohesion parameter Λ made up of a constant, invariant Hildebrand parameter δ and a variable (chameleonic) term Γ which depends on the Walker solute-solvent interaction parameter ${}^{ij}K$ (see Section 5.7):

$$ {}^{j}\Lambda^2 = {}^{j}\delta^2 + {}^{j}\Gamma^2 \qquad (19) $$

The variable cohesion parameter of a solute obtained from observations of interactions with relatively polar liquids would then be designated not ${}^{j}\delta$ but ${}^{j}\Lambda$, and the (invariant) Hildebrand parameter would be determined from vaporization enthalpy as proposed originally, or by interaction with relatively nonpolar liquids with which it formed near-regular solutions.

Kamlet, Doherty, Taft, and Abraham[74] incorporated cohesion parameters in their linear solvation energy approach to self-association in water and alcohols; further details appear in Chapter 8.

5.6 COMBINATION OF COMPONENT COHESION PARAMETERS

One of the assumptions central to the cohesion approach to interactions is that the various contributions to the cohesive pressure of a fluid (either pure or mixed) are additive, so the interchange cohesive pressure or exchange energy density for a system is:

$$ {}^{ij}A = {}^{ij}A_d + {}^{ij}A_o + {}^{ij}A_i + {}^{ij}A_{ab} \tag{20} $$

$$ = ({}^i\delta_d - {}^j\delta_d)^2 + ({}^i\delta_o - {}^j\delta_o)^2 + 2({}^i\delta_d - {}^j\delta_d)({}^i\delta_i - {}^j\delta_i) + 2({}^i\delta_a - {}^j\delta_a)({}^i\delta_b\,{}^j\delta_b) \tag{21} $$

$$ = {}^i\delta_t^2 + {}^j\delta_t^2 - 2\,{}^i\delta_d{}^j\delta_d - 2\,{}^i\delta_o{}^j\delta_o - 2{}^i\delta_i\,{}^j\delta_d - 2\,{}^j\delta_i\,{}^i\delta_d - 2\,{}^i\delta_a{}^j\delta_b - 2\,{}^j\delta_a{}^i\delta_b \tag{22} $$

where ${}^i\delta_t$ is the Hildebrand parameter or total cohesion parameter for component i (evaluated from vaporization energy — see Chapter 7) and related to the component parameters by

$$ {}^i\delta_t^2 = {}^i\delta_d^2 + {}^i\delta_o^2 + 2\,{}^i\delta_i{}^i\delta_d + 2\,{}^i\delta_a{}^i\delta_b \tag{23} $$

or, omitting the superscript i and subscript t for simplicity when there is no ambiguity possible:

$$ \delta^2 = \delta_d^2 + \delta_o^2 + 2\delta_i\delta_d + 2\delta_a\delta_b \tag{24} $$

It is clear that two types of term appear in these equations, quadratic terms (symmetrical) for dispersion and orientation, and double product or cross terms (unsymmetrical) for induction and acid-base interactions. The geometric mean assumption (Equation 12, Chapter 4) is clearly inadequate for the total or Hildebrand parameters in the presence of these cross terms:

$$ {}^{ij}c = ({}^ic\,{}^jc)^{1/2} = {}^i\delta\,{}^j\delta \tag{25} $$

The approach taken by Karger, Snyder, and Eon[4] was to calculate the dispersion term from the refractive index, the induction and orientation terms from the dipole moment and molar volume (or δ_i from retention volumes in gas-liquid chromatography), and 2 δ_a δ_b by difference. The original estimates of these component cohesion parameters for a few common liquids are presented in Table 1. The origins of the various types of cohesion properties of different liquids with similar total cohesion parameters are here apparent, although the precise values of the individual parameters are open to question. A related set of parameters is introduced in Section 5.13.

If hydrogen-bonding capability is absent, it is reasonable to exclude the acid and base components; such an approach was taken by Melder and Ebber,[76-79] allowing an emphasis on dispersion, orientation, and induction components. Ignat and Melder[80,81,] subsequently determined the components of the cohesion parameters of alcohols (Table 2), based on the experimental activity coefficients of alcohols in nonpolar liquids and the partial enthalpies of dissolution in water (for which component values also appear in Table 2). Later, they[161]

TABLE 1
Five-Component Cohesion Parameters of Liquids in Order of Increasing Hildebrand Parameter[1-4]

Liquid	δ/MPa$^{1/2}$						V/cm^{3}mol^{-1}
	δ_t	δ_d	δ_o	δ_i	δ_a	δ_b	
Perfluoroalkanes	~12	~12	—	—	2.1	—	—
Pentane	14.5	14.5	—	—	—	—	115
Diisopropyl ether	14.5	14.1	2.1	0.2	—	6.1	102
Hexane	14.9	14.9	—	—	—	—	131
Diethyl ether	15.3	13.7	4.9	1.0	—	6.1	105
Triethylamine	15.3	15.3	—	—	—	9.2	140
Cyclohexane	16.8	16.8	—	—	—	—	108
Chloropropane, 1-chloropropane	17.2	14.9	5.9	1.2	—	1.4	88
Tetrachloromethane	17.6	17.6	—	—	—	1.0	97
Diethyl sulfide	17.6	16.8	3.5	0.5	—	5.3	108
Ethyl acetate	18.2	14.3	8.2	2.1	—	5.5	98
Propylamine	18.2	14.9	3.5	0.4	3.7	11.3	82
Bromoethane	18.2	16.0	6.3	1.2	—	1.6	77
Toluene	18.2	18.2	—	—	—	1.2	107
Tetrahydrofuran	18.6	15.5	7.2	1.6	—	7.6	82
Benzene	18.8	18.8	—	—	—	1.2	89
Trichloromethane	19.0	16.6	6.1	1.0	13.3	1.0	81
Methyl Ethyl ketone	19.4	14.5	9.6	2.5	—	6.5	90
Acetone, 2-propanone	19.6	13.9	10.4	3.1	—	6.1	74
1,2-Dichloroethane	19.8	16.8	8.6	1.0	—	1.4	79
Anisole, methoxybenzene	19.8	18.6	4.3	0.8	—	3.5	109
Chlorobenzene	19.8	18.8	3.9	0.6	—	2.1	102
Bromobenzene	20.2	19.6	3.1	0.4	—	2.1	105
Iodomethane	20.2	19.0	5.1	0.6	—	1.4	62
1,4-Dioxane	20.7	16.0	10.6	2.1	—	9.4	86
Hexamethylphosphoramide	21.5	17.2	7.0	3.5	—	8.2	176
Pyridine	21.7	18.4	7.8	2.1	—	10.0	81
Acetophenone	21.7	19.6	5.5	1.4	—	6.8	117
Benzonitrile	21.9	18.8	7.0	2.1	—	4.7	103
Propionitrile	22.1	14.1	13.5	3.7	—	4.3	71
Quinoline	22.1	21.1	3.7	0.6	—	8.6	118
N,N-Dimethylacetamide	22.1	16.8	9.6	3.3	—	9.2	92
Nitroethane	22.5	14.9	12.3	4.5	—	2.1	71
Nitrobenzene	22.7	19.4	7.4	2.3	—	2.1	103
Tricresyl phosphate	23.1	19.6	5.1	3.1	—	(?)	316
N,N-Dimethylformamide	24.1	16.2	12.7	4.9	—	9.4	77
1-Propanol	24.5	14.7	5.3	0.8	12.9	12.9	75
Dimethylsulfoxide	24.5	17.2	12.5	4.3	—	10.6	71
Acetonitrile	24.7	13.3	16.8	5.7	—	7.8	53
Phenol	24.7	19.4	4.7	0.8	19.0	4.7	92
Ethanol	26.0	13.9	7.0	1.0	14.1	14.1	59
Nitromethane	26.4	14.9	17.0	6.1	—	2.5	54
γ-Butyrolactone	26.4	16.4	14.7	6.5	—	(?)	77
Propylene carbonate	27.2	20.0	12.1	4.9	—	(?)	85
Diethylene glycol	29.2	16.8	8.2	1.2	10.8	10.8	96
Methanol	29.7	12.7	10.0	1.6	17.0	17.0	41
Ethylene glycol	34.8	16.4	13.9	2.3	12.5	12.5	56
Formamide	39.3	17.0	(?)	(?)	(large)	(large)	40
Water	47.9	12.9	(?)	(?)	(large)	(large)	18

TABLE 2
Five-Component Cohesion Parameters for Alcohols

Liquid	δ/MPa$^{1/2}$ δ_d	δ_o	δ_i	δ_a	δ_b	δ_t
1-Propanol	14.9	6.7	4.5	4.8	23.5	25.0
Isobutanol, 2-methyl-1-propanol	15.0	5.8	3.4	3.8	22.8	23.1
1-Butanol	15.3	5.4	3.1	4.6	22.6	23.8
2-Butanol, *sec*-butanol	15.2	5.4	3.2	3.5	23.3	22.8
1-Pentanol, amyl alcohol	15.7	4.7	2.9	3.7	21.9	22.8
1-Hexanol	15.9	3.7	1.7	4.0	21.5	22.1
Cyclohexanol	17.3	5.4	2.9	2.1	21.2	22.7
1-Octanol	16.2	3.1	1.8	3.6	17.8	21.1
1-Nonanol	16.3	2.8	1.7	3.4	17.2	20.7
1-Decanol	16.4	2.0	0.9	3.8	16.8	20.5
Anisole	18.5	3.5	1.4	0.0	16.5	30.2
Butyl acetate	15.0	4.0	2.4	0.0	19.7	17.7
Nitrobenzene	19.4	7.4	2.2	0.0	15.7	22.7
Water	12.9	31.4	20.8	34.1	22.3	—

Adapted from Ignat, A. V. and Melder, L. I., *Zh. Prikl. Khim. (Leningrad),* 60, 1136, 1987; 62, 419, 1989; *J. Appl. Chem. U.S.S.R.,* 60, 1070, 1987; 62, 376, 1989; *Eesti NSV Tead. Akad. Toim. Keem.,* 36(1), 45, 1987.

TABLE 3
Total and Component Cohesion Parameters for Some Lipophilic Liquids

Liquid	δ/MPa$^{1/2}$ δ_d[a]	δ_o[b]	δ_i[b]	δ_p[c]	δ_h[d]	δ_t[e]	δ_t[f]
Octanoic acid	16.2	0.7	0.0	1.2	11.2	—	19.8
1-Decanol	16.5	2.7	0.6	5.1	11.1	—	20.5
1,2-Propanediyl dinonanoate, propylene glycol dipelargonate	16.6	0.9	0.1	2.3	7.6	18.0	18.4
Dodecane	16.1	0.0	0.0	0.0	0.0	—	16.1
2-Octyldodecanol	16.9	0.5	0.0	1.2	8.7	19.2	19.0
Isopropyl tetradecanoate, isopropyl myristate	16.4	0.9	0.1	2.0	5.7	17.6	17.5
1,2,3-Propanetriyl trioctanodecanoate, glycerol tricaprylocaprate	16.8	0.8	0.1	2.2	8.2	18.0	18.8

[a] Refractive index.
[b] Dipole moment and refractive index.
[c] $\delta_p^2 = \delta_o^2 + 2\delta_i\,\delta_d$
[d] $\delta_h^2 = \delta_t^2 - \delta_d^2 - \delta_p^2$
[e] Gas-liquid chromatography.
[f] Group addition.[82]

Adapted from Munafo, A., Buchmann, M., Hô Nam-Tran, and Kesselring, U. W., *J. Pharm. Sci.,* 77, 169, 1988.

evaluated the component cohesion parameters for polar liquids, from the activity coefficients of alcohols dissolved in them.

Munafo, Buchmann, Hô Nam-Tran, and Kesselring[37] (see Table 3) evaluated for some lipophilic liquids all components, except for separation of δ_a and δ_b, from chromatographic measurements and molecular properties after comparing the various methods, including that of Fedors.[82]

5.7 EMPIRICAL CORRECTIONS FOR GEOMETRIC MEAN DEVIATIONS

Comparison of the equations in Section 4.2 and 5.6 shows that the original Hildebrand-Scatchard equation can be considered as a simplified form of a more general relationship. Many equations of intermediate simplification also have been used.

One of the specific simplifications in the Hildebrand-Scatchard equation was the use of the geometric mean approximation, Equation 25, and there have been several approaches to the accommodation of deviations from geometric mean behavior. Walker[83] proposed the dimensionless specific interchange coefficient, ${}^{ij}K$, which has values between 0.98 and 1.06 for many liquids,

$${}^{ij}c = {}^{ij}K\,({}^{i}c\,{}^{j}c)^{1/2} \tag{26}$$

so

$${}^{ij}A = {}^{i}c + {}^{j}c - 2\,{}^{ij}K\,{}^{i}c^{1/2}\,{}^{j}c^{1/2} \tag{27}$$

$$= {}^{i}\delta^{2} + {}^{j}\delta^{2} - 2\,{}^{ij}K\,{}^{i}\delta\,{}^{j}\delta \tag{28}$$

This parameter was given the symbol f by Reed[84,85] and Sonnich Thomsen[86-88] and also divided into several factors corresponding to different correction terms. Prausnitz and co-workers and others[89-93] used the dimensionless constant ${}^{ij}l$ (with a magnitude on the order of 10^{-2} to 10^{-1}), characteristic of a given pair of materials and almost independent of temperature and composition,

$${}^{ij}c = ({}^{i}c\,{}^{j}c)^{1/2}\,(1 - {}^{ij}l) \tag{29}$$

and of course this is related to ${}^{ij}K$ by

$${}^{ij}l = 1 - {}^{ij}K \tag{30}$$

so the empirical expression for interchange cohesive pressure, Equation 16, becomes

$${}^{ij}A = ({}^{i}\delta - {}^{j}\delta)^{2} + 2\,{}^{ij}l\,{}^{i}\delta\,{}^{j}\delta \tag{31}$$

from which it is clear that the effect of the correction factor is most pronounced when ${}^{i}\delta \approx {}^{j}\delta$. Examples are given in Table 4.

Martin et al.[94-96] in studies of drug solubilities proposed the binary interaction parameter ${}^{ij}W$, which in terms of ${}^{ij}A$ is defined by

$${}^{ij}A = ({}^{i}\delta^{2} + {}^{j}\delta^{2} - 2\,{}^{ij}W) = ({}^{i}\delta - {}^{j}\delta)^{2} + 2\,({}^{i}\delta\,{}^{j}\delta - {}^{ij}W) \tag{32}$$

They also showed[97] that these parameters are simply three different representations of deviations from the geometric mean mixing approximation,

$$1 - {}^{ij}l = {}^{ij}W/({}^{i}\delta\,{}^{j}\delta) = {}^{ij}K \tag{33a}$$

or

TABLE 4
Binary Interaction Parameters for Solutes in Cryogenic Liquids

Solute(j)-solvent(i)	^{ij}l	Recommended range for use of given ^{ij}l, T/K	Maximum error, $100\lvert ^{j}x_{obs} - {}^{j}x_{calc}\rvert / {}^{ij}X_{obs}$
Acetylene-carbon dioxide	0.04	170—188	5
Acetylene-ethane	0.11	150—177	30
Acetylene-ethylene	0.02	100—160	25
Acetylene-methane	0.10	105—135	40
Acetylene-nitrogen	0.07	77	—
Acetylene-oxygen	0.18	90	—
Argon-methane	0.00	72—83	2
Argon-nitrogen	0.00	70—83	1
Argon-oxygen	0.06	63—83	25
Butane-oxygen	0.08	90	—
Carbon dioxide-acetylene	−0.02	175—190	3
Carbon dioxide-butane	0.09	140—200	20
Carbon dioxide-ethane	0.08	110—170	15
Carbon dioxide-ethylene	0.00	140—170	10
Carbon dioxide-methane	−0.02	110—140	10
Carbon dioxide-nitrogen	−0.18	77	—
Carbon dioxide-oxygen	0.03	90	—
Carbon dioxide-propane	0.08	120—200	10
Carbon dioxide-propylene	0.01	130—210	20
Ethane-ethylene	0.01	—	—
Ethane-oxygen	0.03	60—75	20
Ethylene-nitrogen	0.06	70—85	35
Ethylene-oxygen	0.06	75—90	15
Hydrogen sulfide-butane	0.05	140—170	25
Hydrogen sulfide-ethane	0.07	120—160	30
Hydrogen sulfide-ethylene	−0.01	120—185	15
Hydrogen sulfide-methane	0.04	120—150	40
Hydrogen sulfide-propane	0.06	140—170	45
Hydrogen sulfide-propylene	−0.005	130—180	5
Methane-ethane	0.01	—	—
Methane-ethylene	0.01	—	—
Methane-nitrogen	−0.02	70—90	5
Methane-oxygen	0.05	70—90	3
Nitrogen-hydrogen	0	26—33	—
Nitrogen-oxygen	0	—	—
Propane-oxygen	0.02	60—70	115
Propylene-nitrogen	−0.01	65—80	75
Propylene-oxygen	0.05	65—75	30

Adapted from Preston, G. T. and Prausnitz, J. M., *Ind. Eng. Chem. Proc. Des. Dev.*, 9, 264, 1970.

$$^{ij}W = {}^{ij}K\, {}^{i}\delta\, {}^{j}\delta \tag{33b}$$

Although values of ^{ij}K are never very different from unity, these small differences can account for large variations in solubilities.[98] The application by Martin and co-workers of this geometric mean correction to the Hildebrand-Scatchard equation and polynomial regression to fit complex solutes in various solvents is outlined in Chapter 12. More recently, Li et al.[99] proposed the parameter $^{ij}\delta'$ defined such that

$$^{ij}\delta' = {}^{ij}W^{1/2} = ({}^{ij}K\, {}^{i}\delta\, {}^{j}\delta)^{1/2} = (1 - {}^{ij}l)^{1/2}\, ({}^{i}\delta\, {}^{j}\delta)^{1/2} \tag{34}$$

which has the advantage that the relationship of this correction parameter to $^{i}\delta$ and $^{j}\delta$ is clear, and that it retains the cohesion parameter dimensions of $(\text{pressure})^{1/2}$.

Mikos and Peppas[165] employed the geometric mean correction parameter ^{ij}l in the correlation of polymer-liquid interaction parameters (Section 13.5) for hydrophilic copolymers with water, using

$$\chi_H = {}^iV/(RT)[({}^i\delta - {}^j\delta)^2 + 2\,{}^{ij}l\,{}^i\delta\,{}^j\delta]$$

with ^{ij}l ranging from -0.15 to -0.25 depending on copolymer composition and volume fraction. For example, they obtained -0.25 for poly(methacrylic acid) and -0.18 for poly(2-hydroxyethyl methacrylate) extrapolated to infinite dilution in water at 25°C, values which may be compared with the generally lower magnitudes in Table 4. This parameter is negative for the hydrogels, consistent with specific effects such as hydrogen bonding favoring dissolution.

The virtual Hildebrand parameter $^j\Lambda$ (Section 5.5) is related to ^{ij}W by the parabolic relationship[73]

$$-\,2\,{}^{ij}W = {}^j\Lambda^2 - 2{}^i\delta\,{}^j\Lambda - {}^j\delta^2 \tag{35}$$

The geometric mean corrections ^{ij}l for systems of testosterone propionate and unsaturated hydrocarbons were found to fit a rectilinear function of the branching ratio of the hydrocarbon liquids, and for related esters there was a good correlation between ^{ij}l and infrared carbonyl stretching frequencies.[100]

It is apparent that $^{ij}\delta'$, ^{ij}l, ^{ij}K, and ^{ij}W are binary parameters which depend in a complex manner on many properties of i and j, and are usually determined by empirical means, for example, from liquid phase activity coefficients, when Equation 15, Chapter 4, becomes

$$RT \ln {}^jf_x = {}^jV\,{}^i\phi^2\{({}^i\delta - {}^j\delta)^2 + 2\,{}^{ij}l\,{}^i\delta\,{}^j\delta\} \tag{36}$$

They also may be determined from gas phase data[88,92,101-105] where

$$^{ij}T_c = ({}^iT_c\,{}^jT_c)^{1/2}\,(1 - {}^{ij}k) \tag{37}$$

Although the parameters ^{ij}l and ^{ij}k are not identical, because the former reflects some liquid phase aspects absent from gas mixtures, they are often of comparable magnitude.

For the particular case of bonding energy E, for polar interactions, Pauling[106] suggested

$$^{ij}E/\text{ kJ mol}^{-1} = ({}^iE\,{}^jE)^{1/2}/\text{ kJ mol}^{-1} + 125({}^iX - {}^jX)^2 \tag{38}$$

where X is the electronegativity. The resulting expression in terms of cohesion pressures is

$$^{ij}c = ({}^ic\,{}^jc)^{1/2} + k({}^iX - {}^jX)^2/({}^iV\,{}^jV)^{1/2} \tag{39}$$

where k has been taken as 125 kJ mol^{-1} for intermetallic solutions.[107]

The presence of molecular quadrupole moments introduces significant deviations from geometric mean behavior. In analyzing data on the solubility of solid carbon dioxide in liquid hydrocarbons, Myers and Prausnitz[102,108] calculated the ideal solubility (see Section 12.1) and assumed that the activity coefficient of carbon dioxide (referred to its pure subcooled liquid) was given by Equation 15, Chapter 4:

$$RT \ln {}^jf_x = {}^{ij}A\,{}^jV\,{}^i\phi^2 \tag{40}$$

where ^jV is the molar volume of subcooled liquid carbon dioxide, $^i\phi$ is the volume fraction of the liquid hydrocarbon, and ^{ij}A is the interchange cohesive pressure of subcooled liquid

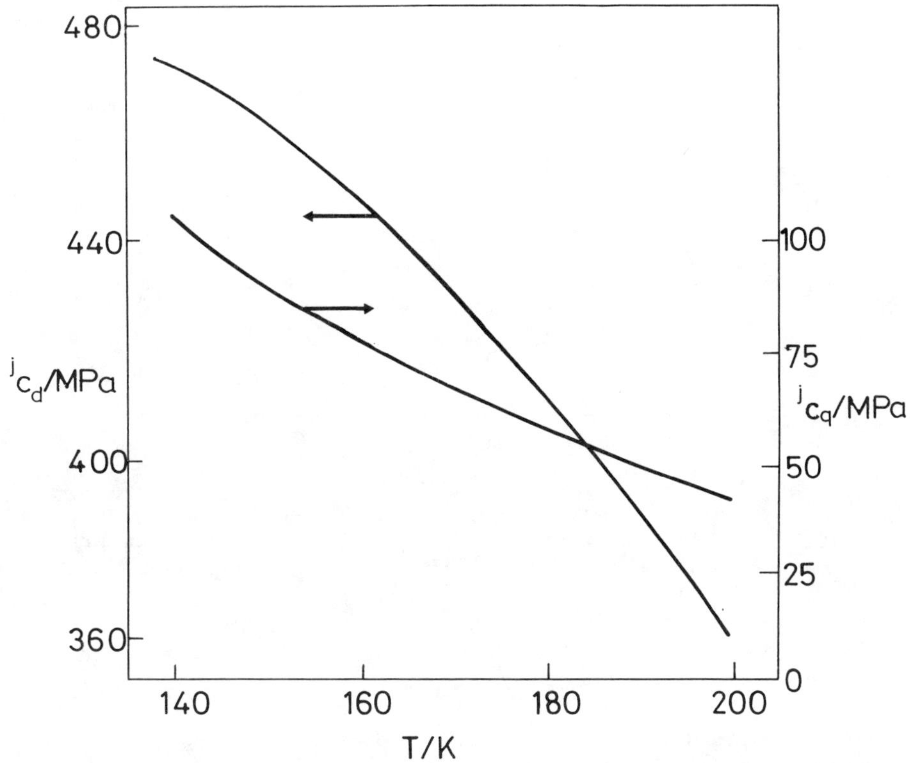

FIGURE 1. Dispersion and quadrupole cohesive pressures of subcooled liquid carbon dioxide as functions of temperature. (Adapted from Myers, A. L. and Prausnitz, J. M., *Ind. Eng. Chem. Fund.*, 4, 209, 1965.)

carbon dioxide with liquid hydrocarbons (Equation 10, Chapter 4). However, since the quadrupole moment of carbon dioxide makes an appreciable contribution to its cohesive pressure, the cohesive pressures were separated into two parts due to dispersion (d) forces and quadrupole (q) forces before the geometric mean approximation was used:

$$ {}^{j}c = {}^{j}c_{\mathrm{d}} + {}^{j}c_{\mathrm{q}} \text{ and } {}^{i}c = {}^{i}c_{\mathrm{d}} + {}^{i}c_{\mathrm{q}} \tag{41} $$

The relative estimated values of dispersion and quadrupole cohesive pressures for subcooled liquid carbon dioxide are shown in Figure 1 (note the different scales) and the sum (Equation 41) is shown in Figure 2. In mixtures, the geometric mean approximation retained for the dispersion interactions, ${}^{ij}c_{\mathrm{q}}$ was evaluated from theoretical expressions, and the total cohesive pressure was given by

$$ {}^{ij}c = ({}^{i}c_{\mathrm{d}}\,{}^{j}c_{\mathrm{d}})^{1/2} + {}^{ij}c_{\mathrm{q}} + {}^{ij}c_{\mathrm{ab}} \tag{42} $$

The last term is a measure of the tendency of carbon dioxide (with Lewis acid properties) to complex with unsaturated hydrocarbons, and was evaluated by difference (Figure 3). Since the complex formation is exothermic, ${}^{ij}c_{\mathrm{ab}}$ rises with falling temperature and becomes relatively more important in such low temperature systems. The full expression for the interchange cohesive pressure is thus

$$ \begin{aligned} {}^{ij}A &= {}^{i}c + {}^{j}c - 2\,{}^{ij}c \\ &= {}^{i}c + {}^{j}c - 2\,[({}^{i}c_{\mathrm{d}}\,{}^{j}c_{\mathrm{d}})^{1/2} + {}^{ij}c_{\mathrm{q}}] \end{aligned} \tag{43} $$

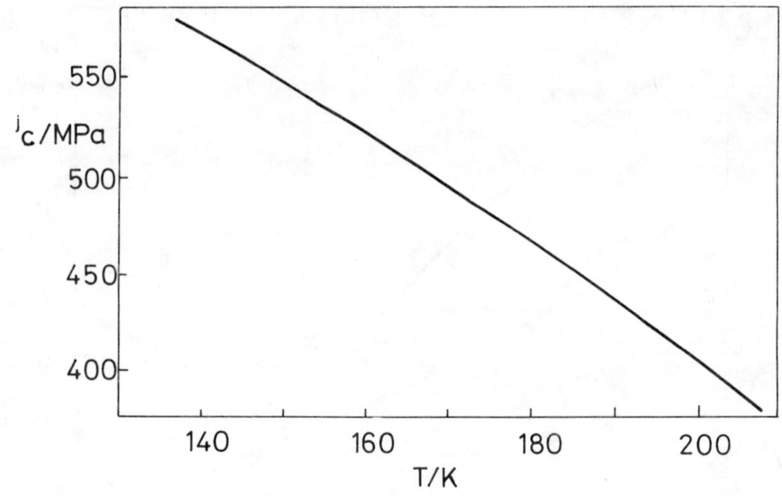

FIGURE 2. Cohesive pressure of subcooled liquid carbon dioxide as a function of temperature. (Adapted from Myers, A. L. and Prausnitz, J. M., *Ind. Eng. Chem. Fund.*, 4, 209, 1965.)

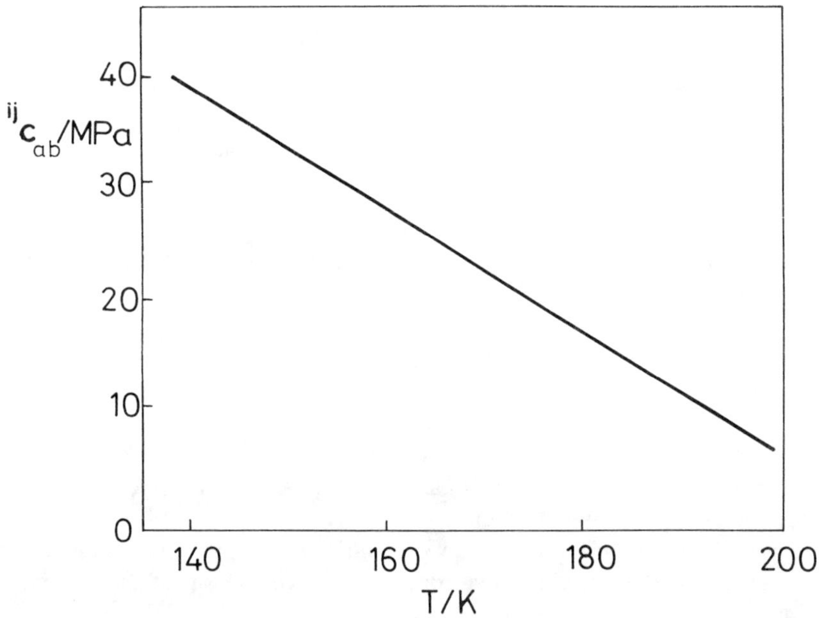

FIGURE 3. Acid-base complex contribution to the cohesive pressure of carbon dioxide-alkene mixtures as a function of temperature. (Adapted from Myers, A. L. and Prausnitz, J. M., *Ind. Eng. Chem. Fund.*, 4, 209, 1965.)

5.8 POLAR-NONPOLAR COHESION PARAMETERS

Van Arkel,[36] Small,[60] and Prausnitz and co-workers[102,109-114] divided the total cohesion parameter into two main components, defining a nonpolar cohesion parameter (δ_λ) and a polar parameter (δ_τ). Although this tends to neglect induction interactions, these may be taken care of by an additional parameter. More of a problem from a practical point of view is the omission of specific interactions. Polar-nonpolar parameters are related to the Hildebrand parameter (total cohesion parameter) by

$$\delta^2 = \delta_\lambda^2 + \delta_\tau^2 \tag{44}$$

and comparison with Equation 24 suggests that δ_λ can be identified with δ_d and that δ_τ corresponds to δ_o.

In their evaluation of δ_λ and δ_τ for polar liquids, Blanks and Prausnitz[110] used the homomorph concept. This is discussed in Section 6.7, but basically it is a method which assumes that the polar energy of vaporization is the difference between the experimentally determined total energy of vaporization and the energy of vaporization of a nonpolar liquid having molecules very nearly the same size and shape as those of the polar liquid. Tables 5a and 5b list those polar and nonpolar cohesion parameter values.

Weimer and Prausnitz[111,113] also published values of polar and nonpolar cohesion parameters for polar liquids; again alcohols and acids were omitted because of their strong hydrogen-bonding properties (Tables 6 and 7). These parameters, evaluated at several temperatures, used homomorph data (Figures 6 to 8, Chapter 6) which were rather different from those of Blanks and Prausnitz, and the numerical values differ slightly. Yet another set of values was presented by Helpinstill and Van Winkle[115] (Tables 8 and 9).

These polar-nonpolar cohesion parameters can be used in a similar way to Hildebrand parameters to evaluate the interchange cohesion pressure, and from this the derived quantities such as energy of mixing. All three sets of data have been tabulated here because each set is still in use, and as stressed in Section 6.10, it is essential to use self-consistent sets of cohesion parameters. Their application to various systems is now summarized, first for nonpolar-nonpolar systems, then for polar-nonpolar and polar-polar systems.

For a mixture of *nonpolar* liquids (i.e., when only dispersion forces occur),

$$^{ij}A = {}^{i}c + {}^{j}c - 2\,{}^{ij}c = {}^{i}\delta_\lambda^2 + {}^{j}\delta_\lambda^2 - 2\,{}^{i}\delta_\lambda{}^{j}\delta_\lambda \tag{45}$$

$$= ({}^{i}\delta_\lambda - {}^{j}\delta_\lambda)^2 = ({}^{i}\delta - {}^{j}\delta)^2 \tag{46}$$

as used in Equation 13, Chapter 4.

For interaction between a *polar* substance (i) and a *nonpolar* substance (j), a term $^{ij}\Psi$ is included for the i-j induction pressure, so the expression for the interchange cohesive pressure is

$$^{ij}A = {}^{i}c + {}^{j}c - 2\,{}^{ij}c = {}^{i}\delta_\tau^2 + {}^{i}\delta_\lambda^2 + {}^{j}\delta_\lambda^2 - 2\,({}^{i}\delta_\lambda\,{}^{j}\delta_\lambda + {}^{ij}\Psi) \tag{47}$$

$$= ({}^{i}\delta_\lambda - {}^{j}\delta_\lambda)^2 + {}^{i}\delta_\tau^2 - 2\,{}^{ij}\Psi \tag{48}$$

As indicated in Equation 4, Chapter 3, the orientation energy (unlike the induction and dispersion energies) has a temperature dependence, so Equation 48 should reflect this, and a more complete expression is

TABLE 5a
Molar Volumes and Hildebrand Parameters of Nonpolar Liquids at 25°C

Liquid	$V/cm^3\ mol^{-1}$	$\delta/MPa^{1/2}$
Alkanes		
Methane	53[a]	9.6[a]
Ethane	69[a]	11.6[a]
Propane	85[a]	12.7[a]
Butane	101.4	13.5
Isobutane, 2-methylpropane	105.5	12.8
Pentane	116.1	14.3
Isopentane, 2-methylbutane	117.4	13.8
Neopentane, 2,2-dimethylpropane	123.3	12.5
Hexane	131.6	14.8
Heptane	147.5	15.2
2,2,3-Trimethylbutane	146.1	14.2
Octane	163.5	15.5
Isooctane, 2,2,4-trimethylpentane	166.1	14.0
Nonane	179.7	15.6
Decane	195.9	15.8
Dodecane	228.6	16.0
Tetradecane	261.3	16.2
Hexadecane	294.1	16.3
Octadecane	326.9	16.4
Eicosane	359.8	16.5
Alkenes		
Ethylene	63[a]	11.3[a]
Propylene	79[a]	12.5[a]
1-Butylene	95.3	13.7
cis-2-Butylene	91.2	14.7
trans-2-Butylene	93.8	14.3
Isobutylene, 2-methylpropene	95.4	13.7
1,3-Butadiene	88.0	14.5
Isoprene	100.8	15.2
Cycloalkanes		
Cyclopentane	94.7	16.6
Cyclohexane	108.7	16.7
Methylcyclohexane	128.3	16.0

TABLE 5a (continued)
Molar Volumes and Hildebrand Parameters of Nonpolar Liquids at 25°C

Liquid	$V/\text{cm}^3\ \text{mol}^{-1}$	$\delta/\text{MPa}^{1/2}$
	Aromatics	
Benzene	89.4	18.7
Toluene, methylbenzene	106.9	18.2
Ethylbenzene	123.1	18.0
o-Xylene, 1,2-dimethylbenzene	121.2	18.4
m-Xylene, 1,3-dimethylbenzene	123.5	18.0
p-Xylene, 1,4-dimethylbenzene	123.9	17.9
Propylbenzene	140.1	17.7
Mesitylene, 1,3,5-trimethylbenzene	140	18.0
Styrene, ethenylbenzene	115.6	19.0
Tetrahydronaphthalene, tetralin	—	19.4
Tetrachloromethane, carbon tetrachloride	97.1	17.6

[a] Estimated from gas solubility data.[114]

Adapted from Blanks, R. F. and Prausnitz, J. M., *Ind. Eng. Chem. Fundam.*, 3, 1, 1964.

$$^{ij}A = (^{i}\delta_{\lambda} - {}^{j}\delta_{\lambda})^2 + (T_*/T)\ {}^{i}\delta_{\tau}^2 - 2\,{}^{ij}\Psi \tag{49}$$

where T_* is the temperature at which the polar parameter ${}^{i}\delta_{\tau}$ was obtained. The tables in this section show that δ_{λ} also varies with temperature, but to a smaller extent, and an alternative (and preferred) procedure is to evaluate all cohesion parameters at the temperatures at which they will be used.

Theory suggests that the induction term, ${}^{ij}\Psi$, should depend on the product ${}^{j}\delta_{\lambda}{}^{i}\delta_{\tau}$, where ${}^{j}\delta_{\lambda}$ is the cohesion parameter of the nonpolar component. When evaluated experimentally by difference[110] the correlation in Figure 4 was observed. Weimer[102,111,113] found that activity coefficient data (Section 7.5) for infinitely dilute solutions of hydrocarbons (*j*) in polar liquids (*i*) could be correlated well by expressing the induction parameter ${}^{ij}\Psi$ as proportional to the square of the polar cohesion parameter ${}^{i}\delta_{\tau}^{2}$:

- For linear and cyclic alkanes, ${}^{ij}\Psi = 0.40\ {}^{i}\delta_{\tau}^{2}$
- For the alkene 1-pentene, ${}^{ij}\Psi = 0.42\ {}^{i}\delta_{\tau}^{2}$
- For the aromatic hydrocarbon benzene, ${}^{ij}\Psi = 0.45\ {}^{i}\delta_{\tau}^{2}$

In other words, this empirical induction cohesive pressure term is related to the polar cohesive pressure of the polar liquid and to the *class* of the hydrocarbon, but is not directly dependent on the cohesive pressure of the hydrocarbon. Weimer and Prausnitz[113] explained this on the basis of the polar molecule interacting with only one carbon-carbon bond during a collision, so the interchange cohesive pressure depends on the polarizability of the carbon-carbon bond, which increases in the order

$$\text{single bond, C–C} < \text{double bond, C = C} < \text{aromatic}$$

Although these correlations were obtained in conditions of high concentrations of the polar component and therefore are expected to be of limited applicability, they provide a simple method of screening polar liquids for use in hydrocarbon separations. For example, the

TABLE 5b
Molar Volumes, Nonpolar Cohesion Parameters, and Polar Cohesion Parameters of Polar Liquids at 25°C

Liquids	V/cm³mol⁻¹	Nonpolar, δ_λ/MPa$^{1/2}$	Polar, δ_τ/MPa$^{1/2}$	Homomorph
Halogen Compounds				
Trichloromethane, chloroform	80.7	15.8	10.0	Figure 5, Chapter 6
Chloroethane, ethyl chloride	73	14.9	9.0	Figure 5, Chapter 6
1,2-Dichloroethane, ethylene dichloride	79.4	16.1	11.9	Figure 5, Chapter 6
1,1-Dichloroethane	84.8	15.5	10.3	Isobutane
1,1,1-Trichloroethane	100.4	15.8	8.9	Neopentane
Trichloroethylene	90.2	16.3	9.6	Isopentane
1,2-Dibromoethane	87	16.9	13.0	Figure 5, Chapter 6
Chlorobenzene	102.1	18.9	6.0	Toluene
Ketones				
Acetone, 2-propanone	74.0	15.5	12.5	Figure 5, Chapter 6
Methyl ethyl ketone, 2-butanone	90.2	15.9	10.3	Figure 5, Chapter 6
Methyl propyl ketone	107.5	15.8	9.1	
Methyl isobutyl ketone	125.8	15.3	8.5	Isohexane
Methyl amyl ketone	140.8	15.9	8.6	Figure 5, Chapter 6
Dipropyl ketone	140.7	15.8	8.8	Figure 5, Chapter 6
Hexamethyl ketone	157.5	15.9	8.4	Figure 5, Chapter 6
Mesityl oxide	115.6	17.1	7.8	2-Methylpentene
Esters				
Ethyl acetate	98.5	15.2	10.6	Figure 5, Chapter 6
Propyl acetate	115.7	15.6	8.8	Figure 5, Chapter 6
Butyl acetate	132.5	15.6	8.7	Figure 5, Chapter 6
Amyl acetate, pentyl acetate	148.9	15.6	8.6	Figure 5, Chapter 6
Ethyl propionate	115.5	15.6	9.2	Figure 5, Chapter 6
Ethers				
Diethyl ether	104.8	14.3	4.6	Figure 5, Chapter 6
Methyl isopropyl ether	103.8	14.4	5.5	Isobutane
1,4-Dioxane	85.7	17.5	9.5	Cyclopentane
Miscellaneous				
Nitropropane	90.4	16.4	13.4	Figure 5, Chapter 6
Acetonitrile	52.9	16.2	18.0	Figure 5, Chapter 6
Propionitrile	70.3	16.3	14.7	Figure 5, Chapter 6

Adapted from Blanks, R. F. and Prausnitz, J. M., *Ind. Eng. Chem. Fundam.*, 3, 1, 1964.

ability of a solvent to ''distinguish'' between a saturated hydrocarbon and an unsaturated hydrocarbon depends primarily on the solvent having a large cohesive pressure (large cohesive energy and small molar volume).

In systems where *both components are polar*, induction interactions may be considered to be negligibly small in comparison with dipole-dipole interactions. If specific interactions are also negligible, the interchange cohesive pressure is

TABLE 6
Molar Volumes and Nonpolar Cohesion Parameters of Hydrocarbons at Various Temperatures

Liquid	t/°C	V/cm^{3}mol^{-1}	δ_λ/MPa$^{1/2}$
Propane	25	89.5	13.4
Butane	25	101.4	14.2
Pentane	0	111.8	15.2
	25	116.2	14.7
	45	120.3	14.2
Hexane	0	127.3	15.6
	25	131.6	15.0
	45	135.4	14.5
	60	138.7	14.2
	100	148.0	13.4
Heptane	25	147.4	15.3
	60	154.3	14.5
	100	164.0	13.6
Decane	25	195.9	16.1
	60	203.4	15.1
	100	213.0	14.2
Hexadecane	25	294.1	17.9
	60	303.6	16.4
	100	315.4	15.1
Cyclopentane	25	94.7	16.8
Methylcyclopentane	25	113.1	16.2
Cyclohexane	0	105.6	17.5
	25	108.7	16.8
	45	111.5	16.4
	60	113.6	16.0
	100	119.4	15.2
Methylcyclohexane	0	124.8	16.7
	25	128.3	16.1
	45	131.3	15.6
	60	133.7	15.3
	100	140.4	14.5
Ethylcyclohexane	25	143.1	16.4
	60	148.5	15.6
1-Butene	25	95.3	14.5
1-Pentene	0	106.1	15.4
	25	110.4	14.8
	45	113.1	14.4
1,3-Butadiene	25	88.0	15.2
Benzene	25	89.4	18.8
	45	91.7	18.3
	60	93.4	17.9
	100	98.8	16.9
Toluene, methylbenzene	25	106.8	18.3
	45	109.2	17.8
	60	111.0	17.4
	100	116.2	16.6
Ethylbenzene	25	123.1	18.4
p-Xylene, 1,4-dimethylbenzene	25	123.9	18.1
Mesitylene, 1,3,5-trimethylbenzene	25	139.6	18.2

Adapted from Weimer, R. F. and Prausnitz, J. M., *Hydrocarbon Process. Petr. Ref.*, 44, 237, 1965.

TABLE 7
Molar Volumes, Nonpolar Cohesion Parameters, and Polar Cohesion Parameters of Polar Liquids at Various Temperatures in Order of Increasing Polarity

Liquid	t/°C	V/cm^{3}mol^{-1}	Nonpolar, δ_λ/MPa$^{1/2}$	Polar, δ_τ/MPa$^{1/2}$
Acetophenone	25	117.4	19.3	7.6
Tetrahydrofuran	0	79.3	17.6	7.9
	25	81.7	17.0	7.6
	45	85.0	16.5	7.1
Pyridine	25	80.9	20.2	7.6
	60	83.9	19.1	7.8
	100	87.4	18.1	8.0
Cyclohexanone	25	104.2	18.1	8.3
Chloroethane	25	74.1	15.1	8.8
Diethyl ketone, 3-pentanone	0	103.3	16.4	9.3
	25	106.4	15.9	9.1
	45	108.9	15.4	8.9
	60	110.7	15.1	8.8
	100	116.1	14.4	8.5
Diethyl carbonate	25	121.9	16.1	9.2
Bromoethane	25	75.1	15.6	9.9
	45	77.3	15.2	9.5
Nitrobenzene	25	102.7	19.8	10.0
Di-(2-chloroethyl) ether	0	109.7	17.8	11.7
	25	117.8	17.1	10.7
Trimethyl phosphate	25	116.2	17.3	10.7
Iodoethane	25	81.1	15.7	10.7
	45	83.0	15.2	10.5
Methyl ethyl ketone, 2-butanone	0	87.3	16.2	11.3
	25	90.1	15.6	10.9
	45	92.6	15.2	10.6
	60	94.5	14.9	10.4
	100	99.9	14.1	10.0
Cyclopentanone	25	89.5	17.8	11.0
2,4-Pentanedione	25	103.0	16.5	11.6
2,5-Hexanedione, acetylacetone	25	117.7	17.3	12.0
Diethyl oxalate	0	132.7	17.8	12.3
	25	136.2	17.1	12.1
	45	139.3	16.6	12.1
2-Nitropropane	25	90.7	16.3	12.3
Methoxyacetone	25	93.2	16.2	12.5
	45	96.0	15.5	12.1
Acetone, 2-propanone	0	72.3	16.2	12.9
	25	74.0	15.7	12.6
	45	76.2	15.2	12.2
	60	78.1	15.0	11.9
	100	83.4	14.1	11.2
Dimethyl carbonate	25	85.0	15.9	12.7
Butyronitrile	0	85.4	16.8	13.3
	25	87.9	16.3	12.8
	45	90.1	15.9	12.6
2,3-Butanedione, diacetyl	25	87.8	15.8	13.0
	45	90.4	15.4	12.6
Aniline	25	91.5	20.1	13.0
	60	94.3	19.2	12.8
	100	97.9	18.3	12.7
1-Nitropropane	25	89.5	16.5	13.1
N-Methyl-2-pyrrolidone	25	96.6	18.7	13.4

TABLE 7 (continued)
Molar Volumes, Nonpolar Cohesion Parameters, and Polar Cohesion Parameters of Polar Liquids at Various Temperatures in Order of Increasing Polarity

Liquid	t/°C	V/cm³mol⁻¹	Nonpolar, δ_λ/MPa$^{1/2}$	Polar, δ_τ/MPa$^{1/2}$
Acetic anhydride	25	95.0	16.1	14.5
Propionitrile	0	68.7	16.9	15.2
	25	70.9	16.3	14.7
	45	72.9	15.9	14.2
	60	74.4	15.6	13.9
	100	79.0	14.9	13.2
Citraconic anhydride	25	89.7	19.3	14.8
Methoxyacetonitrile	25	75.2	16.5	15.0
Furfural	0	81.4	19.2	15.9
	25	83.2	18.5	15.6
	45	84.8	18.0	15.1
	60	85.9	17.7	15.2
	100	89.4	16.9	14.6
Nitroethane	25	72.1	16.4	15.7
	45	73.6	16.1	15.3
N,N-Dimethylacetamide	0	90.7	17.6	16.2
	25	93.2	17.0	15.7
	45	95.3	16.5	15.4
γ-Butyrolactone	25	77.1	19.4	16.4
N,N-Dimethylformamide	0	75.5	17.5	17.0
	25	77.4	17.0	16.5
	45	79.0	16.5	16.2
	60	80.2	16.2	15.9
	100	83.8	15.5	15.3
3-Chloropropionitrile	25	77.7	17.3	17.9
Acetonitrile	0	51.1	16.9	19.1
	25	52.6	16.4	18.4
	45	54.4	16.1	17.7
	60	55.5	15.8	17.3
	100	58.9	15.1	16.3
Ethylenediamine	25	67.3	16.6	19.2
	60	69.7	15.9	18.5
	100	72.9	15.2	17.7
Nitromethane	25	54.3	16.5	19.3
	45	55.3	16.2	18.9
	60	56.3	15.9	18.5
	100	59.3	15.2	17.6
Dimethylsulfoxide	25	71.3	17.5	19.4

Adapted from Weimer, R. F. and Prausnitz, J. M., *Hydrocarbon Process. Petr. Ref.*, 44, 237, 1965.

$$^{ij}A = (^i\delta_\lambda - {}^j\delta_\lambda)^2 + (^i\delta_\tau - {}^j\delta_\tau)^2 \tag{50}$$

or, taking the effect of temperature into account as in Equation 49,

$$^{ij}A = (^i\delta_\lambda - {}^j\delta_\lambda)^2 + (T_*/T)(^i\delta_\tau - {}^i\delta_\tau)^2 \tag{51}$$

Helpinstill and Van Winkle[115] extended the correlation of Weimer and Prausnitz[113] to polar-polar situations, defining the new parameter $^{ij}\psi$ to include both polar-polar and polar-nonpolar systems:

TABLE 8
Molar Volumes, Nonpolar Cohesion Parameters and Polar Cohesion Parameters of Hydrocarbons at Various Temperatures

Liquid	t/°C	V/cm³mol⁻¹	Nonpolar, δ_λ/MPa$^{1/2}$	Polar, δ_τ/MPa$^{1/2}$
Propane	0	83.1	14.2	0.0
	25	89.4	13.4	0.0
Butane	0	96.7	14.8	0.0
	25	101.4	14.2	0.0
	45	106.0	13.8	0.0
1-Butene	0	90.3	14.8	2.9
	25	95.3	14.1	2.7
Pentane	0	111.8	15.3	0.0
	25	116.1	14.6	0.0
	45	120.3	14.2	0.0
	60	122.9	13.9	0.0
	100	131.4	13.5	0.0
1-Pentene	0	106.0	15.4	2.5
	25	110.1	14.8	2.2
	45	113.0	14.4	2.1
	60	115.9	14.0	1.9
	100	121.6	13.5	1.5
Hexane	0	127.3	15.5	0.0
	25	131.6	15.0	0.0
	45	135.4	14.5	0.0
	60	138.7	14.2	0.0
	100	148.0	13.5	0.0
1-Hexene	0	121.6	15.7	2.1
	25	125.9	15.1	1.9
	45	129.5	14.7	1.7
	60	132.4	14.3	1.6
	100	140.7	13.7	1.3
Heptane	0	143.8	15.9	0.0
	25	147.4	15.3	0.0
	45	151.9	14.8	0.0
	60	154.3	14.5	0.0
	100	164.0	13.7	0.0
1-Heptene	45	145.5	14.8	1.4
	60	148.4	14.6	1.3
	100	156.8	13.9	0.9
Decane	25	195.8	15.8	0.0
Cyclopentane	0	91.9	17.4	0.0
	25	94.7	16.8	0.0
	45	97.4	16.1	0.0
	60	99.4	15.8	0.0
	100	105.2	14.9	0.0
Cyclohexane	0	105.5	17.3	0.0
	25	108.7	16.6	0.0
	45	111.5	16.1	0.0
	60	113.6	15.8	0.0
	100	119.4	14.9	0.0
Ethylcyclohexane	25	143.3	16.2	0.7
Butylcyclohexane	25	176.4	16.1	0.6
Benzene	0	86.9	19.6	0.0
	25	89.4	18.8	0.0
	45	91.6	18.2	0.0
	60	93.4	18.3	0.0
	100	98.8	17.0	0.0
Ethylbenzene	25	123.1	18.1	0.4

TABLE 8 (continued)
Molar Volumes, Nonpolar Cohesion Parameters and Polar Cohesion Parameters of Hydrocarbons at Various Temperatures

Liquid	t/°C	V/cm^{3}mol^{-1}	Nonpolar, δ_λ/MPa$^{1/2}$	Polar, δ_τ/MPa$^{1/2}$
Butylbenzene	25	156.8	18.0	0.3
p-Xylene, 1-4-dimethylbenzene	0	121.0	18.8	0.0
	25	123.9	18.1	0.0
	45	126.8	17.5	0.0
	60	128.7	17.1	0.0
	100	134.1	16.4	0.0
Mesitylene, 1,3,5-trimethylbenzene	0	136.5	19.1	0.0
	25	139.8	18.0	0.0
	45	142.7	17.5	0.0
	60	144.8	17.1	0.0
	100	150.8	16.3	0.0
1,4-Diethylbenzene	0	153.1	18.7	0.0
	25	156.9	17.9	0.0
	45	159.8	17.4	0.0
	60	162.0	17.0	0.0
	100	168.6	16.2	0.0
Toluene, methylbenzene	0	104.0	19.0	1.7
	25	106.8	18.2	(1.4)
	45	109.2	17.8	1.3
	60	110.8	17.4	1.2
	100	115.9	16.6	1.0

Adapted from Helpinstill, J. G. and Van Winkle, M., *Ind. Eng. Chem. Proc. Des. Dev.*, 7, 213, 1968.

$$^{ij}A = (^{i}\delta_\tau - {}^{j}\delta_\lambda)^2 + (^{i}\delta_\tau - {}^{j}\delta_\tau)^2 - 2\ {}^{ij}\psi \tag{52a}$$

Following a procedure similar to that of Weimer and Prausnitz, they used activity coefficient data for hydrocarbons in polar systems and evaluated the parameter $^{ij}\psi$ by difference. It was found to be a linear function of $(^{i}\delta_\tau - {}^{j}\delta_\tau)^2$,

- For alkanes, $^{ij}\psi = 0.40\ (^{i}\delta_\tau - {}^{j}\delta_\tau)^2$
- For the alkene, 1-pentene, $^{ij}\psi = 0.22\ (^{i}\delta_\tau - {}^{j}\delta_\tau)^2$
- For aromatic hydrocarbons, $^{ij}\psi = 0.48\ (^{i}\delta_\tau - {}^{j}\delta_\tau)^2$

It follows from this form of the dependence of $^{ij}\psi$ on δ_τ that Equation 52a can be rewritten as[25]

$$^{ij}A = (^{i}\delta_\lambda - {}^{j}\delta_\lambda)^2 + {}^{ij}b\ (^{i}\delta_\tau - {}^{j}\delta_\tau)^2 \tag{52b}$$

Values found for the induction cohesive pressure correction factor ^{ij}b in various situations are listed in Table 10. These examples are not all strictly comparable, but the departure from unity gives an idea of the extent of "correction" necessary as a result of the simple model used.

In terms of the component parameters in Equation 22 (Section 5.6), the induction parameter may be expressed

$$^{ij}\Psi = {}^{i}\delta_i\,{}^{j}\delta_d + {}^{i}\delta_d\,{}^{j}\delta_i + {}^{i}\delta_o\,{}^{j}\delta_o \tag{53}$$

TABLE 9
Molar Volumes, Nonpolar Cohesion Parameters and Polar Cohesion Parameters of Polar Liquids at Various Temperatures

Liquid	t/°C	V/cm^{3}mol^{-1}	Nonpolar, δ_λ/MPa$^{1/2}$	Polar, δ_τ/MPa$^{1/2}$
Acetone, 2-propanone	0	72.3	16.1	12.9
	25	74.0	15.7	12.3
	45	76.2	15.2	11.9
	60	78.1	14.9	12.8
	100	83.4	14.2	10.9
Acetonitrile	0	51.1	16.9	19.0
	25	52.6	16.4	18.2
	45	54.4	16.0	17.5
	60	55.5	15.8	17.3
	100	58.9	15.1	16.3
Acetophenone	0	115.0	20.3	10.1
	25	117.4	19.6	8.1
	45	119.8	19.0	8.0
Aniline	25	91.5	20.1	13.5
	60	94.3	19.1	13.7
	100	97.9	18.2	12.2
1-Butanol	25	92.8	15.9	17.2
Butyl acetate	25	131.5	15.6	7.0
γ-Butyrolactone	25	77.0	19.6	15.4
Butyronitrile	0	85.4	16.9	12.8
	25	87.9	16.3	12.4
	45	90.1	15.9	12.1
Chloropropionitrile	0	75.5	17.8	19.8
	25	77.7	17.2	18.7
	45	79.7	16.8	18.0
Cyclopentanone	25	89.2	18.2	9.4
Diethyl carbonate	0	118.6	16.7	9.6
	25	121.9	16.1	9.0
	45	124.8	15.6	8.8
Diethyl ketone, 3-pentanone	0	103.3	16.4	10.4
	25	106.4	15.8	9.8
	45	108.9	15.4	9.4
Diethyl oxalate	0	132.7	17.9	13.0
	25	136.2	16.6	13.2
	45	139.3	16.1	12.8
N,N-Dimethylacetamide	0	90.7	17.6	16.1
	25	93.2	16.9	14.8
	45	95.3	16.5	14.2
N,N-Dimethylformamide	0	75.5	17.7	16.1
	25	77.4	17.1	15.5
	45	79.0	16.6	14.9
	60	80.2	16.3	15.2
	100	83.8	15.6	14.2
Dimethylsulfoxide	25	71.3	17.3	19.2
	60	73.8	16.6	18.7
	100	77.2	15.9	18.4
Ethanol	25	58.7	16.0	21.9
Ethyl acetate	25	98.5	15.5	9.2
	60	103.3	14.8	8.8
Ethyl butyl ketone	25	139.8	16.1	7.4
	60	146.0	15.3	7.5
Ethylene chlorohydrin	25	66.9	16.5	19.1
	60	68.6	15.9	19.0
	100	77.9	15.1	17.3

TABLE 9 (continued)
Molar Volumes, Nonpolar Cohesion Parameters and Polar Cohesion Parameters of Polar Liquids at Various Temperatures

Liquid	t/°C	V/cm^{3}mol^{-1}	Nonpolar, δ_λ/MPa$^{1/2}$	Polar, δ_τ/MPa$^{1/2}$
Ethylenediamine	0	65.6	17.1	19.9
	25	67.3	16.6	19.1
	45	68.7	16.2	18.4
	60	69.7	15.9	18.1
	100	72.9	15.2	17.4
Ethylene glycol	25	56.0	17.3	31.3
Furfural	0	81.4	19.3	16.1
	25	83.2	18.7	15.4
	45	84.8	18.2	14.9
	60	86.9	17.9	15.0
	100	89.4	16.9	14.4
Methyl Cellosolve®	25	79.4	16.1	15.7
	60	82.3	15.5	15.3
	100	86.0	14.8	16.1
Methyl ethyl ketone, 2-butanone	0	87.3	16.3	11.4
	25	90.1	15.8	11.0
	45	92.6	15.1	10.5
	60	94.5	15.0	10.6
	100	100.0	14.2	10.0
N-Methyl-2-pyrrolidone	0	94.5	19.2	—
	25	96.6	18.5	—
	45	98.6	17.9	7.6
	60	99.6	17.8	13.0
	100	103.3	16.9	12.3
Nitrobenzene	25	102.7	20.1	9.5
	60	105.5	19.3	8.7
	100	109.1	18.2	9.1
Nitromethane	0	53.1	17.3	20.7
	25	54.3	16.8	20.0
	45	55.3	16.4	19.4
	60	56.3	16.1	18.4
	100	59.3	15.4	16.8
Acetylacetone, pentanedione	25	103.1	17.5	10.8
1-Pentanol, amyl alcohol	25	108.7	16.1	14.7
Phenol	25	89.3	20.2	13.0
	100	95.2	18.2	13.9
1-Propanol	25	75.2	15.9	18.9
2-Propanol, isopropanol	25	77.0	15.6	19.0
Propionitrile	0	68.7	16.9	15.1
	25	70.9	16.3	14.7
	45	72.9	15.9	14.4
	60	74.4	15.6	14.1
	100	79.0	14.8	13.7
Pyridine	0	78.7	20.8	9.3
	25	80.9	20.2	7.9
	45	82.9	19.6	6.5
	60	83.9	19.2	6.2
	100	87.4	18.1	6.6
2-Pyrrolidone	25	76.5	20.1	14.1
	45	77.4	20.2	13.3
Tetrahydrofuran	0	79.3	17.7	8.1
	25	81.8	17.1	7.4

Adapted from Helpinstill, J. G. and Van Winkle, M., *Ind. Eng. Chem. Proc. Des. Dev.*, 7, 213, 1968.

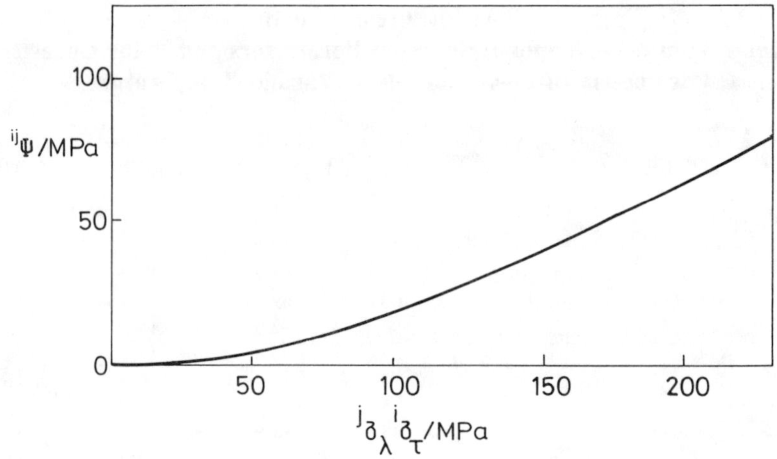

FIGURE 4. Contribution of induction interactions to interchange cohesive pressure at 25°C. (Adapted from Blanks, R. F. and Prausnitz, J. M., *Ind. Eng. Chem. Fund.*, 3, 1, 1964.)

TABLE 10
Values of Induction Cohesive Pressure Correction Factor (${}^{ij}b$)

Component *i* (interacting with liquid *j*)	${}^{ij}b$	Ref.
Induction-free system	1.00	—
Alkane	0.21	113
Alkane	0.20	115
Alkene	0.17	113
Alkene	0.55	115
Aromatic	0.10	113
Aromatic	0.05	115
Vapor of *j*[a]	0.63	120
Vulcanized polyisoprene (natural rubber)	0.19	118
Polymers, surfactants, pigments[b]	0.25	9

[a] From surface free energy, Equation 17, Chapter 17.
[b] Based on Equation 55, and Equation 25, Chapter 17 (the validity of which have been questioned).

Adapted from Hansen, C. M. and Beerbower, A., *Kirk-Othmer Encyclopedia of Chemical Technology,* 2nd ed., Standen, A., Ed., Interscience, New York, 1971, 889. (Related data appear in Table 13.)

Further development of these ideas led to the MOSCED (*mo*dified *s*eparation of *c*ohesive *e*nergy *d*ensity) method (Sections 5.13 and 7.8).

For the special case of a solute in a supercritical solvent (Sections 11.3 and 12.2), Kramer and Thomas[157] neglected the polar and hydrogen-bonding contributions to the cohesive properties of the solvent in its supercritical state while retaining these contributions for the solute and introducing a binary interaction parameter ${}^{ij}\beta$ for solute-solvent interactions so Equation 52b simplified to

$${}^{ij}A = ({}^{i}\delta_d - {}^{j}\delta_d)^2 + {}^{j}\delta_\tau^2 - {}^{ij}\beta \qquad (52c)$$

Seidel and Bittrich[116] from a theoretical model evaluated polar contributions to the enthalpy of vaporization, giving totals which for liquids without hydrogen bonding were in good agreement with experiment.

5.9 HANSEN PARAMETERS

Hansen[9-30] proposed a practical extension of the Hildebrand parameter method to polar and hydrogen-bonding systems, primarily for use in polymer-liquid interactions (Section 14.5). It was assumed that dispersion, polar, and hydrogen-bonding parameters were valid simultaneously, related by Equation 54, with the values of each component being determined empirically on the basis of many experimental observations:

$$\delta_t^2 = \delta_d^2 + \delta_p^2 + \delta_h^2 \tag{54}$$

These parameters are listed in Table 11. Hansen's total cohesion parameter, δ_t, should equal the Hildebrand parameter, although the two quantities may differ for materials with specific interactions when they are determined by different methods. The three component parameters were plotted on a set of three mutually perpendicular axes.

For liquids, the dispersion component, δ_d, obtained by homomorph methods (Section 6.7) was subtracted from the total cohesive property, and the remainder was split into hydrogen-bonding and polar contributions so as to optimize the description of the miscibility behavior of all the polymer-liquid systems investigated (comprising several polymers and many liquids). Both empirical relations (Section 8.2) such as Böttcher's equation[117] and group methods (Section 6.8) were used. Once the three component parameters for each liquid were evaluated, the set of parameters for each polymer could be determined. Burrell[50] pointed out that methods such as this tend to distort the relative magnitudes of the various intermolecular effects, the polar contribution to the cohesive pressure usually being very small in relation to that of hydrogen bonding in those situations where hydrogen bonds exist.

Other Hansen parameter values are presented in Table 12. These data, which draw on earlier results of Hansen,[10,11] as well as on other calculations and observations of elastomer swelling, are based on compilations by Beerbower and Dickey[118] and by Gardon and Teas.[119] Their inclusion is warranted because they provide information on a wide range of pure and mixed liquids. In addition, Table 18 (Section 5.11) presents Hansen parameters for an even wider range of liquids, calculated by the semi-empirical methods described in Sections 5.11 and 7.2.

Occasional negative values of the component parameters have been attributed to the center of maximum interaction lying closer to the axis than the radius of the solubility sphere (see below). The scale on the dispersion axis was doubled with the aim of providing approximately spherical "volumes" of solubility which were drawn up for each solute and then compared with the point locations in this space for each liquid (Figure 5). The distance of the liquid coordinates (${}^i\delta_d$, ${}^i\delta_p$, ${}^i\delta_h$) from the center point (${}^j\delta_d$, ${}^j\delta_p$, ${}^j\delta_h$) of the solute sphere of solubility is

$${}^{ij}R = [4({}^i\delta_d - {}^j\delta_d)^2 + ({}^i\delta_p - {}^j\delta_p)^2 + ({}^i\delta_h - {}^j\delta_h)^2]^{1/2} \tag{55}$$

This may be expressed in a form similar to that of Equation 52b, with $b = 0.25$ and $\delta_\tau^2 = \delta_p^2 + \delta_h^2$:

$${}^{ij}A = ({}^i\delta_d - {}^j\delta_d)^2 + 0.25\,[({}^i\delta_p - {}^j\delta_p)^2 + ({}^i\delta_h - {}^j\delta_h)^2] \tag{56}$$

The distance ${}^{ij}R$ in Equation 55 can be compared with the radius ${}^jR$ of a solute sphere of solubility, and if ${}^{ij}R < {}^jR$ there is a high likelihood of the liquid i dissolving j. This was

TABLE 11
Hansen and Beerbower's 1971 Parameters for Liquids at 25°C

Liquid	$V/cm^3\ mol^{-1}$	$\delta/MPa^{1/2}$			
		δ_d	δ_p	δ_h	δ_t
Alkanes					
Butane	101.4	14.1	0.0	0.0	14.1
Pentane	116.2	14.5	0.0	0.0	14.5
Isopentane, 2-methylbutane	117.4	13.7	0.0	0.0	13.7
Hexane	131.6	14.9	0.0	0.0	14.9
Heptane	147.4	15.3	0.0	0.0	15.3
Octane	163.5	15.5	0.0	0.0	15.5
2,2,4,Trimethylpentane, isooctane	166.1	14.3	0.0	0.0	14.3
Nonane	179.7	15.8	0.0	0.0	15.8
Decane	195.9	15.8	0.0	0.0	15.8
Dodecane	228.6	16.0	0.0	0.0	16.0
Hexadecane	294.1	16.4	0.0	0.0	16.4
Eicosane	359.8	16.6	0.0	0.0	16.6
Cyclohexane	108.7	16.8	0.0	0.2[a]	16.8
Methylcyclohexane	128.3	16.0	0.0	1.0	16.0
cis-Decahydronaphthalene, decalin	156.9	18.8	0.0	0.0	18.8
trans-Decahydronaphthalene	159.9	18.0	0.0	0.0	18.0
Aromatic Hydrocarbons					
Benzene	89.4	18.4	0.0[a]	2.0	18.6
Toluene, methylbenzene	106.8	18.0	1.4	2.0	18.2
Naphthalene[b]	111.5	19.2	2.0	3.9[c]	20.3
Styrene, ethenylbenzene	115.6	18.6	1.0	4.1	19.0
o-Xylene, 1,2-dimethylbenzene	121.2	17.8	1.0	3.1	18.0
Ethylbenzene	123.1	17.8	0.6	1.4	17.8
1-Methylnaphthalene	138.8	20.7	0.8	4.7	21.2
Mesitylene, 1,3,5-trimethylbenzene	139.8	18.0	0.0	0.6	18.0
Tetrahydronaphthalene, tetralin	136.0[a]	19.6[d]	2.0	2.9	20.0
Diphenyl	154.1	21.5	1.0	2.1	21.6
1,4-Diethylbenzene	156.9	18.0	0.0	0.6	18.0

Halohydrocarbons					
Chloromethane, methyl chloride	55.4	15.3[a]	6.1	3.9	17.0
Dichloromethane, methylene dichloride	63.9	18.2	6.3	6.1	20.3
Bromochloromethane	65.0	17.4	5.7	3.5	18.6
Chlorodifluoromethane, Freon 22	72.9	12.3	6.3	5.7	14.9
Dichlorofluoromethane, Freon 21	75.4	15.8	3.1	5.7	17.1
Bromoethane, ethyl bromide	76.9	15.8	3.1	5.7	17.0
1,1-Dichloroethylene	79.0	17.0	6.8	4.5	18.8
1,2-Dichloroethane, ethylene dichloride	79.4	19.0[a]	7.4	4.1	20.9
Diiodomethane, methylene diiodide	80.5	17.8	3.9	5.5	19.0
Trichloromethane, chloroform	80.7	17.8	3.1	5.7	19.0
1,1-Dichloroethane, ethylidene chloride	84.8	16.6	8.2	0.4	18.5
1,1-Dibromoethane, ethylidene bromide	—	18.4[f]	5.1[f]	6.5[f]	20.2[f]
1,2-Dibromoethane, ethylene dibromide	87.0	19.0	6.8	12.1	23.9
Tribromomethane, bromoform	87.5	21.5[a]	4.1	6.1	22.7[a]
1-Chloropropane, propyl chloride	88.1	16.1	7.8	2.0	17.8
Trichloroethylene	90.2	18.0	3.1	5.3	19.0
Dichlorodifluoromethane, Freon 12	92.3	12.3	2.0	0.0	12.5
Trichlorofluoromethane, Freon 11	92.8	15.3	2.1	0.0	15.4
Bromotrifluoromethane	97.0	9.6	2.5	0.0	9.9
Tetrachloromethane, carbon tetrachloride	97.1	17.8	0.0	0.6	17.8
1,1,1-Trichloroethane	100.4	17.0	4.3	2.1	17.7
Tetrachloroethylene, perchloroethylene	101.1	19.0	6.5	2.9[a]	20.3
Chlorobenzene	102.1	19.0	4.3	2.0	19.6
Butyl chloride, 1-chlorobutane	104.9	16.4	5.5	2.1	17.4
1,1,2,2-Tetrachloroethane	105.2[a]	18.8	5.1	9.4	21.6
Bromobenzene	105.3	20.5	5.5	4.1	21.7
o-Dichlorobenzene	112.8	19.2	6.3	3.3	20.5
Benzyl chloride	115.0	18.8[a]	7.2	2.7	20.3
1,1,2,2-Tetrabromoethane[c]	116.8	22.7	5.1	8.2	24.7
1,2-Dichlorotetrafluoroethane, Freon 114[c]	117.0	12.7	1.8	0.0	12.8
1,1,2-Trichlorotrifluoroethane, Freon 113	119.2	14.7	1.6	0.0	14.7
Cyclohexyl chloride	121.3	17.4	5.5	2.1	18.4
1-Bromonaphthalene	140.0	20.3	3.1	4.1	20.9
Trichlorobiphenyl[d]	187.0	19.2	5.3	4.1	20.3
Perfluor(methylcyclohexane)	196.0	12.5	0.0	0.0	12.5
Perfluor(dimethylcyclohexane[d])	217.4	12.5	0.0	0.0	12.5
Perfluoroheptane	227.3	12.5	0.0	0.0	12.1

TABLE 11 (continued)
Hansen and Beerbower's 1971 Parameters for Liquids at 25°C

Liquid	$V/cm^3\ mol^{-1}$	$\delta/MPa^{1/2}$			
		δ_d	δ_p	δ_h	δ_t
Ethers					
Furan	72.5	17.8	1.8	5.3	18.6
Epichlorohydrin	79.9	19.0	10.2	3.7	21.9
Tetrahydrofuran	81.7	16.8	5.7	8.0	19.4
1,4-Dioxane	85.7	19.0	1.8	7.4	20.5
Dimethoxymethane, methylal[c]	88.8	15.1	1.8	8.6	17.5
Diethyl ether	104.8	14.5	2.9	5.1	15.8
Bis(2-chloroethyl) ether	117.6	18.8	9.0	5.7	21.6[a]
Anisole, methoxybenzene[c]	119.1	17.8	4.1	6.8	19.5
Di-(2-methoxyethyl) ether	142.0	15.8	6.1	9.2	19.3
Dibenzyl ether[c]	192.7	17.4	3.7	7.4	19.3
Di-(2-chloroisopropyl) ether[c]	146.0	19.0	8.2	5.1	21.3
Bis-(*m*-phenoxyphenol) ether	373.0	19.6	3.1	5.1	20.5
Ketones					
Acetone, 2-propanone	74.0	15.5	10.4	7.0	20.0
Methyl ethyl ketone, 2-butanone	90.1	16.0	9.0	5.1	19.0
Cyclohexanone	104.0	17.8	6.3	5.1	19.6
Diethyl ketone	106.4	15.8	7.6	4.7	18.1
Mesityl oxide	115.6	16.4	6.1	6.1	18.9
Acetophenone	117.4	19.6[a]	8.6	3.7	21.8
Methyl isobutyl ketone	125.8	15.3	6.1	4.1	17.0
Methyl isoamyl ketone	142.8	16.0	5.7	4.1	17.4
Isophorone	150.5	16.6	8.2	7.4	19.9
Diisobutyl ketone	177.1	16.0	3.7	4.1	16.9

Aldehydes					
Acetaldehyde[c]	57.1	14.7	8.0	11.3	20.2
Furfural, 2-furancarboxyaldehyde	83.2	18.6[a]	14.9	5.1	24.4
Butyraldehyde, butanal	88.5	14.7	5.3[a]	7.0[a]	17.1
Benzaldehyde	101.5	19.4	7.4	5.3	21.5
Esters					
Ethylene carbonate	66.0	19.4	21.7	5.1	29.6
γ-Butyrolactone	76.8	19.0	16.6	7.4	26.3
Methyl acetate	79.7	15.5	7.2	7.6	18.7
Ethyl formate	80.2	15.5	7.2	7.6	18.7
Propylene-1,2-carbonate	85.0	20.0	18.0	4.1	27.3
Ethyl chloroformate	95.6	15.6	10.0	6.8	19.7
Ethyl acetate	98.5	15.8[a]	5.3	7.2[a]	18.1
Trimethyl phosphate	99.9	16.8	16.0	10.2	25.3
Diethyl carbonate	121	16.6	3.1	6.1	17.9
Diethyl sulfate	131.5	15.8	14.7	7.2	22.8
Butyl acetate	132.5	15.8	3.7	6.3	17.4
Isobutyl acetate	133.5	15.1	3.7	6.3	16.8
2-Ethoxyethyl acetate, Cellosolve® acetate	136.2	16.0	4.7	10.6	20.0
Isoamyl acetate	148.8	15.3	3.1	7.0	17.1
Isobutyl isobutyrate	163	15.1	2.9	5.9	16.5
Dimethyl phthalate	163	18.6[a]	10.8[a]	4.9	22.1
Ethyl cinnamate	166.8	18.4	8.2	4.1	20.6
Triethyl phosphate	171.0	16.8	11.5	9.2	22.3
Diethyl phthalate	198	17.6	9.6	4.5	20.6
Dibutyl phthalate	266	17.8[a]	8.6	4.1	20.2
Butyl benzyl phthalate	306	19.0	11.3	3.1	22.3
Tricresyl phosphate	316	19.0	12.3	4.5	23.1
Tributyl phosphate	345	16.4	6.3	4.3	18.1
Isopropyl palmitate[c]	330	14.3	3.9	3.7	15.3
Dibutyl sebacate	339	13.9	4.5	4.1	15.2
Methyl oleate[d]	340	14.5	3.9	3.7	15.5
Dioctyl phthalate	377	16.6	7.0	3.1	18.2
Dibutyl stearate[c]	382	14.5	3.7	3.5	15.4

TABLE 11 (continued)
Hansen and Beerbower's 1971 Parameters for Liquids at 25°C

Liquid	$V/cm^3\ mol^{-1}$	δ/MPa^{1/2} δ_d	δ_p	δ_h	δ_t
	Nitrogen Compounds				
Acetonitrile	52.6	15.3	18.0	6.1	24.4
Acrylonitrile, 2-propenenitrile	67.1	16.4	17.4	6.8	24.8
Propionitrile	70.9	15.3	14.3	5.5	21.7
Butyronitrile	87.0	15.3	12.5	5.1	20.4
Benzonitrile	102.6	17.4	9.0	3.3	19.9
Nitromethane	54.3	15.8	18.8	5.1	25.1
Nitroethane	71.5	16.0	15.5	4.5	22.7
2-Nitropropane	86.9	16.2	12.1	4.1	20.6
Nitrobenzene	102.7	20.0	8.6	4.1	22.2
Ethanolamine, 2-aminoethanol	60.2	17.2	15.6	21.3	31.5
Ethylenediamine	67.3	16.6	8.8	17.0	25.3
1,1-Dimethylhydrazine[c]	76.0	15.3	5.9	11.0	19.8
2-Pyrrolidone	76.4	19.4	17.4	11.3	28.4
Pyridine	80.9	19.0	8.8	5.9	21.8
Propylamine	83.0	17.0	4.9	8.6	19.7
Morpholine	87.1	18.8	4.9	9.2	21.5
Aniline	91.5	19.4	5.1	10.2	22.6
N-Methyl-2-pyrrolidone	96.5	18.0	12.3	7.2	22.9
Butylamine, 1-butanamine	99.0	16.2[a]	4.5[a]	8.0[a]	18.6
Diethylamine	103.2	14.9	2.3	6.1	16.3
Diethylenetriamine	108.0	16.8	13.3	14.3	25.8
Cyclohexylamine	115.2	17.4	3.1	6.6	18.9
Quinoline	118.0	19.4	7.0	7.6	22.0
Dipropylamine	136.9	15.3	1.4	4.1	15.9
Formamide	39.8	17.2	26.2	19.0	36.6
N,N-Dimethylformamide	77.0	17.4	13.7	11.3	24.8
N,N-Dimethylacetamide	92.5	16.8	11.5	10.2	22.7
Tetramethylurea	120.4	16.8	8.2	11.1	21.7
Hexamethyl phosphoramide[c]	175.7	23.2	8.6	11.3	23.2

Sulfur Compounds					
Carbon disulfide	60.0	20.5	0.0	0.6	20.5
Dimethylsulfoxide	71.3	18.4	16.4	10.2	26.7
Ethanethiol[c], ethyl mercaptan	74.3	15.8	6.6	7.2	18.6
Dimethyl sulfone[b]	75	19.0	19.4	12.3	29.8
Diethyl sulfide	108.2	17.0	3.1	2.1	17.4
Acid Halides and Anhydrides					
Acetyl chloride	71.0	15.8	10.6	3.9	19.4
Succinic anhydride[b]	66.8	18.6	19.2	16.6	31.5
Acetic anhydride	94.5	16.0[a]	11.7[a]	10.2[a]	22.3
Alcohols					
Methanol	40.7	15.1	12.3	22.3	29.6
Ethanol	58.5	15.8	8.8	19.4	26.5
Ethylene cyanohydrin, hydracrylonitrile	68.3	17.2	18.8	17.6	31.0
Allyl alcohol, 2-propen-1-ol[c]	68.4	16.2	10.8	16.8	25.7
1-Propanol	75.2	16.0	6.8	17.4	24.5
2-Propanol	76.8	15.8	6.1	16.4	23.5
3-Chloropropanol	84.2	17.6	5.7	14.7	23.6
Furfuryl alcohol	86.5	17.4	7.6	15.1	24.3
1-Butanol	91.5	16.0	5.7	15.8	23.1
2-Butanol	92.0	15.8	5.7	14.5	22.2
Isobutanol, 2-methyl-1-propanol	92.8	15.1	5.7	16.0	22.7
Benzyl alcohol, benzenemethanol	103.6	18.4	6.3	13.7	23.8
Cyclohexanol	106.0	17.4	4.1	13.5	22.4
1-Pentanol, amyl alcohol	109.0	16.0	4.5	13.9	21.7
2-Ethyl-1-butanol	123.2	15.8	4.3	13.5	21.2
Diacetone alcohol	124.2	15.8	8.2	10.8	20.8
1,3-Dimethyl-1-butanol	127.2	15.3	3.3	12.3	19.9
Ethyl lactate	115	16.0	7.6	12.5	21.6
Butyl lactate	149	15.8	6.5	10.2	19.9
Ethylene glycol monomethyl ether, methyl Cellosolve®	79.1	16.2	9.2	16.4	24.8

TABLE 11 (continued)
Hansen and Beerbower's 1971 Parameters for Liquids at 25°C

Liquid	$V/cm^3\ mol^{-1}$	δ/MPa$^{1/2}$ δ_d	δ_p	δ_h	δ_t
Ethylene glycol monoethyl ether, Cellosolve®	97.8	16.2	9.2	14.3	23.5
Diethylene glycol monomethyl ether, methyl Carbitol®	118.0	16.2	7.8	12.7	22.0
Diethylene glycol monoethyl ether, Carbitol®	130.9	16.2	9.2	12.3	22.3
Ethylene glycol monobutyl ether, butyl Cellosolve®	131.6	16.0	5.1	12.3	20.8
2-Ethyl-1-hexanol	157.0	16.0	3.3	11.9	20.2
1-Octanol	157.7	17.0	3.3	11.9	21.0
2-Octanol	159.1	16.2	4.9	11.1	20.2
Diethylene glycol monobutyl ether, butyl Carbitol®	170.6	16.0	7.0	10.6	20.4
1-Decanol	191.8	17.6	2.7	10.0	20.4
'Tridecyl alcohol,' 1-tridecanol[d]	242	14.3	3.1	9.0	17.2
'Nonyl' phenoxy ethanol[d]	275	16.8	10.2	8.4	21.4
Oleyl alcohol, 9-octadecen-1-ol[d]	316	14.3	2.7	8.0	16.6
Triethylene glycol mono-oleyl ether	418.5	13.3	3.1	8.4	16.0
Acids					
Formic acid, methanoic acid	37.8	14.3	11.9	16.6	24.9
Acetic acid, ethanoic acid	57.1	14.5	8.0	13.5	21.4
Benzoic acid[b]	100	18.2, 21.5[f]	7.0, 5.7[f]	9.8	21.8, 24.3[f]
Butyric acid, 1-butanoic acid[c]	110	14.9	4.1	10.6	18.8
Octanoic acid[c]	159	15.1	3.3	8.2	17.5
Oleic acid, 9-octadecenoic acid[d]	320	14.3	3.1	5.5	15.6
Stearic acid, octadecanoic acid[b]	326	16.4	3.3	5.5	17.6
Phenols					
Phenol	87.5	18.0	5.9	14.9	24.1
1,3-Benzenediol, resorcinol[b]	87.5	18.0	8.4	21.1	29.0

m-Cresol, 3-methylphenol	104.7	18.0	5.1	12.9	22.7
o-Methoxyphenol[c]	109.5	18.0	8.2	13.3	23.8
Methyl salicylate	129	16.0	8.0	12.3	21.7
'Nonyl' phenol[d]	231	16.6	4.1	9.2	19.4
Water[c]	18.0	15.6[a]	16.0[a]	42.3[a]	47.8
Polyhydric Alcohols					
Ethylene glycol, 1,2-ethanediol	55.8	17.0	11.0	26.0	32.9
Glycerol, 1,2,3-propanetriol	73.3	17.4	12.1	29.3	36.1
Propylene glycol, propanediol	73.6	16.8	9.4	23.3	30.2
1,3-Butanediol	89.9	16.6	10.0	21.5	28.9
Diethylene glycol	95.3	16.2	14.7	20.5	29.9
Triethylene glycol	114.0	16.0	12.5	18.6	27.5
Hexylene glycol	123.0	15.8	8.4	17.8	25.2
Dipropylene glycol[d] (mixed isomers)	131.3	16.0	20.3	18.4	31.7

[a] Altered from previously published 1967 value; additional values based on 1967 data appear in Table 12.
[b] Solid, treated as subcooled liquid.
[c] Values uncertain.
[d] Impure commercial product of this nominal formula.
[e] Subsequently recalculated by Wu, P. L., Beerbower, A., and Martin, A., *J. Pharm. Sci.,* 71, 1285, 1982.
[f] Subsequently recalculated in 1974 as reported by Beerbower, A., Wu, P. L., and Martin, A., *J. Pharm. Sci.*, 73, 179, 1984.

Adapted from Hansen, C. M. and Beerbower, A., *Kirk-Othmer Encyclopedia of Chemical Technology,* 2nd ed., Standen, A., Ed., Interscience, New York, 1971, Suppl. Vol., 889.

TABLE 12
Hansen Parameters for Liquids, Refrigerants, and Liquid Mixtures at 25°C, Based on 1967 Data

Liquids	$\delta/MPa^{1/2}$		
	δ_d	δ_p	δ_h
Acetic acid, ethanoic acid	14.5	8.0	13.5
Acetone, 2-propanone	15.6	12.3, 10.4[a]	7.0
Acetonitrile	15.3	18.0	6.1
Acetophenone	17.6[a]	8.6[a]	3.7[a]
Acetyl chloride	15.8[a]	10.6[a]	3.9[a]
Acrylonitrile, 2-propenenitrile	16.4	17.4	6.8
Aniline, benzeneamine	19.4, 20.1[a]	5.1, 7.4[a]	10.2, 12.3[a]
Benzaldehyde	19.4	7.4	5.3
Benzene	18.4	1.0	2.1
Benzoic acid	18.2	7.0	9.8
Benzonitrile	17.4[a]	13.3[a]	5.1[a]
Benzyl alcohol, benzenemethanol	18.4	6.3	13.7
Benzyl chloride, (chloromethyl)benzene	17.6, 18.8[a]	7.2	2.7
Biphenyl	21.1[a]	1.0[a]	2.0[a]
Bromobenzene	19.0	5.5	2.1
Bromoethane	16.6[a]	8.0[a]	5.1[a]
1-Bromonaphthalene	20.3[a]	3.1[a]	4.1[a]
Butane	13.7, 14.1[a]	0.0	0.0
1,3-Butanediol	16.6[a]	10.0[a]	21.5[a]
1-Butanol	16.0	5.7	15.8
2-Butanol	15.8	5.7	14.5
Butyl acetate	15.8	3.7	6.3
Butylamine, 1-butanamine	15.6	3.7	7.2
Butyl Carbitol®, 2-(2-butoxyethoxy)ethanol	16.0, 16.4[a]	7.0	10.6
Butyl Cellosolve,® 2-butoxyethanol	16.0	5.1, 6.3[a]	12.3, 12.1[a]
Butyl chloride, 1-chlorobutane	16.2[a]	5.5[a]	2.0[a]
Butyl lactate	15.5[a]	6.5[a]	10.2[a]
Butyraldehyde, butanal	14.7	9.8	5.1, 7.2[a]
Butyronitrile	15.3[a]	12.5[a]	5.1[a]
Carbitol®, 2-(2-ethoxyethoxy)ethanol	16.2	9.2	14.3
Carbon disulfide	20.5, 20.3[a]	0.0	0.6, 0.0[a]
Castor oil	15.6	2.9	9.2
Cellosolve®, 2-ethoxyethanol	16.2	9.2	14.3
Cellosolve® acetate, 2-ethoxyethyl acetate	16.0	4.7	10.6
Chlorobenzene	18.6, 19.0[a]	4.3	4.1, 2.0[a]
Chlorobromomethane	17.4	5.7	3.5
Chlorodifluoromethane	12.3[a]	6.3[a]	5.7[a]
Chlorofluoroethane	11.9[a]	5.7[a]	4.1[a]
Chloromethane	15.3[a]	6.1[a]	3.9[a]
1-Chloropropane	16.0[a]	7.8[a]	2.0[a]
3-Chloropropanol	17.6[a]	5.7[a]	14.7[a]
Cresol, methylphenol	18.0	5.1	12.9
Cyclohexane	16.7	0.0	0.0
Cyclohexanol	17.4	4.1	13.5
Cyclohexanone	17.8	6.3, 8.4[a]	5.1
Cyclohexylamine	17.4[a]	3.1[a]	6.5[a]
Cyclohexyl chloride	19.4[a]	5.5[a]	2.0[a]
1-Decanol	16.0[a]	3.5[a]	8.2[a]
Diacetone alcohol	15.8	8.2	10.8
Dibenzyl ether	17.4	3.7	7.4
Diisobutyl ketone	16.0	3.7	4.1
Dibutyl phthalate	16.8	8.6	4.1
Dibutyl sebacate	13.9	4.5	4.1
o-Dichlorobenzene	19.2	6.3	3.3

TABLE 12 (continued)
Hansen Parameters for Liquids, Refrigerants, and Liquid Mixtures at 25°C, Based on 1967 Data

Liquids	δ/MPa$^{1/2}$		
	δ_d	δ_p	δ_h
2,2-Dichlorodiethyl ether	18.8[a]	9.0[a]	3.1[a]
Dichlorodifluoromethane	12.3[a]	2.0[a]	0.0[a]
1,1-Dichloroethane	16.6[a]	8.2[a]	0.4[a]
1,2-Dichloroethane, ethylene dichloride	18.8	5.3	4.1
Dichloroethylene	17.0	4.7	7.2
Dichloromethane, methylene dichloride	18.2	6.3	6.1
1,2-Dichlorotetrafluoroethane	12.7[a]	1.8[a]	0.0[a]
Diethylamine	14.9[a]	2.3[a]	6.1[a]
1,4-Diethylbenzene	17.8[a]	0.0[a]	0.6[a]
Diethyl carbonate	16.6[a]	3.1[a]	6.1[a]
Diethylene glycol, 2-hydroxyethyl ether	16.2, 15.8[a]	14.7	20.5
Diethylene triamine	16.8[a]	13.3[a]	14.3[a]
Diethyl ether, 1,1′-oxybisethane	14.5	2.9, 4.9[a]	5.1, 2.0[a]
Diethyl ketone	15.8[a]	7.6[a]	4.7[a]
Diethyl sulfate	15.8	14.7	7.2
Diethyl sulfide	16.0[a]	6.3[a]	2.0[a]
Difluoroethane	9.2[a]	10.2[a]	5.7[a]
Dimethoxymethane, formal, methylal	15.1[a]	1.8[a]	8.6[a]
N,N-Dimethylformamide	17.4, 17.0[a]	13.7, 13.3[a]	11.3, 9.2[a]
1,1-Dimethylhydrazine	15.3	5.9	11.1
Dimethyl siloxane	12.1	0.0	0.0
Dimethylsulfone	19.0[a]	19.4[a]	12.3[a]
Dimethylsulfoxide	18.4	16.4	10.2
Dioctyl phthalate	16.6	7.0	3.1
1,4-Dioxane, diethylene oxide	19.0, 17.6[a]	1.8, 8.6[a]	7.4, 4.1[a]
Diphenyl, 1,1′-biphenyl	21.5	1.0	2.1
Dipropylamine	15.3[a]	1.4[a]	4.1[a]
Dipropylene glycol	16.0[a]	13.3[a]	23.9[a]
Epichlorohydrin	19.0[a]	10.2[a]	3.7[a]
Ethanol	15.8	8.8	19.4
Ethanolamine	17.2[a]	15.6[a]	21.3[a]
Ethyl acetate	15.1, 15.3[a]	5.3	9.2
Ethylbenzene	17.8	1.0, 0.6[a]	3.1, 1.4[a]
2-Ethylbutanol	15.8[a]	4.3[a]	13.5[a]
Ethyl carbonate	19.4[a]	21.7[a]	5.1[a]
Ethyl chloroformate	15.6	10.0	6.8
Ethyl cinnamate	18.4	8.2	4.1
Ethylene glycol, 1,2-ethanediol	17.0, 16.8[a]	11.1	26.0
Ethyl formate	15.5[a]	8.4[a]	8.4[a]
2-Ethylhexanol	16.0[a]	3.3[a]	11.9[a]
Ethyl lactate	16.0[a]	7.6[a]	12.5[a]
Formamide	17.2[a]	26.2[a]	19.0[a]
Furan	17.8[a]	1.8[a]	5.3[a]
Furfural, 2-furancarboxyaldehyde	16.8	14.9	5.1
Furfuryl alcohol, 2-furanmethanol	17.4	7.6	15.1
Glycerol, 1,2,3-propanetriol	17.4	12.1	29.3
Heptane	15.3	0.0	0.0
Hexane	14.7, 14.9[a]	0.0	0.0
1-Hexanol	15.8[a]	4.3[a]	13.5[a]
Hexylene glycol	15.8	8.4	17.8
Isoamyl acetate	15.3	3.1	7.0
Isobutyl acetate	15.1[a]	3.7[a]	6.3[a]
Isobutyl isobutyrate	15.1[a]	2.9[a]	5.9[a]

TABLE 12 (continued)
Hansen Parameters for Liquids, Refrigerants, and Liquid Mixtures at 25°C, Based on 1967 Data

Liquids	$\delta/MPa^{1/2}$		
	δ_d	δ_p	δ_h
Isohexanol	15.8	4.3	13.5
Isononyl phenol	16.6	4.1	9.2
Isononyl phenoxyethanol	16.8	10.2	8.4
Isooctane, 2,2,4-trimethylpentane	14.3	0.0	0.0
Isopentane, 2-methylbutane	13.7	0.0	0.0
Isophorone	16.6[a]	8.2[a]	7.4[a]
Mesityl oxide	16.4[a]	7.2[a]	6.1[a]
Methanol, methyl alcohol	15.4, 15.1[a]	12.3	22.3
Methoxymethanol	16.2[a]	9.2[a]	16.4[a]
Methyl acetate	15.6	7.2	7.6
Methyl amyl ketone	16.0	5.7	4.1
Methyl Carbitol®, 2-(2-methoxyethoxy)ethanol	16.2[a]	7.8[a]	12.7[a]
Methyl Cellosolve®, 2-methoxyethanol	16.2[a]	9.2[a]	16.4[a]
Methyl ethyl ketone, 2-butanone	16.0	9.0	5.1
Methyl isoamyl ketone	16.0[a]	5.7[a]	4.1[a]
Methyl isobutyl carbinol, 1,2-dimethylpropanol	15.3[a]	3.3[a]	12.3[a]
Methyl isobutyl ketone, 4-methyl-2-pentanone	15.3	6.1	4.1, 5.7[a]
Methyl oleate	14.5	3.9	3.7
N-Methyl-2-pyrrolidone	18.0[a]	12.3[a]	14.6[a]
Morpholine	18.8[a]	4.9[a]	9.2[a]
Naphthalene	19.2	2.1	5.9
Nitrobenzene	17.6	12.3	4.1
Nitroethane	16.0[a]	15.5[a]	4.5[a]
Nitromethane	15.8	18.8	5.1
2-Nitropropane	16.2[a]	12.1[a]	4.1[a]
Octane	15.3[a]	0.0[a]	0.0[a]
1-Octanol	16.2[a]	7.0[a]	10.6[a]
Oleic acid	14.3	3.1	5.5
Oleyl triethylene glycol ether	13.3	3.1	8.4
Pentane	14.3[a]	0.0[a]	0.0[a]
1-Pentanol	16.0[a]	4.5[a]	13.9[a]
Phenol	18.0	5.9	14.9
Phenoxyethanol	17.8	5.3	12.3
1-Propanol	16.0	6.8, 6.1[a]	17.4, 17.6[a]
2-Propanol	15.8, 15.5[a]	7.2, 9.0[a]	16.0, 16.8[a]
Propionitrile	15.3[a]	12.3[a]	8.2[a]
Propylamine	15.5[a]	4.1[a]	11.3[a]
Propylene carbonate	20.1	18.0	4.1
Propylene glycol, propanediol	16.8	9.4	23.3
Pyridine	19.0	8.8	5.9
2-Pyrrolidone	19.4[a]	17.4[a]	11.3[a]
Resorcinol, 1,3-benzenediol	18.0	8.4	21.1
Stearic acid, octadecanoic acid	16.4	3.3	5.5
Styrene, ethenylbenzene	18.6	1.0	4.1
Succinic anhydride	18.6	19.2	16.6
Tetrachloroethane	18.8	4.1	2.3
Tetrachloroethylene	19.0	6.6	2.9
Tetrachloromethane, carbon tetrachloride	17.8	0.0	0.6
Tetrahydrofuran	16.8[a]	5.7[a]	8.0[a]
Tetrahydronaphthalene	19.2[a]	2.0[a]	2.9[a]
Toluene, methylbenzene	18.0	1.4	2.1
Tribromomethane, bromoform	17.8	4.1	10.8
Tributyl phosphate	16.4	6.3	4.3

TABLE 12 (continued)
Hansen Parameters for Liquids, Refrigerants, and Liquid Mixtures at 25°C, Based on 1967 Data

	$\delta/MPa^{1/2}$		
Liquids	δ_d	δ_p	δ_h
Trichloroethane	17.0	4.3	2.1
Trichloroethylene	18.0	3.1	5.3
Trichloromethane, chloroform	17.8, 17.6[a]	3.1	5.7
Tricresyl phosphate	19.0	12.3	4.5
Triethylene glycol	16.0	10.4	18.6
Water	12.3	31.3	34.2
o-Xylene, 1,2-dimethylbenzene	17.8	1.0	3.1
Refrigerants			
R11, CCl_3F, trichlorofluoromethane	15.3	2.1	0.0
R12, CCl_2F_2 dichlorodifluoromethane	12.3	2.1	0.0
R13, $CClF_3$ chlorotrifluoromethane	4.9	1.8	0.0
R21, CCl_2FH, dichlorofluoromethane	15.8	3.1	5.7
R22, $CClF_2H$, chlorodifluoromethane	12.3	6.3	5.7
R40, $CClH_3$, chloromethane	15.6	6.1	3.9
R113, $C_2Cl_3F_3$, 1,1,2-trichloro-1,2,2-trifluoroethane	14.7	1.6	0.0
R114, $C_2Cl_2F_4$, 1,2-dichlorotetrafluoroethane	12.7	1.8	0.0
R115, C_2ClF_5, chloropentafluoroethane	9.8	1.8	0.0
R13B1, $CBrF_3$, bromotrifluoromethane	9.6	2.5	0.0
R142b, $C_2ClF_2H_3$, chlorodifluoroethane	11.9	5.7	4.1
R152a, $C_2F_2H_4$ difluoroethane	9.2	10.2	5.7
C318, c-C_4F_8, perfluorocyclobutane	11.7	1.2	0.0
Liquid Mixtures			
ASTM fuel 'A', isooctane, 2,2,4-trimethylpentane	14.3	0.0	0.0
ASTM fuel 'B'	15.3	0.4	0.6
ASTM fuel 'C' (calculated)	16.2	0.8	1.0
ASTM oil #1	13.9	0.0	0.0
ASTM oil #2	15.6	0.6	0.2
ASTM oil #3	16.6	1.0	0.4
Auto brake fluid	15.8	6.1	10.2
Auto transmission fluid	14.3	0.4	0.6
Linseed oil	13.9	3.5	3.7
MIL-L-7808 (ester)	14.3	2.9	3.1
MIL-H-8446 (silicate)	14.5	6.1	7.6
MIL-H-5606 (Petr.)	14.7	0.8	0.6
Motor oil-SAE 20W	14.7	0.4	0.4
Neats foot oil	14.3	2.9	3.7
Phosphate hydraulic	14.5	10.4	4.5
Sperm oil	14.3	2.1	2.7
Turpentine	16.4	1.4	0.4
Water-glycol hydraulic fluid	14.3	18.8	22.5

[a] From Gardon and Teas where the values differ from those of Beerbower and Dickey.

Adapted from Beerbower, A. and Dickey, J. R., *Am. Soc. Lubric. Eng. Trans.*, 12, 1, 1969, with alternative and additional values from Gardon, J. L. and Teas, J. P., in *Treatise on Coatings*, Vol. 2, *Characterization of Coatings: Physical Techniques*, Part II, Myers, R. R. and Long, J. S., Eds., Marcel Dekker, New York, 1976, chap. 3.

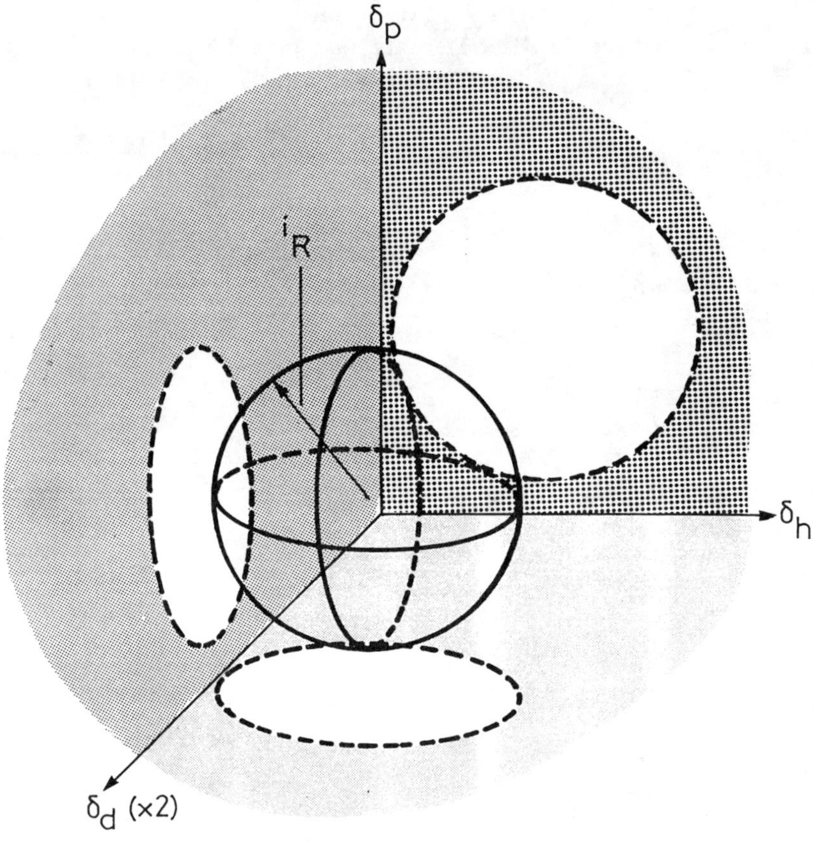

FIGURE 5. Representation of a Hansen parameter solubility sphere and its projections on three axial planes. (Adapted from Beerbower, A. and Dickey, J. R., *Am. Soc. Lubric. Eng. Trans.* 12, 1, 1969.)

found to work well, despite the fact that Equation 55 is very different from the more theoretically justified Equation 21. The "sphere" can be projected onto the three planes passing through two axes and the origin, to provide circles in two-dimensional graphs, as illustrated in Figure 5. Hansen parameters also are related to component free energies of surfaces (Section 17.3).[120]

The incorporation of the numerical factor 4 in Equation 55 does not appear necessary to provide spherical interaction volumes;[121,122] the apparent non-spherical representations which were observed were the result of the restricted range of δ_d values compared with the δ_p and δ_h ranges, and an equation based on Equation 54 is just as satisfactory:

$$^{ij}r = {}^{ij}A^{1/2} = [({}^i\delta_d - {}^j\delta_d)^2 + ({}^i\delta_p - {}^j\delta_p)^2 + ({}^i\delta_h - {}^j\delta_h)^2]^{1/2} \tag{57}$$

Hansen[155] has also defined the *relative energy difference* as the ratio $^{ij}R/^{j}R$, so if this number is less than 1.0, liquid i is predicted to be a solvent for j, with values greater than unity indicating progressively poorer solvents. Applications to several materials of biological interest appear in Table 12a.

Brench[123,124] proposed for component i in a mixture with effective Hansen parameters $\overline{\delta}_d$, $\overline{\delta}_p$, and $\overline{\delta}_h$ the relationship

$$^{ij}A = ({}^i\delta_d - \overline{\delta}_d)^2 + {}^ib\,[({}^i\delta_p - \overline{\delta}_p)^2 + ({}^i\delta_h - \overline{\delta}_h)^2] \tag{58}$$

TABLE 12a
Hansen Parameters for Some Biological Materials, for Use With the 1971 Hansen Liquid Parameters of Table 11

	δ/MPa$^{1/2}$				
	δ_d	δ_p	δ_h	δ_t	^jR
Lard, 23°C	17.8	2.7	4.4	18.4	8.0
Lard, 37°C	15.9	1.2	5.4	16.8	12.0
Water[a]	15.1	20.4	16.5	30.3	18.1
Blood serum	23.2	22.7	30.6	44.6	20.5
Sucrose	21.7	26.3	29.6	45.1	20.4
Urea	20.9	18.7	26.4	38.5	19.4
Keratin (psoriasis scales)	24.6	11.9	12.9	30.3	19.0
Lignin	20.6	13.9	15.3	29.2	11.8

[a] For 1% solutions in water; preferred value, amines not included.

Adapted from Hansen, C. M. and Anderson, B. H., *Am. Ind. Hyg. Assoc.*, 49, 301, 1988.

where ^ib is a Hansen weighting factor, a property of component i, with values (Table 13) comparable to those of ^{ij}b in Table 10. These values have been interpreted physically[124] on the basis that the dispersion forces represent a ''billiard ball'' effect over the whole molecule, while the polar and hydrogen-bonding terms are localized on groups and thus not uniformly distributed.

Numerical results for polymers are quoted in Section 14.5, together with the application of vector methods to the interaction of liquid mixtures with polymers. Hansen parameters are applied to solid solubilities in Chapter 12. In some applications, only two of the three Hansen parameters are used, so that the locations of liquids may be displayed on two-dimensional, projected maps. Figure 6 shows the range of δ_p versus δ_h locations for major liquid groups, regions of overlap predicting mutual miscibility; this technique may be applied to the formulation of extraction systems.[125] Solubility ranges for polymers may be represented in the same way (Section 15.1).

The original ''Hansen parameters'' were determined as outlined above, but very similar parameters have been obtained by alternative calculation methods,[126] as indicated in Section 5.11, with Hoy's major set of values being reported below in Table 17.

There is considerable variation in the Hansen parameters reported for water (Table 14). A study of the solubilities of a variety of *organic compounds in water*[26,155] provides the values in the last three entries of this Table. The δ_h and δ_t values are considerably lower than and inconsistent with those previously reported, which appear to be appropriate for solutions of *water in organic liquids*. (Grunwald[127] discussed aqueous solutions in terms of components described as ''isodelphic'' (unchanged thermodynamic state of solvent during the solution process) and ''lyodelphic'' (partial molar contributions due to solute perturbation of solvent network). Use of the regular solution model gave a water total cohesion parameter of 29 MPa$^{1/2}$, consistent with the ''organics in water'' value.) This variability in Hansen parameter values is a fundamental problem associated with the use of δ_h rather than the more appropriate δ_a and δ_b to represent the hydrogen-bonding cohesion, and attempts to reconcile the divergent δ_h values are futile. The reservations originally expressed by Tawn[128] on Hansen parameters are still valid: this method of describing interactions between liquids and potential solutes is neither as simple as the Burrell method (Sections 4.5, 14.3) nor as complete as a full set of component cohesion parameters. However, the recent values provided by Hansen[155] for ''1% solutions in water'' (Table 12a) acknowledge the problem and provide a guide to behavior in hydrogen-bonded systems. Clearly the use of separate Lewis acid

TABLE 13
Values of Hansen Weighting Factor

Liquid	^{i}b
Hydrocarbons	
Hexane	0.22
Heptane	0.20
Isooctane	0.17
Cyclohexane	0.19
Benzene	0.15
Toluene	0.064
Ethylbenzene	0.12
Styrene	0.13
Chlorinated Hydrocarbons	
Trichloromethane	0.22
Tetrachloromethane	0.24
Trichloroethylene	0.20
Tetrachloroethylene	0.24
1,1,2-Trichloroethane	0.18
Ketones	
Acetone	0.14
Methyl ethyl ketone	0.14
Methyl isobutyl ketone	0.13
Esters	
Methyl acetate	0.16
Ethyl acetate	0.14
Propyl acetate	0.16
Butyl acetate	0.19
Amyl acetate, pentyl acetate	0.19
Ethyl propionate	0.20
Ethyl butyrate	0.19
Alcohols	
Methanol	0.26
Ethanol	0.20
1-Propanol	0.24
2-Propanol	0.20
1-Butanol	0.24
2-Butanol	0.19
Isobutanol	0.23
tert-Butanol	0.20
1-Pentanol	0.19
1-Hexanol	0.28
sec-Octanol	0.23
Ethylene glycol	0.31
Glycerol	0.20
Carboxylic Acid	
Acetic acid	0.12

TABLE 13 (continued)
Values of Hansen Weighting Factor

Liquid	^{i}b
Nitrogen-Containing Liquids	
Ethylenediamine	0.29
Pyridine	0.11
Lipids	
Diolein	0.20
Triolein	0.16
Trilinolein	0.17
Olive oil	0.18
Oleic acid	0.18
Other Liquids	
1,4-Dioxane	0.10
Furfural	0.26
Propylene carbonate	0.41
Water	0.32

Adapted from Ashton, N. F., McDermott, C., and Brench, A., in *Handbook of Solvent Extraction,* Lo, T. C., Baird, M. H. I., and Hanson, C., Eds., John Wiley & Sons-Interscience, New York, 1983, 3.

and Lewis base component parameters would be superior, but more cumbersome, in providing an approximate representation of the "unsymmetrical" interactions.

5.10 FRACTIONAL THREE-COMPONENT COHESION PARAMETERS

Teas[129] showed that for several polymer-liquid systems it is possible to use *fractional cohesive pressures* plotted on a triangular chart to represent miscibility limits:

$$U_d/U = \delta_d^2/\delta_t^2;\ U_p/U = \delta_p^2/\delta_t^2;\ U_h/U = \delta^2{}_h/\delta_t^2 \tag{59}$$

where

$$\delta_t^2 = \delta_d^2 + \delta_p^2 + \delta_h^2 \tag{60}$$

and

$$U = U_d + U_p + U_h \tag{61}$$

This method was used by Vial[130,131] (Table 15), but Teas[119,129,132] chose instead to use *fractional cohesion parameters*, which may be defined

$$f_d = \frac{\delta_d}{\delta_d + \delta_p + \delta_h};\ f_p = \frac{\delta_p}{\delta_d + \delta_p + \delta_h};\ f_h = \frac{\delta_h}{\delta_d + \delta_p + \delta_h} \tag{62}$$

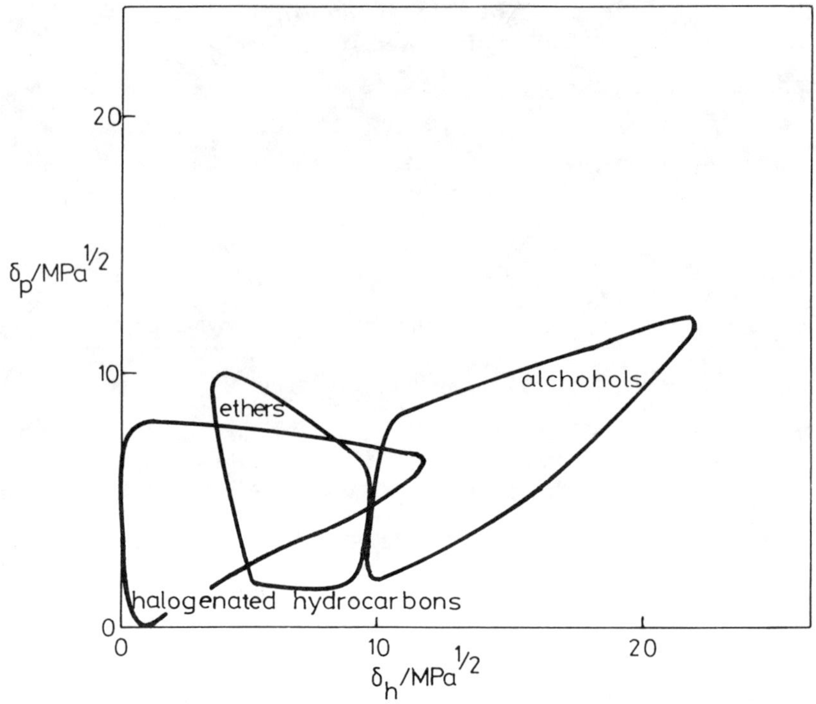

FIGURE 6. Hansen paremeter $\delta_p - \delta_h$ locations for major liquid groups: (A) ethers, halogenated hydrocarbons, and alcohols; (B) esters, aromatic hydrocarbons, ketones, and phenols; (C) aldehydes, polyhydric alcohols, unionized acids, and alkanes; and (D) proton donors (acids, phenols, amines, alcohols, polyhydric alcohols). (Adapted from Klein, E., Eichelberger, J., Eyer, C., and Smith, J., *Water Res.*, 9, 807, 1975.)

The fractional cohesion parameters of Gardon and Teas[119] in Table 16 are calculated from the data published by Hansen and Skaarup in 1967; also included are the values of Teas[28,129] based on Hansen's earlier publication,[10] as these have been widely used. The fractional cohesion parameters defined in Equation 62 have the advantage of spreading the data points more uniformly over the triangular chart, but the disadvantage that they are completely empirical, without even the limited theoretical justification of Hansen parameters. Examples of fractional maps for polymers are shown in Figures 16 to 33, Chapter 15.

The triangular representations make the simplifying assumption that the total cohesion parameter, δ_t, is constant for all materials, and that the *relative* magnitudes of the three contributions (dispersion forces, polar interactions, hydrogen bonding) determine the extent of miscibility. Inspection of tables of Hansen parameter values shows that although there is much greater variation in δ_p and δ_h than in δ_t, the total cohesion parameter is not even approximately constant.

5.11 OTHER TWO- AND THREE-COMPONENT COHESION PARAMETERS

If it is assumed that the cohesive energy $(-U)$ is made up of an additive combination of contributions from nonpolar or dispersive interactions $(-U_d)$, polar interactions $(-U_p)$, and hydrogen-bonding or similar specific association interactions $(-U_h)$,

$$-U = -U_d - U_p - U_h \tag{63}$$

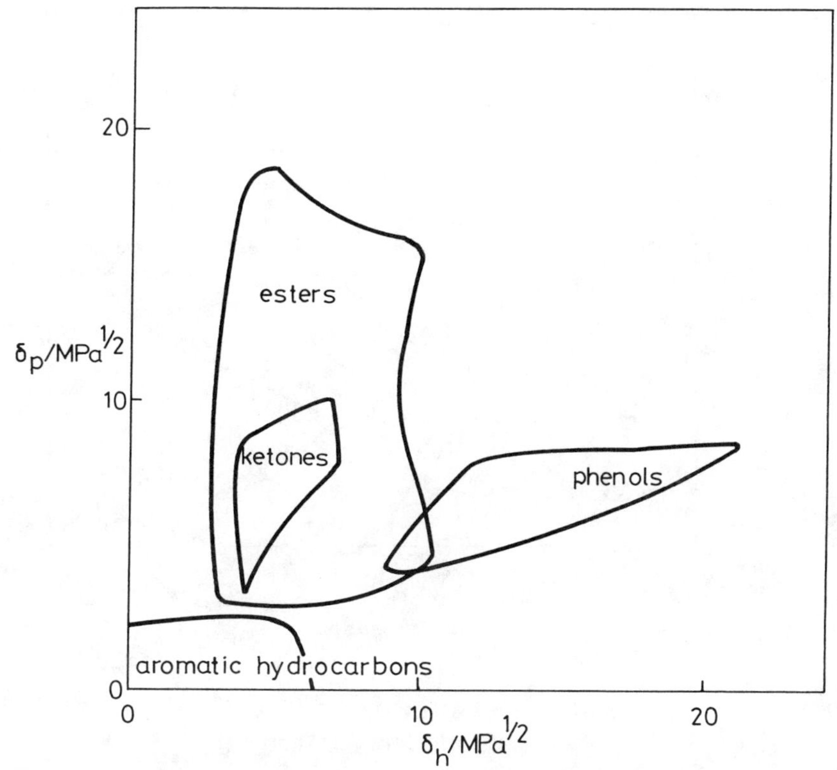

FIGURE 6B.

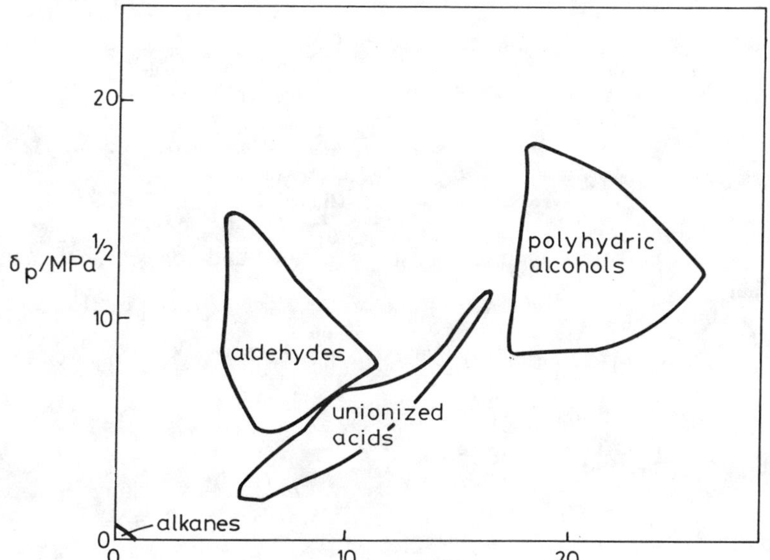

FIGURE 6C.

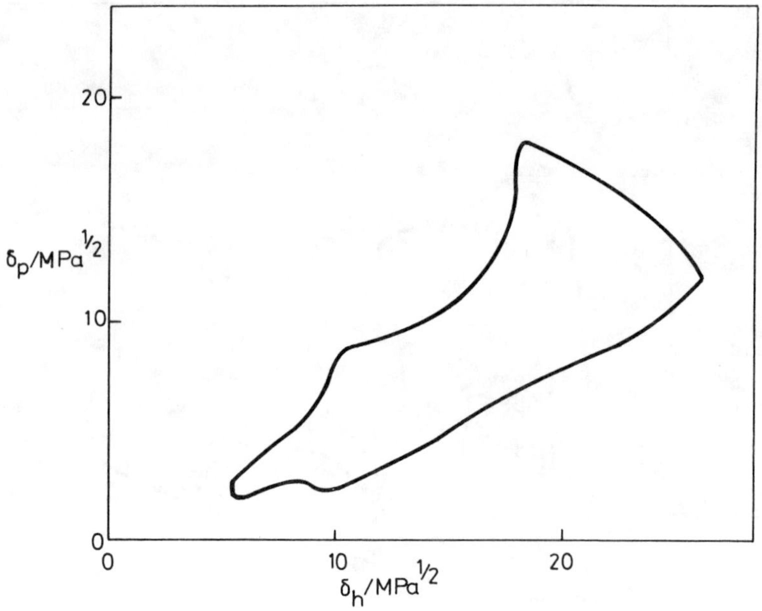

FIGURE 6D.

TABLE 14
Hansen Parameters for Water

δ/MPa$^{1/2}$						
δ_d	δ_p	δ_h	δ_t	^jR	Source	Ref.
12	23	40	48	—	Table 18	126
12	31	34	48	—	Table 12	10, 11, 118
16	16	42	48	—	Table 11	25
20	18	18	32	15	Organic liquids in water	26
15	20	17	30	18	1% solutions in water, Table 12a	155
—	—	—	33—36	—	Aqueous phase in solvent extraction	158
—	—	—	29	—	Grunwald	127

then it follows that the corresponding cohesive pressures and cohesion parameters can be defined so that

$$-U/V = -U_d/V - U_p/V - U_h/V \tag{64}$$

$$\delta_t^2 = \delta_d^2 + \delta_p^2 + \delta_h^2 \tag{65}$$

as seen in Equation 54, and

$$^{ij}A = (^i\delta_d{}^i\delta_d - {}^j\delta_d)^2 + (^i\delta_p - {}^j\delta_p)^2 + (^i\delta_h - {}^j\delta_h)^2 \tag{66}$$

This method was developed by Hansen[9-30,155] (Section 5.10) on an empirical basis and by means of semiempirical equations[28,117,120] but it may also be used for theoretical subdivisions of the Hildebrand parameter with other bases. With the aid of relationships of the type indicated in Section 8.2, the Hansen parameters can be described in terms of molecular parameters related to intermolecular forces and molecular sizes.[133]

Null and Palmer[111,134,135] extended the polar-nonpolar cohesion parameter concept of

TABLE 15
Fractional Cohesive Pressures in Order of Increasing Total Value

Liquid	δ_t/MPa$^{1/2}$	100 U_d/U	100 U_p/U	100 U_h/U
Hexane	14.8	100.0	0.0	0.0
Diethyl ether	15.6	85.8	3.4	10.8
Diisobutyl ketone	16.7	89.3	4.8	5.9
Cyclohexane	16.7	100.0	0.0	0.0
Isoamyl acetate	17.0	80.1	3.2	16.7
Butyl acetate	17.3	82.1	4.5	13.4
1,1,1-Trichloroethane	17.5	92.6	6.0	1.4
Methyl isobutyl ketone	17.5	81.2	13.0	5.8
Tetrachloromethane, carbon tetrachloride	17.7	100.0	0.0	0.0
Toluene, methylbenzene	18.2	98.1	0.6	1.3
Ethyl acetate	18.6	67.2	8.2	24.6
Benzene	18.7	98.5	0.3	1.2
Trichloromethane, chloroform	18.8	88.1	2.6	9.3
Methyl ethyl ketone, 2-butanone	19.0	70.2	22.5	7.3
Trichloroethylene	19.0	89.5	2.6	7.9
Styrene, ethenylbenzene	19.0	95.1	0.3	4.6
Tetrahydronaphthalene, tetralin	19.4	96.7	1.1	2.2
Tetrahydrofuran	19.5	74.6	8.6	16.8
Ethylglycol acetate	19.6	65.2	5.7	29.1
Acetophenone	19.8	77.8	18.8	3.4
Isophorone	19.9	69.4	16.9	13.7
1,4-Dioxane	19.9	86.3	0.8	12.9
1,2-Dichloroethane, ethylene dichloride	20.0	88.7	7.1	4.2
Acetone, 2-propanone	20.0	60.5	27.4	12.1
Cyclohexanone	20.2	76.4	17.2	6.4
Dichloromethane, methylene dichloride	20.3	81.0	9.8	9.2
2-Nitropropane	20.5	61.6	34.4	4.0
Pyridine	21.7	76.1	16.4	7.5
Nitrobenzene	21.7	64.9	31.6	3.5
Cyclohexanol	22.4	60.3	3.3	36.4
Nitroethane	22.7	49.3	46.8	3.9
1-Butanol	23.1	47.6	6.1	46.3
Acetonitrile	24.0	39.4	54.3	6.3
1-Propanol	24.5	41.9	7.6	50.5
Methylglycol	24.7	42.6	13.8	43.6
N,N-Dimethylformamide	24.8	49.1	30.4	20.5
Nitromethane	25.2	39.5	56.3	4.2
Ethanol	26.4	35.5	11.0	53.5
Dimethylsulfoxide	26.4	47.6	37.6	14.8
1,3-Butanediol	28.9	32.8	12.0	55.2
Methanol	29.2	26.2	17.2	56.6
Diethylene glycol	29.9	28.9	24.3	46.8
Ethanolamine	31.7	29.6	24.5	45.9
Glycol	33.3	26.3	11.3	62.4
Formamide	36.4	22.0	51.0	27.0
Water	48.1	6.6	42.6	50.8

Adapted from Vial, J., *C.R. Acad. Sci. Ser. C*, 270, 683, 1970; and Thesis, Faculty of Science, Paris, 1970.

Section 5.8 by defining a parameter described here as δ_ξ to represent the cohesive pressure due to association or hydrogen bonding:

$$\delta_t^2 = \delta_\lambda^2 + \delta_\tau^2 + \delta_\xi^2 \tag{67}$$

Comparison of Equations 65 and 67 with the five-parameter Equation 24 suggests

TABLE 16
Fractional Hansen Parameters of Liquids, Plasticizers and Oils

Compound	δ_t/MPa$^{1/2}$	100f_d	100f_p	100f_h
Liquids				
Acetic acid, ethanoic acid	(21.2)	(40)	(19)	(41)
Acetic anhydride	(22.2)	(36)	(37)	(27)
Acetone, 2-propanone	20.1	47(50)	32(37)	21(13)
Acetonitrile	24.6	39(41)	45(43)	16(16)
Acetophenone, 1-phenylethanone	19.8	59(58)	29(25)	12(17)
Acetyl chloride	19.4	52	35	13
Acrylonitrile, 2-propenenitrile	21.7	41	43	16
Aniline	24.6(22.6)	50(55)	19(21)	31(24)
Benzaldehyde	21.3	61(57)	23(16)	16(27)
Benzene	18.8	78(76)	8(7)	14(17)
Benzonitrile	22.1	48	37	15
Benzyl alcohol, benzenemethanol	24.6	48	16	36
Benzyl chloride	20.3	66	25	9
Biphenyl, diphenyl	21.5	88	4	8
Bromobenzene	20.5	72	21	7
Bromoethane, ethyl bromide	18.2	56	27	17
1-Bromonaphthalene	21.1	74(68)	11(18)	15(14)
Butane	14.1	100	0	0
1,3-Butanediol	28.8	34(35)	21(17)	45(48)
1-Butanol	23.3	43(43)	15(14)	42(43)
Butyl acetate	17.4	60(60)	13(16)	17(24)
Butylamine, 1-butanamine	17.6	59	14	27
Butyl Carbitol®, diethylene glycol monobutyl ether	20.1	46	18	36
Butyl Cellosolve®, ethylene glycol monobutyl ether	20.5(21.0)	46(46)	18(20)	36(34)
Butyl lactate	19.2	48(52)	20(22)	32(26)
Butyraldehyde, butanal	18.4	50	33	17
Butyric acid, butanoic acid	(23.1)	(53)	(13)	(34)
γ-Butyrolactone	26.2	44(43)	39(36)	17(21)
Butyronitrile	20.5	44(48)	41(39)	15(13)
Carbitol®, diethylene glycol monoethyl ether	21.3	48	23	29
Carbitol® acetate, diethylene glycol monoethyl ether acetate	19.4	54	33	13
Carbon disulfide	20.3	88(86)	8(7)	4(7)
Cellosolve®, ethylene glycol monoethyl ether	22.1(24.3)	42(43)	20(20)	38(37)
Cellosolve® acetate, ethylene glycol monomethyl ether acetate	19.2	51(50)	15(34)	34(16)
Chlorobenzene	19.6	65(70)	17(15)	18(15)
1-Chlorobutane	17.2	68(66)	24(23)	8(11)
Chlorodifluoromethane, Freon 22	14.9	50	26	24
Chlorofluoroethane	14.1	55	26	19
Chloromethane, methyl chloride	17.0	61	24	15
1-Chloropropane, propyl chloride	17.4	62	30	8
3-Chloropropanol	23.7	46	15	39
m-Cresol, 3-methylphenol	22.7	50(49)	14(17)	36(34)
Cyclohexane	16.8	94	2	4
Cyclohexanol	22.5	50	12	38
Cyclohexanone	20.3	55(56)	28(22)	17(22)
Cyclohexylamine	(18.5)	(63)	(15)	(22)
Cyclohexyl chloride	19.4(18.4)	70(68)	21(24)	9(8)
Decane	15.8	100	0	0
1-Decanol	19.2	58	13	29

TABLE 16 (continued)
Fractional Hansen Parameters of Liquids, Plasticizers and Oils

Compound	δ_t/MPa$^{1/2}$	100f_d	100f_p	100f_h
Diacetone alcohol	20.3	45(37)	24(29)	31(34)
Dibenzyl ether	19.2	61	13	36
Diisobutyl ketone	16.6	67	16	17
o-Dichlorobenzene	20.5	67(69)	22(15)	11(16)
2,2-Dichlorodiethyl ether	21.1	61(54)	29(37)	10(9)
Dichlorodifluoromethane, Freon 12	12.5	86	14	0
1,1-Dichloroethane	18.4	66	33	2
1,2-Dichloroethane, ethylene dichloride	20.1	67(63)	19(23)	14(14)
Dichloromethane, methylene dichloride	19.8	59	21	20
1,2-Dichloroetetrafluoroethane, Freon 114	12.9	87	13	0
Diethylamine	16.4	64(62)	10(19)	26(19)
1,4-Diethylbenzene	18.0	97	0	3
Diethyl carbonate	18.0	64	12	24
Diethyl ether, 1,1′-oxybisethane	15.6	64(67)	13(23)	23(10)
Diethyl ketone, 3-pentanone	18.2	56	27	17
Diethyl sulfate	22.7	42	39	19
Diethyl sulfide	17.4	77(66)	14(26)	9(8)
Diethylene glycol	29.9	31(25)	29(30)	40(45)
Diethylenetriamine	25.8	38	30	32
Difluoroethane	14.9	37	40	23
Dimethoxymethane, formal, methylal	17.4	59(57)	7(32)	34(11)
N,N-Dimethylformamide	24.8	41	32	27
Dimethylsiloxane	12.1	100	0	0
Dimethyl sulfone	29.9	38	38	24
Dimethylsulfoxide	26.4	41(37)	36(33)	23(30)
1,4-Dioxane, diethylene oxide	20.7	67(58)	7(28)	26(14)
Dipentene	17.4	75	20	5
Dipropylamine	16.2	74(72)	7(12)	19(16)
Dipropylene glycol, oxybispropanol	31.7	35(30)	26(25)	39(45)
Epichlorohydrin, (chloromethyl)oxirane	21.9	58	31	11
Ethanol	26.2	36(36)	18(19)	46(45)
Ethanolamine	31.5	32(31)	29(32)	40(39)
Ethyl acetate	18.4	51	18(37)	31(17)
Ethylbenzene	18.0	87(80)	3(5)	10(15)
2-Ethylbutanol	21.3	48	10	42
Ethyl chloroformate	19.6	48	31	21
Ethyl formate	19.4	48	26	26
2-Ethylhexanol	20.3	50(53)	9(6)	41(41)
Ethyl lactate	21.1	44	21(23)	35(33)
Ethyl carbonate	29.7	42	47	11
Ethylene glycol, 1,2-ethanediol	33.3	30(32)	18(17)	52(51)
Formamide	36.4	28	42	30
Formic acid	(24.9)	(33)	(20)	(47)
Furfural, 2-furancarboxyaldehyde	24.3	46	41	13
Furan	18.6	71(65)	7(12)	22(23)
Furfuryl alcohol, 2-furanmethanol	24.3	43	19	38
Glycerol, 1,2,3-propanetriol	36.2(43.1)	25(26)	23(22)	52(52)
Heptane	15.3	100	0	0
1-Heptanol	21.5	47	14	39
Hexane	14.9	100(96)	0(2)	0(2)
1-Hexanol	21.9	47	13	40
Isoamyl acetate	17.2	60	12(27)	28(13)
Isobutyl acetate	16.6	60	15	25

TABLE 16 (continued)
Fractional Hansen Parameters of Liquids, Plasticizers and Oils

Compound	δ_t/MPa$^{1/2}$	$100f_d$	$100f_p$	$100f_h$
Isobutyl isobutyrate	16.6(16.0)	63(63)	12(22)	25(15)
Isononyl phenol	19.4	55	14	31
Isophorone	19.8	51(52)	25(25)	24(23)
Mesityl oxide	18.6	55	24	21
Methanol	29.3	30(31)	22(23)	48(46)
Methoxymethanol	24.8	39	22	39
Methyl acetate	19.4	45	36	19
Methyl Carbitol®, diethylene glycol monomethyl ether	22.7(22.1)	44(44)	21(22)	35(34)
Methyl Cellosolve®, ethylene glycol monomethyl ether	24.6	39	22	39
Methyl Cellosolve® acetate, ethylene glycol monomethyl ether acetate	20.3	46	17	37
Methyl ethyl ketone, 2-butanone	(19.0)	(53)	(30)	(17)
Methyl formate	20.7	46	22	32
Methyl isoamyl ketone	17.6	62(58)	20(22)	18(20)
Methyl isobutyl carbinol	19.8	50(51)	10(7)	40(42)
Methyl isobutyl ketone	17.6	58(56)	22(23)	20(21)
N-Methyl 2-pyrrolidone	22.9	48	32	20
Morpholine	21.7	57(53)	15(21)	28(26)
Naphthalene	20.3	70	8	22
Nitrobenzene	21.7	52(59)	36(29)	12(12)
Nitroethane	22.7	44(47)	43(42)	13(11)
Nitromethane	25.2	40(41)	47(46)	13(13)
2-Nitropropane	20.5	50(58)	37(33)	13(9)
Octane	15.1	100	0	0
1-Octanol	20.5	53	9	38
1-Pentanol, amyl alcohol	22.7(21.7)	46(47)	13(12)	41(41)
Pentane	14.3	100	0	0
Phenol	24.1	46	15	39
1-Propanol	24.6	40	16	44
Propionitrile	21.9	43	34	23
Propyl acetate	18.0	57	15	28
Propylene carbonate	27.1	48(48)	38(43)	14(9)
Propylene glycol, propanediol	30.3	34(39)	16(15)	50(46)
Pyridine	21.7	56(56)	26(22)	18(22)
2-Pyrrolidone	28.4	41	36	23
Resorcinol, 1,3-benzenediol	29.1	38	18	44
Styrene, ethenylbenzene	19.0	78(76)	4(9)	18(15)
Tetrachloroethane	21.7	56	15	19
Tetrachloroethylene	20.3	67	23	10
Tetrachloromethane, carbon tetrachloride	17.8	85	2	13
Tetrahydrofuran	19.4	55(55)	19(22)	26(23)
Tetrahydronaphthalene, tetralin	19.4	80(83)	8(4)	12(13)
Toluene, methylbenzene	18.2	80(78)	7(6)	13(16)
Tribromomethane, bromoform	21.5	54	13	33
Trichloroethane	17.6	70(68)	19(17)	11(15)
Trichloroethylene	19.0	68	12	20
Trichloromethane, chloroform	17.8	85	2	13
Triethylene glycol	26.6	36	23	41
Water	48.1	18(19)	28(22)	54(58)
Xylene, dimethylbenzene (mixed isomers)	18.0	83(82)	5(6)	12(12)

TABLE 16 (continued)
Fractional Hansen Parameters of Liquids, Plasticizers and Oils

Compound	δ_t/MPa$^{1/2}$	$100f_d$	$100f_p$	$100f_h$
Plasticizers				
Butyl stearate	15.3	67	17	16
Dibutyl maleate	17.2	60	12	28
Dibutyl phthalate	20.1	57	29	14
Dibutyl sebacate	15.1	62	20	18
Dimethyl phthalate	22.1	48	38	14
Dioctyl adipate	18.2	64	23	13
Dioctyl phthalate	18.2	62	26	12
Ethyl cinnamate	20.7	60	27	13
Methyl oleate	15.6	67	17	16
Tricresyl phosphate	24.8	51	30	29
Trimethyl phosphate	25.4	39	37	24
Tributyl phosphate	18.0	45	35	20
Trioctyl adipate	—	62	24	14
Oils and Commercial Solvents				
Castor oil	18.2	56	11	33
Linseed oil (white refined)	14.9	66	17	17
Neats foot oil	14.7	69	14	17
Sperm oil	14.5	75	11	14
Mineral oil (white refined)	14.5	100	0	0
Pine oil	16.6	70	14	16
Cottonseed oil	14.9	67	15	18
Mineral spirits	—	90	4	6
VM & P naphtha	—	94	3	3
Odorless mineral spirits	—	98	1	1
Turpentine	—	77	18	5

Adapted from Gardon, J. L. and Teas, J. P., *Treatise on Coatings,* Vol. 2, *Characterization of Coatings: Physical Techniques,* Part II, Myers, R. R. and Long, J. S., Eds., Marcel Dekker, New York, 1976, chap. 8. Values in parentheses are from the earlier publication of Teas, J. P., *J. Paint Technol.*, 40(516), 19, 1968, for those cases where differences occur.

$$\delta_\lambda^2 \approx \delta_d^2 \tag{68}$$

$$\delta_\tau^2 \approx \delta_p^2 \approx \delta_o^2 + 2\delta_i\delta_d \tag{69}$$

$$\delta_\xi^2 \approx \delta_h^2 \approx 2\delta_a\delta_b \tag{70}$$

The association parameter δ_ξ was obtained from the entropy and enthalpy of association as defined by Wiehe and Bagley,[65,136,137] who had investigated the activity coefficients in mixtures of alcohols with nonpolar liquids and developed multiparameter equations for their correlation (Section 7.5). Wiehe, in unpublished work cited in Reference 156, also developed an alternative two-component formalism:

$$\delta^2 = \delta_c^2 + \delta_f^2$$

where δ_c is a measure of ability to form "complexes" and δ_f is a measure of ability to interact by field forces which do not depend on orientation ($\delta_f \approx \delta_d$). These cohesion parameters, which were used for solvent selection by two-dimensional mapping, are collected in Table 16a.

Hoy[126] determined nonpolar, polar, and hydrogen-bonding parameters by semi-empirical methods which involved:

TABLE 16a
Two-Component Complexing — Field Force Cohesion Parameters for Liquids

Liquid	$\delta/\text{MPa}^{1/2}$		
	δ_t	δ_f	δ_c
Acetone	19.7	15.0	12.8
Acetonitrile	24.8	15.3	19.5
Acetophenone	21.6	17.5	12.7
Benzene	18.7	18.3	4.0
Butyl acetate	17.8	15.7	8.3
Carbon disulfide	20.3	20.3	0.0
Chlorobenzene	19.8	19.2	4.7
Cyclohexane	16.8	16.8	0.0
Cyclohexanone	21.3	17.4	12.3
Dibromomethane	21.3	20.2	6.7
o-Dichlorobenzene	20.5	19.8	5.6
1,2-Dichloroethane	20.2	19.2	6.1
Dichloromethane	20.2	18.2	8.7
Dimethylsulfoxide	26.4	18.4	19.0
N,N-Dimethylacetamide	22.1	17.4	13.6
N,N-Dimethylformamide	24.1	19.1	14.7
1,4-Dioxane	20.7	18.4	9.5
Ethyl acetate	18.2	15.2	10.0
Ethylbenzene	18.1	17.9	2.9
Heptane	15.3	15.3	0.0
Hexane	14.9	14.9	0.0
Iodomethane	20.9	20.1	5.7
Methylcyclohexane	16.0	16.0	0.0
Methyl ethyl ketone	19.3	15.9	11.0
Nitromethane	26.4	16.7	20.5
Styrene	19.1	18.7	4.1
Tetrachloromethane	17.5	17.5	0.0
Tetrahydrofuran	19.5	16.8	9.8
Toluene	18.3	18.1	2.7
Trichloroethylene	18.7	18.2	4.6
Trichloromethane	18.7	17.1	6.2
1,1,1-Trichloroethane	17.5	16.9	4.8
p-Xylene	18.1	18.0	0.8

Adapted from Dickerson, C. G. and Wiehe, I. A., *Pac. Chem. Eng. Congr. (Proc.),* 2(1), 243, 1977.

1. Evaluation of the total cohesion parameter or Hildebrand parameter δ_t as outlined in Section 7.2.
2. Separation of the total cohesion parameter by calculation of the aggregation number (α) from a regression analysis of molar volumes as a function of ratio of boiling point to critical temperature as well as molecular weight, and density

$$\log\alpha = 3.39066\, T_b/ T_c - 0.15848 - \log (M/\rho) \tag{71}$$

The ratio T_b/T_c may be estimated from Lyderson's equation,

$$(T_b/ T_c = 0.567 + \Sigma_z\Delta_T - (\Sigma_z\Delta_T)^2 \tag{72}$$

and the Δ_T values (critical temperature Lyderson group constants) were provided by Hoy,[125] also (Table 17), so it is possible to estimate the component cohesion parameters

TABLE 17
Lyderson Group Constants

Group, z	$10^2\ {}^z\Delta_T$		$10^2\ {}^z\Delta_T^P$
	Aliphatic	Cyclic	
$-CH_3$	2.0	—	2.26
$-CH_2-$	2.0	1.3	2.00
$>CH-$	1.2	1.2	1.31
$>C<$	0.0	−0.7	0.40
$=CH_2$	1.8	—	1.92
$=CH-$	1.8	1.1	1.84
$=C<$	0.0	1.1	1.29
$=CH-$ (aromatic)	—	1.1	1.78
$=C'$ (aromatic)	—	1.1	1.49
$-O-$	2.1	1.4	1.75
$>O$ (epoxide)	—	2.7	2.67
$-COO-$	4.7	—	4.97
$>C=O$	4.0	3.3	4.00
$-CHO$	4.8	—	4.45
$-(CO_2O)$	8.6	—	8.63
$-COOH$	3.9	—	3.90
$-OH\rightarrow$	8.2	—	3.43
$-OH$ (primary)	8.2	—	4.93
$-OH$ (secondary)	8.2	—	4.40
$-OH$ (tertiary)	8.2	—	5.93
$-OH$ (phenolic)	3.5	—	6.00
$-NH_2$	3.1	—	3.45
$-NH-$	3.1	2.4	2.74
$>N-$	1.4	0.7	0.93
$-C=N$	0.6	—	5.39
$-NCO$	5.4	—	5.39
$HCON<$	6.2	—	5.46
$-CONH-$	7.1	—	8.43
$-CON<$	5.4	—	7.29
$-CONH_2$	7.1	—	8.97
$-CONH-$	7.8	—	9.38
$-S-$	1.5	0.8	3.18
$-SH$	1.5	—	1.5
$-Cl$ (primary)	1.7	—	3.11
$-Cl$ (secondary)	1.7	—	3.17
$-Cl_2$ (twin)	3.4	—	5.21
$-Cl$ (aromatic)	1.7	—	2.45
$-Br$	1.0	—	3.92
$-Br$ (aromatic)	1.0	—	3.13
$-F$	1.8	—	0.6
$-I$	1.2	—	—
Conjugation	—	—	0.5
Cis double bond	—	—	−0.10
Trans double bond	—	—	−0.20
4-membered ring	—	—	1.18
5-membered ring	—	—	0.3
6-membered ring	—	—	−0.35
7-membered ring	—	—	0.69

TABLE 17 (continued)
Lydersen Group Constants

Group, z	$10^2\ {}^z\Delta_T$ Aliphatic	$10^2\ {}^z\Delta_T$ Cyclic	$10^2\ {}^z\Delta_T^P$
Ortho	—	—	0.15
Meta	—	—	0.10
Para	—	—	0.60
Bicycloheptyl	—	—	0.34
Tricyclodecane	—	—	0.95

Adapted from Hoy, K. L., *The Hoy Tables of Solubility Parameters,* Union Carbide Corporation, Solvents and Coatings Materials Division, South Charleston, WV, 1985.

knowing only the density and structure, although the accuracy is limited by the approximations made.

3. Calculation of the hydrogen-bonding parameter from

$$\delta_h = \delta_t\,[\,(\alpha - 1)/\alpha]^{1/2} \tag{73}$$

4. Evaluation of the polar parameter by a group molar attraction method based on Equation 34, Chapter 6, with the data of Table 7, Chapter 6:

$$\delta_p = \delta_t\,[\Sigma_z {}^zF_p/(\alpha\Sigma_z\,{}^zF)]^{1/2} \tag{74}$$

5. Calculation of the nonpolar parameter δ_d by difference (Equation 65).

Hoy's values of δ_t, δ_d, δ_p, and δ_h are included in Table 18. Estimates of three-component parameters for phenol, resorcinol, and about 50 alkyl derivatives have been published by Lille, Kundel, and Eisen[138] (Table 19). Stekol'shchikov, Krivtsova, and Ratner[139] have calculated values for hydrocarbons and a few other liquids (Table 20) by various semi-empirical methods based on correlations with physical properties (Section 8.2), and have compared them with literature data. Martin, Wu, and Beerbower[140,141] used three-component cohesion parameters in their study of solubilities of solids in polar and nonpolar liquids (Section 12.2).

Hoy's dispersion components, δ_d, being evaluated by difference, may be considered less reliable than those of Hansen, which were evaluated directly by homomorph methods. On the other hand, Hansen's method introduces small hydrogen-bonding components to the aromatic liquids. The positron method for evaluation of multicomponent cohesion parameters for liquids, mentioned in Section 3.6, may be able to provide more correct separations of the components. The positronium (Ps) state in pure liquids is considered a "bubble" state as a result of strong repulsive interactions with liquid molecules at short distances. It has been suggested by Mogensen[142] that hydrogen bonds in the liquid are not broken when the Ps bubble is formed, so most of the molecules in its vicinity continue to participate in hydrogen bonding to the same extent as in the bulk liquid. Consequently, the Ps "pick off" rates should be able to be used to determine either δ_d or $(\delta_d^2 + \delta_p^2)^{1/2}$, permitting evaluation of δ_h by optimization of solubility behavior. The positron method seems to show correct component values; correlation with the Hansen δ_d values is better than with the Hoy δ_d results, whereas the $(\delta_d^2 + \delta_p^2)^{1/2}$ correlation is better with the Hoy data than with the Hansen figures, and the hydrogen-bonding components in Hansen's values for aromatic liquids are not supported.

TABLE 18
Hoy's Cohesion Parameters for Liquids (and Solids as Subcooled Liquids) at 25°C

Liquid	M/g mol^{-1}	ρ/g cm^{-3}	δ/MPa$^{1/2}$ δ_d	δ_p	δ_h	δ_t
Acetaldehyde	44.1	0.771	11.5	10.6	12.7	20.2
Acetic acid, ethanoic acid	60.1	1.044	13.9	12.2	18.9	26.5
Acetic anhydride	102.1	1.075	10.0	11.3	15.7	21.8
Acetone, 2-propanone	58.1	0.785	13.0	9.8	11.0	19.7
Acetonitrile	41.1	0.776	10.3	11.1	19.6	24.8
Acetophenone, 1-phenylethanone	120.2	1.024	16.1	11.9	8.2	21.6
1-Acetoxy-1,3-butadiene	112.1	0.947	14.2	9.9	8.7	19.4
Acetylacetone, 2,4-pentanedione	100.1	0.968	11.3	11.8	10.7	19.5
Acrolein, 2-propenal	56.1	0.835	11.4	11.1	12.2	20.1
Acrylic acid, 2-propenoic acid	72.1	1.040	13.5	12.8	17.9	25.9
Acrylonitrile, 2-propenenitrile	53.1	0.801	10.6	12.5	14.0	21.6
Allyl acetate	100.1	0.922	14.0	9.5	8.3	18.8
Allyl acetoacetate	142.2	1.032	13.4	11.3	10.5	20.5
Allyl alcohol, 2-propen-1-ol	58.1	0.848	13.0	11.8	18.7	25.7
Allyl chloride, 3-chloropropene	76.5	0.931	13.8	8.9	7.3	18.0
Allyl cyanide, 3-butenenitrile	67.1	0.830	12.3	11.8	13.2	21.6
N-(2-Aminoethyl)ethanolamine	104.2	1.025	13.9	12.8	20.8	28.1
N-(2-Aminoethyl)piperazine	129.2	0.978	15.6	11.7	9.0	21.4
N-(3-Aminopropyl)morpholine	144.2	0.981	16.1	10.2	7.3	20.4
Amyl alcohol, see 1-Pentanol						
sec-Amyl alcohol, see 2-Pentanol	88.2	0.815	14.5	9.1	13.9	22.0
prim. active Amyl alcohol, 2-methyl-1-butanol						
tert-Amyl alcohol, 2-methyl-2-butanol	88.2	0.805	13.8	10.0	12.4	21.1
Benzene	78.1	0.874	16.1	8.6	4.1	18.7
Benzyl alcohol, benzenemethanol	108.1	1.042	14.7	12.2	15.6	24.6
Benzyl Cellosolve®	152.2	1.064	14.2	10.2	13.8	22.3
N,*N*-(Bis(3-aminopropyl))methylamine	145.3	0.897	14.3	9.9	10.1	20.1
Bromobenzene	157.0	1.486	18.4	8.2	0.0	20.1
2-Bromobutane, *sec*-butyl bromide	137.0	1.251	16.9	4.4	3.0	17.7
Bromochloromethane	129.4	1.919	16.9	11.1	6.5	21.2
Bromoethane, ethyl bromide	109.0	1.447	16.2	5.1	6.6	18.2
o-Bromostyrene	183.1	1.408	17.7	9.6	0.0	20.1
o-Bromotoluene	171.0	1.437	17.9	8.8	0.0	20.0
p-Bromotoluene	171.0	1.391	17.7	8.4	0.0	19.6
Bromotrichloromethane	198.3	1.998	12.0	13.8	3.1	18.6
1,3-Butadiene	54.1	0.614	13.0	6.3	4.5	15.1
Butadiene dioxide	86.1	1.106	16.5	13.1	11.6	24.1
Butane	58.1	0.572	13.5	0.0	0.0	13.5
1,4-Butanediol	90.1	1.013	15.0	13.6	27.0	33.7
1-Butanol, butyl alcohol	74.1	0.806	15.0	10.0	15.4	23.7
2-Butanol, *sec*-butyl alcohol	74.1	0.802	14.5	9.1	14.8	22.7
1-Butene	56.1	0.588	12.7	4.2	2.9	13.6
cis-2-Butene	56.1	0.614	12.9	3.8	5.8	14.6
trans-2-Butene	56.1	0.597	13.1	3.6	4.2	14.2
cis-1-Butenyl methyl ether	86.1	0.770	14.2	6.4	6.2	16.8
trans-2-Butenyl methyl ether	86.1	0.780	13.9	6.2	6.4	16.5
3-Butoxybutanol	146.2	0.879	14.2	7.4	11.7	19.9
Butoxydipropylene glycol	190.3	0.911	14.5	8.2	7.0	18.1
Butoxyethoxypropanol	176.3	0.925	13.6	7.8	11.6	19.6
Butyl acetate	116.2	0.876	14.5	7.8	6.8	17.8
sec-Butyl acetate	116.2	0.867	13.7	7.7	5.0	16.5
Butyl acetoacetate	158.2	0.963	13.6	9.5	10.3	19.5
Butyl acrylate	128.2	0.895	14.0	8.3	6.8	17.7
Butylamine	73.1	0.734	13.6	8.1	8.0	17.7

TABLE 18 (continued)
Hoy's Cohesion Parameters for Liquids (and Solids as Subcooled Liquids) at 25°C

Liquid	M/g mol^{-1}	ρ/g cm^{-3}	δ/MPa$^{1/2}$			
			δ_d	δ_p	δ_h	δ_t
Butylaniline	149.2	0.923	16.9	9.1	6.6	20.4
Butylbenzene	134.2	0.855	16.3	6.5	0.0	17.5
Butyl benzoate	178.2	1.001	15.9	9.4	5.9	19.4
Butyl butyrate	144.2	0.865	14.6	6.9	6.2	17.3
Butyl Carbitol®	162.2	0.948	13.3	8.1	12.6	20.0
Butyl Carbitol acetate	204.3	0.974	14.2	8.1	8.7	18.5
Butyl Cellosolve®	118.2	0.896	13.3	7.9	13.0	20.2
Butyl Cellosolve® acetate	160.2	0.936	14.4	8.0	7.8	18.2
Butyl chloride, see chlorobutane						
Butyl cyclohexane	140.3	0.794	16.1	2.4	0.0	16.2
Butylcyclohexylamine	155.3	0.839	16.0	6.3	3.0	17.4
Butylcyclopentane	126.2	0.779	16.1	2.9	0.0	16.4
Butylene glycol, butanediol	90.1	1.001	12.3	12.2	22.2	28.1
Butylene oxide	72.1	0.824	15.0	8.0	7.0	18.4
Butylethanolamine	117.2	0.888	14.4	9.1	15.0	22.7
Butyl ether, 1,1′-oxybisbutane	130.2	0.764	14.6	4.3	4.5	15.9
Butyl ethyl Cellosolve®	146.2	0.833	14.4	6.0	6.2	16.8
Butyl isopropenyl ether	114.2	0.784	15.2	6.2	1.8	16.5
Butyl lactate	146.2	0.972	12.5	9.6	12.3	20.0
Butyl methyl Cellosolve®	132.2	0.841	14.6	6.3	6.7	17.2
Butyl-α-methylbenzylamine	177.3	0.890	15.3	7.9	4.1	17.7
Butyl 6-methyl-3-cyclohexane carboxylate	196.3	0.939	15.4	7.5	2.2	17.3
2-Butyloctanol	186.3	0.831	14.5	6.3	10.0	18.8
Butyl salicylate	194.2	1.069	15.6	11.2	5.4	19.9
o-Butyltoluene	148.2	0.865	16.6	6.1	0.0	17.7
m-Butyltoluene	148.2	0.853	16.5	6.0	0.0	17.5
p-Butyltoluene	148.2	0.851	16.5	5.9	0.0	17.5
Butyraldehyde, butanal	72.1	0.796	13.1	8.9	9.6	18.6
Butyric acid, butanoic acid	88.1	0.953	15.8	10.2	15.8	24.5
Butyric anhydride	158.2	0.962	13.1	10.2	8.9	18.8
γ-Butyrolactone	86.1	1.122	18.6	12.2	14.0	26.3
Butyronitrile	69.1	0.786	13.3	10.6	12.0	20.8
ε-Caprolactone	114.1	1.071	19.1	9.9	14.4	25.9
Carbitol® acetate	176.2	1.004	14.4	9.0	9.4	19.4
Carbitol	134.2	0.983	13.0	8.9	14.1	21.2
Carbon disulfide	76.1	1.256	10.9	16.6	4.3	20.3
Cellosolve® acetate	132.2	0.968	14.4	9.0	8.9	19.1
Cellosolve acrylate	144.2	0.976	14.2	9.4	8.7	19.1
Cellosolve	90.1	0.925	13.0	9.1	15.2	21.9
2-Chloroallylidene diacetate	192.6	1.202	14.9	11.8	7.5	20.4
Chlorobenzene	112.6	1.098	17.4	9.4	0.0	19.7
2-Chloro-1,3-butadiene	88.5	0.949	15.0	7.5	3.5	17.2
1-Chlorobutane, butyl chloride	92.6	0.880	15.3	6.9	3.5	17.1
2-Chlorobutane, *sec*-butyl chloride	92.6	0.867	14.9	6.9	2.4	16.6
2-Chloroethyl acetate	122.6	1.140	15.1	11.4	9.6	21.2
2-Chloroethyl ethyl ether	108.6	0.992	14.4	8.5	7.5	18.4
Chloroform, see trichloromethane						
1-Chloropropane, propyl chloride	78.5	0.885	14.4	7.2	6.0	17.2
2-Chloropropane, isopropyl chloride	78.5	0.856	14.2	7.3	4.3	16.5
o-Chlorostyrene	138.6	1.093	17.1	9.4	0.0	19.6
p-Chlorostyrene	138.6	1.080	17.2	9.2	0.0	19.5
Crotonaldehyde, 2-butenal	70.1	0.847	13.4	10.5	11.5	20.5
Cyclohexane	84.2	0.774	16.5	3.1	0.0	16.8
Cyclohexanol, cyclohexyl alcohol	100.1	0.956	13.8	8.6	15.3	22.3
Cyclohexanone	98.2	0.942	15.6	9.4	11.0	21.3

TABLE 18 (continued)
Hoy's Cohesion Parameters for Liquids (and Solids as Subcooled Liquids) at 25°C

Liquid	$M/g\ mol^{-1}$	$\rho/g\ cm^{-3}$	$\delta/MPa^{1/2}$ δ_d	δ_p	δ_h	δ_t
1-Cyclohexyldecane	224.4	0.814	15.9	1.9	0.0	16.1
1-Cyclohexyldodecane	252.5	0.818	15.9	1.8	0.0	16.0
1-Cyclohexylheptane	182.3	0.806	16.0	2.1	0.0	16.1
1-Cyclohexylnonane	210.4	0.811	16.0	1.9	0.0	16.1
1-Cyclohexyloctane	196.4	0.809	16.0	2.0	0.0	16.1
1-Cyclohexylundecane	238.4	0.815	15.9	1.8	0.0	16.0
Cyclopentane	70.1	0.739	16.1	3.9	0.6	16.5
Cyclopentanone	84.1	0.944	16.2	11.1	8.8	21.5
Cyclopentene	68.1	0.765	15.3	5.8	4.1	16.9
2-Cyclopentenyl alcohol	84.1	0.976	15.4	11.5	16.5	25.3
1-Cyclopentyldecane	210.4	0.806	17.2	2.4	1.3	17.4
1-Cyclopentylheptane	168.3	0.796	16.6	2.6	0.0	16.8
1-Cyclopentylhexane	154.3	0.791	16.5	2.7	0.0	16.7
1-Cyclopentylnonane	196.4	0.803	17.0	2.4	0.0	17.1
1-Cyclopentyloctane	182.3	0.800	16.7	2.5	0.0	16.9
1-Cyclopentylpentane	140.3	0.786	16.2	2.8	0.0	16.5
1-Cyclopentylundecane	224.4	0.808	17.2	2.3	1.2	17.4
Decane	142.3	0.725	15.8	0.0	0.0	15.8
1-Decanol, decyl alcohol	158.3	0.826	15.4	7.0	11.6	20.5
2-Decanol, *sec*-decyl alcohol	158.3	0.821	13.8	6.2	9.8	18.0
1-Decene	140.3	0.736	15.7	3.3	1.3	16.0
Decylbenzene	218.4	0.850	16.5	5.0	0.0	17.3
Diacetone alcohol	116.2	0.934	10.7	11.4	12.6	20.0
Diallylamine	97.2	0.783	13.5	8.5	7.5	17.6
1,1-Diallyloxyethane	142.2	0.871	13.3	8.0	6.2	16.7
3,3′-Diaminodipropylamine	131.2	0.925	13.7	11.5	12.9	22.1
1,3-Diaminopropane	74.1	0.882	13.9	12.9	14.7	24.0
o-Dibromobenzene	235.9	1.973	18.5	10.0	0.0	21.0
Dibutylamine	129.3	0.757	14.6	5.6	5.7	16.7
Dibutyl Carbitol®	218.3	0.880	14.3	6.1	7.1	17.0
Dibutyl Cellosolve®	174.3	0.832	14.6	5.5	6.1	16.8
N,N-Dibutylethanolamine	173.3	0.856	13.2	6.0	9.9	17.6
Dibutyl fumarate	228.3	0.981	14.2	8.6	8.0	18.4
Dibutylisopropanolamine	187.3	0.837	13.2	5.8	8.8	16.9
Dibutyl maleate	228.3	0.990	14.1	8.6	8.4	18.5
Dibutyl phthalate	278.4	1.042	15.9	9.5	8.1	20.2
o-Dichlorobenzene	147.0	0.298	18.0	9.8	0.0	20.5
m-Dichlorobenzene	147.0	0.281	17.6	9.5	0.0	20.0
p-Dichlorobenzene	147.0	0.239	17.5	9.2	0.0	19.8
1,1-Dichloroethane, ethylidene chloride	99.0	1.168	13.8	10.5	5.6	18.3
1,2-Dichloroethane, ethylene dichloride	99.0	1.246	14.2	11.2	9.1	20.2
Di(2-Chloroethoxy)methane	173.0	1.226	13.4	10.2	12.6	21.1
Di(2-Chloroethyl) ether	143.0	1.214	15.4	11.0	9.1	21.0
Dichloroisopropyl ether	171.1	1.106	15.2	9.7	4.6	18.6
Dichloromethane, methylene dichloride	84.9	1.316	13.4	11.7	9.6	20.2
1,3-Dichloropropane, propylene dichloride	113.0	1.150	14.4	10.4	5.4	18.5
2,3-Dichloropropanol	129.0	1.355	12.2	13.6	16.4	24.6
Dicrotylpropional	184.3	0.870	14.6	7.1	6.6	17.5
Di(1,3-dimethylbutyl)amine	185.4	0.781	14.4	4.9	1.5	15.3
1,1-Diethoxybutane	146.2	0.823	13.8	6.2	4.9	15.9
1,1-Diethoxyethane	118.2	0.821	13.5	6.8	5.1	15.9
2,5-Diethoxytetrahydrofuran	161.2	0.962	14.4	7.7	5.9	17.3
Diethoxytriglycol	206.3	0.950	14.2	7.5	8.8	18.3
Diethylamine	73.1	0.701	13.4	7.0	6.3	16.5

TABLE 18 (continued)
Hoy's Cohesion Parameters for Liquids (and Solids as Subcooled Liquids) at 25°C

Liquid	M/g mol^{-1}	ρ/g cm^{-3}	δ/MPa$^{1/2}$			
			δ_d	δ_p	δ_h	δ_t
Diethylaminoethylamine	116.2	0.815	14.0	7.8	7.1	17.5
3-(Diethylamino)propylamine	130.2	0.823	14.4	7.4	7.1	17.7
1,2-Diethylbenzene	134.2	0.874	16.6	6.5	0.0	17.8
1,3-Diethylbenzene	134.2	0.858	16.6	6.3	0.0	17.7
1,4-Diethylbenzene	134.2	0.856	16.6	6.2	0.0	17.7
Diethyl Carbitol®	162.2	0.902	14.1	7.1	8.0	17.7
Diethyl Cellosolve®	118.2	0.835	14.3	6.7	6.7	17.1
Diethylene glycol, 2,2′-oxybisethanol	106.1	1.113	12.4	12.3	23.3	29.1
Diethylenetriamine	103.2	0.948	13.0	13.1	14.7	23.6
N,N-Diethylethanolamine	117.2	0.879	13.1	7.3	12.0	19.2
Diethyl 2-ethylhexanal	202.3	0.835	14.2	5.3	4.9	16.0
Diethyl 2-ethyl-3-methylglutarate	230.2	0.976	13.6	8.0	6.7	17.1
Diethyl fumarate	172.2	1.046	13.8	10.3	9.3	19.5
3,3-Diethylhexane	142.3	0.762	15.7	0.0	0.0	15.7
3,4-Diethylhexane	142.3	0.749	15.4	0.0	0.0	15.4
Di(3-ethylhexyl)amine	241.5	0.801	14.6	4.2	4.8	16.0
Di(2-ethylhexyl) ether	242.5	0.807	14.9	3.3	4.2	15.9
Diethylisopropanolamine	131.2	0.841	12.4	6.7	9.7	17.1
Diethyl ketone, 3-pentanone	86.1	0.809	14.5	8.7	7.6	18.5
Diethyl maleate	172.2	1.063	14.3	10.5	9.9	20.3
3,3-Diethylpentane	128.3	0.749	15.4	0.0	0.0	15.4
2,2-Diethylpentanol	144.3	0.853	14.2	7.3	10.6	19.2
Diethyl phthalate	222.3	1.114	15.0	10.9	8.5	20.4
Diethyl pimelate	216.3	0.988	14.4	8.5	8.3	18.6
Diethyl succinate	172.2	1.035	14.3	9.8	9.4	19.7
3,9-Diethyl-6-tridecanol	256.5	0.843	14.4	5.2	7.5	17.0
Diglycolamine	105.1	1.051	11.9	11.1	19.2	25.1
Diglycol chlorohydrin	124.6	1.168	13.3	12.1	17.1	24.8
Diglycol diacetate	190.2	1.108	14.7	10.7	11.2	21.3
Dihexylamine	185.4	0.785	15.1	4.8	6.1	17.0
Dihexyl ether	186.4	0.789	15.1	3.7	5.2	16.4
1,1-Diisobutoxyethane	174.3	0.815	14.2	5.8	2.6	15.6
Diisobutylene	112.2	0.711	13.9	3.4	0.0	14.3
Diisobutyl ketone	142.2	0.802	14.5	6.8	3.9	16.5
1,1-Diisopropoxyethane	146.2	0.810	13.6	6.2	1.7	15.1
Diisopropylamine	101.2	0.712	13.8	6.2	2.0	15.2
Diisopropylethanolamine	145.3	0.870	14.1	6.9	10.5	18.9
Diisopropyl maleate	200.2	1.005	14.4	9.6	7.2	18.7
Diketene	84.1	1.108	14.5	13.5	12.7	23.6
1,3-Dimethoxybutane	118.2	0.844	14.2	6.7	6.0	16.9
1,1-Dimethoxyethane	90.1	0.845	13.0	7.8	6.8	16.6
1,1-Di(methoxyethoxy)ethane	178.2	0.971	13.8	8.4	8.8	18.4
1,1-Di(methoxyethoxy)methane	164.2	0.991	14.0	8.8	10.0	19.3
1,1-Dimethoxy-2-methylpropane	118.2	0.839	13.5	7.0	4.7	15.9
Dimethoxytetraglycol	222.3	1.007	14.5	8.4	10.3	19.7
3-(Dimethylamino)propionitrile	98.2	0.866	14.3	10.5	11.4	21.1
3-(Dimethylamino)propylamine	102.2	0.812	14.6	8.5	7.9	18.6
Di(α-methylbenzyl) ether	226.3	0.997	16.3	8.9	0.0	18.6
2,2-Dimethylbutane	86.2	0.643	13.7	0.0	0.0	13.7
2,3-Dimethylbutane	86.2	0.656	14.2	0.0	0.0	14.2
2,2-Dimethylbutanol	102.2	0.824	14.6	8.7	12.1	20.8
2,3-Dimethylbutanol	102.2	0.826	14.1	8.5	12.2	20.5
2,4-Dimethylbutanol	102.2	0.809	14.2	8.5	11.4	20.1
2,3-Dimethyl-2-butanol	102.2	0.819	13.4	9.4	10.7	19.6

TABLE 18 (continued)
Hoy's Cohesion Parameters for Liquids (and Solids as Subcooled Liquids) at 25°C

Liquid	$M/g\ mol^{-1}$	$\rho/g\ cm^{-3}$	$\delta/MPa^{1/2}$ δ_d	δ_p	δ_h	δ_t
2,3-Dimethyl-1-butene	84.2	0.672	13.9	3.8	3.0	14.7
3,3-Dimethylbutyl acetate	144.2	0.863	14.9	7.1	3.0	16.8
1,3-Dimethylbutylamine	101.2	0.743	14.2	7.3	3.8	16.3
Dimethyl Cellosolve®	90.1	0.862	13.9	7.7	7.7	17.7
1,1-Dimethylcyclohexane	112.2	0.775	15.4	2.6	0.0	15.7
1,1-Dimethylcyclopentane	98.2	0.749	15.1	3.2	0.0	15.4
2,5-Dimethyl-3,4-dihydro-2*H*-pyran-2-carboxyaldehyde	140.2	0.991	13.9	10.5	4.7	18.1
Dimethyl diketone	86.1	0.976	10.6	12.7	11.7	20.3
2,6-Dimethyl-1,4-dioxane	116.2	0.939	15.7	8.3	4.3	18.2
N,*N*-Dimethylethanolamine	89.1	0.882	13.5	8.5	14.0	21.2
N,*N*-Dimethyl-2-ethylhexylamine	157.3	0.764	15.3	3.1	1.6	15.7
2,5-Dimethylfuran	96.1	0.894	15.8	7.0	4.8	17.9
2,2-Dimethylheptane	128.3	0.705	14.9	0.0	0.0	14.9
2,3-Dimethylheptane	128.3	0.721	15.3	0.0	0.0	15.3
2,4-Dimethylheptane	128.3	0.711	15.0	0.0	0.0	15.0
2,5-Dimethylheptane	128.3	0.710	15.1	0.0	0.0	15.1
2,6-Dimethylheptane	128.3	0.704	15.0	0.0	0.0	15.0
3,3-Dimethylheptane	128.3	0.720	15.1	0.0	0.0	15.1
3,4-Dimethylheptane	128.3	0.726	15.4	0.0	0.0	15.4
3,5-Dimethylheptane	128.3	0.718	15.1	0.0	0.0	15.1
4,4-Dimethylheptane	128.3	0.720	15.0	0.0	0.0	15.0
2,6-Dimethyl-4-heptanol	144.3	0.807	14.3	6.7	9.2	18.3
2,6-Dimethyl-4-heptyl acetate	186.3	0.847	14.4	6.1	2.2	15.8
2,2-Dimethylhexane	114.2	0.690	14.5	0.0	0.0	14.5
2,3-Dimethylhexane	114.2	0.707	15.0	0.0	0.0	15.0
2,4-Dimethylhexane	114.2	0.695	14.7	0.0	0.0	14.7
2,5-Dimethylhexane	114.2	0.688	14.6	0.0	0.0	14.6
3,3-Dimethylhexane	114.2	0.705	14.7	0.0	0.0	14.7
3,4-Dimethylhexane	114.2	0.704	14.9	0.0	0.0	14.9
2,8-Dimethyl-6-isobutyl-4-nonanol	228.4	0.822	13.8	5.3	6.9	16.3
Dimethylisopropanolamine	103.2	0.845	12.6	7.7	11.2	18.5
Dimethyl maleate	144.1	1.146	14.5	12.1	11.8	22.3
2,6-Dimethylmorpholine	115.2	0.928	16.4	10.0	2.3	19.3
2,2-Dimethyloctane	142.3	0.719	14.9	0.0	0.0	14.9
2,3-Dimethyloctane	142.3	0.732	15.3	0.0	0.0	15.3
2,4-Dimethyloctane	142.3	0.721	14.9	0.0	0.0	14.9
2,5-Dimethyloctane	142.3	0.731	15.1	0.0	0.0	15.1
2,6-Dimethyloctane	142.3	0.723	15.1	0.0	0.0	15.1
2,7-Dimethyloctane	142.3	0.719	15.0	0.0	0.0	15.0
3,3-Dimethyloctane	142.3	0.734	15.3	0.0	0.0	15.3
3,4-Dimethyloctane	142.3	0.741	15.4	0.0	0.0	15.4
3,5-Dimethyloctane	142.3	0.731	15.2	0.0	0.0	15.2
3,6-Dimethyloctane	142.3	0.731	15.2	0.0	0.0	15.2
4,4-Dimethyloctane	142.3	0.732	15.2	0.0	0.0	15.2
4,5-Dimethyloctane	142.3	0.742	15.4	0.0	0.0	15.4
2,3-Dimethylpentaldehyde	114.2	0.823	14.5	7.7	6.3	17.6
2,2-Dimethylpentane	100.2	0.668	14.1	0.0	0.0	14.1
2,3-Dimethylpentane	100.2	0.690	14.8	0.0	0.0	14.8
2,4-Dimethylpentane	100.2	0.667	14.2	0.0	0.0	14.2
3,3-Dimethylpentane	100.2	0.688	14.5	0.0	0.0	14.5
2,3-Dimethylpentanol	116.2	0.837	14.8	8.2	12.0	20.8
2,3-Dimethylpentenal	112.2	0.837	14.1	8.7	6.7	17.8
Dimethyl phthalate	194.2	0.119	15.9	12.6	9.7	22.5

TABLE 18 (continued)
Hoy's Cohesion Parameters for Liquids (and Solids as Subcooled Liquids) at 25°C

Liquid	$M/g\ mol^{-1}$	$\rho/g\ cm^{-3}$	$\delta/MPa^{1/2}$			
			δ_d	δ_p	δ_h	δ_t
Dimethyl pimelate	188.2	1.034	14.8	9.4	9.6	20.0
2,2-Dimethylpropane	72.2	0.584	12.5	0.0	0.0	12.5
2,5-Dimethyltetrahydrofuran	100.2	0.825	15.6	6.5	0.0	16.9
1,4-Dioxane, diethylene oxide	88.1	1.028	16.3	10.1	7.9	20.7
1,3-Dioxolane	64.1	1.058	14.8	11.3	13.9	23.2
Dipentene	136.2	0.836	16.3	5.8	0.0	17.3
Diphenyl ether	170.2	1.061	17.4	11.2	0.0	20.7
Diphenylmethane	168.2	1.000	17.2	9.4	0.0	19.5
1,1-Dipropoxyethane	146.2	0.825	13.9	6.2	5.1	16.0
Dipropylamine	101.2	0.732	14.0	6.2	5.8	16.3
Dipropylene glycol, Oxybispropanol	134.2	1.016	12.2	10.3	17.4	23.6
Dipropylenetriamine	131.2	0.898	13.2	11.5	10.6	20.5
Dipropyl ketone	114.2	0.811	14.6	7.5	6.9	17.8
2,3-Dithiabutane	94.2	1.055	13.9	13.4	5.4	20.1
Divinyl Carbitol®	158.2	0.964	14.3	9.0	9.4	19.3
Dodecane	170.3	0.744	16.2	0.0	0.0	16.2
1-Dodecanol, lauryl alcohol	186.3	0.829	15.5	6.5	10.8	20.0
2-Dodecanol, 2-dodecyl alcohol	186.3	0.825	13.7	5.8	9.2	17.4
Epichlorohydrin, (chloromethyl)oxirane	92.5	1.168	15.9	12.3	10.5	22.7
3,4-Epoxy-1-butene	70.1	0.869	14.8	9.8	8.6	19.7
Ethanol	46.1	0.785	12.6	11.2	20.0	26.1
2-Ethoxy-3,4-dihydro-1,2-pyran	128.2	0.963	14.8	8.8	4.9	17.9
1-Ethoxyethoxy-2-propanol	148.2	0.950	13.4	8.5	12.8	20.4
3-Ethoxy-4-ethyloctanol	202.3	0.879	14.4	6.4	9.9	18.5
2-Ethoxy-4-methyl-3,4-dihydropyran	142.2	0.927	14.9	8.3	3.7	17.4
5-Ethoxy-3-methylpentanol	146.2	0.894	14.4	8.6	12.5	20.9
2-Ethoxy-4-methyltetrahydropyran	144.2	0.908	15.4	7.4	2.9	17.3
1-Ethoxy-2-propanol	104.2	0.892	13.0	8.5	12.8	20.1
3-Ethoxypropanol	104.2	0.912	14.0	8.7	14.8	22.1
3-Ethoxypropionaldehyde	102.1	0.911	13.5	9.7	10.5	19.7
3-Ethoxypropionic acid	118.1	1.041	16.5	11.5	16.5	26.0
3-Ethoxypropyl acetate	158.2	0.957	14.5	8.2	8.2	18.6
2-Ethoxy-4-propyl-5-ethyl-3,4-dihydropyran	198.3	0.914	14.9	6.8	4.8	17.0
2-Ethoxytetrahydropyran	130.2	0.933	14.9	7.7	4.2	17.3
Ethoxytriglycol	178.2	1.015	13.1	8.8	13.5	20.8
Ethyl acetate	88.1	0.894	13.4	8.6	8.9	18.2
Ethyl acetoacetate	130.1	1.019	13.3	11.1	10.2	20.1
Ethyl acrylate	100.1	0.915	13.3	9.3	7.9	18.0
Ethyl amyl ketone	128.2	0.816	14.5	7.2	5.7	17.2
N-Ethylaniline	121.2	0.957	17.1	10.5	7.7	21.5
Ethylbenzene	106.2	0.862	16.5	7.4	0.0	18.1
Ethyl benzoate	150.2	1.041	15.7	10.6	6.3	19.9
Ethyl bromide, see Bromoethane						
2-Ethyl-1,3-butadiene	82.2	0.713	14.3	5.6	0.0	15.4
2-Ethylbutanol	102.2	0.829	14.4	8.5	13.0	21.2
2-Ethyl-1-butene	84.2	0.684	14.3	3.9	3.1	15.2
2-Ethylbutyl acetate	144.2	0.874	14.6	7.0	5.3	17.0
N-Ethylbutylamine	101.2	0.735	13.9	6.2	5.9	16.3
2-Ethylbutyl Carbitol®	190.3	0.933	13.3	7.4	11.2	18.9
2-Ethylbutyl Cellosolve®	146.2	0.890	13.7	7.3	11.5	19.4
Ethyl butyl ether	102.2	0.745	14.3	4.8	4.5	15.7
2-Ethylbutyl 2-ethylbutyrate	200.3	0.861	14.3	5.8	4.7	16.1
2-Ethylbutyl 2-ethylhexanoate	228.4	0.853	14.7	5.4	4.9	16.5
2-Ethylbutyl hexanoate	200.3	0.863	14.4	5.8	5.5	16.4

TABLE 18 (continued)
Hoy's Cohesion Parameters for Liquids (and Solids as Subcooled Liquids) at 25°C

Liquid	M/g mol^{-1}	ρ/g cm^{-3}	δ/MPa$^{1/2}$ δ_d	δ_p	δ_h	δ_t
Ethyl butyl ketone	114.2	0.813	14.2	7.4	6.9	17.4
2-Ethylbutyraldehyde	100.2	0.810	14.0	7.9	7.2	17.6
Ethyl butyrate	116.2	0.874	13.8	7.6	6.4	17.0
2-Ethylbutyric acid	116.2	0.919	17.4	9.3	12.9	23.5
2-Ethylcrotonaldehyde	98.2	0.852	14.4	9.5	7.6	18.9
Ethyl crotonate	114.1	0.913	14.0	8.8	7.6	18.2
Ethylcyclohexane	112.1	0.783	16.1	2.6	0.0	16.3
N-Ethylcyclohexylamine	127.2	0.842	15.8	7.0	3.0	17.6
Ethylcyclopentane	98.2	0.761	15.9	3.3	0.0	16.2
N-Ethyldiethanolamine	133.2	1.010	13.2	9.5	18.2	24.4
N-Ethyldiisopropanolamine	161.3	0.940	12.5	8.4	14.1	20.6
4-Ethyl-2,6-dimethylmorpholine	136.2	0.876	15.8	6.6	2.1	17.2
3-Ethyl-2,2-dimethylpentane	128.3	0.730	14.9	0.0	0.0	14.9
3-Ethyl-2,3-dimethylpentane	128.3	0.749	15.3	0.0	0.0	15.3
Ethyl 2,5-endomethylene-3-cyclohexanecarboxylate	166.2	1.021	14.7	7.7	10.9	19.8
Ethylene chlorohydrin, 2-chloroethanol	80.5	1.197	12.0	13.3	19.6	26.5
Ethylene cyanohydrin	71.1	1.041	11.1	15.2	24.7	31.0
Ethylenediamine, 1,2-ethanediamine	60.1	0.903	12.6	14.0	16.9	25.3
Ethylene glycol, 1,2-ethanediol	62.1	1.110	10.1	15.1	29.8	34.9
N-Ethylethanolamine	89.1	0.913	14.3	10.5	17.8	25.1
Ethyl ether, diethyl ether	74.1	0.703	13.4	5.3	5.6	15.4
Ethyl 3-ethoxybutyrate	160.2	0.924	14.1	8.0	6.7	17.5
Ethyl 3-ethoxypropionate	146.2	0.943	14.3	8.5	8.0	18.4
3-Ethylheptane	128.3	0.722	15.4	0.0	0.0	15.4
4-Ethylheptane	128.3	0.725	15.5	0.0	0.0	15.5
5-Ethyl-3-hepten-2-one	140.2	0.841	15.0	7.8	6.6	18.1
2-Ethylhexaldehyde	128.2	0.815	14.3	7.1	6.5	17.2
3-Ethylhexane	114.2	0.708	15.2	0.0	0.0	15.2
2-Ethylhexanoic acid	144.2	0.902	17.4	8.3	11.6	22.5
2-Ethylhexanol	130.2	0.829	15.1	7.7	12.0	20.8
2-Ethyl-1-hexene	112.2	0.722	15.0	3.6	0.0	15.5
2-Ethylhexyl acetate	172.3	0.867	14.8	6.3	5.4	17.0
2-Ethylhexyl acrylate	184.3	0.881	13.9	6.8	4.5	16.1
2-Ethylhexylamine	129.3	0.784	14.8	6.6	5.9	17.3
2-Ethylhexyl Carbitol®	218.3	0.918	13.9	7.1	10.6	18.9
2-Ethylhexyl Cellosolve®	174.3	0.880	13.7	6.7	10.6	18.5
2-Ethylhexyl chloride	148.7	0.877	15.5	5.4	2.4	16.6
2-Ethylhexyl crotonate	198.3	0.881	15.4	6.7	6.2	17.9
2-Ethylhexyl 2-ethylbutyrate	228.4	0.858	14.1	5.3	4.5	15.7
2-Ethylhexyl-2-ethylhexanoate	256.4	0.855	14.7	5.1	4.7	16.2
2-Ethylhexyl hexanoate	228.4	0.859	14.5	5.4	5.5	16.4
3-(2-Ethylhexoxy)propylamine	187.3	0.843	14.2	6.7	6.8	17.1
2-Ethylhexyl 4-pentenoate	212.3	0.874	14.6	6.3	5.8	17.0
Ethylideneacetone	84.1	0.850	14.3	9.9	9.7	19.9
N-Ethylisopropanolamine	103.2	0.882	13.6	9.5	14.6	22.1
Ethyl isopropenyl ether	86.13	0.764	14.1	6.9	2.7	15.9
Ethyl isopropyl ether	100.2	0.806	15.3	5.1	6.2	17.3
N-Ethyl-α-methylbenzylamine	149.2	0.906	15.4	8.8	4.3	18.2
2-Ethyl-2-methyl-1,3-dioxolane	116.2	0.932	15.4	8.5	0.0	17.6
3-Ethyl-2-methylheptane	142.3	0.741	15.4	0.0	0.0	15.4
3-Ethyl-3-methylheptane	142.3	0.745	15.4	0.0	0.0	15.4
3-Ethyl-4-methylheptane	142.3	0.748	15.6	0.0	0.0	15.6
4-Ethyl-2-methylheptane	142.3	0.731	15.2	0.0	0.0	15.2

TABLE 18 (continued)
Hoy's Cohesion Parameters for Liquids (and Solids as Subcooled Liquids) at 25°C

Liquid	$M/g\ mol^{-1}$	$\rho/g\ cm^{-3}$	$\delta/MPa^{1/2}$ δ_d	δ_p	δ_h	δ_t
4-Ethyl-3-methylheptane	142.3	0.748	15.6	0.0	0.0	15.6
4-Ethyl-4-methylheptane	142.3	0.747	15.6	0.0	0.0	15.6
5-Ethyl-2-methylheptane	142.3	0.731	15.1	0.0	0.0	15.1
5-Ethyl-3-methylheptane	142.3	0.738	15.3	0.0	0.0	15.3
3-Ethyl-2-methylhexane	128.3	0.726	15.2	0.0	0.0	15.2
3-Ethyl-3-methylhexane	128.3	0.736	15.2	0.0	0.0	15.2
4-Ethyl-2-methylhexane	128.3	0.718	15.0	0.0	0.0	15.0
4-Ethyl-3-methylhexane	128.3	0.737	15.4	0.0	0.0	15.4
3-Ethyl-2-methylpentane	114.2	0.714	15.0	0.0	0.0	15.0
3-Ethyl-3-methylpentane	114.2	0.722	14.9	0.0	0.0	14.9
2-Ethyl-3-methyl-1,5-pentanediol	146.2	0.962	14.2	10.7	16.9	24.5
5-Ethyl-2-methylpiperidine	127.2	0.834	15.7	6.9	1.6	17.2
7-Ethyl-2-methyl-4-undecanol	214.4	0.831	14.3	5.6	8.2	17.4
N-Ethylmorpholine	115.2	0.908	16.6	7.6	0.7	18.3
5-Ethyl-2-nonanone	170.3	0.838	15.1	6.3	6.3	17.5
5-Ethyl-2-nonanol	172.3	0.831	14.3	6.1	9.6	18.3
3-Ethyloctane	142.3	0.735	15.8	0.0	0.0	15.8
4-Ethyloctane	142.3	0.735	15.8	0.0	0.0	15.8
4-Ethyloctanol	158.3	0.834	14.9	7.0	11.0	19.8
3-Ethylpentane	100.2	0.693	15.0	0.0	0.0	15.0
2-Ethyl-1,5-pentanediol	132.2	0.964	13.8	10.8	18.9	25.8
Ethyl propionate	102.1	0.884	14.0	8.1	7.8	17.9
2-Ethyl-3-propylacrolein	126.2	0.846	14.9	8.4	6.6	18.3
2-Ethyl-3-propylacrylic acid	142.2	0.943	18.9	10.1	12.9	25.1
2-Ethyl-2-propylhexanol	172.3	0.850	13.9	6.6	9.7	18.2
Ethyl propyl ketone	100.2	0.811	13.9	7.9	7.3	17.6
3-Ethyl-4-propyltetrahydropyran	156.3	0.874	16.9	5.2	0.0	17.6
m-Ethylstyrene	132.2	0.889	16.8	7.7	0.0	18.5
m-Ethyltoluene	120.2	0.859	16.4	6.7	1.5	17.8
o-Ethyltoluene	120.2	0.875	16.8	6.9	0.0	18.1
p-Ethyltoluene	120.2	0.859	16.4	6.6	2.1	17.8
Fluorobenzene	96.1	1.015	15.4	9.0	5.2	18.6
2-Formyl-3,4-dihydro-2H-pyran	112.1	1.070	13.8	12.0	9.6	20.6
Glycerol, 1,2,3-propanetriol	92.1	1.258	9.3	15.4	31.4	36.2
Glycerol triacetate	218.2	1.153	14.9	11.6	11.2	22.0
Glycol diacetate	146.1	1.099	14.4	11.2	10.6	21.1
Glycol dipropionate	174.2	1.042	14.1	9.7	9.2	19.4
Heptane	100.2	0.679	15.3	0.0	0.0	15.3
2,4-Heptanediol	132.2	0.926	13.4	9.9	16.8	23.7
1-Heptanol	116.2	0.819	15.1	8.0	13.0	21.5
2-Heptanol	116.2	0.813	14.6	7.4	11.7	20.1
3-Heptanol	116.2	0.817	14.6	7.5	11.8	20.2
1-Heptene	98.2	0.692	14.6	3.6	2.7	15.3
cis-2-Heptene	98.2	0.703	15.0	3.4	2.3	15.6
trans-2-Heptene	98.2	0.696	14.9	3.1	2.9	15.5
cis-3-Heptene	98.2	0.697	14.8	3.3	2.5	15.4
trans-3-Heptene	98.2	0.693	14.8	3.1	2.3	15.3
3-Hepten-2-one	112.2	0.842	15.5	9.0	6.4	19.0
2-Heptyl acetate	158.2	0.857	14.6	6.6	4.9	16.8
3-Heptyl acetate	158.2	0.860	14.5	6.6	5.0	16.7
Heptylbenzene	176.3	0.851	16.5	5.6	0.0	17.4
3-Heptyl Carbitol®	204.3	0.924	13.4	7.2	10.7	18.6
3-Heptyl Cellosolve®	160.3	0.880	13.3	6.9	10.7	18.4
2,4-Hexadienal	96.1	0.888	15.1	10.4	11.2	21.5
Hexaldehyde	100.2	0.809	14.2	7.9	8.2	18.2

TABLE 18 (continued)
Hoy's Cohesion Parameters for Liquids (and Solids as Subcooled Liquids) at 25°C

Liquid	$M/g\ mol^{-1}$	$\rho/g\ cm^{-3}$	$\delta/MPa^{1/2}$			
			δ_d	δ_p	δ_h	δ_t
Hexane	86.2	0.654	14.9	0.0	0.0	14.9
2,5-Hexandeiol	118.2	0.957	14.2	10.7	19.3	26.2
2,5-Hexanedione	114.1	0.967	13.6	11.6	11.6	21.1
Hexanoic acid	116.2	0.923	17.2	9.2	13.8	23.9
1-Hexanol	102.2	0.816	15.0	8.5	13.7	22.0
2-Hexanol	102.2	0.810	14.4	7.9	12.3	20.5
3-Hexanol	102.2	0.814	13.5	7.7	11.6	19.4
1-Hexene	84.2	0.668	14.4	3.9	0.0	15.0
cis-2-Hexene	84.2	0.681	14.6	3.5	3.1	15.3
trans-2-Hexene	84.2	0.673	14.6	3.3	2.9	15.2
cis-3-Hexene	84.2	0.674	14.4	3.5	3.4	15.2
trans-3-Hexene	84.2	0.671	14.5	3.3	3.2	15.2
Hexyl acetate	144.2	0.869	14.9	7.0	6.4	17.7
Hexyl acrylate	156.2	0.883	14.7	7.5	6.6	17.8
Hexylamine, 1-hexanamine	101.2	0.762	14.2	7.1	6.9	17.3
Hexylene glycol	118.2	0.918	12.5	11.4	16.8	23.1
Hexylene glycol diacetate	202.3	0.994	13.4	8.7	7.0	17.4
Hexyl Carbitol®	190.3	0.929	13.9	7.5	12.0	19.8
Hexyl Cellosolve®	146.2	0.883	13.8	7.2	12.1	19.7
Hexyl chloride, 1-chlorohexane	120.6	0.873	15.6	6.0	4.0	17.2
Hexyl 2-ethylbutyrate	200.3	0.856	14.3	5.7	5.2	16.2
Hexyl ethyl Carbitol®	218.3	0.881	14.3	6.1	7.2	17.1
Hexyl 2-ethylhexanoate	228.4	0.858	13.8	5.2	5.2	15.6
Hexyl hexanoate	200.3	0.858	14.7	5.8	6.1	17.0
N-(2-Hydroxyethyl)morpholine	131.2	1.068	15.3	8.7	13.8	22.4
N-(2-Hydroxyethyl)piperidine	130.2	1.056	15.0	11.0	15.3	24.1
N-(2-Hydroxyethyl)propylenediamine	118.2	0.988	15.0	12.3	19.2	27.3
Isoamyl alcohol, 3-methyl-1-butanol	88.2	0.806	15.2	9.3	14.1	22.7
sec-Isoamyl alcohol, 3-methyl-2-butanol	88.2	0.813	14.0	8.4	12.4	20.5
Isobutyl acetate	116.2	0.866	14.5	7.8	5.1	17.2
Isobutyl acrylate	128.2	0.884	14.1	8.4	5.7	17.4
Isobutanol, 2-methyl-1-propanol	74.1	0.798	14.4	9.8	15.0	23.0
Isobutyl benzene	134.2	0.848	15.7	6.3	0.0	16.9
Isobutyl Cellosolve®	118.2	0.887	13.9	7.6	12.6	20.2
Isobutylene oxide	72.1	0.801	15.7	8.4	1.3	17.8
Isobutyl heptyl ketone	184.3	0.813	14.7	6.0	3.6	16.3
Isobutyl isobutyrate	144.2	0.849	14.2	7.0	0.0	15.8
Isobutyraldehyde	72.1	0.784	13.1	9.0	8.4	17.9
Isobutyric acid	88.1	0.943	15.4	9.9	16.2	24.5
Isobutyronitrile	69.1	0.765	13.3	10.6	10.6	20.1
Isodecanol	158.3	0.835	14.4	6.9	10.7	19.3
Isooctanol	130.2	0.827	14.4	7.3	12.9	20.7
Isophorone	138.2	0.917	16.4	9.4	3.2	19.2
Isoprene, 2-methyl-1,3-butadiene	68.1	0.675	13.7	5.8	4.0	15.3
Isopropyl acetate	102.1	0.866	14.3	8.4	5.7	17.6
Isopropyl acetoacetate	144.2	0.984	13.4	10.4	8.5	19.0
Isopropyl alcohol, see 2-propanol						
Isopropylamine, see 2-propanamine	59.1	0.681	12.6	8.6	7.0	16.8
Isopropylbenzene	120.2	0.857	16.2	6.9	0.0	17.6
Isopropyl benzoate	164.2	1.005	15.6	10.0	3.9	18.9
Isopropylcyclohexane	126.2	0.797	16.0	2.5	0.0	16.2
Isopropylcyclopentane	112.2	0.771	15.6	3.1	0.0	15.9
N-Isopropyldiisopropanolamine	175.3	0.935	12.7	8.2	13.2	20.1
Isopropyl ether, 2,2′-oxybispropane	102.2	0.718	13.6	4.7	1.5	14.4

TABLE 18 (continued)
Hoy's Cohesion Parameters for Liquids (and Solids as Subcooled Liquids) at 25°C

Liquid	M/g mol^{-1}	ρ/g cm^{-3}	δ/MPa$^{1/2}$			
			δ_d	δ_p	δ_h	δ_t
N-Isopropylisopropanolamine	117.2	0.868	13.5	9.0	12.8	20.7
4-Isopropylheptane	142.3	1.056	15.2	0.0	0.0	15.2
Isopropyl 6-methyl-3-cyclohexenecarboxylate	182.3	0.939	15.4	7.9	0.0	17.3
N-Isopropylmorpholine	129.2	0.909	16.4	7.1	0.0	17.9
Lauryl alcohol, see 1-Dodecanol						
Mesityl oxide	98.2	0.854	14.7	9.6	6.6	18.8
Methacryladehyde	70.1	0.841	12.5	10.7	8.8	18.7
Methacrylic acid	86.1	1.009	17.0	13.0	16.2	26.8
Methallyl cyanide	81.1	0.829	13.9	11.5	10.5	20.9
Methanol	32.0	0.786	11.6	13.0	24.0	29.7
1-Methoxy-1,3-butadiene	84.1	0.826	14.4	8.0	8.2	18.4
3-Methoxybutanol	104.2	0.917	14.1	8.8	14.3	22.0
3-Methoxybutyl acetate	146.2	0.949	14.3	8.6	7.5	18.3
3-Methoxybutyraldehyde	102.1	0.921	13.7	9.9	9.9	19.6
4-Methoxy-2,6-dipropyl-1,3-dioxane	202.3	0.929	14.9	7.3	5.3	17.4
4-Methoxy-4-methyl-2-pentanone	130.2	0.903	13.4	8.8	6.3	17.2
3-Methoxypropionitrile	85.1	0.934	13.7	12.0	14.5	23.3
Methoxytriglycol	164.2	1.043	13.4	9.4	14.8	22.0
Methyl acetate	74.1	0.927	13.3	9.5	10.4	19.4
Methyl acetoacetate	116.1	1.071	13.6	12.2	12.0	21.8
Methyl acrylate	86.1	0.950	13.2	10.2	9.4	19.2
Methylamyl alcohol	102.2	0.803	13.1	7.6	10.5	18.4
Methyl amyl ketone	114.2	0.811	15.1	7.6	7.2	18.4
Methyl benzoate	136.2	1.081	15.9	11.5	7.2	20.8
α-Methylbenzylamine	121.2	0.948	16.1	11.1	5.8	20.4
α-Methylbenzyl Cellosolve®	166.2	1.034	13.7	9.5	11.9	20.5
α-Methylbenzyldimethylamine	149.2	0.899	15.9	7.3	0.0	17.5
2-Methyl-1,3-butadiene, see Isoprene						
2-Methylbutane	72.2	0.614	13.8	0.0	0.0	13.8
2-Methyl-1-butene	70.1	0.645	13.6	4.1	3.7	14.7
2-Methyl-2-butene	70.1	0.657	14.2	4.1	3.8	15.2
3-Methyl-1-butene	70.1	0.621	12.7	3.8	3.7	13.7
N-Methylbutylamine	87.2	0.728	14.2	6.7	6.6	17.0
Methyl butyl ketone	100.2	0.806	14.1	7.9	7.1	17.7
Methyl *sec*-butyl ketone	100.2	0.808	14.1	8.0	6.2	17.4
Methyl *tert*-butyl ketone	100.2	0.801	14.1	8.2	4.1	16.8
3-Methylbutyraldehyde	86.1	0.795	13.6	8.5	7.5	17.7
Methyl Carbitol®	120.2	1.015	13.4	9.6	15.8	22.8
Methyl Carbitol acetate	162.2	1.034	14.6	9.6	10.3	20.3
Methyl Cellosolve®	76.1	0.960	13.0	10.0	17.4	23.9
Methyl Cellosolve acetate	118.1	0.999	14.7	9.8	9.9	20.3
Methyl chloroform, see 1,1,1-Trichloroethane						
4-Methylcyclohexylmethylamine	127.2	0.888	16.2	8.0	5.1	18.8
Methylcyclopentane	84.2	0.743	15.7	3.5	0.0	16.0
N-Methyldibutylamine	143.3	0.755	15.2	3.2	2.9	15.8
N-Methyldiethanolamine	119.2	1.036	12.8	11.0	20.1	26.2
4-Methyl-3,4-dihydropyran	98.2	0.890	15.5	7.9	2.5	17.5
2,5-*endo*Methylenecyclohexylmethylamine	125.2	0.936	16.5	8.2	6.1	19.4
2,5-*endo*Methylenecyclohexane-1-methanol	140.3	0.986	16.4	9.0	11.1	21.8
2,5-*endo*Methylene-3-cyclohexenyl acetate	152.2	1.045	16.0	9.2	5.4	19.3
2,5-*endo*Methylenecyclohexyl acetate	154.2	1.022	16.7	8.2	5.1	19.3
Methylene dichloride, see Dichloromethane						
N-Methylethanolamine	75.1	0.936	14.1	11.4	20.0	27.0
Methyl ethyl Carbitol®	148.2	0.918	14.5	7.6	8.6	18.5

TABLE 18 (continued)
Hoy's Cohesion Parameters for Liquids (and Solids as Subcooled Liquids) at 25°C

Liquid	$M/g\ mol^{-1}$	$\rho/g\ cm^{-3}$	$\delta/MPa^{1/2}$			
			δ_d	δ_p	δ_h	δ_t
Methyl ethyl ether	60.1	0.692	13.0	5.7	6.2	15.5
Methyl ethyl ketone	72.1	0.800	14.1	9.3	9.5	19.3
2-Methyl-5-ethylpyridine	121.2	0.916	17.7	7.7	2.5	19.5
2-Methylheptane	114.2	0.693	15.0	0.0	0.0	15.0
3-Methylheptane	114.2	0.700	15.1	0.0	0.0	15.1
4-Methylheptane	114.2	0.699	15.1	0.0	0.0	15.1
2-Methyl-1-heptene	112.2	0.715	14.9	3.6	0.0	15.3
2-Methyl-2-heptene	112.2	0.719	15.3	3.5	0.0	15.7
3-Methyl-1-heptene	112.2	0.706	14.4	3.4	1.3	14.9
3-Methyl-2-heptene	112.2	0.724	15.3	3.5	0.0	15.7
4-Methyl-1-heptene	112.2	0.712	14.6	3.5	0.0	15.0
5-Methyl-1-heptene	112.2	0.711	14.5	3.4	2.0	15.0
cis-5-Methyl-2-heptene	112.2	0.718	15.2	3.2	0.0	15.5
6-Methyl-1-heptene	112.2	0.707	14.6	3.4	2.2	15.2
cis-6-Methyl-2-heptene	112.2	0.713	15.1	3.2	0.0	15.4
Methyl heptyl ketone	142.2	0.821	14.6	6.9	4.6	16.8
2-Methylhexane	100.2	0.673	14.8	0.0	0.0	14.8
3-Methylhexane	100.2	0.682	14.9	0.0	0.0	14.9
5-Methyl-2-hexanol	116.2	0.808	14.9	7.6	11.3	20.2
Methyl hexyl ketone	128.2	0.814	15.1	7.2	6.8	18.0
N-Methylhomopiperazine	114.2	0.922	16.3	8.9	5.6	19.4
Methyl 3-hydroxybutyrate	118.1	1.053	11.9	10.9	14.6	21.8
Methyl isoamyl ketone	114.2	0.807	14.9	7.7	5.8	17.7
Methyl isobutyl ketone	100.2	0.796	14.4	8.1	5.9	17.6
Methyl isocyanate	57.1	0.951	11.2	12.2	13.6	21.7
Methyl isopropenyl ketone	84.1	0.847	14.0	10.3	7.1	18.8
Methyl isopropyl ketone	86.1	0.798	14.6	8.8	6.6	18.3
Methyl methacrylate	100.1	0.930	13.7	9.8	6.1	17.9
1-Methyl-2,5-*endo*methylenecyclohexane-1-methanol	126.2	0.998	14.6	7.5	21.6	27.2
1-Methylnaphthalene	142.2	1.014	17.9	9.5	0.0	20.3
2-Methylnonane	142.3	0.723	15.5	0.0	0.0	15.5
3-Methylnonane	142.3	0.728	15.7	0.0	0.0	15.7
4-Methylnonane	142.3	0.727	15.7	0.0	0.0	15.7
5-Methylnonane	142.3	0.728	15.6	0.0	0.0	15.6
2-Methyloctane	128.3	0.708	15.4	0.0	0.0	15.4
3-Methyloctane	128.3	0.716	15.4	0.0	0.0	15.4
4-Methyloctane	128.3	0.715	15.4	0.0	0.0	15.4
2-Methylpentanal	100.2	0.805	14.1	8.0	7.1	17.7
2-Methylpentane	86.2	0.648	14.4	0.0	0.8	14.4
3-Methylpentane	86.2	0.659	14.6	0.0	0.0	14.6
2-Methyl-1,3-pentanediol	118.2	0.969	13.0	10.9	18.6	25.2
2-Methyl-1,5-pentanediol	118.2	0.970	14.2	11.6	20.6	27.6
2-Methylpentanoic acid	116.2	0.918	15.6	8.9	11.8	21.5
2-Methylpentanol	102.2	0.820	14.2	8.5	12.6	20.8
2-Methyl-2-pentanol	102.2	0.811	13.3	9.2	11.1	19.6
2-Methyl-3-pentanol	102.2	0.820	14.7	8.0	12.1	20.6
3-Methylpentanol	102.2	0.819	14.8	8.6	13.0	21.5
3-Methyl-2-pentanol	102.2	0.825	15.2	8.2	12.6	21.4
3-Methyl-3-pentanol	102.2	0.824	11.4	8.7	10.1	17.5
4-Methylpentanol	102.2	0.809	15.6	8.7	13.4	22.3
2-Methyl-2-pentenal	98.2	0.851	14.8	9.6	7.7	19.2
2-Methyl-1-pentene	84.2	0.674	14.1	3.9	3.2	14.9
2-Methyl-2-pentene	84.2	0.680	14.4	3.8	3.4	15.3
3-Methyl-1-pentene	84.2	0.662	13.7	3.7	2.9	14.5

TABLE 18 (continued)
Hoy's Cohesion Parameters for Liquids (and Solids as Subcooled Liquids) at 25°C

Liquid	$M/g\ mol^{-1}$	$\rho/g\ cm^{-3}$	$\delta/MPa^{1/2}$ δ_d	δ_p	δ_h	δ_t
4-Methyl-1-pentene	84.2	0.658	13.6	3.7	3.0	14.4
cis-4-Methyl-2-pentene	84.2	0.664	14.2	3.5	0.0	14.6
trans-4-Methyl-2-pentene	84.2	0.663	14.4	3.3	0.0	14.7
4-Methyl-2-pentyl acetate	144.2	0.853	14.4	7.0	3.0	16.3
4-Methyl-2-pentyl butyrate	172.3	0.849	14.1	6.2	3.6	15.8
2-Methylpentyl Cellosolve®	146.2	0.882	13.6	7.2	11.4	19.2
4-Methyl-2-pentyl Cellosolve	146.2	0.874	13.4	7.3	10.6	18.6
N-Methylpiperazine	100.2	0.898	16.7	9.0	5.2	19.6
2-Methylpropane, isobutane	58.1	0.550	12.5	0.0	0.0	12.5
2-Methylpropene, isobutylene	56.1	0.587	12.6	4.2	3.0	13.6
Methyl 1-propenyl ether	72.1	0.771	14.7	7.3	7.8	18.2
Methyl propionate	88.1	0.909	13.9	8.8	8.9	18.7
Methyl propyl ketone	86.1	0.801	14.6	8.8	7.0	18.4
2-Methyl-2-propylpentanol	144.3	0.838	14.3	7.3	10.2	19.0
N-Methyl-2-pyrrolidone	99.1	1.020	16.5	10.4	13.5	23.7
α-Methylstyrene	118.2	0.905	16.6	8.5	0.0	18.6
2-Methyl-1,2,3,6-tetrahydrobenzaldehyde	124.2	0.942	15.4	9.5	6.5	19.2
4-Methyltetrahydropyran	100.2	0.953	15.8	6.3	0.0	17.0
2-Methylthioethyl acrylate	146.2	1.060	14.4	11.5	7.9	20.1
2-Methylthiophene	98.2	1.012	15.1	12.1	2.6	19.5
3-Methylthiophene	98.2	1.015	15.2	12.1	1.8	19.5
Monoethanolamine, 2-aminoethanol	61.1	1.012	13.9	13.8	25.0	31.8
Monoisopropanolamine, isopropanolamine	75.1	0.956	13.9	12.3	20.1	27.3
Morpholine	87.1	0.995	16.0	11.4	10.1	22.1
Nitrobenzene	123.1	1.190	17.6	14.0	0.0	22.5
Nitroethane	75.1	1.045	19.0	13.0	0.0	23.0
1-Nitropropane	89.1	0.996	18.1	11.2	0.0	21.3
2-Nitropropane	89.1	0.985	16.5	10.4	6.6	20.6
Nonane	128.3	0.713	15.6	0.0	0.0	15.6
1-Nonanol	144.3	0.823	15.3	7.3	12.0	20.7
sec-Nonanol	144.3	0.819	13.9	6.6	10.3	18.5
1-Nonene	126.2	0.724	15.4	3.4	0.0	15.8
Nonylbenzene	204.3	0.851	16.0	6.8	0.0	17.4
Nonyl Cellosolve®	188.3	0.865	13.7	6.4	9.2	17.7
1-Nonylynaphthalene	254.4	0.932	17.4	6.6	0.0	18.6
Nonylphenol	220.4	0.941	15.6	6.7	9.1	19.3
Octane	114.2	0.698	15.4	0.0	0.0	15.4
1-Octanol	130.2	0.822	15.2	7.7	12.4	21.1
2-Octanol	130.2	0.817	14.1	6.9	10.9	19.1
1-Octene	112.2	0.710	15.0	3.5	2.3	15.5
cis-2-Octene	112.2	0.719	15.1	3.2	0.0	15.5
trans-2-Octene	112.2	0.714	15.1	3.0	0.0	15.4
cis-3-Octene	112.2	0.716	15.1	3.2	0.0	15.5
trans-3-Octene	112.2	0.710	15.1	3.0	0.0	15.4
cis-4-Octene	112.2	0.716	15.1	3.2	0.0	15.4
trans-4-Octene	112.2	0.709	15.1	3.0	0.0	15.4
Octylbenzene	190.3	0.851	16.7	5.4	0.0	17.5
1,4-Oxathiane	104.2	1.112	16.8	12.3	5.2	21.5
Pentadecane	212.4	0.764	16.8	0.0	0.0	16.8
trans-1,3-Pentadiene	68.1	0.671	14.1	5.3	4.8	15.8
Pentane	72.2	0.621	14.4	0.0	0.0	14.4
1,5-Pentanediol	104.2	0.987	13.9	12.2	22.8	29.4
1-Pentanol, amyl alcohol	88.2	0.811	14.8	9.1	14.7	22.7
2-Pentanol, *sec*-amyl alcohol	88.2	0.805	15.0	8.5	13.7	22.0

TABLE 18 (continued)
Hoy's Cohesion Parameters for Liquids (and Solids as Subcooled Liquids) at 25°C

Liquid	$M/g\ mol^{-1}$	$\rho/g\ cm^{-3}$	$\delta/MPa^{1/2}$ δ_d	δ_p	δ_h	δ_t
3-Pentanol	88.2	0.816	13.9	8.3	13.0	20.8
1-Pentene	70.1	0.635	13.9	4.1	0.0	14.5
cis-2-Pentene	70.1	0.649	14.5	3.8	0.0	15.0
trans-2-Pentene	70.1	0.642	14.1	3.4	3.6	14.9
Pentyl acetate	130.2	0.869	14.3	7.3	5.7	17.1
Pentylcyclohexane	154.3	0.799	16.0	2.3	0.0	16.1
Perchloroethylene, tetrachloroethylene	165.9	1.611	11.4	15.2	0.0	19.0
Phenoxypropylene oxide	150.2	1.104	17.6	11.8	7.6	22.5
Phenyl Carbitol®	182.2	1.109	14.4	10.4	14.3	22.9
Phenyl Cellosolve®	138.2	1.103	13.7	10.4	16.0	23.5
Phenyl Cellosolve® acetate	180.2	1.101	16.2	11.1	8.8	21.5
N-Phenylethanolamine	137.2	1.091	15.2	11.9	17.0	25.7
Phenylhydrazine	108.1	1.095	15.6	14.9	13.7	25.5
N-Phenylpiperazine	162.2	1.064	18.1	11.4	3.2	21.6
Picoline	93.1	0.950	18.2	7.8	6.8	20.9
Pinacolyl alcohol	102.2	0.815	14.2	8.0	10.6	19.5
α-Pinene	136.2	0.853	15.6	4.3	0.0	16.2
Piperidine	85.2	0.857	16.2	8.7	5.8	19.3
Pivaldehyde	86.1	0.786	13.4	8.5	5.5	16.8
Polyglycolamine H221M	221.3	1.004	13.9	9.2	10.8	19.8
Propanal, propionaldehyde	58.1	0.791	12.6	9.7	11.0	19.3
Propane	44.1	0.492	11.8	0.0	0.0	11.8
1,3-Propanediol, trimethylene glycol	76.1	1.050	12.4	14.1	27.1	33.0
1-Propanol	60.1	0.799	14.1	10.5	17.7	24.9
2-Propanol, isopropanol	60.1	0.781	14.0	9.8	16.0	23.4
Propasol® Solvent B, butoxypropanol	132.2	0.878	13.2	7.5	11.5	19.0
Propasol® Solvent DM	148.2	0.955	13.1	8.6	12.2	19.8
Propasol® Solvent M	90.1	0.917	12.9	9.2	14.2	21.3
Propasol® Solvent P, propoxypropanol	118.2	0.880	13.2	8.0	12.1	19.5
Propionic acid	74.1	0.988	15.3	11.2	17.1	25.5
Propionic anhydride	130.1	1.005	12.6	11.4	10.4	19.9
Propionitrile	55.1	0.777	12.6	11.5	13.8	22.0
Propyl acetate	102.1	0.881	14.1	8.1	7.8	18.0
Propylamine	59.1	0.712	13.9	9.3	7.2	18.2
Propylbenzene	120.2	0.856	16.2	6.9	0.0	17.6
Propyl Carbitol®	148.2	0.963	13.1	8.4	13.2	20.4
Propyl chloride, see chloropropane						
Propylcyclohexane	126.2	0.788	16.0	2.5	0.0	16.2
Propylcyclopentane	112.2	0.771	16.0	3.1	0.0	16.3
Propylene chlorohydrin	94.5	1.106	12.9	11.6	16.0	23.6
Propylene diamine	74.1	0.858	13.5	12.8	12.8	22.6
Propylene dichloride, see 1,3-dichloropropane						
Propylene glycol, 1,2-propanediol	76.1	1.033	11.8	13.3	25.0	30.7
Propylene oxide, methyloxirane	58.1	0.823	13.6	8.2	10.4	18.9
4-Propylheptane	142.3	0.731	15.3	0.0	0.0	15.3
2-Propylheptanol	158.3	0.828	14.5	6.9	10.6	19.2
m-Propyltoluene	134.2	0.855	16.5	6.3	0.0	17.7
o-Propyltoluene	134.2	0.869	16.7	6.5	0.0	17.9
p-Propyltoluene	134.2	0.853	16.5	6.2	0.0	17.6
Pyridine	79.1	0.978	17.6	10.1	7.7	21.7
Styrene, ethenylbenzene	104.2	0.901	16.8	9.1	0.0	19.1
Styrene oxide, phenyloxirane	120.2	1.048	17.5	11.4	5.1	21.5
α-Terpineol	154.2	0.927	13.9	7.9	10.2	19.0

TABLE 18 (continued)
Hoy's Cohesion Parameters for Liquids (and Solids as Subcooled Liquids) at 25°C

Liquid	$M/g\ mol^{-1}$	$\rho/g\ cm^{-3}$	$\delta/MPa^{1/2}$ δ_d	δ_p	δ_h	δ_t
Tetradecane	198.4	0.758	16.6	0.0	0.0	16.6
1,1,3,3-Tetraethoxypropane	220.3	0.913	13.5	7.4	6.7	16.8
1,2,3,6-Tetrahydrobenzaldehyde	110.2	0.966	15.6	10.3	7.6	20.1
1,2,3,6-Tetrahydrobenzoic acid	126.2	1.077	18.6	11.3	13.7	25.7
1,2,3,6-Tetrahydrobenzonitrile	107.2	0.951	15.4	11.5	10.0	21.7
Tetrahydrofuran	72.1	0.882	13.3	11.0	6.7	18.5
1,2,3,4-Tetrahydronaphthalene	132.2	0.964	17.8	7.8	0.0	19.4
Tetrahydropyran-2-methanol	116.2	1.021	14.4	9.6	13.5	22.0
1,1,3,5-Tetramethoxyhexane	206.3	0.952	13.7	7.7	7.1	17.3
2,2,3,3-Tetramethylbutane	114.2	0.820	17.1	0.0	0.0	17.1
N,N,N,N-Tetramethyl-1,3-butanediamine	144.3	0.797	15.2	4.7	2.0	16.0
N,N,N,N-Tetramethyl-1,2-ethylenediamine	116.2	0.772	15.0	5.1	2.7	16.0
2,2,3,3-Tetramethylpentane	128.3	0.746	15.2	0.0	0.0	15.2
N,N,N,N-Tetramethyl-1,3-propanediamine	130.2	0.777	15.1	4.8	2.6	16.1
2-Thiabutane	76.2	0.835	14.6	8.7	5.7	17.9
Thiacyclopentane, tetrahydrothiophene	88.2	0.992	17.4	10.9	0.6	20.5
2-Thiapropane	62.1	0.841	14.2	9.6	6.7	18.4
Thiophene	84.1	1.057	14.0	12.4	7.5	20.0
Toluene, methylbenzene	92.1	0.862	16.4	8.0	1.6	18.3
Tolylene diisocyanate	174.2	1.212	15.3	16.4	7.8	23.7
N-(*o*-Tolyl)ethanolamine	151.2	1.067	14.3	10.9	15.0	23.4
Triallylamine	137.2	0.795	13.8	6.6	4.1	15.8
Tributylamine	185.4	0.775	15.1	2.8	4.0	15.9
1,1,1-Trichloroethane, methyl chloroform	133.4	1.218	10.8	11.5	0.0	15.8
1,1,2-Trichloroethane	133.4	1.432	13.9	12.9	7.0	20.2
1,1,2-Trichloroethylene	131.4	1.455	11.7	14.0	4.4	18.7
Trichloromethane	119.4	1.477	11.0	13.7	6.3	18.7
1,2,3-Trichloropropane	147.4	1.384	14.8	12.7	6.7	20.6
2,2,3-Trichloropropionaldehyde	161.4	1.467	11.0	14.6	8.0	20.0
Tridecane	184.4	0.751	16.4	0.0	0.0	16.4
1,1,3-Triethoxy-4-ethyloctane	275.5	0.866	14.0	5.6	4.8	15.9
Triethylamine	101.2	0.723	14.6	3.7	1.9	15.2
Triethylene glycol	150.2	1.249	15.0	12.2	21.8	29.1
Triethylenetetramine	146.2	0.976	12.7	12.4	14.1	22.8
Triglycol dichloride	187.0	1.190	15.3	10.5	9.6	20.9
1,1,3-Trimethoxybutane	148.2	0.916	13.6	7.8	6.6	17.0
1,2,3-Trimethylbenzene	120.2	0.889	17.3	7.3	0.0	18.7
1,2,4-Trimethylbenzene	120.2	0.870	16.9	7.1	0.0	18.3
1,3,5-Trimethylbenzene, mesitylene	120.2	0.860	16.7	7.0	0.0	18.1
2,2,3-Trimethylbutane	100.2	0.685	14.2	0.0	0.0	14.2
1,1,2-Trimethylcyclopentane	112.2	0.767	15.1	3.0	0.0	15.4
1,1,3-Trimethylcyclopentane	112.2	0.742	14.6	2.9	0.0	14.9
2,2,3-Trimethylheptane	142.3	0.737	15.2	0.0	0.0	15.2
2,2,4-Trimethylheptane	142.3	0.722	14.6	0.0	0.0	14.6
2,2,5-Trimethylheptane	142.3	0.721	14.7	0.0	0.0	14.7
2,2,6-Trimethylheptane	142.3	0.714	14.6	0.0	0.0	14.6
2,3,3-Trimethylheptane	142.3	0.743	15.3	0.0	0.0	15.3
2,3,4-Trimethylheptane	142.3	0.746	15.4	0.0	0.0	15.4
2,3,5-Trimethylheptane	142.3	0.736	15.0	0.0	0.0	15.0
2,3,6-Trimethylheptane	142.3	0.729	14.9	0.0	0.0	14.9
2,4,4-Trimethylheptane	142.3	0.728	14.8	0.0	0.0	14.8
2,4,5-Trimethylheptane	142.3	0.736	15.0	0.0	0.0	15.0
2,4,6-Trimethylheptane	142.3	0.717	14.5	0.0	0.0	14.5
2,5,5-Trimethylheptane	142.3	0.731	14.8	0.0	0.0	14.8

TABLE 18 (continued)
Hoy's Cohesion Parameters for Liquids (and Solids as Subcooled Liquids) at 25°C

Liquid	$M/g\ mol^{-1}$	$\rho/g\ cm^{-3}$	$\delta/MPa^{1/2}$			
			δ_d	δ_p	δ_h	δ_t
3,3,4-Trimethylheptane	142.3	0.752	15.4	0.0	0.0	15.4
3,3,5-Trimethylheptane	142.3	0.737	15.0	0.0	0.0	15.0
3,4,4-Trimethylheptane	142.3	0.752	15.4	0.0	0.0	15.4
3,4,5-Trimethylheptane	142.3	0.754	15.4	0.0	0.0	15.4
2,2,3-Trimethylhexane	128.3	0.724	15.0	0.0	0.0	15.0
2,2,4-Trimethylhexane	128.3	0.710	14.5	0.0	0.0	14.5
2,2,5-Trimethylhexane	128.3	0.702	14.5	0.0	0.0	14.5
2,3,3-Trimethylhexane	128.3	0.733	15.1	0.0	0.0	15.1
2,3,4-Trimethylhexane	128.3	0.734	15.2	0.0	0.0	15.2
2,3,5-Trimethylhexane	128.3	0.717	14.8	0.0	0.0	14.8
2,4,4-Trimethylhexane	128.3	0.719	14.7	0.0	0.0	14.7
3,3,4-Trimethylhexane	128.3	0.740	15.3	0.0	0.0	15.3
2,6,8-Trimethyl-4-nonanol	186.3	0.814	13.9	5.9	7.8	16.9
2,2,3-Trimethylpentane	114.2	0.711	14.7	0.0	0.0	14.7
2,2,4-Trimethylpentane, isooctane	114.2	0.687	14.0	0.0	0.0	14.0
2,3,3-Trimethylpentane	114.2	0.721	14.8	0.0	0.0	14.8
2,3,4-Trimethylpentane	114.2	0.714	14.9	0.0	0.0	14.9
2,2,4-Trimethylpentanol	130.2	0.825	14.3	7.7	9.9	19.0
2,2,4-Trimethyl-3-pentenyl acetate	170.3	0.892	15.2	7.6	0.0	17.0
2,2,4-Trimethylpentyl acetate	172.3	0.861	14.8	6.5	1.1	16.2
2,4,6-Trimethyl-1,2,3,6-tetrahydrobenzaldehyde	152.2	0.914	14.7	8.2	5.3	17.7
Undecane	156.3	0.735	16.0	0.0	0.0	16.0
1-Undecanol	172.3	0.828	15.4	6.7	11.2	20.2
2-Undecanol	172.3	0.823	13.9	6.0	9.8	18.0
Valeraldehyde, 1-pentanal	86.1	0.805	13.8	8.4	8.8	18.4
Valeric acid	102.1	0.934	16.7	9.7	14.3	24.0
Vinyl acetate	86.1	0.926	12.9	10.0	8.7	18.5
Vinyl allyl ether	84.1	0.798	13.2	7.9	7.0	16.8
Vinyl butyl Carbitol®	188.3	0.913	14.2	7.3	8.0	17.9
Vinyl butyl ether	100.2	0.774	13.8	6.1	5.5	16.1
Vinyl *S*-butylmercaptoethyl ether	160.3	0.921	15.0	8.5	5.9	18.2
Vinyl butyrate	114.2	0.895	13.8	8.7	7.0	17.7
Vinyl Carbitol®	132.2	1.023	13.3	9.9	15.3	22.6
Vinyl 2-chloroethyl ether	106.6	1.041	14.2	9.9	8.7	19.4
Vinyl crotonate	112.1	0.937	14.0	10.0	8.4	19.2
Vinyl ethyl Carbitol®	160.2	0.932	14.0	8.0	8.6	18.3
Vinyl ethyl ether	72.1	0.748	13.2	7.0	5.7	16.0
Vinyl 2-ethylhexanoate	170.3	0.869	14.0	7.0	5.2	16.6
Vinyl 2-ethylhexyl ether	156.3	0.805	14.9	5.2	4.5	16.3
Vinyl *S*-ethylmercaptoethyl ether	132.2	0.946	14.9	9.6	6.6	18.9
Vinyl isobutyl ether	100.2	0.763	14.4	6.3	4.0	16.2
Vinyl isopropyl ether	86.1	0.747	13.6	6.6	4.2	15.7
Vinyl methyl Cellosolve®	102.1	0.890	14.0	8.4	8.6	18.4
Vinyl methyl ether	58.1	0.742	10.5	7.0	6.8	14.4
Vinyl propionate	100.1	0.910	13.5	9.4	7.7	18.1
Vinyl propyl ether	86.1	0.762	13.5	6.5	5.9	16.1
m-Vinyltoluene	118.2	0.905	16.8	8.3	0.0	18.7
o-Vinyltoluene	118.2	0.897	16.8	8.4	0.0	18.8
p-Vinyltoluene	118.2	0.904	16.9	8.2	0.0	18.8
Vinyl 2,6,8-trimethyl-4-nonyl ether	212.4	0.802	14.5	4.4	1.5	15.3
Water	18.02	0.997	12.2	22.8	40.4	48.0

TABLE 18 (continued)
Hoy's Cohesion Parameters for Liquids (and Solids as Subcooled Liquids) at 25°C

Liquid	M/g mol^{-1}	ρ/g cm^{-3}	δ/MPa$^{1/2}$			
			δ_d	δ_p	δ_h	δ_t
m-Xylene, 1,3-dimethylbenzene	106.2	0.860	16.5	7.2	2.4	18.2
o-Xylene, 1,2-dimethylbenzene	106.2	0.876	17.0	7.5	0.0	18.5
p-Xylene, 1,4-dimethylbenzene	106.2	0.856	16.5	7.0	2.0	18.1

Adapted from Hoy, K. L., *The Hoy Tables of Solubility Parameters,* Union Carbide Corporation, Solvents and Coatings Materials Division, South Charleston, WV, 1985.

5.12 OTHER MULTICOMPONENT COHESION PARAMETERS

Beerbower, Martin, and Wu[31,32] developed a four-component approach to the solubilities of solids in polar and nonpolar systems which is intermediate in complexity between five-component parameters and the three-component Hansen method. Induction parameters were ignored on the grounds that they did not improve significantly the solubility predictions, so Equation 24 reduces to

$$\delta_t^2 = \delta_d^2 + \delta_o^2 + 2\,\delta_a\delta_b \tag{75}$$

If δ_o is identified with δ_p and $\delta_h^2 = 2\,\delta_a\,\delta_b$, the three-component Hansen parameters are recovered. But, more importantly, Equation 21 simplifies to

$$^{ij}A = ({}^i\delta_d - {}^j\delta_d)^2 + ({}^i\delta_p - {}^j\delta_p)^2 + 2({}^i\delta_a - {}^j\delta_a)({}^i\delta_b - {}^i\delta_b) \tag{76}$$

The value of δ_b can be determined[143] from the spectroscopic proton-accepting parameter β, and δ_a can then be calculated from $(\delta_t^2 - \delta_d^2 - \delta_p^2/2\delta_b)$. The original data set by this method appears in Table 21. The δ_p and δ_d values were based on those of Hansen and Beerbower[25] (Table 11) with a few corrections, and some of the component parameters were adjusted on the basis of solubility observations. If a larger set of data is developed, this four-parameter method may become a generally acceptable improvement on the Hansen method, as it takes into account the unsymmetrical nature of hydrogen-bonding interactions.

The Hansen component parameters have been combined in various alternative ways. Sometimes δ_p and δ_h (the "association" interactions) are described collectively:

$$\delta_{hp}^2 = \delta^2{}_h + \delta_p^2 = \delta_t^2 - \delta_d^2 \tag{77}$$

Beerbower and Jensen[35] developed a semiquantitative approach to the prediction of the hard/soft character of Lewis acid-base species using a sorting map based on the composite association parameter δ_{hp} and the dispersion parameter. On this map, alkanes, aromatics and sulfides were soft, oxygen donor species were hard, and anilines and pyridines were borderline. The ratio δ_{hp}/δ_d was critical, reflected in a softness value (based on a zero value for water) defined as

$$S = 20.1 - (\delta_{hp}/\text{MPa}^{1/2})/((\delta_d/\text{MPa}^{1/2}) - 13.3) \tag{78}$$

Table 22 lists Lewis base liquids according to the hard/borderline/soft classification, together with softness values defined as in Equation 78. Lewis acids such as glycols and carboxylic acids appeared in the hard region; trichloromethane was in the soft region.

TABLE 19
Three-Component Cohesion Parameters for Alkyl Derivatives of Phenol and Resorcinol From Vaporization, Distribution Coefficient, and Gas Chromatographic Data

	20°C		80°C			
	$\delta/MPa^{1/2}$ (vaporization data)	$\delta/MPa^{1/2}$ (distribution data)	$\delta_t/MPa^{1/2}$	$\delta_d/MPa^{1/2}$	$\delta_p/MPa^{1/2}$	$\delta_h/MPa^{1/2}$
			Alkylphenols, Ph(OH)R			
—	24.6	25.8	23.1	17.8	5.5	13.9
2-C_1	22.7	23.1	21.5	17.2	0.8	12.9
2-C_2	21.9	—	20.7	16.4	3.3	11.9
2-C_3	21.3	—	19.8	16.0	3.7	11.3
2-C_4	20.9	—	19.2	15.6	4.3	10.4
2-C_5	20.3	—	18.6	14.9	4.9	10.0
2-C_6	19.8	—	18.2	14.5	5.5	9.4
3- and 4-C_1	23.9	25.0	22.3	17.0	6.8	12.9
3- and 4-C_2	23.3	22.1	21.7	16.6	7.2	11.9
3- and 4-C_3	22.9	—	21.1	16.0	7.6	11.3
3- and 4-C_4	22.3	—	20.5	15.6	8.0	10.4
3- and 4-C_5	21.9	—	20.1	15.3	8.2	10.0
3- and 4-C_6	21.5	—	19.4	14.7	8.6	9.4
2,6-$(C_1)_2$	21.9	—	20.7	16.6	2.9	11.9
2,4-$(C_1)_2$	22.5	22.9	20.7	16.6	2.9	11.9
2,3-$(C_1)_2$	22.3	22.9	20.7	16.6	2.9	11.9
3,5-$(C_1)_2$	23.7	23.1	21.7	16.6	7.2	11.9
3,4-$(C_1)_2$	23.1	23.1	21.7	16.6	7.0	11.9
			Alkylresorcinols, Ph-1,3-$(OH)_2$R			
—	31.5	30.1	29.5	18.4	11.3	20.1
2-C_1	27.8	28.2	26.0	17.6	5.7	18.2
2-C_2	27.2	27.6	25.2	17.0	7.8	17.0
2-C_3	26.6	27.0	24.3	16.6	8.2	15.8
2-C_4	26.2	26.6	23.9	16.2	9.4	14.9
2-C_5	25.8	26.4	23.3	15.6	10.0	14.1
2-C_6	25.6	26.2	22.9	15.1	10.6	13.5
2-C_7	25.4	25.8	22.7	14.7	11.5	13.1
2-C_8	25.0	—	22.5	14.3	12.1	12.5
4-C_1	28.4	28.6	26.4	17.6	7.4	18.2
4-C_2	27.6	27.8	25.6	17.2	8.6	17.0
4-C_3	27.0	27.2	25.0	16.6	9.8	15.8
4-C_4	26.6	26.6	24.3	16.4	10.0	14.9
4-C_5	26.2	26.2	23.9	16.0	10.8	14.1
4-C_6	25.8	25.8	23.7	15.3	12.1	13.5
4-C_7	25.6	25.6	23.3	14.9	12.3	13.1
4-C_8	25.4	—	23.1	14.5	12.9	12.5
5-C_1	30.1	29.1	27.8	18.0	10.8	18.2
5-C_2	29.1	28.2	27.0	17.4	11.9	17.0
5-C_3	28.4	27.6	26.2	17.0	12.1	15.8
5-C_4	27.8	27.0	25.6	16.6	12.5	14.9
5-C_5	27.4	26.8	25.0	16.2	12.7	14.1
5-C_6	27.0	26.4	24.6	15.8	13.1	13.5
5-C_7	27.0	26.2	24.1	15.3	13.3	13.1
5-C_8	26.4	—	23.9	14.9	13.9	12.5

Adapted from Lille, Ü, Kundle, H., and Eisen, O., *J. Chromatogr.*, 116, 1, 1976.

TABLE 20
Calculated Hansen and Hildebrand Parameters

	δ/MPa$^{1/2}$					
Liquid	δ_d	δ_p	δ_h	δ_t[a]	δ_t[b]	δ_t[c]
Hexane	14.9	0.0	0.0	14.9	14.7	14.7
Isohexane	15.0	0.0	0.0	15.0	14.3	14.3
Heptane	15.3	0.0	0.0	15.3	15.0	15.1
Octane	15.5	0.0	0.0	15.5	15.1	15.3
Decane	15.9	0.0	0.0	15.9	15.2	15.2
Dodecane	16.0	0.1	0.0	16.0	15.1	15.6
Cyclohexane	16.2	0.2	0.0	16.2	16.7	17.2
Benzene	17.7	0.3	2.2	17.8	18.5	18.9
Toluene	17.6	0.7	1.8	17.7	18.1	18.2
o-Xylene	17.8	1.0	1.6	17.9	18.2	18.3
Phenol	18.5	5.3	16.2	25.1	22.8	—
Ethylbenzene	17.6	0.8	1.6	17.7	17.8	17.2
Methylcyclohexane	16.1	0.1	0.0	16.1	16.2	—
Acetone	14.7	11.5	5.6	19.5	19.1	—
Methanol	18.6	0.5	1.9	18.7	18.7	—
Nefras S-155/200 (mineral spirits)	16.3	1.0	0.6	16.3	16.3	15.8
Nefras S-80/120 (BR-2)	15.4	0.4	0.2	15.4	14.9	14.9
Nefras AR-120/200	17.8	0.5	1.7	17.9	17.4	17.6

[a] $\delta_d^2 + \delta_p^2 + \delta_h^2$
[b] From vaporization enthalpy.
[c] From surface tension, $\delta_t/\text{MPa}^{1/2} = 4.1\ (\gamma/{}^iV^{1/3})^{0.43}$

Adapted from Stekol'shchikov, M. N., Krivtsova, L. M., and Ratner, M. I., *Khim. Tekhnol. Topl. Masel*, (6), 33, 1987; *Chem. Technol. Fuels Oils*, (6), 292, 1987.

An alternative combination of component cohesion parameters[144,145] uses a cohesion parameter determined from gas-liquid chromatography (Chapter 18) and given by $(\Delta H_{add}/V)^{1/2}$ where ΔH_{add} is the "additional" enthalpy of solution (the enthalpy of solution less the contributions from dispersion and induced dipole interactions).

Wingefors[146,147] extended the polar-nonpolar approach (Section 5.8) for solvent extraction in terms of a proton donor parameter, δ_σ:

$$\delta_t^2 = \delta_\lambda^2 + \delta_\tau^2 + k\delta_\lambda\,\delta_\tau + \delta_\sigma\delta_\tau \tag{79}$$

The terms on the right-hand side represent dispersion, dipole-dipole, dipole-induced dipole, and hydrogen-bonding components, respectively; the combination of the dipole parameter δ_τ with the proton donor parameter δ_σ follows from the assumption that proton acceptor ability and electron donor ability should be correlated in some way. Comparison with other equations suggests:

$$\delta_\lambda^2 \approx \delta_d^2 \tag{80}$$

$$\delta_\tau^2 + k\delta_\lambda\delta_\tau \approx \delta_p^2 \approx 2\delta_i\delta_d \tag{81}$$

$$\delta_\sigma\,\delta_\tau \approx 2\delta_a\delta_b \tag{82}$$

Bagley, Nelson, and Scigliano[148] suggested that their cohesion parameters δ_V and δ_R (based on internal pressure considerations as indicated in Section 7.5) are related to Hansen parameters:

TABLE 21
Four-Component Cohesion Parameters for Common Liquids

Liquid	δ/MPa$^{1/2}$						^iV
	δ_d	δ_p	δ_a	δ_b	δ_h	δ_t	cm^3 mol^{-1}
Pentane	14.5	0.0	0.0	0.0	0.0	14.5	116
Hexane	14.9	0.0	0.0	0.0	0.0	14.9	132
Heptane	15.3	0.0	0.0	0.0	0.0	15.3	148
Nonane	15.8	0.0	0.0	0.0	0.0	15.8	180
Decane	15.8	0.0	0.0	0.0	0.0	15.8	196
Cyclohexane	16.8	0.0	0.0	0.0	0.0	16.8	109
Benzene	18.4	1.0	1.4	1.4	2.0	18.6	89
Toluene	18.0	1.4	1.6	1.2	2.0	18.2	107
Chlorobenzene	19.0	4.3	2.0	1.0	2.0	19.6	102
1,2-Dichloroethane	19.0	7.4	4.1	2.0	4.1	20.8	79
1,1-Dichloroethane	16.6	8.2	0.4	0.2	0.4	18.5	85
Trichloromethane	17.8	3.1	6.1	2.7	5.7	18.9	81
Tetrachloromethane	17.8	0.0	0.1	1.8	0.6	17.8	97
1,2-Dibromoethane	19.0[a]	3.5	22.9	1.6	8.6	21.2	87
1,1-Dibromoethane	18.4[a]	5.1[a]	20.5	1.0	19.6[a]	20.2	93
Diethyl ether	14.5	2.9	1.0	12.9	5.1	15.7	105
Dipropyl ether	14.9	2.3	0.8	10.2	4.1	15.6	139
Dibutyl ether	15.6	1.6	0.6	8.0	3.1	15.9	170
Ethyl acetate	15.1	5.3	10.8	3.9	9.2	18.5	99
Propyl acetate	15.8	4.3	6.8	7.4	3.1	17.7	116
Butyl acetate	15.8	3.9	6.3	5.7	3.5	17.4	133
Hexyl acetate	16.0	3.1	5.9	3.9	4.5	17.3	165
Carbon disulfide	20.5	0.0	0.4	0.4	0.6	20.5	60
Acetone	15.5	10.4	4.9	4.9	7.0	20.0	74
1,4-Dioxane	19.0	1.8	2.1	13.3	7.4	20.5	86
Aniline	19.4	5.1	3.9	13.3	10.2	22.5	92
Nitrobenzene	20.1	8.6	4.1	2.1	4.1	22.2	103
Acetophenone	19.6	8.6	2.3	3.1	3.7	21.7	117
Benzyl alcohol	18.4	6.3	12.1	7.8	13.7	23.8	104
Cyclohexanol	17.4	4.1	14.9	6.1	13.5	22.4	106
Methanol	15.1	12.3	17.2	22.3	22.3	29.6	41
Ethanol	15.8	8.8	17.0	11.3	19.4	26.5	59
1-Propanol	16.0	6.8	15.3	9.8	17.4	24.5	75
2-Propanol	15.8	6.1	14.5	9.2	16.4	23.5	77
1-Butanol	16.0	5.7	13.1	9.4	15.8	23.1	92
Isobutanol	15.1	5.7	12.3	10.4	16.0	22.8	92
2-Butanol	15.8	5.7	13.5	7.8	14.5	22.2	93
tert-Butanol	14.9	5.1	19.6	4.9	13.9	21.0	94
1-Pentanol	16.0	4.5	11.1	8.8	13.9	21.7	109
1-Hexanol	16.4	4.3	11.1	7.2	12.9	21.3	125
1-Heptanol	16.6	4.1	10.8	7.0	12.3	21.0	142
1-Octanol	17.0	3.3	10.6	6.6	11.9	20.9	158
Ethylene glycol	17.0	11.1	36.6	9.0	25.8	32.7	56
1,2-Propanediol	16.8	9.4	28.8	9.4	23.3	30.2	74
1,3-Propanediol	16.6	10.8	22.3	15.1	26.0	32.7	73
Glycerol	17.4	12.1	40.1	10.4	29.3	36.1	73
1,4-Butanediol	16.8	16.6	37.2	7.6	23.7	33.5	89
Acetic acid	14.5	8.0	14.3	6.3	13.5	21.4	58
Propionic acid	14.7	7.8	12.3	6.1	12.3	20.7	75
Butyric acid	14.9	4.1	13.1	4.3	10.6	18.8	92
Dimethylsulfoxide	18.4	16.4	4.5	11.7	10.2	26.7	71
Pyridine	19.0	8.8	2.9	6.6	6.1	21.8	81
Formamide	17.2	26.2	11.7	15.6	19.0	36.7	40
N-Methylformamide	17.2	20.7	9.8	8.0	12.5	29.6	59

TABLE 21 (continued)
Four-Component Cohesion Parameters for Common Liquids

Liquid	$\delta/\text{MPa}^{1/2}$ δ_d	δ_p	δ_a	δ_b	δ_h	δ_t	${}^iV$ $\text{cm}^3\ \text{mol}^{-1}$
N,N-Dimethylformamide	17.4	13.7	7.0	9.0	11.3	24.8	77
N,N-Diethylformamide	16.8	11.5	5.5	7.0	8.8	22.1	112
N,N-Dimethylacetamide	16.8	11.5	5.9	8.8	10.2	22.7	93
N,N-Diethylacetamide	16.8	8.4	4.3	6.6	7.6	20.2	127
Water	15.6	16.0	13.7	65.5	42.3	47.9	18

[a] Recalculated by Beerbower et al.

Adapted from Beerbower, A., Wu, P. L., and Martin, A., *J. Pharm. Sci.*, 73, 179, 1984.

$$\delta_V^2 \approx \delta_d^2 + \delta_p^2 \approx \delta_d^2 + \delta_o^2 + 2\delta_i\delta_d \tag{83}$$

and

$$\delta_R^2 \approx \delta_h^2 \approx 2\delta_a\delta_b \tag{84}$$

As shown above (Section 5.6), Munafo, Buchmann, Hô Nam-Tran and Kesselring[37] used $\delta_h^2 = 2\ \delta_a\ \delta_b$ for δ_h which they evaluated instead of δ_a and δ_b, together with δ_d, δ_i, δ_o, and $\delta_p = (\delta_o^2 + 2\ \delta_i\ \delta_d)^{1/2}$.

Obviously there are many such ways of subdividing the Hildebrand or total cohesion parameter to provide component or composite parameters that are simpler to use than the full set of five parameters, but less general. The choice finally made in each case depends on the particular application.

It is important to remember that cohesion parameter values are derived from many different types of information, both theoretical and experimental, and they will not all be in good agreement with each other. The aim should be to select a cohesion parameter method that is appropriate for the particular task, and to determine the values for all the necessary materials. The greater the extent of polar and specific interactions that occur in the systems investigated, the greater the divergence between different methods is likely to be. Other solubility scales are outlined in Chapter 8.

5.13 MODIFIED SEPARATION OF COHESIVE ENERGY DENSITY

Thomas and Eckert[149] proposed a *mo*dified *s*eparation of *c*ohesive *e*nergy *d*ensity (MOSCED) model to predict limiting activity coefficients (Section 7.8) using only pure component parameters. They achieved an average error of only 9% for over 3000 limiting activity coefficients in aprotic and protic (but non-aqueous) systems over wide temperature ranges, with very few errors greater than 30%.

They first considered an equation of the form

$$\begin{aligned} {}^{ij}A = ({}^i\delta_\chi - {}^j\delta_\chi)^2 + ({}^i\delta_\tau - {}^j\delta_\tau)^2 + \\ ({}^i\delta_\psi - {}^j\delta_\psi)({}^i\delta_\tau - {}^j\delta_\tau) + ({}^i\delta_\alpha - {}^j\delta_\alpha)({}^i\delta_\beta - {}^j\delta_\beta) \end{aligned} \tag{85}$$

TABLE 22
Softness of Lewis-Base Liquids Determined From Hansen Parameters

Liquid	Classification	S
Saturated Hydrocarbons		
Pentane	Soft	20.1
Hexane	Soft	20.1
Heptane	Soft	20.1
Decane	Soft	20.1
Cyclohexane	Soft	20.1
Aromatic Hydrocarbons		
Ethylbenzene	Soft	19.8
o-Xylene, 1,2-dimethylbenzene	Soft	19.4
Toluene, methylbenzene	Soft	19.6
Benzene	Soft	19.7
Naphthalene	Soft	19.3
Sulfides		
Diethyl sulfide	Soft	19.0
Dimethyl sulfide	Soft	18.7
Carbon disulfide	Soft	19.1
Anilines		
Aniline	Borderline	18.2
o-Toluidine, 2-methylaniline	Borderline	18.4
Pyridines		
2,4,6-Trimethylpyridine	Borderline	18.4
2,6-Dimethylpyridine	Borderline	18.3
4-Methylpyridine, γ-picoline	Borderline	18.2
Pyridine	Borderline	18.2
Ethers		
Diethyl ether	Hard	15.3
Dipropyl ether	Hard	17.2
Anisole, methyl phenyl ether	Borderline	18.3
Ketones		
Acetone, 2-propanone	Hard	14.6
3-Pentanone, diethyl ketone	Hard	16.2
2-Butanone, methyl ethyl ketone	Hard	15.4
Cyclohexanone	Hard	17.9
Acetophenone, phenyl methyl ketone	Borderline	18.6
Esters		
Ethyl acetate	Hard	14.7
Propyl acetate	Hard	16.9
Butyl acetate	Hard	17.1
Hexyl acetate	Hard	17.6

TABLE 22 (continued)
Softness of Lewis-Base Liquids Determined From Hansen Parameters

Liquid	Classification	S
Aliphatic Amines		
Butylamine	Hard	17.0
Propylamine	Hard	17.4
Nitriles		
Acetonitrile	Hard	10.8
Propionitrile	Hard	12.8
Butyronitrile	Hard	13.5
Benzonitrile	Hard	17.8
Nitro Compounds		
Nitromethane	Hard	12.1
Nitroethane	Hard	14.1
2-Nitropropane	Hard	15.1
Nitrobenzene	Borderline	18.7
Water		
Water	Hard	0.0
Alcohols		
Methanol	Hard	6.2
Isobutanol, 2-methyl-1-propanol	Hard	10.9
tert-Butanol, 2-methyl-2-propanol	Hard	11.1
Ethanol	Hard	11.4
2-Propanol	Hard	13.0
2-Butanol	Hard	13.8
1-Propanol	Hard	13.1
1-Butanol	Hard	13.8
1-Hexanol	Hard	15.6
1-Octanol	Hard	16.8
Cyclohexanol	Hard	16.4
Benzyl alcohol	Hard	17.1
Amides		
N-Diethylacetamide	Hard	16.9
N,N-Dimethylacetamide	Hard	15.7
N,N-Diethylformamide	Hard	15.9
N-Methylformamide	Hard	13.9
Formamide	Hard	11.8
N,N-Dimethylformamide	Hard	15.8
Sulfoxide		
Dimethylsulfoxide	Hard	11.8

Adapted from Beerbower, A. and Jensen, W. B., *Inorg. Chim. Acta,* 75, 193, 1983.

where δ_λ is a measure of polarizability; δ_τ is a measure of polarity; δ_ψ reflects the induction interactions; δ_α is the acidity parameter; and δ_β is the basicity parameter. The Hildebrand parameter or total cohesion parameter of component i is then

$$^i\delta^2 = {^i\delta_\lambda}^2 + {^i\delta_\tau}^2 + {^i\delta_\psi}{^i\delta_\tau} + {^i\delta_\alpha}{^i\delta_\beta} \tag{86}$$

(This is similar to the formulation of Karger, Snyder, et al., but does not include the factors of 2 associated with the induction and acid-base cross-terms in Equations 21 and 23.) Equation 85 was simplified into a form analogous to that of Weimer and Prausnitz (Section 5.8),

$$\begin{aligned} {^{ij}A} = &({^i\delta_\lambda} - {^j\delta_\lambda})^2 + {^iq^2}{^jq^2}\,({^i\delta_\tau} - {^j\delta_\tau})^2 \\ &+ ({^i\delta_\alpha} - {^j\delta_\alpha})({^i\delta_\beta} - {^j\delta_\beta}) \end{aligned} \tag{87}$$

where the induction parameter was considered solely a function of degree of unsaturation, with $q = 1.0$ for saturated molecules, 0.9 for aromatics, and values between these limits for unsaturated aliphatics. Two factors, $^i\psi$ and $^i\xi$ were included to account for the "asymmetry" in volume fraction dependence due to polarity and hydrogen bonding, respectively. Neglecting the combinatorial term, the ratio of limiting activity coefficients of components i and j according to Equation 85 are

$$(\ln {^if_x^\infty})/(\ln {^jf_x^\infty}) = {^iV}/{^jV}$$

but this was found not to hold in solutions containing polar molecules. For example, pure ethanol is strongly hydrogen-bonded, but at infinite dilution in cyclohexane, ethanol molecules have no opportunity for hydrogen bonding. With this modification, the infinite dilution activity coefficient expression (which can be compared with Equation 21 in Chapter 4) was

$$\begin{aligned} \ln {^jf_x^\infty} = &[{^jV}/(RT)][({^i\delta_\lambda} - {^j\delta_\lambda})^2 + {^iq^2}\,{^jq^2}\,({^i\delta_\tau} - {^j\delta_\tau})^2/{^i\psi} \\ &+ ({^i\delta_\alpha} - {^j\delta_\alpha})({^i\delta_\beta} - {^j\delta_\beta})/\,{^i\xi}] + {^{ij}d} \end{aligned} \tag{88}$$

where ^{ij}d is the Flory-Huggins combinatorial molecular size effect term (Section 13.2)

$$^{ij}d = \ln({^jV} / {^iV}) + 1 - {^jV} / {^iV} \tag{89}$$

although in the correlation of Thomas and Eckert the "overcorrection" often caused by this form was largely avoided by using

$$^{ij}d = \ln\,({^jV} / {^iV})^{\,^ja} + 1 - ({^jV} / {^iV})^{\,^ja} \tag{90}$$

where

$$^ja = 0.953 - 0.00231\,({^j\delta_\tau^2}/\mathrm{MPa} + {^j\delta_\alpha}{^i\delta_\beta}\,/\mathrm{MPa}) \tag{91}$$

The parameters δ_τ, δ_α and δ_β were treated as being adjustable, with δ_λ calculated *a priori* from refractive index (Section 3.4). All the parameters necessary for calculation of limiting activity coefficients (and other thermodynamic properties) are collected in the publication by Thomas and Eckert[149] and its associated Supplementary Material. The cohesion parameters in Table 23 have been rounded to 2 or 3 significant figures, and refer to 20°C. The quantities V, δ_λ and q were treated as temperature-independent; the other parameters were temperature-dependent:

TABLE 23
Cohesion Parameters from the MOSCED Approach at 20°C

Liquid	V cm³ mol⁻¹	δ/MPa^1/2 δ_τ	δ_γ	δ_α	δ_β
Acetone	74.4	15.3	8.4	0.0	10.0
Acetonitrile	52.2	15.2	12.3	1.8	8.1
Acetophenone	116.9	18.0	6.2	1.8	6.4
Aniline	91.1	18.9	8.6	7.8	4.4
Anisole	108.1	17.7	5.7	0.6	3.3
Benzene	89.1	17.4	4.0	0.5	1.2
Benzonitrile	102.6	17.9	6.8	1.3	6.0
Benzyl acetate	142.4	17.8	6.7	0.5	5.1
Benzyl chloride	117.2	18.1	5.5	1.5	1.9
Bromoanisole	128.4	18.5	4.5	1.6	6.3
1-Bromobutane	107.4	17.1	3.1	0.5	0.4
Bromodichloromethane	82.7	18.3	4.1	5.9	0.2
Bromoethane	76.5	16.8	4.2	0.7	0.5
1-Bromohexadecane	305.6	17.6	1.4	0.2	0.1
Bromonaphthalene	149.6	20.0	4.0	1.2	0.8
1-Bromopentane	124.0	17.2	2.8	0.4	0.3
1-Bromopropane	90.9	17.0	3.6	0.6	0.4
Bromotrichloromethane	98.5	18.5	2.0	1.4	0.2
Butane	100.4	14.7	0.0	0.0	0.0
1-Butanol	91.5	16.2	2.1	9.5	9.5
2-Butanone, methyl ethyl ketone	89.6	15.8	6.7	0.0	8.3
Butyl acetate	132.0	16.1	4.5	0.0	5.2
Butylbenzene	156.0	17.2	2.0	0.1	0.8
tert-Butyl chloride	109.9	15.9	3.3	0.5	0.3
Butylcyclohexane	175.5	17.1	0.0	0.0	0.0
Carbon disulfide	72.0	20.0	0.6	0.6	0.3
Chlorobenzene	101.8	17.8	3.8	1.5	0.9
1-Chlorobutane	104.5	16.3	3.4	0.5	0.3
2-Chlorobutane	106.0	16.2	3.3	0.5	0.3
1-Chlorohexadecane	301.5	17.4	1.5	0.2	0.1
1-Chloropropane	88.1	16.0	3.9	0.6	0.4
Cycloheptane	121.3	17.2	0.0	0.0	0.0
Cyclohexane	108.1	16.8	0.0	0.0	0.0
Cyclohexanol	105.4	17.7	1.8	8.2	8.2
Cyclohexanone	103.6	17.3	6.2	0.0	9.9
Cyclohexene	101.4	17.3	0.6	0.0	0.5
Cyclooctane	134.4	17.5	0.0	0.0	0.0
Cyclopentane	94.1	16.4	0.0	0.0	0.0
Cyclopentanol	90.9	17.4	2.3	9.2	9.2
Decahydronaphthalene	154.0	18.0	0.0	0.0	0.0
Decane	194.9	16.5	0.0	0.0	0.0
Decanenitrile	186.9	16.9	4.6	0.3	6.5
1-Decanol	190.8	17.1	1.2	5.6	5.6
1-Decene	189.3	16.7	0.4	0.0	0.4
Dibromomethane	69.6	19.3	5.9	4.1	0.2
Dibutyl ketone	173.1	16.7	7.4	0.0	4.5
1,1-Dichloroethane	84.2	16.6	4.5	2.0	0.6
1,2-Dichloroethane	79.1	17.2	6.4	1.6	1.0
Dichloromethane	64.1	16.8	5.7	5.1	0.8
Diisobutyl ketone	176.6	16.5	3.3	0.0	4.9
1,4-Dioxane	84.2	16.5	6.8	0.0	8.5
2,2-Dimethylbutane	132.9	15.6	0.0	0.0	0.0
2,3-Dimethylbutane	130.3	15.7	0.0	0.0	0.0
N,N-Dimethylformamide	77.0	16.9	9.5	1.3	21.1
2,2-Dimethylpentane	148.7	15.9	0.0	0.0	0.0

TABLE 23 (continued)
Cohesion Parameters from the MOSCED Approach at 20°C

Liquid	V cm^3 mol^{-1}	δ/MPa$^{1/2}$ δ_τ	δ_γ	δ_α	δ_β
2,4-Dimethylpentane	149.0	15.8	0.0	0.0	0.0
Dodecane	227.5	16.7	0.0	0.0	0.0
Dotriacontane	555.0	17.4	0.0	0.0	0.0
Eicosane	358.3	17.2	0.0	0.0	0.0
Ethanol	58.4	15.4	2.8	12.7	12.7
Ethyl acetate	97.8	15.6	5.8	0.0	6.7
Ethylbenzene	122.5	17.4	2.6	0.1	1.0
Ethyl butyl ketone	139.5	16.4	4.4	0.0	5.5
Ethylcyclohexane	142.4	17.0	0.0	0.0	0.0
Ethyl formate	80.4	15.3	6.8	0.0	7.8
Ethyl octanoate	143.6	18.5	3.2	0.0	3.7
Ethyl propionate	114.5	15.9	5.1	0.0	5.9
Heneicosane	374.6	17.2	0.0	0.0	0.0
Heptane	146.6	16.0	0.0	0.0	0.0
Heptanenitrile	137.2	16.7	5.9	0.5	4.1
1-Heptene	140.9	16.2	0.5	0.0	0.5
Hexadecane	292.8	17.0	0.0	0.0	0.0
1-Hexadecanol	288.0	17.3	0.9	4.1	4.1
Hexadecene	287.3	17.1	0.2	0.0	0.2
Hexane	130.8	15.7	0.0	0.0	0.0
1-Hexanol	124.7	16.7	1.7	7.6	7.6
1-Hexene	125.0	16.0	0.0	0.0	0.4
Iodoethane	93.6	18.7	3.4	0.6	0.6
Iodomethane	79.1	19.0	4.0	0.7	0.6
2-Iodopropane	111.6	18.1	2.9	0.3	0.3
Isooctane	165.1	16.0	0.0	0.0	0.0
Isopentane	116.4	15.2	0.0	0.0	0.0
Isopentene	105.9	16.0	0.5	0.0	0.4
Isoprene	100.0	16.7	1.1	0.0	0.7
Methanol	40.5	14.6	5.2	15.2	15.2
Methyl acetate	79.3	15.4	6.8	0.0	7.8
3-Methyl-1-butene[a]	107.8	14.8	0.5	0.0	0.4
3-Methyl-1-butene[a]	111.8	15.4	0.5	0.0	0.4
Methyl butyrate	113.7	16.0	5.1	0.0	5.9
Methylcyclohexane	124.4	16.8	0.0	0.0	0.0
Methylcyclopentane	112.4	16.5	0.0	0.0	0.0
3-Methylhexane	145.8	16.0	0.0	0.0	0.0
Methyl isopropyl ketone	107.3	16.0	5.1	0.0	6.6
2-Methylpentane	131.9	15.6	0.0	0.0	0.0
3-Methylpentane	129.7	15.7	0.0	0.0	0.0
4-Methyl-2-pentanone	125.1	16.1	5.0	0.0	6.1
2-Methyl-1-pentene	126.7	15.9	0.5	0.0	0.4
2-Methyl-2-pentene	122.6	16.3	0.5	0.0	0.4
4-Methyl-1-pentene	123.8	16.1	0.5	0.0	0.4
Methyl propionate	96.3	15.7	5.8	0.0	6.7
Nitrobenzene	102.3	18.3	7.7	1.3	2.8
Nitroethane	72.0	16.1	9.9	0.6	4.5
Nitromethane	53.7	15.8	12.8	2.7	4.9
1-Nitropropane	89.0	16.3	8.5	0.5	3.9
2-Nitropropane	90.1	16.1	8.5	0.5	3.8
Nonane	178.7	16.4	0.0	0.0	0.0
Octacosane	489.4	17.4	0.0	0.0	0.0
Octadecane	327.1	17.1	0.0	0.0	0.0
Octane	162.6	16.2	0.0	0.0	0.0
1-Octanol	157.8	16.9	1.4	6.4	6.4

TABLE 23 (continued)
Cohesion Parameters from the MOSCED Approach at 20°C

Liquid	V cm^3 mol^{-1}	δ/MPa$^{1/2}$ δ_τ	δ_γ	δ_α	δ_β
1-Octene	157.0	16.4	0.5	0.0	0.4
Palmitonitrile	286.0	17.2	3.3	0.2	2.3
trans-1,3-Pentadiene	100.8	16.9	1.2	0.0	0.8
Pentane	115.3	15.3	0.0	0.0	0.0
1-Pentanol	108.1	16.5	1.8	8.4	8.4
2-Pentanone, methy propyl ketone	106.5	16.0	5.7	0.0	7.0
3-Pentanone, diethyl ketone	105.8	16.1	5.7	0.0	7.0
Pentatriacontane	604.4	17.5	0.0	0.0	0.0
1-Pentene	109.6	15.6	0.5	0.0	0.4
2-Pentene	107.0	15.9	0.5	0.0	0.4
Phenol	89.0	18.3	4.4	33.1	3.4
Propane	88.1	13.7	0.0	0.0	0.0
1-Propanol	74.8	15.9	2.4	10.8	10.8
Propionitrile	70.4	15.5	9.9	0.8	6.8
Propiophenone	132.9	17.9	5.6	1.4	6.0
Propyl acetate	115.0	15.9	5.1	0.0	5.9
Propyl formate	97.3	15.7	5.8	0.0	6.7
Pyridine	80.5	17.5	6.5	1.4	13.7
Quinoline	118.1	19.6	5.7	0.7	6.3
Squalene	478.5	18.4	0.6	0.0	1.4
Tetrachloromethane	96.5	17.6	1.8	1.2	0.3
Tetracosane	423.8	17.3	0.0	0.0	0.0
Tetrahydrofuran	81.1	16.4	4.7	0.0	9.4
Toluene	106.3	17.3	3.2	0.3	1.2
Triacontane	522.2	17.4	0.0	0.0	0.0
1,1,1-Trichloroethane	99.6	17.1	2.9	1.0	0.4
1,1,2-Trichloroethane	92.7	17.8	4.9	3.1	0.4
Trichloromethane	80.7	17.2	4.0	6.2	0.1
Triethylamine	139.0	15.4	1.1	0.0	10.2
1,3,5-Trimethylcyclohexane	163.8	16.8	0.0	0.0	0.0
Tripropylamine	189.6	15.6	0.9	0.0	6.8
o-Xylene	120.6	17.6	3.4	0.1	1.8
p-Xylene	123.3	17.3	2.6	0.1	1.5

[a] Two entries reported in original publication.

Adapted from Thomas, E. R. and Eckert, C. A., *Ind. Eng. Chem. Proces. Des. Dev.*, 23, 194, 1984.

$$\delta_\tau(T) = (293/T)^{0.4}\ \delta_\tau(293)$$

$$\delta_\alpha(T) = (293/T)^{0.8}\ \delta_\alpha(293)$$

$$\delta_\beta(T) = (293/T)^{0.8}\ \delta_\beta(293)$$

Park and Carr[150] subsequently proposed that the original MOSCED assumption $\delta_\alpha = \delta_\beta$ for the alcohols be abandoned in favor of a more realistic evaluation. The method has been applied to extractive distillation (Section 7.10).

5.14 LIQUID METALS

Although the Hildebrand parameter concept has been applied to liquid metal miscibilities,[151,159,160,164] it appears[151] that incorporation of electronegativity differences is necessary to predict behavior. Kumar,[152] on the basis that the properties (particularly the degree of order) of liquid metals were closer to those of their solids than to their gaseous states, proposed that the much smaller enthalpies of fusion be used in place of enthalpies of vaporization to evaluate cohesion parameters, "Kumar parameters" (Table 24). As indicated in Section 2.2, the cohesion parameters calculated from the vaporization enthalpies of *solid* metals are not *Hildebrand* parameters, which are liquid-state properties. Similar values for other elements are listed in Table 4, Chapter 2. Table 25 reports internal pressures and cohesion parameters for liquid alkali metals from Shrivastava and Pandey[163] (Section 7.5).

Information on ionic liquids (molten salts) is included in Section 12.3.

TABLE 24
Hildebrand and Kumar Cohesion Parameters for Liquid Metals

Group	Element	Melting point, K	Electronegativity[a]	δ $MPa^{1/2}$	δ_K $MPa^{1/2}$
I A	Li	459	0.95	111	15.2
	Na	371	0.90	68	10.4
	K	336	0.81	43	7.2
	Rb	312	0.78	39	6.5
	Cs	301	0.76	33	5.5
II A	Be	1553	1.42	264	48.6
	Mg	923	1.16	102	25.3
	Ca	1123	1.03	82	18.3
	Sr	1043	0.99	70	17.2
	Ba	977	0.92	68	14.0
III A	Sc	1473	1.27	164	29.9
	Y	1763	1.20	147	32.7
	La	1099	1.17	129	20.5
IVA	Ti	2093	1.62	192	42.0
	Zr	2023	1.48	192	38.3
	Hf	1973	1.48	—	39.3
	Th	2073	1.36	174	30.5
VA	V	2008	1.85	243	55.9
	Nb	2686	1.77	260	49.8
	Ta	3269	1.77	278	47.6
VI A	Cr	2163	2.15	221	53.8
	Mo	2898	2.05	262	54.4
	W	3683	2.05	297	59.5
	U	1403	1.30	205	33.0
VII A	Mn	1518	1.69	194	44.5
VIII	Fe	1812	2.21	239	46.5
	Co	1768	2.26	258	51.0
	Ni	1728	2.24	254	51.7
	Ru	2773	2.12	284	55.8
	Rh	2239	2.12	264	51.3
	Pd	1827	2.08	209	44.5
	Ir	2727	2.10	284	55.4
	Pt	2047	2.07	248	46.5
Rare earths	Ce	1070	1.21	139	21.0
	Gd	1585	1.20	129	27.7
I B	Cu	1356	2.00	219	42.9
	Ag	1234	1.90	168	34.1
	Au	1336	2.30	190	34.8
II B	Zn	693	1.50	119	28.4
	Cd	594	1.55	92	21.6
	Hg	234	1.80	63	12.4
III B	Al	933	1.48	176	10.4
	Ga	303	1.62	151	21.8
	In	429	1.48	123	14.4
	Tl	573	1.46	100	15.7
IV B	Si	1703	1.82	180	65.0
	Ge	1231	1.77	156	48.5
	Sn	505	1.61	133	21.1
	Pb	600	1.56	104	16.2
V B	As	1087	2.04	135	28.2
	Sb	904	2.10	121	32.8
	Bi	544	1.78	98	22.6
VI B	Se	493	2.35	72	18.2
	Te	723	2.08	63	29.2

[a] Data from Mott, B. W., *J. Mater. Sci.*, 3, 424, 1968.

Adapted from Kumar, R., *J. Mater. Sci.*, 7, 1409, 1972.

TABLE 25
Internal Pressures and Cohesion Parameters for Liquid Alkali Metals

Metal	t/°C	π/MPa	$\pi^{1/2}$/MPa$^{1/2}$
Sodium	100	473	21.8
	150	528	23.0
	200	573	23.9
	221	591	24.3
	245	614	24.8
Potassium	70	241	15.5
	113	262	16.2
	163	291	17.0
	191	294	17.2
	229	313	17.7
	263	314	17.7
Cesium	35	139	11.8
	76	154	12.4
	122	168	13.0
	169	180	13.4
	207	189	13.8
	247	196	14.0
Rubidium	43	182	13.5
	84	198	14.1
	120	214	14.6
	179	234	15.3
	208	244	15.6
	245	254	16.0

Adapted from Shrivastava, S. N. and Pandey, J. D., *Acoust. Lett.*, 3, 219, 1980.

REFERENCES

1. **Karger, B. L., Snyder, L. R., and Horvath, C.,** *An Introduction to Separation Science,* John Wiley & Sons, New York, 1973.
2. **Snyder, L. R.,** Solvent selection for separation processes, in *Separation and Purification,* 3rd ed., Perry, E. S. and Weissberger, A., Eds., John Wiley & Sons-Interscience, New York, 1978, 25.
3. **Karger, B. L., Snyder, L. R., and Eon, C.,** An expanded solubility parameter treatment for classification and use of chromatographic solvents and adsorbents. Parameters for dispersion, dipole and hydrogen bonding interactions, *J. Chromatogr.,* 125, 71, 1976.
4. **Karger, B. L., Snyder, L. R., and Eon, C.,** Expanded solubility parameter treatment for classification and use of chromatographic solvents and adsorbents, *Analyt. Chem.,* 50, 2126, 1978.
5. **Keller, R. A., Karger, B. L., and Snyder, L. R.,** Use of the solubility parameter in predicting chromatographic retention and eluotropic strength, *Gas Chromatography 1970* (Proc. Int. Symp., *8*, Dublin, 1970), Stock, R., Ed., The Institute of Petroleum, London, 1971, 125.
6. **Keller, R. A. and Snyder, L. R.,** Relation between the solubility parameter and the liquid-solid solvent strength parameter, *J. Chromatogr. Sci.,* 9, 346, 1971.
7. **Schoenmakers, P. J.,** *Optimization of Chromatographic Selectivity: A Guide to Method Development* (Journal of Chromatography Library, Vol. 35), Elsevier, Amsterdam, 1986.
8. **Schoenmakers, P. J., Billiet, H. A., and De Galan, L.,** Solubility parameter as a tool in understanding liquid chromatography, *Chromatographia,* 15, 205 (English) and 387 (German), 1982.
9. **Hansen, C. M.,** *Three Dimensional Solubility Parameter and Solvent Diffusion Coefficient. Importance in Surface Coating Formulation,* Doctoral Dissentation, Danish Technical Press, Copenhagen, 1967.
10. **Hansen, C. M.,** The three dimensional solubility parameter — key to paint component affinities. I. Solvents, plasticizers, polymers and resins, *J. Paint Technol.,* 39, 104, 1967.

11. **Hansen, C. M.,** The three dimensional solubility parameter — key to paint component affinities. II. Dyes, emulsifiers, mutual solubility and compatibility, and pigments, *J. Paint Technol.*, 39, 505, 1967.
12. **Hansen, C. M.,** On the application of the three dimensional solubility parameter concept to the prediction of mutual solubility and compatibility, *Skand. Tidskr. Färg Lack,* 13, 132, 1967.
13. **Hansen, C. M.,** The three dimensional solubility parameter and its use. I. Introduction and non-pigmented systems, *Skand. Tidskr. Färg Lack,* 14(2), 28, 1968.
14. **Hansen, C. M.,** The universality of the solubility parameter, *Ind. Eng. Chem. Prod. Res. Dev.*, 8, 2, 1969.
15. **Hansen, C. M.,** Characterization of surfaces by spreading liquids, *J. Paint Technol.*, 42, 660, 1970.
16. **Hansen, C. M.,** Solubility in the coatings industry, *Skand. Tidskr. Färg Lack,* 17, 69, 1971.
17. **Hansen, C. M.,** Surface dewetting and coatings performance, *J. Paint Technol.*, 44(570), 57, 1972.
18. **Hansen, C. M.,** Solvents for coatings, *Chemtech,* 2, 547, 1972.
19. **Hansen, C. M.,** Solvent selection by computer, in *Solvents Theory and Practice (Adv. Chem. Ser.* 124), Tess, R. W., Ed., American Chemical Society, Washington, D.C., 1973, chap. 4.
20. **Hansen, C. M.,** Solvents in water-borne coatings, *Ind. Eng. Chem. Prod. Res. Dev.*, 16, 266, 1977.
21. **Hansen, C. M.,** Some aspects of acid/base interactions, *Farbe Lack,* 83, 595, 1977.
22. **Hansen, C. M.,** Advances in the technology of solvents in coatings, *FATIPEC (Fed. Assoc. Tech. Ind. Peint. Vernis, Emaux Encres Impr. Eur. Cont.) Congr.*, 14, 97, 1978.
23. **Hansen, C. M.,** Some answered and unanswered questions about the solubility parameter, in *Macromolecular Solutions: Solvent-Polarity Relationships in Polymers* (Symp. Proc., Am. Chem. Soc., New York, 1981), Seymour, R. B. and Stahl, G. A., Eds., Pergamon, New York, 1982, 1; *Org. Coat. Plast. Chem.*, 45, 227, 1981.
24. **Hansen, C. M.,** Organic solvents in high solids and water-reducible coatings, *Prog. Org. Coat.*, 10, 331, 1982.
25. **Hansen, C. M. and Beerbower, A.,** Solubility parameters, in *Kirk-Othmer Encyclopedia of Chemical Technology,* 2nd ed., Standen, A., Ed., Interscience, New York, Suppl. Vol., 1971, 889.
26. **Hansen, C. M. and Pierce, P. E.,** Surface effects in coating processes, *Ind. Eng. Chem. Prod. Res. Dev.*, 13, 218, 1974.
27. **Skaarup, K.,** The three dimensional solubility parameter and its use. II. Pigmented systems, *Skand. Tidskr. Färg Lack,* 14(3), 45, 1968.
28. **Hansen, C. M. and Skaarup, K.,** The three dimensional solubility parameter — key to paint component affinities. III. Independent calculation of the parameter components, *J. Paint Technol.*, 39, 511, 1967.
29. **Hansen, C. M. and Skaarup, K.,** Aspects of the three-dimensional solubility parameter, *Dan. Kemi,* 48(6), 81, 1967.
30. **Hansen, C. M. and Wallström, E.,** On the use of cohesion parameters to characterize surfaces, *J. Adhesion,* 15, 275, 1983.
31. **Beerbower, A., Wu, P. L., and Martin, A.,** Expanded solubility parameter approach. I. Naphthalene and benzoic acid in individual solvents, *J. Pharm. Sci.*, 73, 179, 1984.
32. **Martin, A., Wu, P. L., and Beerbower, A.,** Expanded solubility parameter approach. II. *p*-Hydroxybenzoic acid and *p*-hydroxybenzoate in individual solvents, *J. Pharm. Sci.*, 73, 188, 1984.
33. **Hildebrand, J. H.,** Absence of outer non-bonding electrons in methyl groups affects solubility parameters, *Ind. Eng. Chem. Fundam.*, 17, 365, 1978.
34. **Hildebrand, J. H.,** An improvement in the theory of regular solutions, *Proc. Natl. Acad. Sci., U.S.A.*, 76, 6040, 1979.
35. **Beerbower, A. and Jensen, W. B.,** The HSAB principle and extended solubility theory, *Inorg. Chim. Acta,* 75, 193, 1983.
36. **van Arkel, A. E.,** Mutual solubility of liquids, *Trans. Faraday Soc.*, 42B, 81, 1946.
37. **Munafo, A., Buchmann, M., Hô Nam-Tran, and Kesselring, U. W.,** Determination of the total and partial cohesion parameters of lipophilic liquids by gas-liquid chromatography and from molecular properties, *J. Pharm. Sci.*, 77, 169, 1988.
38. **Arnett, E. M.,** Quantitative comparison of weak organic bases, *Progr. Phys. Org. Chem.*, 1, 223, 1963.
39. **Bent, H. A.,** Structural chemistry of donor-acceptor interactions, *Chem. Rev.*, 68, 587, 1968.
40. **Bent, H. A.,** The nature of intermolecular donor-acceptor bonds, in *Solutions and Solubilities,* Part II, Dack, M. R. J., Ed., John Wiley & Sons-Interscience, New York, 1976, chap. 10.
41. **Coetzee, J. F. and Ritchie, C. D., Eds.,** *Solute-Solvent Interactions,* Marcel Dekker, New York, 1969.
42. **Drago, R. S.,** Quantitative evaluation and prediction of donor-acceptor interactions, *Struct. Bonding (Berlin),* 15, 73, 1973.
43. **Drago, R. S. and Matwiyoff, N. A.,** *Acids and Bases,* Heath, Lexington, Mass., 1968, 66.
44. **Jensen, W. B.,** The Lewis acid-base definitions: a status report, *Chem. Rev.*, 78, 1, 1978.
45. **Jensen, W. B.,** Lewis acid-base interactions and adhesion theory, *Rubber Chem. Technol.*, 55, 881, 1982.
46. **Jensen, W. B.,** Acids, bases, and adhesion: a synergy, *Chemtech,* 12, 755, 1982.

47. **Lim, Y. Y. and Drago, R. S.,** Lewis acidity of antimony pentachloride, *Inorg. Chem.*, 11, 202, 1972.
48. **Reichardt, C.,** *Solvent Effects in Organic Chemistry,* (Monographs in Modern Chemistry, Vol. 3), Verlag Chemie, Weinheim, 1979.
49. **Wrona, P. K.,** On the correlation between empirical Lewis acid-base solvent parameters and the thermodynamic parameters of ion solvation, *J. Electroanal. Chem.*, 108, 153, 1980.
50. **Burrell, H.,** The challenge of the solubility parameter concept, *J. Paint Technol.*, 40, 197, 1968.
51. **Gardon, J. L.,** The influence of polarity upon the solubility parameter concept, *J. Paint Technol.*, 38(492), 43, 1966.
52. **Makitra, R. G. and Pirig, Ya. N.,** Interdependence between the basicity of donor solvents and heats of their mixing with chloroform, *Ukr. Khim. Zh. (Russ. Ed.),* 46, 83, 1980.
53. **Nelson, R. C., Hemwall, R. W., and Edwards, G. D.,** Treatment of hydrogen bonding in predicting miscibility, *J. Paint Technol.*, 42, 636, 1970.
54. **Sörensen, P.,** New aspects on the application of the solubility parameters, *Skand. Tidskr. Färg Lack,* 20(2), 9, 1974.
55. **Sörensen, P.,** Application of the acid/base concept describing the interaction between pigments, binders and solvents, *J. Paint Technol.*, 47(602), 31, 1975.
56. **Benesi, H. A. and Hildebrand, J. H.,** A spectrophotometric investigation of the interaction of iodine with aromatic hydrocarbons, *J. Am. Chem. Soc.*, 71, 2703, 1949.
57. **Alley, S. K., Jr. and Scott, R. L.,** NMR studies of hydrogen bonding in hydrofluorocarbon solutions, *J. Phys. Chem.*, 67, 1182, 1963.
58. **Pimentel, G. C. and McClellan, A.,** *The Hydrogen Bond,* W. H. Freeman, San Francisco, 1960.
59. **Burrell, H.,** Trends in solvent science and technology, in *Solvents Theory and Practice (Adv. Chem. Ser.* 124), Tess, R. W., Ed., American Chemical Society, Washington, D.C., 1973, chap. 1.
60. **Small, P. A.,** Some factors affecting the solubility of polymers, *J. Appl. Chem.*, 3, 71, 1953.
61. **Rider, P. E.,** A two-parameter model for estimating hydrogen bond enthalpies of reaction, *J. Appl. Polym. Sci.*, 25, 2975, 1980; Hydrogen bonding potential, HBP: A model for predicting resin solubilities in organic solvent, *Polym. Eng. Sci.*, 23, 810, 1983.
62. **Rider, P. E.,** Theories of solvency and solution, in *Applied Polymer Science,* 2nd ed., Vol. 285, Tess, R. W. and Poehlein, G. W., Eds., 1985, chap. 27, 643.
63. **Chastrette, M., Rajzmann, M., Chanou, M., and Purcell, K. F.,** Approach to a general classification of solvents using a multivariate statistical treatment of quantitative solvent parameters, *J. Am. Chem. Soc.*, 107, 1, 1985.
64. **Renon, H. and Prausnitz, J. M.,** On the thermodynamics of alcohol-hydrocarbon solutions, *Chem. Eng. Sci.*, 22, 299, 1967.
65. **Wiehe, I. A. and Bagley, E. B.,** Estimation of dispersion and hydrogen bonding energies in liquids, *Am. Inst. Chem. Eng. J.*, 13, 836, 1967.
66. **Bagley, E. B. and Chen, S.-A.,** Hydrogen bonding effects in non-electrolyte solutions, *J. Paint Technol.*, 41, 494, 1969.
67. **Harris, H. G. and Prausnitz, J. M.,** Thermodynamics of solutions with physical and chemical interactions, *Ind. Eng. Chem. Fundam.*, 8, 180, 1969.
68. **Kirchnerova, J. and Cavé, G. C. B.,** The solubility of water in low-dielectric solvents, *Can. J. Chem.*, 54, 3909, 1978.
69. **Hoy, K. L.,** New values of the solubility parameters from vapor pressure data, *J. Paint Technol.*, 42(541), 76 and 115, 1970.
70. **Kojima, I., Yoshida, M., and Tanaka, M.,** Distribution of carboxylic acids between organic solvents and aqueous perchloric acid solution, *J. Inorg. Nucl. Chem.*, 32, 987, 1970.
71. **Takeda, K., Yamashita, H., and Akiyama, M.,** Dimerization of some carboxylic acids in organic phases, *Solvent Extr. Ion Exch.*, 5, 29, 1987.
72. **Iscan, M.,** Temperature effect on the partition and dimerization constant of benzoic acid, *Thermochim. Acta,* 94, 305, 1985.
73. **Martin, A., Wu, P. L., Liron, Z., and Cohen, S.,** Dependence of solute solubility parameters on solvent polarity, *J. Pharm. Sci.*, 74, 638, 1985.
74. **Kamlet, M. J., Doherty, R., Taft, R. W., and Abraham, M. H.,** Linear solvation energy relationships. 26. Some measures of relative self-association of alcohols and water, *J. Am. Chem. Soc.*, 105, 6741, 1983.
75. **Nisbet, K. D.,** Structure-solubility parameter relationships in alcohols, in *Structure-Solubility Relationships in Polymers* (Proc. Symp., 1976), Harris, F. W. and Seymour, R. B., Eds., Academic Press, New York, 1977, 33.
76. **Melder, L. I. and Ebber, A. V.,** Determination of the orientation and induction components of the solubility parameter of polar compounds without hydrogen bonds, *Zh. Prikl. Khim.*, 55, 1979, 1982; *J. Appl. Chem. U.S.S.R.*, 55, 1823, 1982.

77. **Melder, L. and Ebber, A.,** Use of the theory of regular solutions for the description of solutions with polar components without a hydrogen bond. 1. Selection of the method for the expression of polarizability of solution components in the calculation of induction interaction energy, *Tr. Tallin. Politekhn. In-ta,* No. 509, 3, 1981.
78. **Ebber, A.,** Use of the theory of regular solutions for the description of solutions with polar components without a hydrogen bond. 2. Determination of orientation and induction components of the solubility parameter, *Tr. Tallin. Politekhn. In-ta,* No. 509, 15, 1981.
79. **Melder, L. and Ebber, A.,** Splitting of the overall solubility parameter of polar solvents without hydrogen bonding into dispersion, orientation and induction components, *Eesti NSV Tead. Akad. Toim. Keem.,* 30(1), 39, 1981.
80. **Ignat, A. V. and Melder, L. I.,** Determination of components of the solubility parameter of alcohols, *Zh. Prikl. Khim. (Leningrad),* 60, 1136, 1987; *J. Appl. Chem. U.S.S.R.,* 60, 1070, 1987.
81. **Ignat, A. and Melder, L.,** Splitting the solubility parameter of self-associating compounds into components, *Eesti NSV Tead. Akad. Toim. Keem.,* 36(1), 45, 1987.
82. **Fedors, R. F.,** A method for estimating both the solubility parameters and molar volumes of liquids, *Jet Propul. Lab. Quart. Tech. Rev.,* 3(1), 45, 1973; *Polym. Eng. Sci.,* 14, 147 and 472, 1974.
83. **Walker, E. E.,** Solvent action of organic substances on polyacrylonitrile, *J. Appl. Chem.,* 2, 470, 1952.
84. **Reed, T. M., III,** The theoretical energies of mixing for fluorocarbon-hydrocarbon mixtures, *J. Phys. Chem.,* 59, 425, 1955.
85. **Reed, T. M., III,** The polarizabilities of molecules in liquid mixtures, *J. Phys. Chem.,* 63, 1798, 1959.
86. **Sonnich Thomsen, E.,** *The Energy of Mixing of Non-Polar Liquids. Reflections upon the Hildebrand Theory,* Thesis, Royal Danish School of Pharmacy, Copenhagen, 1965.
87. **Sonnich Thomsen, E.,** The energy of mixing of non-polar liquids. Reflections upon the theories of Prigogine and Hildebrand, *Dansk Kemi,* 47(3), 35, 1966.
88. **Sonnich Thomsen, E.,** Interactions between unlike nonpolar molecules: correction of the geometric mean rule. 1. Basic derivations; 2. Dilute systems; 3. Energies of mixing, *Acta Chem. Scand.,* 25, 260 and 265, 1971; 26, 2100, 1972.
89. **Reid, R. C., Prausnitz, J. M., and Sherwood, T. K.,** *The Properties of Gases and Liquids,* McGraw-Hill, New York, 3rd ed., 1977.
90. **Preston, G. T. and Prausnitz, J. M.,** Thermodynamics of solid solubility in cryogenic solvents, *Ind. Eng. Chem. Process Des. Dev.,* 9, 264, 1970.
91. **Funk, E. W. and Prausnitz, J. M.,** Thermodynamic properties of liquid mixtures: aromatic — saturated hydrocarbon systems, *Ind. Eng. Chem.,* 62(9), 8, 1970.
92. **Bazua, E. R. and Prausnitz, J. M.,** Vapour-liquid equilibria for cryogenic mixtures, *Cryogenics,* 11, 115, 1971.
93. **Utracki, L. A.,** Interactions in poly(vinyl chloride) solutions, *Polym. J.,* 3, 551, 1972.
94. **Martin, A., Newburger, J., and Adjei, A.,** New solubility equation, *J. Pharm. Sci.,* 68(10), iv, 1979.
95. **Martin, A., Newburger, J., and Adjei, A.,** Extended Hildebrand solubility approach. II. Solubility of theophylline in polar binary solvents, *J. Pharm. Sci.,* 69, 487, 1980.
96. **Adjei, A., Newburger, J., and Martin, A.,** Extended Hildebrand approach. III. Solubility of caffeine in dioxane-water mixtures, *J. Pharm. Sci.,* 69, 659, 1980.
97. **Martin, A., Wu, P. L., Adjei, A., Mehdizadeh, M., James, K. C., and Metzler, C.,** Extended Hildebrand solubility approach: testosterone and testosterone propionate in binary solvents, *J. Pharm. Sci.,* 71, 1334, 1982.
98. **Martin, A., Paruta, A. N., and Adjei, A.,** Extended Hildebrand solubility approach: methylxanthines in mixed solvents, *J. Pharm. Sci.,* 70, 1115, 1981.
99. **Li, Y.-G., Teng, T., Lu, J.-F., Chen, G., and Li, J.-D.,** A study on Scatchard-Hildebrand solution theory for metal solvent extraction systems, *Fluid Phase Equil.,* 30, 297, 1986.
100. **James, K. C., Ng, C. T., and Noyce, P. R.,** Solubilities of testosterone propionate and related esters in organic solvents, *J. Pharm. Sci.,* 65, 656, 1976.
101. **Chueh, P. L. and Prausnitz, J. M.,** Vapor-liquid equilibrium at high pressures. Vapor-phase fugacity coefficients in nonpolar and quantum gas mixtures, *Ind. Eng. Chem. Fundam.,* 6, 492, 1967.
102. **Hildebrand, J. H., Prausnitz, J. M., and Scott, R. L.,** *Regular and Related Solutions,* Van Nostrand-Reinhold, Princeton, NJ, 1970.
103. **Knobler, C. M.,** Volumetric properties of gaseous mixtures, *Chem. Soc. Spec. Per. Rep. Chem. Thermodyn.,* 2, 199, 1978.
104. **Prausnitz, J. M. and Chueh, P. L.,** *Computer Calculations for High Pressure Vapor Equilibria,* Prentice-Hall, Englewood Cliffs, NJ, 1966.
105. **Sonnich Thomsen, E.,** Present state of the solubility parameter theory, *Mitt. Chem. Gesell. DDR,* 16, 150, 1969.
106. **Pauling, L.,** *The Nature of the Chemical Bond,* 3rd ed., Cornell University Press, Ithaca, NY, 1960.
107. **Stringfellow, G. B.,** The calculation of regular solution interaction parameters between elements from Groups III, IV and V of the Periodic Table, *Mater. Res. Bull.,* 6, 371 and 639, 1971.

108. **Myers, A. L. and Prausnitz, J. M.,** Thermodynamics of solid carbon dioxide solubility in liquid solvents at low temperatures, *Ind. Eng. Chem. Fundam.*, 4, 209, 1965.
109. **Andersen, R.,** Ph.D. Thesis, Dept. of Chemical Engineering, University of California, Berkeley, 1961.
110. **Blanks, R. F. and Prausnitz, J. M.,** Thermodynamics of polymer solubility in polar and nonpolar systems, *Ind. Eng. Chem. Fundam.*, 3, 1, 1964.
111. **Kumar, R. and Prausnitz, J. M.,** Solvents in chemical technology, in *Solutions and Solubilities,* Part 1, Dack, M. R. J., Ed., John Wiley & Sons-Interscience, New York, 1975, chap. 5.
112. **Prausnitz, J. M. and Anderson, R.,** Thermodynamics of solvent selectivity in extractive distillation of hydrocarbons, *Am. Inst. Chem. Eng. J.*, 7, 96, 1961.
113. **Weimer, R. F. and Prausnitz, J. M.,** Screen extraction solvents this way, *Hydrocarbon Proc. Petr. Ref.*, 44, 237, 1965.
114. **Prausnitz, J. M. and Shair, F. H.,** A thermodynamic correlation of gas solubilities, *Am. Inst. Chem. Eng. J.*, 7, 682, 1961.
115. **Helpinstill, J. G. and Van Winkle, M.,** Prediction of infinite dilution activity coefficients for polar-polar binary systems, *Ind. Eng. Chem. Proc. Des. Dev.*, 7, 213, 1968.
116. **Seidel, P. and Bittrich, H.-J.,** The polar contribution to the heat of vaporization, *Chem. Tech.*, 36, 382, 1984.
117. **Böttcher, C. J. F.,** *Theory of Electric Polarisation,* Elsevier, New York, 1952.
118. **Beerbower, A. and Dickey, J. R.,** Advanced methods for predicting elastomer/fluids interactions, *Am. Soc. Lubric. Eng. Trans.*, 12, 1, 1969.
119. **Gardon, J. L. and Teas, J. P.,** Solubility parameters, in *Treatise on Coatings,* Vol. 2, Part II, Myers, R. R. and Long, J. S., Eds., Marcel Dekker, New York, 1976, chap. 8, 413.
120. **Beerbower, A.,** Surface free energy: a new relationship to bulk energies, *J. Colloid Interface Sci.*, 35, 126, 1971.
121. **Izumi, Y. and Miyake, Y.,** Study of linear poly(*p*-chlorostyrene)-diluent systems. I. Solubilities, phase relationships, and thermodynamic interactions, *Polym. J.*, 3, 647, 1972.
122. **Knox, B. H.,** Bimodal character of polyester-solvent interactions. I. Evaluation of the solubility parameters of the aromatic and the aliphatic ester residues of poly(ethylene terephthalate), *J. Appl. Polym. Sci.*, 21, 225, 1977.
123. **Brench, A.,** unpublished presentation, *Soc. Chem. Ind. Symp.*, 1978 (cited in Ref. 124).
124. **Ashton, N. F., McDermott, C., and Brench, A.,** Chemistry of extraction of nonreacting solutes, in *Handbook of Solvent Extraction,* Lo, T. C., Baird, M. H. I., and Hanson, C., Eds., John Wiley & Sons-Interscience, New York, 1983, 3.
125. **Stephanopoulos, G. and Townsend, D. W.,** Synthesis in process development, *Chem. Eng. Res. Des.*, 64, 160, 1986.
126. **Hoy, K. L.,** *Tables of Solubility Parameters,* Union Carbide Corporation, Chemicals and Plastics Research and Development Dept., South Charleston, WV, 1969, 1975, and 1985.
127. **Grunwald, E. G.,** Thermodynamic properties of nonpolar solutes in water and the structure of hydrophobic hydration shells, *J. Am. Chem. Soc.*, 108, 5726, 1986.
128. **Tawn, A. R. H.,** Discussion following paper by K. Skaarup, *Skand. Tidskr. Färg Lack,* 15, 52, 1968.
129. **Teas, J. P.,** Graphic analysis of resin solubilities, *J. Paint Technol.*, 40(516), 19, 1968.
130. **Vial, J.,** Tridimensional representation of the solubility parameters: study of the solubilization of a vinyl chloride-vinyl acetate copolymer, *Compt. Rend. Acad. Sci. Ser. C,* 270, 683, 1970.
131. **Vial, J.,** *Predicting the Solubility of Polymers: Application to some Vinyl Chloride-Vinyl Acetate Copolymers,* Thesis, Faculty of Science, Paris, 1970.
132. **Teas, J. P.,** *Predicting Resin Solubilities,* Ashland Chemical Co., Tech. Bull. 1206, Ashland Chemical Co., Ashland, KY, 1971.
133. **Peiffer, D.,** The molecular factors affecting the solubility parameter, *J. Appl. Polym. Sci.*, 25, 369, 1980.
134. **Null, H. R.,** *Phase Equilibrium in Process Design,* John Wiley & Sons-Interscience, New York, 1970.
135. **Null, H. R. and Palmer, D. A.,** Predicting phase equilibria, *Chem. Eng. Prog.*, 65(9), 47, 1969.
136. **Wiehe, I. A.,** *Thermodynamics of Alcohol-Inert Solvent Solutions,* Thesis, Washington University, St. Louis, MO, 1967; *Diss. Abstr. B,* 28, 166, 1967.
137. **Wiehe, I. A.,** Thermodynamic properties of solutions of alcohols in inert solvents, *Ind. Eng. Chem. Fundam.*, 6, 209, 1967.
138. **Lille, Ü., Kundel, H., and Eisen, O.,** Solubility parameters of alkyl derivatives of phenol and resorcinol, *J. Chromatogr.*, 116, 1, 1976.
139. **Stelkol'shchikov, M. N., Krivtsova, L. M., and Ratner, M. I.,** Evaluation of solvency of hydrocarbon solvents, *Khim. Tekhnol. Topl. Masel,* No. 6, 33, 1987; *Chem. Technol., Fuels Oils,* No. 6, 292, 1987.
140. **Martin, A., Wu, P. L., Adjei, A., and Beerbower, A.,** Extended Hansen solubility approach — naphthalene in individual solvents, *J. Pharm. Sci.*, 70, 1260, 1981.
141. **Wu, P. L., Beerbower, A., and Martin, A.,** Extended Hansen solubility approach — calculating partial solubility parameters of solid solutes, *J. Pharm. Sci.*, 71, 1285, 1982.

142. **Mogensen, O. E.,** Information on the structure of liquids and solutions obtained by positron annihilation, *Electrochim. Acta,* 33, 1203, 1988.
143. **Kamlet, M. J., Abboud, J. L. M., and Taft, R. W.,** An examination of linear solvation energy relationships, in *Progress in Physical Organic Chemistry,* Vol. 13, Taft, R. W., Ed., John Wiley & Sons, New York, 1981, 485.
144. **Karim, K. A. and Bonner, D. C.,** Thermodynamic interpretation of solute-polymer interactions at infinite dilution, *J. Appl. Polym. Sci.,* 22, 1277, 1978; erratum, *J. Appl. Polym. Sci.,* 23, 2825, 1979.
145. **Karim, K. A. and Bonner, D. C.,** An improved concept in solubility theory applied to amorphous polymers. I. Poly(ethyl methacrylate), *Polym. Eng. Sci.,* 19, 1174, 1979.
146. **Wingefors, S.,** *Studies of the Solvent Extraction Chemistry of Dialkyldithiophosphates and Development of Models for Non-Ideal Behavior in Solvent Extraction,* Thesis, Dept. of Nuclear Chemistry, Chalmers University of Technology, Göteborg, Sweden, 1979.
147. **Wingefors, S., Liljenzin, J. O., and Saalman, E.,** Three-dimensional solubility parameters for the correlation and prediction of liquid-liquid partition constants of metal chelate complexes, *Proc. Int. Solvent Extr. Conf.,* 3, Paper 8-116, 1980.
148. **Bagley, E. B., Nelson, T. P., and Scigliano, J. M.,** Three-dimensional solubility parameters and their relationship to internal pressure measurements in polar and hydrogen bonding solvents, *J. Paint Technol.,* 43(555), 35, 1971.
149. **Thomas, E. R. and Eckert, C. A.,** Prediction of limiting activity coefficients by a modified separation of cohesive energy density model and UNIFAC, *Ind. Eng. Chem. Process Des. Dev.,* 23, 194, 1984.
150. **Park, J. H. and Carr, P. W.,** Predictive ability of the MOSCED and UNIFAC activity coefficient estimation methods, *Analyt. Chem.,* 59, 2596, 1987.
151. **Mott, B. W.,** Liquid immiscibility in metal systems, *Phil. Mag.,* [8], 2, 259, 1957; Immiscibility in liquid metals systems, *J. Mater. Sci.,* 3, 424, 1968.
152. **Kumar, R.,** Clusters in liquid metals, *Contemp. Phys.,* 10, 49, 1969; Immiscibility in binary alloys of group 1B metals (copper, silver and gold) — a semi-empirical approach, *J. Mater. Sci.,* 7, 1409, 1972.
153. **Huyskens, P. L., Haulait-Pirson, M. C., Siegel, G. G., and Kapuku, F.,** Gibbs energy of some binary systems containing hydrogen bonds, *J. Phys. Chem.,* 92, 6841, 1988; **Huyskens, P. L. and Siegel, G. G.,** Fundamental questions about entropy, *Bull. Soc. Chem. Belg.,* 97, 809, 815, 821, and 825, 1988.
154. **Huyskens, P. L.,** Prediction of the dissolving power of pure solvents for crystalline substances and drugs, 3rd Cycle Course in Pharmaceutical Sciences, Champéry, Switzerland, October 1989.
155. **Hansen, C. M. and Andersen, B. H.,** The affinities of organic solvents in biological systems, *Am. Ind. Hyg. Assoc.,* 49, 301, 1988.
156. **Dickerson, C. G. and Wiehe, I. A.,** Spherical encapsulated polymer particles by spray drying, *Pac. Chem. Eng. Congr. [Proc.],* 2(1), 243, 1977.
157. **Kramer, A. and Thodos, G.,** Adaptation of the Flory-Huggins theory for modeling supercritical solubilities of solids, *Ind. Eng. Chem. Res.,* 27, 1506, 1988.
158. **Wakahayashi, T., Oki, S., Omori, T., and Suzuki, N.,** Some applications of the regular solution theory to solvent extraction — I. Distribution of β-diketones, *J. Inorg. Nucl. Chem.,* 26, 2255, 1964.
159. **Bowersox, D. F.,** Thermodynamic properties of selected solutes in liquid metal solutions, *U.S. A. E. C.,* CONF-690801, 565-75, 1969.
160. **Yatsenko, S. P. and Dieva, E. N.,** Solubility of high-melting metals in molten indium, *Zh. Fiz. Khim.,* 47, 2948, 1973; *Russ. J. Phys. Chem.,* 47, 1658, 1973.
161. **Ignat, A. and Melder, L.,** Determination of components of the solubility parameter of polar solvents, *Kh. Prikl. Khim. (Leningrad),* 62, 419, 1989; *J. Appl. Chem. U.S.S.R.,* 62, 376, 1989.
162. **Huyskens, P. L.,** Differences in the structures of highly polar and hydrogen-bonded liquids, *J. Mol. Struct.,* 198, 123, 1989.
163. **Shrivastava, S. N. and Pandey, J. D.,** Excess intermolecular free length and molecular interaction for binary molten metal halide and cryogenic liquid mixtures, *Acoust. Lett.,* 3, 219, 1980.
164. **Martin, B. G.,** Calculation of component partial pressures in the Hg-Te and Cd-Te binary systems using regular associated solution theory, *Bull. Am. Phys. Soc.,* 25, 349, 1980.
165. **Mikos, A. G. and Peppas, N. A.,** Flory interaction parameter χ for hydrophilic copolymers with water, *Biomaterials,* 9, 419, 1988.

Chapter 6

CALCULATED COHESION PARAMETERS

Many properties of materials change in regular ways with increasing chain length in a homologous series, and some properties are conveniently linear. Thus, the logarithms of partition coefficients for some solutes between two liquids change linearly with chain length,[1] and a correlation exists[2] between methylene group increment and liquid Hildebrand parameter in octane-polar liquid systems (see Section 7.11). From the information presented in previous chapters, it is apparent that the miscibility behavior of materials depends to a large extent on two properties, the molar cohesive energy, $-U$, and the molar volume, V. Therefore, it is instructive to consider how $-U$, V, and the ratio and product of these two quantities vary through homologous series of organic compounds as chain lengths change, and how they can be estimated by means of contributions from groups of atoms.

Reference should also be made to Section 14.8 for additional sets of data, not only for polymer systems but also for materials such as long-chain lipophilic liquids with cohesion parameters that can be assessed in similar ways.[3]

6.1 GROUP MOLAR VAPORIZATION ENTHALPY

Dunkel[4] demonstrated 60 years ago that liquid molar vaporozation enthalpies at a given temperature could be estimated by the summation of atomic or group contributions, ${}^{z}\Delta H$:

$$\Delta H = \sum_{z} {}^{z}\Delta H \tag{1}$$

This approach (Table 1) was followed up by several investigators, including Hayes[5] and Fedors,[6] while Bondi[7,8] estimated the contributions to the enthalpy of vaporization extrapolated to absolute zero. Stein[9] also applied additive estimation methods to vaporization properties.

6.2 GROUP MOLAR COHESIVE ENERGY

It follows from the close relationship betwwen ΔH and $-U$ (Section 2.1) that the molar cohesive energy also increases in an approximately linear fashion with chain length in homologous series of compounds, and that group contribution methods may be used.[4,10-13] The lines obtained when $-U$ is plotted against chain length for molecules containing different functional groups are not quite parallel, and for accurate results in the use of group contributions the slopes of the lines have to be considered:[14]

$$-{}^{z}U = n\,{}^{z}\mu + {}^{z}\epsilon \tag{2}$$

where n is the number of carbon atoms, ${}^{z}\epsilon$ is the group molar cohesive energy constant, and ${}^{z}\mu$ is the chain length correction factor. Table 2 lists the values of these parameters for several groups. The molar cohesive energy is

$$-U = -\sum_{z} {}^{z}U \tag{3}$$

and the Hildebrand parameters may be calculated from

TABLE 1
Group Contributions to the Molar Vaporization Enthalpy[4,6]

Atom or group, z	$^z\Delta H$/kJ mol^{-1}
$-CH_3$	7.45
$=CH_2$	7.45
$-CH_2-$	4.14
$=CH-$	4.14
$>CH-$	−1.59
$-O-$	6.82
$-OH$	30.3
$=CO$	17.9
$-CHO-$	19.7
$-COOH$	37.5
$-COOCH_3$	23.4
$-COOC_2H_5$	26.1
$-NH_2$	14.8
$-Cl$	14.2
$-F$	8.6[a]
$-Br$	18.0[a]
$-I$	21.1[a]
$-NO_2$	30.1[a]
$-SH$	17.8[a]

[a] Provisional value.

TABLE 2
Group Molar Cohesive Energies, Molar Volumes, and Attraction Constants at 25°C

Group, z	$^z\epsilon$/kJ mol^{-1}	$^z\mu$/kJ mol^{-1}	^zV/cm^{3}mol^{-1}	^zF/J$^{1/2}$cm$^{3/2}$ mol^{-1}
$>CH-$	4.06	0.0	−0.5	121
$-CH_2-$	5.15	0.0	16.5	286
$-CH_3$	4.14	0.0	34.0	405
$-CH{=}CH_2$	8.37	0.0	44.0	610
$-CH(CH_3)_2$	12.3	0.0	67.5	931
$-NH_2$	10.8	−0.38	19.0	477
$-Cl$	11.7	−0.25	24.0	497
$-CHO$	18.2	−0.71	26.0	710
CH_3COO-	23.2	−0.42	50.5	1105
Cyclohexyl	29.5	−0.92	95.0	1606
Phenyl	31.2	−0.92	75.0	1493
$-OH$	32.8	−0.34	8.7	908
$-COOH$	32.8	+0.92	27.0	1352[a]

[a] More correctly, 1113 + 80n.

Adapted from Rheineck, A. E. and Lin, K. F., *J. Paint Technol.*, 40, 611, 1968.

$$\delta = (-U/V)^{1/2} = \left(-\sum_z {}^zU \Big/ \sum_z {}^zV\right)^{1/2} \tag{4}$$

using group molar volumes (Section 6.5). This method requires knowledge of only the chemical structure, without the molar volume (or density), but it is applicable only at 25°C, and as presented in Table 2 is limited to 1-substituted n-alkanes because the dependence of

${}^z\mu$ on chemical structure is not reported. The Hildebrand parameters of 1-substituted n-alkanes by this procedure are therefore given by

$$\delta = \left(\frac{{}^e\epsilon + {}^z\epsilon + (n - 1 + {}^z\mu)\ {}^u\epsilon}{{}^eV + {}^zV + (n - 1)\ {}^uV} \right)^{1/2} \tag{5}$$

for end-groups e = $-CH_3$ and repeat units u = $-CH_2-$. Also shown in Table 2 are the group molar attraction constants (Section 6.4),

$$\sum_z {}^zF = \left(-\sum_z {}^zU \sum_z {}^zV \right)^{1/2} = \delta V \tag{6}$$

Fedors[6] also has calculated molar vaporization energy contributions, ${}^z\Delta U$, and molar volume contributions, ${}^zV$, for structural components (Table 3). Although it is considered[15] that they give less accurate estimates of cohesive energy than other sets of data, this compilation is noteworthy because of the great number of groups considered, and because of the inclusion of metals for use in calculations on organometallic compounds. Fedors found that it was possible to estimate both U and V (and therefore also δ) for cyclic compounds from the properties of linear compounds having the same structure, by means of a cyclization increment. For example, the liquid N-phenylpiperazine,
is considered as

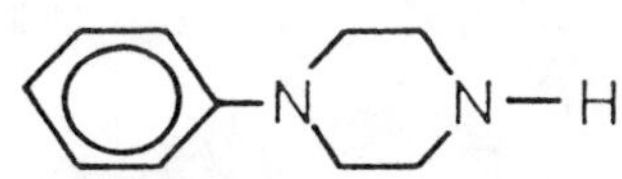

$-CH_2-NH-CH_2-CH_2-N-CH_2-$
|
$-CH{=}CH-CH{=}CH-C{=}CH-$

in which any functional group listed in Table 3 and present in the ring remains intact. Table 4 lists the group contributions, which result in an estimate for the Hildebrand parameter of 21.1 $MPa^{1/2}$. Kreibich and Batzer[16] used Fedors' cohesive energy increment values in an extensive study of polymer glass transition temperatures, while Sammar and Nagaka[103] applied the method to non-ionic surfactants.

It should be noted that when central atoms contain more than one substituent (other than the groups CH_3, CH_2, CH, or C) the contributions of those groups to the cohesive energy in Table 3 are decreased by approximately 20%, while the contributions to the molar volume are increased by about 7%.[6] Thus, in $CHCl_2$ the Cl contributions to $-U$ and V are -17% and $+8\%$, respectively, relative to the Cl contribution in CH_2Cl, while in $CHCl_3$ the figures are -34% and $+14\%$ relative to a single Cl substituent.

The increment of cohesion energy per $-CH_2-$ group would not be expected to be constant unless determined under conditions of constant reduced temperature (Section 7.3) rather than constant temperature. Meyer and co-workers[17-21] have done this in estimating the relative magnitudes of dispersion, induction, and orientation cohesive energies. A plot of cohesive energy against the number, m, of C–H bonds at the same reduced temperature for a homologous series of ketones was compared with a similar plot for alkanes (Figure 1). As the ketone alkyl chain length increases the orientation interactions decrease, the induction interactions are unaltered, and the dispersion interactions increase at the same rate as those for alkanes. At any m value, $-U_d$ is given by the alkane line, $-U_i$ is the vertical separation between the alkane line and the parallel line passing through the high-m ketone points, and

TABLE 3
Group Contributions to the Molar Vaporization Energy and Molar Volume at 25°C

Group, z	${}^{z}\Delta U$/kJ mol^{-1}	${}^{3}V$/cm^{3} mol^{-1}
$-CH_3$	4.71	33.5
$-CH_2-$	4.94	16.1
$>CH-$	3.43	−1.0
$>C<$	1.47	−19.2
$H_2C=$	4.31	28.5
$-CH=$	4.31	13.5
$>C=$	4.31	−5.5
$HC\equiv$	3.85	27.4
$-C\equiv$	7.07	6.5
Phenyl	31.9	71.4
Phenylene (*o*, *m*, *p*)	31.9	52.4
Phenyl (trisubstituted)	31.9	33.4
Phenyl (tetrasubstituted)	31.9	14.4
Phenyl (pentasubstituted)	31.9	−4.6
Phenyl (hexasubstituted)	31.9	−23.6
Ring closure, 5 or more atoms	1.05	16
Ring closure, 3 or 4 atoms	3.14	18
Conjugation in ring, for each double bond	1.67	−2.2
Halogen attached to C atom with double bond	−20% of halogen ${}^{z}\Delta U$	
–F	4.19	18.0
–F (disubstituted)	3.56	20.0
–F (trisubstituted)	2.30	22.0
$-CF_2-$ (for perfluoro compounds)	4.27	23.0
$-CF_3-$ (for perfluoro compounds)	4.27	57.5
–Cl	11.55	24.0
–Cl (disubstituted)	9.63	26.0
–Cl (trisubstituted)	7.53	27.3
–Br	15.49	30.0
–Br (disubstituted)	12.4	31.0
–Br (trisubstituted)	10.7	32.4
–I	19.05	31.5
–I (disubstituted)	16.7	33.5
–I (trisubstituted)	16.3	37.0
–CN	25.5	24.0
–OH	29.8	10.0
–OH (disubstituted or on adjacent C atoms)	21.9	13.0
–O–	3.35	3.8
–CHO (aldehyde)	21.4	22.3
–CO–	17.4	10.8
$-CO_2-$	18.0	18.0
$-CO_3-$ (carbonate)	17.6	22.0
$-C_2O_3-$ (anhydride)	30.6	30.0
HCOO– (formate)	18.0	32.5
$-CO_2CO_2-$ (oxalate)	26.8	37.3
$-HCO_3$	12.6	18.0
–COF	13.4	29.0
–COCl	17.6	38.1
–COBr	24.2	41.6
–COI	29.3	48.7
$-NH_2$	12.6	19.2
–NH–	8.4	4.5
$-N<$	4.2	−9.0
–N=	11.7	5.0

TABLE 3 (continued)
Group Contributions to the Molar Vaporization Energy and Molar Volume at 25°C

Group, z	$^{z}\Delta U$/kJ mol^{-1}	^{3}V/cm^{3} mol^{-1}
$-NHNH_2$	22.0	—
$-NNH_2$	16.7	16
$-NHNH$	16.7	16
$-N_2$(diazo)	8.4	23
$-N{=}N-$	4.2	—
$>C{=}N{-}N{=}C<$	20.1	0
$-N{=}C{=}N-$	11.47	—
$-NC$	18.8	23.1
$-NF_2$	7.66	33.1
$-NF-$	5.07	24.5
$-CONH_2$	41.9	17.5
$-CONH-$	33.5	9.5
$-CON<$	29.5	−7.7
$HCON<$	27.6	11.3
$HCONH-$	44.0	27.0
$-NHCOO-$	26.4	18.5
$-NHCONH-$	50.2	—
$-NHCON<$	41.9	—
$>NCON<$	20.9	−14.5
NH_2COO-	37.0	—
$-NCO$	28.5	35.0
$-ONH_2$	19.1	20.0
$>C{=}NOH$	25.1	11.3
$-CH{=}NOH$	25.1	24.0
$-NO_2$ (aliphatic)	29.3	24.0
$-NO_2$ (aromatic)	15.36	32.0
$-NO_3$	20.9	33.5
$-NO_2$ (nitrite)	11.7	33.5
$-NHNO_2$	39.8	28.7
$-NNO$	27.2	10
$-SH$	14.44	28.0
$-S-$	14.15	12
$-S_2-$	23.9	23.0
$-S_3-$	13.40	47.2
$>SO$	39.1	—
SO_3	18.8	27.6
SO_4	28.5	31.6
$-SO_2Cl$	37.1	43.5
$-SCN$	20.1	37.0
$-NCS$	25.1	40.0
P	9.42	−1.0
PO_3	14.2	22.7
PO_4	20.9	28.0
$PO_3(OH)$	31.8	32.2
Si	3.4	0
SiO_4	21.8	20.0
B	13.8	−2.0
BO_3	0.0	20.4
Al	13.8	−2.0
Ga	13.8	−2.0
In	13.8	−2.0
Tl	13.8	−2.0
Ge	8.1	−1.5
Sn	11.3	1.5

TABLE 3 (continued)
Group Contributions to the Molar Vaporization Energy and Molar Volume at 25°C

Group, z	${}^{z}\Delta U$/kJ mol^{-1}	${}^{z}V$/cm^3 mol^{-1}
Pb	17.2	2.5
As	13.0	7.0
Sb	16.3	8.9
Bi	21.4	9.5
Se	17.2	16.0
Te	20.1	17.4
Zn	14.5	2.5
Cd	17.8	6.5
Hg	22.8	7.5

Adapted from Fedors, R. F., *J. Polym. Sci. C,* 26, 189, 1969, and van Krevelen, D. W., *Properties of Polymers,* 2nd ed., Elsevier, Amsterdam, 1976.

TABLE 4
Calculation of Hildebrand Parameter of N-Phenylpiperazine from Group Contributions

Group, z	$-{}^{z}U$/kJ mol^{-1} (Table 3)	$-\sum_z {}^{z}U$/kJ mol^{-1}	${}^{z}V$/cm^3 mol^{-1} (Table 3)	$\sum_z {}^{z}V$/cm^3 mol^{-1}
4(–CH$_2$–)	4.94	19.8	16.1	64.4
1(–NH–)	8.4	8.4	4.5	4.5
1(–N<)	4.2	4.2	−9.0	−9.0
5(–C=)	4.31	21.6	13.5	67.5
1(>C=)	4.31	4.3	−5.5	−5.5
3(Conjugated double bonds)	1.67	5.0	−2.2	−6.6
2(6-Membered rings)	1.05	2.1	16	32
		65.4		147

$$\delta = (65400/147)^{1/2} = 21.1 \text{ MPa}^{1/2}$$

Adapted from Fedors, R. F., *Polym. Eng. Sci.,* 14, 147 and 472, 1974.

$-U_o$ is given by the separation of the ketone curve from the straight line passing through the high-m points. (Figure 1 was derived on the basis of a –CH$_2$– group molar volume of 19.08 cm^3 mol^{-1}, characteristic of hexane at 0°C. The temperature at which each member of the series displays 19.08 cm^3 per mole of –CH$_2$– groups was calculated from thermal expansion data, and $-U$ was then calculated for each member at its own appropriate temperature. This plot therefore assumes a reduced temperature scale which differs from that based on the critical temperature.)

Lawson[22] has determined a set of group contributions to molar vaporization energy and molar volume for fluoro compounds, based on their boiling points and Equation 19, Chapter 7 (Table 5). Further information on cohesive energies has been presented by Polak,[23] Fried and Schneier,[24] and Maffiolo, Vidal, and Renon.[12] Group partial molal entropies, enthalpies, and Gibbs free energies of solution also have been used.[25-27]

6.3 MOLAR ADDITIVE FUNCTIONS

There is an important group of molar additive properties having the form[28]

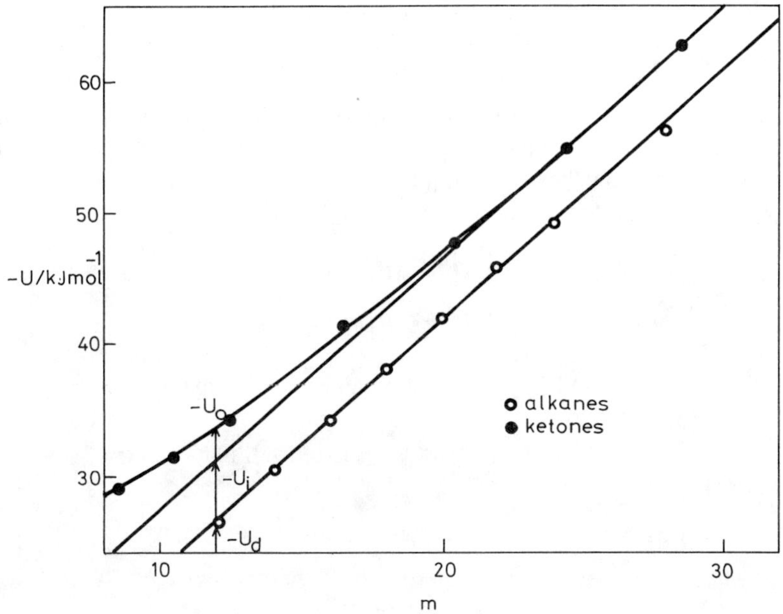

FIGURE 1. Molar cohesive energies as functions of C–H bond number for alkanes and ketones. (Adapted from Meyer, E. F. and Wagner, R. E., *J. Phys. Chem.*, 70, 3162, 1966.)

TABLE 5
Group Contributions to the Molar Vaporization Energy and Molar Volume at 25°C for Fluoro Compounds

Group, z	${}^z\Delta U$/kJ mol^{-1}	${}^zV$/cm^3 mol^{-1}
$-CF_3$	8.09	54.8
$-CF_2-$	3.28	23.1
–CFH–	1.77	18.6
–CF–	−1.66	−15.0
–C–(perfluoro)	−6.34	−38.3
–N–(perfluoro tertiary amine)	−3.82	−16.3
–O– (perfluoro ether)	0.03	19.0
5-Atom ring	8.46	37.7
6-Atom ring	9.51	39.9

Adapted from Lawson, D. D., *Appl. Energ.*, 6, 241, 1980.

$$Y = M\rho^{-1}f(x) = Vf(x) \tag{7}$$

where x is some physical quantity, including the molar parachor (Section 17.1), the lyoparachor (see below), molar attraction (Section 6.4), and molar refraction (Section 3.1).

In general, the group contributions ${}^z\Delta H$ are functions of temperature because ΔH itself is temperature dependent, but Bowden and Jones[29] showed that for a wide variety of both polar and nonpolar liquids it was possible to define constants related to ΔH and ΔU which are independent of temperature:

$$\frac{M}{{}^l\rho - {}^g\rho}\left(\frac{\Delta H}{M}\right)^{4/5} = [L] \tag{8}$$

$$\frac{M}{{}^{l}\rho - {}^{g}\rho}\left(\frac{\Delta U}{M}\right)^{3/4} = [\Lambda] \tag{9}$$

The exponents 4/5 and 3/4 were determined empirically. The constants [L] and [Λ], termed the normal lyoparachor and true lyoparachor, respectively, are independent of temperature. For temperatures at or below the normal liquid boiling point, the vapor density ${}^{g}\rho$ is negligible compared with the liquid density ${}^{l}\rho$, so Equations 8 and 9 can be written

$$\Delta H = M([\mathrm{L}]/V)^{5/4} \tag{10}$$

$$\Delta U = M([\Lambda]/V)^{4/3} \tag{11}$$

It was found that both [L] and [Λ] could be obtained by summation of group constants:

$$[\mathrm{L}] = \sum_{z} {}^{z}[\mathrm{L}] \qquad [\Lambda] = \sum_{z} {}^{z}[\Lambda] \tag{12}$$

so the Hildebrand parameter can be expressed

$$\delta = \rho^{5/6}\left(\sum_{z} {}^{z}[\Lambda]/M\right)^{2/3} \tag{13}$$

Also, the temperature dependence of δ can be estimated from the temperature dependence of the density because the other factor is independent of temperature. Additional values have been provided by Wright[30] and the information on ${}^{z}[\mathrm{L}]$ and ${}^{z}[\Lambda]$ has been collected by Fedors[6] (Table 6).

6.4 GROUP MOLAR ATTRACTION CONSTANTS

Scatchard[31] and Small[32] observed that there are parallel linear relationships among several homologous series when the square root of the product of the molar volume and molar cohesion energy, $(-UV)^{1/2}$, is plotted against the chain length (number of carbon atoms). In other words, $(-UV)^{1/2}$, or the equivalent quantity (δV), has additive properties. It has been described as the molar attraction, F (compare Equation 7),

$$F = \delta V \tag{14}$$

and can be considered to be made up of the sum of the group molar attraction constants ${}^{z}F$ of all the molecular groups z:

$$-U = \left(\sum_{z} {}^{z}F\right)^{2}/V \tag{15}$$

so

$$c = \left(\sum_{z} {}^{z}F/V\right)^{2} \tag{16}$$

and

$$\delta = \sum_{z} {}^{z}F/V = \sum_{z} {}^{z}F/\sum_{z} {}^{z}V \tag{17}$$

Molar cohesive energies and Hildebrand parameters can thus be estimated for any molecular compound once the individual group molar attraction constants are determined. Although

TABLE 6
Group Contributions to the Normal Lyoparachor, [L] and True Parachor, [Λ][6,29,30]

Group or atom, z	${}^{z}[L]/J^{4/5}g^{1/5}$ cm^3 mol^{-1}	${}^{z}[\Lambda]/J^{3/4}$ g$^{-3/4}$ cm^3 mol^{-1}
C	−3751	−2353
H	2655	1736
N	−354.2	−141.3
O (ether)	560.4	428.5
O (ketone)	6576	4403
O (ketone)[a]	6934	4601
O (carboxylate)	2840	1899
O (anhydride)[a]	4182	2796
O (carbonate)[a]	2074	1395
CN (aliphatic)[a]	7870	5294
CN (aromatic)[a]	6704	4543
Cl	2747	1881
Br	682.7	554
I	−54.1	118.5
Branch in a carbon chain	−417.1	−277.0
Benzene ring bonds	18190	11710
Cyclohexane ring bonds	3665	2322
Cyclohexene ring bonds[a]	8990	5754
O (carbonate)[b]	1729	1229
F[b]	1666	1112
Sn[b]	−4526	−2808
NO_2 (nitro)[b]	5280	3598
Double bond[b]	4620	2954
Double bond[a]	5249	3434
Triple bond[b]	11250	7342
Triple bond[a]	10510	6827
Pyridine ring bonds[b]	18540	12080
Furane ring bonds[b]	13390	8629

[a] From Reference 30 and the remainder from Reference 29.
[b] Provisional values.

the additivity of group molar attraction constants on an incremental basis is questionable on theoretical grounds, the method is adequate for many applications.[33,34] It provides acceptable data for hydrocarbons and many other compounds, but it is less satisfactory for situations where several large groups are packed around a central atom (as in tetrachloromethane, for example) and where steric, ring-closure, conjugation, polar, and hydrogen bonding effects are significant. (Small's values for group molar attraction constants at 25°C appear below in Table 11.)

Hoy[35,36] redetermined Small's constants on the basis of new values of Hildebrand parameters calculated from vapor pressure data for a large number of compounds. The Hildebrand parameters may be calculated from

$$\delta/\text{MPa}^{1/2} = \left(\sum_{z} {}^{z}F + 276.3\right)/V \tag{18}$$

In Table 7 are collected values of ${}^{z}F$, together with the corresponding polar constants that are used in the same way to evaluate the polar cohesion parameter, δ_p (Section 5.9 and Section 6.8) and also the group molar volumes at the glass transition temperature.

Numerical values of Hoy's cohesion parameters are presented in Table 18, Chapter 5,

TABLE 7
Hoy Group Molar Attraction Constants at 25°C

Group, z	Bond type	${}^zF/J^{1/2}\ cm^{3/2}\ mol^{-1}$	${}^zF_P/J^{1/2}\ cm^{3/2}\ mol^{-1}$	${}^zV'\ cm^3\ mol^{-1}$
Base value		276.3	—	—
$-CH_3$	Saturated	303.4	0.0	21.548
$-CH_2-$	Saturated	269.0	0.0	15.553
$>CH-$	Saturated	175.9	0.0	9.557
$>C<$	Saturated	65.5	0.0	3.562
$CH_2=$	Alkene	258.8	66.9	19.173
$-CH=$	Alkene	248.5	59.5	13.178
$>C=$	Alkene	172.8	63.0	7.183
$-CH=$	Aromatic	239.9	62.2	13.417
$>C=$	Aromatic	200.7	64.8	7.422
$-O-$	Ether	235.2	216.0	6.462
$-O-$	Acetal	236.3	10.2	6.462
$-O-$	Epoxide	360.4	155.9	6.462
$-COO-$	Ester	668.1	524.1	23.728
$>C=O$	Ketone	538.0	525.7	17.265
$-CHO$	Aldehyde	598.6	531.6	23.261
$-(CO)_2O$	Anhydride	1160.4	1159.8	40.993
$-COOH$	Acid	564.8	415.6	26.102
$-OH\rightarrow$	H-bond OH	485.8	485.8	10.647
$-OH$	Primary (not H-bonded)	673.8	673.8	12.457
$-OH$	Secondary	591.6	591.6	12.457
$-OH$	Tertiary	798.6	798.6	12.457
$-OH$	Phenolic	349.8	349.8	12.457
$-NH_2$	Amino, primary	463.5	463.5	17.012
$-NH-$	Amino, secondary	368.2	368.2	11.017
$>N-$	Amino, tertiary	125.0	125.0	12.569
$-C{=}N$	Nitrile	725.3	724.5	23.066
$-NCO$	Isocyanate	733.7	8.2	25.907
$HCON<$	Formamide	1017.0	724.1	35.830
$-CONH-$	Amide	1134.6	893.9	28.302
$-CONH_2$	Amide	1206.6	989.2	34.297
OCONH	Urethane	1261.9	892.2	34.784
$-S-$	Thioether	428.3	428.3	18.044
Cl	Primary	419.5	306.8	19.504
Cl	Secondary	426.1	315.0	19.504
Cl_2	Twinned	701.0	562.5	39.008
Cl	Aromatic	329.3	81.4	19.504
Br	Primary	527.5	122.7	25.305
Br	Aromatic	420.6	100.2	25.305
F	Primary	84.5	73.2	11.200
Conjugation		47.6	−19.8	—
Cis		−14.6	−14.5	—
Trans		−27.6	−27.6	—
4-Membered ring		159.1	200.5	—
5-Membered ring		42.9	84.9	—
6-Membered ring		−48.0	61.0	—
7-Membered ring		92.3	0.0	—
Bicycloheptane ring		46.2	—	—
Tricyclodecane ring		127.8	—	—
Ortho substitution		19.8	−13.3	—
Meta substitution		13.5	−24.3	—
Para substitution		82.4	−33.8	—

Adapted from Hoy, K. L., *The Hoy Tables of Solubility Parameters*, Union Carbide Corporation, 1975 and 1985; *J. Paint Technol.*, 42, 76, 1970.

which is based on the values in the 1975 and 1985 editions of the Union Carbide publication.[35,36] (The glycol ether values in the later editions have polar and nonpolar cohesion parameters that differ from those published earlier, a result of correcting a group value error.) Hoy found that the only compounds for which the simple molar attraction procedure was unsatisfactory were the acids, alcohols, and other compounds capable of association and "chameleonic" behavior (Section 5.5). He therefore took dimerization into account in such compounds:

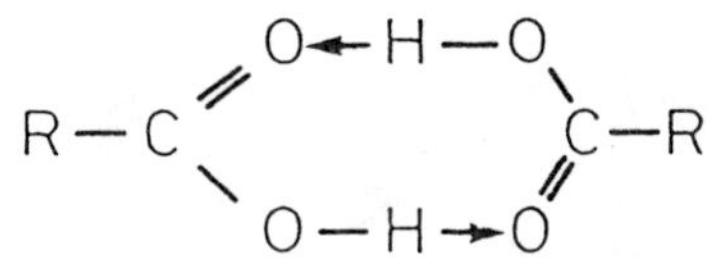

If the cohesion parameter is expressed

$$\delta = (-U\rho/M)^{1/2} \tag{19}$$

then in the case of dimeric molecules the molar mass is $2M$, so

$$\delta = (-U\rho/2M)^{1/2} \tag{20}$$

Matsuura[37] used modified Small constants in a discussion of reverse osmosis separation. Another set of atomic attraction constants was derived by van Krevelen[38,39] based on the contributions of atoms and features such as double bonds (Table 8). This was intended for the investigation of coal (Section 16.11), but it was used by Chavan et al.[40] for anionic dyes (Section 17.8), and derived group attraction constants for polymers appear in Table 12, Chapter 14. In an interesting extension of the concept, molar attraction constants have been applied to the correlation of biological activity with chemical structure (Chapter 19). McGowan[41] estimated molecular attraction constants and cohesion parameters from characteristic atomic volumes.

It is not necessary to fully evaluate the Hildebrand parameter of a compound to benefit from information on group contribution values. Boberski, Seiner, and Petracca[42] have applied molar attraction constants to silicon chemistry, calculating a "working Hildebrand parameter" from a weighted average of individual carbon-based functional groups bonded to silicon (Section 17.5).

6.5 GROUP MOLAR VOLUME

As indicated earlier in this chapter, the molar volumes of homologous series of molecules increase approximately linearly with increasing numbers of carbon atoms. This is true not only among the members of each series, but also between different series of compounds.[6,10,14,15,28,43,44] Even more significantly, the contribution of a polar- or hydrogen-bonding end group changes very little with the chain length — the effect of hydrogen bonding and polarity remains unchanged or undiluted by the increasing nonhydrogen bonding and nonpolar portions of the molecules.

Thus,

$$V = M\rho^{-1} = \sum_z {}^zV \tag{21}$$

where z represents the contributing group. (Group contributions are found to be preferable

TABLE 8
Atomic Attraction Constants at 25°C

Atom or structural feature, *z*	Bond type	$^zF/J^{1/2}\ cm^{3/2}\ mol^{-1}$
C		0.0
H		140.1
O	Ether	256
O	Ester	256
O	Ketone	685
O	Primary alcohol	736
O	Secondary alcohol	614
O	Phenol	511
S	Thioether	460
S	Thiol	(511)
F		164
Cl		471
Br		614
I		859
Double bond (nonaromatic)		164
Double bond (aromatic)		272
Triple bond		440
N	Aliphatic primary amine	205
N	Aliphatic secondary amine	286
N	Aromatic primary amine	133
N	Heterocycle	235
N	Nitrile	982
N + O	Aliphatic nitro	941
N + O	Aromatic nitro	665
N + O	Acid amide, –C(=O)–NH–	1227
N + O	Acid amide, –C(=O)–N	818
Nonaromatic ring		123
Ramification in chain		−92
Conjugation of double bonds		51
2 OHs on adjacent Cs		−389

Adapted from van Krevelen, D. W., *Fuel*, 44, 229, 1965.

to ''atom'' contributions, which tend to be more dependent on the nature of the surrounding atoms.) For an homologous series the additive character can be tested by plotting ρ^{-1} against M^{-1}. If the increments for the repeating units (u) and for the constant ''end'' groups (e) of the molecule are designated ^uM, ^uV and ^eM, ^eV, respectively, for the molar masses and molar volumes, it can be shown[45] that

$$\rho^{-1} = {}^uV/{}^uM + ({}^uM{}^eV - {}^eM{}^uV)M^{-1}/{}^uM \tag{22}$$

so for exact additivity a set of lines is obtained, intersecting at (0, $^uV/^uM$). Exner[28] found that the additivity criterion was better fulfilled at constant temperature than at the individual boiling points.

The degree of molecular flexibility (conformational freedom) can influence the space requirement of groups.[46] Askadskii and co-workers[47-50] have discussed molar volumes in terms of the van der Waals volumes, ^zV_v of the constituent atoms or groups, and the molecular packing coefficient k which they defined

$$k = (N_A/V) \sum_z {}^zV_v = (N_A\rho/M)\sum_z {}^zV_v \tag{23}$$

where N_A is the Avogadro constant. Values of ${}^zV_v$ are given in Table 9; it should be noted that these are molecular constants, in units of Å^3, which must be multiplied by N_A to provide group molar constants. van der Waals volumes and molecular packing factors have been used widely in other ways related to cohesion parameters.[51-53]

Some group contributions to molar volumes at 25°C are included in Tables 2 and 3, and van Krevelen[15] has provided a summary of other values. This information may be used to estimate molar volumes for compounds when experimental values are unavailable, or to check that the additivity in a function of the form of Equation 7 is fulfilled with an accuracy greater than that of the molar volume alone.[28] Information on group molar volumes of polymers is included in Section 13.1, and the significance of molar volume in the activity of drugs in Chapter 19. For the "molar surface" modification of the Hildebrand-Scatchard equation,[54,55] molecular surface increments may be combined in the same way as volume increments.

6.6 GROUP COHESION PARAMETERS

During their search for improved methods of cohesion parameter calculations based on group contributions, Rheineck and Lin[14] plotted Hildebrand parameters against numbers of carbon atoms, n, in homologous series. For the hydrocarbons (Figure 2) a straight line relationship exists above about five carbon atoms; Hildebrand parameters have been evaluated[56] for n-alkanes up to C_{25}. For the other homologous series (Figure 3) no simple relationship is apparent.

In alternative methods of interpolating and extrapolating cohesion parameters, Rätzsch and Krahn[57] found that graphs of log δ against n^{-1} were linear, while Watanabe and Sugiyama[58] plotted log δ against log M to give straght lines for homologous series of saturated aliphatic hydrocarbons, alcohols, alkylbenzenes, ethers, ketones, and monobrominated hydrocarbons, but slightly curved lines for nitriles and esters of acetic acid.

Konstam and Feairheller[59] proposed a modified method for calculating Hildebrand parameters more directly from functional group contributions. In general, straight lines result when the Hildebrand parameters of a homologous series of monofunctional compounds are plotted against the reciprocal of the molar volume, V^{-1} (Figure 4):[59,60]

$$\delta = A + BV^{-1} \tag{24}$$

as discussed further in Section 10.1. This is equivalent to plotting $(-UV)^{1/2}$ against the number, n, of carbon atoms if V is a linear function of n and is more satisfactory for estimating the Hildebrand parameter of high molar mass members of a homologous series because the intercept of the line with the δ axis represents the Hildebrand parameter of an infinitely long member of the series ($V^{-1} = 0$). This means that all higher members of the series fall on a line between this axis and the first datum point and so can be readily obtained by extrapolation. (The lighter members of the homologous series of unsubstituted hydrocarbons are exceptional in not lying on a straight line when plotted against V^{-1}.) The functional group molar volumes may also be estimated from this information.

The analysis can be extended in terms of end groups (e) and repeating units (u).[61]

$$\delta = \sum_z {}^zF/V = [{}^uF(V - {}^eV)/{}^uV + {}^eF]/V \tag{25}$$

TABLE 9
van der Waals Volumes of Atoms and Atomic Groups

Atom or group	Bond lengths/Å	$^{z}V_{w}$/Å^3
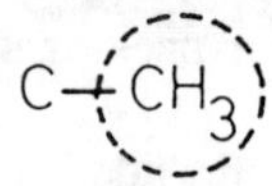	C–C, 1.48	22.7
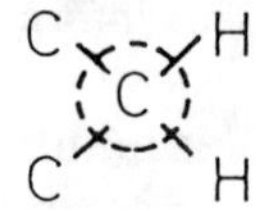	C–C, 1.54, 1.48	12.6
	C–C, 1.40	7.8
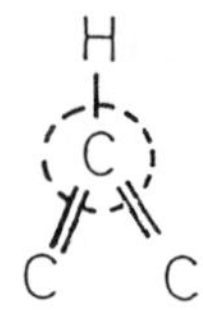	C–C, 1.40; C–H, 1.08	13.9
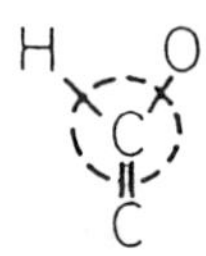	C–O, 1.50; C–H, 1.08; C=O, 1.34	16.2
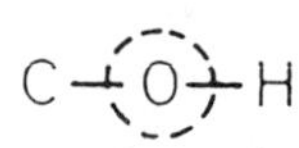	C–O, 1.37; C–H, 1.08	5.2
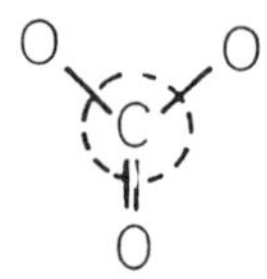	C–O, 1.50; C=O, 1.28	19.0
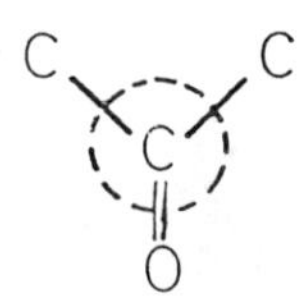	C–C, 1.54; C=O, 1.28	12.8
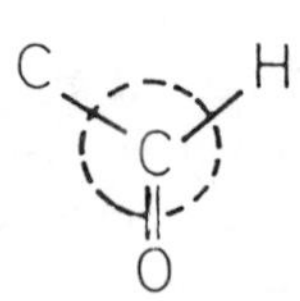	C–C, 1.45; C–H, 1.08; C=O, 1.28	16.9
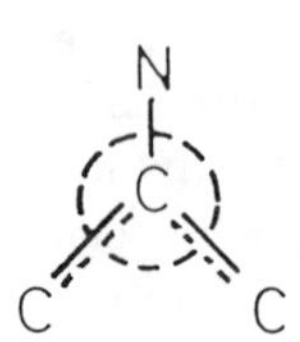	C–C, 1.40; C–N, 1.37	10.2

TABLE 9 (continued)
van der Waals Volumes of Atoms and Atomic Groups

Atom or group	Bond lengths/Å	${}^{z}V_{v}/\text{Å}^{3}$
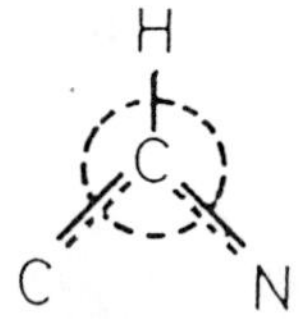	C–C, 1.40; C–H, 1.08; C–N, 1.34	14.8
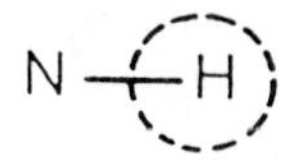	N–H, 1.08	3.5
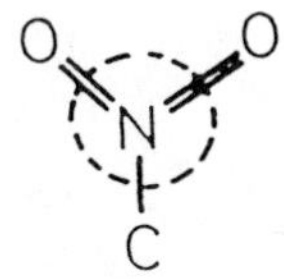	C–N, 1.37; N=O, 1.20	7.0
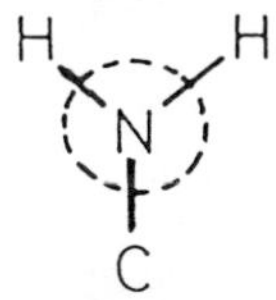	C–N, 1.37; N–H, 1.08	8.7
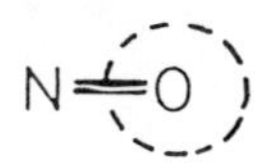	N=O, 1.20	7.2
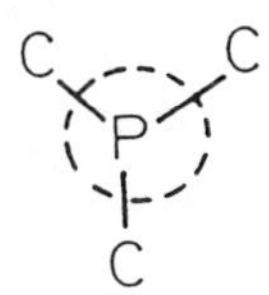	C–P, 1.81	14.7
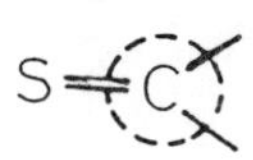	C=S, 1.71	20.2
	C–S, 1.76	16.5
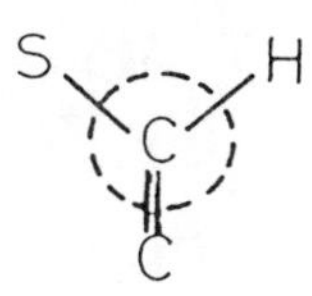	C=C, 1.34; C–S, 1.76, C–H, 1.08	14.2
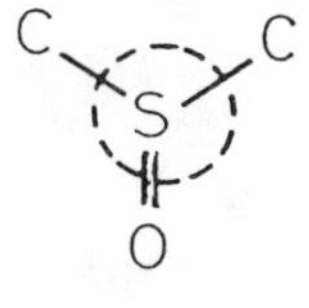	C–S, 1.76; S=O, 1.44	14.7
	C–S, 1.76; S–H, 1.33	19.5
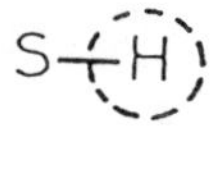	S–H, 1.33	3.2

TABLE 9 (continued)
van der Waals Volumes of Atoms and Atomic Groups

Atom or group	Bond lengths/Å	$^{z}V_{v}$/Å^{3}
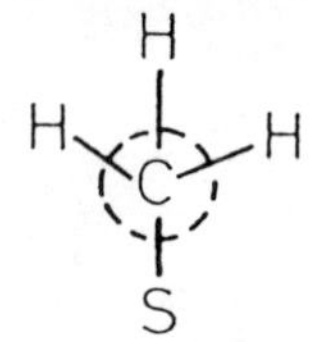	C–H, 1.08; C–S, 1.76	18.4
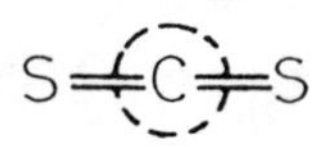	C–S, 1.55	14.9
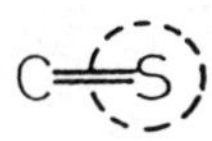	C–S, 1.55	19.7
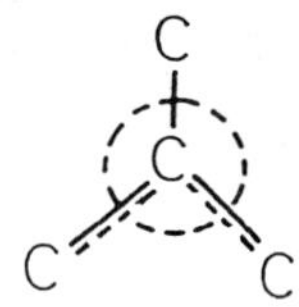	C–C, 1.40; C–Cl, 1.70	10.5
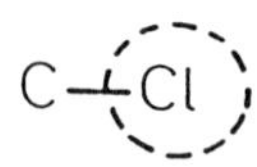	C–Cl, 1.70	19.5
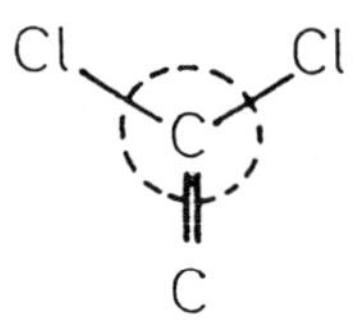	C=C, 1.34; C–Cl, 1.77	11.0
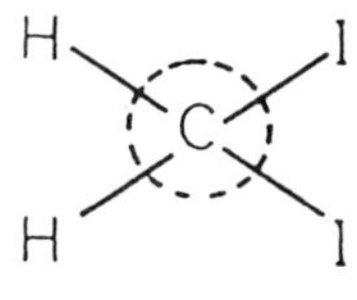	C–H, 1.08; C–I, 2.21	13.5
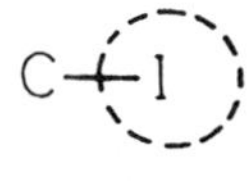	C–I, 2.21	38.4
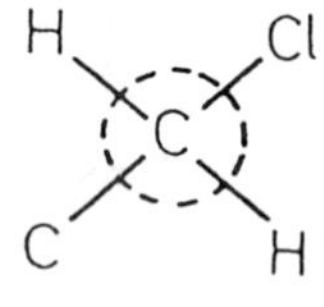	C–C, 1.54; C–H, 1.08: C–Cl, 1.77	14.1
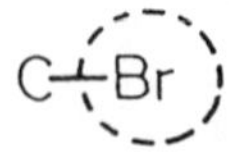	C–Br, 1.85	27.1
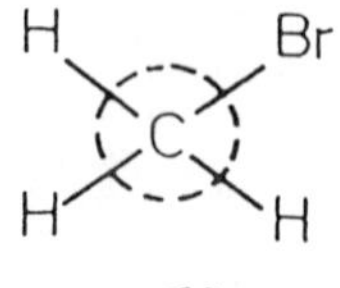	C–H, 1.08; C–Br, 1.94	17.7
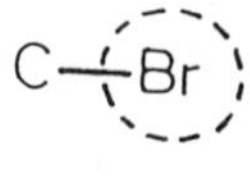	C–Br, 1.94	27.3

TABLE 9 (continued)
van der Waals Volumes of Atoms and Atomic Groups

Atom or group	Bond lengths/Å	${}^zV_v$/Å^3
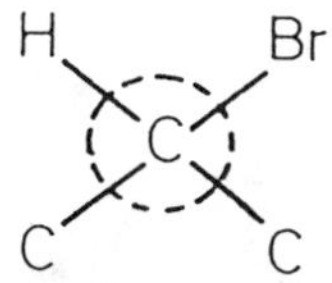	C–C, 1.54; C–H, 1.08; C–Br, 1.94	9.6
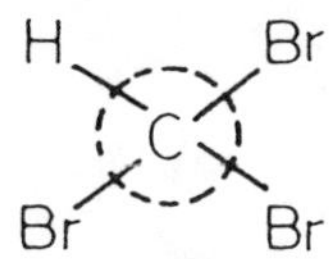	C–H, 1.08; C–Br, 1.94	10.5
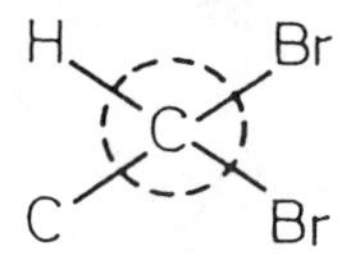	C–C, 1.54; C–H, 1.08; C–Br, 1.94	10.0
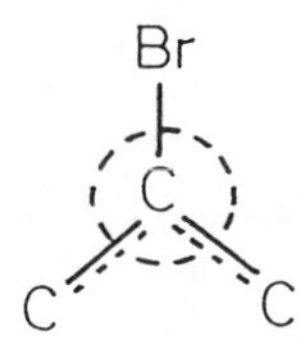	C–C, 1.40; C–Br, 1.85	9.9

Adapted from Askadskii, A. A., Kolmakova, L. K., Tager, A. A., Slonimskii, G. L., and Korshak, V. U., *Dokl. Akad. Nauk. SSSR,* 226, 857, 1976; *Dokl. Phys. Chem.,* 226, 99, 1976; *Vysokomol. Soedin. Ser. A* 19, 1004, 1977; *Polym. Sci. U.S.S.R.,* 19, 1159, 1977.

where $(V - {}^eV)/{}^uV$ is a measure of the number of repeating units. Group Hildebrand parameters may be defined,

$$ {}^z\delta = {}^zF/{}^zV \tag{26} $$

so

$$ {}^u\delta = {}^uF/{}^uV, \qquad {}^e\delta = {}^eF/{}^eV \tag{27} $$

and

$$ \delta = {}^u\delta(1 - {}^eV/V) + {}^e\delta{}^eV/V \tag{28} $$

which has the same form as Equation 24. The calculation of polymer cohesion parameters is discussed further in Section 14.8. Table 10 lists some group cohesion parameters and group molar volumes. The group molar volumes may be compared with the molar volumes of their molecular analogues, that is of molecules consisting of functional group plus hydrogen atom. For example, water (V = 18.0 cm^3 mol^{-1}) and acetic acid (V = 57.9 cm^3 mol^{-1}) have rather smaller molar volumes than –OH (${}^zV$ = 19.7 cm^3 mol^{-1}) and CH_3COO– (${}^zV$ = 58.5 cm^3 mol^{-1}) because of the extensive hydrogen bonding, but most molecules are slightly larger than the sum of the volumes of their functional groups.

The group molar attraction constants also calculated by Konstam and Feairheller[59] are listed in Table 11, along with Small's[32] earlier values. It should be noted that Small's value for the hydroxyl group represents only the cohesive pressure resulting from dispersion, while

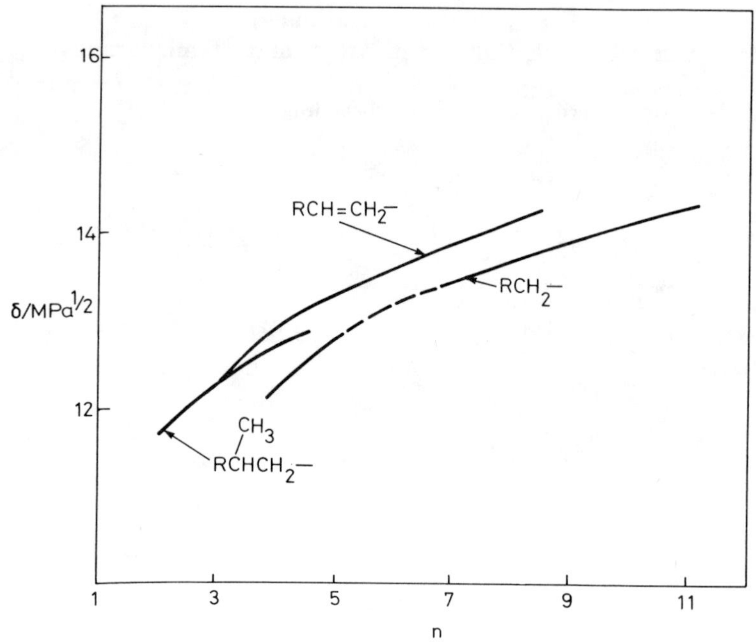

FIGURE 2. Hildebrand parameters of hydrocarbons as functions of numbers of carbon atoms. (Adapted from Rheineck, A. E. and Lin, K. F., *J. Paint Technol.*, 40, 611, 1968.)

the method of calculation of Konstam and Feairheller takes hydrogen bonding into account. (In fact,[62] Small's method provides a reasonable estimate of internal pressure rather than cohesive pressure: see Section 7.5.) Sears and Darby[102] applied Small's method to plasticizers, and many other estimates of Hildebrand parameters and related quantities have been based on these particular attraction constants, despite the publication of several revised sets (Section 6.4).

Askadskii and coworkers[47-50] calculated cohesion parameters by means of the van der Waals volumes described in the previous section. It follows from the definition of the packing coefficient k that

$$\delta^2 = -U/V = -\frac{Uk}{N_a \sum_z {}^z V_v} = \frac{U^*}{N_A \sum_z {}^z V_v} = -\frac{\sum_z {}^z U^*}{N_A \sum_z {}^z V_v} \quad (29)$$

where $-U^*$ is the effective molar cohesive energy (the molar cohesive energy, $-U$, divided by the factor describing the number of times the van der Waals volume is smaller than the molar volume). The effective cohesive energy atomic contributions are listed in Table 12, and the denominator terms may be evaluated from the data in Table 9. These have been used both for liquids, and for polymers (Section 14.8).

Fisher[43,63] developed simple logarithmic equations for relating homolog cohesion parameters to number of carbon atoms (n) of the form $\delta = b + m \log (n + k)$. Cohesion parameters of some moderately polar homologs could be represented by simple linear equations, $\delta = b + m\,n$. The tables of b, m, and k values can be used to estimate Hildebrand parameters for a very large number of compounds. Also, the logarithm expression may be used to relate Hildebrand parameters to other quantities such as molar volume, boiling point, and vaporization enthalpy, which are linearly related to homolog chain length. Fisher has

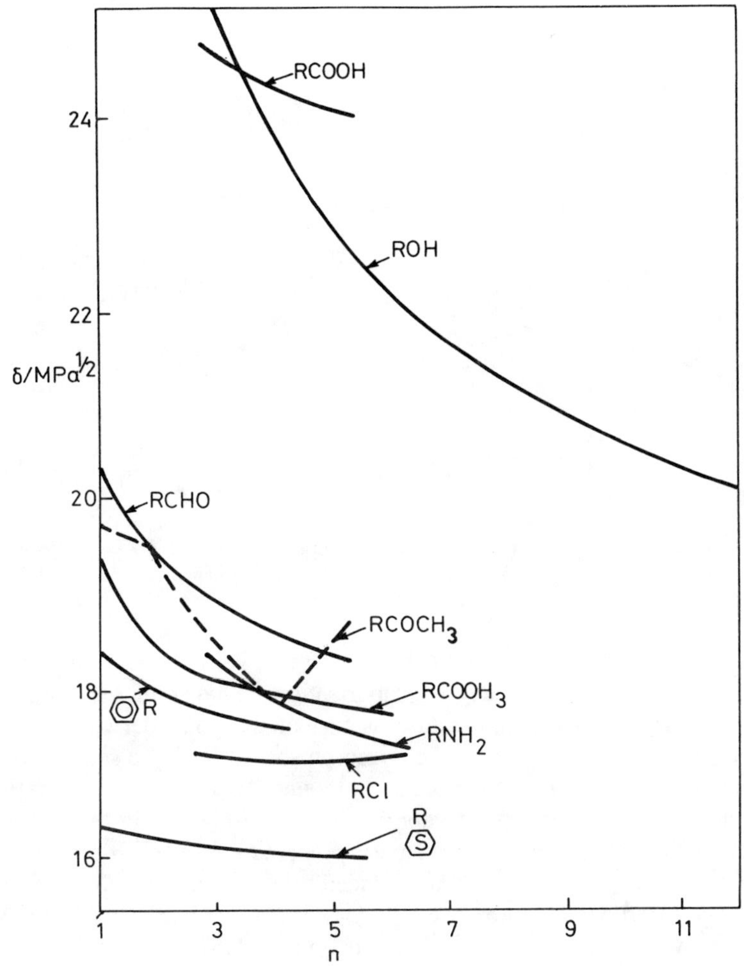

FIGURE 3. Hildebrand parameters of *n*-alkyl derivatives as functions of number of carbon atoms. (Adapted from Rheineck, A. E. and Lin, K. F., *J. Paint Technol.*, 40, 611, 1968.)

proceeded to devise alternative equations correlating Hildebrand parameters and other properties of homologs with chain length.[64,65]

A novel correlation[66] of cohesion energy parameters is based on the ratio of the sum of atomic weights of hetero atoms in the liquid molecule to the total molecular weight, and the ratio of the number of methyl or CF_3 groups per molecule to the total number of carbon atoms per molecule.

6.7 HOMOMORPHS AND DISPERSION COHESION PARAMETERS

In any multicomponent cohesion parameter system there arises the problem of evaluating the various components separately, and one obvious approach is to compare the properties of compounds which differ only in the presence or absence of a certain group. Here the homomorph concept is important:[67] the homomorph of a polar molecule is a nonpolar molecule having very nearly the same size and shape. Bondi and Simkin[68,69] proposed that for the purpose of estimating the dispersion component of cohesive pressure or cohesion

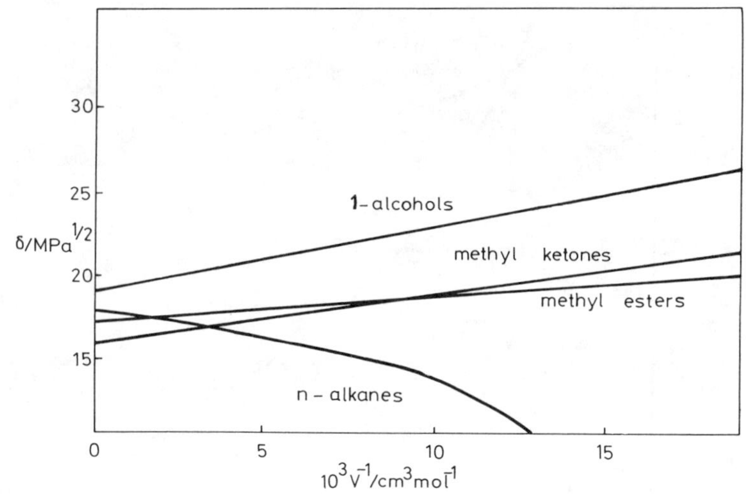

FIGURE 4. Hildebrand parameters of homologous series as functions of reciprocal molar volume. (Adapted from Konstam, A. H. and Feairheller, W. R., Jr., *Am. Inst. Chem. Eng. J.*, 16, 837, 1970.)

parameter, the homomorph of a polar molecule is that hydrocarbon whose structure is the same as that of the polar molecule, the vaporization enthalpy of the homomorph being evaluated at the same reduced temperature as that of the polar derivative. Following the additional recommendation[70,71] that the homomorph should have the same molar volume as well as a structure similar to that of the polar compound, Blanks and Prausnitz[72,73] published a homomorphic plot of the energy of vaporization for straightchain hydrocarbons (Figure 5). On the same basis, Weimer and Prausnitz[73] prepared homomorph plots of the cohesive pressure against molar volume at various reduced temperatures for alkanes, cycloalkanes, and aromatic hydrocarbons (Figures 6 to 8). (The axes in Figures 1 to 3 of Weimer and Prausnitz[74] should be labeled "cal cm^{-3}", and Equation 9 lacks the factor *RT* in the last two terms.[76]) The figures also have been reproduced elsewhere.[17,75] The cohesive pressure properties of a series of similar liquids vary in a smooth, continuous manner with their molar volumes, and Thompson[17,77] has expressed this analytically. Similar homomorphic plots have been presented by Helpinstill and Van Winkle.[78]

Wingefors and Liljenzin[79,80] developed a new homomorph equation for the calculation of dispersion cohesive pressure, based on Antoine constants and vaporization enthalpies at the boiling point and at 25°C. This correlation is depicted here for primary alcohols at 25°C (Figure 9a), *n*-alkanes at 25°C (Figure 9b), and *n*-alkanes at their normal boiling points (Figure 9c). The nonlinearity of the *V* scale should be noted; $\delta_\lambda{}^2$ is a linear function of *V* raised to the power of θ:

$$\delta\lambda^2/\mathrm{J\ cm^{-3}} = (-U/V)/\mathrm{J\ cm^{-3}} = a(T_r) - b(T_r)V^\theta$$

where

$$\theta = 0.124687, \tag{30}$$

$$a(T_r) = 1022.31 - 1586.36T_r + 1082.07T_r^2$$

$$b(T_r) = 277.790 - 531.335\ T_r + 482.039T_r^2$$

The broken lines in Figure 9 indicate a region where Equation 30 has limited accuracy.

Pussemier et al.[104] used the homomorph principle in evaluating Hildebrand parameters

TABLE 10
Group Cohesion Parameters and Group Molar Volumes of Some Functional Groups

Group, z	${}^{z}\delta$/MPa$^{1/2}$	${}^{z}V$/cm^3 mol^{-1}
$-NH_2$	24.03	29.49
–F	13.46	30.30
–Cl	18.18	34.98
–Br	19.33	33.07
–I	19.80	42.30
–COOH	29.60	36.28
–OH	41.50	19.66
–CHO	22.15	37.22
–(CO)–	28.47	32.52
CH_3CO–	23.13	53.49
–O–	20.54	27.96
–CN	30.40	31.23

Adapted from Konstam, A. H. and Feairheller, W. R., Jr., *Am. Inst. Chem. Eng. J.*, 16, 837, 1970.

for pesticides (containing highly polar bioactive functions) from experimental information on compounds corresponding to the aromatic portions of the pesticide molecules.

Varushchenko, Loseva, and Druzhinina[101] used the assumption[19] that a carbon-chlorine bond and three carbon-hydrogen bonds make almost equal contributions to the dispersion energy of a polar liquid to calculate

$$\Delta U_p^o = \Delta U^o(C_nH_{2n}Cl_2) - \Delta U^o(C_{n+2}\,H_{2n+6})$$

and so separated polar and dispersion components. Another experimental approach to information of this kind is gas-liquid chromatography (Section 18.3).

From graphs or functions such as those above, the dispersion cohesive pressure, $-U_d/V$, and dispersion parameter, δ_d, at any desired molar volume and temperature may be determined for a particular type of molecular skeleton. However, there are difficulties. Hansen and Beerbower[76] have pointed out that the homomorph charts of Weimer and Prausnitz[73] and of Helpinstill and Van Winkle[78], which were extended (but not used) by Beerbower and Dickey,[81] yield dispersion parameter values higher than those of Blanks and Prausnitz.[73] This last set was the basis for several subsequent successful correlations by Hansen,[82,83] Beerbower[81,84,85] and Zisman,[86] as well as the data used by Hansen and Beerbower.[76] When using group contribution and cohesion parameter data from different literature sources, it is essential to check that they are based on the same models and assumptions, and that they are self-consistent.

6.8 HANSEN PARAMETER GROUP CONTRIBUTIONS

It would be useful to be able to estimate the Hansen parameters δ_d, δ_p, and δ_h (Section 5.9) directly from molecular physical properties, but although correlations may be found between the cohesion parameter components and these quantities (Section 8.2, and in particular the work of Peiffer,[53] Loev and Lavrishchev,[87] Stekol'schchikov et al.,[88] and Martin, Wu, Adjei, and Beerbower[89,90]), it is usually more convenient to use methods based on structural combinations. If a quick estimate of the dispersion parameter is required, the dispersion contributions of Table 13 can be used:[15,76,91]

TABLE 11
Group Molar Attraction Constants at 25°C

Group, z	Bond type	$^{z}F/J^{1/2}\ cm^{3/2}\ mol^{-1}$	
		Small[32]	Konstam and Feairheller[59]
$-CH_3$		438	—
$-CH_2$	Single	272	—
$>CH-$	Single	57	—
$>C<$	Single	−190	—
$CH_2=$	Double	389	—
$-CH=$	Double	227	—
$>C=$	Double	39	—
$CH\equiv C-$	Triple	583	—
$-C\equiv C-$	Triple	454	—
Phenyl		1503	—
Phenylene (*o*,*m*,*p*)		1346	—
Naphthyl		2344	—
Ring, 5-membered		215-235	—
Ring, 6-membered		195-215	—
Conjugation		40-60	—
S	Sulfide	460	—
SH	Thiol	644	—
ONO_2	Nitrate	~900	—
NO_2	Aliphatic nitro	~900	—
PO_4	Organic phosphate	~1020	—
H	Variable	160-205	—
–CHO		—	824
–O–	Ether	143	575
–CO–	Ketone	563	681
–COO–	Ester	634	—
–COOH		—	1074
CN		839	948
Cl	Single	552	636
Cl	$>CCl_2$	532	636
Cl	$-CCl_3$	511	493[a]
Cl	>C(Cl)–C(Cl)<	—	497
Br	Single	696	736
Br	$>CBr_2$	—	677
Br	$-CBr_3$	—	581[a]
Br	>C(Br)–C(Br)<	—	630
F	Single	—	407
CF_2	*n*-Fluorocarbon	307	311
CF_3	*n*-Fluorocarbon	561	—
I	Single	869	839
OH	Single	348[b]	816[b]
OH	Diol	—	634
NH_{82}		—	710

[a] Values based on data for only one compound.
[b] See text for explanation of discrepancy between values of ^{z}F for –OH group.

$$\delta_d = \left(\sum_z {}^{z}F_d\right)/V \qquad (31)$$

If only one polar group is present, it is possible to evaluate δ_p from Table 7 by means of

$$\delta_p = F_p/V \qquad (32)$$

TABLE 12
Effective Cohesive Energy Atomic Contributions[47-50]

Atom or molecular feature, z	$-^zU^*$/kJ mol^{-1}
Carbon	2.304
Hydrogen	0.200
Oxygen	0.597
Nitrogen	5.04
Fluorine	0.101
Chlorine	−0.93
Bromine	2.44
Iodine	7.11
Sulfur	7.32
Dipole-dipolc coupling in dipolar aprotic solvents	6.79
Dipole-dipole coupling in amide-type solvents	6.79
Dipole-dipole coupling in dimethylsulfoxide type solvents	10.88
Hydrogen bonding	16.44
Aromatic ring (skeleton)	2.98
Double bond	−1.35

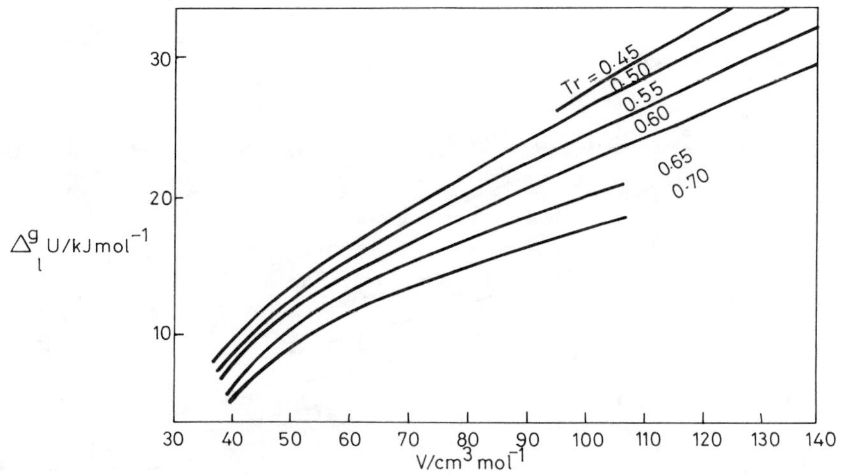

FIGURE 5. Vaporization energy for straight-chain hydrocarbons plotted against molar volume at various reduced temperatures. (Adapted from Blanks, R. F. and Prausnitz, J. M., *Ind. Eng. Chem. Fund.*, 3, 1, 1964.)

but for more than one polar group it is necessary to correct for the interaction of polar groups by using[15]

$$\delta_p = \left(\sum_z {}^zF_p^2\right)^{1/2}/V \qquad (33)$$

This must be reduced by multiplying δ_p calculated by Equation 33 by a symmetry factor if two identical polar groups are present in symmetrical positions:

- 0.5 for one plane of symmetry
- 0.25 for two planes of symmetry
- 0 for more planes of symmetry

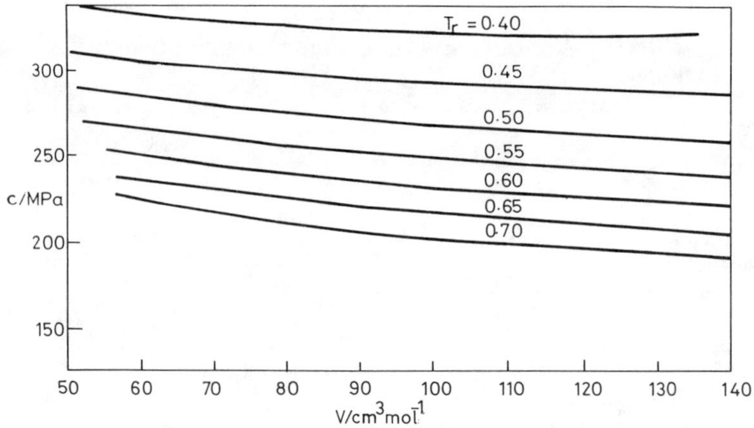

FIGURE 6. Homomorph plots for *n*-alkanes of cohesive pressure against molar volume at various reduced temperatures. (Adapted from Weimer, R. F. and Prausnitz, J. M., *Hydrocarbon Proc. Petr. Ref.*, 44, 237, 1965: see text.)

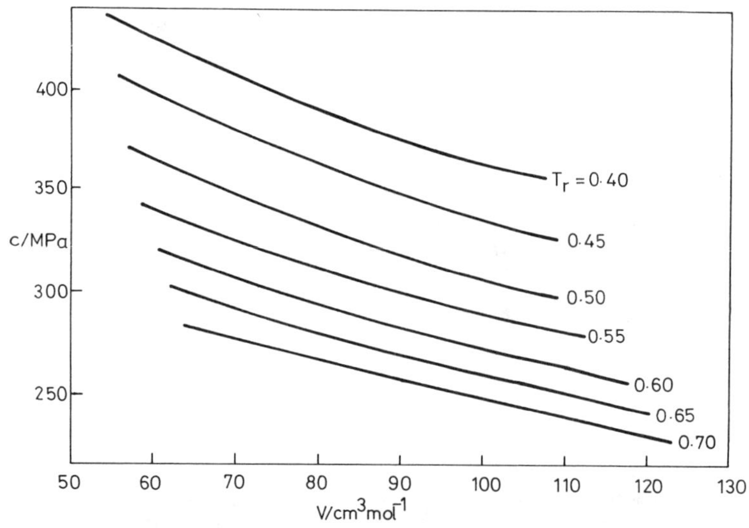

FIGURE 7. Homomorph plots for cycloalkanes of cohesive pressure against molar volume at various reduced temperatures. (Adapted from Weimer, R. F. and Prausnitz, J. M., *Hydrocarbon Proc. Petr. Ref.*, 44, 237, 1965: see text.)

These factors are included with the ${}^zF_p$ values in Table 13. Hansen and Beerbower[76] provided alternative values, warning against indiscriminate addition of functional group contributions and pointing out that the effect of adding a second chlorine could reduce δ_p, but ortho –COO– groups in phthalate could quadruple δ_p.

Hoy[35,36] used a group contribution method for the polar cohesion parameter (see Sections 5.11 and 6.4) based on

$$\delta_p = \delta_t\left(\sum_z {}^zF_p / \sum_z {}^zF\right) \tag{34}$$

with the ${}^zF_p$ values in Table 7.

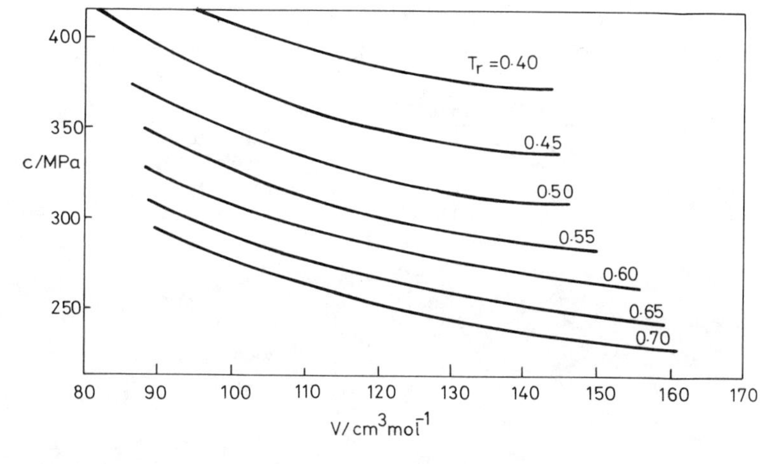

FIGURE 8. Homomorph plots for aromatic hydrocarbons of cohesive pressure against molar volume at various reduced temperatures. (Adapted from Weimer, R. F. and Prausnitz, J. M., *Hydrocarbon Proc. Petr. Ref.*, 44, 237, 1965: see text.)

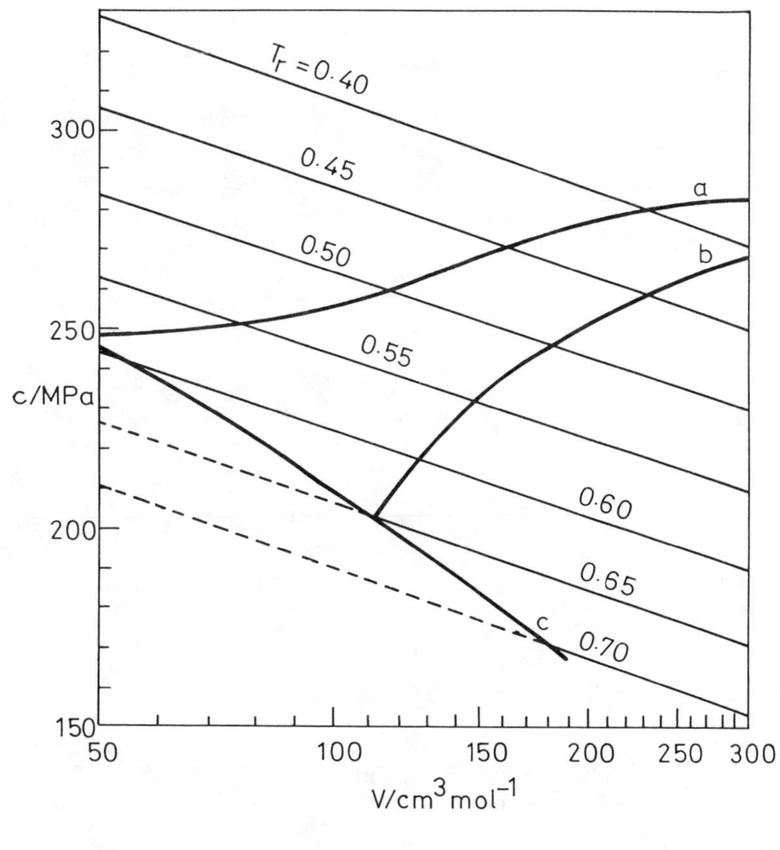

FIGURE 9. Cohesive pressures of *n*-alkanes in terms of molar volume and reduced temperature: a, primary alcohols; b, *n*-alkanes at 25°C; and c, *n*-alkanes at their normal boiling points. (Adapted from Wingefors, S. and Liljenzin, J. O., *J. Chem. Technol. Biotechnol.*, 31, 523, 1981.)

TABLE 13
Dispersion and Polar Group Molar Attraction Contributions

Structural group	${}^{z}F_d$/J$^{1/2}$ cm$^{3/2}$ mol^{-1} Koenhen and Smolders[91]	${}^{z}F_d$/J$^{1/2}$ cm$^{3/2}$ mol^{-1} van Krevelen[15]	${}^{z}F_p$/J$^{1/2}$ cm$^{3/2}$ mol^{-1} van Krevelen[15]	(${}^{z}F_p = {}^{z}V\,{}^{z}\delta_p$)/J$^{1/2}$ cm$^{3/2}$ mol^{-1} Hansen and Beerbower[76]
$-CH_3$	411	420	0	—
$-CH_2-$	284	270	0	—
$>C-$	104	80	0	—
$>C<$	—	−70	0	—
$=CH_2$	—	400	0	—
$=CH-$	—	200	0	—
$=C<$	—	70	0	—
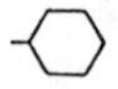	—	1620	0	—
	1640	—	—	—
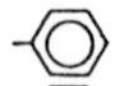	1510	1430	110	—
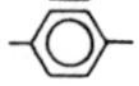 (o, m, p)	1350	1270	110	—
$-F$	—	(220)	—	460 ± 50
$-Cl$	—	450	550	610 ± 200
$>Cl_2$	—	—	—	360 ± 50
$-Br$	—	(550)	—	610 ± 50
$-I$	—	—	—	665 ± 50
$-CN$	446	430	1100	1070 ± 100
$-OH$	202	210	500	510 ± 60
$(-OH)_n$	—	—	—	n(350 ± 50)
$-O-$	—	100	400	410 ± 100
$-COH$	—	470	800	—
$-C=O$	325	—	—	—
$-CO-$	—	290	770	800 ± 30
$-COOH$	409	530	420	450 ± 20
$-COO-$	395	390	490	510 ± 50
$HCOO-$	—	530	—	—
$-NH_2$	—	280	—	610 ± 200
$-NH-$	143	160	210	205 ± 30
$-N<$	—	20	800	—
$-NO_2$	440	500	1070	1020 ± 100
$-S-$	—	440	—	—
$=PO_4^-$	—	740	1890	—
Ring	—	190	—	—
One plane of symmetry	—	—	0.50 ×	—
Two planes of symmetry	—	—	0.25 ×	—
More planes of symmetry	—	—	0 ×	—

Although the F-method is not applicable directly to the calculation of δ_h, Hansen and Beerbower[76] have assumed that hydrogen bonding cohesive energy is additive, leading to

$$\delta_h = \left(\sum_z -{}^{z}U_h/V\right)^{1/2} \tag{35}$$

Two sets of estimates of $-{}^{z}U_h$ are shown in Table 14. Again, extreme caution is needed in adding group contributions in the use of a single hydrogen bonding parameter to describe

TABLE 14
Hydrogen Bonding Parameter Group Contributions

Structural group, z	${}^{z}U_h$/J mol^{-1} van Krevelen[15]	$-{}^{z}U_h = {}^{z}V^{z}\delta_h^2$/J mol^{-1} Hansen and Beerbower[76] Aliphatic	Aromatic
$-CH_3$	0	—	—
$-CH_2-$	0	—	—
$>CH-$	0	—	—
$>C<$	0	—	—
$=CH_2$	0	—	—
$=CH-$	0	—	—
$=C<$	0	—	—
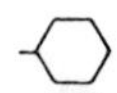	0	—	—
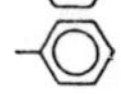	0	—	—
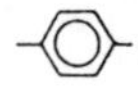 (o, m, p)	0	—	—
$-F$	—	~0	~0
$-Cl$	400	400 ± 80	400 ± 80
$>Cl_2$	—	690 ± 40	750 ± 40
$-Br$	—	2100 ± 400	2100 ± 400
$-I$	—	4000 ± 800	—
$-CN$	2500	2100 ± 800	2300 ± 800
$-OH$	20000	19500 ± 1700	19500 ± 2100
$-(OH)_n$	—	n(19500 ± 1700)	n(19500 ± 1700)
$-O-$	3000	4800 ± 1200	5200 ± 1200
$-COH$	4500	—	—
$-CO-$	2000	3300 ± 600	3300 ± 600
$-COOH$	10000	11500 ± 1000	9400 ± 1000
$-COO-$	7000	5200 ± 600	3300 ± 600
$-NH_2$	8400	5600 ± 800[a]	9400 ± 800[a]
$-NH-$	3100	3100 ± 800	—
$-N<$	5000	—	—
$-NO_2$	1500	1700 ± 200	1700 ± 200
$=PO_4^-$	13000	—	—

[a] Data from Bondi[8] corrected to 25°C and $V\delta_p^2$ subtracted. For important steric shielding effects, see his Tables 7.6 and 7.7.

an interaction really requiring both donor and acceptor components (Section 5.5), where ''chameleonic'' effects can occur.

Meusberger[92] presented the Hansen parameter group contributions in Table 15 compiled by Beerbower as a compromise between the 1971 Hansen-Beerbower data (Table 11, Chapter 5) and Hoy's cohesion parameters (Table 18, Chapter 5), and which at least have the advantage of being the most recently compiled comprehensive set. Examples of calculations of component cohesion parameters are provided in Table 16.

6.9 PARTICIPATING AND NONPARTICIPATING GROUPS

In applying cohesion parameter methods to the solubility of steroids and other drugs, Ostrenga and Steinmetz[34,93,94] formulated a parameter which assessed the fraction of the liquid molecule which could participate in specific interactions, considered to be highly significant in the solubilization process:

TABLE 15
Group Contributions for Molar Volumes and Hansen Parameters Compiled by Beerbower

Structural group	M g mol^{-1}	^zV cm^3 mol^{-1}	^zF_d (= $^zV\ ^z\delta_d$) J$^{1/2}$ cm$^{3/2}$ mol^{-1}	^zF_p (= $^zV\ ^z\delta_p$) J$^{1/2}$ cm$^{3/2}$ mol^{-1}	$^zV\ ^z\delta_h^2$ J mol^{-1}
$-CH_3$	15.04	31.7	419	0	0
$-CH_2-$	14.03	16.6	270	0	0
>CH–	13.02	−1.0	80	0	0
>C<	12.01	−19.2	−70	0	0
$=CH_2$	14.03	32.1	403	94	143
=CH–	13.02	12.4	223	70	143
=C<	12.01	−5.7	45	70	143
5 or 6 member ring	0.00	13.5	190	0	0
Conjugation:chain	0.00	−1.7	0	0	563
Cis configuration	0.00	0.0	−409	0	0
Conjugation:ring	0.00	0.80	43	0	−227
–O–	16.0	3.80	100	401	1467
>C=O	28.0	10.5	291	769	978
–OCO–	44.0	18.96	391	489	3420
$>PO_4^-$	96.0	28.00	741	1000	6353
–CN	26.0	22.40	430	1101	1221
$-NO_2$	46.0	24.00	499	1070	732
$-NH_2$	16.0	17.93	370	419	3220
–NH–	15.0	4.50	160	211	1516
>N–	14.0	−9.00	31	149	366
–N=	14.0	4.00	164	1323	1759
–CONH	43.0	16.80	516	1270	5985
–CON<	42.0	13.20	301	1229	4772
–OH	17.0	10.47	211	499	9773
–COOH	45.0	27.83	530	419	4887
>SiO<	44.1	3.80	266	307	921
–Cl	35.5	25.30	419	614	205
–Cl aromatic	35.5	24.00	329	550	78
$(-Cl)_2$ adjacent	70.9	51.20	900	1101	880
>C=O	28.0	10.00	291	769	978
–F	19.0	18.00	221	542	0
–Br	79.9	29.00	1080	614	1023
–Br aromatic	79.9	30.00	550	401	440
$(-Br)_2$adjacent	159.8	54.00	1101	929	3224
–I	126.9	32.20	655	665	2046
–COO– ester	44.0	8.20	667	511	2557
–COO– ester aromatic	44.0	8.20	667	511	1636
–O–	16.0	3.60	235	409	2352
–O– aromatic	16.0	3.80	100	401	1467
$-NH_2$ amide	15.0	25.10	669	1023	5523
–S–	32.1	8.00	454	362	221
Phenyl–	77.1	75.40	1499	121	205
Phenyl<	76.1	60.40	1319	133	205
$-NO_2$ aromatic	46.0	24.00	499	1070	732
–NH– aromatic	15.0	4.50	160	211	1516
>N–aromatic	14.0	−9.00	31	149	366
$>CO_3$	60.0	22.00	640	221	2246
$>CO_3$ ring	60.0	22.00	761	1528	945
HCOO– formate	44.0	32.50	542	248	3226
–CO–O–CO– anhydride	72.0	30.00	675	1105	4838

Adapted from Meusberger, K. E., *Am. Chem. Soc. Symp. Ser.*, 371 (*Pesticide Formulations*), 151, 1988.

TABLE 16
Group Molar Hansen Parameter Calculations for Diacetone Alcohol at 25°C

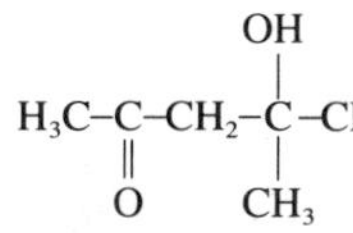

$V = 123.8\ \mathrm{cm^3\ mol^{-1}}$

Group, z	$\sum_z {}^zF_d/\mathrm{J^{1/2}\ cm^{3/2}\ mol^{-1}}$ (Table 13)	$\sum_z {}^zF_p^2/\mathrm{kJ\ cm^3\ mol^{-2}}$ (Table 13)	$-\sum_z {}^zU_h/\mathrm{kJ\ mol^{-1}}$ (Table 14)
3(–CH_3)	1260	0.0	0.0
(–CH_2)	270	0.0	0.0
(>C<)	−70	0.0	0.0
(–CO–)	290	593.0	2.0
(–OH)	210	250.0	20.0
	1960	843.0	22.0

$\delta_d = 1960/123.8 = 15.8\ \mathrm{MPa^{1/2}}$ (Equation 31)

$\delta_p = 843.000^{1/2}/123.8 = 7.4\ \mathrm{MPa^{1/2}}$ (Equation 33)

$\delta_h = (22.000/123.8)^{1/2} = 13.3\ \mathrm{MPa^{1/2}}$ (Equation 34)

$$\delta = (\delta_d^2 + \delta_p^2 + \delta_h^2)^{1/2} = 21.9\ \mathrm{MPa^{1/2}}$$

Adapted from van Krevelen, D. W., *Properties of Polymers: Their Estimation and Correlation with Chemical Structure,* 1976.

$$\delta = \delta_{sp} + \delta_{np} = (F_{sp} + F_{np})/V \tag{36}$$

$$\delta_{sp} = F_{sp}/V;\ \delta_{np} = F_{np}/V \tag{37}$$

$$f_{sp} = \delta_{sp}/\delta = F_{sp}/F = \sum_z {}^zF_{sp}/(\delta V) \tag{38}$$

The partial cohesion parameters, indicated here by δ_{sp} and δ_{np}, are the contributions made to δ by participating and nonparticipating functional groups, respectively. They are defined in terms of the corresponding molar attraction constants, F_{sp} and F_{np}, and the fractional participating molar attraction constant or fractional cohesion parameter is given by f_{sp}. F_{sp} values were calculated from Small's[32] group ${}^zF$ values, while δV values were obtained from whole-molecule sources if available, rather than by summing group contributions. For diols such as ethylene glycol and propylene glycol, the value of ${}^zF$ for a single –OH group was calculated from δV for the whole molecule less the $\epsilon_Z\ {}^zF$ value of the remaining groups. For strictly nonpolar (non-participating) liquids such as heptane, $f_{sp} = 0$, while for strictly polar or completely participating liquids such as water, $f_{sp} = 1$. The values for some other liquids are included in Table 17, rounded to two significant figures. This topic is referred to again in Section 19.1.

6.10 COMPARISON OF METHODS

The values reported by different authors of particular group contributions show considerable variation.[60,95-98] In particular, Jayasri and Yaseen[60,97] compared Hildebrand parameter values calculated by several group contribution methods with those reported on the basis of experimental data. They also reported (Table 18) preferred values, obtained after making linear plots of δ as functions of inverse molar volume (Sections 6.6 and 9.2).

TABLE 17
Participating Fractional Parameters

Liquid	f_{sp}
Water	1.00
Propylene glycol	0.60
Diethylene glycol	0.54
1,4-Butanediol	0.51
Glyceryol triacetate	0.50
Ethylene glycol diacetate	0.47
Propylene carbonate	0.46
Methyl Cellosolve®	0.45
Ucar® solvent LM	0.42
Cellosolve®	0.39
Carbitol®	0.36
Ethyl acetate	0.36
Methyl Cellosolve® acetate	0.35
Poly(ethylene glycol) 400	0.33
Cellosolve® acetate	0.32
Carbitol® acetate	0.29
Butyl Carbitol®	0.29
Pentyl acetate	0.24
Isopropyl myristate	0.11
Heptane	0.00

Adapted from Ostrenga, J. A. and Steinmetz, C., *J. Pharm. Sci.*, 59, 414, 1970.

TABLE 18
Hildebrand Parameter Values at 25°C Suggested by Jayasri and Yaseen[60,97] After Correlation With Inverse Molar Volume Dependence.

		$\delta/MPa^{1/2}$			
Liquid	*V* cm³ mol⁻¹	Jayasri & Yaseen	Hoy	Burrell	Hansen
	Alcohols				
Methanol	40.4	30.7	29.7	29.7	29.2
Ethanol	58.4	26.5	26.1	26.0	26.4
1-Propanol	74.7	24.7	24.9	24.3	24.5
1-Butanol	91.7	23.5	23.7	23.3	23.1
1-Pentanol, amyl alcohol	108.2	22.4	22.8	22.3	21.7
1-Hexanol	124.8	22.1	22.0	21.9	21.9
1-Heptanol	141.9	21.5	21.5	21.7	21.1
1-Octanol	157.9	21.1	21.1	21.1	19.8
1-Nonanol	174.3	20.7	20.7	—	—
1-Decanol	190.9	20.7	20.5	—	—
2-Propanol, isopropanol	60.1	23.7	23.4	23.5	23.6
Isobutanol, 2-methyl-1-propanol	74.1	23.0	23.0	—	22.8
2-Butanol, *sec* Butanol	74.1	22.8	22.7	22.1	22.2
2-Ethylbutanol	102.2	21.7	21.2	21.5	21.2
2-Ethylhexanol	130.2	20.7	20.8	—	20.2
Allyl alcohol, 2-propen-1-ol	58.2	25.5	25.7	—	24.9
	Aliphatic Hydrocarbons				
Methane	38.7	9.5	—	—	9.6
Ethane	53.6	11.6	—	—	11.6

TABLE 18 (continued)
Hildebrand Parameter Values at 25°C Suggested by Jayasri and Yaseen[60,97] After Correlation With Inverse Molar Volume Dependence.

Liquid	V cm^3 mol^{-1}	δ/MPa$^{1/2}$ Jayasri & Yaseen	Hoy	Burrell	Hansen
Propane	75.3	13.2	11.8	—	12.7
Butane	96.9	14.1	13.5	—	13.5
Pentane	115.3	14.5	14.4	14.3	14.4
Hexane	130.5	14.8	14.8	—	14.8
Heptane	146.5	15.1	15.3	15.1	15.2
Octane	162.3	15.3	15.4	15.6	15.4
Nonane	178.8	15.5	15.6	—	15.6
Decane	194.9	15.6	15.8	—	15.8
Cyclopentane	93.4	16.3	16.6	—	15.8
Cyclohexane	108.0	16.8	16.8	16.8	16.7
Aromatic Hydrocarbons					
Benzene	88.9	18.7	18.7	18.8	18.7
Toluene, methylbenzene	106.3	18.3	18.3	18.2	18.2
o-Xylene, 1,2-dimethylbenzene	120.6	18.1	18.5	18.4	18.4
m-Xylene, 1,3-dimethylbenzene	122.9	18.3	18.2	18.0	18.0
p-Xylene, 1,4-dimethylbenzene	123.3	18.0	18.1	17.9	17.9
Ethylbenzene	122.5	18.0	18.1	18.0	18.0
Propylbenzene	129.4	17.6	—	17.7	17.7
Isopropylbenzene	139.5	17.9	17.6	17.4	17.4
Styrene, ethenylbenzene	114.8	18.3	19.1	19.0	19.0
Esters					
Methyl formate	62.1	20.1	20.3	20.9	20.6
Ethyl formate	80.2	18.9	19.0	19.2	19.5
Propyl formate	97.8	18.3	—	18.8	19.6
Isobutyl formate	116.7	17.3	—	—	17.6
Isoamyl formate	133.4	17.1	—	—	17.3
Methyl acetate	79.9	19.1	19.4	19.6	19.4
Ethyl acetate	97.8	18.1	18.2	18.6	18.6
Propyl acetate	115.1	17.6	18.0	18.0	17.9
Butyl acetate	131.7	17.4	17.8	17.4	17.3
Isopropyl acetate	116.5	17.1	17.6	17.2	17.3
Isobutyl acetate	133.3	16.9	17.2	17.0	17.2
Isoamyl acetate	148.1	16.9	—	—	17.0
Methyl propionate	96.3	18.3	18.7	—	18.5
Ethyl propionate	114.0	17.6	17.9	—	17.9
Propyl propionate	131.6	17.3	—	17.4	16.9
Isobutyl propionate	146.7	16.5	—	—	16.5
Isoamyl propionate	165.8	16.7	—	—	16.7
Methyl butyrate	113.7	17.8	—	18.2	18.2
Ethyl butyrate	132.2	17.2	17.0	17.4	17.4
Propyl butyrate	148.1	17.1	—	17.2	16.9
Methyl isobutyrate	114.6	17.1	—	—	17.4
Ethyl isobutyrate	133.6	16.7	—	—	16.8
Propyl isobutyrate	147.3	16.7	—	—	16.3

TABLE 18 (continued)
Hildebrand Parameter Values at 25°C Suggested by Jayasri and Yaseen[60,97] After Correlation With Inverse Molar Volume Dependence.

Liquid	V cm^3 mol^{-1}	δ/MPa$^{1/2}$ Jayasri & Yaseen	Hoy	Burrell	Hansen
	Ketones				
Acetone, 2-propanone	73.3	19.8	19.7	20.5	20.0
Methyl ethyl ketone, 2-butanone	89.6	18.9	19.3	—	19.0
Diethyl ketone, 3-pentanone	105.6	18.4	—	18.0	18.1
Methyl propyl ketone, 2-pentanone	106.1	18.3	—	17.8	18.3
Methyl butyl ketone, 2-hexanone	120.7	18.3	—	—	17.6
Methyl isobutyl ketone	124.9	18.1	—	17.2	17.5
Methyl amyl ketone	138.9	17.7	—	17.4	18.1
Ethyl butyl ketone	139.5	17.8	—	—	18.5
Dipropyl ketone	139.7	17.6	—	—	18.1
Methyl hexyl ketone	196.7	17.7	—	17.2	17.4
Cyclopentanone	88.7	20.6	21.5	21.3	21.6
Cyclohexanone	103.6	20.8	21.3	20.3	20.2
Acetophenone	117.1	21.5	21.6	—	19.8
Isophorone	150.5	19.2	19.2	18.6	19.9
	Ethers				
Methyl isopropyl ether	100.9	15.7	—	—	15.4
Diethyl ether	103.9	15.2	15.4	—	15.6
Ethyl propyl ether	120.4	15.7	—	—	15.9
Ethyl butyl ether	136.6	15.9	15.7	—	16.5
Diisopropyl ether	142.1	14.9	14.4	—	14.4
Dibutyl ether	170.2	16.3	15.9	—	15.9
	Carbitols®				
Methyl Carbitol	118.2	21.7	22.8	—	21.9
Carbitol	136.2	21.1	21.1	19.6	22.5
Butyl Carbitol	170.5	19.4	20.0	18.2	18.3
Diethyl Carbitol	182.0	17.1	17.1	—	18.4
Dibutyl Carbitol	248.0	16.8	17.0	—	17.7
	Cellosolves®				
Methyl Cellosolve	79.1	23.3	23.9	22.1	24.7
Butyl Cellosolve	131.6	19.6	20.2	18.2	21.0
Diethyl Cellosolve	133.7	18.7	17.1	—	20.4
Dibutyl Cellosolve	209.0	16.4	16.8	—	17.1
	Halogen Compounds				
Chloromethane, methyl chloride	57.5	16.4	—	—	—
Chloroethane, ethyl chloride	73.6	16.6	—	—	—
1-Chloropropane, propyl chloride	88.3	16.9	17.2	17.4	17.4
1-Chlorobutane, butyl chloride	104.7	16.9	17.1	—	17.3
2-Chloropropane, isopropyl chloride	90.0	16.4	16.5	—	16.5
2-Chlorobutane, *sec*-butyl chloride	106.1	16.6	16.6	16.6	16.6
2-Ethylhexylchloride	169.1	16.9	16.6	—	17.2
Chlorobenzene	101.7	19.8	19.8	19.4	19.6
Bromobenzene	104.7	20.6	20.2	—	20.0
o-Dichlorobenzene	112.7	21.1	20.5	20.5	20.4

TABLE 18 (continued)
Hildebrand Parameter Values at 25°C Suggested by Jayasri and Yaseen[60,97] After Correlation With Inverse Molar Volume Dependence.

Liquid	V cm^3 mol^{-1}	δ/MPa$^{1/2}$ Jayasri & Yaseen	Hoy	Burrell	Hansen
Tetrachloromethane, carbon tetrachloride	96.5	18.5	17.5	17.6	17.7
Trichloromethane, chloroform	79.7	18.6	18.7	19.0	18.8
1,2-Dichloroethane, ethylene dichloride	79.0	19.2	20.2	20.1	19.0
Trichloroethylene	90.3	18.9	18.7	19.0	19.0
1,3-Dichloropropane, propylene dichloride	97.5	18.8	18.5	19.3	19.3
1,1-Dichloroethane	84.3	19.0	18.3	—	20.0
	Amines				
Propylamine, propanamine	82.2	18.0	18.1	—	18.4
Butylamine, butanamine	98.8	17.6	17.7	—	—
Diethylamine	102.9	17.0	16.5	—	16.3
Hexylamine, hexanamine	132.7	17.3	17.3	—	—
Dipropylamine	137.0	16.2	16.3	—	15.9
Dibutylamine	168.5	16.4	16.7	—	16.7
	Anilines				
Aniline	91.1	21.3	21.9	—	22.6
Methyl aniline	108.7	20.3	20.1	—	21.4
Ethyl aniline	126.8	20.0	21.5	—	—
Dimethyl aniline	126.8	18.2	18.2	—	19.5
Butyl aniline	161.4	19.7	20.4	—	—
	Nitriles				
Acetonitrile	52.4	24.5	26.0	24.3	24.0
Propionitrile	70.3	21.9	22.0	22.1	21.5
Butyronitrile	86.8	20.9	20.8	21.5	20.4
Capronitrile, hexanenitrile	120.1	19.7	—	19.2	18.5

Not unexpectedly, comparison of calculated results with experiment for both liquids and polymers (Section 14.8) suggests that more recent sets of group contribution constants are preferable to Small's original values. However, despite the differences between the values in various sets of group constants, the molar cohesive energies and cohesion parameters for molecules obtained by adding the group contributions usually do not differ too widely, and each system usually provides values acceptable for many practical purposes. It is, of course, important to use self-consistent sets of group contributions. When different techniques are combined (e.g., δ_h contributions added to δ_d evaluated from molar attraction constants[103]) greater variations occur.

Like most other aspects of the cohesion parameter method, even the decision as to the preferred method of defining group contributions or of estimating values is often based on empirical observation. It is essential to be alert for any characteristics of a particular material or combination of materials that are not accounted for in the rather simple models which form the bases of the calculations.

The various cohesion parameter methods described in Chapters 5, 7, and 8 require specific data for each compound, but group methods can predict activity coefficients and other properties in systems containing previously unstudied compounds. Although the cohesion parameter group methods described in this chapter may be used for this purpose, the most popular approach is the UNIFAC (*UNI*QUAC *f*unctional group *ac*tivity) functional group method[99] based on the UNIQUAC (*uni*versal *qua*si *c*hemical) equation.[100] Some comparisons are referred to in Section 7.8.

REFERENCES

1. **Yalkowsky, S. H., Flynn, G. L., and Slunick, T. G.,** Importance of chain length on physicochemical and crystalline properties of organic homologs, *J. Pharm. Sci.*, 61, 852, 1972.
2. **Leshchev, S. M., Starobinets, G. L., Rakhmanko, E. M., and Bubel, O. N.,** Influence of the solvent nature on the solvophobic effect, *Dokl. Akad. Nauk B.S.S.R.*, 31, 725, 1987.
3. **Munafo, A., Buchmann, M., Hô Nam-Tran, and Kesselring, U. W.,** Determination of the total and partial cohesion parameters of lipophilic liquids by gas-liquid chromatography and from molecular properties, *J. Pharm. Sci.*, 77, 169, 1988.
4. **Dunkel, M.,** Calculation of intermolecular forces in organic compounds, *Z. Phys. Chem.*, A138, 42, 1928.
5. **Hayes, R. A.,** The relationship between glass temperature, molar cohesion, and polymer structure, *J. Appl. Polym. Sci.*, 5, 318, 1961.
6. **Fedors, R. F.,** A method for estimating both the solubility parameters and molar volumes of liquids, *Jet Propul. Lab. Quart. Tech. Rev.*, 3(1), 45, 1973; *Polym. Eng. Sci.*, 14, 147, and 472, 1974.
7. **Bondi, A.,** On the solubility of paraffin-chain compounds, *J. Phys. Chem.*, 51, 891, 1947.
8. **Bondi, A. A.,** *Physical Properties of Molecular Crystals, Liquids and Glasses,* John Wiley & Sons, New York, 1968.
9. **Stein, S. E.,** Application of additive estimation methods to vaporization properties of liquids: n-alkanes, *J. Chem. Soc. Faraday Trans. I,* 77, 1457, 1981.
10. **Bunn, C. W.,** The melting points of chain polymers, *J. Polym. Sci.*, 16, 323, 1955.
11. **DiBenedetto, A. T.,** Molecular properties of amorphous high polymers. I. A cell theory for amorphous high polymers, *J. Polym. Sci., A-1*, 3459, 1963.
12. **Maffiolo, G., Vidal, J., and Renon, H.,** Cohesive energy of liquid hydrocarbons, *Ind. Eng. Chem. Fundam.*, 11, 100, 1972.
13. **van Krevelen, D. W. and Hoftyzer, P. J.,** Practical evaluation of the [η]-M relation, *J. Appl. Polym. Sci.*, 11, 2189, 1967.
14. **Rheineck, A. E. and Lin, K. F.,** Solubility parameter calculations based on group contributions, *J. Paint Technol.*, 40, 611, 1968.
15. **van Krevelen, D. W.,** *Properties of Polymers: Their Estimation and Correlation with Chemical Structure,* 2nd ed., Elsevier, Amsterdam, 1976.
16. **Kreibich, U. T. and Batzer, H.,** Influence of the segment structure and cross-linking on the glass transition T_g. Possibilities of predicting T_g using the values of cohesive energy E_{coh}, *Angew. Makromol. Chem.*, 83, 57, 1979.
17. **Kumar, R. and Prausnitz, J. M.,** Solvents in chemical technology, in *Solutions and Solubilities, Part I,* Dack, M. R. J., Ed., John Wiley & Sons-Interscience, New York, 1975, chap. 5.
18. **Meyer, E. F. and Wagner, R. E.,** Cohesive energies in polar organic liquids, *J. Phys. Chem.*, 70, 3162, 1966.
19. **Meyer, E. F., Renner, T. A., and Stec, K. S.,** Cohesive energies in polar organic liquids. 2. The *n*-alkyl nitriles and the 1-chloro-alkanes, *J. Phys. Chem.*, 75, 642, 1971.
20. **Meyer, E. F. and Hotz, C. A.,** Cohesive energies in polar organic liquids. 3. Cyclic ketones, *J. Chem. Eng. Data,* 21, 274, 1976.
21. **Meyer, E. F., Awe, M. J., and Wagner, R. E.,** Cohesive energies in polar organic liquids. 4. *n*-Alkyl acetates, *J. Chem. Eng. Data,* 25, 371, 1980.
22. **Lawson, D. D.,** Methods of calculating engineering parameters for gas separations, *Appl. Energ.*, 6, 241, 1980.
23. **Polak, J.,** Heat of vaporization and cohesion energy of liquids, *Collect. Czech. Chem. Commun.*, 31, 1483, 1966.
24. **Fried, V. and Schneier, G. B.,** Some comments on cohesion energies of liquids, *J. Phys. Chem.*, 72, 4688, 1968.
25. **Figgins, C. E., Risby, T. H., and Jurs, P. C.,** The use of partial molal entropies and free energies in the characterization of gas chromatographic stationary phases, *J. Chromatogr. Sci.*, 14, 453, 1976.

26. **Figgins, C. E., Reinhold, B. L., and Risby, T. H.,** Partial molal and mole fraction free energies as characterizers for stationary liquid phases, *J. Chromatogr. Sci.,* 15, 208, 1977.
27. **Risby, T. H., Jurs, P. C., and Reinhold, B. L.,** Use of thermodynamic solubility parameters for the characterization of liquid phases, *J. Chromatogr.,* 99, 173, 1974.
28. **Exner, O.,** Additive physical properties. II. Molar volume as an additive property, *Collect. Czech. Chem., Commun.,* 32, 1, 1967.
29. **Bowden, S. T. and Jones, W. J.,** Latent heat of vaporization and composition, *Philos. Mag.,* 39, 155, 1948.
30. **Wright, F. J.,** Latent heat of vaporization and composition, *Recl. Trav. Chim. Pays-Bas,* 79, 784, 1960.
31. **Scatchard, G.,** Equilibria in non-electrolyte solutions in relation to the vapor pressures and densities of the components, *Chem. Rev.,* 8, 321, 1931.
32. **Small, P. A.,** Some factors affecting the solubility of polymers, *J. Appl. Chem.,* 3, 71, 1953.
33. **Cammarata, A. and Yau, S. J.,** Molecular estimates for molar attraction constants, *J. Pharm. Sci.,* 61, 723, 1972.
34. **Ostrenga, J. A.,** On the additivity of molar attraction constants, *J. Pharm. Sci.,* 58, 1281, 1969.
35. **Hoy, K. L.,** New values of the solubility parameters from vapor pressure data, *J. Paint Technol.,* 42(541), 76 and 115, 1970.
36. **Hoy, K. L.,** *Tables of Solubility Parameters,* Union Carbide Corporation, Chemicals and Plastics Research and Development Dept., South Charleston, WV, 1969, 1975, and 1985.
37. **Matsuura, T.,** Reverse osmosis separation of solute from aqueous solution, *Yukui Gosei Kagaku Kyokai (J. Syn. Org. Chem. Jpn.),* 31, 717, 1973.
38. **Hoftyzer, P. J. and van Krevelen, D. W.,** Paper IIIa-15, *Int. Symp. Macromol. (IUPAC), Leyden,* 1970.
39. **van Krevelen, D. W.,** Chemical structure and properties of coal. XXVIII — Coal constitution and solvent extraction, *Fuel,* 44, 229, 1965.
40. **Chavan, R. B., Venkata Rao, J., and Ahmad, A.,** Solubility parameters for dyeing water swollen, solvent exchanged cotton with anionic dyes from mixed solvents, *Text. Res. J.,* 53, 347, 1983.
41. **McGowan, J. C.,** The estimation of solubility parameters and related properties of liquids, *J. Chem. Tech. Biotechnol.,* 34A, 38, 1984.
42. **Boberski, W. G., Seiner, J. A., and Petracca, V. G.,** Reducing the viscosity of magnesia cement with organosilicon compounds, *Ind. Eng. Chem. Prod. Res. Dev.,* 21, 531, 1982.
43. **Fisher, C. H.,** Equations correlate *n*-alkane physical properties with chain length, *Chem. Eng.,* 89(19), 111, 1982.
44. **Scatchard, G.,** Equilibrium in nonelectrolyte mixtures, *Chem. Rev.,* 44, 7, 1949.
45. **Exner, O.,** Additive physical properties. I. Relations and problems of statistical nature, *Collect. Czech. Chem. Commun.,* 31, 3222, 1966.
46. **Brower, K. R., Peslak, J., Jr., and Elrod, J.,** A correlation of molecular flexibility with volume and heat of mixing of organic solutes with water and glycol-water mixture, *J. Phys. Chem.,* 73, 207, 1969.
47. **Askadskii, A. A.,** Influence of chemical structure on the properties of polymers, *Pure Appl. Chem.,* 46, 19, 1976.
48. **Askadskii, A. A.,** Computational procedures for the determination of the physical characteristics of polymers, *Usp. Khim.,* 46, 1122, 1977; *Russ. Chem. Rev.,* 46, 589, 1977.
49. **Askadskii, A. A., Kolmakova, L. K., Tager, A. A., Slonimskii, G. L., and Korshak, V. V.,** Evaluation of solubility parameters of low-molecular-weight liquids and polymers, *Dokl. Akad. Nauk SSR,* 226, 857, 1976; *Dokl. Phys. Chem.,* 226, 99, 1976.
50. **Askadskii, A. A., Kolmakova, L. K., Tager, A. A., Slonimskii, G. L., and Korshak, V. V.,** Estimation of cohesion energy of low-molecular-weight liquids and polymers, *Vysokomol. Soedin. Ser. A,* 19, 1004, 1977; *Polym. Sci. U.S.S.R.,* 19, 1159, 1977.
51. **Kaelble, D. H.,** *Physical Chemistry of Adhesion,* John Wiley & Sons-Interscience, New York, 1971.
52. **Moriguchi, I., Kanada, Y., and Komatsu, K.,** Quantitative structure-activity studies. II. van der Waals volume and the related parameters for hydrophobicity in structure-activity studies, *Chem. Pharm. Bull.,* 24, 1799, 1976.
53. **Peiffer, D.,** The molecular factors affecting the solubility parameter, *J. Appl. Polym. Sci.,* 25, 369, 1980.
54. **Drahos, J.,** Liquid-vapor equilibrium. LXXVII. Comparison of the volume and surface modifications of the regular solution model, *Collect. Czech. Chem. Commun.,* 43, 805, 1978; **Drahos, J., Wichterle, I., and Hala, E.,** A generalized method for calculation and prediction of vapour-liquid equilibria at high pressures, *Fluid Phase Equil.,* 1, 173, 1977—1978.
55. **Erdös, E.,** Liquid-vapor equilibriums. XV. Calculation of the constants in the van Laar equation from the properties of the pure components, *Collect. Czech. Chem. Commun.,* 21, 1528, 1956.
56. **Sokolova, S. P. and Pereverzev, A. N.,** Solubility parameters of individual n-alkanes, *Zh. Fiz. Khim.,* 51, 1267, 1977; *Russ. J. Phys. Chem.,* 51, 750, 1977; *Deposited Doc. VINITI,* 472—477, 1977.
57. **Rätzsch, M. and Krahn, G.,** Excess functions in polymer solutions, *J. Polym. Sci. Polym. Symp.,* 42, 1001, 1973.

58. **Watanabe, A. and Sugiyama, S.,** Relation between molecular weight and solubility parameter for pure organic liquids, *Kogyo Kagaku Zasshi,* 72, 1195, 1969.
59. **Konstam, A. H. and Feairheller, W. R., Jr.,** Calculation of solubility parameters of polar compounds, *Am. Inst. Chem. Eng. J.,* 16, 837, 1970.
60. **Jayasri, A. and Yaseen, M.,** Solubility parameter values suggested using the reported and calculated values for organic compounds, *J. Oil Colour. Chem.,* 63(2), 61, 1980.
61. **Wolf, B. A.,** An extrapolation method for the determination of solubility parameters of polymers demonstrated for polyethylene, *Makromol. Chem.,* 178, 1869, 1977.
62. **Wiehe, I. A. and Bagley, E. B.,** Estimation of dispersion and hydrogen bonding energies in liquids, *Am. Inst. Chem. Eng. J.,* 13, 836, 1967.
63. **Fisher, C. H.,** Equations for calculating homolog solubility parameters: The limiting solubility parameter, *J. Coatings Technol.,* 54(688), 67, 1982.
64. **Fisher, C. H.,** Calculating solubility parameters from their molar products and molar quotients, *Polym. Mater. Sci. Eng.,* 51, 588, 1984.
65. **Fisher, C. H.,** Equations for calculating homolog solubility parameters. II. Simple equations based on chain length, *J. Coatings Technol.,* 57(724), 69, 1985.
66. **Rawat, B. S. and Gulati, I. B.,** Cohesive energy density of polar and non-polar liquids, *J. Appl. Chem. Biotechnol.,* 27, 459, 1979.
67. **Brown, H. C., Barbaras, G. K., Berneis, H. L., Bonner, W. H., Johannensen, R. B., Grayson, M., and Nelson, K. L.,** Strained homomorphs. 14. General summary, *J. Am. Chem. Soc.,* 75, 1, 1953.
68. **Bondi, A. and Simkin, D. J.,** On the hydrogen bond contribution to the heat of vaporization of aliphatic alcohols, *J. Chem. Phys.,* 25, 1073, 1956.
69. **Bondi, A. and Simkin, D. J.,** Heats of vaporization of hydrogen-bonded substances, *Am. Inst. Chem. Eng. J.,* 3, 473, 1957.
70. **Anderson, R.,** Ph.D. Thesis, Dept. of Chemical Engineering, University of California, Berkeley, CA, 1961.
71. **Prausnitz, J. M. and Anderson, R.,** Thermodynamics of solvent selectivity in extractive distillation of hydrocarbons, *Am. Inst. Chem. Eng. J.,* 7, 96, 1961.
72. **Blanks, R. F. and Prausnitz, J. M.,** Volumetric properties of gas mixtures containing one or more polar components, *Am. Inst. Chem. Eng. J.,* 8, 86, 1962.
73. **Blanks, R. F. and Prausnitz, J. M.,** Thermodynamics of polymer solubility in polar and nonpolar systems, *Ind. Eng. Chem. Fundam.,* 3, 1, 1964.
74. **Weimer, R. F. and Prausnitz, J. M.,** Screen extraction solvents this way, *Hydrocarbon Process. Petr. Ref.,* 44, 237, 1965.
75. **Hildebrand, J. H., Prausnitz, J. M., and Scott, R. L.,** *Regular and Related Solutions,* Van Nostrand — Reinhold, Princeton, NJ, 1970.
76. **Hansen, C. M. and Beerbower, A.,** Solubility parameters, in *Kirk-Othmer Encyclopedia of Chemical Technology,* 2nd ed., Suppl. Vol., Standen, A., Ed., Interscience, New York, 1971, 889.
77. **Thompson, R. W.,** Ph.D Thesis, Dept. of Chemical Engineering, University of California, Berkeley, CA, 1972.
78. **Helpinstill, J. G. and Van Winkle, M.,** Prediction of infinite dilution activity coefficients for polar-polar binary systems, *Ind. Eng. Chem. Proc. Des. Dev.,* 7, 213, 1968.
79. **Wingefors, S.,** *Studies of the Solvent Extraction Chemistry of Dialkyldithiophosphates and Development of Models for Non-Ideal Behavior in Solvent Extraction,* Thesis, Dept. of Nuclear Chemistry, Chalmers University of Technology, Göteborg, Sweden, 1979.
80. **Wingefors, S. and Liljenzin, J. O.,** Cohesion energy densities and solubility parameters for the n-alkanes expressed as a new homomorph equation, *J. Chem. Technol. Biotechnol.,* 31, 523, 1981.
81. **Beerbower, A. and Dickey, J. R.,** Advanced methods for predicting elastomer/fluids interactions, *Am. Soc. Lubric. Eng. Trans.,* 12, 1, 1969.
82. **Hansen, C. M.,** *Three Dimensional Solubility Parameter and Solvent Diffusion Coefficient. Importance in Surface Coating Formulation,* Doctoral Dissertation, Danish Technical Press, Copenhagen, 1967.
83. **Hansen, C. M.,** The three dimensional solubility parameter — key to paint component affinities. I. Solvents, plasticizers, polymers and resins. II. Dyes, emulsifiers, mutual solubility and compatibility, and pigments, *J. Paint Technol.,* 39, 104 and 505, 1967.
84. **Beerbower, A.,** Surface free energy: a new relationship to bulk energies, *J. Colloid Interface Sci.,* 35, 126, 1971.
85. **Beerbower, A. and Nixon, J.,** Properties of high-internal-phase emulsions. II. Theory of emulsifier selection, *Am. Chem. Soc. Div. Pet. Chem. Prepr.,* 14(1), 62, 1969.
86. **Zisman, W. A.,** Relation of the equilibrium contact angle to liquid and solid constitution, in *Contact Angle, Wettability and Adhesion,* (*Adv. Chem. Ser.* 43), American Chemical Society, Washington, D.C., 1964, chap. 1.

87. **Loev, A. M. and Lavrishchev, L. P.,** Method for the selection of solvents for coating materials, *Lakokras. Mater. Ikh. Primen.,* No. 4, 4, 1977.
88. **Stekol'shchikov, M. N., Krivtsova, L. M., and Ratner, M. I.,** Evaluation of the dissolving ability of hydrocarbon solvents, *Khim. Tekhnol. Topl. Masel,* No. 6, 33, 1987; *Chem. Technol. Fuels Oils,* No. 6, 292, 1987.
89. **Martin, A., Wu, P. L., Adjei, A., and Beerbower, A.,** Extended Hansen solubility approach — naphthalene in individual solvents, *J. Pharm. Sci.,* 70, 1260, 1981.
90. **Wu, P. L., Beerbower, A., and Martin, A.,** Extended Hansen solubility approach — calculating partial solubility parameters of solid solutes, *J. Pharm. Sci.,* 71, 1285, 1982.
91. **Koenhen, D. M., Mulder, M. H. V., and Smolders, C. A.,** Phase separation phenomena during the formation of asymmetric membrane, *J. Appl. Polym. Sci.,* 21, 199, 1977.
92. **Meusberger, K. E.,** Determination of cohesive energy parameters for pesticide formulation development, *Am. Chem. Soc. Symp. Ser.,* 371(Pesticide Formulations), 151, 1988.
93. **Ostrenga, J. A.,** Correlation of biological activity with chemical structure. Use of molar attraction constants, *J. Med. Chem.,* 12, 349, 1969.
94. **Ostrenga, J. A. and Steinmetz, C.,** Estimation of steroid solubility: use of fractional molar attraction constants, *J. Pharm. Sci.,* 59, 414, 1970.
95. **Ahmad, H. and Yaseen, M.,** Estimation of the solubility parameters of low molecular weight compounds by a chemical group contribution technique, *J. Oil Col. Chem. Assoc.,* 60, 99, 1977.
96. **Ahmad, H. and Yaseen, M.,** Principle of additivity of chemical group, atom, and bond used for determination of solubility parameters, *Paintindia,* 30(4), 3, 1980.
97. **Jayasri, A. and Yaseen, M.,** Heat of vaporization: its determination and application in evaluation of solubility parameter: a review, *Prog. Org. Coat.,* 7, 167, 1979.
98. **Qian, X., Lin, Y., and Ren, S.,** Calculation and application of the solubility parameter of disperse dyes, *Huadong Huagong Xueyan Xuebao,* 13, 573, 1987.
99. **Abrams, D. S. and Prausnitz, J. M.,** Statistical thermodynamics of liquid mixtures. New expression for the excess Gibbs energy of partly or completely miscible systems, *Am. Inst. Chem. Eng. J.,* 21, 116, 1975.
100. **Fredenslund, A., Jones, R. L., and Prausnitz, J. M.,** Group-contribution estimation of activity coefficients in nonideal liquid mixtures, *Am. Inst. Chem. Eng. J.,* 21, 1086, 1975.
101. **Varushchenko, R. M., Loseva, O. L., and Druzhinina, A. I.,** Energies of evaporation of dichloroalkanes, *Zh. Fiz. Khim.,* 62, 2329, 1988; *Russ. J. Phys. Chem.,* 62, 1204, 1988.
102. **Sears, J. K. and Darby, J. R.,** *The Technology of Plasticizers,* Wiley-Interscience, New York, 1982.
103. **Samaha, M. W. and Naggar, V. F.,** Micellar properties of non-ionic surfactants in relation to their solubility parameters, *Int. J. Pharm.,* 42, 1, 1988.
104. **Pussemier, L., De Borger, R., Cloos, P., and Van Bladel, R.,** Relation between the molecular structure and the adsorption of arylcarbamate, phenylurea and anilide pesticides in soil and model organic adsorbents, *Chemosphere,* 18, 1871, 1989.

Chapter 7

LIQUIDS

From the definition of the Hildebrand parameter, Equation 5, Chapter 2,

$$\delta = (\Delta U/V)^{1/2} \tag{1}$$

it is apparent that it is necessary to determine both the vaporization energy and the molar volume of a liquid in order to evaluate its Hildebrand parameter. There is rarely much difficulty in finding a reliable value for the molar volume of a liquid, and solids may be treated as subcooled liquids with molar volumes extrapolated to lower temperatures from the liquid state values, but frequently there are problems in obtaining the molar vaporization energy.

7.1 VAPORIZATION ENTHALPY

If the enthalpy of vaporization, ΔH, has been determined calorimetrically (or from the temperature dependence of the vapor pressure) at the required temperature, and if this is well below the boiling point of the liquid, the Hildebrand parameter may be evaluated with the assumption that the vapor is ideal:

$$\delta = (\Delta H - RT)^{1/2}/V^{1/2} \tag{2}$$

(At higher vapor pressures, gas law corrections may be applied, as shown in Section 7.2, but even at the normal boiling point this correction is usually quite small, and unnecessary for many purposes). A typical calculation is shown in Table 1, and values obtained by this method, based on the work of Hildebrand, are listed in Table 3, Chapter 2.

Jayasri and Yaseen[1] reviewed methods of determination of vaporization enthalpies and evaluation of Hildebrand parameters. They also[2] compared experimental values (Table 17, Chapter 6) with those calculated in various ways from the vaporization enthalpies. Majer and Svoboda and co-workers have published several reports[3-8] of experimental vaporization enthalpies and derived thermodynamic properties. In addition, a volume in the IUPAC Chemical Data Series compiled by Majer and Svoboda[9] provides a critical review and data compilation of vaporization enthalpies for organic compounds. (Cohesive energies derived from this source appear in Table 2, Chapter 2.) Wagner, Majer and Svoboda[6] calculated Hildebrand parameters at 25°C and at the normal boiling points for a range of liquids (Table 2).

Other contributions in this area include:

- Sokolova and Pereverzev[10] (higher alkanes)
- Cavé[11] (differential scanning calorimetry method)
- Bollinger, Yvernault, Yvernault, and Houriet[12] (23.4 MPa$^{1/2}$ for 3,3′-oxydipropionitrile)
- Fuchs and Stephenson[13] (vaporization enthalpies, included in the Majer and Svoboda compilation,[9] and a diiodomethane Hildebrand parameter of 24.0 MPa$^{1/2}$)
- Mizerovskii and Vansyatskaya[14] (evaluation of Hildebrand parameters over a wide temperature range for 57 organic compounds from published vaporization enthalpies)

The division of vaporization energy into nonpolar, polar, and hydrogen-bonding components using the homomorph concept is discussed in Chapter 5, particularly Section 5.8. The method may be also applied to liquid mixtures (Section 4.4).

TABLE 1
Calculation of the Hildebrand Parameter of Tetrachloromethane

$t = 20.0°C; T = 293.1$ K
$\Delta H = 33.6$ kJ mol^{-1}
$\rho = 1.60$ g cm^{-3}
$M = 153.8$ g mol^{-1}
$V = M/\rho = 96.1$ cm^3 mol^{-1}
$R = 8.314$ J mol^{-1} K^{-1}
$\Delta H - RT = 31.2$ kJ mol^{-1}
$-U/V = (\Delta H - RT)/V = 325$ J cm^{-3}
$\delta = (-U/V)^{1/2} = 18.0$ MPa$^{1/2}$

When the value of the vaporization enthalpy is known at one temperature (commonly the normal boiling point), it is desirable to be able to evaluate it at another temperature. In such procedures of extrapolating volumes or enthalpies, a *corresponding states* procedure is often used. This is frequently based on the reduced temperature, reduced vapor pressure, and reduced molar volume defined by

$$T_r = T/T_c,\ p_r = p/p_c,\ V_r = V/V_c \tag{3}$$

where the subscript c refers to the values at the critical point. For example, ΔH may be evaluated to within about 2% even within a few degrees of T_c by means of the Watson empirical relationship[15-17]

$$\frac{\Delta H_2}{\Delta H_1} = \left(\frac{T_c - T_2}{T_c - T_1}\right)^{0.38} = \left(\frac{1 - T_{r,2}}{1 - T_{r,1}}\right)^{0.38} \tag{4}$$

Direct experimental information on ΔH, however, is frequently unavailable at any temperature, and many methods have been used to estimate it.[1] Some are thermodynamic, but most are empirical or semiempirical, and the diversity of the methods reflects the fundamental nature of cohesive pressure parameters.

7.2 VAPOR PRESSURE

The most direct thermodynamic method for evaluating Hildebrand parameters is utilization of information on the variation of vapor pressure with temperature and the application of the Clausius-Clapeyron equation:[1,18-21]

$$\mathrm{d}\ \ln p/\mathrm{d}T = \frac{\Delta H}{pT\Delta V} \tag{5}$$

Although exact calorimetric measurements of ΔH are difficult, vapor pressures can be determined accurately as functions of temperature. If it is assumed that the vapor is ideal, from Equation 5 there is obtained

$$\mathrm{d}\ \ln p/\mathrm{d}T = \Delta H/(RT^2) \tag{6}$$

Thus Snow and Barger[357] calculated $\delta = 23.1$ MPa$^{1/2}$ for methanesulfonyl fluoride from the temperature dependence of its vapor pressure and Fujimatsu et al.[374] found Hildebrand parameters at 110°C for liquids forming polyethylene gel adhesives.

If, however, it is not assumed that the vapor is ideal,

TABLE 2
Hildebrand Parameters from Experimental Vaporization Enthalpies

Liquid	δ/MPa$^{1/2}$ 25°C	Normal boiling point (°C)
Aliphatic hydrocarbons		
Pentane	14.5	14.1 (36.0)
Isopentane, 2-methylbutane	13.9	13.9 (27.9)
Hexane	14.9	13.8 (68.7)
2-Methylpentane	14.4	13.5 (60.3)
3-Methylpentane	14.6	13.6 (63.3)
2,2-Dimethylbutane	13.8	12.2 (49.7)
2,3-Dimethylbutane	14.4	13.5 (58.0)
Heptane	15.2	13.4 (98.4)
2-Methylhexane	14.8	13.2 (90.1)
3-Methylhexane	15.1	13.4 (91.9)
2,2,3-Trimethylbutane	14.4	13.1 (80.9)
2,3-Dimethylpentane	15.2	13.6 (89.8)
Octane	15.5	13.0 (125.6)
2-Methylheptane	15.3	13.0 (117.7)
4-Methylheptane	15.3	13.0 (117.7)
Isooctane, 2,2,4-trimethylpentane	14.1	12.4 (99.2)
1-Octene	15.5	13.2 (121.3)
Alicyclic hydrocarbons		
Cyclopentane	17.0	16.3 (49.3)
Cyclohexane	16.6	15.2 (80.7)
Methylcyclopentane	16.1	14.9 (71.8)
Methylcyclohexane	16.0	14.3 (100.9)
cis-1,3-Dimethylcyclopentane	16.0	14.3 (90.8)
Ethylcyclohexane	16.3	13.8 (131.8)
Aromatic hydrocarbons		
Benzene	18.7	17.1 (80.1)
Toluene, methylbenzene	18.2	16.1 (110.6)
Ethylbenzene	18.0	15.2 (136.2)
o-Xylene, *o*-dimethylbenzene	18.3	15.5 (144.4)
m-Xylene, *m*-dimethylbenzene	18.1	15.4 (139.1)
p-Xylene, *p*-dimethylbenzene	18.0	15.3 (138.4)
Ethers		
Diethyl ether	15.5	15.2 (34.5)
Dipropyl ether	15.6	13.8 (90.0)
Diisopropyl ether	14.5	13.3 (68.3)
Dibutyl ether	15.8	13.0 (140.3)
Methyl propyl ether	15.8	15.4 (39.0)
Methyl isopropyl ether	15.5	15.2 (32.5)
Methyl butyl ether	15.9	14.6 (70.2)
Ethyl propyl ether	15.5	14.4 (63.1)
Ethyl isopropyl ether	15.3	14.4 (54.0)
Ethyl butyl ether	15.7	13.9 (92.3)

TABLE 2 (continued)
Hildebrand Parameters from Experimental Vaporization Enthalpies

Liquid	δ/MPa^1/2 25°C	δ/MPa^1/2 Normal boiling point (°C)
Esters		
Ethyl formate	19.2	18.1 (54.3)
Propyl formate	19.0	17.2 (80.9)
Butyl formate	18.3	15.9 (105.6)
Methyl acetate	19.4	18.3 (56.9)
Ethyl acetate	18.4	16.7 (77.2)
Propyl acetate	18.0	15.6 (101.6)
Butyl acetate	17.8	14.8 (126.1)
Methyl propionate	18.6	16.9 (79.8)
Ethyl propionate	17.9	15.6 (99.1)
Propyl propionate	17.6	14.8 (122.4)
Amines		
Propylamine	18.8	17.9 (47.3)
Butylamine	18.4	16.7 (77.1)
Isobutylamine	17.8	16.3 (67.5)
sec-Butylamine	17.3	16.0 (62.7)
Diethylamine	17.0	16.0 (55.6)
Diisopropylamine	14.7	12.3 (83.6)
N-Methylisopropylamine	16.5	15.6 (50.4)
N-Ethylisopropylamine	15.9	14.5 (69.7)
N-Propylisopropylamine	16.3	14.2 (96.9)
N-Butylisopropylamine	15.9	13.3 (123.4)
N-Ethylbutylamine	16.6	14.2 (107.5)
Cyclohexylamine	18.9	16.0 (134.0)
Triethylamine	15.2	13.7 (88.9)
Chlorinated hydrocarbons		
Dichloromethane, methylene dichloride	20.3	19.8 (40.0)
Trichloromethane, chloroform	18.9	17.8 (61.2)
Tetrachloromethane, carbon tetrachloride	17.6	16.2 (76.7)
1,2-Dichloroethane, ethylene dichloride	20.3	18.5 (83.5)
1,1,1-Trichloroethane, methyl chloroform	17.3	16.0 (74.0)
1,1,2-Trichloroethane	20.1	17.4 (113.5)
1,1,2,2-Tetrachloroethane	20.2	16.9 (146.1)
Trichloroethylene	18.9	17.2 (87.3)
Tetrachloroethylene	19.0	16.6 (121.0)
Aldehyde		
Propanal, propionaldehyde	19.4	18.5 (48.8)
Ketones		
Acetone, 2-propanone	19.7	18.6 (56.2)
Methyl ethyl ketone, 2-butanone	19.0	17.2 (79.6)
Methyl propyl ketone, 2-pentanone	18.2	16.1 (102.4)
Diethyl ketone, 3-pentanone	18.4	16.2 (102.0)
Ethyl butyl ketone, 3-hexanone	17.9	15.2 (123.5)
3-Methyl-2-butanone	17.8	15.8 (94.4)

TABLE 2 (continued)
Hildebrand Parameters from Experimental Vaporization Enthalpies

Liquid	δ/MPa$^{1/2}$ 25°C	Normal boiling point (°C)
Alcohols		
Methanol	30.5	29.1 (64.8)
Ethanol	26.5	24.2 (78.3)
2-Propanol, isopropanol	23.9	21.4 (82.2)
1-Butanol	23.8	20.1 (117.8)
1-Pentanol, amyl alcohol	22.5	18.3 (138.1)
Isobutanol, 2-methyl-1-propanol	22.9	19.7 (108.0)

Data selected and adapted from Wágner, Z., Majer, V., and Svoboda, V., *Chem. Listy*, 77, 1150, 1983.

$$d\ \ln p/dT = \Delta H_{ap}/(RT^2) \tag{7}$$

which defines the apparent molar enthalpy of vaporization, ΔH_{ap}. It follows that

$$\Delta H = \Delta H_{ap}\left(\frac{p\Delta V}{\mathrm{RT}}\right) = \Delta H_{ap}\, Z\left(\frac{{}^{g}V - V}{{}^{g}V}\right) \tag{8}$$

where $Z = p\ {}^{g}V/(RT)$ is the compressibility factor, which is unity for an ideal gas but less than unity for real gases. Consequently,

$$\Delta U = \Delta H - p\ \Delta V = \Delta H_{ap}\, Z\left(\frac{{}^{g}V - V}{{}^{g}V}\right) - p({}^{g}V - V) = Z(\Delta H_{ap} - RT) \tag{9}$$

and

$$\delta = [(\Delta H_{ap} - RT)Z/V]^{1/2} \tag{10}$$

This equation reduces to the original definition (Equation 6, Chapter 2) for $Z = 1$. It was used in this form by Sunwoo and Eisen[22] for solvents of pharmaceutical interest. Polak[21] used the alternative nonideality correction coefficient

$$B = {}^{g}V - RT/p \tag{11}$$

in calculating the cohesion energies previously presented (Table 1, Chapter 2).

Several empirical and semiempirical equations have been used for the evaluation of vaporization energies and cohesion parameters of liquids.[20,23-27] In particular, Hoy (Union Carbide Corporation)[26,27] used semiempirical relations in the estimation of Hildebrand parameters and Hansen parameters (Section 5.11) for the wide range of liquids (and solids as subcooled liquids) listed in Table 18, Chapter 5. The Haggenmacher[28] equations are

$$p({}^{g}v - v) = \frac{RT}{M}\left(1 - \frac{pT_c^3}{p_cT^3}\right)^{1/2} \tag{12}$$

$$\Delta H = \frac{dp}{\mathrm{dT}}\frac{RT^2}{Mp}\left(1 - \frac{pT_c^3}{p_cT^3}\right)^{1/2} \tag{13}$$

where ${}^{g}v$ and v are the specific volumes of gas and liquid phases, respectively; the subscript c refers to critical properties and other symbols have their previously defined meanings. The Antoine equation for the vapor pressure has the form

$$\log_{10}p = A - B/(t + C) \tag{14}$$

where A, B, and C are constants. Hoy derived for the Hildebrand or total cohesion parameter

$$\delta_t = \left[\frac{RT\rho}{M}\left(1 - \frac{pT_c^3}{p_cT^3}\right)^{1/2}\frac{2.303BT^2}{(T + C - 273.16)^2} - 1\right]^{1/2} \tag{15}$$

A similar equation was used by Null and Palmer.[29] The data required to calculate the Hildebrand parameter of a pure liquid at any desired absolute temperature T are therefore:

Antoine constants, B, C

critical temperature, T_c

critical pressure, p_c

density, ρ

molar mass, M

The Antoine equation was found to be unacceptably inaccurate at very low pressures in some cases when the vaporization enthalpy at about 25°C was being estimated from data at higher temperatures. The alternative empirical equation used for extrapolation was then

$$\Delta H = \Delta H^\circ \exp(-mt) \tag{16a}$$

or

$$\log_{10}\Delta H = -mt/2.303 + \log_{10}\Delta H^\circ \tag{16b}$$

With this relationship it was possible to estimate ΔH at 25°C after calculating ΔH in the temperature range in which the Antoine constants are valid, and fitting these values to determine the parameters m and ΔH°. Hildebrand parameters may also be estimated from a study of the vapor-liquid properties of multicomponent volatile systems (Section 11.5).

House[30] used the simpler relationship

$$-U = RT\left[\frac{2.303\,BT}{(T + C - 273.16)^2} - 1\right]$$

to determine Hildebrand parameters of 33.3 MPa$^{1/2}$ for xenon difluoride and 30.9 MPa$^{1/2}$ for xenon tetrafluoride at 25°C, decreasing linearly with increasing temperature to 23.6 and 28.9 MPa$^{1/2}$, respectively, at 120°C. House[31] also calculated 15.4 MPa$^{1/2}$ at 20°C for triethylboron and 18.2 MPa$^{1/2}$ at 25°C for diethylzinc. Values are included in Table 3 for tetra-*n*-alkylgermanes[32] and for alkylaluminum compounds.[33]

So far in this section the concern has been the calculation of cohesion parameters from vapor pressures. The process can be reversed, providing a simple method of calculating partial vapor pressure of binary liquid mixtures from the individual vapor pressures of the components by means of cohesion parameters.[18,34] The expressions are

TABLE 3
Hildebrand Parameters as Functions of Temperature for Tetraalkylgermanes and Alkylaluminum Compounds

	$\delta/MPa^{1/2}$			
Compound	**0°C**	**25°C**	**77°C**	**127°C**
$Ge(CH_3)_4$	15.6	13.6	10.9	—
$Ge(C_2H_5)_4$	19.6	17.6	14.9	13.2
$Ge(C_3H_7)_4$	19.9	18.0	15.4	13.7
$Ge(C_4H_9)_4$	23.2	20.3	16.6	14.5
$Ge(C_5H_{11})_4$	25.1	21.5	17.2	14.9
$(CH_3)_3Al$	—	20.8[a]	—	—
$(C_2H_5)_3Al$	—	23.7	—	—
$(C_3H_7)_3Al$	—	17.0[a]	—	—
$(i\text{-}C_4H_9)_3Al$	—	15.7[a]	—	—
$(C_2H_5)_2AlCl$	—	19.9	—	—
$C_2H_5AlCl_2$	—	21.6	—	—
$(CH_3)_3Al_2Cl_3$	—	16.2	—	—
$(C_2H_5)_3Al_2Cl_3$	—	15.2	—	—

[a] 20°C

Adapted from House, J. E., Jr., *Thermochim. Acta*, 69, 361, 1983.

$$\log{}^ip = \log{}^ix + \log{}^ip^\circ + {}^iV{}^j\phi^2 ({}^i\delta - {}^j\delta)^2/(2.303RT) \tag{17a}$$

and

$$\log{}^jp = \log{}^jx + \log{}^jp^\circ + {}^jV{}^i\phi^2 ({}^i\delta - {}^j\delta)^2/(2.303RT) \tag{17b}$$

The value of $({}^i\delta - {}^j\delta)$ is the same in both expressions; if not known it can be estimated, then refined by successive approximation until the sum of the calculated partial vapor pressures, $p_{ca} = {}^ip + {}^jp$, equals the measured total pressure, p_{ob}. This procedure obviates the need for time-consuming analysis of both phases at equilibrium, and is simpler than other procedures.[35] This is illustrated[18,34] for mixtures of benzene ($V = 89.4$ cm^3 mol^{-1}, $p^\circ = 12.7$ kPa, $\delta = 18.7$ MPa$^{1/2}$ at 25°C) with trichlorotrifluoroethane ($V = 120.0$ cm^3 mol^{-1}, $p^\circ = 44.4$ kPa, $\delta = 14.4$ MPa$^{1/2}$). With the quoted values for ${}^i\delta$ and ${}^j\delta$, p_{ca} is higher than p_{ob}, but by reducing $({}^i\delta - {}^j\delta)$ from 4.3 to 4.0 MPa$^{1/2}$ a set of p values is obtained which agrees closely with the measured total pressures (Table 4). The asymmetry in this system (with maximum pressure above ideal at ${}^ix = 0.675$) is fully accounted for by the volume fraction term in Equation 17. Another example is the calculation of component partial pressures in the mercury-tellurium and cadmium-tellurium binary metal systems.[36]

7.3 BOILING POINT

A simple, convenient method of estimating the vaporization enthalpy is the application of Hildebrand and Scott's equation[18,37,38] which expresses ΔH° in terms of the normal boiling point, T_b.

$$\Delta H^\circ/\text{J mol}^{-1} = -12340 + 99.2(T_b/\text{K}) + 0.084(T_b/\text{K})^2 \tag{18}$$

This is based on ''Hildebrand's rule'', which states that the slopes, $(dp/dT)_v$, of liquid-vapor

TABLE 4
Observed and Calculated Total Vapor Pressures at 25°C of Mixtures of Benzene With 1,1,2-Trichloro-1,2,2-Trifluoroethane

Benzene mole fraction	p/kPa			
	p_{ob}	p_{ca}	$^{i}p_{ca}$	$^{j}p_{ca}$
0.260	38.9	38.7	4.7	34.0
0.348	32.1	37.1	6.3	30.8
0.359	36.9	36.9	6.4	30.5
0.510	33.4	33.3	8.1	25.2
0.650	29.6	29.7	9.6	20.1
0.674	28.8	28.8	9.7	19.1
0.776	25.2	24.9	10.1	14.8

Adapted from Hildebrand, J. H., Prausnitz, J. M., and Scott, R. L., *Regular and Related Solutions*, Van Nostrand Reinhold, Princeton, NJ, 1970.

saturation curves of all simple liquids are identical at temperatures where their vapor volumes are equal, the comparison being made at large vapor volume where liquid volumes and departures from gas laws are negligible. Lawson[39] pointed out that no fluorochemical data were included in determining Equation 18 and found for 19 perfluorochemicals the alternative equation

$$\Delta H^{\circ}/\text{J mol}^{-1} = 22210 - 71.8(T_b/\text{K}) + 0.303(T_b/\text{K})^2 \qquad (19)$$

Burrell[40-42] found Equation 18 reasonably accurate only for liquids which are not hydrogen bonded, and recommended that the following corrections be made to the calculated Hildebrand parameter values:

- For alcohols, add 2.9 $\text{MPa}^{1/2}$
- For esters, add 1.2 $\text{MPa}^{1/2}$
- For ketones with boiling points under 100°C, add 1.0 $\text{MPa}^{1/2}$
- For other liquids, no correction

Jayasri and Yaseen[2] suggested adding 3.4 $\text{MPa}^{1/2}$ for alcohols. Thomas[43,44] proposed an alternative empirical relationship between ΔH and T_b for unassociated liquids, and an analogous expression was found by Srivastava[45,46] to relate the internal pressures of 60 liquids to their boiling points. A general relation was found[47] between Hildebrand parameter and temperature, using the normal boiling point:

$$\delta_{298}/\delta_b = 298.15A(\text{K}/T_b) + B \qquad (20)$$

where A and B have characteristic values for noncyclic alkanes and for aromatic hydrocarbons. Kojima, Hamada, Miyahara, and Hosaka[48] used this method for cohesion parameters of oligo(*m*-phenylene)s and oligo(*m*-benzamide)s (Table 5). Other relationships between T_b and Hildebrand parameter have been presented by Fisher and Huddle.[49,50]

Alternative methods proposed by Hildebrand for the estimation of cohesion parameters of nonvolatile materials use fusion enthalpies, and these properties of solids are included in Section 12.2.

TABLE 5
Hildebrand Parameters for Oligo(*m*-Benzamide)s and Oligo(*m*-Phenylene)s from Boiling Points

Compound	$\delta/\text{MPa}^{1/2}$
Diphenyl, $\phi-\phi$	20.5
m-Terphenyl, $\phi-\phi_m-\phi$	17.6
m-Quinquephenyl, $\phi-(\phi_m)_3-\phi$	15.7
Benzanilide, ϕCONHϕ	>19.4
m-Benzamidobenzanilide, ϕCONHϕCONHϕ	>16.1
1,1,3,3-Tetramethylurea	19.6
N,N-Dimethylacetamide	22.0

Adapted from Kojima, T., Hamada, M., Miyahara, M., and Hosaka, Y., *J. Polym. Sci. Polym. Phys.*, 25, 1481, 1987.

7.4 CORRESPONDING STATES

It will be recalled that the complete expression for the cohesive energy of a liquid (Equation 2, Chapter 2) is

$$-U = \Delta H + {}_g\Delta_\infty H - RT + p_s V$$

The terms in this equation can be estimated individually by means of expressions incorporating the reduced quantities in Equation 3. Pitzer and co-workers developed generalized tables for ΔH, ${}_g\Delta_\infty H$ and p_s in terms of T_r, p_r, and the Pitzer acentric factor, ω, defined by the equation[18,51]

$$\omega = -\log p_r - 1.000 \tag{21}$$

where p_r is the reduced vapor pressure at $T_r = 0.7$.

This parameter (Table 6) expresses the effect of a finite molecular core size regardless of the shape of the core, and is a measure of the extra entropy of vaporization above that expected for a simple fluid for which $\omega = 0$. The empirical functions used took the forms

$$\Delta H/(RT) = f_1(T_r,\omega) \qquad {}_g\Delta_\infty H/(RT) = f_2(T_r,p_r,\omega) \qquad \ln p/p_c = f_3(T_r,\omega) \tag{22}$$

where f_1, f_2, and f_3 were functions obtained from experimental data on a variety of nonpolar and slightly polar fluids.

Although Pitzer and co-workers also developed a correlation for liquid volumes, it was based on fluids well above their normal boiling points and with only modest acentric factors. Prausnitz and co-workers[52] considered a wider range of liquids with larger acentric factors. In place of the linear correlation in ω, a quadratic function was used for the reduced volume:

$$V_r = V/V_c = {}^oV_r(T_r) + \omega\,{}^1V_r(T_r) + \omega^2\,{}^2V_r(T_r) \tag{23}$$

where V_c is the critical volume and ${}^oV_r$, ${}^1V_r$, and ${}^2V_r$ are generalized functions evaluated for various liquids from reliable density data for argon, nitrogen, ethylene, propylene, propane, tetrachloromethane, benzene, and heptane. The functions ${}^oV_r$, ${}^1V_r$, and ${}^2V_r$ are given by

$${}^oV_r = 0.11917 + 0.009513\,T_r + 0.21091\,T_r^2 - 0.06922\,T_r^3$$

$$+ 0.07480\,T_r^{-1} - 0.84476\ln(1 - T_r) \tag{24a}$$

TABLE 6
Pitzer Acentric Factors[18,52,54]

Substance	ω
Argon	−0.002
Krypton	−0.002
Xenon	0.002
Methane	0.013
Oxygen	0.021
Nitrogen	0.040
Carbon monoxide	0.041
Ethylene	0.084
Hydrogen sulfide	0.100
Ethane	0.105
Propylene	0.143
Propane	0.152
Tetrafluoromethane, carbon tetrafluoride	0.174
Acetylene, ethyne	0.186
Ethyl mercaptan	0.186
Isobutane, 2-methylpropane	0.189
Tetrachloromethane carbon tetrachloride	0.193
Neopentane, 2,2-dimethylpropane	0.195
1,3-Butadiene	0.197
Butane	0.201
Isopentane, 2-methylbutane	0.206
Dimethyl ether	0.208
Cyclohexane	0.209
Benzene	0.215
Carbon dioxide	0.225
Methylcyclohexane	0.237
Toluene, methylbenzene	0.241
2,3-Dimethylbutane	0.246
Sulfur dioxide	0.246
Ammonia	0.250
Bromobenzene	0.250
Pentane	0.252
Chlorobenzene	0.252
Methyl formate	0.253
2,2,3-Trimethylbutane	0.256
Ethyl ether	0.276
1-Hexene	0.283
p-Xylene, 1,2-dimethylbenzene	0.293
Hexane	0.297
o-Xylene, 1,4-dimethylbenzene	0.300
Isooctane	0.303
Acetone, 2-propanone	0.304
Acetonitrile	0.320
Ethylbenzene	0.342
Heptane	0.352
Ethyl acetate	0.373
Methyl butyrate	0.380
Octane	0.398
Acetic acid, ethanoic acid	0.434
Methanol	0.556
1-Propanol	0.614
Ethanol	0.636

$$^1V_r = 0.98465 - 1.60378\,T_r + 1.82484\,T_r^2 - 0.61432\,T_r^3 - 0.34546\,T_r^{-1} + 0.087037\,\ln(1 - T_r) \quad (24b)$$

$$^2V_r = -0.55314 - 0.15793\,T_r - 1.01601\,T_r^2 + 0.34095\,T_r^3 + 0.46795\,T_r^{-1} - 0.239938\,\ln(1 - T_r) \quad (24c)$$

for $T_r \geqslant 0.56$. For values of $T_r < 0.56$, Prausnitz and Chueh[18,53] proposed

$$V_r = {}^oV_r(T_r) + \omega\,{}^1V_r(T_r) \quad (25)$$

where

$$^0V_r = 0.34037 + 0.009531\,T_r + 0.01131\,T_r^2$$

and

$$^1V_r = -0.16262 + 0.20403\,T_r - 0.64159\,T_r^2 + 0.77987\,T_r^3$$

If reliable experimental values are unobtainable for nonpolar fluids, V_c can be calculated from Pitzer's equation,

$$V_c = Z_c R T_c / p_c \quad (26)$$

where the critical compressibility factor is

$$Z_c = 0.291 - 0.08\omega \quad (27)$$

Lyckmann, Eckert, and Prausnitz[52] calculated $-U$ from Pitzer's tables, and developed a cohesion parameter correlation which is also a quadratic function of the acentric factor, and is reported to have an accuracy of about 1 to 2% for most nonpolar or slightly polar liquids:

$$\delta / p_c^{1/2} = {}^o\delta_r(T_r) + \omega\,{}^1\delta_r(T_r) + \omega^2\,{}^2\delta_r(T_r) \quad (28)$$

The functions $^n\delta_r$ are given by the empirical equation

$$^n\delta_r = {}^n\mathrm{a} + {}^n\mathrm{b}T_r + {}^n\mathrm{c}T_r^2 + {}^n\mathrm{d}T_r^3 \quad (29)$$

with the coefficients in Table 7 for reduced temperatures between 0.5 and 1.0. (These coefficients have been converted to their pure number form from the cal$^{1/2}$ cm$^{-3/2}$ atm$^{-1/2}$ units of the original, using the value from Section 2.5, 1 cal$^{1/2}$ cm$^{-3/2}$ atm$^{-1/2}$ = 6.4260.) The variation of reduced cohesion parameter with reduced temperature and acentric factor is illustrated in Figure 1; at constant acentric factor it decreases with rising reduced temperature until at the critical point it has a small but nonzero value. Preston and Prausnitz[54] extended this work to low reduced temperatures to estimate liquid molar volumes and cohesion parameters for systems of solids in cryogenic solvents (Table 8: again the cohesion parameters have been multiplied by 6.4260). Bazúa and Prausnitz[55] also used reduced Hildebrand parameters when considering cryogenic mixtures.

Deák[56] derived correlations for the determination of the acentric factor and critical pressure of the homomorph of unsaturated normal hydrocarbons from the molar liquid volume at a given temperature.

TABLE 7
Coefficients for Calculating Reduced Cohesion Parameters at Reduced Temperatures Between 0.5 and 1.0

n	^{n}a	^{n}b	^{n}c	^{n}d
0	6.5210759	3.9261171	− 6.9043354	0
1	13.426542	−2.0174752	−24.762043	16.543901
2	−22.896378	52.942741	−17.572308	−13.13696

Adapted from Hildebrand, J. H., Prausnitz, J. M., and Scott, R. L., *Regular and Related Solutions*, Van Nostrand Reinhold, Princeton, NJ, 1970.

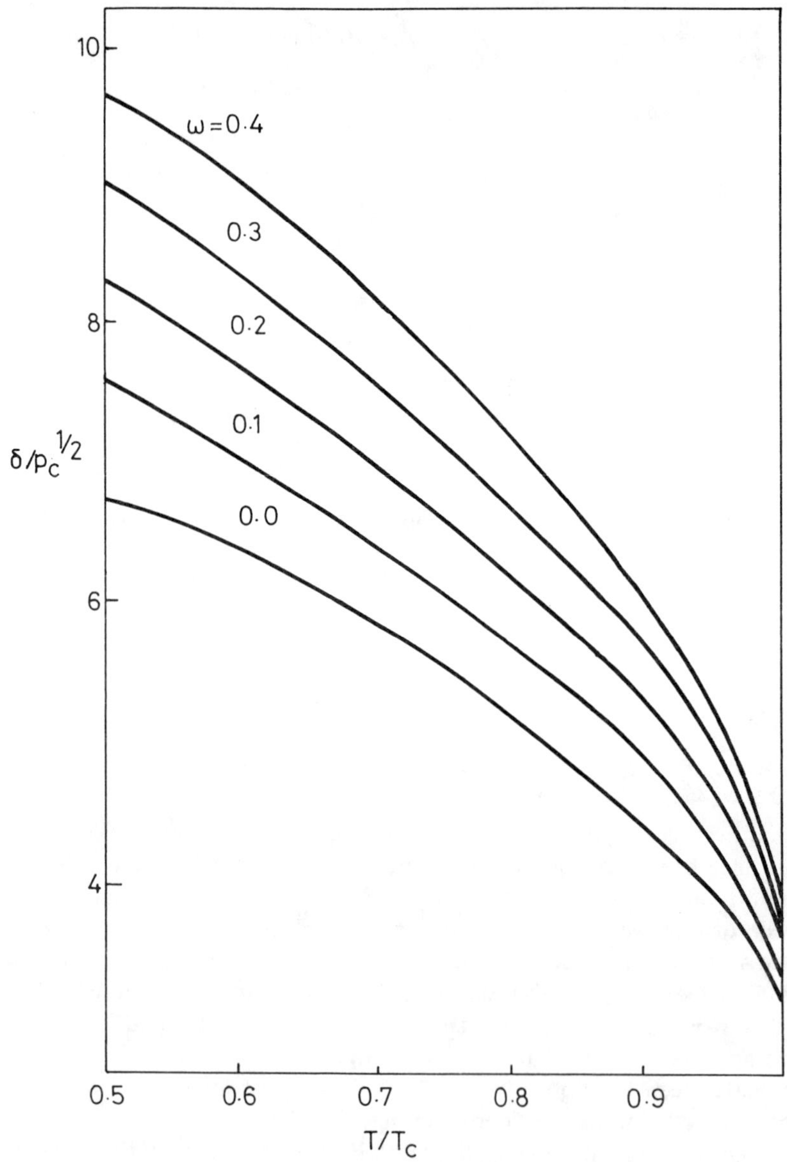

FIGURE 1. Reduced cohesion parameter plotted against reduced temperature at various acentric factor values. (Adapted from Lyckman, E. W., Eckert, C. A., and Prausnitz, J. M., *Chem. Eng. Sci.*, 20, 703, 1965.)

TABLE 8
Reduced Cohesion Parameters[a] (with Reduced Molar Volumes[b] in Parentheses) for Stable and Subcooled Liquids with Various Acentric Factors and at Various Reduced Temperatures

	ω				
T_r	0.0	0.1	0.2	0.3	0.4
0.2	7.711(0.347)	8.771(0.332)	9.639(0.318)	10.410(0.304)	10.892(0.290)
0.3	7.454(0.353)	8.418(0.340)	9.253(0.326)	9.992(0.312)	10.571(0.298)
0.4	7.133(0.362)	8.033(0.349)	8.804(0.336)	9.543(0.322)	10.121(0.309)
0.5	6.747(0.373)	7.551(0.361)	8.322(0.348)	8.996(0.336)	9.639(0.325)
0.6	6.362(0.388)	7.069(0.378)	7.711(0.369)	8.354(0.359)	9.061(0.348)
0.7	5.848(0.414)	6.426(0.408)	7.004(0.400)	7.583(0.390)	8.225(0.379)

[a] Reduced cohesion parameter = $\delta/p_c^{1/2}$.

[b] Reduced volume = V/V_c.

Adapted from Preston, G. T. and Prausnitz, J. M., *Ind. Eng. Chem. Proc. Des. Dev.*, 9, 264, 1970.

TABLE 9
Reduced Temperature Dependence of Hildebrand Parameters for Hydrocarbons

Hydrocarbon	T_c/K	δ_c/MPa$^{1/2}$	k/MPa$^{1/2}$	m	Average % deviation with m = 0.447
Methane	191.1	4.84	13.71	0.448	0.41
Ethane	305.4	5.08	15.44	0.477	1.48
Propane	370.0	4.83	15.24	0.446	0.62
Butane	425.4	4.62	15.18	0.446	0.88
Pentane	469.8	4.44	15.16	0.446	0.72
Hexane	507.9	4.26	15.14	0.446	0.53
Heptane	540.6	4.13	15.14	0.446	0.47
Octane	569.4	3.97	14.73	0.446	1.53
Nonane	593.8	3.88	15.24	0.448	1.31
Decane	616.1	3.78	15.18	0.448	1.06
Undecane	636.0	3.69	15.14	0.446	0.64
Dodecane	653.9	3.60	15.24	0.446	1.26
Ethylene	283.1	5.18	15.55	0.450	1.15
Propylene	365.0	5.08	15.65	0.447	0.50
1,3-Butadiene	425.2	5.00	16.36	0.450	0.90
Cyclohexane	553.0	4.81	15.96	0.446	0.45
Benzene	562.0	5.33	18.00	0.448	0.49
					Mean 0.85

Adapted from Bradford, M. L. and Thodos, G., *Can. J. Chem. Eng.*, 44, 345, 1966.

Bradford and Thodos[57] calculated the Hildebrand parameter from vapor pressures at higher temperatures, up to the critical point (δ_c) and found that the relationship

$$\delta - \delta_c = k(1 - T_r)^m \tag{30}$$

was well obeyed for a wide range of hydrocarbons, with k depending on the class of hydrocarbons and m = 0.447 (Table 9). Corresponding states methods which have been used in calculation of cohesive energy and cohesion parameters include those of Lozada, Monfort, and Del Rio[58] and of Sharma.[59]

Methods of this type were originally applied to substances consisting of spherical or highly symmetrical molecules, but a development of the theory extends it to chain molecules, with three characteristic parameters which have a linear dependence on chain length.[60,61]

7.5 INTERNAL PRESSURE AND COHESIVE PRESSURE

It is clear from Chapter 2 that cohesive pressure is a measure of the *total* molecular cohesion per unit volume, and that on vaporization of a liquid to noninteracting vapor *all* intermolecular bonds or interactions are broken. Cohesive pressure has extremely high values for liquids of high polarity, and very low values for materials such as fluorocarbons where forces of attraction are weak.

On the other hand, the internal pressure is a measure of the change in internal energy of a liquid as it experiences a small isothermal expansion (Equation 7, Chapter 2). Such a differential change in volume does not necessarily disrupt all intermolecular interactions to an equal extent.[46,62-64] The interactions most affected include repulsion, dispersion, and weak dipole processes, and for liquids in which these effects predominate, the values of internal pressure approach those of cohesive pressure (Table 10). In other words, a small volume expansion has a disruptive effect on these interactions which is equivalent to that of vaporization (and of cavity formation:[381] see also Section 8.4).

However, the existence of intermolecular hydrogen bonding in a liquid substantially increases the cohesive pressure, while the internal pressure is comparable to that of liquids without hydrogen bonding. The hydrogen bond possesses a localized, ''chemical'' nature (Section 5.5) which tends not to contribute to the internal pressure, and the hydrogen bonding pressure or energy density contribution can be measured by the difference $(c - \pi)$. The repulsive, dispersive, and dipole interactions can be considered as ''nonchemical'' or ''physical'' interactions.

Mogensen,[65] on this assumption, plotted ortho-positron pick-off rate (Sections 3.6 and 5.11) against $\pi^{1/2}$ and found that the limited number of data points available lay close to a straight line through the points for saturated hydrocarbons, with the exception of 1,4-dioxane, acetonitrile, and carbon disulfide.

Equation 21, Chapter 2, introduced the parameter n as a measure of the ratio of internal pressure to cohesive pressure:

$$(\partial U/\partial V)_T = n(\Delta U/V) \qquad \text{or} \qquad \pi = nc \tag{31}$$

Table 10 lists the values of internal pressure, cohesive pressure, and their ratio n based on those collected by Allen et al.[62] It can be seen that the fluorocarbons show high values of n, and that at the other extreme are the low n, hydrogen-bonded liquids. Values of n for a few polymers appear in Table 6, Chapter 13. Dunlap and Scott[379] showed that n increased from 1.08 to 1.10 between 5°C and 55°C for hexane, but decreased from 1.32 to 1.29 over the same temperature range for perfluorohexane.

Amoros, Solana, and Villar[15] have emphasized that n is temperature-dependent; rounded values from their report appear in Tables 11 and 12. Tiwari and Pandey[66] interpreted the temperature dependence of the internal pressures of mixtures. Bagley, Nelson, Barlow, and Chen[67] suggested

$$\pi = \Delta U + 3\,RT/2 \tag{31a}$$

for nonpolar hydrocarbons, although Amoros et al.[15] found no improvement over $\pi = n\,c$. Another of the alternative equations linking cohesive pressure and internal pressure is the compressibility-solubility relationship of Varadachary and Ramamoorthy.[68] However, a comparison[69] of the internal pressure and cohesive pressure values of carbon disulfide,

TABLE 10
Comparison of Internal Pressures and Cohesive Pressures for Liquids at 20°C[62,72,78]

Liquid	π/MPa	c/MPa	n	Δf
Pentane[a]	229	210	1.09	—
	221	210	1.05	—
Hexane	239	225	1.06	—
Heptane	256	235	1.09	—
Octane	266	242	1.10	—
Nonane	276	249	1.11	—
Decane	280	253	1.11	—
Dodecane	291	260	1.12	—
Tetradecane[a]	305	266	1.15	—
	295	266	1.11	—
Isopentane, 2-methylbutane	229	192	1.19	3.5
2-Methylhexane	249	221	1.12	3.3
2,2-Dimethylpentane	234	203	1.15	3.7
2,3-Dimethylpentane	259	222	1.16	4.3
2,4-Dimethylpentane	236	206	1.14	3.6
2,2,3-Trimethylpentane	238	205	1.16	5.7
3-Methylhexane	252	226	1.11	4.1
3-Ethylpentane	267	231	1.16	4.3
3,3-Dimethylpentane	251	213	1.17	4.4
2-Methylheptane	253	229	1.10	2.4
2,2-Dimethylhexane	251	212	1.18	5.2
2,3-Dimethylhexane	273	229	1.19	5.8
2,5-Dimethylhexane	253	218	1.16	4.6
2,4-Dimethylhexane	252	217	1.16	4.7
2,2,3-Trimethylpentane	255	218	1.18	4.9
Isooctane, 2,2,4-trimethylpentane	236	200	1.18	4.9
2,3,4-Trimethylpentane	264	224	1.18	5.2
2,3,3-Trimethylpentane	262	223	1.18	5.1
3-Methylheptane	266	233	1.14	4.4
3,3-Dimethylhexane	260	218	1.18	5.4
3,4-Dimethylhexane	275	232	1.18	5.6
3-Ethylhexane	271	233	1.16	5.0
2-Methyl-3-ethylpentane	268	229	1.17	—
3-Methyl-3-ethylpentane	280	228	1.22	—
1-Pentene	229	219	1.05	0.9
1-Hexene	245	231	1.06	1.4
1-Octene	267	246	1.09	2.8
1-Decene	283	261	1.09	3.5
Cyclohexane	326	285	1.14	3.6
Methylcyclohexane	297	260	1.14	3.8
Benzene	379	357	1.06	1.6
Toluene, methylbenzene	355	337	1.05	1.5
Ethylbenzene	352	328	1.07	2.5
Isopropylbenzene, cumene	347	308	1.13	—
o-Xylene, 1,3-dimethylbenzene	356	342	1.04	1.4
m-Xylene, 1,3-dimethylbenzene	345	330	1.04	1.5
p-Xylene, 1,4-dimethylbenzene	345	326	1.05	1.9
Diethyl ether	264	251	1.05	—
Dipropyl ether	287	243	1.18	—
1,4-Dioxane	499	402	1.24	—
Acetone, 2-propanone	337	398	0.85	—
Methyl ethyl ketone, 2-butanone	342	383	0.89	—
Cyclohexanone	413	364	1.13	—
Acetophenone	457	456	1.00	—
Methanol	285	887	0.32	—

TABLE 10 (continued)
Comparison of Internal Pressures and Cohesive Pressures for Liquids at 20°C[62,72,78]

Liquid	π/MPa	*c*/MPa	*n*	Δ*f*
Ethanol	291	703	0.41	—
1-Propanol	288	607	0.47	—
1-Butanol	300	485	0.62	—
tert-Butanol	339	473	0.72	—
Ethylene glycol	502	887	0.57	—
Methyl acetate	371	388	0.95	—
Ethyl acetate	354	347	1.02	—
Diisooctyl phthalate	388	331	1.17	—
Diisooctyl sebacate	365	305	1.20	—
Iodomethane	375	414	0.90	—
Dichloromethane	408	414	0.98	0.0
Dibromomethane	470	485	0.97	—
Trichloromethane, chloroform	370	362	1.02	0.5
Tribromomethane	455	473	0.96	—
Tetrachloromethane, carbon tetrachloride	345	312	1.10	2.6
1,2-Dichloroethane	427	416	1.03	0.7
1,2-Dibromoethane	451	418	0.98	—
1,1,2-Trichloro-1,2,2-trifluoroethane	267	205	1.30	—
Trichloroheptafluorobutane	277	189	1.47	—
Perfluoroheptane	220	136	1.62	—
Perfluoro(methylcyclohexane)	231	147	1.58	—
Acetonitrile	379	590	0.64	—
Carbon disulfide	377	410	0.92	—
Hexamethyldisiloxane	188	205	0.92	—

[a] More than one figure is shown for *n*, from different experimental results.

TABLE 11
Variation of the Ratio of Internal Pressure to Cohesive Pressure with Temperature for Gases

	T/K							
Gas	**25**	**40**	**65**	**90**	**130**	**170**	**215**	**255**
Neon	0.79	1.20	—	—	—	—	—	—
Argon	—	—	—	0.89	0.97	—	—	—
Krypton	—	—	—	—	0.85	1.00	—	—
Xenon	—	—	—	—	—	0.83	0.98	1.39
Nitrogen	—	—	0.88	0.99	—	—	—	—
Oxygen	—	—	0.79	0.93	1.03	—	—	—
Methane	—	—	—	—	1.05	1.15	—	—

Adapted from Amoros, J., Solana, J. R., and Villar, E., *Mater, Chem. Phys.*, 10, 557, 1984.

benzene, and cyclohexane emphasizes that one should not expect to find simple relationships between them, or simple trends in their variation with temperature and pressure.

In mixed systems, it can be predicted that[18]

$$\frac{{}^j\overline{V} - {}^jV}{{}^jV} = \frac{({}^i\delta - {}^j\delta)^2}{{}^i(\partial U/\partial V)_T}$$

TABLE 12
Variation of the Ratio of Internal Pressure to Cohesive Pressure with Temperature for Liquids

	T/K							
Liquid	**250**	**280**	**300**	**320**	**340**	**370**	**400**	**450**
Tetrachloromethane	1.10	1.11	1.13	1.13	1.12	—	—	—
Benzene	—	1.08	1.10	1.10	1.09	1.07	1.06	—
Trichlorotrifluoroethane	—	—	1.18	—	—	—	—	—
Perfluorohexane	—	1.32	1.31	1.29	—	—	—	—
Perfluoroheptane	—	1.50	1.48	—	—	—	—	—
Hexane	—	—	1.10	1.11	1.11	1.12	1.12	—
Heptane	—	1.11	1.11	1.12	1.13	1.12	1.11	—
Octane	—	—	1.12	1.13	1.14	1.14	1.14	—
Nonane	—	—	1.11	1.14	1.16	1.12	—	—
Undecane	—	—	1.14	1.15	—	1.11	—	—
Dodecane	—	—	1.20	1.20	1.18	1.18	1.16	—
Hexadecane	—	—	1.28	1.27	1.26	1.26	1.25	—

Adapted from Amoros, J., Solana, J. R., and Villar, E., *Mater. Chem. Phys.*, 10, 557, 1984.

where ${}^j\overline{V}$ is the partial molal volume of the solute j at infinite dilution, ${}^jV$ is its molar volume, and ${}^i(\delta U/\delta V)_T$ is the internal pressure of solvent i. This equation has given reasonably satisfactory estimates of ${}^j\delta$ for drug molecules.[70] Similar calculations have been made for tin(IV) iodide in organic solvents.[71]

An alternative relationship between internal pressure and cohesive pressure has been derived[72] in terms of the change, Δf, in the effective number of classical external degrees of freedom in going from the gas to the liquid state. The parameter Δf was shown to be involved in the general correction term needed to calculate the dispersion cohesion parameter:

$$\delta_d^2 = -U_d/V = \Delta U/V + (RT/2)(\Delta f/V) \tag{32}$$

For a liquid with $n = 1$

$$\Delta f = 2(\pi V - \Delta U)/(RT) \tag{33}$$

implying that $\delta_d{}^2 = \pi$.

For n not equal to unity,

$$\Delta U = \pi V/n - RT\Delta f/2 \tag{34}$$

an equation with two variable parameters. The values of Δf for some liquids are included in Table 10.

Chemical bonding energies are more commonly expressed in units of kJ mol^{-1} than J cm^{-3} or MPa, and these units are used in Table 13 as an alternative way of comparing "non-chemical" or "volume-dependent" interactions, described by

$$-U_V = \pi V \tag{35}$$

with "chemical" or "residual" interactions, described in terms of $-U_R$,

$$-U_R = \pi V - V\Delta U \tag{36}$$

It is clear from Table 13 that $-U_R$ includes other interactions as well as hydrogen bonds,

TABLE 13
Volume-Dependent and Residual Cohesive Energies of Liquids at 25°C

Liquid	$-U_V$/kJ mol^{-1}	$-U_R$/kJ mol^{-1}	$-U$/kJ mol^{-1}
Water	3	39	42
Methanol	13	23	36
Formamide	22	41	63
Acetonitrile	21	10	31
Ethanol	17	24	41
Ethylene glycol	28	22	50
Acetone	25	4	29
Methyl acetate	29	0	29
N,N-Dimethylformamide	36	9	45
Dimethylsulfoxide	36	14	50
Propylene carbonate	46	~19	65
1,4-Dioxane	42	0	42
Ethyl acetate	35	0	35
Benzene	34	0	34
Tetrachloromethane	34	0	34

Adapted from Dack, M. R. J., *Chem. Soc. Rev.*, 4, 211, 1975.

such as strong dipolar interactions, otherwise appreciable hydrogen bonding would be attributed to dipolar aprotic liquids such as *N,N*-dimethylformamide.[46] Hydrocarbons and other liquids of *low dipole moment* (1,4-dioxane, tetrachloromethane) have intermediate $-U_v$ values. *Protic liquids* (notably water) have large $-U_R$ values and a distinct geometrical structure resulting from the hydrogen bonding which hinders the effective operation of nonchemical forces, so they have smaller $-U_v$ terms. *Polar aprotic* liquids, on the other hand, have high values for both terms.

Bagley and co-workers[64,67,73-75] have utilized this chemical bond discriminating property of cohesive pressure and internal pressure to subidivide the Hildebrand parameter in another way. One part corresponding to the physical or nonchemical effects is the *volume-dependent cohesion parameter,* defined by

$$\delta_V^2 = \pi \tag{37}$$

and from Equation 34 it follows that δ_v values are larger than δ_t values calculated from vaporization data. The other component is a *residual parameter,* arising from the chemical effects,

$$\delta_R^2 = (\Delta U - \pi V)/V \tag{38}$$

One advantage of this approach is that for separation of the component parameters it makes use of the thermodynamic quantities π and ΔU which are fairly readily available for most liquids, in contrast to other empirically determined component cohesion parameters. The internal pressure can be measured experimentally by determining the thermal pressure coefficient or the coefficients of expansion and isothermal compressibility (Equation 10, Chapter 2).* Examples of values of δ_V and δ_R as functions of temperature are shown in Table 14.

The responses of materials to ultrasonic energy, which depend on the adiabatic compressibility, may also be used,[89-91] and the internal pressures of liquids and liquid mixtures

* For example: Bagley et al.;[67,75-78] Barton;[79] Chapman and Macdonald[80] Dack;[63] Gibson and Loeffler;[81] Macdonald et al.;[82-84] Malcolm and Ritchie;[85] Suri and Ramakrishna;[86] Watarai and Suzuki;[87] and Westwater et al.[88]

TABLE 14
Volume-Dependent and Residual Cohesion Parameters for Acetone and Alcohols

Liquid	t/°C	V cm^3 mol^{-1}	δ_V/MPa$^{1/2}$	δ_R/MPa$^{1/2}$
Acetone	25	74.0	18.2	8.2
	35	75.1	17.9	8.1
	45	76.3	17.6	7. 8
	55	77.5	17.3	7.6
	65	78.8	17.0	7.3
	75	80.1	16.7	7.1
	85	81.5	16.3	6.8
	95	82.9	16.0	6.6
Methanol	25	40.7	16.9	23.9
	35	41.2	16.7	23.5
	45	41.7	16.6	23.1
	55	42.3	16.5	22.6
	65	42.8	16.4	22.1
	75	43.4	16.3	21.5
	85	44.0	16.2	20.9
	95	44.6	16.2	20.2
Ethanol	30	—	16.7	19.7
	40	—	16.7	19.3
	50	—	16.6	18.9
	60	—	16.6	18.5
	70	—	16.5	18.0
	80	—	16.5	17.4
	90	—	16.5	16.7
1-Propanol	25	—	17.2	17.8
	35	—	17.2	17.3
	45	—	17.1	16.8
	55	—	17.1	16.3
	65	—	17.1	15.7
	75	—	17.1	15.1
	85	—	17.1	14.4
	95	—	17.1	13.6
2-Propanol	25	—	16.7	17.0
	35	—	16.7	16.4
	45	—	16.8	15.7
	55	—	16.9	15.0
	65	—	16.9	14.2
	75	—	16.9	13.4
	85	—	16.9	12.6
	95	—	16.8	11.8
1-Butanol	25	—	17.3	15.0
	35	—	17.3	14.6
	45	—	17.3	14.1
	55	—	17.3	13.6
	65	—	17.1	13.0
	75	—	17.1	12.5
	85	—	17.1	12.5
	85	—	17.1	11.9
	95	—	17.1	11.3

Adapted from Bagley, E. G., Nelson, T. P., and Scigliano, J. M., *J. Paint Tech.*, 43(555), 35, 1971.

can be calculated from other physical properties.[92-96] Paul and co-workers[371] related internal pressures of bisphenol A polymers to gas sorption properties. In the work of Pandey and Shrivastava and co-workers, internal pressures have been identified closely with cohesive energy densities, and physical properties have been used to determine cohesion parameters for liquid metals[97] (Table 25, Chapter 5), liquid metal halides[97] (Table 7, Chapter 12), castor oil ($\delta = 19.8\ \text{MPa}^{1/2}$ at 25°C),[98] and liquid tetrafluoromethane and methane.[99] Sound velocity measurements as probes for determining hydrogen-bonding behavior are discussed in Section 8.1.

Vavruch[100] developed and tested an expression for internal pressure in terms of critical temperature, T_c,

$$\pi/\text{MPa} = 63.0\ (1.5\ T_c - T)/V$$

where V is the molar volume. The mean percentage deviation was 3.8% for 25 selected liquids (excluding alkanes, alkenes, monohydric alcohols, carboxylic acids, water, liquid metals, and probably cryogenic fluids). Previously, Vavruch[101] had explored the relationship between surface energy, internal pressure, and other physical properties. Pant et al.[102] compared the internal pressure and cohesion parameter values of nine hard-sphere fluids; excess internal pressure has been correlated with interactions in binary mixtures of real liquids;[103-106] and Sharma and Das[107] considered internal pressure relationships for slightly non-spherical molecules.

As indicated in Section 5.12 (Equations 83 and 84) it would be expected that

$$\delta_R^2 \approx \delta_h^2 \approx 2\delta_a\delta_b \tag{39}$$

and

$$\delta_V^2 \approx (\delta_d^2 + \delta_p^2) \approx \delta_t^2 - 2\delta_a\delta_b = \delta_d^2 + \delta_o^2 + 2\delta_i\delta_d \tag{40}$$

but because the various types of cohesion parameter have rather different bases, there is no reason to expect these relationships to be exact. Tijssen, Billiet, and Schoenmakers[108] have used a similar set of partial parameters in their discussion of the application of cohesion parameters to the prediction of selectivity and retention in chromatography (Chapter 18):

$$\delta_T^2 = \delta_t^2 + 3RT/(2V) = \delta_d^2 + \delta_o^2 + 2\delta_a\delta_b$$

The term $3RT/(2V)$ is the ideal difference between the external degrees of freedom of vibration and rotation in the gas and liquid phases.

It is clear that the physical/chemical or volume-dependent/residual division is one of the many ways of simplifying the general expression in Equation 22, Chapter 5, with the advantage that it leads to a convenient two-dimensional diagram (such as those used in the presentation of polymer solubility in Section 15.1) and that its components may be evaluated from thermodynamic data.

There is also a densitometric method[18] for Hildebrand parameter determination used by Liron, Cohen, and co-workers[70,109,110] which requires knowledge of liquid internal pressures (see Section 12.1).

7.6 LIQUID-LIQUID MISCIBILITY

This subject was introduced at the end of Section 4.2, where expressions derived from the Hildebrand-Scatchard equation were presented for liquid-liquid phase boundary determination and for the (low) solubilities of systems such as water in low-permittivity liquids.[111,112] Information on mixed liquids appears in Section 4.4.

Limiting low concentration activity coefficients (predictable from cohesion parameters: Section 7.8) can be used to predict binary azeotropes and mutual solubility regions in binary liquid systems. Hildebrand parameter differences may be evaluated from equations such as that developed by Fujishiro and Hildebrand[113] for two-phase regular liquid systems,

$$({}^i\delta - {}^j\delta)^2 = \frac{RT}{2({}^i\phi_2 - {}^i\phi_1)}[{}^iV^{-1}\ln({}^ix_2/{}^ix_1) + {}^jV^{-1}\ln({}^jx_1/{}^ix_2)]$$

for distribution of components *i* and *j* between phases 1 and 2. This has been utilized recently in supercritical fluid systems (Section 11.2) and their application to supercritical extraction (Section 11.4) and supercritical fluid chromatography (Section 18.5).

In addition to Hildebrand, Prausnitz, and co-workers,[18,37,114-119] reviewers including Staverman,[120] Ashton, McDermott and Brench,[121] Williamson,[122] Novák, Matouš, and Pick[123] and Acree[124] have discussed the use of the Scatchard-Hildebrand equation for solubility correlation and prediction.

Enthalpies of solution and related properties in associated systems have been ordered and subdivided on the basis of cohesion parameters.[125-127] In other reports, modeling of liquid-liquid phase equilibria was studied by Bittrich et al.,[128] while Abraham, Kamlet, and Taft[129,130] correlated the standard free energies of solution of nonpolar solutes with the Hildebrand parameter (solvent-solvent interactions), and for polar solutes incorporated the solvatochromic π^* parameter (Chapter 8) which is related to the ability of a bulk solvent to stabilize a dipole through dipole-dipole interactions (solute-solvent interactions). Amidon and Williams[131] developed a solubility equation for nonelectrolytes in water that incorporated the Hildebrand parameter as well as the octanol/water partition coefficient (Section 7.11).

However, not all attempts to correlate properties with cohesion parameters are successful. Grec, Riess, and Devallez[132] found no correlation between the critical solution temperatures of fluorocarbons with hexane and the Hildebrand parameters. Buchtová[133] studied correlation of miscibilities with Hildebrand parameters and π^* parameters (Section 8.4); in most cases the Hildebrand parameter had low significance. Aliev[134] presented a new method of predicting miscibility of liquids that used two mutual solubility parameters; the quality of prediction was similar to that of Hildebrand parameters for liquids without strong hydrogen bonding, and better for strongly hydrogen-bonding liquids.

Liquids of high molecular weight (such as hydrocarbon mineral oils) interact in much the same way as polymers with low molecular weight liquids, and their swelling can be used to determine their Hildebrand parameter values,[135] as described in Section 14.3 for polymers.

7.7 BINARY SOLVENTS

The regular solution method of evaluating the solvent properties of mixed liquids is described in Section 4.4, together with an outline of some of the attempts that have been made to cope with non-regularity. Vapor pressures are discussed in Section 7.2, and activity coefficients are the subject of Section 7.8. Hansen parameters of a few liquid mixtures are collected in Table 12, Chapter 5; methods involving other criteria for solvent formulation are described in Chapter 8; and the behavior of polymers in mixed liquids is discussed in Chapters 13 and 14, particularly Section 13.6.

Acree and Rytting[124,136-143,376,377] used the "Nearly Ideal Binary Solvent" (NIBS) model, extending it to allow for association between the solute and a complexing cosolvent. Excess Gibbs free energies of the mixed solvent systems were estimated from the Scatchard-Hildebrand model, using the liquid Hildebrand parameters. Results reported included the binary systems

- Dibutyl ether with alkanes[138,139]
- Dibutyl ether with 1-chlorohexane[140]
- 1,4-Dioxane with alkanes[142]
- Cyclohexane[136,137,143] and xylene and benzene[377] with alkanes
- Tetrachloromethane with alkanes[136,376]

This approach was linked[141] to that of Huyskens and Haulait-Pirson (Section 8.4).

Hildebrand parameters have been applied to a variety of other binary liquid mixture properties:

- The use of reversed-flow gas chromatography (RFGC) to determine Hildebrand parameter values for ethanol, 1-propanol, acetone, and methyl ethyl ketone assuming a value for water[380]
- Solubilities at the immiscibility critical point of a binary solvent, by Ludmer, Shinnar, and Yakhot[144]
- Excess Gibbs function and vapor pressure values of hydrocarbons with alcohol or phenol, by Książczak and Buchowski[145]
- Binary solvent for drycleaning wool fabrics with a degreasing ability comparable to that of 1,1,1-trichloroethane, by Li et al.[146]
- Volumetric interactions of cyclohexanone with alkanes, by Chawla, Parkash, and Suri[147]
- Coal liquefaction by binary tetralin-polynuclear aromatic solvents, by Chiba, Tagaya, and Saito[148]
- Drug cosolvency[149] (Section 19.2)

The properties of benzene-chlorobenzene and of benzene-toluene mixtures have been discussed by Tiwari and Pandey[66] in terms of the temperature dependence of internal pressure.

7.8 ACTIVITY COEFFICIENTS IN LIQUID-LIQUID SYSTEMS

From the Hildebrand-Scatchard regular solution equation the activity coefficients on the mole fraction scale for components *i* and *j* in an *i-j* mixture are given by Equation 15, Chapter 4:

$$RT\ln {}^{j}f_{x} = {}^{j}V\,{}^{i}\phi^{2}\,({}^{i}\delta - {}^{j}\delta)^{2} \tag{41}$$

$$RT\ln {}^{i}f_{x} = {}^{i}V\,{}^{j}\phi^{2}\,({}^{i}\delta - {}^{j}\delta)^{2} \tag{42}$$

The volume fractions (ϕ) are known from the constitution of the mixture, and the quantities V and δ are available from the properties of the pure components, so that for regular solutions no experimental data on the *mixture* are required. For nonregular solutions the calculated activity coefficients are too high, but at least their values for a number of solvents are usually in the correct rank order.

In infinitely dilute solutions of j in i, ${}^{i}\phi \rightarrow 1$, and for *regular solutions*

$$RT\ln {}^{j}f_{x}^{\infty} = {}^{j}V({}^{i}\delta - {}^{j}\delta)^{2} \tag{43}$$

For systems in which a size effect causes a nonideal entropy of mxing, the Flory-Huggins correction (Section 4.2) is incorporated. With an ideal enthalpy of mixing,

$$\ln {}^{j}f_{x}\,(\Delta_{m}H=0) = \ln({}^{j}\phi/{}^{j}x) + {}^{i}\phi(1 - {}^{j}V/{}^{i}V) \tag{44}$$

Thus, for a solution in which neither $\Delta_{m}H$ nor $\Delta_{m}S$ is ideal,

$$\ln {}^jf_x = \ln({}^j\phi/{}^jx) + {}^i\phi(1 - {}^jV/{}^iV) + ({}^jV/RT){}^i\phi^2({}^i\delta - {}^j\delta)^2 \quad (45)$$

The corresponding expression for infinitely dilute solutions of j (${}^jx \to 0$, ${}^ix \to 1$) is

$$\ln {}^jf_x^\infty = \ln({}^jV/{}^iV) + 1 - {}^jV/{}^iV + ({}^jV/RT)({}^i\delta - {}^j\delta)^2 \quad (46)$$

For multicomponent solutions, expressions involving the volume fraction weighted average Hildebrand parameter (Equation 41, Chapter 4) may be used, for example

$$RT\ln {}^jf_x^\infty = {}^jV({}^j\delta - \bar{\delta})^2 \quad (47)$$

Usually, of course, solutions are not regular because there are specific solute-solvent effects, and activity coefficients are functions of differences not only in cohesive pressure but also of molecular structure.[150] This has been shown clearly by Hildebrand[18,151,152] for the solubility of iodine in different types of solvent (Section 12.1). In these cases, more general expressions are required, for example:

$$RT\ln {}^jf_x = {}^{ij}A\,{}^i\phi^2\,{}^jV \text{ (Equation 15, Chapter 4)}$$

and

$${}^{ij}A = {}^ic + {}^jc - 2\,{}^{ij}c \text{ (Equation 10, Chapter 4)}$$

and

$${}^{ij}c = ({}^ic\,{}^jc)^{1/2}(1 - {}^{ij}l) \text{ (Equation 21, Chapter 5)}$$

leading to

$$RT\ln {}^jf_x = {}^jV\,{}^i\phi^2[({}^i\delta - {}^j\delta)^2 + 2\,{}^{ij}l\,{}^i\delta\,{}^j\delta] \quad (48)$$

or

$$RT\ln {}^jf_x^\infty = {}^jV[({}^i\delta - {}^j\delta)^2 + 2\,{}^{ij}l\,{}^i\delta\,{}^j\delta] \quad (49)$$

A Flory-Huggins size effect term can be incorporated, as shown above. Alternatively, some form of polar-nonpolar approach (Section 5.8) can be used:

$$RT\ln {}^jf_x^\infty = {}^jV[({}^i\delta_\lambda - {}^j\delta_\lambda)^2 + ({}^i\delta_\tau - {}^j\delta_\tau)^2 - 2\,{}^{ij}\psi] + RT\ln({}^jV/{}^iV) + RT(1 - {}^jV/{}^iV) \quad (50)$$

Activity coefficients at infinite dilution are particularly important,[119,153-158] providing predictive information useful for liquid-liquid miscibility, binary liquid-vapor equilibria, and chromatographic retention and selectivity. Monfort, Vidal, and Renon[154] utilized the Weimer-Prausnitz approach[156] (Section 5.8), while Cavé and co-workers[157,158] used limiting activity coefficients to determine cohesion parameters from the expression (Equation 50) of Helpinstill and Van Winkle[159] (also Section 5.8), and thus estimated values of δ_t, δ_d, δ_p, δ_i, and δ_o (Table 15).

A very flexible approach to the correlation and prediction of activity coefficients has been proposed by Null and Palmer,[29,30] assuming that the various contributions, y, to ln ${}^jf_x^\infty$ are additive:

$$\ln {}^jf_x^\infty = \sum_y (\ln {}^jf_x^\infty)_y \quad (51)$$

Contributions may include any or all of the following, together with other effects which may be added as they are developed for any particular system:

TABLE 15
Component Cohesion Parameters from Limiting Activity Coefficients by Cavé and Co-workers[157,158]

Liquid	V/cm^3 mol^{-1}	δ/MPa$^{1/2}$ δ_t	δ_d	δ_p	δ_i	δ_o
Heptane	146.6	15.3	15.3	0.0	0.0	0.0
1-Heptene	—	15.4	15.2	2.7	0.1	1.4
1,6-Heptadiene	—	15.6	15.1	3.8	0.3	2.4
Butyl acetate	131.8	17.8	16.0	7.7	1.4	—
Ethyl acetate	97.8	18.4	15.3	10.2	2.3	5.8
2-Nitropropane	90.1	20.9	16.2	13.1	3.3	8.0
1-Nitropropane	89.0	21.5	16.4	13.9	3.8	—
Nitroethane	71.5	23.5	16.3	17.0	5.5	10.2
Acetonitrile	52.5	24.3	14.5	19.5	6.3	14.1
Nitromethane	53.6	25.9	16.2	20.3	7.0	13.6

- Hildebrand-Scatchard regular enthalpy of mixing, Section 4.2
- Polar orientation and induction effects proposed by Weimer and Prausnitz[156] and Helpinstill and Van Winkle,[159] Section 5.8
- Entropy effects from hydrogen bonding equilibria by the method of Wiehe and Bagley[64,160] (Section 5.5) which reduces to the Flory-Huggins term (Section 13.2) for nonassociating molecules
- Gibbs free energy term for breaking of associating component "bonds" in terms of hydrogen bonding constant K_h

These contributions are summarized in Table 16, and for polar but nonassociating components the expression simplifies to

$$\ln {}^jf_x^\infty = \frac{{}^jV}{RT}\left[({}^i\delta_\lambda - {}^j\delta_\lambda)^2 + {}^{ij}\eta({}^i\delta_\tau - {}^j\delta_\tau)^2\right] + \ln({}^jV/{}^iV) + 1 - {}^jV/{}^iV \qquad (52)$$

The correlating expressions used by Null and Palmer include:

$$^{ij}\eta = {}^jd - {}^je({}^i\delta_\lambda - {}^j\delta_\lambda)^2 \qquad (53)$$

(${}^jd$ and ${}^je$ being parameters characteristic of solute j) and the equilibrium constant defined in terms of the entropy and enthalpy of hydrogen bonding:[64]

$$K_h = \exp[1 - \Delta S_h/R + \Delta H_h/(RT)] \qquad (54)$$

The Hildebrand parameter is correlated by

$$\delta_t^2 = \frac{2.30RB}{V}\left(\frac{T}{t + C}\right)^2 - \frac{RT}{V} \qquad (55)$$

where B and C are constants of the Antoine equation (Section 7.2):

$$\log_{10}(p/\text{mmHg}) = A - \frac{B}{t/°\text{C} + C} \qquad (56)$$

and the polar cohesion parameter is given by an expression similar to that for δ_t:

TABLE 16
Additive Contribution Terms to Infinite Dilution Solute Activity Coefficient

Contribution to $ln\,{}^jf^{\infty}_x$	Include contribution if
Regular enthalpy of mixing	
$({}^jV/RT)[({}^i\delta_\lambda - {}^j\delta_\lambda)^2 + {}^{ij}\eta({}^i\delta_\tau - {}^j\delta_\tau)^2]$	both i and j polar
$({}^jV/RT)[({}^i\delta_\lambda - {}^j\delta_\lambda)^2 + {}^{ij}\eta\,{}^i\delta_\tau^2]$	j nonpolar, i polar
$({}^jV/RT)[({}^i\delta_\lambda - {}^j\delta_\lambda)^2 + {}^{ij}\eta\,{}^j\delta_\tau^2]$	j polar, i nonpolar
$({}^jV/RT)[({}^i\delta_\lambda - {}^j\delta_\lambda)^2]$	both i and j nonpolar
Athermal entropy of mixing	
$1 + \ln({}^jV/{}^iV) + \ln{}^jK_h[\ln(1 + {}^jK_h) - {}^jK_h]/{}^jK_h$	j is associating
$1 + \ln({}^jV/{}^iV)$	j is nonassociating
$-({}^jV/{}^iV)\left\{\frac{1}{1 + {}^iK_h} + \frac{\ln{}^iK_h}{{}^iK_h}\left[\ln(1 + {}^iK_h) - \frac{{}^iK_h}{1 + {}^iK_h}\right]\right\}$	i is associating
$-({}^jV/{}^iV)$	i is nonassociating
Gibbs free energy of association "bonds" broken	
$-\frac{\Delta{}^jG_h}{RT}\left[1 - \frac{\ln(1 + {}^jK_h)}{{}^jK_h}\right]$	j is associating
$-\frac{\Delta{}^jG_h}{RT}\left[\frac{\ln(1 + {}^iK_h)}{{}^jK_h} - \frac{1}{1 + {}^iK_h}\right]\frac{{}^jV}{{}^iV}$	i is associating

Adapted from Null, H. R., *Phase Equilibrium in Process Design,* John Wiley & Sons, New York, 1970.

$$\delta_\lambda^2 = \frac{2.303RB_\lambda}{V}\left(\frac{T}{t + T_L}\right)^2 - \frac{RT}{V} \tag{57}$$

The hydrogen bonding cohesion parameter is given by

$$\delta_\xi^2 = -\frac{\Delta H_h}{V}\left[\frac{K_h - \ln(1 + K_h)}{K_h}\right] \tag{58}$$

and the nonpolar cohesion parameter is obtained by difference:

$$\delta_\tau^2 = \delta_t^2 - \delta_\lambda^2 - \delta_\xi^2 \tag{59}$$

Also used are the equations

$$V = M/(a - bt) \tag{60}$$

to link the molar volume, V, molar mass, M, and density $(a - bt)$; and

$$\Delta G_h = \Delta H_h - T\Delta S_h \tag{61}$$

Each component is therefore characterized by several parameters, and these are listed by Null[29] The overall average error in prediction for 845 literature data points for approximately 300 systems was 25% in ${}^jf^{\infty}_x$ but for all saturated hydrocarbons (including cyclic ones) in all solvents it was less than 9%. This is adequate for screening and for some design purposes.[29]

In *multicomponent* systems for regular solutions

$$RT\ln{}^kf_x = {}^kV({}^k\delta - \bar{\delta})^2$$

where the volume fraction mean Hildebrand parameter is

$$\bar{\delta} = \sum_i {}^i\phi{}^i\delta$$

and it is interesting to note that it follows from regular solution theory that the quantity ${}^kV({}^k\delta - \bar{\delta})^2$ is independent of temperature. However, as pointed out in Section 4.4, it is only if ${}^ix = {}^i\phi$ for all components i that Equation 41, Chapter 4, is obtained from the more general expression, Equation 38, Chapter 4:[18]

$$RT\ln{}^kf_x = {}^kV\sum_i\sum_j ({}^{ik}A - {}^1\!/_2{}^{ij}A){}^i\phi{}^j\phi + RT[\ln({}^k\phi/{}^kx) + 1 - {}^k\phi/{}^kx]$$

The second term (the Flory-Huggins size effect term) is included if there is a large difference in molar volumes of the components. If this full multicomponent equation is used, it is important that the binary coefficients ${}^{ij}A$ are obtained from experimental data using the full binary equation with Flory-Huggins term:

$$RT\ln{}^jf_x = {}^jV{}^i\phi^{2\,ij}A + RT[\ln({}^j\phi/{}^jx) + 1 - {}^j\phi/{}^jx]$$

Alternatively, if the shortened form is used, the ${}^{ij}A$ values must be evaluated from the binary equation without the size effect term:

$$RT\ln{}^jf_x = {}^jV{}^i\phi^2\,{}^{ij}A$$

Activity coefficients may be determined by means of related equations from liquid-liquid distribution data (Sections 7.10 and 7.11 and by chromatographic methods (Chapter 18).

The more recent MOSCED model[153] (Section 5.13) has the form

$$RT\ln{}^jf_x^\infty = {}^jV[({}^i\delta_\lambda - {}^j\delta_\lambda)^2 + {}^iq^2\,{}^jq^2\,({}^i\delta_\tau - {}^j\delta_\tau)^2/{}^i\psi + ({}^i\delta_\alpha - {}^j\delta_\alpha)$$
$$({}^i\delta_\beta - {}^j\delta_\beta)/{}^i\xi] + RT\,{}^{ij}d$$

where ${}^i\psi$ is the polar asymmetry factor, ${}^i\xi$ is the hydrogen-bonding asymmetry factor, and ${}^{ij}d$ is the Flory-Huggins combinatorial size effect term,

$${}^{ij}d = \ln({}^jV/{}^iV)^{ja} + 1 - ({}^jV/{}^iV)^{ja}$$

The degree of unsaturation parameter, q, has values ranging from 1.0 for saturated hydrocarbons to 0.9 for aromatics.

For systems composed of two partially miscible liquids, cohesion parameters thus may be used in combination with other empirical expressions,[161] and although satisfactory data *correlation* is possible, *prediction* is less reliable than one would like. The UNIFAC estimation of activity coefficients by a functional group method (Section 6.10) has been reasonably successful.[124,162,163] Grünbauer and Tomlinson[164] compared the COMMON UNIQUAC method with the much more limited Hildebrand parameter method, while Thomas and Eckert[153] found that their cohesion parameter MOSCED method gave predictive results for limiting activity coefficients which compared well with UNIFAC. The relative predictive abilities have been compared by Cockrem, Meyer, and Lightfoot,[165] and tested further against experimental values by Park and Carr.[155] Hiranuma[368] used aspects of the Hansen parameter approach in a liquid phase activity coefficient method by assuming additive vaporization energies for dispersion, polar, and hydrogen-bonding effects. For fully miscible liquids, excess Gibbs free energies have been estimated from Hildebrand parameters for use in the Nearly Ideal Binary Solvent method (Section 7.7).

Cohesion parameter methods have been used for the study of activity coefficients or of related thermodynamic quantities in many systems in addition to those already mentioned:

- Esters and ketones in a range of solvents, by Aarna, Melder, et al.[166-168]
- Hydrocarbons with phthalate esters and glycols (gas chromatographic stationary phases), by Alessi, Kikic, Tlustos, and Torriano[169,170]
- Long-chain hydrocarbons, carboxylic acids and alcohols by Bondi[150,171]
- Mutual solubilities of phenol, salicylaldehyde and water with and without electrolytes, by Kralj and Sincic[172]
- Donor-acceptor complexes, by Christian, Childs, and Lane[173]
- Organic solutes in molten sulfur, by Dowell and Stewart[174]
- Hydrocarbon systems, by Funk and Prausnitz[175] and Schille and Bittrich[176]
- Long-chain alkylamines and their hydrochlorides, by Kertes[177,178]
- Methyldiphenyl phosphate with hexane and water, by Apelblat[179,180]
- Phase equilibrium studies, by Ghosh[181]
- Solute-solvent interactions of oligo(*m*-benzamide)s with *N,N*-dimethylacetamide, by Kojima et al.[182] (Table 5)
- Resorcinols from liquid-liquid distribution, by Arro, Melder, and Lohmus[183]
- Carbazole in binary alkane-dibutyl ether mixtures, by Acree and McCargar[141] using the model of Huyskens and Haulait-Pirson
- Cryogenic and other vapor-liquid systems (Section 11.3)[365]

7.9 COMPLEXES

One of the early applications of regular solution theory was the discussion of activity coefficient ratios and complex equilibrium constants in terms of Hildebrand parameters. There has been considerable debate as to the correct basis of the equilibrium constant (concentration, mole fraction, volume fraction, or molality), and on activity coefficient values.[184,185] Benesi and Hildebrand[186] pointed out that the equations commonly used for determining donor-acceptor complex equilibrium constants, K_c, from absorbance-concentration data are correct only if the activity coefficient ratio (${}^{DA}f_x/{}^{D}f_x{}^{A}f_x$) remains constant through the composition range. Using Equation 41 they estimated that the cohesion parameter of the complex, ${}^{DA}\delta$, is approximately the arithmetic mean of ${}^{D}\delta$ and ${}^{A}\delta$, and

$$ {}^{DA}f_x/{}^{D}f_x{}^{A}f_x \approx 1 \tag{62} $$

In discussion of a polymeric system with Flory-Huggins entropy of mixing (Sections 4.2, 13.2) Alley and Scott[187] showed that if ${}^{DA}\delta$ is equal to the volume-weighted average of ${}^{D}\delta$ and ${}^{A}\delta$,

$$ {}^{DA}\delta = \frac{{}^{D}V{}^{D}\delta + {}^{A}V{}^{A}\delta}{{}^{D}V + {}^{A}V} \tag{63} $$

then the volume fraction equilibrium constant K_ϕ (and the virtually equivalent concentration constant, K_c, but not the mole fraction constant, K_x) should be approximately independent of composition so that Equation 62 holds.

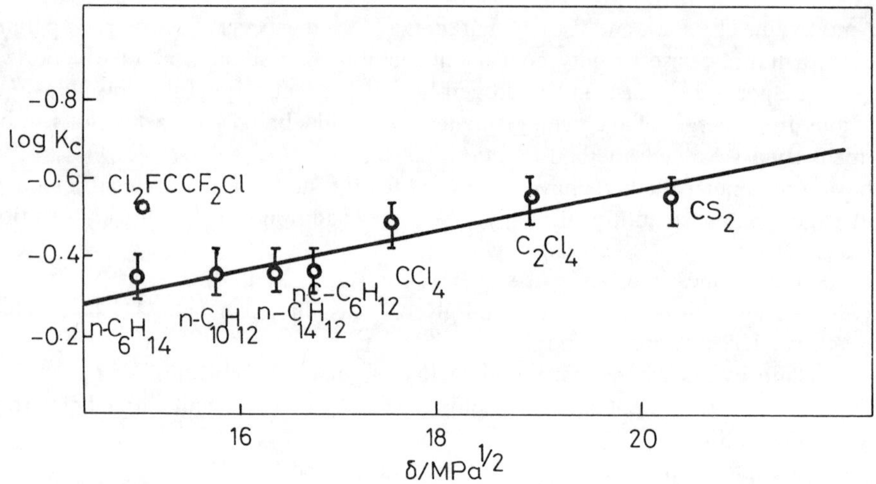

FIGURE 2. Comparison of log K_c for the 1-heptyne-acetone complex with the solvent Hildebrand parameter. (Adapted from Buchowski, H., Devaure, J., Huong, P. V., and Lascombe, J., *Bull. Soc. Chim. Fr.*, 2532, 1966.)

Using Scatchard-Hildebrand theory with no correction for different size effects, Buchowski et al.[188] showed that the mole fraction equilibrium constant, K_x, of a donor-acceptor complex was related to the Hildebrand parameter ${}^i\delta$ of the solvent by

$$\log \frac{{}^{DA}x}{{}^{D}x\,{}^{A}x} = \log K_x = a_1 + b\,{}^i\delta \tag{64}$$

where a_1 and b depend only on the properties of donor and acceptor, including their molar volumes and cohesion parameters. However, no simple correlation of this kind was observed for the 1-heptyne/acetone system in a range of solvents with Hildebrand parameters between 15 and 20 MPa$^{1/2}$. On the other hand, if account is taken of the different molar volumes by a Flory-Huggins term (as discussed in Section 4.3, for example), the resulting expressions are

$$\log \frac{{}^{DA}\phi}{{}^{A}\phi\,{}^{D}\phi} = \log K_\phi = a_2 + b\,{}^i\delta \tag{65}$$

and

$$\log \frac{[DA]}{[A][D]} = \log K_c = a + b\,{}^i\delta \tag{66}$$

which are found to give a reasonably good fit to experimental data for heptyne/acetone[188] (Figure 2), iodine/pyridine[173,189,190] (Figure 3) and phenothiazine/*p*-chloroaniline.[191]

Christian[189] has developed a similar expression by considering the effect of solvation on complex formation equilibria,

$$\log {}^iK_c = \log {}^oK_c + (\alpha - 1)X \tag{67}$$

in which ${}^iK_c$ and ${}^oK_c$ represent 1:1 formation constants for the reaction in solvent i and in a reference medium (either another solvent or the vapor phase), respectively, and α is the ratio of the free energy of transfer of the complex from the reference medium into solvent

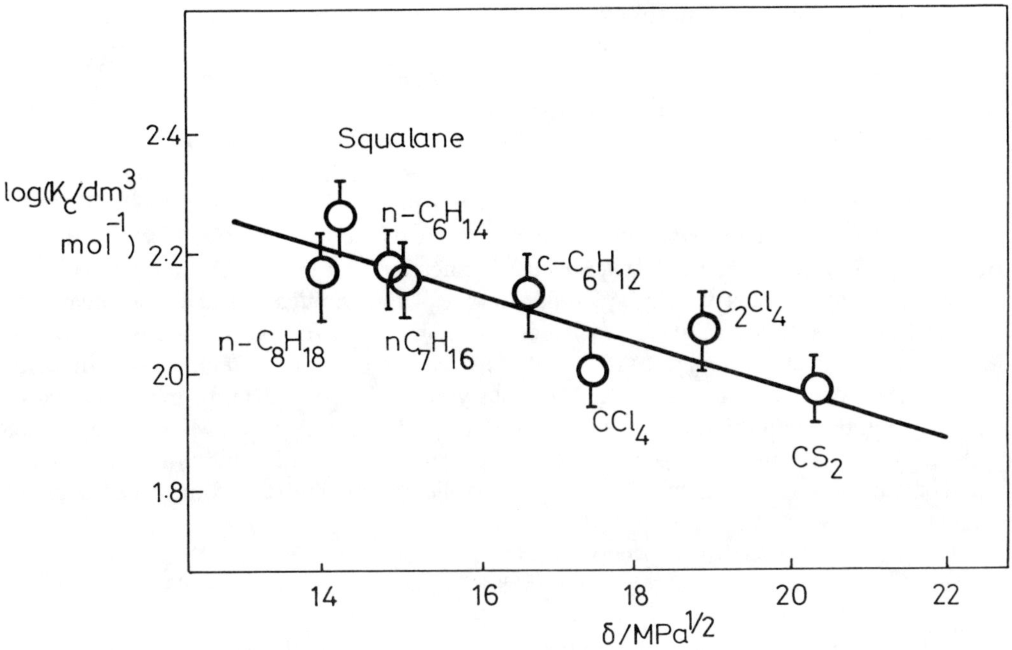

FIGURE 3. Comparison of log K_c for the iodine-pyridine complex with the solvent Hildebrand parameter. (Adapted from Huong, P. V., Platzer, N., and Jasien, M. L., *J. Am. Chem. Soc.*, 91, 3669, 1969.

i to the sum of the free energy of transfer of the uncomplexed donor and acceptor. The parameter *X* is defined

$$X = \log\,({}^{A}k{}^{D}k) \tag{68}$$

where ${}^{A}k$ and ${}^{D}k$ are distribution ratios at infinite dilution for the solutes A and D between solvent *i* and the reference medium. *X* can be expressed in terms of cohesion parameters and molar volumes of the materials involved. Although Equations 65 and 67 are not identical, they are similar, the better performance of the latter being attributed to the inclusion of a term 2log(${}^{o}V/{}^{i}V$) in the variable *X*. Earlier, Alley and Scott[187] had concluded that the use of Flory-Huggins entropy over-corrected for the effect of size differences, and that the actual situation lay somewhere between a constant K_x and a constant K_ϕ or K_c.

In contrast to the linear dependence of log *K* on ${}^{i}\delta$ found by Buchowski and co-workers,[188] Tanaka[192] derived the quadratic function

$$\log K = a + b\,{}^{i}\delta({}^{i}\delta - 2\bar{\delta}) \tag{69}$$

where

$$\bar{\delta} = \frac{{}^{A}V{}^{A}\delta + {}^{D}V{}^{D}\delta - {}^{DA}V{}^{DA}\delta}{{}^{A}V + {}^{D}V - {}^{DA}V} \tag{70}$$

and also illustrated its application to the pyridine/iodine, heptyne/acetone and pyridine hydrate complexes. Tanaka pointed out that when ${}^{i}\delta$ differs markedly from $\bar{\delta}$, (either greater or smaller), the function approximates to a linear dependence of log *K* on either ${}^{i}\delta^2$ or ${}^{i}\delta$. The conclusion from this is that although it may be convenient and useful to use empirical equations to correlate complex equilibrium constants with solvent cohesion parameters, care must be taken in their theoretical interpretation.

Buchtová studied the applicability of the π^* and δ parameters (Section 8.4) for correlation of solvent effects on equilibrium constants and other properties: in most cases the Hildebrand parameter had low significance. Galin and co-workers[193] carried out a similar study of tautomerism of *t*-butyl 2-picolyl and *t*-butyl quinaldyl ketones with δ, π^*, and a hydrogen-bonding parameter.

Cohesion parameter theory can be extended to metal ion complexes and ion-pairs. Tanaka[194,195] has shown that the Hildebrand parameter of a noncharged and unsolvated complex MR_n is equal to or slightly greater than the Hildebrand parameter of the ligand HR. Irving, Smith, and Karmali[196] tested the implication that the Hildebrand parameter of a chelated metal complex should be similar to that of the conjugate ligand acid regardless of the nature of the ''buried'' metal, M. Hildebrand parameters of Be(acetylacetonate)$_2$ (18.6 MPa$^{1/2}$) and Th(acetylacetonate)$_4$ (18.4 MPa$^{1/2}$) from viscosity measurements on dilute solutions in benzene were compared with previously published[197,198] data for trisacetylacetonates of aluminum, chromium, manganese, iron, and cobalt (20.2 $\pm$ 0.4 MPa$^{1/2}$ in benzene, 19.8 $\pm$ 0.6 MPa$^{1/2}$ in trichloromethane). Solvation of the complex by either the organic liquid or water also can be taken into account by consideration of relative values of Hildebrand parameters.

The self-association which occurs in liquids and solutions[29,30,64,160,199], and which can be regarded as a weak form of complexing, is discussed elsewhere, particularly in Sections 5.5 and 2.8.

Complex equilibrium constants and the effects of solvents on them have been determined or correlated with the aid of cohesion parameters for the following systems:

- Tributyl phosphate complexes[200-202]
- Thenoyltrifluoroacetone interactions[200,201,203,204]
- Acetylacetonates of thorium, beryllium, aluminum, chromium, manganese, iron, and cobalt:[196-198,205,358] uranium(inner complex salts);[206,207] thulium;[208] and copper and other metal β-diketone complexes[209-211]
- Metal-porphine complexes[212,213]
- Iodine-solvent interactions[173,189,190,192,214-218]
- γ-Picoline complexes[219-221]
- Quinoline and pyridine complexes[217,222,223]
- Interactions of aromatic donors with arsenic and antimony trihalides[224-226]
- Interactions between tetrabromomethane and aromatic donors by gas chromatography[227]
- Other donor-acceptor complexes[228,229]
- Alpha-beta transition of copper phthalocyanine[369]
- Voltammetric studies of the transition metals manganese (III), iron (III), and copper (II), complexes with *N,N*-disubstituted dithiocarbamates[230]
- Dimerization of benzoic acid and other carboxylic acids[231,232]
- Self-association of alcohols and water[233]
- Acetylene in organic solvents[234]
- Water in organic solvents[111]
- Solvatochromism of bis(bathophenanthroline) dicyanoiron(II) complex[235]

Other complexing sytems which are used in solvent extraction operations are included in Section 7.11.

7.10 SOLVENT-AIDED SEPARATION

A common application of liquids as solvents is to assist the separation of the components of mixtures, either by means of countercurrent liquid extraction, or by extractive distillation (extractive stripping), or by azeotropic distillation. These techniques are known as solvent-aided separation processes.[197,236-239] For example, ethanol cannot be obtained in pure form from water solution by simple distillation because these two liquids form an azeotrope at 96 weight % ethanol, but if benzene is added the binary azeotrope does not form and alcohol may be distilled off at any desired degree of purity. Such vapor-liquid equilibria are considered in Section 11.5.

Another notable example is in the separation of aromatics and naphthenics from petroleum feedstock[239-241] with the aid of a solvent such as nitrobenzene, furfural, phenol, propane, or sulfur dioxide. In this case the aromatics are preferentially dissolved in the added component, enabling the hydrocarbon phase with low aromatic content to be taken from the top of the extractor while the solvent, rich in aromatics together with some light alkanes, is withdrawn from the bottom. The solvent also increases preferentially the volatility of the nonaromatics, allowing extractive distillation. Other applications include the extraction of ionic materials in the form of electrically neutral ion-pairs or complexes (Section 7.11), studies of the lipophilic properties of drug molecules (Chapter 19), and phase-transfer catalysis.[249]

The extent of separation possible at equilibrium in any separation or extraction process may of course be expressed in terms of the product compositions in the two phases.[30,238] The *separation factor* ${}^{ij}\alpha$, is defined as the ratio of distribution ratios,

$$ {}^{ij}\alpha = {}^{i}D_x / {}^{j}D_x \tag{71} $$

where each distribution ratio, D_x, is the ratio of mole fractions in phases 1 and 2 at equilibrium:

$$ {}^{i}D_x = {}^{i}x_1 / {}^{i}x_2 \tag{72} $$

so

$$ {}^{ij}\alpha = \frac{{}^{i}x_1 / {}^{i}x_2}{{}^{j}x_1 / {}^{j}x_2} = \frac{{}^{i}x_1 {}^{j}x_2}{{}^{j}x_1 {}^{i}x_2} \tag{73} $$

Separation can be effected if ${}^{ij}\alpha$ is significantly different from unity. It may be either greater or smaller than unity, but by convention components i and j are usually defined so that ${}^{ij}\alpha$ is greater than one.

For separation processes such as extractive distillation which are based on equilibration between gas and liquid phases, the selectivity or separation factor is a measure of the ability of a solvent to increase the volatility of one component relative to another. It can be shown that in this case, at low to moderate pressures,

$$ {}^{i}p = {}^{i}f_x {}^{i}p_s {}^{i}x \tag{74} $$

where ${}^{i}p$ is the partial pressure, ${}^{i}p_s$ is the pure component saturation pressure and ${}^{i}f_x$ is the activity coefficient of component i. The separation factor (in this context commonly called the relative volatility) is

$$ {}^{ij}\alpha = \frac{{}^{i}f_x {}^{i}p_s}{{}^{j}f_x {}^{j}p_s} \tag{75} $$

This property ${}^{ij}\alpha$, is dependent on composition, because of the composition dependence of ${}^{i}f_x$ and ${}^{j}f_x$, but it is relatively insensitive to temperature and pressure, and can be evaluated with activity coefficients estimated from cohesion parameters.

For equilibrium between two immiscible liquid phases, Equation 74 can be used for both phases, and as these are necessarily in equilibrium with the same vapor phase,

$$^ia_1 = {}^ia_2 \tag{76}$$
$$^iD_x = {}^ix_1/{}^ix_2 = {}^if_{x,2}/{}^if_{x,1}$$

and

$$^{ij}\alpha = \frac{{}^ix_1/{}^jx_2}{{}^ix_2/{}^jx_2} = \frac{{}^if_{x,2}/{}^jf_{x,1}}{{}^if_{x,1}\ {}^jf_{x,2}} \tag{77}$$

This indicates the tendency for component i to be extracted more readily from phase 2 (the feed) into the solvent phase 1 than component j, and only a knowledge of the activity coefficients is necessary to predict selectivity at any particular composition. If the concentration of solvent in the feed (phase 2) is very low, it is sometimes possible to assume that

$$^if_{x,2}/{}^jf_{x,2} \approx 1$$

so the *selectivity* of a solvent toward the i-j mixture will be determined by

$$s_{ij} = {}^if_{x,1}/{}^jf_{x,1} \tag{78}$$

From Equation 76, which gives the solubility of component i in solvent phase 1 at equilibrium compared with its solubility in feed phase 2, it can be seen that as $^if_{x,2}$ (the activity coefficient of the transferring solute in the feed phase) is relatively independent of the nature of the solvent, the capacity of any solvent for the transferring solute i will largely depend on $^if_{x,1}$.

To illustrate the use of cohesion parameters, if in the situation described by Equation 78 the solvent k is present in exess (when the selectivity is a maximum) so that the solution is infinitely dilute with respect to components i and j, an equation such as Equation 16, Chapter 4, can be used to estimate the selectivity,

$$RT \ln {}^if_x^\infty = {}^iV\ {}^{ik}A \tag{79}$$

$$RT \ln {}^jf_x^\infty = {}^jV\ {}^{jk}A \tag{80}$$

$$RT \ln {}^ks_{ij} = {}^iV\ {}^{ik}A - {}^jV\ {}^{jk}A \tag{81}$$

Any of the appropriate prevously described expressions for the cohesive pressure in terms of cohesion parameters can then be substituted for ^{ik}A and ^{jk}A. In general terms, the choice of a noninteracting solvent with a cohesion parameter which differs significantly from the cohesion parameters of the components to be separated enhances the solvent *selectivity*, although it reduces the solvent *capacity*. It is therefore necessary to reach a compromise such that the solvent has an adequate capacity but retains good selectivity and immiscibility between the phases. However, if a solvent can be obtained which specifically interacts with the component to be extracted, then both the selectivity and the solvent capacity are increased. This can be illustrated in terms of component cohesion parameters. Using Equation 21, Chapter 5, results in the expression

$$RT \ln {}^k s_{ij} = {}^iV\, {}^{ik}A - {}^jV\, {}^{jk}A = ({}^i\delta_d - {}^k\delta_d)^2\, {}^iV + ({}^i\delta_o^2 + {}^k\delta_o^2)\, {}^iV - 2\, {}^iV\, ({}^i\delta_i\, {}^k\delta_d + {}^k\delta_i\, {}^i\delta_d) - ({}^j\delta_k - {}^k\delta_d)^2\, {}^jV - ({}^j\delta_o^2 + {}^k\delta_o^2)\, {}^jV + 2{}^jV\, ({}^j\delta_i\, {}^k\delta_d + {}^k\delta_i\, {}^j\delta_d) \tag{82}$$

For the separation of nonpolar hydrocarbons i and j with the aid of a polar liquid k, the component cohesion parameters ${}^i\delta_o$, ${}^j\delta_o$, ${}^i\delta_i$ and ${}^j\delta_i$ are negligible, so

$$RT \ln {}^k s_{ij} = [({}^i\delta_d - {}^k\delta_d)^2\, {}^iV - ({}^j\delta_d - {}^k\delta_d)^2\, {}^jV] + [{}^k\delta_o^2\, ({}^iV - {}^jV)] + [2\, {}^k\delta_i({}^jV\, {}^j\delta_d - {}^iV\, {}^i\delta_d)] \tag{83}$$

The three bracketed terms in this equation indicate the contributions made by the added solvent to the selectivity as a result, respectively, of the dispersion effect (D), the orientation effect (O), and the inductive effect (I), and the equation can be written

$$RT \ln {}^k s_{ij} = D + O + I + E + AB$$

with the addition of an entropy term E[156,159] and an acid-base term AB if desired.

Although equations such as this are not quantitatively reliable for prediction, they provide useful qualitative guides. Hydrocarbons have similar δ_d values, so the dispersion term is significant only if ${}^iV$ and ${}^jV$ are markedly different. The polar effect also is proportional to the difference between the hydrocarbon molar volumes, and usually dominates in Equation 83. The inductive energies of hydrocarbons do not vary significantly, so the third term is usually small. The conclusion in this particular case is that a solvent is most selective if its orientation parameter value is high, and the selectivity depends strongly on the difference in size of the hydrocarbons being separated. (On a molecular level, this implies that the larger hydrocarbon molecules have a higher probability of interacting with the polar solvent molecules).

In addition, of course, the selectivity also depends on the ability of a solvent to form complexes, with the electron donor-acceptor or Lewis acid-base contribution to cohesive pressure depending on the six δ_a and δ_b parameters of components i,j and k. The term to be incorporated in Equation 83 has the form

$$AB = [2\, {}^iV\, ({}^i\delta_a\, {}^i\delta_b + {}^k\delta_a\, {}^k\delta_b - {}^i\delta_a\, {}^k\delta_b - {}^k\delta_a\, {}^i\delta_b) - 2\, {}^jV\, ({}^j\delta_a\, {}^j\delta_b + {}^k\delta_a\, {}^k\delta_b - {}^j\delta_a\, {}^k\delta_b - {}^k\delta_a\, {}^j\delta_b)] \tag{84}$$

In the simplest case of an alkane i (for which ${}^i\delta_a$ and ${}^i\delta_b$ are negligible), an aromatic hydrocarbon j(${}^j\delta_a$ negligible) and an electron-accepting solvent k (${}^k\delta_b$ negligible) this reduces to

$$AB = [2\, {}^jV\, {}^k\delta_a\, {}^j\delta_b] \tag{85}$$

A different but related formalism was used by Buchowski and Pawlowski[243] for the influence of electron donor-acceptor interactions on the partition coefficients of iodine between water and organic liquids. Other cohesion parameter based methods of activity coefficient estimation (Section 7.8), as well as several alternative procedures, have been suggested for determining suitable mixtures for solvent-aided separation processes from a knowledge of pure component properties.[239]

Berg and Yeh[244,245] have used cohesion parameters (Hansen parameter plots and the MOSCED model: see Chapter 5) for investigating the reversal of relative volatilities in binary liquid systems, the breaking of ternary acetate-ethanol-water azeotropes by extractive distillation, and the separation of acetone-methanol mixtures.

7.11 LIQUID EXTRACTION

The discussion in Section 7.10 is concerned with the situation where a solvent k is present (often in excess) during the separation of i and j. Another common situation involves a relatively *small* amount of material k (often an electrically neutral ion-pair or complex used to transport ionic species) distributed between two phases, frequently an organic solvent, o, and water, aq.[197,246] The activity of a solid k is referred to the hypothetical supercooled liquid at the same temperature, and its Hildebrand parameter has the value that the supercooled liquid would have (Section 12.2). Equation 15, Chapter 4, can be written for each phase in turn, relating the activity, a, of the solute in that solvent to its mole fraction, x. The exchange cohesive pressure terms can be expressed in full as in Equation 82, but this is cumbersome, and usually the Hildebrand parameters are used:

$$\ln {}^k a_o = \ln {}^k x_o + {}^k V \, {}^o\phi^2 \, ({}^k\delta - {}^o\delta^2/(RT) \tag{86}$$

$$\ln {}^k a_{aq} = \ln {}^k x_{aq} + {}^k V \, {}^{aq}\phi^2 \, ({}^k\delta - {}^{aq}\delta)^2/(RT) \tag{87}$$

When partition equilibrium is established between the two phases (Equation 76),

$${}^k a_o = {}^k a_{aq}$$

and if ${}^o\phi$ and ${}^{aq}\phi$ can be equated with unity, the mole fraction distribution ratio of distribuend k is

$$\ln {}^k D_x = \ln ({}^k x_o / {}^k x_{aq}) = {}^k V \, [({}^k\delta - {}^{aq}\delta)^2 - ({}^k\delta - {}^o\delta)^2]/(RT) \tag{88}$$

This is sometimes expressed in the equivalent form

$$\ln {}^k D_x = {}^k V \, [({}^o\delta + {}^{aq}\delta - 2\,{}^k\delta)\,({}^{aq}\delta - {}^o\delta)]/(RT) \tag{89}$$

The limiting distribution ratio or limiting distribution coefficient,

$${}^k K_x = \lim_{{}^j x_o \to 0} \left(\frac{{}^k x_o}{{}^k x_{aq}} \right) \tag{90}$$

often replaces ${}^k D_x$ in the above equations; the extraction free energy may also be used. If the Flory-Huggins correction for volume dissimilarity is included (Section 13.2) the analogous expression is in terms of the volume fraction distribution ratio,

$$\ln {}^k D_\phi = \ln ({}^k\phi_o / {}^k\phi_{aq}) = [{}^k V/(RT)][({}^k\delta - {}^{aq}\delta)^2 - ({}^k\delta - {}^o\delta)^2] + {}^k V \, ({}^o V^{-1} - {}^{aq} V^{-1}) \tag{91}$$

An equivalent equation in terms of the molar concentration distribution coefficient is[114]

$$\ln {}^k D_c = {}^k V \, ({}^{aq}\delta + {}^o\delta_{FH} - 2\,{}^k\delta)\,({}^{aq}\delta - {}^o\delta)/(RT) \tag{92}$$

where

$${}^o\delta_{FH} = {}^o\delta + RT \, ({}^o V^{-1} - {}^{aq} V^{-1})\,({}^{aq}\delta - {}^o\delta)^{-1} \tag{93}$$

Equations such as 88 and 91 are not directly applicable to practical extraction systems which are by no means regular. Olszer and Siekierski[224] proposed an alternative way of using such equations by selecting one organic substance (s) as a standard:

$$\ln ({}^kD_x/{}^sD_x) = {}^kV\,[({}^k\delta - {}^s\delta)^2 - ({}^k\delta - {}^o\delta)^2]/(RT) \tag{94}$$

or

$$\ln ({}^kD_\phi/{}^sD_\phi) = [({}^kV/(RT)][({}^k\delta - {}^s\delta)^2 - ({}^k\delta - {}^o\delta)^2] + {}^kV\,({}^oV^{-1} - {}^sV^{-1}) \tag{95}$$

Assumptions implicit in these equations include:

- ${}^ka_o = {}^ka_s = {}^ka_{aq}$;
- ${}^kV$ is the same in all solvents;
- the composition of the aqueous phase does not vary;
- all the solutions are regular.

It follows from Equations 88 and 94 that $\ln {}^kD_x$ and $\ln({}^kD_x/{}^sD_x)$ should be parabolic functions of ${}^o\delta$, and this is observed in some systems.[197]

Provided the ratio of activity coefficients remains constant, the ${}^kD$ values will remain constant right up to the stage where each phase is simultaneously saturated with the distribuend k, so Equation 88 becomes

$$\ln {}^kD_x = \ln ({}^kx_{o,s}/{}^kx_{aq,s}) = {}^kV\,[{}^{aq}\phi^2\,({}^k\delta - {}^{aq}\delta)^2 - {}^o\phi^2\,({}^k\delta - {}^o\delta)^2]/(RT) \tag{96}$$

This equates distribution coefficients with the ratio of solubilities, ${}^kx_{o,s}/{}^kx_{aq,s}$, (Chapter 12).

Srebrenik and Cohen[247] derived a more general form of Equation 91 which differs in not assuming a regular solution model, and in incorporating the partial molal volume ${}^k\overline{V}$ of solute k in organic and aqueous phases in place of the molar volume ${}^kV$. If ${}^k\overline{V}_{aq} = {}^k\overline{V}_o = {}^kV$, this equation reduces to the form of Equation 91. The major barrier to its use is the extremely limited amount of data available on partial molar volumes, but where tested it gives good agreement between theoretically and experimentally derived values.[247]

Cohesion parameters have been used in various other ways in liquid extractions:

- Siekierski[248] has shown that the interaction of extractants (e) with diluents (d) can be described in terms of regular solution theory, with the use of

$$\ln {}^ef_x = {}^eV\,{}^d\phi^2\,({}^e\delta - {}^d\delta)^2/(RT) \tag{97}$$

- A correlation function used for a similar purpose (phase transfer catalysis) but based on hydrophile-lipophile balance ideas (Section 17.7) is[249]

$$D(\delta) = (\delta - {}^{aq}\delta)^2/(\delta - {}^o\delta)^2 \tag{98}$$

- Noel and Meloan[250] correlated distribution constants with molecular refractive index, solvent Hildebrand parameter, and density
- Infinite dilution activity coefficients of solutes such as dihydric phenols[183] have been determined on the basis of partition between two liquid phases with Hildebrand parameter expressions
- Cohesive pressure or internal pressure terms have been included in other treatments of liquid-liquid equilibria, such as that of Irving and Lewis[251] for the distribution ratio of indium halides between halogen acids and various organic solvents, and the study of Watarai and Suzuki[87] on partition between heptane and water-dimethylsulfoxide
- Distribution coefficients (or partition coefficients) and cohesion parameters are used as measures of the lipophilic properties of drug molecules, which limit the magnitude

of a biological response by partitioning a chemical agent betwen aqueous and organic biophases and so governing the penetration of the molecules through tissues (Chapter 19)

- Assessment of cohesion parameter components for a polar substance can be based on its partition between water and a polar liquid incapable of hydrogen bonding[252-255] (Section 5.6)
- Correlation of octanol/water partition coefficients has been carried out with solvatochromic parameters (Section 8.4) including Hildebrand parameters,[256] and by means of other solubility equations[257]
- Buchtová, Koutek, and Bekárek[133] correlated distribution coefficients of various materials between water and other liquids with the solvatochromic π^* and δ parameters; the Hildebrand parameter had low significance
- The methylene group increment effect on extraction in sixteen polar liquid-octane systems has been studied,[258] the liquids falling into three groups with different slopes in the linear relationships between methylene group increment and Hildebrand parameter
- Octanol/water partition for modeling more complex natural polymer systems (Section 16.11)
- Hansen parameter diagrams may be used when designing solvent extraction mixtures[121,259] (see also Section 5.9)

Another aspect of solvent extraction to have been discussed in terms of cohesion parameters is the effect of diluents. An empirical equation has been developed.[372] Alekperov[373] explained the effects of diluents on the extraction of boric acid with 2-hydroxy-5-octylbenzylethanolamine, and Horwitz, Martin, and Diamond[375] considered the influence of diluents on the extractant octyl(phenyl)-*N,N*-diisobutylcarbamoylmethylphosphine oxide.

As in all other applications of cohesion parameters, it should not be expected that distribution ratios can be predicted in detail, and even if the infinite dilution values are approximately correct, these have limited application. Rather, the emphasis should be on formulating broad trends and on making correlations. When extrapolations and predictions are made, these are unlikely to prove precise, although they are useful as estimates.[260,261]

Most extraction systems use an aqueous phase, and one problem in the application of Hildebrand parameters for the prediction or correlation of distribution data is the value of ${}^{aq}\delta$. This is usually quoted as being about 50 $MPa^{1/2}$, and although this seems correct for solutions of water in hydrocarbons, for solutions of hydrocarbons in water a more appropriate value is about 30 $MPa^{1/2}$, the δ_h contribution being lower for this type of system (Table 14, Chapter 5).[161,262-264]

The Hansch lipophilic parameter based on partition coefficient information is described in Section 19.1. Grünbauer, Tomlinson, and co-workers[265] have compared UNIQUAC and cohesion parameter approaches to liquid-liquid distribution. Information on complex formation is included in Section 7.9 and on ion-pair formation in Section 12.4.

The application of cohesion parameters to liquid-liquid extraction has been discussed by several authors.* Equations of the type presented above have been applied to a great variety of systems, including:

- Distribution between aqueous solutions and organic liquids of thenoylacetone, thenoyltrifluoroacetone, thiothenoyltrifluoroacetone, and their metal chelates[200,201,203,204,278-284]
- Organophosphorus extractants including diethyl phosphate, tributyl phosphate, tributyl phosphine oxide, trioctyl phosphine oxide, triphenyl phosphine oxide, and other phosphine oxides[201,202,248,285-289,375,378]

* For example: Dobson and van der Zeeuw;[366] Hampe;[266] Högfeldt;[261] Irving;[197] King;[238] Noel and Meloan;[250] Ritcey;[267] Sierkierski;[248] Tanaka;[192,195] and Wingefors.[268-277]

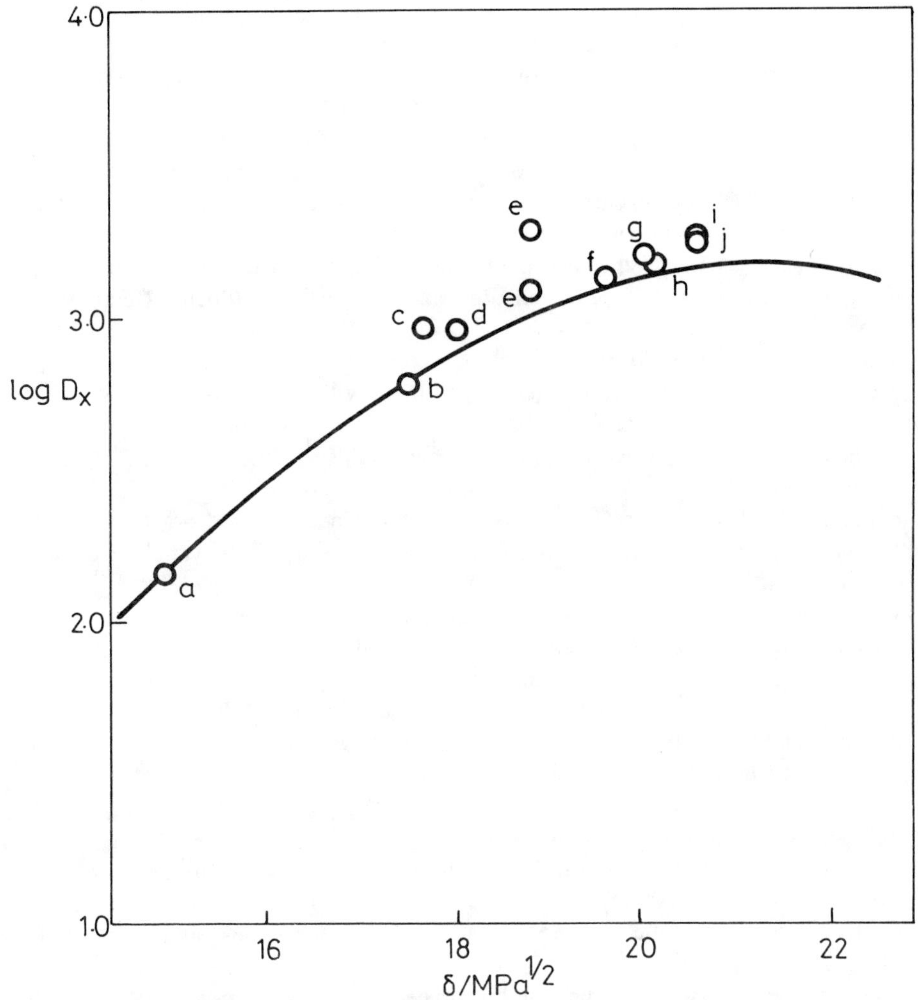

FIGURE 4. Effect of liquid Hildebrand parameter on the distribution ratio of 8-quinolinol between aqueous and organic phases: a, hexane; b, tetrachloromethane; c, bromobutane; d, toluene; e, trichloromethane; f, dichloromethane; g, 1,2-dichloroethane; h, dichlorobenzene; i, bromochloromethane; j, dibromethane. (Adapted from Mottola, H. A. and Freiser, H., *Talanta*, 14, 864, 1967.)

- 8-Hydroxyquinoline (oxine; $\delta = 22\ MPa^{1/2}$; for aqueous soluton $35\ MPa^{1/2}$), alkyl hydroxy quinoline (Kelex® 100), 8-quinolinols, and their metal chelates (Figure 4)[290-292,362,363]
- β-Diketones such as acetylacetone and their complexes with a wide variety of metal ions[196-198,206-211,264,278,293-296]
- Nickel dimethylglyoximate ($\delta = 26\ MPa^{1/2}$; $38\ MPa^{1/2}$ for aqueous solution)[361]
- Baes, Moyer, et al.[359] developed a computer program for interpreting solvent extraction data which used Hildebrand parameters for the organic phase, the solvent Hildebrand parameter being non-adjustable but all solute Hildebrand parameters being adjustable
- Selectivity of sulfolane for aromatic hydrocarbons[240]
- Ion-association extraction systems (trilaurylammonium perrhenate, dextromorphan iodide in cyclohexane-trichloromethane)[297]
- Solute-solvent interactions in ion-pair extraction of tetraalkylammonium iodides (δ from relations of the type indicated above for R_4NI, R = ethyl, propyl, butyl, isoamyl, $22.3 \pm 0.4\ MPa^{1/2}$)[298,299]

- Partition coefficients between organic liquids and water of phenols[300-304,364] and their derivatives,[302,305,306,364] and anisidines[304]
- Partition coefficient between organic and aqueous phase of 1-phenyl-3-methyl-4-benzoyl-5-pyrazolone[307]
- Distribution coefficients of organics between water (and wastewater) and several organic liquids using the polar-nonpolar cohesion parameter model (Section 5.8)[308]
- Partition coefficients of carboxylic acids, nitro-compounds, anilines, pyridines and iodine between organic and aqueous phases[232,234,305,309-312]
- Benzoic acid[231,313] and other carboxylic acids[310] and their extraction of copper(II)[314]
- Copper extraction with capric acid (decanoic acid)[321] and diluent effects with LIX® 64N[322]
- Water in organic liquid mixtures[261,315]
- Extraction of nitric acid from aqueous solutions with butyl acetate[316]
- Amine,[289,317,318] amine salt,[177,178,319] and quinoline[289] extractants
- Extraction of arsenic trihalides into aromatic liquids[225,226]
- Germanium tetrahalides[320]
- γ-Picoline and cobalt(γ-picoline)$_4$(NCS)$_2$ between aqueous and organic phases[219-221]
- Sulfur-containing extractants (dialkyldithiophosphates and dihexyl sulfoxide) and their metal chelates[269-273]
- Quaternary ammonium and phosphonium salts as phase transfer catalysts[249], and for solvent extraction of thiocyanate[323] and metal ions[324]
- Chlorobenzene partition coefficients[325]
- Distribution of benzidine and *o*-tolidine between organic phases[326]
- Aluminum extraction with mononomyl phosphoric acid[327]
- Diluent effects in metal extractions[289,318,324,327]
- Americium(III) chelates[363]
- Extractability of potassium as picrate[328]
- Extraction properties of protonated bis(4-ethyl-3,5-dipropyl-1-pyrazolyl)methane for iron, gallium, and thallium[329]
- Distribution of aliphatic alcohols (C_7 to C_{15}) between octane and solutions of zinc chloride in acetone[330]
- Activity coefficients by gas chromatograph head space analysis for diluents in D2EHPA-hydrocarbon binary systems, with δ for D2EHPA calculated at 17.0 $MPa^{1/2}$ [331]
- Solvent leaching of metal-containing compounds from residual oils[332]
- Solvent extraction of organics from atmospheric particulate matter[333]
- Lipid extraction from biological tissue[334] and from flour[335-340]
- Partition of hexane into lipid bilayer membranes[341]
- Solvent extraction of leaf juice curd[342]
- Distribution of cholesterol in the fractions of milk fat[360]
- Effect of Hildebrand parameter on extraction of thiocyanate[367]
- Liquid-liquid partition of 21H,23H-porphine[370]

Other systems used in solvent extraction are listed in Section 7.10.

7.12 ACIDITY-BASICITY LINEAR ENERGY SCALES

Drago[343-346] proposed for Lewis acid-base interactions (Section 5.4) a double scale based on the equation

$$-\Delta H_{ab} = E_a E_b + C_a C_b \tag{99}$$

to correlate enthalpies of adduct formation between acceptor or Lewis acid (a) and donor or Lewis base (b) in the gas phase and in poorly solvating solvents. The values of the parameters were obtained empirically from directly measurable enthalpy values, E_a and C_a being set equal to 1.0 for iodine. Self-consistent sets of E and C values for a wide range of acids and bases allow the prediction of ΔH_{ab} for many adduct combinations.

This approach to electron pair donor/electron pair acceptor interactions has something in common with the expanded interaction cohesion parameter expressions in that it separates the tendency to undergo electrostatic or ionic interactions (the E term) from the tendency to take part in covalent interactions (the C term), and in that the dimensions of the parameters include $(\text{energy})^{1/2}$. However, comparison with Equation 22, Chapter 5, shows that there is little similarity in the detailed division of the terms, and that no simple correlation between the two approaches is possible.

There are several related systems. Examples of reaction rate correlations appear in Section 10.1, and the *lipophilic parameter* is introduced in Section 19.1. The *proton affinity*[347-349] relates to compounds in the gas phase, without interference from specific and nonspecific interactions in the liquid phase. The *Hammett*[350] *equation* takes the form

$$\text{kog}\ (K/\ K_o)\ =\ \sigma\rho \tag{100}$$

where K is the equilibrium constant or rate coefficient for a side-chain reaction of a *meta-* or *para*-substituted benzene derivative, and K_o is the constant for the ''parent'' compound. The substituent constant, σ, is a measure of the basicity of the substituent (that is, relative to hydrogen) and ideally would be independent of the nature of the reaction. The reaction constant, ρ, is characteristic of the particular reaction and its susceptibility to changes in substituent basicity. The parameter σ is thus a measure of the relative basicity of a homologous series of solvents, and has something in common with δ_b. Rider[351-353] developed a two-parameter method of treating hydrogen-bonding interactions between liquids and polymers in terms of hydrogen-bond donating (b) and accepting (C) tendencies of each component, such that $-\Delta H = b\,C$. This model provides solubility maps which have similarities to Hansen maps (Section 5.9). Abraham, Kamlet, Taft, and co-workers also have developed a system of linear solvation energy relationships (Sections 8.2 and 8.4) with one set of constants characterizing a given solute and another set of parameters characteristic of the solvents, the appropriate products providing energy terms.

More direct measures of liquid basicity include the enthalpy of mixing with trichloromethane and with other halogenated hydrogen-bond donors in basic solvents[349,354] and the enthalpy of coordination with antimony trichloride.[354,355] Most of the other numerous scales of ''basicity'' or ''electron donating ability'' or ''proton affinity'' (Section 8.3) are single-valued, and even more difficult to correlate with each other or with cohesion parameters. Although they agree in their general trends, the detailed orders of solvent basicity depend on the system from which the scale is derived.[349,356]

REFERENCES

1. **Jayasri, A. and Yaseen, M.,** Heat of vaporization: its determination and application in evaluation of the solubility parameter: a review, *Progr. Org. Coat.*, 7, 167, 1979.
2. **Jayasri, A. and Yaseen, M.,** Solubility parameter values suggested using the reported and calculated values for organic compounds, *J. Oil Col. Chem.*, 63(2), 61, 1980.
3. **Majer, V., Svoboda, V., Pošta, A., and Pick, J.,** Enthalpy data of liquids. XXII. Determination of heats of vaporization and some other thermodynamic quantities for three fluorinated halogen ethanes, *Collect. Czech. Chem. Commun.*, 46, 817, 1981.

4. **Svoboda, V., Charvátová, V., Majer, V., and Pick, J.,** Enthalpy data on liquids. XXIV. Determination of heats of vaporization and some other thermodynamic quantities for four alkylcycloparaffins, *Collect. Czech. Chem. Commun.*, 46, 2983, 1981.
5. **Svoboda, V.,** Cohesion energy of liquids, *Chem. Listy,* 76, 389, 1982.
6. **Wágner, Z., Majer, V., and Svoboda, V.,** Calculation of solubility parameters and their temperature dependence for a selected group of substances, *Chem Listy,* 77, 1150, 1983.
7. **Uchytilová, V., Majer, V., and Svoboda, V.,** Enthalpies of vaporization and cohesive energies of 2-chloro-1,1,2-trifluoroethyl ethyl ether, 2-chloro-1,1,2-trifluoroethyl propyl ether, 2-chloro-1,1,2-trifluoroethyl chloromethyl ether, and 2-chloro-1,1,2-trifluoroethyl difluoromethyl ether, *J. Chem. Thermodyn.*, 16, 475, 1984.
8. **Majer, V., Svoboda, V., and Lencka, M.,** Enthalpies of vaporization and cohesive energies of pyridine and isomeric methylpyridines, *J. Chem. Thermodyn.*, 16, 1019, 1984.
9. **Majer, V. and Svoboda, V.,** *Enthalpies of Vaporization of Organic Compounds. A Critical Review and Data Compilation* (IUPAC Chemical Data Series, Vol. 32), Blackwell, Oxford, 1985.
10. **Sokolova, S. P. and Pereverzev, A. N.,** Solubility parameters of individual n-alkanes, *Zh. Fiz. Khim.*, 51, 1267, 1977; *Russ. J. Phys. Chem.*, 51, 750, 1977; *Deposited Doc. VINITI,* 472-77.
11. **Cavé, G., Ceolin, R., and Puisieux, F.,** Demonstration of a method for determining the solubility parameters of liquids by measuring the heat of vaporization, *Pharm. Acta Helv.*, 58, 172, 1983.
12. **Bollinger, J.-C., Yvernault, G., Yvernault, T., and Houriet, R.,** Physical properties and basicity of oxydipropionitrile, *Can. J. Chem.*, 62, 361, 1984.
13. **Fuchs, R. and Stephenson, W. K.,** personal communication, 1986.
14. **Mizerovskii, L. N. and Vansyatskaya, L. N.,** Temperature dependence of the solubility parameter of organic liquids, *Izv. Vyssh. Uchebn. Zaved., Khim. Khim. Tekhnol.*, 31(4), 51, 1988.
15. **Amoros, J., Solana, J. R., and Villar, E.,** Behaviour of the internal pressure of liquids in accordance with variations in temperature, volume, and cohesive energy density, *Mater. Chem. Phys.*, 10, 557, 1984.
16. **Fedors, R. F.,** A method for estimating both the solubility parameters and molar volumes of liquids, *Jet Propul. Lab. Quart. Tech. Rev.*, 3(1), 45, 1973; *Polym. Eng. Sci.*, 14, 147, and 472, 1974.
17. **Watson, K. M.,** Prediction of critical temperature and heats of vaporization, *Ind. Eng. Chem.*, 23, 360, 1931; Thermodynamics of the liquid state. Generalized prediction of properties, *Ind. Eng. Chem.*, 35, 398, 1943.
18. **Hildebrand, J. H., Prausnitz, J. M., and Scott, R. L.,** *Regular and Related Solutions,* Van Nostrand-Reinhold, Princeton, NJ, 1970.
19. **Hobin, T. P.,** Model polyethers. IV. Vapour pressures of oligomers of poly(tetramethylene oxide), *Polymer,* 9, 65, 1968.
20. **Mandik, L. and Korinsky, J.,** Cohesion energy of nonpolar liquids, *Chem. Listy,* 61, 727, 1967.
21. **Polak, J.,** Heat of vaporization and cohesion energy of liquids, *Collect. Czech. Chem. Commun.*, 31, 1483, 1966.
22. **Sunwoo, C. and Eisen, H.,** Solubility parameter of selected sulfonamides, *J. Pharm. Sci.*, 60, 238, 1971.
23. **Christian, S. D. and Yeo, K. O.,** A method for predicting vapor pressures of nonpolar liquids, *Thermochim. Acta,* 5, 95, 1972.
24. **Miller, A. A.,** Vaporization energy and density relations in nonpolar liquids, *J. Phys. Chem.*, 68, 3900, 1964; Volume-energy relations in liquids at 0°K, *J. Phys. Chem.*, 69, 3190, 1965.
25. **Miller, A. A.,** Free volumes in simple and polymeric liquids, *J. Polym. Sci. A-2,* 4, 415, 1966.
26. **Hoy, K. L.,** *Tables of Solubility Parameters,* Union Carbide Corporation, Chemicals and Plastics Research and Development Dept., South Charleston, WV, 1969, 1975, 1985.
27. **Hoy, K. L.,** New values of the solubility parameters from vapor pressure data, *J. Paint Technol.*, 42(541), 76 and 115, 1970.
28. **Haggenmacher, J. E.,** The heat of vaporization as a function of pressure and temperature, *J. Am. Chem. Soc.*, 68, 1633, 1946.
29. **Null, H. R.,** *Phase Equilibrium in Process Design,* John Wiley & Sons-Interscience, New York, 1970; **Null, H. R. and Palmer, D. A.,** Predicting phase equilibria, *Chem. Eng. Progr.*, 65(9), 47, 1969.
30. **House, J. E., Jr.,** Cohesion energies and solubility parameters for xenon difluoride and xenon tetrafluoride, *J. Fluorine Chem.*, 22, 299, 1983.
31. **House, J. E., Jr.,** Cohesion energies and solubility parameters for triethylboron and diethylzinc, *Thermochim. Acta,* 71, 215, 1983.
32. **House, J. E., Jr.,** Cohesion energies and solubility parameters for tetra-*n*-alkylgermanes, *Thermochim. Acta,* 69, 361, 1983.
33. **House, J. E., Jr.,** Molecular association in trialkylaluminium and alkylaluminium compounds determined from solubility parameters and related thermodynamic quantities, *J. Organomet. Chem.*, 263, 267, 1984.
34. **Linford, R. G. and Hildebrand, J. H.,** Calculation of partial from total vapor pressures. System C_6H_6 + CCl_2FClCF_2, *Ind. Eng. Chem. Fundam.*, 8, 846, 1969.
35. **Prausnitz, J. M., Eckert, C. A., Orye, R. V., and O'Connell, J. P.,** *Computer Calculations for Multicomponent Vapor-Liquid Equilibria,* Prentice-Hall, Englewood Cliffs, NJ, 1966.

36. **Martin, B. G.,** Calculation of component partial pressures in the Hg-Te and Cd-Te binary systems using regular associated solution theory, *Bull. Am. Phys. Soc.,* 25, 349, 1980.
37. **Hildebrand, J. H.,** *Solubility of Non-Electrolytes,* 2nd ed., Reinhold, New York, 1936; Russian translation, GONTI, Moscow-Leningrad, 1938.
38. **Scott, R. L.,** The solubility of fluorocarbons, *J. Am. Chem. Soc.,* 70, 4090, 1948.
39. **Lawson, D. D.,** Methods of calculating engineering parameters for gas separations, *Appl. Energ.,* 6, 241, 1980.
40. **Burrell, H.,** Solubility parameters, *Interchem. Rev.,* 14, 3 and 31, 1955.
41. **Burrell, H.,** Solubility parameters for film formers, *Off. Dig. Fed. Paint Varn. Prod. Clubs,* 27, 726, 1955.
42. **Burrell, H.,** Solubility parameter values, in *Polymer Handbook,* Brandrup, J. and Immergut, E. H., Eds., John Wiley & Sons-Interscience, New York, 1966, IV-341; 2nd ed., 1975, IV-337.
43. **Thomas, L. H.,** Viscosity of molecular association, *J. Chem. Soc.,* 1345 and 1349, 1948.
44. **Gopal, R. and Rizvi, S. A.,** Estimation of molecular interaction energies in some amides, *Z. Phys. Chem., (Frankfurt),* 81, 330, 1972.
45. **Srivastava, S. C.,** Relation between ultrasonic velocity and other physical properties of pure organic liquids, *Indian J. Phys.,* 33, 503, 1959.
46. **Dack, M. R. J.,** The importance of solvent internal pressure and cohesion to solution phenomena, *Chem. Soc. Rev.,* 4, 211, 1975.
47. **Watanabe, A. and Sugiyama, S.,** Generalized relations between solubility parameter and temperature for hydrocarbons, *Kagaku Kogaku,* 36, 680, 1972.
48. **Kojima, T., Hamada, M., Miyahara, M., and Hosaka, Y.,** Solute-solvent interactions of oligo (*m*-benzamide)s in *N,N*-dimethylacetamide, *J. Polym. Sci. Polym. Phys.,* 25, 1481, 1987.
49. **Fisher, C. H. and Huddle, B. P., Jr.,** Calculating properties of *n*-alkanes from boiling points, *Chem. Eng.,* 91(7), 102, 1984.
50. **Fisher, C. H.,** Estimating polarities and solubility parameters from T/V and $T/V^{1/2}$ quotients, *Polym. Mater. Sci. Eng.,* 51, 593, 1984.
51. **Lewis, G. N. and Randall, M.,** *Thermodynamics,* 2nd ed., revised by Pitzer, K. S. and Brewer, L., Eds., McGraw-Hill, New York, 1961.
52. **Lyckman, E.W., Eckert, C. A., and Prausnitz, J. M.,** Generalized liquid volumes and solubility parameters for regular solution application, *Chem. Eng. Sci.,* 20, 703, 1965.
53. **Prausnitz, J. M. and Chueh, P. L.,** *Computer Calculations for High Pressure Vapor Equilibria,* Prentice-Hall, Englewood Cliffs, NJ, 1968.
54. **Preston, G. T. and Prausnitz, J. M.,** Thermodynamics of solid solubility in cryogenic solvents, *Ind. Eng. Chem. Proc. Des. Dev.,* 9, 264, 1970.
55. **Bazuá, E. R. and Prausnitz, J. M.,** Vapour-liquid equilibria for cryogenic mixtures, *Cryogenics,* 11, 115, 1971.
56. **Deák, G.,** A method for the calculation of the nonpolar solubility parameter of unsaturated normal hydrocarbons, *Hung. J. Ind. Chem.,* 1, 343, 1973.
57. **Bradford, M. L. and Thodos, G.,** Solubility parameters of hydrocarbons, *Can. J. Chem. Eng.,* 44, 345, 1966.
58. **Lozada, C. M., Monfort, J. P., and Del Rio, F.,** Calculation of solubility parameters from an equation of state, *Chem. Phys. Lett.,* 45, 130, 1977.
59. **Sharma, B. K.,** A new method of estimation of solubility parameter of liquids, *Indian J. Pure Appl. Phys.,* 14, 939, 1976; Solubility parameter and evaluation of some thermodynamic properties of simple liquids, *Indian J. Pure Appl. Phys.,* 15, 633, 1977.
60. **Hijmans, J.,** Phenomenological formulation of the principle of corresponding states for liquids consisting of chain molecules, *Physica,* 27, 433, 1961.
61. **Meyer, E. F. and Wagner, R. E.,** Cohesive energies in polar organic liquids, *J. Phys. Chem.,* 70, 3162, 1966.
62. **Allen, G., Gee, G., and Wilson, G. J.,** Intermolecular forces and chain flexibility in polymers. I. Internal pressures and cohesive energy densities of simple liquids, *Polymer,* 1, 456, 1960.
63. **Dack, M. R. J.,** Solvent structure. The use of internal pressure and cohesive energy density to examine contributions to solvent-solvent interactions, *Aust. J. Chem.,* 28, 1643, 1975.
64. **Wiehe, I. A. and Bagley, E. B.,** Estimation of dispersion and hydrogen bonding energies in liquids, *Am. Inst. Chem. Eng. J.,* 13, 836, 1967.
65. **Mogensen, O. E.,** Information on the structure of liquids and solutions obtained by positron annihilation, *Electrochim. Acta,* 33, 1203, 1988.
66. **Tiwari, V. and Pandey, J. D.,** Interaction studies in benzene-chlorobenzene and benzene-toluene through internal pressure and solubility parameter, *Z. Phys. Chem. (Leipzig),* 262, 53, 1981.
67. **Bagley, E. B., Nelson, T. P., Barlow, J. W., and Chen, S.-A.,** Internal pressure measurements and liquid-state energies, *Ind. Eng. Chem. Fundam.,* 9, 93, 1970.

68. **Varadachary, P. S. and Ramamoorthy, K.,** Relation between compressibility and solubility parameters of some non-polar and nearly non-polar liquids, *Indian J. Pure Appl. Phys.*, 8, 185, 1970.
69. **Renuncio, J. A. R., Breedveld, G. J. F., and Prausnitz, J. M.,** Internal pressures and solubilty parameters for carbon disulfide, benzene and cyclohexane, *J. Phys. Chem.*, 81, 324, 1977.
70. **Cohen, S., Goldschmid, A., Shtacher, G., Srebrenik, S., and Gitter, S.,** The inhalation convulsants: a pharmacodynamic approach, *Molec. Pharmacol.*, 11, 379, 1975.
71. **Suri, S. K. and Ramakrishna, V.,** Partial molar volumes of tin(IV) iodide in organic solvents containing a six-membered ring, *J. Chem. Thermodyn.*, 7, 573, 1975.
72. **Bagley, E. B., Nelson, T. P., Chen, S.-A., and Barlow, J. W.,** Internal pressure measurements and evaluation of external molecular vibrational modes in the liquid state, *Ind. Eng. Chem. Fundam.*, 10, 27, 1971.
73. **Bagley, E. B. and Chen, S.-A.,** Hydrogen bonding effects in non-electrolyte solutions, *J. Paint Technol.*, 41, 494, 1969.
74. **Bagley, E. B., Nelson, T. P., and Scigliano, J. M.,** Three-dimensional solubility parameters and their relationship to internal pressure measurements in polar and hydrogen bonding solvents, *J. Paint Technol.*, 43(555), 35, 1971.
75. **Scigliano, J. M.** *Consideration of Complex Solutions Using Internal Pressure Measurements,* Doctoral Thesis, Washington University, St. Louis, MO, 1972; *Diss. Abstr. B,* 32, 6984, 1972.
76. **Bagley, E. G. and Scigliano, J. M.,** Free volume of liquids and polymers from internal pressure studies, *Polym. Eng. Sci.*, 11, 177, 1971.
77. **Bagley, E. B. and Scigliano, J. M.,** Internal pressure behavior of polymers above and below the glass temperature, *Polym. Eng. Sci.*, 11, 320, 1971.
78. **Bagley, E. B. and Wood, H. H.,** Internal pressure measurements of diisooctylphthalate and dioctylsebacate, *Polym. Eng. Sci.*, 6, 141, 1966.
79. **Barton, A. F. M.,** Internal pressure. A fundamental liquid property, *J. Chem. Educ.*, 48, 156, 1971.
80. **Chapman K. M. and Macdonald, D. D.,** Thermal pressure and energy-volume coefficients for dimethylsulfoxide and methanol, *J. Chem. Thermodyn.*, 8, 675, 1976.
81. **Gibson, R. E. and Loeffler, O. H.,** Pressure-volume-temperature relations in solutions. V. The energy-volume coefficients of carbon tetrachloride, water, and ethylene glycol, *J. Am. Chem. Soc.*, 63, 898, 1941.
82. **Grant-Taylor, D. F. and Macdonald, D. D.,** Thermal pressure and energy-volume coefficients for the acetonitrile + water system, *Can. J. Chem.*, 54, 2813, 1976.
83. **Macdonald, D. D. and Hyne, J. B.,** The thermal pressure and energy-volume coefficients of dimethylsulfoxide-water mixtures, *Can. J. Chem.*, 49, 611, 1971; The thermal pressure and energy-volume coefficients of the methyl alcohol-water and t-butyl alcohol-water systems, *Can. J. Chem.*, 49, 2636, 1971.
84. **Macdonald, D. D., Hyne, J. B., and Swinton, F. L.,** Energy-volume coefficients of alcohol-water mixtures, *J. Am. Chem. Soc.*, 92, 6355, 1970.
85. **Malcolm, G. N. and Ritchie, G. L. D.,** Thermal pressure coefficient and the entropy of melting at constant volume of polyethylene oxide, *J. Phys. Chem.*, 66, 852, 1962.
86. **Suri, S. K. and Ramakrishna, V.,** Partial molal volume of iodine in *o*-, *m*-, and *p*-xylenes, *Indian J. Chem.*, 13, 175, 1975.
87. **Watarai, H. and Suzuki, N.,** Pressure-volume approach to the activity coefficient of nonelectrolyte in aqueous mixed solvent, *Chem. Lett. (Jpn.),* 321, 1977.
88. **Westwater, W., Frantz, H. W., and Hildebrand, J. H.,** The internal pressure of pure and mixed liquids, *Phys. Rev.*, 31, 135, 1928; 34, 649, 1929.
89. **Bhattacharyya, B.,** Application of Henderson's partition function in the computation of sound velocity and solubility parameters, *Indian J. Chem. A,* 14, 796, 1976.
90. **Malakondaiah, K., Khan, V. H., and Subrahmanyam, S. V.,** Internal pressure studies on binary mixtures of benzene with p-xylene, dioxane, toluene and chlorobenzene and of isooctane with cyclohexane, benzene and carbon tetrachloride on the basis of ultrasonic measurements, *Indian J. Chem. A,* 16, 733, 1978.
91. **Mishra, R. A.,** Some thermodynamic properties of dense fluid methane, *Vijnana Parishad Anusandhan Patrika,* 21, 293, 1978.
92. **Joseph, P. S., Srinivasan, G. M., and Savariraj, G. A.,** Role of internal pressure in the study of azeotropic mixtures, *Indian J. Chem. A,* 19, 1005, 1980.
93. **Srinivasan, G. M. and Savariraj, G. A.,** Variation of internal pressure in liquids and liquid mixtures: benzene and triethylamine, *Indian J. Chem.*, 12, 610, 1974; Evaluation of the free-volume of a liquid from vapor pressure, *J. Acoust. Soc. India,* 4(1/2), 10, 1976.
94. **Srinivasan, G. M., Savariraj, G. A., and Balasubramanian, S.,** Relation between entropy and internal pressure in liquids and solutions — carbon tetrachloride-ethyl acetate system, *J. Acoust. Soc. India,* 6(3), 48, 1978.
95. **Srinivasan, G. M., Savariraj, G.A., and Suryanarayana, C. V.,** Relation between vapor pressure and internal pressure in liquids, *Indian J. Phys.*, 52B, 24, 1978.

96. **Suryanarayana, C. V.,** Thermodynamic significance of internal pressure and its role in transport processes, *Indian J. Chem.*, 10, 713, 1972; The nature of internal pressure and its temperature coefficient in liquids and liquid solutions, *Proc. Natl. Acad. Sci. India A*, 43, 6, 1973.
97. **Shrivastava, S. N. and Pandey, J. D.,** Excess intermolecular free length and molecular interaction for binary molten metal halide and cryogenic liquid mixtures, *Acoustics Lett.*, 3, 219, 1980.
98. **Pandey, J. D., Dubey, P., and Saxena, M. C.,** Thermodynamic and related properties of castor oil determined from sound velocity, *Acustica*, 48, 277, 1981.
99. **Pandey, J. D. and Shrivastava, S. N.,** Temperature dependence of some acoustical and thermodynamical properties of tetrafluoromethane and methane, *Acustica*, 51, 66, 1982.
100. **Vavruch, I.,** personal communication, 1988.
101. **Vavruch, I.,** On the predictions of total surface energy of pure liquids, *J. Colloid Interface Sci.*, 73, 267, 1980; Stefan's rule as a consequence of cohesive forces, *Colloids Surfaces*, 15, 57, 1985; On the evaluation of the volume of surface formation of pure liquids, *J. Colloid Interface Sci.*, 114, 290, 1986; On the relation between surface energy, internal pressure and molar volume in pure fluids, *Colloids Surfaces*, 30, 405, 1988.
102. **Pant, N., Chaturvedi, C. V., and Chaturvedi, G. D.,** Thermal pressure coefficient, internal pressure and solubility parameter of hard sphere fluids, *Z. Phys. Chem. (Leipzig)*, 264, 513, 1983.
103. **Pandey, J. D., Pant, N., and Chaturvedi, B. R.,** Evaluation of excess internal pressure and correlation with interactions in binary liquid mixtures, *Chem. Scr.*, 18, 224, 1981.
104. **Srivastava, A. P. and Dubey, S. M.,** Evaluation of excess internal pressure and correlation with interaction in binary liquid mixtures, *Current Sci.*, 53, 74, 1984.
105. **Srivastava, A. P.,** Evaluation of excess internal pressure in binary liquid mixtures of water with associated liquids, *Current Sci.*, 55, 180, 1986.
106. **Srivastava, A. P. and Triparthi, S. N.,** Evaluation of excess internal pressure, excess free length and sound velocity in binary liquid mixtures at 298.15, 313.15, and 323.15 K, *Z. Phys. Chem. (Leipzig)*, 268, 793, 1987.
107. **Sharma, B. K. and Das, N.,** Internal pressure — temperature behavior and solubility parameter of liquids; Modified expressions for characteristic parameters of liquid state and estimation of the Gruneisen parameter, *Indian J. Pure Appl. Phys.*, 19, 668, 1981; 20, 249, 1982.
108. **Tijssen, R., Billiet, H. A. H., and Schoenmakers, P. J.,** Use of the solubility parameter for predicting selectivity and retention in chromatography, *J. Chromatogr.*, 122, 185, 1976.
109. **Liron, Z. and Cohen, S.,** Densitometric determination of the solubility parameter and molal volume of compounds of medicinal relevance, *J. Pharm. Sci.*, 72, 499, 1983.
110. **Martin, A., Wu, P. L., Liron, Z., and Cohen, S.,** Dependence of solute solubility parameters on solvent polarity, *J. Pharm. Sci.*, 74, 638, 1985.
111. **Kirchnerova, J. and Cavé, G. C. B.,** The solubility of water in low-dielectric solvents, *Can. J. Chem.*, 54, 3909, 1976.
112. **Black, C., Joris, G. G., and Taylor, H. S.,** The solubility of water in hydrocarbons, *J. Chem. Phys.*, 16, 537, 1948.
113. **Fujishiro, R. and Hildebrand, J. H.,** The liquid-liquid solubility of cyclohexane and perfluorotributylamine at 25°, *J. Phys. Chem.*, 66, 573, 1962.
114. **Hildebrand, J. H. and Scott, R. L.,** *Solubility of Non-Electrolytes*, 3rd ed., Reinhold, New York, 1950; Dover, New York, 1964.
115. **Hildebrand, J. H. and Scott, R. L.,** *Regular Solutions*, Prentice-Hall, Englewood Cliffs, NJ, 1962.
116. **Jolley, J. E. and Hildebrand, J. H.,** The liquid-liquid solubility of octamethyl-cyclotetrasiloxane with perfluoromethylcyclohexane and perfluoro-*n*-heptane, *J. Phys. Chem.*, 61, 791, 1957.
117. **Prausnitz, J. M.,** *Molecular Thermodynamics of Fluid Phase Equilibria*, Prentice-Hall, Englewood Cliffs, NJ, 1969.
118. **Prausnitz, J. M., Lichtenthaler, R. N., and Gomez de Azevedo, E.,** *Molecular Thermodynamics of Fluid Phase Equilibria*, 2nd ed., Prentice-Hall, Englewood Cliffs, NJ, 1986.
119. **Reid, R. C., Prausnitz, J. M., and Sherwood, T. K.,** *The Properties of Gases and Liquids*, McGraw-Hill, New York, 3rd ed., 1977.
120. **Staverman, A. J.,** Cohesion forces and phase theory of binary liquid systems, *Rec. Trav. Chim. Pays-Bas*, 60, 827, 1941.
121. **Ashton, N. F., McDermott, C., and Brench, A.,** Chemistry of extraction of nonreacting solutes, in *Handbook of Solvent Extraction*, Lo, T. C., Baird, M. H. I., and Hanson, C., Eds., John Wiley & Sons-Interscience, New York, 1983 chap 1, 3; Brench, A., Society of Chemical Industry Symposium, 1978, unpublished.
122. **Williamson, A. G.,** Solutions of nonelectrolytes, *Ann. Rev. Phys. Chem.*, 15, 63, 1964.
123. **Novák, J. P., Matouš, J., and Pick, J.,** *Liquid-Liquid Equilibria*, (Studies in Modern Thermodynamics Vol. 7), Elsevier, Amsterdam, 1987.
124. **Acree, W. E., Jr.,** *Thermodynamic Properties of Nonelectrolyte Solutions*, Academic Press, Orlando, FL, 1984.

125. **Krishnan, C. V. and Friedman, H. L.,** Solvation enthalpies of hydrocarbons and normal alcohols in highly polar solvents, *J. Phys. Chem.,* 75, 3598, 1971.
126. **Rogers, D. W., Kim, J.-H., Patel, S. C., and Wurm, R. A.,** A microcomputer-controlled microcalorimeter calibration circuit. 3. Enthalpies of solution of linear and cyclic alkanes in glacial acetic acid, *Mikrochim. Acta,* 3, 297, 1985.
127. **Góralski, P. and Tkaczyk, M.,** Calorimetric investigations of association in ternary systems. Part 4. The influence of solvation on enthalpy of complex formation in phenol-tetrahydrofuran and 2,6-dimethylphenol-tetrahydrofuran systems, *J. Chem. Soc. Faraday Trans. I,* 83, 3083, 1987.
128. **Bittrich, H.-J., Hammerl, I., Lempe, D., and Sarius, A.,** Modeling of phase equilibrium in the liquid-liquid phase separation. I. Zero and one-parameter models, *Wiss. Zeit. T. H. Leuna-Merseberg,* 23, 407, 1981.
129. **Kamlet, M. J., Carr, P. W., Taft, R. W., and Abraham, M. H.,** Linear solvation energy relationships. 13. Relationship between the Hildebrand solubility parameter, δ_H, and the solvatochromic parameter, π^*, *J. Am. Chem. Soc.,* 103, 6062, 1981.
130. **Abraham, M. H., Kamlet, M. J., and Taft, R. W.,** Linear solvation energy relationships. Part 19. Correlation of the free energies of solution of 41 solutes in select solvents with Hildebrand's solubility parameter, δ_H, and with the solvatochromic parameter, π^*, *J. Chem. Soc. Perkins Trans. II,* 923, 1982.
131. **Amidon, G. L. and Williams, N. A.,** A solubility equation for non-electrolytes in water, *Int. J. Pharm.,* 11, 249, 1982.
132. **Grec, J.-J., Reiss, J. G., and Devallez, B.,** Study of perfluoroalkylated solvents in biomedical use: Higher critical temperatures of solubility of bis(F-alkyl)ethanes in hexanes and rates of excretion: Solubility parameters, excess quantities of carboxylic acid mixtures and perfluoroalkyl compounds, *Nouv. J. Chim.,* 9, 637, 1985.
133. **Buchtová, M., Koutek, B., and Bekárek, V.,** Study of the use of Hildebrand's solubility parameter in the evaluation of media effects, *Acta Univ. Palacki Olomuc., Fac. Rerum Nat.,* 82(Chem.-24), 67, 1985.
134. **Aliev, A. M.,** Semi-empirical method of selecting solvents of varying purpose, *Dep. Doc. VINITI,* 3935-81, 1981; New scientific concept for prediction of the miscibility of liquids, *Sb. Tr.-Akad. Nauk Az. SSR Inst. Neftekhim. Protsessov im Akad. Yu. G. Mamedalieva,* 16, 132, 1987.
135. **Lazurin, E. A., Samorodov, V. T., and Kosmodemyanskii, L. V.,** Determination of solubility parameters of mineral oils, *Prom-st. Sint. Kauch.,* No. 8, 10, 1980; *Int. Polym. Sci. Technol.,* 8(12), T12, 1981.
136. **Acree, W. E., Jr. and Rytting, J. H.,** Solubility in binary solvent systems. I. Specific versus nonspecific interactions, *J. Pharm. Sci.,* 71, 201, 1982.
137. **Acree, W. E., Jr. and Rytting, J. H.,** Solubility in binary solvent systems. II. The importance of nonspecific interactions, *Int. J. Pharm.,* 10, 231, 1982.
138. **McCargar, J. W. and Acree, W. E., Jr.,** Thermochemical investigations of associated solutions. 3. Effect of the inert cosolvent on solute-solvent association constants calculated from solubility measurements, *Phys. Chem. Liq.,* 17, 123, 1987.
139. **McCargar, J. W. and Acree, W. E., Jr.,** Thermochemical investigations of associated solutions. 4. Calculation of carbazole-dibutyl ether association constants from measured solubility in binary solvent mixtures, *J. Pharm. Sci.,* 76, 572, 1987.
140. **Acree, W. E., Jr. and McCargar, J. W.,** Thermochemical investigations of associated solutions. 5. Calculation of solute-solvent equilibrium constants from solubility in mixtures containing two complexing solvents, *J. Pharm. Sci.,* 76, 575, 1987.
141. **Acree, W. E., Jr. and McCargar, J. W.,** Thermochemical investigations of associated solutions. 7. Development of solubility expressions based on the Huyskens and Haulait-Pirson definition of solution ideality, *J. Molec. Liq.,* 37, 251, 1988.
142. **Procyk, A. D., Bissell, M., Street, K. W., Jr., and Acree, W. E., Jr.,** Solubility in binary solvent systems: 8. Estimation of binary alkane plus *p*-dioxane solvent nonideality from measured anthracene solubilities, *J. Pharm. Sci.,* 76, 621, 1987.
143. **Judy, C. L., Pontikos, N. M., and Acree, W. E., Jr.,** Solubilities of pyrene in binary solvent mixtures containing cyclohexane, *J. Chem. Eng. Data,* 32, 60, 1987.
144. **Ludmer, Z., Shinnar, R., and Yakhot, V.,** Solubility in binary mixtures at the immiscibility critical point, *Am. Inst. Chem. Eng. J.,* 33, 1776, 1987.
145. **Książczak, A. and Buchowski, H.,** Prediction of thermodynamic properties of associated systems on the basis of properties of pure liquids. II. Excess Gibbs energy and vapour pressure, *Fluid Phase Equil.,* 16, 361, 1984.
146. **Li, J.-Y., Zhao, Z., He, M., Qiao, Y., and Zhang, S.,** Application of the solubility parameter concept to drycleaning agents for wool fabrics, *Fangzhi Xuebao,* 6, 461, 1985.
147. **Chawla, B., Parkash, R., and Suri, S. K.,** A systematic study on volumetric interactions of cyclohexanone with various aliphatic hydrocarbons, *Can. J. Chem.,* 61, 1647, 1983.
148. **Chiba, K., Tagaya, H., and Saito, N.,** Liquefaction of Yallourn coal by binary system solvent, *Energy Fuels,* 1, 338, 1987.

149. **Rubino, J. T. and Yalkowsky, S. H.,** Cosolvency and cosolvent polarity, *Pharm. Res.*, 4, 220, 1987.
150. **Bondi, A.,** On the solubility of paraffin-chain compounds, *J. Phys. Chem.*, 51, 891, 1947.
151. **Hildebrand, J. H.,** Order from chaos, *Science,* 150, 441, 1965.
152. **Hildebrand, J. H. and Dymond, J. H.,** Effect of methyl groups upon the solvent power of aliphatic liquids, *Proc. Natl. Acad. Sci. U.S.A.*, 54, 1001, 1965.
153. **Thomas, E. R. and Eckert, C. A.,** Prediction of limiting activity coefficients by a modified separation of cohesive energy density model and UNIFAC, *Ind. Eng. Chem. Process Des. Dev.*, 23, 194, 1984.
154. **Monfort, J.-P., Vidal, J., and Renon, H.,** Activity coefficients of hydrocarbons at infinite dilution in two families of solvents. Effects of the solvent structure, *J. Chim. Phys.*, 67, 748, 1970.
155. **Park, J. H. and Carr, P. W.,** Predictive ability of the MOSCED and UNIFAC activity coefficient estimation methods, *Analyt. Chem.*, 59, 2596, 1987.
156. **Weimer, R. F. and Prausnitz, J. M.,** Screen extraction solvents this way, *Hydrocarbon Process Petr. Ref.*, 44, 237, 1965.
157. **Milanová, E. and Cavé, G. C. B.,** Limiting activity coefficients in dilute solutions of nonelectrolytes. I. Determination for polar-nonpolar binary mixtures by a novel apparatus and a solubility-parameter treatment of results, *Can. J. Chem.*, 60, 2697, 1982.
158. **Afrashtehfar, S. and Cavé, G. C. B.,** Limiting activity coefficients in dilute solutions of nonelectrolytes. II. Determination for polar-nonpolar and polar-polar binary mixtures, and the application of some solubility-parameter treatments, *Can. J. Chem.*, 64, 198, 1986.
159. **Helpinstill, J. G. and Van Winkle, M.,** Prediction of infinite dilution activity coefficients for polar-polar binary systems, *Ind. Eng. Chem. Process Des. Dev.*, 7, 213, 1968.
160. **Wiehe, I. A.,** *Thermodynamics of Alcohol-Inert Solvent Solutions,* Thesis, Washington University, St. Louis, MO, 1967; *Diss. Abstr. B*, 28, 166, 1967; Thermodynamic properties of solutions of alcohols in inert solvents, *Ind. Eng. Chem. Fundam.*, 6, 209, 1967.
161. **Koskas, A. and Durandet, J.,** Thermodynamics and prediction of two-component liquid-liquid equilibria, *Chim. Ind.-Gen. Chim.*, 98, 1386, 1967.
162. **Abrams, D. S. and Prausnitz, J. M.,** Statistical thermodynamics of liquid mixtures. New expression for the excess Gibbs energy of partly or completely miscible systems, *Am. Inst. Chem. Eng. J.*, 21, 116, 1975.
163. **Fredenslund, A., Jones, R. L., and Prausnitz, J. M.,** Group-contribution estimation of activity coefficients in nonideal liquid mixtures, *Am. Inst. Chem. Eng. J.*, 21, 1086, 1975; Arbuckle, W. B., Estimating activity coefficients for use in calculating environmental parameters, *Environ. Sci. Technol.*, 17, 537, 1983.
164. **Grünbauer, H. J. M. and Tomlinson, E.,** Local composition models in pharmaceutical chemistry. I. Liquid-liquid distribution coefficients from COMMON UNIQUAC parameters, *Int. J. Pharm.*, 21, 61, 1984.
165. **Cockrem, M. C. M., Meyer, R. A., and Lightfoot, E. N., Jr.,** Recovery of *n*-butanol from dilute solution by extraction, *Sep. Sci. Technol.*, 21, 1059, 1986.
166. **Aarna, A., Arro, J., Melder, L., and Tamvelius, H.,** Dividing the solubility parameter into components and determining the activity coefficients for self-associated substances from the liquid-liquid equilibrium, *Izv. Akad. Nauk Est. SSR Khim. Geol.*, 24, 115, 1975.
167. **Aarna, A., Melder, L. I., and Ebber, A. V.,** Extraction of esters by saturated organic solvents, *Zh. Prikl. Khim,* 50, 2604, 1977; *J. Appl. Chem. U.S.S.R.*, 50, 2476, 1977; Determination of the limiting activity coefficients of complex esters in nonpolar solvents, *zh. Prikl. Khim.*, 51, 1359, 1978; *J. Appl. Chem. U.S.S.R.*, 51, 1295, 1978.
168. **Aarna, A., Melder, L., and Ebber, A.,** Use of a modified theory of regular solutions for describing solutions of ketones in nonpolar solvents, *Zh. Prikl. Khim.*, 53, 156, 1980; *J. Appl. Chem. U.S.S.R.*, 53, 134, 1980.
169. **Alessi, P., Kikic, I., and Tlustos, G.,** Prediction of activity coefficients of hydrocarbons in glycols using solubility parameters, *Chim. Ind. (Milan),* 53, 1148, 1971.
170. **Alessi, P., Kikic, I., and Torriano, G.,** Activity coefficients of hydrocarbons in phthalates, *J. Chromatogr.*, 105, 257, 1975; Correlation of hydrocarbon activity coefficients with solubility parameters for phthalate esters, *J. Chromatogr.*, 106, 17, 1975.
171. **Bondi A.,** Solubility and swelling of alkali soaps in organic solvents, *J. Colloid Sci.*, 5, 458, 1950.
172. **Kralj, F. and Sincic, D.,** Mutual solubilities of phenol, salicylaldehyde, phenol-salicylaldehyde mixture, and water with and without the presence of sodium chloride or sodium chloride plus sodium sulfate, *J. Chem. Eng. Data,* 25, 335, 1980.
173. **Christian, S. D., Childs, J. D., and Lane, E. H.,** Activity coefficient effects in spectral and solubility studies of molecular complex equilibria, *J. Am. Chem. Soc.*, 94, 6861, 1972.
174. **Dowell, F. and Stewart, G. H.,** Solvent characteristics of molten sulfur at 132°C, *Phosphorus and Sulfur,* 1, 135, 1976.
175. **Funk, E. W. and Prausnitz, J. M.,** Thermodynamic properties of liquid mixtures: aromatic-saturated hydrocarbon systems, *Ind. Eng. Chem.*, 62(9), 8, 1970.
176. **Schille, W. and Bittrich, H.-J.,** Calculating free excess enthalpies in hydrocarbon systems, *Chem. Tech. (Leipzig),* 25, 292, 1973.

177. **Kertes, A. S.,** Solubility parameters of long-chain alkylamines and their activities in organic solvents, *J. Inorg. Nucl. Chem.,* 26, 1764, 1964.
178. **Kertes, A. S.,** Solubility and activity of high-molecular amine hydrochlorides in organic solvents, *J. Inorg. Nucl. Chem.,* 27, 209, 1965.
179. **Apelblat, A.,** Mutual solubilities of methyl diphenyl phosphate and some aliphatic hydrocarbons, *J. Chem. Soc. (B),* 1459, 1970.
180. **Apelblat, A.,** Thermodynamic studies of solvent extraction systems. 3. Binary mixtures of methyl diphenyl phosphate with benzene and water, *J. Inorg. Nucl. Chem.,* 40, 1783, 1978.
181. **Ghosh, D. P.,** *Phase Equilibrium Studies. Activity Coefficients, Solubility Parameters, and Apparent Molar Volumes of Solutes in Binary Systems,* Dissertation, University of Idaho, Moscow, ID, 1972; *Diss. Abstr. Int. B,* 33, 2046, 1972.
182. **Kojima, T., Hamada, M., Miyahara, M.-A., and Hosaka, Y.,** Solute-solvent interactions of oligo(*m*-benzamide)s in *N,N*-dimethylacetamide, *J. Polym. Sci. Polym. Phys.,* 25, 1481, 1987.
183. **Arro, Ya. V. and Melder, L. L.,** Activity coefficients of dihydric phenols in water from their distribution between two liquid phases, *Zh. Fiz. Khim.,* 49, 1077, 1975; *Russ. J. Phys. Chem.,* 49, 635, 1975; Arro, Ya. V., Melder, L. L., and Lohmus, H. R., Determination of activity coefficients of 5-ethyl- and 4,5-dimethyl resorcinols from the liquid-liquid equilibrium, *Tr. Tallin. Politekhn. Inst.,* 449, 35, 1978.
184. **Christian, S. D. and Lane, E. H.,** Solvent effects on molecular complex equilibria, in *Solutions and Solubilities,* Part I, Dack, M. R. J., Ed., John Wiley & Sons-Interscience, New York, 1975, chap. 6.
185. **Lane, E. H., Christian, S. D., and Childs, J. D.,** Spectral and solubility studies of molecular complex equilibria. Concentration scale dependence and activity coefficient effects, *J. Am. Chem. Soc.,* 96, 38, 1974.
186. **Benesi, H. A. and Hildebrand, J. H.,** A spectrophotometric investigation of the interaction of iodine with aromatic hydrocarbons, *J. Am. Chem. Soc.,* 71, 2703, 1949.
187. **Alley, S. K., Jr. and Scott, R. L.,** NMR studies of hydrogen bonding in hydrofluorocarbon solutions, *J. Phys. Chem.,* 67, 1182, 1963.
188. **Buchowski, H., Devaure, J., Huong, P. V., and Lascombe, J.,** Influence of solvents on the association constants of heptyne-1 and of deuterochloroform with acetone, *Bull. Soc. Chim. Fr.,* 2532, 1966.
189. **Christian, S. D.,** Solvent effects on molecular complex formation. Pyridine-iodine system, *J. Am. Chem. Soc.,* 91, 6514, 1969.
190. **Huong, P. F., Platzer, N., and Josien, M. L.,** Solvent effects on the iodine-pyridine charge-transfer complex, *J. Am. Chem. Soc.,* 91, 3669, 1969.
191. **Barigand, M., Orszagh, J., and Tondeur, J.-J.,** Molecular complexes. The equilibrium constant of phenothiazine with *p*-chloroaniline in several solvents, *Bull. Soc. Chim. Fr.,* 48, 1973.
192. **Tanaka, M.,** Solvent effect on equilibria and SN2 reactions of non-charged species, *Z. Phys. Chem. (Frankfurt),* 96, 239, 1975.
193. **Roussel, R.,** Oteyza de Guerro, M., Spegt, P., and Galin, J. C., Solvent effects on the tautomerism of t-butyl 2-picolyl and t-butyl quinaldyl ketone, *J. Heterocyclic Chem.,* 19, 785, 1982.
194. **Tanaka, M.,** Partition coefficient and extraction constant of non-charged species as considered from the regular solution theory, *Proc. Int. Solvent Extraction Conf., The Hague, 1971,* Society of Chemical Industry, London, 1971, 18.
195. **Tanaka, M.,** Separation by solvent extraction. A study by regular solution theory, *Bunseki Kiki,* 14, 327, 1976.
196. **Irving, H. M. N. H., Smith, J. S., and Karmali, S.,** The solubility parameters of metal acetylacetonates, *J. Inorg. Nucl. Chem.,* 43, 1139, 1981.
197. **Irving, H. M. N. H.,** Application of the solubility concept in liquid extraction, in *Ion Exchange and Solvent Extraction,* Vol. 6, Marinsky, J. A. and Marcus, Y., Eds., Marcel Dekker, New York, 1974, chap. 3, 139.
198. **Irving, H. M. N. H.,** The theory of regular solutions applied to the viscosity of dilute solutions of chromium(III)acetylacetonate in organic solvents, *J. Inorg. Nucl. Chem.,* 30, 1873, 1968; 31, 1212, 1969.
199. **Anderson, B. D., Rytting, J. H., and Higuchi, T.,** Influence of self-association on the solubility of phenol in isooctane and cyclohexane, *J. Am. Chem. Soc.,* 101, 5194, 1979; Solubility of polar organic solutes in non-aqueous systems — role of specific interactions, *J. Pharm. Sci.,* 69, 676, 1980.
200. **Akiba, K. and Suzuki, N.,** The effect of solvents on the adduct formation of uranyl thenoyltrifluoroacetonate with tributyl phosphate, *Bull. Chem. Soc. Jpn.,* 44, 1043, 1971.
201. **Akiba, K., Suzuki, N., and Kanno, T.,** Role of the solvent in the distribution of the zinc-TTA chelate and the adduct formation with TBP, *Bull. Chem. Soc. Jpn.,* 42, 2537, 1969.
202. **Kang, I. K.,** The activity coefficients of the tri-*n*-butylphosphate and europium complex in the extraction of europium by tri-*n*-butylphosphate in nitric acid solution, *Taehan Hwahak Hoechi (Daehan Hwahak Hwoejee) (J. Korean Chem. Soc.)* 18, 272, 1974.
203. **Shibuya, H., Watarai, H., and Suzuki, N.,** The solution behavior of thenoyltrifluoroacetone in alcoholic solvents, *Bull. Chem. Soc. Jpn.,* 51, 2932, 1978.

204. **Akiba, K., Suzuki, N., Asano, H., and Kanno, T.,** Regularities of extraction of uranyl thenoyltrifluoroacetonate into a number of solvents, *J. Radioanal. Chem.*, 7, 203, 1971.
205. **Allard, B.,** On the formation and properties of tetrakis(acetylacetonato)thorium(IV) hemibenzene, *Acta Chem. Scand.*, A30, 381, 1976.
206. **Jones, M. M., Wood, J. L., and May, W. R.,** Thermodynamics of solution of some inner complex salts, *J. Inorg. Nucl. Chem.*, 23, 305, 1961.
207. **Jones, M. M., Davila, A. T., Hix, J. E., Jr., and Pilts, R. V.,** Thermodynamics of solution of some inner complex salts, II., *J. Inorg. Nucl. Chem.*, 25, 369, 1963.
208. **Nakamura, S., Imura, H., and Suzuki, N.,** Solvent effect on the synergic extraction of thulium(III) with acetylacetone and 1,10-phenanthroline, *Inorg. Chim. Acta,* 109, 157, 1985.
209. **Imura, H. and Suzuki, N.,** Solute-solvent interactions in liquid-liquid partition of bis(acetylacetonato)copper(II), *Kakuriken Kenkyu Hokoku (Tohoku Diagaku)(Res. Rep. Lab. Nucl. Sci. Tohoku Univ.),* 16, 95, 1983; Evaluation of the liquid-liquid partition coefficient of bis(acetylacetonato)copper(II): contribution of specific solute-solvent interaction, *J. Radioanalyt. Nucl. Chem. Art.*, 88, 63, 1985.
210. **Mitchell, J. W. and Banks, C. V.,** Solvent effects on solubilities and partition coefficients of some copper(II)-beta-diketone complexes, *J. Inorg. Nucl. Chem.*, 31, 2105, 1969.
211. **Watarai, H., Oshima, H., and Suzuki, N.,** Regularities of the partition coefficients of bis, tris, and tetrakis(acetylacetonato)metal(II, III and IV) complexes, *Quant. Struct. - Act. Relat.*, 3, 17, 1984.
212. **Kobayashi, M., Saitoh, K., and Suzuki, N.,** Reversed-phase, high-performance, thin-layer chromatography of metal tetrakis(p-tolyl)porphine complexes on octadecylbonded stationary phase, *Chromatographia,* 20, 49, 1985.
213. **Wakui, Y., Saitoh, K., and Suzuki, N.,** Non-aqueous, reversed phase, high-performance, thin-layer chromatography and column liquid chromatography of metal complexes of porphine, *Chromatographia,* 22, 160, 1986.
214. **Irving, H. M. N. H. and Smith, J. S.,** A viscometric study of iodine-solvent interactions, *J. Inorg. Nucl. Chem.*, 31, 3163, 1969.
215. **Nakanishi, K. and Asakura, S.,** Solubility and entropy of solution of iodine in ethanol-cyclohexane solutions, *Chem. Lett.*, 493, 1972.
216. **Uruska, I.,** Solvent effect on electron donor-acceptor complexes. Pyridine-molecular iodine system, in *Proc. Int. Symp. Specific Interact. Mol. Ions 3rd,* Vol. 2, Institute of Chemistry, University of Wroclaw, Poland, 1976, 542.
217. **Uruska, I.,** Solvent effect on thermodynamic stability of iodine complexes with quinoline and pyridine bases, *Spectrochim. Acta,* 36A, 639, 1980.
218. **Uruska, I. and Koschmidder, M.,** A calorimetric study of complex formation between molecular iodine and pyridine or 2-methylpyridine in weakly polar solvents, *J. Chem. Soc. Perkins II,* 1713, 1987.
219. **Narbutt, J.,** Molecular complexes between γ-picoline and chloroform. A liquid-liquid extraction study, *Bull. Acad. Pol. Sci. Ser. Sci. Chim.*, 22, 1095, 1974.
220. **Narbutt, J. and Siekierski, S.,** Effect of intermolecular interactions in the organic phase on the liquid-liquid extraction of γ-picoline and of the $Co(\gamma\text{-picoline})_4(NCS)_2$ complex, *Rocz. Chem.*, 48, 1777, 1974.
221. **Narbutt, J. and Siekierski, S.,** Second-sphere molecular complexes between bis(isothiocyanato)tetrakis(picoline)cobalt $[Co(\gamma\text{-picoline})_4(NCS)_2]$ and chloroform. A liquid-liquid extraction study, *J. Inorg. Nucl. Chem.*, 38, 305, 1976.
222. **Libús, W., Chachulska, K., and Stangret, J.,** Free energies of transfer of $ZnCl_2py_2$ and of related tetrahederal complexes from pyridine to other solvent media, *J. Sol. Chem.*, 12, 529, 1983.
223. **Szpakowska, M. and Uruska, I.,** Solvent effect on u.v.-vis and EPR spectra and on the relative stability of mononuclear and binuclear complexes of Cu(II)acetate with pyridine derivatives, *Spectrochim. Acta,* 43A, 763, 1987.
224. **Olszer, R. and Siekierski, S.,** *An Extraction Study on the Interactions between Arsenic Tribromide and Organic Solvents,* Rep. 707/V/C, Institute of Nuclear Research, Warsaw, 1966.
225. **Olszer, R. and Siekierski, S.,** An extraction study on the interaction between arsenic trichloride and aromatic solvents, *J. Inorg. Nucl. Chem.*, 28, 1991, 1966.
226. **Olszer, R. and Siekierski, S.,** An extraction study on the interaction between antimony trichloride and aromatic solvents, *Bull. Acad. Pol. Sci. Ser. Sci. Chim.*, 14, 325, 1966.
227. **King, J. W. and Quinney, P. R.,** Investigation of charge-transfer interactions between carbon tetrabromide and aromatic donors by gas chromatography, *J. Chromatogr.*, 49, 161, 1970.
228. **Christian, S. D., Yeo, K. O., and Tucker, E. E.,** Effects of solvents on molecular complex formation equilibria. Use of nonpolar analogs for polar solutes. *J. Phys. Chem.*, 75, 2413, 1971.
229. **Christian, S. D., Frech, R., and Yeo, K. O.,** Use of the nonpolar analog model in predicting the effects of solvents on molecular complex formation equilibria, *J. Phys. Chem.*, 77, 813, 1973.
230. **Ichimura, A. and Kitagawa, T.,** Voltammetric studies of tris(1-pyrrolidine carbodithioato)manganese(III) and -iron(III) and bis(1-pyrrolidinecarbodithioata)copper(II) in nonaqueous solvents, *Bull. Chem. Soc. Jpn.*, 53, 2538, 1980.

231. **İşcan, M.,** Temperature effect on the partition and dimerization constant of benzoic acid, *Thermochim. Acta*, 94, 305, 1985.
232. **Takeda, K., Yamashita, H., and Akiyama, M.,** Dimerization of some carboxylic acids in organic phases, *Solv. Extr. Ion Exch.*, 5, 29, 1987.
233. **Kamlet, M. J., Doherty, R., Taft, R. W., and Abraham, M. H.,** Linear solvation energy relationships. 26. Some measures of relative self-association of alcohols and water, *J. Am. Chem. Soc.*, 105, 6741, 1983.
234. **Harris, H. G. and Prausnitz, J. M.,** Thermodynamics of solutions with physical and chemical interactions, *Ind. Eng. Chem. Fundam.*, 8, 180, 1969.
235. **Dubois, P., Levillain, P., and Mariaud, M.,** Solvatochromism of the bis(bathophenanthroline)dicyanoiron(II) complex, *Bull. Soc. Chim. Fr.*, 243, 1978.
236. **Grieger, R. A. and Eckert, C. A.,** Solvent mixtures for separation processes, *Ind. Eng. Chem. Process Des. Dev.*, 6, 250, 1967.
237. **Hampe, M.,** Liquid-liquid extraction: application and choice of solvent, *Chem. Ing. Tech.*, 50, 647, 1978.
238. **King, C. J.,** *Separation Processes*, McGraw-Hill, New York, 1971.
239. **Kumar, R. and Prausnitz, J. M.,** Solvents in chemical technology, in *Solutions and Solubilities, Part I*, Dack M. R. J., Ed., John Wiley & Sons-Interscience, New York, 1975, chap. 5.
240. **Mukhopadhyay, M.,** A thermodynamic method based upon the theory of regular solutions for selection of solvents and process conditions for aromatics extraction, *J. Chem. Technol. Biotechnol.*, 29, 634, 1979.
241. **Prausnitz, J. M. and Anderson, R.,** Thermodynamics of solvent selectivity in extractive distillation of hydrocarbons, *Am. Inst. Chem. Eng. J.*, 7, 96, 1961.
242. **Funk, E. W.,** Study of heavy hydrocarbons by inverse-phase chromatography, *Ind. Eng. Chem. Prod. Res. Dev.*, 16, 115, 1977.
243. **Buchowski, H. and Pawlowski, W.,** Influence of electron donor-acceptor interactions on partition coefficients of iodine, *Chem. Anal. (Warsaw)*, 17, 557, 1972.
244. **Berg, L. and Yeh, A.-I.,** The breaking of ternary acetate-alcohol-water azeotropes by extractive distillation, *Chem. Eng. Commun.*, 48, 93, 1986.
245. **Yeh, A.-I. and Berg, L.,** Reversing the relative volatility by extractive distillation, *Am. Inst. Chem. Eng. Symp. Ser.*, 82(250) (Recent Adv. Sepn. Tech. III), 169, 1986; Yeh, A.-I., Berg, L. and Warren, K. J., The separation of acetone-methanol mixture by extractive distillation, *Chem. Eng. Commun.*, 68, 69, 1988.
246. **Buchowski, H.,** Relation between partition coefficients and properties of solvents, *Nature (London)*, 194, 674, 1962.
247. **Srebrenik, S. and Cohen, S.,** Theoretical derivation of partition coefficients from solubility parameters, *J. Phys. Chem.*, 80, 996, 1976.
248. **Siekierski, S.,** The role of diluent in extraction processes, *Nucleonika*, 9, 601, 1964.
249. **Fukunaga, K., Shirai, H., Ide, S., and Kimura, M.,** Solubility parameters of phase-transfer catalysts and their applications, *Nippon Kagaku Kaishi*, 1148, 1980.
250. **Noel, D. E. and Meloan, C. E.,** Some empirical correlations in solvent extraction, *Separ. Sci.*, 7, 935, 1972.
251. **Irving, H. and Lewis, D. C.,** The solvent extraction of solvated ion-pairs: a theoretical approach, *Proc. Chem. Soc.*, 222, 1960.
252. **Melder, L. I. and Ebber, A. V.,** Determination of the orientation and induction components of the solubility parameter of polar compounds without hydrogen bonds, *Zh. Prikl. Khim.*, 55, 1979, 1982; *J. Appl. Chem. U.S.S.R.*, 55, 1823, 1982.
253. **Melder, L. and Ebber, A.,** Use of the theory of regular solutions for the description of solutions with polar components without a hydrogen bond. 1. Selection of the method for the expression of polarizability of solution components in the calculation of induction interaction energy, *Tr. Tallin. Politekhn. In-ta.*, No. 509, 3, 1981.
254. **Ebber, A.,** Use of the theory of regular solutions for the description of solutions with polar components without a hydrogen bond. 2. Determination of orientation and induction components of the solubility parameter, *Tr. Tallin. Politekhn. In-ta.*, No. 509, 15, 1981.
255. **Melder, L. and Ebber, A.,** Splitting of the overall solubility parameter of polar solvents without hydrogen bonding into dispersion, orientation and induction components, *Eesti NSV Tead. Akad. Toim. Keem.*, 30(1), 39, 1981.
256. **Kamlet, M. J., Abraham, M. H., Doherty, R. M., and Taft, R. W.,** Solubility parameters in polymers and biological media. 4. Correlation of octanol/water partition coefficients with solvatochromic parameters, *J. Am. Chem. Soc.*, 106, 464, 1984.
257. **Amidon, G. L. and Williams, N. A.,** A solubility equation for non-electrolytes in water, *Int. J. Pharm.*, 11, 249, 1982.
258. **Leshchev, S. M., Starobinets, G. L., Rakhman'ko, E. M., and Bubel, O. N.,** Influence of the solvent nature on the solvophobic effect, *Dokl. Akad. Nauk B.S.S.R.*, 31, 725, 1987.
259. **Stephanopoulos, G. and Townsend, D. W.,** Synthesis in process development, *Chem. Eng. Res. Des.*, 64, 160, 1986.

260. **Barford, R. A.,** Modern countercurrent distribution, *Sep. Purif. Meth.,* 4, 351, 1975.
261. **Högfeldt, E.,** The diluent not only a solvent for the extractant but also a participant in the extraction process, *Chem. Ind.,* 184, 1976.
262. **Davis, S. S.,** Molar attraction constants applied to structure activity relationships, *Experientia,* 26, 671, 1970.
263. **Hansen, C. M. and Pierce, P. E.,** Surface effects in coating processes, *Ind. Eng. Chem. Prod. Res. Dev.,* 13, 218, 1974.
264. **Wakahayashi, T., Oki, S., Omori, T., and Suzuki, N.,** Some applications of the regular solution theory to solvent extraction. I. Distribution of β-diketones, *J. Inorg. Nucl. Chem.,* 26, 2255, 1964.
265. **de Meere, A. L. J., Taylor, P. J., Grünbauer, H. J. M., and Tomlinson, E.,** Local composition models in pharmaceutical chemistry. II. Differentiation of hydrophobic fragmental constants, *Int. J. Pharm.,* 31, 209, 1986.
266. **Hampe, M. J.,** Selection of solvents in liquid/liquid extraction according to physico-chemical aspects, *Chem. Ing. Tech.,* 57, 669, 1985; *Ger. Chem. Eng.* 9, 251, 1986.
267. **Ritcey, G. M. and Lucas, B. H.,** Diluents and modifiers — their effect on mass transfer and separation, *Proc. Int. Solvent Extr. Conf. (ISEC '74),* Jeffreys, G. V., Ed., Society of Chemical Industry, London, 1974, Vol. 3, 2437.
268. **Wingefors, S.,** *Studies of the Solvent Extraction Chemistry of Dialkyldithiophosphates and Development of Models for Non-Ideal Behaviour in Solvent Extraction,* Thesis, Dept. of Nuclear Chemistry, Chalmers University of Technology, Göteborg, Sweden, 1979, comprising the following nine publications.
269. **Wingefors, S.,** Solvent extraction chemistry of dialkyldithiophosphates. I. Potentiometric two-phase titration studies of dibutylheptyldithiophosphoric and methylheptyldithiophosphoric acids, *Acta Chem. Scand.* A34, 289, 1980.
270. **Wingefors, S.,** Solvent extraction chemistry of dialkyldithiophosphates. II. Evaluation of two-phase titration data for dibutylheptyldithiophosphoric and methylheptyldithiophosphoric acids according to the solubility parameter concept, *Acta Chem. Scand.,* A34, 297, 1980.
271. **Wingefors, S.,** Solvent extraction chemistry of dialkyldithiophosphates. III. A spectrophotometric study of the protolysis and distribution equilibria of dibutyldithiophosphoric acid in the systems carbon tetrachloride/1 M (Na,H)ClO_4 and carbon tetrachloride/1 M (Na,H)Cl, *Acta Chem. Scand.,* 34, 301, 1980.
272. **Wingefors, S. and Rydberg, J.,** Solvent extraction chemistry of dialkyldithiophosphates. IV. Formation and partition of zinc bis(dibutyldithiophosphate) in the carbon tetrachloride and n-heptane/1 M (Na,H)ClO_4 systems, *Acta Chem. Scand.,* 34, 313, 1980.
273. **Wingefors, S.,** Solvent extraction chemistry of dialkyldithiophosphates. V. Formation and partition of nickel bis(dimethyldithiophosphate) species in the system toluene — 1 M Na(DMDTP,ClO_4), *Acta Chem. Scand.,* 34, 327, 1980.
274. **Wingefors, S., Liljenzin, J. O., and Saalman, E.,** Three-dimensional solubility parameters for the correlation and prediction of liquid-liquid partition constants of metal chelate complexes, *Proc. Int. Solvent Extr. Conf.,* 3, 116, 1980.
275. **Wingefors, S. and Liljenzin, J. O.,** Development of a correlation between the nonpolar solubility parameter, refractive index and molecular structure. I. Aliphatic hydrocarbons, *J. Chem. Technol. Biotechnol.,* 31, 115, 1981.
276. **Wingefors, S. and Liljenzin, J. O.,** Development of a correlation between the nonpolar solubility parameter, refractive index and molecular structure. II. Aliphatic ethers and alcohols, *J. Chem. Technol. Biotechnol.,* 31, 530, 1981.
277. **Wingefors, S. and Liljenzin, J. O.,** Cohesion energy densities and solubility parameters for the n-alkanes expressed as a new homomorph equation, *J. Chem. Technol. Biotechnol.,* 31, 523, 1981.
278. **Omori, T., Wakahayashi, T., Oki, S., and Suzuki, N.,** Some applications of the regular solution theory to solvent extractions-II. Distribution of scandium chelates, *J. Inorg. Nucl. Chem.,* 26, 2265, 1964.
279. **Suzuki, N., Akiba, K., and Kanno, T.,** Regularities in the distribution of TTA and its scandium chelate into a series of ester solvents, *Analyt. Chim. Acta,* 43, 311, 1968.
280. **Suzuki, N., Akiba, K., Kanno, T., and Wakahayashi, T.,** Relation between distribution of TTA and "solubility parameter" of alcohol solvent, *J. Inorg. Nucl. Chem.,* 30, 2521, 1968.
281. **Suzuki, N., Akiba, K., and Asano, H.,** Regularities in the distribution of TTA and its scandium chelate into a series of ether solvents, *Analyt. Chim. Acta,* 51, 115, 1970.
282. **Suzuki, N., Shibuya, H., and Watarai, H.,** The partition mechanism of thenoylacetone and thenoyltrifluoroacetone in water/alcohol systems, *Bull. Chem. Soc. Jpn.,* 51, 3232, 1978.
283. **Suzuki, N., Suzuki, J., and Saitoh, K.,** Liquid chromatography separation of metal(II) chelates of thenoyltrifluoroacetone on polyester beads, *J. Chromatogr.,* 177, 166, 1979.
284. **Wakabayashi, T., Takaizumi, K., Seto, K., Suzuki, N., and Akiba, K.,** Distribution of thenoyltrifluoroacetone and its scandium chelate in ketone solvent system, *Bull. Chem. Soc. Jpn.,* 41, 1854, 1968.
285. **Akiba, K. and Kanno, T.,** Liquid-liquid distribution coefficient of trioctyl phosphine oxide, *Kakuriken Kenkyu Hokoku (Tohoku Daigaku),* 9, 272, 1976.

286. **Akiba, K., Kanno, T., and Wada, M.,** Distribution coefficient of triphenyl phosphine oxide, *Kakuriken Kenkyu Hokoku (Tohoku Daigaku),* 10, 131, 1977.
287. **Akiba, K., Wada, M., and Kanno, T.,** Distribution coefficients of tributylphosphate and tributyl phosphine oxide, *Kakuriken Kenkyu Hokoku (Tohoku Daigaku),* 10, 268, 1977.
288. **Siekierski, S.,** The influence of diluent on extraction of europium and thorium nitrates by tri-*n*-butyl phosphate, *J. Inorg. Nucl. Chem.,* 24, 205, 1962.
289. **Sato, T.,** Solvent extraction of some metals from aqueous solutions, *J. Radioanalyt. Nucl. Chem.,* 101, 77, 1986.
290. **Mottola, H. A. and Freiser, H.,** Distribution of certain 8-quinolinols and their copper(II) chelates in a series of organic solvent-aqueous pairs, *Talanta,* 13, 55, 1966; Some solvent effects on the solvent extraction of 8-quinolinol, *Talanta,* 14, 864, 1967.
291. **Spink, D. R. and Okuhara, D. N.,** The effect of diluents on the extraction of tracer level copper by an alkyl hydroxy quinoline (Kelex® 100), *Proc. Int. Solvent Extr. Conf. (ISEC '74),* Jeffreys, G. V., Ed., Society Chemical Industry, London, 1974, Vol. 3, 2527.
292. **Zharovskii, F. G. and Ryzhenko, V. L.,** Applicability of solubility parameters to the extraction of hydroxyquinolinates, *Khim. Protsessov Ekstr., Mater. Konf. Khim. Ekstr., 3rd, 1969,* Zolotov, Yu. A., Ed., Nauka, Moscow, 1972, 237.
293. **Allard, B.,** Studies of tetravalent acetylacetonato complexes. II. Species formed in two-phase systems as indicated by the use of the regular solution concept, *J. Inorg. Nucl. Chem.,* 39, 694, 1977.
294. **Rydberg, J.,** Thermodynamics of two-phase liquid-liquid distribution of metal complexes studied by AKUFVE technique, *J. Indian Chem. Soc.,* 51, 15, 1974.
295. **Yamamoto, Y., Yamamoto, M., Ebisui, S., Takagi, T., Hashimoto, T., and Izuhara, M.,** Gel chromatography of metal acetylacetonates in organic solvents, *Analyt. Lett.,* 6, 451, 1973.
296. **Yamamoto, M. and Yamamoto, Y.,** Solvent effects in gel chromatography: relationship between distribution coefficients of metal acetylacetonates and solubility parameters of solvents, *Analyt. Chim. Acta,* 87, 375, 1976.
297. **Freiser, H.,** Relevance of solubility parameter in ion association extraction systems, *Analyt. Chem.,* 41, 1354, 1969.
298. **Iwamoto, E., Ito, K., and Yamamoto, Y.,** Solute-solvent interactions in ion-pair extraction of tetraalkylammonium iodides. 1. A new approach to the extraction constant, *J. Phys. Chem.,* 85, 894, 1981.
299. **Iwamoto, E., Ohmori, H., and Yamamoto, Y.,** Solute-solvent interactions in ion-pair extraction of tetraalkylammonium iodides. 3. The role of relative permittivity and solubility parameter in nitrobenzene-tetrachloromethane mixtures at 25°C., *J. Phys. Chem.,* 87, 1776, 1983.
300. **Buchowski, H.,** Effect of electron donor properties of solvents on partition coefficients and infrared absorption of phenol, *Rocz. Chem.,* 41, 627, 1967.
301. **Kiezyk, P. R. and Mackay, D.,** The screening and selection of solvents for the extraction of phenol from water, *Can. J. Chem. Eng.,* 51, 741, 1973.
302. **Korenman, Ya. I. and Pereshein, V. V.,** Extraction of phenol and o-cresol, *Tr. Khim. Khim. Tekhnol.,* No. 1,66,1969.
303. **Korenman, Ya. I.,** Solubility parameter and extraction of phenols, *Zh. Fiz. Khim.,* 46, 578, 1972; *Russ. J. Phys. Chem.,* 46, 334, 1972.
304. **Tamvelius, H. Ya. and Melder, L. I.,** Solubility parameters of phenol and anisidine obtained from interphase distribution data, *Tr. Tallin. Politekhn. Inst.,* No. 356, 47, 1974.
305. **Kojima, I. and Davis, S. S.,** Thermodynamics of distribution of para-substituted phenols between organic solvents and sodium chloride solution, *Int. J. Pharm.,* 20, 247, 1984.
306. **Korenman, Ya. I.,** Relationships of xylenols, *Zh. Prikl. Khim.* (Leningrad), 46, 380, 1973; *J. Appl. Chem. U.S.S.R.,* 46, 394, 1973.
307. **Li, L.-M. and Hsu, K.-H.,** Determination of the partition coefficients of 1-phenyl-3-methyl-4-benzoyl-5-pyrazolone between organic solvents and water and their correlation with the solubility parameters, *Hua Hsueh Hsueh Pao,* 36, 239, 1978.
308. **Campbell, J. R. and Luthy, R. G.,** Measurement and prediction of distribution coefficients for wastewater aromatic solutes, *Environ. Sci. Technol.,* 17, 582, 1983.
309. **Kemula, W., Buchowski, H., Lewanaowski, R., Pawlowski, W., and Teperek, J.,** Distribution studies. V. The effect of solvents on the partition coefficients and the role of solubility parameters, *Bull. Acad. Polon. Sci. Ser. Sci. Chim.,* 14, 395, 1966.
310. **Kojima, I., Yoshida, M., and Tanaka, M.,** Distribution of carboxylic acids between organic solvents and aqueous perchloric acid solution, *J. Inorg. Nucl. Chem.,* 32, 987, 1970.
311. **Hirose, K. and Tanaka, M.,** Partition of pyridine bases between benzene and aqueous solution, *J. Inorg. Nucl. Chem.,* 38, 2285, 1976.
312. **Fujii, Y. and Tanaka, M.,** Partition of carboxylic acids between benzene and aqueous solution as considered from the regular solution theory, *Bull. Chem. Soc. Jpn.,* 54, 3696, 1981.

313. **Fujii, Y., Sobue, K., and Tanaka, M.,** Solvent effect on the dimerization and hydration constant of benzoic acid, *J. Chem. Soc. Faraday I,* 74, 1467, 1978.
314. **Kojima, I., Uchida, M., and Tanaka, M.,** Extraction of copper(II) with aliphatic carboxylic acids, *J. Inorg. Nucl. Chem.,* 32, 1333, 1970.
315. **Högfeldt, E. and Fredlund, F.,** The solubility of water in mixtures or organic solvents, *Acta Chem. Scand.,* 24, 1858, 1970.
316. **Moroto, S. and Watanabe, A.,** Solvent extraction of nitric acid with butyl acetate from aqueous mixture of phosphoric acid, *Nippon Kagaku Kaishi,* 2317, 1972.
317. **Starobinets, G. L., Rakhman'ko, E. M., and Leshchev, S. M.,** Effects of the medium during distribution of primary aliphatic amines between mixtures of water with lower alcohols and octane, *Zh. Fiz. Khim.,* 53, 2720, 1979; *Russ. J. Phys. Chem.,* 53, 1560, 1979.
318. **Sato, T., Nakamura, T., and Kuwahara, M.,** Diluent effect on the extraction of uranium(VI) from hydrochloric acid solutions by trioctylamine, *Solv. Extr. Ion Exch.,* 3, 283, 1985.
319. **Schmidt, V. S., Shesterikov, V. N., and Mezhov, E. Z.,** The solubility of amine salts in solvents of low polarity and the influence of diluents on the extraction properties of amine salts, *Usp. Khim.,* 36, 2167, 1967; *Russ. Chem. Rev.,* 36, 946, 1967.
320. **Siekierski, S. and Olszer, R.,** Relation between the partition coefficient of GeX_4 molecules (X = Cl, Br, I) and the solubility parameter of the solvent, *J. Inorg. Nucl. Chem.,* 25, 1351, 1963.
321. **Yamada, H. and Tanaka, M.,** Solvent effects on extraction of copper with capric acid, *J. Inorg. Nucl. Chem.,* 38, 1501, 1976.
322. **Whewell, R. J., Hughes, M. A., and Hanson, C.,** Role of the diluent in the solvent extraction of copper with LIX 64N, *Advances in Extractive Metallurgy* (Proc. Int. Symp., 3rd), Institute of Mining and Metallurgy, London, 1977.
323. **Reuben, B., Jones, S., and Kaur, N.,** Solvent extraction of thiocyanate ions with quaternary ammonium salts, *Chem. Ind.,* No.1, 14, 1985.
324. **Sato, T., Nakamura, T., and Fujimatsu, T.,** Extraction of bivalent manganese, cobalt, copper, zinc, and cadmium from hydrochloric acid solutions by long-chain alkyl quaternary ammonium chloride in various organic solvents, *Bull. Chem. Soc. Jpn.,* 54, 2656, 1981.
325. **Suzuki, N., Yoshida, T., and Watarai, H.,** Partition coefficients of chlorobenzene in heptane-water and heptane-acetonitrile systems, *Bull. Chem. Soc. Jpn.,* 55, 121, 1982.
326. **Nikolaev, B. A. and Korenman, I. M.,** Application of a theory of regular solutions to the distribution of benzidine and *o*-tolidine, *Dep. Doc.* SPSTL 125Khp-D82, 1982.
327. **Bailey, N. T. and Mahi, P.,** The effect of diluents on the metal extracted and phase separation in the extraction of aluminum with mononenyl phosphoric acid, *Hydrometall.,* 18, 351, 1987.
328. **Iwachido, T., Minami, M., Naito, H., and Tōei, K.,** The extractability of potassium as picrates (KPic and K(L)Pic, L: 18-Crown-6) and its relationship to the properties of the solvents, *Bull. Chem. Soc. Jpn.,* 55, 2378, 1982.
329. **Kremenskaya, I. N. and Evseeva, N. E.,** Study of the extraction properties of bis(4-ethyl-3,5-dipropyl-1-pyrazolyl)methane in acid media, *Radiokhim.,* 18, 835, 1976.
330. **Starobinets, G. L., Rakhman'ko, E. M., and Voronova, G. N.,** *Vesti Akad. Navuk B.S.S.R. Ser. Khim. Navuk,* (4), 33, 1981.
331. **Li, Y. G., Teng, T., Lu, J. F., Chen, G., and Li, J. D.,** A study on Scatchard-Hildebrand solution theory for metal-solvent extraction systems, *Fluid Phase Equil.,* 30, 297, 1986.
332. **Tsukuda, T., Saitoh, K., and Suzuki, N.,** Leaching behavior of metal-containing species in residual oils with different organic solvents, *Analyt. Chim. Acta,* 183, 89, 1986.
333. **Grosjean, D.,** Solvent extraction and organic carbon determination in atmospheric particulate matter: the organic extraction — organic carbon analyzer (OE-OCA) technique, *Analyt. Chem.,* 47, 797, 1975; **Grosjean, D., Van Cauwenberghe, K., Schmid, J. P., Kelley, P. E., and Pitts, J. N., Jr.,** Identification of $C_3 - C_{10}$ aliphatic dicarboxylic acids in airborne particulate matter, *Environ. Sci. Technol.,* 12, 313, 1978.
334. **Schmid, P.,** Extraction and purification of lipids. II. Why is chloroform-methanol such a good lipid solvent?, *Physiol. Chem. Phys.,* 5, 141, 1973.
335. **Chung, O. K., Pomeranz, Y., Finney, K. F., and Sjogren, M. D.,** Defatted and reconstituted wheat flours. I. Effects of solvents and Soxhlet types on functional (breadmaking) properties, *Cereal Chem.,* 54, 454, 1977.
336. **Chung, O. K., Pomeranz, Y., and Jacobs, R. M.,** Extractability of wheat-flour lipids by solvents that vary in solubility parameters, *J. Am. Oil Chem. Soc.,* 56(Abstr.), A194, 1979.
337. **Chung, O. K., Pomeranz, Y., Hwang, E. C., and Dikeman, E.,** Defatted and reconstituted wheat flours. IV. Effects of flour lipids on protein extractability from flours that vary in bread-making quality, *Cereal Chem.,* 56, 220, 1979.
338. **Chung, O. K., Pomeranz, Y., Finney, K. F., Shogren, M. D., and Carville, D.,** Defatted and reconstituted wheat flours. V. Breadmaking response to shortening of flour differentially defatted by varying solvent and temperature, *Cereal Chem.,* 57, 106, 1980.

339. **Chung, O. K., Pomeranz, Y., Jacobs, R. M., and Howard, B. G.,** Extraction conditions to differentiate among hard red winter wheats that vary in breadmaking, *J. Food Sci.*, 45, 1168, 1980.
340. **Chung, O. K., Pomeranz, Y., and Jacobs, R. M.,** Solvent solubility parameter and flour moisture effects on lipid extractability, *J. Am. Oil. Chem. Soc.*, 61, 793, 1984.
341. **Simon, S. A., Stone, W. L., and Bennett, P. B.,** Can regular solution theory be applied to lipid bilayer membranes?, *Biochim. Biophys. Acta*, 550, 38, 1979.
342. **Hernandez, A., Martinez, C., and Gonzalez, G.,** Freezing of alfalfa leaf juice-formation and solvent extraction of freezing curd, *J. Sci. Food Agric.*, 42, 173, 1988.
343. **Drago, R. S. and Wayland, B. B.,** A double-scale equation for correlating enthalpies of Lewis acid-base interactions, *J. Am. Chem. Soc.*, 87, 3571, 1965.
344. **Drago, R. S. and Matwiyoff, N. A.** *Acids and Bases,* Heath, Lexington, MA, 1968, 66.
345. **Drago, R. S., Vogel, G. E., and Needham, T. E.,** A four-parameter equation for predicting enthalpies of adduct formation, *J. Am. Chem. Soc.*, 93, 6014, 1971.
346. **Drago, R. S.,** Quantitative evaluation and prediction of donor-acceptor interactions, *Struct. Bonding (Berlin),* 15, 73, 1973.
347. **Chong, S.-L. and Franklin, J. L.,** Proton affinities of benzene, toluene, and the xylenes, *J. Am. Chem. Soc.*, 94, 6630, 1972.
348. **Long, J. and Munson, B.,** Proton affinities of some oxygenated compounds, *J. Am. Chem. Soc.*, 95, 2427, 1973.
349. **Dack, M. R. J.,** The influence of solvent on chemical reactivity, in *Solutions and Solubilities,* Dack, M. R. J., Ed., Part II, John Wiley & Sons-Interscience, New York, 1976, chap. 11.
350. **Hammett, L. P.,** *Physical Organic Chemistry,* McGraw-Hill, New York, 1940.
351. **Rider, P. E.,** A two-parameter model for estimating hydrogen bond enthalpies of reaction, *J. Appl. Polym. Sci.*, 25, 2975, 1980.
352. **Rider, P. E.,** Hydrogen bonding potential, HBP: a model for predicting resin solubilities in organic solvents, *Polym. Eng. Sci.*, 23, 810, 1983.
353. **Rider, P. E.,** Theories of solvency and solution, in *Applied Polymer Science,* 2nd ed. (*Am. Chem. Soc. Symp. Ser.*, Vol. 285), Tess, R. W. and Poehlein, G. W., Eds., 1985, chap. 27, 643.
354. **Arnett, E. M.,** Quantitative comparison of weak organic bases, *Progr. Phys. Org. Chem.*, 1, 223, 1963.
355. **Gutmann, V.,** Reactions of complex inorganic compounds in nonaqueous solvents, *Rec. Chem. Progr.*, 30, 169, 1969.
356. **Gutmann, V.,** Significance of empirical solvent parameter (donor number, acceptor number) for organic chemists, *Z. Chem.*, 20, 37, 1980.
357. **Snow, A. W. and Barger, W. R.,** A chemical comparison of methanesulfonyl fluoride with organofluorophosphorus ester anticholinesterase compounds, *Chem. Res. Toxicol.*, 1, 379, 1988.
358. **Ito, K., Isono, H., Iwamoto, E., and Yamamoto, Y.,** Solvation and ionic association of bis(β-diketo)(diamine) cobalt(III) perchlorates in alcohols and ketones, *Bull. Chem. Soc. Jpn.*, 61, 4077, 1988.
359. **Moyer, B. A., Caley, C. A., and Baes, C. F., Jr.,** Hydration and aggregation of monofunctional sulfoxide and other neutral oxygen-donor extractants: the di(2-ethylhexyl)sulfoxide, dodecane, water system, *Solv. Extr. Ion Exch.*, 6, 785, 1988; **Baes, C. F., Jr., Moyer, B. A., Case, G. N., and Case, F. I.,** SXLSQA, a computer program for including both complex formation and activity effects in the interpretation of solvent extraction data, *Separation Science and Technology* (Proc. Sixth Symp. Sep. Sci. Technol. Energy Appl., Knoxville, TN, October 1989), to be published.
360. **Arul, J., Boudreau, A., Makhlouf, J., Tardif, R., and Grenier, B.,** Distribution of cholesterol in milk fat fractions, *J. Dairy Res.*, 55, 361, 1988.
361. **Oki, S.,** Distribution coefficient of nickel dimethylglyoximate between an aqueous solution and organic solvents, *Talanta,* 18, 1233, 1971.
362. **Wakabayashi, T.,** Applications of the regular solution theory to solvent extraction. IV. Oxine-inert solvent system, *Bull. Chem. Soc. Jpn.*, 40, 2836, 1967.
363. **Wada, M. and Kanno, T.,** Liquid-liquid partition of americium(III) chelates with 2-thenoyltrifluoroacetone, 8-quinolinol, and cupferron, *Nippon Kagaku Kaishi,* No. 10, 1332, 1979.
364. **Korenman, Ya. I.,** Extraction of phenol and *p*-ethylphenol, *Tr. Khim. Khim. Tekhnol.*, No. 2, 68, 1969.
365. **Iguchi, A.,** Predicting cryogenic vapor-liquid equilibrium, *Kagaku Sochi,* 13(11), 86, 1971.
366. **Dobson, S. and van der Zeeuw, A. J.,** Hydrocarbon solvent diluents in hydroxy oxime solvent extraction processes, *Chem. Ind.*, No. 5, 175, 1976.
367. **Thonchk, N. K., Jones, S. D., Reuben, B. G., and Mahi, P.,** Extraction of thiocyanate ions from coal gasification effluents: Part I — Ion-pair extraction system, *Chem. Eng. Res. Des.*, 66, 503, 1988.
368. **Hiranuma, M.,** Estimation of liquid phase activity coefficients by the use of three-dimensional vaporization energy, *Fluid Phase Equil.*, 41, 81, 1988.
369. **Condorelli, G.,** The alpha→beta transition of phthalocyaninatocopper. Influence of type of vehicle, *Pitture Vernici,* 49, 259, 1973.

370. **Wakui, Y., Imura, H., and Suzuki, N.,** Solvent effect on the liquid-liquid partition of 21H,23H-porphine between various organic solvents and acidic aqueous solutions, *Bull. Chem. Soc. Jpn.*, 62, 2520, 1989.

371. **Barbari, T. A., Koros, W. J., and Paul, D. R.,** Gas sorption in polymers based on bisphenol A, *J. Polym. Sci. B Polym. Phys.*, 26, 729, 1988.

372. **Marcus, Y.,** Diluent effects in solvent extraction, *Solvent Extr. Ion Exch.*, 7, 567, 1989.

373. **Alekperov, E. R.,** Effect of diluents on the extraction of boric acid with 2-hydroxy-5-octylbenzylethanolamine, *Azerb. Khim. Zh.*, No. 1, 109, 1988.

374. **Fujimatsu, H., Ogasawara, S., Satoh, N., Komori, K., and Kuroiwa, S.,** Adhesive effect of high density polyethylene gels on polyethylene moldings, *Colloid Polym. Sci.*, 267, 500, 1989.

375. **Horowitz, E. P., Martin, K. A., and Diamond, H.,** The influence of the diluent on the distribution behavior of octyl(phenyl)-*N,N*-diisobutylcarbamoylmethylphosphine oxide, *Solvent Extr. Ion Exch.*, 42, 261, 1980.

376. **Tucker, S. A. and Acree, W. E., Jr.,** Solubility of anthracene in binary carbon tetrachloride + alkane solvent mixtures, *Phys. Chem. Liq.*, 19, 73, 1989.

377. **Acree, W. E., Jr. and Tucker, S. A.,** Solubility of anthracene in binary *p*-xylene + alkane and benzene + alkane solvent mixtures, *Phys. Chem. Liq.*, 20, 31, 1989.

378. **Akiba, K., Wada, M., and Kanno, T.,** Liquid-liquid partition constants of tri-*n*-octyl, tris(2-ethylhexyl) and triphenyl phosphine oxides, *J. Inorg. Nucl. Chem.*, 42, 261, 1980.

379. **Dunlap, R. D. and Scott, R. L.,** Internal pressures of perfluoro-*n*-hexane, *n*-hexane, and their mixtures, *J. Phys. Chem.*, 66, 631, 1962.

380. **Koliadima, A., Agathonos, P., and Karaiskakis, G.,** Estimation of solubility and interaction parameters in binary liquid mixtures by reversed-flow gas chromatography, *Chromatographia*, 26, 29, 1988.

381. **Moura Ramos, J. J.,** Molecular shape and orientational order. Effects in the energy of cavity formation, *J. Solution Chem.*, 18, 957, 1989.

Chapter 8

SOLVENT SCALES

Parameters describing and correlating the solvent abilities of liquids have been based on a great variety of chemical and physical properties. Some are measures of solvent basicity. Others are direct determinations of the solubility of a representative solute in a range of liquids; the solubility of hydrogen chloride in liquids at 10°C was used in this way by Gerrard[1,2] and the resulting solvent basicity scale was compared with other scales by Arnett[3] and by Dack.[4] Once a solvent scale has been established, it is necessary to determine the position on it of any required liquid. If the solvent scale has a theoretical basis, it may be possible to calculate values from the information available on other properties, but if it is an empirical or semiempirical scale, direct testing is required.

In this chapter, several solvent scales related to cohesion parameters are described, together with some numerical values. Subsequently, in Chapters 14 and 15, some values for polymers are presented.

8.1 HYDROGEN BONDING

Burrell[5-12] was one of the first to attempt to deal with the hydrogen bonding factor in the application of cohesion parameters to practical systems. He did this in a very simple way, dividing liquids into three classes according to their hydrogen bonding capacities, on the assumption that complete miscibility can occur only if the degree of hydrogen bonding is comparable in the components:

- *Weak or poor hydrogen-bonding capacity liquids,* including hydrocarbons, chlorinated hydrocarbons, and nitrohydrocarbons
- *Moderate hydrogen-bonding capacity liquids,* including ketones, esters, ethers, and glycol monoethers
- *Strong hydrogen-bonding capacity liquids*, such as alcohols, amines, acids, amides, and aldehydes

This classification is still used widely in practical applications. A few typical examples are presented in Table 1, Chapter 4, in the form of solubility spectra series, and a fuller listing appears in Table 1 of this Chapter. An alternative classification in a widely used reference work[13] distinguishes between nonpolar liquids, moderately polar liquids, and hydrogen-bonded liquids.

Lieberman et al.[14-16] developed the classification by assigning quantitative values (see, for example, Table 3 below) to the hydrogen-bonding capabilities and plotting them against Hildebrand parameters. Subsequently, spectroscopic hydrogen-bonding parameters were used (Section 8.3).

An alternative quantitative measure of the ability of organic liquids to donate and accept hydrogen bonds is sonic velocity in liquid-paper systems.[4,17] Paper fibers are held together largely by hydrogen bonds, and on wetting paper with a liquid, most disruption of the water-fiber bonds occurs in the presence of those liquids which preferentially form fiber-liquid hydrogen bonds. As a result, the velocity of sound through the paper, which depends on the degree of bonding, decreases as the liquid hydrogen bonding capability increases. Water, as the most common liquid for disrupting fiber-fiber bonds in paper, was chosen as a reference standard, and the hydrogen-bonding parameter was defined as

$$\gamma_w = \frac{100 \times \text{sound velocity in water-soaked paper}}{\text{sound velocity in liquid-soaked paper}} \tag{1}$$

TABLE 1
Hildebrand Parameters and Hydrogen-Bonding Capabilities of Liquids at 25°C

Liquid	$\delta/MPa^{1/2}$	Hydrogen-bonding capability
Acetaldehyde	21.1	Moderate
Acetic acid, ethanoic acid	20.7	Strong
Acetic anhydride	21.1	Strong
Acetone, 2-propanone	20.2	Moderate
Acetonitrile	24.3	Poor
Acetophenone, 1-phenylethanone	21.7	Moderate
Acetyl chloride	19.4	Moderate
N-Acetylmorpholine	23.7	Moderate
N-Acetylpiperidine	22.9	Strong
N-Acetylpyrrolidine	23.3	Strong
Acrolein, 2-propenal	20.0	Strong
Acrylic acid, 2-propenoic acid	24.5	Strong
Acrylonitrile, 2-propenenitrile	21.5	Poor
Allyl acetate	18.8	Moderate
Allyl alcohol, 2-propen-1-ol	24.1	Strong
Allyl chloride, 3-chloropropene	18.0	Moderate
Ammonia	33.3	Strong
sec-Amyl acetate	17.0	Moderate
p-Anethole, *p*-propenylanisole	17.2	Moderate
Aniline, benzeneamine	21.1	Strong
Anthracene	20.2	Poor
Apco® #18 solvent	15.3	Poor
Apco #140 solvent	14.9	Poor
Apco thinner	16.0	Poor
Aroclor® 1248	18.0	Poor
Benzaldehyde	19.2	Moderate
Benzene	18.8	Poor
Benzonitrile	17.2	Poor
Benzyl alcohol, benzenemethanol	24.7	Strong
Benzyl Cellosolve,® ethylene glycol monobenzyl ether	22.3	Moderate
Bromobenzene	20.2	Poor
1-Bromobutane	17.8	Moderate
2-Bromobutane, *sec*-butyl bromide	17.2	Moderate
Bromoethane, ethyl bromide	19.6	Moderate
Bromonaphthalene	21.7	Poor
1-Bromopentane, amyl bromide	15.5	Moderate
1-Bromopropane, propyl bromide	18.2	Moderate
o-Bromostyrene	20.0	Poor
1,3-Butadiene, bivinyl	14.5	Poor
Butane	13.9	Poor
1,3-Butanediol	23.7	Strong
1,4-Butanediol	24.7	Strong
2,3-Butanediol	22.7	Strong
1-Butanol	23.3	Strong
2-Butanol, *sec*-butyl alcohol	22.1	Strong
tert-Butanol	21.7	Strong
Butyl acetate	17.4	Moderate
sec-Butyl acetate	16.8	Moderate
Butyl acrylate	18.0	Moderate
Butylamine, 1-butanamine	17.8	Strong
Butyl butyrate	16.6	Moderate
Butyl Carbitol,® diethylene glycol monobutyl ether	19.4	Moderate
Butyl Cellosolve,® ethylene glycol monobutyl ether	19.4	Moderate
2,3-Butylenecarbonate	24.7	Moderate
Butyl ether, 1,1′-oxybisbutane	16.0	Moderate
Butyl formate	18.2	Moderate
Butyl iodide, 1-iodobutane	17.6	Moderate

TABLE 1 (continued)
Hildebrand Parameters and Hydrogen-Bonding Capabilities of Liquids at 25°C

Liquid	δ/MPa$^{1/2}$	Hydrogen-bonding capability
Butyl lactate	19.2	Moderate
Butyl methacrylate	16.8	Moderate
Butyl propionate	18.0	Moderate
Butyl stearate	15.3	Moderate
Butyraldehyde, butanal	18.4	Moderate
Butyric acid, butanoic acid	21.5	Strong
γ-Butyrolactone	25.8	Moderate
Butyronitrile	21.5	Poor
ε-Caprolactam	26.0	Moderate
ε-Caprolactone, 2-oxepanone	20.6	Moderate
Capronitrile, hexanenitrile	19.2	Poor
Carbitol,® diethylene glycol monoethyl ether	20.9	Moderate
Carbitol® acetate, diethylene glycol monoethyl ether acetate	17.4	Moderate
Carbon disulfide	20.4	Poor
Celanese® solvent 601	18.8	Moderate
Cellosolve,® ethylene glycol monoethyl ether	21.5	Moderate
Cellosolve® acetate, ethylene glycol monoethyl ether acetate	17.8	Moderate
Chloroacetonitrile	25.8	Poor
Chlorobenzene	19.4	Poor
Chloroethane, ethyl chloride	18.8	Moderate
β-Chloroethyl acetate	19.8	Moderate
Chloromethane, methyl chloride	19.8	Moderate
1-Chloropentane, amyl chloride	17.0	Moderate
1-Chloropropane, propyl chloride	17.4	Moderate
o-Chlorostyrene	18.0	Poor
p-Chlorostyrene	18.0	Poor
p-Chlorotoluene	18.0	Poor
m-Cresol, 3-methylphenol	20.9	Strong
Cyclobutanedione	22.5	Moderate
Cyclohexane	16.8	Poor
Cyclohexanol	23.3	Strong
Cyclohexanone	20.3	Moderate
Cyclopentane	17.8	Poor
Cyclopentanone	21.3	Moderate
p-Cymene	16.8	Poor
Decahydronaphthalene, decalin	18.0	Poor
Decane	13.5	Poor
Diacetone alcohol	18.8	Moderate
Diacetone alcohol methyl ether, Pentoxone®	16.8	Moderate
N,N-Diacetylpiperazine	28.0	Moderate
Diamyl ether	14.9	Moderate
Diamyl phthalate	18.6	Moderate
Dibenzyl ether	19.2	Moderate
1,2-Dibromoethane, ethylene dibromide	21.3	Poor
1,2-Dibromoethylene	20.7	Poor
Dibutoxyethyl phthalate, Kronisol®	16.4	Moderate
Dibutylamine	16.6	Strong
Dibutyl fumarate	18.4	Moderate
Dibutyl maleate	18.4	Moderate
Dibutyl phenyl phosphate	17.8	Moderate
Dibutyl phthalate	19.0	Moderate
Dibutyl sebacate	18.8	Moderate
Dichloroacetic acid	22.5	Strong
o-Dichlorobenzene	20.5	Poor
Dichlorodifluoromethane, Freon 12	11.3	Poor
1,1-Dichloroethane, ethylidene chloride	18.2	Poor

TABLE 1 (continued)
Hildebrand Parameters and Hydrogen-Bonding Capabilities of Liquids at 25°C

Liquid	δ/MPa$^{1/2}$	Hydrogen-bonding capability
cis-1,2-Dichloroethylene, dichloroethene	18.6	Poor
trans-1,2-Dichloroethylene	18.4	Poor
Dichloroethyl ether	20.0	Moderate
Dichlorofluoromethane, Freon 21	17.0	Poor
1,2-Dichloropropane	18.4	Poor
2,2-Dichloropropane	16.8	Poor
Dicyclohexyl, bicyclohexyl	17.4	Poor
N,N-Diethylacetamide	20.3	Moderate
Diethylamine	16.4	Strong
Diethyl carbonate	18.0	Moderate
Diethylene glycol, 2,2′-oxybisethanol	24.8	Strong
Diethylene glycol monobutyl ether, butyl Carbitol®	19.4	Moderate
Diethylene glycol monoethyl ether, Carbitol	20.9	Moderate
Diethylene glycol monoethyl ether acetate, Carbitol acetate	17.4	Moderate
Diethylene glycol monolaurate	17.8	Moderate
Diethyl ether, 1,1′-oxybisethane	15.1	Moderate
N,N-Diethylformamide	21.7	Moderate
Diethylketone, 3-pentanone	18.0	Moderate
Diethyl maleate	20.3	Moderate
Diethyl oxalate	17.6	Moderate
Diethyl phthalate	20.5	Moderate
2,2-Diethyl-1,3-propanediol, heptylene glycol	20.3	Strong
Diethyl sulfone	25.4	Moderate
Difluorodichloromethane, Freon 12	11.3	Poor
Difluorotetrachloroethane, Freon 112	16.0	Poor
N,N-Diformylpiperazine	31.5	Moderate
Dihexyl ether	16.4	Moderate
Dihexyl phthalate	18.2	Moderate
Diisobutylene	15.8	Poor
Diisobutyl ketone	16.0	Moderate
Diisodecyl phthalate	14.7	Moderate
Diisopropyl ether	14.1	Moderate
Diisopropyl ketone	16.4	Moderate
Dipropyl ether	16.0	Moderate
N,N-Dimethylacetamide	22.1	Moderate
N,N-Dimethylaniline	19.8	Moderate
2,2-Dimethyl-1,3-butanediol	20.5	Strong
Dimethyl carbonate	20.3	Moderate
Dimethyl ether	18.0	Moderate
N,N-Dimethylformamide	24.8	Moderate
Dimethyl malonate	22.5	Moderate
N,N-Dimethylnitroamine	26.8	Strong
Dimethyl oxalate	22.5	Moderate
Dimethyl phosphite	25.6	Moderate
Dimethyl phthalate	21.9	Moderate
Dimethylsiloxanes	10–12	Poor
Dimethyl sulfide	19.2	Poor
Dimethyl sulfone	29.7	Moderate
Dimethylsulfoxide, sulfinylbismethane	24.5	Moderate
Dimethyltetramethylene sulfone	24.8	Moderate
Dioctyl adipate	17.8	Moderate
Dioctyl phthalate	16.2	Moderate
Dioctyl sebacate	17.6	Moderate
1,4-Dioxane, diethylene oxide	20.5	Moderate
1,3-Dioxolane	20.9	Moderate
Dipentene	17.4	Poor

TABLE 1 (continued)
Hildebrand Parameters and Hydrogen-Bonding Capabilities of Liquids at 25°C

Liquid	$\delta/MPa^{1/2}$	Hydrogen-bonding capability
Diphenyl ether	20.7	Moderate
Diphenyl 2-ethylhexyl phosphate	17.6	Moderate
Dipropylene glycol, oxybispropanol	20.5	Strong
Dipropylene glycol monomethyl ether	19.0	Moderate
Dipropylphthalate	19.8	Moderate
Dipropyl sulfone	23.1	Moderate
Dodecane	16.2	Poor
1-Dodecanol	20.0	Strong
Epichlorohydrin, (chloromethyl)oxirane	22.5	Strong
Ethane	12.3	Poor
Ethanol	26.0	Strong
N-Ethylacetamide	25.2	Strong
Ethyl acetate	18.6	Moderate
Ethyl acrylate	17.6	Moderate
Ethylamine	20.5	Strong
Ethyl amyl ketone	16.8	Moderate
Ethylbenzene	18.0	Poor
Ethyl benzoate	16.8	Moderate
Ethyl bromide, bromoethane	19.6	Moderate
2-Ethyl-1-butanol	21.5	Strong
Ethyl butyrate	17.4	Moderate
Ethyl caprylate	14.9	Moderate
Ethyl chloride, chloroethane	18.8	Moderate
Ethyl cyanoacetate	22.5	Moderate
Ethylene bromide, 1,2-dibromoethane	19.8	Poor
Ethylene carbonate	30.0	Moderate
Ethylene chlorohydrin, 2-chloroethanol	24.9	Strong
Ethylene cyanohydrin, 3-hydroxypropanenitrile	31.1	Strong
Ethylenediamine, 1,2-ethanediamine	25.2	Strong
Ethylene dichloride, 1,2-dichloroethane	20.0	Poor
Ethylene glycol, 1,2-ethanediol	29.9	Strong
Ethylene glycol diacetate	20.5	Moderate
Ethylene glycol diethyl ether, diethyl Cellosolve®	17.0	Moderate
Ethylene glycol dimethyl ether, diethyl Cellosolve®	17.6	Moderate
Ethylene glycol methyl ether acetate, methyl Cellosolve® acetate	18.8	Moderate
Ethylene glycol monobenzyl ether, benzyl Cellosolve	22.3	Moderate
Ethylene glycol monobutyl ether, butyl Cellosolve	19.4	Moderate
Ethylene glycol monoethyl ether, Cellosolve	21.5	Moderate
Ethylene glycol monoethyl ether acetate, Cellosolve acetate	17.8	Moderate
Ethylene glycol monomethyl ether, methyl Cellosolve	23.3	Moderate
Ethylene glycol monophenyl ether, phenyl Cellosolve	23.5	Moderate
Ethylene oxide, oxirane	22.7	Moderate
N-Ethylformamide	28.4	Strong
Ethyl formate	19.2	Moderate
2-Ethyl-1,3-hexanediol, octylene glycol	19.2	Strong
Ethylhexanol	19.4	Strong
Ethylhexyl acrylate	15.9	Moderate
Ethylidene chloride, 1,1-dichloroethane	18.2	Poor
Ethyl iodide, iodoethane	19.2	Moderate
Ethyl isobutyl ether	15.3	Moderate
Ethyl isobutyrate	16.2	Moderate
Ethyl lactate	20.5	Moderate
Ethyl mercaptan, ethanethiol	18.8	Poor
Ethyl methacrylate	17.0	Moderate
N-Ethyl morpholine	18.2	Strong

TABLE 1 (continued)
Hildebrand Parameters and Hydrogen-Bonding Capabilities of Liquids at 25°C

Liquid	δ/MPa$^{1/2}$	Hydrogen-bonding capability
Ethyl orthoformate	17.0	Moderate
Ethyl propionate	17.2	Moderate
Fluorocarbons, aliphatic	11-13	Poor
Fluorocarbons, aromatic	15-17	Poor
Formamide	39.3	Strong
Formic acid, methanoic acid	24.7	Strong
N-Formylmorpholine	26.6	Moderate
N-Formylpiperidine	23.5	Moderate
Furan	19.2	Moderate
Furfural, 2-furancarboxaldehyde	22.9	Moderate
Furfuryl alcohol, 2-furanmethanol	25.6	Strong
Glycerol, 1,2,3-propanetriol	33.7	Strong
Heptane	15.1	Poor
1-Heptanol, *n*-heptyl alcohol	21.7	Strong
Hexamethylphosphoramide	21.5	Strong
Hexane	14.9	Poor
2,5-Hexanediol	21.1	Strong
1-Hexene	15.1	Poor
1-Hexanol, *n*-hexyl alcohol	21.9	Strong
Hydrazine	37.0	Strong
Hydrogen	6.1	Poor
Hydrogenated terphenyl (Monsanto HB-40)	18.4	Poor
Hydrogen cyanide, hydrocyanic acid	24.7	Strong
Iodobenzene	20.7	Poor
Iodoethane, ethyl iodide	19.2	Moderate
1-Iodopentane, amyl iodide	17.2	Moderate
Isoamyl acetate	16.0	Moderate
Isoamyl alcohol, 3-methyl-1-butanol	20.5	Strong
Isoamyl formate	16.4	Moderate
Isobutanol	21.5	Strong
Isobutyl acetate	17.0	Moderate
Isobutyl acrylate	17.4	Moderate
Isobutyl butyrate	16.0	Moderate
Isobutyl chloride	16.6	Moderate
Isobutylene	13.7	Poor
Isobutylene glycol, 2,2,-dimethyl-1,2-butanediol	22.9	Strong
Isobutyl formate	16.8	Moderate
Isobutyric acid	21.1	Strong
Isobutyronitrile	20.0	Poor
Isodecyl acrylate	16.8	Moderate
Isophorone, 3,5,5-trimethyl-2-cyclohexen-1-one	18.6	Moderate
Isoprene, 2-methyl-1,3-butadiene	15.1	Poor
Isopropanol, 1-methylethanol	23.5	Strong
Isopropyl acetate	17.2	Moderate
Isopropyl isobutyrate	16.2	Moderate
Isopropyl chloride, 2-chloropropane	16.6	Moderate
Lauryl alcohol, 1-dodecanol	16.6	Strong
Low odor mineral spirits	14.1	Poor
Maleic anhydride, 2,5-furandione	27.8	Strong
Malonitrile	30.9	Poor
Mesitylene, 1,3,5-trimethylbenzene	18.0	Poor
Mesityl oxide, 4-methyl-3-penten-2-one	18.4	Moderate
Methacrylic acid, 2-methyl-2-propenoic acid	22.9	Strong
Methane	11.0	Poor
Methanol	29.6	Strong
4-Methoxy-4-methyl-2-pentanol	17.4	Strong

TABLE 1 (continued)
Hildebrand Parameters and Hydrogen-Bonding Capabilities of Liquids at 25°C

Liquid	δ/MPa$^{1/2}$	Hydrogen-bonding capability
4-Methoxy-4-methyl-2-pentanone	17.0	Moderate
Methylacetamide	29.9	Strong
Methyl acetate	19.6	Moderate
Methyl acrylate	18.2	Moderate
Methylamine, methanamine	22.9	Strong
Methylamyl acetate	16.4	Moderate
Methyl amyl ketone	17.4	Moderate
Methyl benzoate	21.5	Moderate
Methyl bromide, bromomethane	19.6	Moderate
Methyl butyl ketone	17.0	Moderate
Methyl butyrate	18.2	Moderate
Methyl caprolactone	20.2	Moderate
Methyl Cellosolve,® ethylene glycol monomethyl ether	23.3	Moderate
Methyl Cellosolve® acetate, ethylene glycol methyl ether acetate	18.8	Moderate
Methyl chloride, chloromethane	19.8	Moderate
Methylcyclohexane	15.9	Poor
Methylcyclohexanone	19.0	Moderate
Methylene chloride, dichloromethane	19.8	Poor
Methylene glycolate	25.4	Moderate
Methylene iodide, diiodomethane	24.1	Poor
Methyl ethyl ketone, 2-butanone	19.0	Moderate
Methyl ethyl sulfone	27.4	Moderate
N-Methylformamide	32.9	Strong
Methyl formate	20.9	Moderate
Methyl hexyl ketone	17.0	Moderate
Methyl iodide, iodomethane	20.9	Moderate
Methyl isoamyl ketone	17.2	Moderate
Methylisobutylcarbinol, 1,2-dimethylpropanol	20.5	Strong
Methyl isobutyl ketone	17.2	Moderate
Methyl isobutyrate	17.0	Moderate
Methyl isopropyl ketone	17.4	Moderate
Methyl isovalerate	16.2	Moderate
Methyl methacrylate	18.0	Moderate
Methyl nonyl ketone, 2-undecanone	16.0	Moderate
2-Methyl-1,3-pentanediol	21.1	Strong
2-Methyl-2,4-pentanediol, hexylene glycol	19.8	Strong
2-Methylpentanediol monoethyl ether, Pentoxol®	17.4	Strong
Methyl propionate	18.2	Moderate
Methyl propyl ketone, 2-pentanone	17.8	Moderate
Methyl propyl sulfone	25.6	Moderate
N-Methylpyrrolidone-2	23.1	Moderate
Methyl salicylate	21.7	Moderate
α-Methylstyrene	17.4	Poor
Methyltetramethylene sulfone	26.4	Moderate
Methyl valerate	16.2	Moderate
Monofluorodichloromethane, Freon 21	17.0	Poor
Monofluorotrichloromethane, Freon 11	15.5	Poor
Morpholine	22.1	Strong
Naphthalene	20.3	Poor
Nitrobenzene	20.5	Poor
Neopentane, 2,2-dimethylpropane	12.9	Poor
Neopentyl glycol	22.5	Strong
Nitroethane	22.7	Poor
Nitromethane	26.0	Poor
Nitrooctane	14.3	Poor

TABLE 1 (continued)
Hildebrand Parameters and Hydrogen-Bonding Capabilities of Liquids at 25°C

Liquid	$\delta/MPa^{1/2}$	Hydrogen-bonding capability
1-Nitropropane	21.1	Poor
2-Nitropropane	20.3	Poor
Nonyl phenol	19.2	Strong
Octane	15.6	Poor
1-Octanol, octyl alcohol	21.1	Strong
Pentachloroethane	19.2	Poor
1-Pentanamine, amylamine	17.8	Strong
Pentane	14.3	Poor
1,5-Pentanediol, pentamethylene glycol	23.5	Strong
2,4-Pentanediol	22.1	Strong
1-Pentanol, amyl alcohol	22.3	Strong
1-Pentene, amylene	14.1	Poor
Pentoxone,® diacetone alcohol methyl ether	16.8	Moderate
Pentyl acetate, amyl acetate	17.4	Moderate
Pentyl formate, amyl formate	17.4	Moderate
Perfluoroheptane	11.9	Poor
Perfluoro(methylcyclohexane)	12.3	Poor
Phenanthrene	20.1	Poor
Phenyl Cellosolve,® ethylene glycol monophenyl ether	23.5	Moderate
Phenylhydrazine	25.6	Strong
Pine oil	17.6	Poor
Piperidine	17.8	Strong
Piperidone, piperidinone	27.8	Strong
Propane	13.1	Poor
1-Propanol	24.3	Strong
2-Propanol, isopropanol	23.5	Strong
β-Propiolactone, 2-oxetanone	27.2	Moderate
Propionic acid, propanoic acid	20.3	Strong
Propionic anhydride	20.5	Strong
Propionitrile, propanenitrile	22.1	Poor
Propyl acetate	18.0	Moderate
Propyl alcohol, 1-propanol	24.3	Strong
Propylbenzene	17.6	Poor
Propyl bromide, 1-bromopropane	18.2	Moderate
Propyl butyrate	17.2	Moderate
Propyl chloride, 1-chloropropane	17.4	Moderate
Propylene-1,2-carbonate	27.2	Moderate
Propylene glycol, propanediol	25.8	Strong
Propylene glycol methyl ether	20.7	Moderate
Propylene oxide, methyloxirane	18.8	Moderate
Propyl formate	18.8	Moderate
Propyl propionate	17.4	Moderate
Pyridine	21.9	Strong
γ-Pyrone	27.4	Moderate
α-Pyrrolidone, pyrrolidinone	30.1	Strong
Quinoline	22.1	Strong
Santicizer® 8	24.3	Moderate
Shellsol® 72	14.7	Poor
Shell TS28 solvent	15.1	Poor
Silicon tetrachloride, tetrachlorosilane	15.1	Poor
Socal® Solvent No.1	16.6	Poor
Socal® Solvent No. 2	16.2	Poor
Socal Solvent No. 3	15.8	Poor
Solvesso® 100	17.6	Poor
Solvesso 150	17.4	Poor
Styrene, ethenylbenzene	19.0	Poor

TABLE 1 (continued)
Hildebrand Parameters and Hydrogen-Bonding Capabilities of Liquids at 25°C

Liquid	δ/MPa$^{1/2}$	Hydrogen-bonding capability
Styrene oxide, phenyloxirane	21.5	Moderate
Succinic anhydride	31.5	Strong
Terpene B	17.2	Poor
Tetrachlorodifluoroethane, Freon 112	16.0	Poor
1,1,2,2-Tetrachloroethane	19.8	Poor
Tetrachloroethylene, perchloroethylene	19.0	Poor
Tetrachloromethane, carbon tetrachloride	17.6	Poor
Tetraethylene glycol	20.3	Strong
Tetrahydrofuran	18.6	Moderate
Tetrahydronaphthalene	19.4	Poor
Tetramethylene sulfone	27.4	Moderate
Tetramethyloxamide	23.3	Moderate
Thiophene	20.1	Moderate
Toluene, methylbenzene	18.2	Poor
Tolylene diisocyanate	23.7	Strong
Tributylamine	15.8	Strong
1,1,2-Trichloroethane	19.6	Poor
Trichloroethylene	18.8	Poor
Trichlorofluoromethane, Freon 11	15.5	Poor
Trichloromethane, chloroform	19.0	Poor
Trichlorotrifluoroethane, Freon 113	14.9	Poor
Tricresyl phosphate	17.2	Moderate
Triethylamine, *N,N*-diethylethanamine	15.1	Strong
Triethylene glycol	21.9	Strong
Triethylenetetramine	22.7	Strong
Trifluorotrichloroethane, Freon 113	14.9	Poor
3,5,5-Trimethylhexanol	17.2	Strong
Triphenyl phosphate	17.6	Moderate
Triphenyl phosphite	19.0	Moderate
Tripropylene glycol	18.8	Strong
Tripropylene glycol methyl ether	17.8	Moderate
Turpentine	16.6	Poor
Valeric acid	20.1	Strong
Valeronitrile	19.6	Poor
Varnolene Varsol® #2	15.5	Poor
Vinyl acetate	18.4	Moderate
Vinyl chloride	16.0	Moderate
Vinyltoluene	18.6	Poor
VM & P Naphtha	15.5	Poor
Water	47.9	Strong
Xylene, dimethylbenzene	18.0	Poor

Note: The values were calculated either from Equation 2, Chapter 7, if ΔH was available, or from Hildebrand's Equation 18, Chapter 7, with the corrections described in the text, or in the case of commercial mixed hydrocarbon solvents, from the Kauri-Butanol number (Section 8.5). Each liquid is classified as capable of poor, moderate, or strong hydrogen bonding.

Adapted from Burrell, H., *Polymer Handbook*, Brandrup, J. and Immergut, E. H., Eds., John Wiley & Sons, New York, 1966, IV-341; 2nd ed., 1975, IV-337.

using the longitudinal (compressional) sound velocity in the machine direction of Whatman® No. 2 chromatography paper.

Results are listed in Table 2. It is noteworthy that a 10% solution of urea in dimethyl sulfoxide has a much higher hydrogen bonding capacity than water. Alkanes have no effect

TABLE 2
Sound Velocity Hydrogen Bonding Parameters for Liquids

Liquid	Longitudinal sonic velocity in machine direction, km s^{-1}	γ_w
Dry paper	1.95	23.6
Water (reference)	0.46	100.0
Alcohols		
Methanol	0.64	72
Ethanol	1.21	38
1-Propanol	1.33	35
2-Propanol	1.60	29
1-Butanol	1.41	33
Isobutanol	1.61	29
2-Butanol	1.66	28
Cyclohexanol	1.56	30
1-Octanol	1.82	25
2-Octanol	1.59	28
2-Ethylhexanol	1.45	32
Glycols		
Ethylene glycol, 1,2-ethanediol	0.71	65
Propylene glycol, propanediol	0.95	48
Diethylene glycol, 2,2′-oxybisethanol	0.85	54
Glycerol, 1,2,3-propanetriol	0.70	66
Ethers		
Diethyl ether, 1,1′-oxybisethane	1.9	24
Tetrahydrofuran	1.43	32
1,4-Dioxane, diethylene oxide	1.0	46
Ether-Alcohols		
Cellosolve®, ethyl Cellosolve	1.05	44
Butyl Cellosolve	1.52	30
Hexyl Cellosolve	1.47	31
Esters		
Ethyl acetate	1.61	29
Propyl formate	1.72	27
Methyl propionate	1.75	26
γ-Butyrolactone	1.44	32
Dimethyl phthalate	1.51	31
Phosphorus-Containing Esters		
Trimethyl phosphate	1.11	41
Tributyl phosphate	1.32	35
Trioctyl phosphate	1.82	25
Trioctyl phosphite	1.75	26
Triphenyl phosphite	1.56	30
Dimethyl acid phosphite	0.59	78

TABLE 2 (continued)
Sound Velocity Hydrogen Bonding Parameters for Liquids

Liquid	Longitudinal sonic velocity in machine direction, km s^{-1}	γ_w
Amides and Amines		
Formamide	0.49	94
N,N-Dimethylformamide	0.68	68
N,N-Diethylformamide	1.50	31
N,N-Dimethylacetamide	0.53	87
Ethylenediamine	0.19	240
Diethylenetriamine	0.23	200
Aniline	0.96	94
Cyclohexylamine	1.67	28
Pyridine	0.58	80
Piperidine	0.52	89
Morpholine	0.23	200
Ketones		
Acetone-2-propanone	1.56	29
Methyl ethyl ketone, 2-butanone	1.48	31
Cyclohexanone	1.20	38
Acids		
Acetic acid (glacial), ethanoic acid	1.10	41
Acetic acid + 7% water	0.88	52
Sulfur Compounds		
Carbon disulfide	1.35	33
Dimethylsulfoxide, sulfinylbismethane	0.57	79
Aqueous Solutions		
10% Lithium chloride	0.46	100
10% Potassium chloride	0.46	100
10% Sodium chloride	0.46	100
10% Urea	0.48	95
10% Sucrose	0.44	105
Dimethylsulfoxide Solutions		
50 wt % Trimethyl phosphate	0.41	112
50 wt % Methanol	0.86	54
50 wt % Ethanol	1.07	43
10 wt % Urea	0.26	173
10 wt % Acetic acid	0.33	139
10 wt % Phenol	0.62	74
Hydrocarbons		
Pentane	1.95	24
Hexane	1.95	24
Octane	1.95	24
Benzene	1.64	28
Toluene	1.70	27
Diphenylmethane	1.31	35

Adapted from Craver, J. K., *J. Appl. Polym. Sci.*, 14, 1755, 1970.

TABLE 3
Comparison of the Hydrogen Bonding Parameters as Defined by Craver (y_w), Crowley et al. (Y_C), and Lieberman (γ_L)

Liquid	γ_w	γ_C	γ_L
Water	100	—	2.89
Aniline	94	18.1	0.49
Piperidine	89	24.2	—
N,N-Dimethylacetamide	87	12.3	—
Pyridine	80	18.1	2.89
Dimethylsulfoxide	79	7.7	—
Methanol	72	18.7	2.89
N,N-Dimethylformamide	68	11.7	2.89
Glycerol, 1,2,3-propanetriol	66	—	2.89
Ethylene glycol, 1,2-ethanediol	65	20.6	2.89
Diethylene glycol, 2,2′-oxybis-ethanol	55	—	1.96
1,4-Dioxane	46	9.7	0.64
Cellosolve®, ethyl Cellosolve	44	13.0	1.69
Ethanol	38	18.7	2.89
Cyclohexanone	38	11.7	0.49
1-Propanol	35	18.7	2.89
Carbon disulfide	33	0.0	0.09
1-Butanol	33	18.7	2.89
γ-Butyrolactone	32	9.7	1.0
2-Ethylhexanol	32	18.7	2.89
N,N-Diethylformamide	31	11.7	2.89
Dimethyl phthalate	31	—	1.0
Methyl ethyl ketone, 2-butanone	31	7.7	0.64
Butyl Cellosolve	30	13.0	—
Cyclohexanol	30	18.7	2.89
Acetone, 2-propanone	29	9.7	1.0
2-Propanol	29	—	2.89
Ethyl acetate	29	8.4	1.0
2-Butanol	28	—	2.89
Benzene	28	0.0	0.09
Toluene, methylbenzene	27	4.5	0.09
Propyl formate	27	—	1.0
Methyl propionate	26	8.4	—
1-Octanol	25	18.7	2.89
Diethyl ether, 1,1′-oxybisethane	24	13.0	1.0
Hexane	24	0.0	0.0
Pentane	24	0.0	0.09
Octane	24	—	0.09

Adapted from Craver, J. K., *J. Appl. Polym. Sci.*, 14, 1755, 1970.

on the fiber hydrogen bonding, and have the same γ_w value as air-dried paper. Table 3 shows that these results do not agree in detail with the parameters of Lieberman[14] or Crowley et al.[18] (Section 83). There is no reason for considering them any less correct as single parameter measures of extent of hydrogen bonding, but the limitation of all such single-term parameters is clear (Section 5.5). Dyck and Hoyer[19] have provided another alternative hydrogen bonding parameter, and a semi-empirical method was described by Aliev.[20]

Hydrogen bonding in terms of component cohesion parameters is introduced in Sections 5.4 and 5.5; the simpler but less accurate Hansen approach is outlined in Sections 5.9 and 5.10. The formalism of hydrogen-bond donor (α) and hydrogen-bond acceptor (β) of the

TABLE 4
Classification of Solvents by Electrostatic Factor

Class	Solvent type	10^{30}EF/C m
I.	Hydrocarbon	0–7
II.	Electron donor	7–70
III.	Hydroxylic	50–170
IV.	Dipolar aprotic	≥170

Adapted from Dack, M. R. J., Ed., *Solutions and Solubilities,* John Wiley & Sons, New York, 1976, chap. 11.

solvatochromic equation is discussed in Section 8.4 as well as Section 7.12, together with other approaches based on linear solvation energy relations.

8.2 ELECTROSTATIC PARAMETERS AND EMPIRICAL RELATIONSHIPS

It has been shown in Chapter 3 that cohesion parameter values (and cohesive properties generally) are closely related to the electrostatic properties such as dipole moment, relative permittivity, polarizability, ionization potential, and refractive index.

Both dipole moment (μ) and relative permittivity (ϵ) are of particular importance.[4] In the absence of specific association, the dipole moment tends to determine the orientation of solvent molecules around molecular solutes, while dissolution of ions is promoted by high relative permittivity of the solvent. The *electrostatic factor*, **EF**, which is defined as the product of ϵ and μ, takes both effects into account and provides a basis for the classification of solvents. In Table 4 the **EF** limits have been converted to SI units, 1 D = 3.336×10^{-30} C m. Common liquids classified in this way by Dack[4] are presented in Table 5

There is a great variety of reports* of empirical and semiempirical relationships linking the electrical properties of liquids to solubility, to Hildebrand parameters. or to Hansen parameters. Gardon's fractional polarity is particularly noteworthy.

Gardon[30,31] discussed the nature of molecular interactions in terms of the fraction of total interaction due to dipole-dipole or orientation attractions (p), induction (i), and dispersion (d, effects, such that

$$p + i + d = 1 \tag{2}$$

In particular, the fractional polarity, p, may be calculated from the dipole moment, μ, polarizability, α, and ionization potential, I, once certain assumptions have been made. From Equations 4,5, and 6 in Chapter 3, for the case of identical i and j molecules and assuming additivity of energy components,

* See, for example: Beerbower and Dickey;[21] Böttcher;[22] Bonn and van Aartsen;[23] Cammarata, Yau, and Rogers;[24] Carr;[25] Coomber;[26] Dack;[4,27] Darby, Touchette, and Sears;[28] Djordjevic and Porter;[29] Gardon;[30,31] Gjaldbaek and Anderson;[32] Gordon;[33] Gorman and Hall;[35] Griffiths and Pugh;[36] Hansen and co-workers;[37-39] Herbrandson and Neufeld;[40] Keller, Karger, Snyder, and Eon;[41,42] Kesselring and co-workers;[43] Kirchnerova and Cavé;[44] Koenhen and Smolders;[45] Lawson and Ingham;[46] Liszi;[47] Loev and Lavrishchev[48] Longster and Walker;[49] Martin and co-workers;[50,51] Matsuura and co-workers;[52-56] McGinnis;[57] Moore;[58] Noel and Meloan;[59] Paruta and co-workers;[60-66] Pastor et al.;[67] Peiffer;[68] Sewell;[69] Shetty;[70] Skraga and Wieckiewicz;[71] Stekol'shchikov et al.;[72] Sunwoo and Eisen;[73] Tatsumi;[74] Thomas and Eckert;[75] Tijssen et al.;[76] Turyshev et al.;[77] Watanabe and Sugiyama;[78] Wiesner;[79] Wingefors;[80,81] Wunkelmann;[82] Wolkoff;[83] Xu;[84] Yamamoto et al.;[85] and Yeo.[86]

TABLE 5
Electrostatic Factors of Organic Liquids

	ϵ_r(temperature)	10^{30} μ/C m (in solution unless otherwise stated)	10^{30} EF/C m
Class I, Hydrocarbons			
Benzene	2.28 (25°C)	0.00	0.0
Cyclohexane	2.02 (20°C)	0.00	0.0
Heptane	1.92 (20°C)	0.00	0.0
Naphthalene	2.54 (85°C)	0.00	0.0
Pentane	1.84 (20°C)	0.00	0.0
p-Xylene, 1,4-dimethylbenzene	2.27 (20°C)	0.00	0.0
Hexane	1.89 (20°C)	0.27	0.5
Styrene, ethenylbenzene	2.43 (25°C)	0.40	1.0
Ethylbenzene	2.41 (20°C)	1.17	2.8
m-Xylene, 1,3-dimethylbenzene	2.37 (20°C)	1.23	2.9
Toluene, methylbenzene	2.38 (25°C)	1.30	3.1
1-Pentene	2.10 (20°C)	1.57	3.3
tert-Butylbenzene	2.38 (20°C)	1.77	4.2
o-Xylene, 1,2-dimethylbenzene	2.57 (20°C)	2.06(vapor)	5.3
Class II, Electron-Donors			
Tetrachloromethane[a]	2.24 (20°C)	0.00	0.0
Tetrachloroethylene[a]	2.30 (25°C)	0.00	0.0
Propionic acid	3.44 (40°C)	2.10	7.2
Trichloroethylene	3.42(16°C)	3.0	10.3
Phenetole	4.22 (20°)	3.34	14.1
Bromoform, tribromoethane	4.39 (20°C)	3.30	14.5
Acetic acid, ethanoic acid	6.13 (20°C)	2.77	17.0
Anisole, methoxybenzene	4.33 (25°C)	4.00	17.3
1,2-Dibromoethane, ethylene dibromide	4.78 (25°C)	3.74	17.9
Trichloromethane, chloroform	4.81 (20°C)	3.84	18.5
Diethyl ether, 1,1′-oxybisethane	4.34 (20°C)	4.54	19.7
Butylamine	5.3 (21°C)	4.67	24.8
Fluorobenzene	5.42 (25°C)	4.90	26.6
Chlorobenzene	5.62 (25°C)	5.20	29.3
Pentyl acetate	4.75 (20°C)	6.37	30.3
Bromobenzene	5.40 (25°C)	5.77	31.2
Ethyl propionate	5.65 (19°C)	5.81	32.8
o-Toluidine	6.34 (18°C)	5.34	33.8
Aniline	6.89 (20°C)	5.04	34.7
Propyl acetate	5.69 (19°C)	6.20	35.3
Ethyl acetate	6.02 (25°C)	6.04	36.4
Methyl acetate	6.68 (25°C)	5.84	39.0
Ethyl benzoate	6.02 (20°C)	6.64	40.0
Methyl benzoate	6.59 (20°C)	6.21	40.9
Tetrahydrofuran	7.61 (20°C)	5.44	41.4
Chloroethane	6.29 (170°C)	6.61	41.5
Ethyl formate	7.6 (25°C)	6.47	45.3
Dichloromethane, methylene dichloride	9.08 (20°C)	5.17	46.9
Propyl formate	7.72 (19°C)	6.31	48.7

[a] In this class because of structure.

TABLE 5 (continued)
Electrostatic Factors of Organic Liquids

	ϵ_r(temperature)	10^{30} μ/C m (in solution unless otherwise stated)	10^{30} EF/C m
1-Chloropropane, propyl chloride	7.7 (20°C)	6.57	50.6
1-Chlorobutane, butyl chloride	7.39 (20°C)	7.04(vapor)	52.0
Bromoethane	9.39 (20°C)	6.71	63.0
1,1-Dichloroethane, ethylidene chloride	10.0 (18°C)	6.51	65.1
Class III, Hydroxylic Solvents			
p-Cresol, 4-methylphenol	9.91 (58°C)	5.14	50.9
o-Cresol, 2-methylphenol	11.5 (25°)	4.70	54.1
Phenol	9.78 (60°C)	5.77	56.4
tert-Butanol, 2-methyl-2-propanol	10.9 (30°C)	5.54	60.3
m-Cresol, 3-methylphenol	11.8 (25°C)	5.14	60.7
Benzyl alcohol, benzenemethanol	13.1 (20°C)	5.54	72.6
1-Pentanol, amyl alcohol	13.9 (25°C)	6.0	83.5
2-Methylcyclohexanol	13.3 (20°C)	6.51	86.5
Isoamyl alcohol, 3-methyl-1-butanol	14.7 (25°C)	6.07	89.2
Cyclohexanol	15.0 (25°C)	6.34	95.1
1-Butanol	17.1 (25°C)	5.61	95.8
2-Propanol	18.3 (25°C)	5.61	102.6
Isobutanol, 2-methyl-1-propanol	17.7 (25°C)	5.97	105.7
1-Propanol	20.1 (25°C)	5.54	111.3
Ethanol	24.30 (25°C)	5.61	136.2
Methanol	32.63 (25°C)	5.54	180.7
Water (*in this class because of structure*)	78.5 (25°C)	6.14	481.9
Class IV, Dipolar Aprotic Solvents			
Benzaldehyde	17.8 (20°C)	9.24	165
Acetaldehyde	21.1 (21°C)	8.31	175
Acetone, 2-propanone	20.7 (25°C)	9.61(vapor)	199
Valeronitrile	17.4 (21°C)	11.91	207
Cyanopropane	20.3 (21°C)	11.91	242
Benzonitrile	25.20 (25°C)	13.51	341
Nitromethane	35.87 (30°C)	10.58	379
Nitrobenzene	34.82 (25°C)	13.31	464
N,N-Dimethylformamide	36.7 (25°C)	12.74(vapor)	468
N,N-Dimethylacetamide	37.8 (25°C)	12.64	478
Acetonitrile	37.5 (20°C)	12.81(vapor)	480
Hexamethylphosphoramide	29.6 (30°C)	17.91	530
Dimethylsulfoxide	46.6 (25°C)	14.98(vapor)	698
Formamide (*not aprotic*)	109.5 (25°C)	11.24	1231

Adapted from Dack, M. R. J., Ed., *Solutions and Solubilites,* John Wiley & Sons, New York, 1976, chap. 11.

$$U = U_o + U_i + U_d \tag{3}$$

then

$$-U = \frac{2\mu^4}{3kTr^6} + \frac{2\alpha\mu^2}{r^6} + \frac{3I\alpha^2}{4r^6} \tag{4}$$

and the fractional polarity is

$$p = U_o/U \tag{5}$$

Gardon's values of p for various liquids are collected in Table 6, together with the molar volumes, V, and Hildebrand parameters, δ, taken by Gardon from various sources. Other values have been quoted by Wyart and Dante[87] from a Shell publication,[88] by Nelson et al.[89,90] (see next Section, Table 10) and by Shell[153] (Tables 11 and 14 in next Section). The application of two-dimensional plots of Hildebrand parameter against fractional polarity has been discussed by Wyart and Dante,[87] by Somerville and Lopez,[91] and by Kumar and Prausnitz,[90] and is illustrated in Figure 1. This is a map divided into p-δ regions of solubility and insolubility for a given solute. The criterion for solubility in a pure liquid or liquid mixture is that the liquid (p,δ) point should lie within the solubility region, with volume fraction averages (Equation 31, Chapter 4) being used for the calculation of p and δ in mixed systems. In Figure 1, points A and B represent the low-boiling liquid (acetone) and medium-boiling liquid (methyl isobutyl ketone). It will be assumed that the constraints on the formulation are[87]:

1. A volume ratio of 30:70 for the low to medium boiling liquids.
2. A ratio of 13:87 for high-boiling to (low plus medium)-boiling liquids.
3. A ratio of 75:25 for ''active'' liquid (solvent) to ''latent'' solvent. (The latent solvent, 2-propanol or isopropanol, represented by point F, lies outside the solubility region.)
4. A ratio of 50:50 for hydrocarbon (H) to oxygenated liquids.

Criterion 1 determines point C on line AB. The high-boiling liquid (4-methyl-4-methoxy pentanone) is represented by point D, and point E divides CD according to criterion 2. The line EF, linking the points representing the active and latent solvents, is then divided at G in 25:75 ratio to give the oxygenated liquid blend parameters. Finally, GH is divided at J to meet criterion 4, giving a blend which lies within the solubility region.

On the basis of the relationship between polarizability and refractive index (Chapter 3), Koenhen and Smolders[45] found for a variety of liquids an empirical correlation between the dispersion cohesion parameter δ_d and the refractive index at the sodium ''D'' line, n_D,

$$\delta_d/\text{MPa}^{1/2} = 19.5\, n_D - 11.4 \tag{6}$$

with correlation coefficient 0.90 and standard error of 0.7 MPa$^{1/2}$ in estimation of δ_d. A similar relationship was used by Loev and Lavrishchev[48] and Stekol'shchikov et al.[72] Winkelmann[82] calculated thermodynamic excess functions from dielectric data for systems which included *N,N*-dimethylformamide. Experimental information from 96 hydrocarbons suggested[41,42]

$$\delta_d/\text{MPa}^{1/2} = 63\,(n^2 - 1)/(n^2 + 2) \tag{7}$$

while the assumption that cohesion parameters should be evaluated from internal pressures (Sections 2.4, 7.5) led to [76]

TABLE 6
Liquid Fractional Polarities and Hildebrand Parameters[30,31]

	$V/cm^3\ mol^{-1}$	$\delta/MPa^{1/2}$	p
Alkanes			
Pentane	116	14.4	0.000
Isopentane, 2-methylbutane	117	14.4	0.000
Hexane	132	14.9	0.000
Cyclohexane	109	16.8	0.000
Heptane	147	15.2	0.000
Octane	164	15.4	0.000
Isooctane, 2,2,4-trimethylpentane	166	14.0	0.000
Nonane	180	15.7	0.000
Decane	194	15.9	0.000
Tetradecane	258	16.3	0.000
Aromatic Hydrocarbons			
Benzene	89	18.7	0.000
Toluene, methylbenzene	107	18.2	0.001
o-Xylene, 1,2-dimethylbenzene	121	18.4	0.001
m-Xylene, 1,3-dimethylbenzene	123	18.0	0.001
p-Xylene, 1,4-dimethylbenzene	124	17.9	0.000
Ethylbenzene	123	18.0	0.001
Naphthalene	123	20.3	0.000
Propylbenzene	140	17.7	0.000
Cumene, isopropylbenzene	140	18.1	0.002
Mesitylene, 1,3,5-trimethylbenzene	140	18.0	0.000
p-Cymene, *p*-isopropyltoluene	156	20.2	0.000
Alkenes			
Butadiene	87	13.7	—
Isoprene	100	14.8	—
Styrene, ethenylbenzene	115	17.7	0.000
Perfluoro Compounds			
Perfluorobutane	163	10.6	0.000
Perfluoropentane	168	11.3	0.000
Perfluorohexane	205	11.5	0.000
Perfluoroheptane	227	11.7	0.000
Perfluorononane	279	11.7	0.000
Perfluorocyclobutane	133	11.7	0.000
Perfluorocyclohexbutane	170	12.3	0.000
Perfluorobenzene	115	16.6	0.000
Perfluorotoluene	142	15.8	0.000
Chloroalkanes			
Dichloromethane	65	19.8	0.120
Trichloromethane, chloroform	81	19.0	0.017
Tetrachloromethane, carbon tetrachloride,	97	17.6	0.000
Chloroethane, ethyl chloride	73	17.4	0.319
1,1-Dichloroethane, ethylidene chloride	85	18.6	0.215
1,2-Dichloroethane, ethylene dichloride	79	20.1	0.043
1,1,1-Trichloroethane	100	17.4	0.069
Tetrachloroethane	101	21.3	0.092
Isobutyl chloride	105	16.8	0.171
tert-Butyl chloride	109	16.2	0.189

TABLE 6 (continued)
Liquid Fractional Polarities and Hildebrand Parameters[30,31]

	$V/cm^3\ mol^{-1}$	$\delta/MPa^{1/2}$	p
Chloroalkenes			
cis-Dichloroethylene	76	19.8	0.166
trans-Dichloroethylene	77	18.8	0.000
Trichloroethylene	90	18.8	0.005
Tetrachloroethylene	101	19.1	0.000
Vinyl chloride	68	17.8	—
1,1-Dichloroethylene, vinylidene chloride	80	17.6	—
Chloroprene	91	19.0	0.0006
Aromatic Halogen Compounds			
Chlorobenzene	107	19.4	0.058
Bromobenzene	105	20.5	0.029
Iodobenzene	111	23.1	0.019
o-Bromotoluene	120	20.1	—
Bromoalkanes			
Bromomethane, methyl bromide	57	19.2	0.252
Tribromomethane, bromoform	88	21.5	0.019
Bromoethane, ethyl bromide	76	18.2	0.206
1,2-Dibromoethane, ethylene dibromide	87	21.3	0.012
1,1,2,2-Tetrabromoethane	116	21.1	0.019
1,2,3-Tribromopropane	110	21.9	0.038
Iodoalkanes			
Iodomethane, methyl iodide	63	20.3	0.129
Diiodomethane, methylene diiodide	81	24.1	0.011
Iodoethane, ethyl iodide	81	19.2	0.057
Hydroxyl Compounds			
Water	18	47.5	0.819
Phenol	88	29.7	0.057
Glycol	56	32.1	0.476
Glycerol, 1,2,3-propanetriol	73	33.8	0.468
Cyclohexanol	104	23.3	0.075
Methanol	41	29.7	0.388
Ethanol	58	26.0	0.268
1-Propanol	76	24.3	0.152
1-Butanol	91	23.3	0.096
2-Butanol, 2-methyl-1-propanol	91	21.9	0.111
1-Pentanol, amyl alcohol	108	22.3	0.074
1-Hexanol	125	21.9	0.060
1-Heptanol	142	20.5	0.049
1-Octanol	158	21.1	0.035
Carboxylic Acids			
Acetic acid, ethanoic acid	57	25.8	0.296
Butyric acid	92	23.5	0.108
Isovaleric acid	108	22.3	0.031
Valeric acid	109	22.1	—

TABLE 6 (continued)
Liquid Fractional Polarities and Hildebrand Parameters[30,31]

	$V/cm^3\ mol^{-1}$	$\delta/MPa^{1/2}$	p
Aldehydes			
Acetaldehyde	57	20.1	0.715
Heptaldehyde	134	19.8	0.231
Benzaldehyde	101	22.1	0.282
Ketones			
Acetone, 2-propanone	74	20.5	0.695
Methyl ethyl ketone, 2-butanone	90	19.0	0.510
Diethyl ketone, 3-pentanone	106	18.0	0.286
Methyl propyl ketone	106	17.8	0.398
Methyl butyl ketone	122	17.6	0.353
Methyl pentyl ketone	129	17.4	—
Methyl hexyl ketone	156	17.3	0.232
Cyclohexanone	109	20.3	0.380
Ethers			
Diethyl ether	105	15.1	0.033
Diisopropyl ether	141	14.3	—
β,β-Dichloroethyl ether	117	20.1	0.275
1,4-Dioxane	86	20.5	0.006
Esters			
Methyl formate	61	21.9	0.306
Ethyl formate	80	19.2	0.131
Methyl acetate	80	19.6	0.182
Ethyl acetate	99	18.6	0.167
Propyl acetate	115	17.9	0.129
Isopropyl acetate	117	17.6	—
Ethyl propionate	115	18.2	0.096
Pentyl formate	130	17.7	0.102
Butyl acetate	132	17.5	0.167
Ethyl butyrate	132	16.7	0.071
Pentyl acetate	148	17.3	0.067
Butyl propionate	147	17.4	—
Ethyl isovalerate	149	17.7	0.092
Pentyl propionate	164	17.2	—
Butyl butyrate	166	16.4	—
Isobutyl isobutyrate	165	17.8	—
Isoamyl butyrate	179	17.4	—
Ethyl benzoate	143	19.8	0.057
Diethyl phthalate	198	20.6	0.068
Monomer Esters			
Vinyl acetate	92	17.8	0.052
Methyl acrylate	90	18.2	—
Ethyl acrylate	109	17.2	—
Methyl methacrylate	106	17.8	0.149
Ethyl methacrylate	125	17.2	0.146
Butyl acrylate	143	18.2	—
Butyl methacrylate	160	16.8	0.096

TABLE 6 (continued)
Liquid Fractional Polarities and Hildebrand Parameters[30,31]

	$V/cm^3\ mol^{-1}$	$\delta/MPa^{1/2}$	p
Isobutyl methacrylate	161	16.2	—
Diethyl carbonate	121	18.0	0.002
Diethyl malonate	152	21.1	0.182
Ethylene carbonate	66	29.7	0.924
Amines			
Dibutylamine	168	16.1	—
Aniline	91	22.1	0.063
Pyridine	81	21.9	0.174
Amides			
Formamide	40	36.4	0.88
Acetamide	51	34.2	0.82
N,N-Dimethylformamide	77	24.8	0.772
N,N-Dimethylacetamide	93	22.7	0.682
Nitriles			
Acetonitrile	53	24.3	0.852
Propionitrile	71	21.9	0.753
Butyronitrile	87	21.5	0.719
Valeronitrile	103	20.7	0.690
Capronitrile	120	20.9	0.525
Malonitrile	63	30.9	0.798
Monomer Nitriles			
Acrylonitrile	67	21.4	0.802
Methacrylonitrile	84	18.6	0.746
Nitro Compounds			
Nitromethane	54	25.8	0.780
Nitroethane	76	22.7	0.710
1-Nitropropane	89	20.9	0.710
2-Nitropropane	89	20.3	0.720
Nitrobenzene	103	20.5	0.625
o-Nitrotoluene	118	21.5	0.524
m-Nitrotoluene	118	21.3	0.790
Sulfur Compounds			
Carbon disulfide	62	20.5	0.000
Ethyl mercaptan	74	17.7	0.097
Dimethyl sulfide	74	18.4	—
Diethyl sulfide	110	18.0	0.065
Thiophene	78	20.1	0.071
Dimethyl sulfone	75	29.9	0.782
Diethyl sulfone	109	25.6	—
Dipropyl sulfone	142	23.1	—
Dimethylsulfoxide	71	27.4	0.813

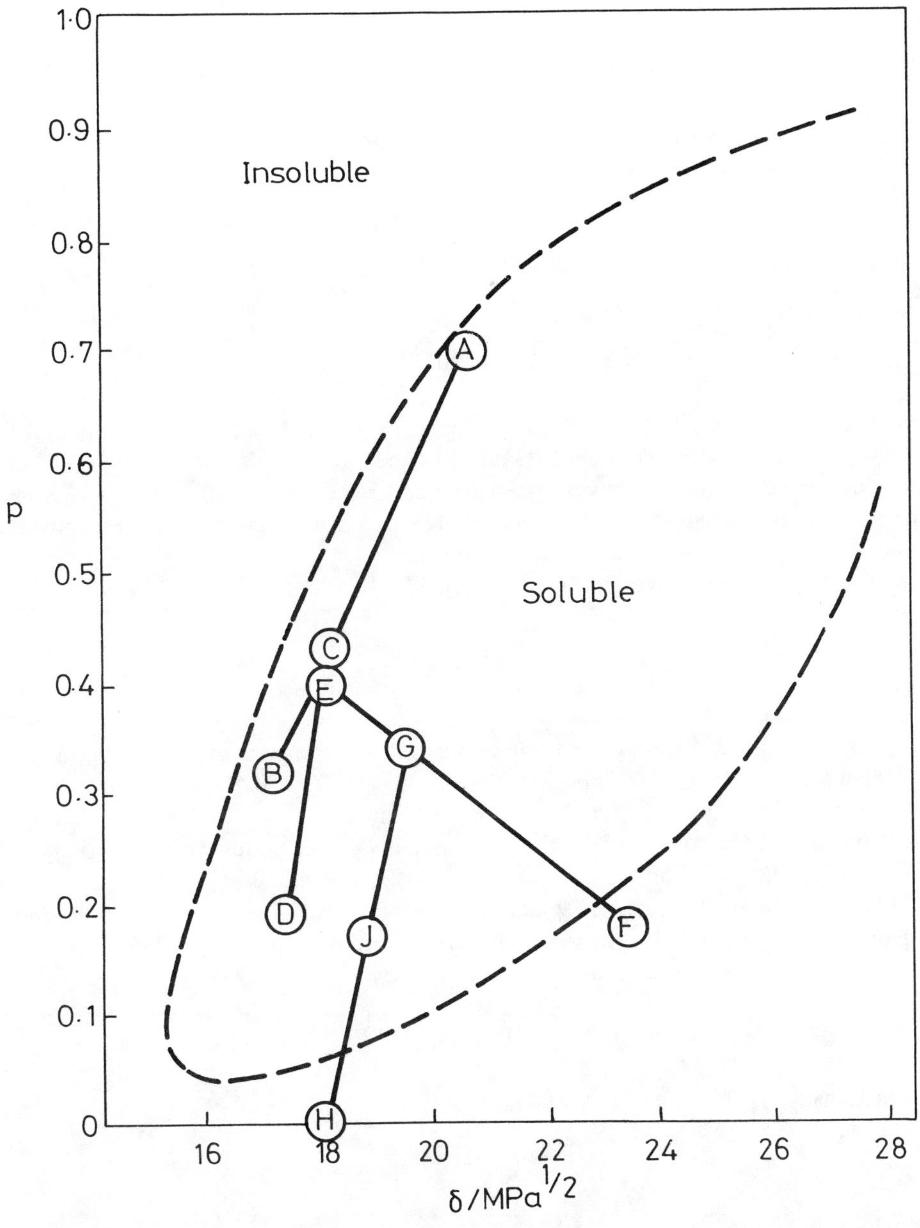

FIGURE 1. Fractional polarity-Hildebrand parameter map used as described in the text to design a blend of low boiling point solvent acetone (A), medium boiling point solvent methyl isobutyl ketone (B), "latent" solvent isopropanol (F), and high boiling point solvent 4-methyl-4-methoxy-2-pentanone (D) for a hydrocarbon (H). (Adapted from Wyart, J. W. and Dante, M. F., *Kirk-Othmer Encyclopedia of Chemical Technology*, 2nd ed., Vol. 18, John Wiley & Sons-Interscience, New York, 1969, 564.)

$$\delta_d^2/\text{MPa} = 954\ (n^2 - 1)/(n^2 + 2) + 21.8 \tag{8}$$

These relationships were used and compared by Poole and co-workers.[70] Thomas and Eckert,[75] in their MOSCED approach (Section 5.13) used for non-aromatics

$$\delta_d/\text{MPa}^{1/2} = 41.5\ (n_D^2 - 1)/(n_D^2 + 2) + 6.17 \tag{9}$$

and for aromatics

$$\delta_d/\text{MPa}^{1/2} = 39.9\,(n_D^2 - 1)/(n_D^2 + 2) + 5.71 \tag{10}$$

Expressions which have been fairly widely used for the polar cohesion parameter δ_p and the orientation cohesion parameter δ_o are

$$\delta_p = A\mu V^{-1/2} \tag{11}$$

and

$$\delta_p = B\mu V^{-3/4} \tag{12}$$

where A and B are constants. For δ_p in cal$^{1/2}$ cm$^{-3/2}$, μ in debye units and V in cm^3 mol^{-1}, Beerbower and Dickey[21] gave $A = 1.83$ and Koenhen and Smolders[45] reported 9.5, though this should have been 18.3 as given by Hansen and Beerbower[37] and Karim and Bonner.[92] In the same units, the constant B is 50.1.[45,92] If required, these values may be converted to SI units with the conversion factors of Section 2.5.

A function of the relative permittivity defined by the equation

$$\Theta(\epsilon) = \frac{3\epsilon_r \ln \epsilon_r}{\epsilon_r \ln \epsilon_r - \epsilon_r + 1} - \frac{6}{\ln \epsilon_r} - 2 \tag{13}$$

has also been used as a solvent polarity function, varying from zero at $\epsilon_r = 1.000$ (a vacuum) to 1.000 when ϵ_r is very high. For water, $\epsilon_r = 80$, $\Theta(\epsilon) = 0.50$. Carr[25] has shown that $\Theta(\epsilon)$ can be used, together with the dipole moment and molar volume, to estimate the orientation cohesion parameter, δ_o. Other relationships involving relative permittivity, dipole moment, and polar cohesion parameters have been developed.[41,42,70,78,198]

Relations similar to Equations 9 and 10 have been used for total cohesion parameters (Hildebrand parameters) as well as the dispersion component alone. Thus Sewell[69] for a range of polymers showed that the molar refraction was given by

$$R/\text{cm}^3 = 0.0163\,(-UV)^{1/2}/\text{J}^{1/2}\,\text{cm}^{3/2} \tag{14}$$

so from Equation 3, Chapter 3,

$$R' = V\,(n^2 - 1)/(n^2 + 2) \tag{15}$$

it follows that

$$\delta/\text{MPa}^{1/2} = 62.0\,(n^2 - 1)/(n^2 + 2) \tag{16}$$

Paruta and coworkers[60-63,65,66] used the empirical relationship

$$\delta/\text{MPa}^{1/2} = 0.45\epsilon_r + 15.3 \tag{17}$$

and this was modified by Skraga and Wieckiewicz.[71]

The "solvatrochromic" π^* polarity scale used by Abboud, Abraham, Carr, Doherty, Kamlet, and Taft[25,93-109,191,195] is an index of liquid dipolarity and polarizability which measures the ability of a liquid to stabilize a charge or a dipole by virtue of its dielectric effect, with values generally proportional to dipole moments when consideration is restricted to a set of aprotic, nonchlorinated aliphatic "select" liquids that contain a single dominant bond dipole. Values and further details appear in Section 8.4.

A rather different way of expressing the degree of polarity is the ratio of the van der Waals volume of polar groups to the total van der Waals volume.[197]

Topics related to the various relationships referred to here appear in Sections 12.2, 14.5 and 16.1, but no attempt is made to provide a comprehensive review of the numerous and diverse empirical correlations between molecular constants and cohesion properties.

8.3 SINGLE-COMPONENT DONOR-ACCEPTOR PARAMETERS

This section is concerned with attempts to describe solute-solvent interactions by means of a single parameter for each component. Spectroscopic studies used for correlating solvent behavior are described in Section 3.5, while Sections 7.12 and 8.4 deal with two-component donor-acceptor parameters.

Small[110] determined the "iodine-bonding number" of a few liquids by visually matching the color of iodine solutions, which range from violet in inert solvents to brown in strongly associating solvents. A plot of Hildebrand parameters against iodine-bonding numbers for solvents and nonsolvents of poly(vinyl chloride) provides an array of points which fall fairly clearly into general regions of solubility and nonsolubility (Figure 2). Although Small's iodine-bonding number (γ_N) has had only limited use,[111] his work demonstrated the importance both of spectroscopic parameters and of multiparameter correlation in the practical description of solubility behavior. Numerical values of γ_N are included in Table 7, Chapter 14.

Lieberman's empirical hydrogen-bonding parameter, γ_L, (Section 8.1) proved useful, and it was made more reliable by Crowley et al.[18] and Beerbower et al,[112] who introduced a direct hydrogen-bonding parameter with a spectroscopic origin from the work of Gordy and Stanford.[3,113,114] This was based on the effect that a solvent had on a small amount of alcohol introduced into the solvent: the greater the extent of hydrogen bonding between the solvent and the alcohol, the weaker the O–H bond and the lower the frequency of infrared radiation absorbed. Deuterated alcohol was used because the O–D stretch band is in a region with little interference, permitting detection at low alcohol concentrations. The extent of the shift to lower frequencies of the O–D stretching infrared absorption of deuterated methanol in the liquid under study thus provides a measure of its hydrogen bond acceptor capability. The spectrum of a solution of deuterated methanol in the test liquid was compared with that of a solution in benzene, and the hydrogen bonding parameter was defined

$$\gamma_C = \Delta\nu/10 \tag{18}$$

where the OD absorption shift was expressed in wave numbers. (The subscript C has been attached to distinguish this from alternative definitions of γ.) Thus, replacing benzene with nitrobenzene as solvent shifts the peak from 2681 cm^{-1} to 2653 cm^{-1}, indicating a γ_C value of 2.8 for nitrobenzene. Some numerical values appear in Tables 7 and 8.

More precise measurements on the deuterated methanol system were made by Nelson et al.,[115] who also discussed the choice of reference solvent. Benzene is not an inert solvent, but is a hydrogen bond acceptor, the aromatic π-electron system having some proton attracting properties. There is even the suggestion of a small amount of hydrogen bonding between methanol and tetrachloromethane, the solvent used by other investigators.[116-118] It appeared that OH absorption bands of methanol in cyclohexane formed a suitable reference system. The hydrogen-bonding parameter values for alcohols or for any other donor/acceptor compounds are subtracted rather than added in the expression of Nelson et al.[115] for the effective or *net hydrogen-bond accepting ability*, θ, of a solvent mixture

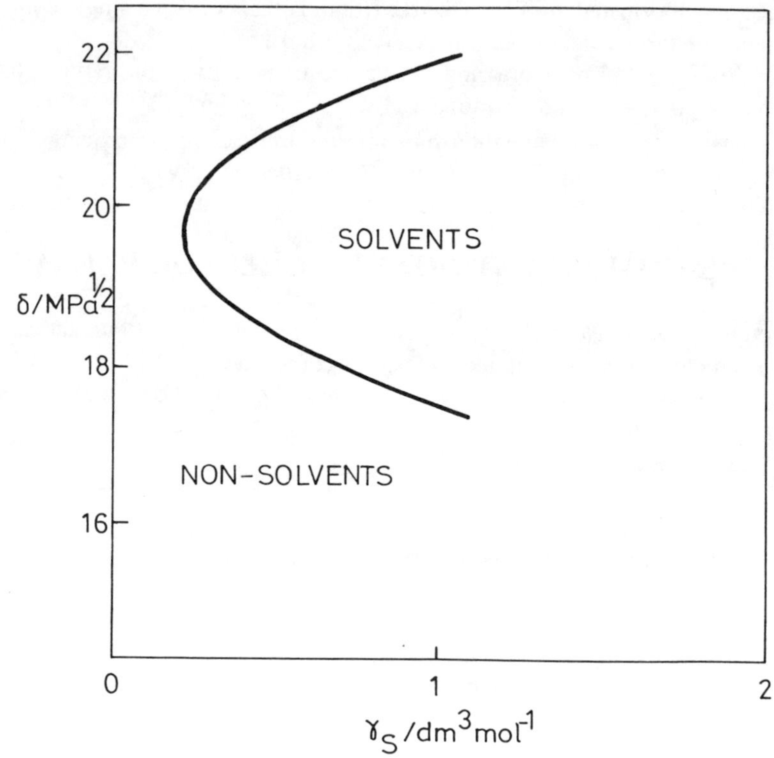

FIGURE 2. Hildebrand parameter-Small's iodine bonding number map for poly(vinyl chloride). (Adapted from Small, P. A., *J. Appl. Chem.*, 3, 71, 1953.

$$\theta = \sum_i {}^i\mathrm{k}\ {}^i\phi\ {}^i\gamma_C \tag{19}$$

For a single liquid, (${}^i\theta = 1$),

$${}^i\theta = {}^i\mathrm{k}\ {}^i\gamma_C \tag{20}$$

Here ${}^i\gamma_C$ is the hydrogen-bonding parameter of component i,${}^i\theta$ is its volume fraction, and ${}^ik$ is a weighting factor taken as -1 for donor-acceptor liquids such as alcohols, zero for ether-alcohols, and $+1$ for all other compounds. The sign and magnitude of ${}^i\theta$ reflect the hydrogen bonding contribution to the solution, systems with large, positive values being "better" solvents. This went some way toward recognizing the fact that hydrogen bonding involves both a donor and an acceptor with different roles, rather than two equivalent, mutually hydrogen bonding species. Values of the parameter ${}^i\theta$ for some pure liquids are included in tables 10, 11 and 14 (next Section).

Another definition of the hydrogen-bonding parameter,[119-121] yielding what is described here as γ_D, is

$$\gamma_D = 0.0359\Delta v + 2.2 \tag{21}$$

where Δv is defined as before. This restricts the range of γ_D to values above 2.2 and below about 10, rather than ranging from 0 to 21 as in the case of γ_c. Both sets of data are included in the ASTM D3132 standard test method[121] "Solubility Range of Resins and Polymers",

TABLE 7
Orr and Gordy Hydrogen-Bonding Parameters

Solvent	Orr (CH_3CH_2OD, isooctane reference) ν_{max}/cm^{-1}	Orr γ_O	Gordy (CH_3OD, benzene reference) ν_{max}/cm^{-1}	Gordy γ_G
Isooctane, 2,2,4-trimethylpentane	2696	0.0	—	—
Decane	2696	0.0	—	—
Naphtha, VM &P	2695	0.0	—	—
Heptane	2695	0.0	—	—
Isopar® H (mixed aliphatic hydrocarbons)	2692	0.0	—	—
Tetrachloromethane	2684	1.1	—	—
Trichloromethane	2677	1.8	—	—
Chlorobenzene	2673	2.2	2666	1.5
Dichloromethane	2673	2.2	—	—
1,1,2-Trichloroethane	2669	2.6	—	—
Nitromethane	2668	2.7	—	—
1,2-Dichloroethane	2664	3.1	2666	1.5
Benzene	2664	3.1	2681	0.0
Styrene	2660	3.5	—	—
Toluene	2659	3.6	—	—
Nitrobenzene	2659	3.6	2653	2.8
Xylol	2658	3.7	—	—
Solvesso® 150 (mixed aromatic hydrocarbons)	2654	4.1	—	—
Isobutyl isobutyrate	2638	5.7	—	—
Ethyl acetate	2630	6.5	2597	8.4
Isopropyl acetate	2630	6.5	—	—
Pentyl acetate	2630	6.5	2591	9.0
Benzyl acetate	2628	6.7	—	—
Butyl acetate	2628	6.7	—	—
Acetonitrile	2625	7.0	2618	6.3
Ethyl amyl ketone	2623	7.2	—	—
Propylene carbonate	2620	7.5	—	—
Acetone	2620	7.5	2584	9.7
Methyl propyl ketone	2620	7.5	—	—
Methyl ethyl ketone	2619	7.6	2604	7.7
Methyl isopropyl ketone	2618	7.7	—	—
Methyl isobutyl ketone	2618	7.7	2604	7.7
Diisobutyl ketone	2610	8.5	2597	8.4
1,4-Dioxane	2610	8.5	2584	9.7
Cellosolve® acetate	2600	9.5	—	—
N,N-Dimethylformamide	2595	10.0	2564	11.7
Carbitol® acetate	2590	10.5	—	—
Diisopropyl ether	2590	10.5	2558	12.3
Diethyl ether	2588	10.7	2551	13.0
Tetrahydrofuran	2580	11.5	—	—
Cellosolve®	2540	15.5	—	—
Butyl Cellosolve®	2540	15.5	—	—
Dimethylsulfoxide	2540	15.5	—	—
Aniline	2512	18.3	2500	18.1
2-Ethyl hexanol	2485	21.0	—	—
Tributylamine	2430	26.5	2463	21.8
Tripropylamine	2430	26.5	2457	22.4
Butyl Carbitol®	band too broad		—	—
1-Butanol	band too broad		—	—

Adapted from Orr, A. A., *J. Paint Technol.*, 47(607), 45, 1975.

TABLE 8
Liquid Hildebrand Parameters, Hydrogen-Bonding Parameters and Dipole Moments[18,148,149]

Liquids	δ/MPa$^{1/2}$	γ_c	$\mu/10^{30}$ C m
Acetone, 2-propanone	20.5	9.7	9.7
Acetonitrile	24.3	6.3	13.0
Acetylacetone, 2,4-pentanedione	22.1	8.4	10.3
Acrylonitrile	21.5	5.7	12.7
Anisole, methoxybenzene	20.2	7.0	4.7
Aniline, benzeneamine	24.1	18.1	5.0
Benzene	18.8	0.0 (ref. pt.)	0.0
Benzonitrile	17.2	7.7	14.7
Benzyl alcohol, benzenemethanol	24.7	18.7	5.7
1-Bromobutane	17.8	1.5	6.7
1-Butanol	23.3	18.7	5.7
Butyl acetate	17.4	8.8	6.3
Butyl butyrate	16.6	8.8	6.3
Butyl Carbitol®, diethylene glycol monobutyl ether	18.2	13.0	5.3
Butyl Cellosolve®	18.2	13.0	5.3
Butyl formate	17.8	7.5	6.3
Butyl lactate	19.2	7.0	6.3
Butyl propionate	17.6	8.4	6.3
Butylene carbonate	24.7	4.9	17.7
γ-Butyrolactone	27.2	9.7	13.7
Butyronitrile	21.5	7.7	13.3
Capronitrile, hexanenitrile	19.2	7.7	13.3
Carbitol®, diethylene glycol monoethyl ether	19.6	13.0	5.3
Carbitol® acetate, diethylene glycol monoethyl ether acetate	17.4	9.4	6.0
Carbon disulfide	20.5	0.0	0.0
Cellosolve®, ethylene glycol monoethyl ether	20.2	13.0	5.3
Cellosolve® acetate, ethylene glycol monoethyl ether acetate	17.8	9.4	6.0
Chlorobenzene	19.4	1.5	5.3
1-Chloropentane	17.0	1.5	7.3
Cyclohexane	16.8	0.0	0.0
Cyclohexanol	23.3	18.7	5.7
Cyclohexanone	19.8	11.7	9.0
Cyclopentanone	21.3	8.4	9.0
Decane	13.5	0.0	0.0
Diacetone alcohol	18.8	13.0	10.7
Dibutyl ether	14.5	11.0	4.0
1,2-Dichloroethane	20.0	1.5	3.7
Dichloromethane	19.8	1.5	5.0
1,2-Dichloropropane	18.4	1.5	5.3
2,2-Dichloropropane	16.8	1.5	7.7
N,N-Diethylacetamide	20.2	12.3	12.7
Diethyl carbonate	18.0	5.0	4.0
Diethyl ether, 1,1′-oxybisethane	15.1	13.0	4.0
N,N-Diethylformamide	21.7	11.7	12.7
Diethyl ketone, 3-pentanone	18.0	7.7	9.0
Diisobutyl ketone	16.0	8.4	9.0
Diisopropyl ketone	16.4	8.4	9.0
N,N-Dimethylacetamide	22.1	12.3	12.7
Dimethyl carbonate	20.2	4.9	3.3
N,N-Dimethylformamide	24.7	11.7	12.7
Dimethylsulfoxide, sulfinylbismethane	26.4	7.7	13.3
1,4-Dioxane, diethylene oxide	20.2	9.7	1.3

TABLE 8 (continued)
Liquid Hildebrand Parameters, Hydrogen-Bonding Parameters and Dipole Moments[18,148,149]

Liquids	$\delta/MPa^{1/2}$	γ_c	$\mu/10^{30}$ C m
Dipentene	17.4	1.5	0.0
Dipropyl ether	14.1	11.7	4.3
Dipropyl sulfone	23.1	7.7	15.0
Ethanol	26.0	18.7	5.7
Ethyl acetate	18.6	8.4	6.0
Ethyl amyl ketone	16.8	7.7	9.0
Ethylbenzene	18.0	1.5	2.0
Ethyl benzoate	16.8	6.3	6.7
Ethyl butyl ether	14.7	11.7	4.0
Ethylene carbonate	30.1	4.9	16.3
Ethylene glycol, 1,2-ethanediol	29.0	20.6	7.7
Ethylene oxide, oxirane	22.7	10.0	6.3
N-Ethylformamide	28.4	12.0	13.3
Ethyl formate	19.2	8.4	6.3
2-Ethylhexanol	19.4	18.7	5.7
Ethyl lactate	20.5	7.0	6.3
Heptane	15.1	0.0	0.0
Hexane	14.9	0.0	0.0
1-Hexanol	21.9	18.7	5.7
Isobutyl acetate	17.0	8.8	6.3
Isobutyl isobutyrate	16.4	8.8	6.3
Isobutyl chloride	16.6	1.5	7.0
Isopropyl acetate	17.2	8.6	6.3
Mesityl oxide	18.4	9.7	9.0
Methanol	29.7	18.7	5.7
Methyl acetate	19.6	8.4	5.7
Methylamyl acetate	16.4	9.0	6.3
Methyl amyl ketone	17.4	7.7	9.0
Methyl isoamyl ketone	17.0	7.4	9.0
Methyl benzoate	21.3	6.3	6.3
Methyl isobutyl carbinol, 1,2-dimethylpropanol	20.5	18.7	5.7
Methyl butyl ketone	17.4	8.4	9.0
Methyl isobutyl ketone	17.2	7.7	9.0
Methylcyclohexane	16.0	0.0	0.0
Methyl ethyl ketone, 2-butanone	19.0	7.7	9.0
Methyl formate	20.9	8.4	6.3
Methyl nonyl ketone, 2-undecanone	16.0	7.7	9.0
Methyl propionate	19.0	8.4	6.3
Methyl propyl ketone, 2-pentanone	18.2	8.0	9.0
Methyl isopropyl ketone	17.4	8.0	9.0
Nitrobenzene	20.1	2.8	14.3
Nitroethane	22.7	2.5	12.0
Nitromethane	26.0	2.5	11.3
Nitropropane	21.9	2.5	12.3
1-Octanol, octyl alcohol	21.1	18.7	5.7
Pentane	14.3	0.0	0.0
Pentyl acetate	17.4	9.0	6.0
Pentyl formate	17.4	7.0	6.3
Phenyl acetate	21.3	7.7	6.3
Phenyl vinyl ether	18.8	6.3	4.0
Piperidine	17.8	24.2	4.0
1-Propanol	24.3	18.7	5.7
β-Propiolactone	27.2	7.7	13.7
Propionitrile	22.1	7.7	13.3
Propyl acetate	18.0	8.6	6.3

TABLE 8 (continued)
Liquid Hildebrand Parameters, Hydrogen-Bonding Parameters and Dipole Moments[18,148,149]

Liquids	δ/MPa$^{1/2}$	γ_c	$\mu/10^{30}$ C m
Propylene carbonate	27.2	4.9	16.7
Propylene oxide	18.8	10.0	6.0
Propyl propionate	17.4	7.0	6.3
Pyridine	21.9	18.1	7.3
Styrene, ethenylbenzene	19.0	1.5	0.0
Tetrachloromethane, carbon tetrachloride	17.6	0.0	0.0
Toluene, methylbenzene	18.2	4.5	1.3
1,1,2-Trichloroethane	19.6	1.5	4.0
Trichloromethane, chloroform	19.0	1.5	4.0
Tributylamine	17.8	21.8	2.3
Tripropylamine	18.0	22.4	2.3
Xylene, dimethylbenzene(commercial)	18.0	4.5	1.3
VM & P naphtha	15.5	0.0	0.0

but only values of γ_C as defined in Equation 18 are listed in Table 9 and used in tables and figures in this book. It should be noted that Equations 18 and 21 may be combined to give:

$$\gamma_D = 0.359\gamma_C + 2.2 \tag{22}$$

$$\gamma_C = 2.79\gamma_D - 6.13 \tag{23}$$

An improved procedure has been developed for the determination of hydrogen bonding parameters, based on subtracting the bond maximum in 0.1 *M* deuterated ethanol in the liquid from the reference value in isooctane or heptane,[122] ethanol being soluble in a range of liquids broader than that for methanol. It is claimed that this method is more accurate and comprehensive, while not invalidating data obtained by the original method. The accuracy with which the band maximum may be determined decreases as the extent of hydrogen bonding increases (Figure 3), the error being given as ± 10% of the bond shift with a minimum of ± 2 cm^{-1} (that is, ± 0.2 in γ). Orr's values (described here as γ_O) are presented in Table 7, together with Gordy's original results (γ_G) for comparison. So far, this procedure does not seem to have been widely adopted.

Hydrogen bonding parameters are special cases of a more general type of electron donor-acceptor parameter. Kagiya et al.[123] also repeated Gordy's type of experiment for many liquids, using benzene as a reference and calculating the *electron-donating power,* $\Delta\upsilon_D$. Some liquids, including tetrachloromethane, have a negative shift, and the data can be corrected to tetrachloromethane as reference for comparison with the results of Nelson et al.[115] In addition, Kagiya and co-workers went further, and the *electron-accepting power,* $\Delta\upsilon_A$, of liquids was measured by determining the shift in the C=O vibrational band of acetophenone. The competitive power to accept or donate electrons could be measured in liquid mixtures.

There are other examples. Koppel and Palm[124] suggested that the O–D stretching frequency in the gas phase was a more logical basis for a Lewis basicity scale, and defined

$$B = \nu_{gas} - \nu_{liquid}$$

Shorter and co-workers[125] measured the O–D stretching band of CH_3OD in many other aprotic liquids. Numerous other parameters of electron donating or accepting power and

TABLE 9
Liquids and Liquid Mixtures Used in the ASTM Standard Test Method for Solubility Ranges of Resins and Polymers, with Hildebrand Parameters Hydrogen-Bonding Parameters, and Dipole Moments[121,150]

Liquid or liquid mixture, % by volume	δ/MPa$^{1/2}$	γ_C	μ/10^{30} C m
Diisopropyl ether	14.1	11.7	4.3
Pentane	14.3	0.0	0.0
Heptane	14.9	0.0	0.0
Diisobutyl ketone/diisopropyl ether, 50:50	15.1	10.1	6.7
Diethyl ether	15.1	13.0	4.0
Heptane/diisobutyl ketone, 50:50	15.6	4.2	4.7
Heptane/diisobutyl ketone, 25:75	15.8	6.3	6.7
Methylcyclohexane	16.0	0.0	0.0
Heptane/butyl acetate, 66.7:33.3	16.0	2.9	2.0
Diisobutyl ketone	16.0	8.4	9.0
Diethyl ether/butyl acetate, 57:43	16.2	11.2	5.0
Diethyl ether/2-ethylhexanol, 66.7:33.3	16.6	14.9	4.7
Cyclohexane	16.8	0.0	0.0
Heptane/butyl acetate, 40:60	16.8	5.3	3.7
Cyclohexane/toluene, 71:29	17.2	1.3	0.3
Butyl acetate	17.4	8.8	6.3
Cyclohexane/toluene, 40:60	17.6	2.7	0.7
Cyclohexane/benzene, 50:50	17.8	0.0	0.0
Butyl acetate/toluene, 50:50	17.8	6.7	4.0
Ethylene glycol monobutyl ether/butyl acetate, 50:50	17.8	10.9	6.0
Diethyl ether/2-ethylhexanol, 33.3:66.7	18.0	16.8	5.0
Toluene (methylbenzene)	18.2	4.5	1.3
Ethylene glycol monobutyl ether (butyl Cellosolve®)	18.2	13.0	5.3
Benzene	18.8	0.0	0.0
Propylene oxide (methyloxirane)	18.8	10.0	6.0
Ethylene glycol monobutyl ether/2-ethylhexanol, 56:44	18.8	15.5	5.3
Methylene chloride/toluene, 50:50	19.0	3.0	3.3
Toluene/dioxane, 50:50	19.2	7.1	1.3
2-Ethylhexanol	19.4	18.7	5.7
Methyl acetate	19.6	8.4	5.7
Dichloromethane(methylene chloride)	19.8	1.5	5.0
Dichloromethane/1,4-dioxane,66.7:33.3	20.1	4.2	3.7
1,4-Dioxane/dichloromethane, 50:50	20.1	5.6	3.3
1,4-Dioxane (diethylene oxide)	20.3	9.7	1.3
Cyclohexanone	20.3	11.7	9.0
Ethylene glycol monoethyl ether (EGMEE, Cellosolve®)	20.3	13.0	5.3
EGMEE/methyl isobutyl carbinol, 75:25	20.3	14.4	5.3
Carbon disulfide	20.5	0.0	0.0
EGMEE/methylisobutylcarbinol, 30:70	20.5	17.0	5.7
Methylisobutylcarbinol (1,2-dimethylpropanol)	20.5	18.7	5.7
Dichloromethane/nitroethane, 66.7:33.3	20.9	1.8	7.3
1,4-Dioxane/nitroethane, 66.7:33.3	21.1	7.3	5.0
Toluene/acetonitrile, 50:50	21.3	5.4	7.3
EGMEE/1-pentanol, 50:50	21.3	15.9	5.7
2-Ethylbutanol	21.5	18.7	5.7
Dichloromethane/nitroethane, 33.3:66.7	21.7	2.2	9.7
Nitroethane/dioxane, 75:25	22.1	4.3	9.3
1,4-Dioxane/propylene carbonate, 75:25	22.1	8.5	5.3
N,N-Dimethylacetamide	22.1	12.3	12.7
EGMEE/methanol, 80:20	22.1	14.1	5.3
EGMEE/1-butanol, 33.3:66.7	22.3	16.8	5.7
1-Pentanol (amyl alcohol)	22.3	18.7	5.7

TABLE 9 (continued)
Liquids and Liquid Mixtures Used in the ASTM Standard Test Method for Solubility Ranges of Resins and Polymers, with Hildebrand Parameters Hydrogen-Bonding Parameters, and Dipole Moments[121,150]

Liquid or liquid mixture, % by volume	$\delta/MPa^{1/2}$	γ_C	$\mu/10^{30}$ C m
Benzene/nitromethane, 48:52	22.5	1.3	6.0
1,4-Dioxane/*N,N*-dimethylformamide, 50:50	22.5	10.7	7.0
Nitroethane	22.7	2.5	12.0
Butyl acetate/acetonitrile, 25:75	22.7	6.9	11.3
1-Butanol	23.3	18.7	5.7
Nitroethane/acetonitrile, 50:50	23.5	4.4	12.7
EGMEE/methanol, 65:35	23.5	15.0	5.3
1-Butanol/*N,N*-dimethylformamide, 66.7:33.3	23.7	16.4	8.0
1,4-Dioxane/dimethylsulfoxide, 50:50	23.9	8.7	7.3
Nitroethane/nitromethane, 50:50	24.3	2.5	11.7
Acetonitrile	24.3	6.3	13.0
1-Butanol/*N,N*-dimethylformamide, 30:70	24.3	13.8	10.7
1-Propanol	24.3	18.7	5.7
Acetonitrile/*N,N*-dimethylformamide, 70:30	24.6	8.0	13.0
N,N-Dimethylformamide/acetonitrile, 68:32	24.6	10.0	12.7
N,N-Dimethylformamide	24.8	11.7	12.7
1-Propanol/ethanol, 50:50	25.2	18.7	5.7
Nitromethane/acetonitrile, 60:40	25.4	4.0	12.0
Ethanol/*N,N*-dimethylformamide, 50:50	25.4	15.2	9.3
1,4-Dioxane/propylene carbonate, 23:77	25.6	6.0	13.0
N,N-Dimethylformamide/dimethylsulfoxide, 40:60	25.8	9.3	13.0
Ethanol/*N,N*-dimethylformamide, 70:30	25.8	16.6	7.7
N,N-Dimethylformamide/methanol, 80:20	25.8	13.1	11.3
Nitromethane	26.0	2.5	11.3
Ethanol	26.2	18.7	5.7
Dimethylsulfoxide/ethanol, 77:28	26.4	10.8	11.3
Dimethylsulfoxide (sulfinylbismethane)	26.6	7.7	13.3
Methanol/*N,N*-dimethylformamide, 46:54	27.0	14.9	9.3
Propylene carbonate	27.2	4.9	16.7
Dimethylsulfoxide/methanol, 80:20	27.2	9.9	11.7
N,N-Dimethylformamide/*N*-methylformamide, 70:30	27.2	11.8	12.7
Ethanol/methanol, 50:50	27.8	18.7	5.7
Methanol/*N,N*-dimethylformamide, 66.7:33.3	28.0	16.4	8.0
Methanol/dimethylsulfoxide, 57:43	28.4	14.0	9.0
Methanol	29.7	18.7	5.7
Propylene glycol (propanediol)	30.7	20.6	7.3
Methanol/*N*-methylformamide, 63:37	30.9	16.2	8.3
Ethylene glycol (1,2-ethanediol)	35.0	20.6	7.7

polarity have been proposed, many based on spectroscopic properties.* The correlation between cohesion parameters and spectroscopically determined charge transfer complex formation constants is described in Section 7.6; correlations of other spectroscopic parameters with cohesion parameters have less theoretical foundation, but do exist in some special cases.[40]

* For example: Abboud, Abraham, Carr, Doherty, Kamlet, and Taft;[93-109] Allerhand and Schleyer;[126] Arnett et al.;[127] Balakrishnan and Easteal;[128] Brooker et al.;[129,130] Cardinal and Mukerjee;[131] Dack;[4,132] Dubois et al.;[133] Gebelein;[134] Gordon;[33,34,135] Griffiths and Pugh;[36] Herbrandson and Neufeld;[40] Kosower;[136,137] Lassau and Jungers;[138] Marcus;[194] Reeves, Maggio, and Costa;[139] Reichardt et al.;[140-145] Smid;[146] and Walther.[147]

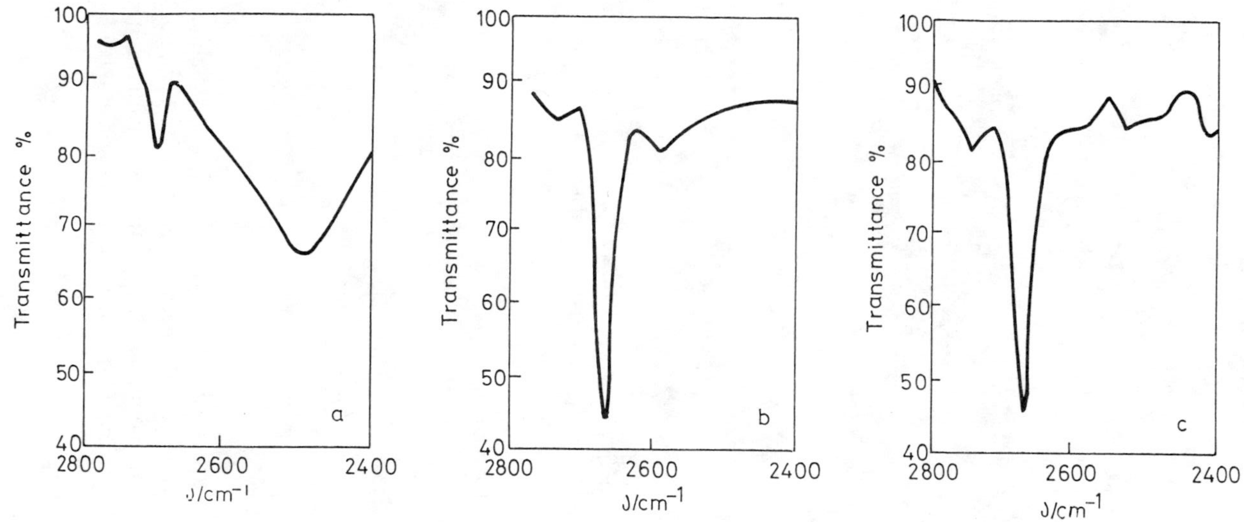

FIGURE 3. Spectra of the O-D stretch band of 0.1 *M* CH_3CH_2OD in nine solvents (0.496 mm cell): a, heptane; b, benzene; c, dichloromethane; d, nitrobenzene; e, ethyl acetate; f, acetone, g, diethyl ether; h, butyl Cellosolve®; i, tripropylamine: (Adapted from Orr, A. A., *J. Paint Technol.*, 47(607), 45, 1975.)

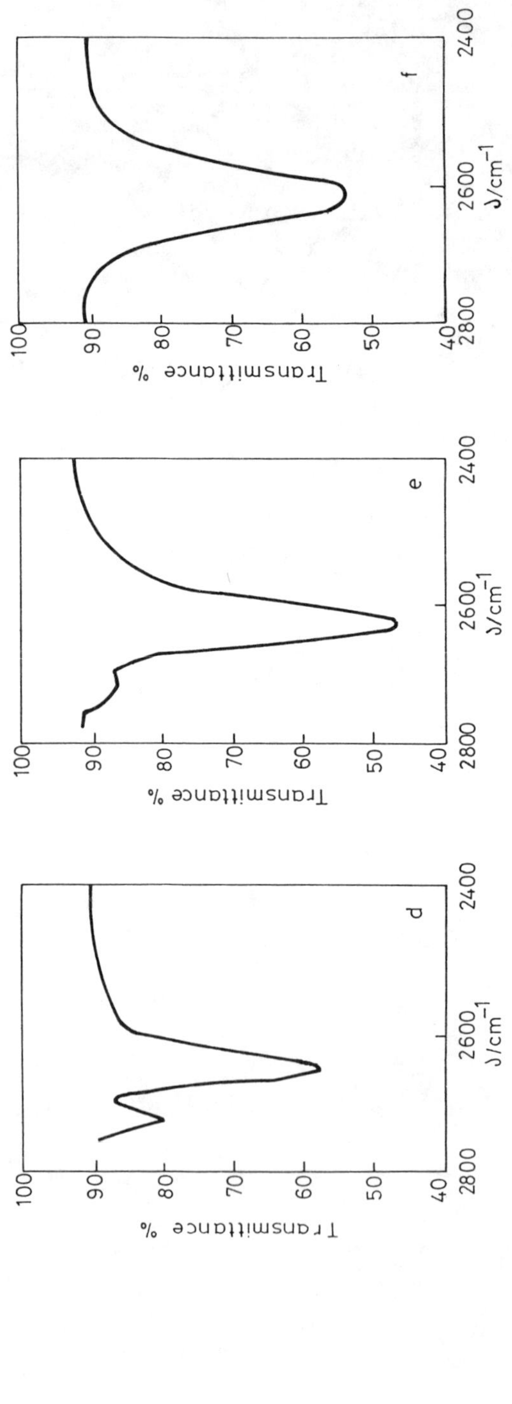

FIGURE 3A.

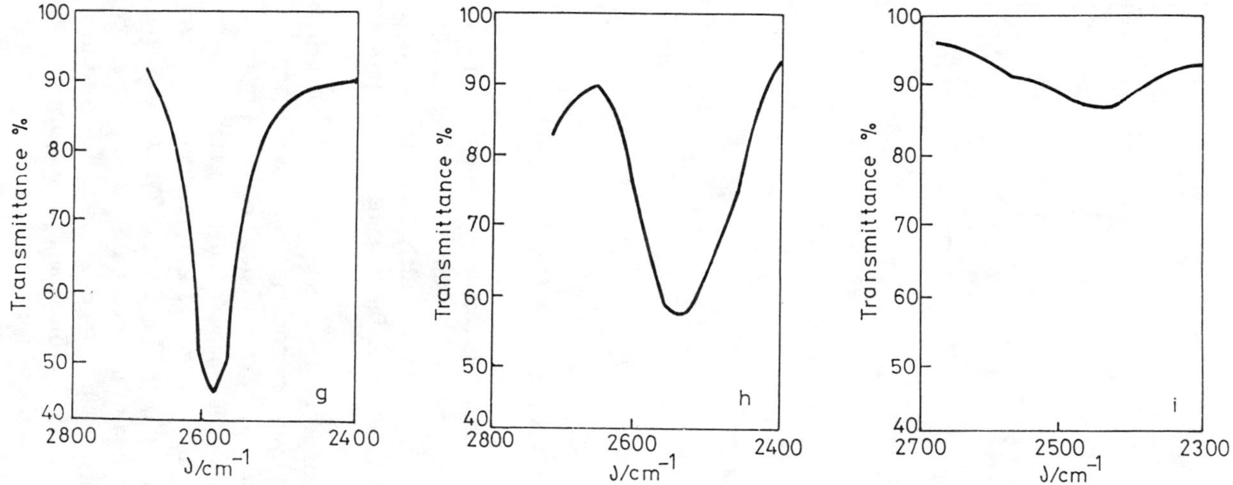

FIGURE 3 (continued)

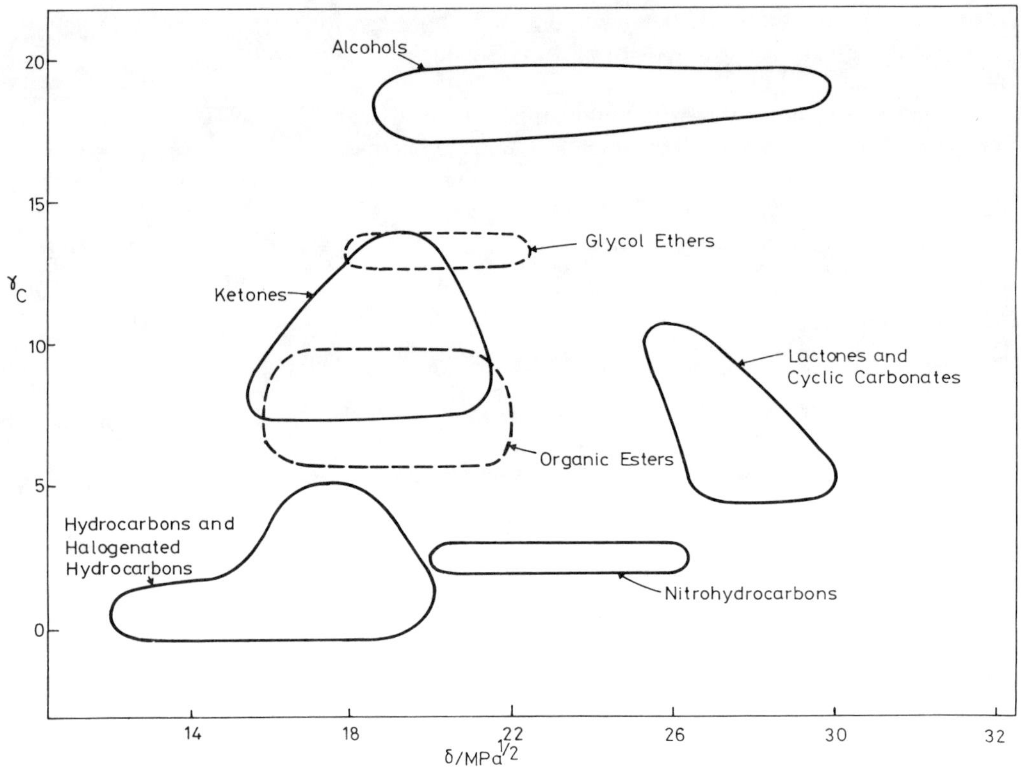

FIGURE 4. Hildebrand parameter-hydrogen bonding index locations for major liquid groups. (Adapted from Cosaert, E., *Chim. Peintures*, 34, 169, 1971.)

8.4 HYBRID COHESION PARAMETER SOLVENT SCALES

The Hildebrand parameter is an adequate measure of interactions for solubilities of nonpolar molecules, but it does not account for polar interactions or specific association. Because Hildebrand parameter values can be determined readily and are widely available, they have been used in conjunction with various other theoretical or empirical parameters to provide more effective predictions of solvent properties.

Crowley et al[18,148,149] used, in addition to the Hildebrand parameter, the dipole moment[150] as a measure of polar forces, together with the hydrogen bonding parameter described in Section 8.3. In this way it is possible to construct three-dimensional models to represent solubility behavior, as well as two-dimensional contour maps for simpler graphical formulation. These are illustrated in Figures 34 to 50, Chapter 15, and the three parameters are listed in Table 8 for a variety of liquids and in Table 9 for the ASTM D3132 liquids and liquid mixtures. Figure 4 provides an indication of the location of liquid groups on a Hildebrand parameter-hydrogen bonding parameter map.[119,120] Equation 23 has been used to convert the original du Pont γ_D data to the γ_C scale; the Hildebrand parameter values are from Burrell (Table 1) or calculated in the same manner, and the original data in $cal^{1/2}\ cm^{-3/2}$ and debye units have been recalculated in SI units (section 2.5). Figure 4 can be used as an "overlay" to Figures 54 to 65, Chapter 15, to assist in solvent selection; this type of representation clearly "separates" different liquid types better than $\delta_p - \delta_h$ maps (Section 5.9).

Gardon's[30,31] fractional polarity and the application of two-dimensional δ-*p* plots were discussed in Section 8.2. Nelson et al.[89,90,151-153] also used the Hildebrand parameter and fractional polarity, but in addition incorporated the hydrogen bonding parameter θ described in Section 8.3 to provide an alternative three-component solubility scale for coatings solvents (Tables 10 and 11).

Henry[154] has shown that with multivariate statistical analysis it is possible to find alternative ways of presenting three-component Hansen parameters and hybrid sets of $\delta - \gamma_c - \mu$ parameters. Using principal component analysis, data were plotted in the $U-V$ plane, with W values indicated numerically on the diagrams, for the following functions, δ being expressed in $\mathrm{cal}^{1/2}\ \mathrm{cm}^{-3/2}$ and μ in debye:

$$U = 0.66\delta + 0.55\gamma_C + 0.51\mu \quad (25a)$$

$$V = -0.03\delta - 0.66\gamma_C + 0.75\mu \quad (25b)$$

$$W = -0.75\delta + 0.52\delta_C + 0.42\mu \quad (25c)$$

$$U = -0.50\delta_d + 0.59\delta_p + 0.63\delta_h \quad (26a)$$

$$V = 0.85\delta_d + 0.49\delta_p + 0.20\delta_h \quad (26b)$$

$$W = 0.19\delta_d - 0.64\delta_p + 0.75\delta_h \quad (26c)$$

This provides a greater separation of typical liquids, but is likely to be less convenient for the practical user as the chemical significance of the relative positions on the diagrams is not as immediately obvious.

The linear free energy relationships introduced in Section 7.12 (and also discussed in Section 10.1) can be thought of as hybrid cohesion parameter solvent scales, and there have been several relatively successful developments.

A Hildebrand parameter term has been included in the solvatochromic "universal solubility" treatment of Abboud, Abraham, Carr, Doherty, Kamlet, and Taft[93-109,191,195,196] (and of related approaches[155,163,199]) as a measure of the solvent-solvent cavity effect, that is as an estimate of the energy necessary to create a cavity in the bulk liquid to accommodate a solute molecule. This contribution is combined with the π^* parameter (Section 8.2) to allow for solute-solvent dipole interactions, the α scale of hydrogen bond donor acidity, the β scale of hydrogen bond acceptor basicity, and an empirical coordinate covalency parameter ξ in expressions for the linear solvation energy of the form

$$g + h^i\delta^2 + s\pi^* + a\alpha + b\beta + e\xi \quad (27)$$

where g, h, s, a, b, and e are constants characterizing a given solute while the other parameters describe the behavior of a particular solvent. In some cases the $s\pi^*$ term is replaced by $s\,(\pi + d\Delta)$ where Δ is a polarizability correction term.

The π^* parameter (and other measures of donor-acceptor interactions and polarity) thus tend not to correlate well with the Hildebrand parameter, but they are complementary, both contributing to the free energy of solution in this approach.

Huyskens, Haulait-Pirson, and coworkers[156-160] developed a related method, defining the dissolving power \$ of a liquid for polymer j in terms of the chemical potential of j in its own pure phase, ${}^j\mu^o$, and in the solution, ${}^j\mu$:

$$^j\$ = ({}^j\mu^o - {}^j\mu)/(RT) \quad (28)$$

The factor RT (gas constant $R = 8.315\ \mathrm{J\ K^{-1}\ mol^{-1}}$, and absolute temperature T) makes \$ conveniently dimensionless. When ${}^j\$$ is positive, the affinity of j for the liquid is greater than that of j for its own phase, and more solute can be dissolved; when it is negative, phase

TABLE 10
Liquid Hildebrand Parameters, Fractional Polarities and Hydrogen-Bonding Parameters[89,90,152]

Liquid	δ/MPa$^{1/2}$	p	θ
Hydrocarbons			
(Additional hydrocarbon values appear in Table 11)			
Shellsol® B	14.8	0.000	0.0
Hexane	14.9	0.000	0.0
Shellsol B-8	15.1	0.000	0.4
Tolusol® 5	14.7	0.000	0.0
Tolusol 10	15.1	0.000	0.5
Tolusol 25	15.5	0.000	3.5
Shellsol 260	14.7	0.000	0.07
Toluene, methylbenzene	18.2	0.001	4.2
Special VM & P naphtha	15.1	0.000	0.32
Super VM & P naphtha	15.2	0.000	0.0
Ethylbenzene	18.0	0.001	4.2
Xylene, dimethylbenzene	18.1	0.001	4.5
Tolusol 28	17.4	0.001	3.4
Shell mineral spirits 135W	15.5	0.000	4.2
Cyclosol® 53	17.9	0.001	5.0
Shellsol 70	14.6	0.000	0.0
Shellsol 360	15.2	0.000	0.5
Shellsol 340	15.3	0.000	0.13
Cyclosol 63	17.8	0.001	5.3
Shellsol 140	15.1	0.000	0.5
Ketones			
Acetone, 2-propanone	20.5	0.695	12.5
Methyl ethyl ketone, 2-butanone	19.0	0.510	10.5
Methyl isobutyl ketone	17.2	0.315	10.5
Methyl isoamyl ketone	17.0	0.255	10.9
Cyclohexanone	20.2	0.380	13.7
Ethyl amyl ketone	16.8	0.223	8.5
Pentoxone®	17.4	0.190	12.5
Diisobutyl ketone	16.0	0.123	9.8
Diacetone alcohol	18.8	0.312	0.0
Isophorone	18.6	0.190	14.9
Mesityl oxide	18.4	0.332	12.0
Esters			
Ethyl acetate (85-88%)	19.6	0.182	4.9
Ethyl acetate (99%)	18.6	0.167	8.4
Isopropyl acetate	17.6	0.100	8.5
Propyl acetate	17.9	0.129	8.5
sec-Butyl acetate	16.8	0.082	8.3
Isobutyl acetate	17.0	0.097	8.7
Butyl acetate (90-92%)	18.4	0.118	5.7
Butyl acetate (99%)	17.6	0.120	8.0
Pentyl acetate	17.3	0.067	8.2
Methylamyl acetate	16.8	0.050	8.3
Isobutyl isobutyrate	15.7	0.042	8.0
Methyl Cellosolve®, ethylene glycol monomethyl ether acetate	18.8	0.095	10.5
Cellosolve® acetate, ethylene glycol monoethyl ether acetate	17.8	0.073	10.1

TABLE 10 (continued)
Liquid Hildebrand Parameters, Fractional Polarities and Hydrogen-Bonding Parameters[89,90,152]

Liquid	δ/MPa$^{1/2}$	p	θ
Butyl Cellosolve acetate, ethylene glycol monobutyl ether acetate	16.8	0.060	10.3
Ethers and Glycol Ethers			
Tetrahydrofuran	20.2	0.075	12.0
1,4-Dioxane	20.5	0.006	14.6
Methyl Oxitol®	22.1	0.126	0.0
Oxitol®	20.2	0.086	0.0
Butyl Oxitol®	18.2	0.048	0.0
Methyl Dioxitol®	20.9	0.058	0.0
Dioxitol®	19.6	0.043	0.0
Butyl Dioxitol®	18.2	0.028	0.0
Alcohols			
Methanol	29.7	0.388	−19.8
Neosol®	27.0	0.296	−17.7
2-Propanol	23.5	0.178	−16.7
1-Propanol	24.3	0.152	−16.5
2-Butanol	22.1	0.123	−17.5
Isobutanol	21.9	0.111	−17.9
1-Butanol	23.3	0.096	−18.0
Methyl isobutyl carbinol, 1,2-dimethylpropanol	20.5	0.066	−18.7
1-Pentanol, *n*-amyl alcohol	22.3	0.074	−18.2
Cyclohexanol	23.3	0.075	−16.5
2-Ethylhexanol	19.4	0.042	−18.7
Miscellaneous			
Dichloromethane, methylene dichloride	19.8	0.12	1.5
Trichloromethane, chloroform	19.0	0.017	1.5
1,2-Dichloroethane, ethylene dichloride	20.0	0.043	1.5
1,1,2,2-Tetrachloroethane	21.3	0.092	1.5
2-Nitropropane	20.2	0.72	4.0
N,N-Dimethylformamide	24.7	0.772	18.9
Chlorobenzene	19.4	0.058	2.7

separation can be expected. The value of ${}^j\mu$ is concentration-dependent, so ${}^j\$$ refers to a particular value of ${}^j\phi$; at the equilibrium situation ${}^j\$ = 0$ the value of ${}^j\phi$ corresponds to the saturation concentration. The dissolving power is made up of a constant contribution A of the solid polymer, a term β based on the influence of the ratio ${}^jV_u/{}^iV$ between the polymer repeat unit molar volume and the liquid molar volume, a contribution D obtained from a partial cohesion parameter (nonspecific dispersion and polar interactions), a term H depending on the extent of hydrogen bonding possible between liquid and polymer, and a term C arising from segment-segment hydrogen bonding:

$$\$ = H + \beta - A - C - D \tag{29}$$

TABLE 11
Solvent Parameters of Shell Hydrocarbons[89,151,153]

Hydrocarbon solvent	Kauri-butanol number (ASTM D1133)	Aniline point, °C (ASTM D611) (M = mixed aniline point)	$\delta/\text{MPa}^{1/2}$	p	θ
White spirit (LAWS)	37	56	16.1	0.000	0.5
Hexane	29	66	14.9	0.000	0.0
SBP 62/82	30	64	~15.0	0.000	0.0
SBP 80/110	32	61	15.1	0.000	0.0
SBP 100/140	32	64	15.3	0.000	0.0
SBP 140/165	30	67	15.5	0.000	0.0
Shellsol® A	90	14 M	18.0	0.001	5.0
Shellsol E	75	25 M	17.2	0.001	4.5
Shellsol AB	91	16 M	17.8	0.001	5.3
Shellsol R	81	23 M	17.2	0.001	5.0
Shellsol RA	90	15 M	17.7	0.001	5.5
Shellsol T	26	85	15.5	0.000	0.0
Shellsol TD	26	84	15.6	0.000	0.0
Shellsol K	27	75	15.8	0.000	0.1
Shell Ink Solvent 2325	31	72	—	—	—
Shell Ink Solvent 2426	32	67	—	—	—
Shell Ink Solvent 2729	28	72	—	—	—
Shell Ink Solvent 2831	26	77	—	—	—
SBP 1	32	66	15.1	0.000	0.0
SBP 2	33	64	15.3	0.000	0.0
SBP 3	32	64	15.8	0.000	0.25
SBP 4	33	61	15.5	0.000	0.1
SBP 5	33	65	15.5	0.000	0.0
SBP 6	36	56	16.0	0.000	0.5
SBP 11	38	53	15.9	0.000	0.5
Rubber solvent	32	64	15.8	0.000	0.3

Further details appear in Sections 12.1 and 14.6, but at this stage it should be noted that Huyskens and Haulait-Pirson defined yet another cohesion parameter, δ', a modified cohesion parameter taking into account only the non-specific dispersion and dipolar cohesion forces, and excluding hydrogen bonds. A reference value of $\delta' = 16.0\ \text{MPa}^{1/2}$ was chosen for octacosane, and δ' values for liquids were determined from the solubility of this solid alkane. The resulting values were influenced by entropy factors, and in particular by the choice of the magnitude of α in

$$\beta = (\alpha + b)({}^{j}V_{u}/{}^{i}V)(1 - {}^{j}\phi) \tag{30}$$

where α is zero for the "classical" Hildebrand-Scatchard relation, unity for the Flory-Huggins equation, and with the intermediate value of 0.5 chosen for use. Some typical values of ${}^{i}\delta'$ are listed in Table 12. This cohesion parameter is related to δ_{dp} and δ_{V}, but of course the exact values depend on the particular choice for α.

Amidon and Williams[161] developed a solubility equation based on the processes solid → subcooled liquid → octanol solution → water solution, the estimation of the standard Gibbs function change for the process of subcooled liquid solute mixing with octanol including a cohesion parameter term using a δ value for octanol of $21.1\ \text{MPa}^{1/2}$.

Chernyi and coworkers[162] presented a five-parameter linear equation for the solubility of anthracene and carbazole with a term in ${}^{i}\delta^{2}$.

Chastrette and coworkers[163] used a multivariate statistical treatment of liquids with a variety of parameters, including Hildebrand parameters as a measure of liquid cohesion: the energy required to create a cavity to accommodate a solute. The original representation of

TABLE 12
Modified Nonspecific Cohesion Parameters[a] for Liquids at 25°C

Liquid	δ/MPa$^{1/2}$			
	δ'[a]	δ_d	δ_p	δ_h
Aliphatic hydrocarbons				
Pentane	15.5	14.5	0.0	0.0
Hexane	15.7	14.9	0.0	0.0
Heptane	15.8	15.3	0.0	0.0
Octane	15.8	15.5	0.0	0.0
Decane	15.9	15.8	0.0	0.0
Dodecane	15.9	16.0	0.0	0.0
Pentadecane	15.9	16.3	0.0	0.0
Hexadecane	16.0	16.4	0.0	0.0
Cyclohexane	16.0	16.8	0.0	0.2
Aromatic hydrocabons				
Benzene	19.0,18.9[b]	18.4	0.0	2.0
Toluene	18.5,18.1[b]	18.0	1.4	2.0
Xylene	18.0,17.3[b]	17.8	1.0	3.1
Mesitylene	16.9[b]	—	—	—
Ethylbenzene	18.0[b]	—	—	—
Chlorocarbons				
Dichloromethane	20.5,20.6[b]	18.2	6.3	6.1
Trichloromethane	18.8	17.8	3.1	5.7
Tetrachloromethane	17.1	17.8	0 .0	0.6
o-Dichlorobenzene	18.7[b]	—	—	—
Tetrachloroethane	17.0[b]	—	—	—
Esters				
Methyl formate	22.9,23.0[b]	—	—	—
Methyl acetate	21.7,21.8[b]	15.5	7.2	7.6
Ethyl acetate	20.8,20.9[b]	15.8	5.3	7.2
Butyl acetate	19.7,19.8[b]	15.8	3.7	6.3
Hexyl acetate	19.2,19.3[b]	—	—	—
Ethyl propionate	20.1,20.2[b]	—	—	—
Ketones				
Acetone	22.1,22.2[b]	15.5	10.4	7.0
Methyl ethyl ketone	21.0,21.1[b]	16.0	9.0	5.1
Diethyl ketone	20.3	15.8	7.6	4.7
Methyl isobutyl ketone	20.0,20.1[b]	15.3	6.1	4.1
Diisopropyl ketone	19.5	—	—	—
Alcohols and Water				
Methanol	19.3	15.1	12.3	22.3
Ethanol	17.8	15.8	8.8	19.4
1-Propanol	17.3	16.0	6.8	17.4
2-Propanol	17.5	15.8	6.1	16.4
1-Butanol	17.0	16.0	5.7	15.8
2-Butanol	16.5	15.8	5.7	14.5
1-Pentanol	16.8	16.0	4.5	13.9

TABLE 12 (continued)
Modified Nonspecific Cohesion Parameters[a] for Liquids at 25°C

Liquid	$\delta/MPa^{1/2}$			
	δ'^{a}	δ_d	δ_p	δ_h
1-Hexanol	16.7	—	—	—
1-Octanol	16.5	17.0	3.3	11.9
1-Decanol	16.4	17.6	16.0	10.0
Water	20.5	15.6	16.0	42.3

[a] Based on $\delta' = 16.0$ MPa$^{1/2}$ for octacosane and $\alpha = 0.5$.
[b] From Haulait-Pirson, M. C., Huys, G., and Vanstraelen, E., *Ind. Eng. Chem. Res.*,, 26, 447, 1987

Adapted from Huyskens, P. L., Haulait-Pirson, M. C., Brandts Buys, L. D., and Van der Borght, X., *J. Coatings Technol.*, 57(724), 57, 1985.

83 liquids in eight-dimensional space could be reduced to a three-dimensional subspace with only an 18% loss of information. The first principal component was correlated with properties associated with liquid polarizability, the second was representative of the liquid polarity, and the third was linked to electron acceptor properties. The Hildebrand parameter accounted for ~ 25% of the variance of the four first principal components, without being significantly correlated to any of them. Chastrette and Carretto[164] also examined and interpreted the Reichardt parameter E_T,[141,142] the Kamlet-Abboud-Taft parameter π^*,[93] and the Kosower Z parameter.[136,137]

8.5 PRACTICAL SOLVENT SCALES

There are many empirical tests commonly used to quantify the behavior of solvents. One group of tests describes the "power" or "strength" of hydrocarbon solvents.[1-4,21,87,153,166-168]

The *aniline point* or *aniline cloud point* (ACP), which is based on the fact that aniline is a poor solvent for aliphatic hydrocarbons and an excellent one for aromatics, is defined as the minimum equilibrium solution temperature for equal volumes of aniline and solvent, with heptane a standard at 69.3°C. Because aromatic hydrocarbons yield temperatures for the aniline point which are so low that aniline can crystallize from the mixture, the *mixed aniline point* can be used: the minimum equilibrium solution temperature of a mixture of two volumes of aniline, one volume of solvent, and one volume of heptane. The procedure is described in ASTM D611, and representative values are listed in Table 11. A rough quantitative correlation exists between Hildebrand parameters and the aniline point,[21,167,168] but is of limited value.

The *Kauri-butanol number*, KB (ASTM D1133, also included in Table 11) is a measure of the tolerance of a standard solution of kauri resin in 1-butanol to added hydrocarbon diluent, being the volume in milliliters at 25°C of the solvent required to produce a certain defined degree of turbidity when added to 20 g of a standard kauri-1-butanol solution. Standard values are toluene (KB = 105) and 75% by volume of heptane with 25% toluene (KB = 40). Special methods must be used for the high volatility solvents used as propellants[169,170] (Table 13). There is an approximately linear relationship between the Hildebrand parameter and the KB number for hydrocarbons with KB > 35:[153,167,171]

$$\delta/MPa^{1/2} = 0.04\ KB + 14.2 \tag{31}$$

TABLE 13
Hildebrand Parameters and Kauri-Butanol Numbers for Isotron Propellants[169,170]

Propellent	$\delta/MPa^{1/2}$	KB
Vinyl chloride	16.0	60
Propellant 133a	15.8	23.0
Propellant 11	15.6	60
Dimethyl ether	14.9	91.0
Propellant A	13.7	26.5
Butane	13.5	19.5
Propellant P	13.4	29.3
Propellant 114	13.0	12
Isobutane, 2-methylpropane	12.6	17.5
Propellant 12	12.5	18
Propane	11.9	15.2

$$KB = 25(\delta/MPa^{1/2}) - 348 \quad (32)$$

For the aliphatic hydrocarbons with KB < 35, there is a different relationship which includes a correction for molecular size:

$$\left(\frac{\delta/V}{MPa^{1/2}/cm^3\ mol^{-1}}\right) = 0.00986KB - 0.163 \quad (33)$$

$$KB = 101.5\left(\frac{\delta/V}{MPa^{1/2}/cm^3\ mol^{-1}}\right) + 16.52 \quad (34)$$

These relationships are illustrated in Figures 5 and 6.

Lin[172] presented a relationship between the Hildebrand parameter and the UOP characterization factor, K, for petroleum solvents, an additive parameter which is a function of many physical properties such as boiling point, density, molecular weight, viscosity and critical properties, and which is useful for solvent formulation:

$$\delta/MPa^{1/2} = 32.7 - 1.416\,K \quad (35)$$

The solvent ''strength'' of chemical or oxygenated solvents (such as alcohols, ketones, esters, and glycol ethers) is much greater than that of hydrocarbons, and the techniques just described for hydrocarbons are inadequate. The *dilution ratio* is widely used since it is a direct measure of the tolerance of a solvent-resin mixture to added diluent. The standard definition and method for its determination, described in terms of ASTM 1720 for cellulose nitrate as an example, is ''the maximum number of unit volumes of a diluent that can be added to a unit volume of solvent to cause the first persistent heterogeneity in a solution of 8 g of cellulose nitrate per 100 mL of combined solvent plus diluent and at a temperature of 25 ± 3°C.'' Dilution ratios must specify the diluent and, as can be seen from Table 14, considerably more aromatic diluent can be tolerated than aliphatic. Qualitative correlations exist between dilution ratios and Hildebrand parameters;[6,171] an example of the application of this method is given in Section 14.4. The *heptane number* of hydrocarbons is a measure of the relative solvent power of high solvency hydrocarbons in the presence of resins not soluble in heptane, and is the ratio of the volume of heptane required to produce a particular degree of turbidity in the resin-solvent solution to that in a resin-commercial 2° toluene

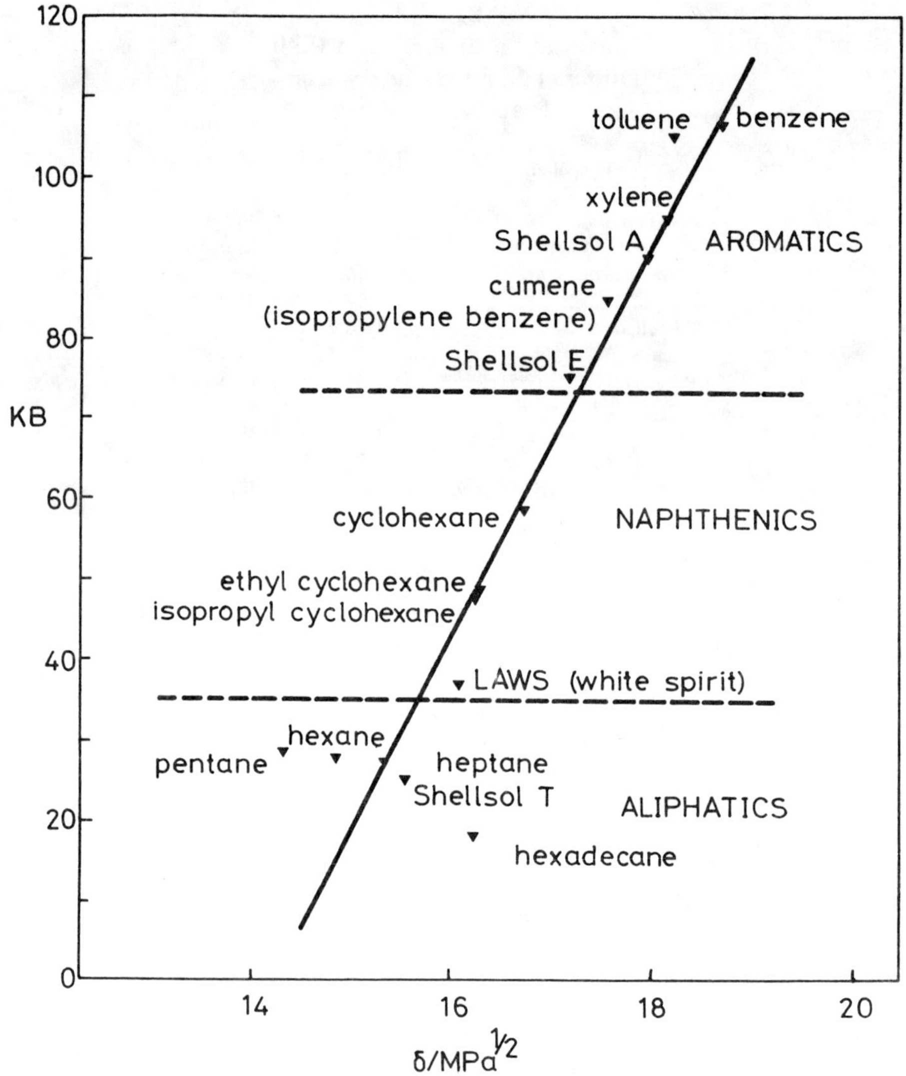

FIGURE 5. Hildebrand parameter—kauri-butanol number relationship for hydrocarbons and Shell solvents. (Adapted from Shell Chemicals, *Solvent Power,* 1979.)

solution. There are other related procedures, such as the heptane miscibility of lacquer solvents. The *wax number*, which is the number of milliliters of a liquid required to produce a turbidity when added to 10 mL of trichloroethylene saturated with Montan wax, or to 10 mL of benzene saturated with beeswax, may also be correlated with liquid Hildebrand parameter.[173]

Godfrey,[174-176] following Rothmund[177] and Francis,[178] developed a single-parameter system for organic liquids which he called the *miscibility number* (*M*-number, with values between 1 and 31) based on experimentally observed critical solution temperatures (CST) and similar in principle to the dilution ratio or cloud point method. In general, pairs of liquids with *M*-numbers differing by 15 units or less are miscible in all proportions at 25°C; each pair with an *M*-number difference of 16 has a CST between 25°C and 75°C (partially miscible); and a difference of 17 or more corresponds to immiscibility or a CST above 75°C. Liquids with $M = 16$ must generate *M*-number differences less than 15 when combined with other liquids, and so are predicted to be "universal" solvents. A set of "standard"

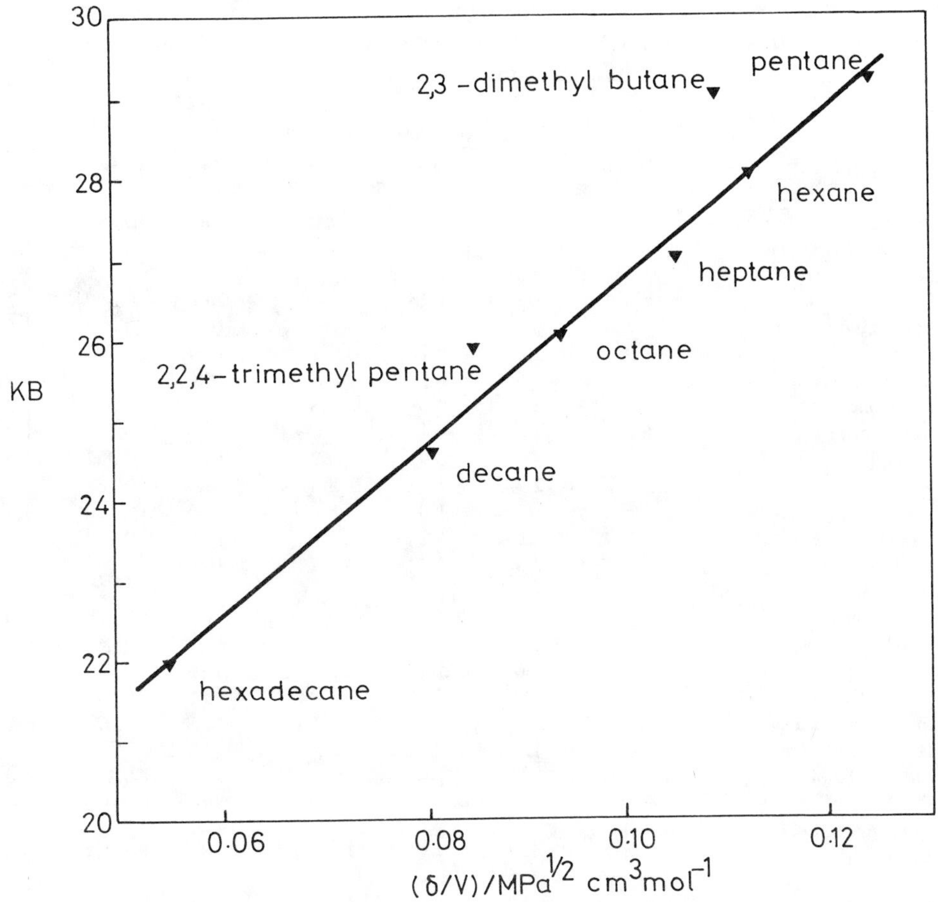

FIGURE 6. Hildebrand parameter/molar volume—kàuri-butanol number relationship for alkanes. (Adapted from Shell Chemicals, *Solvent Power,* 1979.)

liquids selected to conform to these criteria appears in Table 15. It can be seen that there is a close correlation between M and δ.

Another type of empirical solvent classification scheme has been developed in connection with chromatography (Chapter 18) where it is useful to be able to distinguish solvent strength or polarity from solvent selectivity. Gas-liquid chromatographic methods are particularly convenient for quantitative characterization of the solvent properties of the stationary liquid.[143,179] This may involve a "retention index", or more rigorously, take into account the nature of the test sample. Rohrschneider[180-183] introduced the concept of measuring solvent polarity and selectivity through the use of model solutes. Thus, nitromethane was considered to exemplify dipole interactions, pyridine to reveal acceptor interactions, and ethanol to show donor behavior. Snyder[184-186,195] transformed Rohrschneider's solubility data for six test solutes in many commonly used solvents into a set of solvent characterization parameters, approximately corrected for molecular weights of solvent and solute, and for dispersion interactions. The resulting *polarity index, P'*, is a measure of the strength of a solvent or its ability to dissolve or to interact with various polar test solutes. The selectivity parameters, x_e, x_d, and x_n represent the fraction of P' contributed by interactions associated with ethanol, dioxane, and nitromethane, respectively, and reflect the relative ability of the solvent to function as proton acceptor (x_e), or proton donor (x_d), or to interact strongly with dipoles (x_n). The selectivity parameters may be plotted on triangular diagrams or "solvent triangles"

TABLE 14
Dilution Ratios and Other Solvent Parameters for Organic Liquids and Shell Chemical Solvents at 25°C[89,153]

Solvent	ASTM dilution ratio[a] Toluene	Xylene	SBP[b]	1-Butanol	δ/MPa	*p*	θ
Acetone	4.1	3.7	0.8	7.0	20.5	0.695	12.5
Methyl ethyl ketone	4.1	3.0	0.9	—	19.0	0.510	10.5
Methyl isobutyl ketone	3.6	2.9	1.0	—	17.2	0.315	10.5
Cyclohexanone	7.3	7.3	1.25	—	20.2	0.380	13.7
Ethyl amyl ketone	2.2	—	0.9	—	16.8	0.223	8.5
Pentoxone® [c]	2.9	2.6	0.7	—	17.4	0.190	12.5
Diisobutyl ketone	1.5	1.5	0.6	—	16.0	0.123	9.8
Diacetone alcohol	3.4	2.0	0.6	7.8	18.8	0.312	0.0
Isophorone	6.2	5.1	—	—	18.6	0.190	14.9
Mesityl oxide	3.6	—	0.9	9.2	18.4	0.332	12.0
Ethyl acetate	2.9	2.3	1.0	8.4	18.6	0.167	8.4
Isopropyl acetate	2.6	2.4	1.0	8.8	17.6	0.100	8.5
Propyl acetate	3.0	2.0	1.2	—	17.9	0.129	8.5
sec-Butyl acetate	2.6	2.5	1.2	8.2	16.8	0.082	8.3
Isobutyl acetate	2.6	—	1.3	—	17.0	0.097	8.7
Butyl acetate	2.7	2.7	1.3	8.2	17.6	0.120	8.0
Amyl acetate	2.2	2.2	1.2	7.3	17.3	0.067	8.2
Methyl amyl acetate	1.7	1.6	1.0	—	16.8	0.050	8.3
Methyl Oxitol® acetate	2.3	1.9	0.6	—	18.8	0.095	10.5
Ethyl Oxitol acetate	2.6	2.3	0.9	7.5	17.8	0.073	10.1
Butyl Dioxitol® acetate	—	1.8	0.6	—	—	—	—
Methyl Oxitol	4.0	2.9	—	—	22.1	0.126	0.0
Ethyl Oxitol	3.1	4.1	6.1	6.9	20.2	0.086	0.0
Isopropyl Oxitol	4.4	3.8	1.7	—	—	—	—
Butyl Oxitol	3.3	3.2	1.8	—	18.2	0.048	0.0

[a] ASTM D1720, 1/4 second N/C cellulose.
[b] Aliphatic hydrocarbons.
[c] 4-Methoxy-4-methyl-2-pentanone.

in the same way as fractional cohesion parameters (section 5.10) and, as expected, it is found that liquids with the same chemical functional group tend to appear in the same region of the triangle. Table 16 reports Snyder's most recent data[195] modified from that originally published.[184,185] There is, in general, a good correlation between the polarity index values and Hildebrand parameter values (Figure 7), but liquids such as diethyl ether and triethylamine, which are strong proton acceptors but which have no proton donor ability, have δ values similar to those of alkanes although they show up as moderately polar on the P' scale. This is because Hildebrand parameters are based on pure liquid properties, while the polarity index is based on the properties of *mixtures*. For mixed solvents, the ${}^{ij}P'$ value is determined on a volume fraction basis, in the same way as ${}^{ij}\delta$:

$$ {}^{ij}P' = {}^{i}\phi\,{}^{i}P' + {}^{j}\phi\,{}^{j}P' \tag{42}$$

Spychaj and Hamielec[187,188] used both P' and δ in discussing fractional separation of oligomers in binary liquid mixtures.

Szymanowski et al.[192] defined another chromatographic polarity index, and correlated it with Hildebrand parameters. The description of the selectivity in high pressure liquid chromatography solvents (methanol, acetonitrile, and tetrahydrofuran) by component cohesion parameters and by selectivity parameters has been compared by Bakalyar et al.[189] This publication also proposed another characterization method, based on interactions with ni-

TABLE 15
Miscibility Numbers of 'Standard' Solvents, with Hildebrand Parameters

M	Liquid	$\delta/MPa^{1/2}$
1.	Glycerol, 1,2,3-propanetriol	36
2.	Ethylene glycol, 1,2-ethanediol	35
3.	1,4-Butanediol, butylene glycol	34
4.	2,2′-Thiodiethanol	—
5.	Diethylene glycol	29
6.	Triethylene glycol	29
7.	Tetraethylene glycol	20
8.	Methoxyacetic acid	—
9.	Dimethylsulfoxide	25
10.	*N*-Formylmorpholine	27
11.	Furfuryl alcohol	26
12.	Methyl Carbitol,® 2-(2-methoxyethoxy)ethanol	23
13.	Methyl Cellosolve,® 2-methoxyethanol	25
14.	Cellosolve,® 2-ethoxyethanol	24
15.	Butyl Carbitol,® 2-(2-butoxyethoxy)ethanol	20
16.	Butyl Cellosolve,® 2-butoxyethanol	21
17.	1,4-Dioxane	21
18.	Diethyl ketone, 3-pentanone	18
19.	1,1,2,2-Tetrachloroethane	20
20.	1,2-Dichloroethane, ethylene dichloride	20
21.	Chlorobenzene	20
22.	1,2-Dibromobutane, ethylene dibromide	21
23.	1-Bromobutane, butyl bromide	18
24.	Isoamyl bromide, 1-bromo-3-methylbutane	—
25.	*sec*-Amylbenzene	—
26.	4-Vinylcyclohexene	—
27.	1-Methylcyclohexene	—
28.	Cyclohexane	17
29.	Heptane	15
30.	Tetradecane	17
31.	Petrolatum	15

Adapted from Godfrey, N. B., *Chemtech,* 2, 359, 1972.

trobenzene, anisole, and phenol. Vavruch[176,190] and Pytela[193] are among those to have reviewed these simpler solvent scales.

TABLE 16
Snyder's Polarity Index and Selectivity Parameters in Order of Increasing P′

Liquid	P'	x_e	x_d	x_n
Perfluorohexane	−0.84	—	—	—
Hexane	−0.14	—	—	—
Isooctane	−0.03	—	—	—
Decane	0.06	—	—	—
Cyclohexane	0.17	—	—	—
Hexadecane	0.20	—	—	—
Squalane	0.44	—	—	—
Carbon disulfide	1.07	0.22	0.39	0.39
Tetrachloromethane	1.56	0.26	0.40	0.34
Dibutyl ether	1.65	0.48	0.14	0.38
Diisopropyl ether	1.83	0.51	0.10	0.39
Triethylamine	2.19	0.66	0.08	0.26
p-Xylene, 1,4-dimethylbenzene	2.55	0.28	0.26	0.45
Toluene, methylbenzene	2.68	0.28	0.27	0.45
Pentadecanenitrile	2.84	0.36	0.21	0.43
Chlorobenzene	3.02	0.23	0.33	0.44
Fluorobenzene	3.03	0.26	0.29	0.44
Bromobenzene	3.14	0.28	0.31	0.42
Diethyl ether	3.15	0.53	0.13	0.34
1-Octanol	3.23	0.58	0.17	0.25
Ethoxybenzene	3.34	0.30	0.27	0.43
Isoamyl alcohol, isopentanol	3.46	0.58	0.17	0.25
Nonanenitrile	3.66	0.36	0.22	0.42
Anisole	3.87	0.30	0.28	0.42
2-Propanol	3.92	0.57	0.17	0.26
tert-Butanol	4.03	0.56	0.20	0.24
1-Butanol	4.11	0.54	0.19	0.27
1-Propanol	4.13	0.54	0.19	0.27
Bis-(2-ethoxyethyl ether)	4.13	0.38	0.20	0.42
Ethyl acetate	4.24	0.36	0.22	0.42
Tetrahydrofuran	4.28	0.41	0.19	0.40
Dichloromethane, methylene dichloride	4.29	0.27	0.33	0.40
Trichloromethane, chloroform	4.31	0.31	0.35	0.34
Diiodomethane, methylene diiodide	4.31	0.28	0.37	0.35
Ethanol	4.40	0.52	0.19	0.29
Butyronitrile	4.60	0.35	0.23	0.42
Methyl ethyl ketone, 2-butanaone	4.62	0.36	0.20	0.43
Cyclohexanone	4.72	0.37	0.21	0.41
Nitrobenzene	4.74	0.29	0.29	0.43
Benzonitrile	4.77	0.32	0.26	0.41
Acetophenone	4.95	0.35	0.25	0.40
2-Picoline	5.15	0.44	0.20	0.36
1,4-Dioxane	5.27	0.37	0.23	0.40
Pyridine	5.53	0.42	0.22	0.36
Acetonitrile	5.64	0.33	0.25	0.42
Methoxyethanol	5.71	0.41	0.22	0.36
Nitromethane	5.78	0.29	0.30	0.41
Benzyl alcohol	6.06	0.40	0.29	0.31
Acetic acid, ethanoic acid	6.13	0.41	0.30	0.30
γ-Butyrolactone	6.27	0.34	0.26	0.40
Aniline, benzeneamine	6.31	0.33	0.31	0.36
N,N-Dimethylformamide	6.31	0.40	0.21	0.39
N,N-Dimethylacetamide	6.45	0.41	0.20	0.38
N-Methyl-2-pyrrolidone	6.45	0.41	0.21	0.39
Hexamethylphosphoramide	7.10	0.47	0.16	0.37

TABLE 16 (continued) Snyder's Polarity Index and Selectivity Parameters in Order of Increasing P′

Liquid	P'	x_e	x_d	x_n
Dimethylsulfoxide	7.29	0.40	0.22	0.37
Trifluoroethanol	7.55	0.40	0.33	0.27
Hexafluoroisopropanol	8.68	0.45	0.27	0.27

Adapted from Rutan, S. C., Carr, P. W., Cheong, W. J., Park, J. H., and Snyder, L. R., *J. Chromatogr.*, 463, 21, 1989.

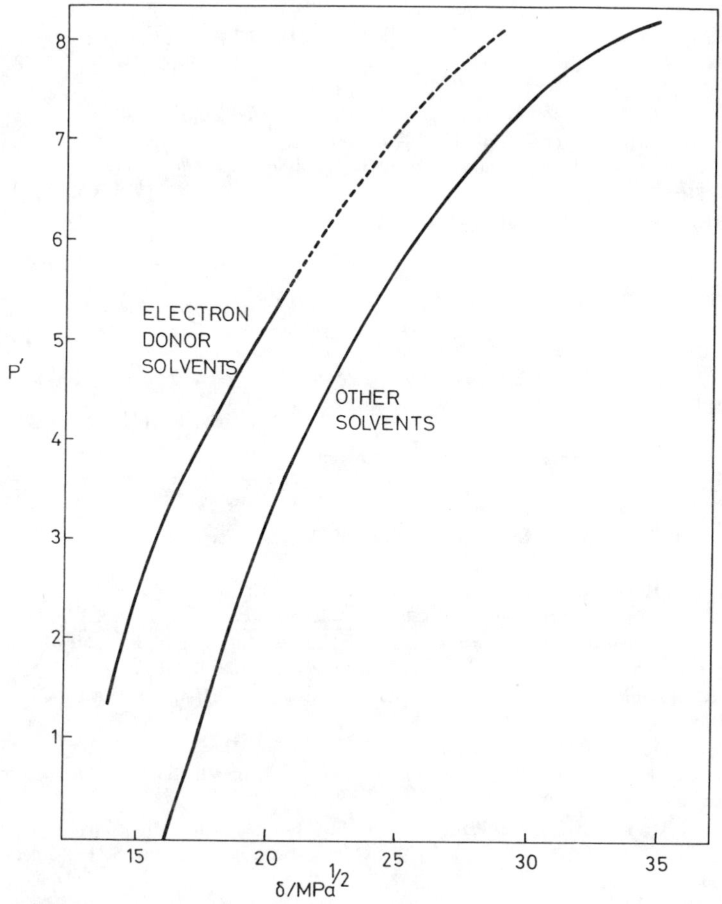

FIGURE 7. Relationship of polarity index to Hildebrand parameter. (Adapted from Snyder, L. R., *Separation and Purification*, 3rd ed., Perry, E. S. and Weissberger, A., Eds., John Wiley & Sons, New York, 1978, 25.)

REFERENCES

1. **Gerrard, W., Mincer, A. M. A., and Wyvill, P. L.,** Solubility of hydrogen halides in organic compounds containing oxygen. V. Solubility of hydrogen chloride and bromide in some esters and ethers and vapor pressures of hydrogen chloride in certain solutions, *J. Appl. Chem.*, 10, 115, 1960.
2. **Gerrard, W.,** *Solubility of Gases and Liquids. A Graphic Approach,* Plenum Press, New York, 1976.
3. **Arnett, E. M.,** Quantitative comparison of weak organic bases, *Progr. Phys. Org. Chem.*, 1, 223, 1963.
4. **Dack, M. R. J.,** The influence of solvent on chemical reactivity, in *Solutions and Solubilities,* Dack, M. R. J., Ed., Part II, John Wiley & Sons-Interscience, New York, 1976, chap. 11.
5. **Burrell, H.,** Solubility parameters, *Interchem. Rev.*, 14, 3 and 31, 1955.
6. **Burrell, H.,** Solubility parameters for film formers; Solubility parameters of resins; A solvent formulating chart, *Off. Dig. Fed. Paint. Varn. Prod. Clubs,* 27, 726, 1955; 29, 1069 and 1159, 1957.
7. **Burrell, H.,** The use of the solubility parameter concept in the United States, *FATIPEC (Fed. Assoc. Tech. Ind. Peint. Vernis, Emaux Encres Impr. Eur. Cont.) Congr.*, 6, 21, 1962.
8. **Burrell, H.,** Solubility parameter values, in *Polymer Handbook,* Brandrup, J. and Immergut, E. H., Eds., John Wiley & Sons-Interscience, New York, 1966, IV-341; 2nd ed., 1975, IV-337.
9. **Burrell, H.,** The challenge of the solubility parameter concept; Entropy: the hidden ingredient, *J. Paint Technol.*, 40, 197, 1968; 42(540), 3, 1970.
10. **Burrell, H.,** Solubility of polymers, in *Kirk-Othmer Encyclopedia of Science and Technology,* 2nd ed., John Wiley & Sons, New York, Vol. 12, 1970, 618.
11. **Burrell, H.,** Trends in solvent science and technology, in *Solvents, Theory and Practice (Adv. Chem. Ser.* 124), Tess, R. W., Ed., American Chemical Society, Washington, DC, 1973, chap. 1.
12. **Burrell, H.,** Pigment wetting parameter; Pigment dispersion parameter, *Am. Chem. Soc. Div. Org. Coat. Plast. Chem. Pap.*, 35(2), 18 and 25, 1975.
13. **Weast, R. C., Ed.,** Solubility parameters of organic compounds, in *CRC Handbook of Chemistry and Physics,* 69th ed., Boca Raton, FL, 1988.
14. **Lieberman, E. P.,** Quantification of the hydrogen bonding parameter, *Off. Dig. Fed. Soc. Paint Technol.*, 34(444), 30, 1962.
15. **Hagman, J. F.,** *Solvent Systems for Neoprene — Predicting Solvent Strength,* A35117, du Pont Elastomer Chemicals Dept., Wilmington, DE, March 1964.
16. **du Pont de Nemours, E. I. and Co.,** *A New Dimension in Solvent Formulation — Elvacite Acrylic Resins,* A44369, du Pont Electrochemicals Dept., Wilmington, DE, 1965.
17. **Craver, J. K.,** A classification of hydrogen-bonding solvents by sonic velocity measurements, *J. Appl. Polym. Sci.*, 14, 1755, 1970.
18. **Crowley, J. D., Teague, G. S., Jr., and Lowe, J. W.,** A three-dimensional approach to solubility, *J. Paint Technol.*, 38, 269, 1966.
19. **Dyck, M. and Hoyer, P.,** Solubility and hydrogen-bridge parameters, *Farbe Lack,* 70, 522, 1964.
20. **Aliev, A. M.,** Semi-empirical method of selecting solvents of varying purpose, *VINITI Dep. Doc.*, 3935-81, 1981; New Scientific concept for prediction of the miscibility of liquids, *Sb. Tr.-Akad. Nauk Az. SSR Inst. Neftekhim. Protsessov im. Akad. Yu. G. Mamedalieva,* 16, 132, 1987.
21. **Beerbower, A. and Dickey, J. R.,** Advanced methods for predicting elastomer/fluids interactions, *Am. Soc. Lubric, Eng. Trans.*, 12, 1, 1969.
22. **Böttcher, C. J. F.,** *Theory of Electric Polarisation,* Elsevier, New York, 1952.
23. **Bonn, R. and van Aartsen, J. J.,** Solubility of polymers in relation to surface tension and index of refraction, *Eur. Polym. J.*, 8, 1055, 1972.
24. **Cammarata, A., Yau, S. J., and Rogers, K. S.,** Structure-activity relationships having a basis in regular solution theory, *J. Med. Chem.*, 14, 211, 1971; **Cammarata, A. and Yau, S. J.,** Molecular estimates for molar attraction constants, *J. Pharm. Sci.*, 61, 723, 1972.
25. **Carr, P. W.,** Fundamental, predictive approach to dipole-dipole interactions based on the dipole moment of the solute and the dielectric constant of the solvent, *J. Chromatogr.*, 194, 105, 1980.
26. **Coomber, D. I.,** Relation between dielectric constants and internal pressures of non-polar liquids, *Trans. Faraday Soc.*, 35, 304, 1939.
27. **Dack, M. R. J.,** The importance of solvent internal pressure and cohesion to solution phenomena, *Chem. Soc. Rev.*, 4, 211, 1975.
28. **Darby, J. R., Touchette, N. W., and Sears, K.,** Dielectric constants of plasticizers as predictors of compatibility with polyvinyl chloride, *Polym. Eng. Sci.*, 7, 295, 1967.
29. **Djordjevic, M. B. and Porter, R. S.,** New empirical relation between cohesion energy density and Onsager reaction field for several classes of organic compounds, in *Macromolecular Solutions: Solvent-Polarity Relationships in Polymers* (Symp. Proc. Am. Chem. Soc., New York, 1981), Seymour, R. B. and Stahl, G. A., Eds., Pergamon, New York, 1982, 198; **Djordjevic, M. B.,** *Effect of Polymer Structure on Solubility, Compatibility and Thermal Stability,* Ph. D. Thesis, University of Massachusetts, 1983; *Diss. Abstr. Int. B,* 43, 4006, 1983.

30. **Gardon, J. L.,** Cohesive-energy density, in *Encyclopedia of Polymer Science and Technology,* Vol. 3, John Wiley & Sons, New York, 1964, 833.
31. **Gardon, J. L.,** The influence of polarity upon the solubility parameter concept, *J. Paint Technol.,* 38(492), 43, 1966.
32. **Gjaldbaek, J. C. and Anderson, E. K.,** The solubility of carbon dioxide, oxygen, carbon monoxide and nitrogen in polar solvents, *Acta Chem. Scand.,* 8, 1398, 1954.
33. **Gordon, J. E.,** Transition energies for a merocyanine dye in aqueous electrolyte solutions. Solvent polarity indicator transition energy-internal pressure relations, *J. Phys. Chem.,* 70, 2413, 1966.
34. **Gordon, J. E.,** Application of fused salts in organic chemistry, *Tech. Meth. Org. Organomet. Chem.,* 1, 51, 1969.
35. **Gorman, W. G. and Hall, G. D.,** Dielectric constant correlations with solubility and solubility parameters, *J. Pharm. Sci.,* 53, 1017, 1964.
36. **Griffiths, T. R. and Pugh, D. C.,** Solvent polarity studies. Part 1. New *Z* values and relationships with other solvent polarity scales, *J. Solution Chem.,* 8, 247, 1979; Correlations among solvent polarity scales, dielectric constant and dipole moment, and a means to reliable predictions of polarity scale values from current data, *Coord. Chem. Rev.,* 29, 129, 1979.
37. **Hansen, C. M. and Beerbower, A.,** Solubility parameters, in *Kirk-Othmer Encyclopedia of Chemical Technology,* 2nd ed., Suppl. Vol., Standen, A., Ed., Interscience, New York, 1971, 889.
38. **Hansen, C. M.,** *Three Dimensional Solubility Parameter and Solvent Diffusion Coefficient. Importance in Surface Coating Formulation,* Doctoral Dissertation, Danish Technical Press, Copenhagen, 1967.
39. **Hansen, C. M. and Skaarup, K.,** The three dimensional solubility parameter — key to paint component affinities. III. Independent calculation of the parameter components, *J. Paint Technol.,* 39, 511, 1967.
40. **Herbrandson, H. F. and Neufeld, F. R.,** Organic reactions and the critical energy density of the solvent. The solubility parameter, δ, as a new solvent parameter, *J. Org. Chem.,* 31, 1140, 1966.
41. **Keller, R. A., Karger, B. L., and Snyder, L. R.,** Use of the solubility parameter in predicting chromatographic retention and eluotropic strength, *Gas Chromatography 1970* (Proc. Int. Symp., 1970, 8, Dublin), Stock, R., Ed., The Institute of Petroleum, London, 1971, 125.
42. **Karger, B. L., Snyder, L. R., and Eon, C.,** An expanded solubility parameter treatment for classification and use of chromatographic solvents and adsorbents. Parameters for dispersion, dipole and hydrogen bonding interactions, *J. Chromatogr.,* 125, 71, 1976.
43. **Munafo, A., Buchmann, M., Hô Nam-Tran, and Kesselring, U. W.,** Determination of the total and partial cohesion parameters of lipophilic liquids by gas-liquid chromatography and from molecular properties, *J. Pharm. Sci.,* 77, 169, 1988.
44. **Kirchnerova, J. and Cavé, G. C. B.,** The solubility of water in low-dielectric solvents, *Can. J. Chem.,* 54, 3909, 1976.
45. **Koenhen, D. M. and Smolders, C. A.,** The determination of solubility parameters of solvents and polymers by means of correlation with other physical quantities, *J. Appl. Polym. Sci.,* 19, 1163, 1975.
46. **Lawson, D. D. and Ingham, J. P.,** Estimation of solubility parameters from refractive index data, *Nature(London),* 223, 614, 1969.
47. **Liszi, J.,** Cohesion energy of a liquid due to permanent dipoles, *Acta Chim. Acad. Sci. Hung.,* 98, 93, 1978.
48. **Loev, A. M. and Lavrishchev, L. P.,** Method for the selection of solvents for coating materials, *Lakokras. Mater. Ikh. Primen.,* No. 4, 4, 1977.
49. **Longster, G. F. and Walker, E. E.,** Effect of permanent dipoles on the cohesive energy and solvent action of some highly polar substances, *Trans. Faraday Soc.,* 49, 228, 1953.
50. **Martin, A., Wu, P. L., Adjei, A., Beerbower, A., and Prausnitz, J. M.,** Extended Hansen solubility approach: Naphthalene in individual solvents, *J. Pharm. Sci.,* 70, 1260, 1981.
51. **Wu, P. L., Beerbower, A., and Martin, A.,** Extended Hansen solubility approach — calculating partial solubility parameters of solid solutes, *J. Pharm. Sci.,* 71, 1285, 1982.
52. **Matsuura, T.,** Reverse osmosis separation of solute from aqueous solution, *Yukui Gosei Kagaku Kyokai (J. Syn. Org. Chem. Jpn.),* 31, 717, 1973.
53. **Matsuura, T.,** Transport processes in membranes for reverse osmosis, *Chem.-Ing. Tech.,* 50, 565, 1978.
54. **Matsuura, T., Blais, P., and Sourirajan, S.,** Polar and nonpolar parameters for polymeric reverse osmosis membrane materials from liquid chromatographic data, *J. Appl. Polym. Sci.,* 20, 1515, 1976.
55. **Matsuura, T. and Sourirajan, S.,** Reverse osmosis separation of hydrocarbons in aqueous solutions using porous cellulose acetate membranes, *J. Appl. Polym. Sci.,* 17, 3683, 1973.
56. **Matsuura, T. and Sourirajan, S.,** Characterization of reverse osmosis membrane materials by liquid chromatography data, *Proc. 6th Int. Symp., Fresh Water from the Sea,* 3, 227, 1978.
57. **McGinniss, V. D.,** Prediction of solvent and polymer characteristics through the use of easy to measure properties, *Abstr. Pap. Am. Chem. Soc.,* 176, ORPL 102, 1978; *Org. Coat. Plast. Chem.,* 39, 529, 1978.
58. **Moore, W. E.,** The use of an approximate dielectric constant to blend solvent systems, *J. Am. Pharm. Assoc. Sci. Ed.,* 47, 855, 1958; Use of an approximate dielectric constant in solubility studies, *J. Pharm. Sci.,* 51, 391, 1962.

59. **Noel, D. E. and Meloan, C. E.,** Some empirical correlations in solvent extraction, *Separ. Sci.*, 7, 935, 1972.
60. **Paruta, A. N.,** *A Dielectric Constant Approach to Solubility phenomena,* Thesis, Rutgers - The State University, Brunswick, NJ, 1963; *Diss. Abstr.*, 24, 2701, 1964.
61. **Paruta, A. N.,** Solubility of parabens in alcohols, *J. Pharm. Sci.*, 58, 216, 1969.
62. **Paruta, A. N. and Mauger, J. W.,** Solubility of sodium salicylate in mixed solvent systems, *J. Pharm. Sci.*, 60, 432 1971.
63. **Paruta, A. N., Sciarrone, B. J., and Lordi, N. G.,** Correlation between solubility parameters and dielectric constants, *J. Pharm. Sci.*, 51, 704, 1962.
64. **Lordi, N. G., Sciarrone, B. J., Ambrosio, T. J., and Paruta, A. N.,** Dielectric constants and solubility, *J. Pharm. Sci.*, 53, 463, 1964.
65. **Alexander, K. S., Mauger, J. W., Peterson, H., Jr., and Paruta, A. N.,** Solubility profiles and thermodynamics of parabens in aliphatic alcohols, *J. Pharm. Sci.*, 66, 42, 1977.
66. **Breon, T. L., Mauger, J. W., and Paruta, A. N.,** Determination of solubility parameter values for pure solvents and binary mixtures, *Drug. Dev. Ind. Pharm.*, 6, 87, 1980.
67. **Pastor, G. R., Angelovich, J. M., and Silver, H. F.,** Study of the interaction of solvent and catalyst in coal hydrogenation, *Ind. Eng. Chem. Proc. Des. Dev.*, 9, 609, 1970.
68. **Peiffer, D.,** The molecular factors affecting the solubility parameter, *J. Appl. Polym. Sci.*, 25, 369, 1980.
69. **Sewell, J. H.,** *A Study of Cohesive Forces, Refractive Indexes and Glass Transition Temperature Data of Polymers,* Royal Aircraft Establishment Technical Report No. 66185, Farnborough, Hampshire, 1966, AD 641308.
70. **Shetty, P. H., Youngberg, P. J., Kersten, B. R., and Poole, C. F.,** Solvent properties of liquid organic salts used as mobile phases in microcolumn reversed-phase liquid chromatography, *J. Chromatogr.*, 411, 61, 1987.
71. **Skraga, J. and Wieckiewicz, Z.,** Method for determination of the Hildebrand solubility parameter of plasticizers and their blends, *Polimery(Warsaw),* 26(2), 45, 1981.
72. **Stekol'shchikov, M. N., Krivtsova, L. M., and Ratner, M. I.,** Evaluation of the dissolving ability of hydrocarbon solvents, *Khim. Tekhnol. Topl. Masel,* No. 6, 33, 1987; *Chem. Technol. Fuels Oils,* No. 6, 292, 1987.
73. **Sunwoo, C. and Eisen, H.,** Solubility parameter of selected sulfonamides, *J. Pharm. Sci.*, 60, 238, 1971.
74. **Tatsumi, M., Yamaguchi, Y., Moriwaki, S., Usami, S., and Yamamoto, S.,** Determination of solubility parameter of polymer by microwave absorption in waveguide, *Kobunshi Ronbunshu,* 45, 161, 1988.
75. **Thomas, E. R. and Eckert, C. A.,** Prediction of limiting activity coefficients by a modified separation of cohesive energy density model and UNIFAC, *Ind. Eng. Chem. Proc. Des. Dev.*, 23, 194, 1984.
76. **Tijssen, R., Billiet, H. A. H., and Schoenmakers, P. J.,** Use of the solubility parameter for predicting selectivity and retention in chromatography, *J. Chromatogr.*, 122, 185, 1976.
77. **Turyshev, B. I., Shuvaev, V. P., Goldenberg, A. D., and Sazhin, B. I.,** Electric conductivity of polymer dielectrics in contact with liquid media, *Plast. Massy,* No. 7, 69, 1978; *Int. Polym. Sci. Technol.*, 6(4), 25, 1979.
78. **Watanabe, A. and Sugiyama, S.,** Relation between molecular weight and solubility parameter for pure organic liquids, *Kogyo Kagaku Zasshi,* 72, 1195, 1969.
79. **Wiesner, I.,** Solubility parameter of epoxy resins, *Chem. Prumysl.*, 26, 532, 1976.
80. **Wingefors, S.,** *Studies of the Solvent Extraction Chemistry of Dialkyldithiophosphates and Development of Models for Non-Ideal Behavior in Solvent Extraction,* Thesis, Department of Nuclear Chemistry, Chalmers University of Technology, Göteborg, Sweden, 1979.
81. **Wingefors, S. and Liljenzin, J. O.,** Development of a correlation between the nonpolar solubility parameter, refractive index and molecular structure. I. Aliphatic hydrocarbons; II. Aliphatic ethers and alcohols, *J. Chem. Technol. Biotechnol.*, 31, 115 and 530, 1981.
82. **Winkelmann, J.,** Calculation of the thermodynamic excess functions from dielectric data for systems with dimethylformamide (DMF), *Z. Phys. Chem. (Leipzig),* 263, 1185, 1982.
83. **Wolkoff, H. N.,** Liquids, in *The Theory and Practice of Industrial Pharmacy,* Lachman, L., Lieberman, H. A., and Kanig, J. L., Eds., Lea & Febiger, Philadelphia, PA, 1970, chap. 15.
84. **Xu, C. and Song, X.,** Correlation between solubility parameter and dielectric constant of polymer, *Hecheng Xiangjiao Gongye,* 10, 352, 1987.
85. **Yamamoto, T., Isshiki, S., and Nakayama, S.,** Synthetic paper for extra high voltage cable, *IEEE Trans. Power Appl. Syst.*, PAS 91, 2415, 1972.
86. **Yeo, R. S.,** Ion clustering in Nafion and its relationship to membrane conductivity, in *Symposium on Ion Exchange,* Yeo, R. S. and Buck, R. P., Eds., Electrochemical Society, Princeton, NJ, 1981.
87. **Wyart, J. W. and Dante, M. F.,** Solvents, industrial, in *Kirk-Othmer Encyclopedia of Chemical Technology,* 2nd ed., Vol. 18, John Wiley & Sons-Interscience, New York, 1969, 564.
88. Shell Chemical Co., *Solvent Notes,* Bulletin IC:67-64-SN, 1967.

89. **Nelson, R. C., Figurelli, V. F., Walsham, J. G., and Edwards, G. D.,** Solution theory and the computer. Effective tools for the coatings chemist, *J. Paint Technol.*, 42, 644, 1970.
90. **Kumar, R. and Prausnitz, J. M.,** Solvents in chemical technology, in *Solutions and Solubilities, Part I*, Dack, M. R. J., Ed., John Wiley & Sons-Interscience, New York, 1975, chap. 5.
91. **Somerville, G. R. and Lopez, J. A.,** Technique for reformulating solvent mixtures in epoxy resin coatings, in *Solvents, Theory and Practice* (Adv. Chem. Ser. 124), Tess, R. W., Ed., American Chemical Society, Washington, DC, 1973, chap. 12, 175.
92. **Karim, K. A. and Bonner, D. C.,** An improved concept in solubility parameter theory applied to amorphous polymers. I. Poly(ethyl methacrylate), *Polym. Eng. Sci.*, 19, 1174, 1979.
93. **Kamlet, M. J., Abboud, J.-L. M., and Taft, R. W.,** The solvatochromic comparison method. 6. The π^* scale of solvent polarities, *J. Am. Chem. Soc.*, 99, 6027, 1977.
94. **Abboud, J.-L., Kamlet, M. J., and Taft, R. W.,** Regarding a generalized scale of solvent polarities, *J. Am. Chem. Soc.*, 99, 8325, 1977.
95. **Abboud, J.-L. M. and Taft, R. W.,** An interpretation of a general scale of solvent polarities. A simplified reaction theory modification, *J. Phys. Chem.*, 83, 412, 1979.
96. **Kamlet, M. J., Hall, T. N., Boykin, J., and Taft, R. W.,** Linear solvation energy relationships. 6. Addition to and correlations with the π^* scale of solvent polarities, *J. Org. Chem.*, 44, 2599, 1979.
97. **Kamlet, M. J., Abboud, J.-L. M., and Taft, R. W.,** An examination of linear solvation energy relationships, in *Progress in Physical Organic Chemistry,* Vol. 13, Taft, R. W., Ed., John Wiley & Sons, New York, 1981, 485.
98. **Abraham, M. H., Taft, R. W., and Kamlet, M. J.,** Linear solvation energy relationships. 15. Heterolytic decomposition of the tert-butyl halides, *J. Org. Chem.*, 46, 3053, 1981.
99. **Kamlet, M. J., Carr, P. W., Taft, R. W., and Abraham, M. H.,** Linear solvation energy relationships. 13. Relationship between the Hildebrand solubility parameter, δ_H, and the solvatochromic parameter, π^*, *J. Am. Chem. Soc.*, 103, 6062, 1981.
100. **Abraham, M. H., Kamlet, M. J., and Taft, R. W.,** Linear solvation energy relationships. Part 19. Correlation of the free energies of solution of 41 solutes in select solvents with Hildebrand's solubility parameter, δ_H, and with the solvatochromic parameter, π^*, *J. Chem. Soc. Perkin Trans. II*, 923, 1982.
101. **Kamlet, M. J., Taft, R. W., Carr, P. W., and Abraham, M. H.,** Linear solvation energy relationships. Part 9. Correlations of gas/liquid partition coefficients with the solvatochromic parameters, π^*, α and β, *J. Chem. Soc. Faraday Trans. I*, 78, 1689, 1982.
102. **Kamlet, M. J., Aboud, J.-L. M., Abraham, M. H., and Taft, R. W.,** Linear solvation energy relationships. 23. A comprehensive collection of the solvatochromic parameters, π^*, α, and β, and some methods for simplifying the generalized solvatochromic equation, *J. Org. Chem.*, 48, 2877, 1983.
103. **Kamlet, M. J., Doherty, R., Taft, R. W., and Abraham, M. H.,** Linear solvation energy relationships. 26. Some measures of relative self-association of alcohols and water, *J. Am. Chem. Soc.*, 105, 6741, 1983.
104. **Abraham, M. H., Kamlet, M. J., Taft, R. W., and Weathersby, P. K.,** Solubility properties in polymers and biological media. 1. Correlations in various media of the solubility of gaseous anesthetics and other solutes with a number of solute parameters, *J. Am. Chem. Soc.*, 105, 6797, 1983.
105. **Kamlet, M. J., Abraham, M. H., Doherty, R. M., and Taft, R. W.,** Solubility properties in polymers and biological media. 4. Correlation of octanol/water partition coefficients with solvatochromic parameters, *J. Am. Chem. Soc.*, 106, 464, 1984.
106. **Taft, R. W., Abraham, M. H., Doherty, R. M., and Kamlet, M. J.,** Linear solvation energy relationships. 29. Solution properties of some tetraalkylammonium halide ion pairs and dissociated ions, *J. Am. Chem. Soc.*, 107, 3105, 1985.
107. **Kamlet, M. J., Gal, J.-F., Maria, P.-C., and Taft, R. W.,** Linear solvation energy relationships. Part 32. A co-ordinate covalency parameter, ξ, which, in combination with the hydrogen bond acceptor basicity parameter, β, permits correlation of many properties of neutral oxygen and nitrogen bases (including aqueous pK_a), *J. Chem. Soc. Perkin Trans. II*, 1583, 1985.
108. **Abraham, M. H., Doherty, R. M., Kamlet, M. J., Harris, J. M., and Taft, R. W.,** Linear solvation energy relationships. Part 37. An analysis of contributions of dipolarity-polarisability, nucleophilic assistance, electrophilic assistance, and cavity terms to solvent effects on t-butyl halide solvolysis rates; Part 38. An analysis of the use of solvent parameters in the correlation of rate constants, with special reference to the solvolysis of t-butyl chloride, *J. Chem. Soc. Perkin Trans. II*, 913 and 1097, 1987.
109. **Rawes, R.,** Chemists propose universal solubility equation, *Chem. Eng. News*, (March 18), 20, 1985; **Taft, R. W., Abraham, M. H., Doherty, R. M., and Kamlet, M. J.,** The molecular properties governing solubilities or organic nonelectrolytes in water, *Nature*, 313, 384, 1985.
110. **Small, P. A.,** Some factors affecting the solubility of polymers, *J. Appl. Chem.*, 3, 71, 1953.
111. **Izumi, Y. and Miyake, Y.,** Study of linear poly(*p*-chlorostyrene)-diluent systems. I. Solubilities, phase relationships, and thermodynamic interactions, *Polym. J.*, 3, 647, 1972.
112. **Beerbower, A., Kaye, L. A., and Pattison, D. A.,** Picking the right elastomer to fit your fluids, *Chem. Eng. (NY)*, 74(26), 118, 1967.

113. **Gordy, W.**, Spectroscopic comparison of the proton-attracting properties of liquids, *J. Chem. Phys.*, 7, 93, 1939.
114. **Gordy, W. and Stanford, S. C.**, Spectroscopic evidence for hydrogen bonds: comparison of proton-attracting properties of liquids, II and III, *J. Chem. Phys.*, 8, 170, 1940; 9, 204, 1941.
115. **Nelson, R. C., Hemwall, R. W., and Edwards, G. D.**, Treatment of hydrogen bonding in predicting miscibility, *J. Paint Technol.*, 42, 636, 1970.
116. **Searles, S. and Tamres, M.**, Hydrogen bond formation with saturated cyclic ethers, *J. Am. Chem. Soc.*, 73, 3704, 1951.
117. **Tamres, M.**, Aromatic compounds as donor molecules in hydrogen bonding, *J. Am. Chem. Soc.*, 74, 3375, 1952.
118. **Searles, S., Tamres, M., and Barrow, G. M.**, Hydrogen-bonding of esters and lactones. Site of bonding and effect of ring sizes, *J. Am. Chem. Soc.*, 75, 71, 1953.
119. **du Pont de Nemours, E. I. and Co.**, *Solvent Formulating Maps for Elvacite Acrylic Resins*, Bulletin PA-12-770, A70562, Wilmington, DE, 1970; PA-12-174, 1974.
120. **Cosaert, E.**, Solubility parameters allowing the prediction of the solubility of a solid in a solvent system, *Chim. Peint.*, 34, 169, 1971.
121. **ASTM** *Standard Test Method for Solubility Range of Resins and Polymers*, ANSI/ASTM D3132-72, American Society for Testing and Materials, Philadelphia, and American National Standards Institute, New York, reapproved 1976.
122. **Orr, A. A.**, Improved infrared technique for measurement of hydrogen bonding parameter of solvents, *J. Paint Technol.*, 47(607), 45, 1975.
123. **Kagiya, T., Sumida, T., and Inoue, T.**, A measure of the electron-donating power and electron-accepting power of liquid organic compounds, *Bull. Chem. Soc. Jpn.*, 41, 767, 1968.
124. **Koppel, I. A. and Palm, V. A.**, The influence of the solvent on organic reactivity, in *Advances in Linear Free Energy Relationships*, Chapman, N. B. and Shorter, J., Eds., Plenum Press, London, 1972, chap. 5.
125. **Burden, A. G., Collier, G., and Shorter, J.**, Influence of aprotic solvents on the O-D stretching band of methan[^{2}H]ol, *J. Chem. Soc. Perkin Trans. II*, 1627, 1976.
126. **Allerhand, A. and Schleyer, P. v. R.**, Solvent effects in infrared spectroscopic studies of hydrogen bonding, *J. Am. Chem. Soc.*, 85, 371, 1963.
127. **Arnett, E. M., Joris, L., Mitchell, E., Murty, T. S. S. R., Gorric, T. M., and Schleyer, P. v. R.**, Studies of hydrogen-bonded complex formation. III. Thermodynamics of complexes by infrared spectroscopy and calorimetry, *J. Am. Chem. Soc.*, 92, 2365, 1970.
128. **Balakrishnan, S. and Easteal, A. J.**, Empirical polarity parameters for some binary solvent mixtures; Intermolecular interactions in water + acetonitrile mixtures: evidence from the composition variation of solvent polarity parameters, *Aust. J. Chem.*, 34, 933 and 943, 1981.
129. **Brooker, L. G. S., Keyes, G. H., and Heseltine, D. W.**, Color and constitution. XI. Anhydronium bases of *p*-hydroxystyryl dyes as solvent polarity indicators, *J. Am. Chem. Soc.*, 73, 5350, 1951.
130. **Brooker, L. G. S., Craig, A. C., Heseltine, D. W., Jenkins, P. W., and Lincoln, L. L.**, Color and constitution. 13. Merocyanines as solvent property indicators, *J. Am. Chem. Soc.*, 87, 2443, 1965.
131. **Cardinal, J. R. and Mukerjee, P.**, Solvent effects on the ultraviolet spectra of benzene derivatives and naphthalene. Identification of polarity sensitive spectral characteristics, *J. Phys. Chem.*, 82, 1614, 1978.
132. **Dack, M. R. J.**, Organic solvents — their effect on chemical reactivity, *Chem. Br.*, 6, 347, 1970; Solvent effects and reactivities, *Chemtech*, 108, 1971; The influence of solvent on chemical reactivity. An alternative approach, *J. Chem. Educ.*, 51, 231, 1974.
133. **Dubois, J. E., Goetz, E., and Bienvenue, A.**, Effect of solvents on the $n \rightarrow \pi^*$ transition of saturated ketones, *Spectrochim. Acta*, 20, 1815, 1964.
134. **Gebelein, C. G.**, The correlation of the solubility parameter with some solvent polarity parameters. I. *Abstr. Pap. Am. Chem. Soc.*, 176, ORPL 101, 1978; *Org. Coat. Plast. Chem.*, 39, 524, 1978.
135. **Gordon, J. E.**, Fused organic salts. IV. Characterization of low melting quaternary ammonium salts. Phase equilibria for salt-salt and salt-nonelectrolyte systems. Properties of the liquid salt medium, *J. Am. Chem. Soc.*, 87, 4347, 1965.
136. **Kosower, E. M.**, The effect of solvent on spectra. I. A new empirical measure of solvent polarity: Z-values; II. Correlation of spectral absorption data with Z-values; III. The use of Z-values in connection with kinetic data; *J. Am. Chem. Soc.*, 80, 3253 and 3261 and 3267, 1958.
137. **Kosower, E. M.**, *An Introduction to Physical Organic Chemistry*, John Wiley & Sons, New York, 1968.
138. **Lassau, C. and Jungers, J. C.**, Effect of solvent on chemical reactions. The quaternization of tertiary amines by methyl iodide, *Bull. Soc. Chim. Fr.*, 2678, 1968.
139. **Reeves, R. L., Maggio, M. S., and Costa, L. F.**, Importance of solvent cohesion and structure in solvent effect on binding site probes, *J. Am. Chem. Soc.*, 96, 5917, 1974.
140. **Dimroth, K., Reichardt, C., Siepmann, T., and Bohlmann, F.**, Pyridinium *N*-phenolbetaines and their use for the characterization of the polarity of solvents, *Ann. Chem.*, 661, 1, 1963.

141. **Maksimovic, Z. B., Reichardt, C., and Spiric, A.,** Determination of empirical parameters of solvent polarity E_T in binary mixtures by solvatochromic pyridinium-*N*-phenol betaine dyes, *Z. Analyt. Chem.*, 270, 100, 1974.
142. **Reichardt, C.,** Solvent scales and chemical reactivity, in *Organic Liquids: Structure, Dynamics, Chemical Properties,* Buckingham, A. D., Lippert, E., and Bratos, S., Eds., John Wiley & Sons, Chichester, U.K., 1978, chap. 16.
143. **Reichardt, C.,** *Solvent Effects in Organic Chemistry* (Monographs in Modern Chemistry, Vol. 3), Verlag Chemie, Weinheim, 1979.
144. **Reichardt, C.,** Empirical parameters of solvent polarity as linear free energy relationships, *Angew, Chem. Int. Ed.*, 18, 98, 1979.
145. **Reichardt, C.,** Empirical parameters of solvent polarity and chemical reactivity, in *Molecular Interactions,* Vol. 3, Ratajczak, H. and Orville-Thomas, W. J., Eds., John Wiley & Sons, Chichester, 1982, chap. 5, 241.
146. **Smid, J.,** Spectrophotometric studies of ion-pair equilibria, in *Ions and Ion Pairs in Organic Reactions,* Vol. 1, Szwarc, M., Ed., John Wiley & Sons, New York, 1972, chap. 3.
147. **Walther, D.,** Polarity of organic solvents and film-forming polymers by solvatochromism of 1,2-diimine complexes of molybdenum(O), *J. Prakt. Chem.*, 316, 604, 1974.
148. **Crowley, J. D., Teague, G. S., Jr., and Lowe, J. W.,** A three-dimensional approach to solubility. II. *J. Paint Technol.*, 39, 19, 1967.
149. **Eastman Kodak,** *Three Dimensional Approach to Solubility,* Rochester, NY (undated).
150. **McLellan, A. L.,** *Tables of Experimental Dipole Moments,* W. H. Freeman, San Francisco, 1963.
151. **Ramsbotham, J.,** Solvent formulation for surface coatings, *Prog. Org. Coatings,* 8, 113, 1980.
152. **Shell Chemicals,** *Solvent System Design,* Tech. Bull. ICS(X)/78/2, 1978.
153. **Shell Chemicals,** *Solvent Power,* Tech. Bull. ICS(X)/79/2, 1979.
154. **Henry, F.,** The solubility parameter in multivariate analysis, FATIPEC *(Fed. Assoc. Tech. Ind. Peint. Vernis Emaux Encres Impr. Eur. Cont.) Congr.*, 14, 263, 1978.
155. **Roussel, R., Oteyza de Guerrero, M., Spegt, P., and Galin, J. C.,** Solvent effects on the tautomerism of *t*-butyl 2-picolyl and *t*-butyl quinaldyl ketones, *J. Heterocyc. Chem.*, 19, 785, 1982.
156. **Huyskens, P. L. and Haulait-Pirson, M. C.,** A new approach for the quantitative prediction of the solubilities of polymers in solvents and solvent mixtures, *Tenth Int. Conf. Org. Coatings Sci. Technol.*, Athens, July 1984.
157. **Haulait-Pirson, M. C. and Huyskens, P. L.,** New possibilities in the quantitative prediction of the solubility of polymers, *FATIPEC (Fed. Assoc. Tech. Ind. Peint Vernis Emaux Encres Impr. Eur. Cont.) Congr.*, 17(Lugano), 235, 1984.
158. **Huyskens, P. L., Haulait-Pirson, M. C., Brandts Buys, L. D., and Van der Borght, X.,** Dissolving power of solvents and solvent blends for polymers, *J. Coatings Technol.*, 57(724), 57, 1985.
159. **Huyskens, P., Haulait-Pirson, M. C., and van der Borght, X.,** Calculation of the solvent power of solvents for resins, *Farbe Lack,* 92, 1145, 1986.
160. **Haulait-Pirson, M.-C., Huys, G., and Vanstraelen, E.,** New predictive equation for solubility of *n*-alkanes in organic solvents, *Ind. Eng. Chem. Res.*, 26, 447, 1987.
161. **Amidon, G. L. and Williams, N. A.,** A solubility equation for non-electrolytes in water, *Int. J. Pharm.*, 11, 249, 1982.
162. **Chernyi, A. V., Rubtsov, V. I., Aleksandrov, V. V., Nazarov, V. N., and Vail, E. I.,** Dependence of anthracene and carbazole solubility on the nature of the solvent, *Ukr. Khim. Zh.*, 52(6), 597, 1986; *Sov. Progr. Chem.*, 52(6), 41, 1986.
163. **Chastrette, M., Rajzmann, M., Chanou, M., and Purcell, K. F.,** Approach to a general classification of solvents using a multivariate statistical treatment of quantitative solvent parameters, *J. Am. Chem. Soc.*, 107, 1, 1985.
164. **Chastrette, M. and Carretto, J.,** Statistical study of solvent effects. III. Calculation and interpretation of empirical polarity parameters from physicochemical properties of pure solvents, *Can. J. Chem.*, 63, 3492, 1985.
165. **Mandik, L.,** Correlation between the KB (kauri-butanol) value and the solubility parameter, *Chem. Prum.*, 15, 283, 1965.
166. **Mandik, L.,** Additivity of the critical solution temperatures of hydrocarbons with aniline, *Collect. Czech. Chem. Commun.*, 31, 3869, 1966.
167. **Reynolds, W. W. and Larson, E. C.,** The fundamentals of solvent selection, *Off. Dig. Fed. Soc. Paint Technol.*, 34, 311, 1962.
168. **Mandik, L.,** Evaluation of dissolving capacity from the aniline point, *Ropa Uhlie,* 8, 143, 1966.
169. **Isotron Chemicals Department,** *Propellent Solvency Values,* Isotron Tech. Inf. File, 3(2), Pennwalt Chemical Corp., Philadelphia, PA, 1964.
170. **Reed, A. B., Jr.,** Solubility parameters and their application to aerosols, *Chem. Spec. Manuf. Assoc. Proc., 46th Ann. Meet.*, Washington, DC, 1959, 94.

171. **Sevestre, J.,** On some correlations involving the solubility parameter "δ", *Double Liaison,* No. 109, 67, 1964.
172. **Lin, J.-S.,** The relationship between the characterization factor "K" and the solubility parameter "δ" of petroleum solvents, *Proc. Natl. Sci. Counc. Repub. China,* 2, 314, 1978.
173. **Stawiszynski, A. and Zawadzki, J.,** Use of the solubility parameter δ for evaluation and selection of organic solvents, *Powloki Ochr.,* 7, 26, 1979.
174. **Godfrey, N. B.,** Solvent selection via miscibility number, *Chemtech,* 2, 359, 1972.
175. **Jefferson Chemical,** *Miscibility Predictor,*© Houston, TX.
176. **Vavruch, I.,** Miscibility of solvents, *Chem. Labor. Betr.,* 35, 385, 1984.
177. **Rothmund, V.,** *Z. Phys. Chem.,* 26, 433, 1898.
178. **Francis, A. W.,** *Critical Solution Temperatures (Adv. Chem. Ser.,* 31), American Chemical Society, Washington, DC, 1961.
179. **Vernon, F.,** The characterization of solute-solvent interactions in gas-liquid chromatography, *Dev. Chromatogr.,* 1, 1, 1978.
180. **Rohrschneider, L.,** A method of characterization for gas chromatographic separation of liquids, *J. Chromatogr.,* 22, 6, 1966.
181. **Rohrschneider, L.,** The polarity of stationary liquid phases in gas chromatography, *Adv. Chromatogr.,* 4, 333, 1967.
182. **Rohrschneider, L.,** Gas chromatographic separation in regular solutions, *J. Gas Chromatogr.,* 6, 5, 1968.
183. **Rohrschneider, L.,** Solvent characterization by gas-liquid partition coefficients of selected solutes, *Analyt. Chem.,* 45, 1241, 1973.
184. **Snyder, L. R.,** Classification of the solvent properties of common liquids, *J. Chromatogr.,* 92, 223, 1974.
185. **Snyder, L. R.,** Solvent selection for separation processes, in *Separation and Purification,* 3rd ed., Perry, E. S. and Weissberger, A., Eds., John Wiley & Sons-Interscience, New York, 1978, 25.
186. **Snyder, L. R.,** Classification of the solvent properties of common liquids, *J. Chromatogr. Sci.,* 16, 223, 1978.
187. **Spychaj, T.,** Solvent/nonsolvent solubility and polarity parameters in fractional separation of oligomers, 1. Epoxy resins, *Angew. Makromol. Chem.,* 149, 127, 1987.
188. **Spychaj, T. and Hamielec, A. E.,** Solvent/nonsolvent solubility and polarity parameters in fractional separation of oligomers. 2. Styrene-acrylic acid copolymers, *Angew. Makromol. Chem.,* 157, 137, 1988.
189. **Bakalyar, S. R., McIlwrick, R., and Roggendorf, E.,** Solvent selectivity in reversed-phase high-pressure liquid chromatography, *J. Chromatogr.,* 142, 353, 1977.
190. **Vavruch, I.,** Solvent effects and solvent selection, *Chem. Labor. Betr.,* 35, 536, 1984.
191. **Kamlet, M. J., Doherty, R. M., Carr, P. W., Mackay, D., Abraham, M. H., and Taft, R. W.,** Linear solvation energy relationships. 44. Parameter estimation rules that allow accurate prediction of octanol/water partition coefficients and other solubility and toxicity properties of polychlorinated biphenyls and polycyclic aromatic hydrocarbons, *Environ. Sci. Technol.,* 22, 503, 1988.
192. **Szymanowski, J., Sobzyńska, A., and Voelkel, A.,** Solubility and polarity parameters of polyoxyethylene glycol dialkyl ethers, *J. Pharm. Sci.,* 77, 893, 1988.
193. **Pytela, O.,** Empirical approach to description of solvent effect on processes in solutions: a review, *Collect. Czech. Chem. Commun.,* 53, 1333, 1988.
194. **Marcus, Y.,** Diluent effects in solvent extraction, *Solvent Extr. Ion Exch.,* 7, 567, 1989.
195. **Rutan, S. C., Carr, P. W., Cheong, W. J., Park, J. H., and Snyder, L. R.,** Re-evaluation of the solvent triangle and comparison to solvatochromic based scales of solvent strength and selectivity, *J. Chromatogr.,* 463, 21, 1989.
196. **Rutan, S. C., Carr, P. W., and Taft, R. W.,** Solvatochromic linear solvation energy relationships for gas-liquid partition coefficients, *J. Phys. Chem.,* 93, 4292, 1989.
197. **Mizerovskii, L. N., Vansyatskaya, L. N., and Smurova, G. I.,** Investigation of the reaction between a ternary ethylene-propylene copolymer and organic liquids, *Vysokomol. Soedin. Ser. A,* 29, 1512, 1987; *Polym. Sci. U.S.S.R.* 29, 1663, 1987.
198. **Meeten, G. H.,** Theoretical basis for Coomber's relationship, *Nature (London),* 223, 827, 1969.
199. **Moura, Ramos, J. J.,** Molecular shape and orientational order. Effects in the energy of cavity formation in liquids, *J. Solution Chem.,* 18, 957, 1989.

Chapter 9

EFFECTS OF PHYSICAL CONDITIONS

In previous chapters no emphasis has been placed on the variation in the cohesion parameter values which occur when such factors as temperature, pressure, and concentration are changed. Although these variations may be safely ignored in many cases, there are some situations where they are significant.

9.1 TEMPERATURE

By definition, the enthalpy of mixing of *regular* solutions is independent of temperature, and for mixtures of hydrocarbons i and j the temperature dependence of the enthalpy of mxing is in fact observed to be small; in terms of Hildebrand parameters this corresponds to ${}^{i}\delta$ and ${}^{j}\delta$ having similar temperature dependences.

It follows from its definition that the Hildebrand parameter decreases with increasing temperature, and to a good approximation over a moderate temperature range for the simpler, nonpolar liquids

$$\mathrm{d} \ln \delta/\mathrm{d}T = -k \tag{1}$$

where k is a positive constant. Hildebrand parameters for hydrocarbons may be fitted also to an equation of the form

$$\delta = \delta_o + mT \tag{2}$$

where m is of the order of -0.03 MPa$^{1/2}$ K^{-1}. Mizerovskii and Vansyatskaya[1] used literature data on enthalpy of vaporization and density to calculate the Hildebrand parameters of 57 organic compounds over a wide temperature range. The information was fitted to Equation 2, with m being a linear function of the reciprocal of the van der Waals molar volume,

$$-10^3\ m/\mathrm{MPa}^{1/2} = 5.4 + 1250\ \mathrm{cm}^3\ \mathrm{mol}^{-1}/V_o \tag{3}$$

Values of these parameters for hydrocarbons are presented in Table 1.

There is a variety of other empirical and semiempirical equations describing the temperature dependence of Hildebrand parameters,[2-5] some of which were mentioned in Chapter 7. They include for two temperatures, 1 and 2, the expression

$$(\delta_2/\delta_1)^2 = (V_1/V_2)^{2.27} + (R/V_2)[T_1(V_1/V_2)^{1.27} - T_2] \tag{4}$$

If $|T_1 - T_2|$ is less than or are equal to 150 K, and if both T_1 and T_2 are below T_b, the second term can be neglected:

$$(\delta_2/\delta_1)^2 = (V_1/V_2)^{2.27}$$

or

$$(\delta_2/\delta_1) = (V_1/V_2)^{1.13} = (\rho_2/\rho_1)^{1.13} \tag{5}$$

If it is assumed that

$$V_2 = V_1[1 + \alpha\ (T_2 - T_1)] \tag{6}$$

TABLE 1
Variation with Temperature of Hildebrand Parameters for Hydrocarbons

Liquid	δ_o/MPa$^{1/2}$	$-10^3\ m$/K^{-1}	T/K range
	Alkanes		
Pentane	22.4	27.0	213 – 313
Hexane	22.0	24.1	213 – 323
Heptane	21.5	21.2	213 – 323
Octane	21.6	20.8	273 – 323
Nonane	21.0	18.3	223 – 323
Decane	21.6	19.8	—
Isobutane, 2-methylpropane	22.4	32.4	213 – 263
Isopentane	22.2	28.0	213 – 298
Isooctane	19.7	19.0	213 – 323
	Cycloalkanes		
Cyclopentane	24.9	27.8	263 – 313
Cyclohexane	24.2	25.0	283 – 333
Methylcyclopentane	23.2	23.8	263 – 313
1,1-Dimethylcyclopentane	23.4	23.3	283 – 313
Ethylcyclopentane	22.8	22.0	283 – 313
Methylcyclohexane	22.2	21.0	273 – 333
	Alkenes		
Propylene	23.9	37.0	113 – 223
1-Butene	23.6	32.6	193 – 263
cis-2-Butene	24.9	33.8	223 – 273
trans-2-Butene	24.1	32.8	223 – 273
Isobutene	23.5	32.0	203 – 263
1-Pentene	23.7	31.0	283 – 303
cis-2-Pentene	23.5	27.7	213 – 303
trans-2-Pentene	23.1	27.2	273 – 303
1-Hexene	22.8	26.4	273 – 323
1-Heptene	22.1	23.0	283 – 313
1-Octene	21.6	20.8	263 – 333
	Cycloalkenes		
Cyclopentene	26.0	20.0	283 – 313
Cyclohexene	24.9	25.0	283 – 313
	Dienes		
Propadiene	26.8	39.0	203 – 233
1,3-Butadiene	24.8	34.0	223 – 268
2-Methyl-1,3-butadiene	25.0	32.5	283 – 303
	Aromatic hydrocarbons		
Benzene	26.2	25.2	280 – 330
Toluene	24.7	21.7	213 – 323
Ethylbenzene	23.9	19.8	213 – 323
Isopropylbenzene	23.3	19.8	273 – 333
Butylbenzene	22.3	18.3	283 – 313
o-Xylene	24.8	21.5	253 – 323
m-Xylene	23.5	18.3	233 – 323
p-Xylene	24.5	22.0	293 – 323

Adapted from Mizerovskii, L. N. and Vansyatskaya, L. N., *Izv. Vyssh. Uchebn. Zaved. Khim. Khim. Tekhnol.*, 31(4), 51, 1988.

TABLE 2
Calculated Variation with Temperature of Hansen Parameters

Liquid	$(d\delta/dT)/MPa^{1/2}\ K^{-1}$		
	$(d\delta_d/dT)$	$(d\delta_p/dT)$	$(d\delta_h/dT)$
Acetone, 2-propanone	−0.023	−0.006	−0.013
1-Propanol	−0.019	−0.003	−0.030

Adapted from Hansen, C. M. and Beerbower, A., *Kirk-Othmer Encyclopedia of Chemical Technology*, 2nd ed., Standen, A., Ed., Interscience, New York, 1971, 889.

where α is the coefficient of expansion, then

$$\delta_2 = \delta_1 [1 + 1.13\alpha(T_1 - T_2)] \tag{7}$$

Alternatively, Watson's relationship (Equation 4, Chapter 7) can be used, and when combined with equations of the above type, yields[5]

$$(\delta_2/\delta_1) = [(1 - T_{r,2})/(1 - T_{r,1})]^{0.34} \tag{8}$$

Jayasri and Yaseen[5] have presented $\delta - T$ nomograms based on this expression. For a variety of hydrocarbons, Bradford and Thodos[6] determined Hildebrand parameters for temperatures up to the critical point and found that they fitted Equation 30, Chapter 7:

$$\delta - \delta_c = k(1 - T_r)^m$$

where m is close to 0.447 for all the hydrocarbons (Table 11, Chapter 7). Calculated values deviated from observed values by an average of only 0.85% for these compounds. The above expressions provide different forms of fitting equations for the temperature dependence of hydrocarbon Hildebrand parameters, and they are also suitable for the dispersion cohesion parameters of all materials.

Tiwari and Pandey[7] considered the temperature dependence of internal pressures of two liquid mixtures (benzene-chlorobenzene and benzene-toluene).

It was shown in Section 5.8 that the orientation cohesion parameter is expected to have a temperature dependence,

$$(\delta_{0,1}/\delta_{0,2})^2 = T_2/T_1 \tag{9}$$

of the same sign and of comparable magnitude to or smaller than that of δ_d. Hansen and Beerbower[8] have reviewed methods of calculating the variation with temperature of δ_o, δ_p, δ_h and their results for two liquids are summarized in Table 2. Experimental values of polar and nonpolar cohesion parameters as functions of temperature appear in Tables 6 to 9, Chapter 5.

For some practical purposes, for processes carried at or near ambient temperatures, the 25°C cohesion parameters can be used quite successfully because functions such as ${}^jV\ {}^i\phi^2\ ({}^i\delta - {}^j\delta)$ remain almost independent of temperature, as predicted by regular solution theory. Also, the entropy-temperature factor in the free energy equation $\Delta G = \Delta H - T\Delta S$ is of much greater importance than the small temperature dependence of the enthalpy (cohesion parameter) term.

However, it should be noted that the effect on cohesion parameters of changing tem-

TABLE 3
The Dependence of Hildebrand Parameter on Inverse Molar Volume

Class of compound	Slope/MPa$^{1/2}$ mol cm^{-3}	Intercept/MPa$^{1/2}$
Aliphatic hydrocarbons	−294.6	17.1
Ethers	−208.1	17.6
Halogenated liquids	−77.7	17.7
Polychlorinated liquids	118.8	17.5
Aromatic hydrocarbons	225.7	16.2
Ketones	304.1	15.6
Formates	310.2	15.1
Acetates	351.0	14.6
Butyrates	344.3	14.7
Mixed alcohols	332.4	18.3
Anilines	409.1	18.8
Nitriles	497.4	15.0
Aliphatic 1-alcohols	515.2	17.9
Cellosolves®	864.3	12.5
Carbitols®	1228.5	11.6

Adapted from Jayasri, A. and Yaseen, M., *J. Coatings Tech.*, 52 (667), 41, 1980.

perature can be utilized to provide a cohesion parameter spectrum (Sections 4.5 and 14.2), and also that there are other cases, particularly in liquid/vapor situations (Chapter 11) where the temperature dependence of δ values is even more important. For example, in considering the factors involved in the choice of mixed refrigerants, Slinn[9] plotted Hildebrand parameter-temperature data through the normal boiling points and pointed out that azeotropes are expected where the Hildebrand parameters are different but boiling points are similar. Values of cohesion parameters at higher temperatures are now becoming available from gas chromatographic studies (Section 18.3) exemplified by the Hansen parameters at 80°C of the alkyl derivatives of phenol and resorcinol (Table 19, Chapter 5).

9.2 DENSITY, VOLUME AND PRESSURE

By the way they are defined, cohesion parameters depend strongly on molar volume. This is particularly noticeable in a comparison between aliphatic hydrocarbons and fluorocarbons, the latter having larger molar volumes and consequently smaller cohesion parameters.[10] The effect of molar volume can be significant, but is often neglected, partly because molar volumes of common solvents tend to be rather similar, and partly to reduce the number of parameters for simplicity.

In view of the relationships (such as Equation 17) in Chapter 6, it is not surprising that plots of cohesion parameter against inverse molar volume are found to be approximately linear for particular classes of compound. Jayasri and Yaseen[11] have used this fact to assist in the determination of each liquid of a single recommended Hildebrand parameter from the wide range of reported values (Table 18, Chapter 6). The slopes are positive, that is, the Hildebrand parameters decrease with increasing molar volume in all series except aliphatic hydrocarbons, ethers, and halogenated solvents (Table 3). Peiffer[12] was able to correlate Hansen parameters with van der Waals volumes (which are discussed in Section 6.5).

Also significant in the use of cohesion parameters to evaluate energy and activity terms is the fact that they must be multiplied by a volume, to give a term of the form ${}^jV({}^i\delta - {}^j\delta)^2$. When ${}^i\delta$ and ${}^j\delta$ are similar, the effects of terms such as this are often unimportant, but in other cases where ${}^i\delta$ and ${}^j\delta$ are significantly different, such terms may predominate.

This shows up well in Equations 82 to 85, Chapter 7, and is illustrated by the variation in the extent of interactions of lower aliphatic hydrocarbons (which have similar δ values but different molar volumes) with polar materials.[13]

The effect of volume difference also appears in the entropy term, and is most noticeable for polymer-liquid systems. For this purpose the Flory-Huggins theory provides the appropriate entropy of mixing term (Sections 4.2 and 13.2), although it tends to over-correct for volume differences. Helpinstill and Van Winkle[14] found it necessary in predicting activity coefficients to include the Flory-Huggins correction if the molar volumes differed by more than 5%; but for the less demanding application of solvent selection criteria it is not required unless the molar volumes differ substantially.

The most noticeable effect of an increase in *pressure* is the indirect effect of a decrease in molar volume, which has the result described above. The direct effect of pressure is relatively unimportant and has received little attention. Compressibility measurements can provide values of internal pressure (Section 7.5).

As applied to supercritical phases, the regular solution concept leads to[15-18]

$$\delta = 1.25\ \rho_c^{1/2}\ (\rho/\rho_1) \tag{10}$$

where p_l is the reference density in the liquid state; properties of supercritical systems are considered further in Sections 11.3, 11.4, and 18.5.

9.3 CONCENTRATION

One of the important inconsistencies in the conditions under which different sets of solute cohesion parameters are determined and used is the concentration at which the observation is made, and this is particularly noticeable in the case of polymers.[19]

Formally, from Equation 31, Chapter 4, it is clear that if account is not taken of ${}^j\phi$, the apparent measured value of ${}^j\delta$ is incorrect. In practice there are also other effects. Equations such as 14 or 15, Chapter 4, contain additional volume fraction terms, introducing other concentration contributions. The energy of mixing has a maximum value when ${}^i\phi = {}^xj\phi = 0.5$, and at high ${}^j\phi$ values (which tend to be those commercially used) the energy of mixing is much smaller. However, entropy effects in polymer systems may cause concentration to produce phase separation at dilute concentrations during clean-up operations.

The composition dependence of cohesion parameters for mixed liquids is discussed elsewhere, particularly in Sections 4.4 and 7.7. Mizerovskii and Belova[20] used the equation of Wrewsky[21] for the composition dependence of the vaporization enthalpy of binary liquid mixtures (and of random copolymers). Current equations were shown to result in unacceptable errors in calculation when applied to binary mixtures involving benzene, tetrachloromethane, acetone, hexane, and trichloromethane.

REFERENCES

1. **Mizerovskii, L. N. and Vansyatskaya, L. N.,** Temperature dependence of the solubility parameter of organic liquids, *Izv. Vyssh. Uchebn. Zaved., Khim. Khim. Tekhnol,* 31(4), 51, 1988.
2. **Bowden, S. T. and Jones, W. J.,** Latent heat of vaporization and composition, *Philos. Mag.,* 39, 155, 1948.
3. **Fedors, R. F.,** A method for estimating both the solubility parameters and molar volumes of liquids, *Jet Propul. Lab. Quart. Tech. Rev.,* 3(1), 45, 1973; *Polym. Eng. Sci.,* 14, 147, and 472, 1974.
4. **Jayasri, A. and Yaseen, M.,** Heat of vaporization: its determination and application in evaluation of the solubility parameter: a review, *Progr. Org. Coatings,* 7, 167, 1979.

5. **Jayasri, A. and Yaseen, M.,** Nomograms for solubility parameter at 25°C and at different temperatures, *J. Coat. Technol,* 52(667), 41, 1980.
6. **Bradford, M. L. and Thodos, G.,** Solubility parameters of hydrocarbons, *Can. J. Chem. Eng.*, 44, 345, 1966.
7. **Tiwari, V. and Pandey, J. D.,** Interaction studies in benzene-chlorobenzene and benzene-toluene through internal pressure and solubility parameter, *Z. Phys. Chem. (Leipzig),* 262, 53, 1981.
8. **Hansen, C. M. and Beerbower, A.,** Solubility parameters, in *Kirk-Othmer Encyclopedia of Chemical Technology,* 2nd ed., Suppl. Vol., Standen, A., Ed., Interscience, New York, 1971, 889.
9. **Slinn, D. S. L.,** Factors in the choice of refrigerants for air conditioning, *Refrig. Air Cond.*, 74(880), 49, 1971.
10. **Reed, T. M., III,** The theoretical energies of mixing for fluorocarbon-hydrocarbon mixtures, *J. Phys. Chem.*, 59, 425, 1955.
11. **Jayasri, A. and Yaseen, M.,** Solubility parameter values suggested using the reported and calculated values for organic compounds, *J. Oil Col. Chem. Assoc.* 63(2), 61, 1980.
12. **Peiffer, D.,** The molecular factors affecting the solubility parameter, *J. Appl. Polym. Sci.*, 25, 369, 1980.
13. **Reynolds, W. W. and Larson, E. C.,** The fundamentals of solvent selection, *Off. Dig. Fed. Soc. Paint Technol.*, 34, 311, 1962.
14. **Helpinstill, J. G. and Van Winkle, M.,** Prediction of infinite dilution activity coefficients for polar-polar binary systems, *Ind. Eng. Chem. Proc. Des. Dev.*, 7, 213, 1968.
15. **Hildebrand, J. H. and Scott, R. L.,** *Solubility of Non-Electrolytes,* 3rd ed., Reinhold, New York, 1950; Dover, New York, 1964.
16. **Giddings, J. C., Myers, M. N., McLaren, L., and Keller, R. A.,** High pressure gas chromatography of nonvolatile species, *Science,* 162, 67, 1968.
17. **Giddings, J. C., Myers, M. N., and King, J. W.,** Dense gas chromatography at pressures to 2000 atmospheres, *J. Chromatogr. Sci.*, 7, 276, 1969.
18. **Leyendecker, D.,** Selection of conditions for an SFC (Supercritical Fluid Chromatography) separation, in *Supercritical Fluid Chromatography* (RSC Chromatography Monograph), Smith, R. M., Ed., Royal Society of Chemistry, London, 1988, chap. 3, 53.
19. **Burrell, H.,** The challenge of the solubility parameter concept, *J. Paint Technol.*, 40, 197, 1968.
20. **Mizerovskii, L. N. and Belova, O. V.,** Calculation of the solubility parameter of binary mixtures of liquids and copolymers, *Vysokomol. Soedin. Ser. A.*, 29, 1236, 1987.
21. **Wrewsky (Vrevskii), M. S.,** Theoretical and experimental investigations on the vaporization of binary mixtures. II. Theory of vaporization, *Z. Phys. Chem. A.*, 144, 359, 1929.

Chapter 10

KINETIC AND TRANSPORT PROCESSES

10.1 KINETIC SOLVENT EFFECTS

Many varied empirical scales are available for the description of the effect of solvents on chemical processes in solution. Qualitative similarities exist between these scales, which are based on electrostatic, thermodynamic, kinetic, or spectroscopic properties, but simple quantitative relationships between them should not be expected as each scale is the resultant of many molecular factors combined in a unique way.

Probably most well-known are the linear free energy relationships (see also Section 7.12), which have the form

$$\log {}^{B}k_i = m\log {}^{A}k_i + c \tag{1}$$

where ${}^{A}k$ and ${}^{B}k$ are the rate coefficients (or equilibrium constants) of two reaction series, A and B, which are subject to the same variation (designated by i) in the structure of the reactants or the nature of the surrounding medium. Because the logarithm of the rate coefficient is proportional to the Gibbs free energy of activation, Equation 1 describes a linear relationship between free energies. The Hammett equation[1-3] is

$$\log^{RX} k - \log^{RH} k = \sigma\rho \tag{2}$$

in which ${}^{RX}k$ is the rate constant of substituted aromatic substrates in a particular reaction. This is compared with the unsubstituted rate constant ${}^{RH}k$ by means of the substituent constant σ and the reaction constant ρ. The same type of equation can be applied to changes in reaction medium, with σ being the medium constant, and it is this factor which can be discussed in terms of cohesion parameters.

There are numerous reports and reviews* of correlations between cohesion parameter solvent scales, other solvent parameters (such as those described in Chapter 8) and kinetic data, and examples are listed in Section 10.3.

10.2 TRANSITION STATE THEORY

The transition state theory of reaction rates assumes the transitory formation of an activated complex (usually designated by the symbol ‡) from the reactants, and its subsequent rearrangement into products. Pseudo-thermodynamic properties such as volume, enthalpy, and Hildebrand parameter can be attributed to this activated state, and an expression may be derived** for the rate constant of a reaction involving one or two reactants r in solvent i relative to the rate in a standard solvent o; specific solvent effects are neglected. This is obtained by combining the Brönsted-Bjerrum equation for the rate constant's terms of activity coefficients,

* For example: Abraham;[4,5] Buchtova;[6] Carr;[7] Chapman, Dack, Shorter and co-workers;[2,3,8-16] Eckert and co-workers;[17-21] Gardon;[22,23] Griffiths and Pugh;[24,25] Grunwald and Winstein;[26] Herbrandson and Neufeld;[27] Kamachi;[28] Kamlet, Abboud, and Taft;[29] Koppel and Palm;[30] Kosower;[31-32] Lay;[34] Mancini et al.;[35] Pearson;[36,37] Reichardt;[38-42] and Winstein et al.[43]

** For example: Abraham;[4] Dack;[13,16] Eckert and co-workers;[19,20] Frost and Pearson;[36,44] Glasstone;[45] Laidler and Eyring;[46,47] Reichardt;[40] and Tanaka.[48]

$$k_x = \frac{{}^{o}k_x \, \Pi_r \, {}^{r}f_x}{{}^{+}f_x} \tag{3}$$

with the regular solution expression for activity coefficients,

$$RT \ln f_x = V(\delta - \bar{\delta})^2 \tag{4}$$

$$RT \ln ({}^{i}k_x/{}^{o}k_x) = \Sigma_r \, {}^{r}V({}^{r}\delta - {}^{i}\delta)^2 - {}^{+}V({}^{+}\delta - {}^{i}\delta)^2 \tag{5}$$

or

$$RT \ln {}^{i}k_x = \text{constant} + \Sigma_r \, {}^{r}V({}^{r}\delta - {}^{i}\delta)^2 - {}^{+}V({}^{+}\delta - {}^{i}\delta)^2 \tag{6}$$

The rate constants should be expressed in mole fraction units rather than the more common concentration units. The volume of activation, defined by

$$\Delta V^{+} = {}^{+}V - \Sigma_r \, {}^{r}V \tag{7}$$

may be determined experimentally by the effect of external pressure on the reaction rate constant,

$$(\delta \ln k_x/\delta p)_T = -\Delta V^{+}/(RT) \tag{8}$$

The activation volume is positive for a reaction where there is an increase of volume and loosening of chemical bonds in achieving the activated complex structure, and here it is reasonable to expect $\Sigma_r \, {}^{r}\delta > {}^{\ddagger}\delta$. For $\Delta V^{\ddagger}$ negative, there is a compression of intermolecular bonds in the reactants, and $\Sigma_r \, {}^{r}\delta < {}^{\ddagger}\delta$.

In one of the limiting cases, that of nonionic reactions involving molecules of low polarity, ${}^{\ddagger}V$ does not differ significantly from the molar volumes of the reagents, and a linear relationship is predicted between $\ln {}^{i}k_x$ and ${}^{i}\delta$. For example, in the case of unimolecular reactions, Equation 5 for two solvents i and j yields

$$\ln {}^{i}k_x/{}^{j}k_x = P({}^{i}\delta^2 - {}^{j}\delta^2) + Q({}^{i}\delta - {}^{j}\delta) \tag{9}$$

where

$$P = ({}^{r}V - {}^{+}V)/RT = -\Delta V^{+}/(RT) \tag{10}$$

and

$$Q = (2 \, {}^{+}V \, {}^{+}\delta - 2 \, {}^{r}V \, {}^{r}\delta)/(RT) \tag{11}$$

If $\Delta V^{\ddagger}$ is small, this leads to a linear relationship between $\ln {}^{i}k^{x}$ and ${}^{i}\delta$ for different i. (For the other limiting case of ionic or very polar reactants, electrostriction effects are great, the dominant factor is the difference between $\Sigma_r \, {}^{r}V$ and ${}^{\ddagger}V$, and the contribution of cohesion energy terms is small.)

The correlation between Hildebrand parameters and volumes of activation can also be understood by recalling that the activation volume is determined by varying the *external* pressure, while the Hildebrand parameter is closely related to the *internal* pressure of the system. For nonpolar reactions, and for polar reactions in nonpolar solvents, the internal pressure of solvents influence reaction rates in the *same direction* as external pressure.

However, for polar or ionic processes the effect of solvent is more complex, and the cohesion parameter concept is less useful.

Relations such as Equation 9 can be used to estimate $^{\ddagger}\delta$, the Hildebrand parameters of activated complexes. To some extent the solvent is being used as a probe in the region surrounding the activated complex, which is the aim of most kinetic and mechanistic studies. The activated complex tends to have properties which approach those of the products of the reactions. The general rule[45,49] is that if the reaction is one in which the products are of higher cohesion (higher internal pressure or cohesive pressure) than the reactants, it is accelerated by solvents of high cohesion. This is justified by a consideration of Equation 5. Conversely, if the solvent is similar to the reactants in cohesion properties, the rate tends to be lower.

The activation energy, E, defined by the empirical expression for the temperature dependence of reaction rate constants,

$$k = A \exp(-E/RT) \tag{12}$$

or arising from transition state theory, is related to the Hildebrand parameters by

$$^{i}E = \text{constant} + \Sigma_r \, ^{r}V(^{r}\delta - {}^{i}\delta)^{1/2} \tag{13}$$

and in some cases this simplifies to a linear relation between ^{i}E and $^{i}\delta$. Such a relationship was observed by Halpern et al.[50] for the unimolecular, nonionic *cis-trans* isomerization of azobenzene in 16 solvents. The more polar solvents showed greater scatter, but only water deviated significantly. This equation was used for ethyl radical reactions by Stefani[51] and for growth kinetics of environmental crazing in polycarbonate.[149]

Correlations based on regular solution theory seem to work surprisingly well, even in relatively polar systems, and this appears to be because the *ratio* of activity coefficients $^{r}f_x/^{\ddagger}f_x$ is nearly constant, although the individual activity coefficient values vary greatly from solvent to solvent.[17,20] However, rates in mixed solvents tend to show considerable deviations from what is expected on cohesion parameter considerations, and one of the reasons is that preferential solvation of the reactants and/or the activated complex by one solvent is likely to occur.

10.3 RATES AND MECHANISMS

Systems in which rate studies in different solvents have been correlated with cohesion parameters (usually Hildebrand parameters) include the Menschutkin reactions between: pyridine and benzyl bromide;[1,52] pyridine and iodomethane;[51] triethylamine and iodoethane;[18,51] and trimethylamine with iodoethane.[48] In these reactions, which are of the type

$$R_3N + R'X \longrightarrow [R_3N...R'...X]^{\ddagger} \longrightarrow R_3R'N^+X^- \tag{14}$$

the transition states are very polar, but not ionic, and the activated complexes have estimated Hildebrand parameters of the order of 30 $MPa^{1/2}$. Murai and Kimura[53] found a better correlation of log k with the Hildebrand parameter than with $(\epsilon - 1)/(2\epsilon + 1)$ for aliphatic alcohols and aprotic liquids except tetrahydrofuran,[1,4] 1,4-dioxane and benzene for a Menschutkin reaction (quarternization of pyridine with dodecyl bromide, $C_{12}H_{25}Br$).

Other systems considered in terms of cohesion parameters are

- Amination of benzyl chloride[54]
- Heterogeneous hydrogenation of toluene[55]
- Racemization of 1,1′-binaphthyl[56]

- Thermal decomposition of di-tert-butyl peroxide[144], in which non-regular behavior was observed
- Combination and disproportionation of ethyl radicals in solution and cycloaddition reactions of methyl acrylate with cyclopentadiene[51]
- *cis-trans* Isomerization of azobenzene[50]
- Diels-Alder addition reactions of maleic anhydride with chloroprene,[20,57,58] 1,3-butadiene,[4,7,57] and isoprene;[18,57-59] of acrylonitrile with isoprene;[17,20,57,61] and of tetracyanoethylene with anthracene derivatives[62]
- Activation enthalpies of solvent exchange at bivalent transition metal ions and the formation and dissociation of nickel complexes[63]
- Micellar catalysis[143]
- Catalysis by covalently immobilized colloidal metals[145]
- Application of the Kamlet-Abboud-Abraham-Taft approach (Section 8.4), including a Hildebrand parameter term, to *tert*-butyl halide solvolysis rates[5,64] electron transfer thermodynamics and kinetics[34] and other media effects,[6] the contribution of the δ term being frequently of minor importance
- Aromatic nucleophilic substitution (2,4-dinitrochlorobenzene with piperidine in aprotic solvents): low correlation with Hildebrand parameter[35]
- Quaternary ammonium and phosphonium salts as phase transfer catalysts[65] (Equation 98, Chapter 7, and Section 12.3)
- Alkylation of KCN by solid-liquid phase transfer catalysis[66]
- *N*-Acetylation of chitosan[67]
- The anion role in the dissociation of sodium dibenzo-24-crown-8 in nitromethane[146]
- The effect of solvent Hildebrand parameter on retention of immobilized-cell activity in microbial epoxidation of propylene and 1-butene;[68] in choosing a water-immiscible organic solvent for multiphase biocatalysis, a low Hildebrand parameter and high molecular weight tend to lead to a high activity retention of the free or immobilized cells
- Biochemical structure-activity relationships (Chapter 19)

10.4 POLYMERIZATION AND POLYMER REACTIONS

Control can be exerted on the rates and products of polymerization reactions by varying the cohesion parameter of the solvent.[69-72] Some polymerizations and copolymerizations are faster in poor solvents than in good solvents, and the cohesion parameters of initiators may be important if there is preferential solvation.

An additional contribution can be made by cohesion parameters in planning the synthesis of block and graft copolymers: the nature of the solvents used for preparation and recovery can have a considerable effect on the structure, properties, and rate of formation of the copolymer.

This section lists examples of applications of cohesion parameters to polymerization, copolymerization, grafting, crosslinking, and other polymer reactions; polymer cohesion parameters themselves and their implications are considered in detail in Chapters 13 to 16. The effects of liquids on polymer surface properties are discussed in Chapter 17.

- Production of long-lived macroradicals during heterogeneous polymerization in poor solvents[72-74]
- Copolymerization from binary mixtures of monomers[139]
- Microlatex formation by emulsion polymerization[148]
- Acrylonitrile-butadiene-styrene copolymerization: preferential solvation[75]
- Acrylonitrile-styrene emulsion copolymerization[76,139]
- Porosity in divinylbenzene-styrene copolymerization[142]

- Acrylonitrile-vinyl bromide[77]
- Polybutadiene double bond reactivity[78]
- Butadiyne polymer[79]
- Butyl acrylate[80] and dibutyl itaconate[147] radical polymerization
- Acetylation of cellulose[81] and the reaction of cellulose with epoxy compounds[82]
- Decamethylenediamine polyamides[83]
- 3-Hydroxybenzoates, linear and branched[84]
- Maleic anhydride-vinyl monomer[85,86]
- Methyl methacrylate-acrylonitrile copolymerization[73]
- Epoxidation of silk[87]
- Styrene and styrene-divinylbenzene[88-93]
- Styrene-acrylic[94,95]
- Styrene triblock and star copolymers[96,97,136]
- Styrene[98-100] and trifluorostyrene[137] graft polymerization
- Trimethylolpropane trimethacrylate (TRIM) (effect on macroporosity and residual unreacted C = C)[101]
- Polyurethane[102,103]
- Vinyl acetate heterogeneous solution polymerization[104,105]
- Vinyl acetate-alkyl acrylate ester emulsion copolymers[106]
- Vinyl acetate-ethylene and vinyl alcohol-ethylene copolymerization[107,108]
- Vinyl esters[109]
- Vinyl bromide polymerization[77]
- Vinyl fluoride radical polymerization[110]
- Group-transfer polymerization of methacrylates and acrylates[138]
- Epoxidation of wool fiber[111]
- Hydrogenation reactions catalyzed by polymer-anchored rhodium[112]
- The effect of solvents on radiation grafting[113-118] and cross-linking: swelling solvents tend to promote efficient grafting[119]
- Radiation grafting onto wool[120]
- Dispersion polymerization[91,92,94] (See also dispersions in Section 17.8)
- Rate of soft polymer gelation[121]
- Polypentenamers from cyclopentene and isobutane[133]
- Spray-drying for polymer particle encapsulation[134]
- Polymerization in slurries of poly(ethyl methacrylate) in methyl methacrylate[141]

10.5 VISCOSITY

Viscous flow can be regarded as a rate process in which molecules move into holes or voids in the liquid. On this model, the activation energy for viscous flow, E_η, is related to the energy required to form such a hole, that is, to the vaporization energy, and the value of the ratio $(-U/E_\eta)$ is a useful parameter.[122-124] Both transition state theory and regular solution theory may be applied to the transport process, and it is also possible to relate internal pressure to transport properties.[125,126]

Gordon[22,23] has used the viscous properties of molten salts in a study of their cohesive nature, along with Hildebrand parameters and other measures of cohesion. The quantity (E_η/V) can be correlated to some extent with $(-U/V)$, although discrepancies occur, of course. Cohesion parameters also have been used in the description of the viscous properties of solutions. One successful application was the relationship deduced between Hildebrand parameters and the initial slopes of plots of dilute solution relative viscosities against mole fraction. This was applied to chromium (III) acetylacetonate and iodine[127] in several organic solvents, the Hildebrand parameters of the solutes being evaluated and providing evidence

TABLE 1
Coefficients in the Linear Relation Between Hildebrand Parameter and the Natural Logarithm of the Viscosity Activation Energy

Class of substance	α/MPa	β/MPa
Alkane	−193	−109
Bromoalkane	641	81
Nitrile	6080	1431
Carboxylic acid	4176	898
Alcohol	6051	1293

Adapted from Rätzsch, M. T. and Friedel, B., *Z. Phys. Chem. (Leipzig)*, 256, 829, 1975.

on the nature of the solute and the degree of solvation in different solvents. Makitra and Vasyutin[132] found that the linear correlation between E_η and cohesive energy density held only for *n*-alkanes and alkyl derivatives. Röthlisberger et al.[135] considered viscosities of binary water-aprotic and water-protic polar systems in terms of various parameters, including Hildebrand parameters.

Rätzsch and Friedel[128] correlated liquid viscosity coefficients with thermodynamic properties and calculated Hildebrand parameters on the basis of viscosity activation energies from the linear relationship

$$\delta = (\alpha - \beta \ln E)_\eta^{1/2} \tag{15}$$

Values of α and β Appear in Table 1.

Correlations between viscosity and Hildebrand parameters also have been widely used for solutions of polymers; these are presented in Section 14.7.

In a very different kind of system, the dropping points of greases made from lithium stearate soap and oil have been shown to be a function of the Hildebrand parameter of the oil used in the formulation.[129] Apparent viscosities of Carbopol® 941 dispersions in hydroalcoholic mixtures have been investigated;[130] δ was found to be 38 $MPa^{1/2}$. Hildebrand parameters have been applied to the prediction of viscosities in other solid-liquid systems, such as cements.[131]

Other aspects of wetting, adhesion, and lubrication are dealt with in Section 17.5.

10.6 OTHER TIME-DEPENDENT PROPERTIES

Other transport properties may be correlated with cohesion parameters in the same way as viscosity:

- Gas, liquid, and solid diffusion and permeation in polymers (Section 16.6)
- Reverse osmosis membranes (Sections 16.6 and 18.6)
- Swelling kinetics of cellulose and other natural fibers and lignin (Section 16.11)
- Time-dependent mechanical properties of polymers (Section 16.2)
- Biological transmission (Section 19.2)
- Spray-drying processes (Section 10.4)
- Initial solvent extraction rates[140]

For polymeric systems, Flory-Huggins terms are required for the nonideal entropy associated with the dissimilar molecular sizes, in addition to the cohesion parameter enthalpy terms. In some cases, the rate of chemical reaction is determined by a time-dependent transport property, often diffusion rate. Thus, heterogeneous polymerization, where the polymer is insoluble in the monomer or solvent, has its rate controlled by the polymer swelling rate, which can be correlated by means of Hildebrand parameters.

REFERENCES

1. **Hammett, L. P.,** *Physical Organic Chemistry,* McGraw-Hill, New York, 1940.
2. **Shorter, J.,** The separation of polar, steric and resonance effects in organic reactions by the use of linear free energy relationships, *Quart Rev.,* 24, 433, 1970.
3. **Chapman, N. B. and Shorter, J., Eds.,** *Advances in Linear Free Energy Relationships,* Plenum Press, London, 1972.
4. **Abraham, M. H.,** Solvent effects on transition states and reaction rates, *Prog. Phy. Org. Chem.,* 11, 1, 1974.
5. **Abraham, M. H., Doherty, R. M., Kamlet, M. J., Harris, J. M., and Taft, R. W.,** Linear solvation energy relationships. Parts 37 and 38. An analysis of contributions of dipolarity-polarisability, nucleophilic assistance, electrophilic assistance, and cavity terms to solvent effects on t-butyl halide solvolysis rates; An analysis of the use of solvent parameters in the correlation of rate constants, with special reference to the solvolysis of t-butyl chloride, *J. Chem. Soc. Perkin Trans. II,* 913 and 1097, 1987.
6. **Buchtová, M., Koutek, B., and Bekárek, V.,** Study of the use of Hildebrand's solubility parameter in the evaluation of media effects, *Acta Univ. Palacki. Olomuc. Fac. Rerum Nat.,* 82 (Chem.-24), 67, 1985.
7. **Carr, P. W.,** Fundamental, predictive approach to dipole-dipole interactions based on the dipole moment of the solute and the dielectric constant of the solvent, *J. Chromatogr.,* 194, 105, 1980.
8. **Chapman, N. B., Dack, M. R. J., and Shorter, J.,** The separation of polar and steric effects. XII. The influence of protic and aprotic solvents on the kinetics of the reactions of benzoic acid with diazodiphenylmethane, *J. Chem. Soc. B,* 834, 1971.
9. **Chapman, N. B., Dack, M. R. J., Newman, D. J., Shorter, J., and Wilkinson, R.,** The influence of solvent on organic reactions. I. Kinetics and mechanism of the reactions of carboxylic acids with diazodiphenylmethane in donor aprotic solvents, *J. Chem. Soc. Perkin Trans. II,* 962, 1974.
10. **Chapman, N. B., Dack, M. R. J., Newman, D. J., Shorter, J., and Wilkinson, R.,** Kinetics and mechanism of the reactions of carboxylic acids with diazodiphenylmethane in aprotic solvents in which the acids may be dimeric, *J. Chem. Soc. Perkin Trans. II,* 971, 1974.
11. **Dack, M. R. J.,** Organic solvents — their effect on chemical reactivity, *Chem. Br.* 6, 347, 1970.
12. **Dack, M. R. J.,** Solvent effects and reactivities, *Chemtech,* 1, 108, 1971.
13. **Dack, M. R. J.,** The influence of solvent on chemical reactivity. An alternative approach, *J. Chem. Educ.,* 51, 231, 1974.
14. **Dack, M. R. J.,** The importance of solvent internal pressure and cohesion to solution phenomena, *Chem. Soc. Rev.,* 4, 211, 1975.
15. **Dack, M. R. J.,** Solvent structure. The use of internal pressure and cohesive energy density to examine contributions to solvent-solvent interactions, *Aust. J. Chem.,* 28, 1643, 1975.
16. **Dack, M. R. J.,** The influence of solvent on chemical reactivity, in *Solutions and Solubilities,* Dack, M. R. J., Ed., Part II, John Wiley & Sons-Interscience, New York, 1976, chap. 11.
17. **Wong, K. F. and Eckert, C. A.,** Solution thermodynamics and kinetic solvent effects on a Diels-Alder reaction, *Trans. Faraday Soc.,* 66, 2313, 1970.
18. **Snyder, R. B. and Eckert, C. A.,** Chemical kinetics at a critical point, *Am. Inst. Chem. Eng. J.,* 19, 1126, 1973.
19. **Eckert, C. A., Hsieh, C. K., and McCabe, J. R.,** Molecular thermodynamics for chemical reaction design, *Am. Inst. Chem. Eng. J.,* 20, 20, 1974.
20. **Eckert, C. A.,** Molecular thermodynamics of reactions in solution, in *Solutions and solubilities,* Part I, Dack, M. R. J., Ed., John Wiley & Sons-Interscience, New York, 1975, chap. 1.
21. **Eckert, C. A.,** Introduction to theories of solution and application to phase equilibria, in *Solutions and Solubilities,* Part II, Dack, M. R. J., Ed., John Wiley & Sons-Interscience, New York, 1976, chap. 9.

22. **Gordon, J. E.,** Fused organic salts. IV. Characterization of low melting point quarternary ammonium salts. Phase equilibria for salt-salt and salt-nonelectrolyte systems. Properties of the liquid salt medium, *J. Am. Chem. Soc.,* 87, 4347, 1965; Applications of fused salts in organic chemistry, *Tech. Meth. Org. Organomet. Chem.,* 1, 51, 1969.
23. **Gordon, J. E.,** Transition energies for a merocyanine dye in aqueous electrolyte solutions. Solvent polarity indicator transition — internal pressure relations, *J. Phys. Chem.,* 70, 2413, 1966.
24. **Griffiths, T. R. and Pugh, D. C.,** Solvent polarity studies. Part 1 New *Z* values and relationships with other solvent polarity scales, *J. Solution Chem.,* 8, 247, 1979.
25. **Griffiths, T. R. and Pugh, D. C.,** Correlations among solvent polarity scales, dielectric constant and dipole moment, and a means to reliable predictions of polarity scale values from current data, *Coord. Chem. Rev.,* 29, 129, 1979.
26. **Grunwald, E. and Winstein, S.,** The correlation of solvolysis rates, *J. Am. Chem. Soc.,* 70, 846, 1948.
27. **Herbrandson, H. F. and Neufeld, F. R.,** Organic reactions and the critical energy density of the solvent. The solubility parameter, δ, as a new solvent parameter, *J. Org. Chem.,* 31, 1140, 1966.
28. **Kamachi, M.,** Influence of solvent on free radical polymerization of vinyl compounds, *Adv. Polym. Sci.,* 38, 55, 1981.
29. **Kamlet, M. J., Abboud, J.-L. M., and Taft, R. W.,** An examination of linear solvation energy relationships, in *Progress in Physical Organic Chemistry,* Vol. 13, Taft, R. W., Ed., John Wiley & Sons, New York, 1981, 485.
30. **Koppel, I. A. and Palm, V. A.,** The influence of the solvent on organic reactivity, in *Advances in Linear Free Energy Relationships,* Chapman, N. B. and Shorter, J. Eds., Plenum Press, London, 1972, chap. 5.
31. **Kosower, E. M.,** The effect of solvent on spectra. I. A new empirical measure of solvent polarity: *Z*-values; II. Correlation of spectral absorption data with *Z*-values; III. The use of *Z*-values in connection with kinetic data, *J. Am. Chem. Soc.,* 80, 3253, 3261, and 3267, 1958.
32. **Kosower, E. M.,** Reactions through charge-transfer complexes, *Prog. Phys. Org. Chem.,* 3, 81, 1965.
33. **Kosower, E. M.,** *An Introduction to Physical Organic Chemistry,* John Wiley & Sons, New York, 1968.
34. **Lay, P. A.,** Specific hydrogen-bonding contributions to the thermodynamics and kinetics of electron transfer. Modifications to the classical and semiclassical theories of electron transfer, *J. Phys. Chem.,* 90, 878, 1986.
35. **Mancini, P. M. E., Martinez, R. D., Vottero, L. R., and Nudelman, N. S.,** Solvent effects on aromatic nucleophilic substitution. Part 3. The kinetics of the reaction of 2,4-dinitrochlorobenzene with piperidine in aprotic solvents, *J. Chem. Soc. Perkin Trans. II,* 1133, 1984.
36. **Pearson, R. G.,** Influence of the solvent on rates of ionization, entropies, and heats of activation, *J. Chem. Phys.,* 20, 1478, 1952.
37. **Pearson, R. G.,** The influence of the reagent on organic reactivity, in *Advances in Linear Free Energy Relationships,* Chapman, N. B. and Shorter, J., Eds., Plenum Press, London, 1972, 281.
38. **Reichardt, C.,** Solvent scales and chemical reactivity, in *Organic Liquids: Structure, Dynamics, Chemical Properties,* Buckingham, A. D., Lippert, E., and Bratos, S., Eds., John Wiley & Sons, Chichester, 1978, chap. 16.
39. **Reichardt, C.,** Empirical parameters of solvent polarity as linear free-energy relationships, *Angew. Chem.,* 91, 119, 1979; *Angew. Chem. Int. Ed.,* 18, 98, 1979.
40. **Reichardt, C.,** *Solvent Effects in Organic Chemistry* (Monographs in Modern Chemistry, Vol. 3), Verlag Chemie, Weinheim, 1979.
41. **Reichardt, C.,** Empirical parameters of solvent polarity and chemical reactivity, in *Molecular Interactions,* Ratajczak, H. and Orville-Thomas, W. J., Eds., John Wiley & Sons, Chichester, 1982, chap. 5, 241.
42. **Reichardt, C.,** Solvent effects on chemical reactivity, *Pure Appl. Chem.,* 54, 1867, 1982.
43. **Smith, S. G., Fainberg, A. H., and Winstein, S.,** Correlation of solvolysis rates. IX. *p*-Methoxyneophyl toluenesulfonate in a variety of solvents. Ionizing power of hydroxylic and nonhydroxylic solvents, *J. Am. Chem. Soc.,* 83, 618, 1961.
44. **Frost, A. A. and Pearson, R. G.,** *Kinetics and Mechanism,* 2nd ed., John Wiley & Sons, New York, 1961.
45. **Glasstone, S.,** The "activated complex" theory and the influence of solvents on reaction velocity, *J. Chem. Soc.,* 723, 1936.
46. **Laidler, K. J. and Eyring, H.,** Effect of solvents on reaction rates, *Ann. N.Y. Acad. Sci.,* 39, 303, 1940.
47. **Laidler, K. J.,** *Chemical Kinetics,* 2nd ed., McGraw-Hill, New York, 1965.
48. **Tanaka, M.,** Solvent effect on equilibria and SN2 reactions of non-charged species, *Z. Phys. Chem. (Frankfurt),* 96, 239, 1975.
49. **Richardson, M. and Soper, F. G.,** The effect of the cohesion of the medium on reaction velocity. The velocity of interaction of *N*-chloroacetanilide and hydrobromic acid in aqueous solutions, *J. Chem. Soc.,* 1873, 1929.
50. **Halpern, J., Brady, G. W., and Windler, C. A.,** The *cis-trans* isomerization of azobenzene in solution, *Can. J. Res.,* 28B, 140, 1950.

51. **Stefani, A. P.,** Solvent effects in electroneutral reactions. I. Combination and disproportionation of ethyl radicals in solutions, *J. Am. Chem. Soc.*, 90, 1694, 1968.
52. **Kondo, Y., Ohnishi, M., and Tokura, N.,** Solvent effects under high pressure. IV. Procedure for the estimation of solvation in the transition state, *Bull. Chem. Soc. Jpn.*, 45, 3579, 1972; and earlier publications cited there.
53. **Murai, K. and Kimura, C.,** Kinetic studies on quaternization of tertiary amines. Consideration of solubility parameter in Menschutkin reaction, *Asahi Garasu Kogyo Gijutsu Shoreikai Kenkyu Hokoku,* 15, 481, 1969; IX. Solvent polarity parameters in the quaternization of pyridine with dodecyl bromide, *Nippon Kagaku Kaishi,* No. 11, 2209, 1972.
54. **Shinoda, K.,** The solvent effect on the amination of benzyl chloride, *Toyama Kogyo Koto Semmon Gakko Kiyo,* 10(1), 9, 1976.
55. **Yoshida, T. and Keii, T.,** Solvent effect on heterogeneous hydrogenation of toluene, *Kogyo Kagaku Zasshi,* 74, 2100, 1971.
56. **Colter, A. K. and Clemens, L. M.,** Solvent effects in the racemization of 1,1-binaphthyl. A note on the influence of internal pressure on reaction rates, *J. Phys. Chem.*, 68, 651, 1964.
57. **McCabe, J. R.,** *High Pressure Kinetic Studies of Solvent and Substituent Effects on a Diels-Alder Reaction,* Ph.D. Thesis, University of Illinois, Urbana-Champaign, 1973; *Diss. Abstr. B,* 34, 4351B, 1974.
58. **McCabe, J. R. and Eckert, C. A.,** High-pressure kinetic studies of solvent and substituent effects on Diels-Alder reactions, *Ind. Eng. Chem. Fundam.*, 13, 168, 1974.
59. **Dewar, M. J. S. and Pyron, R. S.,** Nature of the transition state in some Diels-Alder reactions, *Trans. Faraday Soc.*, 66, 2579, 1970.
60. **Grieger, R. A. and Eckert, C. A.,** Solvent effects on the activation volume of a Diels-Alder reaction, *Trans. Faraday Soc.*, 66, 2579, 1970.
61. **Soula, J. C., Lumbroso, D., Hellin, M., and Coussemant, F.,** Kinetic investigation of the diene condensation of isoprene with acrylonitrile and methyl acrylate. I. Thermal reaction. II. Catalyzed reaction, *Bull. Soc. Chem. Fr.*, 2059, and 2065, 1966.
62. **Lotfi, M. and Roberts, R. M. G.,** Kinetics and mechanisms of the Diels-Alder addition of tetracyanoethylene to anthracene derivatives-II, *Tetrahedron,* 35, 2137, 1979.
63. **Tanaka, M.,** Reactivity of metal ions and metal complexes in different solvents, in *Ions and Molecules and Solution* (Studies in Physical and Theoretical Chemistry, Vol. 27; from VI Int. Symp. Solute-Solute-Solvent Interactions, Minoo, Japan, 1982), Tanaka, N., Ohtaki, H., and Tamamushi, R., Eds., Elsevier, Amsterdam, 1983, 267.
64. **Abraham, M. H., Taft, R. W., and Kamlet, M. J.,** Linear solvation energy relationships. 15. Heterolytic decomposition of the tert-butyl halides, *J. Org. Chem.*, 46, 3053, 1981.
65. **Fukunaga, K., Shirai, H., Ide, S., and Kimura, M.,** Solubility parameters of phase-transfer catalysts and their applications, *Nippon Kagaku Kaishi,* 1148, 1980.
66. **Bram, G., Loupy, A., and Pédoussant, M.,** The effect of water and other additives on the alkylation of KCN in solid-liquid phase transfer without solvents, *Tetrahedron Lett.*, 27, 4171, 1986.
67. **Moore, G. K. and Roberts, G. A. F.,** Studies on the acetylation of chitosan, *Proc. First Int. Conf. Chitin Chitosan,* Muzzarelli, R. A. A. and Pariser, E. R., Eds., MIT Sea Grant Report MITSG78-7, 1978, PB 285640, p. 421; Reactions of chitosan. 2. Preparation and reactivity of *N*-acyl derivatives of chitosan, *Int. J. Biol. Macromol.*, 3, 292, 1981.
68. **Brink, L. E. S. and Tramper, J.,** Optimization of organic solvent in multiphase biocatalysis, *Biotech. Bioeng.*, 27, 1258, 1985.
69. **Seymour, R. B., Kincaid, P. D., and Owen, D. R.,** New block copolymer from macroradicals, in *Polymerization Reactions and New Polymers (Adv. Chem. Ser.* 129), Platzer, N. A. J., Ed., American Chemical Society, Washington, D.C., 1973, chap. 14.
70. **Seymour, R. B. and Owen, D. R.,** Role of solubility parameters and other fundamental concepts in the synthesis and characterization of block copolymers, *Austin Paint J.*, 18(4), 4, 1973.
71. **Seymour, R. B. and Owen, D. R.,** Recent advances in water-based polymer systems, *Am. Paint J.*, 58(8), 14, 1973.
72. **Seymour, R. B., Owen, D. R., and Kincaid, P. D.,** Macroradical reactions, *Chemtech,* 3, 549, 1973.
73. **Seymour, R. B., Owen, D. R., and Stahl, G. A.,** Preparation of block copolymers from macroradicals *Am. Chem. Soc. Div. Org. Coat. Plast. Chem. Pap,* 33(1), 670, 1973; Poly(methyl methacrylate-*co*-acrylonitrile), *Polymer*, 14, 324, 1973.
74. **Seymour, R. B. and Patel, V.,** Chain transfer of telogens with styrene in bulk and emulsion systems, *J. Macromol. Sci. Chem.*, A7, 961, 1973.
75. **Locatelli, J. L. and Reiss, G.,** ABS (acrylonitrile-butadiene-sytrene) copolymers. Preferential solvation of polybutadiene by the initiator, *J. Polym. Sci. Polym. Chem.*, 11, 3309, 1973.
76. **Seymour, R. B., Owen, D. R., McGee, M. L., Losado, C. J., and Clark, M.,** Acrylonitrile-styrene copolymer emulsions, *Ind. Eng. Chem. Prod. Res. Dev.*, 12, 198, 1973.
77. **Chapiro, A. and Mankowski, Z.,** Polymerization and radiochemical copolymerization of vinyl bromide, *Eur. Polym. J.*, 18, 707, 1982.

78. **Zuchowska, D.,** Structure and properties of chemically modified low-molecular weight polybutadienes, *Pr. Nauk. Inst. Elektron Politekh. Wroclaw,* 33, Monograph 7, 1979; Polybutadiene modified by epoxidation. 1. Effect of polybutadiene microstructure on the reactivity of double bounds, *Polymer,* 21, 514, 1980.
79. **Snow, A. W.,** Thermal characterization of polybutadiene, *J. Macromol. Sci. Chem.,* A22, 1429, 1985.
80. **Kamachi, M., Fujuii, M., Ninomiya, S.-I., Katsuki, S., and Nozakua, S.-I.,** Solvent effect on radical polymerization of butyl acrylate, *J. Polym. Sci. Polym. Chem. Ed.,* 20, 1489, 1982.
81. **Sakai, I., Nakamae, K., Okawa, T., Inubushi, H., and Matsumoto, T.,** Studies on suspension and emulsion. 33. Adsorption behavior of polybutadiene having carboxyl end groups at water-oil interface, *Kobunshi Ronbunshu,* 34, 661, 1977.
82. **Tanaka, Y., Shimura, Y., and Shiozaki, H.,** Effects of salts, solvents and substituents on the reaction of cellulose with epoxy compounds, *Makromol. Chem.,* 177, 1301, 1976.
83. **Mansour, E. S. M. E., Khalifa, A. M. I., and Rateb, L.,** Synthesis and properties of some new polyamides. Part II, *J. Macromol. Sci. Chem.,* A16, 669, 1981.
84. **Kricheldorf, H. R., Zang, Q.-Z., and Schwarz, G.,** New polymer syntheses: 6. Linear and branched poly(3-hydroxy-benzoates), *Polymer,* 23, 1821, 1982.
85. **Seymour, R. B., Tatum, S. D., Boriack, G. J., and Tsang, H.-S.,** Role of solubility parameters in the copolymerization of maleic anhydride and vinyl monomer, *Tex. J. Sci.,* 21(1), 13, 1969.
86. **Seymour, R. B., Tsang, H. S., Jones, E. E., Kincaid, P. D., and Patel, A. K.,** Multicomponent systems from copolymers of maleic anhydride and vinyl monomers, in *Multicomponent Polymer Systems (Adv. Chem. Ser.* 99), Platzer, N. A., Ed., American Chemical Society, Washington, D.C., 1971, chap. 26.
87. **Tanaka, Y. and Shiozaki, H.,** Salt-catalyzed reaction of phenyl glycidyl ether with silk fibroin, *Makromol. Chem.,* 129, 12, 1969.
88. **Owen, D. R., McLemore, D., Liu, W.-L., Seymour, R. B., and Tinnerman, W. N.,** Effect of viscosity and solubility parameter of a nonreactive liquid additive on the emulsion polymerization of styrene, in *Emulsion Polymerization (ACS Symp. Ser.* 24), Piirma, I. and Gardon, J. L., Eds., American Chemical Society, Washington, D.C., 1975, 299.
89. **Seymour, R. B., Owen, D. R., Tinnerman, W. N., and Losada, C. J.,** The effect of liquid additives on the emulsion polymerization of styrene, *Polym. Prepr. Am. Chem. Soc. Div. Polym. Chem.,* 16, 261, 1975.
90. **Gan, L. M., Chew, C. H., and Friberg, S. E.,** Polymerization of styrene in water-alcohol-ionic surfactant solutions, *J. Macromol. Sci. Chem.,* A19, 739, 1983.
91. **Lok, K. P. and Ober, C. K.,** Particle size control in dispersion polymerization of polystyrene, *Can. J. Chem.,* 63, 209, 1985.
92. **Ober, C. K., Lok, K. P., and Hair, M. L.,** Monodispersed, micron-sized polystyrene particles by dispersion polymerization, *J. Polym. Sci. Polym. Lett. Ed.,* 23, 103, 1985.
93. **Takeda, K., Akiyama, M., and Yamamizu, T.,** Shape-memory pore structure in porous crosslinked polystyrene, *Angew. Makromol. Chem.,* 157, 123, 1988.
94. **Ober, C. K. and Lok, K. P.,** Formation of large monodisperse copolymer particles by dispersion polymerization, *Macromolecules,* 20, 268, 1987.
95. **Dale, J. A. and Millar, J. R.,** Cross-linker effectiveness in styrene copolymerization, *Macromolecules,* 14, 1515, 1981.
96. **Cowie, J. M. G., Lath, D., and McEwen, I. J.,** Dynamic mechanical behavior of polystyrene-(ethene-*co*-butene)-polystyrene triblock copolymer films cast from various solvents, *Macromolecules,* 12, 52, 1979.
97. **Wilkes, G. L., Bagrodia, S., Ophir, Z., and Emerson, J. A.,** A systematic study of the effect of casting solvent on the bulk properties of linear SBS block copolymers, *J. Appl. Phys.,* 49, 5060, 1978.
98. **Omichi, H. and Araki, K.,** Solvent effects on the radiation-induced graft polymerization of styrene onto poly(isobutylene oxide), *J. Polym. Sci. Polym. Chem.,* 14, 2773, 1976.
99. **Omichi, H. and Araki, K.,** Kinetics of radiation-induced graft polymerization of styrene onto poly(ethylene oxide), *J. Polym. Sci. Polym. Chem.,* 17, 1401, 1979.
100. **Omichi, H. and Araki, K.,** Kinetics of radiation-induced graft copolymerization of styrene with ethyl acrylate, *J. Polym. Sci. Polym. Chem.,* 17, 1701, 1979.
101. **Rosenberg, J.-E. and Flodin, P.,** Macroporous gels. 2. Polymerization of trimethylolpropane trimethacrylate in various solvents, *Macromolecules,* 20, 1518, 1987.
102. **Millich, F. W.,** The polymerization of isocyanides, *Chem. Rev.,* 72, 101, 1972.
103. **Matsunaga, K. and Yamashita, T.,** Solvent effects in urethane formation, *Nippon Kagaku Kaishi,* No. 5, 993, 1985.
104. **Seymour, R. B. and Stahl, G. A.,** The effect of solvent on polymerization rate and particle size in the heterogeneous solution polymerization of vinyl acetate, *Abstr. Pap. Am. Chem.* Soc., 1972, POLY134, 1976; Preparation of block copolymers of vinyl acetate in viscous poor solvents, *J. Macromol. Sci. Chem.,* A11, 53, 1977.

105. **Stahl, G. A.,** The effect of solvents on polymerization rate, particle size, and molecular weight distribution in the heterogeneous solution polymerization of vinyl acetate, in *Structure-Solubility Relationships in Polymers,* Harris, F. W. and Seymour, R. B., Eds., Academic Press, New York, 1977, 259.
106. **Jayasuriya, R. M., Klein, A., El-Aasser, M. S., and Vanderhoff, J. W.,** Prediction of the microstructure of vinyl acetate-alkyl acrylate ester emulsion copolymers based on solubility parameters, *Abs. Pap. Am. Chem. Soc.,* 188(Aug.), 154, 1984; *Polym. Mater. Sci. Eng.,* 51, 766, 1984.
107. **Van der Meer, R., Aarts, M. W. A. M., and German, A. L.,** Effect of solvent on the copolymerization of ethylene and vinyl acetate, *J. Polym. Sci. Polym. Chem.,* 18, 1347, 1980.
108. **Yamashita, S., Nagata, S., and Takabura, K.,** Ethylene-vinyl alcohol copolymer membranes for hemodialysis. 1. Studies on preparation conditions and properties of ethylene-vinyl alcohol copolymer membranes for hemodialysis, *Kobunshi Ronbunshu,* 36, 249, 1979.
109. **Nazakura, S., Sumi, M., Uoi, M., Okamoto, T., and Murahashi, S.,** Stereoregulation in the radical polymerization of vinyl esters, *J. Polym. Sci. Polym. Chem.,* 11, 279, 1973.
110. **Raucher, D. and Levy, M.,** Kinetics of vinyl fluoride polymerization, *J. Polym. Sci. Polym. Chem.,* 17, 2663, 1979.
111. **Tanaka, Y. and Shiozaki, H.,** Reaction of wool keratin with epoxides, in *Proc. Int. Wool Text. Res. Conf. Aachen,* Vol. III, Schriftenreihe Deutsches Wollforschungsinstut an der Technischen Hochschule Aachen e.V., 1975, p. 192; *Sen'i Kobunshi Zairyo Kenkynsho Kenkyu Hokoku,* 109, 19, 1975; **Shiozaki, H. and Tanaka, Y.,** Solvent effect on the epoxide-wool reactions, *Sen'i Gakkaishi,* 28, 174, 1972.
112. **Innorta, G., Modelli, A., Scagnolari, F., and Foffani, A.,** Kinetic studies in various media on some hydrogenation reactions catalyzed by polymer-anchored $RhCl(PPh_3)_3$, *J. Organomet. Chem.,* 185, 403, 1980.
113. **Chapiro, A. and Jendrychowska-Bonamour, A. M.,** Properties of some graft copolymers. 1. Preparation of samples. Solubility and swelling, *J. Chim. Phys.,* 60, 1029, 1963.
114. **Wilson, J. E.,** Solvent effects on the rate of radiation-induced grafting of vinyl monomers on polymeric films, *J. Macromol. Sci. Chem.,* A8, 733, 1974; Polymer/solvent interaction in the radiation-induced grafting of vinyl monomers on polymeric films, *J. Macromol. Sci. Chem.,* A9, 607, 1975; Reaction kinetics of the radiation-induced grafting of styrene on polyethylene film, *J. Macromol. Sci. Chem.,* A10, 1441, 1976.
115. **Hartley, F. R., Murray, S. G., and Nicholson, P. N.,** γ-Radiation-produced supported metal complex catalysts. 3. Phosphinated polypropylene supports, *J. Polym. Sci. Polym. Chem. Ed.,* 20, 2395, 1982.
116. **Hartley, F. R., Murray, S. G., and Sayer, A. T.,** γ-Radiation-produced supported metal complex catalysts. Part VII. The influence of the solvent used during radiation grafting on the catalytic properties of supported rhodium (I)-phosphine hydroformylation catalysts, *J. Mol. Catal.,* 38, 295, 1986.
117. **Omichi, H. and Stannett, V. T.,** Radiation-induced grafting of methacrylic acid onto polyurethane in the presence of solvents, *J. Appl. Polym. Sci.,* 30, 3059, 1985.
118. **Omichi, H. and Okamoto, J.,** Solvent effects on the synthesis of ion-exchange membranes by radiation-induced graft polymerization of methyl α,β,β-trifluoroacrylate, *J. Polym. Sci. Polym. Chem. Ed.,* 20, 1559, 1982.
119. **Zurawski, D. E. and Sperling, L. H.,** Characterization of electron beam crosslinked poly(styrene-block-ethylene-co-butylene-block-styrene), *Polym. Eng. Sci.,* 23, 510, 1983.
120. **Mullerschule, D.,** Gamma radiation induced grafting onto wool. 1. Fundamental studies on loose wool. 2. Studies on fabrics and practical implication, *Radiat. Phys. Chem.,* 16, 149, and 159, 1980.
121. **Jones, D. W.,** Influence of plasticizer on soft polymer gelation, *J. Dent. Res.,* 65, 634, 1986.
122. **Gopal, R. and Rizvi, S. A.,** Estimation of molecular interaction energies in some amides, *Z. Phys. Chem. (Frankfurt),* 81, 330, 1972.
123. **Grunberg, L. and Nissan, A. H.,** The energies of vaporisation, viscosity and cohesion and the structure of liquids, *Trans. Faraday Soc.,* 45, 125, 1949.
124. **Van Zeggeren, F.,** Relation between activation energy of the viscosity and heat of vaporization for molten salts, *Can. J. Chem.,* 34, 1512, 1956.
125. **Rao, R. V. G. and Seshaiah, V. V.,** The equation of state and calculation of viscosity of liquids through internal pressure, *Z. Phys. Chem. (Frankfurt),* 78, 26, 1972.
126. **Suryanarayana, C. V.,** Thermodynamic significance of internal pressure and its role in transport processes, *Indian J. Chem.,* 10, 713, 1972.
127. **Irving, H. M. N. H. and Smith, J. S.,** The theory of regular solutions applied to the viscosity of dilute solutions of chromium (III) acetylacetonate in organic solvents, *J. Inorg. Nucl. Chem.,* 30, 1873, 1968; 31, 1212, 1969; A viscometric study of iodine-solvent interactions, *J. Inorg. Nucl. Chem.,* 31, 3163, 1969.
128. **Rätzsch, M. T. and Friedel, B.,** Correlation of viscosity coefficients with thermodynamic properties, *Z. Phys. Chem. (Leipzig),* 256, 829, 1975.
129. **Bolster, R. N. and Little, R. C.,** Dependence of lithium grease dropping points on solubility parameter of oil component, *Ind. Eng. Chem. Prod. Res. Dev.,* 5, 198, 1966.
130. **Contreras, M. D., Ruiz, M. A., Cerezo, A., and Parero, A.,** Determination of Carbopol 941 solubility parameters in hydroalcoholic mixtures using apparent viscosity, *Farm. Clin.,* 3, 607, 1986.

131. **Boberski, W. G., Seiner, J. A., and Petracca, V. G.,** Reducing the viscosity of magnesia cements with organosilicon compounds, *Ind. Eng. Chem. Prod. Res. Dev.*, 21, 531, 1982.
132. **Makitra, R. G. and Vasyutin, Y. M.,** Relationship of the activation energy of viscous flow of hydrocarbons with cohesive energy density, *Ukr. Khim. Zh.*, 54, 276, 1988.
133. **Hara, T., Horino, H., and Kita, R.,** Curable trans-rich polypentenamines, *Jpn. Pat.* 49/2900, 1974.
134. **Dickerson, C. G. and Wiehe, I. A.,** Spherical encapsulated polymer particles by spray drying, *Pac. Chem. Eng. Congr.* [*Proc.*], 2(1), 243, 1977.
135. **Röthlisberger, T., Testa, B., Carrupt, P.-A., Mayer, J. M., and Etter, J.-C.,** The influence of hydration and structure-promoting effects on the viscosity of binary solvent mixtures, *Int. J. Pharm.*, 44, 141, 1988.
136. **Hoover, J. M., Ward, T. C., and McGrath, J. E.,** Hydrogenated star block copolymers of butylstyrene-isoprene and divinyl benzene, *Adv. Polyolefins* (Proc. Am. Chem. Soc. Int. Symp., 1985), Seymour, R. B. and Chen, T., Eds., Plenum, New York, 1987, 89.
137. **Momose, T., Tomiie, K., Ishigaki, I., and Okamoto, J.,** Radiation grafting of α, β, β-trifluorostyrene onto various polymer films by preirradiation method; **Momose, T., Yoshioka, H., Ishigaki, I., and Okamoto, J.,** Radiation grafting of α,β,β-trifluorostyrene onto poly(ehtylene-tetrafluoroethylene) film by preirradiation method. 1. Effects of preirradiation dose, monomer concentration, reaction temperature and film thickness, *J. Appl. Polym. Sci.*, 37, 2165 and 2817, 1989.
138. **Zou, Y. and Pan, R.,** Steric effect and solvent effect in group-transfer polymerization, *Xiamen Daxue Xuebao, Ziran Kexueban*, 28(1), 70, 1989.
139. **Egorochkin, G. A., Smirnova, L. A., Semchikov, Yu. D., and Tikhonova, Z. A.,** Solubility parameters and copolymerization features of a series of monomer mixtures, *Vysokomol. Soedin., Ser. B*, 31, 433, 1989.
140. **Szymanowski, J. and Prochaska, K.,** Interfacial activity of model 2-hydroxy-5-alkylbenzophenone oximes and their intermediates, *J. Colloid Interface Sci.*, 123, 456, 1988.
141. **Singh, D. P., Kalachandra, S., and Turner, D. T.,** Glass transition of polymers made from slurries of poly(ethyl methacrylate) and methyl methacrylate, *J. Polym. Sci. Polym. Phys.*, 26, 1795, 1988.
142. **Poinescu, I. C. and Beldie, C.,** Styrene divinylbenzene copolymers — influence of diluent mixture on matrix structure, *Angew. Mak.*, 164, 45, 1988; **Poinescu, I. C. and Carpov, A.,** Relationship between crosslinked copolymer synthesis and their features, *Rev. Roum. Chim.*, 34, 1061, 1989.
143. **Haddad-Fahed, O., Lelievre, J., and Goriaud, R.,** Micellar catalysis and substituent effects in the case of charged substituents: nucleophilic attack of 1-methoxy-2,4,6-trinitrobenzene by aminobenzoate anions, *J. Chim. Phys. Phys-Chim. Biol.*, 85, 637, 1988.
144. **Cafferata, L. F. R. and Quintans, M. T.,** Kinetics and mechanism of di-tert-butyl peroxide thermal decomposition in N,N-dimethylformamide solution, *An. Asoc. Quim. Argen.*, 75, 461, 1987.
145. **Komiyama, M., Ohtaki, M., and Hirai, H.,** Covalently immobilized colloidal metal particles as selective catalysts for olefin hydrogenation, *J. Coord Chem.*, 18, 185, 1988.
146. **Stöver, H. D. H. and Detellier, C.,** Investigation by Na-23 NMR of the anion role in the dissociation kinetics of sodium dibenzo-24-crown-8 in nitromethane, *J. Phys. Chem.*, 93, 3174, 1989.
147. **Sato, T., Morita, N., Tanaka, A., and Ota, T.,** Solvent effect on the radical polymerization of di-*n*-butyl itaconate, *J. Polym. Sci. Polym. Chem.*, 27, 2497, 1989.
148. **Cuirassier, F., Baradji, Ch. H., and Riess, G.,** Preparation of microlatex with functionalized polyesters as surfactants, in *Polymer Association Structures — Microemulsions and Liquid Crystals* (Am. Chem. Soc. Symp. Ser. 384), El Nokaly, M. A., Ed., Americal Chemical Society, Washington, D.C., 1989, 100.
149. **Priori, A., Nicolais, L., and DiBenedetto, A. T.,** The kinetics of surface craze growth in polycarbonate exposed to normal hydrocarbons, *J. Mater. Sci.*, 18, 1466, 1983; **Lee, W., Nobile, M. R., DiBenedetto, A. T., and Nicolais, L.,** Growth kinetics of environmental crazing in polycarbonate, *Int J. Polym. Mater.*, 12, 185, 1989.

Chapter 11

GASES

Liquid-gas equilibrium plays an important role in a wide range of natural and industrial processes. Although "chemical bond" effects are important in some cases, such as the transport of oxygen in blood, in other situations (like the dissolution of oxygen in water) there are negligible specific effects and it is reasonable to expect cohesion parameters to provide correlation and prediction methods. They have been used for this purpose in the design of liquid membranes for the production of oxygen-enriched air,[1] and in the solubility of oxygen and alkene oxides in solvents used for multiphase biocatalysis.[2] Most gas mixtures can be separated by contact with a suitable liquid, if it is chosen to preferentially dissolve one component. Analysis of one phase in an equilibrium situation can provide information on the other phase, as in the application of Hildebrand parameters in the choice of model compounds for head space analysis for volatile trace organics in water.[3] Gas solubility data have been correlated and reviewed by Battino and Clever,[4,5] Hildebrand et al.,[6-10] Gerrard,[12] Linford and Thornhill[13], Wilhelm,[14] and Abraham, Kamlet, Taft et al.[15-17] Applications of cohesion parameters to gas scrubbing have been reviewed by Sweeney, Ritter, and Mc Ginley.[110]

11.1 NOMENCLATURE

Consider a binary system with component i ("solvent" or denser component) and j (gaseous solute or lighter component) in phases "1" (liquid) and "g" (gas). The sum of partial pressures of i and j in the gas phase constitutes the total gas pressure, p,

$$^{i}p + {}^{j}p = p \tag{1}$$

The mole fraction of each component in liquid and gas phases is denoted by x and y, respectively:

$$^{i}x + {}^{j}x = 1 \tag{2}$$

$$^{i}y + {}^{j}y = 1 \tag{3}$$

The fugacity, ^{j}f, of substance j in the gaseous mixture is a measure of its escaping tendency, with dimensions of pressure. It is an effective gas pressure, corrected for deviations from the ideal gas laws, and from its definition in terms of the chemical potential it follows that

$$\lim_{p \to 0}({}^{j}f/{}^{i}yp) = 1 \tag{4}$$

and that

$$RT \ln({}^{j}f/{}^{j}yp) = \int_{0}^{p} ({}^{j}V_{1} - RT/p)\mathrm{d}p \tag{5}$$

If Raoult's law (Equation 28, Chapter 2) holds, the solubility of an ideal gas j in the liquid phase is

$$^{j}x_{s} = {}^{j}yp/{}^{j}\mathrm{p}_{s} \tag{6}$$

where ${}^{j}p_s$ is the (saturation) vapor pressure of the pure liquid. Observed gas solubilities often conform approximately to this relationship, but, quantitatively, the ideal solubility is usually higher than the measured value. If ${}^{j}x \ll 1$, Henry's law is often used, based on ideal gas behavior in the gas phase and ideal dilute solution behavior in the liquid phase,

$$ {}^{j}x = {}^{j}yp/{}^{ij}H \tag{7} $$

where Henry's constant, ${}^{ij}H$, depends on temperature as well as on the chemical nature of both i and j. When solute and solvent differ substantially in properties, ${}^{ij}H$ tends to be large; if there is significant interaction its value is smaller. Correlation of ${}^{ij}H$ with ${}^{i}\delta$ and ${}^{j}\delta$ is discussed in Section 11.5.

The activity, ${}^{j}a$, of the component j in a particular solution is the ratio of its fugacity to its fugacity in some arbitrary but specified standard state. In regular solution theory, the standard state is usually the pure liquid at the same temperature (extrapolated below the melting point in the case of a solid, and to a hypothetical condensed liquid in the case of a gas[8,11].) The activity of j is therefore

$$ {}^{j}a = {}^{j}x\,{}^{j}f_x = {}^{j}f/{}^{j}f_l \tag{8} $$

where ${}^{j}f_l$ is the hypothetical liquid fugacity and ${}^{j}f_x$ is the mole fraction activity coefficient of j in solution. The fugacity coefficient, ${}^{i}\varphi$, of component j in the gas phase is defined by the ratio of the fugacity ${}^{j}f$ of the component relative to that of an ideal gas, ${}^{j}yp$,

$$ {}^{j}\varphi = {}^{j}f/{}^{j}yp $$

$$ {}^{j}f = {}^{j}\varphi\,{}^{j}yp \tag{9} $$

The corresponding fugacity coefficient ${}^{j}\nu$ of the reference "liquid" j is

$$ {}^{j}\nu = {}^{j}f_l/p $$

so

$$ {}^{i}f_l = {}^{j}\nu p \tag{10} $$

It follows that

$$ {}^{j}f/{}^{j}f_l = {}^{j}\varphi\,{}^{j}y/{}^{j}\nu \tag{11} $$

and from Equation 8

$$ {}^{j}y/{}^{j}x = {}^{j}f_x\,{}^{j}\nu/{}^{j}\varphi \tag{12} $$

11.2 HILDEBRAND PARAMETERS AND GAS SOLUBILITIES

Hildebrand[6,9,18] and others[19-22,110] have pointed out that a close relationship exists between the logarithm of the gas solubility, log ${}^{j}x_s$, and the Hildebrand parameter, ${}^{i}\delta$, of the liquid. A linear relationship was assumed at first, and this has been utilized subsequently.[23-25] For example, log ${}^{j}x_a$ for nitrogen, argon, methane, ethylene, and ethane were found to be linear functions of ${}^{i}\delta$ for the normal primary alcohols as solvents. However, it was later shown that some curved lines became straight when ${}^{j}\delta^2$ was used in place of ${}^{j}\delta$ (see Equation 24).

TABLE 1
Volumes and Hildebrand Parameters as Hypothetical Dissolved Liquids at 25°C for Gaseous Solutes[11,21]

Gas	${}_jV_l$/cm^3 mol^{-1}	${}^j\delta$/MPa$^{1/2}$
Nitrogen	32.4	5.3
Carbon monoxide	32.1	6.4
Oxygen	33.0	8.2
Argon	57.1	10.9
Methane	52.0	11.6
Carbon dioxide	55.0	12.3
Krypton	65.0	13.1
Ethylene	65.0	13.5
Ethane	70.0	13.5
Radon	70.0	14.0
Chlorine	74.0	17.8

Prausnitz and Shair[11,21] considered a gaseous component j being dissolved isothermally in a liquid, the process being accompanied not only by changes in enthalpy and entropy, as in the case of the mixing of two liquids, but also by a large decrease in volume of the gas because the molar volume of j in the liquid phase, ${}^jV_l$, is much smaller than that in the gas phase, ${}^jV_g$. In order to apply regular solution theory, which assumes no volume change, the gas is considered to be condensed to a liquid-like volume before being mixed with i:

gas $j \rightarrow$ hypothetical liquid j

hypothetical liquid j + liquid $i \rightarrow$ mixture

The resulting equation is equivalent to that deduced in Section 4.2 for liquid-liquid mixtures,

$$\ln {}^jf_x = \ln ({}^ja/{}^jx) = \ln ({}^jf/{}^jf_l\,{}^jx) = {}^jV_l\,{}^i\phi^2({}^i\delta - {}^j\delta)^2/(RT) \tag{13}$$

where $({}^jf_l/{}^jf)$ is the ratio of the fugacities of the gaseous solute in its "liquid" and pure gaseous states and, ${}^jV_l$ is its hypothetical liquid molar volume. This means that, provided the reference state of the gas is chosen appropriately, liquid-liquid cohesion parameter methods may be applied to the calculation of the gas solubility, ${}^jx_s$,

$$-\ln {}^jx_s = \ln ({}^jf_l/{}^jf) + {}^jV_l\,{}^i\phi^2({}^i\delta - {}^j\delta)^2/RT \tag{14}$$

These equations involve three parameters for the gaseous component as a hypothetical liquid (${}^jf_l$, ${}^j\delta$, and ${}^jV_l$), and although all are individually temperature dependent, regular solution theory predicts that the term ${}^jV_l\,{}^i\phi^2\,({}^i\delta - {}^j\delta)^2$ is not temperature dependent. These three hypothetical liquid parameters may be obtained from solubility data; the values of ${}^jV_l$ and ${}^j\delta$ as listed in Table 1 are not unique, but are chosen to reflect physical significance as well as to optimize the representation of available solubility data. Values of the fugacity of the hypothetical liquid at one atmosphere pressure can be obtained from the plot of corresponding states in Figure 1,[11,21,22] for which the critical temperature and pressure of the solute gas are required. At a considerably higher pressure, p, the "liquid" phase fugacity as read from Figure 1, may be corrected by means of

$${}^jf_l\ (\text{at } p) = {}^jf_l \exp[{}^jV(p - 1)/(RT)] \tag{15}$$

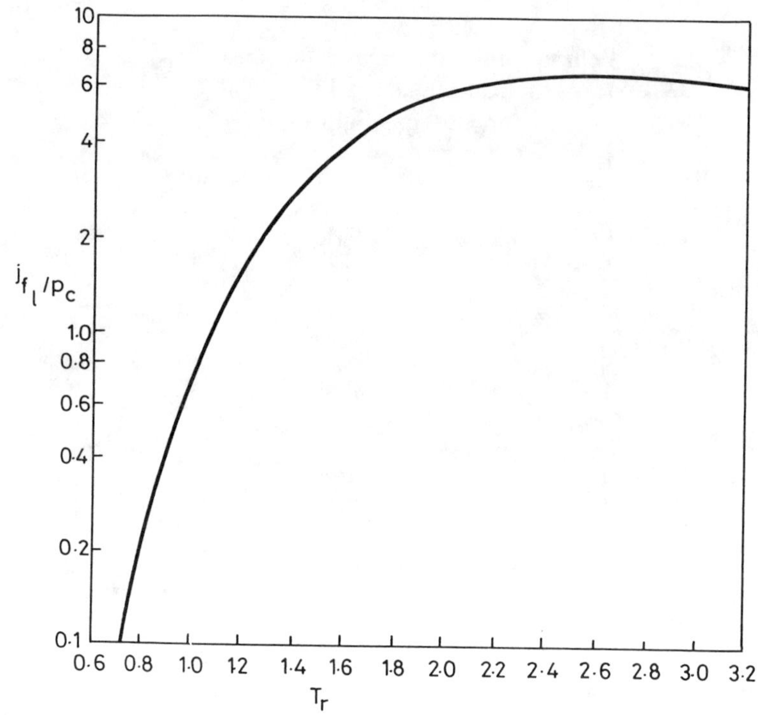

FIGURE 1. Fugacity of hypothetical liquefied gases at atmospheric pressure in terms of critical pressure and reduced temperature. (Adapted from Prausnitz, J. M. and Shair, F. H., *Am. Inst. Chem. Eng. J.*, 7, 682, 1961.)

and the pure gas fugacity is

$$^jf = {}^j\varphi {}^jyp \tag{16}$$

(The gas phase fugacity coefficient ${}^j\varphi$ can be calculated from an equation of state for the gas.) An example may be found in the work of Shaw and Vosper[26] on the solubility of nitrogen oxide in various liquids, but in this case a plot of the logarithm of the vapor pressure of liquid nitrogen oxide was extrapolated from the critical temperature down to 25°C. The solubility of oxygen in aromatic liquids has also been correlated with cohesive pressure.[27]

It is found that the fugacities of the hypothetical liquid solutes j at atmospheric pressure are essentially independent of the solvent i, so it is possible to use these fugacities to calculate solute activity coefficients ${}^jf_x$, which are applicable in both polar and nonpolar solvents:

$${}^jx_s^{-1} = ({}^jf_l/{}^jf){}^jf_x = {}^jf_l f_x/({}^j\varphi {}^j\mathrm{yp}) \tag{17}$$

Variations on Equations 13 and 14 are possible. For mixed solvents (multicomponent systems) ${}^i\delta$ is replaced by $\overline{\delta}$ as defined in Equation 41, Chapter 4, in terms of volume fractions:

$$\overline{\delta} = \Sigma_i\, {}^i\phi\, {}^i\delta$$

Using the more general formalism of Section 4.2, an alternative to Equation 13 is

$$\ln {}^jf_x = {}^jV_1\, {}^i\phi^2\, {}^{ij}A/(RT) \tag{18}$$

TABLE 2
Hildebrand Parameters for Hydrogen from Solubility[11,32]

Solvent	$^{j}\delta/\mathrm{MPa}^{1/2}$
Diethyl ether	4.3
m-Xylene	4.1
Tetrachloromethane	4.3
Toluene	4.3
Benzene	4.4
Carbon disulfide	4.1[a]

[a] Recalculated by Hildebrand et al.[11] from corrected solubility information.

an equation that can make use of the component cohesion parameters of Section 5.6 ($^{j}\delta_d$, $^{j}\delta_i$, $^{j}\delta_a$, and $^{j}\delta_b$) for the hypothetical liquid j as well as its molar volume, and a reliable estimation of their values obviously presents many problems. The correlation of Yen and McKetta[28] for nonpolar gases in polar, nonassociated liquids is an example of a model of intermediate complexity, introducing an additional parameter Δ for polar, non-associated liquids. Another is the use of geometric mean correction factors described in Section 5.7, values obtained from gas-phase virial data being applied subsequently to liquid-liquid or liquid-vapor systems.

The Hildebrand-Scatchard equation had already been applied in a slightly different way to solutions of argon, nitrogen, hydrogen, and other gases, with a relationship still in wide use,[11,29-35]

$$-\ln {}^{j}x_s = -\ln {}^{j}x_{id} + [{}^{j}\overline{V}/(RT)]({}^{i}\delta - {}^{j}\delta)^2 \tag{19}$$

where $^{j}\overline{V}$ is the average partial molar volume of the gas in a range of solvents, sometimes taken as equal to $_{j}V$ at the normal boiling point if there is only a small volume charge on mixing. For hydrogen, $^{j}\overline{V} = 28.6\ \mathrm{cm}^3\ \mathrm{mol}^{-1}$ was used, and the cohesion parameter values obtained in different solvents are in good agreement (Table 2). The "ideal" gas solubility, ($^{j}x_{id}$) the solubility of a gas that does not interact energetically with the liquid at temperature T, is calculated from

$$\ln {}^{j}x_{id} = (\Delta H/R)(T^{-1} - T_b^{-1}) \tag{20}$$

where ΔH is the enthalpy of vaporization of the gas j at its normal boiling point, T_b. If the molar volumes of solute and solvent differ greatly, a correction of the Flory-Huggins type (Section 13.2) is made:

$$-\ln {}^{j}x_s = -\ln {}^{j}x_{id} + ({}^{j}\overline{V}/RT)({}^{i}\delta - {}^{j}\delta)^2 + \ln ({}^{j}\overline{V}/{}^{i}V) + (1 - {}^{j}\overline{V}/{}^{i}V) \tag{21}$$

or

$$-\log {}^{j}x_s = -\log {}^{j}x_{id} + ({}^{j}\overline{V}/2.30RT)({}^{i}\delta - {}^{j}\delta)^2 + \log ({}^{j}\overline{V}/{}^{i}V) + 0.434\,(1 - {}^{j}V/{}^{i}V) \tag{22}$$

Lawson used these equations for the estimation of the solubilities of oxygen, nitrogen, and

TABLE 3
Parameters for Gas Solubility Calculation with Flory-Huggins Correction[36,37]

Gas, j	$\log {}^jx_{id}$ (25°C)	${}^j\bar{V}$/cm³ mol⁻¹	${}^j\delta$/MPa$^{1/2}$
Hydrogen	−3.2596	37	10.4
Nitrogen	−2.7959	53	10.6
Oxygen	−2.7536	46	11.7
Carbon dioxide	−1.4209	105	6.8
Carbon monoxide	−2.8100	52	11.9
Methane	−2.4001	52	12.7
Ethane	−1.4855	66	13.5
Propane	−0.9208	89	13.6

carbon dioxide in battery electrolyte solvents (thionyl chloride, sulfuryl chloride, phosphorous oxychloride, 2-methyltetrahydrofuran, propylene carbonate),[46] and for a variety of gases in hydrocarbons and perfluorocarbons for thermochemical cycles.[37] The appropriate values of $\log {}^jx_{id}$, ${}^j\bar{V}$ and Hildebrand parameter ${}^j\delta$ for the gases, appear in Table 3. This form of the Hildebrand-Scatchard equation and this set of self-consistent parameters are an alternative to Equation 14 and Table 1. They may be considered less satisfactory from some points of view, but provide adequate correlations and predictions.

Both the Flory-Huggins correction and related correction terms have been incorporated to allow for molar volume differences in systems such as carbon dioxide, hydrogen sulfide, and propane in alkanes.[38] Special methods of calculation with quantum corrections are required for the light gases hydrogen,[21,39] helium,[40] and neon. Solutions of chemically reactive gases like hydrogen chloride, sulfur dioxide, and ammonia may not be able to be quantitatively fitted by the above equations. In this situation, and in others where values of ${}^j\bar{V}$ or ${}^jV_1$ and ${}^i\delta$ are not available, the form of the equations may still be valuable:

$$-\log {}^jx_s = a + b\,({}^i\delta - {}^j\delta)^2 \tag{23}$$

or more simply,

$$-\log {}^jx_s = a + b\,{}^i\delta^2 \tag{24}$$

Hildebrand[7,11] also provided graphical correlations of gas solubilities with liquid Hildebrand parameters (Figure 2), which are widely used. The slopes are similar for many of the simpler gas solutes. Points for tetrafluoromethane and sulfur hexafluoride do not lie on straight lines and have been omitted. The points for carbon dioxide in some liquids that provide electron donor-acceptor interactions (tetrachloromethane, toluene, benzene) lie well above the line for nonbasic liquids, and also have been omitted.

This method also provides good linear correlations for the solubilities of hydrogen, argon, ethane, and perfluoroethane in mixtures of benzene with trichlorotrifluoroethane[41] with $\bar{\delta}$ evaluated as the volume fraction average of the Hildebrand parameters of the pure liquids. Estimates of gas-solubilities in liquid mixtures may be made with this technique, but it is preferable to have at least some data obtained directly from mixtures.[7,11,13,42]

The closely related linear solvation energy relationships (Section 7.12) have been used by Abraham, Kamlet, Taft, and co-workers[15-17] for correlation of solubilities and Gibbs energies of solution with various solution parameters, including cohesion parameters.

Hayduk and co-workers[24,25] provided a semi-empirical method for predicting the temperature coefficient of gas solubility for regular solutions that requires a knowledge of the Hildebrand parameter and critical temperature of the liquid, as well as at least one solubility

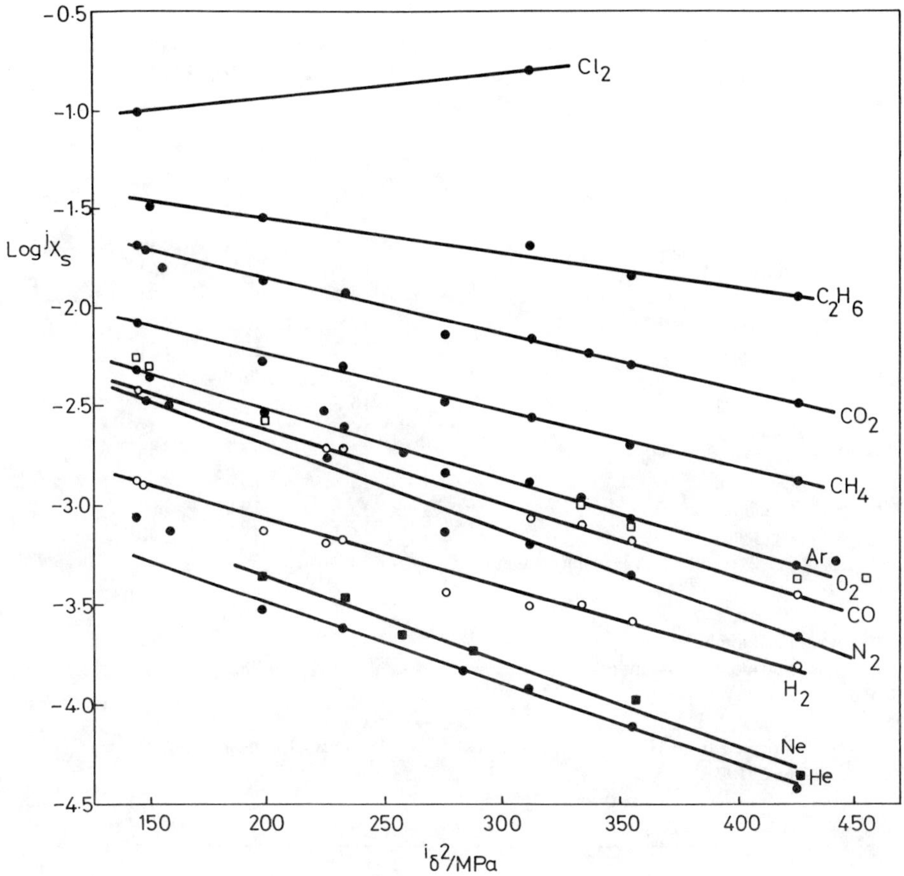

FIGURE 2. Linear correlation of logarithm of gas solubility with square of liquid Hildebrand parameter. (Adapted from Hildebrand, J. H., Prausnitz, J. M., and Scott, R. L., *Regular and Related Solutions,* Van Nostrand-Reinhold, New York, 1970.)

measurement (at any temperature) of the gas in that liquid. An average reference solubility (x_o) at the liquid critical temperature was used and correlated with the Hildebrand parameter:

$$\log (x_2/x_1) = \log (x_o/x_1) \log (T_2/T_1)/\log (T_c/T_1) \qquad (25)$$

$$\log (10^4 x_o) = 2.265 - 0.0655 (\delta/\text{MPa}^{1/2}) \qquad (26)$$

Numerical values are listed in Table 4.

Iwai, Nagafui, and Arai[43] used cohesion parameters and reduced cohesion parameters (Section 7.4) in their prediction of gas solubilities in molten polymers, and Devallez et al.[108] used cohesion parameters in correlations for the solubilities of respiratory gases (oxygen and carbon dioxide) in 1,2-bis(fluoroalkyl)ethenes.

It is possible to extrapolate measured liquid-liquid data to provide liquid-vapor data at higher temperatures by means of cohesion parameters[44] (see Section 11.5).

Another simple application of cohesion parameters to gas-liquid solubility problems is to plot gas solubility directly against liquid Hildebrand parameters, as in the results reported by Steinberg and Manowitz,[45] who wished to separate xenon from a nuclear reactor by absorption in a liquid solvent (Figure 3). A relatively sharp maximum is obtained at a Hildebrand parameter value close to that estimated for ''liquid'' xenon by other methods.

TABLE 4
Average Reference Solubilities, Critical Temperatures and Hildebrand Parameters used by Hayduk for Gas Solvents

Liquid	$\delta/MPa^{1/2}$ (25°C)	10^4x_o	T_c/K
Carbon disulfide	20.5	7.4	546
Benzene	18.8	12.3	563
Toluene	18.2	14.5	594
Tetrachloromethane	17.6	11.9	556
Docosane	17.2	15.5	801
Eicosane	17.0	12.8	775
Cyclohexane	16.8	13.5	553
Octadecane	16.6	13.7	748
Methylcyclohexane	16.0	17.3	572
Dodecane	15.8	19.1	658
Octane	15.3	19.7	569
Heptane	15.1	17.3	540
1,1,2-Trichloro-1,2,-trifluoroethane (Freon 113)	14.9	20.4	487
Isooctane	14.1	16.6	544
Perfluoroheptane	12.5	30.7	487
Perfluorotributylamine	12.1	28.9	553
Perfluoroheptane	11.9	33.0	476

Adapted from Hayduk, W. and Buckley, W. D., *Can. J. Chem. Eng.*, 49, 667, 1971.

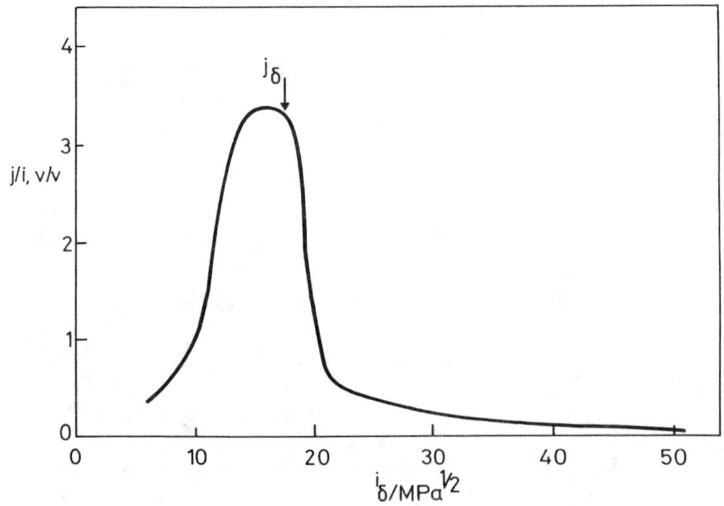

FIGURE 3. Variation of xenon solubility (volume/volume at constant temperature and low pressure) with liquid Hildebrand parameter. (Adapted from Steinberg, M. and Manowitz, B., *Ind. Eng. Chem.*, 51(1), 47, 1959.)

Lewis, Hopke, and Stukel[46] found a Hildebrand parameter of (17.2 ± 0.2) $MPa^{1/2}$ for radon from solubilities in a series of perfluorocarbons by this method. Hayduk carried out a similar procedure for the solubilities of ethane in nonpolar[47] and polar[48] liquids. England[45] published curves for carbon dioxide solubility as functions of cohesion parameter for various molar volumes: solubility was highest in liquids of low molar volume.

Snyder's polarity index (Section 8.5), which is related to cohesion parameters, also has been used for gas solubilities, particularly in chromatographic studies.[49]

11.3 COMPRESSED GASES

In addition to the situation described above of a gas possessing a hypothetical liquid-like molar volume when it dissolves in a liquid at low pressures, there is that of gases that become liquids or that achieve liquid-like molar volumes because of low temperatures and/or high pressures. The calculation of cohesion parameters of cryogenic solvents (gases liquefied by low temperatures) by corresponding states methods is described in Section 7.4, and regular solution theory has been used in the prediction of the vapor-liquid equilibrium properties in mixed cryogenic fluids including argon, nitrogen, oxygen, carbon monoxide, carbon dioxide, methane, ethylene, ethane, and propane. Cohesion parameters of refrigerants and other cryogenic liquids are included in many of the data tables in previous chapters, particularly Table 13 of Chapter 8 (Hildebrand parameters) and also Table 12 of Chapter 5 (Hansen parameters). Activity coefficient methods of Section 7.4 are also applicable.[104]

The solvent properties of compressed gases, especially carbon dioxide, for biochemicals and polymers have been receiving more attention recently because of their application in high pressure gas chromatography and supercritical fluid chromatography (Chapter 18). It is reported[50] that even at 25 K above their critical temperatures, a mixture of helium ($\delta \approx$ 8MPa)$^{1/2}$ and xenon ($\delta \approx$ 16MPa)$^{1/2}$ at 200 MPa separates into two phases because of the divergent cohesion properties. Lagowski[51] has commented on the solvent ability of liquid ammonia in terms of Hildebrand parameters, dense fluid methane has been studied[52] and aerosol propellants have been discussed on the basis of their cohesion parameters[53-55] (Table 13, Chapter 8).

The equation[9,56-58] introduced in Section 9.2 for the Hildebrand parameter of the supercritical fluid introduced in Section 9.2 is

$$\delta = (1.25\, p_c^{1/2})(\rho_r/\rho_{r,l}) = \delta_l(\rho/\rho_{r,l}) \tag{27}$$

where p_c is the critical pressure, ρ_r is the reduced density of the supercritical fluid, and $\rho_{r,l}$ is the reduced density of the liquid, normally about 2.7.

It has been suggested that the solvent power of compressed gases and supercritical fluids depends on both a state or physical effect, and a chemical effect. Equation 27 indicates that the Hildebrand parameter of a gas reaches the value characteristic of the parent liquid when the gas is compressed to the liquid density; the first factor is associated with the chemical effect and the second with the state effect. Thodos and Kramer[106] used this expression, together with Equation 53, Chapter 5, and the Flory-Huggins entropy of mixing for infinitely dilute solutions with solutes and solvents of disparate molecular size (Equation 21, Chapter 4) to model the solubilities of solids in supercritical fluids and applied it to the solubility of 1-hexadecanol and palmitic acid in carbon dioxide.

In chromatography (Chapter 18), the chemical effect of the mobile phase determines its position in the eluotropic series, which orders the elution strengths of various fluids. The elution strengths and the Hildebrand parameters of liquids show a similar ordering, and this order is preserved in compressed gases if the state effect is considered.

Allada[59] defined an alternative generalized cohesion parameter for use in supercritical conditions,

$$\delta = [(U^\infty - U)/V]^{1/2} \tag{28}$$

where U and V are the internal energy and volume of the supercritical fluid at the specified

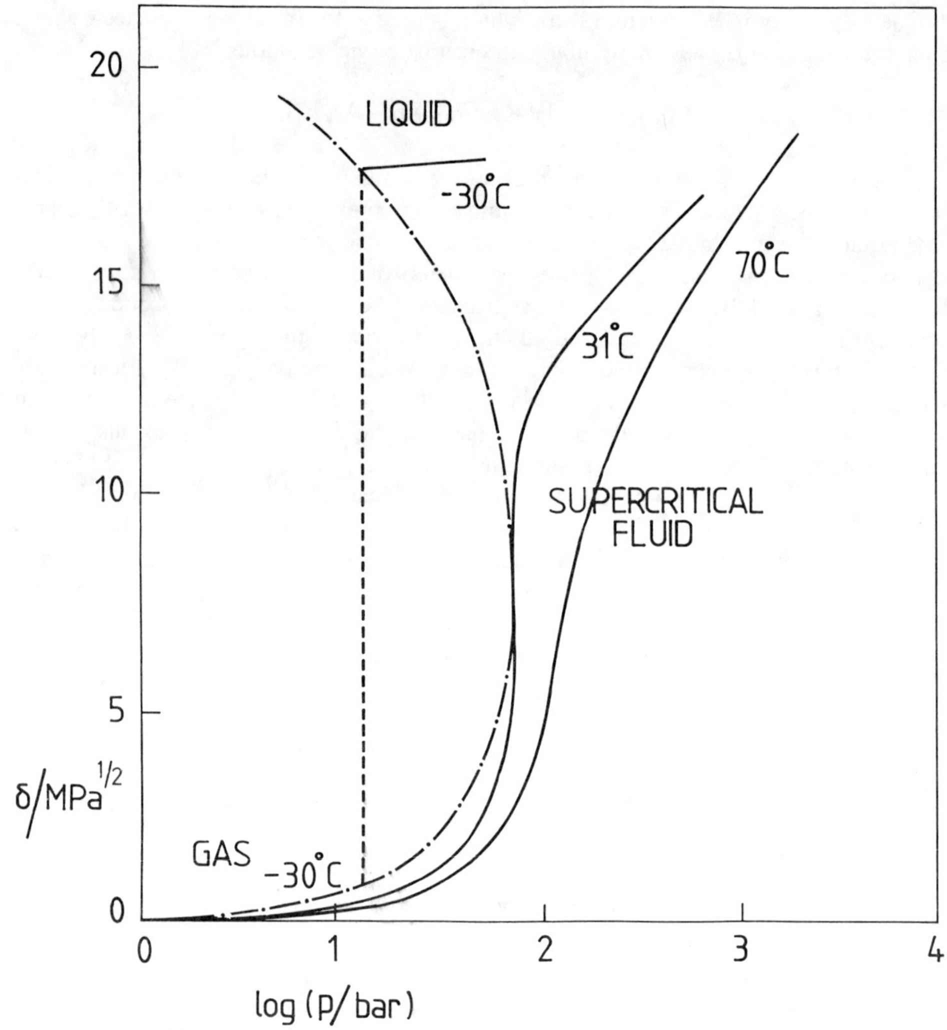

FIGURE 4. Hildebrand parameter (defined by Equation 28) of gaseous, liquid, and supercritical fluid carbon dioxide as a function of pressure at various temperatures. (Adapted from Johnston, K. P., *Am. Chem. Soc. Symp. Ser.*, 406, 1, 1989.)

temperature and pressure, and U^{∞} is the internal energy of the gas isothermally expanded to zero pressure where intermolecular separation is infinite.[60] The Hildebrand parameter also was modified for supercritical fluids by defining

$$\delta = [(\Delta H - RT(1 - Z)/V]^{1/2} \tag{29}$$

where Z is the compressibility factor.

The Hildebrand parameter of a supercritical fluid (defined by Equation 28, for example) may be varied over a wide range with small changes of pressure or temperature. Figure 4 shows[112] that at $-30°C$ δ for carbon dioxide rises discontinuously from near zero in the gas to about 17 $MPa^{1/2}$ in the liquid, but above the critical temperature δ varies continuously and may be readily adjusted to the desired value.

Relations between Hildebrand parameters, internal pressure (Section 7.5)[61] and equations of state[62] have been used to estimate parameter values for supercritical fluids, particularly carbon dioxide.

TABLE 5
Hildebrand Parameter (δ/MPa$^{1/2}$) Differences Between Supercritical Carbon Dioxide and Liquids, at 71—72°C for Oils and at 52°C for Polar Liquids

Liquid	p/MPa 34.5	55.2	69.0	110	140
Soy oil	3.9	3.5	3.3	3.3	3.3
Jojoba oil	4.3	3.9	3.9	—	—
Castor oil	3.1	3.7	3.7	—	—
Glycerol	12.3	12.7	12.5	—	—
Ethylene glycol	11.7	12.9	13.7	—	—
Triethylene glycol	11.1	11.1	10.4	—	—
1-Butanol	9.2	9.2	9.4	—	—

Adapted from Eissler, R. L. and Friedrich, J. P., *J. Am. Oil Chem. Soc.*, 65, 764, 1988.

Determination or calculation of reliable cohesion parameter values of oils and other liquids under conditions in which the other component is supercritical are not simple, and an alternative approach has been to estimate Hildebrand parameter *differences* between supercritical carbon dioxide and the liquid solute (often a natural oil) from the compositions of the two fluid phases by making use of the relationship of Fujishiro and Hildebrand[63] (Section 7.6). Such studies have been made by Eissler and Friedrich[64] (Table 5).

Other applications of cohesion parameters to supercritical systems have been the modification of supercritical fluids by polar cosolvents,[65] and a consideration of clustering in supercritical fluid mixtures.[66] Johnston[112] has reviewed recent applications of supercritical fluids.

11.4 SUPERCRITICAL EXTRACTION

The application of supercritical fluids to extraction is becoming increasingly important,[67] and cohesion parameters are providing a means of assessing extraction conditions, particularly for natural oils.[68-71,109,111]

Pang and McLaughlin[72] applied cohesion parameter assessment to the supercritical extraction of the aromatic solid fluorene with ethylene and carbon dioxide and to extraction of tar sands at 300°C with supercritical carbon dioxide and a range of other fluids. But Amestica and Wolf[73] found no linear relation between Hildebrand parameter and coal extraction yields with supercritical fluids such as toluene and ethanol. Yonker and Smith[74] considered Hildebrand parameters when discussing the solubilities of caffeine and theophylline in supercritical ammonia. Supercritical carbon dioxide extraction of cholesterol in milk fat fractions has been discussed in terms of Hildebrand parameters and molar volumes.[103] The extraction and separation of the methyl esters of oleic, linolenic, and linolenic acids from their mixture by means of supercritical carbon dioxide and entrainers (Table 6) also has been described by a model incorporating Hildebrand parameters.[107]

11.5 VAPOR-LIQUID EQUILIBRIA IN MIXTURES

Vapor-liquid equilibria, particularly of hydrocarbon mixtures, may be expressed in terms of the vaporization ratio, ${}^iK$ of component i in gas and liquid phases (Equation 12):

$$ {}^iK = {}^iy/{}^ix = {}^if_x\,{}^i\nu/{}^i\varphi \tag{30}$$

TABLE 6
Hildebrand Parameters and Molar Volumes Used in the Selective Extraction of Methyl Esters of Higher Aliphatic Acids with Supercritical Carbon Dioxide and Entrainers

Compound	$V/cm^3\ mol^{-1}$	$\delta/MPa^{1/2}$
Methyl esters		
Methyl oleate	344	16.4
Methyl linoleate	337	16.3
Methyl linolenate	331	16.2
Entrainers		
Octane	164	15.4
Heptane	147	15.3
Hexane	132	14.9
Ethyl valerate	148	17.4
Ethyl butyrate	132	17.5
Ethyl propionate	116	17.7
Ethyl acetate	99	18.3
Supercritical carbon dioxide		
9.3 MPa, 40°C	90	8.2
10.8 MPa, 32°C	59	12.1

Adapted from Ikushima, Y., Saito, N., and Goto, T., *Ind. Eng. Chem. Res.*, 28, 1364, 1989.

K values are functions of composition as well as of pressure and temperature, although the composition dependence is frequently neglected for mixtures involving similar liquids, such as aliphatic hydrocarbons. The relative volatility of component i with respect to j is[75]

$$^{ij}\alpha = {}^iK/{}^jK = ({}^iy\,{}^jx)/({}^ix\,{}^jy) \tag{31}$$

The composition dependence of K or α may be expressed in terms of Hildebrand parameters. From Equations 13 and 30 it is predicted that log ^iK or log $(^iy/^ix)$ is a linear function of $(^i\delta - \overline{\delta})^2$, where $\overline{\delta}$ is the effective Hildebrand parameter of the mixture, and this can be observed (Figure 5).

Barton and co-workers[76-78] have used Hildebrand parameters for the liquid phase activity in the development of a correlation system for C_3 to C_5 hydrocarbons that can predict K values with average deviations of 2%. This involves a set of Hildebrand parameters that are self-consistent although slightly different from those previously published. They are presented in Table 7 in the form

$$\delta = \delta_o + mT \tag{32}$$

which may be compared with Table 1, Chapter 9. There is also excellent agreement between experimental values of K, f_x, and α for multicomponent systems with those calculated using

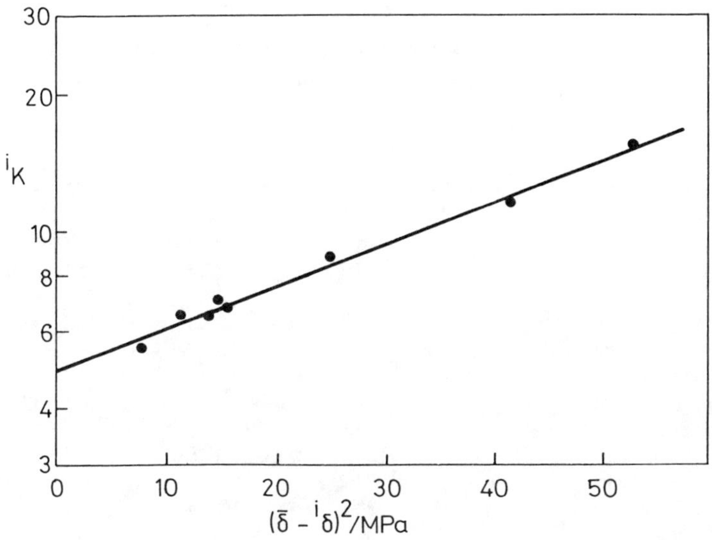

FIGURE 5. Effect of solvent mixture Hildebrand parameter ($\bar{\delta}$) on methane vapor-liquid equilibrium at 38°C and 3.5 MPa. (Adapted from Prausnitz, J. M., Edmister, W. C., and Chao, K. C., *Am. Inst. Chem. Eng. J.*, 6, 214, 1960.)

Equation 41, Chapter 4. Similar correlations have been made by other workers for mixtures covering a wider range of compounds.*

Yeo and Christian[87] have presented a general method for the prediction of the temperature dependence of solubilities of solutes distributed between nonpolar solvents and vapor. For extractive distillation and other separation processes where a polar solvent is used to facilitate the separation of nonpolar hydrocarbon components, it is necessary to use multicomponent cohesion parameters.[84,88-92]

The volatility of component *i* relative to that of component *j* can be expressed in terms of the selectivity ${}^{k}s_{ij}$ of a solvent *k* with respect to *i* and *j*:

$$ {}^{ij}\alpha = {}^{k}s_{ij}\, {}^{j}p_s/{}^{j}p_s = {}^{i}f_x\, {}^{i}p_s/({}^{j}f_x\, {}^{j}p_s) \tag{33} $$

where p_s is the (saturation) vapor pressure of the pure component. For a good extractive solvent, ${}^{k}s_{ij}$ must be significantly different from unity, and its value for extractive distillations may be predicted or correlated in terms of cohesion parameters, just as described in Section 7.7 for solvent extraction. The selectivity is a maximum when the solvent *k* is present in excess, and in the limiting case, at infinite dilution of components *i* and *j* (Equation 81, Chapter 7),

$$ RT \ln {}^{k}s_{ij} = {}^{i}V\, {}^{ik}A - {}^{j}V\, {}^{jk}A \tag{34} $$

For the separation of nonpolar hydrocarbons *i* and *j* with a polar liquid *k* (Equation 83, Chapter 7).

$$ \begin{aligned} RT\ ln\ {}^{k}s_{ij} = {} & [({}^{i}\delta_d - {}^{k}\delta_d)^2\, {}^{i}V - ({}^{j}\delta_d - {}^{k}\delta_d)^2\, {}^{j}V] + \\ & [{}^{k}\delta_o^2({}^{i}V - {}^{j}V)] + [2\,{}^{k}\delta_i({}^{j}V\, {}^{i}\delta_d - {}^{i}V\, {}^{i}\delta_d)] \end{aligned} \tag{35} $$

* For example: Chao and Seader;[79] Connolly;[80] Drahos et al.,[81] Gothard et al.,[82] Hayduik and co-workers;[24,25] Nitta and Katayama;[105] Prausnitz and co-workers;[22,83,84] and Saito, Sagara et al.[85,86]

TABLE 7
Coefficients of the Hildebrand Parameter Temperature Dependence Equation[76,77]

Compound	δ_o/MPA$^{1/2}$	-10^3 m/K^{-1}	Range, T/K
Propane	18.8	18.0	255 – 370
Propene, propylene	22.8	27.2	283 – 333
Propadiene	28.5	40.8	283 – 333
Propyne	29.3	37.2	283 – 333
Butane	20.4	20.5	311 – 378
Isobutane, 2-methylpropane	20.3	20.5	266 – 378
1-Butene, 1-butylene	24.9	30.8	311 – 378
Isobutene, 2-methylpropane	24.0	29.1	328 – 378
trans-2-Butene	23.6	27.3	311 – 378
cis-2-Butene	26.9	36.0	328 – 378
1,3-Butadiene	26.0	31.8	311 – 367
1,2-Butadiene	27.2	34.3	311 – 367
1-Butyne	28.5	33.4	311 – 367
1-Buten-3-yne	30.0	36.8	311 – 367
2-Butyne	27.6	31.6	311 – 367
Pentane	22.4	24.1	325 – 336
Isopentane, 2-methylbutane	22.3	24.0	273 – 348
3-Methyl-2-butene	23.4	24.7	325 – 336
2-Methyl-1,3,-butadiene	23.8	23.4	325 – 336

The significance of the dispersion, orientation, and induction terms in this expression is discussed in Section 7.10.

The methods of Chapter 7 for activity coefficient estimation can be extended to vapor-liquid equilibria; corrections to the geometric mean assumption of the regular solution theory are usually necessary, an area that has been explored in detail by Prausnitz and co-workers.[21,22,84,88,93-96]

Prediction of vapor-liquid equilibrium data for completely miscible systems often is approached by the evaluation of activity coefficients at infinite dilution,[97,98] and these can be estimated by alternative cohesion parameter methods such as MOSCED (Section 5.13) as well as group methods such as UNIFAC (Section 7.8). Kamlet, Taft, Carr, and Abraham[17] found that when gas/liquid partition coefficients were referenced to an alkane solute of similar molar volume they were well correlated by the solvatochromic parameters (Section 8.4) that include the Hildebrand parameter.

Multicomponent liquid-vapor equilibrium data are sometimes correlated by Wilson equations[22,99] (Section 4.6) because only the binary parameters are required. Often, however, the parameters of all possible binary systems are not known, and it is desirable to be able to estimate the parameters in the Wilson equations from the properties of the pure components. Gothard and co-workers[82] have shown that this can be done in terms of Hildebrand parameters by relating the absolute values of the Hildebrand parameter differences to the Wilson interaction energy differences.

Graboski and Daubert[100] correlated Soave equation of state interaction coefficients from vapor-liquid equilibria with Hildebrand parameter differences. Lee[101] went back to the Hildebrand-Scatchard theory (Section 4.2), reformulated it in terms of statistical mechanical quantities, incorporated entropic effects into the regular solution model, and applied it to vapor-liquid equilibrium properties of mixtures including carbon dioxide, hydrogen sulfide, propane, pentane, and acetone.

11.6 HENRY'S LAW

From the definition of ${}^{ij}H$ (Equation 7) and Equation 13 it follows that for a dilute solution (${}^{j}\phi \approx 1$),

$$\ln {}^{ij}H = \text{constant} + {}^{i}V_1\,({}^{i}\delta - {}^{j}\delta)^2/(RT) \tag{36}$$

and that for any solute j

$$\ln {}^{ij}H = \text{constant} + {}^{j}V_1\,{}^{i}\delta^2/(RT) \tag{37}$$

${}^{j}V_1$ and ${}^{j}\delta$ refer to a hypothetical liquid and can be determined in a variety of ways, as discussed in Section 11.2 and as illustrated by Sagara et al.[86] for hydrogen, nitrogen, methane, ethane, and propane in hydrocarbons.

An alternative correlation that has been used for the Henry's constant for helium[40] is

$$\ln {}^{ij}H = a_o + a_1/T + a_2\,{}^{i}\delta \tag{38}$$

where the temperature dependence of ${}^{i}\delta$ is expressed in the form of Equations 28 and 29, Chapter 7. Another suggestion in terms of Hildebrand parameters is the reduced Henry's constant, (${}^{ij}H/{}^{i}\delta^2$), which has been plotted against temperature with the aim of providing a universal curve for hydrogen in all solvents.[39]

Tassios and co-workers[102] have devised a method for the prediction of Henry's constant values for single gases in mixed liquids from the corresponding binary data and a one-parameter form of the Wilson equation, making use of Hildebrand parameters.

It should be noted that Henry's law assumes that the gas-phase fugacity is equal to the partial pressure; this is not necessary and it is possible to define ${}^{ij}H$ as

$${}^{j}f = {}^{j}\varphi\,{}^{j}yp = {}^{ij}H\,{}^{j}x \tag{39}$$

in which the fugacity coefficient ${}^{j}\phi$ (Equation 9) is incorporated. Alternative and more rigorous forms of Henry's law are also used.

REFERENCES

1. **Baker, R. W., Roman, I. C., and Lonsdale, H. K.,** Liquid membranes for the production of oxygen-enriched air. I. Introduction and passive liquid membranes, *J. Membrane Sci.*, 31, 15, 1987.
2. **Brink, L. E. S. and Tramper, J.,** Optimization of organic solvent in multiphase biocatalysis, *Biotechnol. Bioeng.*, 27, 1258, 1985.
3. **Friant, S. L. and Suffet, I. H.,** Interactive effects of temperature, salt concentration and pH on head space analysis for isolating volatile trace organics in aqueous environmental samples, *Analyt. Chem.*, 51, 2167, 1979.
4. **Battino, R. and Clever, H. L.,** The solubility of gases in liquids, *Chem. Rev.*, 66, 395, 1966.
5. **Clever, H. L. and Battino, R.,** The solubility of gases in liquids, in *Solutions and Solubilities,* Part I, Dack, M. R. J., Ed., John Wiley & Sons-Interscience, New York, 1975, chap. 7.
6. **Hildebrand, J. H.,** A simple correlation of gas solubilities, *J. Phy. Chem.*, 58, 671, 1954.
7. **Hildebrand, J. H. and Lamoreaux, R. H.,** Solubility of gases in liquids, *Ind. Eng. Chem. Fundam.*, 13, 110, 1974.
8. **Hildebrand, J. H.,** *Solubility of Non-Electrolytes,* 2nd ed., Reinhold, New York, 1936; Russian translation, GONTI, Moscow-Leningrad, 1938.
9. **Hildebrand, J. H. and Scott, R. L.,** *Solubility of Non-Electrolytes,* 3rd ed., Reinhold, New York, 1950; Dover, Princeton, NJ, 1970.

10. **Hildebrand, J. H. and Scott, R. L.,** *Regular Solutions,* Prentice-Hall, Englewood Cliffs, NJ, 1962.
11. **Hildebrand, J. H., Prausnitz, J. M., and Scott, R. L.,** *Regular and Related Solutions,* Van Nostrand-Reinhold, Princeton, NJ, 1970.
12. **Gerrard, W.,** *Solubility of Gases and Liquids. A Graphic Approach,* Plenum Press, New York, 1976.
13. **Linford, R. G. and Thornhill, D. G. T.,** Use of Hildebrand and Lamoreaux's theory to predict solubility and related properties in gas-liquid systems, *J. Appl. Chem. Biotechnol.,* 27, 479, 1977.
14. **Wilhelm, E.,** Dilute solutions of gases in liquids, *Fluid Phase Equil.,* 27, 233, 1986.
15. **Abraham, M. H., Kamlet, M. J., and Taft, R. W.,** Linear solvation energy relationships. Part 19. Correlation of the free energies of solution of 41 solutes in select solvents with Hildebrand's solubility parameter, δ_H, and with the solvatochromic parameter, π^*, *J. Chem. Soc. Perkin Trans. II,* 923, 1982.
16. **Abraham, M. H., Kamlet, M. J., Taft, R. W., and Weathersby, P. K.,** Solubility properties in polymers and biological media. 1. Correlations in various media of the solubility of gaseous anesthetics and other solutes with a number of solute parameters, *J. Am. Chem. Soc.,* 105, 6797, 1983.
17. **Kamlet, M. J., Taft, R. W., Carr, P. W., and Abraham, M. H.,** Linear solvation energy relationships. Part 9. Correlations of gas/liquid partition coefficients with the solvatochromic parameters π^*, α and β, *J. Chem. Soc. Faraday Trans. I,* 78, 1689, 1982.
18. **Jolley, J. E. and Hildebrand, J. H.,** Solubility, entropy and partial molal volumes in solutions of gases in non-polar solvents, *J. Am. Chem. Soc.,* 80, 1050, 1958.
19. **Lachowicz, S. K. and Weale, K. E.,** Prediction of gas solubility in nonpolar liquids, *J. Chem. Eng. Data,* 3, 162, 1958.
20. **Smith, E. B. and Walkley, J.,** The thermodynamic properties of gases in solution. I. The partial molal volume, *J. Phys. Chem.,* 66, 597, 1962.
21. **Prausnitz, J. M. and Shair, F. H.,** A thermodynamic correlation of gas solubilities, *Am. Inst. Chem. Eng. J.,* 7, 682, 1961.
22. **Prausnitz, J. M.,** *Molecular Thermodynamics of Fluid Phase Equilibria,* Prentice-Hall, Englewood Cliffs, NJ, 1968; **Prausnitz, J. M., Lichtenthaler, R. N., and Gomez de Azevedo, E.,** *Molecular Thermodynamics of Fluid Phase Equilibria,* 2nd ed., Prentice-Hall, Englewood Cliffs, NJ, 1986.
23. **Boyer, F. L. and Bircher, L. J.,** The solubility of nitrogen, argon, methane, ethylene and ethane in normal primary alcohols, *J. Phys. Chem.,* 64, 1330, 1960.
24. **Hayduk, W. and Buckley, W. D.,** Temperature coefficient of gas solubility for regular solutions, *Can. J. Chem. Eng.,* 49, 667, 1971.
25. **Hayduk, W. and Castaneda, R.,** Solubilities of the highly soluble gases, propane and butane, in normal paraffin and polar solvents, *Can. J. Chem. Eng.,* 51, 353, 1973.
26. **Shaw, A. W. and Vosper, A. J.,** Solubility of nitric oxide in aqueous and nonaqueous solvents, *J. Chem. Soc. Faraday Trans. I,* 73, 1239, 1977.
27. **Naumenko, N. K., Mukhin, N. N., and Aleskovskii, V. B.,** Investigations of the solubility of oxygen in liquid derivatives of the aromatic series, *Zh. Prikhl. Khim.,* 42, 2522, 1969; *J. Appl. Chem. U.S.S.R.,* 42, 2376, 1969.
28. **Yen, L. C. and McKetta, J. J., Jr.,** A thermodynamic correlation of nonpolar gas solubilities in polar, nonassociated liquids, *Am. Inst. Chem. Eng. J.,* 8, 501, 1962.
29. **Gjaldbaek, J. C. and Anderson, E. K.,** The solubility of carbon dioxide, oxygen, carbon monoxide and nitrogen in polar solvents, *Acta Chem. Scand.,* 8, 1938, 1954.
30. **Gjaldbaek, J. C. and Hildebrand, J. H.,** The solubility of nitrogen in carbon disulfide, benzene, normal- and cyclohexane, and in three fluorocarbons, *J. Am. Chem. Soc.,* 71, 3147, 1949.
31. **Sonnich Thomsen, E. and Gjaldbaek, J. C.,** The solubility of hydrogen, nitrogen, oxygen and ethane in normal hydrocarbons; The solubility of propane in nonpolar solvents, *Acta Chem. Scand.,* 17, 127, and 134, 1963.
32. **Gonikberg, M. G.,** Regular solutions of gases in liquids. I. Regular solutions of hydrogen, *J. Phys. Chem.,* 14, 582, 1940; *Acta Physicochim. U.R.S.S.,* 12, 489, 1940.
33. **Gonikberg, M. G.,** Regular solutions of gases in liquids. II. More concentrated solutions of hydrogen at high pressure, *Acta Physicochim. U.R.S.S.,* 12, 921, 1940.
34. **Williamson, A. G.,** Solutions of nonelectrolytes, *Ann. Rev. Phys. Chem.,* 15, 63, 1964.
35. **England, C.,** Gas solubilities in physical solvents, *Chem. Eng.,* 93(8), 63, 1986.
36. **Frank, H. A. and Lawson, D. D.,** Methods for estimating the solubility of gases in electrolytes used in lithium batteries, Electrochemical Society Meeting, Hollywood, FL, October 1980.
37. **Lawson, D. D.,** Methods of calculating engineering parameters for gas separations, *Appl. Energ.,* 6, 241, 1980.
38. **King, M. B., Kassim, K., and Al-Najjar, H.,** The solubilities of carbon dioxide, hydrogen sulfide and propane in some normal alkane solvents. II. Correlation of data at 25°C in terms of solubility parameters and regular solution theory, *Chem. Eng. Sci.,* 32, 1247, 1977.
39. **Orentlicher, M. and Prausnitz, J. M.,** Thermodynamics of hydrogen solubility in cryogenic solvents at high pressures, *Chem. Eng. Sci.,* 19, 775, 1964.

40. **Marathe, V. V. and Prausnitz, J. M.,** Correlation for helium solubilities in liquids, *J. Appl. Chem. Biotechnol.*, 21, 173, 1971.
41. **Linford, R. G. and Hildebrand, J. H.,** Solubilities of gases in mixtures of nonpolar liquids, *J. Phys. Chem.*, 73, 4410, 1969.
42. **Vosmansky, J.,** The methods of predicting the solubility of gases in mixed solvents, *Chem. Listy*, 81, 592, 1987.
43. **Iwai, Y., Nagafui, M., and Arai, Y.,** Prediction of solubilities for gases in molten polymers using a solution model, *Technol. Rep. Kyushu Univ.*, 53, 387, 1980.
44. **Koskas, A. and Durandet, J.,** Thermodynamics and prediction of two-component liquid-liquid equilibria, *Chim. Ind. Gen. Chim.*, 98, 1386, 1967.
45. **Steinberg, M. and Manowitz, B.,** Recovery of fission product noble gases, *Ind. Eng. Chem.*, 51(1), 47, 1959.
46. **Lewis, C., Hopke, P. K., and Stukel, J. J.,** Solubility of radon in selected perfluorocarbon compounds and water, *Ind. Eng. Chem. Res.*, 26, 356, 1987.
47. **Hayduk, W.,** Ethane/Non-polar, non-paraffin solvents, in *Ethane,* Hayduk, W., Ed., Solubility Data Series, Vol. 9, Pergamon, Oxford, 1982, 138.
48. **Hayduk, W.,** Ethane/Polar solvents excluding alcohols, water, and aqueous solutions, in *Ethane,* Hayduk, W., Ed., Solubility Data Series Vol. 9, Pergamon, Oxford, 1982, 196.
49. **Bakalyar, S. R., Bradley, M. P. T., and Honganen, R.,** The role of dissolved gases in high-performance liquid chromatography, *J. Chromatogr.*, 158, 277, 1978.
50. **Hansen, C. M.,** The universality of the solubility parameter, *Ind. Eng. Chem. Prod. Res. Dev.*, 8, 2, 1969.
51. **Lagowski, J. J.,** The chemistry of liquid ammonia, *J. Chem. Educ.*, 55, 752, 1978.
52. **Mishra, R. A.,** Some thermodynamic properties of dense fluid methane, *Vijnana Parishad Anusandhan Patrika*, 21, 293, 1978.
53. **Hobin, T. P.,** Surface tension in relation to cohesive energy with particular reference to hydrocarbon polymers, *J. Adhes.*, 3, 237, 1972.
54. **Isotron Chemicals Department,** *Propellent Solvency Values,* Isotron Tech. Inf. File, 3(2), Pennwalt Chemical Corporation, Philadelphia, PA, 1964.
55. **Reed, A. B., Jr.,** Solubility parameters and their application to aerosols, *Chemical Specialities Manufacturers Association, Proc. 46th Ann. Meet.*, Washington, DC, 1959, 94.
56. **Giddings, J. C., Myers, M. N., McLaren, L., and Keller, R. A.,** High pressure gas chromatography of nonvolatile species, *Science,* 162, 67, 1968.
57. **Giddings, J. C., Myers, M. N., and King, J. W.,** Dense gas chromatography at pressures to 2000 atmospheres, *J. Chromatogr. Sci.*, 7, 276, 1969.
58. **Leyendecker, D.,** Selection of conditions for an SFC separation, in *Supercritical Fluid Chromatography* (RSC Chromatography Monograph), Smith, R. M., Ed., Royal Society of Chemistry, London, 1988, chap. 3, 53.
59. **Allada, S. R.,** Solubility parameters of supercritical fluids, *Ind. Eng. Chem. Proc. Des. Dev.*, 23, 344, 1984.
60. **Ikushima, Y., Goto, T., and Arai, M.,** Modified solubility parameter as an index to correlate the solubility in supercritical fluids, *Bull. Chem. Soc. Jpn.*, 60, 4145, 1987.
61. **Eissler, R. L., Ritter, J. M., and Friedrich, J. P.,** Solubility parameters of supercritical fluids from pressure-volume-temperature data, *J. Am. Oil Chem. Soc.*, 60, 710, 1983.
62. **Bente, P. F., III and Weaver, H. E.,** Hildebrand parameters of liquid and supercritical CO_2, *Abs. Pap. Am. Chem. Soc.*, 186, PHYS 23, 1983.
63. **Fujishiro, R. and Hildebrand, J. H.,** The liquid-liquid solubility of cyclohexane and perfluorbutylamine at 25°, *J. Phys. Chem.*, 66, 573, 1962.
64. **Eisslen, R. L. and Friedrich, J. P.,** Estimation of supercritical fluid-liquid solubility parameter differences from stirred reactor data, *Prepr. Am. Chem. Soc. Div. Fuel Chem.*, 30(3), 62, 1985; Estimation of supercritical fluid-liquid solubility parameter differences for vegetable oils and other liquids from data taken with a stirred autoclave, *J. Am. Oil Chem. Soc.*, 65, 764, 1988.
65. **Dobbs, J. M., Wong, J. M., Lahiere, R. J., and Johnston, K. P.,** Modification of supercritical fluid behavior using polar cosolvents, *Ind. Eng. Chem. Res.*, 26, 56, 1987.
66. **Kim, S. and Johnston, K. P.,** Clustering in supercritical fluid mixtures, *Am. Inst. Chem. Eng. J.*, 33, 1603, 1987.
67. **Randall, L. G.,** The present status of dense (supercritical) gas extraction and dense gas chromatography. Impetus for DGC-MS development: Review, *Sep. Sci. Technol.*, 17, 1, 1982.
68. **King, J. W.,** Application of the solubility parameter concept to critical fluid extraction, 16th Great Lakes Regional Am. Chem. Soc. Meeting, Normal, IL, 1982.
69. **King, J. W.,** Generalized extraction conditions for the critical fluid processing of oils and oleophilic compounds, 74th National Meeting, American Oil Chemists Society, Chicago, IL, 1983; *J. Am. Chem. Soc.*, 60, 711, 1983.

70. **King, J. W.,** Supercritical fluid extraction of polymers and solvents: utilization of the solubility parameter concept, 188th National Am. Chem. Soc. Meeting, Philadelphia, PA, 1984; *Polym. Mater. Sci. Eng.*, 51, 707, 1984.
71. **King, J. W.,** Prediction of the threshold pressure in supercritical fluid extraction utilizing the Flory-Huggins interaction parameter, 20th Great Lakes Regional Am. Chem. Soc. Meeting, Milwaukee, WI, 1986.
72. **Pang, T.-H. and McLaughlin, E.,** Supercritical extraction of aromatic hydrocarbon solids and tar sand bitumens, *Ind. Eng. Chem. Process Des. Dev.*, 24, 1027, 1985.
73. **Amestica, L. and Wolf, E. E.,** Supercritical liquefaction of an Illinois No. 6 coal, *Proc. Int. Conf. Coal Sci., (Cent. Conf. Manage.: Pittsburgh, PA)*, 3-5, 1983.
74. **Yonker, C. R. and Smith, R. D.,** Solubility of caffeine and theophylline in supercritical ammonia, *Fluid Phase Equilibria*, 22, 175, 1985.
75. **King, C. J.,** *Separation Processes*, McGraw-Hill, New York, 1971.
76. **Barton, P., Holland, R. E., and McCormick, R. H.,** Correlation of vapor-liquid equilibria of C_3-C_5 hydrocarbons using solubility parameters, *Ind. Eng. Chem. Proc. Des. Dev.*, 13, 378, 1974.
77. **Barton, P., Campbell, M. L., McCormick, R. H., and Holland, R. E.,** Vapor-liquid equilibria for multicomponent C_4 hydrocarbons, Am. Chem. Soc. Meet., Washington, DC, 1979.
78. **Holland, R. E.,** *Correlation of Vapor-Liquid Equilibriums Using Solubility Parameters*, Ph.D. Dissertation, Pennsylvania State University, 1971; *Diss. Abstr. Int. B.*, 33, 173, 1972.
79. **Chao, K. C. and Seader, J. D.,** A general correlation of vapor-liquid equilibria in hydrocarbon mixtures, *Am. Inst. Chem. Eng. J.*, 7, 598, 1961.
80. **Conolly, J. F.,** Thermodynamic properties of hydrogen in benzene solutions, *J. Chem. Phys.*, 36, 2897, 1962.
81. **Drahos, J.,** Liquid-vapor equilibrium. LXXVII. Comparison of the volume and surface modifications of the regular solution model, *Collect. Czech. Chem. Commun.*, 43, 805, 1978; **Drahos, J., Wichterle, I., and Hala, E.,** A generalized method for calculation and prediction of vapour-liquid equilibria at high pressures, *Fluid Phase Equil.*, 1, 173, 1977/78.
82. **Gothard, F. A., Codrea Ciobanu, M. F., Breban, D. G., Bucur, C. I., and Sorescu, G. V.,** Predicting the parameters in the Wilson equations for activity coefficients in binary hydrocarbon systems, *Ind. Eng. Chem. Proc. Des. Dev.*, 15, 333, 1976; 16, 156, 1977.
83. **Funk, E. W. and Prausnitz, J. M.,** Thermodynamic properties of liquid mixtures: aromatic-saturated hydrocarbon systems, *Ind. Eng. Chem.*, 62(9), 8, 1970.
84. **Prausnitz, J. M., Edmister, W. C., and Chao, K. C.,** Hydrocarbon vapor-liquid equilibria and solubility parameter, *Am. Inst. Chem. Eng. J.*, 6, 214, 1960.
85. **Sagara, H., Yokobori, Y., and Saito, S.,** Application of regular solution theory to vapor-liquid equilibriums of cryogenic fluids, *J. Chem. Eng. Jpn.*, 7, 8, 1974.
86. **Sagara, H., Arai, Y., and Saito, S.,** Calculation of Henry's constant of gases in hydrocarbon solvent by regular solution theory, *J. Chem. Eng. Jpn.*, 8, 93, 1975.
87. **Yeo, K. O. and Christian, S. D.,** A thermodynamic method for predicting solubilities of solutes in nonpolar solvents, *Ind. Eng. Chem. Fundam.*, 13, 196, 1974.
88. **Anderson, R.,** Ph.D. Thesis, Dept. of Chemical Engineering, University of California, Berkeley, 1961.
89. **Finch, R. N. and Van Winkle, M.,** Prediction of vapor-liquid equilibrium for polar-nonpolar binary systems, *Am. Inst. Chem. Eng. J.*, 8, 455, 1962.
90. **Grieger, R. A. and Eckert, C. A.,** Solvent mixtures for separation processes, *Ind. Eng. Chem. Process Des. Dev.*, 6, 250, 1967.
91. **Sanada, Y. and Honda, H.,** Swelling equilibrium of coal by pyridine at 25°C, *Fuel*, 45, 295, 1966.
92. **Spelyng, W. A. and Tassios, D. P.,** Solvent selectivity for hydrocarbons with close molar volumes, *Ind. Eng. Chem. Process Des. Dev.*, 13, 328, 1974.
93. **Bazua, E. R. and Prausnitz, J. M.,** Vapour-liquid equilibria for cryogenic mixtures, *Cryogenics*, 11, 115, 1971.
94. **Prausnitz, J. M. and Anderson, R.,** Thermodynamics of solvent selectivity in extractive distillation of hydrocarbons, *Am. Inst. Chem. Eng. J.*, 7, 96, 1961.
95. **Prausnitz, J. M. and Chueh, P. L.,** *Computer Calculations for High Pressure Vapor Equilibria*, Prentice-Hall, Englewood Cliffs, NJ, 1968.
96. **Prausnitz, J. M., Eckert, C. A., Orye, R. V., and O'Connell, J. P.,** *Computer Calculations for Multicomponent Vapor-Liquid Equilibria*, Prentice-Hall, Englewood Cliffs, NJ, 1966.
97. **Thomas, E. R. and Eckert, C. A.,** Prediction of limiting activity coefficients by a modified separation of cohesive energy density model and UNIFAC, *Ind. Eng. Chem. Process Des. Dev.*, 23, 194, 1984.
98. **Park, J. H. and Carr, P. W.,** Predictive ability of the MOSCED and UNIFAC activity coefficient estimation methods, *Analyt. Chem.*, 59, 2596, 1987.
99. **Wilson, G. M.,** Vapor-liquid equilibrium. XI. A new expression for the excess free energy of mixing, *J. Am. Chem. Soc.*, 86, 127, 1964.
100. **Graboski, M. S. and Daubert, T. E.,** A modified Soave equation of state for phase equilibrium calculations, *Ind. Eng. Chem. Process Des. Dev.*, 17, 448, 1978.

101. **Lee, L. L.,** A molecular theory of solubility parameters: generalization to polar fluid mixtures, *Fluid Phase Equil.,* 35, 77, 1987.
102. **Kung, J. K., Nazario, F. N., Joffe, J., and Tassios, D.,** Prediction of Henry's constants in mixed solvents from binary data, *Ind. Eng. Chem. Process Des. Dev.,* 23, 170, 1984.
103. **Arul, J., Boudreau, A., Makhlouf, J., Tardif, R., and Gressier, B.,** Distribution of cholesterol in milk fat fractions, *J. Dairy Res.,* 55, 361, 1988.
104. **Iguchi, A.,** Predicting cryogenic vapor-liquid equilibrium, *Kagaku Sochi,* 13(11), 86, 1971.
105. **Nitta, T. and Katayama, T.,** Correlation of excess thermodynamic properties for alcohol-nonpolar solvent systems by a continuous linear association model, *J. Chem. Eng. Jpn.,* 6, 224, 1973.
106. **Kramer, A. and Thodos, G.,** Adaptation of the Flory-Huggins theory for modeling supercritical solubilities of solids, *Ind. Eng. Chem. Res.,* 27, 1506, 1988; Solubility of 1-hexadecanol and palmitic acid in supercritical carbon dioxide, *J. Chem. Eng. Data,* 33, 230, 1988.
107. **Ikushima, Y., Saito, N., and Goto, T.,** Selective extraction of oleic, linoleic, and linolenic acid methyl esters from their mixture with supercritical carbon dioxide-entrainer systems and a correlation of the extraction efficiency with a solubility parameter, *Ind. Eng. Chem. Res.,* 28, 1364, 1989.
108. **Devallez, B., Gauffeetau, Y., Le Blanc, M., and Riess, J. G.,** Solubility of respiratory gases in the 1,2-bis(fluoroalkyl) ethenes, *J. Chim. Phys. Phys. Chim. Biol.,* 85, 947, 1988.
109. **Ikushima, Y., Saito, N., Hatakeda, K., Ito, S., and Goto, T.,** Development of extraction and separation monitoring in situ with an on-line SFE-SFC/FTIR system, *Chem. Lett.,* No. 10, 1707, 1989.
110. **Sweeney, C. W., Ritter, T. J., and McGinley, E. B.,** A strategy for screening physical solvents, *Chem. Eng.,* 95(9), 119, 1988.
111. **King, J. W.,** Fundamentals and applications of supercritical fluid extraction in chromatographic science, *J. Chromatogr. Sci.,* 27, 355, 1989.
112. **Johnston, K. P.,** New directions in supercritical fluid science and technology, *Am. Chem. Soc. Symp. Ser.,* 406, 1, 1989.

Chapter 12

SOLIDS

As emphasized in Chapter 2, the Hildebrand parameter has a liquid state basis, and a hypothetical subcooled liquid reference state is necessary when the method is extended to solids. Use of the vaporization or sublimation enthalpy of a solid at 25°C less the expansion work *RT* but uncorrected for the crystal-liquid transition does not yield a Hildebrand parameter, although this cohesion parameter value may be valuable for other purposes. An example of such an application is the comment[1] that diamond ($c^{1/2} \approx 430$ MPa$^{1/2}$) is prepared from graphite ($c^{1/2} \approx 364$ MPa$^{1/2}$) using metals of high cohesion pressure such as iron (239 MPa$^{1/2}$) to promote the process. Values of cohesion parameters of the solid elements evaluated from $[(\Delta H - RT)/V]^{1/2}$, where ΔH is the enthalpy of vaporization or sublimation, are included in Table 4, Chapter 2.

Most of the general accounts of cohesion properties referred to in earlier chapters consider solids as well as liquids; for example, the monographs by Hildebrand, Prausnitz and Scott[2] and by Martin, Swarbrick and Cammarata.[3] The cohesive properties of solids have been emphasized by Kendall,[4] while Acree[5] has reviewed the use of the Hildebrand-Scatchard model (among others) to predict activity coefficients and solubilities for solid solutes in liquids.

An alternative approach is the calculation of the cohesive energies of solids from their structures using the appropriate atomic parameters, such as that carried out for Ti_4O_7 by Iguchi and Matsushima.[6,7] Wright[177] used the poly(ethylene oxide) Hildebrand parameter value in estimating lattice energy of its complexes with alkali halides which are of interest as polymer electrolytes.

As well as cohesion parameter values for solids in Table 4, Chapter 2 and Table 1 of this chapter, the cohesion parameters for many other organic compounds that are solids at room temperature are included among the liquids throughout the book. This chapter is concerned with the interactions of solids with liquids. Sections 12.1 and 12.2 deal with covalent solids, organic solids used as drugs having received particular attention. Sections 12.3 and 12.4 are concerned with ionic systems. Crystalline polymers and the incorporation in polymers of solid plasticizers and fillers are discussed in Chapter 16. Solid particles small enough for adsorption properties to be significant are considered in Chapter 17; powders in Section 17.5, colloids in Section 17.7, and dispersions in Section 17.8. Systems of large monomeric hydrocarbon-chain molecules have properties in common with polymer systems, and reference may be made to Sections 13.5, 14.1, and 18.3.

12.1 ACTIVITIES AND SOLUBILITIES

The regular solution theory and the Hildebrand parameter are based on the enthalpy changes occurring when liquids are mixed. In order to extend these concepts to solutions of crystalline solids in liquids, it is necessary to calculate the activity of the solid referred to a hypothetical liquid subcooled below its melting point. The solid state activity can then be used in conjunction with expressions relating cohesion parameter-activity expressions such as Equation 15, Chapter 4,

$$RT \ln ({}^{j}a/{}^{j}x) = RT \ln {}^{j}f_{x} = {}^{j}V\,{}^{i}\phi^{2}\,{}^{ij}A = {}^{j}V\,{}^{i}\phi^{2}\,({}^{i}\delta - {}^{j}\delta)^{2} \quad (1)$$

or

$$RT \ln {}^j a = RT \ln {}^j x + {}^j V \, {}^i \phi^2 \, ({}^i \delta - {}^j \delta)^2 \tag{2}$$

to evaluate cohesion parameters of nonvolatile solutes or to estimate their solubilities.

Solubilities of solids in hydrocarbon liquids can be accounted for by athermal or entropic factors, while for many other liquids Hildebrand parameters are adequate to predict behavior, but specific interactions must be provided for by geometric mean corrections, multicomponent cohesion parameters, or association constants, as explained in previous chapters. This has been discussed by various authors such as Fung and Higuchi.[8]

The solubility (${}^j x_s$ on the mole fraction scale) of a solid j in a solvent i has been shown to vary regularly with the Hildebrand parameter of the solvent. In some cases,[169] ${}^j x_s$ versus ${}^i \delta$ plots give a maximum used to estimate ${}^j \delta$. In others, plots of log ${}^j x_s$ against ${}^i \delta$ are nearly linear or only slightly curved, as is illustrated by the behavior of several types of ion-pairs in organic solvents: complexes of ruthenium with *o*-phenanthroline,[9] potassium halides in alcohols,[10] and alkali picrates in alcohols and esters.[11] As well as solid solubilities being expressed in terms of Hildebrand parameters in this way, other correlations also exist, such as the linear relationship between the ratio (mass solid/mass saturated solution) and the number of carbon atoms in a homologous series.[12] It is useful to explore the thermodynamics of solutions in rather more detail in order to seek more precise and useful expressions.

The equation for the variation of the activity (${}^c a$) of a crystalline solid c with temperature (T) at constant pressure (p) relative to the corresponding subcooled liquid under the same conditions is[2]

$$\begin{aligned}(\partial \ln {}^c a / \partial T)_p &= \{\partial [{}^c G - {}^1 G)/RT]/\partial T\}_p = ({}^c H - {}^1 H)/RT^2 \\ &= \Delta H / RT^2\end{aligned} \tag{3}$$

This can be integrated when the difference between the molar heat capacities of liquid and solid is known:

$$\Delta C_p = {}^1 C_p - {}^c C_p \tag{4}$$

The enthalpy of fusion at the melting point, ΔH_m is introduced:

$$\Delta H = \Delta H_m - \Delta C_p \, (T_m - T) \tag{5}$$

Integration of Equation 3 with the assumption that ΔC_p is independent of temperature then gives

$$\ln {}^c a = \frac{\Delta H_m}{R} \left\{ \frac{T - T_m}{T_m T} \right\} + \frac{\Delta C_p}{R} \left\{ \frac{T_m - T}{T} \right\} - \frac{\Delta C_p}{R} \ln \left\{ \frac{T_m}{T} \right\} \tag{6}$$

The activity of the solid is here referred to the same standard state as the activity, ${}^j a$, of the solute in solution, so the condition

$${}^j a = {}^c a \tag{7}$$

is sufficient to determine the equilibrium solubility of the solid, which on the mole fraction scale is ${}^j x_s$:

$${}^c a = {}^j a = {}^j f_x \, {}^j x_s \tag{8}$$

For an *ideal* solution,

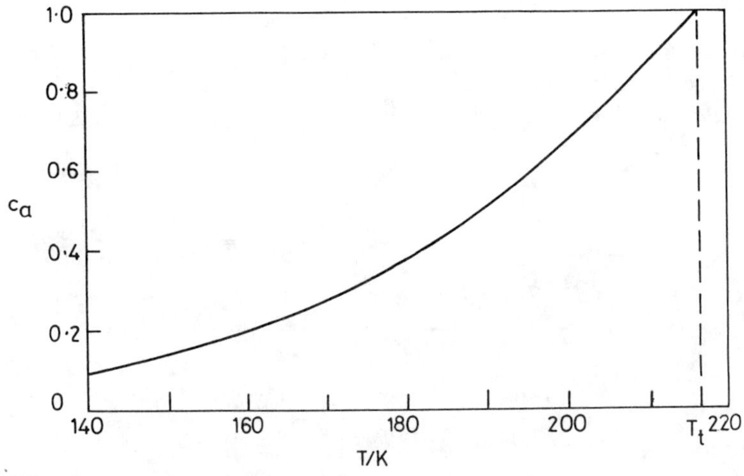

FIGURE 1. Activity of crystalline solid carbon dioxide relative to the subcooled liquid as a function of temperature. At the triple point, ${}^{c}a = 1$. (Adapted from Myers, A. L. and Prausnitz, J. M., *Ind. Eng. Chem. Fund.*, 4, 209, 1965.)

$$ {}^{c}a = {}^{j}a = {}^{j}x_{\mathrm{id}} \tag{9} $$

so the solubility can be estimated from Equation 6.

The use of this equation is illustrated by the work of Myers and Prausnitz[13] on the solubility of solid carbon dioxide in liquids at low temperatures. A more general form of Equation 6 was used to allow for the slight temperature dependence of ΔC_p:

$$ RT\ln {}^{c}a = \Delta H_t(T/T_t - 1) + \int_T^{T_t} \Delta C_p \, \mathrm{d}T - T\int_T^{T_t} (\Delta C_p/T)dT \tag{10} $$

where the subscript t refers to the triple point temperature (which is close to the melting point for many common materials). This equation is depicted in Figure 1. For an ideal solution (${}^{j}f = 1$) the solubility ${}^{j}x_s$ can be read directly from the graph, falling with decreasing temperature from complete miscibility at the triple point temperature to ${}^{j}x_s = 0.09$ at 140 K. The form of cohesive pressure relationship used in the case of carbon dioxide-hydrocarbon mixtures is discussed in Section 5.7.

If ΔC_p is assumed to be zero, Equation 6 reduces to

$$ \ln {}^{c}a = \frac{\Delta H^{m}}{R}\left\{\frac{T - T_m}{T_m T}\right\} \tag{11} $$

and from Equation 9,

$$ \ln {}^{j}x_s = \frac{\Delta H_m}{\mathrm{R}}\left\{\frac{T - T_m}{T_m T}\right\} = \frac{\Delta H_m}{R}\left\{\frac{1}{T_m} - \frac{1}{T}\right\} \tag{12} $$

an equation predicting a linear relationship between the logarithm of solubility and the reciprocal of the absolute temperature; Mortimer[14] suggested using the slopes of such lines as aids to solvent selection. Another common assumption is to equate the heat capacity to the entropy of fusion, $(\Delta\ H)/T_m$. Neau, Flynn, and Yalkowsky[171] showed for *n*-alkyl *p*-

aminobenzoates as test solutes in hydrocarbons and tetrachloromethane that such simplifying heat capacity assumptions had no greater effect on Hildebrand parameter estimation than the intrinsic experimental error.

For a *regular* solution, combination of Equations 1, 6, and 8 predicts that mole fraction solubility as a function of temperature T is

$$\log {}^{j}x_{s} = \frac{\Delta H_{m}}{2.303R}\left\{\frac{T - T_{m}}{T_{m}T}\right\} + \frac{\Delta C_{p}}{2.303R}\left\{\frac{T_{m} - T}{T}\right\} - \frac{\Delta C_{p}}{R}\log\frac{T_{m}}{T} \qquad (13)$$

$$= \frac{\Delta H_{m}}{2.303R}\left\{\frac{T - T_{m}}{T_{m}T}\right\} - \frac{{}^{j}V\,{}^{i}\phi^{2}\,{}^{ij}A}{2.303RT}$$

$$= \frac{\Delta H_{m}}{2.303R}\left\{\frac{1}{T_{m}} - \frac{1}{T}\right\} - \frac{{}^{j}V\,{}^{i}\phi^{2}\,({}^{i}\delta - {}^{i}\delta)^{2}}{2.303RT} \qquad (14)$$

Equations such as these provide predictions which are quite good for nonpolar solvents, fair for polar aprotic solvents, but unreliable for highly protic solvents.[15,16] The method has been employed for correlating or predicting the solubilities of:

- Long-chain hydrocarbons (solubilities estimated by UNIFAC superior to those by regular solution theory)[17]
- Alkanes in binary liquids[18]
- Pitch and other polynuclear aromatics[19-21]
- Benzoic acid[22] and maleic anhydride[23] (See Table 1)
- *p*-Hydroxybenzoic esters (a multiple regression method from alcohol solvent data[12] to determine the solute Hildebrand parameters in Table 1 which were self-consistent, although the solute molar volumes were anomalous)[24]
- *n*-Alkyl *p*-aminobenzoates[25]
- Eicosanoic acid from vapor-liquid-solid equilibria (see Table 1)[26] although the Hildebrand-Scatchard theory is not always successful even if association of acids is allowed for)[27,28]
- Tetrasulfur tetranitride[29] and heptasulfur imide[30] (Table 1)
- Nitrogen compounds in low dielectric liquids[31] (See Table 1)
- 1,12-Dodecamethylenediamine[32] (Table 1)
- Pyridine monocarboxamides (picolinamide, nicotinamide, isonicotinamide) and pyrazine carboxamide[33]
- Cholesteryl myristate, cholesteryl palmitate, and cholesteryl oleate[34] (Table 1)
- Mestanolone, methandienone, methyltestosterone, nandrolone and testosterone in alkanes and alcohols[35]
- Morphine: [176] Solubility showing a regular solution parabola for 8 = 27.0 MPa$^{1/2}$
- 1,3- and 1,4-Dicyanobenzenes[174]
- Hydrocortisone: a reliable 25.4 MPa$^{1/2}$ for the Hildebrand parameter from solubility in a wide variety of liquids; 23.3 MPa$^{1/2}$ calculated by the group contribution method of Hoy (Section 7.2)[36]
- Bromides[37] and iodides[38] of arsenic, antimony, bismuth and tin; iodides of tin and germanium[39] (covalent nonelectrolytes; see Table 1)
- Sulfur[40,41]

As an alternative to the assumption of zero ΔC_p, Equation 3 can be rewritten[2] as

TABLE 1
Hildebrand Parameters and Apparent Molar Volumes for Predominantly Covalent Crystalline Solids as Subcooled Liquids at ~25°C

Material (liquid, or solid as subcooled liquid)	${}^{j}V$/cm³ mol⁻¹	${}^{j}\delta$/ MPa$^{1/2}$ (2 significant figures)	References
Alcohols			
1-Hexadecanol	290	20	173
1-Octadecanol	323	19	173
Aliphatic acids			
Acetic	60	22	135
Propionic, propanoic	76	20	135
Butanoic	93	20	135
Pentanoic	110	19	135
Hexanoic	127	19	135
Heptanoic	143	19	135
Octanoic	160	19	135
Nonanoic	177	19	135
Lauric, dodecanoic	227	19	135
Palmitic, hexadecanoic	293	18	135, 173
Stearic, octadecanoic	327	18-24	101, 173
Eicosanoic	—	18	26
2-Methylbutyric	110	19	135
3-Methylbutyric	111	20	135
2-Ethylbutyric	126	19	135
3,3-Dimethylbutyric	128	19	135
Pivalic	110	21	135
2-Phenylbutyric	151	16	135
Cyclopropane carboxylic	80	19	135
α-Hydroxystearic	—	19-29	101
9,10,16-Trihydroxypalmitic	—	19	101
Oleic	—	17	125
Maleic anhydride	—	24	23
Amines, Amides, Nitriles (See also Tables 3 & 4)			
1,12-Dodecamethyleneamine	—	23	32
2,4-Dinitrodiphenylamine	182	18-23	31
2,4,6-Trinitrodiphenylamine	197	20-25	31
3-Nitroaniline	106	25-27	31
4-Nitroaniline	101	27-29	31
Tolbutamide	210	23	70
Acetohexamide	234	24	70
1,3-Dicyanobenzene	118	22	174
Analgesics (narcotic)			
Morphine	216	24-27	176
Hydromorphone	216	23-24	176
Codeine	227	22-23	176
Fentanyl	303	20	176
Sufentanil	345	19-20	176
Meperidine	222	20	176
Aromatic hydrocarbons			
Naphthalene	123	20-22	63-65, 96, 97
Phenanthrene	150	20	50

TABLE 1 (continued)
Hildebrand Parameters and Apparent Molar Volumes for Predominantly Covalent Crystalline Solids as Subcooled Liquids at ~25°C

Material (liquid, or solid as subcooled liquid)	^iV/cm³ mol⁻¹	$^i\delta$/ MPa$^{1/2}$ (2 significant figures)	References
Barbituric Acid Derivatives			
Barbital (barbitone)	—	28	104-106, 134, 162
Phenobarbital	—	27-28	105, 106, 134, 162, 167
Allobarbital	—	26	105
Aprobarbital	—	25	105
Cyclobarbital	—	25	105
Butalbarbital	—	24	105
Butobarbital	—	24	105, 106
Pentobarbital	—	24	105, 162
Secobarbital	—	23, 30	105, 162
Hexobarbital	—	23	105, 106
Methylbarbital	—	26	106, 162
Amylobarbital	—	24	106, 162
5-(Anisylidene)-2-thiobarbituric acid	—	25	162
Benzoic Acid and Derivatives			
Benzoic acid	101, 104	23, 24	12, 22, 65
p-Hydroxybenzoic acid	—	26, 31	67, 73
Salicylic acid	—	29-32	110
Methyl *p*-hydroxybenzoate	121	23-25	12, 60, 61, 67
Ethyl *p*-hydroxybenzoate	140	23-25	12, 60, 61
Propyl *p*-hydroxybenzoate	158	23-25	12, 60, 61
Butyl *p*-hydroxybenzoate	176	23-27	12, 60, 61
Methyl *p*-aminobenzoate	128	25-27	25, 171
Ethyl *p*-aminobenzoate	144	25	25, 171
Propyl *p*-aminobenzoate	160	24	25, 171
Butyl *p*-aminobenzoate	177	24	25, 171
Pentyl *p*-aminobenzoate	193	23	25, 171
Hexyl *p*-aminobenzoate	209	22	25
Heptyl *p*-aminobenzoate	225	22	25
Octyl *p*-aminobenzoate	241	21	25
Nonyl *p*-aminobenzoate	258	21	25
Dodecyl *p*-aminobenzoate	306	21	25
Hexadecyl *p*-aminobenzoate	371	20	25
Cholesterol and Derivatives			
Cholesterol	391	19	135
Cholesteryl acetate	428	17	135
Cholesteryl propionate	440	16	135
Cholesteryl butyrate	455	16	135
Cholesteryl hexanoate	494	17	135
Cholesteryl heptanoate	513	19	135
Cholesteryl laurate	597	18	135
Cholesteryl myristate	627	18	34
Cholesteryl palmitate	663	18	34
Cholesteryl stearate	692	17	135
Cholesteryl oleate	671	15	34
Cholesteryl chloride	398	16	135

TABLE 1 (continued)
Hildebrand Parameters and Apparent Molar Volumes for Predominantly Covalent Crystalline Solids as Subcooled Liquids at ~25°C

Material (liquid, or solid as subcooled liquid)	${}^{j}V$/cm³ mol⁻¹	${}^{j}\delta$/ MPa$^{1/2}$ (2 significant figures)	References
Cortisone and related compounds			
Cortisone	—	27	131
Corticosterone	—	27	131
Cortisol	—	28	131
Aldosterone	—	30	131
Griseofulvin			
Griseofulvin	—	21	107
Halogens and Halides			
Iodine, I_2	59	29	2, 39
Arsenic (III) bromide	93	23-29	37
Antimony (III) bromide	93	27-34	37
Bismuth (III) bromide	87	36-38	37
Tin (IV) bromide	131	30	37
Arsenic (III) iodide	104	27-29	38
Antimony (III) iodide	110	27-30	38
Bismuth (III) iodide	115	23-24	38
Tin (IV) iodide	151	23-24	38, 39
Germanium (IV) iodide	—	25	39
Silicon (IV) iodide	—	23	39
Lipids			
Dodecylstearate	—	25	101
Tristearin	—	18-21	101, 131, 132
Tiolein	—	19	131
Tricaprin	—	18	132
Trilaurin	—	18	132
Trimyristin	—	18	132
Tripalmitin	—	18	132
Distearin	—	18-23	101, 131
Monostearin	—	18-27	101
Mercaptopurine			
6-Mercaptopurine	—	33	130
Metal Soaps			
Lead stearate	—	18	119-121
Zinc stearate	—	18	119-121
Lithium stearate	—	18, 19	119-121
Sodium stearate	—	19	116, 119
Magnesium stearate	—	19	123
Methyl xanthines			
Theophylline	124	24-29	68, 69, 164, 165
Caffeine	144	24-29	68, 69, 71, 164, 165
Theobromine	124	25-29	68, 69, 164, 165

TABLE 1 (continued)
Hildebrand Parameters and Apparent Molar Volumes for Predominantly Covalent Crystalline Solids as Subcooled Liquids at ~25°C

Material (liquid, or solid as subcooled liquid)	^{j}V/cm^3 mol^{-1}	$^{j}\delta$/ MPa$^{1/2}$ (2 significant figures)	References
		Norethindrone and Derivatives	
Norethindrone	—	22	160
Norethindrone acetate	—	22	160
Norethindrone heptanoate	—	22	160
Norethindrone dimethyl propionate	—	22	160
Norethindrone *trans*-4-hexylcyclohex-ylcarboxylate	—	22	160
Norethindrone *trans*-3-(4-butylcycloh-exyl)propionate		21	160
Norethindrone benzoate	—	22	160
Norethindrone 4-biphenylcarboxylate	—	22	160
Norethindrone 4-cyclohexyl-benzoate	—	21	160
Norethindrone-6-(4-chloro-phenyl)hexanoate	—	23	160
Norethindrone pentamethylidisiloxyl ether	—	20	160
Norethindrone 4-bicyclo-hexylcarbox-ylate	—	21	160
Norethindrone 4-phenoxybenzoate	—	22	160
		Phenazones	
Aminophenazone	200	23-30	165
Phenylbutazone	274	21-28	163, 165
		Phenolic Antioxidants	
Topanol CA	—	19	102
Irganox 1010	—	18	102
Goodrite 3125	—	19	102
Ionox 330	—	17	102
Plastonox	—	19	102
Santowhite P	—	19	102
		Phenols	
2,5-Dimethylphenol	124	19	168
3,4-Dimethylphenol	121	19	168
2-Nitrophenol	—	21-22	166
2-Nitro-5-methylphenol	—	22	100
		Phosphorous	
Phosphorus, P_4	70	27	2
		Porphine	
21H,23H-Porphine	—	22	170
		Plasticizers	

Plasticizers
(see Table 21, Chapter 14; Tables 4 and 5, Chapter 18)

TABLE 1 (continued)
Hildebrand Parameters and Apparent Molar Volumes for Predominantly Covalent Crystalline Solids as Subcooled Liquids at ~25°C

Material (liquid, or solid as subcooled liquid)	^jV/cm^3 mol^{-1}	$^j\delta$/ MPa$^{1/2}$ (2 significant figures)	References
Quaternary Ammonium Salts (see Table 5, and "Surfactants" in this table)			
Sulfonamides			
Sulfathiazole	162	23, 27	74, 161
Sulfamerazine	186	27	161
Sulfamethazine	197	26	161
Sulfisomidine	182, 195	26, 40	70, 74, 161
Sulfadiazine	156	27, 55	74, 161
Sulfameter	183	38	161
Sulfamethoxazole	176	24, 34	74, 161
Sulfisoxazole	188	31	161
Sulfamethoxypyridazine	173	25, 28	172
Sulfur and Sulfur Compounds			
Sulfur, S_8	135	25, 26	2, 41
Tetrasulfur tetranitride	86	28	29
Heptasulfur amide	127	26	30
Surfactants (See also Table 7, Chapter 17, and Table 4 in this chapter)			
Glycerol monostearate	—	17	125
Glycerol monolaurate	—	17	124
Sorbitan monolaurate	—	17-18	124,125
Polyoxyethylene(10)oleyl ether	—	18	125
Polyoxyethylene(20)cetyl ether	—	19	125
Sodium octadecanoate	—	19	125
Sodium hexadecanoate	—	19	125
Sodium dodecanoate	—	20	125
Sodium octanoate	—	21	125
Sodium dinonylnaphthalene sulfonate	—	22	124, 125
Dioctyl sodium sulfosuccinate	—	25	125
Sodium heptadecyl-1-sulfate	—	26	125
Sodium dodecyl-1-sulfate	—	29	125
Sodium decyl-1-sulfate	—	30	125
Sodium octyl-1-sulfate	—	32	125
Di-(2-ethylhexyl)sodium succinate	—	17, 21	124
Tridodecylammonium chloride	—	21, 22	124, 128
Propyldodecyldimethylammonium chloride	—	20	124
Propyldodecylidimethylammonium bromide	—	19	124
Didodecyldimethyl ammonium chloride	—	20	124
Didodecylidimethylammonium bromide	—	18	124
Didodecylidimethylammonium iodide	—	16	124
Polyoxyethylene nonylphenyl ether	—	19	124
Octaethylene glycol mono-dodecyl ether	—	18	124

TABLE 1 (continued)
Hildebrand Parameters and Apparent Molar Volumes for Predominantly Covalent Crystalline Solids as Subcooled Liquids at ~25°C

Material (liquid, or solid as subcooled liquid)	^jV/cm^3 mol^{-1}	$^j\delta$/ MPa$^{1/2}$ (2 significant figures)	References
Testosterone and Derivatives			
Testosterone	—	19, 22	72, 114
Testosterone formate	—	19	114
Testosterone acetate	—	19	114
Testosterone propionate	—	19-20	72, 112-114
Testosterone butyrate	—	19	114
Testosterone valerate	—	19	114

$$\left[\frac{\partial \ln {}^c a}{\partial \ln T}\right]_p = \frac{\Delta H}{RT} \tag{15}$$

On taking the second derivative there is obtained the relationship:

$$\left[\frac{\partial^2 \ln {}^c a}{(\partial \ln T)^2}\right]_p = T\left[\frac{\partial}{\partial T}\left\{\frac{\Delta H}{RT}\right\}\right]_p = \frac{\Delta C_p}{\mathrm{R}} - \frac{\Delta H}{RT} \tag{16}$$

At the melting point, T_m

$$\Delta H_m = T_m \, \Delta S_m \tag{17}$$

so the second derivative would be zero at this point if $\Delta S_m = \Delta C_p$. If this assumption is made,

$$\Delta H_m \approx T_m \Delta S_m \approx T_m \Delta C_p \tag{18}$$

$$\left[\frac{\partial^2 \ln {}^c a}{(\partial \ln T)^2}\right]_p \approx \frac{\Delta H_m}{RT_m} = \frac{\Delta S_m}{R} \tag{19}$$

$$\ln {}^c a \approx \frac{\Delta S_m}{R} \ln \frac{T}{T_m} \tag{20}$$

This is at least as good an approximation as Equation 11.

With the same argument as that used in deriving Equation 14, it follows from Equation 20 for *regular* solutions that

$$\ln {}^j x_s = \frac{\Delta S_m}{R} \ln \frac{T}{T_m} - \frac{{}^jV \, {}^i\phi^2 \, {}^{ij}A}{RT} \tag{21}$$

or

$$-\log {}^j x_s = \frac{\Delta S_m}{R} \log \frac{T_m}{T} + \frac{{}^jV \, {}^i\phi^2}{2.303RT} ({}^i\delta - {}^j\delta)^2 \tag{22}$$

Therefore, a plot of $\log {}^j x_s$ against $\log T$ is expected to be linear (for regular solutions) with the intercept and slope depending on enthalpy and entropy terms, respectively. One of the most fully investigated solid solutes is iodine.[2,42,43] The lines for all violet ("non-associated") solutions in the region ${}_j x_s < 0.1$ are quite straight, allowing accurate determination of ΔS_m. The lines for nonviolet solutions of iodine have smaller slopes, corresponding to smaller entropies of mixing as a result of complexing (which reduces the number of independent molecules). However, the lines curve upward (tending to greater slopes) as complexing decreases with rising temperature.

It follows from Equation 41, Chapter 7,

$$RT \ln {}^j x_s = {}^j V \, {}^i\phi^2 ({}^i\delta - {}^j\delta)^2$$

that a line of unit slope fits data for many regular solutions when $RT \ln ({}^j a / {}^j x_s) = RT \ln {}^i f_x$ is plotted against ${}^j V \, {}^i\phi^2 \, ({}^i\delta - {}^j\delta)$.[22,42,44] Hildebrand et al.[2] have summarized the information on solubilities of other solids, including stannic iodide, white phosphorous, naphthalene, aluminum bromide, and sulfur in terms of regular solutions and deviations from regularity.

Both Equations 14 and 22 suggest that for solutions approaching regular behavior, plots of $\log {}^j x_s$ against $({}^i\delta - {}^j\delta)^2$ should be approximately linear, and this relatively simple procedure is probably the most widely used method of correlating solid solubilities by means of Hildebrand parameters.[10,11,45,46]

It should be noted that straight lines for plots based on Equation 22 may be obtained fortuitously from very irregular systems; an example from the work of A. Adjei is theobromine in dioxane-water mixtures.[47]

Martin and co-workers[48] also showed that the log-linear equation of Yalkowsky, Flynn and Amidon,[49] which relates the solubilities of some drugs in binary aqueous systems to those in water, can be derived from the Hildebrand-Scatchard expression corrected for geometric mean deviations (Section 5.7).

If ${}^i\delta$, ${}^j\delta$, and ${}^j V$ (for the supercooled liquid j corresponding to the solid of interest) as well as T_m and ΔS_m are known for approximately regular systems, it is possible by means of Equations 14 or 22 to calculate phase diagrams. These may be T vs. ${}^j x_s$ plots for binary systems, or triangular compositon diagrams with T contours for ternary systems. Examples are

- Phenanthrene-cyclohexane-methylene iodide[50]
- Vanadium-chromium, vanadium-niobium, chromium-niobium[51]
- Scandium oxide-lanthanum oxide[52]
- Elements from Groups III, IV, including III-V semiconductors[53]
- First transition metals[54]
- Met-aluminous silicate liquids[55]
- Molten salts, including binary alkali halides[56]
- Gallium-phosphorous-germanium, gallium-arsenic-germanium, and gallium-arsenic-zinc[57]
- Liquid crystal materials exhibiting azeotrope-like behavior in liquid two-phase regions[58,59]

The activity coefficient part of the expression for solubility,

$$-\log {}^j x_s = \frac{{}^j V \, {}^i\phi^2 \, {}^{ij}A}{2.303RT} \tag{23}$$

has been expanded for solid solubilities in terms of binary interaction parameters instead of assuming that ${}^{ij}A = ({}^i\delta - {}^j\delta)^2$. Thus, ${}^{ij}W$ (Equation 32 Chapter 5), defined by

$$ {}^{ij}A = {}^{i}\delta^2 + {}^{j}\delta^2 - 2\,{}^{ij}W \tag{24}$$

has been applied to a variety of systems.

These geometric mean correction methods have been applied also to mixed solvent systems. Martin and co-workers investigated methylxanthines (theophylline, caffeine, theobromine);[68,69] as well as tolbutamide, acetohexamide, and sulfisomidine;[70] and caffeine, *p*-aminobenzoates and sulfadiazine in water-cosolvent systems.[48,71] Values of the Hildebrand parameters used, in some cases calculated by the method of Fedors (Section 6.2), are included in Table 1. Also studied were testosterone and testosterone propionate,[72] *p*-hydroxybenzoic acid,[73] and sulfonamides.[74] Sulfisomidine showed a Hildebrand parameter of 25.8 $MPa^{1/2}$ by the method of Fedors, 26.0 $MPa^{1/2}$ from peak solubility.

The geometric mean correction factor ${}^{ij}l$ (Equation 29, Chapter 5), defined by

$$ {}^{ij}A = ({}^{i}\delta - {}^{j}\delta)^2 + 2\;{}^{ij}l\;{}^{i}\delta\;{}^{j}\delta \tag{25}$$

has been used by Prausnitz and co-workers[75,76] for solids dissolved in cryogenic liquid solvents (Table 4, Chapter 5). An important system is carbon dioxide in liquid methane. (The case of solid carbon dioxide,[76,77] which has significant quadrupole interactions, is discussed in Section 5.7) Ksiazczak and Anderko[78] also used a method incorporating geometric mean correction factors for systems of polar solids with hydrocarbons. Any of these methods, even with the binary interaction parameters selected by intelligent guesswork, usually provide improved accuracy in solubility predictions from Hildebrand parameters.

Component cohesion parameters have not been used widely in the correlation of solid-liquid solubilities. The work of Martin and co-workers is described in Section 12.2, and another example is the use of polar fractional molar attraction constants, $F_p/F = \delta_p/\delta$, for the steroids fluocinolone acetonide and fluocinolide.[79] The logarithm of this quantity plotted against the logarithm of the solubility exhibits linearity separately on each side of a maximum for each steroid at $\delta_p/\delta = 0.4$. Where there are specific interactions, Hildebrand parameters are inadequate to predict solubilities, but the assumption of solute-solvent complexes allows the rationalization of observations.[80]

In correlating the solubilities of pharmaceutical solids, Paruta and co-workers[61,81-85] have developed an equation linking Hildebrand parameters with relative permittivity ϵ_r (see also Section 8.2):

$$\delta/\text{MPa}^{1/2} = 0.45\epsilon_r + 15.3 \tag{26}$$

This was modified and used by Skraga and Wieckiewicz[86] for the determination of Hildebrand parameters of a series of poly(vinyl chloride) plasticizers.

Following publications of Yalkowsky, Valvani, and Roseman,[87,88] a solubility equation was developed by Amidon and Williams,[89] which was based on Gibbs energy changes for the processes solid → subcooled liquid → solution in octanol → aqueous solution, and which included the Hildebrand parameters of octanol and subcooled liquid for the second step with values determined by the method of Hoy (Section 7.2).

Haulait-Pirson, Huys, and Vanstraelen[90] presented an equation based on the fusion and transition properties of solid alkanes, their molar volumes and the molar volumes of liquids to predict solubilities of *n*-alkanes in nonpolar liquids, and incorporated a term in the differences of modified non-specific cohesion parameters, δ' (Section 8.4) to take into account changes in cohesion forces due to different chemical natures.

Cohesion parameters have been used in various ways:

- Saturated vapor pressures of binary three-phase solid-liquid-vapor systems incorporating highly polar compounds (particularly phenols) have been considered by Ksiazczak[91]

- Iscan reported the effect of temperature on partition and dimerization of benzoic acid[92]
- Equilibria of 2-nitrophenol in hydrocarbons[166]
- Properties in binary solvents have been investigated for the aromatic hydrocarbons pyrene and anthracene[93-95] and naphthalene[96-97] and also for carbazole,[98-99] 2-nitro-5-methylphenol[100] and dimethylphenols[168] (see also Section 7.7)
- Schmid solvents such as trichloromethane-methanol for lipids, long chain aliphatic acids and esters[101]
- Billingham, Calvert and Manke[102] studied the solubility of phenolic antioxidants in polyalkenes
- Drug activity of barbiturates[103-106]
- Grant and Abougela[107] predicted the solubility of griseofulvin
- Infinite dilution partial molar volumes of atropine and hyoscine in relation to their respective potencies[108]
- King and Friedrich considered the solubility of caffeine in supercritical fluids[109]
- Solubilities of nalidixic and salicylic acids in binary liquids[110]
- DDT insecticides in petroleum fractions[111] and other pesticides[169]
- Testosterone and derivatives[72,112-114]
- For a vanadium pentoxide xerogel (V_2O_5. $1.6H_2O$) no molecular intercalation was observed when the Hildebrand parameter of the liquid was less than 27 $MPa^{1/2}$ [115]
- Properties of metal soaps: lithium grease dropping points;[116] solubility of potassium acid (lauric, stearic, and oleic) soaps;[117] and it has been shown that the critical solution temperatures of oil-metal soap systems can be used to estimate the Hildebrand parameter of an oil by means of a calibrating curve established for oils with known properties. Zinc stearate was chosen for its low usable temperature range (70 to 120°C) and low water sensitivity.[116,118-122] Similar information is available from lithium grease dropping points, and other solubility properties of alkali soaps can be correlated by means of Hildebrand parameters.[186]
- Compression of binary powder mixtures, yielding a Hildebrand parameter of 19 $MPa^{1/2}$ for magnesium stearate based on 23 $MPa^{1/2}$ for caffeine[123]
- Surfactants[124-128] (See Section 17.7)
- Drug transmission rates[129,130] (See Section 19.2)
- Lipids[131,132]

12.2 COHESION PARAMETERS FOR NONIONIC SOLID SOLUTES

Hildebrand parameters and other cohesion parameters are available in data tables elsewhere in this Handbook for solids considered as subcooled liquids at or near 25°C. For example, Table 18 in Chapter 5 gives Hildebrand parameters and Hansen parameters of condensed materials regardless of their melting points. Further values of Hildebrand parameters for solids are summarized in Table 1. It is clear that the Hildebrand parameter is a relatively non-discriminating method of characterizing the properties of solids, particularly pharmaceutical solids.

When expressions such as Equations 12 and 14 are used to estimate the solubilities of solids, it is necessary to have information which may not be readily available: most solids are non-volatile and there is uncertainty in the molar volume when extrapolation is made from the molten substance to the subcooled liquid. For solids with significant vapor pressures the Antoine equation method (Section 7.2) can be used.[166]

Both Hildebrand parameters and apparent molar volumes of crystalline solids (considered as subcooled liquids at 25°C) may be calculated by the group contribution methods of Section 6.5 just as if they were liquids. In some cases the temperature dependence of the density of liquid solutes is extrapolated down past the freezing point to the subcooled liquid at 25°C. The enthalpy of fusion may be determined by Walden's method,[133]

$$\Delta H_m = T_m K \tag{27}$$

where K is approximately constant at 56.5 J K^{-1}, although it does vary appreciably with the class of compound,[14]

The Hildebrand parameters, such as those in Table 1, can be obtained from equations based on relationships in Section 12.1, or from the maximum of the solubility spectrum (compare Sections 8.1 and 14.1) where the ${}^i\delta$ value should equal ${}^j\delta$, or from the methods of Section 12.1. Sometimes binary liquid mixtures are used, with the individual Hildebrand parameters being combined by Equation 31, Chapter 4:

$${}^{ij}\delta = \frac{{}^i\phi\,{}^i\delta + {}^j\phi\,{}^j\delta}{{}^i\phi + {}^j\phi} \tag{28}$$

The resulting solute Hildebrand parameter values tend to vary when measured in different pairs of liquids, and there is evidence in the case of phenobarbital[134,167] for two distinct values, depending on the hydrogen-bonding capability of the liquid or liquid mixture. The use of gas-liquid chromatography to characterize plasticizers in terms of cohesion parameters is discussed in Section 18.2.

Liron and Cohen[135] used partial molar volumes from high precision density measurements of dilute solutions of solids (alkanoic acids, cholesterol, and cholesteryl esters) in reference solvents (Hildebrand parameters from 17.6 $MPa^{1/2}$ for tetrachloromethane to 20.5 $MPa^{1/2}$ for nitrobenzene) to estimate molar volumes and Hildebrand parameters as subcooled liquids; these values are included in Table 1. Group contributions were also considered, the incremental molar volume in solution per CH_2 group for all compounds tested being 16.7 cm^3 mol^{-1}.

Martin and Carstensen[60] correlated solubilities and Hildebrand parameters from Restaino and Martin[12] and Alexander et al.[61] (Table 1).

The determination of Hansen parameters for solid materials has not been pursued to as great an extent as in the case of polymers. Martin, Wu, Beerbower, and co-workers[62-65] used a multiple regression method with Hansen parameters[66] (Section 5.9) and geometric mean correlated Hildebrand parameters (Section 5.7) to correlate and predict the solubilities of naphthalene. The regression gave

$$\delta_d = 20.6\ MPa^{1/2}$$

$$\delta_p = 4.0\ MPa^{1/2}$$

$$\delta_h = 1.9\ MPa^{1/2}$$

$$\delta_t = 21.1\ MPa^{1/2}$$

while maximum solubility gave $\delta_t = 19.6\ MPa^{1/2}$.

Hansen parameters for benzoic acid were

$$\delta_d = 17.3\ MPa^{1/2}$$

$$\delta_p = 16.1\ MPa^{1/2}$$

$$\delta_h = 4.2\ MPa^{1/2}$$

$$\delta_t = 24.0\ MPa^{1/2}$$

Beerbower, Wu and Martin[65] incorporated acid and base cohesion parameters in an expanded cohesion parameter system (Section 5.12) for the solubilities of naphthalene and benzoic acid by a regression method.

Triangular Hansen parameter solubility diagrams were provided also. With a four-parameter equation, for benzoic acid

$$\delta_d = 17.2 \text{ MPa}^{1/2}$$

$$\delta_p = 15.1 \text{ MPa}^{1/2}$$

$$\delta_a = 9.0 \text{ MPa}^{1/2}$$

$$\delta_b = 3.3 \text{ MPa}^{1/2}$$

$$\delta_t = 24.1 \text{ MPa}^{1/2}$$

while for naphthalene

$$\delta_d = 21.5 \text{ MPa}^{1/2}$$

$$\delta_p = 3.6 \text{ MPa}^{1/2}$$

$$\delta_a = -5.9 \text{ MPa}^{1/2}$$

$$\delta_b = 1.1 \text{ MPa}^{1/2}$$

$$\delta_t = 22.1 \text{ MPa}^{1/2}$$

(Negative signs occurred occasionally also in Hansen's original method.) For *p*-hydroxybenzoic acid,[67]

$$\delta_d = 17.0 \text{ MPa}^{1/2}$$

$$\delta_p = 14.5 \text{ MPa}^{1/2}$$

$$\delta_a = 3.7 \text{ MPa}^{1/2}$$

$$\delta_b = 8.0 \text{ MPa}^{1/2}$$

$$\delta_t = 26.3 \text{ MPa}^{1/2}$$

The total cohesion parameter calculated by the method of Fedors (Section 6.2) was 31 $\text{MPa}^{1/2}$, but it was noted that this calculation method tends to overestimate Hildebrand parameter values. For methyl *p*-hydroxybenzoate

$$\delta_d = 18.6 \text{ MPa}^{1/2}$$

$$\delta_p = 14.9 \text{ MPa}^{1/2}$$

$$\delta_a = 9.6 \text{ MPa}^{1/2}$$

$$\delta_b = 1.5 \text{ MPa}^{1/2}$$

$$\delta_t = 24.4 \text{ MPa}^{1/2}$$

Subsequently,[172] the solubility of sulfamethoxypyridazine was investigated.

Kesselring and co-workers[163,165] evaluated phenylbutazone Hansen parameters by means

TABLE 1a
Reported Ranges of Hansen Parameters for Some Pharmaceutical Solids, from the Review by Rey-Mermet[165]

Solid	Method	$\delta/MPa^{1/2}$		
		δ_d	δ_p	δ_h
Caffeine	Partition	21	4	8
	Solubility	19-21	7-14	10-19
	Calorimetry	17	14	13
	Gas chromatography	17	12	17
Theophylline	Solubility	24	13	12
	Calorimetry	17	11	14
Theobromine	Solubility	25	15	11
	Calorimetry	17	12	13
Methyl *p*-hydroxybenzoate	Solubility	18-28	10-15	5-25
(methyl paraben)	Calorimetry	17	14	13
	Gas chromatography	21	18	24
Aminophenazone	Partition	19	2	9
	Solubility	9-21	7-13	7-11
	Calorimetry	17	14	13
Phenylbutazone	Solubility	14-25	9-14	8-11
	Calorimetry	18	13	11

of a new expression for the Gibbs function of solution ($\delta_d = 25 \pm 2$, $\delta_p = 9 \pm 1$, $\delta_h = 7.6 \pm 0.6$, $\delta_t = 25 \pm 2$ $MPa^{1/2}$, results which compared favorably with gas-solid chromatography results ($\delta_d = 22.5 \pm 0.6$, $\delta_p = 14 \pm 2$, $\delta_h = 9 \pm 2$, $\delta_t = 28 \pm 1$ $MPa^{1/2}$). Hansen parameters were determined also[164,165] from solution calorimetry for caffeine ($\delta_d = 17$, $\delta_p = 14$, $\delta_h = 13$, $\delta_t = 26$ $MPa^{1/2}$), theophylline ($\delta_d = 17$, $\delta_p = 11$, $\delta_h = 12$ to 13, $\delta_t = 24$ $MPa^{1/2}$) and theobromine ($\delta_d = 17$, $\delta_p = 12$, $\delta_h = 13$, $\delta_t = 25$ $MPa^{1/2}$). However, as Rey-Mermet[165] showed, Hansen parameters vary widely according to the experimental method and form of calculation (Table 1a). In other studies, Kramer and Thodos[173] evaluated δ_t (Table 1) and δ_d (16-17 $MPa^{1/2}$ at 25 °C for 1-hexadecanol, 1-octadecanol, palmitic (hexadecanoic) acid and stearic (octadecanoic) acid using supercritical carbon dioxide and the approach described in Section 5.8, while Durbut and Broze[175] found for glycerol tripalmitate (tripalmitin) an ellipsoidal Hansen volume centered on $\delta_d = 18.0$, $\delta_p = 4.3$, and $\delta_h = 4.3$ $MPa^{1/2}$.

Fractional cohesive pressures (Section 5.10) were used by James et al.[113] to estimate Hansen parameters for testosterone propionate ($\delta_d = 14.9$ $MPa^{1/2}$, $\delta_p = 8.6$ $MPa^{1/2}$, $\delta_h = 7.6$ $MPa^{1/2}$, $\delta_t = 19.4$ $MPa^{1/2}$). The binary interaction parameter ^{ij}l (Equation 25) was also used in an attempt to improve the correlation between cohesion parameters and solubilities.

12.3 IONIC SYSTEMS

Most inorganic systems investigated in terms of Hildebrand parameters have had significant covalent bonding character, although Hansen[1] (Section 5.9) extended his polymer-liquid technique to a range of ionic solids, mainly hydrated salts, and Meusburger[136] applied Hansen parameters to ionic emulsifiers. England, Lawson,and Hrubes[137] showed that superconductivity above 8K occurred in alloys and metallic compounds over a relatively narrow range of cohesion parameters (see Table 4 in Chapter 2 for cohesion parameters of the elements).

In some cases (Table 2),[9-11] the cohesive energy density has been equated with the lattice energy density and the resulting cohesion parameters are not strictly comparable with the liquid state Hildebrand parameters. However, other work on ionic systems does conform to

TABLE 2
Molar Volumes and Cohesion Parameters of Ion Pairs from Lattice Energy Densities

Ion pair	${}^{j}V$/cm^3 mol^{-1}	$c^{1/2}$/MPa$^{1/2}$
Potassium chloride	37	135
Potassium bromide	43	124
Potassium iodide	51	110
Potassium picrate	98	72
Rubidium picrate	108	68
Cesium picrate	124	62

Adapted from Takamatsu, T., *Bull. Chem. Soc. Jpn.*, 47, 1287 and 2647, 1974.

TABLE 3
Hildebrand Parameters of Long-Chain *n*-Alkylamines

Number of atoms per chain	${}^{j}\delta$/MPa$^{1/2}$		
	Alkylamine	Dialkylamine	Trialkylamine
6	18	17	16
7	18	17	16
8	18	17	16
10	18	17	16
12	18	17	16
13	18	17	—
14	18	17	15
15	18	17	—
16	18	17	15
18	18	17	15
20	18	17	15

Adapted from Kertes, A. S., *J. Inorg. Nucl. Chem.*, 26, 1764, 1964.

the liquid state approach, and Hildebrand parameters of complexes are discussed in Sections 7.9 and 7.10.

If different molar volumes are taken into account with the Flory-Huggins correction (Equation 44, Chapter 7 and Section 13.2) the expression for an ideal saturated solution is

$$\ln {}^{j}x_{s} + \ln {}^{j}f_{x} = \ln {}^{j}x_{id} = \ln {}^{j}\phi + {}^{i}\phi\,(1 - {}^{j}V/{}^{i}V) + ({}^{j}V/RT)\,{}^{i}\phi^{2}\,({}^{i}\delta - {}^{i}\delta)^{2} \tag{29}$$

This has been used by Kertes[127,128] and others[138] for long chain alkylamines (Table 3) and their hydrochlorides (Table 4). The terms for $\ln {}^{j}x_{id}$ for a given amine with two liquids were equated and the resulting values of ${}^{j}\delta$ were in agreement with Hildebrand parameters calculated from ${}^{j}\delta = ({}_{l}\Delta_{g} H - RT/{}^{j}V)^{1/2}$ for the subcooled liquids. When Equation 29 was applied to alkylammonium chlorides in different liquids, similar values of ${}^{j}\delta$ were obtained (Table 4). The diminishing Hildebrand parameter values from primary through secondary to tertiary alkylammonium chlorides are in agreement with the higher solubilities of tertiary amine salts in solvents of low polarity. Quaternary ammonium salts (Table 5) are discussed below.

Poole and co-workers[139] reported dispersion and orientation (polar) cohesion parameters for some liquid organic salts, in connection with their use in liquid chromatography (Chapter

TABLE 4
Hildebrand Parameters of Long-Chain *n*-Alkylamine Hydrochlorides

Amine hydrochloride	Solvent	$^{j}\delta$/MPa$^{1/2}$ (in various solvents)	$^{j}\delta$/MPa$^{1/2}$ (preferred value)
Dihexyl-	Tetrachloromethane	25.0	26
	Benzene	27.0	
	Heptane	25.8	
	m-Xylene	25.2	
Diheptyl-	Tetrachloromethane	24.6	25
	Benzene	26.8	
	Heptane	25.0	
Triheptyl-	Heptane	21.7	22
Octyl-	Benzene	27.6	28
Dioctyl-	Tetrachloromethane	24.3	25
	Benzene	26.4	
	Heptane	24.8	
Trioctyl-	Heptane	21.7	22
Decyl-	Benzene	28.2	28
Didecyl-	Tetrachloromethane	25.4	25
	Benzene	27.0	
	Heptane	23.9	
Dodecyl-	Benzene	28.8	29
	Trichloromethane	26.2	
Didodecyl-	Tetrachloromethane	25.4	25
	Benzene	26.8	
	Heptane	23.9	

Adapted from Kertes, A. S., *J. Inorg. Nucl. Chem.*, 27, 209, 1965.

TABLE 5
Hildebrand Parameters at 100°C for Quaternary Ammonium and Phosphonium Salt Phase-Transfer Catalysts

Organic salt	t_m/°C	$^{j}\delta$/MPa$^{1/2}$
$C_4H_9P(C_6H_5)_3Br$	249-251	40
$(C_4H_9)_4NBr$	87-89	30
$C_{12}H_{25}N(CH_3)_3Cl$	paste	29
$C_{16}H_{33}N(CH_3)_3Cl$	paste	28
$C_{18}H_{37}N(CH_3)_3Cl$	84-100	26
$(C_4H_9)_4PBr$	99-101	26
$(C_{18}H_{37})_2N(CH_3)_2Cl$	100-117	26
$C_6H_5CH_2N(C_2H_5)_3Cl$	102-103	23
$(C_8H_{17})_4NBr$	70-72	22
$C_{16}H_{33}P(C_4H_9)_3Br$	55-57	22
$C_{18}H_{37}N(C_8H_{17})_3Br$	paste	21
$C_6H_5CH_2P(C_4H_9)_3Cl$	160<	18

Adapted from Fukunaga, K., Shirai, H., Ide, S., and Kimura, M., *Nippon Kagaka Naishi,* 1148, 1980.

TABLE 6
Dispersion and Orientation Cohesion Parameters for some Liquid Organic Salts

	V/cm³ mol⁻¹	δ/MPa$^{1/2}$ δ_d	$\delta_o = \delta_p$
Ethylammonium nitrate	98	16 – 17	7 – 9
Propylammonium nitrate	116	17	7 – 10
Tributylammonium nitrate	270	17	10 – 12
Butylammonium thiocyanate	144	18 – 19	8 – 12
sec-Butylammonium thiocyanate	134	18 – 19	9 – 12
Dipropylammonium thiocyanate	171	18 – 20	8 – 11

Adapted from Shetty, P. H., Youngberg, P. J., Kersten, B. R., and Poole, C. F., *J. Chromatogr.*, 411, 61, 1987.

18). Values were calculated by several of the methods outlined in Section 3.7, the ranges being indicated in Table 6.

Taft, Abraham, Doherty, and Kamlet[140] extended the application of their solvatochromic equation (Section 8.4) (which included a term in δ or in δ^2 corresponding to the free energy required to disrupt solvent/solvent interactions) to the solution properties of tetraalkylammonium halide ion-pairs and disassociated ions. A different equation also incorporating δ^2 (but with either negative or positive coefficients) was used by Glikberg and Marcus[141] to correlate single ion Gibbs function values. Iwamoto, Ito, and Yamamoto,[142] following Sierkierski and Olszer,[143] Wakahayashi et al.[144] and Irving[145] applied regular solution theory and Hildebrand parameters to the extraction of ion-pairs (Section 7.11).

The mixing of hydrocarbons with water is exothermic ($\Delta H < 0$) but the low miscibility shows that $\Delta G > 0$. Consequently the entropy of the system must be decreasing during mixing. It is energetically advantageous for dissolved molecules in such systems to cluster together, the effect being known as solvophobic interaction.[146] Solvent dipolar molecules also pack around ions in solution, but in this case the solvation is strongly favored ($\Delta G < 0$), and ions are stabilized in solution. Neither of these types of specific solute-solvent interaction is taken into account by conventional cohesion parameter methods that exclude both entropy effects and ionic interactions.

In systems containing ions, ion-dipole and ion-induced dipole interactions which are long range ($U = -\,{}^iq\,{}^j\mu/r^2$ and $-\,{}^iq^2\,{}^j\alpha/r^4$, respectively), dominate all other effects (Chapter 3). It is possible to develop ionic or electrostatic cohesive pressure terms and cohesion parameters, and compared with these contributions the other interaction cohesion parameters would be negligible. This has not often been done, although Tsutsui and Tanaka[147] have correlated the glass transition temperatures (Section 16.3) of ionic polymers such as ionene polymers, polyphosphates, and polyacrylates with the assumption that the cohesive pressure could be approximated by the electrostatic contribution. Takamatsu[9-11] has made a similar assumption in equating the cohesive energy density of ion-pairs in solution with the electrostatic lattice energy density. Shrivastava and Pandey[178] have reported cohesion parameters for liquid metal halides (Table 7) defined as the square root of internal pressure (Section 7.5).

Although it might seem that non-ionic cohesion parameters would not be appropriate measures of miscibility in ionic systems, they remain useful criteria in some cases. Related polarity scales have been used for the range of liquids from monatomic liquids through molten salts to molten metals,[148,149] and there has been a correlation of cohesion energy with activation energy of viscosity of molten salts.[150] The relatively recent interest[151-153] in elec-

TABLE 7
Internal Pressures and Cohesion Parameters for Liquid Alkali Metal Halides

Salt	t/°C	π/MPa	$\pi^{1/2}$/MPa$^{1/2}$
Lithium chloride	610	1823	42.7
	800	2094	45.8
Sodium bromide	800	1849	43.0
	900	2163	46.5
Potassium chloride	800	1823	42.7
	900	1831	42.8
Potassium iodide	681	1314	36.3
	800	1468	38.3
Sodium chloride	800	2055	45.3
	900	2283	47.8
Potassium bromide	800	1180	34.4
	900	2646	51.4
Cesium bromide	700	1211	34.8
	800	1338	36.6
Cesium chloride	700	2671	51.7
	800	1611	40.1
Sodium iodide	700	1522	39.0
	800	1410	37.6
Lithium bromide	700	1475	38.4
	800	1214	34.8

Adapted from Shrivastava, S. N. and Pandey, J. D., *Acoust. Lett.*, 3, 219, 1980.

trolyte systems extending in the liquid phase from molecular liquids to molten salts should enourage a more unified view of liquid properties. In particular, molten organic salts seem to span a range of solvent properties from those comparable to the highly cohesive molten inorganic salts to those of the very low cohesion pressure hydrocarbons. At this lower end of the scale reside the quaternary ammonium salts, $R_4N^+X^-$, and phosphonium salts which owe their small cohesion pressures to their large ionic volumes. Hildebrand parameters of some quaternary ammonium and phosphonium salts used as phase transfer catalysts[154] are shown in Table 5. It is significant that even in some of these ionic systems (such as the thiocyanates and trialkylammonium picrate), hydrogen bonding energy is still significant. Another example of the tendency toward unified treatments of ionic and covalent systems is the correlation between empirical Lewis acid-base solvent parameters and the thermodynamics of ion solvation.[155]

In liquids of relative permittivity below about 30 (see Table 5, Chapter 8), there is evidence that univalent ions exist largely as undissociated ion-pairs, or as solvent-shared ion-pairs,[84,156,157] and it is reasonable to expect that cohesion parameter concepts might be used directly in such cases. The solubility of sodium salicylate in a series of pure alcohols shows a maximum at 1-pentanol (relative permittivity 13.9) similar to the corresponding figure for salicylic acid (15), indicating that sodium salicylate is behaving essentially as a nonelectrolyte.[84] The solvent extraction and solubility of electrically neutral ion-pairs and complexes may be discussed in terms of cohesion parameter and relative permittivity (Equation 27).

There are many situations where the presence of salts in high relative permittivity solvents (notably water) dramatically changes their characteristics. One of the clearest examples is the use of added salts to bring about phase separation, for example, the "salting out" of organic liquids from water by the dissolution of sodium chloride, attributed to electrostriction or increase of cohesive energy density or internal pressure.[158,159] The properties of the aqueous ionic solutions in principle can still be described in terms of conventional cohesion parameters as far as their interactions with other liquids are concerned, but no attempts have been made to interpret the results in terms of electrostatic cohesion parameters. Extension of cohesion parameters to these ionic systems may prove as instructive as the identification of orientation, induction, and Lewis acid-base component parameters in the original Hildebrand parameter.

REFERENCES

1. **Hansen, C. M.,** The universality of the solubility parameter, *Ind. Eng. Chem. Prod. Res. Dev.*, 8, 2, 1969.
2. **Hildebrand, J. H., Prausnitz, J. M., and Scott, R. L.,** *Regular and Related Solutions,* Van Nostrand-Reinhold, Princeton, NJ, 1970.
3. **Martin, A., Swarbrick, J., and Cammarata, A.,** *Physical Pharmacy,* 3rd ed., Lea & Febiger, Philadelphia, PA, 1983, chap. 12, "Solubility and distribution phenomena."
4. **Kendall, K.,** Sticky solids, *Contemp. Phys.*, 21, 277, 1980.
5. **Acree, W. E., Jr.,** *Thermodynamic Properties of Nonelectrolyte Solutions,* Academic Press, Orlando, FL, 1984, 205.
6. **Iguchi, E. and Matsushima, F.,** Calculation of cohesive energy of Ti_4O_7, *J. Mater. Sci.*, 21, 1046, 1986.
7. **Matsushima, F. and Iguchi, E.,** Lattice energy calculations for Ti_4O_7 using a shell model, *J. Phys. Chem. Solids,* 47, 45, 1986.
8. **Fung, H.-O. and Higuchi, T.,** Molecular interactions and solubility of polar nonelectrolytes in nonpolar solvents, *J. Pharm. Sci.*, 60, 1782, 1971.
9. **Takamatsu, T.,** Solubility studies of ion-pairs in organic solvents, *J. Pharm. Sci.*, 60, 1782, 1971.
10. **Takamatsu, T.,** The application of the regular solution theory to the ion-pair systems, *Bull. Chem. Soc. Jpn.*, 47, 1287, 1974.
11. **Takamatsu, T.,** The solubility study of ion-pairs in organic solvents, *Bull. Chem. Soc. Jpn.*, 47, 2647, 1974.
12. **Restaino, F. A. and Martin, A. N.,** Solubility of benzoic acid and related compounds in a series of *n*-alkanols, *J. Pharm. Sci.*, 53, 636, 1964.
13. **Myers, A. L. and Prausnitz, J. M.,** Thermodynamics of solid carbon dioxide solubility in liquid solvents at low temperatures, *Ind. Eng. Chem. Fundam.*, 4, 209, 1965.
14. **Mortimer, F. S.,** Melting point, latent heat of fusion and solubility, *J. Am. Chem. Soc.*, 44, 1415, 1922.
15. **Eckert, C. A.,** Introduction to theories of solutions and application to phase equilibria, in *Solutions and Solubilities,* Part II, Dack, M. R. J., Ed., John Wiley & Sons-Interscience, New York, 1976, chap. 9.
16. **Scatchard, G.,** Equilibria in non-electrolyte solutions in relation to the vapor pressures and densities of the components, *Chem. Rev.*, 8, 321, 1931.
17. **Basu, R. S., Pham, H., and Wilson, D. P.,** Prediction of solubility of long-chain hydrocarbons in various solvents using UNIFAC, *Int. J. Thermophys.*, 7, 319, 1986.
18. **Domanska, U.,** Solubilities of *n*-paraffin hydrocarbons in binary solvent mixtures, *Fluid Phase Equil.*, 35, 217, 1987.
19. **Riggs, D. M. and Diefendorf, R. J.,** The solubility of aromatic compounds, *Ext. Abstr. Program. — Bienn. Conf. Carbon,* 14, 407, 1979.
20. **Riggs, D. M.,** Carbon fiber from solvent extracted pitch, *Prepr. Am. Chem. Soc. Div. Pet. Chem.*, 29, 480, 1984.
21. **Riggs, D. M., Chen, S.-H., and Diefendorf, R. J.,** Solubility and solubility parameters of polynuclear aromatics systems, *Proc. Gov. Conf. Expanding Use Coal N.Y. State: Probl. Issues,* Tress, M. H. and Dawson, J. C., Eds., Res. Found. State Univ. New York, Albany, NY, 1981, 539.
22. **Chertkoff, M. J. and Martin, A. N.,** The solubility of benzoic acid in mixed solvents, *J. Am. Pharm. Assoc. Sci. Ed.*, 49, 444, 1960.
23. **Gaivoronski, A. N. and Granzhan, V. A.,** Solubility of maleic anhydride, *Zh. Prikl. Khim. (Leningrad),* 53, 801, 1980.
24. **Cavé, G., Kothari, R., Puisieux, F., Martin, A. N. and Carstensen, J. T.,** Solubility parameters from maxima in solubility/solvent plot, *J. Pharm.*, 5, 267, 1980.
25. **Yalkowsky, S. H., Flynn, G. L., and Slunick, T. G.,** Importance of chain length on physicochemical and crystalline properties of organic homologs, *J. Pharm. Sci.*, 61, 852, 1972.
26. **Domanska, U.,** Vapour-liquid-solid equilibrium of eicosanoic acid in one- and two-component solvents, *Fluid Phase Equil.*, 26, 201, 1986.

27. **Domanska, U.,** Solid-liquid phase relations of some normal long-chain fatty acids in selected one-component and two-component solvents, *Ind. Eng. Chem. Res.*, 26, 1153, 1987.
28. **Domanska, U.,** Solubility of *o*-toluic acid in some organic solvents. Part II, *Polish J. Chem.*, 60, 847, 1986.
29. **Hamada, S., Kudo, Y., and Kawano, M.,** The solubility parameter of tetrasulfur tetranitride, *Bull. Chem. Soc. Jpn.*, 48, 719, 1975.
30. **Hamada, S., Kudo, Y., and Kawano, M.,** The solubilities and the solubility parameter of heptasulfur imide, *Bull. Chem. Soc. Jpn.*, 48, 2963, 1975.
31. **Lemordant, D. and Gaoriaud, R.,** Prediction of solubilities of molecular compounds in nonpolar liquids, *J. Chim. Phys.*, 78, 427, 1981.
32. **Komarov, V. M., Polyakova, L. P., and Antipenko, S. G.,** Solubility of 1,12-dodecamethylenediamine, *Zh. Prikl. Khim.*, 49, 1606, 1976; *J. Appl. Chem. U.S.S.R.*, 49, 1631, 1976.
33. **Negoro, H.,** Solubility phenomena of pyridine and pyrazine monocarboxamides. III. Solubilities of picolinamide, nicotinamide, isonicotinamide and pyrazine carboxamide, *Yakugaku Zasshi,* 80, 674, 1960.
34. **Kishimoto, H., Iwasaki, T., and Yonese, M.,** The solubility temperature relationship of cholesteryl myristate, palmitate and oleate in various solvents, *Yakugaku Zasshi,* 106, 286, 1986.
35. **Gharavi, M., James, K. C., and Sanders, L. M.,** Solubilities of mestanolone, methandienone, methyltestosterone, nandrolone and testosterone in homologous series of alkanes and alkanols, *Int. J. Pharm.*, 14, 333, 1983.
36. **Hagen, T. A. and Flynn,** Solubility of hydrocortisone in organic and aqueous media: evidence for regular solution behavior in apolar solvents, *J. Pharm. Sci.*, 72, 409, 1983.
37. **Maheshwari, R. C., Suri, S. K., and Ramakrishna, V.,** Solubility of As, Sb, Bi and Sn bromides in organic solvents; Partial molar volume of As, Sb, Bi and Sn bromides in some non-polar organic solvents, *J. Inorg. Nucl. Chem.*, 36, 1809, 1974; 37, 1937, 1975.
38. **Suri, S. K. and Ramakrishna, V.,** Solubility of some iodides in non-polar solvents, *Can. J. Chem.*, 47, 3049, 1969.
39. **Burgess, J. and Kijowski, J.,** Enthalpies of solution and solubilities of tin (IV) iodide and of germanium (IV) iodide, *Inorg. Chim. Acta,* 58, 135, 1981.
40. **Boulegue, J.,** Solubility of elemental sulfur in water at 298 K, *Phosphorous and Sulfur,* 5, 127, 1978.
41. **Scott, R. L.,** Liquid solutions of sulfur, in *Elemental Sulfur, Chemistry and Physics,* Meyer, M., Ed., Interscience, New York, 1965, 337.
42. **Hildebrand, J. H.,** Order from chaos, *Science,* 150, 441, 1965.
43. **Shinoda, K. and Hildebrand, J. H.,** The solubility and entropy of solution of iodine in octamethylcyclotetrasiloxane and tetraethoxysiloxane; The solubility and entropy of solution of iodine in *n*-C_7F_{16}, *c*-$C_6F_{11}CF_3$, $(C_3F_7COOCH_2)_4C$, *c*-$C_4Cl_2F_6$, CCl_2FCClF_2 and $CHBr_3$; Partial molal volumes of iodine in various complexing and non-complexing solvents, *J. Phys. Chem.*, 61, 789, 1957; 62, 292 and 295, 1958.
44. **Hildebrand, J. H. and Dymond, J. H.,** Effect of methyl groups upon the solvent power of aliphatic liquids, *Proc. Natl. Acad. Sci. U.S.A.*, 54, 1001, 1965.
45. **Hildebrand, J. H. and Scott, R. L.,** *Solubility of Non-Electrolytes,* 3rd ed., Reinhold, New York, 1950; Dover, New York, 1964.
46. **Hildebrand, J. H. and Scott, R. L.,** *Regular Solutions,* Prentice Hall, Englewood Cliffs, NJ, 1962.
47. **Martin, A.,** personal communication, 1985.
48. **Martin, A., Wu, P. L., Adjei, A., Lindstrom, R. E., and Elsworthy, P. H.,** Extended Hildebrand solubility approach and the log linear solubility equation, *J. Pharm. Sci.*, 71, 849, 1982.
49. **Yalkowsky, S. H., Flynn, G. L., and Amidon, G. L.,** Solubility of nonelectrolytes in polar solvents, *J. Pharm. Sci.*, 61, 983, 1972.
50. **Gordon, L. J. and Scott, R. T.,** Enhanced solubility in solvent mixtures. I. The system phenanthrene-cyclohexane-methylene iodide, *J. Am. Chem. Soc.*, 74, 4138, 1952.
51. **Molokanov, V. V., Chernov, D. B., and Budberg, P. B.,** Calculation of phase diagrams for the vanadium-chromium, vanadium-niobium, and chromium-niobium systems in terms of the regular solution approximation, *Zh. Fiz. Khim.*, 51, 2022, 1977; *Russ. J. Phys. Chem.*, 51, 1181, 1977.
52. **Badie, J. M. and Foex, M.,** Experimental determination, calculation, and prediction of certain scandium oxide-lanthanum oxide diagrams, *J. Solid State Chem.*, 26, 311, 1978.
53. **Stringfellow, G. B.,** The calculation of regular solution interaction parameters between elements from Groups III, IV and V of the Periodic Table, *Mater. Res. Bull.*, 6, 371, and 639, 1971; Calculation of regular solution interaction parameters in semiconductor solid solutions, *J. Phys. Chem. Solids,* 34, 1749, 1973; Calculation of distribution coefficients of donors in III-V semiconductors, *J. Phys. Chem. Solids,* 35, 775, 1974; Calculation of ternary and quaternary III-V phase diagrams, *J. Cryst. Growth,* 27, 21, 1974.
54. **Suzuki, T. and Hagiwara, M.,** A survey of regular solution parameters for the systems among the first transition metals, *Metall. Trans. A,* 8A, 371, 1977.
55. **Ghiorso, M. S. and Carmichael, I. S. E.,** A regular solution model for met-aluminous silicate liquids: application to geothermometry, immiscibility, and the source regions of basic magmas, *Contrib. Mineral. Petrol.*, 71, 323, 1980.

56. **Nakanishi, K.,** Thermodynamic studies of phase equilibria in molten salt systems. Prediction of liquidus and eutectic behaviour in mixtures having a common ion, *Ind. Eng. Chem. Fund.*, 9, 449, 1970.
57. **Nebauer, E. and Schneider, M.,** Regular solution calculations of ternary phase diagrams for Ga-P-Ge and similar systems using electronegativity data, *Phys. Stat. Solidi A.*, 23, 485, 1974.
58. **Van Hecke, G. R.,** Use of regular solution theory for calculating binary mesogenic phase diagrams exhibiting azeotropic-like behavior for liquid two-phase regions. 1. Simple minimum forming systems, *J. Phys. Chem.*, 83, 2344, 1979.
59. **Van Hecke, G. R., Cantu, T. S., Domon, M., and Billard, J.,** Use of regular solution theory for calculating binary mesogenic phase diagrams exhibiting azeotrope-like behavior. 2. Maxima forming systems, *J. Phys. Chem.*, 84, 263, 1980.
60. **Martin, A. and Cartensen, J.,** Extended solubility approach: solubility parameters for crystalline solid compounds, *J. Pharm. Sci.*, 70, 170, 1981.
61. **Alexander, K. S., Mauger, J. W., Peterson, H., Jr., and Paruta, A. N.,** Solubility profiles and thermodynamics of parabens in aliphatic alcohols, *J. Pharm. Sci.*, 66, 42, 1977.
62. **Martin, A., Newburger, J., and Adjei, A.,** New solubility equation, *J. Pharm. Sci.*, 68(10), iv, 1979.
63. **Martin, A., Wu, P. L., Adjei, A., Beerbower, A., and Prausnitz, J. M.,** Extended Hansen solubility approach: naphthalene in individual solvents, *J. Pharm. Sci.*, 70, 1260, 1981.
64. **Wu, P. L., Beerbower, A., and Martin, A.,** Extended Hansen solubility approach: calculating partial solubility parameters of solid solutes, *J. Pharm. Sci.*, 71, 1285, 1982.
65. **Beerbower, A., Wu, P. L., and Martin, A.,** Expanded solubility parameter approach I: naphthalene and benzoic acid in individual solvents, *J. Pharm. Sci.*, 73, 179, 1984.
66. **Hansen, C. M. and Beerbower, A.,** Solubility parameters, in *Kirk-Othmer Encyclopedia of Chemical Technology*, 2nd ed., Standen, A., Ed., Interscience, New York, Suppl. Vol., 1971, 889.
67. **Martin, A., Wu, P. L., and Beerbower, A.,** Expanded solubility parameter approach II: *p*-hydroxybenzoic acid and methyl *p*-hydroxybenzoate in individual solvents, *J. Pharm. Sci.*, 73, 188, 1984.
68. **Martin, A., Newburger, J., and Adjei, A.,** Extended Hildebrand solubility appraoch. II. Solubility of theophylline in polar binary solvents, *J. Pharm. Sci.*, 69, 487, 1980.
69. **Martin, A., Paruta, A. N., and Adjei, A.,** Extended Hildebrand solubility approach: methylxanthines in mixed solvents, *J. Pharm. Sci.*, 70, 1115, 1981.
70. **Martin, A. and Miralles, M. J.,** Extended Hildebrand solubility approach: solubility of tolbutamide, acetohexamide, and sulfisomidine in binary solvent mixtures, *J. Pharm. Sci.*, 71, 439, 1982.
71. **Adjei, A., Newburger, J., and Martin, A.,** Extended Hildebrand approach. III. Solubility of caffeine in dioxane-water mixtures, *J. Pharm. Sci.*, 69, 659, 1980.
72. **Martin, A., Wu, P. L., Adjei, A., Mehdizadeh, M., James, K. C., and Metzler, C.,** Extended Hildebrand solubility approach: testosterone and testosterone propionate in binary solvents, *J. Pharm. Sci.*, 71, 1334, 1982.
73. **Wu, P. L. and Martin, A.,** Extended Hildebrand solubility approach: *p*-hydroxybenzoic acid in mixtures of dioxane and water, *J. Pharm. Sci.*, 72, 587, 1983.
74. **Martin, A., Wu, P. L., and Velasquez, T.,** Extended Hildebrand solubility approach: sulfonamides in binary and ternary solvents including dioxane-water, *J. Pharm. Sci.*, 74, 277, 1985.
75. **Preston, G. T. and Prausnitz, J. M.,** Thermodynamics of solid solubility in cryogenic solvents, *Ind. Eng. Chem. Process Des. Dev.*, 9, 264, 1970.
76. **Preston, G. T., Funk, E. W., and Prausnitz, J. M.,** Solubility of hydrocarbons and carbon dioxide in liquid methane, *J. Phys. Chem.*, 75, 2345, 1971.
77. **Bukhari, M. A. I.,** Saturated phase enthalpy relationships for solid CO_2 in CH_4 at cryogenic phase behavior, *Acta Tech. Acad. Sci. Hung.*, 91, 373, 1980.
78. **Ksiazczak, A. and Anderko, A.,** Prediction of solid-liquid equilibria on the basis of pure compound properties, *Fluid Phase Equil.*, 35, 127, 1987.
79. **Ostrenga, J. A. and Steinmetz, C.,** Estimation of steroid solubility: use of fractional molar attraction constants, *J. Pharm. Sci.*, 59, 414, 1970.
80. **Anderson, B. D., Rytting, J. H., and Higuchi, T.,** Solubility of polar organic solutes in non-aqueous systems — role of specific interactions, *J. Pharm. Sci.*, 69, 676, 1980.
81. **Breon, T. L., Mauger, J. W., and Paruta, A. N.,** Determination of solubility parameter values for pure solvents and binary mixtures, *Drug. Dev. Ind. Pharm.*, 6, 87, 1980.
82. **Paruta, A. N.,** *A Dielectric Constant Approach to Solubility Phenomena*, Thesis, Rutgers — The State University, Brunswick, NJ, 1963; *Diss. Abstr.*, 24, 2701, 1964.
83. **Paruta, A. N.,** Solubility of parabens in alcohols, *J. Pharm. Sci.*, 58, 216, 1969.
84. **Paruta, A. N. and Mauger, J. W.,** Solubility of sodium salicylate in mixed solvent systems, *J. Pharm. Sci.*, 60, 432, 1971.
85. **Paruta, A. N., Sciarrone, B. J., and Lordi, N. G.,** Correlation between solubility parameters and dielectric constants, *J. Pharm. Sci.*, 51, 704, 1962.

86. **Skraga, J. and Wieckiewicz, Z.,** Method for determination of the Hildebrand solubility parameter of plasticizers and their blends, *Polimery (Warsaw),* 26(2), 45, 1981.
87. **Yalkowsky, S. H. and Valvani, S.,** Solubility and partitioning I. Solubility of nonelectrolytes in water, *J. Pharm. Sci.,* 69, 912, 1980.
88. **Valvani, S. C., Yalkowsky, S. H., and Roseman, T. J.,** Solubility and partitioning IV. Aqueous solubility and octanol-water partition coefficients of liquid non-electrolytes, *J. Pharm. Sci.,* 70, 502, 1981.
89. **Amidon, G. L. and Williams, N. A.,** A solubility equation for non-electrolytes in water, *Int. J. Pharm.,* 11, 249, 1982.
90. **Haulait-Pirson, M.-C., Huys, G., and Vanstraelen, E.,** New predictive equation for the solubility of solid n-alkanes in organic solvents, *Ind. Eng. Chem. Res.,* 26, 447, 1987.
91. **Ksiazczak, A.,** The effect of auto-association on solid-liquid equilibrium; Solid-liquid equilibrium and the structure of solutions of highly polar aromatic compounds, *Fluid Phase Equil.,* 28, 39 and 57, 1986.
92. **Iscan, M.,** Temperature effect on the partition and dimerization constant of benzoic acid, *Thermochim. Acta,* 94, 305, 1985.
93. **Judy, C. L., Pontikos, N. M., and Acree, W. E., Jr.,** Solubility of pyrene in binary solvent mixtures, containing cyclohexane, *J. Chem. Eng. Data,* 32, 60, 1987.
94. **Marthandan, M. V. and Acree, W. E., Jr.,** Solubility of anthracene in binary solvent mixtures containing dibutyl ether, *J. Chem. Eng. Data,* 32, 301, 1987.
95. **Procyck, A. D., Bissell, M., Street, K. W., Jr., and Acree, W. E., Jr.,** Solubility in binary solvent systems: 8. Estimations of binary alkane plus *p*-dioxane solvent nonideality from measured anthracene solubilities, *J. Pharm. Sci.,* 76, 621, 1987.
96. **Kikic, I. and Longo, V.,** Application of theory to solubility parameters through the correlation and prediction of naphthalene solubility in pure and mixed solvents. I. Pure solvents, *Univ. Studi Trieste Fac. Ing. Inst. Chim. Appl.,* 1974, 47.
97. **Domanska, U.,** Solubility and hydrogen bonding. Part VII. Synergic effect of naphthalene in mixed solvents, *Polish J. Chem.,* 55, 1715, 1981.
98. **McCargar, J. W. and Acree, W. E., Jr.,** Thermochemical investigations of associated solutions: 3. Effect of the inert cosolvent on solute-solvent association constants calculated from solubility measurements, *Phys. Chem. Liq.,* 17, 123, 1987.
99. **McCargar, J. W. and Acree, W. E., Jr.,** Thermochemical investigations of associated solutions: 4. Calculation of carbazole-dibutyl ether association constants from measured solubility in binary solvent mixtures; 5. Calculation of solute-solvent equilibrium constants from solubility in mixtures containing two complexing solvents, *J. Pharm. Sci.,* 76, 572, and 575, 1987.
100. **Buchowski, H., Domanska, U., and Ksiazczak, A.,** Solubility and hydrogen bondi ıg. Part 5. Synergic effect of solubility in mixed solvents, *Pol. J. Chem.,* 53, 1127, 1979.
101. **Schmid, P.,** Extraction and purification of lipids. II. Why is chloroform-methanol such a good lipid solvent? *Physiol. Chem. Phys.,* 5, 141, 1973.
102. **Billingham, N. C., Calvert, P. D., and Manke, A. S.,** Solubility of phenolic antioxidants in polyolefins, *J. App. Polym. Sci.,* 26, 3543, 1981.
103. **Khalil, S. A., Abdallah, O. Y., and Moustafa, M. A.,** Absorption of some barbiturates by gambusia fish and its correlation to solubility paramater, *Can. J. Pharm. Sci.,* 11, 26, 1976.
104. **Khalil, S. A. and Martin, A. N.,** Drug transport through model membranes and its correlation with solubility parameters, *J. Pharm. Sci.,* 56, 1225, 1967.
105. **Khalil, S. A., Moustafa, M. A., and Abdallah, O. Y.,** The use of the solubility parameter as an index of drug activity, *Can. J. Pharm. Sci.,* 11, 121, 1976.
106. **Ghosal, S. K. and Gupta, B. K.,** Determination of the solubility parameters of several barbiturates: their significance in drug activity, *J. Inst. Chem. (India),* 51, 117, 1979.
107. **Grant, D. J. W. and Abougela, I. K. A.,** Prediction of the solubility of griseofulvin in glycerides and other solvents of relatively low polarity from simple regular solution theory, *Int. J. Pharm.,* 17, 77, 1983; Solubility behavior of griseofulvin in solvents of relatively low polarity, *Expo.-Congr. Int. Technol. Pharm. 3rd,* Vol. 4, Assoc. Pharm. Galenique Ind., Chatenay-Malabry, France, 1983, 118.
108. **Cohen, S. and Haberman, F.,** The molal volumes of atropine and hyosine in relation to their respective potencies, *Br. J. Pharmac.,* 83, 807, 1984.
109. **King, J. W. and Friedrich, J. P.,** Quantitative correlations between solute molecular structure and solubility in supercritical fluids, Institute of Food Technologists Meeting, Las Vegas, NV, 1987, Paper 202; *J. Chromat.,* 517, 449, 1990.
110. **Gadalla, M. A. F., Ghaly, G. M., and Samaha, M. W.,** The effect of the composition of binary systems on the solubility and solubility parameter estimation of nalidixic and salicyclic acids, *Int. J. Pharm.,* 38, 71, 1987.
111. **Thomas, A. M.,** Petroleum fractions as DDT solvents, *Agric. Food Chem.,* 12, 442, 1964.
112. **Bowen, D. B. and James, K. C.,** Effect of temperature on the solubilities of testosterone propionate in low polarity solvents, *J. Pharm. Pharmacol.,* 22, 104S, 1970.

113. **James, K. C., Ng, C. T., and Noyce, P. R.,** Solubilities of testosterone propionate and related esters in organic solvents, *J. Pharm. Sci.*, 65, 656, 1976.
114. **James, K. C. and Roberts, M.,** The solubilities of the lower testosterone esters, *J. Pharm. Pharmacol.*, 20, 709, 1968.
115. **Lemordant, D., Bouhaouss, A., Aldebert, P., and Baffier N.,** Intercalation of organic solvents in the lamellar structure of V_2O_5 xerogels, *J. Chem. Phys.*, 83, 105, 1986.
116. **Bolster, R. N. and Little, R. C.,** Dependence of lithium grease dropping points on solubility parameter of oil component, *Ind. Eng. Chem. Prod. Res. Dev.*, 5, 198, 1966.
117. **Schott, H. and Chang, S.-L.,** The solubility of potassium acid soaps in organic solvents, *J. Colloid Interface Sci.*, 117, 94, 1987.
118. **Bright, G. S.,** The relationship of the Hildebrand solubility parameter of various oils to the properties of lithium soap greases made therefrom, *Nat. Lubric. Grease Inst. Spokesman*, 34, 156, 1970; *Tex. J. Sci.*, 22(Abstr.), 304, 1971.
119. **Little, R. C.,** Some physical properties of soap/solvent systems and their relation to the solubility parameter of the solvent, *J. Colloid Interface Sci.*, 21, 266, 1966.
120. **Little, R. C.,** Oil solubility parameter as a factor in boundary lubrication effects, *Wear*, 30, 385, 1974.
121. **Little, R. C.,** Some observations on the zinc stearate-toluene system, *J. Colloid Interface Sci.*, 51, 200, 1975.
122. **Bondi, A.,** Solubility and swelling of alkali soaps in organic solvents, *J. Colloid Sci.*, 5, 458, 1950.
123. **Leuenberger, H.,** Compression of binary powder mixtures and solubility parameters of solids, *Int. J. Pharm.*, 27, 127, 1985.
124. **Kon-no, K., Sakiyama, M., Suzuki, T., and Kitahara, A.,** The determination of solubility parameter of surfactants by gas chromatography, *Nippon Kagaku Kaishi*, 2211, 1974.
125. **Little, R. C.,** Correlation of surfactant hydrophile-lipophile balance (HLB) with solubility parameter, *J. Colloid Interface Sci.*, 65, 587, 1978.
126. **Little, R. C. and Singleterry, C. R.,** The solubility of alkali dinonylnaphthalenesulfonates in different solvents and a theory for the solubility of oil-soluble soaps, *J. Phys. Chem.*, 68, 3453, 1964.
127. **Kertes, A. S.,** Solubility parameters of long-chain alkylamines and their activities in organic solvents, *J. Inorg. Nucl. Chem.*, 26, 1764, 1964.
128. **Kertes, A. S.,** Solubility and activity of high-molecular amine hydrochlorides in organic solvents, *J. Inorg. Nucl. Chem.*, 27, 209, 1965.
129. **Sherertz, E. F., Sloan, K. B., and McTiernan, R. G.,** Use of theoretical partition coefficient determined from solubility parameters to predict permeability coefficients for 5-fluorouracil, *J. Invest. Dermatol.*, 86, 506, 1986 (abstract); 89, 147, 1987.
130. **Waranis, R. P., Siver, K. G., and Sloan, K. B.,** The solubility parameter of vehicles as a predictor of relative vehicle effects on the diffusion of 6-mercaptopurine, *Int. J. Pharm.*, 36, 211, 1987.
131. **Gerrits, P. O., Horobin, R. W., Wright, D. J., van Goor, H., and Grond, J.,** Retention, localization and staining of lipids in tissues embedded in water-miscible methacrylates: using the solubility parameter to analyze the influences of resins, plasticizers and other reagents, *Acta Morphol. Neerl.-Scand.*, 25, 29, 1987.
132. **Liversidge, G. G. and Grant, D. J. W.,** Influence of physicochemical interactions on the properties of suppositories. II. Interactions between the constituents of fatty suppository bases and ketoprogen or metronidazole, *Drug. Dev. Ind. Pharm.*, 9, 223, 1983.
133. **Walden, P.,** Heat of fusion, specific cohesion, and molecular complexity at the melting point, *Z. Elektrochem.*, 14, 713, 1908.
134. **Maher, N. and Sirois, G.,** Correlation between the solubility of phenobarbital and the polarity of various solvent systems, *Can. J. Pharm. Sci.*, 6(3), 55, 1971.
135. **Liron, Z. and Cohen, S.,** Densitometric determination of the solubility parameter and molal volume of compounds of medicinal relevance, *J. Pharm. Sci.*, 72, 499, 1983.
136. **Meusberger, K. E.,** Computerized optimization of emulsifiers for pesticide emulsifiable concentrates, in *Advances in Pesticide Formulation Technology*, Scher, H., Ed., American Chemical Society, Washington, D.C., 1984, 121; Determination of cohesive energy density parameters for pesticide formulation development, *Am. Chem. Soc. Symp. Ser.*, 371 (*Pesticide Formulations*), 151, 1988.
137. **England, C., Lawson, D. D., and Hrubes, J. D.,** Superconductivity, cohesive energy density and electron-atom ratio in metals, *J. Appl. Phys.*, 52, 2923, 1981.
138. **Shmidt, V. S., Shesterikov, V. N., and Mezhov, E. A.,** The solubility of amine salts in solvents of low polarity and the influence of diluents on the extraction properties of amine salts, *Usp. Khim.*, 36, 2167, 1967; *Russ. Chem. Rev.*, 36, 946, 1967.
139. **Shetty, P. H., Youngberg, P. J., Kersten, B. R., and Poole, C. F.,** Solvent properties of liquid organic salts used as mobile phases in microcolumn reversed-phase liquid chromatography, *J. Chromatgr.*, 411, 61, 1987.

140. **Taft, R. W., Abraham, M. H., Doherty, R. M., and Kamlet, M. J.,** Linear solvation energy relationships. 29. Solution properties of some tetraalkylammonium halide ion pairs and dissociated ions, *J. Am. Chem. Soc.,* 107, 3105, 1975.
141. **Glikberg, S. and Marcus, Y.,** Relation of the Gibbs free energy of transfer of ions from water to polar solvents to the properties of the solvents and the ions, *J. Soln. Chem.,* 12, 255, 1983.
142. **Iwamoto, E., Ito, K., and Yamamoto, Y.,** Solute-solvent interactions in ion-pair extraction of tetraalkylammonium iodides. 1. A new approach to the extraction constant; 3. The role of relative permittivity and solubility parameter in nitrobenzene-tetrachloromethane mixtures at 25°C, *J. Phys. Chem.,* 85, 894, 1981; 87, 1776, 1983.
143. **Sierkierski, S. and Olszer, R.,** Relation between the partition coefficient of GeX_4 molecules (X = Cl, Br, I) and the solubility parameter of the solvent, *J. Inorg. Nucl. Chem.,* 25, 1351, 1963.
144. **Wakahayashi, T., Oki, S., Omori, T., and Suzuki, N.,** Some applications of the regular solution theory to solvent extraction. I. Distribution of β-diketones, *J. Inorg. Nucl. Chem.,* 26, 2255, 1964.
145. **Irving, H. M. N. H.,** Application of the solubility concept in liquid-liquid extraction, in *Ion Exchange and Solvent Extraction,* Vol. 6, Marinsky, J. A., and Marcus, Y., Eds., Marcel Dekker, New York, 1974, chap. 3, 139.
146. **Reichardt, C.,** *Solvent Effects in Organic Chemistry* (Monographs in Modern Chemistry Vol. 3), Verlag Chemie, Weinheim, 1979.
147. **Tsutsui, T. and Tanaka, T.,** Role of electrostatic forces in the glass transition temperature of ionic polymers, *Polymer,* 18, 817, 1977.
148. **Gardon, J. L.,** The influence of polarity upon the solubility concept, *J. Paint Technol.,* 38(492), 43, 1966.
149. **Gardon, J. L.,** Critical review of concepts common to cohesive energy density, surface tension and butt joint strength, *J. Colloid Interface Sci.,* 59, 582, 1977; *Prog. Org. Coat.,* 5, 1, 1977.
150. **Van Zeggeren, F.,** Relation between activation energy of the viscosity and heat of vaporization for molten salts, *Can. J. Chem.,* 34, 1512, 1956.
151. **Petrucci, S., Ed.,** *Ionic Interactions: From Dilute Solutions to Molten Salts,* 2 vols, Academic Press, New York, 1971.
152. **Pitzer, K. S.,** Electrolytes. From dilute solutions to fused salts, *J. Am. Chem. Soc.,* 102, 2901, 1980.
153. **Smedley, S. I.,** *The Interpretation of Ionic Conductivity in Liquids,* Plenum Press, New York, 1980.
154. **Fukunaga, K., Shirai, H., Ide, S., and Kimura, M.,** Solubility parameters of phase-transfer catalysts and their applications, *Nippon Kagaku Kaishi,* 1148, 1980.
155. **Wrona, P. K.,** On the correlations between empirical Lewis acid-base solvent parameters and the thermodynamic parameters of ion solvation, *J. Electroanal. Chem.,* 108, 153, 1980.
156. **Larson, R. G. and Hunt, H.,** Molecular forces and solvent power, *J. Phys. Chem.,* 43, 417, 1939.
157. **Small, P. A.,** Some factors affecting the solubility of polymers, *J. Appl. Chem.,* 3, 71, 1953.
158. **Dack, M. R. J.,** The importance of solvent internal pressure and cohesion to solution phenomena, *Chem. Soc. Rev.,* 4, 211, 1975.
159. **McDevit, W. F. and Long, F. A.,** The activity coefficient of benzene in aqueous salt solutions, *J. Am. Chem. Soc.,* 74, 1773, 1952.
160. **Lewis, G. A. and Enever, R. P.,** Solution thermodynamics of some potentially long-acting norethindrone derivatives. II. Solutions in 2,2,4-trimethylpentane (iso-octane), *Int. J. Pharm.,* 3, 275, 1979.
161. **Sunwoo, C. and Eisen, H.,** Solubility parameter of selected sulfonamides, *J. Pharm. Sci.,* 60, 238, 1971.
162. **Guillén Sans, R. and Guzmán Chozas, M.,** Determination of dielectric requirement and solubility parameter of 5-(anisylidene)-2-thiobarbituric acid, *Pharm. Acta Helv.,* 64(2), 51, 1989.
163. **Reuteler-Faoro, D., Ruelle, P., Hô Nam-Trân, de Reyff, C., Buchmann, M., Nègre, J. C., and Kesselring, U. W.,** A new equation for calculating partial cohesion parameters of solid substances from solubilities, *J. Phys. Chem.,* 92, 6144, 1988.
164. **Rey-Mermet, C., Ruelle, P., Hô Nam-Trân, Nègre, J. C., Buchmann, M., and Kesselring, U. W.,** Determination of cohesion parameter of three xanthines by heat-flow calorimetry, to be published.
165. **Rey-Mermet, C.,** *Determination of Partial Cohesion Parameter from Binary Thermodynamic Quantities,* Doctoral Thesis, School of Pharmacy, University of Lausanne, Switzerland, 1989.
166. **Ksiazczak, A. and Jodzewicz, W.,** Liquid-liquid and liquid-solid equilibria of 2-nitrophenol in hydrocarbons, *Polish J. Chem.,* 55, 1889, 1981.
167. **Martin, A. and Mauger, J.,** The curious solubility of phenobarbitol: how to use solubility parameters, *Am. J. Pharm. Ed.,* 52, 68, 1988.
168. **Domanska, U.,** Solubility of 2,5-dimethylphenol and 3,4-dimethylphenol in binary solvent mixtures containing alcohols, *Fluid Phase Equil.,* 40, 259, 1988.
169. **Bozzay, J., Dávid, Á., Ékes, G., and Rusznák, I.,** On the solubility of pesticides and compounds of pesticide type, *Period. Polytech. Chem. Eng.,* 23(2), 87, 1979.
170. **Wakui, Y., Imura, H., and Suzuki, N.,** Solvent effect on the liquid-liquid partition of 21H,23H-porphine between various organic solvents and acidic aqueous solutions, *Bull. Chem. Soc. Jpn,* 62, 2520, 1989.

171. **Neau, S. H., Flynn, G. L., and Yalkowsky, S. H.,** The influence of heat capacity assumptions on the estimation of solubility parameters from solubility data, *Int. J. Pharm,* 49, 223, 1989; **Neau, S. H.,** *On the Application of Regular Solution Theory to the Estimation of Solubility Parameters of Crystalline Organic Compounds from Solubility Data,* Thesis, University of Michigan, Ann Arbor, MI, *Diss. Abstr. Int. B,* 49, 5319, 1989.
172. **Bustamante, P., Escalera, B., Martin, A., and Sellés, E.,** Predicting the solubility of sulfamethoxypyridazine in individual solvents. I. Calculating partial solubility parameters, *J. Pharm. Sci.,* 78, 567, 1989; **Martin, A., Bustamante, P., Escalera, B., and Selleés, E.,** Predicting the solubility of sulfamethoxypyridazine in individual solvents. II. Relationship between solute-solvent interaction terms and partial solubility parameters, *J. Pharm. Sci.,* 78, 672, 1989.
173. **Kramer, A. and Thodos, G.,** Solubility of 1-hexadecanol and palmitic acid in supercritical carbon dioxide; Solubility of 1-octadecanol and stearic acid in supercritical carbon dioxide, *J. Chem. Eng. Data,* 33, 230, 1988; 34, 184, 1989.
174. **Gal'perin, L. B. and Yakushkin, M. I.,** Solubility of tere- and isophthalonitriles, *Zh. Prikl. Khim.,* 41, 1829, 1968; *J. Appl. Chem. U.S.S.R.,* 41, 1723, 1968.
175. **Durbut, P. and Broze, G.,** Application of the solubility parameter theory to fatty soil removal, *Comun. Jorn. Com. Esp. Deterg.,* 20, 383, 1989.
176. **Roy, S. D. and Flynn, G. L.,** Solubility and related physicochemical properties of narcotic analgesics, *Pharm. Res.,* 5, 580, 1988.
177. **Wright, P. V.,** The lattice energies of complexes of poly(ethylene oxide) with sodium halides, *Polymer,* 30, 1179, 1989.
178. **Shrivastava, S. N. and Pandey, J. D.,** Excess intermolecular free length and molecular interaction for binary molten metal halide and cryogenic liquid mixtures, *Acoust. Lett.,* 3, 219, 1980.

Chapter 13

POLYMER SOLUTIONS

Because many of the polymers of practical importance are amorphous and, therefore, ''liquid-like'' in some of their properties, their solubility behaviors can be treated by the methods used for liquid-liquid mixtures if allowances are made for the differences due to one of the components being made up of very long molecules and the other of small molecules.

It is found that solutions of polymers deviate from ideality even at very low concentrations, whereas mixtures of liquids obey Raoult's law (Equation 28, Chapter 2) if the enthalpy of mixing is negligible. Consider, for example, a mixture of 100 g of polystyrene (number-average molar mass 290,000 g mol^{-1}) with 3.4 g of toluene.[1] This corresponds to a solvent mole fraction ${}^{i}x = 0.99$, and an ideal mixture of this composition would be expected to have a vapor pressure of 3.7 kPa. But at 25°C it is, in fact, an order of magnitude lower, 0.32 kPa.

The entropy of the liquid state is positive relative to that of the amorphous polymeric state because liquid molecules are free to translate and rotate in a random manner, whereas polymer molecules have little freedom for either translation or rotation, and even segmental motion is limited. Consequently, polymer molecules undergo an increase in entropy when dispersed in a liquid, and when mixing takes place it is usually because there is an associated increase in entropy and not because of any particular forces of ''attraction''. The increase in entropy that occurs when two liquids of equal-sized molecules mix (the combinatorial entropy due to the extra degrees of freedom) is given by Equation 24, Chapter 2. For molecules of different size the number of possible combinations (and therefore the extent of entropy increase) is less.

Any theory or description of polymer solutions has to also explain the fact that even when a polymer-liquid pair exhibits complete miscibility at high polymer concentrations, two phases may be formed on dilution. In other words, if a miscibility gap occurs in a system including a polymer, this miscibility gap is at low solute concentrations, leading to marked asymmetry in the phase diagram. Another complication is the phenomenon of ''co-nonsolvency'', the formation of a non-solvent by the mixing of two solvents (Section 13.6).

Mixtures of polymers and low molar mass liquids are defined frequently in terms of mass or weight fraction (w), volume fraction (ϕ), mole fractions (x) being inconvenient because ${}^{j}x$ is usually a very small quantity even when ${}^{j}w$ and ${}^{j}\phi$ are close to unity.

13.1 PHYSICAL PROPERTIES OF POLYMERS

The polymers that have the greatest similarities with liquids are those in the rubbery amorphous state (Table 1) and above their glass transition temperatures at ambient temperature. Other amorphous polymers (Table 2) have higher glass transition temperatures and are in the glassy state at room temperature. Some polymers are observed in both amorphous and crystalline states that differ substantially in their densities or molar volumes (Table 3).

The molar volume per repeating structural unit of a polymer is defined as

$$ {}^{u}V = {}^{u}M\nu = {}^{u}M/\rho \qquad (1) $$

where ν is the specific volume, ${}^{u}M$ is the molar mass of the structural unit and ρ is the density. van Krevelen and Hoftyzer[2] have compiled experimental information on densities and molar volumes of rubbery amorphous polymers (subscript ru), glassy amorphous polymers (subscript gl), and crystalline polymers (subscript cr) (Tables 1 to 3). Finding that the molar group volume contributions for monomers (Chapter 6) were unsuitable for the

TABLE 1
Densities and Structural Unit Molar Volumes of Rubbery Polymers at 25°C

Polymer	ρ_{ru}/g cm^{-3}	V_{ru}/cm^3 mol^{-1}
Polyacetaldehyde	1.07	41
Poly(3,3-bis(chloromethyl)oxacyclobutane)	1.39	112
Polybutadiene	0.89	61
Polybutylene	0.86	65
Poly(butyl methacrylate)	1.05	135
Poly(butyl vinyl ether)	0.93	93
Poly(*sec*-butyl vinyl ether)	0.92	108
Poly(chlorobutadiene)	1.24	71
Poly(decyl vinyl ether)	0.88	209
Poly(dodecyl methacrylate)	0.93	204
Poly(dodecyl vinyl ether)	0.89	238
Poly(ethyl acrylate)	1.12	89
Poly(2-ethylbutyl methacrylate)	1.04	164
Polyethylene	0.86	33
Poly(ethylene oxide)	1.13	39
Poly(2-ethylhexyl vinyl ether)	0.90	173
Polyformaldehyde	1.25	24
Polyhexene	0.86	98
Poly(hexyl methacrylate)	1.01	169
Poly(hexyl vinyl ether)	0.93	139
Polyisobutylene	0.84	67
Poly(isobutyl vinyl ether)	0.93	108
Poly(isopropyl vinyl ether)	0.92	93
Poly(methyl acrylate)	1.22	71
Poly(2-methylbutadiene)	0.91	75
Poly(1-methylpentyl methacrylate)	1.01	168
Poly(octyl methacrylate)	0.97	204
Poly(octyl vinyl ether)	0.91	171
Polypentadiene	0.89	77
Polypentene	0.85	83
Poly(pentyl vinyl ether)	0.92	124
Poly(5-phenylpentene)	1.05	139
Polypropylene	0.85	50
Poly(propylene oxide)	1.00	58
Poly(tetramethylene oxide)	0.98	74
Poly(tetrafluoroethylene)	2.00	50
Poly(vinyl butyl sulfide)	0.98	119
Poly(vinylidene chloride)	1.66	58
Poly(vinyl methyl sulfide)	1.18	63
Poly(vinyl propionate)	1.02	98

Adapted from van Krevelen, D. W., *Properties of Polymers: Their Estimation and Correlation with Chemical Structure,* 2nd ed., Elsevier, Amsterdam, 1976.

prediction of polymer properties, they also developed new group contributions specifically for rubbery (${}^zV_{ru}$) and glassy (${}^zV_{gl}$) polymers. These are shown in Table 4, together with the range of recommended liquid group contributions, ${}^zV_1$. In addition, van Krevelen and Hoftyzer compiled a set of improved atomic contributions to V_{ru} and V_{gl} for use with polymers containing structural groups not appearing in Table 4. From Table 3 it can be seen that the average value of the density ratio ρ_{cr}/ρ_{am} is 1.13, with $\rho_{cr}/\rho_{ru} \approx 1.15$ and $\rho_{cr}/\rho_{gl} \approx 1.07$ for amorphous polymers with glass transition temperatures below 25°C and above 25°C, respectively. For semicrystalline polymers (subscript sc) with degree of crystallinity x_{cr}, the resulting approximate relationship between ρ_{sc} and ρ_{am} is

TABLE 2
Densities and Structural Unit Molar Volumes of Glassy Polymers at or Near 25°C

Polymers	ρ_{gl}/g cm^{-3}	V_{gl}/cm^3 mol^{-1}
Polyacrylonitrile	1.18	45
Polyamide 6	1.08	104
Polyamide 6,6	1.07	212
Polyamide 6,10	1.04	212
Polyamide 8	1.04	136
Polyamide 11	1.01	182
Polyamide 12	0.99	199
Poly(benzyl methacrylate)	1.18	150
Poly(*tert*-butyl acrylate)	1.00	128
Poly(butyl chloroacrylate)	1.24	131
Poly(*sec*-butyl chloroacrylate)	1.24	131
Poly(*sec*-butyl methacrylate)	1.05	135
Poly(*tert*-butyl methacrylate)	1.02	139
Poly(*p*-*tert*-butylstyrene)	0.95	169
Poly(2-chloroethyl methacrylate)	1.32	113
Poly(1-*o*-chlorophenylethyl methacrylate)	1.27	177
Poly(chlorotrifluoroethylene)	1.92	61
Poly(cyclohexyl chloroacrylate)	1.25	151
Poly(cyclohexylenedimethylene terephthalate)	1.19	231
Poly(cyclohexyl methacrylate)	1.10	153
Poly(*p*-cyclohexylphenyl methacrylate)	1.12	219
Poly(1,3-dimethylbutyl methacrylate)	1.01	170
Poly(3,3-dimethylbutyl methacrylate)	1.00	170
Poly(dimethyl phenylene oxide)	1.07	112
Poly(diphenyl phenylene oxide)	1.14	214
Poly(diphenylmethyl methacrylate)	1.17	216
Poly(1,2-diphenylethyl methacrylate)	1.15	232
Poly(ethyl chloroacrylate)	1.39	97
Poly(ethylene isophthalate)	1.34	144
Poly(ethylene phthalate)	1.34	144
Poly(ethylene terephthalate)	1.33	145
Poly(ethyl methacrylate)	1.12	102
Poly(4-fluoro-2-trifluoromethylstyrene)	1.43	133
Poly(glycolic acid)	1.60	36
Poly(isopentyl methacrylate)	1.03	151
Poly(isopropyl chloroacrylate)	1.27	117
Poly(4,4-isopropylenediphenylene carbonate)	1.20	212
Poly(isopropyl methacrylate)	1.03	124
Polymethacrylonitrile	1.10	61
Poly(1-methylbutyl methacrylate)	1.03	152
Poly(methyl cyanoacrylate)	1.30	85
Poly(4,4-methylenediphenylene carbonate)	1.24	182
Poly(methyl methacrylate)	1.17	86
Poly(4-methylpentene)	0.84	100
Poly(α-methylstyrene)	1.07	111
Poly(*o*-methylstyrene)	1.03	115
Poly(*p*-methylstyrene)	1.04	114
Poly(neopentyl methacrylate)	0.99	157
Poly(1-phenylethyl methacrylate)	1.13	169
Poly(phenyl methacrylate)	1.21	134
Poly(3-phenylpropene)	1.05	113
Polypivalolactone	1.10	91
Poly(propyl chloroacrylate)	1.30	114
Poly(propyl methacrylate)	1.08	119
Polystyrene	1.05	99

TABLE 2 (continued)
Densities and Structural Unit Molar Volumes of Glassy Polymers at or Near 25°C

Polymers	ρ_{gl}/g cm^{-3}	V_{gl}/cm^3 mol^{-1}
Poly(tetramethylene isophthalate)	1.27	174
Poly(thiodiphenylene carbonate)	1.36	180
Poly(2,2,2-trifluoro-1-methylethyl methacrylate)	1.34	134
Poly(*m*-trifluoromethylstyrene)	1.43	133
Poly(3,3,3-trifluoropropylene)	1.58	61
Poly(1,2,2-trimethylpropyl methacrylate)	0.99	172
Poly(vinyl acetate)	1.19	72
Poly(vinyl alcohol)	1.26	35
Poly(vinyl chloride)	1.39	45
Poly(vinyl chloroacetate)	1.45	83
Poly(vinylcyclohexane)	0.95	116
Poly(vinyl methyl ketone)	1.12	63

Adapted from van Krevelen, D. W., *Properties of Polymers: Their Estimation and Correlation with Chemical Structure,* 2nd ed., Elsevier, Amsterdam, 1976.

TABLE 3
Densities of Amorphous and Crystalline States of Polymers

Polymer	ρ_{cr}/g cm^{-3}	ρ_{am}/g cm^{-3}	ρ_{cr}/ρ_{am}
Polyacetylene	1.15	1.00	1.15
Polyamide 6	1.23	1.08	1.14
Polyamide 6,6	1.24	1.07	1.16
Polyamide 6,10	1.19	1.04	1.14
Poly(bisphenol A carbonate)	1.31	1.20	1.09
Polybutadiene	1.01	0.89	1.14
Polybutylene	0.95	0.86	1.10
Polyethylene	1.00	0.85	1.18
Poly(ethylene oxide)	1.33	1.12	1.19
Poly(ethylene terephthalate)	1.50	1.33	1.13
Polyisobutylene	0.94	0.84	1.12
Poly(*cis*-isoprene)	1.00	0.91	1.10
Poly(*trans*-isoprene)	1.05	0.90	1.16
Poly(methylene oxide)	1.54	1.25	1.25
Poly(methyl methacrylate)	1.23	1.17	1.05
Polypentene	0.92	0.85	1.08
Polypivalolactone	1.23	1.08	1.13
Polypropylene	0.95	0.85	1.12
Poly(propylene oxide)	1.15	1.00	1.15
Polystyrene	1.13	1.05	1.09
Poly(tetrafluoroethylene)	2.35	2.00	1.17
Poly(tetramethylene oxide)	1.18	0.98	1.20
Poly(trifluorochloroethylene)	2.19	1.92	1.14
Poly(vinyl alcohol)	1.35	1.26	1.07
Poly(vinyl chloride)	1.52	1.39	1.10
Poly(vinylidene chloride)	1.95	1.66	1.17
Poly(vinylidene fluoride)	2.00	1.74	1.15

Adapted from van Krevelen, D. W., *Properties of Polymers: Their Estimation and Correlation with Chemical Structure,* 2nd ed., Elsevier, Amsterdam, 1976.

TABLE 4
Molar Volume Group Contributions for Glassy and Rubbery Polymers Compared with the Corresponding Liquid Group Contributions

Group, z	${}^{z}V/\mathrm{cm^3\ mol^{-1}}$ Glassy	Rubbery	Liquid
	Bivalent		
$-CH_2-$	15.9	16.5	16—17
$-CH(CH_3)-$	33.4	32.7	32—34
$-C(CH_3)_2-$	52.4	50.4	48—50
$-CH{=}CH-$	—	27.8	24—27
$-CH{=}C(CH_3)-$	—	42.8	40—43
$-C_6H_4-$	65.5	61.4	59—62
$-CH(C_6H_5)-$	82.2	74.5	75—79
$-C_6H_3(CH_3)-$	83.4	—	—
$-C_6H_2(CH_3)_2-$	104.1	—	—
$-C_6H_{10}-$	87.8	—	—
$-CH(C_6H_{11})-$	100.2	—	—
$-CHF-$	20.4	19.9	16—19
$-CHCl-$	29.4	28.3	24—27
$-CH(CN)-$	29.0	—	—
$-O-$	10.0	8.5	6—7
$-CO-$	13.4	—	—
$-COO-$ (general)	23.0	24.6	16—24
$-COO-$ (acrylic)	18.3	21.0	—
$-O-CO-O-$	31.4	—	—
$-CH(OH)-$	19.2	—	—
$-CONH-$	24.9	—	—
$-S-$	17.8	15.0	11—16
	Tetravalent		
$>C<$	4.6	4.8	—
$>C_6H_2<$	56.3	—	—
	Trivalent		
$>CH-$	9.5	9.9	—
$>C_6H_3-$	59.5	—	—
	Monovalent		
$-CH_3$	23.9	22.8	—
$-C_6H_5$	72.7	64.7	—
$-C_6H_{11}$	90.7	—	—
$-F$	10.9	10.0	—
$-Cl$	19.9	18.4	—
$-CN$	19.5	—	—
$-OH$	9.7	—	—

Adapted from van Krevelen, D. W., *Properties of Polymers: Their Estimation and Correlation with Chemical Structure,* 2nd ed., Elsevier, Amsterdam, 1976.

TABLE 5
Estimation of Densities at 25°C of Glassy and Crystalline Poly(Ethylene Terephthalate)

Group, z	${}^zV_{gl}$/cm^3 mol^{-1} (Table 4)	$\Sigma_z\, {}^zV_{gl}$/cm^3 mol^{-1}
$-C_6H_4-$	65.5	65.5
2 ($-$COO$-$)	23.0	46.0
2 ($-CH_2-$)	15.9	31.8
		V_{gl} 143.3

ρ_{gl} = 192.2/143.3 = 1.34 g cm^{-3}
Equation 2: ρ_{cr} = 1.13 ρ_{gl} = 1.52 g cm^{-3}

Adapted from van Krevelen, D. W., *Properties of Polymers: Their Estimation and Correlation with Chemical Structure,* 2nd ed., Elsevier, Amsterdam, 1976.

$$\rho_{sc}/\rho_{am} = V_{am}/V_{sc} \approx 1 + 0.13x_{cr} \qquad (2)$$

As an example,[2] in Table 5 are estimated the densities of glassy and crystalline forms of poly(ethylene terephthalate), with repeat unit $-O-CO-\phi_p-CO-O-CH_2-CH_2-$, ϕ_p being the *p*-phenylene group. The calculated ρ_{gl} is in good agreement with the experimental 1.33 g cm^{-3}, and ρ_{cr} is within the experimental range of 1.46 to 1.52 g cm^{-3}.

When applied to polymers, Askadskii's van der Waals' volume method[3-5] (Section 6.6) shows that the packing coefficient,

$$k = (N_A\rho/M) \sum_z {}^zV_v \qquad (3)$$

is approximately constant for all amorphous polymers, regardless of chemical structure, with an average value of 0.681 at 20°C. The distribution of packing coefficients for crystalline polymers, on the other hand, is very broad.[5] Polarity may be expressed in terms of the van der Waals volume of polar groups.[187] Good examples of the estimation of polymeric chain flexibility, of molar volumes of polymer repeat units, and of polymeric residues (portions of repeat units) are the works of Sewell,[6] Knox[7,8] on poly(ethylene terephthalate), Becker[9] on polyurethane, and Lavygin.[10]

Most of the group contribution methods (Chapter 6) and thermodynamic expressions (Chapter 2) for liquids can be applied to polymers. Equations of state for polymer melts and for amorphous polymers below their melting points have been used in the same way as that described in Section 2.4 for liquids.[11] The zero point cohesive pressure has been evaluated,[12,13] while internal pressures, π, and cohesive pressures for polymers have been determined and correlated.* It can be seen from Table 6 that the correlation expression of Voeks,[20]

$$\pi/c = n \approx 1.3 \qquad (4)$$

is approximately correct. Eskin and Nesterov[23] evaluated cohesive energies.

In liquid-liquid systems, considerable use is made of the concepts of ideal solutions and

* For example: Allen et al.;[14,15] Bagley and Scigliano;[16,17] Booth, Price, et al.;[18,19] Nanda and Simha;[11] Sears and Darby;[180] van Krevelen;[2] Voeks;[20] and Walsh and Rostami.[21,22]

TABLE 6
Comparison of Internal Pressures and Cohesive Pressures for Polymers at 20°C

Polymer	π/MPA	c/MPA	$n = \pi/c$
Poly(chlorotrifluoroethylene)	370	220—260	1.7—1.4
Poly(dimethyl siloxane)	240	220—240	1.1—1.0
Poly(ethyl acrylate)	440	350—370	1.3—1.2
Polyethylene	320	250—290	1.3—1.1
Polyisobutylene	330	250—270	1.3—1.2
Poly(methyl methacrylate)	380	340—690	1.1—0.6
Poly(propylene oxide)	370	230—420	1.6—0.9
Polystyrene	460	300—360	1.5—1.3

Adapted from van Krevelen, D. W., *Properties of Polymers: Their Estimation and Correlation with Chemical Structure,* 2nd ed., Elsevier, Amsterdam, 1976.

regular solutions. An important advance in the development of an understanding of polymer-liquid systems was the formulation of a corresponding polymer solution concept, a hypothetical reference solution with which real polymer solutions may be compared.

13.2 THE POLYMER-LIQUID REFERENCE SOLUTION

One of the first and best-known attempts to formulate a model explaining the deviations arising in polymer solution properties due to the large differences in size between solute and solvent molecules was the Flory-Huggins lattice model. This theory, with its various subsequent variations, has been reviewed by many authors,* and the results presented here are limited to those required to show the relationship between Flory-Huggins and cohesion parameter concepts.

Flory and co-workers[46-49] and Huggins[50-52] calculated the entropy of mixing of long-chain molecules with monomeric liquids using the assumption that each polymer segment and each single monomeric molecule occupied a lattice site. They applied a statistical mechanical treatment, taking into account the fact that the polymer segments must occupy adjacent sites, and derived for the molar combinatorial entropy of mixing:

$$\Delta_m S_{co} = -R({}^i x \ln {}^i\phi + {}^j x \ln {}^j\phi) \qquad (5)$$

where the volume fraction ϕ replaces the mole fraction x in the logarithm term of the ideal entropy of mixing expression for equal-sized small molecules (Equation 24, Chapter 2):

$$\Delta_m S_{id} = -R({}^i x \ln {}^i x + {}^j x \ln {}^j x)$$

The reference state for a mixture of i and j components, of which j is polymeric, can be considered one for which the entropy change on mixing is given by the Flory-Huggins lattice model and for which the enthalpy of mixing is zero. The difference between the entropy of mixing for the production of a real mixture and of this reference mixture is frequently

* For example: Acree;[24] Bagley and Scigliano;[25] Blanks and Prausnitz;[26-28] Burrell;[29] Carpenter;[30] Casassa and Sedlacek;[31] Dauantis;[32] Delmas and Patterson;[33] Flory;[34] Gardon[35,36] Hildebrand et al.,[37] Holten-Anderson and Eng;[179] Kaelble;[38] Kalz;[39] Magat;[40] Mark and Tobolsky[41] Olabisi et al.;[42] Prausnitz et al.;[43,44] and Sheehan and Bisio.[45]

designated the "residual" entropy of mixing (as opposed to the term "excess", used for nonpolymeric mixtures),

$$^{R}\Delta_m S = \Delta_m S - \Delta_m S_{co} \tag{6}$$

and the "correction term" for the size effect in an otherwise ideal mixture is

$$\Delta_m S_{co} - \Delta_m S_{id} = -R\,{}^{i}x\ \ln({}^{i}\phi/{}^{i}x) - R\,{}^{j}x\ \ln({}^{j}\phi/{}^{j}x) \tag{7}$$

Since the enthalpy of mixing is zero for the reference mixture, the residual Gibbs free energy of mixing is

$$\begin{aligned} ^{R}\Delta_m G &= \Delta_m G - (-T\Delta_m S_{co}) \\ &= \Delta_m G - RT({}^{i}x\ \ln {}^{i}\phi + {}^{j}x\ \ln {}^{j}\phi) \end{aligned} \tag{8}$$

The derivative of $^{R}\Delta_m G$ with respect to the amount of component i provides the residual chemical potential of the liquid,

$$({}^{i}\mu - {}^{i}\mu^{o})^{R} = ({}^{i}\mu - {}^{i}\mu^{o}) - RT[\ln {}^{i}\phi + (1 - {}^{i}V/{}^{j}V){}^{j}\phi] \tag{9}$$

The expression within square brackets replaces $\ln {}^{i}x$, which would be the appropriate term for a process with an ideal entropy of mixing.

It should be noted that the use of the Flory-Huggins entropy expression for large non-polymer molecules (such as long-chain surfactant and plasticizer molecules) does not always provide satisfactory results.[53,54] (Section 14.8 includes an example where *calculated* Hildebrand parameters for large plasticizer molecules are more correct than experimental ones.)

13.3 POLYMER-LIQUID INTERACTION PARAMETERS

Equation 9, when rearranged, provides a formal definition[1] of the polymer-liquid interaction parameter, χ

$$\begin{aligned} &({}^{i}\mu - {}^{i}\mu^{o})^{R}/(RT\,{}^{j}\phi^2) = ({}^{i}\mu - {}^{i}\mu^{o})/(RT\,{}^{j}\phi^2) - \\ &\quad [\ln {}^{i}\phi + (1 - {}^{i}V/{}^{j}V){}^{j}\phi]/{}^{j}\phi^2 = \chi \end{aligned} \tag{10}$$

Defined empirically but unambiguously in this way, χ (previously denoted μ by some authors) is a general, dimensionless interaction parameter reflecting the intermolecular forces between the molecules in a particular polymer-liquid system. Its use need not be restricted to any particular theory or model, a preferred approach in that χ has been associated with several rather different theoretical treatments since the original proposal of Flory and Huggins. It was first introduced as an enthalpy of dilution term without an entropy component, but has been considered subsequently as a Gibbs free energy parameter. (More recently it has been redefined in terms of segment fractions in place of volume fractions.)

Because of its definition in terms of chemical potential, χ can be related to several important thermodynamic properties of polymer solutions. For example, it follows from Equation 10 that

$$\ln {}^{i}a = ({}^{i}\mu - {}^{i}\mu^{o})/(RT) = \ln {}^{i}\phi + (1 - {}^{i}V/{}^{j}V){}^{j}\phi + \chi\,{}^{j}\phi^2 \tag{11}$$

and many other relationships may be derived. As the ratio of molar volumes of liquid and polymer, ${}^{i}V/{}^{j}V$, is approximately equal to the reciprocal of the degree of polymerization and is usually negligibly small compared with unity, Equation 11 simplifies to

$$\ln {}^{i}a \approx \ln {}^{i}\phi + {}^{j}\phi + \chi\, {}^{j}\phi^{2} \tag{12}$$

The parameter χ may be obtained experimentally by measuring the activity of the solvent in a binary polymer-liquid mixture. This can be done by a variety of techniques, such as solvent vapor pressure lowering, freezing point depression, boiling point elevation, osmotic pressure, light scattering, viscosity, equilibrium swelling of polymers, stress-strain behavior of swollen polymers, or gas-liquid chromatography.* A few of these methods (swelling and viscosity in Chapter 14, chromatography in Chapter 18) are discussed here, but in general neither the methods nor the results are detailed in this handbook. A compilation of values (*CRC Handbook of Polymer-Liquid Interaction Parameters and Solubility Parameters*) has been published recently.[56]

The Gibbs free energy of mixing expression corresponding to Equation 10 is

$$\Delta_{m}G/(RT) = {}^{i}x \ln {}^{i}\phi + {}^{j}x \ln {}^{j}\phi + \chi\, {}^{i}x\, {}^{j}\phi \tag{13}$$

The third term can be rearranged into various alternative forms which are more convenient for some purposes. For example, from Equation 8, Chapter 4,

$${}^{i}x = {}^{i}\phi({}^{j}V\, {}^{j}x + {}^{i}V\, {}^{i}x)/{}^{i}V \tag{14}$$

so

$$\chi\, {}^{i}x\, {}^{j}\phi = \chi\, {}^{i}\phi\, {}^{j}\phi({}^{i}x + {}^{j}V\, {}^{j}x/{}^{i}V) \tag{15}$$

and

$$\Delta_{m}G/(RT) = {}^{i}x \ln {}^{i}\phi + {}^{j}x \ln {}^{j}\phi + \chi\, {}^{i}\phi\, {}^{j}\phi({}^{i}x + {}^{j}V\, {}^{j}x/{}^{i}V) \tag{16}$$

The first two terms result from the (positive) combinatorial entropy of polymer-liquid mixing, and always make a negative contribution to $\Delta_{m}G$, promoting the mixing process. According to the original Flory-Huggins theory, the third term arises only from the enthalpy of mixing, which in the van Laar form is[91,92]

$$\Delta_{m}H = RT\, \chi\, {}^{i}x\, {}^{j}\phi \tag{17}$$

However, this third term is now considered to account more generally for the residual Gibbs free energy of mixing which is in excess of that arising from the Flory-Huggins combinatorial entropy of the reference polymer-liquid system. In order for the polymer to be miscible with the liquid, $\Delta_{m}G$ (Equation 13) must be negative, so the interaction parameter χ must be either negative, or positive but sufficiently small that this term is less than the sum of the combinatorial entropy terms.

Flory[46,47] showed that a polymer *j* and a liquid *i* are completely miscible through the entire composition range provided that

* For example: Barton;[55,56] Blanks;[26,27] Bristow and Watson;[57] Coran and Anagnostopoulos;[58] Cowie and co-workers;[59-62] DiPaola-Baranyi, Guillet, and co-workers;[63,64] Dort;[65] Eskin and Nesterov;[66,67] Fernandez-Sanchez et al.;[182] Ghosh;[68] Karim and Bonner;[69,70] Koenhen and Smolders;[71] Kruse;[72] Leung;[73] Mizerovskii et al.,[187] Nilsson et al.;[74] Olabisi;[69,70] Orwoll;[1] Prausnitz et al.;[76,77] Price;[78] Rehner;[79,80] Richards and Disselhoff;[81,82] Scott and Magat;[83] Sears and Darby;[180] Sheehan and Bisio;[45] Shvarts et al.;[84-86] Vilcu and Leca;[181] Wolf and co-workers;[87-89] and Yamashita et al;[90] Koningsveld, Klientjens, and Nies;[192]

$$\chi \leq {}^1/_2[1 + ({}^iV/{}^jV)^{1/2}]^2 \tag{18}$$

That is, the critical polymer-liquid interaction parameter corresponding to the situation portrayed in Figure 5b, Chapter 2, is

$$\chi_c = {}^1/_2[1 + ({}^iV/{}^jV)^{1/2}]^2 \tag{19}$$

The associated critical composition of the mixture is

$${}^j\phi_c = [1 + ({}^jV/{}^iV)^{1/2}]^{-1} \tag{20}$$

For χ defined as in Equation 10 (independent of any theoretical model), these expressions are strictly valid only if χ is independent of concentration, or if the molar mass of the polymer is so large that $({}^iV/{}^jV) \approx 0$ and

$$\chi_c = 0.5 \tag{21}$$

$${}^j\phi_c = 0 \tag{22}$$

Experimentally, it is certainly observed that the liquid predominates at the critical composition, as illustrated in Figure 1, where χ^{-1} is plotted against ${}^i\phi$ for a typical polymer-liquid system. The marked asymmetry of the two-phase region for values of $\chi > 0.5$ ($\chi^{-1} < 2$) is apparent, but if for any polymer-liquid system χ is less than 0.5 ($\chi^{-1} > 2$) the components are predicted to be completely miscible.

As originally formulated, the polymer-liquid interaction parameter was expected to be inversely dependent on absolute temperature and independent of polymer concentration, but as now empirically defined it depends in an unspecified way on temperature, composition, and polymer chain length.

Although the definition of the polymer-liquid interaction parameter by means of Equation 10 tacitly assumes that it is not concentration-dependent, χ should be reported as a function of concentration for each polymer-liquid combination, and recent compilations do so.[1,56,89]

In some cases (polymers in "good" solvents) χ is approximately independent of concentration (in accordance with the original Flory-Huggins model). More often (polymers in "poor" solvents) χ increases with polymer concentration increase. Occasionally (in strongly interacting systems) χ decreases with increasing polymer concentration. As can be seen from Figure 2, considerable variation of χ with concentration is possible, but in all cases an empirical expression relating them may be used:

$$\chi = \chi_0 + \chi_1 {}^j\phi + \chi_2 {}^j\phi^2 + \ldots\ldots \tag{23}$$

The polymer-liquid interaction parameter tends to rise gradually to reach a plateau value with increasing polymer molar mass, ${}^jM$, and the predicted model behavior of χ_c is to approach 0.5 at infinite polymer length:

$$\chi_c = {}^1/_2(1 + {}^jM^{-1/2})^2$$

Values of χ tend to decrease slightly but not predictably with increasing temperature, over a limited range the dependence fitting the relationship

$$\chi = a + b/T \tag{24}$$

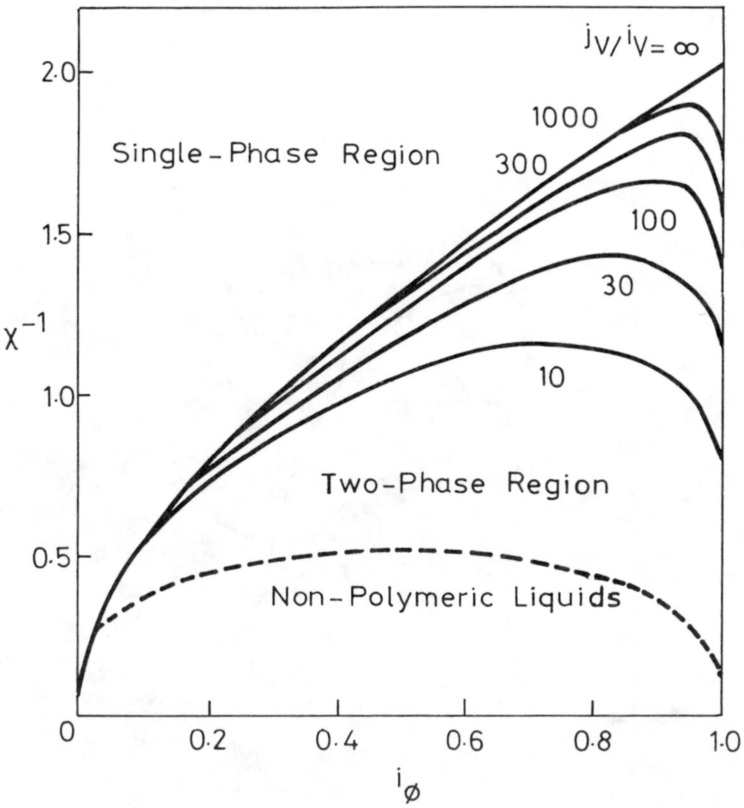

FIGURE 1. Phase equilibrium in polymer-liquid systems in terms of the reciprocal of the polymer-liquid interaction parameter as a function of liquid volume fraction for various molar volume ratios. (Adapted from Small, P. A., *J. Appl. Chem.*, 3, 71, 1953.)

(as illustrated, for example, by Guillet and co-workers[63,64]), supporting the assumption that χ is best treated as a Gibbs free energy parameter that can be divided into entropy and enthalpy components,

$$\chi = \chi_S + \chi_H \tag{25}$$

although in the original Flory-Huggins theory it was a pure enthalpy term. From basic thermodynamic equations it can be shown that

$$\chi_H = -T(\delta\chi/\delta T) \tag{25a}$$

$$\chi_S = d(\chi T)/dT) \tag{25b}$$

In the terminology of Equation 25, from the point of view of regular solution theory, χ_S would be expected to be a small negative correction to the Gibbs free energy change resulting from the mixing. However, χ_S is found to be positive and much larger than χ_H. χ_S is sometimes identified with the reciprocal of the coordination number (which is the number of nearest neighbors of a typical molecule) in polymer solutions, and in some cases has been assumed to have a constant value of about 0.3 for all systems (Section 13.5), although it can vary from zero to 2.5 or more. Thus, if χ must be less than 0.5 for full polymer-

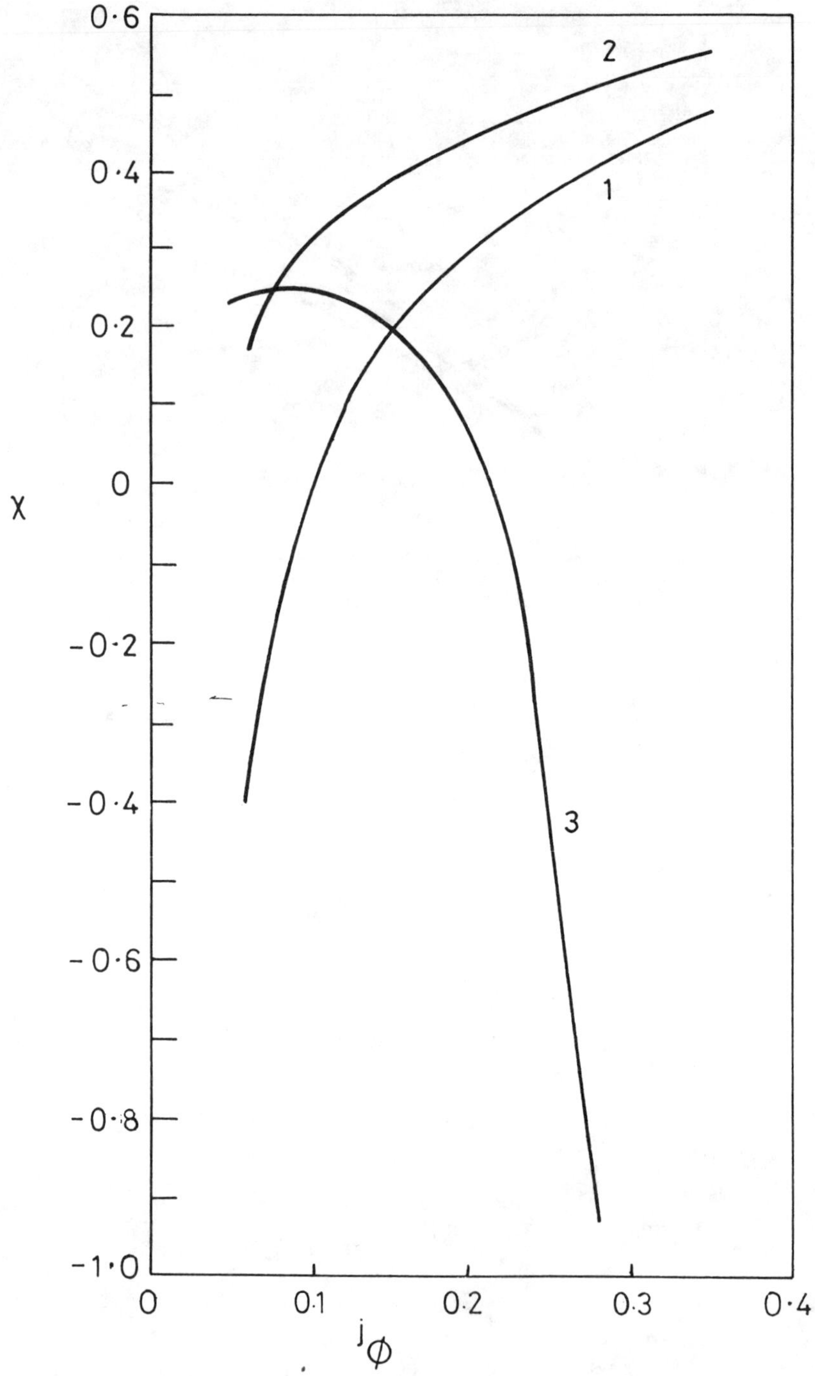

FIGURE 2. Dependence of the polymer-liquid interaction parameter on the polymer volume fraction for solutions of cellulose derivatives: 1, cellulose acetate-acetone, 25°C; 2, cellulose acetate-methyl acetate, 25°C; 3, cellulose nitrate-acetone, 35°C. (Adapted from Heil, J. F. and Prausnitz, J. M., *Am. Inst. Chem. Eng. J.*, 12, 678, 1966.)

liquid miscibility, and χ_S is 0.3 or more, it follows that χ_H must be positive and very small, or negative, to meet the miscibiity criterion.

For nonpolar or only slightly polar systems, χ_H is calculated frequently from the Hildebrand-Scatchard expression, Equation 14, Chapter 4,

$$\Delta_m H = ({}^i x\, {}^i V + {}^j x\, {}^j V)\, {}^{ij}A\, {}^i\phi\, {}^j\phi \tag{26}$$

On equating this with

$$\Delta_m H = RT\chi_H\, {}^i\phi\, {}^j\phi({}^i x + {}^j V\, {}^j x/{}^i V) \tag{27}$$

from Equation 16, it follows that

$$\chi_H = {}^{ij}A\, {}^i V/(RT) \tag{28}$$

which suggests that χ_H can be interpreted as the ratio of the exchange energy (${}^{ij}A\,{}^iV$) to the thermal energy (RT).

In the absence of polar or specific interactions, the Hildebrand parameter can be introduced into Equation 28. It will be recalled (Equation 10, Chapter 4) that

$${}^{ij}A = {}^i c + {}^j c - 2\, {}^{ij}c$$

where

$${}^i c = {}^i\delta^2 \quad \text{and} \quad {}^j c = {}^j\delta^2 \tag{29}$$

and in the special "regular solution" case where ${}^{ij}c = {}^i\delta\,{}^j\delta$ (the geometric mean approximation, Sections 4.1 and 5.7),

$${}^{ij}A = ({}^i\delta - {}^j\delta)^2 \tag{30}$$

and

$$\chi_H = {}^i V({}^i\delta - {}^j\delta)^2/(RT) \tag{31}$$

It follows that in order for polymer-liquid miscibility to be possible (χ_H small), ${}^i\delta$ and ${}^j\delta$ must have very similar values. Specific interactions (such as hydrogen bonding between molecules of type i and type j) add positive terms to ${}^{ij}c$ and so decrease ${}^{ij}A$, with the result that χ_H also decreases and the mutual solubility is enhanced.

Any of the alternative expanded cohesion parameter expressions of Chapter 5 can be used also in the evaluation of ${}^{ij}A$ and χ_H. For example, the corresponding χ_H expression in terms of the three-component Hansen parameters (Section 5.9) is

$$\chi_H = {}^i V[({}^i\delta_d - {}^j\delta_d)^2 + ({}^i\delta_p - {}^j\delta_p)^2 + ({}^i\delta_h - {}^j\delta_h)^2]/(RT) \tag{32}$$

Another energy density parameter introduced by Flory[34] is

$$B = \chi RT/{}^i V \tag{33}$$

which represents the interaction energy density characteristic of the polymer-liquid pair. It follows from Equations 28 and 31 that B corresponds to ${}^{ij}A$ if χ_S is neglected. Although the parameter B has been used in discussing polymer-liquid interactions[93,94] it has not achieved widespread acceptance.

Mikos and Peppas[190] applied the geometric mean correction parameter ^{ij}l (Section 5.7) in the estimation of χ_H from Hildebrand parameters for water with hydrophobic copolymers such as the biomedical hydrogel poly(2-hydroxyethyl methacrylate-*co*-methacrylic acid), rather than using multicomponent parameters to accommodate hydrogen-bonding interactions.

Roe and Lin[95-97] defined a polymer-polymer interaction energy density, Λ, such that when both Λ and χ are independent of concentration

$$\Lambda = \chi RT/^{j}V$$

and

$$^{ij}A = \Lambda = (^{i}\delta - {}^{j}\delta)^2$$

When χ is concentration-dependent (the usual situation) the relation between Λ and χ is more complex, of course. Although Λ has some advantages over χ in the description of polymer-polymer interactions (Section 16.9), and a study[178] of the miscibility of epoxy monomers with carboxyl-terminated butadiene-acrylonitrile random copolymers yielded a correlation of the type

$$\Lambda = \Lambda_o + \Lambda_T^{T}$$

this parameter is not used widely and will not be explored further here.

It has been assumed so far in this discussion of polymer-liquid interactions that the polymers are in the non-crystalline, amorphous, or "liquid-like" form. If a polymer is in the crystalline state, the additional energy necessary to "melt" the crystals is incorporated in the energy of mixing, and such polymers are usually soluble only in those liquids for which specific interactions increase ^{ij}c to such an extent that ^{ij}A becomes negative.

13.4 LIMITATIONS OF INTERACTION PARAMETERS

The Flory-Huggins theory is useful in considering the thermodynamics of dilute polymer solutions, but it has theoretical limitations and is not necessarily of great value in all solvent formulation problems. The shortcomings of the Flory interaction parameter include:[29,98]

- It is not a constant, but depends on polymer concentration and molecular weight as well as temperature
- It is not able to be evaluated simply for every polymer-liquid combination
- It characterizes a polymer-liquid pair, and so is inconvenient for multicomponent systems
- It is a composite term influenced by factors such as hydrogen bonding

There is an additional problem associated with the larger-than-expected χ_S value. The theory offers no explanation for the fact that polymer solutions separate into two phases on increasing the temperature as well as on lowering the temperature.[99-101] This is illustrated in Figure 3. The two critical phase-separation temperatures are called the lower critical solution temperature (LCST) and upper critical solution temperature (UCST), respectively. (The lower critical solution temperature is so named because it lies *below* a two-phase region, despite the fact that it occurs at a higher temperature than the UCST.) The LCST apparently is a general phenomenon in all polymer-liquid systems, although some must be studied in sealed vessels if the LCST lies above the normal boiling point of the liquid. LCST behavior

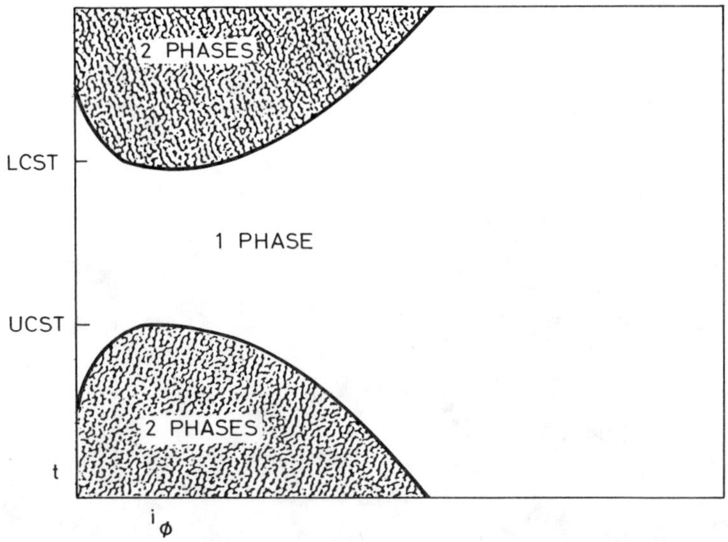

FIGURE 3. Diagrammatic polymer solution volume fraction phase diagram showing upper critical solution temperature (UCST) and lower critical solution temperature (LCST). (After Delmas, G. and Patterson, D., *Off. Dig. Fed. Soc. Paint Technol.*, 34, 677, 1962 and Patterson, D., *Macromolecules*, 2, 672, 1969.)

is observed also in some polymer-polymer systems which are completely miscible at lower temperatures.[102] Equation 19 holds for the LCST as well as the UCST, and because χ_c increases as the polymer molecular size decreases, both critical solution temperatures may be used for polymer fractionation (Section 14.9).

An UCST is due to a positive enthalpy of mixing (endothermic process) but the thermodynamic conditions for a LCST to exist are that both enthalpy and entropy of dilution must be negative, so that from the relationship (Equation 22, Chapter 2)

$$\Delta_m G = \Delta_m H - T\Delta_m S \tag{34}$$

during an increase in temperature $\Delta_m G$ changes sign from negative to positive. This means that as the solutions are diluted, they become *more* ordered, even for systems involving molecules that normally interact with simple van der Waals force fields. This situation is unusual in small-molecule mixtures (although it does sometimes occur), but it is the normal behavior in polymer solutions. In terms of χ_S, this means that the normal, positive combinatorial $\Delta_m S$ terms that make a negative contribution to $\Delta_m G$ have been outweighed by the large, positive χ_S term.

"Free volume" concepts may be invoked to explain the LCST and the large value of χ_S.[32,44,103-105] A dissimilarity in free volume exists between the denser polymer and the more expanded solvent, and although the resulting solution free volume is intermediate between that of the polymer and that of the solvent, there is an overall contraction. Associated with the negative volume change is a negative entropy contribution resulting in a positive component in χ_S and in the Gibbs free energy of mixing. Recalling Equations 16 and 34, the polymer-liquid interaction parameter may be expressed

$$\chi = \chi_H + \chi_S = \chi_H + \chi_{S,co} + \chi_{S,fv} \tag{35}$$

which separates χ_S into a combinatory component (a small negative correction to the first two combinatory terms in Equation 16) and a free volume (large positive) component.

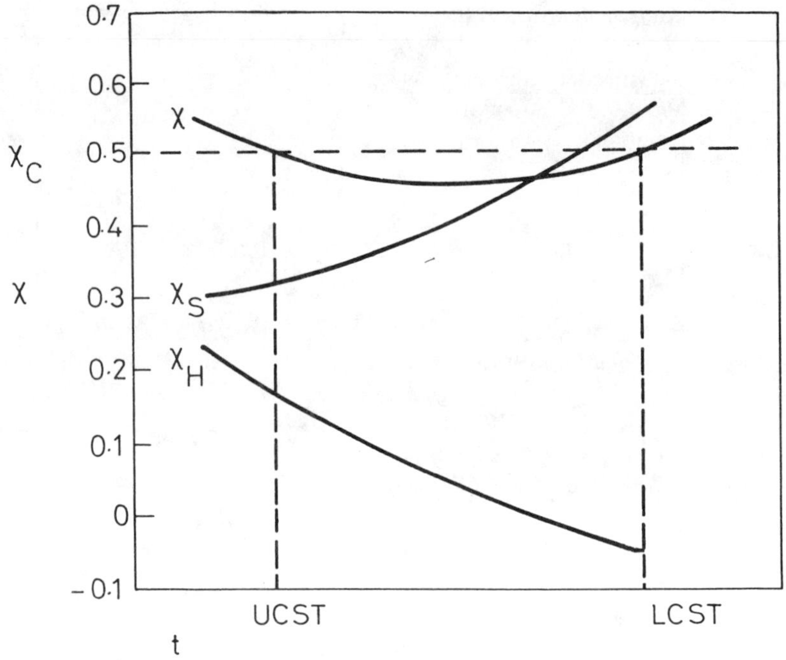

FIGURE 4. Diagrammatic temperature dependence of the polymer-liquid interaction parameter and its enthalpy and entropy components. At the upper critical solution temperature (UCST) and lower critical solution temperature (LCST) the polymer-liquid interaction parameter has its critical value, 0.5. (Adapted from Dayantis, J., *Plast. Mod. Elastomeres,* 29(2), 58, 1977.)

Both the cohesive pressure dissimilarity (χ_H) contribution and the free volume dissimilarity ($\chi_{S,fv}$) contribution to χ are positive (unfavorable to mixing), but whereas the former decreases with increasing temperature, the free volume dissimilarity between polymer and liquid increases (Figure 4). The result is a minimum in the variation of total χ with temperature, so there is a higher temperature LCST as well as a lower temperature UCST at the critical temperatures where $\chi = 0.5$. Evidence for this type of analysis comes from solution properties such as molecular dimensions and intrinsic viscosity (Section 14.7). The values of these properties tend either

1. To decrease with increasing temperature toward the LCST, following the χ_S trend if χ_H is small (as in polyisobutylene-pentane where the cohesive energy χ_H term is negligibly small), or
2. To exhibit a maximum as a function of temperature, following the trend of the total χ curve if there is cohesive energy dissimilarity (as in polyisobutylene-dibutyl ether).[101]

An alternative division of the interaction parameter on a free volume basis has been made by Scigliano.[25,106]

Reviewing these ideas in even simpler terms, it can be seen from Figure 4 that the phase separation at the UCST occurs for reasons associated with enthalpy properties, to some extent predictable by means of exchange cohesive energy and cohesion parameters, while the LCST occurs for reasons related to entropy properties, which cannot be predicted from cohesion parameter considerations.

As well as several variations on theories of the Flory-Huggins type,[107-114] there are many other polymer solution theories including the more generally applicable liquid state partition function and corresponding states approaches.[25,99,104,115-120] However, any increased accuracy available in property prediction frequently does not justify the increased complexity of calculation, or the data needed are not available or not readily available for practical applications. For many purposes, especially at low polymer concentrations, the simple Flory-Huggins treatment, with the incorporation of cohesion parameters, remains particularly valuable.[26,104]

13.5 INTERACTION PARAMETERS AND COHESION PARAMETERS

The close relationship between the polymer-liquid interaction parameter and cohesion parameters, based on Equations 25 and 28, is clear:

$$\chi = \chi_S + {}^{ij}A\,{}^{i}V/(RT) \tag{36}$$

Combination of Equations 13 and 36 yields

$$\Delta_m G/(RT) = {}^{i}x \ln {}^{i}\phi + {}^{j}x \ln {}^{j}\phi + {}^{i}x\,{}^{j}\phi[\chi_S + {}^{ij}A\,{}^{i}V/(RT)] \tag{37}$$

Similarly, from Equations 11 and 36,

$$\ln {}^{i}a = \ln {}^{i}\phi + (1 - {}^{i}V/{}^{j}V){}^{j}\phi + {}^{j}\phi^2[\chi_S + {}^{ij}A\,{}^{i}V/(RT)] \tag{38}$$

These are useful equations in which, if necessary, χ_S may be regarded as an empirical parameter with an approximate value of 0.34, and in which ${}^{ij}A$ may be expressed in terms of cohesion parameters. It may be noted that the middle term on the right-hand-side of Equation 38 can be rearranged to $(1 - {}^{i}\phi/{}^{i}x)$, so that if ${}^{i}x = {}^{i}\phi$ and $\chi_S = 0$, the equation simplifies to

$$\ln {}^{i}a = \ln {}^{i}x + {}^{ij}A\,{}^{i}V\,{}^{j}\phi^2/(RT) \tag{39}$$

$$= \ln {}^{i}x + {}^{i}V\,{}^{j}\phi^2\,({}^{i}\delta - {}^{j}\delta)^2 \tag{40}$$

in agreement with Equation 16, Chapter 2, as expected. Equation 36 can be used in several ways to evaluate cohesion parameters from experimental χ values:

The Hildebrand parameter can be evaluated from a plot of χ against ${}^{i}\delta$[58] by means of the "van Laar-Hildebrand equation",

$$\chi_H = \chi - \chi_S = {}^{i}V\,({}^{i}\delta - {}^{j}\delta)^2/(RT) \tag{41}$$

The polymer Hildebrand parameter is given by

$${}^{j}\delta = {}^{i}\delta \pm (RT|\chi_H|/{}^{i}V)^{1/2} \tag{42}$$

The choice of the correct root of this equation cannot be made on the basis of measurements in only one liquid, but graphical or numerical comparison of the results from several liquids reveals the physically reasonable value. This procedure is illustrated in Table 7 for polystyrene, with enthalpy interaction parameters evaluated in several liquids. In this way, Hildebrand parameters may be obtained by all the methods mentioned in Section 13.3.

Alternatively, some form of multicomponent cohesion parameter, such as those employed by Blanks and Prausnitz[28] (Chapter 5) may be used. Typical values of nonpolar and polar

TABLE 7
Enthalpy Interaction Parameters and Hildebrand Parameters for Polystyrene in Various Liquids

Liquid	χ_H	${}^j\delta$/MPa$^{1/2}$
Toluene	−0.019	18.9 (17.6)[a]
Tetrachloromethane	+0.050	18.7 (16.5)
1,2-Dichloroethane	+0.087	18.4 (20.5)
Butyl acetate	−0.071	18.6 (16.3)
Methyl propyl ketone, 2-pentanone	+0.061	18.9 (16.7)
Cyclohexane	+0.197	18.9 (17.4)
Chlorobenzene	−0.006	19.0 (19.8)
Thiophene	+0.079	18.6 (21.8)
Tetrahydrofuran	−0.031	18.5 (20.4)

[a] In parentheses, the other root of Equation 42.

Adapted from Cowie, J. M. G., Ranson, R. J., and Burchard, W., *Br. Polym. J.*, 1, 187, 1969.

TABLE 8
Nonpolar and Polar Cohesion Parameters for Some Polymers at 25°C

Polymers	δ_λ/MPa$^{1/2}$	δ_τ/MPa$^{1/2}$
Rubber	16.7	0.0
Polyisobutylene	15.7	0.0
Polystyrene	18.6	0.0
Polybutadiene	17.0	0.0
Polyethylene	16.6	0.0
Poly(vinyl chloride)	16.7	10.1
Poly(vinyl acetate)	15.8	11.0
Poly(methyl methacrylate)	15.7	10.5
Poly(propylene oxide)	14.4	5.5

Adapted from Blanks, R. F. and Prausnitz, J. M., *Ind. Eng. Chem. Fundam.*, 3, 1, 1964.

cohesion parameters for polymers are shown in Table 8. Values of χ_S were calculated from Equation 41 for three nonpolar polymers in a large number of nonpolar liquids. The mean values from 23 calculations was 0.34, with a standard deviation from the mean of 0.08 and the largest deviation being 0.15. This value of $\chi_S = 0.34$ was then used for the calculation of χ for polymers in polar liquids, and for 19 polar-polar systems in the absence of specific polymer-liquid interactions the average difference between calculated and experimental values was 0.07 (Table 9).[27,28,104] For the ethylene-propylene terpolymer SKEPT-50, Mizerovskii et al.[187] confirmed that χ_s was approximately 0.35 for alkanes, haloalkanes, and compounds with a degree of polarity below a critical value. For ethers, esters, ketones and alcohols χ_s is proportional to the liquid molar volume and to the degree of polarity expressed as the ratio of the van der Waals volumes of the polar groups to the overall van der Waals volume, rising to 1.6 for acetone and 1.4 for ethanol. Bueche[121] correlated the swelling properties of poly(dimethylsiloxane) by plotting $(\chi - 0.39)/{}^iV$ against $({}^j\delta - {}^i\delta)^2$. Chiou et al.[191] used $\chi_S = 0.25$ for interactions involving humic substances. Polar-nonpolar cohesion parameters were used by Cernia, Mancini and Saini[122] for the thermodynamic properties of branched polyethylene, and for elastomer swelling prediction by De Broutelles and Opigez.[123]

TABLE 9
Calculated and Experimental χ Interaction Parameters for Several Polymer-Polar Liquid Systems[28,37]

System	t/°C	^{ij}A/MPa (Equation 51, Chapter 5)	χ Calculated, $\chi_S = 0.34$	χ Experimental, sources cited in Ref. 28
		Poly(vinyl chloride)		
1,2-Dichloroethane	53	3.76	0.45	0.46
	76	3.55	0.45	0.43
1,2-Dibromoethane	25	8.68	0.65	0.47
Nitropropane	53	10.93	0.71	0.44
		Poly(vinyl acetate)		
1,4-Dioxane	25	5.03	0.51	0.41
Acetone	25	2.53	0.42	0.44
Methyl ethyl ketone	25	0.40	0.35	0.43
		Poly(methyl methacrylate)		
Ethyl acetate	25	0.27	0.35	0.48
Propyl acetate	25	3.12	0.49	0.49
Butyl acetate	25	3.34	0.52	0.50
Pentyl acetate	25	3.62	0.56	0.51
Methyl isobutyrate	25	1.23	0.40	0.48
Acetone	25	4.07	0.46	0.48
Methyl ethyl ketone	25	0.06	0.34	0.47
Methyl propyl ketone	25	1.92	0.42	0.48
Methyl isobutyl ketone	25	4.33	0.56	0.50
Methyl pentyl ketone	25	3.63	0.55	0.49
Diisopropyl ketone	25	3.08	0.51	0.50
Hexamethyl ketone	25	4.53	0.63	0.52

Equation 41 can be rearranged to

$$^{i}\delta = {}^{j}\delta \pm [RT(\chi - \chi_S)/{}^{i}V]^{1/2} \tag{43}$$

and a closely related empirical equation incorporating a correction factor K is[83]

$$^{i}\delta = {}^{j}\delta \pm (RT/K)^{1/2}[(\chi - \chi_S)/{}^{i}V]^{1/2} \tag{44}$$

Shvarts,[84] Sheehan and Bisio,[45] and many subsequent scientists have used this equation graphically by plotting the liquid cohesion parameter $^{i}\delta$ against $[(\chi - \chi_S)/{}^{i}V]^{1/2}$, a "Shvarts plot".

Some results of these three methods of calculation are shown in Table 10. There appears to be an improvement in progressing from Method 1 to Method 3, but all three are suitable for only approximate correlations between χ and δ. If multicomponent parameters (Chapter 5) or empirical corrections for the geometric mean approximation (Section 5.7) are incorporated, better fits are obtained.[124]

Another form of Equation 41 (suggested by Bristow and Watson[57]) has been used[57,63,64,125] to calculate Hildebrand parameters:

$$[{}^{i}\delta^2/(RT) - \chi/{}^{i}V] = [2\,{}^{j}\delta/(RT)]{}^{i}\delta - {}^{j}\delta^2/(RT) - \chi_S/{}^{i}V \tag{45}$$

TABLE 10
Polymer-Liquid Interaction Parameters from Experiment and Different Methods of Calculation

		χ			
			Calculation (with % error in parentheses)		
System	***t*/°C**	**Experimental**	**Method 1 Hildebrand (Equation 41)**	**Method 2 Blanks-Prausnitz (Table 9)**	**Method 3 Shvarts (Equation 43)**
			Poly(vinyl chloride)		
1,2-Dichloroethane	53	0.46	0.34(26%)	0.45(1%)	0.40(13%)
	76	0.43	0.34(21%)	0.45(4%)	0.39(11%)
1,2-Dibromoethane	25	0.47	0.34(27%)	0.65(38%)	0.38(20%)
Nitropropane	25	0.44	0.43(2%)	0.71(61%)	—
			Poly(vinyl acetate)		
Acetone	25	0.44	0.35(19%)	0.42(5%)	0.43(2%)
Methyl ethyl ketone	25	0.43	0.34(22%)	0.35(18%)	0.36(15%)
1,4-Dioxane	25	0.41	0.36(10%)	0.51(26%)	0.43(6%)
			Poly(methyl methacrylate)		
Ethyl acetate	25	0.48	0.35(27%)	0.35(26%)	—
Propyl acetate	25	0.49	0.42(15%)	0.49(0%)	—
Butyl acetate	25	0.50	0.48(3%)	0.52(4%)	—
Pentyl acetate	25	0.51	0.50(1%)	0.56(10%)	—
Acetone	25	0.48	0.36(24%)	0.46(4%)	0.54(13%)
Methyl ethyl ketone	25	0.47	0.35(25%)	0.34(27%)	0.40(15%)
Methyl propyl ketone	25	0.48	0.35(27%)	0.42(11%)	0.43(10%)
Methyl isobutyl ketone	25	0.50	0.39(22%)	0.56(13%)	0.49(1%)
Methyl pentyl ketone	25	0.49	0.34(31%)	0.55(11%)	0.50(1%)
Diisopropyl ketone	25	0.50	0.34(32%)	0.63(25%)	0.50(1%)
Average error			20%	17%	9%

Adapted from Sheehan, C. J. and Bisio, A. L., *Rubber Chem. Technol.*, 39, 149, 1966.

The contribution of χ_S to the right-hand side of Equation 45 is neglected, and a plot of$[{}^i\delta^2/(RT) - \chi/{}^iV]$ against ${}^i\delta$ yields the predicted straight line with slope of $2\,{}^j\delta/(RT)$ and intercept of $[-{}^j\delta^2/(RT)]$, as illustrated in Table 11. Bristow-Watson plots are linear because[57] the first term of the left-hand side of Equation 45 predominates, with $\chi/{}^iV$ providing a relatively small correction effect, smallest in the region of ${}^j\delta = {}^i\delta$, so the weaknesses in the assumption that χ can be identified with $({}^i\delta - {}^j\delta)^2$ are not very significant. Consequently, the data are usually consistent with this equation for a wide range of liquids interacting with any particular polymer. The reverse procedure, the estimation of polymer-liquid interaction parameters from cohesion parameters by means of Equation 41, is far less reliable but is often useful.* Polymer-polymer interaction parameters may be estimated in this way also (Section 16.9).

* Examples may be found in the publications of: Aharoni;[126] Baughan, Jones, and Stewart;[127,128] Bindschaedler, Gurny, Doelker, and Peppas;[129] Bradner, Shapiro, Bowler, Gerrard, and Maddams;[186] Camberlin and pascault;[130] Ceccorulli, Pizzoli, Scandola, and Pezzin;[131] Coleman, Hu, Park, and Painter;[185] Fujioka, Noethiger, Beatty, Baba, and Kagemoto;[132] Fukuda, Nagata, and Inagaki;[133-135] Gomez-Anton, Masegosa, and Horta;[136] Kanbour et al;[188] Katime and Ochoa;[137] Kawaguchi et al;[189] Kelley, Glasser, Rials, and Ward;[183] Lovell and Still;[138] Moore;[139] Oberoi, Lal, and Bhatnagar;[140] Ohde and Okamoto;[141,142] Painter, Park and Coleman;[177] Pascault;[143] Peppas, Lucht and Mikos;[144,190] Shoji and Takeda;[145] Tseng, El-Aasser, and Vanderhoff;[146] Verkholantsev;[147] and Zhigotsky.[148]

TABLE 11
Hildebrand Parameters for Polymers from the Bristow-Watson Equation [63,64]

Polymer	t/°C	${}^{j}\delta$/MPa$^{1/2}$ Slope	Intercept
Polystyrene	19	15.6 ± 0.4	15.6 ± 0.4
Poly(methyl acrylate)	100	17.4 ± 0.6	18.0 ± 0.4
Poly(vinyl acetate)	25	20.7 ± 1.0	21.1 ± 0.8
	135	17.2 ± 0.8	17.6 ± 0.6

If only one polymer-liquid interaction parameter is known, so the graphical methods are unavailable, χ can be substituted into Equation 41 together with an assumed value of χ_S (often the reciprocal of the lattice number, 0.25 or 0.3) and the measured polymer Hildebrand parameter.

Blanks and Prausnitz[28] provided examples of their methods of calculation of interaction parameter values using Equation 36 and the equations of Section 5.8. In systems of *polar polymers and nonpolar liquids* (Equation 49, Chapter 5, with $T_* = T$),

$$ {}^{ij}A = ({}^{i}\delta_\lambda - {}^{j}\delta_\lambda)^2 + {}^{j}\delta_\tau^2 - 2\,{}^{ij}\psi \tag{46} $$

For poly(methyl methacrylate) and hexane at 25°C:

$$ {}^{i}\delta_\lambda = 14.8 \text{ MPa}^{1/2},\ {}^{i}V = 132 \text{ cm}^3 \text{ mol}^{-1} \text{ (Table 5a, Chapter 5)} $$

$$ {}^{j}\delta_\lambda = 15.7 \text{ MPa}^{1/2},\ {}^{j}\delta_\tau = 10.5 \text{ MPa}^{1/2} \text{ (Table 8)} $$

The product ${}^{i}\delta_\tau {}^{j}\delta_\tau = 155.4$ MPa, so from Figure 4, Chapter 5, ${}^{ij}\psi = 40.8$ MPa, and:

$$ {}^{ij}A = (15.7 - 14.8)^2 + (10.5)^2 - 2(40.8) = 29.5 \text{ MPa} $$

$$ \chi_H = {}^{i}V\,{}^{ij}A/(RT) = 132(29.5)/(8.314)(298) = 1.57 $$

$$ \chi = 0.34 + 1.57 = 1.9 $$

This is much greater than the critical value 0.5, indicating that hexane is not a solvent for poly(methyl methacrylate) at 25°C, in agreement with experimental observation. In systems of *polar polymers and polar liquids* without hydrogen bonding (Equation 37),

$$ {}^{ij}A = ({}^{i}\delta_\lambda - {}^{j}\delta_\lambda)^2 + ({}^{i}\delta_\tau - {}^{j}\delta_\tau)^2 \tag{47} $$

For poly(methyl methacrylate) and acetone at 25°C:

$$ {}^{i}\delta_\lambda = 15.5 \text{ MPa}^{1/2},\ {}^{i}\delta_\tau = 12.5 \text{ MPa}^{1/2}, $$

$$ {}^{i}V = 74 \text{ cm}^3 \text{ mol}^{-1} \text{ (Table 5b, Chapter 5)} $$

$$ {}^{j}\delta_\lambda = 15.7 \text{ MPa}^{1/2},\ {}^{j}\delta_\tau = 10.5 \text{ MPa}^{1/2} \text{ (Table 8)} $$

Consequently:

$$ {}^{ij}A = (15.5 - 15.7)^2 + (12.5 - 10.5)^2 = 4.0 \text{ MPa} $$

$$\chi_H = {}^iV\, {}^{ij}A/(RT) = 74(4.0)/(8.314)(298) = 0.12$$

$$\chi = 0.34 + 0.12 = 0.46$$

This may be compared with the experimental value of 0.48 (Table 9), less than 0.5, in agreement with the experimental observation that acetone is a solvent for poly(methyl methacrylate) at 25°C.

Balcar and Bohackova[149] applied this method to viscosities of solutions of poly(1-chloro-2,3-epoxypropane) in 19 solvents. The polar-nonpolar cohesion parameter approach, with allowance for different molar volumes, has been illustrated by Schille and Bittrich[150] for the excess Gibbs free energies of hydrocarbon systems.

Other forms of cohesion parameter have been used in evaluating Flory interaction parameters. Following the principle outlined in Section 5.5 for self-association in alcohols, Chen[151,152] separated "physical" contributions (polar and dispersion interactions) from "chemical" or hydrogen-bonding contributions in polymer-liquid systems by using a parameter related to χ_H for the former, and Hansen's δ_h for the latter. Primes are here attached to the interaction parameter symbol as a reminder that these are not identical with the usual polymer-liquid χ interaction parameters. The general expression is

$$\Delta_m G/(RT) = f({}^i\phi\, {}^j\delta_h\, {}^i\delta_h) + \chi'\, {}^ix\, {}^j\phi \tag{48}$$

where

$$\chi' = \chi'_S + \chi'_H \tag{49}$$

$$\chi'_H = {}^{ij}A\, {}^iV/(RT) \tag{50}$$

and

$$^{ij}A = ({}^i\delta_d - {}^j\delta_d)^2 + ({}^i\delta_p - {}^j\delta_p)^2 \tag{51}$$

It follows that χ'_H can be calculated from the dispersion and polar Hansen parameter values. Chen adopted a "two-dimensional" graphical approach, plotting δ_h against χ'_H for all liquids interacting with a particular polymer. Each liquid was represented by a point (χ'_H, ${}^i\delta_h$). The unit length of the χ'_H scale was twice that of the unit length on the δ_h scale in cal$^{1/2}$ cm$^{-3/2}$ units, and the radius of the circle centered on (O, ${}^j\delta_h$) for the polymer was equal to that of the Hansen solubility sphere (Section 5.9). Because Equation 48 arose from an associated solution theory, the parameters χ'_H, χ'_S and δ_h were considered concentration-independent. This is in contrast to the Flory parameters χ_H and χ_S, which are concentration dependent and, in particular, χ_H can even be negative in associated solutions while χ'_H must always be positive. This type of analysis has been extended to mixed liquids[153-155] (see next Section) and to liquids undergoing acid-base interactions.[155]

13.6 COSOLVENCY AND CONONSOLVENCY

The phenomenon of cosolvency of liquids for polymers (miscibility of polymers with mixed liquids which are individually nonsolvents) is well established and can be discussed in terms of cohesion parameters and polymer-liquid interaction parameters.* Less common

* For example: Horta and Fernańdez-Piérola;[156-158] Kinzer, Lloyd, McGrath, and co-workers;[159,160] Kuhn;[161,162] Narender and Yaseen;[184] Ochoa, Caballero, Valenciano, and Katime;[163] Piccarolo and Titomanlio;[164-166] Trzebicka, Tekely, and Turska;[167] Vidyarthi and Palit;[168] and Wolf and Willms.[87]

(and in some ways more interesting) is cononsolvency (coprecipitation or demixing): polymer insolubility in mixtures of two liquids which individually are solvents for the polymer. For example, in the system *N,N*-dimethylformamide-cyclohexane-polystyrene at some compositions and temperatures the polymer solution becomes two-phase, while the pure components of the mixed solvent exhibit complete miscibility with the polymer. In this particular case the behavior is due to a lack of cohesive interactions between the liquid components, but in other cases it is due to preferential cohesive interactions between the liquids at the expense of polymer-liquid cohesion. Another example is poly(methyl methacrylate) with chlorobutane and pentyl acetate. Other examples are discussed in Section 14.4 (determination of Hildebrand parameters by turbidimetric titration) and Section 16.6.

Blanks and Prausnitz[28,37] extended their prediction of interaction parameters to solutions of polymers(*k*) with mixed nonpolar(*i*)-polar(*j*) liquids. First, the effective cohesion parameters and molar volume of the mixed liquid *ij* were determined on a volume fraction basis:

$$^{ij}\delta_\lambda = {}^{i}\delta_\lambda\,{}^{i}\phi + {}^{j}\delta_\lambda\,{}^{j}\phi \tag{52}$$

$$^{ij}\delta_\tau = {}^{i}\delta_\tau\,{}^{i}\phi + {}^{j}\delta_\tau\,{}^{i}\phi = {}^{i}\delta_\tau\,{}^{j}\phi \tag{53}$$

$$^{ij}V \quad {}^{i}V\,{}^{i}\phi + {}^{j}V\,{}^{j}\phi \tag{54}$$

Then

$$^{ijk}A = ({}^{ij}\delta_\lambda - {}^{k}\delta_\lambda)^2 + ({}^{ij}\delta_\tau - {}^{k}\delta_\tau)^2 + 2({}^{i}\phi\,{}^{j}\phi\,{}^{ij}\Psi - {}^{i}\phi\,{}^{ik}\Psi) \tag{55}$$

The results of one calculation are shown in Figure 5, which predicts that at 25°C the best acetone-hexane mixture as a solvent for poly(methyl methacrylate) has an acetone volume fraction of 0.7 on a solute-free basis, and that the volume fraction of acetone must be greater than about 0.55 for miscibility, in agreement with experimental observation.

13.7 THETA TEMPERATURES AND THETA SOLVENTS

There is in common use another important parameter devised by Flory for polymer-liquid mixtures.[34] In Flory's terminology, the enthalpy and entropy interaction parameters (χ_H and χ_S) are called κ and $(^1/_2 - \psi)$ respectively, and are related to χ by

$$\kappa - \psi = \chi - {}^1/_2 \tag{56}$$

The Flory theta temperature, Θ, is defined

$$\Theta = \kappa T/\psi \tag{57}$$

so that

$$\psi - \kappa = \psi(1 - \Theta/T) \tag{58}$$

In a poor solvent, where both κ and ψ are usually positive, at the theta temperature $\kappa = \psi$ and $\chi = 0.5$, and the partial molar Gibbs free energy of polymer-liquid interaction is zero. This means that the molecular dimensions of the polymer are unperturbed by interactions and the molecules can mutually interpenetrate freely: this is described as *pseudo-ideal* solution

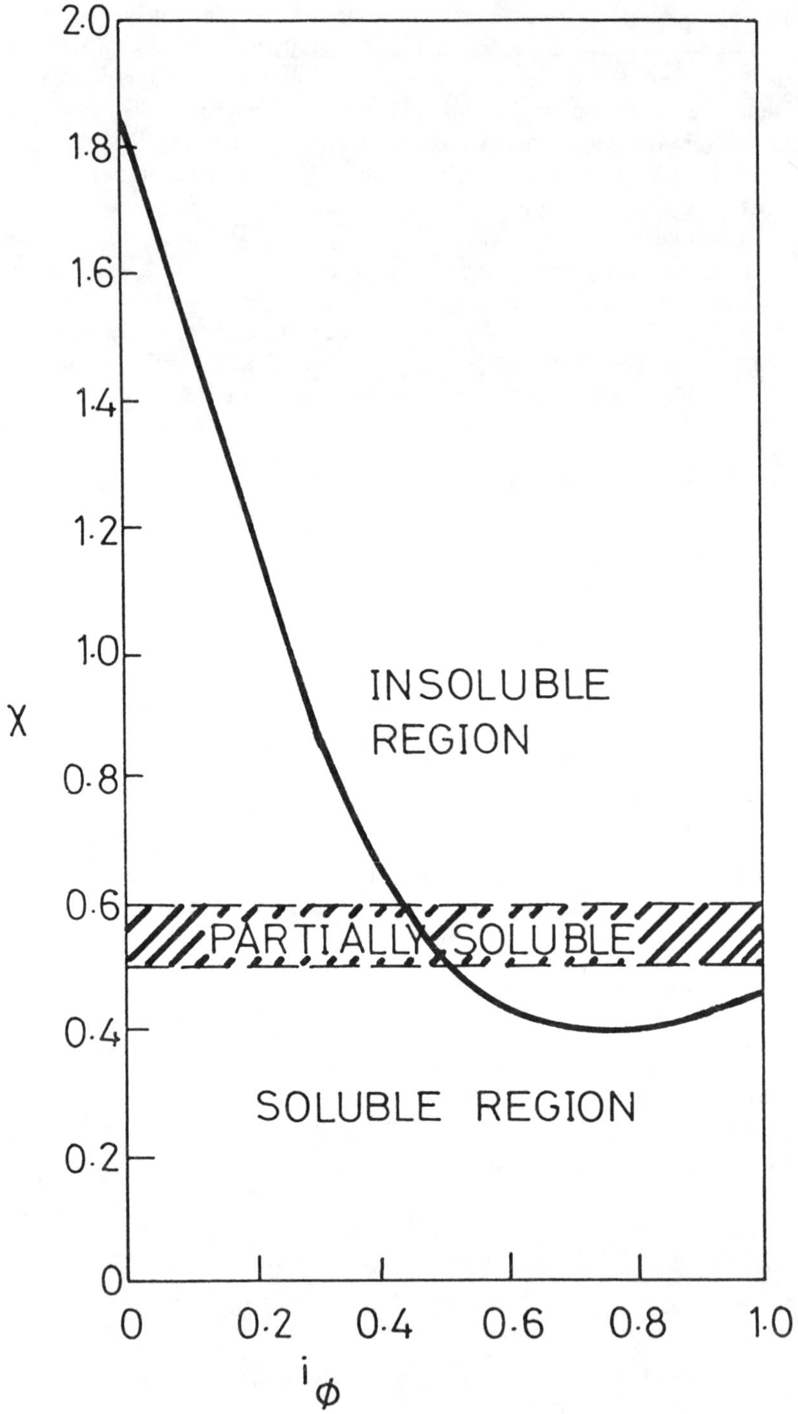

FIGURE 5. Calculated polymer-liquid interaction parameter for poly(methyl methacrylate) in acetone-hexane mixtures as a function of volume fraction of acetone on a polymer-free basis. Solubility is predicted for $\chi < 0.5$. (After Blanks, R. F. and Prausnitz, J. M., *Ind. Eng. Chem. Fund.*, 3, 1, 1964, and Hildebrand, J. H., Prausnitz, J. M., and Scott, R. L., *Regular and Related Solutions,* Van Nostrand-Reinhold, Princeton, NJ, 1970.)

TABLE 12
Theta Temperatures for Some Polymer-Liquid Systems[34,169]

Polymer	Liquid	Θ/°C
Poly(acrylic acid)	1,4-Dioxane	30
Poly(dimethylsiloxane)	Methyl ethyl ketone	20
	Phenetole	83
Polyethylene	Diphenyl ether	161
Poly(hexamethylene adipamide)	0.3 *M* KCl in 90% formic acid	25
Polyisobutylene	Benzene	24
	Phenetole	86
	Anisole	106
Poly(methyl methacrylate)	Acetone	−50
Poly(propylene)	Isoamyl acetate	34
Polystyrene	Cyclohexane	34
	Decahydronaphthalene	31
	Ethylcyclohexane	70
	Methylcyclohexane	71
	Cyclohexane-tetrachloromethane (86.9:13.1)	15
	Methyl ethyl ketone-methanol (88.9:11.1)	30
	Toluene-heptane (47.6:52.4)	30
Poly(vinyl acetate)	3(Heptanone	29
Poly(vinyl chloride)	Benzyl alcohol	155

behavior. Flory theta temperatures are associated with both the upper and lower critical solution temperatures exhibited by all polymer solutions. At temperatures just below the UCST ($T < \Theta$) the molecules attract one another, and if the temperature falls much below Θ, precipitation of the polymer occurs. It has been shown that the critical temperature for phase separation is

$$T_c \approx \Theta/(1 + CM^{-1/2}) \tag{59}$$

where M is the molar mass of the polymer and C is a constant related to ψ for the polymer-liquid system. As M approaches infinity, T_c approaches Θ; that is, the theta temperature is the lowest temperature at which a polymer molecule of infinite length would be completely miscible with the liquid. Below the theta temperature, the polymer exists as a ball or tight coil, with intramolecular interactions between groups in the same molecule. When the temperature is raised to the theta temperature, these interactions disappear and the polymer behaves as an ideal "statistical" coil, the liquid is called a *theta solvent*, and the polymer is said to be in the *theta state*. At temperatures above Θ in "poor" solvents there is little change in polymer chain dimensions, but in "good" solvents the chains tend to uncoil. Some theta temperatures are listed in Table 12, a more comprehensive list appears in the *Polymer Handbook,*[170] and this information is included also in the *CRC Handbook of Polymer-Liquid Interaction Parameters and Solubility Parameters.*[56]

Theta temperatures of polymer-liquid systems have been correlated with Hildebrand parameters[2,171-173] as shown in Figure 6. The lines indicate the range of Hildebrand parameters a liquid should have in order to be a solvent for the polymer at any particular temperature. In theta notation, Equations 28 and 31 become

$$^{ij}A = \psi\Theta R/{}^{i}V \tag{60}$$

and

$$^{i}\delta = {}^{j}\delta \pm (R\Theta\psi/{}^{i}V)^{1/2} \tag{61}$$

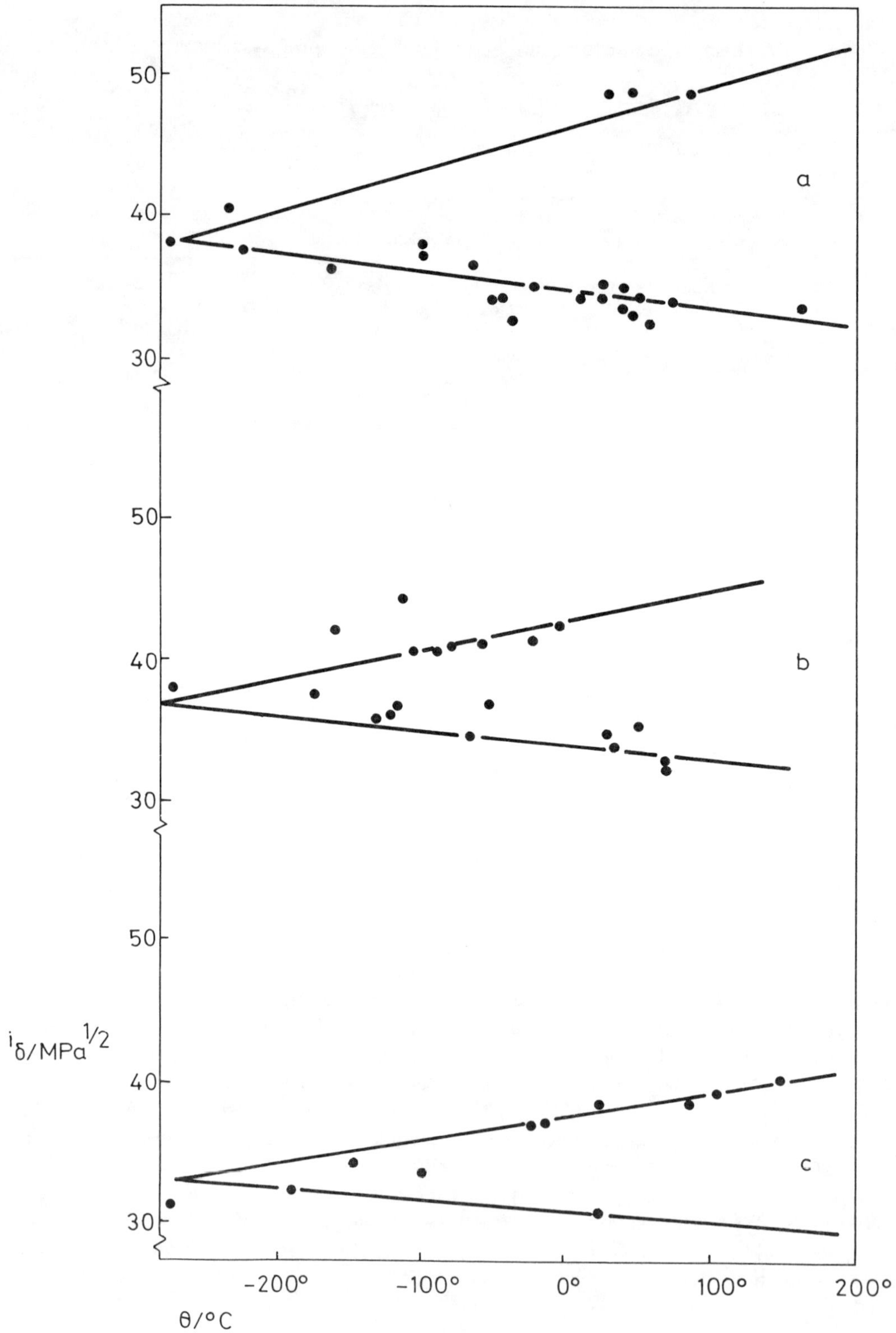

FIGURE 6. Polymer-liquid theta temperatures and liquid Hildebrand parameters: a, poly(methyl methacrylate); b, polystyrene; c, polyisobutylene. (Adapted from van Krevelen, D. W., *Properties of Polymers: Their Estimation and Correlation with Chemical Structure,* 2nd ed., Elsevier, Amsterdam, 1976.)

predicting a linear relationship between the liquid Hildebrand parameter and $\Theta^{1/2}$ if ψ (equal to $^1/_2 - \chi_S$) is constant, consistent with the data of Figure 6. Intrinsic viscosities and other solution properties at theta conditions may be explained also in terms of cohesion parameters.[174,175]

13.8 SEGMENT INTERACTION EQUATION

Heil and Prausnitz[116,117] proposed a segment interaction equation using parameters that have something in common with both cohesion parameters and the polymer-liquid χ interaction parameter. The equation incorporates the local volume fraction concept of Wilson[176] (Section 4.6), with $^{ii}\xi$ representing the local volume fraction of liquid molecules about a central liquid molecule and $^{jj}\xi$ representing the local volume fraction of polymer segments about a central segment. The corresponding local volume fractions involving polymer segments and liquid molecules are $^{ij}\xi$ and $^{ji}\xi$. These are defined in terms of the component mole fractions and molar volumes and the segment interaction energies, and obey the conservation equations

$$^{ij}\xi + {}^{jj}\xi = 1, \; {}^{ji}\xi + {}^{ii}\xi = 1$$

For a binary solution consisting of a polymer (j) of ^jn segments and a liquid i, the molar Gibbs free energy of mixing is

$$\begin{aligned} \Delta_m G/(RT) &= {}^ix \ln {}^{ii}\xi + {}^jn\,{}^jx \ln {}^{jj}\xi + (1 - {}^jn)^jx \ln {}^j\phi \\ &+ {}^ix\,{}^{ji}\xi({}^{ij}g - {}^{ii}g)/(RT) \\ &+ {}^jn\,{}^jx\,{}^{ij}\xi({}^{ij}g - {}^{ji}g)/(RT) \end{aligned} \tag{62}$$

The parameters ^{ii}g, ^{ij}g, and ^{jj}g are the energies of interaction between $i-i$, $i-j$, and $j-j$ pairs (liquid molecules i and polymer segments j). They were assumed to be independent of polymer molecular weight and were expected to be only weak functions of temperature. Although Equation 62 for the binary system involves three different segment interaction energies, there are only two adjustable parameters: the differences $({}^{ij}g - {}^{ii}g)$ and $({}^{ij}g - {}^{jj}g)$. As these differences become small compared with thermal energy RT, Equation 62 reduces to the Flory-Huggins Equation 13, with

$$\chi = (2\,{}^{ij}g - {}^{ii}g - {}^{jj}g)/(RT) \tag{63}$$

This limiting value of χ may be compared with Equation 28 for the enthalpic component of χ,

$$\chi = {}^{ij}A\,{}^iV/(RT) \approx ({}^ic + {}^jc - 2\,{}^ic\,{}^jc)^iV/(RT) \tag{64}$$

The pair of parameters $({}^{ij}g - {}^{ii}g)$ and $({}^{ij}g - {}^{jj}g)$ may be evaluated from thermodynamic activity data just as the Flory interaction parameter (Section 13.3), but this requires reliable vapor pressure data and a considerable amount of calculation. Values (which fit liquid activity over a wide polymer concentration range to an average deviation of well below 1% in most cases) have been listed for several binary polymer-liquid systems (Table 13) and have been used in some excellent predictions of phase diagrams for ternary systems of one polymer in two liquids. The equation works well even for those systems in which strong specific interactions (such as hydrogen bonding) exist.

The parameters in Table 13 follow the expected qualitative trends; thus a positive value

TABLE 13
Segment Interaction Parameters for Some Polymer-Liquid Systems

Polymer, j	Liquid, i	$10^{-3}\,{}^{j}M$ g mol^{-1}	t/°C	$({}^{ij}g - {}^{ii}g)$ J mol^{-1}	$({}^{ij}g - {}^{jj}g)$ J mol^{-1}
Polystyrene	Acetone	290	25	1660	−84
	Benzene	3.33	25	−209	618
	Trichloromethane	290	25	−609	1336
	Cyclohexane	25.9	34	1598	−69
		440	34	1616	−84
	Propyl acetate	290	25	733	159
	Toluene	3.33	40	−163	493
		290	25	−222	662
Poly(propylene glycol)	Methanol	1.96	25	2289	−88
		3.35	25	1200	−61
Poly(ethylene oxide)	Trichloromethane	6	25	1188	−1470
	Water	3	55	3512	−1432
		5	60	3502	−1374
Cellulose	Acetone	104	25	3512	−1046
acetate	1,4-Dioxane	104	25	2003	−814
	Methyl acetate	104	25	1965	−545
	Pyridine	104	25	40	125
Cellulose nitrate	Acetone	94	35	516	−5887
	Methyl acetate	94	35	389	−6389
Rubber	Acetone	—	0	2119	204
	Ethyl acetate	—	25	1449	29
	Methyl ethyl ketone	—	25	1725	−14
Polypropylene	Diethyl ketone	—	25	2390	−14
	Diisobutyl ketone	—	25	1399	−21

Adapted from Heil, J. F. and Prausnitz, J. M., *Am. Inst. Chem. Eng. J.*, 12, 678, 1966.

of $({}^{ij}g - {}^{ii}g)$ for methanol-poly(propylene glycol) indicates that the energy of a methanol-methanol interaction is greater than that of interaction between methanol and an ether segment. (Both ${}^{ij}g$ and ${}^{ii}g$ are negative and $|{}^{ii}g| > |{}^{ij}g|$.) This method is of considerable interest, although, like the Flory-Huggins theory, it does not lead to predictions based on pure component properties.

REFERENCES

1. **Orwoll, R. A.,** The polymer-solvent interaction parameter χ, *Rubber Chem. Technol.*, 50, 451, 1977.
2. **van Krevelen, D. W.,** *Properties of Polymers: Their Estimation and Correlation with Chemical Structure*, 2nd ed., Elsevier, Amsterdam, 1976.
3. **Askadskii, A. A.,** Computational procedures for the determination of the physical characteristics of polymers, *Usp. Khim.*, 46, 1122, 1977; *Russ. Chem. Rev.*, 46, 589, 1977.
4. **Askadskii, A. A, Kolmakova, L. K., Tager, A.A., Slonimskii, G. L., and Korshak, V. V.,** Evaluation of solubility parameters of low-molecular-weight liquids and polymers, *Dokl. Acad. Nauk SSSR*, 226, 857, 1976; *Dokl. Phys. Chem.*, 226, 99, 1976.
5. **Askadskii, A. A., Kolmakova, L. K., Tager, A. A., Slonimskii, G. L., and Korshak, V. V.,** Estimation of cohesion energy of low-molecular-weight liquids and polymers, *Vysokomol. Soedin. Ser. A*, 19, 1004, 1977; *Poly. Sci. U.S.S.R.*, 19, 1159, 1977.
6. **Sewell, J. H.,** Method of calculating densities of polymers, *J. Appl. Polym. Sci.*, 17, 1741, 1973.

7. **Knox, B. H.,** Bimodal character of polyester-solvent interactions. I. Evaluation of the solubility parameters of the aromatic and the aliphatic ester residues of poly(ethylene terephthalate); II. Evaluation of the chemical structures of the aromatic and aliphatic ester residues of poly(ethylene terephthalate); III. The effects of morphology on the nature of the interaction of nonaqueous solvents with the aromatic and the aliphatic ester residues of poly(ethylene terephthalate), *J. Appl. Polym. Sci.,* 21, 225, 249, and 267, 1977.
8. **Knox, B. H., Weigmann, H.-D., and Scott, M. G.,** Interactions of nonaqueous solvents with textile fibers. Part V. Application of the solubility parameter concept to polyester fiber-solvent interactions, *Text. Res. J.,* 45, 203, 1975.
9. **Becker, R.,** Packing density and preliminary calculation of the densities of polyurethanes, *Plaste Kautsch.,* 21, 915, 1974.
10. **Lavygin, I. A.,** Evaluation of the polymeric chain flexibility according to the cohesion energy density and the surface tension of the polymer, *Vysokomol. Soedin. Ser. B,* 18, 426, 1976.
11. **Nanda, V. S. and Simha, R.,** Equation of state and related properties of polymer and oligomer liquids, *J. Phys. Chem.,* 68, 3158, 1964.
12. **Miller, A. A.,** Free volumes in simple and polymeric liquids, *J. Polym. Sci. A-2,* 4, 415, 1966.
13. **Miller, A. A.,** Polymer-melt viscosity and the glass transition: an evaluation of the Adam-Gibbs and the free-volume models, *J. Chem. Phys.,* 49, 1393, 1968; General Electric Reprint 6110.
14. **Allen, G., Gee, G., Mangaraj, D., Sims, D., and Wilson, G. J.,** Intermolecular forces and chain flexibilities in polymers. II. Internal pressures of polymers, *Polymer,* 1, 467, 1960.
15. **Allen, G. and Sims, D.,** Intermolecular forces and chain flexibility. IV. Internal pressures of polyethylene glycol in the region of its melting point, *Polymer,* 4, 105, 1963.
16. **Bagley, E. B. and Scigliano, J. M.,** Free volume of liquids and polymers from internal pressure studies, *Polym. Eng. Sci.,* 11, 177, 1971.
17. **Bagley, E. B. and Scigliano, J. M.,** Internal pressure behavior of polymers above and below the glass temperature, *Polym. Eng. Sci.,* 11, 320, 1971.
18. **Gan, S. T., Leung, Y.-K., Booth, C., and Price, C.,** Internal pressure of solid poly(ethylene oxide), *Eur. Polym. J.,* 13, 971, 1977.
19. **Chan, K. S., Booth, C., and Price, C.,** Internal pressure of solid poly(ethylene oxide) 6000 swollen with liquid poly(ethylene oxide) 200, *Eur. Polym. J.,* 15, 721, 1979.
20. **Voeks, J. F.,** Cohesive energy density and internal pressure of high polymers, *J. Polym. Sci. A,* 2, 5319, 1964.
21. **Rostami, S. and Walsh, D. J.,** Miscibility of ethylene-vinyl acetate copolymers with chlorinated polyethylenes. 3. Simulation of the spinodal using the equation of state theory, *Macromolecules,* 17, 315, 1984.
22. **Rostami, S. and Walsh, D. J.,** Simulation of upper and lower critical phase diagrams for polymer mixtures at various pressures, *Macromolecules,* 18, 1228, 1985.
23. **Eskin, V.E. and Nesterov, A. E.,** Values of cohesive energy for some polymers, *Dokl. Akad. Nauk,* 165, 623, 1965; *Dokl. Phys. Chem.,* 165, 832, 1965.
24. **Acree, W. E., Jr.,** *Thermodynamic Properties of Nonelectrolyte Solutions,* Academic Press, Orlando, FL, 1984.
25. **Bagley, E. B. and Scigliano, J. M.,** Polymer solutions, in *Solutions and Solubilities,* Part II, Dack, M. R. J., Ed., John Wiley & Sons-Interscience, New York, 1976, chap. 16.
26. **Blanks, R. F.,** Engineering applications of polymer solution thermodynamics, *Chemtech.,* 6, 396, 1976.
27. **Blanks, R. F.,** Thermodynamics of polymer solutions, *Polym. Plast. Technol. Eng.,* 8, 13, 1977.
28. **Blanks, R. F. and Prausnitz, J. M.,** Thermodynamics of polymer solubility in polar and nonpolar systems, *Ind. Eng. Chem. Fundam.,* 3, 1, 1964.
29. **Burrell, H.,** The challenge of the solubility parameter concept, *J. Paint Technol.,* 40, 197, 1968.
30. **Carpenter, D. K.,** Solution properties, in *Encyclopedia of Polymer Science and Technology,* Vol. 12, Bikales, N. M., Ed., John Wiley & Sons-Interscience, New York, 1970, 627.
31. **Casassa, E. F. and Sedlacek, B., Eds.,** *Thermodynamic Interactions in Polymer Solutions* (*J. Polym. Sci. C Polym. Symp.,* Vol. 39), 1972.
32. **Dayantis, J.,** The Hildebrand parameter. Solubility of polymers, *Plast. Mod. Elastomeres,* 29(2), 58, 1977.
33. **Delmas, G. and Patterson, D.,** New aspects of polymer solution thermodynamics, *Off. Dig. Fed. Soc. Paint Technol.,* 34, 677, 1962.
34. **Flory, P. J.,** *Principles of Polymer Chemistry,* Cornell University Press, Ithaca, NY, 1953.
35. **Gardon, J. L.,** Cohesive-energy density, in *Encyclopedia of Polymer Science and Technology,* Vol. 3, John Wiley & Sons, New York, 1964, 833.
36. **Gardon, J. L.,** The influence of polarity upon the solubility parameter concept, *J. Paint Technol.,* 38(492), 43, 1966.
37. **Hildebrand, J. H., Prausnitz, J. M., and Scott, R. L.,** *Regular and Related Solutions,* Van Nostrand-Reinhold,Princeton, NJ, 1970.
38. **Kaelble, D. H.,** *Physical Chemistry of Adhesion,* John Wiley & Sons-Interscience, New York, 1971.

39. **Kalz, G.,** Thermodynamics of polymer solutions and conclusions on precipitations, *Plaste Kautsch.*, 29, 401, 1982.
40. **Magat, M.,** Thermodynamics of solutions of high polymers. II. Solubility of high polymers, *J. Chim. Phys.*, 46, 344, 1949.
41. **Mark, H. F. and Tobolsky, A. V.,** Thermodynamics of solutions, *Physical Chemistry of High Polymeric Systems,* 2nd ed., John Wiley & Sons-Interscience, New York, 1950, chap. 8.
42. **Olabisi, O., Robeson, L. M., and Shaw, M. T.,** *Polymer-Polymer Miscibility,* Academic Press, New York, 1979.
43. **Prausnitz, J. M.,** *Molecular Thermodynamics of Fluid Phase Equilibria,* Prentice-Hall, Englewood Cliffs, NJ, 1969; Prausnitz, J. M., Lichtenthaler, R. N., and Gomez de Azevedo, E., *Molecular Thermodynamics of Fluid Phase Equilibria*, 2nd ed., Prentice-Hall, Englewood Cliffs, NJ, 1986.
44. **von Tapavicza, S. and Prausnitz, J. M.,** Thermodynamics of polymer solutions: an introduction, *Chem. Ing. Tech.*, 47, 552, 1975; *Internat. Chem. Eng.*, 16, 329, 1976.
45. **Sheehan, C. J. and Bisio, A. L.,** Polymer/solvent interaction parameters, *Rubber Chem. Technol.*, 39, 149, 1966.
46. **Flory, P. J.,** Thermodynamics of high polymer solutions, *J. Chem. Phys.*, 10, 51, 1942.
47. **Flory, P. J.,** Statistical mechanics of swelling of network structures, *J. Chem. Phys.*, 18, 108, 1950.
48. **Flory, P. J. and Rehner, J., Jr.,** Statistical mechanics of crosslinked polymer networks. I. Rubberlike elasticity; II. Swelling, *J. Chem. Phys.*, 11, 512, and 521, 1943.
49. **Wall, F. T.,** Statistical thermodynamics of rubber. III, *J. Chem. Phys.*, 11, 527, 1943.
50. **Huggins, M. L.,** Solutions of long chain compounds, *J. Chem. Phys.*, 9, 440, 1941.
51. **Huggins, M. L.,** Some properties of long-chain compounds, *J. Phys. Chem.*, 46, 151, 1942; *Ann. N.Y. Acad. Sci.*, 43, Art. 1, 1, 1942.
52. **Huggins, M. L.,** Theory of solutions of high polymers, *J. Am. Chem. Soc.*, 64, 1712, 1942.
53. **Shinoda, K. and Hildebrand, J. H.,** The solubility and entropy of solution of iodine in octamethylcyclotetrasiloxane and tetraethoxysiloxane, *J. Phys. Chem.*, 61, 789, 1957.
54. **Fowkes, F. M.,** The interactions of polar molecules, micelles, and polymers in nonaqueous media, in *Solvent Properties of Surfactant Solutions,* Shinoda, K., Ed., Marcel Dekker, New York, 1967; Edward Arnold, London, 1968, chap. 3.
55. **Barton, A. F. M.,** Cohesion parameters, in *Polymer Yearbook,* Elias, H.-G. and Pethrick, R. A., Eds., Harwood, Chur, Switzerland, 1983, 149; Applications of solubility parameters and other cohesion parameters in polymer science and technology, *Pure Appl. Chem.*, 57, 905, 1985; Cohesion parameters, in *Encyclopedia of Physical Science and Technology,* Academic Press, Vol. 3, 1987, 140; Polymer interaction parameters and cohesion parameters, *Polymer Yearbook 6,* Pethrick, R. A., Ed., Harwood, Chur, Switzerland, 1990, 87.
56. **Barton, A. F. M.,** *Handbook of Polymer-Liquid Interaction Parameters and Solubility Parameters,* CRC Press, Boca Raton, FL, 1990.
57. **Bristow, G. M. and Watson, W. F.,** Cohesive energy densities of polymers. I. Cohesive energy densities of rubbers by swelling measurements; II. Cohesive energy densities from viscosity measurements, *Trans. Faraday Soc.*, 54, 1731, and 1742, 1958.
58. **Coran, A. Y. and Anagnostopoulos, C. E.,** Polymer-diluent interactions. III. Polyethylene-diluent interactions, *J. Polym. Sci.*, 57, 13, 1962.
59. **Cowie, J. M. G.,** The effect of solvent on the unperturbed dimensions of polystyrene, *J. Polym. Sci. C,* 23, 267, 1968.
60. **Cowie, J. M. G.,** Estimation of the cohesive energy density of a polymer from critical opalescence measurements, *Can. J. Chem.*, 46, 3919, 1968.
61. **Cowie, J. M. G., Dey, R., and McCrindle, J. T.,** A comparative study of preferential adsorption in bromoform(1), benzene(2), and polymer(3) systems by light scattering and density gradient ultracentrifugation, *Polym. J.*, 2, 88, 1971.
62. **Cowie, J. M. G. and Ranson, R. J.,** The dissolution and stability of cellulose triacetate in acetone, *Makromol. Chem.*, 143, 105, 1971.
63. **DiPaola-Baranyi, G. and Guillet, J. E.,** Estimation of polymer solubility parameters by gas chromatography, *Macromolecules,* 11, 228, 1978.
64. **DiPaola-Baranyi, G., Guillet, J. E., Klein J., and Jeberien, H. E.,** Estimation of solubility parameters for poly(vinyl acetate) by inverse gas chromatography, *J. Chromatgr.*, 166, 349, 1978.
65. **Dort, I.,** On the possibility of quantitative evaluation of the polymer-solvent interaction from Huggins viscosity constant, *Polymer,* 29, 490, 1988.
66. **Eskin, V. E. and Nesterov, A. E.,** The critical opalescence of solutions of poly β-naphthyl methacrylate and the properties of polymeric coils close to the precipitation point; The critical opalescence and chohesive energy of polymers, *Vysokomol. Soedin.*, 8, 1045, and 1051, 1966; *Polym. Sci. U.S.S.R.*, 8, 1146, and 1153, 1966.

67. **Eskin, V. E. and Nesterov, A. E.,** Critical opalescence of polymer solutions, *J. Polym. Sci. C,* 16, 1619, 1967.
68. **Ghosh, S. K.,** Solubility parameter and hydrocarbon sorption of low density polyethylene, *Makromol. Chem.,* 143, 181, 1971.
69. **Karim, K. A. and Bonner, D. C.,** Thermodynamic interpretation of solute-polymer interactions at infinite dilution, *J. Appl. Polym. Sci.,* 22, 1277, 1978; erratum, 23, 2126, 1978.
70. **Karim, K. A. and Bonner, D. C.,** An improved concept in solubility parameter theory applied to amorphous polymers, I. Poly(ethyl methacrylate), *Polym. Eng. Sci.,* 19, 1174, 1979.
71. **Koenhen, D. M. and Smolders, C. A.,** The determination of solubility parameters of solvents and polymers by means of correlation with other physical quantities, *J. Appl. Polym. Sci.,* 19, 1163, 1975.
72. **Kruse, R. L.,** Styrene-polymer interaction parameters in high impact polystyrene, *Adv. Chem. Ser.,* 142 (Copolym. Polyblends, Compos., Symp., 1974), 141, 1975.
73. **Leung, Y.-K.,** Thermodynamics of α,ω-methoxypoly(ethylene oxide)/*n*-alkane systems by gas-liquid chromatography, *Polymer,* 17, 374, 1976.
74. **Nilsson, H., Silvergren, C., and Törnell, B.,** Swelling of PVC latex particles by VCM, *Eur. Polym. J.,* 14, 737, 1978.
75. **Olabisi, O.,** Polymer compatibility by gas-liquid chromatography, *Macromolecules,* 8, 316, 1975.
76. **Newman, R. D. and Prausnitz, J. M.,** Polymer-solvent interactions from gas-liquid partition chromatography, *J. Phys. Chem.,* 76, 1492, 1972; Polymer-solvent interactions from gas-liquid chromatography, *J. Paint Technol.,* 45(585), 33, 1973.
77. **Merk, W., Lichtenthaler, R. N., and Prausnitz, J. M.,** Solubilities of fifteen solvents in copolymers of poly(vinyl acetate) and poly(vinyl chloride) from gas-liquid chromatgraphy. Estimation of polymer solubility parameter, *J. Phys. Chem.,* 84, 1694, 1980.
78. **Price, G. J.,** Calculation of solubility parameters by inverse gas chromatography, *Polym. Mater. Sci. Eng.,* 58, 1009, 1988; in *Inverse Gas Chromatography Characterization of Polymers and Other Materials (Am. Chem. Soc. Symp. Ser.* 391), Lloyd, D. R., Schreiber, H. P., and Ward, T. C., Eds., American Chemical Society, Washington, DC, 1988, 48.
79. **Rehner, J., Jr.,** Literature references to values of the polymer-solvent interaction parameter, *J. Polym. Sci.,* 46, 550, 1960.
80. **Rehner, J., Jr.,** Polymer-solvent interaction parameter, in *Polymer Handbook,* Brandrup, J. and Immergut, E. H., Eds., John Wiley & Sons-Interscience, New York, 1966, IV-157.
81. **Richards, R. W.,** Dilute solution properties of poly(benzyl methacrylate), *Polymer,* 18, 114, 1977.
82. **Richards, R. W. and Disselhoff, G.,** Second virial coefficients of polybenzyl methacrylate near the theta point, *Angew. Makromol. Chem.,* 66, 221, 1978.
83. **Scott, R. L. and Magat, M.,** Thermodynamics of high-polymer solutions. III. Swelling of cross-linked rubber, *J. Polym. Sci.,* 4, 555, 1949.
84. **Shvarts, A. G.,** Evaluation of rubber-solvent interaction, *Kolloid Zh.,* 19, 376, 1957; *Colloid J. U.S.S.R.,* 19, 375, 1957; *Rubber Chem. Technol.,* 31, 691, 1958.
85. **Shvarts, A. G.,** Comparative determination of [specific] cohesion energy of natural and polyisoprene rubber, *Zh. Fiz. Khim.,* 32, 718, 1958.
86. **Shvarts, A. G., Chefranova, E. K., and Iotkovskaya, L. A.,** Solubility parameters of resins based on dimethylvinylethynyl-*p*-hydroxyphenylmethane, *Kolloid Zh.,* 32, 603, 1970; *Colloid J. U.S.S.R.,* 32, 506, 1970.
87. **Wolf, B. A. and Wilms, M. M.,** Measured and calculated solubility of polymers in mixed solvents: co-nonsolvency, *Makromol. Chem.,* 179, 2265, 1978.
88. **Wolf, B. A.,** Polymer-solvent interaction parameters, in *Polymer Handbook,* Brandrup, J. and Immergut, E. H., Eds., John Wiley & Sons-Interscience, New York, 2nd ed., 1975, IV-131.
89. **Gundert, F. and Wolf, B. A.,** Polymer-solvent interaction parameters, in *Polymer Handbook,* Brandrup, J. and Innergut, E. H., Eds., John Wiley & Sons - Interscience, New York, 3rd. ed., 1989, VII-173.
90. **Yamashita, S., Minoura, Y., and Maekawa, E.,** Solubility and polymer-solvent interaction parameters of 1,2-polybutadiene, *Makromol. Chem.,* 180, 705, 1979.
91. **van Laar, J. J.,** Relation between the deviation of the vapor pressure of binary mixtures of normal compounds from the straight line, and the heat of mixing, *Z. Phys. Chem.,* A137, 421, 1928.
92. **Scatchard, G.,** Equilibria in non-electrolyte solutions in relation to the vapor pressures and densities of the components, *Chem. Rev.,* 8, 321, 1931.
93. **Anagnostopoulos, C. E., Coran, A. Y., and Gamrath, H. R.,** Polymer-diluent interactions. I. A new micromethod for determining poly(vinyl chloride)-diluent interactions, *J. Appl. Polym. Sci.,* 4, 181, 1960.
94. **Anagnostopoulos, C. E. and Coran, A. Y.,** Polymer-diluent interactions. II. Polyvinyl chloride-diluent interactions, *J. Polym. Sci.,* 57, 1, 1962.
95. **Roe, R.-J.,** Evaluation of the polymer-polymer interaction parameter from the equation of state thermodynamic theory, in *Multiphase Polymers* (*Adv. Chem. Ser.,* 176), Cooper, S. L. and Estes, G. M., Eds., American Chemical Society, New York, 1979, chap. 31, 599.

96. **Roe, R.-J. and Zin, W.-C.,** Determination of the polymer-polymer interaction parameter for the polystyrene-polybutadiene pair, *Macromolecules,* 19, 1221, 1980.
97. **Lin, J.-L. and Roe, R.-J.,** Relationship among polymer-polymer interaction energy densities and the deuterium isotope effect, *Macromolecules,* 20, 2168, 1987.
98. **Huglin, M. B. and Pass, D. J.,** Cohesive energy density of polytetrahydrofuran, *J. Appl. Polym. Sci.,* 12, 473, 1968.
99. **Patterson, D.,** Thermodynamics of non-dilute polymer solutions, *Rubber Chem. Technol.,* 40, 1, 1967.
100. **Patterson, D., Delmas, G., and Somcynsky, T.,** A comparison of lower critical solution temperatures of some polymer solutions, *Polymer,* 8, 503, 1967.
101. **Patterson, D.,** Free volume and polymer solubility. A qualitative view, *Macromolecules,* 2, 672, 1969.
102. **Bernstein, R. E., Cruz, C. A., Paul, D. R., and Barlow, J. W.,** LCST behavior in polymer blends, *Macromolecules,* 10, 681, 1977.
103. **Patterson, D. and Robard, A.,** Thermodynamics of polymer compatibility, *Macromolecules,* 11, 690, 1978.
104. **Biros, J., Zeman, L., and Patterson, D.,** Prediction of chi parameter by solubility parameter and corresponding states theories, *Macromolecules,* 4, 30, 1971.
105. **Dayantis, J.,** Free volume concept and polymer solutions thermodynamics, *Makromol. Chem.,* 176, 2437, 1975.
106. **Scigliano, J. M.,** *Consideration of Complex Solutions Using Internal Pressure Measurements,* Doctoral Dissertation, Washington University, St. Louis, MO; *Diss. Abstr. B,* 32, 6984, 1972.
107. **Eichinger, B. E. and Flory, P. J.,** Thermodynamics of polymer solutions. Part 1 — Natural rubber and benzene; Part 2 — Polyisobutylene and benzene; Part 3 — Polyisobutylene and cyclohexane.; Part 4 — Polyisobutylene and *n*-pentane, *Trans. Faraday Soc.,* 64, 2035, 2053, 2061, and 2066, 1968.
108. **Eichinger, B. E. and Flory, P. J.,** Determination of the equation of state of polyisobutylene, *Macromolecules,* 1, 285, 1968.
109. **Huggins, M. L.,** Thermodynamic properties of liquids, including solutions. VI. A new approach to the prediction of solubilities, *J. Paint Technol.,* 44(567), 55, 1972.
110. **Huggins, M. L.,** Thermodynamic properties of liquids, including solutions. IX. Thermodynamic properties of polymer solutions; X. Prediction of polymer solution properties, *Polym. J.,* 4, 502, and 511, 1973.
111. **Lacombe, R. H. and Sanchez, I. C.,** Statistical thermodynamics of fluid mixtures, *J. Phys. Chem.,* 80, 2568, 1976.
112. **Sanchez, I. C.,** Statistical thermodynamics of polymer blends, in *Polymer Blends,* Vol. 1, Paul, D. R. and Newman, S., Eds., Academic Press, New York, 1978, chap. 3.
113. **Sanchez, I. C.,** Equilibrium distribution of a minor constituent between a polymer and its environment, in *Durability of Macromolecular Materials* (*Am. Chem. Soc. Symp. Ser.,* Vol. 95), Eby, R. K., Ed., American Chemical Society, Washington, DC, 1979, chap. 12.
114. **Holten-Andersen, J., Rasmussen, P., and Fredenslund, A.,** Phase equilibria of polymer solutions by group contribution. 1. Vapor-liquid equilibria, *Ind. Eng. Chem. Research,* 26, 1382, 1987.
115. **Bonner, D. C. and Prausnitz, J. M.,** Vapor-liquid equilibrium calculations for concentrated polymer solutions, *Am. Inst. Chem. Eng. J.,* 19, 943, 1973.
116. **Heil, J. F.,** *Phase Equilibria in Multicomponent Polymer Solutions,* Doctoral dissertation, University of California, Berkeley, CA, 1965; *Diss. Abstr.,* 26, 7196, 1966.
117. **Heil, J. F. and Prausnitz, J. M.,** Phase equilibria in polymer solutions, *Am. Inst. Chem. Eng. J.,* 12, 678, 1966.
118. **Patterson, D.,** Role of free volume changes in polymer solution thermodynamics, *J. Polym. Sci. C,* 16, 3379, 1968.
119. **Burrell, H.,** Trends in solvent science and technology, in *Solvents Theory and Practice* (Adv. Chem. Ser., 124), Tess, R. W., Ed., American Chemical Society, Washington, DC, 1973, chap. 1.
120. **Koningsveld, R., Kleintjens, L. A., and Schoffeleers, H. M.,** Thermodynamic aspects of polymer compatibility, *Pure Appl. Chem.,* 39, 1, 1974.
121. **Bueche, A. M.,** Interaction of polydimethylsiloxanes with swelling agents, *J. Polym. Sci.,* 15, 97, 1955.
122. **Cernia, E. M., Mancini, C., and Saini, A.,** Thermodynamic interactions of branched polyethylene in polar and nonpolar systems, *J. Appl. Polym. Sci.,* 12, 789, 1968.
123. **De Broutelles, R. and Opigez, D.,** Quantitative prediction of elastomer swelling in solvents, *Rev. Gen. Caoutch. Plast.,* 52, 264, 1975.
124. **Conder, J. R. and Young, C. L.,** *Physicochemical Measurement by Gas Chromatography,* John Wiley & Sons-Interscience, New York, 1979, Section 5.5.
125. **Rami Reddy, C. and Kalpagam, V.,** A rough estimation of the solubility parameters of styrene-acrylonitrile copolymers, *Current Sci.,* 43, 681, 1974.
126. **Aharoni, S. M., Cilurso, F. E., and Hanrahan, J. M.,** Trifluoroethanol/chloroalkane mixtures: excellent novel solvents for aliphatic polyamides, *J. Appl. Polym. Sci.,* 30, 2505, 1985.
127. **Baughan, E. C.,** The absorption of organic vapours by thin films of polystyrene, *Trans. Faraday Soc.,* 44, 495, 1948.

128. **Baughan, E. C., Jones, A. L., and Stewart, K.,** The absorption of organic vapours by thin films of nitrocellulose, *Proc. Roy. Soc. London Ser. A,* 225, 478, 1954.

129. **Bindschaedler, C., Gurny, R., Doelker, E., and Peppas, N. A.,** Thermodynamics of swelling of polymer latex particles by a water-soluble solvent. II. Experimental results on cellulose acetate latices, *J. Colloid Interface Sci.,* 108, 83, 1985.

130. **Camberlin, Y. and Pascault, J. P.,** Phase segregation kinetics in segmented linear polyurethanes — relations between equilibrium time and chain mobility and between equilibrium degrees of segregation and interaction parameter, *J. Polym. Sci., Polym. Phys.,* 22, 1835, 1984.

131. **Ceccorulli, G., Pizzoli, M., Scandola, M., and Pezzin, G.,** Solubility predictions for poly(vinyl chloride)-ester systems, *Polym. Commun.,* 24, 107, 1983.

132. **Fujioka, K., Noethiger, N., Beatty, C. L., Baba, Y., and Kagemoto, A.,** Compatibility of a random copolymer of varying composition with each homopolymer, in *Polymer Blends and Composites in Multiphase Systems* (*Adv. Chem. Ser.,* 206), Han, C. D., Ed., 1984, 149.

133. **Fukuda, T. and Inagaki, H.,** Interactions between unlike polymers versus dilute solution properties of copolymers, *Pure Appl. Chem.,* 55, 1541, 1983.

134. **Nagata, M., Fukuda, T., and Inagaki, H.,** Light scattering from polymer blend solutions. 5. Characterization of systems of relatively high incompatibility, *Macromolecules,* 20, 2173, 1987.

135. **Fukuda, T. and Inagaki, H.,** Light scattering characterization of polymer blends in solution, in *Proc. Second Rolduc Polymer Meeting* (Limburg, Netherlands), Elsevier, Amsterdam, 1987.

136. **Gomez-Anton, M. R., Masegosa, R. M., and Horta, A.,** Polymethyltrifluoropropylsiloxane: polymer-solvent interaction, solubility and network swelling, *Polymer,* 28, 2116, 1987.

137. **Katime, I. and Ochoa, J. R.,** Polymer cosolvent systems. VII. PMMA-CCl_4-chloroalkane preferential solvation and second virial coefficient, *Eur. Polym. J.,* 20, 99, 1984.

138. **Lovell, P. A. and Still, R. H.,** Synthesis and characterisation of poly(arylene sulfides), 7. Transitional behaviour of random and block copoly(arylene sulfides), *Makromol. Chem.,* 188, 1551, 1987.

139. **Moore, W. R.,** Some aspects of mixing theory, *Trans. J. Plast. Inst.,* 32, 247, 1964.

140. **Oberoi, R. S., Lal, K., and Bhatnagar, H. L.,** Estimation of unperturbed dimensions of GR-S from intrinsic viscosity measurements in good solvents, *Indian J. Chem.,* 9, 1128, 1971.

141. **Okamoto, H. and Ohde, Y.,** Environmental stress cracking and crazing in polymers in terms of irreversible thermodynamics, *Polymer,* 23, 1204, 1982.

142. **Ohde, Y. and Okamoto, H.,** Growth rate *versus* temperature relations characteristic of environmental stress cracks in low-density polyethylenes and the underlying thermodynamic mechanism, *Polymer,* 24, 783, 1983.

143. **Pascault, J. P. and Camberein, Y.,** Quantitative determination of the phase segregation degree in a poly(dimethylsiloxane)-based urethane, *Polym. Commun.,* 27, 230, 1986.

144. **Peppas, N. A. and Lucht, L. M.,** Macromolecular structure of coals. X. Thermodynamic interaction parameter for solvents and coal networks, *Polym. Bull. (Berlin),* 16, 375, 1986.

145. **Shoji, K. and Takeda, M.,** Study on interaction of organic vapor and polyethyleneglycol by gas-chromatography, *Kobunshi Ronbunshi,* 31, 420, 1974; *English Ed.,* 3, 1606, 1974.

146. **Tseng, C.-M., El-Aasser, M. S., and Vanderhoff, J. W.,** Modeling the equilibrium swelling of latex particles with monomers, *Am. Chem. Soc. Symp. Ser.,* 197, 197, 1982.

147. **Verkholantsev, V. V.,** Nonhomogeneous-in-layer coatings, *Progr. Org. Coatings,* 13, 71, 1985.

148. **Zhigotsky, A. G., Dronov, S. V., and Shvarts, T. M.,** Effect of the solvent nature on rheological properties of the solution vinyl chloride copolymer with vinyl acetate, *Ukr. Khim. Zh.,* 46, 841, 1980.

149. **Balcar, H. and Bohackova, V.,** Solution properties of poly(1-chloro-2,3-epoxypropane), *Collect. Czech. Chem. Commun.,* 42, 2145, 1977.

150. **Schille, W. and Bittrich, H.-J.,** Calculating free excess enthalpies in hydrocarbon systems, *Chem. Tech. (Leipzig),* 25, 292, 1973.

151. **Chen, S.-A.,** Polymer miscibility in organic solvents and in plasticizers — a two-dimensional approach, *J. Appl. Polym. Sci.,* 15, 1247, 1971.

152. **Chen, S.-A.,** Polymer miscibility in mixed organic liquids. An extension of the two-dimensional approach, *J. Appl. Polym. Sci.,* 16, 1603, 1972.

153. **Chen, S.-A.,** Estimation of the miscibility of solutes in mixed solvents using a pseudo-binary approach, *Chem. Eng. Sci.,* 30, 575, 1975.

154. **Chen, S.-A. and Lai, W.-J.,** Applicability of the two-dimensional approach to polymer miscibility in pure and mixed organic liquids, *J. Appl. Polym. Sci.,* 23, 317, 1979.

155. **Chen, S.-A. and Lai, W.-J.,** Polymer miscibility in mixed organic liquids having acid-base interactions — a modified two-dimensional approach, *J. Appl. Polym. Sci.,* 23, 319, 1979.

156. **Fernańdez-Piérola, I. and Horta, A.,** Co-nonsolvency of PMMA, *Polym. Bull.,* 3, 273, 1980.

157. **Fernańdez-Piérola, I. and Horta, A.,** Cosolvents of PMMA, *Makromol. Chem.,* 182, 1705, 1981.

158. **Horta, A. and Fernańdez-Piérola, I.,** Depression of the critical temperature in PMMA-cosolvent systems, *Polymer,* 22, 783, 1981.

159. **Kinzer, K. E., Lloyd, D. R., Wightman, J. P., and McGrath, J. E.,** Asymmetric membrane preparation from nonsolvent casting systems, *Desalination,* 46, 327, 1983.
160. **Kinser, K. E., Lloyd, D. R., Gay, M. S., Wightman, J. P., Johnson, B. C., and McGrath, J. E.,** Phase inversion sulfonated polysulfone membranes, *J. Membrane Sci.,* 22, 1, 1985.
161. **Kuhn, R.,** Fractionation of polymers with demixing solvents, *Makromol. Chem.,* 177, 1525, 1976.
162. **Kuhn, R.,** Characterization of polymer blends, block copolymers, and graft copolymers by fractionation procedures using demixing solvents, in *Polym. Sci. Technol.,* 20 (*Polymer Alloys 3: Blends, Blocks, Grafts, Interpenetrating Networks,* Klempner, D. and Frisch, K. C., Eds., Plenum), 45, 1983.
163. **Ochoa, J. R., Caballero, B., Valenciano, R., and Katime, I.,** Polymer cosolvent systems. VIII. PMMA/CCl_4/(iso)amyl acetate. Study by laser-light scattering and viscometry, *Mater. Chem. Phys.,* 9, 477, 1983.
164. **Piccarolo, S. and Titomanlio, G.,** A co-solvation case: possibility of prediction using solubility parameters, *Conv. Ital. Sci. Macromol.,* [Atti], 5th, Assoc. Ital. Sci. Technol. Macromol., Milan, Italy, 1981, 273.
165. **Piccarolo, S. and Titomanlio, G.,** Synergism in the swelling and solubiity of poly(methyl methacrylate) in presence of ethanol/water mixtures, *Makromol. Chem., Rapid Commun.,* 3, 383, 1982.
166. **Piccarolo, S. and Titomanlio, G.,** Thermodynamic behavior of single polymer-binary solvent systems. Qualitative comparison with solubility parameter approach, *Ind. Eng. Chem. Prod. Res. Dev.,* 22, 146, 1983.
167. **Trzebicka, B., Tekely, P., and Turska, E.,** Phenomena observed on swelling poly(2,6-dimethyl-phenylene-oxide) in benzene-nitromethane mixtures, *Acta Polymerica,* 33, 83, 1982.
168. **Vidyarthi, N. and Palit, S. R.,** Cosolvency studies of polyethylene. Part I: Lowering of cloud temperatures in cosolvent mixtures; Part II: Viscometric studies in different cosolvent media, *J. Indian Chem. Soc.,* 59, 748, and 753, 1982.
169. **Seymour, R. B.,** *Introduction to Polymer Chemistry,* McGraw-Hill, New York, 1971, section 2.2.
170. **Elias, H.-G., Adank, G., Dietschy, Hj., Etter, O., Gruber, U., and Ibrahim, F. W.,** Theta-solvents, in *Polymer Handbook,* Brandrup, J. and Immergut, E. H., Eds., John Wiley & Sons-Interscience, New York, 1966, IV-163; Elias, H.-G. Theta-solvents, in *Polymer Handbook,* Brandrup, J. and Immergut, E. H., Eds., John Wiley & Sons-Interscience, New York, 3rd ed., 1989, VII-205.
171. **Fox, T. G.,** Properties of dilute polymer solutions. III. Intrinsic viscosity/temperature relationships for conventional polymethyl methacrylate, *Polymer,* 3, 111, 1962.
172. **Hayduk, W. and Bromfield, H. A.,** Theta-cosolvent solutions for poly(vinyl chloride) by cloud point and osmotic pressure measurements, *J. Appl. Polym. Soc.,* 22, 149, 1978.
173. **van Krevelen, D. W. and Hoftyzer, P. J.,** Practical evaluation of the $[\eta]$-M relation, *J. Appl. Polym. Sci.,* 11, 2189, 1967.
174. **Crescenzi, V. and Flory, P. J.,** Configuration of the poly-(dimethylsiloxane) chain. II. Unperturbed dimensions and specific solvent effects, *J. Am. Chem. Soc.,* 86, 141, 1964.
175. **Izumi, Y. and Miyake, Y.,** Study of linear poly(*p*chlorostyrene)-diluent systems. I. Solubilities, phase relationships, and thermodynamic interactions. *Polymer J.,* 3, 647, 1972.
176. **Wilson, G. M.,** Vapor-liquid equilibrium. XI. A new expression for the excess free energy of mixing, *J. Am. Chem. Soc.,* 86, 127, 1964.
177. **Painter, P. C., Park, Y., and Coleman, M. M.,** Thermodynamics of coal solutions, *Energy Fuels,* 2, 693, 1988.
178. **Verchère, D., Sautereau, H., Pascault, J. P., Moschiar, S. M., Riccardi, C. C. and Williams, R. J. J.,** Miscibility of epoxy monomers with carboxyl-terminated butadiene-acrylonitrile random copolymers, *Polymer,* 30, 107, 1989.
179. **Holten-Andersen, J. and Eng, K.,** Activity coefficients in polymer solutions, *Progr. Org. Coatings,* 16, 77, 1988.
180. **Sears, J. K. and Darby, J. R.,** *The Technology of Plasticizers,* Wiley-Interscience, New York, 1982.
181. **Vilcu, R. and Leca, M.,** Thermodynamic characterization of polymers by chromatography with gas mobile phase, *Rev. Roum. Chim.,* 34, 387, 1989; *Polymer Thermodynamics by Gas Chromatography,* Elsevier, Amsterdam, 1990.
182. **Fernández-Sánchez, E., Fernández-Torres, A., Garcia-Domínguez, J. A., and Santiuste, J. M.,** Solubility parameters of gas chromatographic mixed stationary phases, *J. Chromatogr.,* 457, 55, 1988.
183. **Kelley, S. S., Glasser, W. G., and Ward, T. C.,** Engineering plastics from lignin. XV. Polyurethane films from chain-extended hydroxypropyl lignin, *J. Appl. Polym. Sci.,* 36, 759, 1988; **Kelley, S. S., Ward, T. C., Rials, T. G., and Glasser, W. G.,** Engineering plastics from lignin. XVII. Effect of molecular weight on polyurethane film properties, *J. Appl. Polym. Sci.,* 37, 2961, 1989.
184. **Narender, G. and Yaseen, M.,** Phenomenon of cosolvency and solution parameters of polyamide resins, *J. Coat. Technol.,* 61(773), 41, 1989.
185. **Coleman, M. M., Hu, J., Park, Y., and Painter, P. C.,** Amorphous polyurethane-polyether blends, *Polymer,* 29, 1659, 1988.
186. **Bradner, F. P., Shapiro, J. S., Bowley, H. J., Gerrard, D. L., and Maddams, W. F.,** Some insights into the microstructure of polypyrrole, *Polymer,* 30, 914, 1989.

187. **Mizerovskii, L. N., Vansyatskaya, L. N., and Smurova, G. I.,** Investigation of the reaction between a ternary ethylene-propylene copolymer and organic liquids, *Vysokomol. Soedin. Ser. A,* 29, 1512, 1987; *Polym. Sci. U.S.S.R.,* 29, 1663, 1987.

188. **Kambour, R. P., Gundlach, P. E., Want, I. C. W., White, D. M., and Yeager, G. W.,** Miscibility of poly(2,6-dimethyl-1,4-phenylene oxide) with several styrenic homopolymers — dependence of interaction parameters from critical molecular weights on cohesive energy densities, *Polym. Commun.,* 29(6), 170, 1988.

189. **Kawaguchi, M., Chikazawa, M., and Takahashi, A.,** Displacement of polymers by displacers. 1. Polystyrene at the silica surface, *Macromolecules,* 22, 2195, 1989; **Kawaguchi, M., Hada, T., and Takahashi, A.,** Displacement of polymers by displacers. 2. Poly(ethylene oxide) at the silica surface, *Macromolecules,* 22, 4045, 1989.

190. **Mikos, A. G. and Peppas, N. A.,** Flory interaction parameter χ for hydrophilic copolymers with water, *Biomaterials,* 9, 419, 1988.

191. **Chiou, C. T., Porter, P. E., and Schmedding, D. W.,** Partition equilibria of nonionic organic compounds between soil organic matter and water, *Environ. Sci. Technol.,* 17, 227, 1983.

192. **Koningsveld, R., Klentjens, L. A., and Nies, E.,** Polymers and Thermodynamics, *Croat. Chem. Acta,* 60, 53, 1987.

Chapter 14

DETERMINATION OF POLYMER COHESION PARAMETERS

14.1 HILDEBRAND PARAMETER VALUES

If account is taken of the major deviations from regular behavior in polymer solutions due to the substantial size difference between polymer and solvent molecules, as described in Section 13.5, cohesion parameters may be applied to them just as they are to mixtures of liquids. Because polymers are more often encountered as crystalline or amorphous solid-like materials than as liquids, and because polymers can act as "solvents" as well as "solutes", the use of the more general term "cohesion parameter" is particularly appropriate in place of the rather restrictive "solubility parameter". The polymer cohesion parameter can be thought of as the square root of the cohesive enthalpy per molar structural unit divided by the volume per molar structural unit. Table 1 in Chapter 1 includes the general reviews on polymer cohesion parameters.

The cohesive pressures of polymers can be evaluated directly by thermodynamic means only for vaporizable materials, although some of the indirect thermodynamic methods of Chapter 7 can be applied in correlations and estimations.[1-3] Also, the cohesion parameter concept is directly applicable only to amorphous polymers, and when considering crystalline polymers it is in principle necessary to calculate the activity of the crystalline solid relative to the real or hypothetical amorphous material at the same temperature, as described in Chapter 12. In practice the distinction between amorphous and crystalline polymers is not so clear-cut (Section 16.3), and non-thermodynamic empirical methods are used frequently in the estimation of cohesion parameters for polymers. Preferred single cohesion parameter or Hildebrand parameter values for a few commonly encountered polymers are listed in Table 1. The numerous reviews in Chapter 1 include other data compilations.[4-9] In one of the original applications of cohesion parameters to polymer solutions[4,5] the Hildebrand parameter was combined with a classification into strong, moderate, or poor hydrogen-bonding capability, as described in Sections 4.5 and 8.2, and ranges of values for representative polymers are given in Table 2 in order of increasing Hildebrand parameters. A more complete list appears in Table 3, classified by polymer type.

The best solvent for a polymer is not necessarily the corresponding monomer or oligomer made up of the same repeat units. (An oligomer is a low molecular weight polymer, often a liquid.) These liquids usually have cohesion parameters lower than those of the polymers derived from them, for reasons explained in Section 14.8. The best solvent is usually another liquid, with a cohesive energy rather higher than that of the monomer but with a cohesive energy *density* the same as that of the polymer. It can be noted that hydrogenation has the effect of decreasing the Hildebrand parameter of a polymer.[10]

There is a great variety of indirect methods for estimating polymer cohesion parameters, because they are related to so many physical properties. Once determined the cohesion parameters may be used to correlate or estimate values of other physical properties. This chapter deals with some of the relationships used to evaluate cohesion parameters from other properties of polymers, but consideration of chromatographic properties is deferred until Chapter 18. The properties of oligomers are considered in Section 14.9 and copolymers in Section 16.10. Polymer cohesion and adhesion have much in common,[11] and adhesion is described in cohesion parameter terms in Section 17.5.

TABLE 1
Preferred Hildebrand Parameter Values for Selected Polymers

Polymer	$\delta/MPa^{1/2}$
Polyacrylonitrile	26
Polybutadiene	17.0
Poly(butyl acrylate)	18.5
Cellulose acetate	24
Cellulose nitrate	21
Polychloroprene	18.5
Poly(dimethylsiloxane)	15.5
Ethyl cellulose	20
Polyethylene	17.0
Poly(ethylene oxide)	24
Poly(ethyl methacrylate)	18.5
Polyisobutylene	16.5
Polyisoprene	17.0
Poly(methyl acrylate)	20.5
Poly(methyl methacrylate)	19.0
Polypropylene	16.5
Polystyrene	18.5
Poly(tetrafluoroethylene)	13
Poly(vinyl acetate)	20
Poly(vinyl chloride)	19.5

Selected from *Handbook of Polymer-Liquid Interaction Parameters and Solubility Parameters*, Barton, A. F. M., CRC Press, Boca Raton, FL, 1990.

14.2 HILDEBRAND PARAMETERS FROM POLYMER-LIQUID INTERACTION PARAMETERS

Hildebrand parameters for polymers may be evaluated from polymer-liquid interaction parameters, χ, as described in Section 13.5. For example, χ may be evaluated from swelling measurements by the Flory-Rehner equation[12] and the use of a Bristow-Watson plot (Equation 45, Chapter 13):

$$^{i}\delta^2/(RT) - \chi/^{i}V = (2\,^{j}\delta/(\mathrm{RT}))^{i}\delta - {}^{j}\delta^2/(\mathrm{RT}) - \chi_S/^{i}V \tag{1}$$

The third term on the right-hand side is small compared with the second, so a plot of the left-hand side against the liquid Hildebrand parameter $^{i}\delta$ provides values of the polymer Hildebrand parameter $^{j}\delta$ from both slope and intercept (see Table 4).[13-16] Also useful is the closely related empirical equation 44, Chapter 13,

$$^{i}\delta = {}^{j}\delta \pm (RT/K)^{1/2}[(\chi - \chi_S)/^{i}V]^{1/2} \tag{2}$$

in which $^{i}\delta$ is plotted against $[(\chi - \chi_S)/^{i}V]$, with $^{j}\delta$ being determined from the intercept. This has been used for several natural and synthetic rubbers by Soviet scientists.[17-21] More simply, χ shows a minimum when plotted as a function of $^{i}\delta$, and this minimum can be identified with the polymer Hildebrand parameter.[22] Other variations are also possible.[23]

TABLE 2
Approximate Hildebrand Parameter Ranges for Some Common Polymeric Materials, Classified by Hydrogen Bonding Capability and in Order of Increasing Values[6,7]

Polymer	Hildebrand parameter ranges ($\delta/MPa^{1/2}$) in environments with hydrogen-bonding capability which is		
	Poor	Moderate	Strong
Polytetrafluorocarbons	12—13	—	—
Ester gum	14—22	15—22	19—22
Alkyd 45% soy oil	14—22	15—22	19—24
Silicone DC-1107	14—19	19—22	19—24
Poly(vinyl ethyl ether)	14—23	15—22	19—29
Poly(butyl acrylate)	14—26	15—24	—
Poly(butyl methacrylate)	15—23	15—20	19—23
Silicone DC-23	15—17	15—16	19—21
Polyisobutylene	15—16	—	—
Polyethylene	16—17	—	—
Gilsonite®	16—19	16—17	—
Poly(vinyl butyl ether)	16—22	15—21	19—23
Natural rubber	17	—	—
Hypalon® 20 [chlorosulfonated PE]	17—20	17—18	—
Ethyl cellulose N-22	16—23	15—22	19—30
Chlorinated rubber	17—22	16—22	—
Dammar gum	17—22	16—21	19—22
Versamid® 100 [polyamide]	17—22	17—18	19—23
Polystyrene	17—22	19	—
Poly(vinyl acetate)	17—19	—	—
Poly(vinyl chloride)	17—23	16—22	—
Phenolic resins	17—24	16—27	19—28
Buna N (butadiene-acrylonitrile copolymer)	18—19	—	—
Poly(methyl methacrylate)	18—26	17—27	—
Carbowax® 4000 [poly(ethylene oxide)]	18—26	17—30	19—30
Thiokol® [poly(ethylene sulfide)]	18—21	—	—
Polycarbonate	19—22	19—21	—
Pliolite® P-1230	19—22	—	—
Mylar® [poly(ethylene terephthalate)]	19—22	19—20	—
Vinyl chloride-acetate copolymer	19—23	16—27	—
Polyurethane	20—21	—	—
Styrene-acrylonitrile copolymer	22—23	19—20	—
Vinsol® [rosin derivative]	22—24	16—27	19—26
Epon® 1001 [epoxy]	22—24	17—27	—
Shellac	—	21—23	19—29
Polymethacrylonitrile	—	22—23	—
Cellulose acetate	23—26	21—30	—
Cellulose nitrate	23—26	16—30	26—30
Polyacrylonitrile	—	25—29	—
Poly(vinyl alcohol)	—	—	25—27
Polyamide 6,6	—	—	28—31
Cellulose	—	—	30—33

TABLE 3
Hildebrand Parameter Ranges of Polymers in Environments with Poor, Moderate, or Strong Hydrogen-Bonding Ability

Polymer (with manufacturer)	δ/MPa$^{1/2}$ Ranges in environments with hydrogen-bonding capability which is: Poor	Moderate	Strong
Acrylics			
Acrylic solution polymers (Rohm & Haas)			
Acryloid® B-44	18—24	17—27	—
Acryloid B-66	17—23	16—25	—
Acryloid B-72	17—26	18—27	—
Acryloid B-82	17—23	18—25	—
Acryloid K-120N	17—26	17—27	—
Poly(butyl acrylate) (Rohm & Haas)	14—26	15—25	19—26
Poly(butyl methacrylate) (Rohm & Haas)	15—23	15—20	19—23
Poly(isobutyl methacrylate) (Rohm & Haas)	17—23	17—20	19—23
Poly(ethyl methacrylate) (Rohm & Haas)	17—23	16—27	19—23
Poly(methacrylic acid) (Rohm & Haas)	—	20	26—30
Poly(methyl methacrylate) (Rohm & Haas)	18—26	17—27	—
Butyl methacrylate(BMA)/Acrylonitrile(AN), 80:20	17—26	16—25	—
Isobornyl maleic anhydride/Cyclol®, [a]75:25	15—22	15—17	19—22
Methacrylic acid (MAA)/Blown linseed oil, 20:80	17	—	—
MAA/Ethyl acrylate(EA)/Styrene, 15:38:47	19	17—18	—
MAA/Maleic anhydride(MA)/Vinyl acetate (VA), 15:27.5:57.5	—	20—30	26—30
MAA/MA/VA, 15:17.5:67.5	—	20—30	26—30
Methyl methacrylate (MMA)/Cyclol®, [a] 58:42	19	17—25	—
MMA/EA, 50:50	17—26	16—27	—
MMA/EA, 25:75	—	18—22	19—30
MMA/EA/Allyl glycidyl ether, 40:40:20	17—26	16—27	21
MMA/EA/Acrylamide, 45:45:10	23—26	18—27	26—30
MMA/EA/Acrylonitrile (AN), 55:30:15	22—26	17—20	—
MMA/EA/AN, 40:40:20	22—26	19—30	—
MMA/EA/*tert*-Butylaminomethacrylate, 40:40:20	17—26	16—27	19—30
MMA/EA/Cyclol®, [a] 40:40:20	17—26	16—27	21—22
MMA/EA/MAA, 40:40:20	—	18—22	19—30
MMA/EA/Methylol acrylamide, 45:45:10	17—26	17—27	26—30
MMA/EA/Vinyl butyl ether, 40:40:20	17—26	16—27	26
Pacrosir 9055, 50% (SIR)			
In 50% solutions	16—23	15—22	19—26
In 1% solutions	17—23	15—20	19—25
Alkyd Resins			
40% Adipic, glycerol phthalate	23—26	20—30	26—30
45% Linseed, glycerol phthalate	14—24	15—22	19—24
28% Soy, glycerol phthalate	17—23	17—27	—
30% Soy, glycerol phthalate	17—26	17—30	—
45% Soy, glycerol phthalate	14—23	15—22	19—24
38% Soy, glycerol oil-modified alkyd	17—23	16—25	21—23
45% Soy, pentaerythritol phthalate	14—23	15—22	19—24
Methacrylated dehydrated castor oil, pentaerythritol alkyd	17—23	17—27	—
Pelargonic alkyd	17—23	16—25	—
Pentaerythritol phthalate benzoate	19—26	17—30	—
Pentaerythritol tri-*p*-*tert*-butyl mono-acid phthalate	17—26	15—20	19—23
Pentaerythritol *p*-*tert*-butyl benzoate phthalate	17—23	16—27	21—22
Styrenated linseed alkyd	17—24	15—22	—

TABLE 3 (continued)
Hildebrand Parameter Ranges of Polymers in Environments with Poor, Moderate, or Strong Hydrogen-Bonding Ability

Polymer (with manufacturer)	δ/MPa$^{1/2}$ Ranges in environments with hydrogen-bonding capability which is Poor	Moderate	Strong
Vinyltoluene linseed alkyd	16—22	19—22	—
Siralkid® 2240, 100% (SIR)			
In 50% solutions	17—22	19	—
In 1% solutions	17—22	—	—
Siralkid 2256, 100%			
In 50% solutions	14—23	15—20	19—22
In 1% solutions	16—23	16—20	—
Siralkid 2287/AV, 100%			
In 50% solutions	16—23	18—20	19—23
In 1% solutions	17—22	18—20	22
Siralkid 2550, 100%			
In 50% solutions	15—23	15—22	19—22
In 1% solutions	17—23	16—20	—
Siralkid 2669, 100%			
In 50% solutions	14—23	15—20	19—24
In 1% solutions	14—22	15—20	19—22
Restiroid® 3751, 50% (SIR)			
In 50% solutions	17—23	15—22	22—23
In 1% solutions	17—23	16—20	—
Amine Resins			
Beckamine® P-196 Urea-Formaldehyde (Reichhold)	18—23	17—22	19—26
Beetle® 227-8 Butylated Urea-Formaldehyde (American Cyanamid)	—	—	18—23
Cyzak® 1006 Melamine-Formaldehyde	17—23	16—27	22
Cyzak 1007 Melamine-Formaldehyde	17—23	17—27	—
Cyzak 1013 Melamine-Formaldehyde	17—23	16—25	—
Cyzak 1026 Melamine-Formaldehyde	—	18—27	—
Hexa(methoxy methyl)melamine	17—24	17—30	19—34
Resimene® 888 Butylated melamine-Formaldehyde (Monsanto)	17—22	15—25	19—26
Uformite® MX-61 Benzoguanamine-Formaldehyde (Rohm & Haas)	17—23	15—23	19—23
Versalon® 1112 Dimer acid-Polyamine condensate (General Mills)	—	—	19—23
Versalon 1175	—	—	19—23
Versamid® 100 Dimer acid-Polyamine condensate (General Mills)	17—22	17—18	19—23
Versamid 115	17—22	16—20	19—26
Versamid 900	—	—	—
Versamid 930	—	—	19—23
Versamid 940	—	—	19—23
Siramin® 4147, 67% (SIR)			
In 50% solutions	15—23	15—22	19—30
In 1% solutions	17—23	15—22	19—26
Siramin 6150, 60%			
In 50% solutions	14—26	15—27	19—30
In 1% solutions	14—22	16—22	19—30
Cellulose Derivatives			
Alcohol-soluble butyrate	24	17—27	23—30
Cellulose acetate, LL-1	23—26	20—30	—

TABLE 3 (continued)
Hildebrand Parameter Ranges of Polymers in Environments with Poor, Moderate, or Strong Hydrogen-Bonding Ability

Polymer (with manufacturer)	δ/MPa$^{1/2}$ Ranges in environments with hydrogen-bonding capability which is: Poor	Moderate	Strong
Cellulose acetate-Butyrate	23—26	17—30	26—30
Cellulose butyrate, 0.5 sec	23—26	17—30	26—30
Cyanoethyl cellulose	23—26	25—30	—
Ethyl cellulose, K-200	—	17—22	19—23
Ethyl cellulose, N-22	17—23	15—22	19—30
Ethyl cellulose, T-10	17—19	16—20	19—23
Ethyl hydroxyethyl cellulose	17—21	16—21	—
Nitrocellulose (cellulose nitrate), RS, 25 cps	23—26	16—30	30
Nitrocellulose (cellulose nitrate), SS, 0.5 sec	23—26	16—30	26—30
Epoxy Resins			
DEN 438	18—26	18—30	—
Bisphenol A[b] — Epichlorohydrin condensates (Shell)			
Epon® E-72	17—22	15—20	19—23
Epon 812	18—26	16—30	21—30
Epon 864	19—26	17—30	—
Epon 1001	22—23	17—27	—
Epon 1004	—	17—27	—
Epon 1007	—	17—27	—
Epon 1009	—	17—20	—
Epon 1004 Modified soy ester	17—23	16—20	—
Epon 1004 DHC ester (dehydrated castor oil)	17—23	16—20	—
Eposir® 7120, 100% (SIR)			
In 50% solutions	17—26	15—30	—
In 1% solutions	19—23	17—27	
Eposir 7161, 100%			
In 50% solutions	18—23	17—30	—
In 1% solutions	22	18—27	—
Eposir 7170, 100%			
In 50% solutions	21—23	17—30	—
In 1% solutions	—	17—27	—
Eposir 7180, 100%			
In 50% solutions	22	17—27	—
In 1% solutions	—	18—22	—
Eposir 7190, 100%			
In 50% solutions	—	17—27	—
In 1% solutions	—	18—22	—
Epoester® 7842, 100% (SIR)			
In 50% solutions	17—23	15—22	22
In 1% solutions	17—22	16—20	—
Rubbers			
Alpex® cyclized rubber (Reichhold)	15—22	16	—
Pliolite® NR natural rubber (Goodyear)	17—22	—	—

TABLE 3 (continued)
Hildebrand Parameter Ranges of Polymers in Environments with Poor, Moderate, or Strong Hydrogen-Bonding Ability

Polymer (with manufacturer)	δ/MPa$^{1/2}$ Ranges in environments with hydrogen-bonding capability which is		
	Poor	Moderate	Strong
Hydrocarbon Resins			
Gilsonite Brilliant Black (American Gilsonite)	16—22	16—17	—
Gilsonite Selects	16—19	16—17	19
Hydroxyethylated Nebony® 100 (Neville Chemical)	17—22	19—20	—
Hydroxylated Nebony 100	17—22	19	—
Nebony 100	17—22	16—20	—
Nebony 100/Paraformaldehyde, 90:10	17—22	15—20	—
Nebony 100/Resin, 50:50	16—23	15—20	19—23
Nevchem® 100 (Neville Chemical)	15—22	15—20	—
Nevchem 140	15—23	15—22	—
Nevex® 100 (Neville Chemical)	14—23	15—22	—
Nevillac® 10° (Neville Chemical)	14—26	15—32	19—30
Nevillac Hard	16—26	15—32	19—30
Nevillac Soft	16—26	15—32	19—30
Neville LX685	15—22	19—20	—
Nevindene® R-1 (Neville Chemical)	16—23	15—22	—
Nevindene R-7	16—22	15—22	—
Nevinol® A (Neville Chemical)	14—26	15—24	19—23
Nevpene® (Neville Chemical)	14—22	15—18	—
Panarez® 3-210 (Amoco Chemical)	17—22	—	—
Petrolatum 125 HMP	17—18	—	—
Phenolic Resins			
Amberol® F-7 rosin-modified Phenol-Formaldehyde(Rohm & Haas)	17—22	16—20	19—22
Amberol M-82 resin-modified Phenol-Formaldehyde	15—21	15—20	19—22
Bakelite® BKR2620 Phenolic resin (Union Carbide)	—	17—30	19—30
Bakelite CKR-2400 Phenolic resin	18—24	16—27	19—30
Bakelite CKR-5254 *p*-Phenylphenolic resin	17—21	16—27	19—22
Bakelite CKR-5360 Phenolic resin	17—23	16—27	19—23
Durez® 220 Terpene-Phenol (Hooker Chemical)	17—22	16—20	19—23
Durez 550 Phenolic resin	14—24	15—20	19—30
Methylon® 75202 Phenol ether resin (General Electric)	—	18—25	—
Sirfen® S 1248, 49% (SIR)			
In 50% solutions	24—26	20—30	22—30
In 1% solutions	—	20—30	22—30
Sirfen S 1828, 60%			
In 50% solutions	18—26	15—30	22—30
In 1% solutions	24	17—30	19—30
Sirfen S 4551, 59%			
In 50% solutions	18—26	15—30	19—30
In 1% solutions	—	16—30	19—26
Sirfen F 2620, 41%			
In 50% solutions	—	22	30—48
In 1% solutions	—	22	30—48
Sirfen 0 1100, 100%			
In 50% solutions	23—26	15—30	19—30
In 1% solutions	—	16—30	19—30

TABLE 3 (continued)
Hildebrand Parameter Ranges of Polymers in Environments with Poor, Moderate, or Strong Hydrogen-Bonding Ability

Polymer (with manufacturer)	δ/MPa$^{1/2}$ Ranges in environments with hydrogen-bonding capability which is: Poor	Moderate	Strong
Sirfenol® 5112, 100% (SIR)			
In 50% solutions	15—22	15—20	19—23
In 1% solutions	16—22	15—20	19—22
Sirfenol 5470, 100%			
In 50% solutions	17—22	18—20	21—22
In 1% solutions	17—21	18—20	21—22
Polyesters			
Acid diethylene glycol maleate-phthalate	23—26	15—30	19—30
Carboxyl-terminated diethylene glycol phthalate-isophthalate	23—26	18—30	—
Cryplex resin 1473-5	18—23	17—27	—
Diethylene glycol isophthalate	22—26	25—30	—
Diethylene glycol phthalate	23—26	20—30	—
Dipropylene glycol phthalate	19—26	17—30	—
Dow X-2635 adipate-terephthalate	19—26	17—27	—
Dow X-2635 maleate	—	18—20	—
Vitel® PE 100-X linear polyester resin (Goodyear)	23	20	—
Vitel PE 101-X	19—22	20—22	—
Hydrogenated bisphenol A[b] fumarate — isophthalate	—	16—20	—
Hydrogenated bisphenol A[b] — 1,2-propane glycol fumarate-isophthalate	18—23	17—20	—
Pentaerythritol benzoate-maleate glycol	23	17—30	—
Soluble Mylar® 49000 poly(ethylene glycol terephthalate) (du Pont)	22—23	22—23	—
Soluble Mylar 49001 poly(ethylene glycol terephthalate) (du Pont)	18—22	19—20	—
Soluble Mylar 49002 poly(ethylene glycol terephthalate) (du Pont)	19—22	19—20	—
Triethylene glycol-ethylene glycol maleate-terephthalate	22—24	17—29	—
Triethylene glycol maleate	23—26	20—27	—
Varez® 123	23—24	17—27	19—26
Siral® 5849, 100% (SIR)			
In 50% solutions	17—22	15—20	19—22
In 1% solutions	17—22	16—20	19
Polyamides			
Nylon 8, methylolpolyamide (du Pont)	—	—	24—30
Versalon® 1112 dimer acid-polyamine condensate (General Mills)	—	—	19—23
Versalon 1175	—	—	19—23
Versamid® 100 dimer acid-polyamine condensate (General Mills)	17—22	17—18	19—23
Versamid 115	17—22	16—20	19—26
Versamid 900	—	—	—
Versamid 930	—	—	19—23
Versamid 940	—	—	19—23
Rosin Derivatives			
Abalyn®	14—23	15—22	21—26
Abitol®	14—22	15—22	21—30
Alkydol® 160 ester gum (Reichhold)	19	17—22	19—26

TABLE 3 (continued)
Hildebrand Parameter Ranges of Polymers in Environments with Poor, Moderate, or Strong Hydrogen-Bonding Ability

Polymer (with manufacturer)	δ/MPa$^{1/2}$ Ranges in environments with hydrogen-bonding capability which is		
	Poor	Moderate	Strong
Amberol® F-7 rosin-modified Phenol-Formaldehyde (Rohm & Haas)	17—22	16—20	19—22
Amberol 750 maleic rosin	—	18—22	19—26
Amberol 801 maleic rosin	17—23	15—20	—
Arochem® 455 (U.S. Industrial Chemicals)	—	16—27	19—30
Cellolyn® 21	14—23	15—20	21—24
Cellolyn 95-80T	14—23	15—20	21—24
Cellolyn 102	16—21	17—22	21—24
Cellolyn 104	16—21	16—22	21—23
Cellolyn 502-60X	17—23	15—20	21—24
Cellolyn 582-60X	16—23	15—20	21—24
Dymerex® polymerized rosin (Hercules)	15—22	16—20	19—23
Ester Gum	14—22	15—22	19—22
17% Fumarated rosin	17—22	15—27	19—30
22% Fumarated rosin	18—22	15—27	19—30
Glidden B952	19—22	15—22	19—26
Glidden VBR757	—	17—22	19—30
Hercolyn® D (Hercules)	14—24	15—22	21—26
Lewisol® 28	16—21	15—20	21—22
Neolyn® 23 rosin-derived alkyd (Hercules)	17—23	17—27	—
Neolyn 40	18—26	17—27	21—26
Newport V-40 α-pinene resin (Tenneco)	17—23	15—25	19—30
Pentalyn® A pentaerythritol ester of rosin (Hercules)	17—22	15—20	19—23
Pentalyn C	15—22	15—20	21—23
Pentalyn G maleic-modified pentaerythritol ester of rosin	17—22	16—20	19—22
Pentalyn H	14—22	15—20	21—23
Pentalyn K pentaerythritol ester of dimerized rosin	17—22	16—20	19—23
Pentalyn X	16—21	15—20	21
Pentalyn B25	17—23	15—20	21—24
Pentalyn 255	18—21	15—22	21—30
Penatlyn 802A	16—21	15—20	—
Pentalyn 830	17—19	16—22	19—23
Pentalyn 833	17—22	15—20	22
Pentalyn 856	17—23	15—22	19—23
Pentalyn 954	16—22	15—20	—
Petrex® 7-75T	18—26	17—27	23—30
Petrex 130H	16—23	15—16	21—24
Poly-pale® Ester 1	14—22	15—20	21—24
Poly-pale Ester 10	14—23	15—20	21—24
Staybelite® Ester 5	14—22	15—20	21—24
Staybelite Ester 10	14—22	15—20	21—24
Vinsol® (Hercules)	22—24	16—27	19—26
WW gum rosin	17—23	15—22	19—23
Wood rosin M grade	15—22	15—22	19—30
Styrene Polymers and Copolymers			
Amoco® 18-290	17—23	17—20	—
Bakelite® RMD 4511 Styrene-Acrylonitrile (Union Carbide)	22—23	19	—

TABLE 3 (continued)
Hildebrand Parameter Ranges of Polymers in Environments with Poor, Moderate, or Strong Hydrogen-Bonding Ability

	$\delta/MPa^{1/2}$		
	Ranges in environments with hydrogen-bonding capability which is		
Polymer (with manufacturer)	Poor	Moderate	Strong
Buton® 100 Butadiene-Styrene (Enjay)	14—22	15—20	—
Buton 300	17—22	15—20	19—22
Koppers KTPL-A	16—22	17—20	—
Lustrex® High Test 88 rubber-modified polystyrene (Monsanto)	17—22	19	—
Lytron® 810 Styrene-Maleic anhydride (Monsanto)	24	20—30	—
Lytron 820	19	18—30	22—30
Marbon 9200 styrene copolymer	17—22	19—20	—
Parapol® S-50	19	16	—
Parapol S-60	19	16	—
Piccoflex® 120 Styrene-Acrylonitrile-Indene terpolymer (Pennsylvania Industrial)	17—23	16—20	—
Shell Polyaldehyde Resin EX-39	18—23	19—27	—
Shell Polyaldehyde Resin EX40	18—23	19—27	—
Shell X-450	19—22	17—25	19—26
SMA 1430A	24	17—27	19—30
Styrene/Acrylamide, 85:15	—	—	—
Styrene/Acrylonitrile, 85:15	18—23	19—25	—
Styrene/Butenol, 85:15	17—23	16—20	—
Styrene/Cyclol,[a] 82:18	19	17—22	—
Styrene/2-Ethylhexyl acrylate/Acrylic acid, 81:11:8	18—22	16—20	—
Styrene/Methacrylic acid, 90:10	19—22	17—22	—
Styrene/Methyl acrylate, 85:15	17—23	16—22	—
Styrene/Methyl half ester of maleic, 60:40	—	17—27	22—23
Styrene/Propyl half ester of maleic, 57:43	18—24	17—27	19—23
Styrene/Vinyl butyl ether	17—23	16—20	—
Styron® 440M-27 modified polystyrene (Dow)	17—22	19	—
Styron 475M-27	17—22	19	—
Styron 480-27	19—22	19	—
Vinyl Resins			
Acryloid® K 120N	17—26	17—27	—
DQDA 6225	17—19	—	—
DQDA 3457	17—19	—	—
Elvax® 150 Vinyl acetate/Ethylene copolymer (du Pont)	16—22	—	—
Elvax 250	17—19	—	—
Elvax EOD 3206-1	16—23	16—17	—
Exon® 470 poly(vinyl chloride) (Firestone)	17—23	16—20	—
Exon 471	17—23	16—25	—
Exon 473	17—23	16—20	—
Geon® 121 (Goodrich)	22—23	19—20	—
Polycyclol[a]	—	18—22	19—30
Poly(vinyl butyl ether)	16—22	15—20	19—23
Poly(vinyl ethyl ether)	14—23	15—22	19—30
Formvar® 7/70E poly(vinyl formal) (Shawnigan)	—	20—27	—
Formvar 15/95E	—	20—27	—
Poly(vinyl isobutyl ether)	14—22	15—20	19—23
Saran® F-120 Vinylidene chloride-Acrylonitrile (Dow)	19—23	25—30	—

TABLE 3 (continued)
Hildebrand Parameter Ranges of Polymers in Environments with Poor, Moderate, or Strong Hydrogen-Bonding Ability

Polymer (with manufacturer)	δ/MPa$^{1/2}$ Ranges in environments with hydrogen-bonding capability which is: Poor	Moderate	Strong
Saran F-220	19—23	22—30	—
Shawnigan RS3512, 30% hydrolyzed poly(vinyl acetate)	—	—	—
Shawnigan RS3648, 30% hydrolyzed poly(vinyl acetate)	—	—	—
Sinclair 3840A	18—19	16—20	21—24
Tedlar®	—	—	—
Vinyl acetate(VA)/Ethylhexyl acrylate (EHA)/Methacrylic acid(MAA), 63:33:4	—	17—20	—
VA/EHA/Cyclol[a]/MAA, 76:12:8:4	—	19—27	—
VA/Ethyl acrylate/Cyclol,[a] 70:20:10	18—26	16—30	19—30
Vinyl butyl ether(VBE)/Acrylonitrile/Methylol acrylamide, 46:27:27	24	18—22	—
VBE/Maleic anhydride/Methylol acrylamide, 46:27:27	23—26	18—28	22—30
Vinylidene chloride/Acrylic acid, 75:25	—	17—25	19—30
Vinylite® AYAA poly(vinyl acetate) (Union Carbide)	18—26	17—30	30
Vinylite VAGH, partially hydrolyzed Vinyl chloride-Vinyl acetate copolymer	22—23	16—20	—
Vinylite VMCH Vinyl chloride-Vinyl acetate-Maleic acid terpolymer	22—23	16—25	—
Vinylite VXCC Vinyl chloride copolymer	19—23	16—27	—
Vinylite VYHH Vinyl chloride-Vinyl acetate copolymer	19—23	16—27	—
Vinylite VYLF Vinyl chloride-Vinyl acetate copolymer	19—23	16—27	—
Vinylite XYHL poly(vinyl butyral)	—	18—22	19—30
Vinylite XYSG poly(vinyl butyral)	—	18—22	19—30
Vyset® 69	18—19	17—20	19
Miscellaneous			
Bakelite® sulfone resin P-47	2	20	—
Beckolin® #27 modified oil (Reichhold)	14—22	15—20	19—23
Carbowax® 4000 poly(oxyethylene glycol) (Union Carbide)	18—26	15—20	19—23
Chlorinated rubber	17—22	16—22	—
Conoco H-35 hydrocarbon mixture	14—23	15—20	19—23
Dammar gum, dewaxed	17—22	16—20	19—22
Epocryl® E-11	26	20	22
Estane® X-7	—	20	—
Hexadecyl monoester of trimellitic anhydride	—	18—27	19—30
Hydrogenated sperm oil WX135	18—22	—	—
Hypalon® 20 chlorosulfonated polyethylene (du Pont)	17—20	17—18	—
Hypalon 30	17—22	16—17	—
Ketone resin S588	22—26	17—27	19—30
Lexan® 100 Phosgene — Bisphenol A condensate polycarbonate (General Electric)	19—22	19—20	—
Lexan 105	19—22	19—20	—
Modaflow®	15—23	15—20	19—23
Petrex® acid-dibasic terpene	19—23	17—25	—
Resin #510, fossil coal resin	15—22	15—19	19
Santolite® MHP Arylsulfonamide — Formaldehyde condensate (Monsanto)	22—26	16—30	19
Shellac, pale-pale	—	20—22	19—30
Silicone DC-23 (Dow Corning)	15—17	15—16	19—21

TABLE 3 (continued)
Hildebrand Parameter Ranges of Polymers in Environments with Poor, Moderate, or Strong Hydrogen-Bonding Ability

Polymer (with manufacturer)	$\delta/MPa^{1/2}$ Ranges in environments with hydrogen-bonding capability which is: Poor	Moderate	Strong
Silicone DC-1107	14—19	19—22	19—23
Sylkyd® 50 methoxylated partial hydrolyzate of phenyl- and phenylmethyl-silanes (Dow Corning)	14—26	16—26	19—30
Silicone intermediate Z6018, monophenyl polysiloxanol (Dow Corning)	17—23	16—25	21—23
Soy oil	14—23	15—22	19—24
Soy oil, blown	14—26	16—26	19—30
p-Toluene sulfonamide-Formaldehyde	24	20—30	26—30

Note: Values were determined by Burrell[5,8] as described in Section 14.2, with the exception of SIR (Societa Italiana Resine) resins which are from Dernini and Vargiu.[9] Burrell's liquid values appear in Table 1, Chapter 8.

[a] Cyclol® is bicyclo[2.2.1]hept-5-ene-2methanol.
[b] Bisphenol A is 4,4′-(1-methylethylidene)bisphenol.

There are many examples* of the calculation of Hildebrand parameters from polymer-liquid interaction parameters, particularly δ_∞ from χ_∞ data originating from gas-liquid chromatography and the application of the Bristow-Watson equation. Alternatively,[28,439] if zero volume change on mixing is assumed, the partial molar Gibbs free energy of mixing (obtained from the gas chromatography weight fraction activity coefficient, $RT \ln {}^i\Omega_\infty$) equals ${}^iV({}^i\delta - {}^j\delta)^2$, so when the free energy change is zero ${}^i\delta = {}^j\delta$, and also when $[{}^i\delta^2 - (RT/{}^iV)\ln {}^i\Omega_\infty]$ is plotted as a function of ${}^i\delta$, the slope is $2\,{}^j\delta$.

14.3 SOLUBILITY/SWELLING SPECTRUM METHODS

Polymer Hildebrand parameter ranges may be determined experimentally by observation of their dissolution behavior, degree of swelling, or other polymer property in a ''spectrum'' of liquids with a range of Hildebrand parameters.

In one approach the *solubilities* of polymers in liquids are used. The Hildebrand parameter ranges listed in Table 2 were determined by Burrell[5,8] with commercial grades of liquids and polymers for paint laboratory use. The classification into strong, moderate, or poor hydrogen-bonding capability was employed, with the liquids listed in Table 1, Chapter 4, and this simple procedure is still widely used, although the hydrogen-bonding parameter and dipole moment are incorporated into the ASTM test method (Table 9, Chapter 8). The selected liquid is added to a 1 or 2 g sample of solid polymer in a test-tube such that the final solution has about the correct solids content for the expected use (for example, 50%

* See, for example: Belousov et al.,[24] Castells and Mazza;[25] Cervenka et al.;[26] Chee;[27] DiPaola-Baranyi et al.;[28-30] Dudek and Bueche;[31,32] Eguiazabal, Fernández-Berridi, and co-workers;[33-35] Fernández-Sanchez et al.;[439] Freytag and Radusch;[36] Galin;[37] Gräter et al.;[38] Green;[39] Gregory and Huglin;[40] Guillet and co-workers;[41-45] Hippe et al.;[46] Katz and Salee;[47] Kok and Rudin;[48,49] Lawandy and Darwish;[50] Lazcano, Sánchez, Bello, Marco and Fatou;[420,421] Leca et al.;[51,52,439] Mangaraj;[53] Merk, Lichtenthaler and Prausnitz;[54] Narender and Yaseen;[444] Price;[55] RamiReddy and Kalpagam;[56] Roe;[57] Sanetra et al.;[58] Siow et al.;[59] Takahashi;[60] Wu;[61] Yamashita et al.;[62] and Yeo.[63]

TABLE 4
Values of Polymer Hildebrand Parameters from Swelling Measurements and the Bristow-Watson Equation

Polymer[a]	$^{j}\delta$/MPa$^{1/2}$ from Equation 1 Slope	Intercept	$^{j}\delta$/MPa$^{1/2}$ from Ref. 15	$^{j}\delta$/MPa$^{1/2}$ from Ref. 16
Natural rubber	16.6	16.5—16.6	16.2	17.1
Butyl 400	16.0	16.1—15.9		16.5
Neoprene®	17.6	17.7—17.5	16.7	18.9
Butadiene-styrene, 96/4	16.6	16.6—16.5		
Butadiene-styrene, 87.5/12.5	16.6	16.6—16.4	16.6	17.6
Butadiene-styrene, 71.5/28.5	16.6	16.7—16.6		17.7(60/40)
Butadiene-acrylonitrile, 82/18	17.8	17.9—17.7	19.2	19.4(75/25)
Butadiene-acrylonitrile, 70/30	20.3	20.3—20.1		
Butadiene-acrylonitrile, 61/39	21.2	21.4—21.3		

[a] Processed and vulcanized as described in the source.

Adapted from Bristow, G. M. and Watson, W. F., *Trans. Faraday Soc.*, 54, 1731, 1958.

for alkyds, 20% for vinyl polymers). The mixture may be warmed and stirred to increase the rate of dissolution, but it is cooled and observed at room temperature, the polymer being judged insoluble if there are gel particles or cloudiness present. By successive choices, upper and lower pairs of two liquids are found, one of which dissolves the polymer and one of which does not, thus defining the solubility range in each hydrogen-bonding class. A multiple regression method of handling the results has been presented,[64] and although this yields self-consistent Hildebrand parameter values, there are anomalies in the calculated liquid molar volumes as a result of the limitations in the Hildebrand parameter concept when used to describe systems that involve polar and specific interactions. The polymer Hildebrand parameter ranges of Table 3 are consistent with the liquid Hildebrand parameters of Table 1, Chapter 8. The use of mixed liquids allows small δ increments (typically 0.5 MPa$^{1/2}$) to be made: "cohesion parameter spectroscopy" for polymer characterization.[65]

Cohesion parameters also may be assessed by comparing the temperatures at which dilute polymer suspensions (often 1% or 0.1% by volume) become homogeneous, and the dissolution rates of polymers have been used for the same purpose.[66] Indirect indications of polymer-liquid interactions also are possible, such as plasticization[433] or the peel strength adhesion of polymer films cast from various solvents;[67] these too may exhibit curves with maxima when plotted against liquid Hildebrand parameter.

Notable determinations of cohesion parameters by solubility methods have been made for:

- Polystyrene[68-71]
- Poly(vinylidene chloride)[72]
- Ethylene-acrylic acid copolymers[73]
- Styrene-acrylonitrile copolymers[74]
- Adamantane derivatives[75]
- Shellac[76,77]
- Cellulose[78]
- Poly(vinylpyrrolidone) and poly(vinyl chloride) by partial specific volume[79]
- Poly(ethylene terephthalate)[80,433]

Closely related to the solubility of a polymer in a liquid is the corresponding solubility

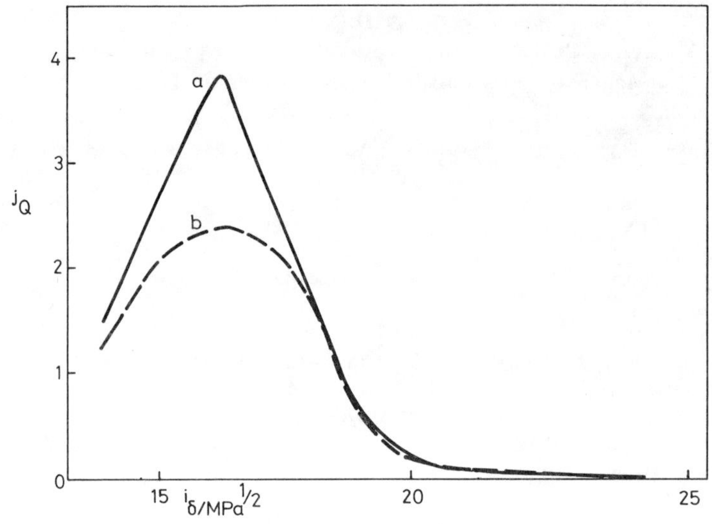

FIGURE 1. Swelling of natural rubber as a function of liquid Hildebrand parameter: a, pure gum; b, tire tread. (Adapted from Mark, H. F. and Tobolsky, A. V., *Physical Chemistry of High Polymeric Systems,* Interscience, New York, 1950, chap. 8.)

of the liquid in the polymer, with resulting *polymer swelling*.[81] A method that may be applied to cross-linked polymers (to polymers that can be cross-linked before testing, for example, by irradiation) and semicrystalline material such as poly(vinyl chloride), assumes that the degree of interaction with a liquid and swelling of a polymer in the liquid is a maximum when the cohesion parameters match. A slightly cross-linked polymer is exposed to a series of liquids (as either liquids or vapors). The polymer is swollen (but does not dissolve because of the cross-links) and the extent of swelling can be plotted against the Hildebrand parameters of the liquids. Figure 1 is a schematic diagram of swelling (in terms of specific volume of liquid imbibed) as a function of liquid Hildebrand parameter. In practice, variation in polar and hydrogen-bonding properties of both polymer and liquids causes considerable scatter in many of these plots, which have been reported for a large number of polymers. Examples (covering both Hildebrand parameter determinations and correlations in the behaviors of polymers of known Hildebrand parameter) are:

- Polyacrylamide terpolymer gels[437]
- Polyacrylates[82,83]
- Polyacrylonitrile[84]
- Polyamide[85]
- Bisphenol A polycarbonate[86] and dicyanobisphenol polycarbonate[87]
- Polybutadiene[88,89]
- Cardopolyheteroarylenes[90]
- Cellulose acetate[91]
- Polychloroprene[50,440]
- Epoxy resins[92-94]
- Oligoethylene glycols[47]
- Polyesters[95]
- Polyester urethanes (Estane®)[96]
- Polyethers, including poly(isobutylene oxide),[97] poly(ethylene oxide),[98,99] poly(propylene oxide),[100] and poly(2,6-dimethyl-1,4-phenylene oxide)[101,102]

- Ethylene-propylene copolymers and terpolymers[103-105]
- Poly(ethylene terephthalate)[106-110]
- Poly(ethylene maleate) and polyfluoropolyol[446,447]
- Indanic polymers[111]
- Ion exchange polymers[112,113] and perfluorinated ionomers including perfluorosulfonic acid (Nafion®)[114-119]
- Polyisobutylene[120]
- Linseed oil films[121]
- Polymethacrylates[83,122-124]
- Polynorbornene[125]
- Polypropylene[22,126-130]
- Polypropylene sulfide[131]
- Shellac[76,132]
- Silicones[53,133-136]
- Silk fibroin[137-139]
- Polystyrene and styrene-divinylbenzene copolymers[140-146]
- Styrene-acrylonitrile copolymers[87,147,148]
- Polysulfone[149]
- Poly(tetrafluoroethylene)[150-152]
- Tetrahydrofuran copolymer[153]
- Polyurethane[53,154]
- Poly(vinyl acetate)[155]
- Poly(vinyl fluoride) and poly(vinylidene fluoride)[156,157]
- Poly(vinyl pyrrolidinone)[158,159]
- Natural rubber[53]
- Peroxide-containing nitrile and butadiene-styrene rubbers[451]
- Coal, shale, and kerogen[160,161]
- Dielectric and insulating properties[162,163]
- Polymer-plasticizer interactions[436]

Despite scatter, the general trend is usually clear and, if required, an approximate single value or a range of Hildebrand parameter values can be obtained from the position of the maximum.

Cohesion parameters can be calculated in other ways from swelling measurements. Thus, Eskin and co-workers[164] have used the fact that the relative differences between the partial specific volumes of a swollen polymer in a given liquid and in one for which ${}^i\delta = {}^j\delta$ are a linear function of $({}^i\delta - {}^j\delta)^2$, deducing information on the thermodynamics of mixing.

Instead of relying on a simple plot of swelling against Hildebrand parameter, some investigators have employed polymer-liquid interaction parameter (χ) relationships. Using Flory's equation linking equilibrium swelling and χ interaction parameter, Gee[15,53,165] showed that the swelling coefficient Q (the volume of liquid imbibed per unit volume of polymer) when plotted against ${}^iV\ ({}^i\delta - {}^j\delta)^2$ yields a Gaussian curve:

$$Q = Q_{max} \exp[-K\ {}^iV({}^i\delta - {}^j\delta)^2] \tag{3}$$

and a plot of $({}^iV^{-1} \ln Q_{max}/Q)^{1/2}$ against ${}^i\delta$ gives a straight line with an intercept equal to ${}^j\delta$. Figure 2 is an example of a Gee plot, for the sorption of liquids by poly(tetrafluoroethylene),[150] where Q is the swelling coefficient expressed in volume of liquid per unit mass of polymer. Gee plots have been used for:

- Natural rubber[53]

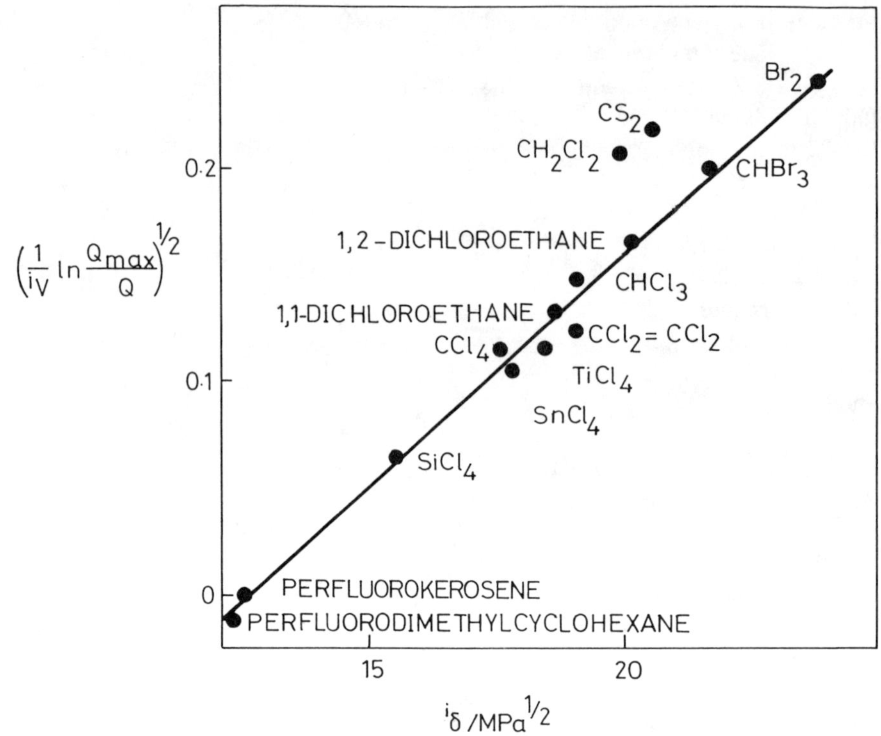

FIGURE 2. Gee plot for the swelling of poly(tetrafluoroethylene) by liquids. (Adapted from Starkweather, H. W., Jr., *Macromolecules,* 10, 1161, 1977.)

- Butyl acrylate-acrylonitrile copolymers[166]
- Butyl and chlorobutyl rubbers[167]
- Poly(dimethylsiloxane)[53]
- Epoxy resins[92]
- Unsaturated polyester resins[60]
- Ethylene-propylene copolymers[103] and terpolymers[457]
- Hydropolybutadiene[53]
- Methacrylate copolymers[168]
- Poly(tetrahydrofuran)[169]
- Trimethylcellulose-poly(tetrahydrofuran) block copolymers[170]
- Polyurethanes[171,172] including polybutadiene urethane and the related polybutadiene ionene[435]

Swelling can be used in conjunction with other physical properties such as stress-strain relationships[169,173] for the evaluation of polymer cohesion parameters. Giuliani[446] showed that it is possible to determine the Hildebrand parameter of a polymer film using an optical waveguide device that probes changes in the dielectric properties of the polymer at the film-vapor interface: a maximum in the solvent spectrum was attributed to reversible variations in surface adhesion to the roughened glass surface caused by softening of the polymer and resulting in changes in the optical contact.

In a reversal of the usual type of polymer-liquid investigation, low cohesive pressure liquids such as iron pentacarbonyl ($\delta \approx 16$ MPa$^{1/2}$) were investigated by their effect on poly(tetrafluoroethylene).[151,152] Extending the swelling method to oligomers (low molecular weight resins) of dimethylvinylethynyl-*p*-hydroxyphenyl methane, Shvarts and co-workers[21] assumed that to determine the Hildebrand parameter ${}^{j}\delta$ of the resin it was sufficient to determine the Hildebrand parameter ${}^{ij}\delta$ of its solution (by the swelling effect it had on rubber) and to calculate ${}^{j}\delta$ from the equation

$$ {}^{ij}\delta^2 = {}^{i}\phi\ {}^{i}\delta^2 + {}^{j}\phi\ {}^{j}\delta^2 \quad (4)$$

which was used in place of Equation 32, Chapter 4, usually employed for mixed liquids:

$$ {}^{ij}\delta = {}^{i}\phi\ {}^{i}\delta + {}^{j}\phi\ {}^{j}\delta \quad (5)$$

In the case of polymers with more than one type of functional group, and particularly for copolymers (Section 16.10) and polymer blends (Section 16.9), more than one maximum may be observed in the Hildebrand parameter swelling spectrum. It is sometimes claimed that three or four distinct peaks can be identified, but the scatter resulting from the imprecision of using a single cohesion parameter for a liquid undergoing many types of interaction means that in these circumstances this approach is of limited value. Polymer blends and copolymer systems that have been studied by solvent spectrum techniques are mainly natural and synthetic rubbers, including styrene-butadiene copolymers.*

Hildebrand parameters for non-polymeric, high molecular weight materials such as mineral oils[183] and plasticizers[184] may be determined by the same method. Alternative solvent spectrum techniques for polymers and non-polymers use a variety of other methods, including solution viscosity (Section 14.7) and surface free energy[185] (Section 17.2). A variation on the procedure of swelling in pure liquids is the use of liquid mixtures, with effective Hildebrand parameters taken as the volume fraction means of the individual liquid values (Equation 5).[96,109,110,154,166,178,186-188] The maxima fall in the same range of Hildebrand parameter values as those estimated by other methods, but the effects of specific interactions can be significant. Additional information on mixed liquid systems appears in Section 4.4, 8.5, and 13.6.

14.4 TURBIDIMETRIC TITRATIONS

Taken further, the mixed liquid approach of the last section becomes the method of dilution ratios or turbidimetric titrations[189,190] that can be used to construct solubility maps in terms of cohesion parameters (Chapter 15). The polymer or resin is dissolved in a "true solvent" at a given weight concentration, and then titrated with a diluent or nonsolvent until turbidity just appears. Usually this cloud point is sharply defined, and if the titration and calculation of effective Hildebrand parameter is carried out for a variety of solvents and diluents, the polymer Hildebrand parameter can be determined or the solubility map drawn up. The procedure is to first determine which liquids are definitely solvents or nonsolvents at the particular concentration, and to note any borderline solvents. This gives an indication of the range of solvency so that suitable solvent-diluent pairs can be selected. In the "Suh and Clarke method," solutions of the polymer are titrated in turn with two nonsolvents, one with a Hildebrand parameter value lower (${}^{l}\delta$) and one higher (${}^{h}\delta$) than the estimated polymer value (${}^{j}\delta$) and solvent value (${}^{i}\delta$).[69,191] Equation 41, Chapter 13, is

$$\chi_H = \chi - \chi_S = {}^{i}V({}^{i}\delta - {}^{j}\delta)^2/(RT) \quad (6)$$

* Reports on these systems, in addition to those already cited, include: Baranwal;[174] Basin and Shvarts;[175] Bhatnagar and Banerjee;[176] Bristow and Watson;[13,177,178] Dudek and Bueche;[14] Gabris;[179] Gee;[15,165,180] Grishin et al.;[89] Samuel and Wilkes;[181] Schneier;[23] Scott and Magat;[16] and Treloar.[182]

For conditions at the cloud point,

$$^{ilj}\chi = {}^{ilj}\chi_S + ({}^{il}V/RT)({}^{j}\delta_{ap} - {}^{il}\delta)^2 \tag{7}$$

and

$$^{ihj}\chi = {}^{ihj}\chi_S + ({}^{ih}V/RT)({}^{ih}\delta - {}^{j}\delta_{ap})^2 \tag{8}$$

where ${}^{j}\delta_{ap}$ is the apparent polymer Hildebrand parameter. Equation these expressions, assuming ${}^{ilj}\chi_S = {}^{ihj}\chi_S$ and using the equations, ${}^{il}\delta = {}^{i}\phi^{i}\delta + {}^{l}\phi^{l}\delta$, ${}^{ih}\delta = {}^{i}\phi^{i}\delta + {}^{h}\phi^{h}\delta$, it follows that:

$$^{j}\delta_{ap} = ({}^{il}V^{1/2}\,{}^{il}\delta + {}^{ih}V^{1/2}\,{}^{ih}\delta)/({}^{il}V^{1/2} + {}^{ih}V^{1/2}) \tag{9}$$

When the nonsolvent content becomes zero, the equation reduces to

$$^{j}\delta = {}^{i}\delta \tag{10}$$

and the intersection of the line representing Equation 9 with that of Equation 10 (which is a 45° line) provides ${}^{j}\delta$. The procedure may be extended to the determination of Hansen parameters.[192] Methods of this type have been applied to:

- Copolymers of acrylonitrile with styrene[192] and with vinyl acetate[193]
- Polyamides[186] and *N*-substituted polymaleimides[194]
- Polybutadiene[88]
- Chlorinated poly(vinyl chloride)[195,196]
- Epoxy resins[197]
- Ethylene copolymers[198]
- Poly(ferrocenylene-methylmethylene)[199]
- Poly(methyl methacrylate)[197,200]
- Poly(oxymethylene)s and related polymers[201-203]
- Polystyrene, poly(α-methylstyrene), poly(2,4-dimethylstyrene), poly(*o*-chlorostyrene),[69,204,205] poly(4-acetoxystyrene),[445] and poly(4-hydroxystyrene)[445]

14.5 HANSEN PARAMETERS

Hansen[206-227] combined an elaborate solubility/swelling spectrum procedure with other methods of evaluation in deriving the three-component cohesion parameters that were introduced in Section 5.9. Solubility data were obtained by visual inspection of polymer mixtures at concentrations of about 0.5 g of polymer per 5 cm^3 of liquid (10% w/v) and classification as

- Soluble
- Almost soluble
- Strongly swollen, slightly soluble
- Swollen
- Little swelling
- No visible effect

Viscous liquids in which the polymer did not appear to dissolve at room temperature were heated, the degree of solubility being assessed after the mixture had cooled to room temperature. Each solute, j, was characterized by a set of Hansen parameters (defining the center of the solubility sphere, coordinates which could be used in Equation 55, Chapter 5) and an interaction radius ${}^{j}R$. For solubility to be predicted, the distance ${}^{ij}R$ of the liquid i at position (${}^{i}\delta_d$, ${}^{i}\delta_p$, ${}^{i}\delta_h$) from the center of the sphere of solubility (${}^{j}\delta_d$, ${}^{j}\delta_p$, ${}^{j}\delta_h$),

$$ {}^{ij}R = [4({}^{i}\delta_d - {}^{j}\delta_d)^2 + ({}^{i}\delta_p - {}^{j}\delta_p)^2 + ({}^{i}\delta_h - {}^{j}\delta_h)^2]^{1/2} \quad (11) $$

must be less than the interaction radius ${}^{j}R$. (The doubling of the δ_d scale, leading to the factor '4' in the first term, was intended to make the "volume of solubility" approximately spherical, although this may not be necessary, as discussed in Section 5.9 and also below.) In a recent publication,[431] Hansen defined the *relative energy difference* as the ratio ${}^{ij}R/{}^{j}R$ (Section 5.9) and also reviewed the application of Hansen parameters to biological polymers (Table 12a, Chapter 5 and Section 16.11). Hansen's data for polymers and resins appear in Table 5; these are 1971 values[213] obtained by computer fitting and differing slightly from the manually calculated results[207,208] of 1967.

An example[228,229] will clarify the procedure for using Hansen parameters to predict polymer solubility. In order to determine if polystyrene is expected to dissolve in a mixture of 60:40 v/v methyl ethyl ketone/hexane, the Hansen parameters are obtained from Table 11, Chapter 5:

$$ \text{Methyl ethyl ketone: } {}^{i}\delta_d = 16.0,\ {}^{i}\delta_p = 9.0,\ {}^{i}\delta_h = 5.1 \text{ MPa}^{1/2} $$

$$ \text{Hexane: } {}^{j}\delta_d = 14.9,\ {}^{j}\delta_p = 0.0,\ {}^{j}\delta_h = 0.0 \text{ MPa}^{1/2} $$

The Hansen parameters are combined on a 60/40 volume fraction basis:

$$ {}^{ij}\delta_d/\text{MPa}^{1/2} = 0.6 \times 16.0 + 0.4 \times 14.9 = 15.6 $$

$$ {}^{ij}\delta_p/\text{MPa}^{1/2} = 0.6 \times 9.0 + 0.4 \times 0.0 = 5.4 $$

$$ {}^{ij}\delta_h/\text{MPa}^{1/2} = 0.6 \times 5.1 + 0.4 \times 0.0 = 3.1 $$

The center-point coordinates and radius of the solubility sphere for polystyrene are (Table 5):

$$ {}^{k}\delta_d = 21.3,\ {}^{k}\delta_p = 5.8,\ {}^{k}\delta_h = 4.3,\ {}^{k}R = 12.7 \text{ MPa}^{1/2} $$

The distance ${}^{ijk}R$ of the point representing the liquid mixture from the center of the polymer solubility sphere (Equation 11) is

$$ [4(21.3 - 15.6)^2 + (5.8 - 5.4)^2 + (4.3 - 3.1)^2]^{1/2} = 11.5 \text{ MPa}^{1/2} $$

As this is less than the radius of the polystyrene solubility sphere, ${}^{k}R = 12.7$ MPa$^{1/2}$, polystyrene is expected to be soluble in a 60/40 methyl ethyl ketone/hexane mixture. Experimentally, this is found to be the case for a 10% w/w solution.

Nunn[230] has provided an example of partial solubility to illustrate Hansen's data. The epoxy resin Epikote® 1001 shows limited solubility in many common liquids, so there is no clear distinction between solvent and nonsolvent in many cases. Table 6 compares the distance of the liquid Hansen parameter coordinates from the center of the resin solubility sphere with the observed miscibility limits. This shows that although Hansen parameters provide useful information, the surfaces of the solubility spheres are not clear-cut boundaries between solvents and nonsolvents. Particular caution should be exercised when there are

TABLE 5
Hansen Parameter and Interaction Radius Values of Some Polymers and Resins (1971 Data) for use with the Liquid Data of Table 11, Chapter 5

Polymer (trade name, supplier)	δ/MPa$^{1/2}$			
	$^j\delta_d$	$^j\delta_p$	$^j\delta_h$	^jR/MPa$^{1/2}$
Acrylonitrile-butadiene elastomer (Hycar® 1052, B. F. Goodrich)	18.6	8.8	4.2	9.6
Alcohol soluble resin (Pentalyn® 255, Hercules)	17.6	9.4	14.3	10.6
Alcohol soluble rosin resin (Pentalyn 830, Hercules)	20.0	5.8	10.9	11.7
Alkyd, long oil (66% oil length, Plexal® P65, Polyplex)	20.4	3.4	4.6	13.7
Alkyd, short oil (coconut oil 34% phthalic anhydride; Plexal C34, Polyplex)	18.5	9.2	4.9	10.6
Blocked isocyanate (phenol) (Suprasec® F5100, ICI)	20.2	13.2	13.1	11.7
Cellulose acetate (Cellidra® A, Bayer)	18.6	12.7	11.0	7.6
Chlorinated polypropylene (Parlon® P-10, Hercules)	20.3	6.3	5.4	10.6
Coumarone-indene resin (Piccoumarone® 450-L, Pennsylvania Industrial Chemical)	19.4	5.5	5.8	9.6
Epoxy (Epikote® 1001, Shell)	20.4	12.0	11.5	12.7
Ester gum (Ester gum BL, Hercules)	19.6	4.7	7.8	10.6
Furfuryl alcohol resin (Durez® 14383, Hooker Chemical)	21.2	13.6	12.8	13.7
Hexamethoxymethyl melamine (Cymel® 300, American Cyanamid)	20.4	8.5	10.6	14.7
Isoprene elastomer (Cariflex® IR305, Shell)	16.6	1.4	−0.8	9.6
Cellulose nitrate ($^1/_2$ sec; H-23, Hagedorn)	15.4	14.7	8.8	11.5
Pentaerythritol ester of rosin, modified (Cellolyn® 102, Hercules)	21.7	0.9	8.5	15.8
Petroleum hydrocarbon resin (Piccopale® 110, Pennsylvania Industrial Chemical)	17.6	1.2	3.6	6.6
Phenolic resin (resole, Phenodur® 373U, Chemische Werke Albert)	19.7	11.6	14.6	12.7
Phenolic resin, pure (Super Beckacite® 1001, Reichhold)	23.3	6.6	8.3	19.8
Polyamide, thermoplastic (Versamid® 930, General Mills)	17.4	−1.9	14.9	9.6
cis-Polybutadiene elastomer (Bunahuls CB10, Chemische Werke Huels)	17.5	2.3	3.4	6.5
Poly(isobutylene) (Lutonal® IC/123, BASF)	14.5	2.5	4.7	12.7
Poly(ethyl methacrylate) (Lucite® 2042, Du Pont)	17.6	9.7	4.0	10.6
Poly(methyl methacrylate) (Rohm & Haas)	18.6	10.5	7.5	8.6
Polystyrene (Polystyrene LG, BASF)	21.3	5.8	4.3	12.7
Poly(vinyl acetate) (Mowilith® 50, Hoechst)	20.9	11.3	9.6	13.7
Poly(vinyl butyral) (Butvar® B76, Shawnigan)	18.6	4.4	13.0	10.6
Poly(vinyl chloride) (Vipla® KR, K = 50, Montecatini)	18.2	7.5	8.3	3.5
Saturated polyester (Desmophen® 850, Bayer)	21.5	14.9	12.3	16.8
Styrene-butadiene (SBR) raw elastomer (Polysar® 5630, Polymer Corp.)	17.6	3.4	2.7	6.6
Terpene resin (Piccolyte® S-1000, Pennsylvania Industrial Chemical)	16.5	0.4	2.8	8.6
Urea-formaldehyde resin (Plastopal® H, BASF)	20.8	8.3	15.0	12.7

Adapted from Hansen, C. M., *Skand. Tidskr. Färg Lack,* 17, 69, 1971.

significant donor-acceptor interactions. Recently, the Hansen technique has been applied to the prediction of the barrier properties of protective clothing[430,431] (Section 16.7). One set of estimates of the resulting Hansen parameters is provided in Table 7.

Izumi and Miyake[231] compared methods based on Equations 55 and 57, Chapter 5,

$$^{ij}r^2 = {}^{ij}A = ({}^i\delta_d - {}^j\delta_d)^2 + ({}^i\delta_p - {}^j\delta_p)^2 + ({}^i\delta_h - {}^j\delta_h)^2 \qquad (12)$$

$$^{ij}R^2 = 4({}^i\delta_d - {}^j\delta_d)^2 + ({}^i\delta_p - {}^j\delta_p)^2 + ({}^i\delta_h - {}^j\delta_h)^2 \qquad (13)$$

TABLE 6
Solubility of Epikote® 1001 Epoxy Resin and 1967 Hansen Parameter Data

Solvent	δ/MPa$^{1/2}$				Miscibility limit, wt % Epikote 1001 in solution
	$^i\delta_d$	$^i\delta_p$	$^i\delta_h$	^{ij}R/MPa$^{1/2}$	
Acetone	15.5	10.4	7.0	5.7	18
Oxitol®	16.1	9.2	14.3	6.1	complete
Methyl ethyl ketone	15.9	9.2	5.1	7.2	complete
Oxitol acetate	15.9	4.7	10.6	7.2	complete
Ethyl acetate	15.2	5.3	9.2	7.6	18
Butyl acetate	15.7	3.7	6.3	9.6	20
Methyl isobutyl ketone	15.3	6.1	4.1	9.6	23
Xylene	17.7	1.0	3.1	13.1	65
	$^j\delta_d$	$^j\delta_p$	$^j\delta_h$	^jR	
Epikote 1001 resin	17.4	11.3	11.3	9.6	—

Adapted from Nunn, C. J., *Chim. Peint.*, 34, 215, 1971.

TABLE 7
Hansen Parameters of Protective Suit Materials as Estimated by Bentz and Billing[430,431]

Polymer	δ/MPa$^{1/2}$				
	δ_d[a]	δ_p	δ_h	δ_t	^{ij}R
Butyl rubber	18	0.8	0.8	18	7.6
Chlorobutyl rubber	18	1.8	0.6	18	7.2
Chlorinated polyethylene	18	8.2	4.3	20	9.6
Viton A copolymer	18	11.5	6.8	22	9.0
Viton B terpolymer	18	9.8	9.6	22	7.4

[a] Mid-range, fixed value used.

for one polar polymer, poly(*p*-chlorostyrene), with over 90 liquids. The two equations were found to be equivalent for solubility predictions, almost all solvents lying within the spherical regions corresponding to

$$^jr = 6.3,\ ^j\delta_d = 18.4,\ ^j\delta_p = 6.1,\ ^j\delta_h = 4.5 \text{ MPa}^{1/2}$$

or

$$^jR = 7.2,\ ^j\delta_d = 17.6,\ ^j\delta_p = 6.1,\ ^j\delta_h = 4.1 \text{ MPa}^{1/2}$$

These values are based on Izumi and Miyake's own set of liquid Hansen parameters (see also Figure 76, Chapter 15).

Hansen parameters have been determined and used in a similar way for other polymers, including:

- Polyacrylonitrile[188]
- Polyamides[188,232]
- Polybutadiene[188]

FIGURE 3. Rectangular "box" representing interaction of polymers with liquids in terms of Hansen parameters. (After Beerbower, A. and Dickey, J. R., *Am. Soc. Lubric. Eng. Trans.*, 12, 1, 1969.)

- Cellulose acetate[233] and cellulose nitrate[232]
- Poly(1-chloro-2,3-epoxypropane)[234]
- Epoxy-based simultaneous interpenetrating networks (SINs)[454]
- Amine-crosslinked aromatic epoxy thermosets[429] (see also Table 15a)
- Poly(ethylene oxide-propylene oxide)[51,52,235]
- Poly(ethylene terephthalate)[106,107,188]
- Polymethacrylates[188,239]
- Furfuryl alcohol resin[232]
- Poly(alkyl siloxanes)[236]
- Polystyrene[51,52,70,235] and poly(4-acetoxystyrene) and poly(4-hydroxystyrene)[445]
- Poly(vinyl alcohol)[187,188,232,237]
- Poly(vinylidene fluoride)[238]

Powders and pharmaceutical tablet formulations are considered in Section 17.5.

Instead of assuming spherical Hansen volumes for polymer solubilities, Beerbower and Dickey[240] considered constant percentage swell contours as rectangular "boxes" (Figure 3). Thus, all liquids represented by Hansen coordinates inside the 25% swell box cause the polymer to swell by more than 25%; those outside the box do not. It is possible to draw "nests" of boxes for each polymer, with copolymers having two nests of boxes. The information may be presented diagrammatically by projecting the boxes on the coordinate planes, or numerically by listing the coordinates of the corners and the centroid (Table 8). One great advantage of these standardized boxes, like the "spheres", is that they can be described easily in numerical terms, unlike the irregular shapes of the maps and models in Chapter 15.

In another variation of the Hansen parameter approach to the correlation of polymer solubilities, Teas[241,242] derived fractional Hansen parameters for polymers in the same way

TABLE 8
Hansen Parameter Ranges and Centroids for 25% Swell of Polymers, for use with the Liquid Data of Table 12, Chapter 5 and the Closely Related Data of Table 11, Chapter 5

Polymer	25% Swell range/MPa$^{1/2}$			Centroid/MPa$^{1/2}$		
	${}^j\delta_d$	${}^j\delta_p$	${}^j\delta_h$	${}^j\delta_d$	${}^j\delta_p$	${}^j\delta_h$
Natural rubber, polyisoprene	14—20	0—12	0—11	18.4	2.1	7.2
Styrene-butadiene	13—23	0—12	0—8	17.4	2.9	6.8
Isobutylene-isoprene	14—21	0—6	0—7	16.0	2.3	3.3
Nitrile-butadiene						
Large block	14—22	0—19	0—9	18.0	4.1	4.1
Small block	14—20	4—19	10—18	18.4	5.1	12.7
Chloroprene						
Block	14—22	0—13	0—8	19.4	3.1	2.7
Small block	15—20	5—8	10—15			
Vinyl silicone						
Main block	14—21	0—12	0—4	16.4	3.7	4.5
Second block	14—18	0—12	4—10			
Vinylidene fluoride-hexafluoropropylene copolymer						
Main block	9—19	3—19	4—7	15.3	7.2	5.3
Small block	9—18	8—19	10—15	15.3	10.2	11.3
Polysulfide						
Block	15—22	0—19	0—8	17.4	9.2	5.1
Small block	16—21	5—17	10—15	18.0	6.8	14.3
Fluorosilicone	15—19	1—15	4—8	15.3	7.2	6.1
Ethylene-propylene copolymer	14—19	0—9	0—6	18.0	0.8	2.1
Chlorosulfonyl polyethylene	15—22	0—19	0—10	18.2	3.1	3.1
Acrylate-acrylic acid						
Main block	14—22	0—19	0—8	18.4	3.1	3.5
Small block	14—20	4—11	10—16	18.4	6.1	13.3
Polyether urethane	14—21	0—19	0—23	17.4	5.1	10.2
Ethylene oxide-epichlorohydrin copolymer	14—21	0—18	0—11	18.6	8.2	7.2

Adapted from Beerbower, A. and Dickey, J. R., *Am. Soc. Lubric. Eng. Trans.*, 12, 1, 1969.

as described in Section 5.10 for liquids. The values listed in Table 9 (based on Hansen's 1967 cohesion parameters) may be used for assessing behaviors of liquids, plasticizers, and oils (Table 16, Chapter 8).

Hansen parameters may be represented in three-dimensional space as vectors[243] (see also Section 4.4 and 16.5), and this approach is particularly useful for the prediction of polymer swelling in mixtures[187,188] and for solvent formulation. For a mixture, *m*, of liquids *i* and *j*, the resultant of the three sets of components due to dispersion, polar, and hydrogen-bonding forces is the endpoint of the radius vector when they are combined by means of the equations

$$ {}^m\delta_x = {}^i\phi\, {}^i\delta_x + {}^j\phi\, {}^j\delta_x \quad (x = d,p,h) \tag{14} $$

The composition of the mixture to give the maximum swelling of a polymer, *p*, corresponds to a minimum distance between the coordinates of the polymer (point P in Figure 4) and of the mixture (point M), and this occurs at a mixture such that PM and IJ are perpendicular. Since PM and IJ are orthogonal, the scalar or dot product of the vectors represented by these lines vanish, so simplifying the analysis. The length of the line PM may be calculated readily from the coordinates ${}^m\delta_x$ and ${}^p\delta_x$, this being an inverse measure of the extent of interaction between the liquid mixture and the polymer. (No prediction is provided by this analysis about the proportions of the liquid components actually absorbed by the polymer

TABLE 9
Fractional Hansen Parameters of Some Polymers Based on 1967 Data

Polymer (brand name, supplier)	δ_t/MPa$^{1/2}$	$100f_d$	$100f_p$	$100f_h$
Poly(ethyl methacrylate), (Lucite® 2042, du Pont)	22.1	55	32	13
Poly(methyl methacrylate) (Rohm & Haas)	23.1	50	27	23
Poly(vinyl acetate) (Mowilith® 50, Hoechst)	23.1	51	27	22
Poly(vinyl butyral) (Butvar® B-76, Shawnigan)	22.5	46	24	30
Poly(vinyl chloride) (Vipla® KR, Montecatini)	22.5	54	26	20
Poly(vinyl pyrrolidone)	—	35	25	40
Epoxy (Epikote® 1001, Shell)	23.5	44	28	28
Polystyrene (Polystyrene LG, Badische)	20.1	63	22	15
Chlorinated polypropylene (Parlon® P-10, Hercules)	20.9	62	24	14
Phenolic resin (Super Beckacite® 1001, Reichhold)	23.1	49	22	29
Polyamide (Versamid® 930, General Mills)	20.9	56	5	39
Petroleum hydrocarbon resin (Piccopale® 110, Pennsylvania Industrial Chemical	18.0	72	13	15
Long oil alkyd resin (Plexal® P65, Polyplex)	19.2	61	18	21
Short oil alkyd resin (Plexal C34, Polyplex)	21.5	56	28	16
Cellulose acetate-butyrate	—	53	22	25
Cellulose nitrate ($^1/_2$ sec, H23, Hagedorn)	22.1	38	38	24
Ethyl cellulose	—	47	28	35
Urea formaldehyde resin (Plastopal® H, Badische)	24.8	48	20	32
Rosin ester (Cellolyn® 102, Hercules)	20.7	57	19	24
Hydrocarbon resin (LX-685, Neville)	—	73	14	13
Furfuryl alcohol resin (Durez® 14383)	—	48	30	22
Styrene-butadiene elastomer	—	70	15	15
Gilsonite®	—	68	16	16

Adapted from Gardon, J. L. and Teas, J. P., *Treatise on Coatings, Vol. 2, Characterization of Coatings: Physical Techniques,* Part II, Myers, R. R. and Long, J. S., Eds., Marcel Dekker, New York, 1976, chap. 8.

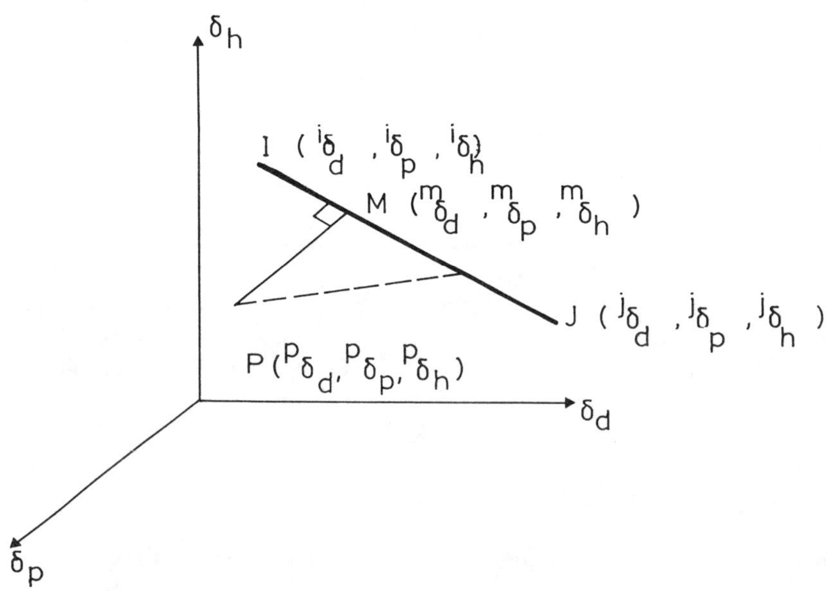

FIGURE 4. Representation in Hansen space of the interaction of a polymer P with a two-component mixture M of liquids I and J. (After Rigbi, Z., *Polymer,* 19, 1229, 1978.)

TABLE 10
Modified Non-Specific Cohesion Parameters and Related Parameters for Three Polymers

Modified parameters	Poly (vinyl acetate)	Poly (methyl methacrylate)	Poly (ethyl methacrylate)
A	0.234	0.247	0.217
$\delta'/\text{MPa}^{1/2}$	19.5	20.3	19.2
$K/\text{cm}^3\ \text{mol}^{-1}$			
$CHCl_3$	40	40	40
ROH	95	50	100
H_2O	1400	750	900
	Hansen parameters, MPa$^{1/2}$		
δ_d	20.9	18.6	17.6
δ_p	11.3	10.5	9.7
δ_h	9.6	7.5	4.0
^jR	13.7	8.6	10.6

Adapted from Huyskens, P. L., Haulait-Pirson, M.-C., Brandts-Buys, L. D., and Van der Borght, X. M., *J. Coatings Technol.*, 57(724), 57, 1985.

during the swelling process.) Wernick[233] used computer simulation and stereographics to rotate image points for polymer solubilities to desired viewing angles and to print stereo pairs for three-dimensional viewing.

As discussed in Chapter 5, although Lewis acid-Lewis base interactions are not adequately treated by Hansen parameter formalism, the three-parameter approach is often adequate for some practical purposes, particularly in systems not involving water or alchols.

14.6 OTHER COHESION PARAMETERS

Lewis acid-base interactions (Section 5.4) are just as important in many polymer-liquid systems as they are in liquid-liquid systems, often dominating the cohesion (and adhesion) properties. Fowkes[244-246] has concentrated on this aspect of polymer interactions, using acid-base enthalpy changes in conjunction with dispersion cohesion parameters.

Division of the total Hildebrand parameter into dispersion, polar, and hydrogen-bonding parts (Hansen parameters, previous section) is clearly only a crude approximation in systems with significant acid-base interactions, but the incorporation of acid and base cohesion parameters in the manner described for liquids in Section 5.6 is not widely practiced except in the characterization of chromatographic materials (Chapter 18). Various other ways of coping with specific interactions in terms of cohesion parameters are described in Chapter 8, some specifically designed for the description of polymer interactions. In particular, the approach of Brandts-Buys, Haulait-Pirson, Huyskens, and Van der Borght incorporates a modified Hildebrand parameter within a comprehensive model.[247-251] For polymers, it is necessary to know values of parameter A (a constant for each polymer at a particular temperature, with a function similar to that of the Hansen parameter ^jR), K (the hydrogen-bonding stability constant for a particular polymer with a particular liquid), and $^j\delta'$ (a modified, non-specific cohesion parameter incorporating only dispersion and dipolar forces). Typical values of these parameters[250] appear in Table 10, together with the corresponding Hansen parameters for comparison.

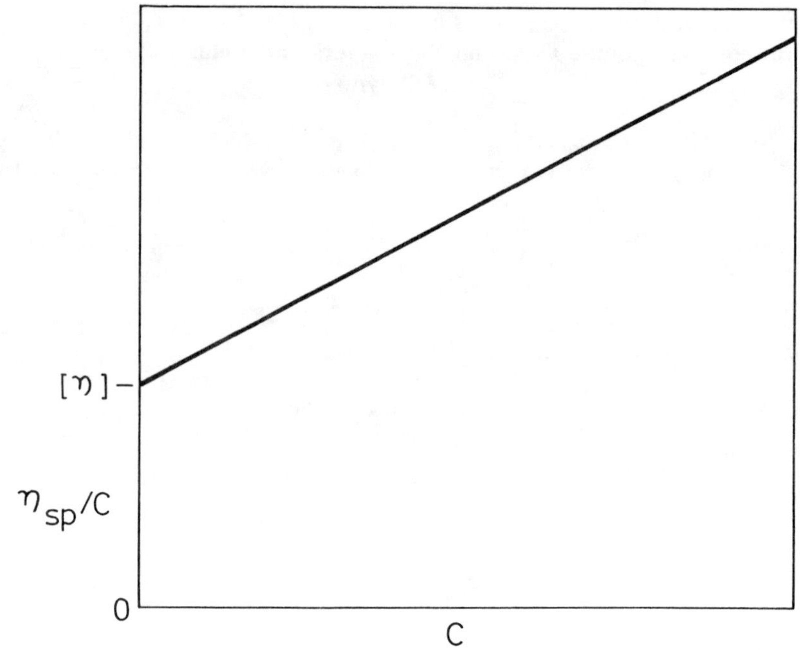

FIGURE 5. Illustration of the determination of intrinsic viscosity [η] in terms of specific viscosity (η_{sp}) and concentration (c).

14.7 SOLUTION VISCOSITY

The viscosity of a *dilute* solution of a polymer is a maximum in the "best" solvent, that is, in one where the cohesion parameters of solvent and polymer are comparable. Equilibrium configuration of molecular chains, both in swelling situations and in polymer solution viscosity studies, depend on the balance between the Gibbs free energy change due to mixing and that due to elastic deformation. The extent of deformation depends on the relative strengths of intramolecular (segment-segment) and intermolecular (segment-solvent) interactions. In a "good" solvent the polymer is unfolded, obtaining to the maximum extent the more favorable polymer-solvent interactions (and therefore the greatest viscosity), while in a "poor" solvent the polymer molecules remain folded because of more favorable intramolecular interactions.

The viscosities of polymer solutions at the limit of infinite dilution have received considerable attention.[252,253] The limiting viscosity number (frequently called the intrinsic viscosity) of a polymer solution is defined (Figure 5):

$$[\eta] = \lim_{c\to 0}\left(\frac{\eta - {}^{i}\eta}{c\,{}^{i}\eta}\right) = \lim_{c\to 0}\left(\frac{\eta_{sp}}{c}\right) \qquad (15)$$

where c is the polymer concentration, ${}^{i}\eta$ is the viscosity of the pure solvent i, and (η_{sp}/c) is known as the reduced viscosity. The specific viscosity of the solution is

$$\eta_{sp} = \eta_{rel} - 1 = \eta/{}^{i}\eta - 1 = (\eta - {}^{i}\eta/{}^{i}\eta) \qquad (16)$$

where η_{rel} is the relative viscosity or viscosity ratio. Also useful is the weight intrinsic viscosity, which is dimensionless:

$$[\eta]_w = {}^{i}\rho[\eta] \qquad (17)$$

Heller[254] presented a detailed analysis of limiting viscosity equations.

The intrinsic viscosity value depends on the properties of isolated polymer molecules and may be related to the polymer viscosity average molecular weight $\overline{M}$ by the empirical Mark-Houwink equation

$$[\eta] = K\overline{M}^{\alpha} \tag{18}$$

The parameters K and α in this equation are evaluated by double logarithmic plots of $[\eta]$ against $\overline{M}$, as presented by Izumi and Miyake[231] for poly(*p*-chlorostyrene), for example. Various other theoretical and empirical expressions have been used, including those predicting the $[\eta]$ value under theta conditions (Section 13.7) where = 0.5, so

$$[\eta]_{\Theta} = K_{\Theta}\overline{M}^{1/2} \tag{19}$$

The Mark-Houwink exponent α is predicted theoretically to be zero for spheres, 2 for rigid rods, and unity for random polymer chains. It clearly depends on the nature of the polymer-solvent interaction, ranging from 0.5 in theta solvents, through 0.65 in ''intermediate'' solvents, to ~0.8 in good solvents. Values of α may be estimated from cohesion parameter differences.[255,256] Other empirical equations such as the Burchard-Stockmayer-Fixman equation can be used to relate intrinsic viscosity and molecular weight to polymer-liquid interaction parameter, and thus to the Hildebrand parameter. Both cohesion parameters and group contribution methods have been used for correlating and predicting intrinsic viscosities,[253] which have been measured as functions of the Hildebrand spectrum of a series of liquids, or occasionally as functions of relative composition ratios of a pair of mixed solvents,[178] or as a function of the Hildebrand parameter of a liquid as it changes with temperature.[257] Viscosity studies and the Mark-Houwink constants have been used in assessing the structural changes (rigid rod to random coil transformations) of polymer solutions,[258] although the cohesion parameters do not seem to provide information on these phenomena.

Polymer systems investigated by means of correlations between polymer-liquid interaction or Hildebrand parameters and viscosity include:

- Alkyds,[259,260] including vegetable oil (linseed, castor) alkyds[261-263]
- Polyamide[444]
- Amylose derivatives[264]
- Aromatic cardo polyesters and polyimides[265,266]
- Polybutadiene[88,267] and polybutadiene modified by epoxidation[268,269] and grafting with maleic anhydride-styrene[270]
- Butyl rubber[271]
- Carbopol® 940 and 934[272]
- Cellulose derivatives[91,185,273]
- Chlorinated poly(vinyl chloride)[195,196]
- Copolymers,[274] including siloxanes,[275] ethylene-vinyl acetate,[276,424] ethylene-diethyl fumarate,[277] and low molecular weight copolymers in high solids coatings[278,279]
- Cyclopolyisoprene[280]
- Polyester urethane[281]
- Polyethylene,[276] polypropylene,[164] and polyisobutylene[164]
- Glycol polyesters[283,284,453]
- Ion-containing polymers[285]
- Itaconate polymers[286-290]
- *N*-substituted polymaleimides[194]
- Poly(methyl methacrylate),[159,289-296] poly(methacrylic acid piperidide) and other polyacrylates[122,289,290,297-299,418,425] and copolymers with organophosphorus compounds[300]

- Polyoxymethylenes[201-203,301-303]
- Silicones[164,275]
- Polystyrene,[292,294,304-309] poly(α-methylstyrene)[164,291] poly(4-acetoxystyrene)[445] and poly(4-hydroxystyrene)[445]
- Styrene-acrylic,[310] styrene-butadiene,[176] styrene-isoprene,[311] styrene-methyl methacrylate,[312] and styrene terpolymers[313]
- Poly(tetrahydrofuran)[169]
- Liquid Thiokols® (δ = 18.0 MPa$^{1/2}$)[422]
- Poly(trimethylene sulfide)[420] and poly(3,3′-dimethylthietane)[421]
- Poly(vinyl acetate),[276,292,295] poly(vinyl chloride)[195,314] poly(vinyl pyrrolidone),[159] and other vinyl polymers and copolymers[315-317]

By analogy with the equilibrium swelling expression (Equation 3) the results of viscosity measurements can be interpreted more readily when rearranged to yield a linear graph and to take the solvent molar volume into account by means of the Flory-Huggins model, the "Mangaraj method":

$$[{}^iV^{-1}\ln([\eta]_{max}/[\eta])]^{1/2} = K^{1/2}({}^i\delta - {}^j\delta) \tag{20}$$

Closely related equations are

$$[(1 - [\eta]/[\eta]_{max}){}^iV^{-1}]^{1/2} = K^{1/2}({}^i\delta - {}^j\delta) \tag{21}$$

$$\log[\eta] = B\log|{}^i\delta - {}^j\delta| + A \tag{22}$$

These equations, which provide more reliable ${}^j\delta$ values, have been used widely.* A more general form of Equation 20 in terms of the interchange cohesive pressure ${}^{ij}A$ is

$$\ln([\eta]_{max}/[\eta]) = K\,{}^{ij}A\,{}^iV \tag{23}$$

With this expression it is possible to correlate polymer solution viscosities with other forms of cohesion parameter expression, for example, by using Equation 12,

$${}^{ij}A = ({}^i\delta_d - {}^j\delta_d)^2 + ({}^i\delta_p - {}^j\delta_p)^2 + ({}^i\delta_h - {}^j\delta_h)^2 \tag{24}$$

or Hansen's interaction radius (Equation 13),

$${}^{ij}R^2 = 4({}^i\delta_d - {}^j\delta_d)^2 + ({}^i\delta_p - {}^j\delta_p)^2 + ({}^i\delta_h - {}^j\delta_h)^2 \tag{25}$$

Bagda[319,320] used as alternatives the equations

$$\eta_{sp} = B({}^i\delta - {}^j\delta)^2 + A \tag{26}$$

and

$$\eta_{sp} = B({}^i\delta_d - {}^j\delta_d)^2 + ({}^i\delta_p - {}^j\delta_p)^2 + {}^i\delta_h - {}^j\delta_p)^2 + A \tag{27}$$

* For example: Arichi and Himura;[445] Ceausescu et al.;[103] Huglin and Pass;[169] Marco, Bello, Fatou, Pérez, and co-workers;[201-203,302] Mangaraj;[83,298,299,315,318] Matsuo;[294] Sosa et al.;[309] Velickovic and Petrovic-Djakov;[286-288,290,425] and Yamashita.[62]

where B and A are dependent on the type of polymer and its concentration, but independent of the nature of the solvent. Erickson[321-324] used an interaction variable, defined as ${}^{j}R\ {}^{i}V^{1/2}$, for concentrated oligomer solutions. Van Dyk, Frisch, and Wu[239] investigated other relationships between $[\eta]$ and δ, including

$$[\eta] = K_{I} - K_{II}\ {}^{i}V\ {}^{ij}R \tag{28}$$

and

$$[\eta] = K_{I} - K_{II}\ {}^{i}V^{1/2}\ {}^{ij}R \tag{29}$$

Several of these methods have the advantage of employing Hansen parameters rather than Hildebrand parameters, and polymers whose solution viscosities have been correlated with such multicomponent parameters include:

- Poly(1-chloro-2,3-epoxypropane)[234]
- Ion-containing (sulfonate, carboxylate) polymers[285]
- Polymethacrylates[239,319,320]
- Polyoxetanes[302,303]
- Poly(vinyl acetate) and vinyl acetate-vinyl chloride copolymers[319,320,325]
- Poly(vinylidene fluoride)[157]
- Other polymers including epoxy, urethane, and acrylic oligomers[320-324,326]

Although the viscosities of *dilute* polymer solutions show minima in poor solvents because the polymer molecules remain coiled as the polymer concentration is increased, there is a ''changeover'' in behavior: the viscosity exhibits a *maximum* in poor solvents at higher concentration.[327,328] There is an aggregation or clustering of polymer molecules, a preliminary to phase separation, when solvents are used that have cohesion parameters near the limits of the miscibility range for the polymer. It is probable that the clusters are not completely broken down to individual molecules even in good solvents with cohesion parameters in the middle of the miscibility range.

This rather complex behavior can be described in terms of entropy and free volume effects which are not present in liquid-liquid systems, and it should not be expected that it can be fully correlated by means of cohesion parameters providing direct information only on the enthalpy aspect of the interactions. Attempts are being made[329,330] to relate concentrated polymer solution behavior to dilute solution parameters such as the Mark-Houwink K and α.

Another approach to polymer viscosity is the use of cohesive energy density as a measure of the cavitation energy necessary to disrupt liquid-liquid interactions in order to accommodate the polymer.[432]

The above discussion relates to flexible polymers that exist in solution as random coils. ''Stiff'' polymers such as cellulose show less change in intrinsic viscosity as the solvent is changed, but the limited solubility in a variety of liquids restricts investigation. It appears that a poly-yne polymer containing a transition metal in the main chain, such as poly[*trans*-bis(tributylphosphine)platinum 1,4-butadiynediyl], retains a rod-like shape in solution, as it exhibits an intrinsic viscosity independent of solvent Hildebrand parameter in the range 13.5 to 20 $MPa^{1/2}$, and a uniformly high solubility in these liquids.[331] This type of polymer is receiving attention because of special properties in both solid state and solution, including the formation of anisotropic (liquid crystal) polymer solutions.

It should be noted that wall adsorption of polymers (Section 17.6) can have an effect on observed polymer solution flow properties.[332] Flow properties of polymers in the absence of solvents are considered in Section 16.2 and dispersions in Section 17.8.

14.8 CALCULATION OF POLYMER COHESION PARAMETERS

The group contribution methods described in Chapter 6 are generally applicable to oligomers and polymers as well as to low molecular weight liquids, the molar mass of the repeating unit, ${}^{u}M$, being used in place of the monomer molar mass. The molar enthalpy of vaporization of an oligomer or polymer made up of n repeating units may be subdivided according to

$$\Delta H = 2\,{}^{e}\Delta H + (n - 2){}^{u}\Delta \mathrm{H} \tag{30}$$

where ${}^{e}\Delta H$ and ${}^{u}\Delta H$ are the contributions of the end groups and repeat units, respectively, to the molar enthalpy of vaporization. The corresponding molar mass is

$$M = 2\,{}^{e}M + (n - 2){}^{u}M \tag{31}$$

where ${}^{e}M$ is the molar mass of the end group and ${}^{u}M$ the molar mass of the repeat unit. Equation 30 can be rewritten:

$$\Delta H/M = 2{}^{e}\Delta H/M + \frac{(n - 2){}^{u}\Delta H}{2\,{}^{e}M + (n - 2){}^{u}M} \tag{32}$$

In polymers where M is large, $2\,{}^{e}\Delta H/M$ and $2\,{}^{e}M$ are negligible, so

$$\Delta H/M = {}^{u}\Delta H/{}^{u}M \tag{33}$$

If the density, $\rho = M/V$, is inserted,

$$\Delta H/V = \rho\,{}^{u}\Delta H/{}^{u}M \tag{34}$$

From the definition of the Hildebrand parameter (Equation 6, Chapter 2),

$$\delta = [(\Delta H - RT)/V]^{1/2} \tag{35}$$

and with RT negligible compared with ΔH for polymers,

$$\delta = (\rho\,{}^{u}\Delta H/{}^{u}M)^{1/2} \tag{36}$$

If this concept is extended[333,334] in terms of group molar attraction constants (Equation 25, Chapter 6),

$$\delta = \sum_{z} {}^{z}F/V \tag{37}$$

then the Hildebrand parameter of an oligomer is

$$\delta = [{}^{u}F(V - {}^{e}V)/{}^{u}V + {}^{e}F]/V \tag{38}$$

where the relative number of repeat units is given by $(V - {}^{e}V/{}^{u}V$. Using ${}^{u}F/{}^{u}V = {}^{u}\delta$ and ${}^{e}F/{}^{e}V = {}^{e}\delta$,

$$\delta = {}^{u}\delta(1 - {}^{e}V/V) + {}^{e}\delta\,{}^{e}V/V \tag{39}$$

as indicated in Section 6.6, and for ${}^{e}V \ll V$ (high polymers),

$$\delta = {}^{u}\delta = {}^{u}F/{}^{u}V \tag{40}$$

From Equation 39 it can be seen that it is possible to evaluate ${}^{u}\delta$ from directly observable δ and V values of liquid and volatile monomers and oligomers, using the molar volume of the smallest member of the homologous series as ${}^{e}V$. Wolf[333] has performed this extrapolation for unbranched alkanes and their 1-chloro and 1-hydroxy derivatives (Figure 6). Even for the homologous alcohols (for which the geometric mean rule is not expected to be applicable) the data (including those for water, not shown in the figure) fall on satisfactory straight lines and give ${}^{u}\delta$ values within the limits of those obtained by other methods. In another case,[335] cohesive pressures have been obtained from the vapor pressures of oligomers for poly(tetramethylene oxide) and related compounds, resulting in a Hildebrand parameter for the liquid polymer of 17.0 $MPa^{1/2}$.

It is also possible to extrapolate Hildebrand parameter values of homologous series to high molar mass by means of relationships such as

$$\log \delta = \log \delta_{\infty} + Bn^{-1} \tag{41}$$

where the coefficient B depends on temperature but δ_{∞} is independent of temperature[336] (Figure 7).

As pointed out in Section 13.1 for the group molar volumes of polymers, the group contributions to cohesive energies of polymers are not necessarily identical to those of the corresponding low molar mass compounds. van Krevelen[253] has compiled a set of $(-{}^{z}U)$ values from polymer data (Table 11). Also, on the basis of the molar attraction constants of Table 8, Chapter 6, for atoms and structural features they derived group molar attraction constants and presented them in the form shown in Table 12 (along with those of Small[134] and of Hoy[337,338] for comparison). Examples have been published[253,339] of the application of the group molar attraction method using these data, for example, for poly(butyl methacrylate), $-CH_2-C(CH_3)(COOCH_2CH_2CH_2CH_3)-$, as indicated in Table 13. The calculated Hildebrand parameter values by both methods are within the range of those reported experimentally (Table 3).

Hoy[338] evaluated cohesion parameters for polymers (in the amorphous state) as well as for monomers (Section 5.11), using the group constant values in Table 7, Chapter 6, and Table 17, Chapter 5. The number (n) of monomer units was estimated for a segment of the polymer chain from

$$n = 1/\left(2 \sum_{z} \Delta_{T}^{P}\right) \tag{42}$$

where $\Sigma\Delta_{T}^{P}$ is the sum of the aggregation constants for the repeat unit of the segment in the polymer chain. The Hildebrand parameter for the polymer was then evaluated from

$$\delta/\mathrm{MPa}^{1/2} = \left[n \sum_{z} ({}^{z}F/\mathrm{J}^{1/2}\ \mathrm{cm}^{3/2}\ \mathrm{mol}^{-1}) + 276.3\right] \Big/ n \sum_{z} ({}^{z}V_{g}/\mathrm{cm}^{3}\ \mathrm{mol}^{-1}) \tag{43}$$

where $\Sigma^{z}V_{g}$ is the sum of the group molar volume constants for the repeat unit. The chain aggregation number α^{*} was estimated from

$$\alpha^{*} = 777.4 \sum_{z} \Delta_{T}^{P} \Big/ \sum_{z} ({}^{z}V_{g}/\mathrm{cm}^{3}\ \mathrm{mol}^{-1}) \tag{44}$$

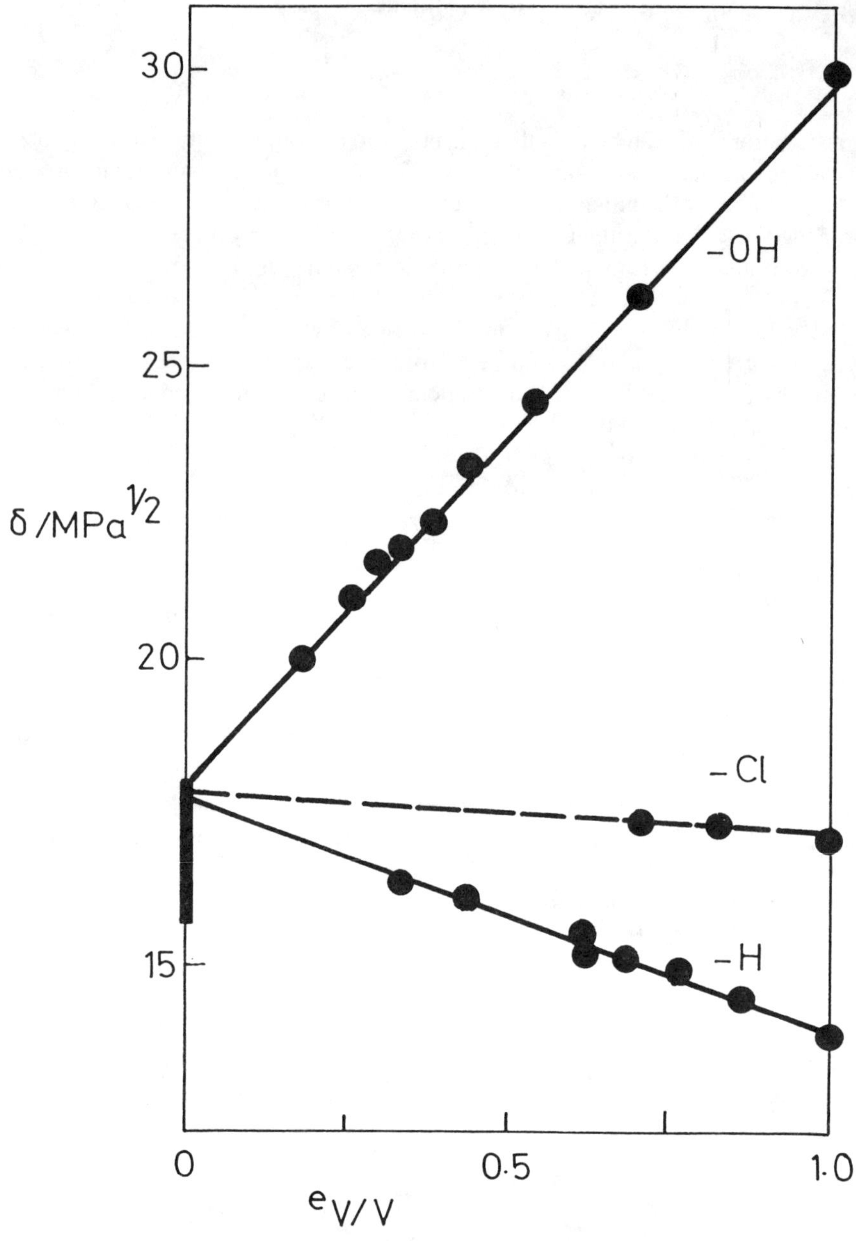

FIGURE 6. Variation of the Hildebrand parameter at 25°C of unbranched alkanes (lowest line) and their 1-hydroxy- (upper line) and 1-chloro- derivatives with the ratio of the functional end group volume to the total volume. The bar on the ordinate indicates the range of polyethylene Hildebrand parameter values by other methods. (Adapted from Wolf, B. A., *Makromol. Chem.*, 178, 1869, 1977.)

Values of δ_h, δ_p, and δ_d were then estimated in the same way as those of monomers (Section 5.11). These values are included in the *Handbook of Polymer-Liquid Interaction Parameters and Cohesion Parameters*.[340]

The method of Fedors[426] (Table 3, Chapter 6) and the other methods described in Chapter 6 have also been applied to polymers. Thus, Sheffer[427] used the Fedors group contributions

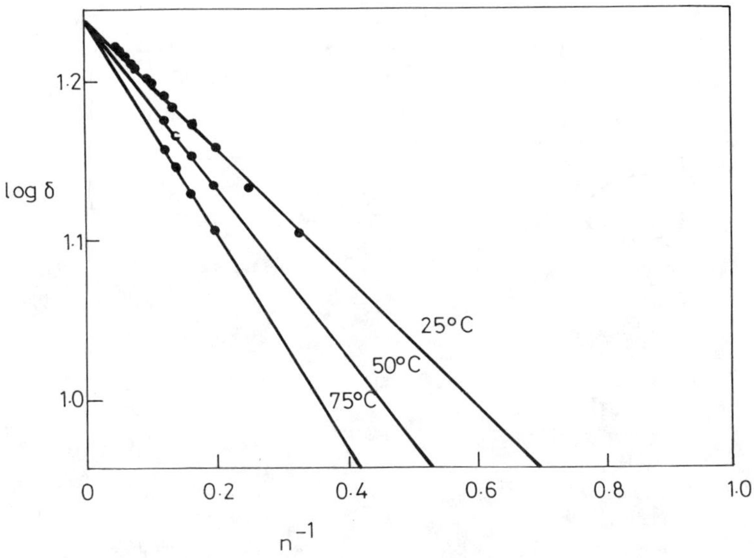

FIGURE 7. The logarithm of the Hildebrand parameter of a homologous series as a function of the reciprocal of the number of carbon atoms, at three temperatures. (Adapted from Rätzsch, M. and Krahn, G., *J. Polym. Sci. Polym. Symp.*, 42, 1001, 1973.)

TABLE 11
Molar Group Contributions to Polymer Cohesive Energies at 25°C

Group, z	^{z}U/kJ mol^{-1}
$-CH_3$	9.64
$-CH_2-$	4.19
$>CH-$	0.42
$>C<$	−5.58
$-CH(CH_3)-$	(10.06)
$-C(CH_3)_2-$	(13.70)
$-CH{=}CH-$	10.20
$>C{=}CH-$	4.86
$-C(CH_3){=}CH-$	(14.50)
Phenyl	31.00
p-Phenylene	25.14
–F	4.47
–Cl	12.99
–Br	15.50
–CN	25.00
–CHCN	25.42
–O–	6.29
–COO–	13.41
–C(=O)–N(H)–	60.76
–S–	8.80

Adapted from van Krevelen, D. W., *Properties of Polymers: Their Estimation and Correlation with Chemical Structure*, Elsevier, Amsterdam, 1976.

TABLE 12
Group Molar Attraction Constants of van Krevelen at 25°C

	$^{z}F/J^{1/2}\ cm^{3/2}\ mol^{-1}$		
Group, *z*	**van Krevelen**	**Hoy**[337]	**Small**[134]
$-CH_3-$	420	303	438
$-CH_2-$	280	269	272
$>C-$	140	176	57
$-C-$	0	66	−190
$-CH(CH_3)-$	560	479	495
$-C(CH_3)_2-$	840	672	686
$-CH{=}CH_2-$	444	497	454
$>C{=}CH-$	304	422	266
$-C(CH_3){=}CH-$	724	(725)	(704)
Cyclopentyl	1384	1295	—
Cyclohexyl	1664	1473	—
Phenyl	1517	1398	1504
p-Phenylene	1377	1442	1346
–F	164	85	(250)
–Cl	471	420	552
–Br	614	528	696
–CN	982	723	839
–CHCN–	1122	(902)	(896)
–OH	754	462	—
–O–	256	235	143
–CO–	685	538	563
–COOH	652	(1000)	—
–COO–	512	668	634
–O–C–O– ‖ O	767	(904)	—
–C–O–C– ‖ ‖ O O	767	1161	—
–C–N– ‖ | O H	1228	(906)	—
–O–C–N– ‖ | O H	1483	(1037)	—
–S–	460	428	460

Adapted from van Krevelen, D. W., *Properties of Polymers: Their Estimation and Correlation with Chemical Structure*, Elsevier, Amsterdam, 1976.

for polyimides prepared by reacting dianhydrides with aromatic diamines (δ = 24 to 29 $MPa^{1/2}$).

Voeks[341] and, more recently, Ahmad, Yaseen and co-workers[342-352] have compared the polymer Hildebrand parameters calculated from various sets of group constants. Table 14 gives an indication of the variation in calculated values, and a comparison with experimental results for a few polymers. Askadskii's van der Waals method[353-355] (Section 6.6) is seen from this table to provide cohesion parameter estimates for polymers that are in good agreement with those of other methods. Sakellariou, Rowe, and White[356] compared Hildebrand parameters for cellulose derivatives and poly(ethylene glycol)s calculated by the methods of Small,[134] Hoy,[337,338] and van Krevelen[253] with literature values; those of Hoy and of van Krevelen were acceptably close to experiment (Table 15).

TABLE 13
Calculation of Hildebrand Parameter of Poly(Butyl Methacrylate) at 25°C from Group Molar Attraction Constants

Group, z	^zU/kJ mol^{-1}	$-\Sigma_z{}^zU$/kJ mol^{-1}	^zF/J$^{1/2}$cm$^{3/2}$ mol^{-1}	$\Sigma_z{}^zF$/J$^{1/2}$cm$^{3/2}$ mol^{-1}
	(Values of van Krevelen, Table 11)		(Values of Small, Table 12)	
4(–CH$_2$ –)	4.19	16.76	272	1088
2(–CH$_3$)	9.64	19.28	438	876
1(>C<)	−5.58	−5.58	−190	−190
1(–COO–)	13.41	13.41		
		43.87		2408

Group, z	^zV/cm^3 mol^{-1}	$\Sigma_z{}^zV$/cm^3 mol^{-1}
	(Values of van Krevelen Table 4, Chapter 13)	
4(–CH$_2$–)	16.45	65.80
2(–CH$_3$)	50.35	50.35
1(>C<)		
1(–COO–)	21.0	21.0
		137.15

$\delta = (-\Sigma_z{}^zU/\Sigma_z{}^zV)^{1/2} = 17.9$ MPa$^{1/2}$
$\delta = (\Sigma_z{}^zF/\Sigma_z{}^zV) = 17.6$ MPa$^{1/2}$

Adapted from van Krevelen, D. W., *Properties of Polymers: Their Estimation and Correlation with Chemical Structure,* Elsevier, Amsterdam, 1976.

TABLE 14
Comparison of Polymer Hildebrand Parameters (δ/MPa$^{1/2}$) from Various Sources

Polymer	Calculation					Experimental	
	Small	Hoy	van Krevelen	Ahmad and Yaseen	Askadskii	Swelling	Intrinsic viscosity
Poly(vinyl acetate)	18.9	19.1	18.2	19.3	—	19.2	—
Poly(methyl acrylate)	18.9	19.1	18.2	19.3	19.6	20.7	20.7
Poly(ethyl acrylate)	18.7	18.9	18.3	18.9	19.0	19.2	19.2
Poly(butyl acrylate)	18.6	18.6	20.4	18.3	—	17.9	18.0
Poly(methyl methacrylate)	17.8	18.0	18.3	20.5	19.0	19.4	19.2
Poly(ethyl methacrylate)	16.8	16.9	17.2	17.6	18.6	18.3	—
Poly(butyl methacrylate)	17.8	18.0	18.4	17.8	18.0	17.9	—
Polyethylene	16.5	16.3	17.0	15.9	—	16.2	—
Polybutadiene	16.6	16.9	17.1	16.2	—	17.3 17.6	—
Polystyrene	18.4	18.5	19.4	18.2	18.6	18.6	17.6 18.6
Polyalkyd (linseed-glycerol phthalate)	19.2	19.5	19.0	18.3	—	—	18.1 18.3

Adapted from Ahmad, H. and Yaseen, M., *Polym. Eng. Sci.*, 19, 858, 1979.

TABLE 15
Calculated and Experimental Hildebrand Parameters for Cellulose Derivatives and Poly(Ethylene Glycol)s at or Near 25°C

	$\delta/MPa^{1/2}$			
	Calculated			
Polymer	Small[134]	Hoy[337,338]	van Krevelen[253]	Experimental range
Ethyl cellulose	16.1	19.0	20.6	19—21
Hydroxypropylmethylcellulose	16.1	21.5	24.4	21—26
Hydroxypropylcellulose	17.0	21.1	24.9	22—24
Hydroxypropylmethylcellulose phthalate	17.2	22.4	26.4	—
Cellulose acetate phthalate	21.7	24.1	27.2	—
Polyethylene glycol 200	19.1	20.2	26.1	24
Polyethylene glycol 400	18.4	20.2	23.1	23
Polyethylene glycol 600	18.0	19.7	22.5	21[a]
Polyethylene glycol 1000	19.1	21.5	23.4	20[a]
Polyethylene glycol 4000	18.9	21.1	22.5	19[a]—25
Polyethylene glycol 6000	18.8	21.2	22.4	24

[a] 100°C

Adapted from Sakellariou, P., Rowe, R. C., and White, E. F. T., *Int. J. Pharm.*, 31, 175, 1986.

Bellenger, Morel, and Verdu[429] calculated Hildebrand parameters for amine-crosslinked aromatic epoxy thermosets by several methods (Table 15a); this was possible for these thermosets because a stoichiometric crosslinking was definable. In a study of blends of poly(alkylene oxide)s with poly(vinyl acetate), Kalfoglou, Sotiropoulou, and Margaritis[434] found that the group molar attraction constants of Hoy were more consistent with observed polymer-polymer interaction parameter values and behaviors than those of Small. Arichi and Himuro[445] carried out a detailed comparison of cohesion parameter determination for polystyrene, poly(4-acetoxystyrene), and poly(4-hydroxystyrene)[445] (Table 15b).

Alternative methods of estimating polymer cohesion parameters are based on theoretical or empirical relationships with other molecular properties (Section 8.2). For example, Sewell[357] used molar refraction, R (Section 8.2)

$$R = (n^2 - 1)V/(n^2 + 2) \tag{45}$$

and showed for a range of polymers that

$$\delta/\text{MPa}^{1/2} = 62.0(n^2 - 1)/(n^2 + 2) \tag{46}$$

Many of the polymer reviews referred to in Table 1, Chapter 1, include descriptions of calculation methods for polymer cohesion parameters. In addition to the work already described, Hildebrand parameters for polymers have been calculated on many occasions.*

Applying to polymeric systems the relationships and data used by Hansen[206,222,225] and others (see Section 8.2), Koenhen and Smolders[379] calculated Hansen parameters for polymers (Table 16). In particular,

$$\delta_d = \sum_z {}^zF_d/{}^uV \tag{47}$$

* For example: Chen;[358] Coleman et al.;[455] Huglin and Zakaria;[359] Kalz;[448,449] Kambour et al.;[452] Ling and Gao;[360] Maiti, Ray, Das, and co-workers;[361-369] Miller;[198] Miroshnichenko;[370] Panda, Patra, Lenka and Nayak;[371,443] Petrovic-Dakov, Velickovic et al.;[286-290,425] Ryan, Stanford, and Still;[450] Saito and Yamaguchi;[372] Sanchez et al.;[373] Shin and Lee;[428] Shtilman et al.;[374] Takeda et al.;[375] Tong, Saenger, and Durning;[456] Tortorello, Kinsella, and Gearon;[376,377] Watanabe and Miyauchi;[378] and Yakovlev et al.[194]

TABLE 15A
Calculated and Experimental Hildebrand Parameters and Experimental Hansen Parameters for Amine-Crosslinked Aromatic Epoxies

System[a]	Hildebrand parameters/$MPa^{1/2}$					Hansen parameters/$MPa^{1/2}$		
	Fedors[426]	Small[134]	Hoy[337,338]	van Krevelen[253]	Experiment	δ_d	δ_p	δ_h
DGEBA-DDME	22.6	24.3	21.5	23.1	20—22	18	10	7
DGEBA-DDM	23.5	24.4	22.1	23.3	22	19	10	7
TGAP-DDME	23.9	25.9	22.2	23.9	22	18	11	8
TGAP-ANI	25.5	27.0	23.0	24.9	24—25	17	11	14
TGAP-DDM	25.7	26.7	23.5	24.7	25—26	17	17	8
TGAP-DDS	27.4	27.7	25.6	25.6	27	16	19	10

[a] DGEBA: diglycidyl ether of bisphenol A.
TGAP: triglycidyl ether of aminophenol.
DDM: diaminodiphenylmethane.
DDME: tetraethyl derivative of DDM.
DDS: diaminodiphenylsulfone.
ANI: aniline.

Adapted from Bellenger, V., Morel, E., and Verdu, J., *J. Appl. Polym. Sci.*, 37, 2563, 1989.

TABLE 15B
Calculated and Experimental Hildebrand Parameters and Experimental Hansen Parameters for Polystyrene, Poly(4-Acetoxystyrene) and Poly(4-Hydroxystyrene)

Method	δ/MPa$^{1/2}$ Polystyrene	Poly(4-acetoxystyrene)	Poly(4-hydroxystyrene)
Viscosity			
Mangaraj method	17.8	22.7	23.9
Turbidimetry			
Suh & Clarke method	18.2	21.9	19.4
Hansen method	18.6	21.7	24.6
Calculation			
Small method[134]	18.6	17.4	19.4
Hoy method[337,338]	21.1	22.9	23.5
Fedors method[426]	21.7	22.9	30.1
Hansen parameters			
δ_d	18.4	17.8	17.6
δ_p	1.0	9.0	10.0
δ_h	3.3	8.4	13.7

Adapted from Arichi, S. and Himuro, S., *Polymer*, 30, 686, 1989.

and

$$\delta_h = \left(-\sum_z {}^zU_h/{}^uV\right)^{1/2} \tag{48}$$

while δ_p values were estimated from the correlations in Equation 14, Chapter 8, with the dipole moment established from liquid data. Matsuura et al.[380-382] utilized the work of Koenhen and Smolders and extended it to provide the group contributions shown in Table 17. Examples of the application of this information to δ_d and δ_h calculations are given in Tables 18 and 19, while Table 20 presents the results of the calculations for a variety of polymers used on chromatographic columns. Ahmad reported component cohesion parameters for poly(ethylene sulfonamide)s,[343] hydroxy polymers,[344] methacrylamide polymers,[351] and acrylamide polymers.[352] Loev and Lavrishchev[383] calculated Hansen parameters for resins used in commercial paint systems; Mishra[384] calculated Hansen parameters for shellac; Sanchez, Bello, Marco, and Fatou[419] estimated Hansen parameters for poly(trimethylene sulfide); Szymanowski, Sobzyńska, and Voelkel[441] calculated Hansen parameters for polyoxyethylene glycol dialkyl ethers; and Kesselring and co-workers[385] calculated component cohesion parameters for large lipophilic molecules.

Thermodynamic properties of polymers, including cohesive pressures and cohesion parameters, may be calculated also by various equation-of-state methods,[341,386] and agreement with other values appears reasonably good. The calculated Hildebrand parameter of liquid polyethylene over wide temperature and pressure ranges is depicted in Figure 8.

In an interesting comparison of experimental and calculation methods of determining Hildebrand parameters for moderately large molecules, the energies of vaporization were determined for several organic plasticizers.[387] They were found to be one half (or less) of the values estimated by Small's additive factor method, so Hildebrand parameters based on these experimental values were low by about 30% (Table 21). It was concluded that large, flexible, linear molecules vaporize in a compact, coiled form and are nearly spherical in the vapor phase, so the energy required for extending the molecules is not included in the measured ΔU. In this case, of course, the ''calculated'' Hildebrand parameter is more appropriate than the ''experimental'' one for describing liquid-state properties.

TABLE 16
Calculated Hansen Parameters for Polymers

Polymer	^{a}V/cm^3 mol^{-1}	δ/MPa$^{1/2}$			
		δ_d(from group contributions, Equation 47)	δ_d (from correlation with refractive index)	δ_p (from dipole moment)	δ_h (from group contributions, Equation 48)
Polyethylene	32.8	17.3	17.6	0.0	0.0
Polystyrene	99	19.2	19.7	0.9	2.1
Poly(tetrafluoroethylene)	50	12.7	14.0	0.0	0.0
Polyacrylonitrile	44.8	18.6	18.2	16.2	6.8
Poly(2,6-dimethyl-1-4,phenylene oxide)	111.7	—	12.3	2.7	4.9
Poly(methyl methacrylate)	85.6	17.5	17.8	5.7	7.8
Polyamide 6,6	211.5	—	18.5	5.1	12.3
Poly(vinyl chloride)	45.1	—	18.7	10.0	3.1
Poly(ethylene terephthalate)	144.5	—	19.5	3.5	8.6

Adapted from Koenhen, D. M. and Smolders, C. A., *J. Appl. Polym. Sci.*, 19, 1163, 1975.

TABLE 17
Group Contributions to Hansen Parameters for Polymers

Group, Z	$-^{Z}U_h$/kJ mol^{-1}	$^{Z}F_d$/J$^{1/2}$ cm$^{3/2}$ mol^{-1}
$-CH_3$	—	411
$>CH_2$	—	284
$>CH-$	—	104
–OH	20.9	203
–C(=O)NH– aliphatic	16.3	468
–C(=O)NH aromatic	27.2	468
–C(=O)NHNHC(=O)–	37.2	—
–C(=O)–O–	5.2	395
–O–	2.3	143
–S(=O)(=O)–	11.3	606
–C(=O)–OH	—	409
–CN	2.1	446
$-NO_2$	—	440
$-CF_2-$	—	327
–Cl	0.4	—
C_6H_6	—	1640
$-C_6H_5$	—	1510
$-C_6H_4-$	0.4	1350

Adapted from Matsuura, T., Blais, P., and Sourirajan, S., *J. Appl. Polym. Sci.* 20, 1515, 1976.

TABLE 18
Calculation of Hansen Parameters for Cellulose Acetate[a] from Group Constants

Group, z	$^{z}F_d$/J$^{1/2}$ cm$^{3/2}$ mol^{-1}	$\Sigma_z\,^{z}F_d$ J$^{1/2}$ cm$^{3/2}$ mol^{-1}	$-\,^{z}U_h$ kJ mol^{-1}	$-\Sigma_z\,^{z}U_h$ kJ mol^{-1}
9.81 ($-CH_3$)	411	4032	—	—
4 ($-CH_2-$)	284	1136	—	—
20 (–CH–)	104	2080	—	—
2.19 (–OH)	203	445	20.9	45.8
9.81 (–COO–)	395	3875	5.2	51.0
8 (–O–)	143	1144	2.3	18.4
Total		12712[b]		115.2[c]

[a] $-(CH_2)_4(CH)_{20}(O)_8(OH)_{2.19}(OCOCH_3)_{9.81};\ ^{u}V = 816.2$ cm^3 mol^{-1}

[b] $\delta_d = (12712/816.2) = 15.6$ MPa$^{1/2}$

[c] $\delta_h = (115200/816.2)^{1/2} = 11.9$ MPa$^{1/2}$

Adapted from Matsuura, T., Blais, P., and Sourirajan, S., *J. Appl. Polym. Scl.*, 20, 1515, 1976.

TABLE 19
Calculation of Hansen Parameters for Aromatic Polyamide-Hydrazide Cop (0.893: 0.107) from Group Constants

$^{ij}M = 238.26 \times 0.893 + 324 \times 0.107 = 247.5 \text{ g mol}^{-1}$
$^{ij}\rho = 1.2 \text{ g cm}^{-3}$
$^{ij}V = 247.5/1.2 = 206 \text{ cm}^3 \text{ mol}^{-1}$
$-\Sigma_z\, {}^zU_h = 55.2 \times 0.893 + 75.3 \times 0.107 = 57.3 \text{ kJ mol}^{-1}$
$\Sigma_z\, {}^zF_d = 3633 \times 0.893 + 4570 \times 0.107 = 3733 \text{ J}^{1/2} \text{ cm}^{3/2} \text{ mol}^{-1}$
$\delta_d = (3733/206) = 18.1 \text{ MPa}^{1/2}$
$\delta_h = (57300/206)^{1/2} = 16.7 \text{ MPa}^{1/2}$

Adapted from Matsuura, T., Blais, P., and Sourirajan, S., *J. Appl. Polym. Sci.*, 20, 1515, 1976.

TABLE 20
Hansen Parameters of Polymers and Copolymers used in Membranes and Chromatographic Columns

Polymer	Column name, trade name	^{u}V/cm^3 mol^{-1}	δ_h/MPa$^{1/2}$	δ_d/MPa$^{1/2}$
Cellulose acetate	CA-398-3-C	816	11.9	15.6
Cellulose acetate	CA-398-3	816	11.9	15.6
Cellulose acetate	CA-383-40	797	12.5	15.5
Cellulose acetate	CA-394-45	811	12.1	15.6
Cellulose triacetate	CTA	861	10.6	15.6
Cellulose acetate propionate	CAP	914	10.1	15.8
Cellulose propionate	CP	1023	9.3	16.7
Cellulose acetate butyrate	CAB 171-2	953	10.1	15.8
Cellulose acetate butyrate	CAB-171-40	953	10.1	15.8
Cellulose acetate butyrate	CAB-381-2	1064	9.9	15.9
Cellulose acetate butyrate	CAB-500-5	1192	8.7	16.0
Cellulose tridecanoate	CTD	2083	6.2	19.8
Ethyl cellulose	ECG	750	11.7	16.7
Ethyl cellulose	ECT	783	9.9	17.1
Cellulose	C	135	22.3	12.6
Copolyamide	31B	199	16.7	18.3
Copolyamide	69B	199	16.7	18.3
Copolyamidohydrazide	80B	206	16.7	18.1
Copolyamidohydrazide	81B	215	16.7	17.9
Copolyamidohydrazide	79B	219	16.7	17.8
Copolyhydrazide	117B	270	16.7	16.9
Copolyamidohydrazide	107B	234	16.7	17.5
Polyamide 4	49B	83	14.1	19.4
Polyamide 6,6	N66, Zytel® (du Pont)	202	12.7	18.6
Bisphenol A polysulfone	PS, Udel® (Union Carbide)	357	7.0	19.0
Polypropylene, isotactic	PP, Profax® 6701 (Hercules)	46	0.0	17.3

Adapted from Matsuura, T., Blais, P., and Sourirajan, S., *J. Appl. Polym. Sci.*, 20, 1515, 1976.

Calculation of Hildebrand parameters is assisted by modern computation methods; for polymers, Fic and Fic[388] described a BASIC computer program for determination of δ values, Van den Berg[389] described a procedure for polymer-liquid interaction parameter calculation from UNIFAC computer programs, Kaelble[390] devised an interactive computer program for estimating cohesive energy density, and Dante et al.[442] described a program for resin-liquid interactions.

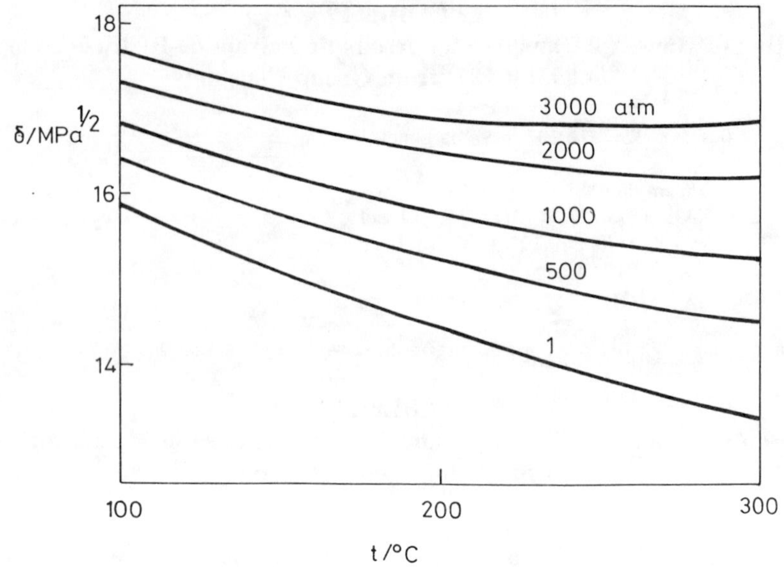

FIGURE 8. Calculated Hildebrand parameter of liquid polyethylene as a function of temperature at various pressures (1 atm = 0.101325 MPa). (Adapted from Maloney, D. P. and Prausnitz, J. M., *J. Appl. Polym. Sci.*, 18, 2703, 1974.)

TABLE 21
Comparison of Experimental and Calculated Hildebrand Parameters of Organic Plasticizers

Plasticizer	Formula	ρ/g cm^{-3} (t/°C)	M/g mol^{-1}	δ/MPa$^{1/2}$ Experimental	δ/MPa$^{1/2}$ Small[134] method
O,O,O-Tributylphosphorothionate	$C_{12}H_{27}O_3PS$	0.933/25	283	12.3	18.6
Triethyleneglycoldipelargonate	$C_{24}H_{46}O_6$	0.964/20	431	14.0	17.6
Diisooctylsebacate	$C_{26}H_{50}O_4$	0.916/20	425	10.8	17.8
2,2′-(2-Ethylhexamidodiethyldi-2-ethylhexoate)	$C_{28}H_{53}NO_5$	0.956/20	484	12.5	18.2
Didecylphthalate	$C_{28}H_{46}O_4$	0.964/25	446	13.4	18.0
Diisodecyl-4,5-epoxytetrahydrophthalate	$C_{28}H_{50}O_5$	0.987/20	466	12.6	16.9
Acetyltri(2-ethylhexyl)citrate	$C_{32}H_{58}O_8$	0.983/25	570	10.7	18.5
Ditridecylphthalate	$C_{34}H_{66}O_4$	0.950/25	538	12.2	17.8
Trioctyltrimellitate	$C_{30}H_{57}O_6$	0.987/25	546	14.8	19.5
Glyceryltriacetoxystearate	$C_{63}H_{116}O_{12}$	0.955/25	1066	12.4	17.6
Glyceryltriacetylricinoleate	$C_{63}H_{110}O_{12}$	0.965/25	1060	13.7	17.7

Adapted from Wall, L. A., Flynn, J. H., and Straus, S., *Polym. Eng. Sci.*, 10, 19, 1970.

14.9 EFFECTS OF TEMPERATURE, CONCENTRATION, MOLECULAR WEIGHT

In addition to the effects on cohesion parameters of the physical variables outlined in Chapter 9, there are particular factors relevant to polymers.

Bianchi[391] and others[392,393] have measured the temperature dependence of internal pressures for polymers and compared the results with those calculated on the assumption that $n = 1$ in Equation 20 and 21, Chapter 2:

$$\overline{U} = -a/V^n \tag{49}$$

$$\pi = (\delta U/\delta V)_T = n\ {}_1\Delta_g\ \overline{U}/V \tag{50}$$

$$(\delta \ln \pi/\delta T)_p = -(n + 1)(\delta \ln V/\delta T)_p = -(n + 1)\alpha \tag{51}$$

As might be expected, agreement is not as good as that for a simple liquid,[394] and in some cases even the sign of the predicted $(\partial \ln\pi/\partial T)_p$ is wrong. More realistic equations of state are available to describe polymer behavior.[395] However, as indicated in Section 14.8, Rätzsch and Krahn[336] showed that for long polymer chains the Hildebrand parameter is independent of both temperature and chain length. It follows that in circumstances where the temperature dependence of δ can be neglected for liquids, the same assumption can be made with some confidence for polymers. Over greater temperature ranges, gas-liquid chromatography (Section 18.2) can now provide experimental information on the temperature dependence of the cohesion parameters of polymers. A consideration of the effects of thermal transitions (glass transition temperature and melting point) are included in Section 16.3.

As explained in Sections 13.4 and 14.7, the temperature and concentration dependences of polymer-liquid interactions are associated largely with the entropy term in the Gibbs free energy criterion for solubility, $T\,\Delta S \geqslant \Delta H$, and the cohesion parameter approach is based on the enthalpy term (to be precise, on the internal energy portion of ΔH). Increasing the entropy change in a solution process permits a greater unfavorable ΔH (greater disparity in cohesion parameter values) and this may often be achieved by raising the polymer concentration. In using the polymer maps of Chapter 15, for example, it should be kept in mind that they can be used at higher solids levels than those for which they were determined: as a solvent evaporates and the resin concentration in the residual solution increases, the system usually remains homogeneous. Conversely, of course, care must be taken when diluting a polymer solution.

The subject of aggregation or association of polymer molecules in systems where phase separation is imminent is discussed in terms of polymer solution viscosity in Section 14.7. This phenomenon can be fully explained only if the associated entropy changes are considered, but near the phase separation condition where $T\,\Delta S \approx \Delta H$, correlation in terms of ΔH and cohesion parameters is possible. Thus, Beevers[396-398] has shown that δ for various nonsolvents is inversely related to the weight percentage of the nonsolvent incorporated at the aggregation point in solutions of polyacrylonitrile in *N,N*-dimethylformamide, γ-butyrolactone and dimethylacetamide.

In most cases a polymer consists of chains of different length or molar mass, and in principle these molecules of different size should be considered as different substances with different cohesion parameters, as shown by the calculations of Section 14.8. In the series of *n*-alkanes and their chloro derivatives (Figure 6) the Hildebrand parameter increases with increasing molecular weight, but in both poly(dimethylsiloxane)[399] and poly(ethylene glycol)s[400] the trend is to lower values at higher molecular weight. In the former case, δ decreases from about 11.5 $MPa^{1/2}$ at a molecular weight of 250 to level out at about 10.3 $MPa^{1/2}$ at molecular weight 900. In the latter, a gas chromatographic determination yielded Hildebrand parameters of 20.5, 20.1, and 19.4 $MPa^{1/2}$, respectively, for PEG 600, PEG 1000, and PEG 4000. Polymer fractionation is possible because a polymer of a given molecular weight has a certain solubility limit in a particular liquid system, and molecular weight fractionation curves can be predicted by means of polymer-liquid interaction parameters[231,316,402-405] and cohesion parameters.[406,407] Wijmans and Smolders[408] explained the accumulation of oligomers in the polysulfone-poor phase of a liquid-liquid phase-separated system in terms of Hildebrand parameters. However, Dyck and Hays[401] reported only a small effect on the Hildebrand solubility diagram over a wide range of poly(methyl methacrylate) molecular weights, and in many other systems the variation of cohesion parameters with molecular weight has been ignored.

Generally, phase separation has been studied either in systems of high polymers and low molecular weight liquids or in polymer-polymer systems, and there are gaps in the information relating to oligomer-oligomer and oligomer-liquid systems which only recently have begun to be adequately bridged, despite the fact that these interactions are of great practical importance.[326,333,334,392,409-413] Equation 39 indicates the relative effects of repeating units and end groups on Hildebrand parameters and miscibility; this concept has been applied to oligo(propylene glycol)s and their acetyl derivatives.[334] Other studies have been made of:

- Mixing of ethylene oxide and propylene oxide oligomers[414]
- Viscosity of oligomer solutions for coatings in terms of Hansen parameters[321-324]
- Vapor pressures and cohesive energies of oligomers of poly(tetramethylene oxide)[335]
- Designing of oligoesteramides based on cohesive pressure calculations[418]
- Hildebrand parameters of resins based on dimethylvinylethynyl-*p*-hydroxyphenylmethane (Equation 2)[21]
- Solvent systems for hydrocarbon resins[416]
- Oligoester chain effects on titanium dioxide dispersions[417]
- The trend in cohesion pressures of polyoxides with increasing methylene chain length, from poly(methylene oxide) through poly(tetramethylene oxide) and poly(hexamethylene oxide) to polyethylene[201]

REFERENCES

1. **Bianchi, U., Cuniberti, C., Pedmonte, E., and Rossi, C.,** Conformational energy contributions to the heat of solution in polymer-solvent systems. I. Theory. II. Experimental results, *J. Polym. Sci., A-2,* 7, 845 and 855, 1969.
2. **Bleha, T. and Berek, D.,** Evaluation of the solubility parameter of amorphous polyethylene, *Sb. Prednasek, "Macrotest 1973"*, 2, 143, 1973.
3. **Cuniberti, C. and Bianchi, U.,** Thermodynamics of mixing: excess volumes in polyisobutylene solvent systems, *Polymer,* 7, 151, 1966.
4. **Burrell, H.,** Solubility parameters, *Interchem. Rev.,* 14, 3 and 31, 1955.
5. **Burrell, H.,** Solubility parameter values, in *Polymer Handbook,* Brandrup, J. and Immergut, E. H., Eds., John Wiley & Sons-Interscience, New York, 1966, IV-341; 2nd ed., 1975, IV-337; **Grulke, E. A.,** Solubility parameter values, 3rd ed., 1989, VII-519.
6. **Seymour, R. B.,** Plastics vs. solvents, *Mod. Plast.,* 48, 150, 1971.
7. **Seymour, R. B.,** *Introduction to Polymer Chemistry,* McGraw-Hill, New York, 1971, section 2.2.
8. **Burrell, H.,** Solubility of polymers, in *Kirk-Othmer Encyclopedia of Science and Technology,* 2nd ed., Vol. 12, John Wiley & Sons, New York, 1970, 618.
9. **Dernini, R. and Vargiu, S.,** Solubility parameters of SIR [Societa Italiana Resine] resins for paints, *Pitture Vernici,* 45, 452, 1969.
10. **Zotteri, L. and Giuliani, G. P.,** Effect of hydrogenation on the elastic properties of poly(styrene-*b*-diene-*b*-styrene) copolymers, *Polymer,* 19, 476, 1978.
11. **Mark, H. F.,** Future improvements in the cohesive and adhesive strength of polymers, *Adhes. Age,* 22(7), 35; 22(9), 45, 1979.
12. **Flory, P. J. and Rehner, J., Jr.,** Statistical mechanics of crosslinked polymer networks. II. Swelling, *J. Chem. Phys.,* 11, 521, 1943.
13. **Bristow, G. M. and Watson, W. F.,** Cohesive energy densities of polymers. I. Cohesive energy densities of rubbers by swelling measurements, *Trans. Faraday Soc.,* 54, 1731, 1958.

14. **Dudek, T. J. and Bueche, F.,** Polymer-solvent interaction parameters and creep behaviour of ethylene-propylene rubbers, *J. Polym. Sci. A,* 2, 811, 1964.
15. **Gee, G.,** Interaction between rubbers and liquids. IV. Factors governing the absorption of oil by rubber, *Trans. Inst. Rubber Ind.,* 18, 266, 1943.
16. **Scott, R. L. and Magat, M.,** Thermodynamics of high polymer solutions. III. Swelling of cross-linked rubber, *J. Polym. Sci.,* 4, 555, 1949.
17. **Grigorovskaya, V. A., Shvarts, A. G., and Bychkova, L. P.,** Determination of cohesive energy density of stereoregular rubbers SKI-3 and SKD, *Vysokomol. Soedin Adgeziya Polim., Sb. Statei,* 41, 1963.
18. **Makarova, L. V., Shvarts, A. G., Zakharov, N. D., and Priborets, A. M.,** Determination of the cohesive energy density of some synthetic rubbers with functional groups, *Vysokomol. Soedin.,* 7, 1056, 1965; *Polym. Sci. U.S.S.R.,* 7, 1168, 1965.
19. **Shvarts, A. G.,** Evaluation of rubber-solvent interaction, *Kolloid Zh.,* 19, 376, 1957; *Colloid J. U.S.S.R.,* 19, 375, 1957; *Rubber Chem. Technol.,* 31, 691, 1958.
20. **Shvarts, A. G.,** Comparative determination of [specific] cohesion energy of natural and polyisoprene rubber, *Zh. Fiz. Khim.,* 32, 718, 1958.
21. **Shvarts, A. G., Chefranova, E. K., and Iotkovskaya, L. A.,** Solubility parameters of resins based on dimethylvinylethynyl-*p*-hydroxyphenylmethane, *Kolloid Zh.,* 32, 603, 1970; *Colloid J. U.S.S.R.,* 32, 506, 1970.
22. **Moore, W. R. and Boden, G. F.,** Heptane-soluble material from atactic polypropylene. II. Interaction with liquids, *J. Appl. Polym. Sci.,* 10, 1121, 1966.
23. **Schneier, B.,** Use of a Gee relationship in a diffusion study, *J. Appl. Polym. Sci.,* 16, 2343, 1972.
24. **Belousov, V. N., Kotsev, B. Kn., and Mikitaev, A. K.,** Study of polycarbonate by inverse gas chromatography, *Vysokomol. Soedin. Ser. A,* 27, 416, 1985; *Polym. Sci. U.S.S.R.,* 27, 467, 1985.
25. **Castells, R. C. and Mazza, G. D.,** Study of polymer-solvent interactions by gas chromatography. Copolymers of vinyl acetate and vinyl alcohol with hydrocarbons and alcohols, *J. Appl. Polym. Sci.,* 32, 5917, 1986; *An. — CIDEPINT,* 123, 1987.
26. **Cervenka, A., Batt, A. M., and Grieveson, B. M.,** Interaction of polyurethane foams with water and its effect on material properties, *Trans. Inst. Mar. Eng.,* (C), Vol. 97, Conf. 2, Paper 15, 103, 1985.
27. **Chee, K. K.,** Novel approach to Mark-Houwink-Sakurada constants and related parameters of polystyrene solutions, *Polymer,* 28, 977, 1987.
28. **DiPaola-Baranyi, G. and Guillet, J. E.,** Estimation of polymer solubility parameters by gas chromatography, *Macromolecules,* 11, 228, 1978.
29. **DiPaola-Baranyi, G., Guillet, J. E., Klein, J., and Jeberien, H. E.,** Estimation of solubility parameters for poly(vinyl acetate) by inverse gas chromatography, *J. Chromatogr.,* 166, 349, 1978.
30. **DiPaola-Baranyi, G.,** Estimation of polymer solubility parameters by inverse gas chromatography, *Macromolecules,* 15, 622, 1982.
31. **Dudek, T. J. and Bueche, F.,** Polymer-solvent interaction parameter and creep of ethylene-propylene rubber, *Rubber Chem. Technol.,* 37, 894, 1964.
32. **Dudek, T. J. and Bueche, F.,** Polymer-solvent interaction parameters and creep behaviour of ethylene-propylene rubbers, *J. Polym. Sci.,* A, 2, 811, 1964.
33. **Fernández-Berridi, M. J., Otero, T. F., Guzmán, G. M., and Elorza, J. M.,** Determination of the solubility parameter of poly(ethylene oxide) at 25°C by gas-liquid chromatography, *Polymer,* 23, 1361, 1982.
34. **Fernández-Berridi, M. J., Guzmán, G. M., Iruin, J. J., and Elorza, J. M.,** Determination of the interaction parameter of poly(ethylene oxide) by gas-liquid chromatography below its melting point, *Polymer,* 24, 417, 1983.
35. **Eguiazabal, J. I., Fernández-Berridi, M. J., Iruin, J. J., and Elorza, J. M.,** Chromatographic determination of polymer solubility parameters, *Polym. Bull. (Berlin),* 13, 463, 1985.
36. **Freytag, U. and Radusch, H. J.,** Study of the thermodynamic compatibility of high-density polyethylene and low-density polyethylene in the melt state. I. Determination of solubility parameters, *Angew. Makromol. Chem.,* 152, 1, 1987.
37. **Galin, M.,** Gas-liquid chromatography study of poly(ethylene oxide)-solvent interactions — estimation of polymer solubility parameter, *Polymer,* 24, 865, 1983.
38. **Gräter, H., Schuster, R. H., and Cantow, H. J.,** Thermodynamics of polymer systems. V. Sorption and mixing properties of selected solutes and solubility parameters of atactic polystyrene by gas chromatography, *Polym. Bull. (Berlin),* 14, 379, 1985.
39. **Green, J. H. S.,** Cohesive density of polystyrene, *Nature (London),* 183, 818, 1959.
40. **Gregory, P. and Huglin, M. B.,** Viscosity of aqueous and alkaline solutions of poly(ethylene oxide), *Makromol. Chem.,* 187, 1745, 1986.
41. **Ito, K. and Guillet, J. E.,** Estimation of solubility parameters for some olefin polymers and copolymers by inverse gas chromatography, *Macromolecules,* 12, 1163, 1979.

42. **Lipson, J. E. G. and Guillet, J. E.,** Studies of polar and non-polar probes in the determination of infinite dilution solubility parameters, *J. Polym. Sci. Polym. Phys. Ed.*, 19, 1199, 1981.
43. **Lipson, J. E. G. and Guillet, J. E.,** Solubilities and solubility parameters of polymers from inverse gas chromatography, in *Macromolecular Solutions: Solvent-Property Relationships in Polymers* (Symp. Proc., Am. Chem. Soc., New York, 1981), Seymour, R. B. and Stahl, G. A., Eds., Pergamon, New York, 1982, 14.
44. **Lipson, J. E. G. and Guillet, J. E.,** Measurement of solubility and solubility parameters for small organic solutes in polymer films by gas chromatography, *J. Coatings Technol.*, 54(684), 89, 1982.
45. **Price, G. J., Guillet, J. E., and Purnell, J. H.,** Measurement of solubility parameters by gas-liquid chromatography, *J. Chromatogr.*, 369, 273, 1986.
46. **Hippe, Z., Balutowska, E., and Lipinski, J.,** Estimation of some thermodynamic properties of polymers in solution from osmotic pressure measurements by means of digital computer programs, *Bull. Acad. Pol. Sci. Ser. Sci. Chim.*, 23, 783, 1975.
47. **Katz, D. and Salee, G.,** Crosslinked oligoethylene glycols: correlation between their structure and some rheological properties, *J. Polym. Sci. A-2*, 6, 801, 1968.
48. **Kok, C. M. and Rudin, A.,** Estimation of solubility parameters from solution properties of polymers, *J. Coatings Technol.*, 55(704), 57, 1983.
49. **Kok, C. M. and Rudin, A.,** An engineering approach to polymer solution properties, *J. Macromol. Sci. Rev. Macromol. Chem. Phys. C*, 24, 49, 1984.
50. **Lawandy, S. N. and Darwish, N. A.,** Interaction constants, solubility parameter and lattice constant of polychloroprene rubber, *Egypt. J. Phys.*, 16, 143, 1985 (pub. 1986).
51. **Leca, M., Groswald-Cremer, L., and Surmeian, M.,** Determination of solubility parameters using gas-liquid chromatography, *Mater. Plast. (Bucharest)*, 16, 114, 1979.
52. **Leca, M. and Enache, M.,** Determination of polymer partial solubility parameters by gas-liquid chromatography, *Rev. Roum. Chim.*, 28, 287, 1983.
53. **Mangaraj, D.,** Cohesive energy densities of high polymers. I. Determination of cohesive energy density from swelling measurements, *Makromol. Chem.*, 65, 29, 1963.
54. **Merk, W., Lichenthaler, R. N., and Prausnitz, J. M.,** Solubilities of fifteen solvents in copolymers of poly(vinyl acetate) and poly(vinyl chloride) from gas-liquid chromatography. Estimation of polymer solubility parameter, *J. Phys. Chem.*, 84, 1694, 1980.
55. **Price, G. J.,** Calculation of solubility parameters by inverse gas chromatography, in *Inverse Gas Chromatography Characterization of Polymers and Other Materials (Am. Chem. Soc. Symp. Ser.* 391), Lloyd, D. R., Schreiber, H. P., and Ward, T. C., Eds., American Chemical Society, Washington, D.C., 1989, 33; *Polym. Mater. Sci. Eng.*, 58, 1009, 1988.
56. **RamiReddy, C. and Kalpagam, V.,** A rough estimation of the solubility parameters of styrene-acrylonitrile copolymers, *Current Sci.*, 43, 681, 1974.
57. **Roe, R.-J.,** Evaluation of the polymer-polymer interaction parameter from the equation of state thermodynamic theory, *Adv. Chem. Ser.*, 176 (Multiphase Polym.), 599, 1979.
58. **Sanetra, R., Kolarz, B. N., and Wlochowicz, A.,** Determination of thermodynamic data for the interaction of aliphatic alcohols with poly(styrene-*co*-divinylbenzene) using inverse gas chromatography, *Polymer*, 28, 1753, 1987.
59. **Siow, K. S., Goh, S. H., and Yap, K. S.,** Solubility parameters of poly(α-methylstyrene-co-acrylonitrile) from gas-liquid chromatography, *J. Chromatogr.*, 354, 75, 1986; *Polym. Mater. Sci. Eng.*, 51, 532, 1984.
60. **Takahashi, S.,** Determination of cohesive energy densities of unsaturated polyester resins from swelling measurements, *J. Appl. Polym. Sci.*, 28, 2847, 1983.
61. **Wu, Y.,** Direct determination of thermodynamic quantities and solubility parameter for poly(tetramethyl-*p*-silphenylene) siloxane by gas-liquid chromatography, *Sepu*, 4, 232, 1986.
62. **Yamashita, S., Minoura, Y., and Maekawa, E.,** Solubility and polymer-solvent interaction parameters of 1,2-polybutadiene, *Makromol. Chem.*, 180, 705, 1979.
63. **Yeo, R. S.,** Solubility parameter of perfluorosulfonated polymer, *Am. Chem. Soc. Symp. Ser.*, 180, 65, 1982.
64. **Cavé, G., Kothari, R., Puisieux, F., Martin, A. N., and Carstensen, T. J.,** Solubility parameters from maxima in solubility solvent plots, *Int. J. Pharm.*, 5, 267, 1980.
65. **Lawson, D. D., Ingham, J. D., and Moacanin, J.,** Solubility parameter spectroscopy, *Eighteenth Canadian High Polymer Forum,* McMaster University, Hamilton, Ontario, Can., 1975.
66. **Bagda, E.,** Determination of solution rate and solubility parameters of polymers, *Colloid Polym. Sci.*, 255, 384, 1977.
67. **Engel, J. H., Jr. and Fitzwater, R. N.,** Adhesion of surface coatings as determined by the peel method, in *Adhesion and Cohesion,* Weiss, P. N. Y., Ed., Elsevier, Amsterdam, 1962, 89.
68. **Boyer, R. F. and Keskkula, H.,** Styrene polymers: characterization, in *Encyclopedia of Polymer Science and Technology,* Vol. 13, Bikales, N. M., Ed., John Wiley & Sons, New York, 1970, 375.

69. **Suh, K. W. and Clark, D. H.,** Cohesive energy densities of polymers from turbidimetric titrations, *J. Polym. Sci. A-1,* 5, 1671, 1967; **Suh, K. W. and Corbett, J. M.,** Solubility parameters of polymers from turbidimetric titrations, *J. Appl. Polym. Sci.,* 12, 2359, 1968.
70. **Mieczkowski, R. and Huppenthal, L.,** Determination of the solubility parameter of polystyrene by partial specific volume measurements, *Polimery (Warsaw),* 32, 397, 1987; The determination of the solubility parameter components of polystyrene by partial specific volume measurements; The determination of the solubility parameter components of polystyrene, *Eur. Polym. J.,* 24, 1185, 1988; 25, 1055, 1989.
71. **Cowie, J. M. G. and McCrindle, J. T.,** Polymer co-solvent systems. II. An asymmetric solubility parameter system — acetone(1), cyclohexanol(2), polystyrene(3), *Eur. Poly. J.,* 8, 1325, 1972.
72. **Wessling, R. A.,** The solubility of poly(vinylidene chloride), *J. Appl. Polym. Sci.,* 14, 1531, 1970.
73. **Tachi, Y., Takahashi, A., and Kagawa, I.,** Solubility of quasi-polyelectrolytes. I. Solubility of ethylene-acrylic acid copolymer in organic solvents, *Kogyo Kagaku Zasshi,* 73, 398, 1970.
74. **Blanks, R. F. and Shah, B. N.,** Solubility of styrene-acrylonitrile copolymers, in *Structure-Solubility Relationships in Polymers,* Harris, F. W. and Seymour, R. B., Eds., Academic Press, New York, 1977, 111.
75. **Hoagland, M. E. and Duling, I. N.,** Polymers and copolymers of 3,5-dimethyl-1-adamantyl acrylate and 3,5-dimethyl-1-adamantyl methacrylate, *Am. Chem. Soc. Div. Petrol. Chem. Prep.,* 15, B85, 1970.
76. **Khanna, B. B.,** Solubility parameter of shellac, *Indian J. Technol.,* 8, 274, 1970.
77. **Banerjee, P. K., Srivastava, B. C., and Kumar, S.,** Cohesive-energy density of shellac, *Polymer,* 23, 417, 1982.
78. **Seymour, R. B., Johnson, E. L., and Stahl, G. A.,** Preparation and properties of cellulose blends in dimethylsulfoxide solution, in *Macromolecular Solutions: Solvent Property Relationships in Polymers* (Symp. Proc., Am. Chem. Soc., New York, 1981), Seymour, R. B. and Stahl, G. A., Eds., Pergamon, New York, 1982, 90.
79. **Mieczkowski, R. and Huppenthal, L.,** Determination of the solubility parameters of poly(vinylpyrrolidone) and PVC by measurement of partial specific volume, *Polimery (Warsaw),* 32, 192, 1987.
80. **Sono, J.,** On the solubility characteristics and physical properties of linear terephthalic polyester coating, *Shikizai Kyokaishi,* 42, 501, 1966.
81. **Schindler, G.,** Swelling behavior of plastics. II. Experimental swelling determination methods; Choice of plastic swelling agent, *Mitteil. Chem. Forschungsinst. Wirt. Oesterr. Oesterr. Kunststoffinstit.,* 25, 127, 1971.
82. **Gur-Arieh, Z. and Ingamells, W.,** The plasticisation and swelling of Acrilan under dyebath conditions, *J. Soc. Dyers Colour.,* 90, 12, 1974.
83. **Mangaraj, D., Patra, S., and Rashid, S.,** Cohesive energy densities of high polymers. II. Cohesive energy densities of polyacrylates and polymethacrylates from swelling measurements, *Makromol. Chem.,* 65, 39, 1963.
84. **Takahashi, M. and Watanabe, M.,** Studies on acrylic fibre. (XVI). Interaction of polyacrylonitrile with solvents, *Sen-i Gakkaishi,* 16, 711, 1960.
85. **Beevers, R. B.,** The swelling of Hercosett® resin, *Text. Res. J.,* 46, 636, 1976.
86. **Kambour, R. P., Gruner, C. L., and Romagosa, E. E.,** Bisphenol-A polycarbonate immersed in organic media. Swelling and response to stress, *Macromolecules,* 7, 248, 1974 (General Electric Reprint 7514).
87. **Kambour, R. P. and Gruner, C. L.,** Effects of polar group incorporation on crazing of glassy polymers: styrene-acrylonitrile copolymer and a dicyano bisphenol polycarbonate, *J. Polym. Sci. Polym. Phys.,* 16, 703, 1978 (General Electric Reprint 8346).
88. **Bhatnagar, S. K.,** Cohesive energy density of cis-polybutadiene, *Makromol. Chem.,* 122, 82, 1969.
89. **Grishin, B. S., Tutorskii, I. A., and Yurovskaya, I. S.,** Application of Gee's equation for assessment of solubility of solid low-molecular substances in polymers, *Vysokomol. Soedin. Ser. B,* 19, 387, 1977.
90. **Tager, A. A., Bessonov, Yu. S., Sivkova, Ye. Ya., Anufriyev, V. A., Vygodskii, Ya. S., and Salazkin, S. N.,** Heats of solution of cardopolyheteroarylenes and the effect of individual macromolecular groups, *Vysokomol. Soedin. Ser. A,* 21, 2136, 1979; *Polym. Sci. U.S.S.R.,* 21, 2358, 1979.
91. **Howard, P. and Parikh, R. S.,** Solution properties of cellulose triacetate. II. Solubility and viscosity studies, *J. Polym. Sci. A-1,* 6, 537, 1968.
92. **Bansal, R. K. and Sahoo, J. C.,** Determination of cohesive energy densities of bisphenol-A and epichlorohydrin epoxy resins from swelling measurements, *Angew. Makromol. Chem.,* 82, 149, 1979.
93. **Sickfeld, J.,** Swelling investigations of technical reaction resins, *Farbe Lacke,* 81, 1113, 1975.
94. **Sickfeld, J.,** Swelling in industrial reactive resins, *DEFAZET — Deutsche Farben.-Z.,* 29, 437, 1975.
95. **Avny, Y., Rebenfeld, L., and Weigmann, H.-D.,** The in situ polymerization of vinyl monomers in polyester yarns, *J. Appl. Polym. Sci.,* 22, 125, 1978.
96. **Cooper, S. L.,** Properties of swollen block copolymer elastomers, *J. Polym. Sci. A-1,* 7, 1765, 1976.
97. **Omichi, H. and Araki, K.,** Solvent effects on the radiation-induced graft polymerization of styrene onto poly(isobutylene oxide), *J. Polym. Sci. Polym. Chem.,* 14, 2773, 1976.

98. **Graham, N. B., Nwahuku, N. E., and Walsh, D. J.,** Interaction of poly(ethylene oxide) with solvents. I. Preparation and swelling of a cross-linked poly(ethylene oxide) hydrogel, *Polymer,* 23, 1345, 1982.
99. **Omichi, H. and Araki, K.,** Kinetics of radiation-induced graft polymerization of styrene onto poly(ethylene oxide), *J. Polym. Sci. Polym. Chem.,* 17, 1401, 1979.
100. **Allen, G. and Crossley, G. H.,** Polypropylene oxide. IV. Preparation and properties of polyether networks, *Polymer,* 5, 553, 1964.
101. **Bernier, G. A. and Kambour, R. P.,** The role of organic agents in the stress crazing and cracking of poly(2,6-dimenthyl-1,4-phenylene oxide), *Macromolecules,* 1, 393, 1968 (General Electric Reprint 6125).
102. **Broens, L., Koenhen, D. M., and Smolders, C. A.,** On the mechanism of formation of asymmetric ultra- and hyperfiltration membranes, *Desalination,* 22, 205, 1977.
103. **Ceausescu, E., Lebada, P., and Petcu, L.,** Determining the solubility parameters of ethylene-propylene co- and ter-polymers, *Mater. Plast. (Bucharest),* 9, 411, 1972; *Lucr. Conf. Nat. Chim. Anal.,* 3rd, Vol. 4, Inst. Cent. Cercet. Chim., Bucharest, Romania, 1971, 7.
104. **Frensdorff, H. K.,** Diffusion and sorption of vapors in ethylene-propylene copolymers. I. Equilibrium sorption, *J. Polym. Sci. A,* 2, 333, 1964.
105. **Kirkham, M. C.,** Properties and microstructures of ethylene-propylene terpolymers, *J. Appl. Polym. Sci.,* 17, 1101, 1973.
106. **Knox, B. H.,** Bimodal character of polyester-solvent interactions. I. Evaluation of the solubility parameters of the aromatic and aliphatic ester residues of poly(ethylene terephthalate); II. Evaluation of the chemical structures of the aromatic and aliphatic ester residues of poly(ethylene terephalate); III. The effects of the morphology on the nature of the interaction of nonaqueous solvents with the aromatic and the aliphatic ester residues of poly(ethylene terephthalate), *J. Appl. Polym. Sci.,* 21, 225, 249, and 267, 1977.
107. **Knox, B. H., Weigmann, H.-D., and Scott, M. G.,** Interactions of nonaqueous solvents with textile fibres. Part V: Application of the solubility parameter concept to polyester fiber-solvent interactions, *Text. Res. J.,* 45, 203, 1975.
108. **Rickert, S. E., Balik, C. M., and Hopfinger, A. J.,** Organization and assembly of macromolecules on inorganic substrates, *Adv. Colloid Interface Sci.,* 11, 149, 1979.
109. **Starobinets, G. L., Levdanskaya, V. V., Rakhman'ko, E. M., and Karabanovich, V. S.,** Interactions of polyethylene terephthalate with binary mixtures of organic solvents, *Zh. Prikl. Khim.,* 51, 922, 1978; *J. Appl. Chem. U.S.S.R.,* 51, 886, 1978.
110. **Starobinets, G. L., Rakhman'ko, E. M., Levdanskaya, V. V., and Karabanovich, V. S.,** Swelling of poly(ethylene terephalate) in binary mixtures of organic solvents, *Vestsi Akad. Navuk BSSR Ser. Khim. Navuk,* No. 1, 12, 1977.
111. **Nishimura, J., Tanaka, N., and Yamashita, S.,** Thermostability of polymers containing units in the main chain, *J. Polym. Sci. Polym. Chem.,* 18, 1203, 1980.
112. **Kamata, S. and Gohara, M.,** Solvent sorption of ion-exchange resins on non-aqueous solvents, *Nippon Kagaku Kaishi,* 532, 1977.
113. **Szentirmay, M. N. and Martin, C. R.,** Chemical properties and film casting of radiation-grafted ion containing polymers, *J. Electrochem. Soc.,* 131, 1652, 1984.
114. **Covitch, M. J.,** Solution processing of perfluorinated ionomers: recent developments, American Chemical Society Meeting, Philadelphia, August, 1984.
115. **McCain, G. H. and Covitch, M. J.,** Solubility characteristics of perfluorinated polymers with sulfonyl fluoride functionality, *J. Electrochem. Soc.,* 131, 1350, 1984.
116. **Yeo, R. S.,** Dual cohesive energy densities of perfluorosulfonic acid (Nafion) membrane, *Polymer,* 21, 432, 1980.
117. **Yeo, R. S.,** Ion clustering in Nafion and its relationship to membrane conductivity, in *Proceedings, Symposium on Ion Exchange,* Yeo, R. S. and Buck, R. P., Eds., Electrochemical Society, Princeton, NJ, 1981.
118. **Yeo, R. S., Chan, S. F., and Lee, J.,** Swelling behavior of Nafion and radiation-grafted cation-exchange membranes, *J. Membrane Sci.,* 9, 273, 1981.
119. **Lee, J., Chan, S. F., and Yeo, R. S.,** Swelling behavior of ion exchange membranes in non aqueous solvents, in *Proceedings, Symposium on Ion Exchange,* Yeo, R. S. and Buck, R. P., Electrochemical Society, Princeton, NJ, 1981.
120. **Nekrasova, T. N. and Eskin, V. Ye.,** The partial specific volume of polyisobutylene in various solvents, *Vysokomol. Soedin. Ser. A,* 15, 2429, 1973; *Polym. Sci. U.S.S.R.,* 15, 2749, 1973.
121. **Eissler, R. L. and Princen, L. H.,** Effect of some pigments on tensile and swelling properties of linseed oil films, *J. Paint Technol.,* 40(518), 105, 1968.
122. **Polowinski, S. and Jacon, E.,** Solubility parameters of aromatic polymethacrylates, *Polimery (Warsaw),* 24, 344, 1979.
123. **Ore, S., Kuluik, E., Ohr, W., and Zapffe, J. W.,** Studies on the swelling of cross-linked polymers of mono-, di-, tri-, and poly(ethylene glycol)monomethacrylates in water and normal alcohols, *Makromol. Chem.,* 148, 211, 1971.

124. **Varma, I. K. and Patnaik, S.,** Determination of cohesive energy densities of 2-hydroxyethylmethacrylate and methacrylate copolymers from swelling measurements, *Angew. Makromol. Chem.,* 55, 109, 1976.
125. **Ross, J. H.,** Demonstration of solvent differences by visible polymer swelling, *J. Chem. Ed.,* 60, 169, 1983.
126. **Michaels, A. S., Vieth, W. R., and Alcalay, H. H.,** The solubility parameter of polypropylene, *J. Appl. Polym. Sci.,* 12, 1621, 1968.
127. **Michaels, A. S., Vieth, W. R., Hoffman, A. S., and Alcalay, H. A.,** Structure-property relations for liquid transport in modified polypropylene membranes, *J. Appl. Polym. Sci.,* 13, 577, 1969.
128. **Shawali, A. S.,** *Absorption of Organic Liquids by Polypropylene,* M.S. Thesis, Lowell Technological Institute, Lowell, MA, 1962.
129. **Valko, E. I. and Shawali, A. S.,** Absorption of organic liquids by polypropylene, *Abstr. Pap. 160th Am. Chem. Soc. Nat. Meeting,* Chicago, Sept 1970.
130. **Valko, E. I.,** Penetration of fibers, in *Chemical Aftertreatment of Textiles,* Mark, H., Wooding, N. S., and Atlas, S. M., Eds., John Wiley & Sons-Interscience, New York, 1971, chap. 2.
131. **De Chirico, A. and Raggi, E.,** Cohesive energy density of polypropylensulphide, *Chim. Ind. (Milan),* 55, 512, 1973.
132. **Mishra, M. K.,** Solution properties of shellac. IV. Solubility parameter of shellac, *Angew. Makromol. Chem.,* 147, 107, 1987.
133. **Yerrick, K. B. and Beck, H. N.,** Solvent resistance of silicone elastomers: solvent-polymer interactions, *Rubber Chem. Technol.,* 37, 261, 1964.
134. **Small, P. A.,** Some factors affecting the solubility of polymers, *J. Appl. Chem.,* 3, 71, 1953.
135. **Hauser, R. L., Walker, C. A., and Kilbourne, F. L., Jr.,** Swelling of silicone elastomers, *Ind. Eng. Chem.,* 48, 1202, 1956.
136. **Humcke-Bogner, R., Liu, J.-C., and Chien, Y. W.,** Methods for determining partial solubility parameters of potential film-coating polymers, *Int. J. Pharm.,* 42, 199, 1988.
137. **Hirabayashi, K.,** Structural formation of silk fibroin (*Bombyx mori L*) by fiber dimensional change, *Zoku Kenshi no Koso,* Nobumasa, H., Ed., Shinshu Daigaku, Sed'igakubu, Ueda, Japan, 1980, 225.
138. **Hirabayashi, K., Oki, R., Shigematsu, M., and Arai, M.,** Studies on the mechanochemical reactivity of silk fibroin. I. Contraction behavior of silk fibers in organic solvents, *Nippon Sanshigaku Zasshi,* 49(1), 27, 1980.
139. **Oki, R., Shigematsu, M., Hirabayashi, K., and Arai, M.,** Studies on the mechanochemical reactivity of silk fibroin. II. Mechanical properties of silk fibers in organic solvents, *Nippon Sanshigaku Zasshi,* 49(1), 39, 1980.
140. **Boyer, R. F. and Soencer, R. S.,** Some thermodynamic properties of slightly cross-linked styrene-divinylbenzene gels, *J. Polym. Sci.,* 3, 97, 1948.
141. **Kambour, R. P., Gruner, C. L., and Romagosa, E. E.,** Solvent crazing of "dry" polystyrene and "dry" crazing of plasticized polystyrene, *J. Polym. Sci. Polym. Phys.,* 11, 1879, 1973 (General Electric Reprint 7435).
142. **Wagner, E. W. and Cotter, R. J.,** Ultrasonic degradation of the gel phase in impact polystyrenes, *J. Appl. Polym. Sci.,* 15, 3043, 1971.
143. **Sieglaff, C. L.,** Viscosity and swelling behaviour of lightly crosslinked microgels, *Polymer,* 4, 281, 1963.
144. **Errede, L. A.,** Correlation of polymer swelling with solubility parameter, *Abstr. Pap. Am. Chem. Soc.,* 191(Apr.), 133, 1986; *Polym. Mater. Sci. Eng.,* 54, 561, 1986.
145. **Errede, L. A.,** Polymer swelling. II. A restudy of poly(styrene-*co*-divinylbenzene) swelling in terms of cross-link density, *J. Appl. Polym. Sci.,* 31, 1749, 1986.
146. **Errede, L. A.,** Polymer swelling. V. Correlation of relative swelling of poly(styrene-*co*-divinylbenzene) with the Hildebrand solubility parameter of the swelling liquid; VI. A study of poly(styrene-*co*-divinylbenzene) swelling in liquid of the homologous series $Ph(CH_2)_nH$, *Macromolecules,* 19, 1522 and 1525, 1986.
147. **Chapiro, A. and Jendrychowska-Bonamour, A. M.,** Properties of some graft copolymers. I. Preparation of samples. Solubility and swelling, *J. Chim. Phys.,* 60, 1029, 1963.
148. **Glöckner, G.,** Interactions between organic solvents and acrylonitrile copolymers, *Wiss. Z. Tech. Hochsch. "Carl Schorlemmer" Leuna-Merseburg,* 17, 279, 1975.
149. **Kambour, R. P., Romagosa, E. E., and Gruner, C. L.,** Swelling, crazing and cracking of an aromatic copolyether-sulfone in organic media, *Macromolecules,* 5, 335, 1972 (General Electric Reprint 7171).
150. **Starkweather, H. W., Jr.,** The sorption of chemicals by perfluorocarbon polymers, *Macromolecules,* 10, 1161, 1977.
151. **Galembeck, F.,** Sorption of iron penacarbonyl in Teflon, *J. Polym. Sci. Polym. Chem.,* 16, 3015, 1978.
152. **Galembeck, F., Galembeck, S. E., and Vicente, E. J.,** Investigation of the interaction of certain low cohesive energy liquids with Teflon and polymeric surfaces, in *Proc. Intl. Symp. Contam. Control,* 4th, Institute of Environmental Science, Mt. Prospect, IL, 1978, 329.

153. **Gorin, Yu. A., Rodina, E. I., and Rapoport, L. Ya.,** Swelling of copolymers of tetrahydrofuran with 3-ethyl-3-(chloromethyl)oxacyclobutane, *Vysokomol. Soedin. Ser. B,* 11, 512, 1969.
154. **Chapiro, A., Lamothe, M., and LeDoan, T.,** Solubility parameter of a polyurethane of defined structure, *Eur. Polym.,* 14, 674, 1978.
155. **Saitoh, K., Ozawa, T., and Suzuki, N.,** Solvent dependence of swelling of Merkogel OR-2000 and various organic solvents, *J. Chromatogr.,* 124, 231, 1976.
156. **Chapiro, A., Mankowski, Z., and Schmitt, N.,** Unusual swelling behavior of films of polyvinyl- and polyvinylidene/fluorides in various solvents, *J. Polym. Sci. Polym. Chem.,* 20, 1791, 1982.
157. **Bottino, A., Capannelli, G., Munari, S., and Turturro, A.,** Solubility parameter of poly(vinylidene fluoride), *J. Polym. Sci. Polym. Phys.,* 26, 785, 1988.
158. **Tashmukhamedov, S. A., Sagdullaev, B. U., Karabaev, A. Sh., and Tillaev, R. S.,** Solubility parameter of poly(vinylpyrrolidone), *Uzb. Khim. Zh.,* 16(6), 47, 1972.
159. **Huppenthal, L., Staszewska, D. U., and Nowacki, J.,** Solubility parameters of poly(vinylpyrrolidone), *Polimery (Warsaw),* 27, 427, 1982.
160. **Vasilakos, N. P. and Lawson, D. D.,** Solubility parameter spectroscopy in coal extraction, *Liq. Fuels Technol.,* 3, 305, 1985.
161. **Kwan, C. L., Chow, E., and Yen, T. F.,** Study of interaction of organic solvents with shale and kerogen using the method of solubility parameter spectrum, *Abstr. Pap. Am. Chem. Soc.,* 177, GEOC 19, 1979.
162. **Yamamoto, T., Isshiki, S., and Nakayamma, S.,** Synthetic paper for extra high voltage cable, *IEEE Trans. Power Appl. Syst.,* PAS 91, 2415, 1972.
163. **Turyshev, B. I., Shuvaev, V. P., Goldenberg, A. D., and Sazhin, B. I.,** Electrical conductivity of polymer dielectrics in contact with liquid media, *Plast. Massy,* No. 7, 69, 1978; *Int. Polym. Sci. Technol.,* 6(4), 25, 1979.
164. **Eskin, V. E., Zhurayev, U., and Nekrosova, T. N.,** On some features of intermolecular interaction exhibited in polymer solutions, *Vysokomol. Soedin. Ser. A,* 18, 2529, 1976; *Polym. Sci. U.S.S.R.,* 18, 2890, 1976.
165. **Gee, G.,** The interaction between rubber and liquids. III. The swelling of vulcanized rubber in various liquids, *Trans. Faraday Soc.,* 38, 418, 1942.
166. **Piglowski, J. and Skowronski, T.,** Compatibility of poly(vinylchloride) with butyl acrylate-acrylonitrile copolymers, *Angew. Makromol. Chem.,* 88, 165, 1980.
167. **Krokhina, L. S., Zyateva, T. V., Shelest, I. I., and Kuleznev, V. N.,** On the energy density of cohesion of butyl and chlorobutyl rubbers and their vulcanizates, *Vysokomol. Soedin. Ser. B.,* 23, 147, 1981.
168. **Choudhary, M. S. and Varma, I. K.,** Copolymers of 2-hydroxyethylmethacrylate and alkyl methacrylates. I. Synthesis and characterization, *J. Polym. Sci. Polym. Chem.,* 23, 1917, 1985.
169. **Huglin, M. B. and Pass, D. J.,** Cohesive energy density of polytetrahydrofuran, *J. Appl. Polym. Sci.,* 12, 473, 1968.
170. **Feger, C. and Cantow, H. J.,** Cellulose containing block copolymers. II. Molecular weights and solution properties of trimethylcellulose poly(oxytetramethylene) block copolymers, *Polym. Bull.,* 6, 321, 1982.
171. **Apukhtina, N. P., Erenburg, Ye. G., and Rappaport, L. Ya.,** Study of the swelling of polyurethanes, *Vysokomol. Soedin.,* 8, 1057, 1966; *Polym. Sci. U.S.S.R.,* 8, 1160, 1966.
172. **Piglowski, J. and Laskawski, W.,** Compatibility of poly(vinylchloride)/polyurethane blends, *Angew. Macromol. Chem.,* 82, 157, 1979.
173. **Lindner, W. L.,** Interaction between organic solvents and poly(ethylene terephthalate) fibers. I; II. The maximal shrinking rate, maximal shrinkage and swelling, *Colloid Polym. Sci.,* 255, 213 and 433, 1977.
174. **Baranwal, K. C.,** Cohesive energy densities of elastomers, *Macromol. Chem.,* 100, 242, 1967.
175. **Basin, V. E. and Shvarts, A. G.,** Determination of cohesive energy densities of some synthetic rubbers, *Vysokomol. Soedin.,* 2, 1470, 1960.
176. **Bhatnager, S. K. and Banerjee, S.,** Cross-linking efficiency of styrene-butadiene rubber with dicumyl peroxide and evaluation of interaction parameter, *Makromol. Chem.,* 109, 217, 1967.
177. **Bristow, G. M. and Watson, W. F.,** Viscosity-equilibrium swelling correlations for natural rubber, *Trans. Faraday Soc.,* 54, 1567, 1958.
178. **Bristow, G. M. and Watson, W. F.,** Swelling and viscosity of rubbers in mixed solvents, *Trans. Inst. Rubber Ind.,* 35, 73, 1959.
179. **Gabris, T.,** Interaction between elastomers and solvents, *Gummi Asbest. Kunstst.,* 24, 694, 1971.
180. **Gee, G.,** Thermodynamic study of rubber solutions and gels, in *Adv. Colloid Sci.,* Vol. II, Interscience, New York, 1946, 145.
181. **Samuels, S. L. and Wilkes, G. L.,** Effects of gamma radiation on the dynamic mechanical and swelling behaviour of styrene butadiene block and random copolymers, *Polym. Eng. Sci.,* 13, 280, 1973.
182. **Treloar, L. R. G.,** *The Physics of Rubber Elasticity,* 2nd ed., Clarendon, Oxford, 1958; 3rd ed., 1975.
183. **Lazurin, E. A., Samorodov, V. T., and Kosmodemyanskii, L. V.,** Determination of solubility parameters of mineral oils, *Prom-st. Sint. Kauch,* No. 8, 10, 1980; *Int. Polym. Sci. Technol.,* 8(12), T12, 1981.

184. **Lychkin, I. P., Fedotova, L. V., Litvinova, T. V., Kopytina, T. D., and Mazyrina, L. A.,** Determination of solubility parameters of plasticizers produced from hydroformylation products, *Izv. Vyssh. Uchebn. Zaved. Khim. Khim. Tekhnol.*, 29(5), 92, 1986.
185. **Roberts, G. A. F. and Thomas, I. M.,** Determination of the solubility parameter of o-hydroxypropylcellulose, *Polymer,* 19, 459, 1978.
186. **Aharoni, S. M., Cilurso, F. E., and Hanrahan, J. M.,** Trifluoroethanol/chloroalkane mixtures: excellent novel solvents for aliphatic polyamides, *J. Appl. Polym. Sci.*, 30, 2505, 1985.
187. **Froehling, P. E., Koenhen, D. M., Bantjes, A., and Smolders, C. A.,** Swelling of linear polymers in mixed swelling agents; predictability by means of solubility parameters, *Polymer,* 17, 835, 1976.
188. **Rigbi, Z.,** Prediction of swelling of polymers in 2 and 3 component mixtures, *Polymer,* 19, 1229, 1978.
189. **Seymour, R. B., Wood, H., and Owen, D. R.,** The characterization of polymeric systems by turbidimetric titration, *Pop. Plast.*, 22(3), 21, 1977.
190. **Riess, G. and Callot, P.,** Fractionation of copolymers, in *Fractionation of Synthetic Polymers,* Tung, L. H., Ed., Marcel Dekker, New York, 1977, Chap. 5.
191. **Elias, H. G.,** Cloud-Point and turbidity titration, in *Fractionation of Synthetic Polymers,* Tung, L. H., Ed., Marcel Dekker, New York, 1977, chap. 4.
192. **Seymour, R. B. and Wood, H. A.,** Determination of solubility parameters of copolymers of styrene and acrylonitrile, in *Structure-Solubility Relationships in Polymers* (Proc. Symp., 1976), Harris, F. W. and Seymour, R. B., Eds., Academic Press, New York, 1977, 99.
193. **Wu, C.-C. and Hsieh, W.-C.,** Solubility parameter, IR spectrometric composition analysis and the flow testing of acrylonitrile and vinyl acetate copolymers, *Chemistry (Hua Hsueh Taiwan),* No. 3, 85, 1974.
194. **Yakovlev, S. A., Hoering, Z., and Ulbricht, I.,** Solubility parameters of some N-substituted polymaleimides, *Vysokomol. Soedin. Ser. B,* 26, 403, 1984.
195. **Tashmukhamedov, S. A., Khasankhanova, M., Fuzailov, Sh., Tillayev, R. S., and Usmanov, Kh. U.,** Turbidimetric and viscometric determination of the solubility parameter of chlorinated polyvinylchloride, *Vysokomol. Soedin. Ser. A,* 15, 699, 1973; *Polym. Sci. U.S.S.R.*, 15, 790, 1973.
196. **Tillaev, R. S., Khasankhanova, M., Tashmukhamedov, S. A., and Usmanov, Kh. U.,** Determination of solubility parameters of chlorinated poly(vinyl chloride) by turbidimetry and viscosimetry, *J. Polym. Sci. Polym. Symp.*, 39, 107, 1972.
197. **Jayasri, A. and Yaseen, M.,** Solubility parameters of polymeric materials by turbidimetric titration, *Proc. Semin. Polym. Surf. Coat.: Recent Dev.*, 21, 1982.
198. **Miller, R. F.,** Evaluation of ethylene copolymers as pour depressants, in *Macromolecular Solutions: Solvent-Property Relationships in Polymers* (Symp. Proc., Am. Chem. Soc., New York, 1981), Seymour, R. B. and Stahl, G. A., Eds., Pergamon, New York, 1982, 84.
199. **Gal, A., Cais, M., and Kohn, D. H.,** Organometallic polymers, II. Crosslinking of ferrocenylene-methylene-type polymers, *J. Appl. Polym. Sci.*, 15, 1571, 1971.
200. **Arndt, K. F. and Häusler, K. G.,** Determining the features of polymers by turbidimetric titrations using the SPEKOL spectrocalorimeter, *Jena Rev.*, 23, 238, 1978.
201. **Marco, C., Bello, J. G., and Fatou, J. G.,** Cohesive energy densities of polyethers: poly(hexamethylene oxide), *Eur. Polym. J.*, 17, 41, 1981.
202. **Alamo, R., Fatou, J. G., and Bello, A.,** Solubility parameter and random coil dimensions of poly (1,3-dioxolane), *Polym. J.*, 15, 491, 1983.
203. **Marco, C., Bello, A., Fatou, J. G., and Garza, J.,** Solubility parameters of poly(1,3-dioxepane), *Makromol. Chem.*, 187, 177, 1986.
204. **Volkova, L. A., Andreeva, N. A., Podol'skii, A. F., Taran, A. A., and Eskin, V. E.,** Critical opalescence of poly(α-methylstyrene) solutions in cyclohexane, *Vysokomol. Soedin. Ser. A.*, 19, 475, 1975, *Polym. Sci. U.S.S.R.*, 19, 543, 1979.
205. **Volkova, L. A., Andreeva, N. A., and Erskin, V. E.,** Critical opalescence of poly-2,4-dimethylstyrene solutions in *n*-heptane, *Vysokomol. Soedin. Ser. A.* 20, 874, 1978; *Polym. Sci. U.S.S.R.*, 20, 986, 1978.
206. **Hansen, C. M.,** *Three Dimensional Solubility Parameter and Solvent Diffusion Coefficient. Importance in Surface Coating Formulation,* Doctoral Dissertation, Danish Technical Press, Copenhagen, 1967.
207. **Hansen, C. M.,** The three dimensional solubility parameter — key to paint component affinities. I. Solvents, plasticizers, polymers and resins, *J. Paint Technol.*, 39, 104, 1967.
208. **Hansen, C. M.,** The three dimensional solubility parameter — key to paint component affinities. II. Dyes, emulsifiers, mutual solubility and compatibility, and pigments, *J. Paint Technol.*, 39, 505, 1967.
209. **Hansen, C. M.,** On the application of the three dimensional solubility parameter concept to the prediction of mutual solubility and compatibility, *Skand. Tidskr. Färg Lack,* 13, 132, 1967.
210. **Hansen, C. M.,** The three dimensional solubility parameter and its use. I. Introduction and non-pigmented systems, *Skand. Tidskr. Färg Lack,* 14(2), 28; 14(3), 45, 1968.
211. **Hansen, C. M.,** The universality of the solubility parameter, *Ind. Eng. Chem. Prod. Res. Dev.*, 8, 2, 1969.
212. **Hansen, C. M.,** Characterization of surfaces by spreading liquids, *J. Paint Technol.*, 42, 660, 1970.

213. **Hansen, C. M.,** Solubility in the coatings industry, *Skand. Tidskr. Färg Lack,* 17, 69, 1971.
214. **Hansen, C. M.,** Surface dewetting and coatings performance, *J. Paint Technol.,* 44(570), 57, 1972.
215. **Hansen, C. M.,** Solvents for coatings, *Chemtech,* 2, 547, 1972.
216. **Hansen, C. M.,** Solvent selection by computer, in *Solvents Theory and Practice (Adv. Chem. Ser.* 124), Tess, R. W., Ed., American Chemical Society, Washington, D.C., 1973, chap. 4.
217. **Hansen, C. M.,** Solvents in water-borne coatings, *Ind. Eng. Chem. Prod. Res. Dev.,* 16, 266, 1977.
218. **Hansen, C. M.,** Some aspects of acid/base interactions, *Farbe Lack,* 83, 595, 1977.
219. **Hansen, C. M.,** Advances in the technology of solvents in coatings, *FATIPEC (Fed. Assoc. Tech. Ind. Peint. Vernis Emoux Encres Impr. Eur. Cont.) Congr.,* 14, 97, 1978.
220. **Hansen, C. M.,** Some answered and unanswered questions about the solubility parameter, in *Macromolecular Solutions: Solvent-Polarity Relationships in Polymers* (Symp. Proc., Am. Chem. Soc., New York, 1981), Seymour, R. B. and Stahl, G. A., Eds., Pergamon, New York, 1982, 1; *Org. Coat. Plast. Chem.,* 45, 227, 1981.
221. **Hansen, C. M.,** Organic solvents in high solids and water-reducible coatings, *Prog. Org. Coat.,* 10, 331, 1982.
222. **Hansen, C. M. and Beerbower, A.,** Solubility parameters, in *Kirk-Othmer Encyclopedia of Chemical Technology,* Suppl. Vol., 2nd ed., Standen, A., Ed., Interscience, New York, 1971, 889.
223. **Hansen, C. M. and Pierce, P. E.,** Surface effects in coating processes, *Ind. Eng. Chem. Prod. Res. Dev.,* 13, 218, 1974.
224. **Skarrup, K.,** The three dimensional solubility parameter and its use. II. Pigmented systems, *Skand. Tidskr. Färg Lack,* 14(2), 28; 14(3), 45, 1968.
225. **Hansen, C. M. and Skaarup, K.,** The three dimensional solubility parameter — key to paint component affinities. III. Independent calculation of the parameter components, *J. Paint Technol.,* 39, 511, 1967.
226. **Hansen, C. M. and Skaarup, K.,** Aspects of the three-dimensional solubility parameter, *Dan. Kemi,* 48(6), 81, 1967.
227. **Hansen, C. M. and Wallström, E.,** On the use of cohesion parameters to characterize surfaces, *J. Adhesion,* 15, 275, 1983.
228. **Ramsbotham, J.,** Solvent formulation for surface coatings, *Prog. Org. Coatings,* 8, 113, 1980.
229. **Shell Chemicals,** *Solubility Parameters,* 2nd ed., Tech. Bull. ICS(X)/78/1, 1978.
230. **Nunn, C. J.,** Solubility parameters: some practical considerations, *Chim. Peintures,* 34, 215, 1971.
231. **Izumi, Y. and Miyake, Y.,** Study of linear poly(*p*-chlorostyrene)-diluent systems. I. Solubilities, phase relationships, and thermodynamic interactions, *Polym. J.,* 3, 647, 1972.
232. **Rider, P. E.,** Hydrogen bonding potential, HBP: A model for predicting resin solubilities in organic solvents, *Polym. Eng. Sci.,* 23, 810, 1983.
233. **Wernick, D. L.,** Stereographic display of three-dimensional solubility parameter corerlations, *Ind. Eng. Chem. Prod. Res. Dev.,* 23, 240, 1984.
234. **Balcar, H. and Bohackova, V.,** Solution properties of poly (1-chloro-2, 3-epoxypropane), *Collect. Czech. Chem. Commun.,* 42, 2145, 1977.
235. **Leca, M.,** A new route for determining the solubility parameters of organic solvents by gas liquid chromatography, *Rev. Roum. Chim.,* 27, 987, 1982.
236. **Baney, R. H., Voigt, C. E., and Mentele, J. W.,** Partial solubility parameters for polydimethylsiloxane and polyethylmethylsiloxanes, in *Structure-Solubility Relationships in Polymers* (Proc. Symp., 1976), Harris, F. W. and Seymour, R. B., Eds., Academic Press, New York, 1977, 225.
237. **Hayduk, W. and Bromfield, H. A.,** Theta-cosolvent solutions for poly(vinyl chloride) by cloud point and osmotic pressure measurements, *J. Appl. Polym. Sci.,* 22, 149, 1978.
238. **Bottino, A., Capannelli, G., Munari, S., and Turturro, A.,** Thermodynamic approach to prepare asymmetric membranes of poly(vinylidene fluoride, in *Synthetic Polymeric Membranes* (Proc. Microsymp. Macomol., 29th, 1986), Sedláček, B. and Kahovec, J., Eds., Walter de Gruyter & Co., Berlin, 1987, 141.
239. **Van Dyk, J. W., Frisch, H. L., and Wu, D. T.,** Solubility, solvency and solubility parameters, *Ind. Eng. Chem. Prod. Res. Dev.,* 24, 473, 1985; *Abs. Pap. Am. Chem. Soc.,* 188, 108, 1984.
240. **Beerbower, A. and Dickey, J. R.,** Advanced methods for predicting elastomer/fluids interactions, *Am. Soc. Lubric. Eng. Trans.,* 12, 1, 1969.
241. **Gardon, J. L. and Teas, J. P.,** Solubility parameters, in *Treatise on Coatings,* Vol. 2, *Characterization of Coatings: Physical Techniques,* Part II, Myers, R. R. and Long, J. S., Eds., Marcel Dekker, New York, 1976, chap. 8, 413.
242. **Teas, J. P.,** Graphic analysis of resin solubilities, *J. Paint Technol.,* 40(516), 19, 1968; *Predicting Resin Solubilities,* Ashland Chemical Co., Tech. Bull. 1206, Ashland, KY, 1971.
243. **Stephanopoulos, G. and Townsend, D. W.,** Synthesis of process development, *Chem. Eng. Res. Des.,* 64, 160, 1986.
244. **Fowkes, F. M.,** Acid-base interactions in polymer adhesion, in *Physicochemical Aspects of Polymer Surfaces,* Vol. 2, Mittal, K. L., Ed., Plenum, New York, 1983, 583.

245. **Fowkes, F. M., Tischler, D. O., Wolfe, J. A., Lannigan, L. A., Ademu-John, C. M., and Halliwell, M. J.,** Acid-base complexes of polymers, *J. Polym. Sci. Polym. Chem.*, 22, 547, 1984.
246. **Fowkes, F. M.,** Acid-base contributions to polymer-filler interactions, *Rubber Chem. Technol.*, 57, 328, 1984.
247. **Brandts-Buys, L. and Huyskens, P.,** Limitations in the applicability of the concept of "solubility parameter," in *Proc. 16th FATIPEC Congress,* Vol. 2, 1981, 191.
248. **Huyskens, P. L. and Haulait-Pirson, M.-C.,** A new approach for the quantitative prediction of the solubility of polymers in solvents and solvent mixtures, 10th Intl. Conf. Organic Coatings Science and Technology, Athens, July 1984.
249. **Haulait-Pirson, M. C. and Huyskens, P. L.,** New possibilities in the quantitative prediction of the solubility of polymers, 27th FATIPEC Congress, September 1984.
250. **Huyskens, P. L., Haulait-Pirson, M. C., Brandts-Buys, L. D., and Van der Borght, X. M.,** *J. Coatings Technol.*, 57(724), 57, 1985.
251. **Huyskens, P., Haulait-Pirson, M.-C., and Van der Borght, X.,** Calculation of the solvent power of solvents for resins, *Farbe Lack,* 92, 1145, 1986.
252. **Eskin, V. E. and Nesterov, A. E.,** Critical opalescence of polymer solutions, *J. Polym. Sci.*, C, 16, 1619, 1967.
253. **van Krevelen, D. W.,** *Properties of Polymers: Their Estimation and Correlation with Chemical Structure,* 2nd ed., Elsevier, Amsterdam, 1976.
254. **Heller, W.,** Treatment of viscosity data on polymer solutions (an analysis of equations and procedures). 1. Intrinsic viscosity and limiting slope constants, *J. Colloid Sci.*, 9, 547, 1954.
255. **Li, H.,** An empirical formula for calculation of the α index in the Mark-Houwink equation, *Ziran Zazhi,* 10, 635, 1987.
256. **van Krevelen, D. W. and Hoftyzer, P. J.,** Practical evaluation of the $[\eta]$-M relation, *J. Appl. Polym. Sci.*, 11, 2189, 1967.
257. **Song, W. R. and Brownawell, D. W.,** Polymer solubility parameters determined from measurements in a single solvent, *Polym. Eng. Sci.*, 10, 22, 1970.
258. **Millich, F.,** Rigid rods and the characterization of polyisocyanides, *Adv. Polym. Sci.*, 19, 117, 1975.
259. **Ahmad, H. and Yaseen, M.,** Application of intrinsic viscosity data for determination of solubility parameters and molecular weights of alkyds, *J. Coatings Technol.*, 50(640), 86, 1978.
260. **Rao, M. V. R. and Yaseen, M.,** Determination of the solubility parameters of alkyd resins. Effect of the solubility parameters of solvents, *Farbe Lack,* 91, 810, 1985.
261. **Rao, M. V. R. and Yaseen, M.,** Viscosity average molecular weights of resins: their dependences on the nature and solubility parameter of solvents, *Polym. Mater. Sci. Eng.*, 51, 601, 1984; *Abstr. Am. Chem. Soc.*, 188, 122, 1985.
262. **Rao, M. V. R. and Yaseen, M.,** Solution parameters of linseed oil alkyd: their dependence on solute-solvent interactions, *J. Coatings Technol.*, 58(743), 49, 1986.
263. **Shareef, K. M. A. and Yaseen, M.,** Rheological properties of alkyd solutions and their relation with solubility parameters of solvents, *J. Coatings Technol.*, 55(701), 43, 1983.
264. **Cowie, J. M. G.,** Studies on amylose and its derivatives. VI. Estimation of cohesive energy densities of some amylose derivatives, *Biopolymers,* 3, 69, 1965.
265. **Tager, A. A. and Kolmakova, L. K.,** The solubility parameter, methods of its determination and its relationship with the solubility of polymers, *Vysokomol. Soedin Ser. A,* 22, 483, 1980; *Polym. Sci. U.S.S.R.*, 22, 533, 1980.
266. **Tager, A. A., Kolmakova, L. K., Shemyakina, G. Ya., Vygodskii, Ya. S., and Salazkin, S. N.,** Cohesion energy density and intrinsic viscosity of aromatic cardo polymers, *Vysokomol. Soedin Ser. B,* 18, 569, 1976.
267. **Yamashita, S., Minoura, Y., and Maekawa, E.,** Solubility and polymer-solvent interaction parameters of 1,2-polybutadiene, *Makromol. Chem.*, 180, 705, 1979.
268. **Zuchowska, D.,** Structure and properties of chemically modified low-molecular weight polybutadienes, *Pr. Nauk. Inst. Elektron. Politech. Wroclaw,* 33, Monograph 7, 1979.
269. **Zuchowska, D.,** Polybutadiene modified by epoxidation. 1. Effect of polybutadiene microstructure on the reactivity of double bonds, *Polymer,* 21, 514, 1980.
270. **Gancarz, I. and Laskawskii, W.,** Peroxide-initiated crosslinking of maleic anhydride-modified low-molecular weight polybutadiene. II. Crosslinking degree and thermal degradation of cured polymers, *J. Polym. Sci. Polym. Chem.*, 17, 1523, 1979.
271. **Lazurin, E. A., Kosmodem'yanskii, L. V., Tret'yakova, N. M., and Samorodov, V. T.,** Viscosity of butyl rubber solutions in mixtures of hydrocarbon solvents with ethanol, *Kauch. Rezina,* No. 7, 7, 1981.
272. **Ruiz, M., Contreras, M., Parera, A., and Cerezo, A.,** Determination of the solubility parameter of Carbopol 940 in ethanol-water, methanol-water mixtures, using apparent viscosity measurements; Determination of the solubility parameter of Carbopol 934 in water-alcohol medium by measuring the apparent viscosity, *Farm. Clin.*, 3, 257 and 344, 1986.

273. **Kent, D. J. and Rowe, R. C.,** Solubility studies on ethyl cellulose used in film coating, *J. Pharm. Pharmacol.*, 30, 808, 1978.
274. **Marti, S.,** Thesis, 1977, cited in personal communication from G. Riess.
275. **Fritzsche, A. K. and Price, F. P.,** Solvent interactions of poly(diphenylsiloxane-dimethylsiloxane-diphenylsiloxane) block copolymers, in *Colloidal and Morphological Behavior of Block and Graft Copolymers*, Molau, G. E., Ed., Plenum Press, New York, 1971, 85.
276. **Rätzsch, M., Freydank, H., and Göbel, G.,** Solubility parameters of polyethylene, poly(vinyl acetate) and ethylene-vinyl acetate copolymers, *Faserforsch. Textiltech.*, 27, 425, 1976.
277. **Yu, A. J., Gobran, R. H., and Foerster, R. F.,** Solubility parameter of ethylene-diethyl fumarate copolymer, *J. Polym. Sci. A-1*, 8, 3039, 1970.
278. **Pfau, J. P. and Sharpe, R. G.,** Solvent absorption as a means to examine factors important in solvent viscosity reduction for high solids oligomer, *J. Coatings Technol.*, 59(750), 33, 1987.
279. **Sherwin, M. A., Koleske, J. V., and Taller, R. A.,** Solution properties of high solids intermediates, *J. Coatings Technol.*, 53(683), 35, 1981.
280. **Vohlidal, J., Matyska, B., and Bohackova, V.,** Relation between the cohesive energy density and the structure of a cyclopolyisoprene, *J. Chim. Phys. Physicochim. Biol.*, 69, 1348, 1972.
281. **Khakimullin, Yu, N., Averko-Antonovich, Yu, O., Kirpichnikov, P. A., and Gasnikova, M. A.,** Determination of the cohesive energy density of polyester urethane rubber, *Vysokomol. Soedin, Ser. B*, 17, 287, 1975.
282. **Lath, D. and Berek, D.,** Viscometric determination of the solubility parameter of atactic polypropylene, *Chem. Zvesti*, 21, 450, 1967.
283. **Guzman, J., Alamo, R., and Fatou, J. G.,** Glass transitions and solubility parameter for poly(diethylene glycol terephthalate), *An. Quim. Ser. A.*, 76, 214, 1980.
284. **Nakatsuka, R.,** The viscometric properties of polyesters. II. Viscosity in the range of low temperatures near the glass transition point, *Bull. Chem. Soc. Jpn.*, 37, 403, 1964.
285. **Peiffer, D. G. and Lundberg, R. D.,** Solution properties of ion-containing polymers in polar solvents, *J. Polym. Sci. Polym. Chem.*, 22, 1757, 1984.
286. **Velickovic, J. and Filipovic, J.,** Cohesive energy densities of poly(di-*n*-alkyl itaconates), *Angew. Makromol. Chem.*, 57, 139, 1977.
287. **Velickovic, J. S., Petrovic-Djakov, D. M., and Filipovic, J.,** Solubility parameters of poly(di-*n*-alkyl cyclohexyl-) and poly(di-*n*-alkyl phenyl itaconates), *Angew. Makromol.*, 113, 21, 1983.
288. **Velickovic, J. S. and Petrovic-Djakov, D. M.,** Solubility parameters of poly(alkyl phenyl methacrylates), *Struct. Prop. Abstr. IUPAC Macro '83*, 545, 1983.
289. **Petrovic-Djakov, D. M. and Velickovic, J. S.,** Solubility parameter of some poly(xylenyl methacrylates) and poly(dixylenyl itaconates), *VIIIth Jugoslav Symp. Chem. Technol. Macromolecules*, Bled, April 1984.
290. **Petrovic-Djakov, D. M., Velickovic, J. S., and Plavsic, M. B.,** Solubility parameters of poly(ditolyl itconates) and poly(tolyl methacrylates), *J. Serb. Chem. Soc.*, 50, 31, 1985.
291. **Cowie, J. M. G.,** The dependence of cohesive energy density on polymer stereostructure, *Polymer*, 10, 708, 1969.
292. **Csuros, Z., Geczy, I., Gara, M., and Gyurkovics, I.,** Rheology of macromolecular substances. VIII. Interaction between dissolved polymers and solvents, determined by viscometry, *Magy. Tud. Akad. Kem. Tud. Oszt. Kozl.*, 13, 139, 1960.
293. **Fox, T. G.,** Properties of dilute polymer solutions. III. Intrinsic viscosity/temperature relationships for conventional polymethyl methacrylate, *Polymer*, 3, 111, 1962.
294. **Matsuo, T.,** Viscometric method for evaluation of the solubility parameter of polymers, *Abstr. Pap. ACS/CSJ Chem. Congr., POLY* 265, 1979.
295. **Utracki, L. A.,** On the solubility parameter of polar polymers, *J. Appl. Polym. Sci.*, 16, 1167, 1972.
296. **Ferguson, J., Hudson, N. E., and Warren, B. C. H.,** Structural changes during elongation of polymer solutions, *J. Non-Newtonian Fluid Mech.*, 23, 49, 1987.
297. **Gargallo, L. and Russo, M.,** Effect of structure on glass transition temperatures of poly(methacrylic ester)s with bulky side groups, *Makromol. Chem.*, 176, 2735, 1975.
298. **Mangaraj, D., Patra, S., and Rath, S. B.,** Cohesive energy densities of high polymers. IV. C.E.D. of polyacrylates, *Makromol. Chem.*, 67, 84, 1963.
299. **Mangaraj, D., Patra, S., and Roy, P. C.,** Cohesive energy density of high polymers. V. C.E.D. of polymethacrylates, *Makromol. Chem.*, 81, 173, 1965.
300. **Gubanov, E. F., Shulyndin, S. V., Gurskaya, V. Sh., Palikhov, N. A., Zuyer, B. M., and Ivanov, B. Ye.,** Effect of molecular interaction on properties of copolymers containing phosphorus, *Vysokomol. Soedin. Ser. A.*, 18, 653, 1976; *Polym. Sci. U.S.S.R.*, 18, 744, 1976.
301. **Sumitomo, H., Kobayashi, K., and Saji T.,** Polymerization of β-carbomethoxypropionaldehyde, *J. Polym. Sci. Polym. Chem.*, 10, 3421, 1972.
302. **Pérez, E., Gomez, M. A., Bello, A., and Fatou, J. G.,** Solubility parameters in polyoxetanes: poly(oxetane), poly(3,3-dimethyl oxetane) and poly(3,3-diethyl oxetane), *J. Appl. Polym. Sci.*, 27, 3721, 1982.

303. **Pérez, E., Bello, A., Fatou, J. G., and Sanchez, R.,** Solubility parameters in polyoxetanes: poly(3-ethyl-3-methyloxetane) and poly(3-tert-butyloxetane), *An. Quim., Ser. A,* 82, 179, 1986.
304. **Bezuaye, A. and Huglin, M. B.,** Effect of temperature and solvent composition on the unperturbed dimensions of polystyrene, *Polymer,* 20, 44. 1972.
305. **Cowie, J. M. G.,** The effect of solvent on the unperturbed dimensions of polystyrene, *J. Polym. Sci. C.,* 23, 267, 1968.
306. **Cowie, J. M. G., Ranson, R. J., and Burchard, W.,** Influence of solvent on measurements of cohesive energy density of polystyrene, *Br. Polym. J.,* 1, 187, 1969.
307. **Green, J. H. S.,** Cohesive energy density of polystyrene, *Nature (London),* 183, 818, 1959.
308. **Sevestre, J.,** Solubility of polymers, *Double Liaison,* No. 105, 81, 1964; On some correlations involving the solubility parameter "δ", *Double Liaison,* No. 109, 67, 1964.
309. **Sosa, J. M., Rubio, A., and Pérez, B. J.,** Viscosities of polystyrene solutions in *Structure-Solubility Relationships in Polymers,* Harris, F. W. and Seymour, R. B., Eds., Academic Press, New York, 1977, 89; Dilute solution viscosities of polystyrene solutions, *Abstr. Pap. Am. Chem. Soc.,* 172, POLY 101, 1976.
310. **Lin, O. C. C.,** The change of structure and viscosity of styrene-acrylate copolymers in aqueous media, in *Structure-Solubility Relationships in Polymers,* Harris, F. W. and Seymour, R. B., Eds., Academic Press, New York, 1977, 165.
311. **Riess, G.,** Synthesis and chemical modification of polystyrene-polydiene copolymer sequences, *Angew. Makromol. Chem.,* 60/61, 21, 1971.
312. **Alfrey, T., Jr., Goldberg, A. I., and Price, J. A.,** Dilute-solution viscosity of polymethyl methacrylate and a methyl methacrylate-styrene copolymer, *J. Colloid Sci.,* 5, 251, 1950.
313. **Vasilova, O. I., Zaitseva, V. V., and Kucher, R. G.,** Solubility parameter of styrene-acrylonitrile-methyl methacrylate terpolymers, *Vysokomol. Soedin. Ser. B,* 29, 912, 1987.
314. **Utrachi, L. A.,** Interactions in poly(vinyl chloride) solutions, *Polym. J.,* 3, 551, 1972.
315. **Mangaraj, D., Patra, S., Roy, P. C., and Bhatnagar, S. K.,** Cohesive-energy-densities of high polymers. VI. C.E.D. of polyvinyl esters, *Makromol. Chem.,* 84, 225, 1965.
316. **Manning, A. J. and Rodriguez, F.,** Predicting solubilities of vinyl polymers, *J. Appl. Polym. Sci.,* 17, 1651, 1973.
317. **Zhigotsky, A. G., Dronov, S. V., and Shvets, T. M.,** Effect of the solvent nature on rheological properties of the solution vinyl chloride copolymer with vinyl acetate, *Ukr. Khim. Zh.,* 46, 841, 1980.
318. **Mangaraj, D., Bhatnagar, S. K., and Rath, S. B.,** Cohesive energy densities of high polymers. III. Estimation of C.E.D. by viscosity measurements, *Makromol. Chem.,* 67, 75, 1963.
319. **Bagda, E.,** The possibility of the prediction of the viscosity of polymer solutions, *FATIPEC (Fed. Assoc. Tech. Ind. Peint. Vernis Emaux Encres Impr. Eur. Cont.) Congr.,* 14, 137, 1978.
320. **Bagda, E.,** Prediction of viscosity of concentrated polymer solutions with the solubility parameter concept, *DEFAZET (Dtsch. Farben Z.),* 32, 372, 1978.
321. **Erickson, J. R.,** Viscosity of oligomer solutions for UV curable coatings, *J. Radiat. Curing,* 3, 12, 1976.
322. **Erickson, J. R.,** Viscosity of oligomer solutions for high solids and UV curable coatings, *J. Coatings Technol.,* 48(620), 58, 1976.
323. **Erickson, J. R.,** personal communications, 1979.
324. **Erickson, J. R. and Garner, A. W.,** Viscosity of concentrated oligomer solutions, *Am. Chem. Soc. Org. Coat. Plast. Prep.,* 37, 447, 1977.
325. **Drinberg, S. A., Frost, E. I., Rozhkov, Yu. P., and Shreiner, S. A.,** Use of the concept of a three-dimensional solubility parameter in selecting a solvent for copolymers of vinyl chloride with vinyl acetate, *Lakokrasoch. Mater. Ikh. Primen.,* No. 1, 13, 1972.
326. **Schoff, C. K.,** Recent advances in the rheology of high solids coatings, *Progr. Org. Coatings,* 4, 189, 1976.
327. **Burrell, H.,** Trends in solvent science and technology, in *Solvents Theory and Practice* (*Adv. Chem. Ser.* 124), Tess, R. W., Ed., American Chemical Society, Washington, D.C., 1973, 1.
328. **Hoernschemeyer, D.,** The influence of solvent type on the viscosity of concentrated polymer solutions, *J. Appl. Polym. Sci.,* 18, 61, 1974.
329. **Rudin, A. and Johnston, H. K.,** Stability of polymer mixtures. I. Coagulation of polymer dispersions, *J. Paint Technol.,* 44(568), 41, 1972.
330. **Rudin, A., Johnston, H. K., and Nicholson, R. E.,** Stability of polymer mixtures. II. Application of kinetic and solubility parameter models to polystyrene dispersions, *J. Paint Technol.,* 44(568), 54, 1972.
331. **Takahashi, S., Kariya, M., Yatake, T., Sonogashira, K., and Hagihara, N.,** Studies of poly-yne polymers containing transition metals in the main chain. 2. Synthesis of poly[*trans-bis*(tri-*n*-butyl(phosphine)platinum 1,4-butadiynediyl] and evidence of a rodlike structure, *Macromolecules,* 11, 1063, 1978.
332. **Lee, K. W. and Boger, D. V.,** The importance of solvent composition and adsorption effects on the flow properties of polybutadiene solutions, *Chem. Eng. Commun.,* 53, 97, 1987.

333. **Wolf, B. A.,** An extrapolation method for the determination of solubility parameters of polymers demonstrated for polyethylene, *Makromol. Chem.*, 178, 1869, 1977.
334. **Wolf, B. A. and Schuch, W.,** Oligomer-oligomer incompatibility. 3. End-group effects, *Makromol. Chem.*, 182, 1801, 1981.
335. **Hobin, T. P.,** Model polyethers. IV. Vapour pressures of oligomers of poly(tetramethylene oxide), *Polymer*, 9, 65, 1968.
336. **Rätzsch, M. and Krahn, G.,** Excess functions in polymer solutions, *J. Polym. Sci. Polym. Symp.*, 42, 1001, 1973.
337. **Hoy, K. L.,** New values of the solubility parameters from vapour pressure data, *J. Paint Technol.*, 42(541), 76 and 115, 1970.
338. **Hoy, K. L.,** *Tables of Solubility Parameters,* Union Carbide Corporation, Chemicals and Plastics Research and Development Dept., South Charleston, WV, 1969, 1975, and 1985.
339. **Hirt, P. and Herlinger, H.,** New elastomer fiber system, *Angew. Makromol. Chem.*, 40/41, 71, 1974.
340. **Barton, A. F. M.,** *Handbook of Polymer-Liquid Interaction Parameters and Solubility Parameters,* CRC Press, Boca Raton, FL, 1990.
341. **Voeks, J. F.,** Cohesive energy density and internal pressure of high polymers, *J. Poly. Sci. A,* 2, 5319, 1964.
342. **Ahmad, H. and Yaseen, M.,** Solubility parameter. Its concept and determination, *J. Colour Sci.*, 13(2), 7, 1974.
343. **Ahmad, H.,** Solubility parameters of poly(ethylene sulfonamide)s by the use of group contribution technique, *Farbe Lack,* 86, 797, 1980.
344. **Ahmad, H.,** Solubility parameter values of hydroxy polymers through its components and chemical group contribution technique, *J. Oil Colour Chem. Assoc.*, 63, 263, 1980.
345. **Ahmad, H. and Yaseen, M.,** Determination of solubility parameters of polymers by a group contribution technique, *J. Oil Colour Chem. Assoc.*, 60, 488, 1977.
346. **Ahmad, H. and Yaseen, M.,** Determination of solubility parameters of low and high molecular weight compounds, *J. Colour Soc.*, 16(2), 19, 1977.
347. **Ahmad, H. and Yaseen, M.,** Determination of solubility parameters of halogenated polymers by a group contribution technique, *Paint Manuf.*, 48(5), 28, 1978.
348. **Ahmad, H. and Yaseen, M.,** Application of a chemical group contribution technique for calculating solubility parameters of polymers, *Polym. Eng. Sci.*, 19, 858, 1979.
349. **Ahmad, H. and Yaseen, M.,** Solubility parameter determination from chemical group constituents. Solubility parameter values for poly(ethylenesulfonates) and poly(ethylenesulfonyl halides) determined by chemical group contribution technique, *Farbe Lack,* 85, 356, 1979.
350. **Ahmad, H. and Yaseen, M.,** Principle of additivity of chemical group, atom, and bond used for determination of solubility parameters, *Paintindia,* 30(4), 3, 1980.
351. **Ahmad, H.,** Determination of solubility parameter of methacrylamide series polymers by the use of its components and group contribution, *J. Colour Soc.*, 20(3-4), 108, 1981.
352. **Ahmad, H.,** Solubility parameter of acrylamide series polymers through its components and group contribution technique, *J. Macromol. Sci. Chem.*, A17, 585, 1982.
353. **Askadskii, A. A.,** Influence of chemical structure on the properties of polymers, *Pure Applied Chem.*, 46, 19, 1976.
354. **Askadskii, A. A.,** Computational procedures for the determination of the physical characteristics of polymers, *Usp. Khim.*, 46, 1122, 1977; *Russ. Chem. Rev.*, 46, 589, 1977.
355. **Askadskii, A. A., Kolmakova, L. K., Tager, A. A., Slonimskii, G. L., and Korshak, V. V.,** Estimation of cohesion energy of low-molecular weight liquids and polymers, *Vysokomol. Soedin. Ser. A,* 19, 1004, 1977; *Polymer Sci. U.S.S.R.*, 19, 1159, 1977; Evaluation of solubility parameters of low-molecular weight liquids and polymers, *Dokl. Akad. Nauk. SSR,* 226, 857, 1976; *Dokl. Phys. Chem.*, 226, 99, 1976.
356. **Sakellariou, P., Rowe, R. C., and White, E. F. T.,** The solubility parameters of some cellulose derivatives and polyethylene glycols used in tablet film coating, *Int. J. Pharm.*, 31, 175, 1986.
357. **Sewell, J. H.,** *A Study of Cohesive Forces, Refractive Indexes and Glass Transition Temperature Data of Polymers,* Royal Aircraft Establishment Technical Report No. 66185, Farnborough, Hampshire, 1966, AD 641308.
358. **Chen, J.,** Calculation and application of solubility parameters of alkyd resins, *Tuliao Gongye,* 84, 21-4, 18, 1985.
359. **Huglin, M. B. and Zakaria, M. B.,** Swelling properties of copolymer hydrogels prepared by gamma irradiation, *J. Appl. Polym. Sci.*, 31, 457, 1986.
360. **Ling, A.-L. and Gao, Y.-Y.,** The estimation of the solubility parameter of polysulfonamide, *Beijing Gongye Daxue Xuebao,* 9(3), 145, 1983.

361. **Maiti, S. and Das, S.,** Synthesis and properties of a new polyesterimide, *Angew. Macromol.*, 86, 181, 1980; Synthesis and properties of polyesterimides and their isomers, *J. Appl. Polym. Sci.*, 26, 957, 1981; A new polyblend: Polyesterimide with phenol-formaldehyde resins, *Polym. Sci. Technol.*, 21 (*Modification of Polymers,* Carraher, C. E. and Morre, J. A., Eds., Plenum, New York), 1983, 247.
362. **Maiti, S. and Ray, A.,** Processable heat resistant polymers. VII. Synthesis and characterization of polyamideimide from *N*-(*p*-carboxyphenyl)trimellitimide and *p,p'*-di(aminocyclohexyl)methane, *J. Appl. Polym. Sci.*, 27, 4345, 1982; IX. Synthesis and properties of polyamideimide from *N*-(*p*-carboxyphenyl)trimellitimide and *p,p'*-di(aminophenyl)sulfone, *J. Macromol. Sci. Chem.* A18, 955, 1982; X. Synthesis and characterization of an unsaturated polyesterimide containing 1,5-dioxoisoindoline-2,5-diyl units, *Macromol. Chem.*, 183, 2949, 1982; XI. Synthesis and characterization of unsaturated polyamideimides, *J. Polym. Sci. Polym. Chem.*, 21, 999, 1983. XII. Poly-amideimide from *N*-(*p*-carboxyphenyl)trimellitimide, *J. Macromol. Sci. Chem.*, A19, 439, 1983; XIII. Structure property relationship in polyamideimides, *J. Appl. Polymer. Sci.*, 28, 225, 1983.
363. **Maiti, S. and Ray, A.,** Unsaturated polyamides. Studies on synthesis, properties and crosslinking, *Angew. Makromol. Chem.*, 116, 175, 1983.
364. **Maiti, M. and Maiti, S.,** Polymers from renewable sources. V. Synthesis and properties of a polyamideimide from rosin, *J. Macromol. Sci. Chem.*, 20, 109, 1983.
365. **Maiti, S., Ray, A., and Maiti, M.,** Processable heat resistant polymers. VIII. Synthesis and characterization of polyamideimide from *N*-(4-carboxyphenyl)trimellitimide and sulfanilamide, *Angew. Makromol.*, 107, 87, 1982.
366. **Guchait, P. K., Ray, A., and Maiti, S.,** Processable heat resistant polymers. XV. Synthesis and properties of polyamideimides from *N*-(*p*-carboxyphenyl)trimellitimide and *p,p'*-di(aminophenyl)methane, *J. Macromol. Sci. Chem.*, A20, 93, 1983.
367. **Ray, A., Bhattacharya, V. K., Rao, Y. V., and Maiti, S.,** Processable heat-resistant polymers. XVI. Polyamideimidesulphonamides, *Eur. Polym. J.*, 19, 1195, 1983.
368. **Ray, S. S., Kundu, A. K., Maiti, M., Ghosh, M., and Maiti, S.,** Polymers from renewable sources. VII. Synthesis and properties of polyamideimides from rosin-maleic anhydride adduct, *Angew. Makromol. Chem.*, 122, 153, 1984.
369. **Ray, S. S., Kundu, A. K., and Maiti, S.,** Polymers from renewable sources. IX. Synthesis and properties of a polyamideimide from rosin, *J. Macromol. Sci. Chem.*, 23, 271, 1986.
370. **Miroshnichenko, A. A.,** Thermodynamics of liquid-vapour equilibrium in solutions. VII. Calculation of excess enthalpy of mixing and Gibbs energy of low molecular weight substances with polymer melts from molecular data, *Zh. Fiz. Khim.*, 56, 2670, 1982; *Russ. J. Phys. Chem.*, 56, 1644, 1982.
371. **Patra, S., Lenka, S., and Nayak, P. L.,** Synthetic resins. II. Preparation and characterization of resins prepared from aminoacetophenone, *J. Appl. Polym. Sci.*, 32, 5071, 1986; IV. Preparation and characterization of resins prepared from hydroxyacetophenone, *Angew. Makromol.*, 144, 23, 1986.
372. **Saito, T. and Yamaguchi, K.,** Cohesive energy density of chlorinated polyethylenes, *Kobunshi Kagaku,* 29, 537, 1972.
373. **Sanchez, A., Bello, A., Marco, C., and Fatou, J. G.,** Solution properties and chain dimensions of poly(trimethylenesulfide), *Makromol. Chem.*, 189, 399, 1988.
374. **Shtilman, M. I., Lyalyushko, N. S., Maiorova, N. V., and Korshak, V. V.,** Structure and properties of polyvinylchloromethylketone, *Vysokomol. Soedin. Ser. B.*, 21, 528, 1979.
375. **Takeda, K., Akiyama, M., and Yamamizu, T.,** Shape-memory pore structure in porous cross-linked polystyrenes, *Angew. Makromol.*, 157, 123, 1988.
376. **Tortorello, A. L., Kinsella, M. A., and Gearon, R. T.,** The solubility parameter concept in the design of polymers for high performance coatings, in *Macromolecular Solutions: Solvent-Property Relationships in Polymers,* Seymour, R. B. and Stahl, G. A., Eds., Pergamon, New York, 1982, 26.
377. **Tortorello, A. J. and Kinsella, M. A.,** Solubility parameter concept in the design of polymers for high performance coatings. I. Acrylic copolymers; II. Urethane copolymers, *J. Coatings Technol.*, 55(696), 99, 1983; 55(697), 29, 1983.
378. **Watanabe, H. and Miyauchi, T.,** Determination of solubility parameter for siloxane segment. Solubility of iodine in liquid poly(dimethylsiloxane), *J. Chem. Eng. Jpn.*, 6, 109, 1973.
379. **Koenhen, D. M. and Smolders, C. A.,** The determination of solubility parameters of solvents and polymers by means of correlation with other physical qualities, *J. Appl. Polym. Sci.*, 19, 1163, 1975.
380. **Matsuura, T.,** Reverse osmosis separation of solute from aqueous solution, *Yukui Gosei Kagaku Kyokai (J. Syn. Org. Chem. Jpn.),* 31, 717, 1973; Transport processes in membranes for reverse osmosis, *Chem.-Ing. Tech.*, 50, 565, 1978.
381. **Matsuura, T. and Sourirajan, S.,** Reverse osmosis separation of hydrocarbons in aqueous solutions using porous cellulose acetate membranes, *J. Appl. Polym. Sci.*, 17, 3683, 1973; Characterization of reverse osmosis membrane materials by liquid chromatography data, *Proc. 6th Intl. Symp., Fresh Water from the Sea,* 3, 227, 1978.

382. **Matsuura, T., Blais, P., and Sourirajan, S.,** Polar and nonpolar parameters for polymeric reverse osmosis membrane materials from liquid chromatographic data, *J. Appl. Polym. Sci.*, 20, 1515, 1976.
383. **Loev, A. M. and Lavrishchev, L. P.,** Method for the selection of solvents for coating materials, *Lakokras. Mater. Ikh. Primen.*, No. 4, 4, 1977.
384. **Mishra, M. K.,** Solution properties of shellac. IV. Solubility parameter of shellac, *Angew. Makromol. Chem.*, 147, 107, 1987.
385. **Munafo, A., Buchmann, M., Hô Nam-Tran, and Kesselring, U. W.,** Determination of the total and partial cohesion parameters of lipophilic liquids by gas-liquid chromatography and from molecular properties, *J. Pharm. Sci.*, 77, 169, 1988.
386. **Maloney, D. P. and Prausnitz, J. M.,** Thermodynamic properties of liquid polyethylene, *J. Appl. Polym. Sci.*, 18, 2730, 1974.
387. **Wall, L. A., Flynn, J. H., and Straus, S.,** Vaporization of organic plasticizers, *Tech. Pap. Reg. Tech. Conf. Soc. Plast. Eng. Binghamton Sect.*, 27, 1968; Rates of molecular vaporization of organic plasticizers, *Polym. Eng. Sci.*, 10, 19, 1970.
388. **Fic, J. and Fic, G.,** Computer programmes for interpreting the results of studies on the field of polymer solutions, *Polimery (Warsaw)*, 24, 155, 1979.
389. **Van den Berg, J. W.,** Direct application of UNIFAC activity coefficient computer programmes to the calculation of solvent activities and chi parameters for polymer solutions, *Ind. Eng. Chem. Prod. Res. Dev.*, 23, 321, 1984.
390. **Kaelble, D. H.,** Relations between polymer chemistry and physical properties, *Org. Coat. Appl. Polym. Sci. Proc.*, 46, 241, 1981.
391. **Bianchi, U.,** Temperature coefficients of the internal pressure of high polymers, *Ric. Sci.*, 32(Ser. 2) IIA, 2, 651, 1962.
392. **Allen, G., Gee, G., Mangaraj, D., Sims, D., and Wilson, G. J.,** Intermolecular forces and chain flexibilities in polymers. II. Internal pressures of polymers, *Polymer*, 1, 467, 1960.
393. **Allen, G., Sims, D., and Wilson, G. J.,** Intermolecular forces and chain flexibilities. III. Internal pressures of polymers below their glass transition temperatures, *Polymer*, 2, 375, 1961.
394. **Bianchi, U., Agabio, G., and Turturro, A.,** Internal pressure of simple liquids, *J. Phys. Chem.*, 69, 4362, 1965.
395. **Olabisi, O. and Simha, R.,** A semiempirical equation of state for polymer melts, *J. Appl. Polym. Sci.*, 21, 149, 1977.
396. **Beevers, R. B.,** The physical properties of polyacrylonitrile and its copolymers, *Macromol. Rev.*, 3, 113, 1968.
397. **Beevers, R. B.,** The effect of non-solvent on association in polyacrylonitrile solutions; The effect of solvent on association in polyacrylonitrile solutions, *Polymer*, 9, 535, 1968; 10, 791, 1969.
398. **Beevers, R. B.,** Dependence of association in polyacrylonitrile solutions on the molecular volume of the nonsolvent, *J. Appl. Polym. Sci.*, 14, 505, 1970.
399. **Bueche, A. M.,** Interaction of polydimethylsiloxanes with swelling agents, *J. Polym. Sci.*, 15, 97, 1955.
400. **Kawakami, M., Egashira, M., and Kagawa, S.,** Measurements of the interactions between polyethylene glycol and organic compounds by gas chromatographic technique, *Bull. Chem. Soc. Jpn.*, 49, 3449, 1976.
401. **Dyck, M. F. and Hays, W. P.,** Influence of the polymer weight on the solubility parameter γ, *Skand. Tidskr. Färg Lack*, 15, 227, 1969.
402. **Koningsveld, R. and Staverman, A. J.,** Determination of critical points in multicomponent polymer solutions, *J. Polym. Sci. C*, 16, 1775, 1967.
403. **Konoingsveld, R., Kleintjens, L. A., and Schoffeleers, H. M.,** Thermodynamic aspects of polymer compatibility, *Pure Appl. Chem.*, 39, 1, 1974.
404. **Koningsveld, R. and Kleintjens, L. A.,** Thermodynamics of polymer mixtures, *J. Polym. Sci. Polym. Symp.*, 61, 221, 1977.
405. **Wolf, B. A. and Willms, M. M.,** Measured and calculated solubility of polymers in mixed solvents: consolvency, *Makromol. Chem.*, 179, 2265, 1978.
406. **Spychaj, T.,** Solvent/nonsolvent solubility and polarity parameters in fractional separation of oligomers. 1. Epoxy resins, *Angew. Makromol. Chem.*, 149, 127, 1987.
407. **Spychaj, T. and Hamielec, A. E.,** Solvent/nonsolvent solubility and polarity parameters in fractional separation of oligomers. 2. Styrene-acrylic copolymers, *Angew. Makromol. Chem.*, 157, 137, 1988.
408. **Wijmans, J. E. and Smolders, C. A.,** Removal of a low molecular-weight crystallizable polymer fraction from solutions of poly[oxy-1,4-phenylene-sulphonyl-1,4-phenyleneoxy-1,4-phenylen(1-methylethylidene)-1,4-phenylene] (polysulfone) *Eur. Polym. J.*, 19, 1143, 1983.
409. **Allen, G., Gee, G., and Wilson, G. J.,** Intermolecular forces and chain flexibility in polymers. 1. Internal pressures and cohesive energy densities of simple liquids, *Polymer*, 1, 456, 1960.
410. **Nanda, V. S. and Simha, R.,** Equation of state and related properties of polymer and oligomer liquids, *J. Phys. Chem.*, 68, 3158, 1964.

411. **Reynolds, W. W. and Larson, E. C.,** The fundamentals of solvent selection, *Off. Dig. Fed. Soc. Paint Technol.*, 34, 311, 1962.
412. **Wolf, B. A. and Blaum, G.,** Dependence of oligomer-oligomer incompatibility on chain length and pressure. 1. Oligo-isobutene/oligo-propylene glycol and oligo-styrene/oligo-ethylene glycol, *J. Polym. Sci. Polym. Symp.*, 61, 251, 1977.
413. **Wolf, B. A. and Blaum, G.,** Oligomer-oligomer incompatibility, 2. Oligo(dimethylsiloxane)/oligo(propylene glycol), *Makromol. Chem.*, 180, 2591, 1979.
414. **Friday, A., Cooper, D. R., and Booth, C.,** Mixing of ethylene and propylene oxide oligomers: 2. Phase separation, *Polymer*, 18, 171, 1977.
415. **Huczkowski, P. and Kapko, J.,** Designing of oligoesteramides based on cohesive energy density calculations, *Makromol. Chem.*, 179, 571, 1979.
416. **Powers, P. O.,** Solvent systems for hydrocarbon resins, *Am. Chem. Soc. Div. Org. Coatings Plast. Chem. Pap.*, 31, 469, 1971; in *Solvents Theory and Practice (Adv. Chem. Ser.* 124), Tess, R. W., Ed., American Chemical Society, Washington, D.C., 1973, chap. 9.
417. **Doroszkowski, A. and Lambourne, R.,** Effect of molecular architecture of long chain fatty acids in the dispersion properties of titanium dioxide in non-aqueous liquids, *Faraday Discuss. Chem. Soc.*, 65, 252, 1978.
418. **Song, B. G. and Lee, D. C.,** Dilute solution properties on molecular conformation of poly(n-octadecyl acrylate), *Polymer (Korea)*, 10, 588, 1986.
419. **Sánchez, A., Bello, A., Carlos, M., and Fatou, J. G.,** Solution properties and chain dimensions of poly(trimethylene sulphide), *Makromol. Chem.*, 189, 399, 1988.
420. **Sánchez, A., Bello, A., Marco, C., and Fatou, J. G.,** Solution properties and chain dimensions of poly(trimethylene sulphide), *Makromol. Chem.*, 189, 399, 1988.
421. **Lazcano, S., Bello, A., Marco, C., and Fatou, J. G.,** Solution properties and random coil dimensions of poly (3,3′-dimethylthietane), *Eur. Polym. J.*, 24, 985, 1988.
422. **Mukhutdinova, T. Z., Averko-Antonovich, L. A., Prokudina, K. N., and Kirpichnikov, P. A.,** Cohesive energy density of liquid thiokols, *Tr. Kazan. Khim.-Tekhnol. Inst.*, No. 50, 141, 1972.
423. **Tashmukhamedov, S. A., Nurglieva, F. F., and Tillaev, R. S.,** Properties of solutions of poly(methacrylic acid piperidide), *Uzb. Khim. Zh.*, 16(4), 44, 1972.
424. **Lathova, E., Lath, D., Florian, S., and Juranicova, V.,** Solution properties of statistical ethylene-vinyl acetate copolymers, *Makrotest Sb. Prednasek Celostatni Konf.*, 5, 91, 1978.
425. **Petrović-Dakov, D. M. and Veličkovič, J. S.,** Solubility parameters of poly(xylenyl methacrylates), *J. Serb. Chem. Soc.*, 54, 285, 1989.
426. **Fedors, R. F.,** A method for estimating both the solubility parameters and molar volumes of liquids, *Jet Propul. Lab. Quart. Tech. Rev.*, 3(1), 45, 1973; *Polym. Eng. Sci.*, 14, 147 and 472, 1974.
427. **Sheffer, H.,** Cresylic soluble polyimides, *J. Appl. Polym. Sci.*, 26, 3837, 1981.
428. **Shin, S. M. and Lee, D. C.,** Conformational study on the thermoplastic urethane polymer in dilute solution, *Nonmunjip — Sanop Kwahak Kisul Yonguso (Inha Taehakkyo)*, 13, 285, 1985.
429. **Bellenger, V., Morel, E., and Verdu, J.,** Solubility parameters of amine-crosslinked aromatic epoxies, *J. Appl. Polym. Sci.*, 37, 2563, 1989.
430. **Hansen, C. M. and Hansen, K. M.,** Solubility parameter prediction of the barrier properties of chemical protective clothing, *ASTM Spec. Tech. Publ.*, 989 (Perform. Protect. Clothing, Second Symp.), 197, 1988; **Hansen, C. M. and Andersen, B. H.,** The affinities of organic solvents in biological systems, *Am. Ind. Hyg. Assoc.*, 49, 301, 1988.
431. **Bentz, A. P. and Billing, C. B., Jr.,** Determination of solubility parameters of new suit materials, *ASTM Spec. Tech. Publ.*, 989 (Perform Protective Clothing, Second Symp.), 209, 1988; U.S. Coast Guard protective clothing research since Solna (NOKOBETEF II), presented at Third Int. Scand. Symp. (NOKOBETEF III), Norway, 1989.
432. **Galin, J. C., Lutringer, G., and Galin, M.,** Solubility properties of poly(1,1-difluoroethylene) in dipolar aprotic solvents, *J. Appl. Polym. Sci.*, 37, 487, 1989.
433. **Ingamells, W. and Yanumet, N.,** Solvent-polymer interaction in the plasticisation of poly(ethylene terephthalate), *Brit. Polym. J.*, 12(1), 12, 1980.
434. **Kalfoglou, N. K., Sotiropoulou, D. D., and Margaritis, A. G.,** Thermal and morphology characterization of blends of poly(ethylene oxide) with poly(vinyl acetate), *Eur. Polym. J.*, 24, 389, 1988; **Margaritis, A. G. and Kalfoglou, N. K.,** Compatibility of poly(vinyl chloride) with polyalkyleneoxides. II. Poly(propylene oxide) and poly(tetramethylene oxide), *J. Polym. Sci. Polym. Phys.*, 27, 1767, 1989.
435. **Yamashita, S., Itoi, M., Kohjiya, S., and Kidera, A.,** Synthesis and properties of elastomeric ionenes. I. Polybutadiene ionenes, *J. Appl. Polym. Sci.*, 35, 1927, 1988.
436. **Sears, J. K. and Darby, J. R.,** *The Technology of Plasticizers*, Wiley-Interscience, New York, 1982.
437. **Nakazato, K. and Suzuki, K.,** Correlation between solubility parameter and exclusion limit on polyacrylamide terpolymer, *Macromolecules*, 22, 1497, 1989.

438. **Vilcu, R. and Leca, M.,** Thermodynamic characterization of polymers by chromatography with gas mobile phase, *Rev. Roum. Chim.*, 34, 387, 1989; *Polymer Thermodynamics by Gas Chromatography,* Elsevier, Amsterdam, 1990.
439. **Fernández-Sánchez, E., Fernández-Torres, A., García-Domínguez, J. A., and Santiuste, J. M.,** Solubility parameters of gas chromatographic mixed phases, *J. Chromatogr.*, 457, 55, 1988.
440. **Balayan, L. I. and Gevorkyan, S. S.,** Multicomponent solvent for chloroprene rubber, *Kauch. Rezina,* No. 12, 48, 1981.
441. **Szymanowski, J., Sobzynska, A., and Voelkel, A.,** Solubility and polarity parameters of polyoxyethylene glycol dialkyl ethers, *J. Pharm. Sci.*, 77, 893, 1988.
442. **Dante, M. F., Bittar, A. D., and Caillault, J. J.,** Program calculates solvent properties and solubility parameters, *Mod. Paint Coat.*, 79(9), 46 and 51, 1989.
443. **Panda, J., Lenka, S., and Nayak, P. L.,** Synthetic resins. VIII. Preparation and characterization of *p*-aminoacetophenone/substituted benzoic acid/formaldehyde resins, *Angew. Makromol. Chem.*, 170, 29, 1980.
444. **Narender, G. and Yaseen, M.,** Phenomenon of cosolvency and solution parameters of polyamide resins, *J. Coatings Technol.*, 61(773), 41, 1989.
445. **Arichi, S. and Himuro, S.,** Solubility parameters of poly(4-acetoxystyrene) and poly(4-hydroxystyrene), *Polymer,* 30, 686, 1989.
446. **Giuliani, J. F.,** An investigation of the solubility of organic vapors in polymer films using an optical wave guide interfacial probe, *J. Polym. Sci. Polym. Phys.*, 26, 2197, 1988.
447. **Snow, A. and Wohltjen, H.,** Poly(ethylene maleate)-cyclopentadiene: a model reactive polymer-vapor system for evaluation of a SAW [surface acoustic wave] microsensor, *Analyt. Chem.*, 56, 1411, 1984.
448. **Kalz, G.,** Solubility and interfacial parameters of chlorinated olefinic and vinyl polymers, *Plaste Kautsch.*, 35, 154, 1988.
449. **Kalz, G.,** Intra- and intermolecular energies of moist dispersed solids, *Plaste Kautsch.*, 36, 23, 1989.
450. **Ryan, A. J., Stanford, J. L., and Still, R. H.,** Calculation of solubility parameters of typical structural units present in segmented polyurethanes, poly(urethane urea)s and polyureas formed by reaction injection molding, *Polym. Commun.*, 29, 196, 1988.
451. **Chuiko, L. S., Petrovskii, V. L., Grinenko, N. M., and Mel'nik, N. M.,** Study of thermodynamic characteristics of peroxidate rubbers, *Visn. L'viv. Politekh. Inst.*, 149, 53, 1981.
452. **Kambour, R. P., Gundlach, P. E., Wang, I. C. W., White, D. M., and Yeager, G. W.,** Miscibility of poly(2,6-dimethyl-1,4-phenylene oxide) with several styrenic homopolymers — dependence of interaction parameters from critical molecular weights on cohesive energy densities, *Polym. Commun.*, 29(6), 170, 1988.
453. **Pérez, C., Guzman, J., Riande, E., Delacamp, J. G., and Deabajo, J.,** Influence of chemical structure on glass transition temperatures of poly(propylene terephthalate) and poly(propylene isophthalate), *Makromol. Chem.*, 189, 691, 1988.
454. **Fukami, A., Iisaka, K., and Etoh, S.,** Simultaneous interpenetrating polymer networks of epoxy resin/bismaleimide-allylester copolymer. I. Dynamic mechanical properties, *J. Makromol. Sci. Phys.*, B28, 251, 1989.
455. **Coleman, M. M., Lee, J. Y., Serman, C.-J., Wang, Z.-Q., and Painter, P. C.,** Poly(ethylene-*co*-methacrylic acid)-polyether blends, *Polymer*, 30, 1298, 1989; **Serman, C. J., Xu, Y., Painter, P. C., and Coleman, M. M.,** On the phase behavior and the limits of miscibility of styrene-*co*-vinylphenol blends with poly(alkyl methacrylates), *Macromolecules,* 22, 2015, 1989.
456. **Tong, H. M., Saenger, K. L., and Durning, C. J.,** A study of solvent diffusion in thin polyimide films using laser interferometry, *J. Polym. Sci. Polym. Phys.*, 27, 689, 1989.
457. **Mizerovskii, L. N., Vansyatskaya, L. N., and Smurova, G. I.,** Investigation of the reaction between a ternary ethylene-propylene copolymer and organic liquids, *Vysokomol. Soedin, Ser. A,* 29, 1512, 1987; *Polym. Sci U.S.S.R.*, 29, 1663, 1987.

Chapter 15

POLYMER MAPS AND MODELS

Solubility maps may be as simple as plots of maximum polymer concentration as a function of Hildebrand parameter (Figure 1), but more often they are used to represent two or more parameters. Section 15.1 is concerned with those involving only some type of cohesion parameter, and examples including other forms of parameter, are described in Section 15.2.

This type of mapping is receiving increasing application in the field of microlithography for the semiconductor industry (Section 16.5).

15.1 COHESION PARAMETER MAPS AND MODELS

Instead of assuming spherical or rectangular volumes of polymer-liquid interaction that allow characterization of polymers in terms of a few numerical quantities (Sections 5.9 and 14.5), it is possible to present this information in the form of irregular volumes. Three-dimensional models, stereographs, sets of projections, and contours have all been used. (Some of these figures are based on Hansen's 1967 solvent data, part of which appears in Table 12, Chapter 5. But use of the 1971 data in Table 11, Chapter 5 and Figure 6, Chapter 5, should not cause major inconsistencies.)

- Three-dimensional Hansen models (often represented by two-dimensional projections) in systems[2-5] such as poly(methyl methacrylate),[6,7] sulfonated polysulfone membranes,[8] maleic anhydride-α-methylstryene copolymer,[9] poly(vinylidene fluoride),[10] and polymer-plasticizer systems[104]
- Hansen contour maps for synthetic resins and elastomers used in the plastics and resin industries[11,12]
- Hansen contour maps (associated with viscosity information) used in the choice of high-solids coating systems,[13] including phthalic anhydride polymers[13,101] (Figures 2 to 4)
- Hansen contour maps for Hoechst alkyd and melamine resins[14] (Figures 5 and 6)
- Hansen contour map for polyepichlorohydrin elastomer used in adhesives[15] (Figure 7)
- Hansen three-dimensional models and contour maps for the explanation of the properties of copolymers[16-19] (Section 16.10)
- Three-dimensional Hansen model in the characterization of polyisocyanides such as poly (α-phenylethyl isocyanide)[20]
- Two-dimensional Hansen projections in the study of polymers containing indanic units in the main chain[21]
- Hansen contour maps in the study of poly(1-chloro-2,3-epoxypropane)[22]

In many cases, however, the solubility behavior of a polymer can be adequately represented by just one plot in two dimensions, such as:

- The δ-δ_h map for cellulose acetate[23] in Figure 8
- Hansen parameter δ_p -δ_h maps for cellulose acetate and ethyl cellulose in the preparation and use of dialysis and reverse osmosis membranes[24-28] and capillary membranes[31] (Section 6.11 and Figures 9 and 10)
- Hansen parameter δ_p-δ_h maps for resins and pigments[29,30]
- Hansen parameter δ_p-δ_h maps of dyes[32] (Figures 11 to 13)

- Hansen parameter δ_p-δ_h maps to show progress of sulfonation of polysulfones[33]
- A δ_p-δ_h map for reinforced reaction injection molded polyurethane[34]
- δ_p-δ_h maps for protective glove materials[35] (Section 16.7) and for extractive distillation agents[103]
- Hansen parameter δ_d-δ_h maps for plasticizers with cellulose derivatives[36,37]
- A δ_d-δ_p map for poly(vinyl pyrrolidone)[38]
- Hansen parameter δ_d-δ_p maps for polymer-polymer miscibility[39] (Section 16.9)
- Polar-nonpolar cohesion parameter maps (δ_λ-δ_τ or δ_d-δ_{hp})[7,40-41]
- δ_c-δ_f maps (Section 5.10)[100]

An interesting example of the application of δ_p-δ_h Hansen maps is shown in Figure 14. The coordinates of the centers of the interaction circles for poly(ethylene terephthalate) with solvents correspond with the Hansen parameters of the aromatic (a) and aliphatic ester (B) residues of the polymer:

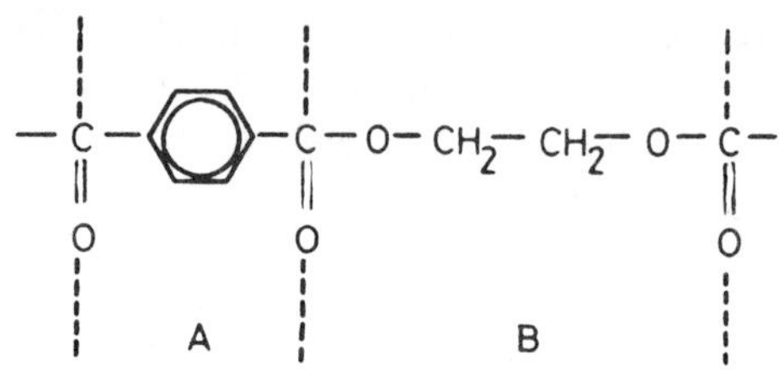

The point marked PET in Figure 14 denotes the center of the interaction sphere for the monomer repeat unit (Table 1). The carbonyl groups of the 1,4-benzenoid residue (-OC-C_6H_4-CO-) are "shared" by both residues, and the observed Hansen parameters provide a measure of the fractional carbonyl content of each residue, found to be approximately one-third in A, two-thirds in B. It appears that the description in terms of polar and hydrogen-bonding Hansen parameters is successful because these effects outweigh the Lewis acid (aromatic residue)-Lewis base (ester residue) effects.[42,83] Similar "bimodal" cohesion parameter behavior is observed in copolymers (Section 16.10).

The most generally useful two-dimensional plots appear to be those based on the subdivision of the Hildebrand parameter into a "volume dependent" part, δ_v, and a "residual" or hydrogen bonding part, δ_R (Section 7.5):

$$\delta_V^2 \approx (\delta_d^2 + \delta_p^2); \quad \delta_R^2 \approx \delta_h^2 \tag{1}$$

Two-dimensional solubility diagrams using δ_v and δ_R for several polymers have been prepared by Scigliano[44] and published by Bagley and Scigliano.[45] In their diagrams the δ_v axis is doubled to provide a circular zone of solubility, a procedure introduced by Hansen and Skaarup (Section 5.9). The center of the circle defines the ${}^j\delta_v$ and ${}^j\delta_R$ values to be assigned to the polymer, *j*. The radius ${}^jR$ of the circle is chosen to include as many solvents and exclude as many nonsolvents as possible (Table 2), and can be used in a manner analogous to that described in Section 14.5, with the estimation of

$${}^{ij}R = [4({}^i\delta_V - {}^j\delta_V)^2 + ({}^i\delta_R - {}^j\delta_R)^2]^{1/2} \tag{2}$$

and comparison of ${}^{ij}R$ with ${}^jR$. van Krevelen[46] used the same method, but evaluated instead:

$${}^{ij}r = [({}^i\delta_V - {}^j\delta_V)^2 + ({}^i\delta_h - {}^j\delta_h)^2]^{1/2} \tag{3}$$

and found for polystyrene ${}^j\delta_v = 18\ \text{MPa}^{1/2}$, ${}^j\delta_h = 5\ \text{MPa}^{1/2}$, ${}^jR = 5\ \text{MPa}^{1/2}$.

In an alternative procedure[47] a circular field of interaction when δ_v is plotted against δ_h is not assumed. Instead, Figure 15 is an example of the kind of map that can be rendered. This shows the effect of liquids on a commercially cured epoxy paint. It should be noted that the δ_v scale is twice that of δ_h, and that the area of intermediate behavior (discoloration, slight softening or minor blistering) was not fully explored because of the unavailability of suitable liquids.

Hansen parameter δ_v-δ_h maps have been used for:

- Chlorine-containing, styrene-based negative acting resists[48]
- Methyl α-chloroacrylate copolymers as electron-sensitive positive resists (Section 16.5)[49]
- Neoprene adhesives[50]
- Polythioamides[51]
- Dyeing of poly(ethylene terephthalate)[52]
- Amine-crosslinked aromatic epoxy thermosets[99] (also δ_d vs. δ_d maps)

Cohesion parameter maps can be used to illustrate properties other than solubility, for example:

- Theta temperatures[46] (Section 13.7)
- Stress-strain properties[53] (Section 16.2) and environmental stress cracking (Section 16.4)
- Surface properties (Section 17.3) such as frictional coefficients,[54] in surface wetting and cleaning (Section 17.9)
- Viscosity[22] (Section 14.7)

A noteworthy example[55-57] is the liquid interaction of liquids with *Halobacterium halobium* bacteriorhodops in purple membrane, described by the use of Equation 3 with ${}^j\delta_v$ = 24.9 MPa$^{1/2}$, ${}^j\delta_h$ = 15.4 MPa$^{1/2}$, ${}^jR$ = 10.3 MPa$^{1/2}$. This can also be represented on a triangular plot, but with a larger circle of interaction.

Triangular diagrams provide a method of conveying information on three properties in two dimensions, as illustrated[58] for polyamide, poly(methyl methacrylate), cellulose nitrate, and short oil alkyd resin in Figure 16. Either cohesion parameters or fractional parameters (Section 5.10) may be used; the latter provide a more even distribution of data. Other examples are:

- Resins such as Epikote® 101 epoxy (Figure 17), Vipla® K poly(vinyl chloride) (Figure 18), Piccopale® 110 (Figure 19), Pentalyn® 225 (Figure 20) and Parlon® P 10 (Figure 21)[59]
- Lignin[59] (Figure 22)
- Epikote® 1007 resin[60] (Figure 23)
- Solubility or swelling[61] of triacetate (Figure 24), diacetate (Figure 25), polypropylene (Figure 26), polyamide 6 (Figure 27), and poly(ethylene terephthalate) (Figure 28) (in this case the regions are bounded by straight lines parallel to the sides of the triangle rather than by curves)
- Alkyd resin (Figure 29)[62,97], formaldehyde-melamine copolymers (Figures 30 and 31), epoxy resin[62] (Figure 32) and phenol-formaldehyde resin (Figure 33)[62]
- Properties of polymethacrylate positive resists[63]
- Dielectric properties of para-substituted poly(α-acetoxystyrene)s[64]
- Sulphonated polysulfones[33]
- Extent of α→β transition of phthalocyaninatocopper[102]
- Poly(DL-lactide)[105]

15.2 HYBRID MAPS AND MODELS

In Section 8.4, two three-component hybrid cohesion parameter schemes for characterizing polymer interactions were described. One used Hildebrand parameter (δ), hydrogen bonding parameter (γ_c), and dipole moment (μ), and is the basis of the ASTM D3132 test method for the ''Solubility Range of Resins and Polymers.'' The other used Hildebrand parameter, hydrogen bonding index (θ) and fractional polarity (p). In either case, one or more countour solubility maps are necessary to describe the behavior of each polymer.

Three-dimensional δ-γ_c-μ representations and two-dimensional solubility maps[65,66] are shown here for cellulose nitrate (Figures 34 and 35), cellulose acetate (Figures 36 and 37), cellulose acetate butyrate (Figures 38 and 39), polymethyl methacrylate (Figures 40 and 41), and a vinyl chloride — acetate copolymer (Figures 42 and 43). Solubility maps are included for alkyds (Figures 44 to 47), an alcohol-soluble butyrate (Figure 48), and other resins (Figures 49 and 50). All these solubility maps may be used with the liquid data of Tables 8 and 9, Chapter 8.

In many cases, two-component maps (without contours) have proved adequate, such as the Hildebrand parameter (δ)-fractional polarity (p) maps shown [60,65-68] for a vinyl copolymer (Figure 51), a chlorinated rubber (Figure 52), and cellulose nitrate (Figure 53). These maps should be used in conjunction with the solvent data of Table 6, 10, 11, and 13, Chapter 8.

Alternatively, Hildebrand parameter—hydrogen-bonding parameter maps can be used. Care should be taken here, as some diagrams are expressed in terms of the Lieberman hydrogen bonding index γ_L(Table 3, Chapter 8), or the du Pont hydrogen bonding index, (γ_D), defined as in Equation 19, Chapter 8), instead of the more usual γ_c of Equation 16, Chapter 8. This is the case for the sources[69,70] of Figures 54 to 62 for Elvacite® acrylic resins and Figures 63 and 64 for Elvax® polyethylene-poly(vinyl acetate) copolymers, but as depicted here they have been converted by means of Equation 21, Chapter 8, to make them consistent with the liquid data of Figure 6 and Tables 8 and 9, Chapter 8. Other examples of two- and three-dimensional presentations are:

- Cellulose nitrate/acrylic polyblend aqueous emulsion[71] (Figure 65)
- Bakelite® vinyl resin VYHH[72]
- Polyepichlorohydrin[15,73]
- Polystyrene[74]
- Polyimide[75]
- Epoxy coatings[76]
- Poly(methyl methacrylate),[77] poly(vinyl chloride), and polysulfone dissolution, crazing and cracking (Figures 3 to 5, Chapter 16)
- Polychloroprene rubber adhesives[78]
- Vinyl chloride-vinyl acetate-maleic acid paints[96]
- Resins[79,98]
- Various polymers in conjunction with calculations in Section 14.8

In the case of three liquid constituents, the triangle formed by the lines joining the points representing the three pure liquids on a δ-γ_c diagram such as Figure 66 provides the entire array of Hildebrand parameter—hydrogen-bonding parameter combinations that can be obtained by blending.[80] When combined with a polymer solubility map, this information provides a good starting point for solvent formulation. Consider a vinyl resin, known to be insoluble in mineral spirits (M). It is found by experiment that cyclohexanone (C) is a solvent, but ethylene glycol monethyl ether (EGMEE), or Cellosolve® (E), is not, and it is desired to formulate a solvent blend made up of these three components. If the resin solubility curve (the broken line in Figure 66) is known, the desired region is clear. If not, a solution of resin in cyclohexane is titrated with mineral spirits until the resin precipitates. (Alternatively, a series of solutions can be prepared with varying ratios of the two liquids.) Point A is placed on the graph so that it divides the cyclohexane-mineral spirits line in the same ratio as the volumes at the solubility limit, in this case CA:AM = 25:75. In a similar manner point B is found to be at EB:BC = 14:86, and in the case of the third side of the triangle, two solubility limits are observed: as mineral spirits is added to EGMEE the resin first dissolves, then precipitates as more mineral spirits is added. These points, A, B, D, and F, indicate the position of the region of resin solubility, and if additional combinations of solvents and nonsolvents are used the whole resin solubility map can be determined.

The hydrogen-bonding index, θ, has advantages over the hydrogen-bonding parameter (Sections 8.3 and 8.4), and in additon contours may be included in Hildebrand parameter—hydrogen-bonding index maps to denote the maximum permissible fractional polarity for a solvent.[81-85] This is illustrated here for Epikote® resins (Figures 67 to 70), acrylic resin (Figure 71), phenolic resin (Figure 72), alkyd resin (Figure 73), and a polyisoprene/polystyrene block copolymer (Figure 74). These figures are used in conjunction with the data of Tables 10, 11 and 14, Chapter 8. As pointed out in Chapter 13, polymer-liquid systems that are completely miscible at high solids concentrations may form two phases on dilution. This means that if a desired formulation has a solids content *lower* than that for which a solubility map has been determined, care must be taken and a generous safety margin allowed. Conversely, for substantially higher solids contents, the solvency requirements of the maps can be relaxed to some extent.

Figure 67 has been annotated to illustrate the method of using the maps.[84] The glycol ether, ethyl Oxitol® (E), with a fractional polarity of 0.086 (Table 10, Chapter 8) is a good solvent for Epikote® 1004 resin, lying on a map contour which indicates that true solvents for the resin at that point can have fractional polarities up to 0.2. In contrast, methyl isobutyl ketone (M) ($p = 0.315$) lies very close to the 0.3 contour, so although the Hildebrand parameter and the hydrogen bonding index are favorable, its fractional polarity is not, and this liquid exhibits only partial solubility for the solid Epikote resins. Consider now the formulation of a solvent blend from the two nonsolvents, xylene (X) and 1-butanol (B), as indicated in Figure 67. As the parameters may be combined in a linear fashion by volume for mixtures, the values for all blends of xylene and 1-butanol lie on the line XB, and the ratio of lengths BC:CX represents the volume ratio of xylene:1-butanol at C. Points between C and D represent potential solvents for the resin, and as BC/CX = 85/15 and BD/DX = 25/75, suitable blends contain 85 to 25% xylene. The same proportionality is applied to fractional polarity, so at C,

$$p_C = p_A + 0.15(p_B - p_X) \tag{4}$$

and

$$p_D = p_A + 0.75(p_B - p_X) \tag{5}$$

Insertion of the values $p_B = 0.096$ and $p_x = 0.001$ (Table 10, Chapter 8) shows that at C the polarity is 0.014 (well below the 0.05 value of the contour on which C and D lie), so solvency is expected. At D, however, the calculated fractional polarity is 0.072, and only limited solvency is expected. Clearly, the greater the xylene content the closer one moves toward C from D, and the lower the polarity. Further components can be added in order to conform to additional solvent formulation constraints.

Two-dimensional Hildebrand parameter vs. fractional polarity solubility maps have also been used, as described in Section 8.8. Figure 75 illustrates such a map for cellulose nitrate.[86]

Izumi and Miyake[87] used their own set of cohesion parameter values, combined with dipole moment and Small's iodine bonding number, γ_s (Section 8.3) for characterizing the solubility behavior of poly(*p*-chlorostyrene). Figure 76 shows the three-dimensional stereograph, the δ-γ_s projection corresponding to Small's original diagram (Figure 2, Chapter 8) and the other two projections.

Hybrid cohesion parameter maps may be used to describe and correlate polymer swelling behavior as well as polymer solubility. Beerbower et al.[88] have used this procedure for a variety of elastomers, based on the hydrogen-bonding parameter data of Crowley et al.[65] (Table 8, Chapter 8). This is illustrated here for styrene-butadiene (Figure 77), isobutylene-isoprene (Figure 78), nitrile-butadiene (Figure 79), chloroprene (Figure 80), and silicone (Figure 81) rubbers, acrylate (Figure 82), and a vinylidene fluoride-hexafluoropropylene copolymer (Figure 83). Liquids in the >100% swell region cause the polymer to swell to at least twice its original volume; from this region either increasing or decreasing the hydrogen-bonding parameter at the same Hildebrand parameter value causes the extent of swelling to decrease. Similar maps have also been published for linseed oil films.[89] Yoshioka[90] reported polymer-tackifier compatibility.

Other forms of hybrid solubility map have been described in other chapters. In Section 13.5 is Chen's δ_h vs. χ_H diagram, where χ_H is a measure of the "physical" interactions.

$$\chi_H = ({}^iV/RT)[({}^i\delta_d - {}^j\delta_d)^2 + ({}^i\delta_p - {}^j\delta_p)^2] \tag{6}$$

and Jackson[91-93] used Hildebrand parameter vs. dipole moment maps in the characterization of contaminants on surfaces (Section 17.9). Plots of polarity index[94] (Section 8.5) and dielectric constant[95] against Hildebrand parameter have also been used.

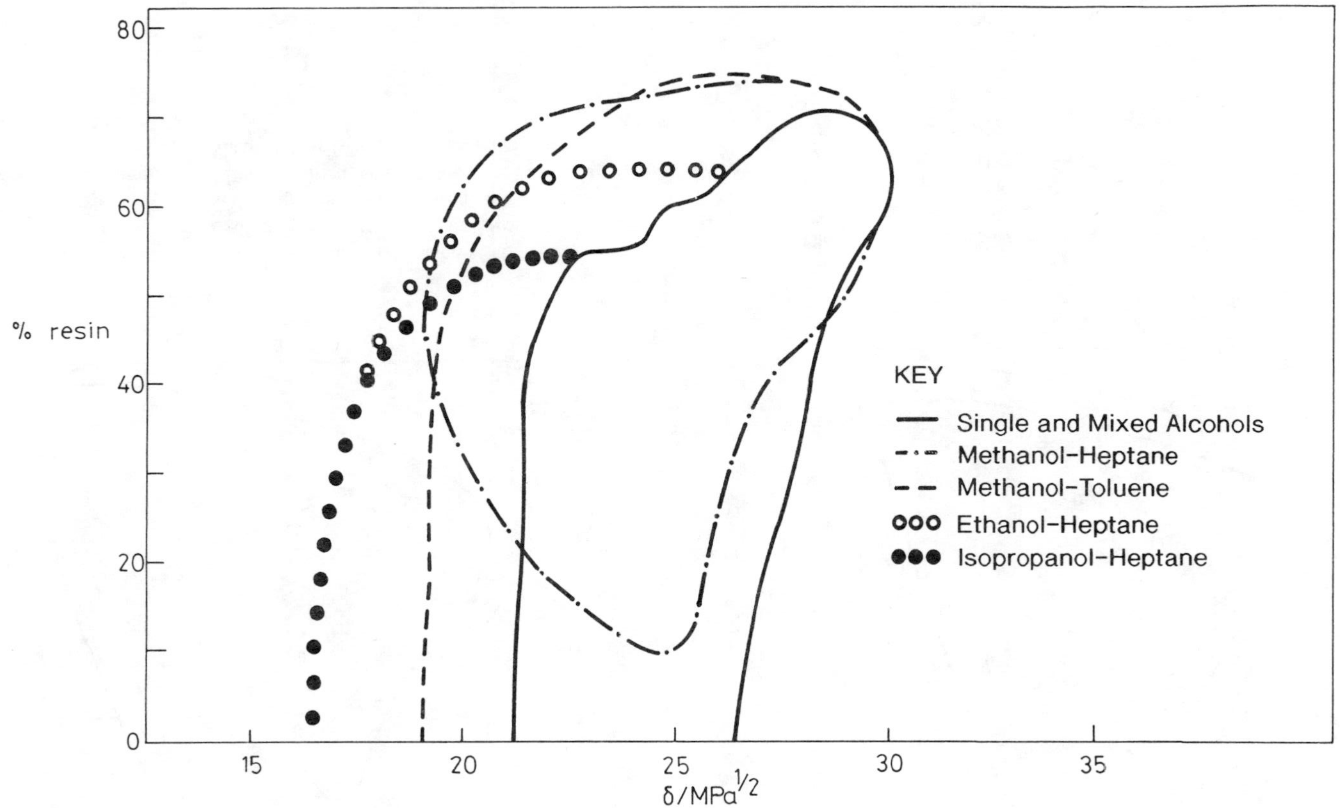

FIGURE 1. Solubility map for a linear polyamide resin (prepared from dimer aliphatic acid and diamine) in terms of Hildebrand parameter values for pure and mixed alcohols and alcohol-hydrocarbon mixtures. Broken lines show solubility extension at lower Hildebrand parameters resulting from the incorporation of hydrocarbons with alcohols. (Adapted from Hinden, E. R. and Whyzmusis, P. D., *Solvents Theory and Practice,* Tess, R. W., Ed., American Chemical Society, Washington, D.C., 1973, chap. 11.)

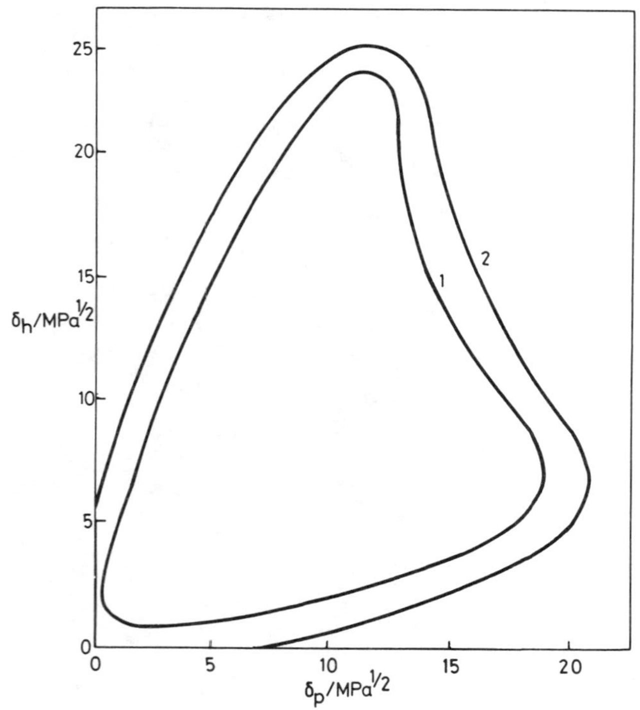

FIGURE 2. Hansen parameter polarity—hydrogen-bonding solubility map with dispersion contours for a 70:30 by weight mixture of 35 weight % *o*-phthalic anhydride-47 weight % synthetic aliphatic acid resin with a highly reactive melamine resin 16% esterified with 1-butanol: 1, δ_d = 15 MPa$^{1/2}$; 2, δ_d = 17 MPa$^{1/2}$. (Adapted from Walz, G., *Kunstharz Nachr. Hoechst,* 33(6), 30, 1974.

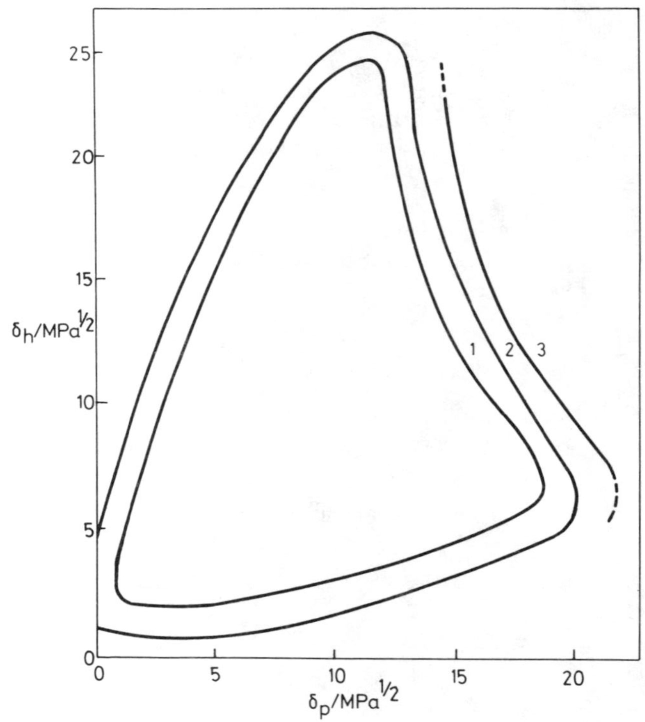

FIGURE 3. Hansen parameter polarity—hydrogen-bonding solubility map with dispersion contours for a 70:30 by weight mixture of 35 weight % *o*-phthalic anhydride-47 weight % synthetic aliphatic acid resin with hexakis(methoxymethyl)melamine resin: 1, δ_d = 15 MPa$^{1/2}$; 2, δ_d = 17 MPa$^{1/2}$; 3, δ_d = 18 MPa$^{1/2}$. (Adapted from Walz, G., *Kunstharz Nachr. Hoechst,* 33(6), 30, 1974.)

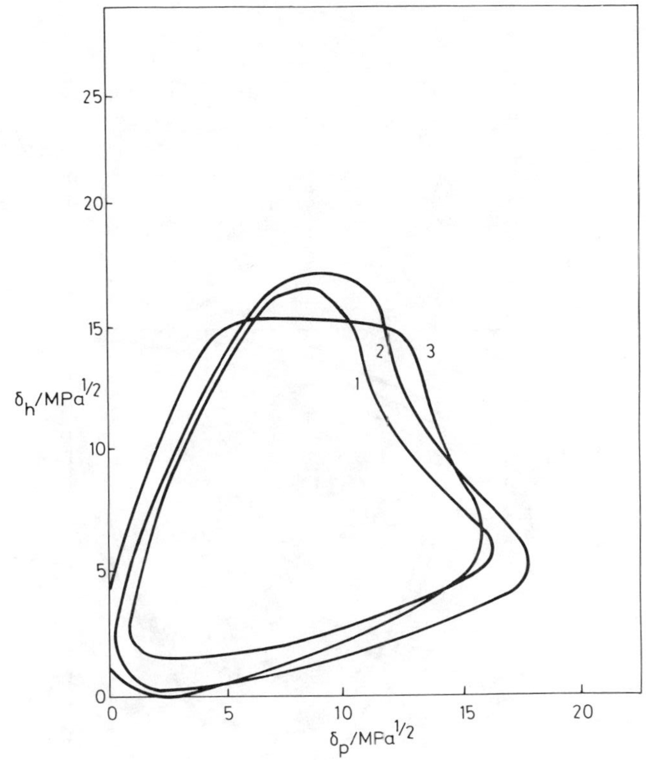

FIGURE 4. Hansen parameter polarity—hydrogen-bonding solubility map with dispersion contours for a colophonium resin (20% solution, 23°C): 1, δ_d = 14 MPa$^{1/2}$; 2, δ_d = 15 MPa$^{1/2}$; 3, δ_d = 16 MPa$^{1/2}$. (Adapted from Walz, G., *Kunstharz Nachr. Hoechst*, 33(6), 30, 1974.)

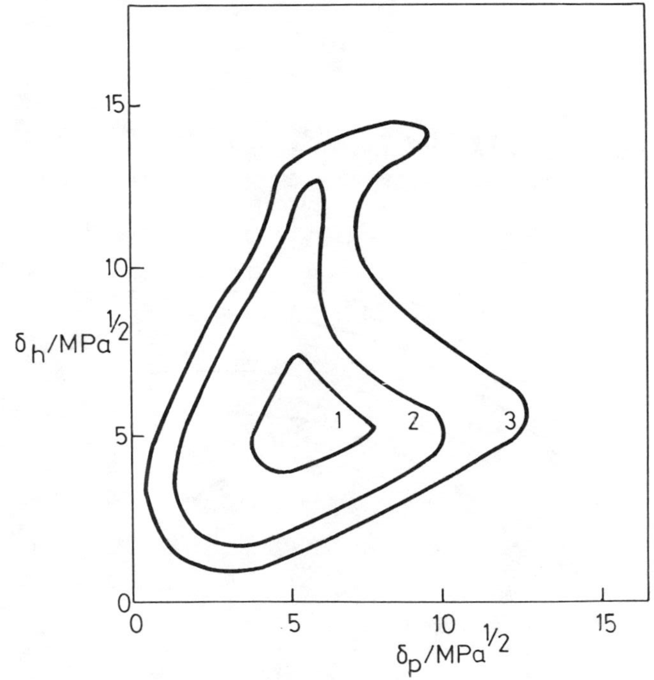

FIGURE 5. Hansen parameter polarity—hydrogen-bonding solubility map with dispersion contours for Hoechst Alftalat® AF342 alkyd resin (20% solution, 23°C): 1, δ_d = 15 MPa$^{1/2}$; 2, δ_d = 16 MPa$^{1/2}$; 3, δ_d = 17 MPa$^{1/2}$. (Adapted from Walz, G. and Emrich, G., *Kunstharz Nachr. Hoechst*, 34(8), 19, 1975.)

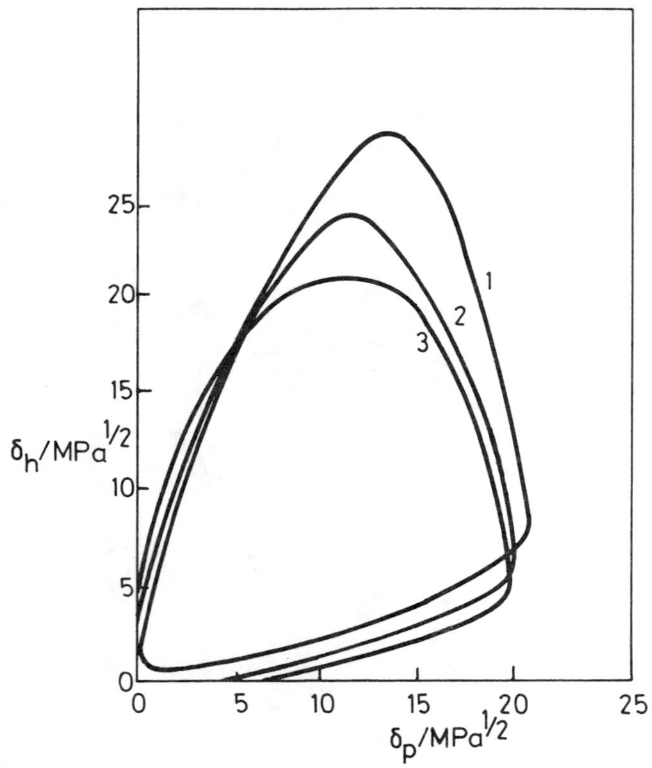

FIGURE 6. Hansen parameter polarity—hydrogen-bonding solubility map with dispersion contours for Hoechst Maprenal® MF800 melamine/formaldehyde resin (20% solution, 23°C): 1, δ_d = 15 MPa$^{1/2}$; 2, δ_d = 16 MPa$^{1/2}$; 3, δ_d = 17 MPa$^{1/2}$. (Adapted from Walz, G. and Emrich, G., *Kunstharz Nachr. Hoechst*, 34(8), 19, 1975.)

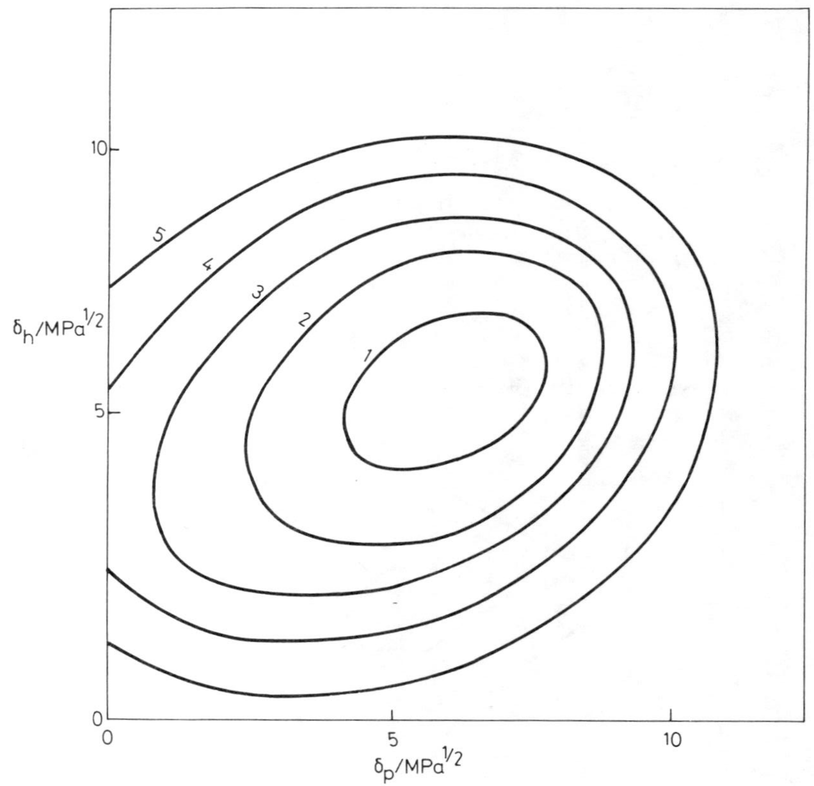

FIGURE 7. Hansen parameter polarity—hydrogen-bonding solubility map with dispersion contours for polyepichlorohydrin: 1, δ_d = 16 MPa$^{1/2}$; 2, δ_d = 16.6 MPa$^{1/2}$; 3, δ_d = 17.4 MPa$^{1/2}$; 4, δ_d = 18 MPa$^{1/2}$; 5, δ_d = 19 MPa$^{1/2}$. (Adapted from Kozakiewicz, J. and Penczek, P., *Angew. Makromol. Chem.*, 50, 67, 1976.)

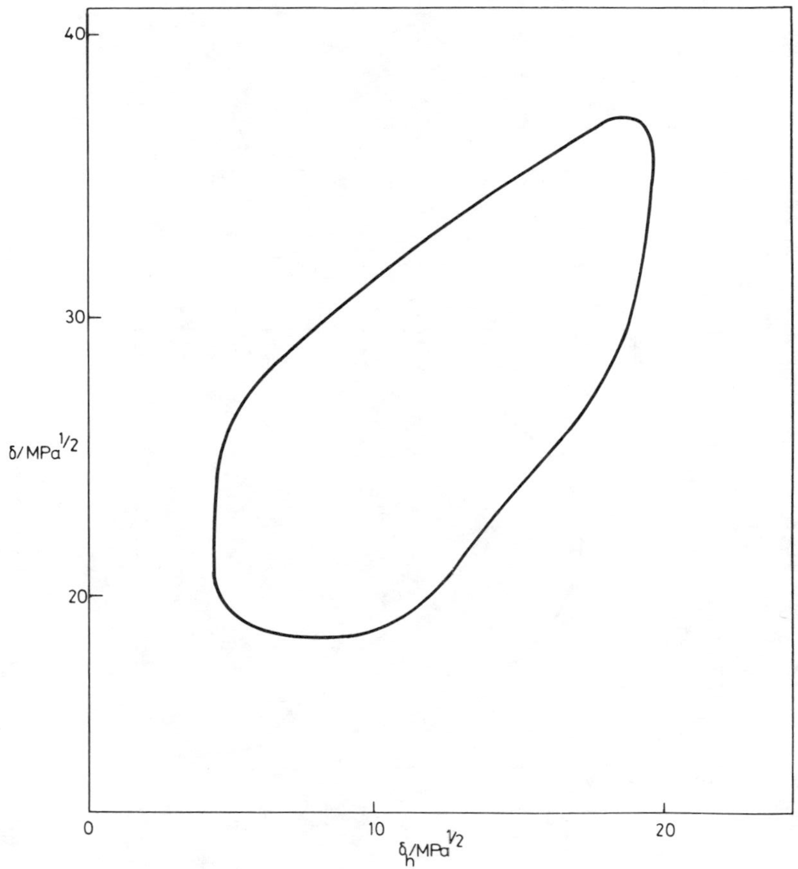

FIGURE 8. Hildebrand parameter—Hansen hydrogen-bonding parameter solubility map for cellulose acetate, for use with Hansen and Beerbower's 1971 liquids data, Table 11, Chapter 5. (Adapted from Hoernschemeyer, D. J., *J. Appl. Polym. Sci.*, 18, 61, 1974.)

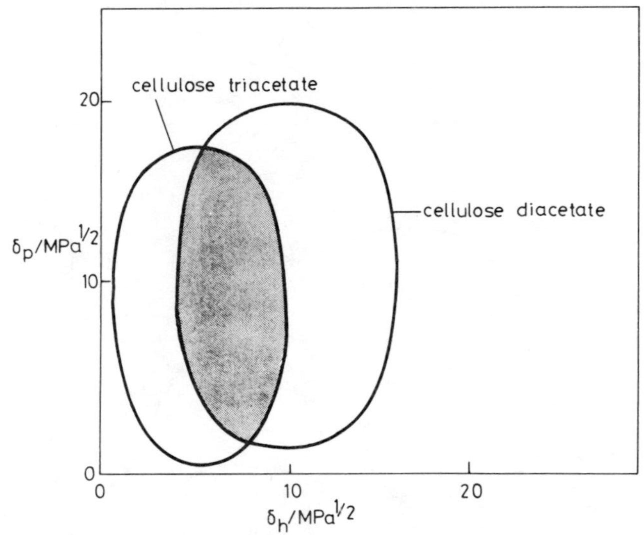

FIGURE 9. Hansen parameter polarity — hydrogen-bonding solubility maps for cellulose triacetate, cellulose diacetate, and (hatched area) a 2.75 DS cellulose acetate blend membrane, for use with Hansen and Beerbower's 1971 liquids data, Table 11, Chapter 5 and Figure 6, Chapter 5. (Adapted from Klein, E., Eichelberger, J., Eyer, C., and Smith, J., *Water Res.*, 9, 807, 1975.)

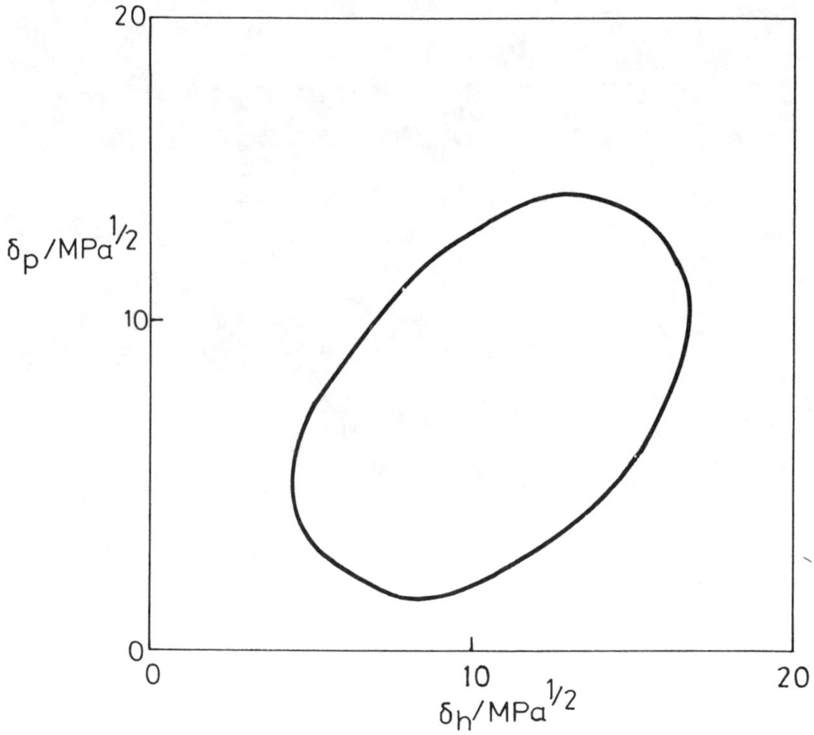

FIGURE 10. Hansen parameter polarity—hydrogen-bonding solubility map for 2.25 DS ethyl cellulose, for use with Hansen and Beerbower's 1971 liquid data, Table 11, Chapter 5 and Figure 6, Chapter 5. (Adapted from Klein, E., Eichelberger, J., Eyer, C., and Smith, J., *Water Res.*, 9, 807, 1975.)

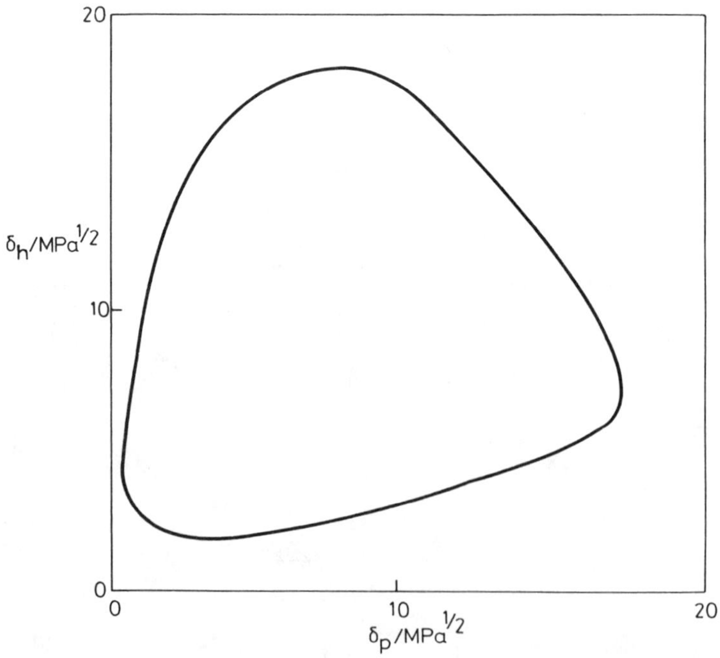

FIGURE 11. Hansen parameter polarity—hydrogen-bonding map for C.I. Solvent Blue 70. (Adapted from Riedel, G., *Farbe Lack*, 82, 281, 1976.)

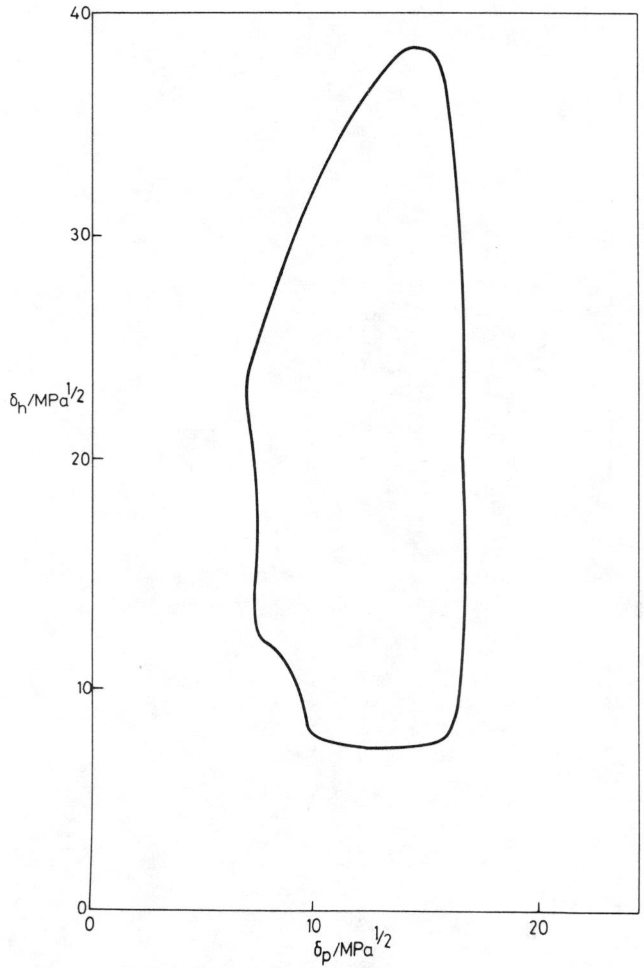

FIGURE 12. Hansen parameter polarity— hydrogen-bonding solubility map for Rhodamin FB (C.I. Basic Violet 10). (Adapted from Riedel, G., *Farbe Lack*, 82, 281, 1976.)

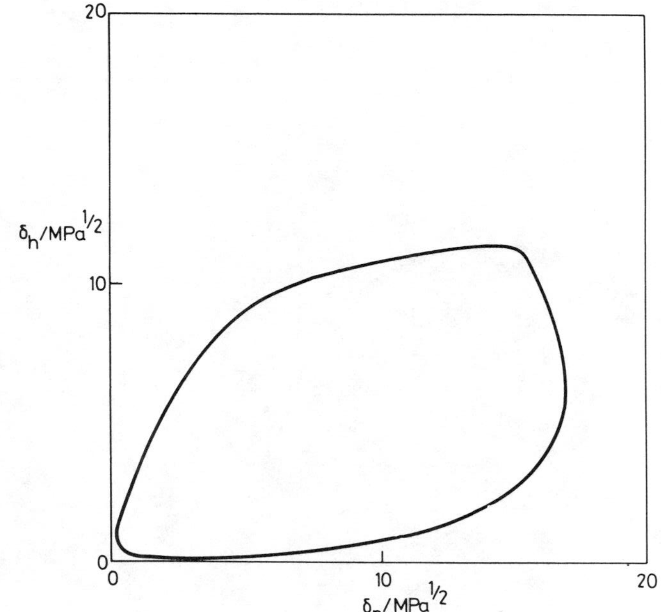

FIGURE 13. Hansen parameter polarity—hydrogen-bonding solubility map for C.I. Solvent Yellow. (Adapted from Riedel, G., *Farbe Lack*, 82, 281, 1976.)

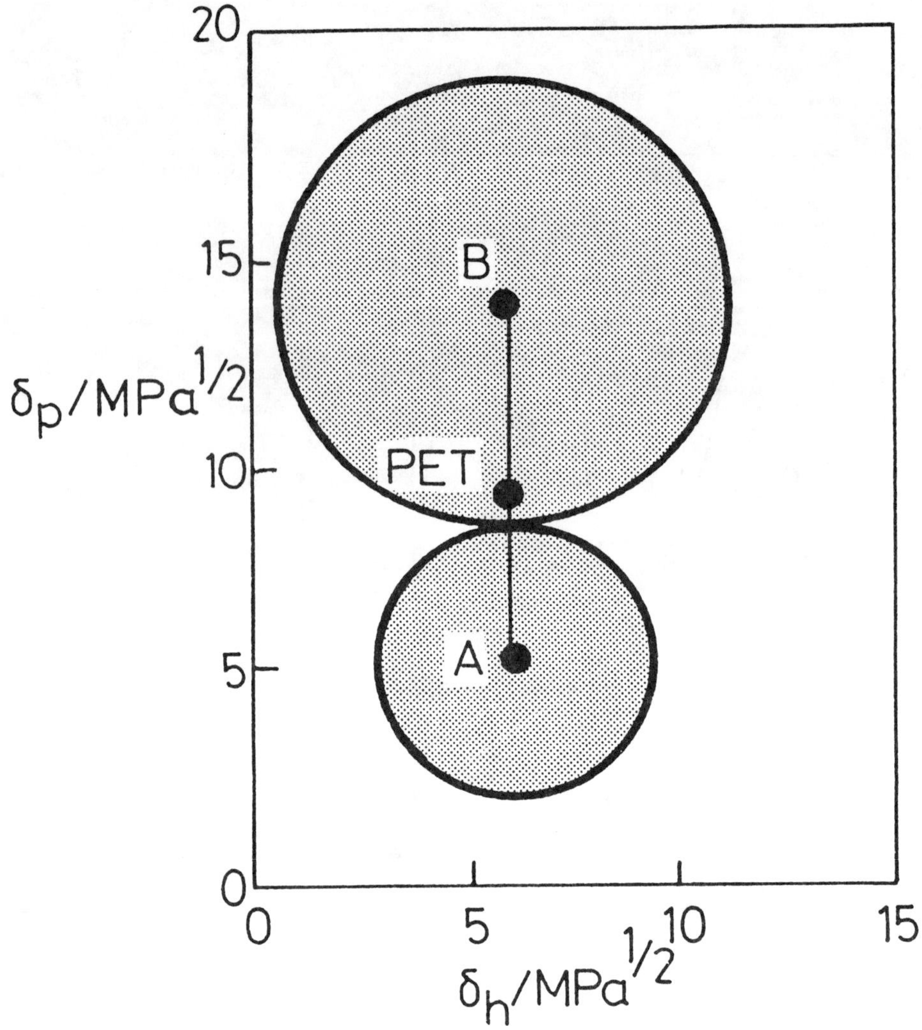

FIGURE 14. Hansen parameter polarity—hydrogen-bonding plot of areas of extensive interaction (>3% shrinkage) between liquids and poly(ethylene terephthalate). The coordinates of the centers of the interaction circles are the δ_p and δ_h values of the aromatic (A) and aliphatic ester (B) residues; the volume fraction average is used for the center of the interaction circle for the monomer repeat unit (PET). (Adapted from Knox, B. H., Weigmann, H.-D., and Scott, M. G., *Text. Res. J.*, 45, 203, 1975.)

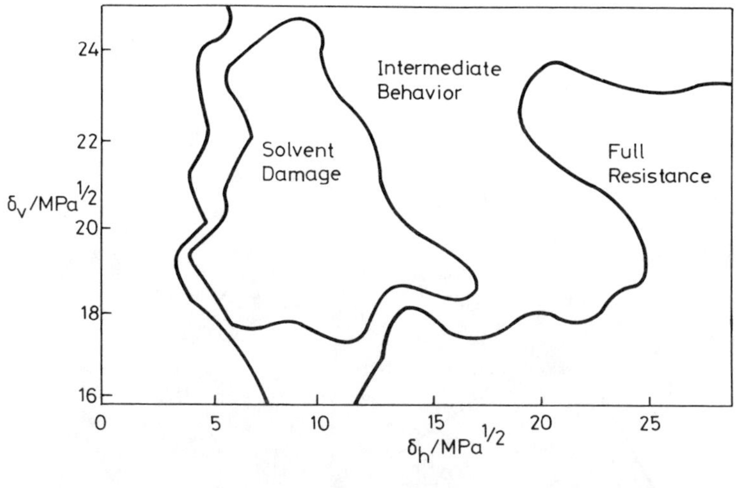

FIGURE 15. Solvent resistance of an epoxy paint portrayed on a map of volume-dependent cohesion parameter versus Hansen hydrogen-bonding parameter. (Adapted from Papo, A., Sciarrone, U., and Torriano, G., *Proc. Intl. Conf. Org. Coatings Sci. Technol.*, 4, 585, 1978.)

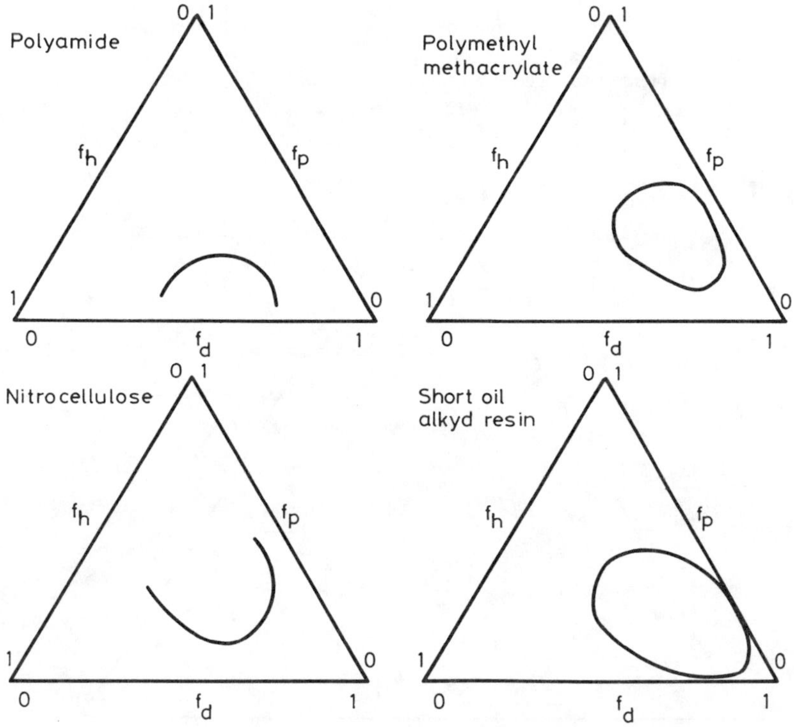

FIGURE 16. Limiting boundaries of solubilities for polyamide, poly(methyl methacrylate), cellulose nitrate, and short oil alkyd resin in terms of fractional Hansen parameters for use with the liquids data of Table 16, Chapter 5. (Adapted from Gardon, J. L. and Teas, J. P., *Treatise on Coatings, Vol. 2, Characterization of Coatings: Physical Techniques,* Myers, R. R. and Long, J. S., Eds., Marcel Dekker, New York, 1976, chap. 8, 413.)

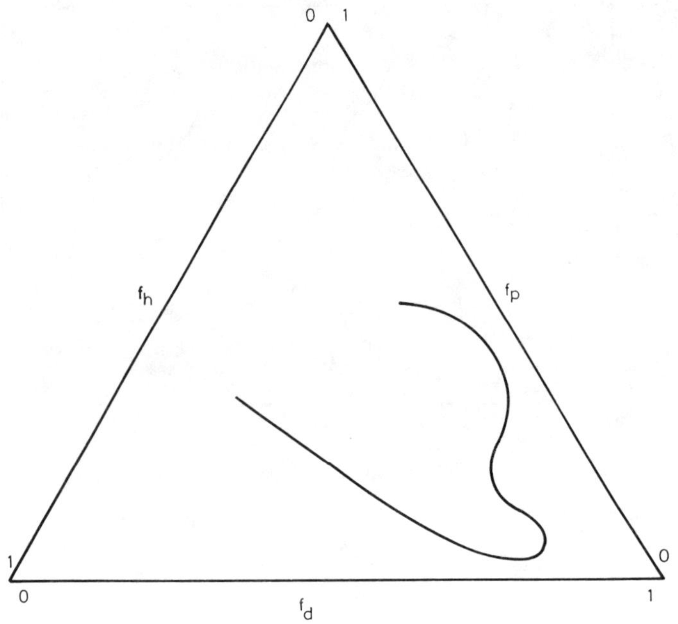

FIGURE 17. Triangular fractional Hansen parameter solubility diagram for Epikote® 101 epoxy resin (Shell), for use with the liquids data of Table 16, Chapter 5. (Adapted from Teas, J. P., *J. Paint Technol.*, 40(516), 19, 1968.)

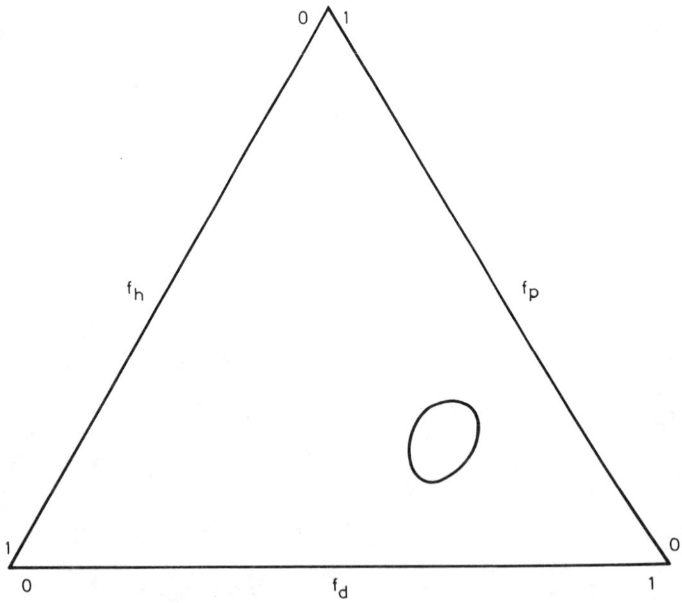

FIGURE 18. Triangular fractional Hansen parameter solubility diagram for Vipla® KR poly(vinyl chloride) resin (Montecatini), for use with the liquids data of Table 16, Chapter 5. (Adapted from Teas, J. P., *J. Paint Technol.*, 40(516), 19, 1968.)

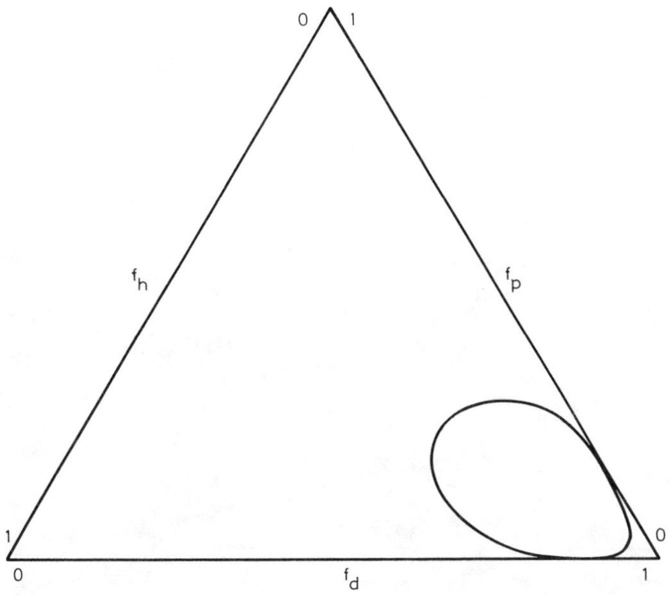

FIGURE 19. Triangular fractional Hansen parameter solubility diagram for Piccopale® 110 resin (Pennsylvania Industrial Chemicals), for use with the liquids data of Table 16, Chapter 5. (Adapted from Teas, J. P., *J. Paint Technol.*, 40(516), 19, 1968.)

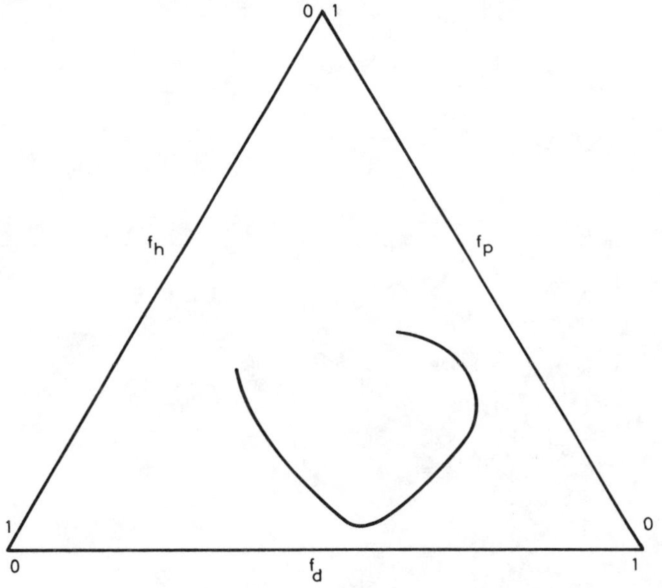

FIGURE 20. Triangular fractional Hansen parameter solubility diagram for Pentalyn® 225 pentaerythritol-rosin resin (Hercules Inc.), for use with the liquids data of Table 16, Chapter 5. (Adapted from Teas, J. P., *J. Paint Technol.*, 40(516), 19, 1968.)

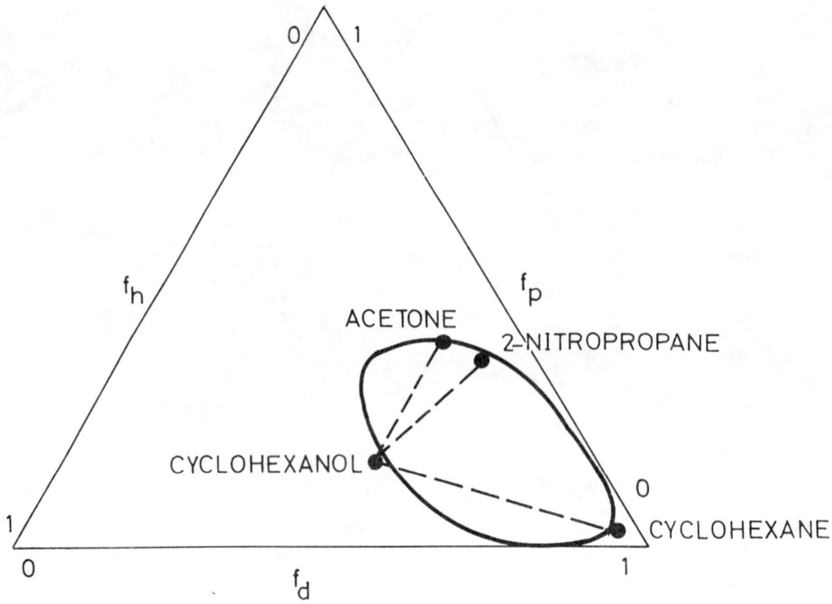

FIGURE 21. Triangular fractional Hansen parameter solubility diagram for Parlon® P10 resin (Hercules Inc.) indicating suitable mixtures of nonsolvents, for use with the liquids data of Table 16, Chapter 5. (Adapted from Teas, J. P., *J. Paint Technol.*, 40(516), 19, 1968.)

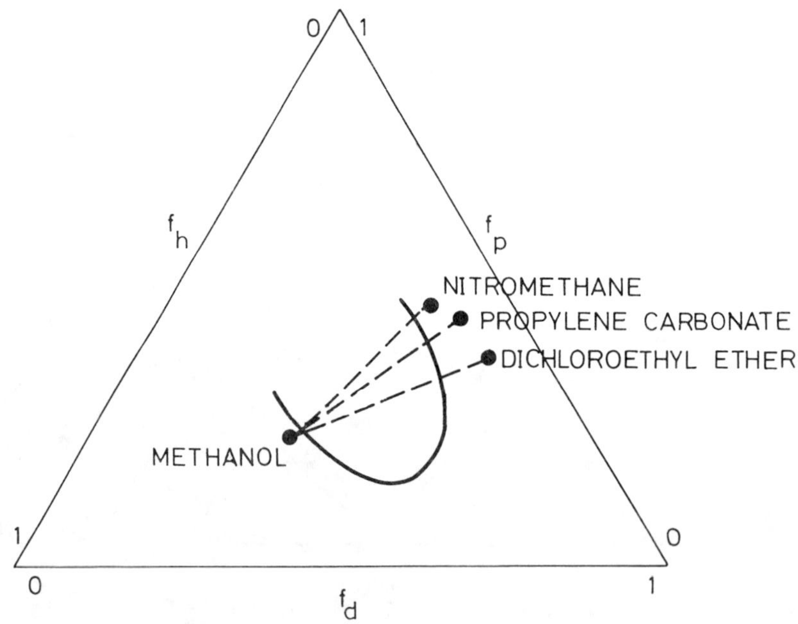

FIGURE 22. Triangular fractional Hansen parameter diagram for the solubility of lignin in mixtures of nonsolvents, for use with the liquids data of Table 16, Chapter 5. (Adapted from Teas, J. P., *J. Paint Technol.*, 40(516), 19, 1968.)

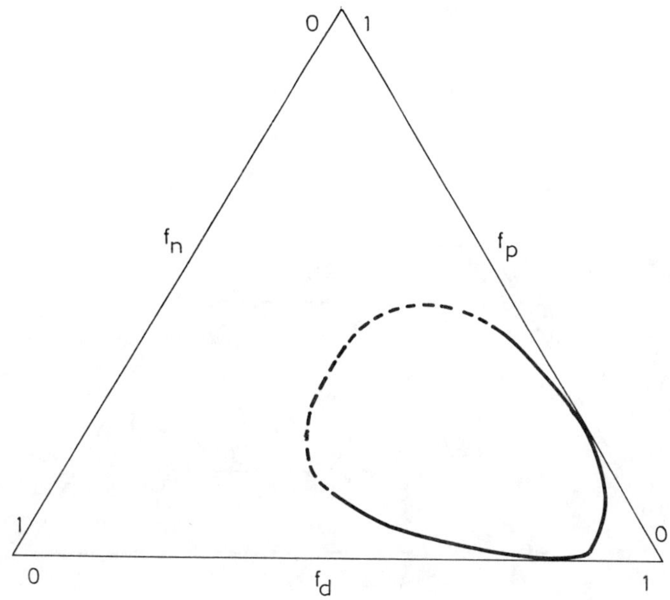

FIGURE 23. Triangular fractional Hansen parameter solubility diagram for Epikote® 1007 (Shell), for use with the liquids data of Table 11, Chapter 5. (Adapted from Shell Chemicals, *Solubility Parameters,* 1978.)

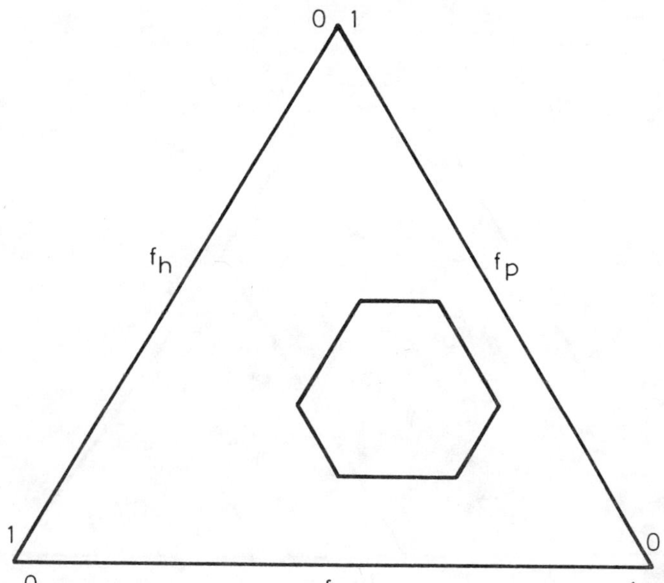

FIGURE 24. Triangular fractional Hansen parameter solubility diagram for triacetate, for use with the liquids data of Table 16, Chapter 5. (Adapted from Hashimoto, I., *Sen-i Gakkaishi,* 34, T469, 1978.)

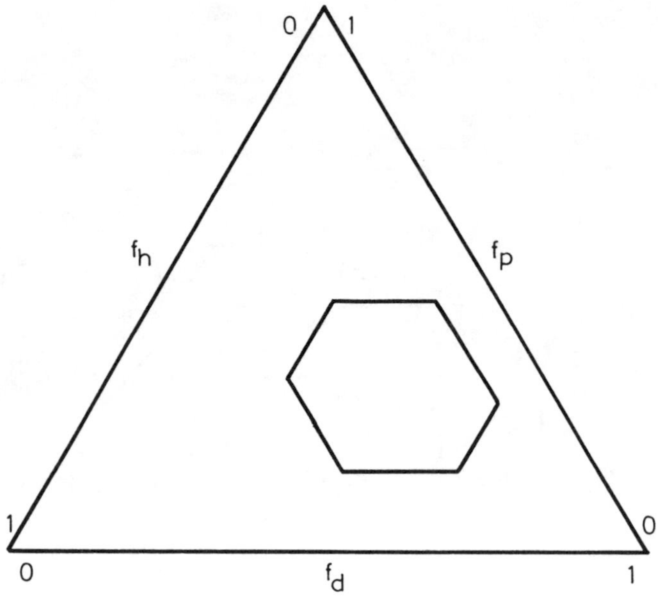

FIGURE 25. Triangular fractional Hansen parameter solubility diagram for diacetate, for use with the liquids data of Table 16, Chapter 5. (Adapted from Hashimoto, I., *Sen-i Gakkaishi,* 34, T469, 1978.)

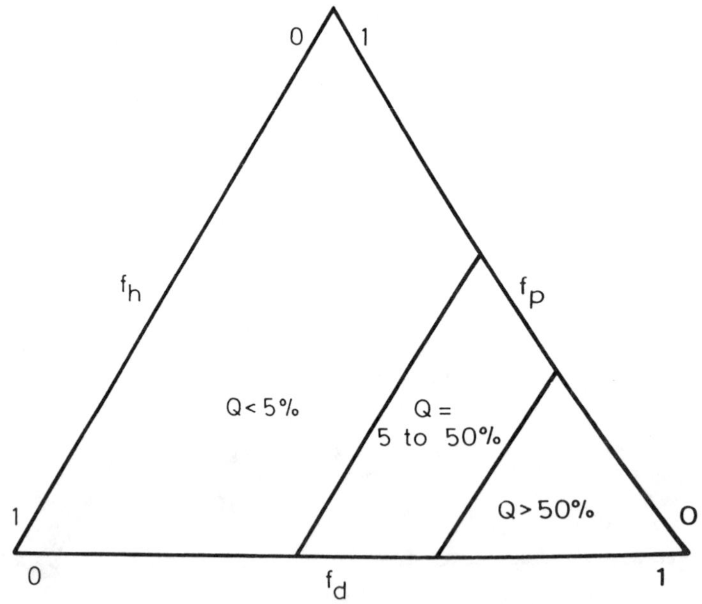

FIGURE 26. Triangular fractional Hansen parameter diagram of the extent of swelling, *Q*, of polypropylene, for use with the liquids data of Table 16, Chapter 5. (Adapted from Hashimoto, I., *Sen-i Gakkaishi,* 34, T469, 1978.)

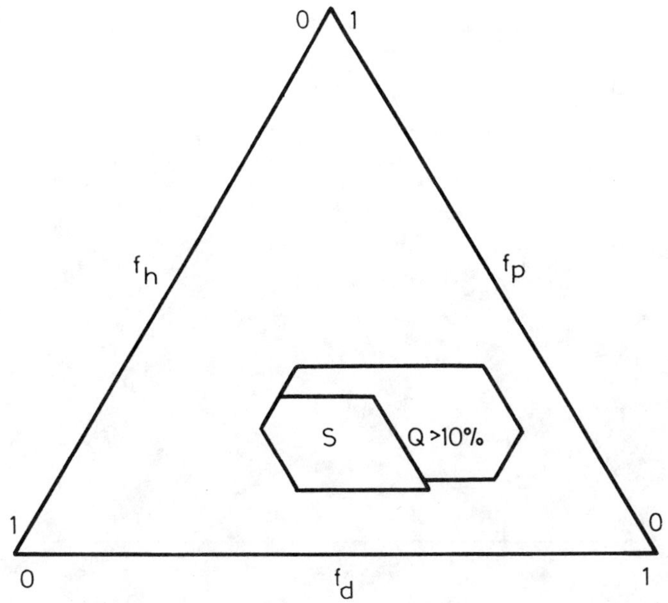

FIGURE 27. Triangular fractional Hansen parameter diagram of the solubility (*S*) and swelling (*Q*) of polyamide-6, for use with the liquids data of Table 16, Chapter 5. (Adapted from Hashimoto, I., *Sen-i Gakkaishi,* 34, T469, 1978.)

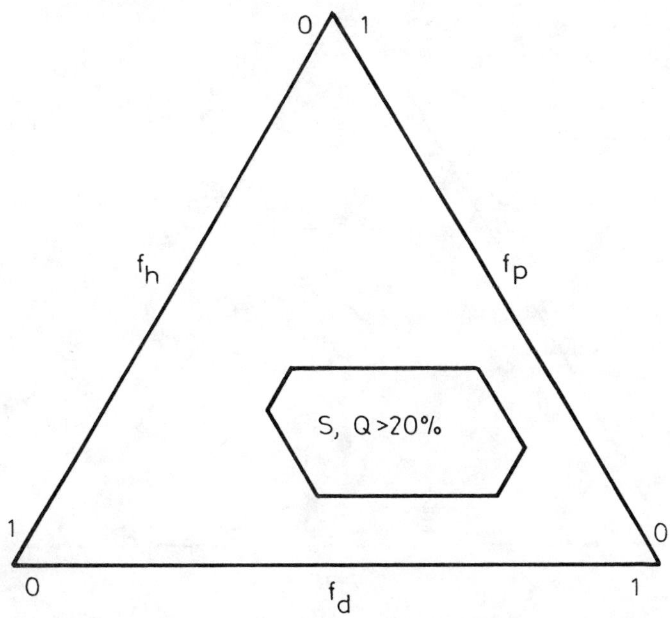

FIGURE 28. Triangular fractional Hansen parameter diagram of the solubility (*S*) and swelling (*Q*) of poly(ethylene terephthalate), for use with the liquids data of Table 16, Chapter 5. (Adapted from Hashimoto, I., *Sen-I Gakkaishi,* 34, T469, 1978.)

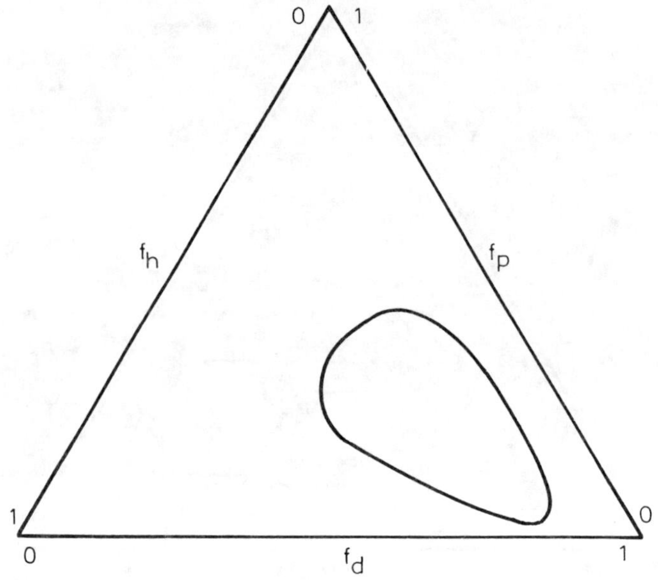

FIGURE 29. Triangular fractional Hansen parameter solubility diagram for an alkyd resin, for use with the liquids data of Table 16, Chapter 5. (Adapted from Loev, A. M. and Lavrishchev, L. P., *Lakokras. Mater. Ikh. Primen.*, No. 4, 4, 1977.)

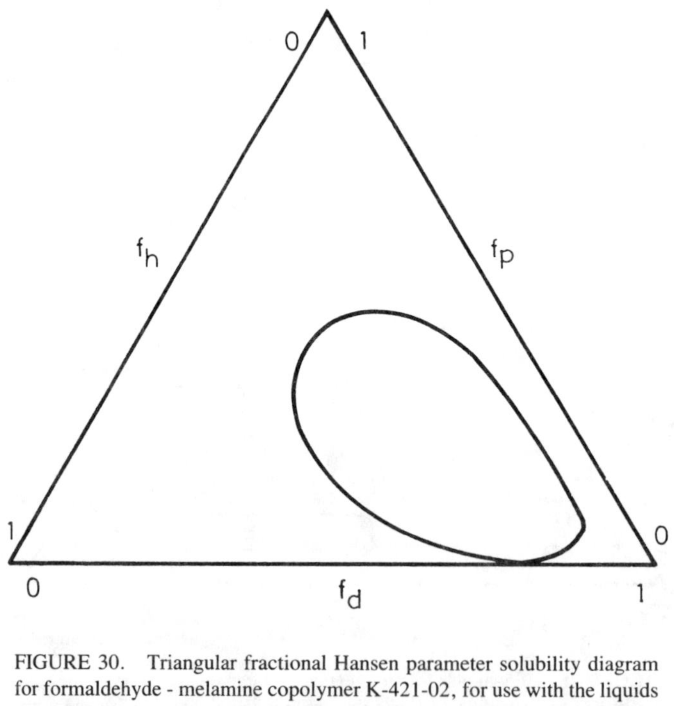

FIGURE 30. Triangular fractional Hansen parameter solubility diagram for formaldehyde - melamine copolymer K-421-02, for use with the liquids data of Table 16, Chapter 5. (Adapted from Loev, A. M. and Lavrishchev, L. P., *Lakokras. Mater. Ikh. Primen.*, No. 4, 4, 1977.)

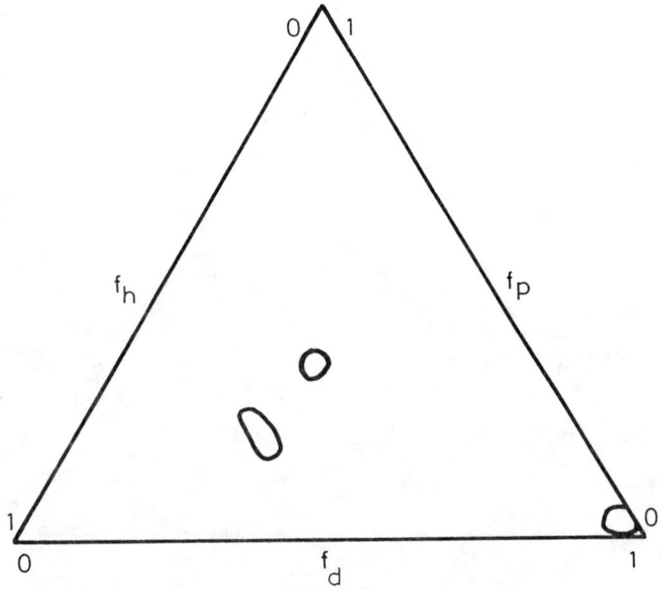

FIGURE 31. Triangular fractional Hansen parameter solubility diagram for formaldehyde-melamine copolymer GM-3, for use with the liquids data of Table 16, Chapter 5. (Adapted from Loev, A. M. and Lavrishchev, L. P., *Lakokras. Mater. Ikh, Primen.*, No. 4, 4, 1977.)

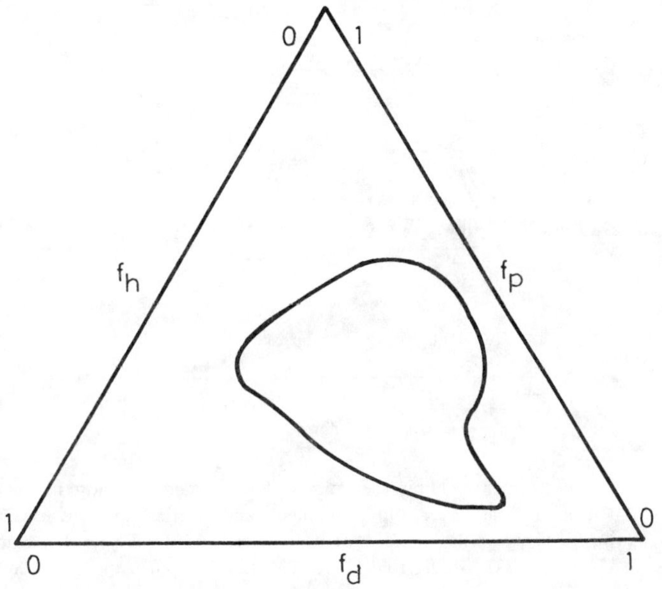

FIGURE 32. Triangular fractional Hansen parameter solubility diagram for E-05K epoxy resin, for use with the liquids data of Table 16, Chapter 5. (Adapted from Loev, A. M. and Lavrishchev, L. P., *Lakokras. Mater. Ikh. Primen.*, No. 4, 4, 1977.)

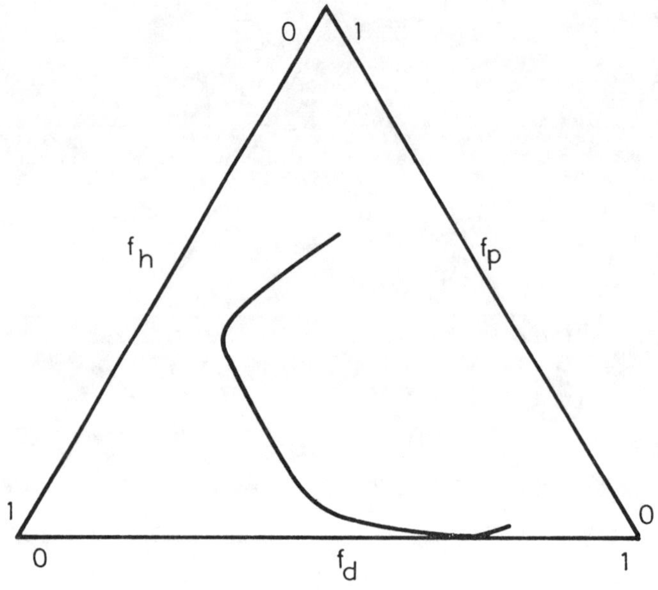

FIGURE 33. Triangular fractional Hansen parameter solubility diagram for phenol-formaldehyde FPF-1F resin, for use with the liquids data of Table 16, Chapter 5. (Adapted from Loev, A. M., and Lavrishchev, L. P., *Lakokras. Mater. Ikh. Primen.*, No. 4, 4, 1977.)

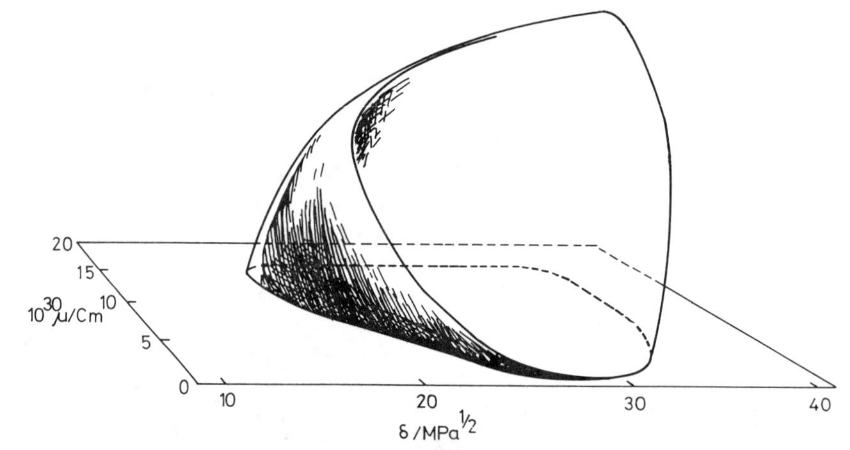

FIGURE 34. Three-dimensional representation of RS-$^1/_2$ second cellulose nitrate solubility in terms of Hildebrand parameter, dipole moment, and (vertical axis) hydrogen-bonding parameter (γ_C), for use with the liquids data of Tables 8 and 9, Chapter 8. (Adapted from Crowley, J. D., Teague, G. S., Jr., and Lowe, J. W., *J. Paint Technol.*, 38, 269, 1966.)

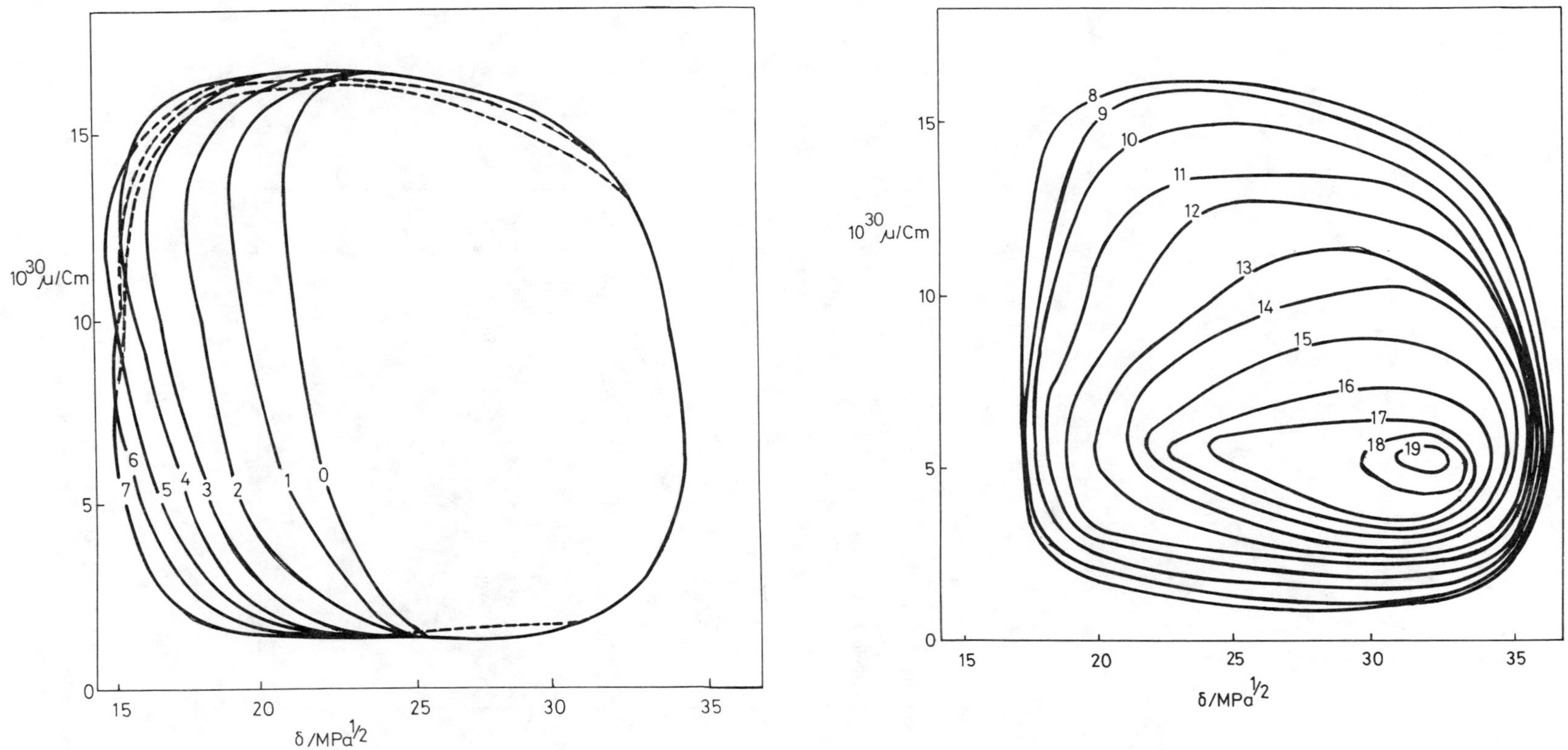

A B

FIGURE 35. Hildebrand parameter—dipole moment solubility map with hydrogen-bonding parameter (γ_C) contours for RS-½ second cellulose nitrate, for use with the liquids data of Tables 8 and 9, Chapter 8: A, γ_C = 0 to 7; B, γ_D = 7 to 19. (Adapted from Crowley, J. D., Teague, G. S., Jr., and Lowe, J. W., *J. Paint Technol.*, 38, 269, 1966.)

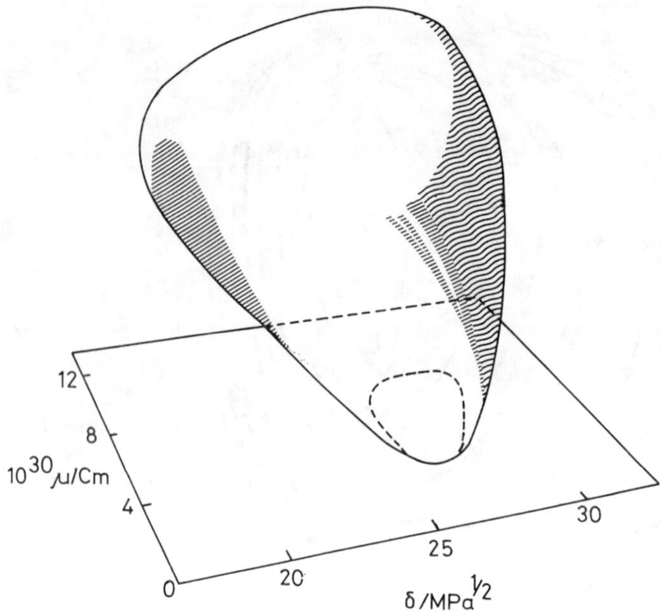

FIGURE 36. Three-dimensional representation of cellulose acetate (E-398-3) solubility in terms of Hildebrand parameter, dipole moment, and (vertical axis) hydrogen-bonding parameter (γ_C), for use with the liquids data of Tables 8 and 9, Chapter 8. (Adapted from Crowley, J. D., Teague, G. S., Jr., and Lowe, J. W., *J. Paint Technol.*, 38, 269, 1966.)

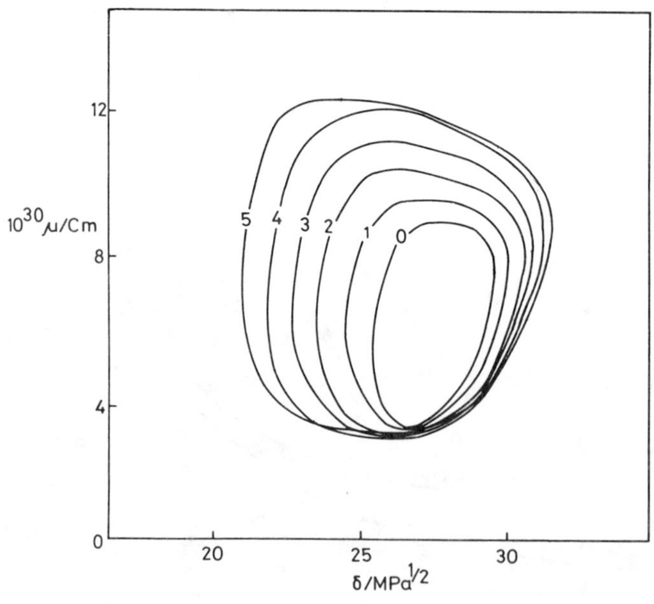

A

FIGURE 37. Hildebrand parameter—dipole moment solubility map with hydrogen-bonding parameter (γ_C) contours for cellulose acetate (E-398-3), for use with the liquids data of Tables 8 and 9, Chapter 8 A, γ_C = 0 to 5; B, γ_C = 5 to 8; C, γ_C = 9 to 13. (Adapted from Crowley, J. D., Teague, G. S., Jr., and Lowe, J. W., *J. Paint Technol.*, 38, 269, 1966.)

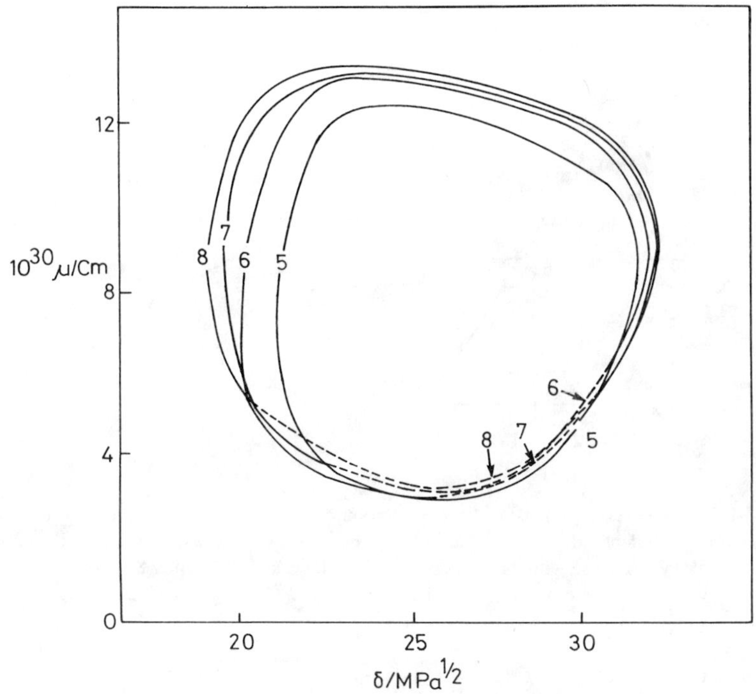

FIGURE 37B

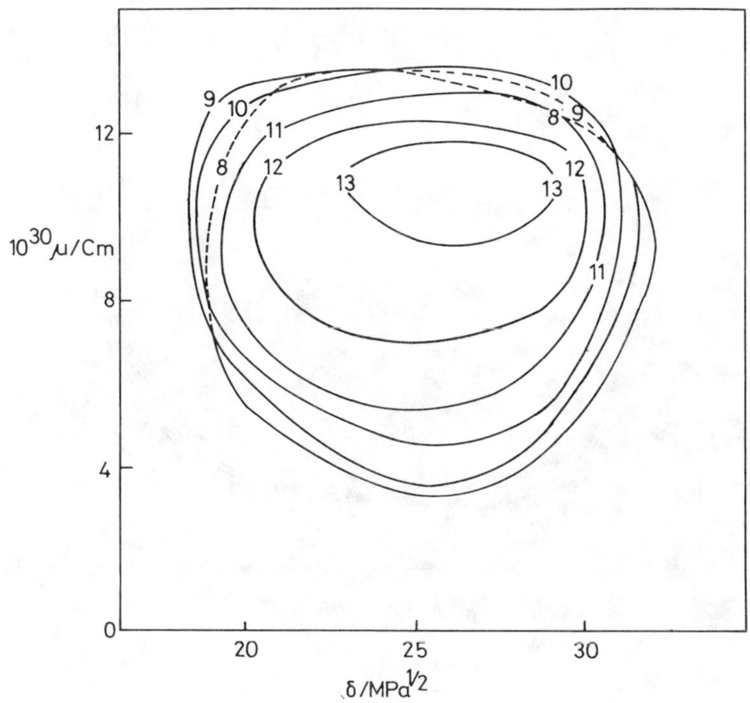

FIGURE 37C

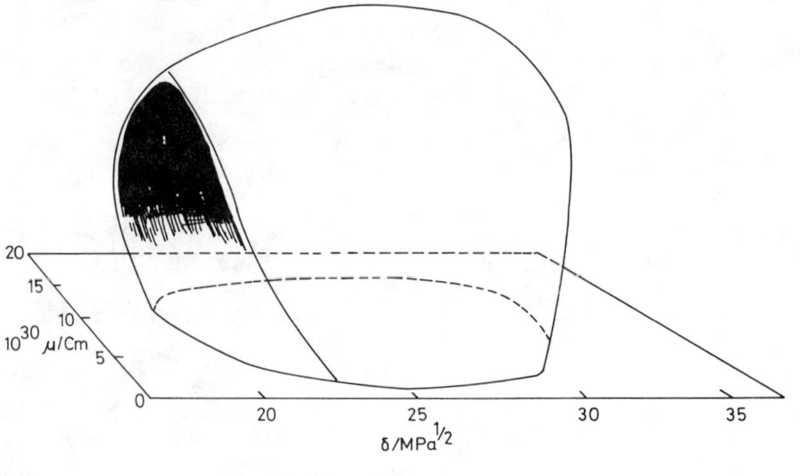

FIGURE 38. Three-dimensional representation of cellulose acetate butyrate solubility in terms of Hildebrand parameter, dipole moment, and (vertical axis) hydrogen-bonding parameter (γ_C), for use with the liquids data of Tables 8 and 9, Chapter 8. (Adapted from Crowley, J. D., Teague, G. S., Jr., and Lowe, J. W., *J. Paint Technol.*, 38, 269, 1966.)

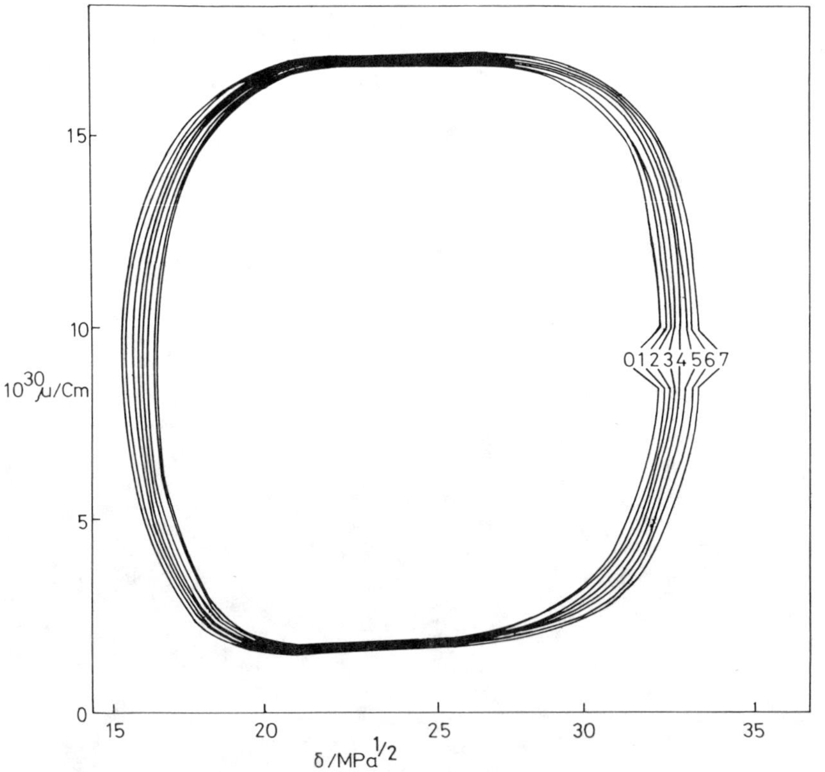

A

FIGURE 39. Hildebrand parameter—dipole moment solubility map with hydrogen-bonding parameter (γ_C) contours for $^1/_2$-second cellulose acetate butyrate, for use with the liquids data of Tables 8 and 9, Chapter 8: A, γ_C = 0 to 7; B, γ_C = 7 to 16.5. (Adapted from Crowley, J. D., Teague, G. S., Jr., and Lowe, J. W., *J. Paint Technol.*, 38, 269, 1966.)

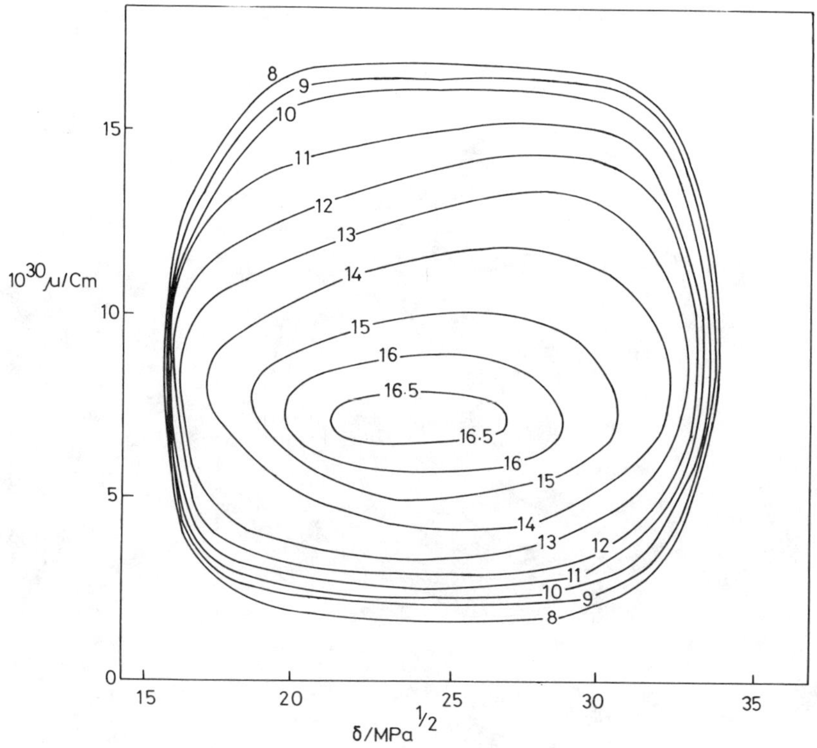

FIGURE 39B

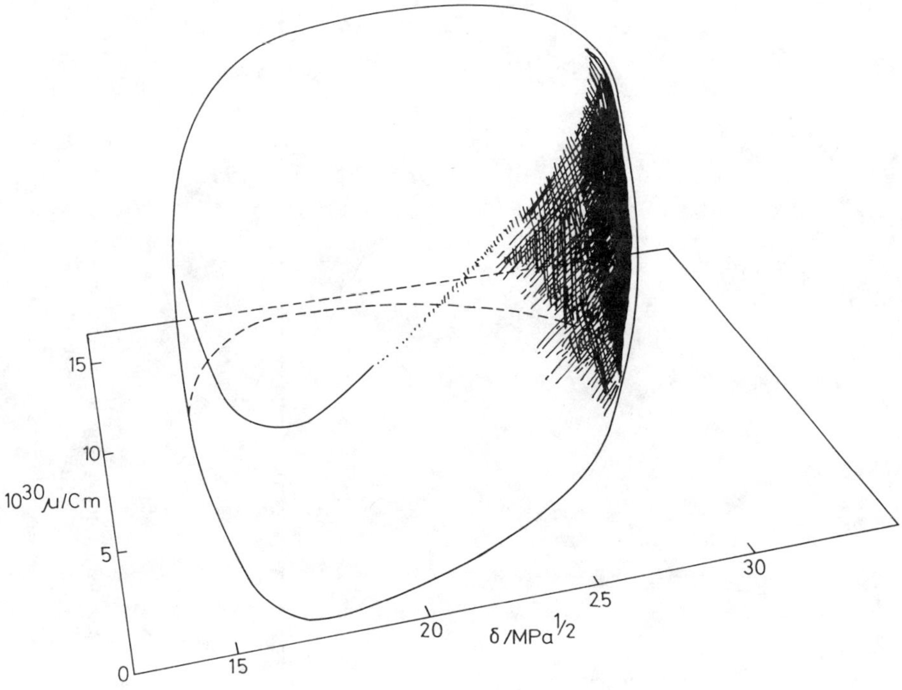

FIGURE 40. Three-dimensional representation of poly(methyl methacrylate) (Elvacite® 2010) solubility in terms of Hildebrand parameter, dipole moment, and (vertical axis) hydrogen-bonding parameter (γ_C), for use with the liquids data of Tables 8 and 9, Chapter 8. (Adapted from Crowley, J. D., Teague, G. S., Jr., and Lowe, J. W., *J. Paint Technol.*, 38, 269, 1966.)

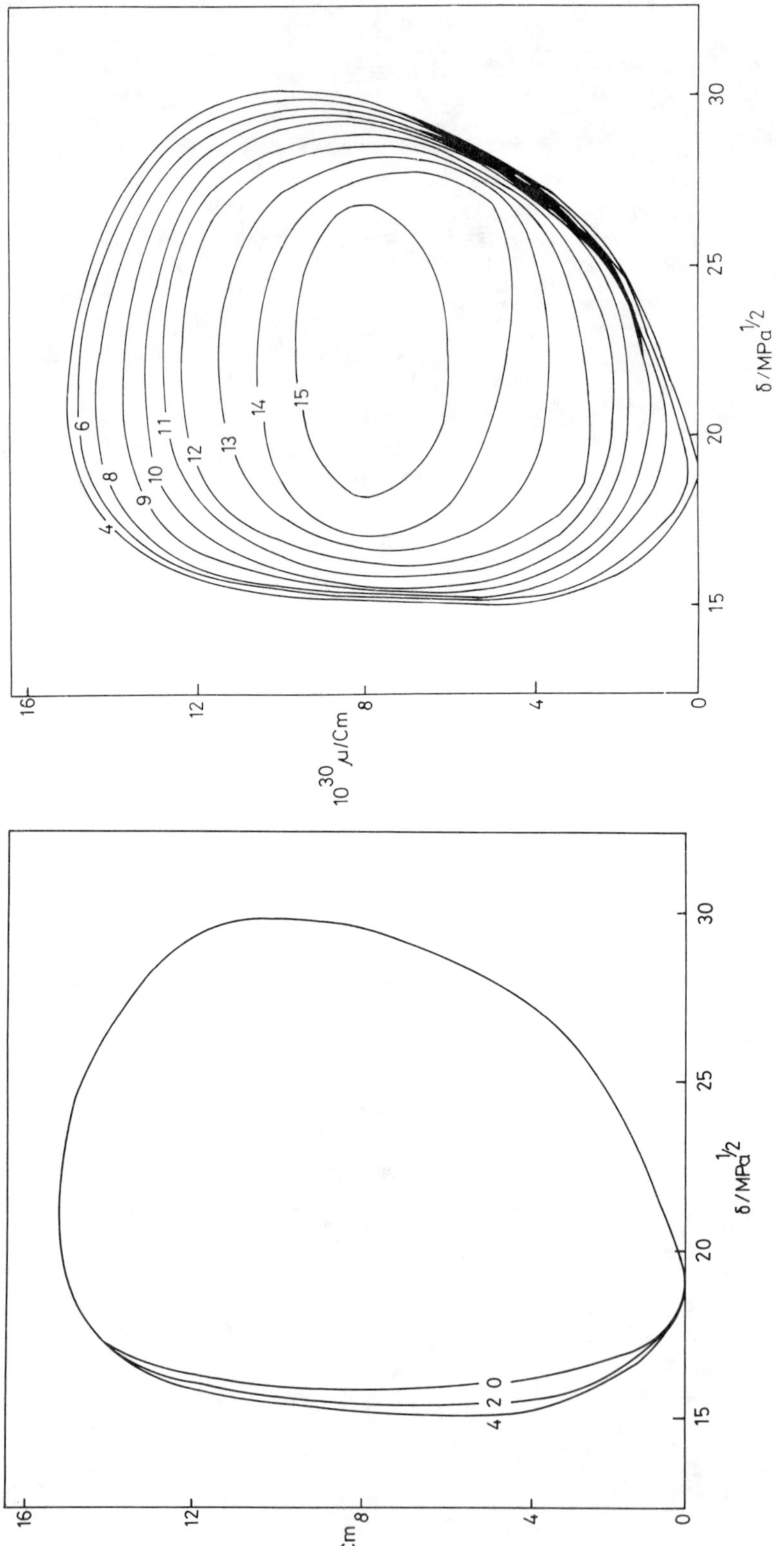

FIGURE 41. Hildebrand parameter—dipole moment solubility map with hydrogen-bonding parameter (γ_c) contours for poly(methyl methacrylate) (Elvacite® 2010), for use with the liquids data of Tables 8 and 9, Chapter 8: A, $\gamma_c = 0$ to 4; B, $\gamma_c = 4$ to 15. (Adapted from Crowley, J. D., Teague, G. S., Jr., and Lowe, J. W., *J. Paint Technol.*, 38, 269, 1966.)

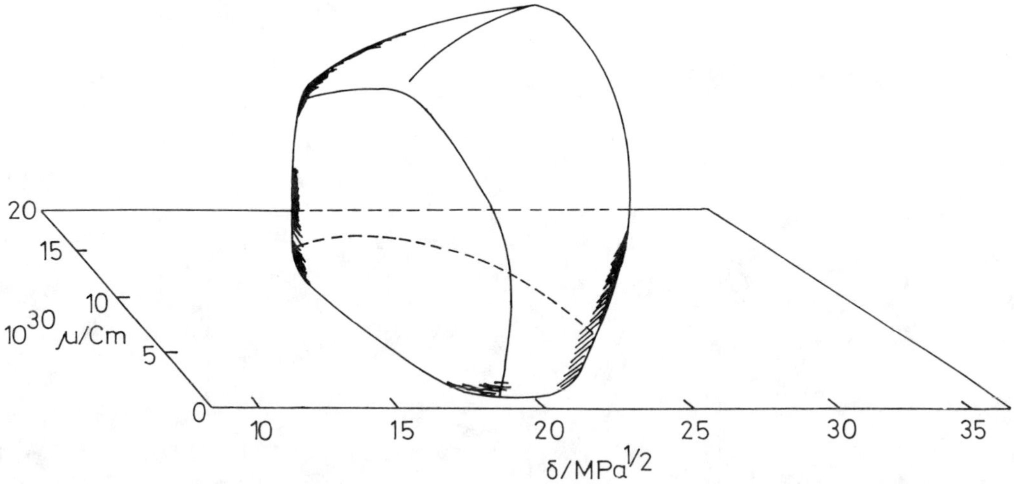

FIGURE 42. Three-dimensional representation of the solubility of Union Carbide Vinylite® VYHH resin (a vinyl chloride-acetate copolymer) in terms of Hildebrand parameter, dipole moment, and (vertical axis) hydrogen-bonding parameter (γ_C), for use with the liquids data of Tables 8 and 9, Chapter 8. (Adapted from Crowley, J. D., Teague, G. S., Jr., and Lowe, J. W., *J. Paint Technol.*, 38, 269, 1966.)

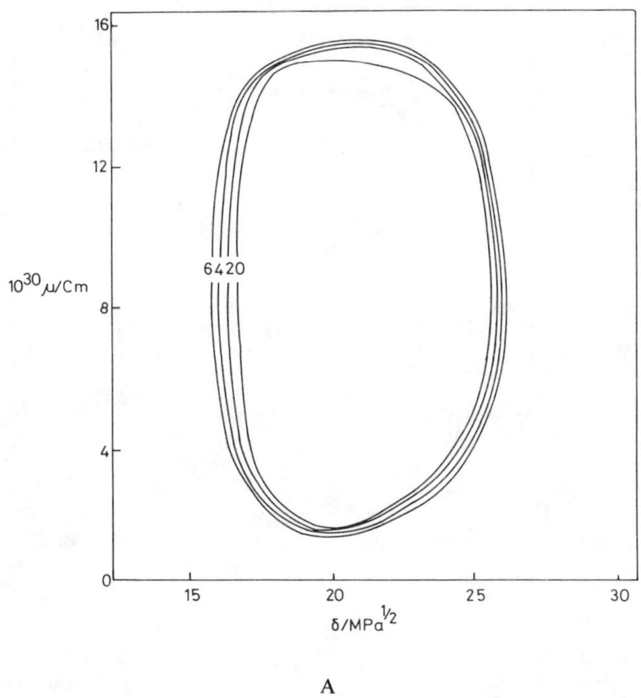

FIGURE 43. Hildebrand parameter—dipole moment solubility map with hydrogen-bonding parameter (γ_C) contours for Union Carbide Vinylite® VYHH resin (a vinyl chloride-vinyl acetate copolymer) for use with the liquids data of Tables 8 and 9, Chapter 8: A, γ_C = 0 to 6; B, γ_C = 6 to 13. (Adapted from Crowley, J. D., Teague, G. S., Jr., and Lowe, J. W., *J. Paint Technol.*, 38, 269, 1966.)

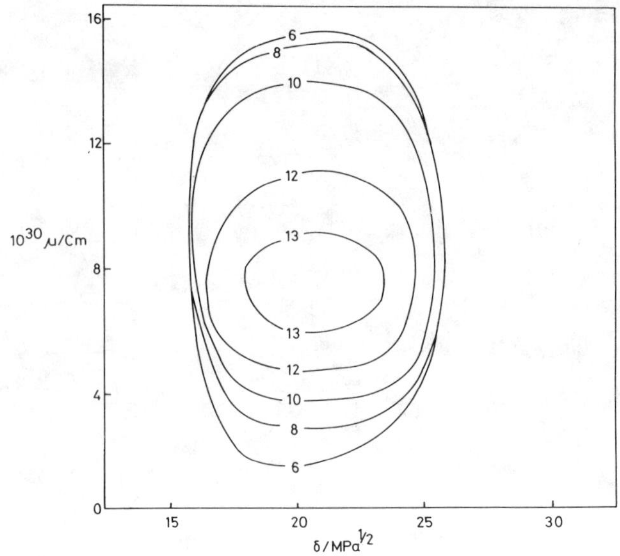

FIGURE 43B

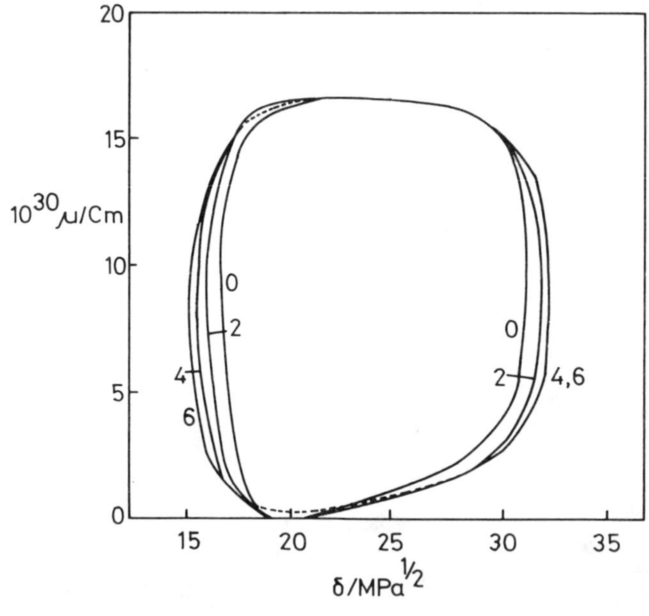

A

FIGURE 44. Hildebrand parameter—dipole moment solubility map with hydrogen-bonding parameter (γ_C) contours for oil-free alkyd Chevron® 1114, for use with the liquids data of Tables 8 and 9, Chapter 8: A, γ_C = 0 to 6; B, γ_C = 8 to 16. (Adapted from Eastman Kodak, *Three Dimensional Approach to Solubility,* Rochester, NY.)

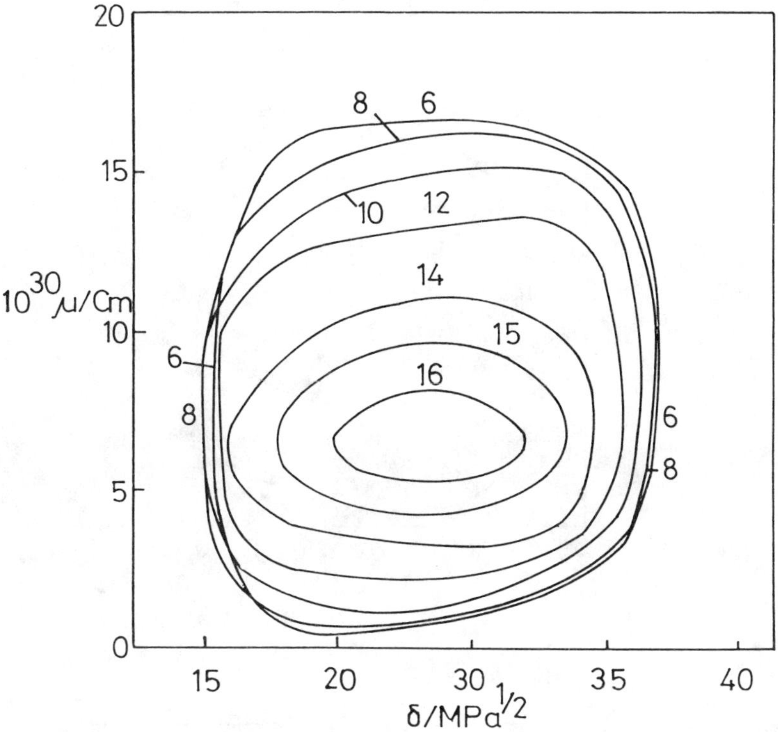

FIGURE 44B

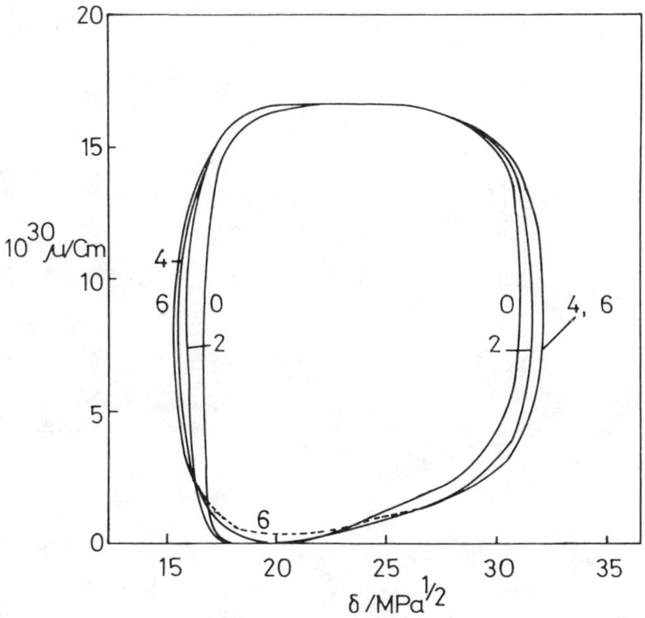

A

FIGURE 45. Hildebrand parameter—dipole moment solubility map with hydrogen-bonding parameter (γ_C) contours for short-oil alkyd, for use with the liquids data of Tables 8 and 9, Chapter 8: A, γ_C = 0 to 6; B, γ_C = 6 to 17. (Adapted from Eastman Kodak, *Three Dimensional Approach to Solubility,* Rochester, NY.)

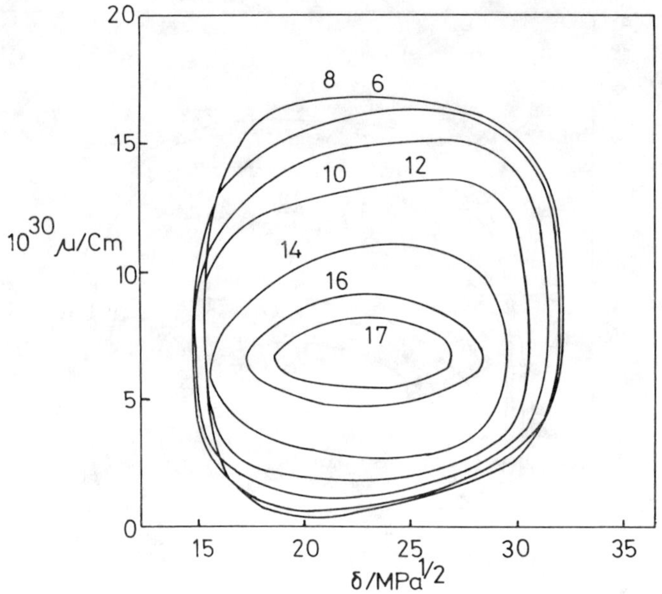

FIGURE 45B

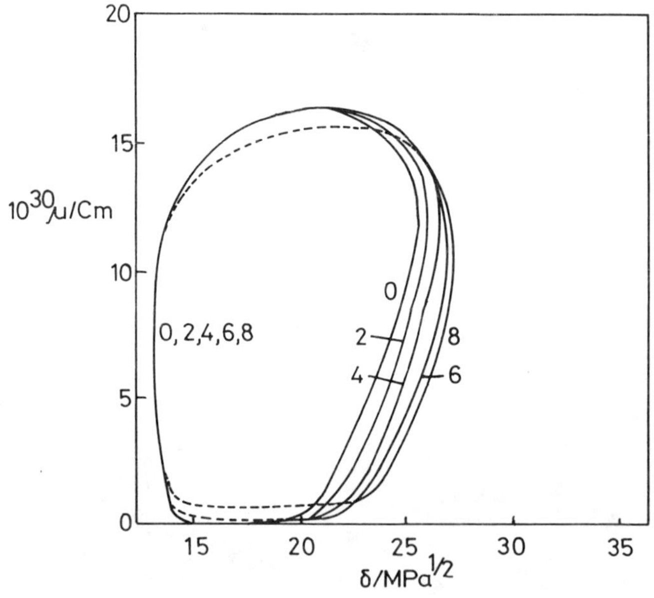

A

FIGURE 46. Hildebrand parameter—dipole moment solubility map with hydrogen-bonding parameter (γ_C) contours for medium-oil alkyd, for use with the liquids data of Tables 8 and 9, Chapter 8: A, γ_C = 0 to 8; B, γ_C = 8 to 17. (Adapted from Eastman Kodak, *Three Dimensional Approach to Solubility,* Rochester, NY.)

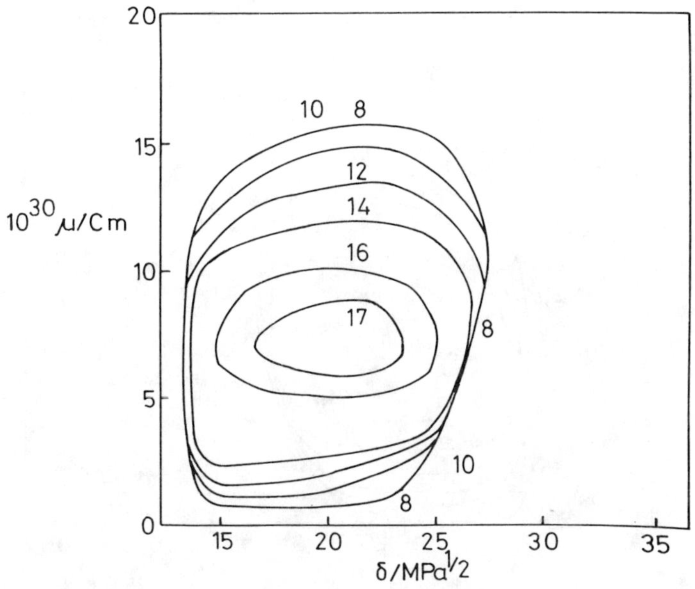

FIGURE 46B

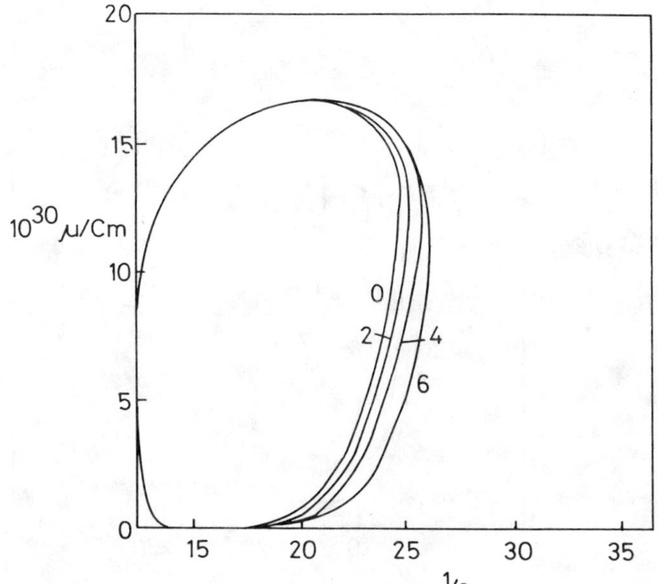

A

FIGURE 47. Hildebrand parameter—dipole moment solubility map with hydrogen-bonding parameter (γ_C) contours for long-oil alkyd, for use with the liquids data of Tables 8 and 9, Chapter 8: A, γ_C = 0 to 6; B, γ_C = 6 to 18. (Adapted from Eastman Kodak, *Three Dimensional Approach to Solubility,* Rochester, NY.)

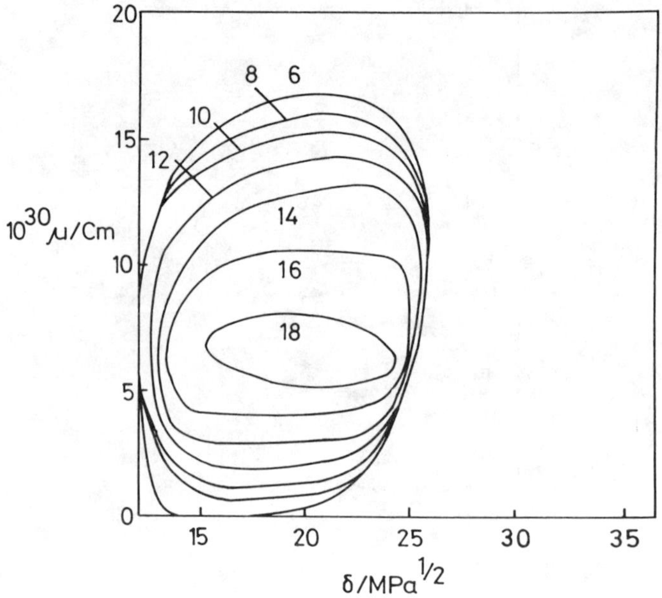

FIGURE 47B

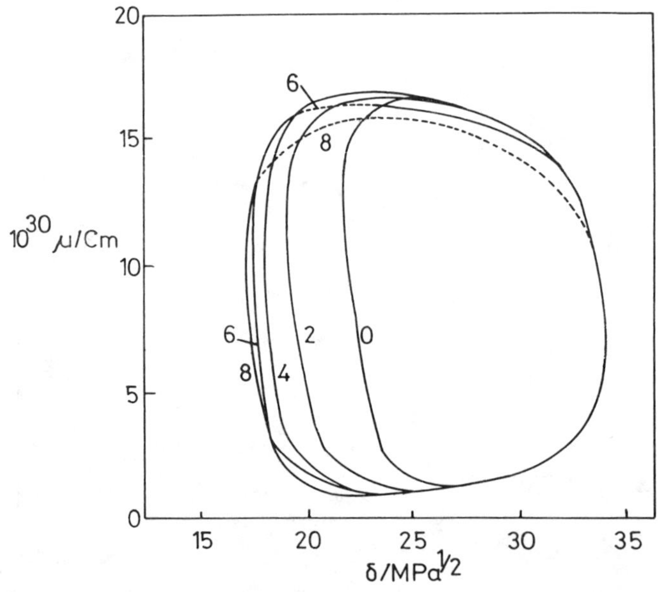

A

FIGURE 48. Hildebrand parameter—dipole moment solubility map with hydrogen-bonding parameter (γ_C) contours for the cellulose derivative alcohol-soluble butyrate, for use with the liquids data of Tables 8 and 9, Chapter 8: A, γ_C = 0 to 8; B, γ_C = 8 to 18.5. (Adapted from Eastman Kodak, *Three Dimensional Approach to Solubility,* Rochester, NY.)

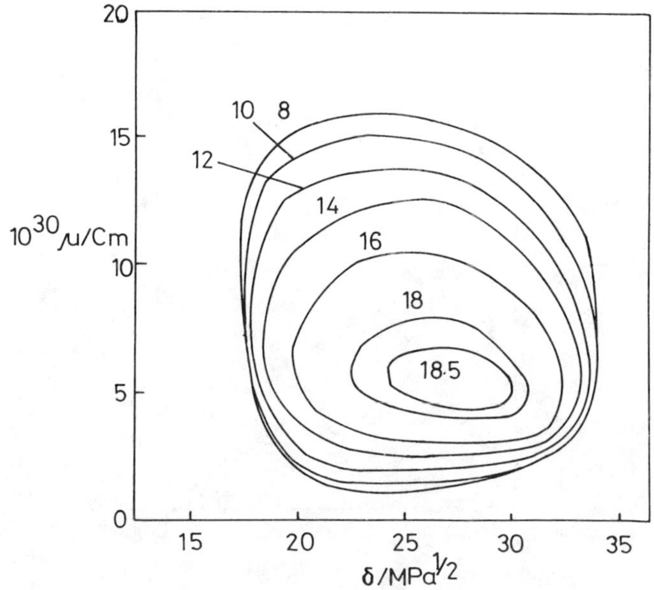

FIGURE 48B

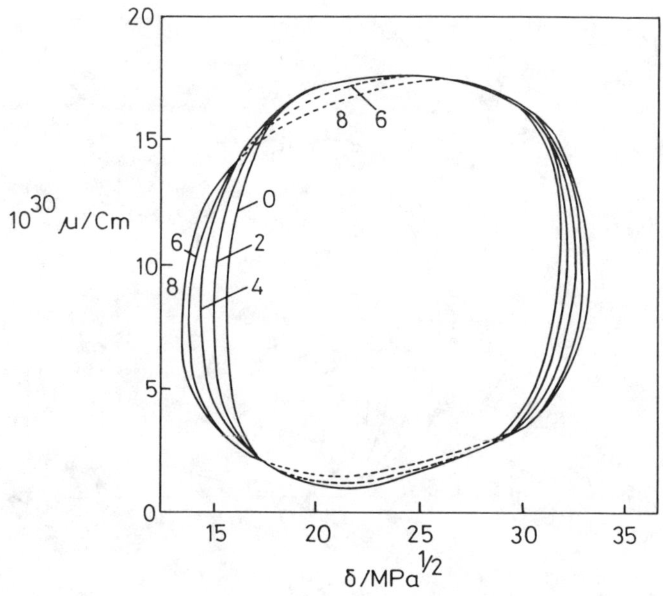

A

FIGURE 49. Hildebrand parameter—dipole moment solubility map with hydrogen-bonding parameter (γ_C) contours for the cellulose derivative EAB-552-0.2, for use with the liquids data of Tables 8 and 9, Chapter 8: A, γ_C = 0 to 8; B, γ_C = 8 to 16. (Adapted from Eastman Kodak, *Three Dimensional Approach to Solubility,* Rochester, NY.)

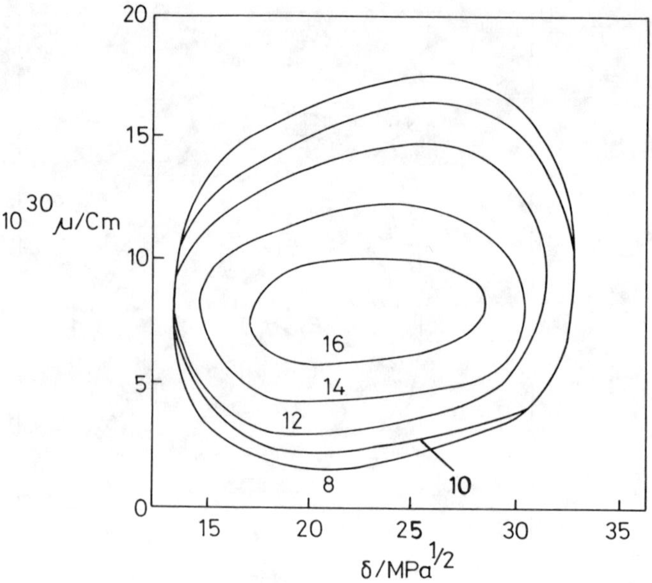

FIGURE 49B

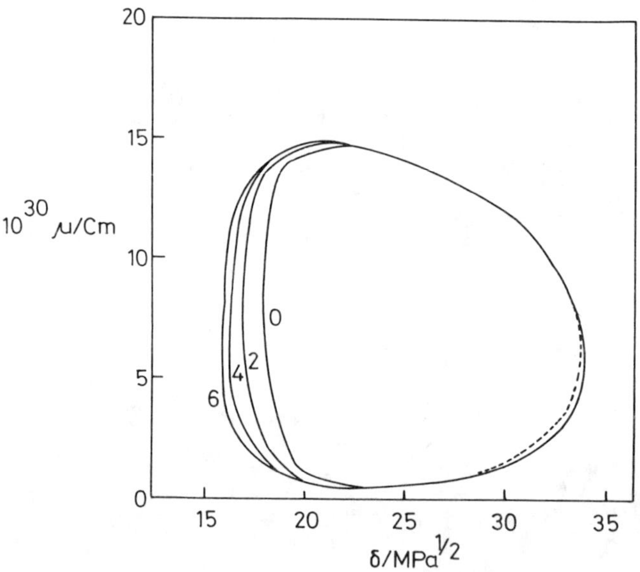

A

FIGURE 50. Hildebrand parameter—dipole moment solubility map with hydrogen-bonding parameter (γ_C) contours for bisphenol A-epichlorohydrin condensate Epon® 1001 epoxy resin (Shell), for use with the liquids data of Tables 8 and 9, Chapter 8: A, γ_C = 0 to 6; B, γ_C = 6 to 18. (Adapted from Eastman Kodak, *Three Dimensional Approach to Solubility*, Rochester, NY.)

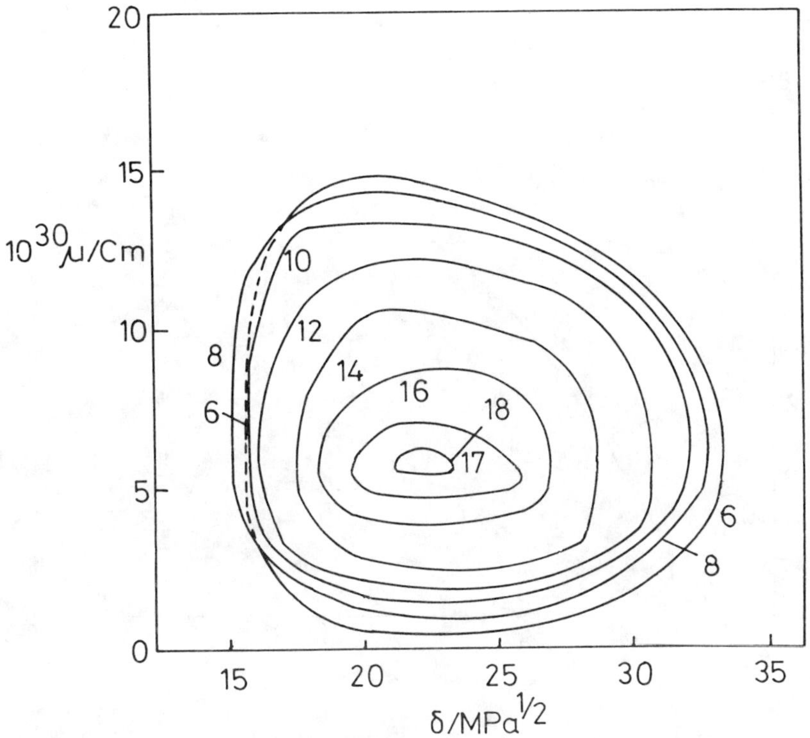

FIGURE 50B

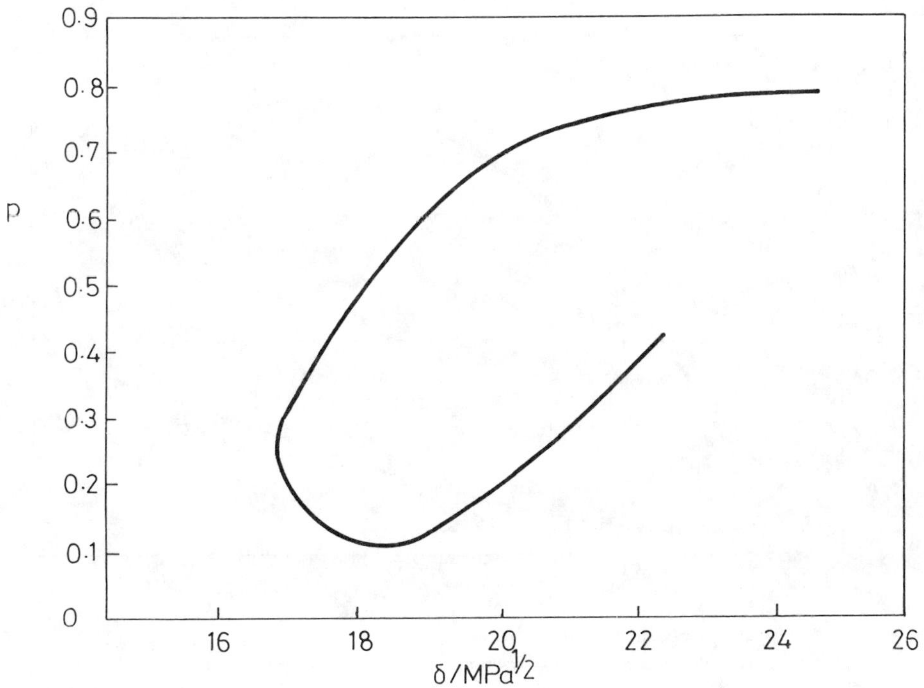

FIGURE 51. Hildebrand parameter—fractional polarity solubility map for 20% by weight Vinylite® VYHH resin (a Union Carbide Corp. vinyl chloride-vinyl acetate copolymer) at 25°C, for use with the liquids data of Tables 6, 10, 11, and 13, Chapter 8. (Adapted from Shell Chemicals, *Solubility Parameters*, 1978.)

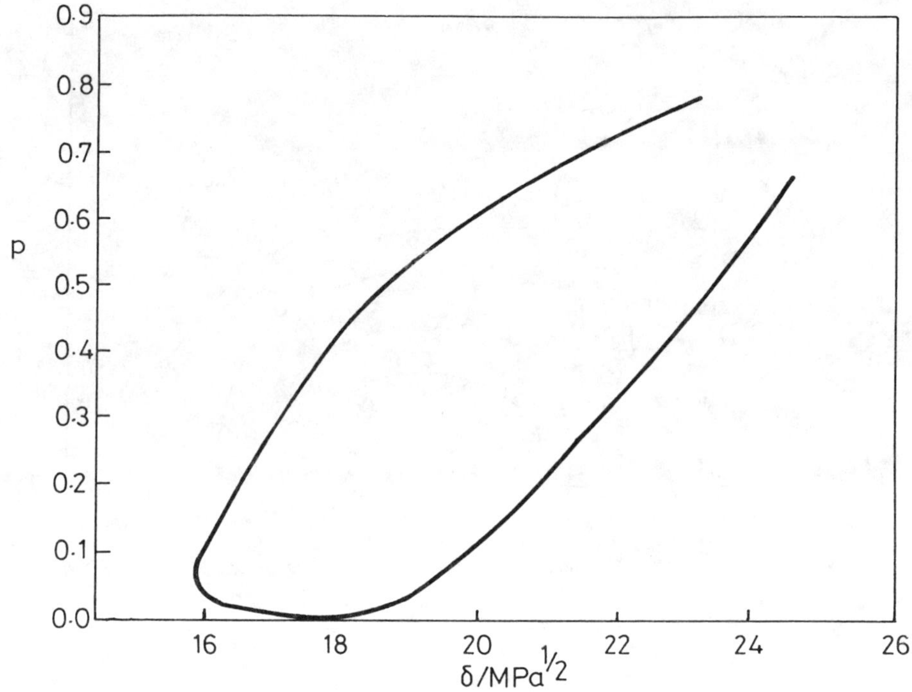

FIGURE 52. Hildebrand parameter—fractional polarity solubility map for 25% by weight Parlon® S20 (a Hercules Inc. chlorinated rubber) at 25°C, for use with the liquids data of Tables 6, 10, 11, and 13, Chapter 8. (Adapted from Shell Chemicals, *Solubility Parameters*, 1978.)

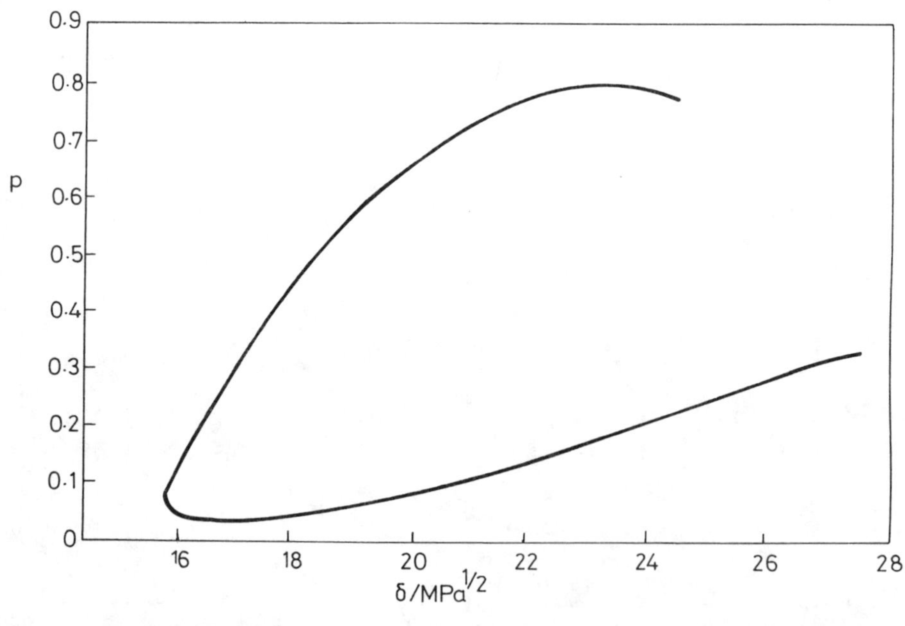

FIGURE 53. Hildebrand parameter—fractional polarity solubility map for 25% by weight SS $^1/_4$, $^1/_2$, and $^5/_6$ second cellulose nitrate at 25°C, for use with the liquids data of Tables 6, 10, 11, and 13, Chapter 8. (Adapted from Nelson, R. C., Figurelli, V. F., Walsham, J. G., and Edwards, G. D., *J. Paint Technol.*, 42, 644, 1970.)

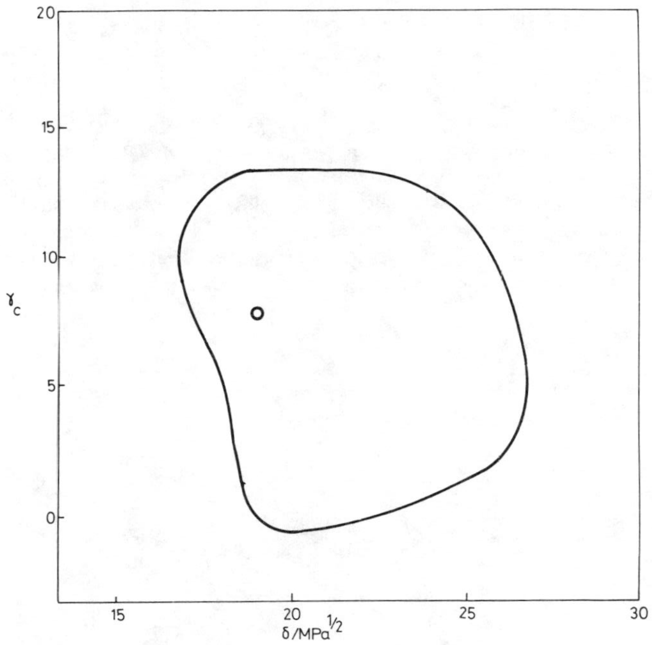

FIGURE 54. Hildebrand parameter—hydrogen-bonding parameter solubility map for Elvacite® 6010 and 2010 poly(methyl methacrylate) acrylic resins, for use with the liquids data of Figure 4, Chapter 8 and Tables 8 and 9, Chapter 8. The circle marks the position of the parameters of Elvacite 6010 solvent, methyl ethyl ketone. (Adapted from du Pont, *Solvent Formulating Maps for Elvacite Acrylic Resins,* Bulletins PA-12-770 and PA-12-174, Wilmington, DE, 1970 and 1974.)

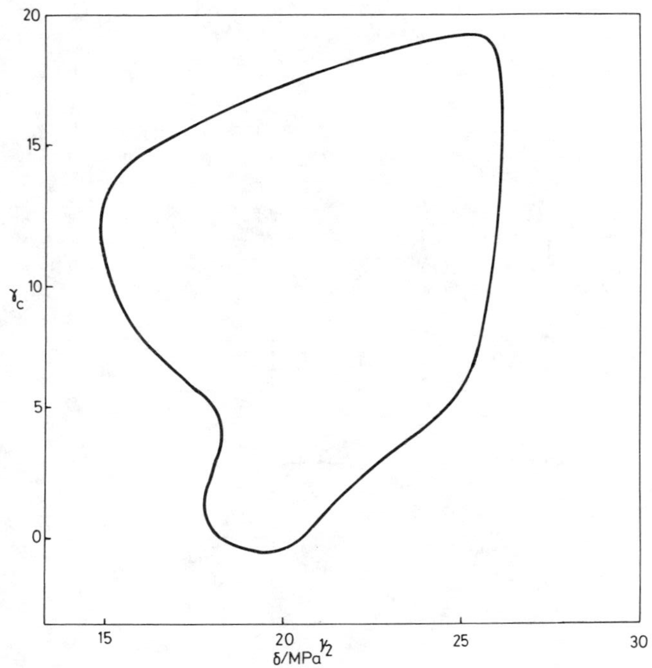

FIGURE 55. Hildebrand parameter—hydrogen-bonding parameter solubility map for Elvacite® 2043 acrylic resin, for use with the liquids data of Figure 4, Chapter 8 and Tables 8 and 9, Chapter 8. (Adapted from du Pont, *Solvent Formulating Maps for Elvacite Acrylic Resins,* Bulletins PA-12-770 and PA-12-174, Wilmington, DE, 1970 and 1974.)

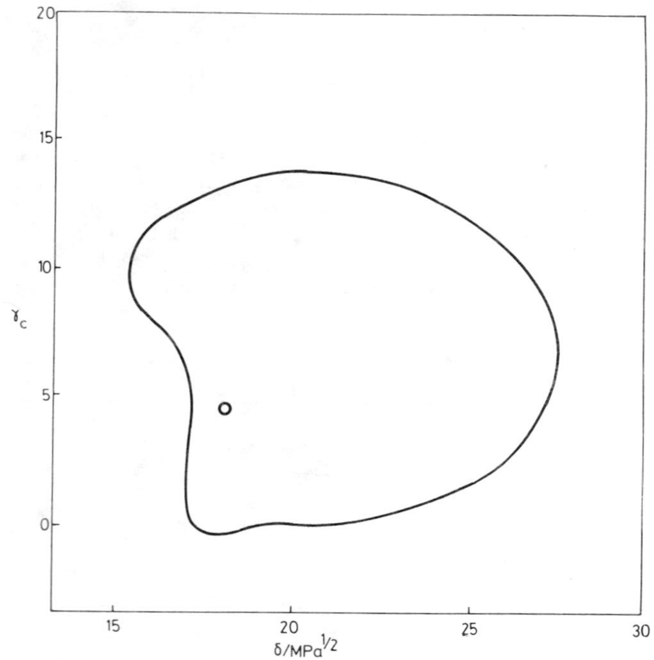

FIGURE 56. Hildebrand parameter—hydrogen-bonding parameter solubility map for Elvacite® 2013, 6013, 6024, and 2042 acrylic resins for use with the liquids data of Figure 4, Chapter 8 and Tables 8 and 9, Chapter 8. The circle marks the position of the parameters of Elvacite 6013 and 6024 solvent, toluene. (Adapted from du Pont, *Solvent Formulating Maps for Elvacite Acrylic Resins,* Bulletins PA-12-770 and PA-12-174, Wilmington, DE., 1970 and 1974.)

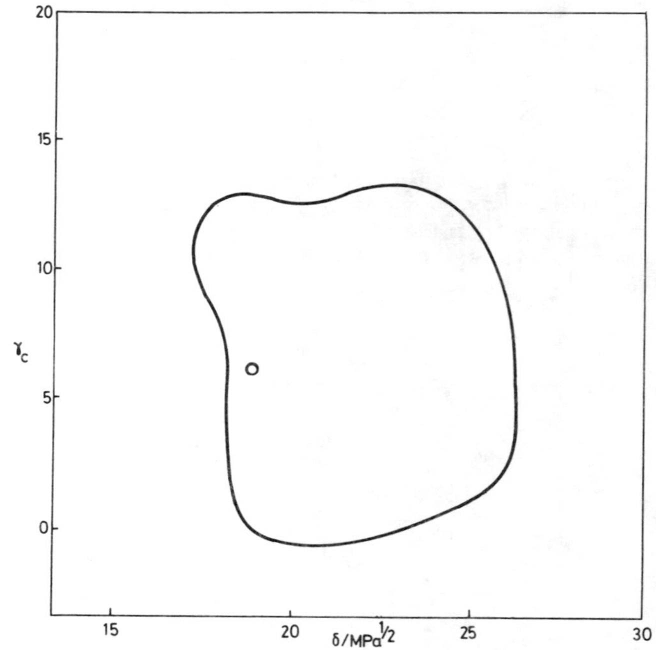

FIGURE 57. Hildebrand parameter—hydrogen-bonding parameter solubility map for Elvacite® 2008, 2041, 6011, and 6012 acrylic resins, for use with the liquids data of Figure 4, Chapter 8 and Tables 8 and 9, Chapter 8. The circle marks the position of the parameters of Elvacite 6011 and 6012 solvent, 70/30 by weight toluene/acetone. (Adapted from du Pont, *Solvent Formulating Maps for Elvacite Acrylic Resins,* Bulletins PA-12-770 and PA-12-174, Wilmington, DE, 1970 and 1974.)

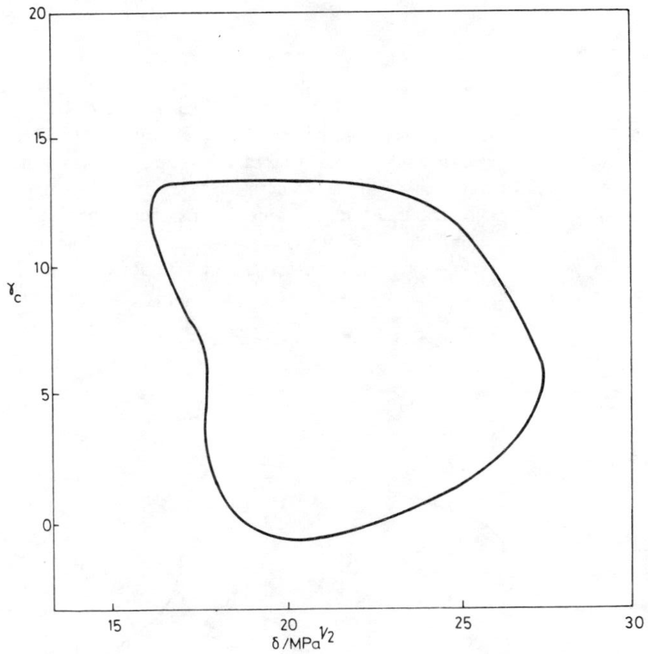

FIGURE 58. Hildebrand parameter—hydrogen-bonding parameter solubility map for Elvacite® 2009 acrylic resin, for use with the liquids data of Figure 4, Chapter 8 and Tables 8 and 9, Chapter 8. (Adapted from du Pont, *Solvent Formulating Maps for Elvacite Acrylic Resins,* Bulletins PA-12-770 and PA-12-174, Wilmington, DE, 1970 and 1974.)

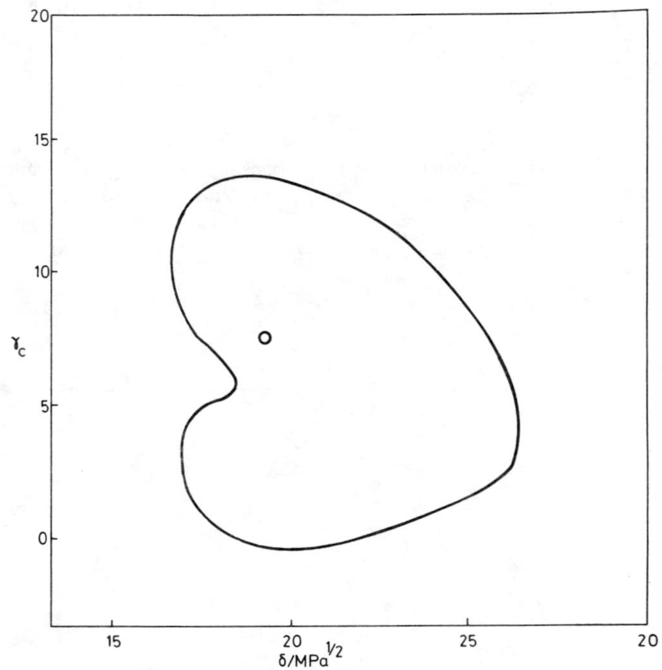

FIGURE 59. Hildebrand parameter—hydrogen-bonding parameter solubility map for Elvacite® 6014 acrylic resin, for use with the liquids data of Figure 4, Chapter 8 and Tables 8 and 9, Chapter 8. The circle marks the position of the parameters of Elvacite 6014 solvent, 80/20 by weight toluene/isopropanol. (Adapted from du Pont, *Solvent Formulating Maps for Elvacite Acrylic Resins,* Bulletins PA-12-770 and PA-12-174, Wilmington, DE, 1970 and 1974.)

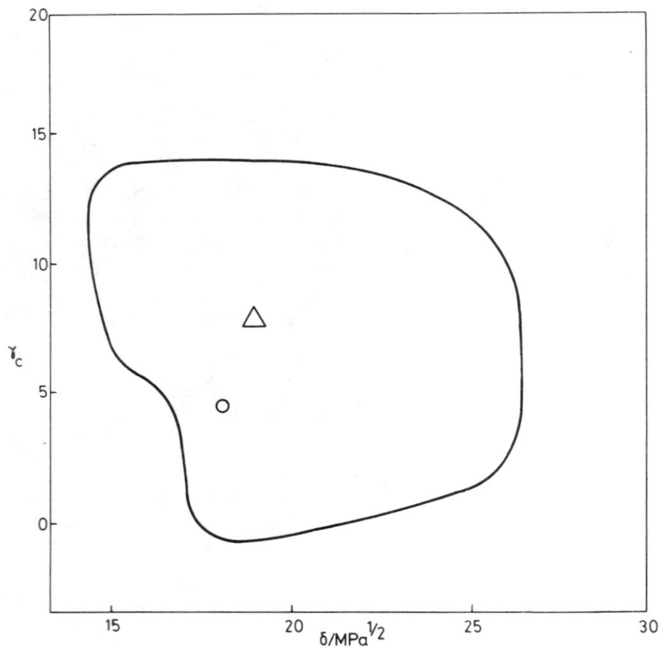

FIGURE 60. Hildebrand parameter—hydrogen-bonding parameter solubility map for Elvacite® 6015, 6106, and 6026 acrylic resins, for use with the liquids data of Figure 4, Chapter 8 and Tables 8 and 9, Chapter 8. The circle marks the position of the parameters of Elvacite 6016 and 6026 solvent, toluene; the triangle that of Elvacite 6015 solvent, methyl ethyl ketone. (Adapted from du Pont, *Solvent Formulating Maps for Elvacite Acrylic Resins,* Bulletins PA-120-770 and PA-12-174, Wilmington, DE, 1970 and 1974.)

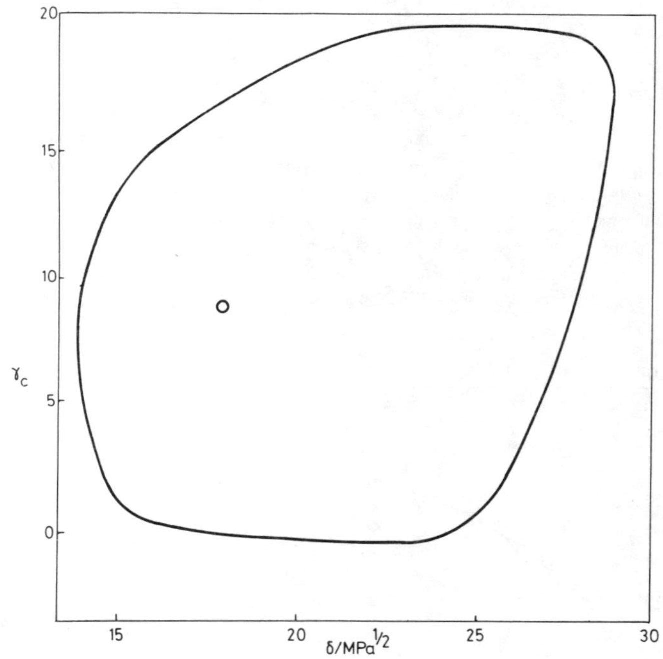

FIGURE 61. Hildebrand parameter—hydrogen-bonding parameter solubility map for Elvacite® 6028 acrylic resin, for use with the liquids data of Figure 4, Chapter 8 and Tables 8 and 9, Chapter 8. The circle marks the positions of the parameters of Elvacite 6028 solvent, propyl acetate. (Adapted from du Pont, *Solvent Formulating Maps for Elvacite Acrylic Resins,* Bulletins PA-12-770 and PA-12-174, Wilmington, DE, 1970 and 1974.)

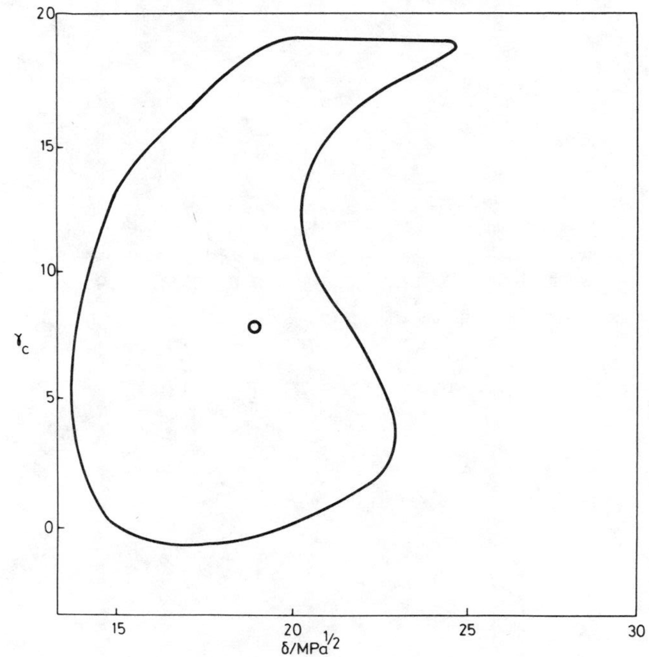

FIGURE 62. Hildebrand parameter—hydrogen-bonding parameter solubility map for Elvacite® 2044, 2045, 2046, and 2047 acrylic resins, for use with the liquids data of Figure 4, Chapter 8 and Tables 8 and 9, Figure 8. The circle marks the position of the parameters of Elvacite 6017 solvent, methyl ethyl ketone. (Adapted from du Pont, *Solvent Formulating Maps for Elvacite Acrylic Resins,* Bulletins PA-12-770 and PA-12-174, Wilmington, DE, 1970 and 1974.)

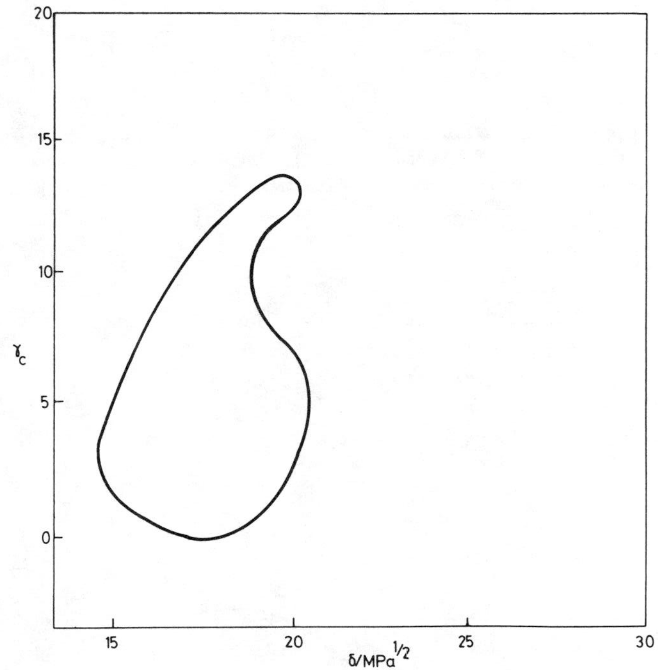

FIGURE 63. Hildebrand parameter—hydrogen-bonding parameter solubility map for the du Pont polyethylene-poly(vinyl acetate) copolymer Elvax® 40, for use with the liquids data of Figure 4, Chapter 8 and Tables 8 and 9, Chapter 8. (Adapted from Cosaert, E., *Chim. Peintures,* 34, 169, 1971.)

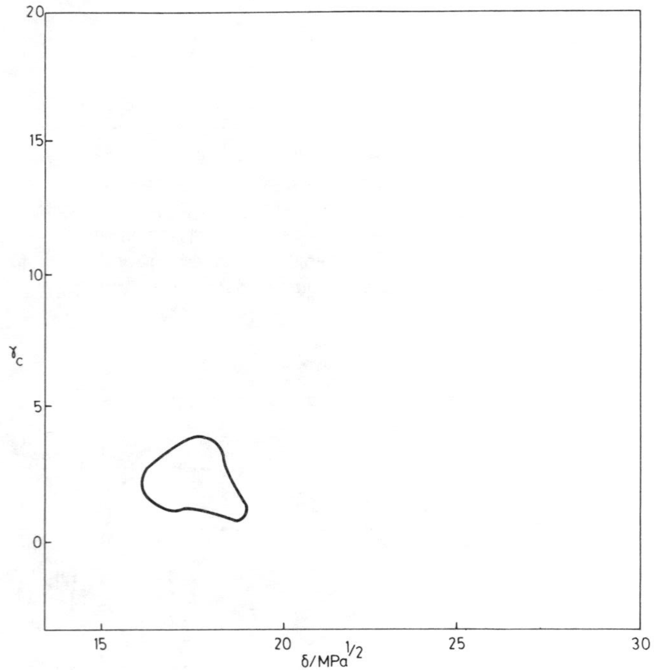

FIGURE 64. Hildebrand parameter—hydrogen-bonding parameter solubility map for the du Pont polyethylene-poly(vinyl acetate) copolymer Elvax® 260, for use with the liquids data of Figure 4, Chapter 8 and Tables 8 and 9, Chapter 8. (Adapted from Cosaert, E., *Chim. Peintures*, 34, 169, 1971.)

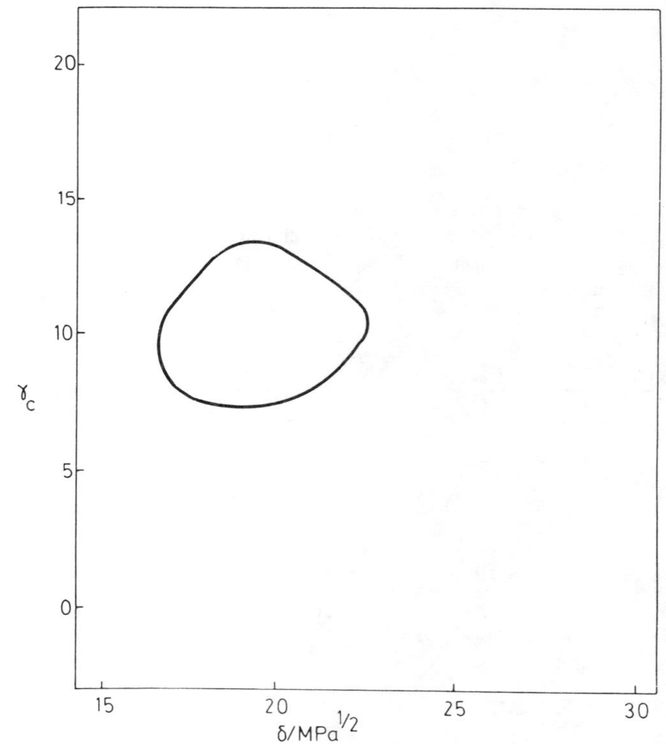

FIGURE 65. Hildebrand parameter—hydrogen-bonding parameter solubility map for cellulose nitrate/acrylic polyblend base emulsion solids, for use with the liquids data of Figure 4, Chapter 8 and Tables 8 and 9, Chapter 8. (Adapted from Zunker, D. W., *J. Coatings Technol.*, 48(616), 37, 1976.)

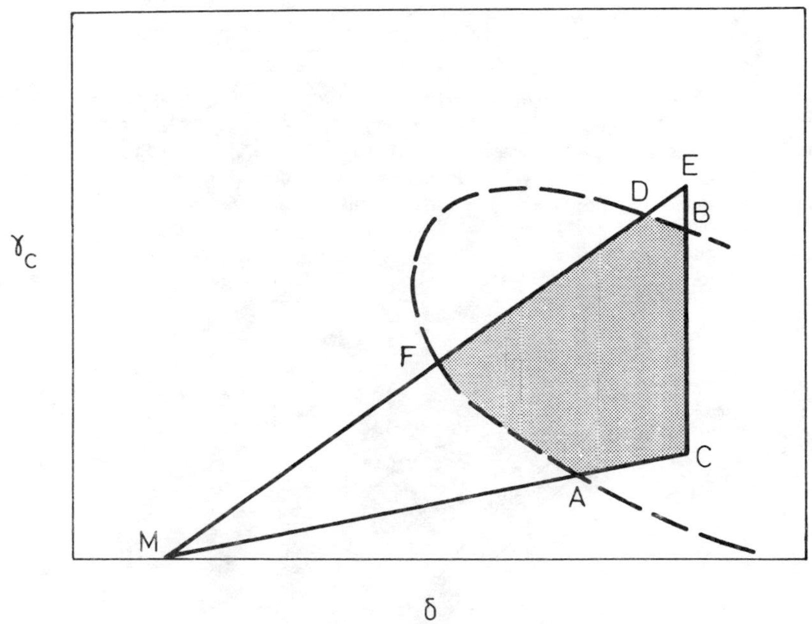

FIGURE 66. Schematic diagram illustrating the use of a Hildebrand parameter—hydrogen-bonding parameter solubility map for solvent formulation: E, ethylene glycol monoethyl ether; C, cyclohexanone; M, mineral spirits (see text). (Adapted from Ellis, W. H., Saary, Z., and Lesnini, D. G., *J. Paint Technol.*, 41, 249, 1969.)

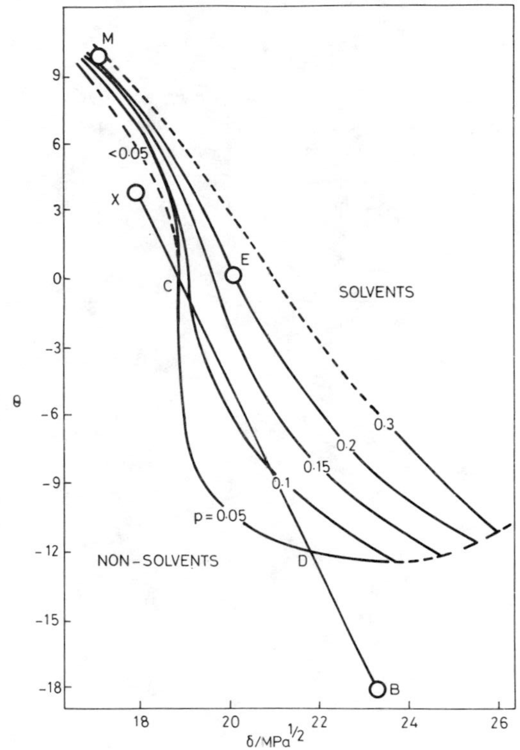

FIGURE 67. Hildebrand parameter—hydrogen-bonding index solubility map, with contours indicating maximum permissible fractional polarities for Epikote® 1004 resin (solids greater than about 30% by weight), for use with the liquids data of Tables 10, 11, and 14, Chapter 8: M, methyl isobutyl ketone; E, ethyl Oxitol®; B, 1-butanol; X, xylene. (Adapted from Shell Chemicals, *Solvent System Design*, 1978.)

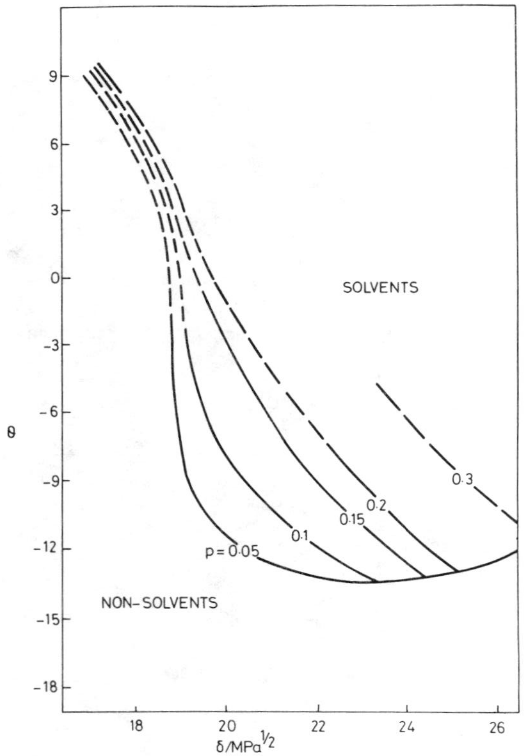

FIGURE 68. Hildebrand parameter—hydrogen-bonding index solubility map, with contours indicating maximum permissible fractional polarities, for Epikote® 1001 resin (solids greater than about 30% by weight), for use with the liquids data of Tables 10, 11, and 14, Chapter 8. (Adapted from Shell Chemicals, *Solvent System Design,* 1978.)

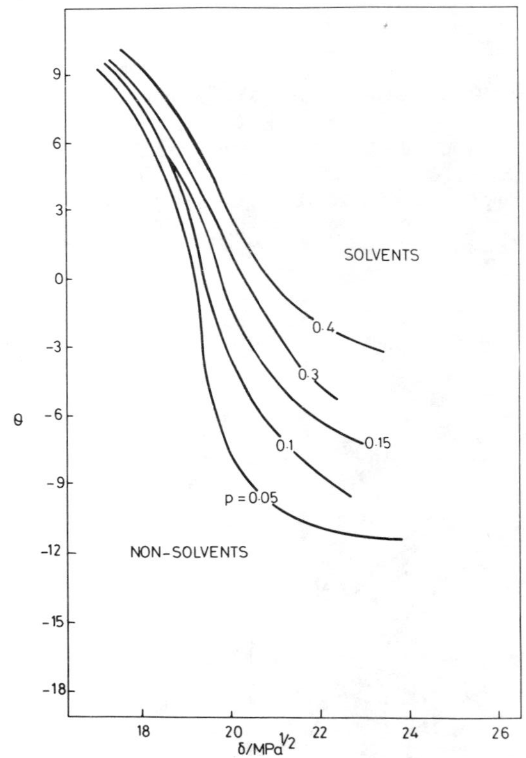

FIGURE 69. Hildebrand parameter—hydrogen-bonding index solubility map, with contours, indicating maximum permissible fractional polarities, for Epikote® 1007 resin (solids greater than about 30% by weight), for use with the liquids data of Tables 10, 11, and 14, Chapter 8. (Adapted from Shell Chemicals, *Solvent System Design,* 1978.)

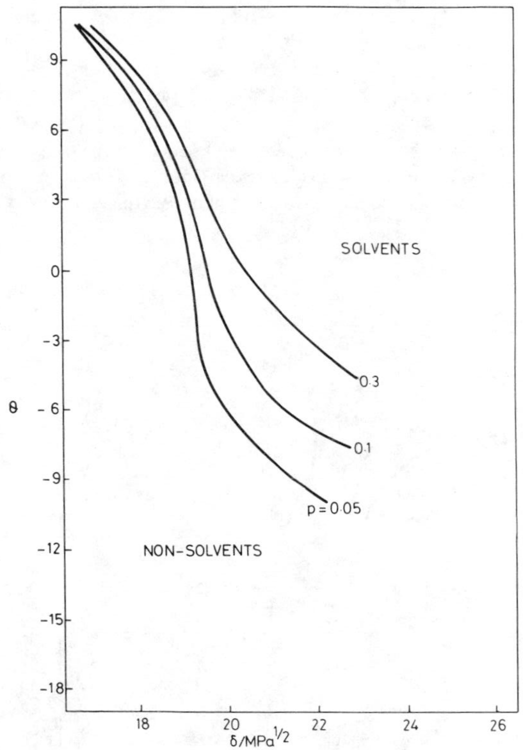

FIGURE 70. Hildebrand parameter—hydrogen-bonding index solubility map, with contours indicating maximum permissible fractional polarities, for Epikote® 1009 resin (solids greater than about 30% by weight), for use with the liquids data of Tables 10, 11, and 14, Chapter 8. (Adapted from Shell Chemicals, *Solvent System Design,* 1978.)

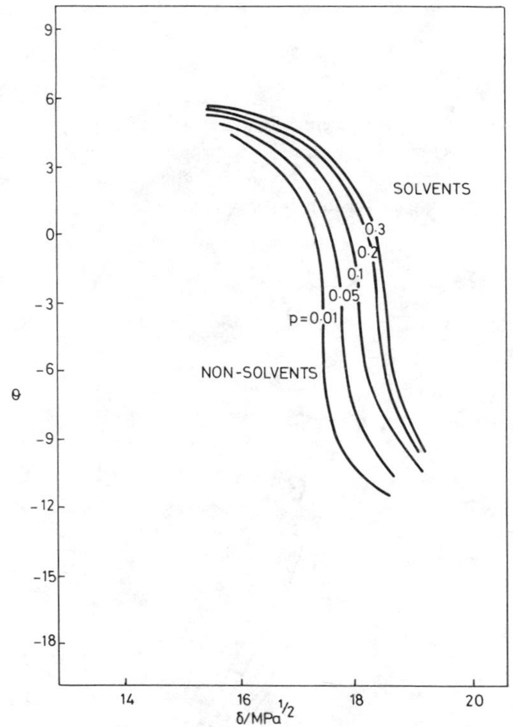

FIGURE 71. Hildebrand parameter—hydrogen-bonding index solubility map, with contours indicating maximum permissible fractional polarities, for acrylic resin (10% solids), for use with the liquids data of Tables 10, 11, and 14, Chapter 8. (Adapted from Shell Chemicals, *Solvent System Design,* 1978.)

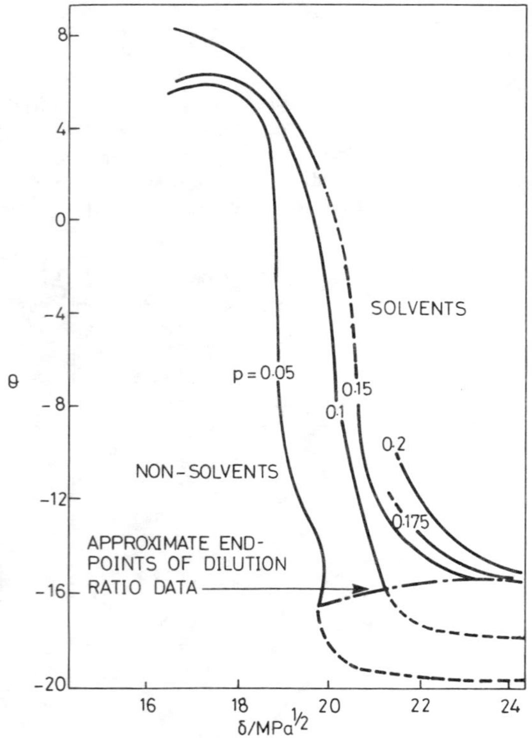

FIGURE 72. Hildebrand parameter—hydrogen-bonding index solubility map, with contours indicating maximum permissible fractional polarities, for Varcum® 6820 phenolic resin (Reichhold) (solids greater than about 10% by weight), for use with the liquids data of Tables 10, 11, and 14, Chapter 8. (Adapted from Ramsbotham, J., *Prog. Org. Coatings*, 8, 113, 1980.)

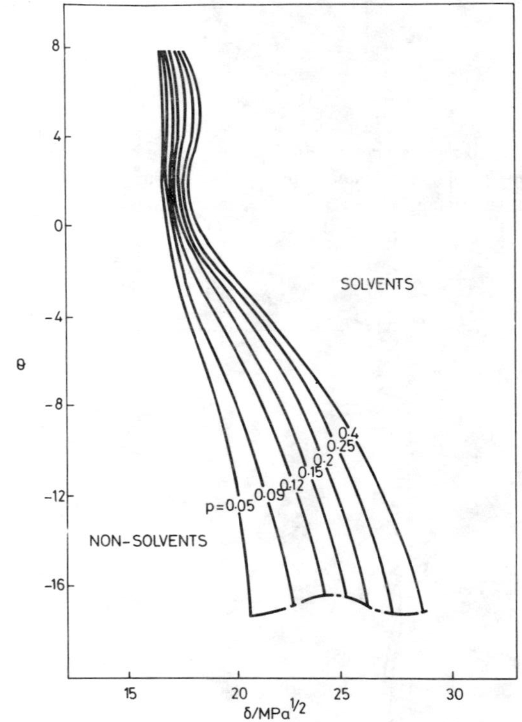

FIGURE 73. Hildebrand parameter—hydrogen-bonding index solubility map, with contours indicating maximum permissible fractional polarities, for Setal® 84 short oil alkyd resin, for use with the liquids data of Tables 10, 11, and 14, Chapter 8. (Adapted from Shell Chemicals, *Solvent System Design*, 1978.)

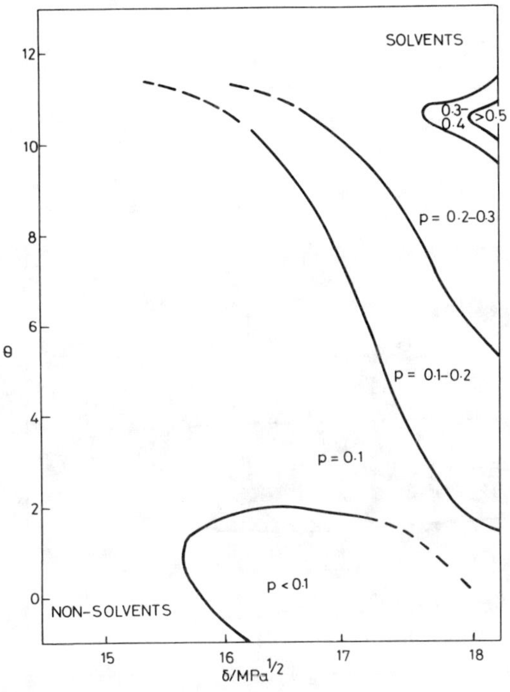

FIGURE 74. Hildebrand parameter—hydrogen-bonding index solubility map, with contours indicating maximum permissible fractional polarities, for polyisoprene-polystyrene block copolymer, for use with the liquids data of Tables 10, 11, and 14, Chapter 8. (Adapted from Shell Chemicals, *Solvent System Design,* 1978.)

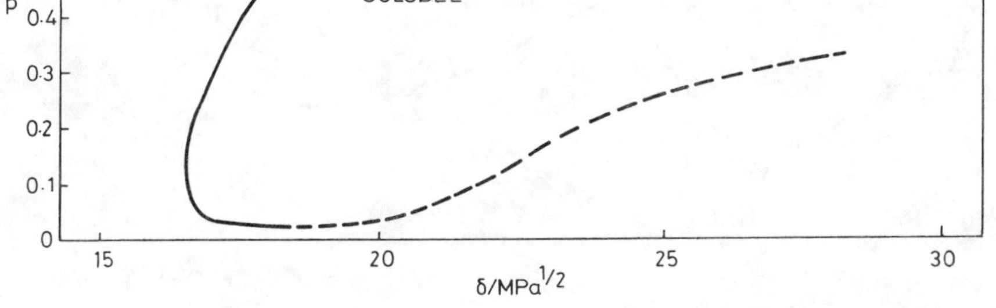

FIGURE 75. Hildebrand parameter—fractional polarity map for RS $^1/_2$ second cellulose nitrate, 8% weight/volume at 25°C, for use with the liquids data of Tables 6, 10, 11, and 14, Chapter 8. (Adapted from Tess, R. W., *Applied Polymer Science,* Craver, J. K. and Tess, R. W., Eds., American Chemical Society, Washington, D.C., 1975, chap. 44.)

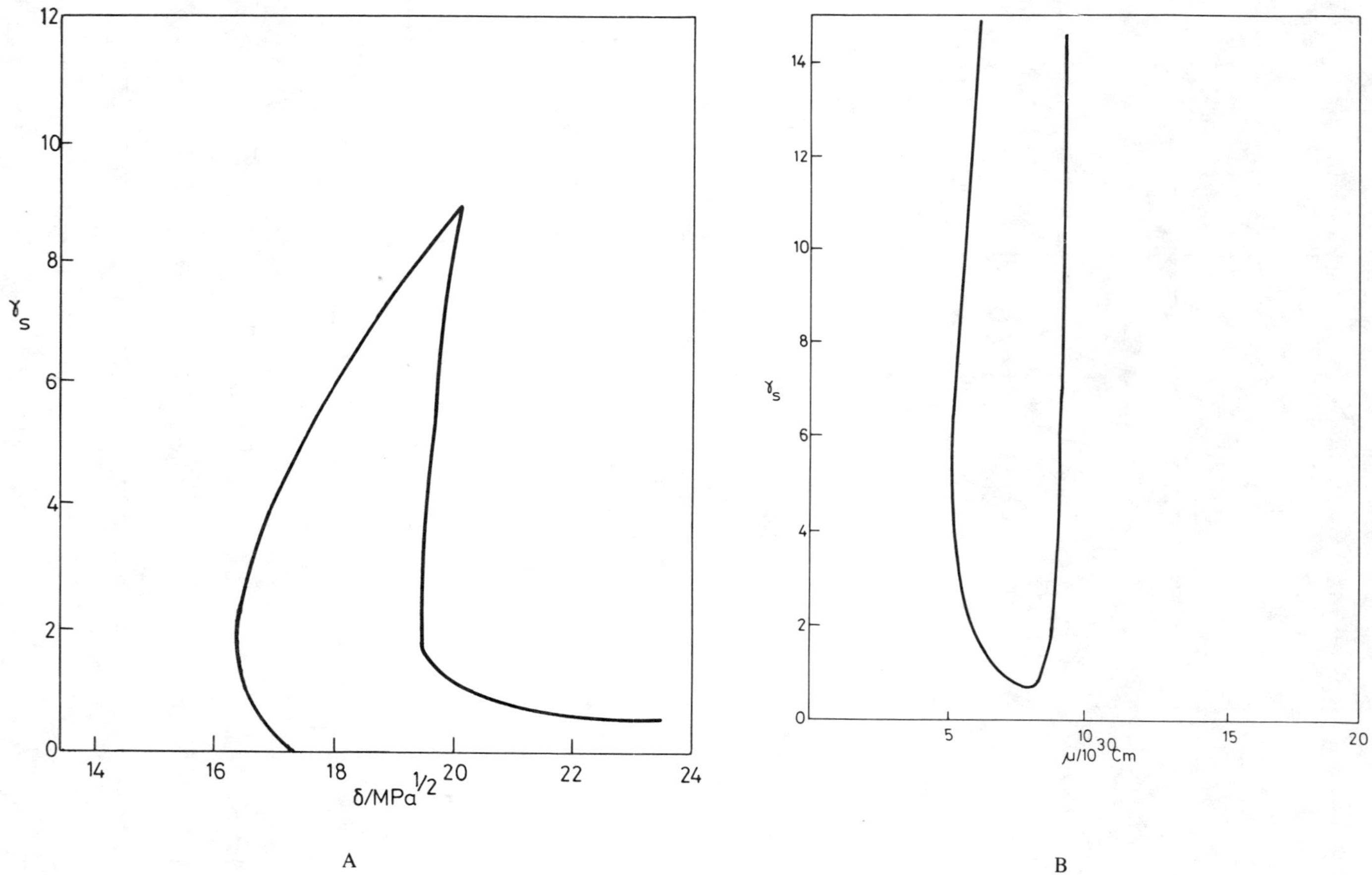

FIGURE 76. Solubility of poly(*p*-chlorostyrene) in terms of cohesion parameter (δ), dipole moment (μ), and Small's iodine-bonding number (γ_S), for use with the liquids data of Table 7, Chapter 14: A, (γ_S, δ) projection; B, (γ_S, μ) projection; C, (δ, μ) projection; D, (δ, μ, γ_S) stereograph. (Adapted from Izumi, Y. and Miyake, Y., *Polym. J.*, 3, 647, 1972.)

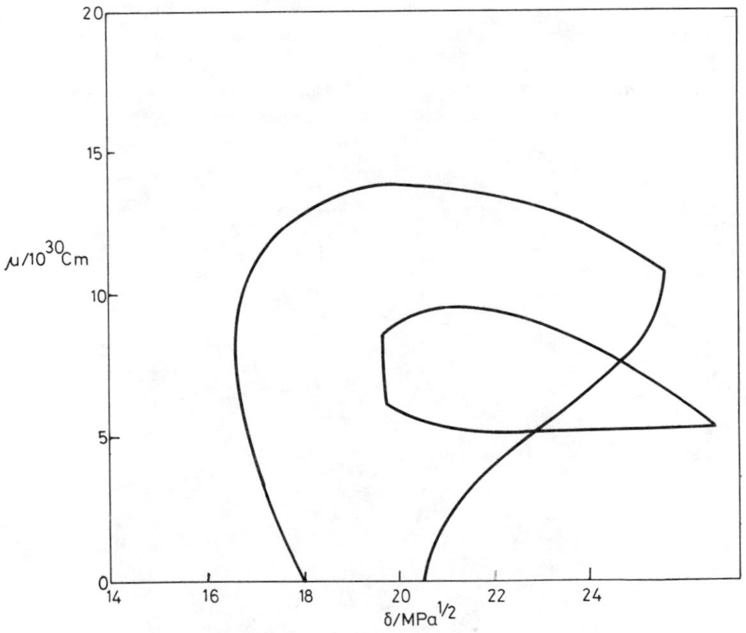

FIGURE 76C

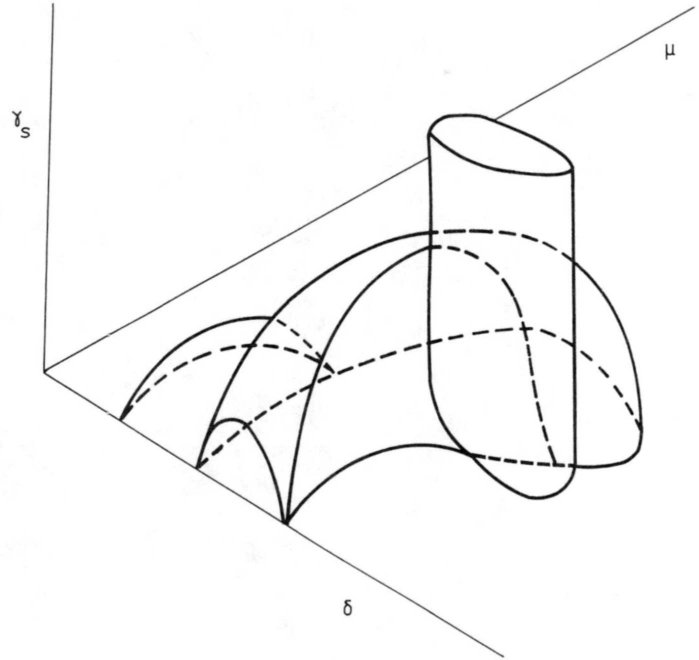

FIGURE 76D

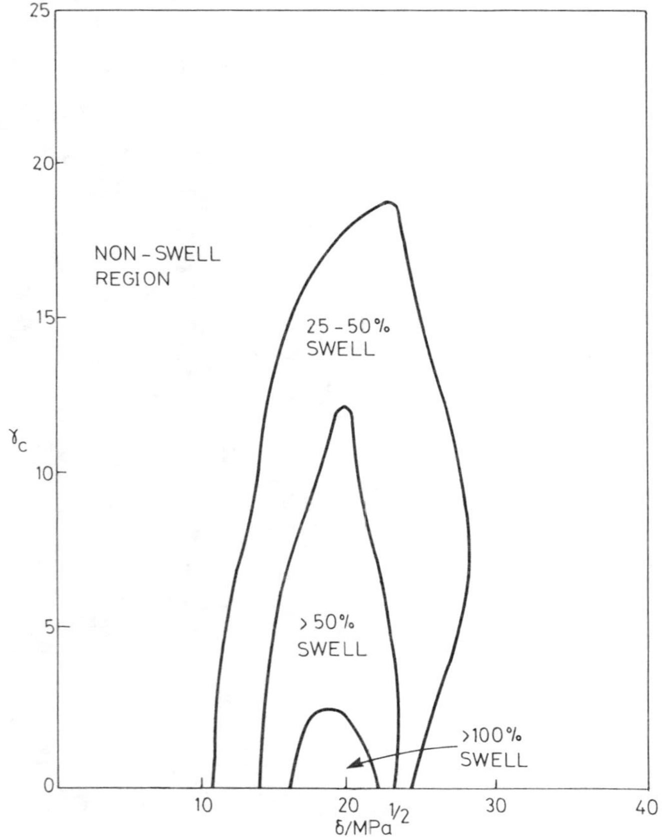

FIGURE 77. Hildebrand parameter—hydrogen-bonding parameter map with % swell contours for styrene-butadiene rubber, for use with the liquids data of Figure 4, Chapter 8 and tables 8 and 9, Chapter 8. (Adapted from Beerbower, A., Kaye, L. A., and Pattison, D. A., *Chem. Eng.* (NY), 74(26), 118, 1967.)

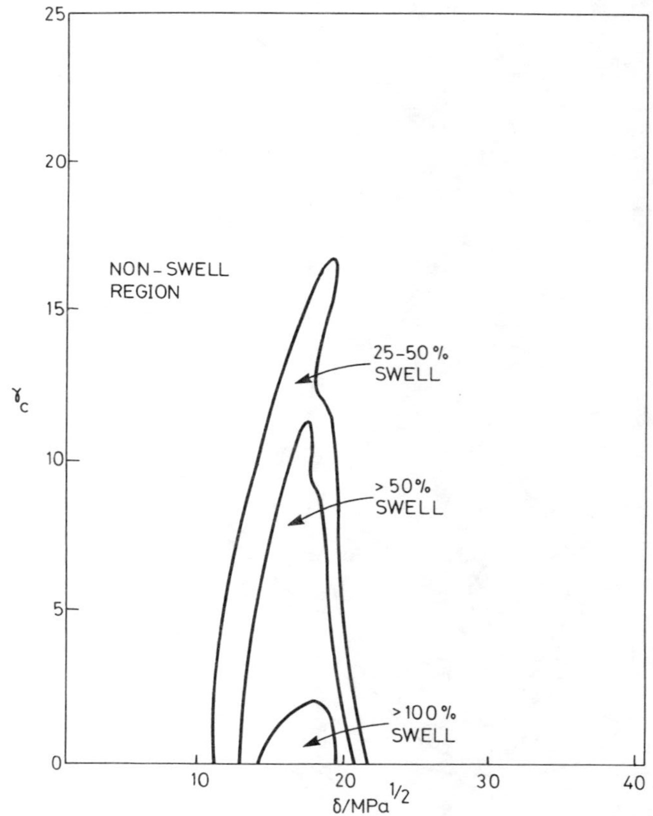

FIGURE 78. Hildebrand parameter—hydrogen-bonding parameter map with % swell contours for isobutylene-isoprene rubber, for use with the liquids data of Figure 4, Chapter 8 and Tables 8 and 9, Chapter 8. (Adapted from Beerbower, A., Kaye, L. A., and Pattison, D. A, *Chem. Eng. (N.Y.)*., 74(26), 118, 1967.)

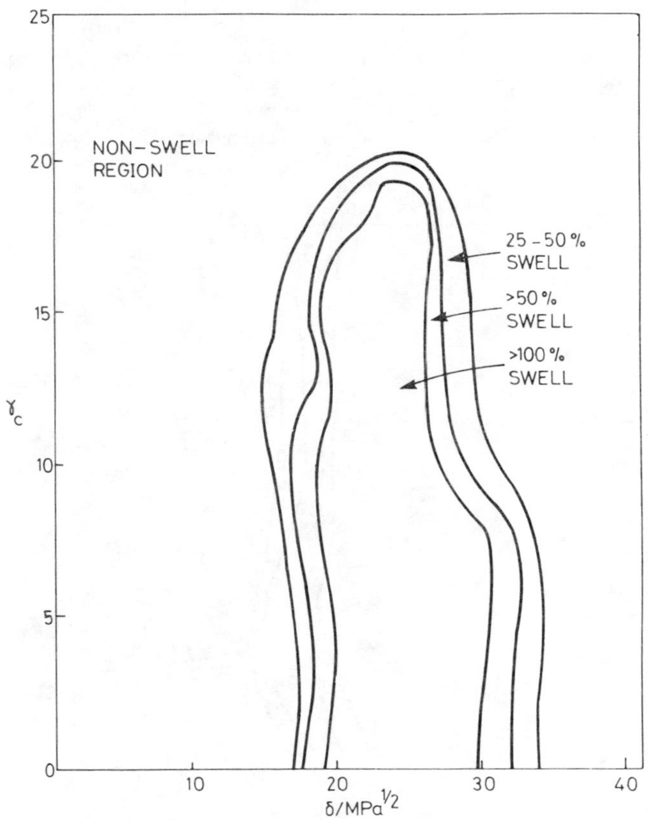

FIGURE 79. Hildebrand parameter—hydrogen-bonding parameter map with % swell contours for nitrile-butadiene rubber, for use with the liquids data of Figure 4, Chapter 8 and Tables 8 and 9, Chapter 8. (Adapted from Beerbower, A., Kaye, L. A., and Pattison, D. A., *Chem. Eng. (N.Y.)*, 74(26), 118, 1967.)

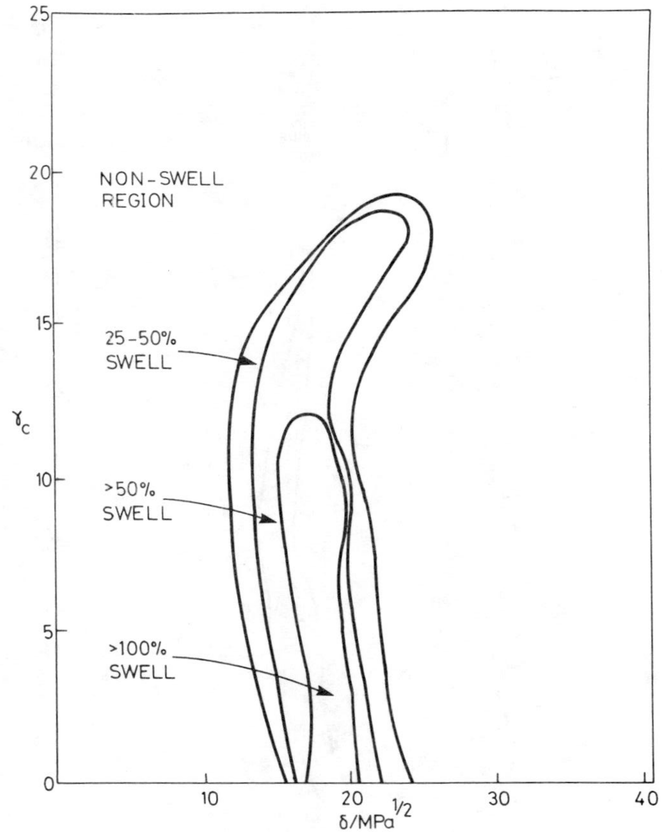

FIGURE 80. Hildebrand parameter—hydrogen-bonding parameter map with % swell contours for chloroprene rubber, for use with the liquids data of Figure 4, Chapter 8 and Tables 8 and 9, Chapter 8. (Adapted from Beerbower, A., Kaye, L. A., and Pattison, D. A., *Chem. Eng. (NY)*, 74(26), 118, 1967.)

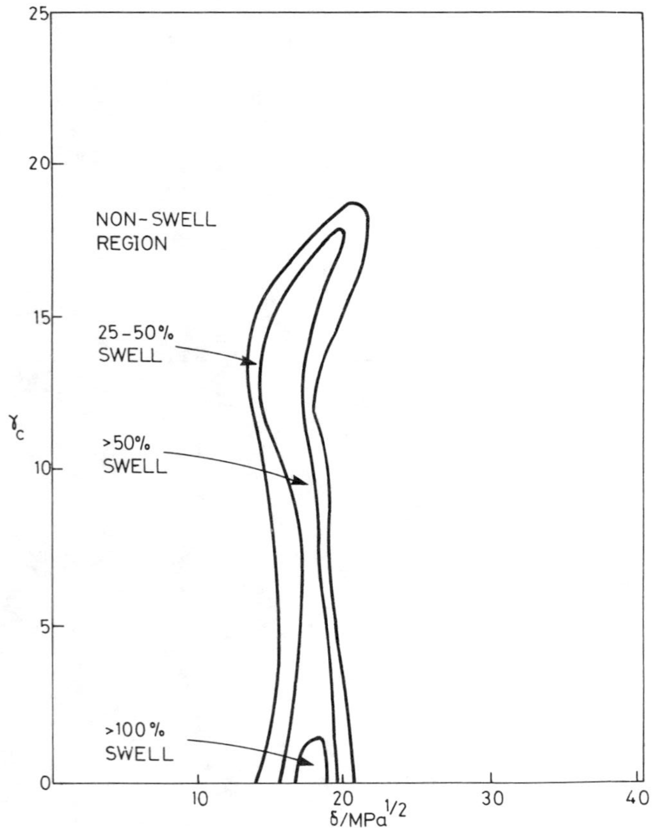

FIGURE 81. Hildebrand parameter—hydrogen-bonding parameter map with % swell contours for silicone rubber, for use with the liquids data of Figure 4, Chapter 8 and Tables 8 and 9, Chapter 8. (Adapted from Beerbower, A., Kaye, L. A., and Pattison, D. A., *Chem. Eng. (NY)*, 74(26), 118, 1967.)

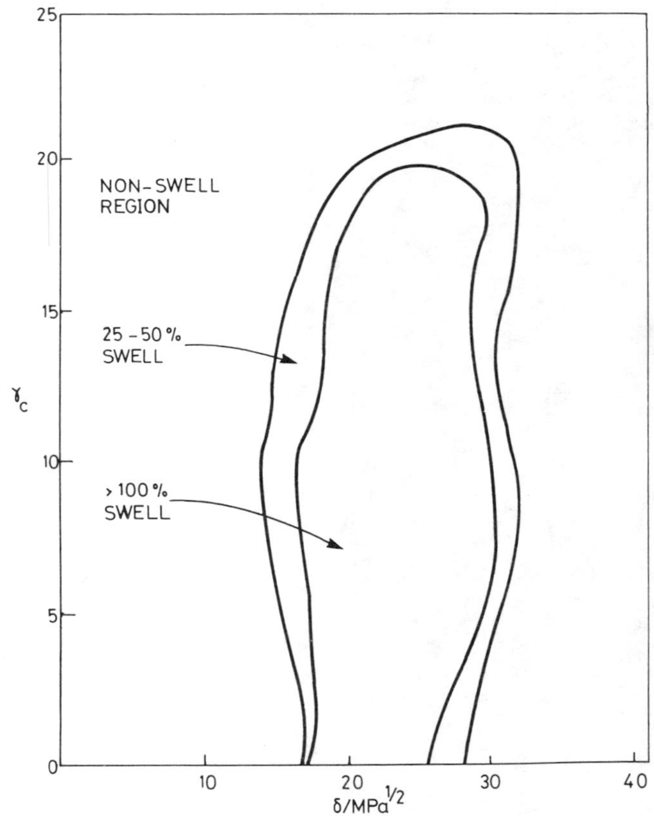

FIGURE 82. Hildebrand parameter—hydrogen-bonding parameter map with % swell contours for acrylate, for use with the liquids data of Figure 4, Chapter 8 and Tables 8 and 9, Chapter 8. (Adapted from Beerbower, A., Kaye, L. A., and Pattison, D. A., *Chem. Eng. (NY)*, 74(26), 118, 1967.)

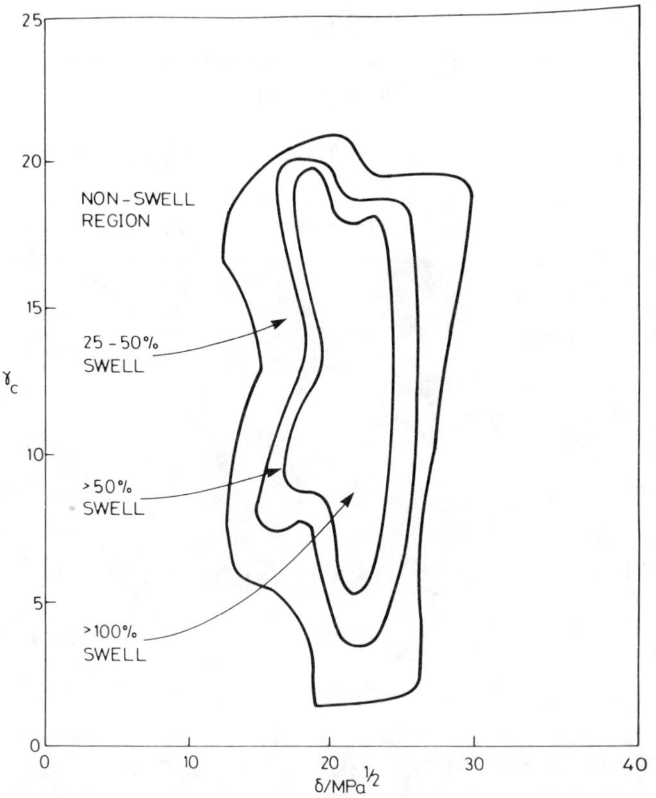

FIGURE 83. Hildebrand parameter—hydrogen-bonding parameter map with % swell contours for vinylidene fluoride-hexafluoropropylene copolymer, for use with the liquids data of Figure 4, Chapter 8 and Tables 8 and 9, Chapter 8. (Adapted from Beerbower, A., Kaye, L. A., and Pattison, D. A., *Chem. Eng. (NY)*, 74(26), 118, 1967.)

TABLE 1
Experimental Hansen Parameters of the Aromatic Residue, the Aliphaic Ester Residue, and the Monomer Repeat Unit of Poly(Ethylene Terephthalate)

	$\delta/MPa^{1/2}$			
	δ_t	δ_d	δ_p	δ_h
Aromatic	20.2	18.5	5.1	6.1
Aliphatic	24.8	19.8	13.7	5.8
PET	22.1	19.0	8.7	6.0

Adapted from Knox, B. H., *J. Appl. Polym. Sci.*, 21, 225, 1977.

TABLE 2
Volume-Dependent and Residual Cohesion Parameters and Interaction Radius for Polymers

Polymer	${}^{j}\delta_v$/MPa$^{1/2}$	${}^{j}\delta_R$/MPa$^{1/2}$	${}^{j}R$/MPa$^{1/2}$
Polystyrene	20.9	0.0	14.3
Poly(methyl methacrylate)	21.3	8.6	11.1
Poly(vinyl chloride)	22.7	7.4	7.8
Poly(vinyl acetate)	21.5	8.4	11.5
Rubber	19.4	0.0	10.6
Polyisobutylene	18.2	0.0	9.4

Adapted from Bagley, E. B. and Scigliano, J. M., *Solutions and Solubilities,* Part II, Dack, M. R. J., Ed., John Wiley & Sons, New York, 1976, chap. 16.

REFERENCES

1. **Hinden, E. R. and Whyzmusis, P. D.,** Polyamide resin solubility, in *Solvents Theory and Practice (Adv. Chem. Ser.* 124), Tess, R. W., Ed., American Chemical Society, Washington, D.C., 1973, chap. 11.
2. **Hansen, C. M.,** The universality of the solubility parameter, *Ind. Eng. Chem. Prod. Res. Dev.,* 8, 2, 1969.
3. **Brandts-Buys, L. and Lhoest, G.,** Solubility parameters and graphic representation of contours of the solubility of polymers, *Chim. Peintures,* 34, 261, 1971.
4. **Unishi, T.,** Numerical evaluation of the solubility for polymers, *Nippon Kagaku Kaishi,* 8, 1342, 1981.
5. **Wernick, D. L.,** Stereographic display of three-dimensional solubility parameter correlations, *Ind. Eng. Chem. Prod. Res. Dev.,* 23, 240, 1984.
6. **Cooper, W. J., Krasicky, P. D., and Rodriguez, F.,** Dissolution rates of poly(methyl methacrylate) films in mixed solvents, *J. Appl. Polym. Sci.,* 31, 65, 1986.
7. **Katime, I., Quintara, J. R., and Straziel, C.,** Relationship between physical parameters of the solvent and its influence in stereocomplex formation, *Makromol. Chem.,* 187, 1441, 1986.
8. **Friedrich, C., Driancourt, A., Noel, C., and Monnerie, L.,** Asymmetric reverse osmosis and ultrafiltration membranes prepared from sulfonated polysulphone, *Desalination,* 36, 39, 1981.
9. **Rodriguez, F., Groele, R. J., and Krasicky, P. D.,** Dissolution rates of thin polymer films using laser interferometry, *Proc. Soc. Photo-Opt. Instrum. Eng.,* 539 (Santa Clara Symp. Microlithography), 14, 1985.
10. **Bottino, A., Capanelli, G., Munari, S., and Turturro, A.,** Solubility parameter of poly(vinylidene fluoride), *J. Polym. Sci. B. Polym. Phys.,* 26, 785, 1988.
11. **Meyer zu Bexten, J. H.,** Solubility parameters and their practical application, *Farbe Lack,* 78, 813, 1972.
12. **Sattelmeyer, R.,** The solubility parameters of synthetic resins and elastomers for the plastics sector, *Adhäsion,* 20, 278, 1976.
13. **Walz, G.,** The choice of solvents for high-solid systems with the aid of the solubility parameter model, *Kunstharz Nachr. (Hoechst),* 33(6), 30, 1974.
14. **Walz, G. and Emrich, G.,** Solubility parameter diagrams of selected alkyd and melamine resins, *Kunstharz Nachr. (Hoechst),* 34(8), 19, 1975.
15. **Kozakiewicz, J. and Penczek, P.,** Adhesives from polyepichlorohydrin elastomers. I. Studies on the properties of adhesive solutions of polyepichlorohydrin, *Angew. Makromol. Chem.,* 50, 67, 1976.
16. **Cabasso, I., Klein, E., and Smith, J. K.,** Polysulfone hollow fibers. I. Spinning and properties, *J. Appl. Polym. Sci.,* 20, 2377, 1976.
17. **Cabasso, I.,** A study of compatible polymer systems employing solubility parameter diagrams, *Abstr. Pap. Am. Chem. Soc.,* 173, ORPL-21, 1977.
18. **Cabasso, I., Jagur-Grodzinski, J., and Vofsi, D.,** Advanced polymer alloys containing polyphosonate components, in *Polymer Science and Technology,* Vol. 10, *Polymer Alloys: Blends, Blocks, Grafts, and Interpenetrating Networks,* Klempner, D. and Frisch, K. C., Eds., Plenum Press, New York, 1977, 1.
19. **Namasté, Y. M. N., Obendorf, S. K., Anderson, C. C., Krasicky, P. D., Rodriguez, F., and Tiberio, R.,** Copolymers of itaconic acid and methyl methacrylate as positive electron beam resists, *J. Vac. Sci. Ser. B,* 1(4), 1160, 1983.

20. **Millich, F.,** Rigid rods and the characterization of polyisocyanides, *Adv. Polym. Sci.,* 19, 117, 1975.
21. **Nishimura, J., Tanaka, N., and Yamashita, S.,** Thermostability of polymers containing indanic units in the main chain, *J. Polym. Sci. Polym. Chem.,* 18, 1203, 1980.
22. **Balcar, H. and Bohackova, V.,** Solution properties of poly(1-chloro-2,3-epoxypropane), *Collect. Czech. Chem. Commun.,* 42, 2145, 1977.
23. **Hoernschemeyer, D.,** The influence of solvent type on the viscosity of concentrated polymer solutions, *J. Appl. Polym. Sci.,* 18, 61, 1974.
24. **Chawla, A. S. and Chang, T. M. S.,** Use of the solubility parameters for the preparation of hemodialysis membranes, *J. Appl. Polym. Sci.,* 19, 1723, 1975.
25. **Deinzer, M., Melton, R., and Mitchell, D.,** Trace organic contaminants in drinking water; their concentration by reverse osmosis, *Water Res.,* 9, 799, 1975.
26. **Klein, E. and Smith, J. K.,** The use of solubility parameters for solvent selection in asymmetric membrane formation, in *Reverse Osmosis Membrane Research,* Lonsdale, H. K. and Podall, H. E., Eds., Plenum Press, New York, 1972, 61; Asymmetric membrane formation. Solubility parameters for solvent selection, *Ind. Eng. Chem. Prod. Res. Dev.,* 11, 207, 1972.
27. **Klein, E., Eichelberger, J., Eyer, C., and Smith, J.,** Evaluation of semipermeable membranes for determination of organic contaminants in drinking water, *Water Res.,* 9, 807, 1975.
28. **Shen, T. C. and Cabasso, I.,** Ethyl cellulose anisotropic membranes, in *Macromolecular Soltutions: Solvent-Property Relationships in Polymers* (Symp. Proc., Am. Chem. Soc., New York, 1981), Seymour, R. B. and Stahl, G. A., Eds., Pergamon, New York, 1982, 108.
29. **Vinther, A.,** Application of the concepts solubility parameter and pigment charge, *Chim. Peintures Engl.,* 34, 363, 1971.
30. **Vinther, A. and Sörensen, P.,** Use of the concepts of solubility parameter and pigment charge in the formulation of pigmented systems, *Farbe Lack,* 77, 317, 1971.
31. **Perepechkin, L. P., Sobakin, V. G., Budnitskii, G. A., Biber, B. L., Baibakova, Z. V., and Rud'ko, L. V.,** Study of formation process of capillaries with selectivity permeable walls, *Vysokomol. Soedin. Ser. A,* 25, 1549, 1983; *Polym. Sci. U.S.S.R.,* 25, 1794, 1983.
32. **Riedel, G.,** Solubility of dyes in organic solvents, *Farbe Lack,* 82, 281, 1976.
33. **Drzewinski, M. and MacKnight, W. J.,** Structure and properties of sulfonated polysulfone ionomers, *J. Appl. Polym. Sci.,* 30, 4753, 1985.
34. **Hart, D. J.,** Factors affecting the distinction of image (DOI) of painted fiberglass reinforced reaction injection molded (RRIM) urethane, *J. Coatings Technol.,* 54(685), 47, 1982.
35. **Christensen, U. L. and Banke, O.,** Gloves and solubility parameters, *Dan. Kem.* 63(3), 70, 1982; *Gloves: Safe/Unsafe,* Institute for Arbejdsmilijo, Danmarks Tekniske Hojshole, Lyngby, Denmark, 1983.
36. **Rowe, R. C.,** The prediction of compatibility/incompatibility in blends of ethyl cellulose with hydroxypropyl methylcellulose or hydroxypropyl cellulose using two-dimensional solubility parameter maps, *J. Pharm. Sci.,* 38, 214, 1986.
37. **Sakellariou, P., Rowe, R. C., and White, E. F. T.,** The interaction and partitioning of low molecule weight polyethylene glycols and diethyl phthalate in ethylcellulose hydroxypropylmethylcellulose blends, *J. Appl. Polym. Sci.,* 34, 2507, 1987.
38. **Huppenthal, L., Staszewska, D. U., and Nowacki, J.,** Solubility parameters of poly(vinylpyrrolidone), *Polimery (Warsaw),* 27, 427, 1982.
39. **Shaw, M. T.,** Studies of polymer-polymer solubility using a two-dimensional solubility parameter approach, *J. Appl. Polym. Sci.,* 18, 449, 1974.
40. **Blanks, R. F. and Prausnitz, J. M.,** Thermodynamics of polymer solubility in polar and nonpolar systems, *Ind. Eng. Chem. Fundam.,* 3, 1, 1964.
41. **Schlegel, L. and Schnabel, W.,** Polymethacrylonitrile as a resist in x-ray lithography, *J. Vac. Sci. Technol. B,* 6(1), 82, 1988.
42. **Knox, B. H.,** Bimodal character of polyester-solvent interactions. I. Evaluation of the solubility parameters of the aromatic and the aliphatic ester residues of poly(ethylene terephthalate); II. Evaluation of the chemical structures of the aromatic and aliphatic ester residues of poly(ethylene terephthalate); III. The effects of morphology on the nature of the interaction of nonaqueous solvents with the aromatic and aliphatic ester residues of poly(ethylene terephthalate), *J. Appl. Polym. Sci.,* 21, 225, 249, and 267, 1977.
43. **Knox, B. H., Weigmann, H.-D., and Scott, M. G.,** Interactions of nonaqueous solvents with textile fibers. Part V: Application of the solubility parameter concept to polyester fiber-solvent interactions, *Text. Res. J.,* 45, 203, 1975.
44. **Scigliano, J. M.,** *Consideration of Complex Solutions Using Internal Pressure Measurements,* Doctoral Dissertation, Washington University, St. Louis, MO, 1972; *Diss. Abstr. B,* 32, 6984, 1972.
45. **Bagley, E. B. and Scigliano, J. M.,** Polymer solutions, in *Solutions and Solubilities,* Part II, Dack, M. R. J., Ed., John Wiley & Sons-Interscience, New York, 1976, chap. 16.

46. **van Krevelen, D. W.,** *Properties of Polymers: Their Estimation and Correlation with Chemical Structure,* 2nd ed., Elsevier, Amsterdam, 1976.
47. **Papo, A., Sciarrone, U., and Torriano, G.,** The solubility parameter. A new tool for the evaluation of the solvent resistance of organic coatings, *Proc. Intl. Conf. Org. Coatings Sci. Technol.,* 4th, State University of New York, New Paltz, NY, 1978, 585.
48. **Novembre, A. G., Masakowski, L. M., and Hartney, M. A.,** Optimal developer selection for negative acting resists, *Polym. Eng. Sci.,* 26, 1158, 1986.
49. **Lai, J. H.,** Application of methyl α-chloroacrylate copolymers as electron sensitive positive resists, *Proc. Soc. Photo-Opt. Instrum. Eng.,* 333 (Submicron Lithography), 8, 1982.
50. **Kim, K. U., Park, T. S., and Lee, K. C.,** Solubility parameter and solvent selection of adhesives, *Polymer (Korea),* 7, 155, 1983.
51. **Gressier, J. C. and Levesque, G.,** Aliphatic polythioamides. III. Study of some physicochemical properties, *Eur. Polym. J.,* 16, 1167, 1980.
52. **Lim, Y. J., Choi, H. J., Yoon, N. S., and Lee, H. K.,** Dyeability of poly(ethylene terephthalate) pretreated with organic solvent, *Han'guk Somyu Konghahhoechi,* 21, 281, 1984.
53. **Henry, L. F.,** Prediction and evaluation of the susceptibilities of glass thermoplastics to environmental stress cracking, in *Predicting Plastics Performance,* (Div. Tech. Conf.), Society of Plastics Engineers, Greenwich, Conn., 1973; *Polym. Eng. Sci.,* 14, 167, 1974.
54. **Beerbower, A. and Panzer, J.,** Comments on "Interaction between lubricants and plastic bearing surfaces", *Wear,* 18, 481, 1971.
55. **Eisenbach, M., Kuppermann, B., Fleischer, N., Robinson, T., and Caplan, S. R.,** Purple membrane in solvents: solubility parameter mapping, in *Energetics and Structure of Halophilic Microorganisms (Dev. Halophilic Microorg.,* 1), Caplan, S. R. and Ginzburg, M., Eds., Elsevier/North-Holland, Amsterdam, 1978, 239.
56. **Eisenbach, M., Caplan, S. R., and Tanny, G.,** Interaction of purple membrane with solvents. I. Applicability of solubility parameter mapping, *Biochim. Biophys. Acta,* 554, 269, 1979.
57. **Eisenbach, M. and Caplan, S. R.,** Interaction of purple membrane with solvents. II. Mode of interaction, *Biochim. Biophys. Acta,* 554, 269, 1979.
58. **Gardon, J. L. and Teas, J. P.,** Solubility parameters, in *Treatise on Coatings,* Vol. 2, *Characterization of Coatings: Physical Techniques,* Part II, Myers, R. R. and Long, J. S., Eds., Marcel Dekker, New York, 19767, chap. 8, 413.
59. **Teas, J. P.,** Graphic analysis of resin solubilities, *J. Paint Technol.,* 40(516), 19, 1968.
60. **Shell Chemicals,** *Solubility Parameters,* 2nd ed., Tech. Bull. ICS(X)/78/1, 1978.
61. **Hashimoto, I.,** Studies of solvent dyeing, 10. Relation between the swelling and solution behaviors and the three dimensional solubility parameters for various polymers in organic solvents, *Sen'i Gakkaishi,* 34, T469, 1978.
62. **Loev, A. M. and Lavrishchev, L. P.,** Method for the selection of solvents for coating materials, *Lakokras. Mater. Ikh. Primen.,* No. 4, 4, 1977.
63. **Parsonage, E. E. and Peppas, N. A.,** Properties of positive resists. I. Irradiation-induced degradation and sensitivity of certain methyl methacrylate copolymers, *Brit. Polym. J.,* 19, 469, 1987.
64. **Belfkira, A., Camps, M., Monthéard, J. P., and Boiteux, G.,** Dielectric study of thin films of para-substituted poly α-acetoxystyrene, *Eur. Polym. J.,* 23, 979, 1987.
65. **Crowley, J. D., Teague, G. S., Jr., and Lowe, J. W.,** A three-dimensional approach to solubility, *J. Paint Technol.,* 38, 268, 1966; 39, 19, 1967.
66. **Eastman Kodak,** *Three Dimensional Approach to Solubility,* Rochester, NY (undated).
67. **Tess, R. W.,** Solvents, *Am. Chem. Soc. Symp. Ser.,* 285 (Applied Polymer Science, 2nd ed., Tess, R. W., and Poehlein, G. W., Eds.), 661, 1985.
68. **Nelson, R. C., Figurelli, V. F., Walsham, J. G., and Edwards, G. D.,** Solution theory and the computer. Effective tools for the coatings chemist, *J. Paint Technol.,* 42, 644, 1970.
69. **Cosaert, E.,** Solubility parameters allowing the prediction of the solubility of a solid in a solvent system, *Chim. Peintures,* 34, 169, 1971.
70. **du Pont de Nemours, E. I. and Co.,** *Solvent Formulating Maps for Elvacite Acrylic Resins,* Bulletin PA-12-770, A.70562, Wilmington, DE, 1970; PA-12-174, 1974.
71. **Zunker, D. W.,** Unique aqueous nitrocellulose/acrylic air-dry systems for industrial finishing, *J. Coatings Technol.,* 48(616), 37, 1976.
72. **Sausaman, D. K.,** A practical approach to solvent applications in coatings and inks, in *Solvents Theory and Practice* (*Adv. Chem. Ser.* 124), Tess, R. W., Ed., American Chemical Society, Washington, D.C., 1973, chap. 8.
73. **Adank, G. and Goshorn, T. R.,** Epichlorohydrin elastomers, rubbers for the seventies, *Angew. Makromol. Chem.,* 16/17, 103, 1971.
74. **Dyck, M. and Hoyer, P.,** Solubility and hydrogen-bridge parameters, *Farbe Lack,* 70, 522, 1964.

75. **Charbonneu, L. F.,** Synthesis and properties of alkylene-linked aromatic polyimides, *J. Polym. Sci. Polym. Chem.,* 16, 197, and 1459, 1978.
76. **Vargiu, S., Parma, G., and Olcese, T.,** The solubility parameter: an approach to optimize the performance of protective coatings, *Pitture Vernice,* 51, 203, 1975; FATIPEC *(Fed. Assoc. Tech. Ind. Peint. Vernis Emaux Encres Impr. Eur. Cont.) Congr.,* 12, 411, 1974.
77. **Dyck, M. F. and Hays, W. P.,** Influence of the polymer weight on the solubility parameter, γ, *Skand. Tidskr. Färg Lack,* 15, 227, 1969.
78. **Penczek, P., Plocharski, S., and Warchol, W.,** Effect of the composition of solvent mixture on the freezing resistance of polychloroproprene rubber based adhesives on storage, *Polimery,* 20, 186, 1975.
79. **DeWalt, C.,** Factors in tackification and suggestions for designing new resins, *Adhes. Age,* 13(3), 38, 1970.
80. **Ellis, W. H., Saary, Z., and Lesnini, D. G.,** Formulation of exempt replacements for aromatic solvents, *J. Paint Technol.,* 41, 249, 1969.
81. **Ramsbotham, J.,** Quantitative polymer solubility maps, *FATIPEC (Fed. Assoc. Tech. Ind. Peint. Vernis Emaux Encres Impr. Eur. Cont.) Congr.,* 14, 567, 1978.
82. **Ramsbotham, J.,** Solvent system design for epoxy resin, *Paint Manuf. Resin News,* 49, (March), 26; (April), 18, 1979.
83. **Ramsbotham, J.,** Solvent formulation for surface coatings, *Progr. Org. Coatings,* 8, 113, 1980.
84. **Shell Chemicals,** *Solvent System Design,* Tech. Bull. ICS(X)/78/2, 1978.
85. **Dangel, B.,** Round table on solubility parameters. Solvent formulation by a graphical method, *Double Liaison Chim. Peintures,* 28(306), 75, 1981.
86. **Tess, R. W.,** Chemistry and technology of solvents, in *Applied Polymer Science,* Craver, J. K. and Tess, R. W., Eds., American Chemical Society, Washington, D.C., 1975, chap. 44.
87. **Izumi, Y. and Miyake, Y.,** Study of linear poly(*p*-chlorostyrene)-diluent systems. I. Solubilities, phase reltionships, and thermodynamic interactions, *Polym. J.,* 3, 647, 1972.
88. **Beerbower, A., Kaye, L. A., and Pattison, D. A.,** Picking the right elastomer to fit your fluids, *Chem. Eng.* (NY), 74(25), 118, 1967.
89. **Eissler, R. L. and Princen, L. H.,** Effect of some pigments on tensile and swelling properties of linseed oil films, *J. Paint Technol.,* 40(518), 105, 1968.
90. **Yoshioka, A.,** Compatibility of polymers with tackifiers from the viewpoint of solubility parameter and hydrogen bonding index, *Nippon Setchakam Kyokaishi,* 19, 504, 1983.
91. **Jackson, L. C.,** *Solubility Parameter Technology and Evaporative Rate Analysis Used to Evaluate Contaminants; Solubility Parameters and Evaporation Rate Analysis in Organic Residue Characterization,* Bendix Reports BDX-613-859(Rev.) and BDX-613-1099(Rev.), Bendix Corporation, Kansas City, MO, 1973 and 1974.
92. **Jackson, L. C.,** How to select a substrate cleaning solvent — contaminant removal using solubility parameter technology; A scientific technique for adhesive development; Surface characterization based on solubility parameters; Solvent cleaning process efficiency — contaminant removal in the electronics industry; Removal of silicone grease and oil contaminants, *Adhes. Age,* 17(12), 22, 1974; 18(3), 30, 1975; 19(10), 17, 1976; 19(7), 31, 1976; 20(4), 29, 1977.
93. **Jackson, L. C.,** Contaminant detection, characterization and removal based on solubility parameters, in *Surface Contamination: Its Genesis, Detection and Control,* Vol. 2, Mittal, K. L., Ed., Plenum Press, New York, 1979, 527; *Proc. Int. Symp. Contam. Control,* 4th, Inst. Environ. Sci., Mt. Prospect, IL, 1978, 226.
94. **Spychaj, T.,** Solvent/nonsolvent solubility and polarity parameters in fractional separation of oligomers. I. Epoxy resins, *Angew. Makromol. Chem.,* 149, 127, 1987.
95. **Vaughan, C. D.,** Using solubility parameters in cosmetics formulation, *J. Soc. Cosmet. Chem.,* 36, 319, 1985.
96. **Pila, S.,** Optimum solvent mixtures for thick vinyl paints, *Pitture Vernici,* 49, 393, 1973.
97. **Petrovic-Dakov, D., Velickovic, J., Batalovic, Z., and Velanac, J.,** Optimization of solvent selection for binders in paints and varnishes, *Polimeri (Zagreb),* 6, 315, 1985.
98. **Verneret, H.,** Solvents to be used in the paints, varnishes, adhesives and printing inks industries, *Afinidad,* 27(274), 87, 1970.
99. **Bellenger, V., Morel, E., and Verdu, J.,** Solubility parameters of amine-crosslinked aromatic epoxies, *J. Appl. Polym. Sci.,* 37, 2563, 1989.
100. **Dickerson, C. G. and Wiehe, I. A.,** Spherical encapsulated polymer particles by spray drying, *Pac. Chem. Eng. Congr.* [*Proc.*], 2(1), 243, 1977.
101. **Kostylev, V. A., Grigor'ev, V. Yu., and Kunin, A. M.,** Use of the three-dimensional solubility parameter of the starting components to calculate the phase diagram of solvent-thinner-polymer systems, *Lakokras. Mater. Ikh. Primen.,* No. 4, 12, 1988.

102. **Condorelli, G.,** The alpha→beta transition of phthalocyaninatocopper. Influence of type of vehicle, *Pitture Vernici,* 49, 259, 1973.
103. **Yeh, A.-I., Berg, L., and Warren, K. J.,** The separation of acetone-methanol mixture by extractive distillation, *Chem. Eng. Commun.,* 68, 69, 1988.
104. **Sears, J. K. and Darby, J. R.,** *The Technology of Plasticizers,* Wiley-Interscience, New York, 1982.
105. **Bodmeier, R. and McGinity, J. W.,** Solvent selection in the preparation of poly (DL-lactide) microspheres prepared by the solvent evaporation method, *Int. J. Pharm.,* 43, 179, 1988.

Chapter 16

CORRELATION OF POLYMER PROPERTIES WITH COHESION PARAMETERS

This chapter further explores polymer properties that have been discussed in terms of cohesion parameters. Solution viscosity is covered in an earlier chapter (Section 14.7), while sorption and surface properties are referred to in Chapter 17. Applications of cohesion parameters in chromatography are discussed in Chapter 18.

16.1 ELECTRICAL PROPERTIES

Molecular properties are outlined in Chapter 3 and empirical correlations between molecular properties and cohesion parameters are discussed in Section 8.2. Koenhen and Smolders[1] examined several of these correlations for polymers, using relationships obtained from studies of nonpolymeric liquids, such as Equation 6, Chapter 8,

$$\delta_d/\text{MPa}^{1/2} = 19.5n_D - 11.4 \tag{1}$$

and Equation 14, Chapter 8,

$$\delta_p = B\mu V^{-3/4} \tag{2}$$

where n_D is the refractive index, μ is the dipole moment, and B is a constant whose numerical value is discussed in Section 8.2. Dipole moments of polymers, where not available,[2] were estimated at 80% of the monomer values. The results are shown in Table 1. Other attempts at the prediction and correlation of polymer cohesion parameters in terms of electrical properties have used molar refraction,[3] refractive index,[4,5] relative permittivity,[6-10] dielectric loss and membrane conductivity,[11] microwave absorption,[12] electrical conductivity in contact with liquds,[13,14] and an optical waveguide device to probe dielectric properties of the film-vapor interface.[749]

16.2 MECHANICAL PROPERTIES

Cohesive pressure or cohesive energy density (δ^2), surface free energy (γ), and ideal tensile strength (σ_{max}) are all manifestations of the energy content of materials and all may, in principle, be determined experimentally. The problem of determining the cohesive pressures of polymers is explained in Chapter 14, the indirect method of calculating the surface free energy of solids by means of their wetting characteristics is described in Chapter 17, and there are difficulties in determining ideal tensile strength because of imperfections and changes in the material during tests. Gardon[15-17] has discussed the relationships between these three properties, and some of his results are shown in Table 2, Chapter 17. The highest reported ideal tensile strengths of polymers and metal whiskers are close to the values predicted by the equation

$$\sigma_{max} = 0.25\delta^2 \tag{3}$$

It is therefore reasonable to expect correlations to exist between the observed mechanical properties of polymers and their cohesion parameters. In particular, stress-strain studies are made on polymers to obtain values of a characteristic constant required for the calculation of the polymer-liquid interaction parameter from the Flory-Rehner equation (Chapter 14).

TABLE 1
Calculation of Polymer Hansen Dispersion and Polar Parameters from Electrical Properties

Polymer	^{n}D	$10^{30}\mu$/C m	δ_d/MPa$^{1/2}$ (Eq. 1)	δ_p/MPa$^{1/2}$ (Eq. 2)
Polyethylene	1.483	0.0	17.6	0.0
Polystyrene	1.591	0.9	19.7	0.9
Poly(tetrafluoroethylene)	1.298	0.0[a]	14.0	0.0
Polyacrylonitrile	1.514	9.2	18.2	16.2
Poly(2,6-dimethyl-1,4-phenylene oxide)	1.567	2.9[a]	19.2	2.7
Poly(methyl methacrylate)	1.490	4—6	17.8	5.7
Polyamide 6,6	1.530	9.3	18.5	5.1
Poly(vinyl chloride)	1.539	5.6—5.8	18.7	10.0
Poly(ethylene terephthalate)	1.580	4.8[a]	19.5	3.5

[a] Estimated.

Adapted from Koenhen, D. M. and Smolders, C. A., *J. Appl. Polym. Sci.*, 19, 1163, 1975.

Polymers are viscoelastic, having the characteristics of either elastic materials or of viscous flowing materials, depending on the time-scale of the perturbation and on the temperature. The mechanical properties of importance include flexibility, tensile strength, glass transition temperature, creep, hardness, and elasticity.[18,19] The tests commonly carried out on polymers are

1. *Stress-strain* test, comparable to those carried out on metals, where the specimen is stretched at a uniform rate while the force on it is monitored. In the elastic region, the specimen stretches uniformly and the strain is directly proportional to the stress. At the yield point there is stretching with little change in stress, and the polymer flows. Eventually, there is some work-hardening as the polymer molecules become oriented and the stress increases until the ultimate strength is reached and the specimen breaks.
2. *Stress relaxation* tests, where the specimen is deformed a given amount, and the stress required to maintain this deformation is measured as a function of time.
3. *Creep* tests, carried out in tension, compression, or shear, with monitoring of deformation with time as a constant load is applied.
4. *Dry crazing* tests for glassy polymers by observation of the surfaces of samples in strain jigs.

The incorporation of plasticizers into polymers modifies these mechanical properties in a way similar to an increase in the working or testing temperature. They lower the glass transition temperature of the polymer (Section 16.3) and modify the mechanical properties that are dependent on the difference between the working temperature and the glass transition temperature. In general, plasticizer addition results in decrease in 100% modulus, decrease in tensile strength, increase in ultimate elongation, faster decay of stress, and faster creep rate. The nature of interaction with polymers of plasticizers, whether ''primary'' (solvent type) or ''secondary'' (nonsolvent type), may be characterized by their cohesion parameters (Section 16.4).

Determination of polymer mechanical properties associated with cohesion parameter studies include:

- Time dependence of the thermal and mechanical properties of segmented urethanes explained in terms of incompatibility between segments and quantified by Hildebrand or Hansen parameters[20]

- Freeze-resistance of polyurethanes[743]
- Stress-strain relationships in poly(tetrahydrofuran)[21]
- Effect of casting solvent on the mechanical properties of styrene-butadiene[22] and polystyrene-(ethylene-*co*-butylene)-polystyrene[23] triblock copolymers
- Maximum tensile strength of epoxy resins from bisphenol A and epichlorohydrin[24]
- Stress-strain, creep, and swelling behavior of du Pont Nordel® ECD-330 ethylene-propylene terpolymer rubber and gum vulcanizate[25]
- Effects of pigments on tensile and swelling properties of linseed oil films[26]
- Rate of abrasive wear of thermoplastic polymers, where a correlation with cohesion parameters is observed if plastic deformation predominates during the wear process[27]
- Dynamic mechanical properties and sonic velocity of polystyrene-poly(vinyl methyl ether)[28] and other blends, including acrylates, methacrylates, and poly(vinyl chloride)[29]
- Mechanical properties of poly(vinyl chloride) and chlorinated poly(vinyl chloride)[30,31]
- Activation energy (temperature dependence) of the creep of glassy polymers under compressive loads[32]
- Mechanical properties and processibility of organo-siloxane block copolymers[33,34]
- Dynamic mechanical properties of ethylene-propylene copolymer[35] and polystyrene(ethylene-*co*-butylene)-polystyrene triblock copolymer[23]
- Dynamic mechanical and swelling properties of styrene-butadiene block and random copolymers as functions of gamma radiation dosage[36]
- Dynamic mechanical properties of blends of poly(vinyl chloride) and ethoxymethylmelamine resin,[37] and of polycarbonate-polystyrene blends and copolymers[38]
- Mechanical behavior of polystyrene and polyisobutylene in the liquid state[39]
- Bulk modulus and dynamic Young's modulus of polymers[40,41]
- Viscoelastic properties of epoxy resins[42]
- Flow properties of acrylonitrile-vinyl acetate copolymers[43]
- Glassy state elasticity modulus[40,44,45]
- Diisodecyl phthalate and tridecyl phthalate (δ = 17.8, 18.0 $MPa^{1/2}$, respectively) as compatible mixed plasticizers for poly(vinyl chloride)[46]
- Broadening of mechanical damping peaks in multiblock polymers due to phase separation[47]
- Ethylene-vinyl chloride copolymers[48,49]
- Dry-crazing behavior of glassy polymers[50-53]
- Acrylamide polymers[54]
- Corresponding states model for prediction of glass temperatures[55]
- Correlation of cohesive energy density, tensile properties and gas sorption by polymers[734]

Frequently more important than the mechanical properties of the polymers alone are the effects on the mechanical performance of polymers of liquids with which they are in contact (Section 16.4).

16.3 THERMAL TRANSITIONS AND PLASTICISATION

The melting points of polymers (as well as of monomers) show a broad correlation with cohesive energy,[56] and once molten, polymers can be described by an equation of state in terms of internal pressure and cohesive pressure,[57,58] while the molten polymer-liquid interaction paramater χ can be related to cohesion parameters and to the volatility of liquids from mixtures with molten polymers.[59,60] However, studies of properties of polymers in the amorphous state (rubbery or glassy) or the crystalline state are usually of greater practical value than those of polymer melts.

The glass transition temperature, T_g, the temperature below which a polymeric substance (or any other substance) becomes elastic, brittle, and glass-like rather than rubbery and viscous, is the temperature at which the thermal energy is insufficient for long-range molecular motion of the main chain to occur. The internal pressure of a glassy polymer (a polymer below T_g) is about half the value of that just above T_g,[61] even when considered at constant volume,[62,63] so cohesive energy and cohesion parameters are likely to play a significant role in the description of the transition, and have received considerable attention.* Kaelble[85] has described an interactive computer model for estimating cohesive energy density, the glass transition temperature, and other properties of acrylic polymers.

The general observation is that there is a broad linear dependence of T_g on cohesive pressure, so that polymers with high Hildebrand parameter values tend to have high values of T_g. Thus, silicone rubber, with a very low δ (~ 15 MPa$^{1/2}$), has one of the lowest glass transition temperatures recorded. In addition, polymers having high internal mobility (low chain stiffness) have lower T_g values. For polymer blends, the nature of the glass transition (whether one or two values of T_g) is a test of compatibility (Section 16.9). For copolymers styrene-methyl methacrylate and acrylonitrile-methyl methacrylate, Ham[691] demonstrated a correlation between cohesive energy densities of homopolymers and the copolymer composition for minimum T_g,

$$j_n/k_n \approx k_c/j_c$$

where $^jn/^kn$ is the mole ratio at minimum T_g.

Plasticization of a polymer is associated with the lowering of the glass transition temperature. More generally and practically,[86] plasticization refers to a change in thermal and mechanical properties involving

- Lowering of rigidity at room temperature
- Increase of elongation to break at room temperature
- Increase of toughness

achieved either by compounding with a lower molecular weight compound (external plasticization) or by modifying the polymer molecules by the introduction of a comonomer to increase chain flexibility (internal plasticization). Cohesion parameters have proved useful in the selection of external plasticizers, an approximate guide[87] to compatibility being Hildebrand parameters of polymer and plasticizer differing by no more than 2.6 MPa$^{1/2}$.

Many of the reviews** listed in Table 1, Chapter 1 deal with polymer plasticization from the cohesion parameter point of view.

For hydrocarbon rubbers, which are produced in large quantities, petroleum-based oils are used widely as extenders and processing aids, with compatibility often being assessed in terms of Hildebrand parameters.[94] Values for poly(isobutylene) and butyl rubber match those of paraffinic oils; natural rubber and styrene-butadiene rubber have Hildebrand parameters comparable to those of aromatic oils, while nitriles match highly aromatic oils or esters. Ahagon, Kobayashi and Misawa[694] found that ice-skid resistance of tire rubber (70NR:30BR) increases with decreasing Hildebrand parameter of both plasticizer and process oil. Group contribution methods (Chapter 6) were used for calculation of Hildebrand parameters of plasticizers (18.3 to 20.2 MPa$^{1/2}$) and oils (16.8 to 19.5 MPa$^{1/2}$).

* For example: Becker;[64-67] Berry;[68] Chiantore;[54] Deanin;[69] DiBenedetto;[55] Gargallo and Russo;[70] Guzman, Alamo and Fatou;[71] Ham;[691] He;[72,73] Kreibich and Batzer;[74-76] Lee and Sewell;[3,77] Miller;[78] Pérez et al.;[79] Rostiashvili and Irzhak;[80] Singh, Kalachandra, and Turner;[737] Tobolsky;[40] Tsutsui and Tanaka;[81] Uematsu;[82] Vicini et al.;[83] and Wymann.[84]

** For example: Brydson;[88,89] Grossman;[714] Immergut and Mark;[86] Li and Long;[90] Sears, Touchette, and Darby;[91,92] and Yoshioka.[93]

Low-volatility plasticizers used in commercial plastics have a long-lasting effect on polymer properties, but volatile liquids can have a similar effect on a short-term basis.[90] In both cases, the cohesion parameter largely determines the efficiency of the plasticizer. Polymers for which such correlations with plasticization and rheology have been made include:

- Poly(2,6-dimethyl-1,4-phenylene oxide)[95]
- Acrilan® (Monsanto) acrylic fiber[96,97] and other acrylics[98,699]
- Polyesters, including poly(ethylene terephthalate)[99,100,716] (see Section 15.1)
- Ionic copolymers[81,770]
- Poly(vinyl chloride)[101-104] and poly(vinyl butyral)[742]
- Polyurethanes[105]
- Polyamides[75]
- Natural polymers (cellulose, lignocellulose, collagen, elastin, keratin, silk)[75,106-111]
- Elastomers[112,113]
- Pharmaceutical tablet binders[114]
- Gel-forming polymers[115]
- Effect of nitrobenzene on nitrated styrene ionomer[770]

In some situations it is convenient to use a slurry of a polymer powder in a monomer which acts as a plasticizer prior to being polymerized to form the required product. Cohesion parameters and glass transition temperatures have been used[737] to explore the compatibility of the initial polymer with the new polymer.

Most of the discussions on polymers so far in Chapters 13 to 16 have been concerned with amorphous polymers which, because of their liquid-like state of disorder, can be treated as subcooled liquids where cohesion parameters are concerned. Crystalline polymers tend to be insoluble in all liquids, which is explained in cohesion parameter terms by attributing an additional enthalpy term associated with fusion or disordering to the dissolution process. Only if there are strong specific interactions such as hydrogen bonding between polymer and liquid molecules can this additional stabilizing energy term be overcome so that dissolution can take place, as illustrated by cellulose in 70% sulfuric acid and by polyamide 6,6 in phenol. However, the cohesion parameter predictions are followed for swelling, with ther crystalline regions appearing to act in the same way as molecular cross-links between polymer chains. As the temperature approaches the melting point ($T > 0.9\ T_m$), cohesion parameters may also be used to discuss dissolution. The degree of cyrstallinity, of course can affect the observed cohesion parameter values.[116]

Amorphous polymers, being metastable with respect to their crystalline forms, can be induced to crystallize in several ways:

1. Thermally induced crystallization, by increasing the temperature above T_g to decrease the viscosity and allow rearrangement
2. Strain-induced crystallization
3. Liquid-induced crystallization

It is this third process which is of particular interest from the point of view of cohesion parameters.[117] Most studies have used poly(ethylene terephthalate),* since it can be quenched easily from the melt to a glass without the complication of the presence of crystallinity before exposure to a liquid. Although some discrepancies are noted in the correlation of liquid-induced crystallization with liquid Hildebrand parameters (particularly because of the

* For example: Cates et al.;[118,119] Desai and Wilkes;[120] Knox, Weigmann et al.;[99,117,121] Kulshreshtha et al.;[122] Liepins et al.;[123] Lindner;[124] Moore and Sheldon;[125] and Vicini.[83]

bimodal distribution of cohesion properties of poly(ethylene terephthalate)[121]: see Section 14.3), when component cohesion parameters such as Hansen parameters are used the agreement improves. This effect of liquid-induced crystallization is to be expected from the previous discussion of plasticization and the lowering of T_g by liquids: crystallization must occur at a temperature between T_g and T_m, and although liquids depress T_m, this melting point lowering is not as great as the depression of T_g.

Studies of interest on systems other than poly(ethylene terephthalate) include:

- Effect of the cohesion parameter value of the solvent mixture on the freezing resistance of polychloroprene rubber adhesives[126]
- Effect of solvent on crystallization from dilute polyethylene solutions[127]
- Liquid-induced crystallization of poly(2,6-dimethyl-1,4-phenylene oxide)[95,128]
- Melting temperatures of mixtures containing polychlorotrifluoroethylene[129]
- Vapor-induced crystallization of polycarbonate[130,131] and of soluble vanadyl phthalocyanine dyes in vinyl chloride-vinyl acetate polymer[132]

Liquid-induced crystallization of polymer blends is included in Section 16.9, examples of plasticization of natural polymers appear in Section 16.11, and interactions of polymers with solids are dealt with in Section 16.8

Anisotropic or liquid crystal behavior of polymers may also be discussed in terms of cohesion parameters.[133-135]

16.4 ENVIRONMENT-INDUCED DEGRADATION

The most obvious effect of a liquid or wetting agent on the mechanical properties of a polymer is deterioration during use, so such phenomena are generally described as *environmental stress cracking* (failure of the material by breaking when exposed to mechanical stress in the medium) or *environmental crazing* (failure of the material by the development of a multitude of fine cracks when exposed to the medium, even as a result of internal stresses in the absence of applied mechanical stress). These effects, which are particularly significant in glassy polymers and less so in semicrystalline polymers, are closely related to the dissolution and swelling phenomena described in Section 14.3.

Explanations of the causes of these processes are varied,[95,136-140] including both the concept of internal surface stabilization or lowering of surface energy causing a change in the mode of failure from the normal necking and flowing to craze and fracture, and the hypothesis of plasticization causing reduction in the stress level at which void propagation takes place. Local reduction in glass transition temperature by liquid plasticization is a readily understandable model, but the surface modification theory is supported by the observation that a short treatment of an acrylic polymer with hot liquids such as ethyl acetate and pentanol modifies the surface and results in a decrease in creep rate. This surface hardening or stabilization effect is considered to be due not only to thermal orientation and elimination of defects in the surface, but also to an extraction of the low molecular weight polymer fraction and completion of the polymerization of residual monomers.[141] For liquids acting in this way, the correlation of the effect with the liquid Hildebrand parameter is not unexpected.

Although the tensile strengths of some polymers show only an approximate correlation with cohesive pressure, the values of the critical flexural strain, ϵ_c, below which neither crazing nor cracking occurs, correlate well with the liquid Hildebrand parameters for poly(2,6-dimethyl-1,4-phenylene oxide), which has a Hildebrand parameter of $\sim$ 18.5 MPa$^{1/2}$. The liquids were of several types, including perfluoroalkanes, alkanes, aliphatic alcohols and formamide (Figure 1). It can be seen that small δ differences give rise to cracking; larger

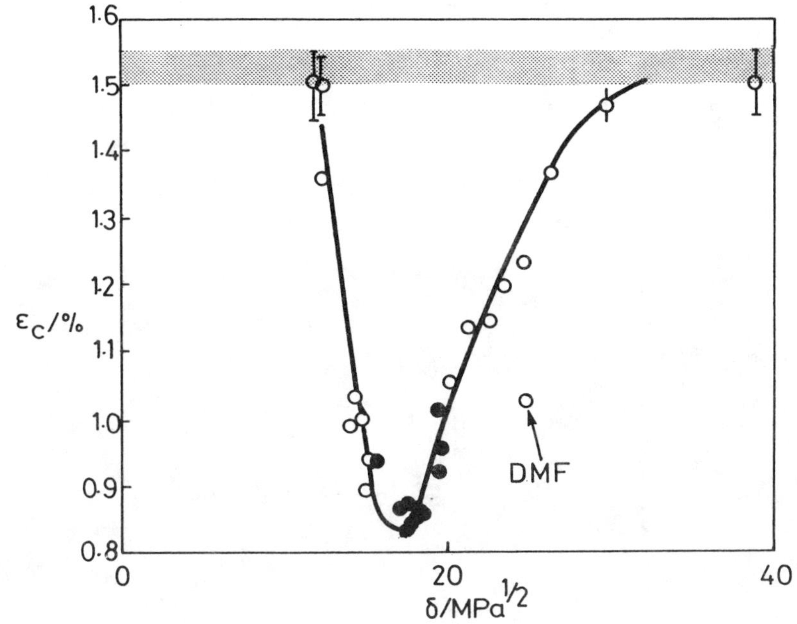

FIGURE 1. Critical strain (ϵ_c) for crazing or cracking of poly(2,6-dimethyl-1,4-phenylene oxide) as a function of the Hildebrand parameter of the environmental liquid. The shaded band indicates the range of ϵ_c values in air; the open circles denote the positions of crazing agents and the filled circles those of the cracking agents. *N,N*-Dimethylformamide (DMF) is the only point showing significant deviation from the curve. (Adapted from Bernier, G. A. and Kambour, R. P., *Macromolecules*, 1, 393, 1968.)

differences cause crazing.[95] Figure 2 shows the critical stress needed to cause crazing of polystyrene[142] as a function of the Hildebrand parameter difference between polymer and liquid, using ${}^j\delta = 18.6$ MPa$^{1/2}$.

When strong polar or hydrogen-bonding interactions occur, the Hildebrand parameter alone is inadquate to correlate the data, as indicated by the critical strain behavior of other polymers investigated by Kambour and co-workers:

- Copolyether-sulfone (polysulfone)[143]
- Bisphenol A polycarbonate[164]
- Polystyrene[51]
- Styrene-acrylonitrile copolymer[52]
- Dicyano bisphenol polycarbonate[52]

Vincent and Raha[140,145] plotted the results for cracking and crazing of the amorphous thermoplastics poly(methyl methacrylate), poly(vinyl chloride), and polysulfone as functions of both δ and the hydrogen-bonding parameter γ_C of liquids to which they are exposed (Figures 3 to 5). In each graph there is an area of solubility, with liquids that cause cracking close to this region and crazing liquids at a greater distance. This is consistent with Figure 1 when hydrogen bonding is taken into account.

Hansen parameters may also be used. Henry[146] correlated the critical strain behavior of Commercial Plastics P-1700 polysulfone and Lexan® polycarbonate by plotting δ_p against δ_h and defining a region of maximum susceptibility, where both ${}^i\delta_p$ and ${}^i\delta_h$ were in the range 3 to 15 MPa$^{1/2}$. Alternatively, the data for several polymers can be plotted on a diagram using Hansen parameters. For example, with $({}^i\delta_h - {}^j\delta_h)$ against $({}^i\delta_{dp} - {}^j\delta_{dp})$, where $\delta_{dp} =$

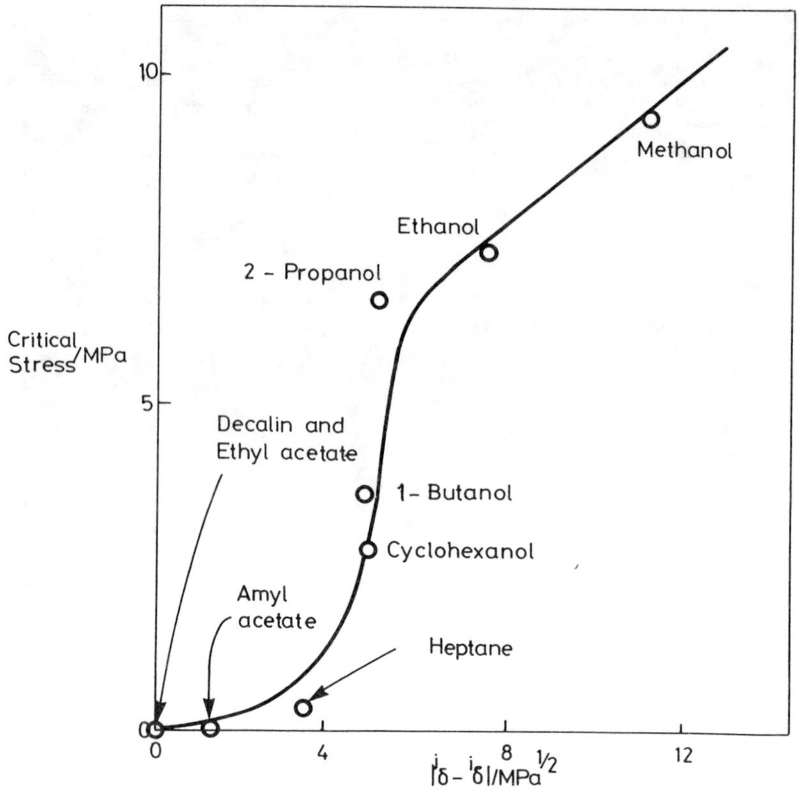

FIGURE 2. Dependence of critical stress for crazing of polystyrene on the difference between the Hildebrand parameter values of polymer and liquid. (Adapted from Boyer, R. F. and Keskkula, H., in *Encyclopedia of Polymer Science and Technology*, Vol. 13, Bikales, N. M., Ed., John Wiley & Sons, New York, 1970, 375.)

$(\delta_d^2 + \delta_p^2)^{1/2}$, van Krevelen[140] reported that the additional polysulfone data of Henry[146] and MacNulty[147] were consistent with those of Vincent and Raha.[145] There is a small central zone in these plots for dissolution, a boundary region of crazing, and an intermediate zone of predominantly cracking with some dissolution, swelling, and crazing.

A full explanation of these phenomena may involve a combination of effects, with surface stabilization being significant at liquid δ values far from that of the polymer, while the plasticizing effect is strongly dependent on δ values near that of the polymer. In any case, cohesion parameter correlations will continue to be important.

It has been shown by Jacques and Wyzgoski[148] that, not unexpectedly, the molar volumes of the liquids are important when cohesion parameters are used to correlate environmental stress cracking effects. They found that although alkanes are nonpolar and are incapable of hydrogen bonding, the Hildebrand parameter does not correlate well with the critical strain of Lexan LS2-111 polycarbonate in linear aliphatic hydrocabons. However, there is good correlation between critical strain and ${}^iV({}^i\delta - {}^j\delta)^2$ for the polymer (*j*) in these liquids (*i*).

Although less critically important, the volume correction is also significant when plotting Hansen parameter terms for polar and hydrogen-bonding liquids. Jacques and Wyzgoski[148] presented their data, together with those of previous workers[144,146] on critical strains for polycarbonate (*j*) in a wide range of liquids (*i*) in the form of plots of ${}^iV\,({}^j\delta_d - {}^i\delta_d)^2$ against ${}^iV\,({}^j\delta_{hp} - {}^i\delta_{hp})^2$, where $\delta_{hp} = (\delta_h^2 + \delta_p^2)^{1/2}$, choosing ${}^j\delta_d = 19.4$ MPa$^{1/2}$ and ${}^j\delta_{hp} = 6.1$ MPa$^{1/2}$ to give the best fit. Most common liquids have molar volumes near 100 cm^3 mol^{-1},

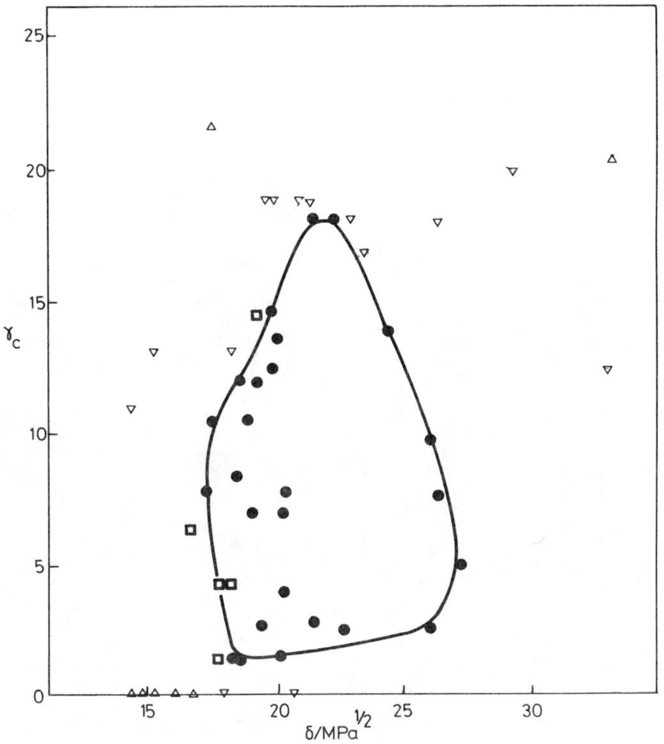

FIGURE 3. Solvents (●), cracking agents (□), low-strain crazing agents (▽), and high-strain crazing agents (△) for poly(methyl methacrylate) plotted on a Hildebrand parameter — hydrogen-bonding parameter solubility map. (Adapted from Vincent, P. I. and Raha, S., *Polymer*, 13, 283, 1972.)

but if ${}^{i}V$ exceeds 150 cm^3 mol^{-1} inclusion of the volume factor is necessary to obtain a good correlation. This procedure was also used by Iisaka et al.[721] for the liquid crazing of bisphenol A polycarbonate.

Rains and Bailey[149] correlated tensile properties of polyester urethanes exposed to various liquids by means of Hansen parameters with the expression

$$\%\ \text{original tensile} - K + A\delta_d + B\delta_p + C\delta_h + D_t + Et$$

where $K = 100$ and E = coefficient of time dependence. The parameters A, B, and C were found to be negative; D was positive, with a value of the same order of magnitude as A; E was negligible. Mai[150] investigated environmental stress cracking of several glassy polymers — poly(methyl methacrylate), poly(phenylene oxide), polystyrene, poly(vinyl chloride), polysulfone, polycarbonate — in terms of Hildebrand parameters, polymer-liquid interaction parameters χ_H, and polar-nonpolar cohesion parameters, concluding that although correlations existed it was not possible to predict the behavior *a priori* from empirical parameters. Numerous other studies have been made of the effects of liquids on the mechanical properties of polymers in terms of cohesion parameters:

- Correlation of both Hildebrand parameters and environmental stress cracking by a semiempirical relationship based on refractive index and hetero-atom content[5]
- Correlation with liquid Hildebrand parameters of critical strains of copolymers with

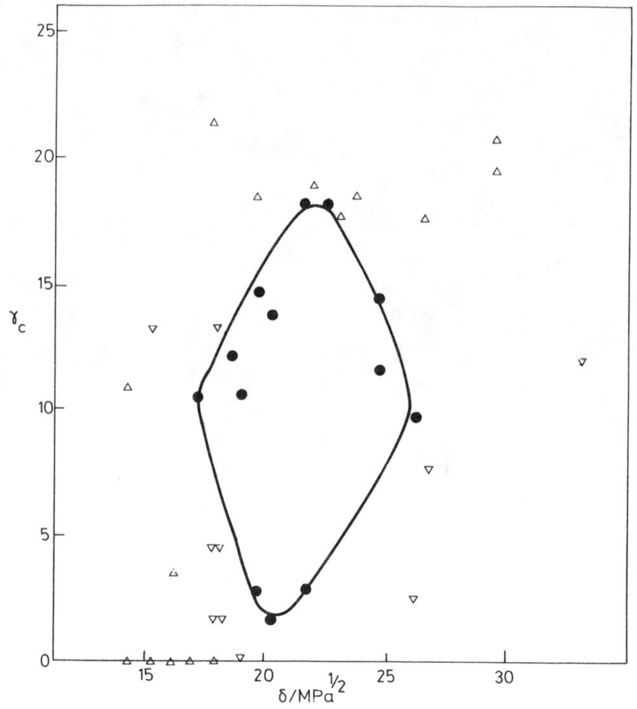

FIGURE 4. Solvents and swelling agents (●), low-strain crazing agents (▽), and high-strain crazing agents (△) for poly-vinyl chloride plotted on a Hildebrand parameter — hydrogen-bonding parameter solubility map. (Adapted from Vincent, P. I. and Raha, S., *Polymer*, 13, 283, 1972.)

different styrene/acrylonitrile ratios and of rubber-modified styrene-acrylonitrile copolymers[146]

- Effect of perfluorokerosene, FC-75, on the tensile modulus of polytetrafluoroethylene[151]
- Environmental stability of polymer systems in terms of component surface free energies[152-157] (Section 17.3)
- Nitrogen and oxygen at liquid nitrogen temperatures as plasticizers and crazing agents for polycarbonate, poly(ethylene terephthalate), and poly(methyl methacrylate)[50]
- Retention of flexural modulus of Epocryl® 480 resin exposed to liquids at room temperature for 120 days[136]
- Environmental crazing of glassy poly(methyl methacrylate) in both pure and mixed liquids[137,138,158-161]
- Craze and creep resistance of high-impact polystyrene in alcohols[162,163]
- Effect of environment and surface coatings on the fatigue properties of polystyrene[164,165]
- Decrease in initial modulus of rayon, polyamide 6,6, polypropylene, cellulose acetate, and polyester yarns in organic liquids as functions of their Hildebrand parameters[99,166-168]
- Correlations with liquid Hildebrand parameter of the fracture behavior of acrylonitrile-butadiene-styrene,[169] plain and glass-filled polystyrene,[170] and poly(methyl methacrylate[171,172]
- Tensile properties of swollen polyester urethane[149,173]
- Dependence of rate of environmental crazing of polycarbonate on liquid Hildebrand parameter[148,174-176,755]
- Effect of liquids on the degradation of polymers by high-speed stirring (lower scission rate observed for liquids in which the polymer molecules are more extended, but no direct relationship with δ)[177,178]

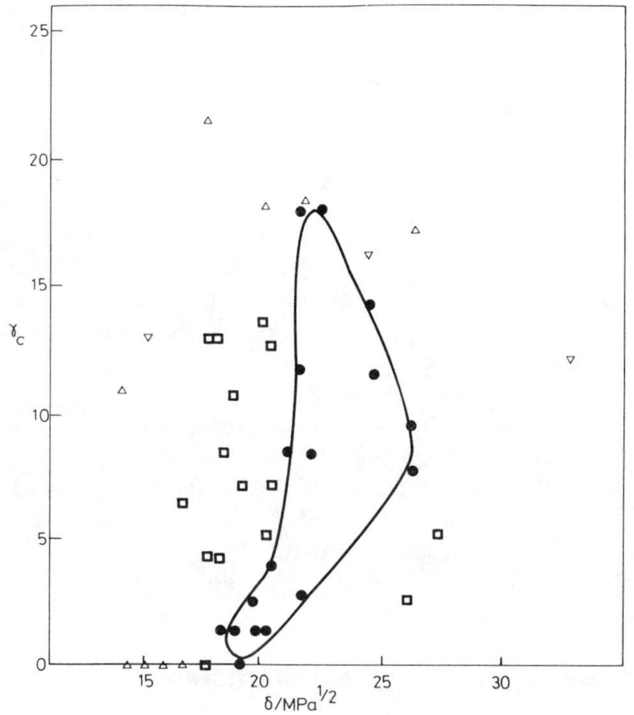

FIGURE 5. Solvents (●), cracking agents (□), low-strain crazing agents (▽), and high-strain crazing agents (△) for polysulfone plotted on a Hildebrand parameter — hydrogen-bonding parameter solubility map. (Adapted from Vincent P. T. and Raha, S., *Polymer*, 13, 283, 1972.)

- Environmental stress crazing of styrene polymers[142] and rigid poly(vinyl chloride)[179]
- Stress-cracking of polyamide induced by zinc chloride solutions[180]
- Environmental influence on the fracture strength of coated glass fibers[181] and effects on polybutadiene glass fiber-reinforced plastics[182]
- Mechanical properties of plasticized cellulose ethers[183]
- Effect of liquid on retained solvent and residual strain in clear epoxy coatings (no definite correlation observed with cohesion parameters)[184]
- Effect of organic liquids on microcrack formation in polypropylene and high-density polyethylene[185] and crack speed in low-density polyethylene[186]
- Environmental stress cracking and crazing of Ultem® polyetherimide and a polyethersulfone in terms of plots[187] similar to Figure 1.
- Whitening (optically opaque regions in highly oriented glassy polymer) and other phenomena in poly(methyl methacrylate)[188,189]
- Heat resistance of polymers[190]

Interactions of polymers with solids are summarized in Section 16.8.

Most of these studies have not used any form of multicomponent cohesion parameter or corrected for liquid volume, which would undoubtedly result in much better correlations.

16.5 RADIATION

Section 3.5 summarizes the use of electromagnetic radiation for spectroscopic studies of materials, including polymers, and Chapter 8 includes related information where the radiation is acting as a "probe" of cohesive properties. An excellent example is the use by Li, Winnick and Guillet of a fluorescence method to determine the Hildebrand parameter of polystyrene;[191] polymer-polymer blend applications are included in Section 16.9.

The interaction of radiation with polymers also brings about chemical or structural changes. The best known of these processes is radiation grafting (Section 10.4). Another application is the use of high energy electrons, ultraviolet, X-ray, or ion-beam radiation in systems with polymers acting as positive resists or negative resists in lithography[192-194] prior to selective dissolution (see also Section 16.6). In negative resists, the exposure to radiation yields an insoluble, cross-linked polymer. However, because the cross-linked portions are liable to swell and distort during the dissolution of the uncrosslinked material, positive resists are sometimes preferred, in which the polymers exposed to radiation degrade and become more soluble as a result of chain scission. Some examples follow:

- Poly(methyl α-chloroacrylate) and other positive resists[195-197]
- Dissolution kinetics of deep UV and electron-beam resists[194,740]
- Chlorinated polymers including chlorinated poly(methylstyrene) as negative resists[198-200]
- Poly(methacrylonitrile) as an X-ray resist[201]
- Deep-etch synchrotron radiation lithography with poly(methyl methacrylate)[202]
- Irradiation-induced degradation of poly(methyl methacrylate)[203]
- Spin coating of electron resists[204]
- Adhesion of lithographic materials (Section 17.5)

16.6 INTERACTIONS WITH GASES AND LIQUIDS

The basis of the explanation of polymer-fluid interactions in terms of cohesion parameters is described in Chapter 13, which concentrates on polymer-liquid interaction parameters; in Sections 14.3 to 14.7 (determination of polymer cohesion parameters); and in Chapter 15 (maps and models). Mixed liquids are dealt with in Sections 4.4 (regular solutions), 13.6 (cosolvency and co-nonsolvency), and 14.4 (turbidimetric titrations). The application of multiple regression to the determination of optimum multicomponent solvent compositions for polymers[205-208,739] is referred to in Sections 4.4 and 14.3. Copolymers are considered in Section 16.10.

A word commonly used in the context of interactions between polymers and liquids is "compatibility", but this is a rather ambiguous term. If one wishes a polymer to dissolve in a liquid, and it does, the two materials are said to be compatible; while if one does *not* wish them to interact, and they do not, they are also described as compatible. More specific terms in common usage are "solubility" or "miscibility" or "swelling" if interaction occurs, and "solvent resistance" for non-interaction.

It is clear that cohesion parameters may be used to aid the selection both of swelling or dissolving agents and of non-interacting agents for a particular polymer. In the former case the aim is to find materials with similar Hildebrand parameters or sets of component cohesion parameters, and in the latter to maximize the differences between the parameters. For polymeric systems, it is important to realise that in some cases it is possible to maintain metastable systems which are thermodynamically unstable. Poly(vinyl chloride), for example, forms a more stable dispersion in methyl ethyl ketone under intensive shear than by conventional slow stirring and heating.[109] Cohesion parameters never provide the complete answer to polymer-liquid interaction problems, but they are always a good guide and have been used widely.[210] Mizerovskii et al.[697] recently derived equations based on the Flory-Rehner equation for prediction of the swelling of vulcanizates.

The range of applications may be illustrated by quoting some published examples; in addition, permeability is covered in Section 16.7, cellulose-based membranes receive attention in Section 16.11, and Chapter 17 includes information on the effect of organic liquids on polymer surface areas and other surface properties.

- Phase separation or coagulation in the production of polyurethane leather substitute[211]
- Asymmetric membranes for hemodialysis, ultrafiltration and reverse osmosis from poly(2,6-dimethyl-1,4-phenylene oxide),[212] polyurethane,[213] ethylene-vinyl alcohol copolymer,[214] and other polymers[215]
- Calculation of the "hydrophilicity" of polymers[216]
- Effects on rubbers of organic liquids,[217-219,756] oils,[220,221] fuels,[222] and hydraulic systems[223,224]
- Solvent resistance of cured hydantoin epoxy resins,[225] polynorbornene,[233] and styrene-maleic anhydride copolymers[234]
- Incorporation of extending oils in elastomers[94] and thickening agents and dispersion media for ureate (polyurea) lubricating greases[226]
- Turbidities of polystyrene dispersions,[227] shape-memory pore structure in porous cross-linked polystyrenes[228] and dynamic mechanical properties of polycarbonate-polystyrene[38]
- Gelation of polystyrene,[229,230] poly(ethyl methacrylate) copolymers,[115] poly(vinyl chloride)[231] and other polymers,[232] including the sol-gel transition of 1,3;2,4-*p,p'*-ditoluylidene sorbitol[731]
- Solubility of Carbowax®, 1-octadecanol and stearic acid in dense carbon dioxide gas[235]
- Interactions of polymer melts with low molecular weight materials[60,236-239]
- Effects of liquids on poly(vinyl chloride)[240-242] and polypropylene[243]
- Properties of vulcanized natural rubber before and after desulfurization[244]
- Anion exchange properties of resins[245-247]
- Liquid interaction with sulfonated and unsulfonated ion exchange resins[248] and with partially metalated polyhydroxystyrenes[249]
- Study of polymer-cosolvent systems by light scattering, ultracentrifugation, and viscosity[250-252]
- Phase equilibria[253,254,719] and swelling[205,206,255,697] in polymer-cosolvent systems
- Washing of propylene film copolymer[256]
- Solution properties of polymer blends[257-259]
- Hansen parameters of synthetic resins and elastomers in the paint and plastics industry[260,261,696,698] including asymmetric membrane preparation and characterization[262-267]
- Organic liquids in textile processing[268] and dyeing[269] (see also Section 17.8)
- Interaction of *Halobacterium halobium* purple membrane with liquids[270,271]
- Light chlorofluorocarbons as solvents for film-forming polymers[272]
- Solubilities of gases in silicones[273] and other polymers[237,274,275]
- Swelling of polyethylene and fluorinated hydrocarbon polymers by trifluorostyrene[751]
- Propylene-based glycol ethers and acetates as solvents in paint stripping and polyurethane foam clean-up[759]
- Effect of casting solvent on polymer properties[38,276,277,713,724,739,750]
- Choice of release agent for polyurethane molding[708]
- Polymer solvent selectivity by gas-liquid chromatography[278] and solvent effects in gel chromatography[279,280] (see also Chapter 18)
- Solubility and swelling (Section 14.3) and viscosity (Section 14.7) of homopolymers and copolymers in organic liquids, including application to lithography(Section 16.5) and photography[758]
- Interactions of liquids with cellulose, lignin, coal, and bitumen (Section 16.11)

- Absorption of blood lipids and other effects of the aqueous physiological environment on biomaterials (Chapter 19) such as hydrogels[773]

An important area of polymer-liquid technological study is concerned with polymeric coatings. Here the nature of the effects associated with the loss of solvent during drying and the long-term resistance to subsequent attack are just as important as the correct formulation for preparation and application. Thus, in water-borne coatings

- The polymer must be soluble in the organic component
- The organic liquid must be miscible with water
- Evaporation of the water-organic blend must be such that two-phase behavior does not occur during drying
- Additional water should not cause phase separation during clean-up

so a detailed knowledge of polymer-liquid interactions is required. Inverse gas chromatography (Section 18.2) can provide information on the situation as the final traces of solvent leave a drying polymer film, yielding infinite dilution activity coefficients or polymer-liquid interaction parameters which can be expressed in terms of cohesion parameters (Section 13.5). ''Good'' solvents favor polymer molecules being left in an extended form, while ''poor'' solvents encourage their coiling. An example is the study reported by Alessi et al.[774] which uses fractional cohesion parameters (Sections 5.10 and 14.5). Other aspects of polymeric coatings studied with the aid of cohesion parameters include:

- Recovery of poly(vinyl chloride) coatings from scrap[241]
- Selection of solvents for coatings, including computer modeling[281-288,727]
- Effect of solvent during amine-cured epoxy film formation[276]
- Design of high performance coatings[289]
- Water-soluble polymers[290-291]
- Solvent system[292] and performance[293,294] of protective epoxy coatings
- Aqueous polymer emulsion formulation[295-298]
- Water-dispersed and water-soluble polymers for coatings[299-301]
- Solvent loss and paint film appearance[302,303]
- Resin-solvent interactions in paint films[304,305,696]
- Solvent effects in spin coating[204]
- Polyamide resin solubility[306,307]
- High solids coatings[308]

16.7 PERMEATION

Permeation rates of gases, liquids, and solutes depend not only on the bulk solubilities and surface adsorption (and therefore on relative cohesion parameter values) but also on diffusion coefficients. In such processes the *thermodynamics* of the interactions determine the equilibrium distribution of the lower molecular weight materials in the polymers, but the *rates* of approach to equilibrium are governed by diffusion rates, and under normal conditions equilibrium or steady state situations may take a very long time (or a high temprature) to attain.

Correlation and prediction methods may be based on polymer-liquid interaction parameters (Chapter 13) and Fick's diffusion equations, combined with cohesion parameters and empirical relations.* Hansen solubility diagrams and similar methods have been widely used

* For example: Einax et al.;[309] Gul';[310] Kadir, Stempler, Liron, and Cohen;[722] König and Schuch;[311] Lloyd;[265] Lonsdale;[312] Mangaraj;[313] Michaels et al.;[314] Mulder;[315] Paul;[316] Plucinski and Szust;[761] Schneier;[317] Senich;[318] and van Krevelen.[140]

in the choice of polymer materials (if organic solutes lie outside the Hansen solubility region they tend not to be permeable) and of solvents for membrane preparation.*

Permeation is important in situations such as the transfer of mobile components between foods and their polymer packaging. Antioxidants, plasticizers, processing aids and residues from the polymerization process (monomers, oligomers, initiators, catalysts) can migrate into the products or between polymers.[102,318,327,328,702] Another situation is the deliberately engineered slow release of materials from polymers.[329,330] Organic liquids can be used to simulate food products, allowing application of higher temperatures and techniques such as inverse gas chromatography[318] (Chapter 17) in the testing processes.

Permeation in protective materials (such as gloves) must be assessed, and cohesion parameters (as well as other group contribution methods)[754] have been used extensively for this purpose.** The selective permeability of membranes can be applied to the reverse osmosis separation of liquids,[265,336] with asymmetric films having received particular attention.[262-265,337] A development of the permeation technique is pervaporation, where components of liquid mixtures are separated by partial vaporization through a non-porous permselective film.[309,315,338-340,728,730]

Such separation processes incorporating use of cohesion parameter principles have been applied to:

- Ions in ion-exchange membranes and ion-selective electrodes[343-347]
- Organonitrogen compounds[348]
- Water-alcohol[340-342,753,767] and other organic-water[349,350,729,767] systems
- Hydrocarbon-alcohol systems[351,352]
- Isomeric xylenes[322] and other hydrocarbons[128,326,730]
- Other organic liquids[319,352]

Membranes are also used in gas-phase separations:[90,353-357]

- Hydrocarbons[358,359]
- Ethylene oxide[275]
- Noble gases[316,360]
- Oxygen and nitrogen[361,362]

Membranes may be liquid films;[363] reference may be made to Chapter 11 for liquid-gas solubility correlations in terms of cohesion parameters.

Permeability is also important in adhesive interactions (Section 17.5) and dyeing (Section 17.8) of polymers,[364] these processes being strongly influenced by the depths of diffusion of the various components, as illustrated by studies of poly(ethylene terephthalate) films.[269,365] Other applications include:

- Retention of lipids in tissue[366]
- Skin permeation (Chapter 19)
- Drug transport[314,704] (Chapter 19)
- Preparation and characterization of cellulose-derived membranes (Section 16.11), including cellulose esters[322,352,367]
- Microencapsulation[368,724]

* Examples are provided in the work of Belfort et al.;[767] Benecki;[128] Cabasso;[319] Christensen and Banke;[320] Henriksen;[321] Mulder;[322] Perkins et al.;[323,324] Perepechkin;[325] Sferrazza and Gooding;[728] and Vahdat.[326]

** For example: Bentz and Billing;[709,710] Christensen and Banke;[320] Derrickson et al.;[331] Forsberg and Faniadis;[332] Hansen and Hansen;[711] Henriksen;[321] Mangaraj;[313] Mikatavage et al.;[333] Perkins et al.;[323,324] Spence;[334] Sumitomo;[335] and Vahdat.[326]

Materials considered from the point of view of cohesion parameters in permeation studies include:

- Polypropylene[369]
- Polypropylene—poly(ethylene terephthalate) mixture[370]
- Polyethylene[328,355,359,371,372]
- Ethylene-propylene copolymer[373]
- Nitrile[333,374]
- Polyamides[335,375-379]
- Poly(vinyl chloride)[329,330,730,753]
- Polystyrene[380,753]
- Fluoropolymers and fluorosulfonate polymers[324,333,381-383]
- Poly(methyl methacrylate)[362,380]
- Poly(ethylene terephthalate)[124,269,361,365]
- Chlorinated poly(vinyltrimethylsilane)[354]
- Other polymers and polymer blends[353,370,384,729]

16.8 INTERACTIONS WITH LOW MOLECULAR WEIGHT SOLIDS

Often solids are incorporated into polymers as plasticizers, antioxidants, fillers, or pigments, and the extent of interaction between the components may be estimated by means of cohesion parameters. In the case of miscibility, allowance for the enthalpy of sublimation of crystalline solids should be made as explained in Chapter 12, but in practice quantitative studies are rare and the simple treatments already described for polymer-liquid systems are usually adequate for qualitative or semi-quantitative correlations. In any case, these are frequently not homogeneous systems. Examples of applications of cohesion parameters to polymer-solid systems are:

- Solubility of sulfur in synthetic rubbers[385]
- Poly(vinyl chloride)-plasticizer systems[6,386-392]
- Characterization of plasticizers and of plasticizer-resin and solvent-resin interactions by chromatography (Chapter 18)
- Mechanical properties and glass transition temperatures of polymer-plasticizer systems[96,97,183,393]
- Adsorption on polymer fillers such as carbon black and microbarite (Section 17.6), including rubber-carbon black interactions,[394] hardened epoxy resins[395] and acid-base interactions[396]
- Pigment-resin interactions[283,397,398] (see also Section 17.8)
- Polymer-plasticizer-solvent interactions in scrap poly(vinyl chloride) recovery[241]
- Organization of macromolecules on inorganic substrates[399]
- Interaction between poly(methyl methacrylate) and silane-treated glass flakes[362]
- Polymer adhesives (Section 17.5)
- Phenolic[400] and hydroxybenzophenone[401] antioxidants in polypropylene and polyethylene

16.9 POLYMER-POLYMER INTERACTIONS

The increasing number of reviews on the subject of polymer blends or alloys* indicates the importance of blending in providing a material with adequate performance at minimum cost, or in achieving a combination of physical properties not found in a single polymer. Polyblends of lower-priced polystyrene or poly(vinyl chloride) with high-performance poly(phenylene oxide) or polycarbonate provide products intermediate in price and performance, while a dispersion of rubber particles in brittle polystyrene or styrene-acrylonitrile polymer gives both high impact resistance and high tensile strength.

Polymers must be ''compatible'' if there is not to be a loss of these desirable properties on blending. Polymer-polymer immiscibility and near-miscibility (microheterogeneous systems) occur very commonly and have been studied widely.[402,403,428-433] Phase separation can occur even on the microdomain scale between the backbone and the grafted side-chains of a graft copolymer or in a block copolymer. For example, a natural rubber-methyl methacrylate graft can exist either as a continuous rubber phase (imparting rubbery properties) in which the side-chains form a dispersed glassy phase, or as a poly(methyl methacrylate) glassy continuous phase with a dispersed rubber phase (having glass-like properties). Despite the high viscosities and long chain lengths in polymer-polymer systems, and the fact that equilibrium may not be achieved in practical situations, the thermodynamics of polymer miscibility are of great importance.[1,58,405,422,430,434-440,757] The entropy of mxing term is rarely sufficiently favorable to overcome an endothermic enthalpy of mixing of polymers.

It should not be assumed, however, that ''compatibility'' is synonymous with ''miscibility.'' Often, true miscibility in polymer blends is not necessary, but good adhesion between the dispersed components is necessary. Just as compatibility is an inadequate term for describing polymer-liquid interactions (Section 16.6), it is also an unfortunate choice when discussing polymer blends. Sometimes, in fact, complete miscibility is undesirable, as in the modification of a glassy polymer by a rubbery component, although adhesion between the phases is still important. It appears, however, that in the case of polymer blends there is no satisfactory alternative to the general term compatibility with its diverse meanings and implications, and it is used here to denote blends which are either homogeneous on a molecular level, or which are microheterogeneous with a satisfactory level of adhesion or cohesion.

It should also be remembered that useful blends can be produced from polymers that are normally classed as incompatible, if the ingredients can be mixed on a fine scale initially(in a melt, for example) and then prevented from demixing. It is possible to modify the mechanical properties of an otherwise incompatible mixture by the addition of a ''compatibilizing agent'', often a block or graft copolymer,[442-444] or by covulcanization.[445,446] In particular, when plastics are toughened by the incorporation of a rubber in finely dispersed (but not molecularly dispersed) form the materials must be incompatible, but not too incompatible, and cohesion parameters provide a convenient measure of the degree of compatibility.[88] Thus, polystyrene (Hildebrand parameter 18.5 $MPa^{1/2}$) may be compounded with styrene-butadiene rubber (16.5 $MPa^{1/2}$), polyisobutylene (16.5 $MPa^{1/2}$) or polyisoprene (17.0 $MPa^{1/2}$).

Another aspect of polymer-polymer interactions is the selection of release agents (natural and synthetic waxes) for molding of polymers such as polyurethane.[708] An area of recent activity is the synthesis of interpenetrating polymer networks (IPNs), with tangling on a

* For example: Bohn;[402-404] Casper and Morbitzer;[405] Cooper and co-workers;[406,407] Corish;[408,409] Gaylord;[747] Gesner;[410] Heidel and Stürzenhofecker;[411] Koningsveld, Kleintjens, and Nies;[757] Sanchez;[422] Schmitt and Kirste;[423,424] Shvarts;[425] Slepakova, Geller et al.,;[426] and Tlusta and Zelinger,[427] Krause;[412,413] Olabisi;[414] Paul and co-workers;[415-418] Platzer;[419,420] Roover;[421] Sanchez;[422] Schmitt and Kirste;[423,424] Shvartz;[425] Slepakova, Geller et al;[426] and Tlusta and Zelinger.[427]

molecular scale of polymers (which would otherwise be incompatible) resulting from mixing at lower molecular weight levels, then polymerizing and cross-linking.[447-452] Compositional quenching (a new process in which two or more polymers are dissolved in a common solvent which is then flashed off) involves knowledge of polymer-liquid interaction parameters and cohesion parameters.[701]

The criteria for compatibility include the existence of only one glass transition temperature in the blend, the observation of lower critical solution temperature behavior[453] (a clear blend becoming cloudy on heating: Section 13.4), and the clarity under magnification of a film cast from homogeneous melt. Incipient phase separation has been studied by such techniques as observing dispersions of individual and aggregated "tagged" molecules.[454]

Various methods[746] may be used in the description and discussion of polymer-polymer phenomena, but Hildebrand parameters provide one of the simplest guides to compatibility, compatible blends of polymers not usually showing a difference greater than 2 or 3 $MPa^{1/2}$ in δ values.[402-404,408,426] Put another way, for mixtures of two polymers to form compatible blends the interaction parameter χ must be very small, and the enthalpy of mixing must be close to zero or negative (compare polymer-liquid interaction parameters, Section 13.5),

$$^{ij}\chi = {}^{u}V\,({}^{i}\delta - {}^{j}\delta)^2/(RT)$$

where ${}^{u}V$ may be taken as the smallest repeat unit volume. Polymer-polymer immiscibility is the rule rather than the exception, and in practice it is necessary to find pairs of materials between which there are specific interactions leading to negative enthalpies of mixing, and component cohesion parameters such as the Hansen δ_h are important in preliminary screening for compatible polymer pairs.

Two-dimensional maps or three-dimensional models can be used, indicating the location of polymers in Hansen parameter space,[455] or cohesion parameters may be combined with other measures of polarity or hydrogen bonding, as illustrated in Chapter 15. Figure 6 shows δ_p and δ_d points for some common homopolymers. More often, in the case of polymers, ranges of cohesion parameters are quoted, and one of the most convenient ways of doing this is to describe the Hansen volume of interaction, i.e. the set of coordinates of the center and the radius of the interaction sphere or ellipsoid (Section 14.5).[456] Figure 7 illustrates the extent of overlap between a poly(phenylene oxide phosphonate ester) and cellulose acetate, a compatible polymer couple.

Expressions based on Hildebrand parameters may be used for polymer-polymer systems to predict thermodynamic functions such as Gibbs free energy of mixing, activity, and chemical potential, just as described in earlier chapters for liquid-liquid and polymer-liquid systems, for example

$$\Delta_m G = RT[{}^{i}x \ln {}^{i}\phi + {}^{j}x \ln {}^{j}\phi + ({}^{i}\delta - {}^{j}\delta)^2\, {}^{i}\phi\, {}^{j}\phi] \tag{4}$$

and the differential energy expression for polymer i in phases 1 and 2 is

$$\begin{aligned} &\ln {}^{i}\phi_1 + (1 - {}^{i}V/{}^{j}V)\,{}^{j}\phi_1 + {}^{i}V({}^{i}\delta - {}^{j}\delta)^2\, {}^{j}\phi_1^2/(\mathrm{RT}) \\ &= \ln {}^{i}\phi_2 + (1 - {}^{i}V/{}^{j}V)\,{}^{j}\phi_2 + {}^{i}V({}^{i}\delta - {}^{j}\delta)^2\, {}^{j}\phi_2^2/(RT) \end{aligned} \tag{5}$$

The first and second terms in each expression are regular mixing entropies, and the third is the interaction enthalpy term.

The problem with the *quantitative* application of Hildebrand parameters (and also Hansen parameters) is that it is not possible to predict negative enthalpies of mixing, which appear to be necessary in polymer blends to achieve miscibility. It is only by extending the formalism

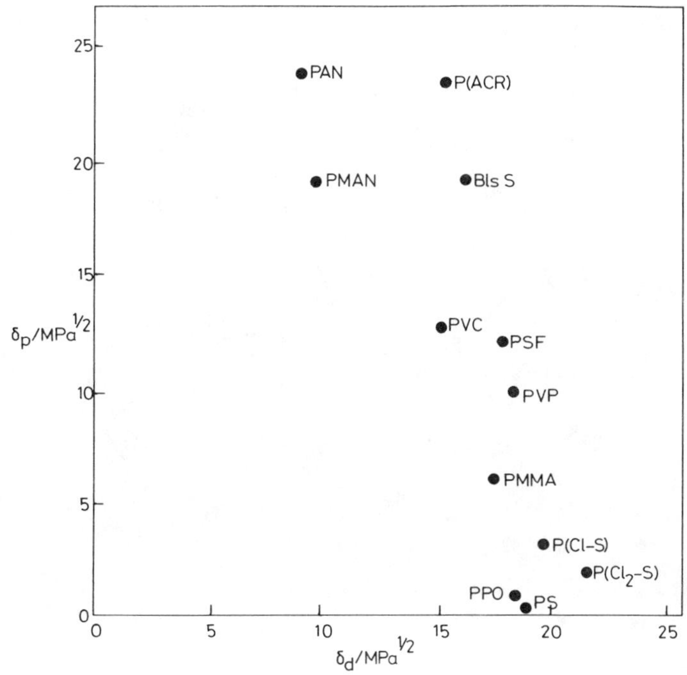

FIGURE 6. Calculated Hansen parameters for some homopolymers plotted on a polar parameter-dispersion parameter map: PAN, polyacrylonitrile; P(ACR), polyacrylamide; PMAN, poly(methacrylonitrile); BisS, poly(oxy-1,4-phenylenesulfonyl-1,4-phenylene); PVC, poly(vinyl chloride); PSF, polysulfone; PVP, poly(2-vinylpyridine); PMMA, poly(methyl methacrylate); P(C1-S), poly(chlorostyrene); P($C1_2$-S), poly(dichlorostyrene); PPO, poly(oxy(2,6-dimethyl-1,4-phenylene)); PS, polystyrene. (Adapted from Shaw, M. T., *J. Appl. Polym. Sci.*, 18, 449, 1974.)

to component cohesion parameters with separate Lewis acid and base terms that it is possible to achieve negative enthalpies of mixing. This involves replacing $({}^i\delta - {}^j\delta)^2$ with ${}^{ij}A$, where (Equation 22, Chapter 5),

$$
\begin{aligned}
{}^{ij}A = {}^i\delta_t^2 + {}^j\delta_t^2 - 2\,{}^i\delta_d\,{}^j\delta_d - 2\,{}^i\delta_o\,{}^i\delta_o - 2\,{}^i\delta_i\,{}^j\delta_d \\
- 2\,{}^j\delta_i\,{}^i\delta_d - 2\,{}^i\delta_a\,{}^j\delta_b - 2\,{}^j\delta_a\,{}^i\delta_b \qquad (6)
\end{aligned}
$$

For polymer-polymer systems, ${}^{ij}A$ is equivalent to Roe's[457,458] interaction parameter Λ (Section 13.3), which can be evaluated and used in a manner equivalent to the polymer-liquid interaction parameter χ. The extended cohesion parameter technique has not yet been applied to polymer blends to any great extent, but there is no reason to expect that it would not be successful. In the meantime, qualitative predictions based on Hansen parameters are still valuable.

Another important group of interactions involves polymer-polymer-liquid systems, such as a polymer alloy used as a membrane for the separation of organic liquid mixtures, or attempts to achieve homogeneous solutions of mixed polymers. Here the plotting of Hansen volumes of interaction for each polymer makes a good starting point.[456,459] A particularly significant application is the solvent separation of mixed plastic scrap prior to recycling.[241,460-463] It would be possible to separate polymers (both homopolymers and copolymers) if there were miscibility at higher temperatures but phase separation at lower temperatures.

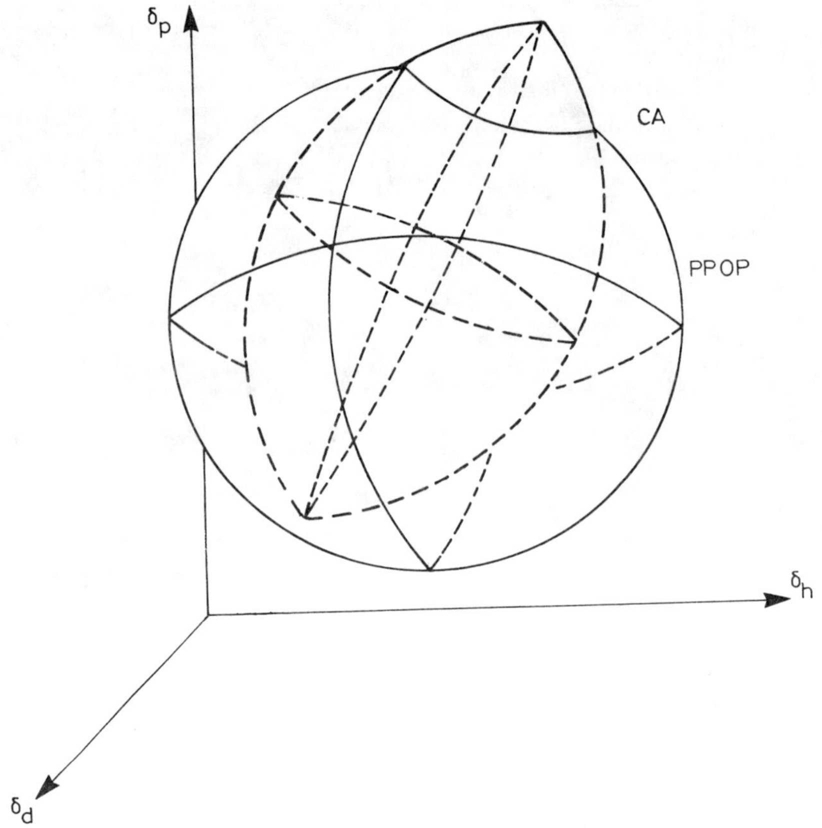

FIGURE 7. Hansen parameter three-dimensional diagram showing substantial overlap between a poly(phenylene oxide phosphonate ester) (PPOP) and cellulose acetate (CA). (Adapted from Cabasso, I., Jagur-Grodzinski, J., and Vofsi, D., in *Polymer Science and Technology*, Vol. 10, *Polymer Alloys: Blends, Blocks, Grafts, and Interpenetrating Networks*, Klempner, D. and Frisch, K. C., Eds., Plenum Press, New York, 1977, 1.)

However, unexpected phenomena may occur. For example, selective specific interactions between the liquid and one member of a pair of compatible polymers may cause phase separation even though the polymers are individually miscible with the lquid.[464] Hildebrand parameters may be used to predict phase diagrams, or simply to justify the neglect of the possibility of phase separation during liquid-induced crystallization of a compatible polymer blend.[465-467]

Polymer compatibility can be investigated by:

- Chromatography (Chapter 18), for example molten polymer miscibility by GLC[468] or molecular probes of polymer blends by inverse gas chromatography[748,768]
- Morphological, thermogravimetric and mechanical studies[405,469-471,765,766]
- Melt viscosity[472]
- Neutron scattering[473-475]
- Excimer fluorescence and infrared spectroscopy[470,476-481,752]

There are many examples of applications of cohesion parameters to polymer-polymer compatibilities:

- Propylene-ethylene polymer blends,[482] including high-impact blends,[483] propylene-isobutylene and ethylene-propylene-diene polymers[484]

- Methacrylates and other acrylic blends[353,470-472,479,480,485-491,763] and dispersions of poly(2-vinylnaphthalene)[752]
- Styrene and substituted styrene polymer blends[444,457,458,492,493,760,765]
- Ethylene-vinyl chloride[494], ethylene-vinyl acetate copolymer blends,[353,465] and ethylene-polyamide 6,6 mixtures[703]
- Blends with epoxies[488,495-497,766] and poly(ethylene oxide)-poly(vinyl acetate) blends[718]
- Poly(vinyl chloride) blends[37,353,432,470,489,498,745,772]
- Polystyrene-poly(vinyl methyl ether) blends[28,464,499,735]
- Polystyrene-acrylic blends[471,472,500,501,738]
- Poly(ethylene terephthalate) blends[502,503]
- Polybutadiene blends[505,506,765]
- Polyϵ-caprolactone blends[505,506,748]
- Other rubbers[507] and coatings[508] and tackifier resins[723]
- Blends with rosin polyamideimide[509,510] and other polyimides[772]
- Cellulose,[511] cellulose derivatives[512,513] and lignin derivatives[777]
- Adhesion at rubber-resin[514] and rubber-tire cord[515] interfaces (see also Section 17.5)
- Polyacrylate textile pigment adhesion[485]
- Polymer blends as high explosive binders[516]
- Polymer-bitumen blends (Section 16.11)
- Behavior of sheet molding compounds[725]
- Polymeric surfactants as compatibilizing agents[747]
- Compatibility in pressure-sensitive adhesives[772]

16.10 COPOLYMERS

The combination of properties achieved by polymer blending sometimes can be obtained more satisfactorily by forming a copolymer (random, block, or graft) (Figure 8).* Random copolymers can be considered as polymer mixtures made up of constituent molecules of different types organized in a random way, but block and graft copolymers with sufficiently long sequences of each type of repeating unit are in many ways analogous to polymer blends, and the components can be either compatible or incompatible in the sense described in Section 16.9.[402,403,424,431,521] Phase separation into microdomains (the limit of separation possible because the sections are linked by intramolecular chemical bonds) in block copolymers is not expected to occur as readily as phase separation in a mixture of the corresponding homopolymers,[492] but the same cohesion factors are involved. It is clear that cohesion parameters are just as important in the design and in the property correlations of copolymers as they are in those of polymer solutions and polymer blends. Reports on the processes of copolymerization, grafting, and crosslinking from a cohesion parameter viewpoint are collected in Section 10.4, and viscosities of copolymer solutions are included in Section 14.7. The structures of macroporous networks, which depend on solvent-polymer interactions during copolymerization,[744] are important in many applications of copolymers.

The solubility behavior of a random copolymer is in general intermediate between the solubility properties of the homopolymer constituents, sometimes enabling easier solvent formulation,[522] or providing "internal plasticization" (Section 16.3).[523] However, incompatible components of block and graft copolymers may be so different in their cohesion properties that a common solvent cannot be found and emulsions or colloidal sols may be formed, resulting in modification of all solution physical properties such as viscosity and thermodynamic functions. The larger value of the entropic part of the polymer-liquid interaction parameter, χ_S, in some block copolymers is an indication of repulsive interactions

* Reviews on this subject include those of: Battaerd and Tregear;[429] Burke and Weiss;[517] Danon *et al.*;[518] Helfand;[431] Lee;[514] Platzer;[420] Riess and Callot;[519] Schmitt;[424] and Shibuya et al.[520]

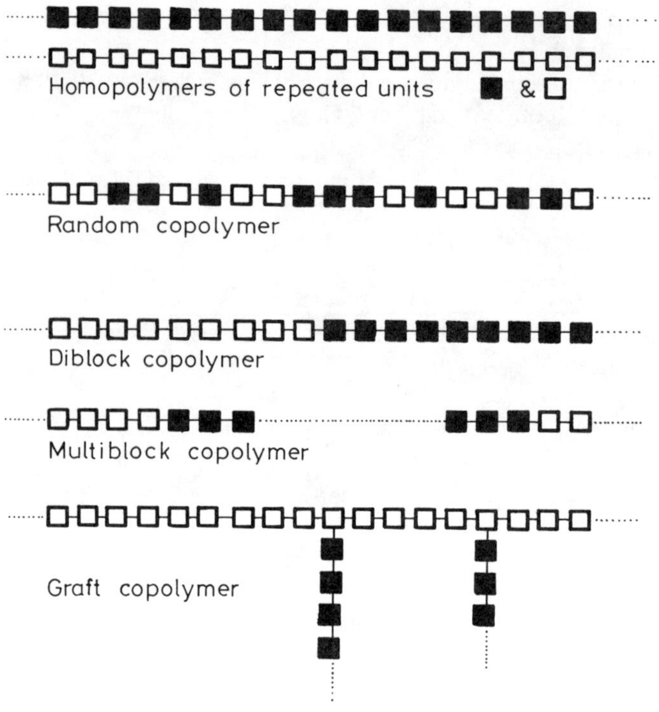

FIGURE 8. Schematic representation of homopolymer blends and types of copolymer. (Adapted from Schmitt, B. J., *Angew. Chem.*, 91, 286, 1979; *Angew. Chem. Int. Ed.*, 18, 273, 1979.

causing intramolecular phase separation. There is a corresponding increase in the enthalpic part, χ_H, due to the incompatibility, and this can be quantified in terms of Hildebrand parameters,[257,524] or more often in terms of component cohesion parameters.[20] Mikos and Peppas[773] have demonstrated the determination of polymer-liquid interaction parameters for hydrophilic copolymers with water, in particular the biomedical hydrogel poly(2-hydroxyethyl methacrylate-*co*-methacrylic acid).

If a copolymer containing both rubbery and glassy material is in a liquid with a cohesion parameter similar to that of the rubber, those parts will be extended and the glassy parts will be tightly coiled so the product will have the properties of a filled rubber. Alternatively, in a liquid which interacts more strongly with the glassy portions, the resulting copolymer will act as a rubber-modified glass.

Cohesion parameters may be determined experimentally for copolymers just as for homopolymers (Chapter 14). Hildebrand parameters for some copolymers are included in Table 3, Chapter 14, and a more complete compilation appears in the *CRC Handbook of Polymer-Liquid Interaction Parameters and Solubility Parameters.*[598] The Hildebrand parameters for random copolymers and some nonrandom copolymers can be calculated with some confidence from data available for homopolymers.[525-528,773] It would appear that this could be done on the assumption that the Hildebrand parameter of a mixture is volume-wise proportional to the Hildebrand parameters of the components in terms of the homopolymer volume fractions (Equation 32, Chapter 4), but an alternative equation has been used[526] which does not assume that

$$^{ij}V = {}^{j}V\,{}^{j}n + {}^{i}V\,{}^{i}n$$

and leads to

$$^{ij}\delta = \frac{(^in\ ^iM^2/^i\rho)^i\delta + (^jn\ ^jM^2/^j\rho)^j\delta}{[^in\ ^iM + ^jn\ ^jM][(^in\ ^iM^2/^i\rho) + (^jn\ ^jM^2/^j\rho)]^{1/2}\ ^{ij}\rho^{-1/2}} \tag{7}$$

where n is the amount (number of moles) and M is the monomer molecular weight, i and j referring to the components and ij to the copolymer. Alternatively, group molar attraction constants (Sections 6.4 and 14.8) may be used, or correlations developed with physical properties.[289,529-532]

Block copolymrs frequently show two clear-cut peaks when the solubility or extent of swelling by liquids is plotted as a function of liquid Hildebrand parameter, reminiscent of the behavior of homopolymers containing functional groups with very different cohesive properties such as poly(ethylene terephthalate) (Table 1, Chapter 15), although not necessarily at the values corresponding exactly to the homopolymers. For example,[223] butadiene-acrylonitrile copolymer shows a polyacrylonitrile peak at 28 $MPa^{1/2}$ and another peak at 19 $MPa^{1/2}$ which corresponds to a copolymer of about 85% butadiene, 15% acrylonitrile. Such shifting of Hildebrand parameter maxima on copolymerization can be used to provide resistance to environmental stress crazing and cracking (Section 16.4), as in styrene-acrylonitrile, for example.[52,143]

Hansen parameters for copolymers have been determined,[533,534,598] and some are included in Tables 8, 15, and 20, Chapter 14.

Studies of copolymer properties of particular relevance because of the use of cohesion parameters include:

- Polyacrylonitrile and its copolymers[43,353,489,535,536]
- Use of PolySAC (polyfunctional surface-active comonomer) in adhesive bonding to hard tooth tissue[537]
- Physical properties and microstructure of ethylene-propylene copolymers and terpolymers[373,538,539]
- Interactions of vinyl acetate-vinyl chloride,[540-542] vinyl acetate-ethylene,[353,543-546] and ethylene-acrylic acid[547] copolymers
- Comparison of Neoprene® W graft copolymers having flexible and rigid polyether branches[548]
- Triblock copolymers, such as styrene-butadiene-styrene[23,549-556]
- Copolymers based on phenylene oxide and *m*-phenylene isophthalamide[133]
- Copolymers and terpolymers of styrene with acrylate, methacrylate and acrylonitrile[283,550,557-560]
- Other styrene copolymers[36,142,246,549,555,561-565,726,738,740,744]
- Butadiene copolymer[566]
- Hydrogenated star block copolymers[726]
- Hydrophilic copolymers (hydrogels)[773]
- Siloxanes[33,34,567-569,732,740]
- Polyurethanes[20,568]
- Effect of copolymer composition on glass transition temperature[691]
- Internal pressures of block copolymers[570]
- Polysulfone-polycarbonate block copolymers[571]
- Miscibilities in blends of poly(acrylonitrile-*co*-styrene) with styrene-maleic anhydride and styrene-*N*-phenylmaleimide copolymers have been discussed[693,733] in terms of Hildebrand parameters.

16.11 NATURAL POLYMERS

Cellulose is one of the materials whose interactions should be described in terms of a full set of cohesion parameters, with both Lewis acid and base terms, as cellulose can act as proton donor, proton acceptor, or both.[572,573,720] This often has not been done in the past, and on its own the Hildebrand parameter is no more than a convenient basis for a first comparison of properties.[574,575] In some cases the hydrogen-bonding parameter γ or the Hansen δ_h parameter has been used as well, and this improves the correlation. Larsson and Johns[720] reported that the liquid Hildebrand parameters and all Hansen parameters except δ_d showed some degree of correlation with tensile energy absorption by paper, but the donor number was superior. Kesselring et al.[576] determined for microcrystalline cellulose from adsorption measurements by gas-solid chromatography the Hansen parameters:

$$\delta_d = 19 \pm 1 \text{ MPa}^{1/2}$$

$$\delta_p = 13 \pm 2 \text{ MPa}^{1/2}$$

$$\delta_h = 31 \pm 1 \text{ MPa}^{1/2}$$

$$\delta_t = 39 + 1 \text{ MPa}^{1/2}$$

Calculated Hansen parameters of cellulose used in chromatographic columns[534] are included in Table 20, Chapter 14. Dambis et al.[690] calculated $\delta_d = 15$, $\delta_t = 27$ MPa$^{1/2}$. Methocel HG90 was reported [695] to have a Hildebrand parameter of 36 MPa$^{1/2}$. Properties to have been studied with the aid of cohesion parameters include:

- Mechanical properties of paper in liquids[574,577]
- Sorption of polymers[511] and liquids[769] by cellulose
- Cellulose accessibility by the thallous ethylate method[578]
- Solubility of cellulose[572,573,580-584]
- Mechanochemical properties of silk[110,111,585]
- Swelling of cotton and other cellulose materials[574,575,588-590]
- Plasticization of cellulose with water[106,107] and other cellulose wetting and plasticizing properties[75,108,591,592]
- Effects of liquids on the reactions of silk and other forms of cellulose with epoxy compounds[594-596] and on the acetylation of viscose rayon[597]
- Polyacrylonitrile textile pigment binder adhesion[485]

Cellulose derivatives have also been studied in particular with the Hildebrand parameter combined with a hydrogen-bonding parameter and dipole moment, and with Hansen δ_p and δ_h parameters. Values of the interaction parameter χ of cellulose derivatives with a variety of liquids are included in the *CRC Handbook of Polymer-Liquid Interaction Parameters and Solubility Parameters*.[598] Most of the studies to have utilized cohesion parameters have been concerned with commercial membranes:

- Interactions of cellulose derivatives with liquids[599]
- Solubility of cellulose acetate[600,601]
- Physicochemical properties of cellulose acetate and amylose acetate[602] and of ethyl cellulose[603]
- Cellulose acetate, ethyl cellulose, and cellulose acetate phthalate reverse osmosis and dialysis membranes for rejection of organics from aqueous solution[349,604-618,729] (Figures 9 and 10, Chapter 15, may be compared with Figure 2, Chapter 8, to indicate the extent of rejection expected for different classes of organic compounds; δ_d and δ_h for these materials are listed in Tables 18 and 20, Chapter 14)
- Three-dimensional models and contour maps[619,620] with Hildebrand parameters, dipole moments, and hydrogen-bonding parameters for cellulose nitrate (Figures 34 and 35,

Chapter 15), cellulose acetate (Figures 36 and 37, Chapter 15), and cellulose acetate butyrate (Figures 38 and 39, Chapter 15)

- Triangular representation of fractional cohesion parameters for triacetate and diacetate[621] (Figures 24 and 25, Chapter 15)
- Dependence of Hildebrand parameter on degree of substitution in the interesterification of starch with methyl palmitate[622]
- Polymer-liquid interaction parameter, χ, for cellulose acetate-acetone, cellulose acetate-methyl acetate, and cellulose nitrate-acetone[253] (Figure 2, Chapter 13)
- Viscosity, polymer-liquid interaction parameters and Hildebrand parameter (25 $MPa^{1/2}$) for cellulose tricarbanilate in dilute solution,[623] and the thermodynamics of cellulose tricarbanilate and poly(vinyl acetate) solutions[624]
- Viscosity, surface free energy, and Hildebrand parameter (22 $MPa^{1/2}$) of *o*-hydroxypropyl cellulose[625]
- Viscosity of ethyl cellulose and hydroxypropyl methylcellulose[183]
- Compatibility in blends of cellulose derivatives[512,513]

Solvents for *chitin and chitosan* have been reviewed in terms of Hildebrand parameters.[626,627] For β-chitin from squid (*Loligo pealei*) the Hildebrand parameter range of effective solvents (formic acid-chlorinated liquid mixtures) was 23—26 $MPa^{1/2}$, compared with 25—26 $MPa^{1/2}$ for α-chitin. Effective solvents for chitin include some tertiary amides ($RCONR_2$) with 5-7% LiCl, notably *N,N*-dimethylacetamide, *N,N*-dimethylpropionamide, *N*-methyl-2-pyrrolidinone and 1,3-dimethyl-2-imidazolidinone.

Hansen parameters have been reported[712] for *keratin* from the swelling of skin scales from psoriasis patients (Table 12a, Chapter 5). The reaction of *wool keratin* with epoxides has been correlated with Hildebrand parameters,[628] γ radiation-induced grafting has been investigated,[629] and the effect of water on thermal transitions has been studied.[75]

Properties of resins such as *shellac*[630-632] derived from natural products are included elsewhere in Chapters 13-16. Properties of *vegetable oils* determined in supercritical fluids, including Hildebrand parameter differences, are discussed in Section 11.3 and for refined *lard* (a semisolid, melting at 42°C, and human depot *fat,* with approximately the same composition), Hansen parameter values[712] are included in Table 12a, Chapter 5.

Lignin is observed to have a Hildebrand parameter between 20 and 29 $MPa^{1/2}$, but polar and hydrogen bonding tends to dominate the interactions so an additional parameter is usually included.[633-641,690] Hansen parameter values[712] appear in Table 12a, Chapter 5. Relative permittivities and dipole moments have been used, and the polar-nonpolar cohesion parameter method has been employed[636] with estimated lignin values δ_λ = 18.6 $MPa^{1/2}$ and δ_τ = 10.7 $MPa^{1/2}$. It is also possible to use the triangular representation of fractional cohesion parameters for lignin (Figure 22, Chapter 15).[642] The extents of dissolution of what may be considered related to lignin decomposition products, *soil colloids, leonardite, humin-kerogen and sediments*, have also been correlated with cohesion parameters.[643-646,762,764,775] The swelling of *wood and lignocellulose* have been discussed in terms of Hildebrand parameters,[646,647,717,720] and Kelley et al.[109] investigated the effects of liquids of appropriate cohesive energy density on the glass transitions of wood. Group contribution methods have been employed[690] for evaluation of the Hildebrand parameters of lignin, hemicellulose and cellulose, used for discussion of the compatibilities of wood cell-wall biopolymers. The cohesive properties of polyurethanes based on chain-extended hydroxypropyl lignin have been investigated,[736] as well as of polymer blends with hydroxypropyl lignin.[777]

For *coal,* the set of atomic attraction constants in Table 8, Chapter 6, was devised for the estimation of Hildebrand parameters as a function of the rank or carbon content; values of δ calculated with these constants for liquids and polymers agree well with other determinations.[648] In calculating δ values for coal, use is made of: the atom ratio, ${}^z r$ (denoting

TABLE 2
Calculation of Hildebrand Parameter of Coalification Series

C, %	H_r	O_r	N_r	S_r	f_a max.	f_a min.	$^{C}M/\rho$ $cm^3\ mol^{-1}$	$\delta/MPa^{1/2}$ max.	$\delta/MPa^{1/2}$ min
70.5	0.862	0.247	0.012	0.002	0.70	0.60	12.0	31.1	29.1
75.5	0.789	0.181	0.012	0.002	0.78	0.63	11.4	29.1	27.1
81.5	0.753	0.108	0.013	0.002	0.83	0.68	11.2	26.0	24.1
85.0	0.757	0.071	0.013	0.002	0.85	0.72	11.0	24.3	23.1
87.0	0.732	0.051	0.013	0.002	0.86	0.75	10.8	23.5	22.3
89.0	0.689	0.034	0.013	0.002	0.88	0.79	10.4	23.0	21.8
90.0	0.655	0.027	0.013	0.002	0.90	0.80	10.1	23.1	21.8
91.2	0.594	0.021	0.011	0.002	0.92	0.83	9.7	23.1	21.8
92.5	0.509	0.016	0.013	0.002	0.96	0.86	9.3	23.3	21.8
93.4	0.440	0.013	0.011	0.003	0.99	0.88	8.9	23.5	21.8
94.2	0.379	0.011	0.009	0.002	1.00	0.91	8.4	23.5	22.1
95.0	0.307	0.009	0.010	0.003	1.00	0.93	8.0	23.5	22.2
96.0	0.223	0.007	0.006	0.002	1.00	0.96	7.3	23.7	22.9
(100.0)	0.000	0.000	0.000	0.000	1.00	1.00	5.4	25.2	25.2

Adapted from van Krevelen, D. W., *Fuel*, 44, 229, 1965.

H/C, O/C, N/C, and S/C for z = H, O, N, and S); the molar volume per carbon atom, ^{C}V; and the molar mass per carbon atom, ^{C}M:

$$\delta = {}^{C}V^{-1} \sum_{z} {}^{z}r\, {}^{z}F = ({}^{C}M/\rho) \sum_{z} {}^{z}r\, {}^{z}F \tag{8}$$

where

$$^{C}M = 12.00 + 1.008\ {}^{H}r + 16.00\ {}^{O}r + \ldots . \tag{9}$$

Also required is the aromaticity of the coal, f_a, which is the proportion of aromatic double bonds. Table 2 shows the composition of a coalification series and the resulting Hildebrand parameters for upper and lower limiting values of f_a, using

$$\sum_{z} {}^{z}r\, {}^{z}F/J^{1/2}\ cm^{3/2}\ mol^{-1} = 140\ {}^{H}r + 614\ {}^{O}r + 233\ {}^{N}r + 460\ {}^{S}r + 136 f_a \tag{10}$$

The average value of the oxygen contributions was thus taken as 614 $J^{1/2}\ cm^{3/2}\ mol^{-1}$; the contributions of the nonaromatic ring (123 $J^{1/2}\ cm^{3/2}\ mol^{-1}$) and of branching in the aliphatic chain (-92 $J^{1/2}\ cm^{3/2}\ mol^{-1}$) were assumed to balance each other; and the aromatic double bond contribution was evaluated from $(3 \times 272 f_a/6)$ $J^{1/2}\ cm^{3/2}\ mol^{-1}$. The result is shown in Figure 9, with a distinct minimum in δ at about 90% carbon content. Also marked is the observed limit of solubility: the increasing network formed as a result of coalification means that liquids such as pyridine and ethylenediamine do not dissolve coal completely, although their Hildebrand parameter values are comparable to those of high rank coals. Of course, this is only a first approximation to solubility prediction, and application of component cohesion parameters is possible (Section 6.8). A similar minimum as a function of rank for six German coals was observed by Hombach,[649] but the range of values was lower (22—23 $MPa^{1/2}$). Results from Hildebrand parameter calculations have been combined with direct swelling observations as well as on residues after extraction and the recovered extract.[650,741] Both simple plots of the volume ratio of swelling, Q, against the liquid Hildebrand parameter (solvent spectra) and the Bristow-Watson method (Equation 1, Chapter 14) yielded δ values

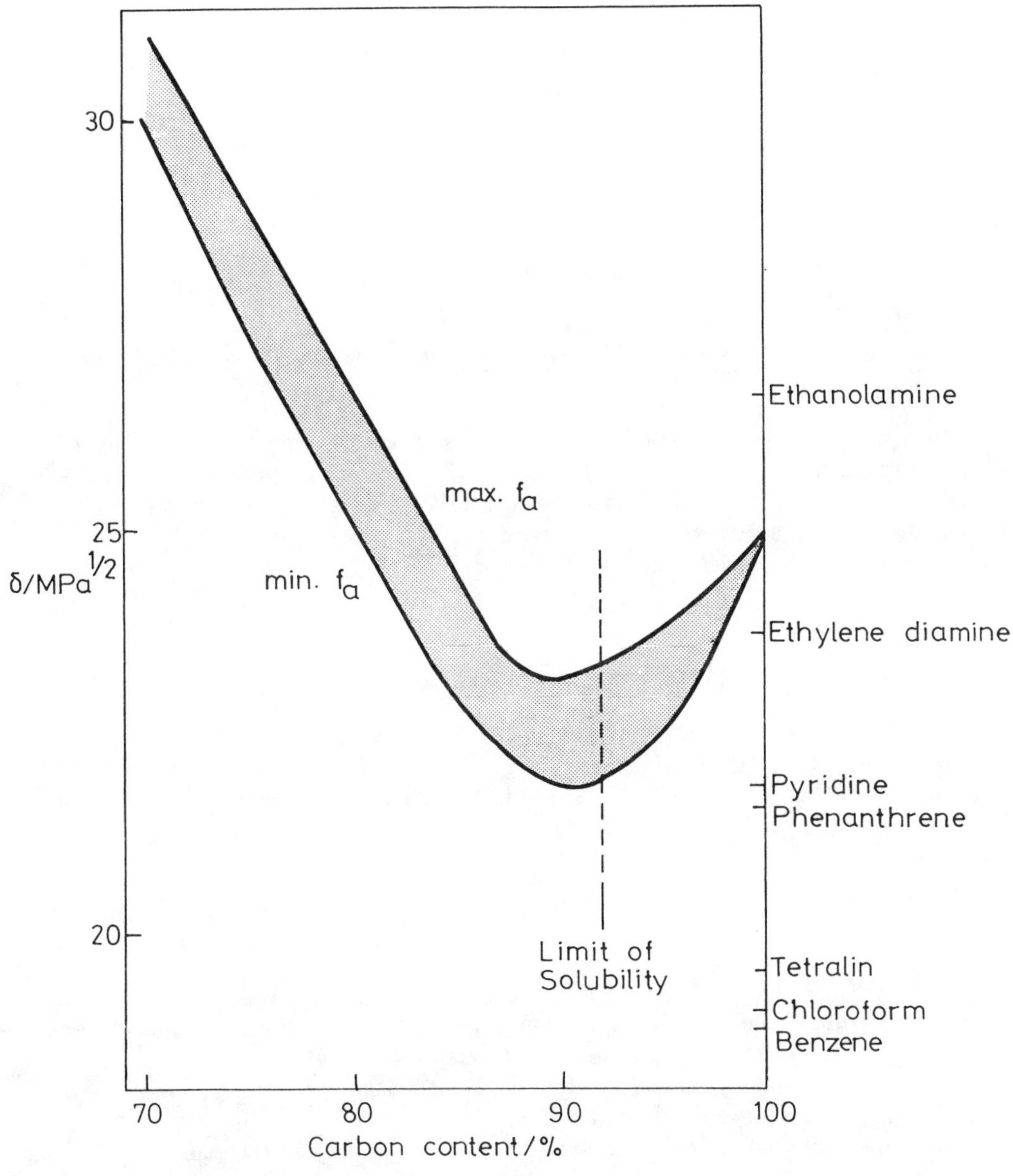

FIGURE 9. Hildebrand parameters of coals of varying ranks. (Adapted from van Krevelen, D. W., *Fuel*, 44, 229, 1965.)

which were approximately independent of rank, in contrast to the calculated values (Table 3).[650] Sanada and Honda[651,652] obtained similar Hildebrand parameter values (22 $MPa^{1/2}$) from the swelling of two Japanese coals and swelling results were interpreted in terms of Hildebrand parameters by Nelson.[653,654] Painter, Park, and Coleman[689] calculated Hildebrand parameters for a series of model coal structures, and showed that only the relative proportions of aliphatic and aromatic constituents and the fraction of oxygen-containing functional groups affected the Hildebrand parameter; the precise distribution of functional groups was unimportant. They then evaluated χ values from the equation (Section 13.5)

$$\chi = 0.3 + ({}^{i}V/RT)({}^{j}\delta - {}^{i}\delta)^2 \qquad (11)$$

TABLE 3
Hildebrand Parameters of Three Australian Coals

	δ/MPa$^{1/2}$		
	Hebe	**Greta**	**Bulli**
Experimental, Bristow-Watson plot			
Slope	22.1	21.1	22.3
Intercept	22.3	21.3	22.3
Experimental, Q vs δ plot maximum	22.1	22.1	22.7
Calculation, van Krevelen method	30.1	23.9	22.1

Adapted from Kirov, N.Y., O'Shea, J. M., and Sergeant, G. D., *Fuel*, 46, 415, 1967.

Hydrogen bonding was accommodated by separate Gibbs function terms. This approach predicted that non-crosslinked coal molecules of low molecular weight and low aromaticity would be completely soluble in pyridine. Peppas and Lucht[658] also evaluated polymer-liquid interaction parameters from Hildebrand parameters.

Hombach[655,656] used Hansen parameters

$$\delta_d = 16 \text{ MPa}^{1/2}$$

$$\delta_p = 11 \text{ MPa}^{1/2}$$

$$\delta_h = 8 \text{ MPa}^{1/2}$$

for a coal with 23.6% VM; Whitehurst et al.[657] reported a fractional Hansen parameter solubility region for SRC centered at

$$f_d = 0.54$$
$$f_p = 0.21$$
$$f_n = 0.25$$

Of the two maxima observed by Bodily, Wann and Kopp[692] as a function of liquid Hildebrand parameter for the swelling of bituminous coal, one was attributed to the solvation of the macromolecular network by a liquid and the other to the breaking by polar liquids of hydrogen bonds serving as crosslinks.

The aims in the study of coal solubility and exploration of these correlations are to find better coal extraction solvents,[649,659-663,741] including easily dehydrogenated liquids,[664] and liquids suitable for use in the conversion of coal to liquid hydrocarbons[665-667] or to ashless solid fuels,[668] although Iino et al.[715] found no correlation with Hildebrand parameter in coal extraction while investigating the synergistic effects of carbon disulfide and *N*-methyl-2-pyrrolidinone. It has been observed that the cohesion properties of coal extraction products and coal liquefaction products are similar, suggesting that molecules similar to those found in liquefaction products already exist in the coal before hydrogenation.[665] Altgelt[670] found that Snyder's selectivity parameters (Section 8.5) were more successful in correlating the results of coal extraction and swelling than Hansen parameters, and it is likely that the set of four or five component cohesion parameters would also be preferable to Hansen parameters. In other studies, Reucroft and co-workers[671-673] interpreted pore volumes, surface areas and gas-induced swelling of coal in terms of cohesion parameters, and desulfurization of coal by chlorination was found to be enhanced if the liquid Hildebrand parameter was close to that providing maximum coal swelling.[700]

Petroleum products usually can be characterized more easily in terms of cohesion parameters than coal or coal derivatives, and values of these parameters for various fractions are included in Section 8.5, while the individual values are included in most of the data tables. *Asphaltenes* (and polynuclear aromatic systems in general, whether or not derived from petroleum or coal) have also received attention.[674-680,771] Cohesion parameters are utilized in the description of chromatographic separation of petroleum fractions (Chapter 18), and this approach can be extended to the study of asphaltenes and whole bitumen (oil plus asphaletenes) from tar-sands by making them the stationary phase.[681,682] Tar sands, oil shale and kerogen may be characterized (and solvents chosen) by a cohesion spectrum method.[683] Hildebrand parameters are also used in assessing the compatibility of added polymers as binders in paving asphalt[684-686] and the plasticizing effectiveness of hydrocarbons in bitumen.[687] But'ko et al.[688] determined Hildebrand parameters for systems containing petroleum resin.

One approach[762,776] to the investigation of natural polymers, which tend to be difficult to characterize, is the study of ''surrogates'' having well-defined physicochemical properties, and the use of cohesion parameters to make predictions based on these model compounds. These model systems may be as simple as octanol-water partitioning being considered as a first approximation to the behavior of water with humic substances[644,775] (see also Section 19.1).

REFERENCES

1. **Koenhen, D. M. and Smolders, C. A.,** The determination of solubility parameters of solvents and polymers by means of correlation with other physical quantities, *J. Appl. Polym. Sci.*, 19, 1163, 1975; Koenhen, D. M. *Phase-Separation in Polymer Solutions,* Thesis, Technische Hogeschool Twente, Enschede, Netherlands, 1977.
2. **McLellan, A. L.,** *Tables of Experimental Dipole Moments,* W. H. Freeman, San Francisco, 1963.
3. **Sewell, J. H.,** *A Study of Cohesive Forces, Refractive Indexes and Glass Transition Temperature Data of Polymers,* Royal Aircraft Establishment Technical Report No. 66185, Farnborough, Hampshire, 1966, AD 641308.
4. **Bonn, R. and van Aartsen, J. J.,** Solubility of polymers in relation to surface tension and index of refraction, *Eur. Polym. J.*, 8, 1055, 1972.
5. **McGinniss, V. D.,** Prediction of solvent and polymer characteristics through the use ot easy to measure properies, *Abstr. Pap. Am. Chem. Sc.*, 176, ORPL 102, 1978; *Org. Coat. Plast. Chem.*, 39, 529, 1978.
6. **Darby, J. R., Touchette, N. W., and Sears, K.,** Dielectric constants of plasticizers as predictors of compatibility with polyvinyl chloride, *Polym. Eng. Sci.*, 7, 295, 1967.
7. **Paruta, A. N., Sciarrone, B. J., and Lordi, N. G.,** Correlation between solubility parameters and dielectric constants, *J. Pharm. Sci.*, 51, 704, 1962.
8. **Pastor, G. R., Angelovich, J. M., and Silver, H. F.,** Study of the interaction of solvent and catalyst in coal hydrogenation, *Ind. Eng. Chem. Proc. Des. Dev.*, 9, 609, 1970.
9. **Wiesner, I.,** Solubility parameter of epoxy resins, *Chem. Prumysl.*, 26, 532, 1976.
10. **Xu, C. and Song, X.,** Correlation between solubility parameter and dielectric constant of polymer, *Hecheng Xiang Xiangjiao Gongye,* 10, 352, 1987.
11. **Yeo, R. S.,** Ion clustering in Nafion and its relationship to membrane conductivity, in *Symposium on Ion Exchange,* Yeo, R. S. and Buck, R. P., Eds., Electrochemical Society, Princteon, NJ, 1981.
12. **Tatsumi, M., Yamaguchi, Y., Moriwaki, S., Usami, S., and Yamamoto, S.,** Determination of solubility parameter of polymer by microwave absorption in waveguide, *Kobunshi Ronbunshu,* 45, 161, 1988.
13. **Turyshev, B. I., Shuvaev, V. P., Goldenburg, A. D., and Sazhin, B. I.,** Electric conductivity of polymer dielectrics in contact with liquid media, *Plast. Massy,* No. 7, 69, 1978; *Int. Polym. Sci. Technol.*, 6(4), 25, 1979.
14. **Yamamoto, T., Isshiki, S., and Nakayama, S.,** Synthetic paper for extra high voltage cable, *IEEE Trans. Power Appl. Sys.*, PAS91, 2415, 1972.
15. **Gardon, J. L.,** Cohesive-energy density, in *Encyclopedia of Polymer Science Technology,* Vol. 3, John Wiley & Sons, New York, 1964, 833.

16. **Gardon, J. L.,** Critical review of concepts common to cohesive energy density, surface tension, tensile strength, heat of mixing, interfacial tension and butt joint strength, *J. Colloid Interface Sci.*, 59, 582, 1977; *Progr. Org. Coatings*, 5, 1, 1977.
17. **Gardon, J. L. and Teas, J. P.,** Solubility parameters, in *Treatise on Coatings*, Vol. 2, *Characterization of Coatings: Physical Techniques*, Part II, Myers, R. R. and Long, J. S., Eds., Marcel Dekker, New York, 1976, chap. 8, 413.
18. **Bernardo, J. J. and Burrell, H.,** Plasticization, in *Polymer Science*, Jenkins, A. D., Ed., North-Holland, Amsterdam, 1972, chap. 8.
19. **Zosel, A.,** Mechanical behavior of coating films, *Progr. Org. Coatings*, 8, 47, 1980.
20. **Wilkes, G. L., Bagrodia, S., Humphries, W., and Wildnauer, R.,** The time dependence of the thermal and mechanical properties of segmented urethanes following thermal treatment, *J. Polym. Sci. Polym. Lett.*, 13, 321, 1975.
21. **Huglin, M. B. and Pass, D. J.,** Cohesive energy density of polytetrahydrofuran, *J. Appl. Polym. Sci.*, 12, 473, 1968.
22. **Bagrodia, S. and Wilkes, G. L.,** Effects of solvent casting copolymer materials as related to mechanical properties, *J. Biomed. Mater. Res.*, 10, 101, 1976.
23. **Cowie, J. M. G., Lath, D., and McEwen, I. J.,** Dynamic mechanical behavior of polystyrene-(ethene-*co*-butene)-polystyrene triblock copolymer films cast from various solvents, *Macromolecules*, 12, 52, 1979.
24. **Bansal, R. K. and Sahoo, J. C.,** Determination of cohesive energy densities of bisphenol-A and epichlorohydrin epoxy resins from swelling measurements, *Angew. Makromol. Chem.*, 82, 149, 1979.
25. **Dudek, T. J. and Bueche, F.,** Polymer-solvent interaction parameters and creep behaviour of ethylene-propylene rubbers, *J. Polym. Sci. A*, 2, 811, 1964.
26. **Eissler, R. L. and Princen, L. H.,** Effect of some pigments on tensile and swelling properties of linseed oil films, *J. Paint Technol.*, 40(518), 105, 1968.
27. **Giltrow, J. P.,** A relationship between abrasive wear and the cohesive energy of materials, *Wear*, 15, 71, 1970.
28. **Hourston, D. J. and Hughes, I. D.,** Dynamic mechanical and sonic velocity behaviour of polystyrene-poly(vinyl methyl ether) blends, *Polymer*, 19, 1181, 1978.
29. **Hourston, D. J.,** personal communication, 1979.
30. **Krozer, S.,** Solubility parameters and some structural characteristics of chlorinated poly(vinyl chloride), *J. Appl. Polym. Sci.*, 15, 1769, 1971.
31. **Krozer, S. and Schneier, B.,** Correlation between some solution and mechanical properties of poly(vinyl chloride) and chlorinated poly(vinyl chloride), *J. Appl. Polym. Sci.*, 15, 1777, 1971.
32. **Peschanskaya, N. N., Bershtein, V. A., and Stepanov, V. A.,** Relation of the activation energy of the creep of glassy polymers with cohesive energy, *Fiz. Tverd. Tela (Leningrad)*, 20, 3311, 1978; *Sov. Phys. Sol. State*, 20, 1945, 1978.
33. **Matzner, M., Noshay, A., and McGrath, J. E.,** Structure-processibility-property relations for organo-siloxane block copolymers, *Polym. Prepr. Am. Chem. Soc. Div. Polym. Chem.*, 14, 68, 1973.
34. **Matzner, M., Noshay, A., and McGrath, J. E.,** Effect of hard segment chemical structure on the processibility of organo-siloxane block copolymers, *Trans. Soc. Rheol.*, 21, 273, 1977.
35. **Romanov, A. and Marcincin, K.,** Molecular motion in chlorinated ethylene-propylene copolymer, *Polymer*, 11, 290, 1970.
36. **Samuels, S. L. and Wilkes, G. L.,** Effects of gamma radiation on the dynamic mechanical and swelling behaviour of styrene butadiene block and random copolymers, *Polym. Eng. Sci.*, 13, 280, 1973.
37. **Tanaka, K., Ogawa, H., and Tanaka, Y.,** Dynamic mechanical properties of blends of polyvinyl chloride and ethoxymethylmelamine resin, *Kobunshi Ronbunshu*, 36, 321, 1979.
38. **Eastmond, G. C. and Haraguchi, K.,** Dynamic mechanical properties of a polycarbonate/polystyrene copolymer: effects of casting solvent, *Polymer*, 24, 1171, 1983.
39. **Vinogradov, G. V.,** Comparison of polymer behaviour in the liquid state under shear and strain, *Plaste Kautsch.*, 17, 241, 1970.
40. **Tobolsky, A. V.,** *Properties and Structure of Polymers*, John Wiley & Sons, New York, 1960.
41. **Willbourn, A. H.,** Molecular design of polymers, *Polymer*, 17, 965, 1976.
42. **Hata, N. and Kumanotoni, J.,** Viscoelastic properties of epoxy resin. I. Effect of prepolymer structure on viscoelastic properties, *J. Appl. Polym. Sci.*, 15, 2371, 1971.
43. **Wu, C.-C. and Hsieh, W.-C.,** Solubility parameter, IR spectrometric composition analysis and flow testing of acrylonitrile and vinyl acetate copolymers, *Chemistry (Hua Hsueh, Taiwan)*, No. 3, 85, 1974.
44. **Kamon, T. and Furukawa, H.,** Curing mechanisms and mechanical properties of cured epoxy resins, *Adv. Polym. Sci.*, 80, 172, 1986.
45. **Holliday, L. and White, J. W.,** Stiffness of polymers in relation to their structure, *Pure Appl. Chem.*, 26, 545, 1971.
46. **Tse, M.-F.,** Rheological and mechanical properties of plasticized poly(vinyl chloride), *Polym. Eng. Sci.*, 21, 1037, 1981.

47. **Watts, M. P. C. and White, E. F. T.,** *Polym. Prepr. Am. Chem. Soc. Div. Polym. Chem.*, 19(1), 38, 1978.
48. **Bower, T. N. and Tonelli, A. E.,** Thermal transitions of ethylene-vinyl chloride copolymers, *Polymer,* 26, 1195, 1985.
49. **Bower, T. N. and Tonelli, A. E.,** Blends of homo- and copolymers of ethylene and vinyl chloride: a compatibility study, *Macromolecules,* 19, 498, 1986.
50. **Kambour, R. P.,** A review of crazing and fracture in thermoplastics, *J. Polym. Sci. D Macromol. Rev.*, 7, 1, 1973.
51. **Kambour, R. P., Gruner, C. L., and Romagosa, E. E.,** Solvent crazing of dry polystyrene and dry crazing of plasticized polystyrene, *J. Polym. Sci. Polym. Phys.*, 11, 1879, 1973 (General Electric Reprint 7435).
52. **Kambour R. P. and Gruner, C. L.,** Effects of polar group incorporation on crazing of glassy polymers: styrene-acrylonitrile copolymer and a dicyano bisphenol polycarbonate, *J. Polym. Sci. Polym. Phys.*, 16, 703, 1978 (General Electric Reprint 8346).
53. **Kambour, R. P.,** Correlations of the dry crazing resistance of glassy polymers with other physical properties, *Polym. Commun.*, 24, 292, 1983.
54. **Chiantore, O., Costa, L., and Guaita, M.,** Glass temperatures of acrylamide polymers, *Makromol. Chem. Rapid. Commun.*, 3, 303, 1982.
55. **DiBenedetto, A. T.,** Prediction of the glass transition temperature of polymers: a model based on the principle of corresponding states, *J. Polym. Sci. Polym. Phys.*, 25, 1949, 1987.
56. **Bunn, C. W.,** The melting points of chain polymers, *J. Polym. Sci.*, 16, 323, 1955.
57. **Maloney, D. P. and Prausnitz, J. M.,** Thermodynamic properties of liquid polyethylene, *J. Appl. Polym. Sci.*, 18, 2703, 1974.
58. **Olabisi, O. and Simha, R.,** A semiempirical equation of state for polymer melts, *J. Appl. Polym. Sci.*, 21, 149, 1977.
59. **Blanks, R. F.,** Engineering applications of polymer solution thermodynamics, *Chemtech,* 6, 396, 1976.
60. **Olabisi, O.,** Volatility of solvents from polysulfone melt, *J. Appl. Polym. Sci.*, 22, 1021, 1978.
61. **Bagley, E. B. and Scigliano, J. M.,** Internal pressure behavior of polymers above and below the glass temperature, *Polym. Eng. Sci.*, 11, 320, 1971.
62. **Bianchi, U.,** Pressure effects on glass transition in polymers, *J. Phys. Chem.*, 69, 1497, 1965.
63. **Turturro, A. and Bianchi, U.,** Measurement of internal pressure applied to glass transitions of polyvinyl chloride, *Chim. Ind. (Milan),* 49, 362, 1967.
64. **Becker, R. and Neumann, G.,** Precalculation of glass temperatures of polymers, *Plaste Kautsch.*, 20, 809, 1973.
65. **Becker, R.,** On the nature of glass temperature of side chain-bearing polymers, *Plaste Kautsch.*, 23, 108, 1976.
66. **Becker, R.,** Quantitive relations between the glass transition temperature of polymers and the energy value, *Z. Phys. Chem. (Leipzig),* 258, 953, 1977.
67. **Becker, R.,** Limiting quantities of the glass transition temperature of polymers, *Plaste Kautsch.*, 25, 1, 1978; Royal Aircraft Establishment Library Translation No. 1992, Farnborough, Hampshire, 1978.
68. **Berry, G. C.,** Remarks on the energy parameters in lattice treatments of the glassy state, *Macromolecules,* 13, 550, 1980.
69. **Deanin, R. D.,** Solubility parameter and glass transition temperature: an example of practical structure-property relationships, *Abstr. Pap. Am. Chem. Soc.*, 174, ORPL 76, 1977.
70. **Gargallo, L. and Russo, M.,** Effect of structure on glass transition temperatures of poly(methacrylic ester)s with bulky side groups, *Macromol. Chem.*, 176, 2735, 1975.
71. **Guzman, J., Alamo, R., and Fatou, J. G.,** Glass transitions and solubility parameter for poly(diethylene glycol terephthalate), *An. Quim. Ser. A,* 76, 214, 1980.
72. **He, T.-B.,** On the chain cross-sectional area and motion of macromolecular chains; Estimating cross-sectional area of a polymer chain by additive method, *J. Appl. Polym. Sci.*, 30, 4319, 1985; 31, 1521, 1986.
73. **He, T.-B. and Ren, S.-J.,** Glass transition temperature and chain flexibility of 1,2-polybutadiene, *J. Appl. Polym. Sci.*, 31, 873, 1986; **He, T.-B., Ren, S.-J., Gao, X.-F., and Li, B.-C.,** Cohesion energy density and glass transition temperature of 1,2-polybutadiene, *Gaofenzi Tongxun,* No. 5, 321, 1982.
74. **Kreibich, U. T. and Batzer, H.,** Influence of the segment structure and cross-linking on the glass transition T_g. Possibilities of predicting T_g using the values of cohesive energy E_{coh}, *Angew. Macromol. Chem.*, 83, 57, 1979.
75. **Batzer, H. and Kreibich, U. T.,** Influence of water in thermal transitions in natural polymers and synthetic polyamides, *Polym. Bull.*, 5, 585, 1981.
76. **Batzer, H. and Kreibich, U. T.,** Connections between glass transition temperature, thermodynamic and mechanical values for predicting material properties from the chemical structure, *Angew. Makromol. Chem.*, 105, 113, 1982.

77. **Lee, W. A. and Sewell, J. H.,** Influence of cohesive forces on the glass transition temperatures of polymers, *J. Appl. Polym. Sci.*, 12, 1397, 1968.
78. **Miller, A. A.,** Polymer-melt viscosity and the glass transition: an evaluation of the Adam-Gibbs and the free volume models, *J. Chem. Phys.*, 49, 1393, 1968; General Electric Reprint 6110.
79. **Pérez, C., Guzman, J., Riande, E., Delacamp, J. G., and de abajo, J.,** Influence of chemical structure on glass transition temperature of poly(propylene terephthalate) and poly(propylene isophthalate), *Makromol. Chem.*, 189, 691, 1988.
80. **Rostiashvili, V. G. and Irzhak, V. I.,** Effect of cohesive energy and chain rigidity on the structural glass-transition temperature, *Vysokomol. Soedin Ser. B*, 20, 837, 1978.
81. **Tsutsui, T. and Tanaka, T.,** Role of electrostatic forces in the glass transition temperatures of ionic polymers, *Polymer*, 18, 817, 1977.
82. **Uematsu, Y.,** Specific cohesive heat of high polymers at the glass transition, *Kobayashi Rigaku Kenkyushu Hokoku*, 12, 101, 1962.
83. **Vincini, L., Seves, A., Canetti, M., Sadocco, P., and Galli, R.,** Influence of an aqueous acetone medium on the thermal behavior of solvent treated poly(ethylene terephthalate), *Polymer*, 25, 1774, 1984.
84. **Wymann, D. P.,** Method for separating intramolecular and intermolecular interactions which influence polymer glass temperatures, *J. Appl. Polym. Sci.*, 11, 1439, 1967.
85. **Kaelble, D. H.,** Relations between polymer chemistry and physical properties, *Org. Coat. Appl. Polym. Sci. Proc.*, 46, 241, 1981.
86. **Immergut, E. H. and Mark, H. F.,** Principles of plasticization, *Adv. Chem. Ser.*, 48, 1, 1965.
87. **Morre, W. R.,** Some aspects of mixing theory, *Trans. J. Plast. Inst.*, 32, 247, 1964.
88. **Brydson, J. A.,** Selection of compounding ingredients, *Plastics (London)*, 26, 107, 1961.
89. **Brydson, J. A.,** *Plastics Materials*, Van Nostrand, Princeton, NJ, 1966; 4th ed., Butterworth, London, 1982, Chap. 5, Relation of structure to chemical properties, 70.
90. **Li, N. N. and Long, R. B.,** Plasticizing effect of permeates on membrane permeation and separation, in *Progress in Separation and Purification*, Vol, 3, Perry, E. S. and Van Oss, C. J., Eds., John Wiley & Sons - Interscience, 1970, 153.
91. **Seats, J. K. and Touchette,** Plasticizers, in *Kirk-Othmer Encyclopedia of Chemical Technology*, 3rd ed., Vol. 18, John Wiley & Sons, New York, 1982, 111.
92. **Sears, J. K., Touchette, N. W., and Darby, J. R.,** Plasticizers, in *Applied Polymer Science*, 2nd ed., Tess, R. W. and Poehlein, G. W., Eds. (*Am. Chem. Soc. Symp. Ser.* 285), 1985, 611; **Sears, J. K. and Darby, J. R.,** *The Technology of Plasticizers*, Wiley-Interscience, NY, 1982.
93. **Yoshioka, A.,** Compatibility of polymers with tackifiers from the viewpoint of solubility parameter and hydrogen bonding index, *Nippon Setchaku Kyokaishi*, 19, 504, 1983.
94. **Studebaker, M. L. and Beatty, J. R.,** The rubber compound and its composition, in *Science and Technology of Rubber*, Eirich, F. R., Ed., Academic Press, New York, 1978, chap. 9.
95. **Bernier, G. A. and Kambour, R. P.,** The role of organic agents in the stress crazing and cracking of poly(2,6-dimethyl-1,4-phenylene oxide), *Macromolecules*, 1, 393, 1968; General Electric Reprint 6125.
96. **Gur-Arieh, Z. and Ingamells, W.,** The plasticisation and swelling of Acrilan under dyebath conditions, *J. Soc. Dyers Colour.*, 90, 12, 1974.
97. **Gur-Arieh, Z., Ingamells, W., and Peters, R. H.,** The effect of plasticizing compounds on acrylic fibers, *J. Appl. Polym. Sci.*, 20, 41, 1976.
98. **Sorokin, G. A., Kutsenko, A. I., Abramova, R. A., and Maksimenko, Ye. G.,** Thermographic investigation of the selective character of the plasticization of copolymers, *Vysokomol. Soedin. Ser. A*, 16, 402, 1974; *Polym. Sci. U.S.S.R.*, 16, 468, 1974.
99. **Ribnick, A. S., Weigmann, H.-D., and Rebenfeld, L.,** Interactions of nonaqueous solvents with textile fibres. I. Effects of solvents on the mechanical properties of a polyester yarn; II. Isothermal shrinkage kinetics of a polyester yarn, *Text. Res. J.*, 42, 720, 1972; 43, 176, 1973; **Ribnick, A. S. and Weigmann, H.-D.,** Interactions of nonaqueous solvents with textile fibers. III. The dynamic shrinkage of polyester yarns in organic solvents, *Text. Res. J.*, 43, 316, 1973.
100. **Nakatsuka, R.,** The viscometric properties of polyesters. II. Viscosity in the range of low temperatures near the glass transition point, *Bull. Chem. Soc. Jpn.*, 37, 403, 1964.
101. **Merzlikina, V. P., Barashkov, O. K., Kirilovich, V. I., Chekhovskaya, N. V., and Ryzhakova, L. V.,** Plasticization of poly(vinyl chloride) by ethers of 1,10-decanedicarboxylic acid, *Plast. Massy*, No. 8, 42, 1985.
102. **Barashkov, O. K. and Barshtein, R. S.,** Effect of thermodynamic factors in the migration of plasticizers from plasticised poly(vinyl chloride) to polyethylene, *Plast. Massy*, No. 11, 26, 1984.
103. **Barashkov, O. K. and Barshtein, R. S.,** Effect of enthalpy and entropy factors on the value of the thermodynamic interaction parameter in poly(vinyl chloride)-plasticizer systems, *Vysokomol. Soedin. Ser. B*, 28, 306, 1986.
104. **Maláč, J., Altman, V., and Zelinger, J.,** Properties of PVC: II. Properties of PVC compounds with solvents, *J. Appl. Polym. Sci.*, 14, 161, 1970.

105. **Chi, M. S.,** Compatibility of cross-linked polymers with plasticizers by glass transition temperature measurement and swelling tests, *J. Polym. Sci. Polym. Ed.*, 19, 1767, 1981.
106. **Salmen, N. L. and Back, E. L.,** The influence of water on the glass transition temperature of cellulose, *Tappi*, 60, 137, 1977.
107. **Salmen, N. L. and Back, E. L.,** The influence of water in the glass transition temperature of cellulose, in *Fibre-Water Interaction in Paper Making*, (Trans. Symp., Oxford, 1977), Technical Division, British Paper and Board Industry Federation, London, 1978, 683.
108. **Naimark, N. I., Kosobutskaya, A. A., and Tarakanov, O. G.,** Interaction of cellulose with plasticizing media, *Zh. Prikl. Khim.*, 57, 1588, 1984; *J. Appl. Chem. U.S.S.R.*, 57, 1473, 1984.
109. **Kelly, S. S., Rials, T. G., and Glasser, W. G.,** Relaxation behaviour of the amorphous components of wood, *J. Mater. Sci.*, 22, 617, 1987.
110. **Hirabayashi, K.,** Structural formation of silk fibroin (*Bombyx mori* L) by fiber dimensional change, in *Zoku Kenshi no Koso*, Nobumasa, H., Ed., Shinshu Daigaku, Sen'igakubu, Ueda, Japan, 1980, 225.
111. **Hirabayashi, K., Oki, R., Shigematsu, M., and Arai, M.,** Studies on the mechanochemical reactivity of silk fibroin. I. Contraction behavior of silk fibers in organic solvents, *Nippon Sanshigaku Zasshi*, 49(1), 27, 1980.
112. **Abu-Isa, I. A.,** Elastomer-gasoline blends interactions. I. Effects of methanol-gasoline mixtures on elastomers, in *Macromolecular Solutions: Solvent-Property Relationships in Polymers*, Seymour, R. B. and Stahl, A. G., Eds., Pergamon Press, New York, 1982, 166: *Rubber Chem. Technol.*, 56, 135, 1983.
113. **Abu-Isa, I. A.,** Elastomer-gasoline blends interactions. II. Effects of ethanol/gasoline and methyl-t-butyl ether/gasoline mixtures in elastomers, *Rubber Chem. Technol.*, 56, 169, 1983; *Am. Chem. Soc. Symp.*, 229, 225, 1983.
114. **Aulton, M. E., Houghton, R. J., and Wells, J. I.,** Compatibility of polymeric binders and potential plasticizers, *J. Pharm. Pharmacol.*, 37, Supp., 113P, 1985.
115. **Jones, D. W., Sutow, E. J., Graham, B. S., Milne, E. L., and Johnston, D. E.,** Influence of plasticizer on soft polymer gelation, *J. Dental Res.*, 65, 634, 1986.
116. **Galembeck, F., Galembeck, S. E., and Vincente, E. J.,** Investigation of the interaction of certain low cohesive energy liquids with Teflon and polyethylene and its implications in the contamination of polymeric surfaces, in *Proc. Int. Symp. Contam. Control*, 4th, Institute of Environmental Science, Mt. Prospect, IL, 1978, 329.
117. **Knox, B. H., Weigmann, H.-D., and Scott, M. G.,** Interactions of nonaqueous solvents with textile fibres. Part V: Application of the solubility parameter concept to polyester fiber-solvent interactions, *Text. Res. J.*, 45, 203, 1975.
118. **Lawton, E. L. and Cates, D. M.,** Liquid induced crystallization of poly(ethylene terephthalate), *J. Appl. Polym. Sci.*, 13, 899, 1969.
119. **Lemons, J. H., Katar, S. K., and Cates, D. M.,** The effect of various organic compounds on poly(ethylene terephthalate) fibers. II. Plasticization and acceleration of dyeing, *Am. Dyestuff Reptr.*, 55(3), 76, 1966.
120. **De Sai, A. B. and Wilkes, G. L.,** Solvent-induced crystallization of polyethylene terephthalate, *J. Polym. Sci. Polym. Symp.*, 46, 291, 1974.
121. **Knox, B. H.,** Bimodal character of polyester-solvent interactions. I. Evaluation of the solubility parameters of the aromatic and the aliphatic ester residues of poly(ethylene terephthalate); II. Evaluation of the chemical structures of the aromatic and aliphatic ester residues of poly(ethylene terephthalate); III. The effects of morphology on the interaction of nonaqueous solvents with the aromatic and the aliphatic ester residues of poly(ethylene terephthalate), *J. Appl. Polym. Sci.*, 21, 225, 249, and 267, 1977.
122. **Kulshreshtha, A. K., Khan, A. H., and Madan, G. L.,** X-ray diffraction study of solvent-induced crystallization in polyester filaments, *Polymer*, 19, 819, 1978.
123. **Liepins, R., Surles, J. R., Morosoff, N., and Stannett, V. T.,** Localized radiation grafting of flame retardants to poly(ethylene terephthalate). I. Bromine-containing monomers, *J. Appl. Polym. Sci.*, 21, 2529, 1977.
124. **Lindner, W. L.,** Interaction between organic solvents and poly(ethylene terephthalate) fibers. I; II. The maximal shrinking rate, maximal shrinking and swelling, *Colloid Polym. Sci.*, 255, 213, 433, 1977.
125. **Moore, W. R. and Sheldon, R. P.,** The crystallization of polyethylene terephalate by organic liquids, *Polymer*, 2, 315, 1961.
126. **Penczek, P., Plocharski, S., and Warchol, W.,** Effect of the composition of solvent mixture on the freezing resistance of polychloroprene rubber based adhesives on storage, *Polimery (Warsaw)*, 20, 186, 1975.
127. **Riande, E. and Fatou, J. M. G.,** Effect of solvent on the crystallization from dilute polyethylene solutions, *Polymer*, 17, 795, 1976.
128. **Benecki, W., Pelszbau, Z., and Turska, E.,** Microscopic investigations of morphological changes during the liquid-induced crystallization of poly(2,6-dimethyl-1,4-phenylene oxide), *Acta Polym.*, 35, 196, 1984.
129. **Bueche, A. M.,** Melting temperature and polymer-solvent interaction: polychlorotrifluoroethylene, *J. Am. Chem. Soc.*, 74, 65, 1952.

130. **Daniewska, I., Kocinska, E., and Dobkowski, Z.,** Morphology of polycarbonate crystallized under the influence of solvent vapours, in *Morphology of Polymers,* Sedláček, B., Ed., Walter de Gruyter & Co., Berlin, 1986, 449.
131. **Daniewska, I., Dobkowski, Z., and Cztonkowska-Kohutnicka, Z.,** Solvent-vapor-induced crystallization of branched polydisperse polycarbonate, *J. Appl. Polym. Sci.,* 31, 2401, 1986.
132. **Law, K.-Y.,** An investigation of solvent-vapor-induced crystallization of soluble vanadyl phthalocyanine dyes in a polymer matrix, *J. Phys. Chem.,* 89, 2652, 1985.
133. **Fellers, J. F., Lee, Y.-D., and Zdrahala, R. J.,** Morphologies and tensile properties of triblock poly(*m*-phenylene isophthalamide-*b*-ethylene oxide) and poly(*m*-phenylene isophthalamide-*b*-dimethylsiloxane), *J. Polym. Sci. Polym. Symp.,* 60, 59, 1977.
134. **Onogi, Y., White, J. L., and Fellers, J. F.,** *Structural Investigations of Polymer Liquid Crystalline Solutions: Aromatic Polyamides Hydroxypropylcellulose and Poly-benzyl-L-glutamate,* Polymer Science and Engineering Report 122, University of Tennessee, Knoxville, TN, (undated, ~ 1978).
135. **Takahasi, S., Kariya, M., Yatake, T., Sonogashira, K., and Hagihara, N.,** Studies of poly-yne polymers containing transition metals in the main chain. 2.Synthesis of poly[*trans*-bis(tri-*n*-butyl(phosphine)platinum 1,4-butadiynediyl] and evidence of a rodlike structure, *Macromolecules,* 11, 1063, 1978.
136. **Allen, R. C.,** Some corrosion mechanisms in attack of resin and resin-glass laminates, *Polym. Eng. Sci.,* 19, 329, 1979.
137. **Andrews, E. H.,** Fracture of polymers, in *MTP Int. Rev. Sci. Phys. Chem. Ser. 1,* Vol. 8 (Macromol. Sci.), 1972, chap. 6.
138. **Andrews, E. H.,** Cracking and crazing in polymeric glasses, in *The Physics of Glassy Polymers,* Haward, R. N., Ed., John Wiley & Sons, New York, 1973, chap. 7.
139. **Kambour, R. P.,** Environmental stress cracking of thermoplastics, in *Corrosion Fatigue: Chemistry, Mechanics and Microstructure,* Devereux, O. F., McEvily, A. J., and Staehle, R. W., Eds., National Association of Corrosion Engineers, Houston, TX, 1972, 681.
140. **van Krevelen, D. W.,** *Properties of Polymers: Their Estimation and Correlation with Chemical Structure,* 2nd. ed., Elsevier, Amsterdam, 1976.
141. **Avramenko, V. L., Novik, E. Yu., and Shturman, A. A.,** Mechanism of the strengthened effect of solvents on polymers, *Fiz.-Khim. Mekh. Mater.,* 9, 114, 1973; *Sov. Mater. Sci.,* 9, 607, 1973.
142. **Boyer, R. F. and Keskkula, H.,** Styrene polymers: crazing and stress cracking, in *Encyclopedia of Polymer Science and Technology,* Vol. 13, Bikales, N. M., Ed., John Wiley & Sons, New York, 1970, 375.
143. **Kambour, R. P., Romagosa, E. E., and Gruner, C. L.,** Swelling, crazing and cracking of an aromatic copolyether-sulfone in organic media, *Macromolecules,* 5, 335, 1972 (General Electric Reprint 7171).
144. **Kambour, R. P., Gruner, C. L., and Romagosa, E. E.,** Bisphenol-A polycarbonate immersed in organic media. Swelling and response to stress, *Macromolecules,* 7, 248, 1974 (General Electric Reprint 7514).
145. **Vincent, P. I. and Raha, S.,** Influence of hydrogen bonding in crazing and cracking of amorphorous thermoplastics, *Polymer,* 13, 283, 1972.
146. **Henry, L. F.,** Prediction and evaluation of the susceptibilities of glassy thermoplastics to environmental stress cracking, in *Predicting Plastics Performance,* (Div. Tech. Conf.), Society of Plastics Engineers, Greenwich, CT, 1973, 10; *Polymer. Eng. Sci.,* 14, 167, 1974.
147. **MacNulty, B. J.,** Critical stress cracking of polymers with phenylene groups in the main chain, *Br. Polym. J.,* 6, 39, 1974.
148. **Jacques, C. H. M. and Wyzgoski, M. G.,** Prediction of environmental stress cracking of polycarbonate from solubility considerations, *J. Appl. Polym. Sci.,* 23, 1153, 1979.
149. **Rains, R. C. and Bailey, R. J.,** Solvent aging studies on polyester urethane elastomers, *J. Elastomers Plast.,* 15, 124, 1983; *Proc. SPI Annu. Tech./Mark. Conf.,* 27th (Meet. Minds: Mark. Polyurethane Technol.), 181, 1982.
150. **Mai, Y.-W.,** Environmental stress cracking of glassy polymers and solubility parameters, *J. Mater. Sci.,* 21, 904, 1986.
151. **Starkweather, H. W., Jr.,** The sorption of chemicals by perfluorocarbon polymers, *Macromolecules,* 10, 1161, 1977.
152. **Kaelble, D. H., Dynes, P. J., and Cirlin, E. H.,** Interfacial bonding and environmental stability of polymer matrix composites, *J. Adhes.,* 6, 23, 1974.
153. **Kaelble, D. H., Dynes, P. J., Crane, L. W., and Maus, L.,** Kinetics of environmental degradation in graphite-epoxy laminates, in *Composite Reliability* (ASTM Spec. Tech. Publ. 580), American Society for Testing and Materials, Philadelphia, PA, 1975, 247.
154. **Kaelble, D. H., Dynes, P. J., and Maus, L.,** Surface energy analysis of treated graphite fibers, *J. Adhes.,* 6, 239, 1974.
155. **Kaelble, D. H., Dynes, P. J., and Pav, D.,** Surface energetics analysis of lithography, in *Polymer Science and Technology,* Vol. 9B, Lee, L.-H., Ed., Plenum Press, New York, 1975, 735.
156. **Kaelble, D. H., and Moacanin, J.,** A surface energy analysis of bioadhesion, *Polymer,* 18, 475, 1977.
157. **Kaelble, D. H. and Moacanin, J.,** Surface energetics analysis of artificial blood substitutes, *Med. Biol. Eng. Comput.,* 17, 593, 1979.

158. **Andrews, E. H. and Bevan, S.,** Mechanics and mechanism of environmental crazing in a polymeric glass, *Polymer,* 13, 337, 1972.
159. **Andrews, E. H., Levy, G. M., and Willis, J.,** Environmental crazing in a glassy polymer: the role of solvent absorption, *J. Mater, Sci.,* 8, 1000, 1973.
160. **Andrews, E. H. and Levy, G. M.,** Solvent stress crazing in PMMA: 1. Geometrical effects, *Polymer,* 15, 599, 1974.
161. **Kawagoe, M. and Kitagawa, M.,** Craze initiation in poly(methyl methacrylate) exposed to *n*-alkanes, *J. Mater. Sci.,* 22, 3000, 1987.
162. **Hoa, S. V.,** Relative influence of the mobility and the solubility parameters of fluids on the mechanical behavior of high impact polystyrene, *Soc. Plast. Eng. Tech. Pap.,* 25, 614, 1979; *Polym. Eng. Sci.,* 20, 1157, 1980.
163. **McCammond, D. and Hoa, V. S.,** Craze and creep resistance of high-impact polystyrene in alcohols, *Polym. Eng. Sci.,* 17, 869, 1977.
164. **Sauer, J. A., Warty, S., and Kaka, S.,** Environmental and surface effects in polymer fatigue, *Fourth Int. Conf. Deformation, Yield and Fracture of Polymers,* Cambridge, England, 1979, paper 26.
165. **Warty, S., Morrow, D. R., and Sauer, J. A.,** Effects of environmental and surface coatings on the fatigue properties of polystyrene, *Polymer,* 19, 1465, 1978.
166. **Weigmann, H.-D. and Ribnick, A. S.,** The effects of non-aqueous solvents on the mechanical properties of textile yarns, in *Textile Solvent Technology — Update '73* (Am. Assoc. Text. Chem. Col. Symp., 1973), Research Triangle Park, NC, 1973, 24.
167. **Weigmann, H.-D. and Ribnick, A. S.,** Interactions of non-aqueous solvents with textile fibers. 4. Effects of solvents on mechanical properties of various textile yarns, *Text. Res. J.,* 44, 165, 1974.
168. **Wyzgoski, M. G.,** The role of solubility in stress cracking of nylon 6,6, in *Macromolecular Solutions: Solvent-Property Relationship in Polymers* (Symp. Proc., Am. Chem. Soc., New York, 1981), Seymour, R. B. and Stahl, G. A., Eds., Pergamon, New York, 1982, 41 (General Motors Research Publication 3779).
169. **Mai, Y. W.,** Fracture initiation and crack propagation of acrylonitrile-butadiene-styrene (ABS) in organic solvents, *J. Mater. Sci.,* 11, 303, 1976.
170. **Mai, Y. W. and Atkins, A. G.,** Effects of rate, temperature and absorption of organic solvents on the fracture of plain and glass-filled polystyrene, *J. Mater. Sci.,* 11, 677, 1976.
171. **Leete, N. B.,** *Environmental Stress Cracking of PMMA — the Solubility Parameter Approach,* B.E. Thesis, University of Sydney, Australia, 1978.
172. **Mai, Y. W. and Leete, N. B.,** Discussion of 'Effect of environment on stability of cracking in brittle polymers,' *J. Mater. Sci.,* 14, 2264, 1979.
173. **Cooper, S. L.,** Properties of swollen block copolymer elastomers, *J. Polym. Sci. A-1,* 7, 1765, 1969.
174. **Miltz, J., DiBenedetto, A. T., and Petrie, S.,** The effect of the environment on the stress crazing of polycarbonate, *J. Mater. Sci.,* 13, 2037, 1978.
175. **Priori, A., Nicolais, L., and DiBenedetto, A. T.,** The kinetics of surface craze growth in polycarbonate exposed to normal hydrocarbons, *J. Mater. Sci.,* 18, 1466, 1983.
176. **Faulkner, D. L.,** The role of surfactant structure and monoethanolamine in the environmental stress cracking of polycarbonate, *J. Appl. Polym. Sci.,* 32, 4909, 1986.
177. **Nakano, A. and Minoura, Y.,** Chemical reactions of polymers studied using mechanical force. 3. Mechanical degradation of polymers with high-speed stirring; 4. Effect of solvent on degradation of polymers by high stirring, *Kogyo Kagaku Zasshi,* 74, 101 and 1452, 1971.
178. **Nakano, A. and Minoura, Y.,** Effect of solvents on the degradation of polymers by high-speed stirring, *J. Appl. Polym. Sci.,* 16, 627, 1972.
179. **Okuda, S. and Iguchi, T.,** Environmental stress crazing of rigid poly(vinyl chloride), *Zairyo,* 24, 961, 1975.
180. **Reimschussel, A. C. and Kim, Y. J.,** Stress-cracking of nylons induced by zinc chloride solutions, *J. Mater Sci.,* 13, 245, 1978.
181. **Schonhorn, H.,** Environmental influence on the fracture strength of coated glass fibers, *J. Appl. Polym. Sci.,* 23, 687, 1979.
182. **Tao, W., Zhao, Y., and Shi, M.,** Corrosion behavior of polybutadiene glass fiber-reinforced plastics in sodium hydroxide solution and some organic solvents, *Huadong Huagong Xueyuan Xuebao,* No. 3, 353, 1983.
183. **Entwistle, C. A. and Rowe, R. C.,** Plasticization of cellulose ethers used in the film coating of tablets, *J. Pharm. Pharmacol.,* 31, 269, 1979.
184. **Croll, S. G.,** Effect of solvent on residual strain in clear epoxy coatings, *J. Oil Col. Chem.,* 63, 230, 1980.
185. **Yefimov, A. V., Bondarev, V. V., Kozlov, P. V., and Bakeyev, N. F.,** Effect of the chemical structure of liquid media on microcrack formation in crystalline polyolefins undergoing deformation, *Vysokomol. Spedin. Ser. A,* 25, 642, 1983; *Polym. Sci. U.S.S.R.,* 25, 755, 1983.

186. **Tonyali, K., Rogers, C. E., and Brown, H. R.,** Stress-cracking of polyethylene in organic liquids, *Polymer,* 28, 1472, 1987.
187. **White, S. A., Weissman, S. R., and Kambour, R. P.,** Resistance of a polyetherimide to environmental stress crazing and cracking, *J. Appl. Polym. Sci.,* 27, 2675, 1982.
188. **Haven, R. E. and Sung, N.-H.,** Non-destructive, quality control testing of polymethylmethacrylate parts, *Soc. Plastic. Eng. Tech. Pap.,* 24, 406, 1978.
189. **Sung, N.-H., Gahan, R. E., and Haven, R. E.,** Studies on whitening phenomena induced by some non-solvents on highly oriented glassy polymers, *Polym. Eng. Sci.,* 23, 328, 1983.
190. **Taova, A. Zh. and Zhelikhazheva, M. V.,** Correlation between the solubility parameters and heat resistance of polymers, *Polikondensats. Protsessy i Polimery,* Natl'chik, 113, 1984.
191. **Li, X.-B., Winnick, M. A., and Guillet, J. E.,** A fluorescence method to determine the solubility parameters δ_H of soluble polymers at infinite dilution. Cyclization dynamics of polymers. II. *Macromolecules,* 16, 992, 1983.
192. **Ritchie, K. and Hughes, H. G.,** Comparative studies of photoresist processing of microelectronics regarding chemical pollution, shortage, and cost, *Solid State Technol.,* 19, 45, 1976.
193. **Turner, S. R. and Daly, R. C.,** Polymers in microlithography, *J. Chem Educ.,* 65, 322, 1988.
194. **Soong, D. S.,** Dissolution kinetics of E-beam resists, *Proc. Soc. Phot.-Opt. Instrum. Eng.,* 539, (Adv. Resist Technol. Process II), 2, 1985.
195. **Lai, J. H.,** Application of methyl α-chloroacrylate copolymers as electron sensitive positive resists, *Proc. Soc. Phot-Opt. Instrum. Eng.,* 333 (Submicron Lithography), 8, 1982.
196. **Rodriguez, F., Krasicky, P. D., and Groele, R. J.,** Dissolution rate measurements, *Solid State Technol.,* 28(5), 125, 1985.
197. **Rodriguez, F., Groele, R. J., and Krasicky, P. D.,** Dissolution rates of thin polymer films using laser interferometry, *Proc. Soc. Phot.-Opt. Instrum. Eng.,* 539 (Microlithography), 14, 1985.
198. **Suzuki, Y., Ishio, N., Eimori, T., Yoshioka, N., Natakake, Y., and Yamazuki, T.,** X-ray lithography using chlorinated polymethylstyrene as a negative X-ray resist, *Jpn. J. Appl. Phys.,* 26, 130, 1987.
199. **Kamoshida, Y., Koshiba, M., Yoshimoto, H., Harita, Y., and Harada, K.,** Application of chlorinated polymethylstyrene, CPMS, to electron beam lithography, *J. Vac. Sci. Technol. B,* 1, 1156, 1985.
200. **Novembre, A. G., Masakowski, L. M., and Hartney, M. A.,** Optimal developer selection for negative acting resists, *Polym. Eng. Sci.,* 26, 1158, 1986.
201. **Schlegel, L. and Schnabel, W.,** Polymethacrylonitrile as a resist in x-ray lithography, *J. Vac. Sci. Technol. B.,* 6(1), 82, 1988.
202. **Ehrfeld, W., Bley, P., Götz, F., Mohr, J., Münchmeyer, D., Schelb, W., Baving, H. J., and Beets, D.,** Progress in deep-etch synchrotron radiation lithography, *J. Vac. Sci. Technol. B.,* 6(1), 178, 1988.
203. **Parsonage, E. E. and Peppas, N. A.,** Properties of positive resists. I. Irradiation-induced degradation and sensitivity of certain methyl methacrylate copolymers, *Brit. Polym. J.,* 19, 469, 1987.
204. **Lai, J. H.,** An investigation of spin coating of electron resists, *Polym. Eng. Sci.,* 19, 1117, 1979.
205. **Rigbi, Z.,** Prediction of swelling of polymers in 2 and 3 component mixtures, *Polymer,* 19, 1229, 1978.
206. **Froehling, P. E., Koenhen, D. M., Bantjes, A., and Smolders, C. A.,** Swelling of linear polymers in mixed swelling agents; predictability by means of solubility parameters, *Polymer,* 17, 835, 1976.
207. **Froehling, P. E. and Hillegers, L. T.,** Solubility parameters of ternary solvent mixtures. Calculation of the solvent composition with maximum polymer interaction, *Polymer,* 22, 261, 1981.
208. **Behnken, D. W.,** Calculation of optimum multicomponent solvent compositions for polymers using a multiple regression program, *Polymer,* 24, 193, 1983.
209. **Budin, A. and Johnston, H. K.,** Stability of polymer mixtures. I. Coagulation of polymer dispersions, *J. Paint Technol.,* 44(568), 41, 1972.
210. **Mellan, I.,** *Compatibility and Solubility,* Noyes, Park Ridge, NJ, 1968.
211. **Iwamoto, T.,** Polyurethane leather substitutes having smooth surfaces, *Jpn. Pat.* 75/39363, 1975.
212. **Broens, L., Koenhen, D. M., and Smolders, C. A.,** On the mechanism of formation of asymmetric ultra- and hyper-filtration membranes, *Desalination,* 22, 205, 1977.
213. **Koenhen, D. M., Mulder, M. H. V., and Smolders, C. A.,** Phase separation phenomena during the formation of asymmetric membranes, *J. Appl. Polym. Sci.,* 21, 199, 1977.
214. **Yamashita, S., Nagata, S., and Takakura, K.,** Ethylene-vinyl alcohol copolymer membranes for hemodialysis. 1. Studies on preparation conditions and properties of ethylene-vinyl alcohol copolymer membranes for hemodialysis, *Kobunshi Ronbunshu,* 36, 249, 1979.
215. **So, M. T., Eirich, F. R., Strathmann, H., and Baker, R. W.,** Preparation of asymmetric Loeb-Sourirajan membranes, *J. Polym. Sci. Polym. Lett.,* 11, 201, 1973.
216. **Chalykh, T. I., Grivtsova, L. A., and Chuikova, L. F.,** Methods for calculating the hydrophilicity of polymers, *Kozh.-Obuvn. Prom-st.,* 21(9), 50, 1979.
217. **Whitby, G. A., Evans, A. B. A., and Pasternack, P. S.,** The influence of chemical structure on the inhibition of liquids by rubber. I. *Trans. Faraday Soc.,* 38, 269, 1942.

218. **Avtushko, V. I., Kedrinskii, I. A., Morozova, O. G., and Raikhelson, L. B.,** Study of the corrosion resistance of some rubbers in solutions based on organic solvents, *Khim. Tekh. Polim.*, 4, 151, 1975.
219. **de Broutelles, R. and Opigez, D.,** Quantitative prediction of elastomer swelling in solvents, *Rev. Gen. Caoutch Plast.*, 52, 264, 1975.
220. **Maeda, S.,** Oil resistance of rubber. Treatment of solubility parameters, *Gosei Gomu,* 19 (1), 21, 1978.
221. **Gurarii, M. G.,** Effect of lubricants on the physical and technological properties of polymers, *Plast. Massy,* No. 4, 33, 1959.
222. **Zavattini, O. and Zotteri, L.,** New fuels and plastics: a developmental stage, *Technopolim. Resine,* No. 2, 13, 1982.
223. **Beerbower, A., Pattison, D. A., and Staffin, G. D.,** Predicting elastomer-fluid compatibility for hydraulic systems, *Am. Soc. Lubric. Eng. Trans.*, 6, 80, 1963.
224. **Eleftherakis, J. G.,** Hydraulic fluid/elastomer compatibility as estimated by solubility parameters, *Rubber Chem. Technol,* 59, 673, 1986 (abstract only).
225. **Catsiff, E. H., Coulehan, R. E., DiPrima, J. F., Gordon, D. A., and Seltzer, R.,** The effect of alkyl substituents on the properties of cured hydantoin epoxy resins, in *Epoxy Resin Chemistry (Am. Chem. Soc. Symp. Ser.* 114), Bauer, R. S., Ed., American Chemical Society, Washington, D.C., 1979, 115.
226. **Danilov, A. M. and Sergeeva, A. V.,** Evaluation of compatibility of thickening agents and dispersion media for ureate lubricating greases, *Khim. Teknol. Topliv Masel,* No. 7, 17, 1985; *Chem. Technol. Fuels Oils,* No. 7, 353, 1985.
227. **Rudin, A., Johnston, H. K., and Nicholson, R. E.,** Stability of polymer mixtures. II. Application of kinetic and solubility parameter models to polystyrene dispersions, *J. Paint Technol.*, 44(568), 54, 1972.
228. **Takeda, K., Akiyama, M., and Yamamizu, T.,** Shape-memory pore structure in porous crosslinked polystyrene, *Angew. Makromol. Chem.*, 157, 123, 1988.
229. **Boyer, R. F., Baer, E., and Hiltner, A.,** Concerning gelation effects in atactic polystyrene solutions, *Macromolecules,* 18, 427, 1985.
230. **Tan, H.-M., Moet, A., Hiltner, A., and Baer, E.,** Thermoreversible gelation of atactic polystyrene solutions, *Macromolecules,* 16, 28, 1983.
231. **Guerrero, S. J. and Keller, A.,** The gelation of PVC: characterization and control, *J. Macromol. Sci. Phys.*, B20, 167, 1981.
232. **Larsen, H. O. and Kindt-Larsen, T.,** Shaped hydrogel articles, *Eur. Pat.* 182659, 1986.
233. **Nelson, P. D. and Ohm, R. F.,** Fluid resistance of Norsorex® polynorbornene (PNR), *Elastomerics,* 111(1), 28, 1979; Paper No. 34, Polymer and Chemicals Dept., American Cyanamid Co., Bound Brook, NJ, 1979.
234. **Miller, R. F.,** Chemical and solvent resistance of selected styrene-maleic anhydride copolymers, *Am. Chem. Soc. Symp. Ser.*, 229, 205, 1983.
235. **Czubryt, J. J., Myers, M. N., and Giddings, J. C.,** Solubility phenomena in dense carbon dioxide gas in the range 270-1900 atmospheres, *J. Phys. Chem.*, 74, 4260, 1970.
236. **Miroshnichenko, A. A.,** Thermodynamics of liquid-vapor equilibrium in solutions. VII. Calculations of excess enthalpy of mixing and Gibbs energy of low molecular weight substances with polymer melts from molecular data, *Zh. Fiz. Khim.*, 56, 2670, 1982; *Russ. J. Phys. Chem.*, 56, 1644, 1982.
237. **Iwai, Y., Nagafui, M., and Arai, Y.,** Prediction of solubilities for gases in molten polymers using a solution model, *Technol. Rep. Kyushu Univ.*, 53, 387, 1980.
238. **Stiel, L. I., Chang, D.-K., Chu, H.-H, and Han, C. D.,** The solubility of gases and volatile liquids in polyethylene and polyisobutylene at elevated temperatures, *J. Appl. Polym. Sci.*, 30, 1145, 1985.
239. **Mikhailov, M. and Bogdanov, B.,** Phase states of liquid mixtures of styrene polymers and copolymers with low molecular weight solvents. I. Liquid mixtures of styrene polymers and a single low-molecular-weight liquid, *God. Vissh. Khim.-Tekhnol. Inst. Burgas Bulg.* (pub. 1982), 16(2), 69, 1980.
240. **Moore, W. R. and Hutchinson, R. J.,** Solvent power of solvents for poly(vinyl chloride), *J. Appl. Polym. Sci.*, 8, 2619, 1964.
241. **Schwenkedel, S. and Koschwitz, D.,** Recovery of PVC from PVC coated scrap, *Kunststoffe,* 70, 313, 1980; *German Plast.*, 70, 3, 1980.
242. **Miller, R. F. and Rider, P. E.,** Hostile effects of oxygenated solvent interaction with Type II rigid poly(vinyl chloride), *Am. Chem. Soc. Symp. Ser.*, 229, 93, 1983.
243. **Moore, W. R. and Boden, G. F.,** Heptane-soluble material from atactic polypropylene. II. Interaction with liquids, *J. Appl. Polym. Sci.*, 10, 1121, 1966.
244. **Mirtscheva-Tabakova, V.,** Physical-chemical and physical-mechanical properties of vulcanizates, obtained by Raney nickel desulfurized rubber networks, *Kautsch. Gummi Kunstst.*, 32, 245, 1979.
245. **Gozdz, A. S. and Kolarz, B. N.,** Exchange equilibria of anion resins, I. Effect of physical structure of macroporous styrene-divinylbenzene copolymers, *Makromol. Chem.*, 180, 2473, 1979.
246. **Gustafson, R. L., Albright, R. L., Heisler, J., Lirio, J. A., and Reid, O. T., Jr.,** Adsorption of organic species by high-surface-area styrene-divinylbenzene copolymers, *Ind. Eng. Chem. Proc. Res. Dev.*, 7, 107, 1968.

247. **Gustafson, R. L. and Lirio, J. A.,** Adsorption of organic ions by anion exchange resins, *Ind. Eng. Chem. Proc. Res. Dev.* 7, 116, 1968.
248. **Kamata, S. and Gohara, M.,** Solvent sorption of ion-exchange resins in non-aqueous solvents, *Nippon Kagaku Kaishi,* 532, 1977.
249. **Takaki, M., Asami, R., and Tsuzuki, S.,** Partial alkali metalation of poly(*p*-hydroxystyrene) and poly(*p*-hydroxystyrene-co-styrene) and the solubility of the resulting polymer in tetrahydrofuran, *Polymer,* 13, 135, 1981.
250. **Cowie, J. M. G.,** The effect of solvent on the unperturbed dimensions of polystyrene, *J. Polym. Sci. C.,* 23, 267, 1968.
251. **Cowie, J. M. G., Dey, R., and McCrindle, J. T.,** A comparative study of preferential adsorption in bromoform (1), benzene (2), and polymer (3) systems by light scattering and density gradient ultracentrifugation, *Polym. J.,* 2, 88, 1971.
252. **Cowie, J. M. G. and McCrindle, J. T.,** Polymer co-solvent systems. II. An asymmetric solubility parameter system — acetone (1), cyclohexanol (2), polystyrene (3), *Eur. Polym. J.,* 8, 1325, 1972.
253. **Heil, J. F.,** *Phase Equilibria in Multi-Component Polymer Solutions,* Doctoral dissertation, University of California, Berkeley, 1965; *Diss. Abstr.,* 26, 7196, 1966; **Heil, J. F. and Prausnitz, J. M.,** Phase equilibria in polymer solutions, *Am. Inst. Chem. Eng. J.,* 12, 678, 1966.
254. **Tseng, C.-M., El-Aasser, M. S., and Vanderhoff, J. W.,** Optimization of polymer-solvent interaction in a multi-solvent system by the UNIFAC group contribution method, *J. Appl. Polym. Sci.,* 32, 5007, 1986.
255. **Kang, S. J. and Ha, W. S.,** Swelling behavior of poly(ethylene terephthalate) by two component solvent mixture, *Han'guk Somyu Konghakhoechi (J. Korean Soc. Text. Eng. Chem.),* 21, 40, 1984.
256. **Chisso Corp.,** Purification of propylene random copolymers for films, *Jpn. Pat.,* 60/47010, 1985.
257. **Fukuda, T. and Inagaki, H.,** Interactions between unlike polymers versus dilute solution properties of copolymers, *Pure Appl. Chem.,* 55, 1541, 1983.
258. **Nagata, M., Fukuda, T., and Inagaki, H.,** Light scattering from polymer blend solutions. 5. Characterization of systems of relatively high incompatibility, *Macromolecules,* 20, 2173, 1987.
259. **Fukuda, T. and Inagaki, H.,** Light scattering characterization of polymer blends in solution, in *Proc. Second Rolduc Polymer Meeting* (Limburg, Netherlands), Elsevier, Amsterdam, 1987.
260. **Sattelmeyer, R.,** The solubility parameters of synthetic resins and elastomers for the plastics sector, *Adhäsion,* 20, 278, 1976.
261. **Van Dyck, J. W., Frisch, H. L., and Wu, D. T.,** Solubility, solvency, solubility parameters, *Abstr. Pap. Nat. Meet. Am. Chem. Soc.,* 188, 108, 1984; *Ind. Eng. Chem. Prod. Res. Nat. Meet. Dev.,* 24, 473, 1985.
262. **Lloyd, D. R., Gerlowski, L. E., Sunderland, C. D., Wightman, J. P., McGrath, J. E., Iqbal, M., and Kang, Y.,** Ploly(aryl ether) membranes for reverse osmosis, *Am. Chem. Soc. Symp. Ser.,* 153 (*Synthetic Membranes,* Vol. 1), 327, 1981.
263. **Ventoza, T. P., Gouye, S. A., and Lloyd, D. R.,** Use of the three component solubility parameter in asymmetric phase inversion membrane preparation, *Polym. Mater Sci Eng.,* 51, 713, 1984; *Abstr. Pap. Nat. Meet. Am. Chem. Soc.,* 188, POLYM 143, 1985.
264. **Ventoza, T. P. and Lloyd, D. R.,** Poly(ether sulfone) membranes for desalination: membrane preparation and characterization, *Desalination,* 56, 381, 1985.
265. **Lloyd, D. R. and Meluch, T. B.,** Selection and evaluation of membrane materials for liquid separations, *Am Chem. Soc. Symp. Ser.,* 269 *(Material Science of Synthetic Membranes),* 47, 1985.
266. **Kinzer, K. E., Lloyd, D. R., Wightman, J. P., and McGrath, J. E.,** Asymmetric membrane preparation from nonsolvent casting systems, *Desalination,* 46, 327, 1983.
267. **Kinzer, K. E., Lloyd, D. R., Gay, M. S., Wightman, J. P., Johnson, B. C., and McGrath, J. E.,** Phase inversion sulfonated polysulfone membranes, *J. Membr. Sci.,* 22, 1, 1985.
268. **Skelly, J. K.,** Alternatives to water in textile processing, *J. Soc. Dye Col.,* 91, 177, 1975.
269. **Lim, Y. J., Choi, H. J., Yoon, N. S., and Lee, H. K.,** Dyeability of poly(ethylene terephthalate) pretreated with organic solvent, *Han'guk Somyu Konghahoechi (J. Korean Soc. Text. Eng. Chem.),* 21, 281, 1984.
270. **Eisenbach, M., Kuppermann, B., Fleischer, N., Robinson, T., and Caplan, S. R.,** Purple membrane in solvents: solubility parameter mapping, in *Energetics and Structure of Halophilic Microorganisms (Dev. Halophilic Microorg.,* 1), Caplan, S. R. and Ginzburg, M., Eds., Elsevier/North Holland, Amsterdam, 1978, 239.
271. **Eisenbach, M., Caplan, S. R., and Tanny, G.,** Interaction of purple membrane with solvents. I. Applicability of solubility parameter mapping; **Eisenbach, M. and Caplan, S. R.,** II. Mode of interaction, *Biochim. Biophys. Acta,* 554, 269 and 281, 1979.
272. **Sevèstre, J.,** The solvent power of light chlorofluorohydrocarbons toward certain film-forming polymers, *Peint. Pigments Vernis,* 42, 838, 1966; Comparison of the solvent powers of trichloroethane and trichloroethylene, *Pitture Vernici,* 49, 227, 1973.

273. **Shah, V. M., Hardy, B. J., and Stern, S. A.,** Solubility of carbon dioxide, methane, and propane in silicone polymers: effect of polymer side chain, *J. Polym. Sci. Polym. Phys.,* 24, 2033, 1986.
274. **Fukuchi, K. and Arai, Y.,** Molar enthalpy of solution of a gas in a polymer near glass transition point, *Netsu Sokutei,* 7(1), 2, 1980.
275. **Schott, H.,** Solubility parameter, specific molar cohesion, and the solubility of ethylene oxide in polymers, *Biomaterials,* 3, 195, 1982.
276. **Cinti, G., Schromek, N., and Torriano, G.,** Amine cured epoxy systems for ballast tanks. The influence of the nature of the solvent mixture on film formation, properties and performance, *J. Oil Col. Chem. Assoc.,* 65, 171, 1982.
277. **Cowie, J. M. G. and McEwen, I. J.,** Observation of a super-glass-transition event in the damping spectra of amorphous block copolymers, *Macromolecules,* 12, 56, 1979.
278. **Karim, K. A. and Bonner, D. C.,** Determination of solvent selectivity for polymers using gas-liquid chromatography, *Soc. Plast. Eng. Tech. Pap.,* 22, 3, 1976.
279. **Saitoh, K. and Suzuki, M.,** Solvent effects in the gel chromatography of n-alkanes on Merckogel OR-2000, *J. Chromatogr.,* 111, 29, 1975; Gel chromatography of β-diketones and their metal complexes. IV. The partition concept in gel chromatography, *Bull. Chem. Soc. Jpn.,* 51, 116, 1978.
280. **Saitoh, K., Ozawa, T., and Suzuki, N.,** Solvent dependence of swelling of Merckogel OR-2000 and various organic solvents, *J. Chromatogr.,* 124, 231, 1976.
281. **Bennett, R. R.,** Solvent balance, *Mod. Paint Coatings,* 67(10), 93, 1977.
282. **Ramsbotham, J.,** Solvent formulation for surface coatings, *Prog. Org. Coatings,* 8, 113, 1980.
283. **Schreiber, H. P.,** Solvent balance, dispersion and rheological properties of pigmented polymer compositions, *J. Paint Technol.,* 46(598), 35, 1974.
284. **Tess, R. W.,** Solvents, in *Applied Polymer Science,* 2nd ed. (*Am. Chem. Soc. Symp. Ser.* 285), Tess, R. W. and Poehlein, G. W., Eds., American Chemical Society, Washington, D.C., 1985, 661.
285. **Wu, D. T.,** Modeling and simulation in the coatings industry, *Chemtech,* 17, 26, 1987.
286. **Ashton, N. F., McDermott, C., and Brench, A.,** Chemistry of extraction of nonreacting solutes, in *Handbook of Solvent Extraction,* Lo, T. C., Baird, M. H. I., and Hanson, C., Ed., John Wiley & Sons, New York, 1983, 3.
287. **Dintenfass, L.,** The nature and effects of solvent power in paint systems, *Paint J. Aust. N.Z.,* 2(8), 9, 1957.
288. **Loev, A. M. and Lavrishchev, L. P.,** Method for the selection of solvents for coating materials, *Lakokras. Mater. Ikh. Primen.,* No. 4, 4, 1977.
289. **Tortorello, A. J., Kinsella, M. A., and Gearon, R. T.,** The solubility parameter concept in the design of polymers for high performance coatings, in *Macromolecular Solutions: Solvent-Property Relationships in Polymers,* Seymour, R. B. and Stahl, G. A., Eds., Pergamon, New York, 1982, 26; **Tortorello, A. J. and Kinsella, M. A.,** Solubility parameter concept in the design of polymers for high performance coatings I. Acrylic polymers; II. Urethane copolymers, *J. Coatings Technol.,* 55 (696), 99, 1983; 55(697), 29, 1983.
290. **Glass, J. E.,** Adsorption characteristics of water-soluble polymers. I. Poly(vinyl chloride) and poly(vinylpyrrolidine) at the aqueous-air interface; II. Poly(ethylene oxide) at the aqueous-air interface, *J. Phys. Chem.,* 72, 4450 and 4459, 1968.
291. **Glass, J. E.,** The role played by water-soluble polymers in paint performance. Part II. Chemical modeling studies, *J. Oil Col. Chem. Assoc.,* 59(3), 86, 1976.
292. **Ramsbotham, J.,** Solvent system design for epoxy resin, *Paint Manuf. Resin News,* 49 (March), 26; (April), 18, 1979.
293. **Papo, A., Sciarrone, U., and Torriano, G.,** The solubility parameter. A new tool for the evaluation of the solvent resistance of organic coatings, *Proc. Int. Conf. Org. Coatings Sci. Technol.,* 4th, State University of New York, New Paltz, NY, 1978, 585.
294. **Vargiu, S., Parma, G., and Olcese, T.,** The solubility parameter: an approach to optimize the performance of protective coatings, *Pitture Vernici,* 51, 203, 1975; *FATIPEC (Fed. Assoc. Tech. Ind. Peint. Vernis Emaux Encres Impr. Eur. Cont.) Congr.,* 12, 411, 1974.
295. **Hansen, C. M.,** Solvents in water-borne coatings, *Ind. Eng. Chem. Prod. Res. Dev.,* 16, 266, 1977.
296. **Paul, S.,** Water-borne acrylic emulsion paints, *Prog. Org. Coatings,* 5, 79, 1977.
297. **Seymour, R. B. and Owen, D. R.,** Recent advances in water-based polymer systems, *Am. Paint J.,* 58(8), 14, 1973.
298. **Zunker, D. W.,** Unique aqueous nitrocellulose/acrylic air-dry systems for industrial finishing, *J. Coatings Technol.,* 48(616), 37, 1976.
299. **Brendley, W. H.,** Fundamentals of acrylic polymers, *Paint Varnish Prod.,* 63(7), 19, 1973.
300. **Brendley, W. H. and Haag, T. H.,** Physical characterization of water dispersed and soluble acrylic polymers, in *Nonpolluting Coatings and Coating Processes,* Gardon, J. L. and Prane, J. W., Eds., Plenum Press, New York, 1973, 5.

301. **Hoy, K. L.,** Solubility parameter as a design parameter for water-borne polymers and coatings, *Polym. Mater. Sci. Eng.*, 51, 566, 1984.
302. **Ramsbotham, J.,** Solvent loss from paint films, *J. Oil Col. Chem. Assoc.*, 62, 359, 1979.
303. **Schärer, D. H. and Tysall, L. A.,** Film appearance and its dependence on solvent formulation: a more systematic approach, *J. Oil Col. Chem. Assoc.*, 54, 927, 1971.
304. **Reichert, K.-H.,** Gas chromatographic study of the interactions forces between binder and solvent in paint films, *J. Oil Col. Chem. Assoc.*, 54, 887, 1971.
305. **Schoff, C. K.,** Recent advances in the rheology of high solids coatings, *Progr. Org. Coatings,* 4, 189, 1976.
306. **Hinden, E. R. and Whyzmusis, P. D.,** Polyamide resin solubility parameters, *Abstr. Pap. Am. Chem. Soc.*, 162, ORPL 65, 1971; *Am. Chem. Soc. Div. Org. Coatings Plast. Chem. Pap.*, 31, 496, 1971.
307. **Hinden, E. R. and Whyzmusis, P. D.,** Polyamide resin solubility, in *Solvents Theory and Practice* (*Adv. Chem. Ser.* 124), Tess, R. W., Ed., American Chemical Society, Washington, D.C., 1973, chap 11.
308. **Kamath, V. R. and Sargent, J. D., Jr.,** Production of high solids acrylic coating resins with t-amyl peroxides: a new way to meet VOC requirements, *J. Coatings Technol.*, 59(746), 51, 1987.
309. **Einax, J., Wolfe, F., and Rödicker, H.,** Relationship between pervaporation, diffusion and solubility of organic solvents on pore-free high polymer membranes, *Plaste Kautsch.*, 28, 202, 1981.
310. **Gul', V. E. and Balavintseva, E. K.,** Influence of specific cohesive energies of polymer and penetrant on the separative properties of polymeric membranes, *Vysokomol. Soedin. Ser B,* 24, 716, 1982; *Int. Polym. Sci. Technol.*, 10, T/66, 1983.
311. **König, U. and Schuch, H.,** Constitution and permeability of plastics, *Kunststoffe,* 67(1), 27, 1977; *Germ. Plast.*, 67(1), 11, 1977.
312. **Lonsdale, H. K.,** The growth of membrane technology, *J. Membr. Sci.*, 10, 81, 1982.
313. **Mangaraj, D., Pfau, J. P., Luttinger, M., Block, F., and Magee, W. S.,** An approach for estimating the permeation of phosphonofluoridates into a high polymer matrix Part 1. Development of the approach, *Polym. Eng. Sci.*, 27, 1338, 1987.
314. **Michaels, A. S., Wong, P. S. L., Prather, R., and Gale, R. M.,** A thermodynamic method of predicting the transport of steroids in polymer matrices, *Am. Inst. Chem. Eng. J.*, 21, 1073, 1975.
315. **Mulder, M. H. V. and Smolders, C. A.,** Pervaporation, solubility aspects of the solubility-diffusion model, *Sep. Purif. Meth.*, 15, 1, 1986.
316. **Paul, D. R.,** Transport properties of polymers, in *Applied Polymer Science,* 2nd ed. (*Am. Chem. Soc. Symp. Ser.* 285), Tess, R. W. and Poehlein, G. W., Eds., American Chemical Society, Washington, D.C., 1985, 253.
317. **Schneier, B.,** Use of a Gee relationship in a diffusion study, *J. Appl. Polym. Sci.*, 16, 2343, 1972.
318. **Senich, G. A.,** Migration to and from plastics, *Chemtech.* 11, 360, 1981.
319. **Cabasso, I.,** Organic liquid mixtures separation by permselective polymer membranes. I. Selection and characteristics of dense isotropic membranes employed in the pervaporation process, *Ind. Eng. Chem. Prod. Res. Dev.*, 22, 313, 1983.
320. **Christensen, U. L. and Banke, O.,** Gloves and solubility parameters, *Dan. Kem.*, 63(3), 70, 1982; *Gloves: safe/unsafe,* Institute for Arbejdsmiljo, Danmarks Tekniske Hojskole, Lyngby, Denmark, 1983.
321. **Henriksen, H. R.,** *Selection of Materials for Protective Gloves. Polymer Membranes to Protect Against Contact with Epoxy Products,* translated from *Materials for Protective Gloves,* Danish Directorate of Labor Inspection Services, Copenhagen, 1982; **Henriksen, H. R., Styr Petersen, H. J., and Madsen, J. O.,** Solubility parameters and permeation resistance (tightness), *Report on Project 1982-39,* Institute for Chemical Industries, Technical University of Denmark, Lyngby, Denmark, 1986.
322. **Mulder, M. H. V., Kruitz, F., and Smolders, C. A.,** Separation of isomeric xylenes by pervaporation through cellulose ester membranes, *J. Membr. Sci.*, 11, 349, 1982.
323. **Perkins, J. L. and Tippit, A. D.,** Use of three-dimensional solubility parameter to predict glove permeation, *Am. Ind. Hyg. Assoc. J.*, 46, 455, 1985.
324. **Perkins, J. L., Ridge, M. C., Holcombe, A. B., and Wang, M. K.,** Skin protection, Viton, and solubility parameters, *Am. Ind. Hyg. Assoc. J.*, 47, 803, 1986.
325. **Perepechkin, L. P., Sobakin, V. G., Budnitskii, G. A., Biber, B. L., Baibakova, Z. V., Rud'ko, L. V., Kutnova, R. K., Yefimova, S. G., and Sokira, A. N.,** Study of the formation of capillaries with selectively permeable walls, *Vysokomol. Soedin. Ser. A,* 25, 1549, 1983; *Polym. Sci. U.S.S.R.*, 27, 1794, 1983.
326. **Vahdat, N.,** Permeation of polymeric materials by toluene, *Am. Ind. Hyg. Assoc. J,* 48, 155, 1987.
327. **Sanchez, I. C.,** Equilibrium distribution of a minor constituent between a polymer and its environment, in *Durability of Macromolecular Materials* (*Am. Chem. Soc. Symp. Ser.* 95), Eby, R. K., Ed., American Chemical Society, Washington, D.C., 1979, chap. 12.
328. **Roe, R.-J., Bair, H. E., and Gieniewski, C.,** Solubility and diffusion coefficient of antioxidants in polyethylene, *J. Appl. Polym. Sci.*, 18, 843, 1974.

329. **Higashide, F., Nozawa, Y., and Ohmura, U.,** Naphthalene release from poly(vinyl chloride) containing glass fillers, *Makromol. Chem. Rapid Commun.,* 2, 531, 1981.
330. **Higashide, F., Ohmura, Y., and Nozawa, Y.,** Effect of glass flakes in poly(vinyl chloride) on the permeability to gases and on the releasing behavior of naphthalene, *Kobunshi Ronbunshu,* 38, 611, 1981.
331. **Derrickson, J. L., Billing, C. B., Jr., Man, V. L., Bastecki, V. A., and Bentz, A. P.,** Permeation of protective clothing materials. Comparison of methods to determine solubility parameter: examination of variables affecting breakthrough time, *Pittsburg Ann. Conf. Analyt. Chem. Appl. Spectr.,* Atlantic City, NJ, 1986.
332. **Forsberg, K. and Faniadis, S.,** The permeation of multi-component liquids through new and pre-exposed glove materials, *Am. Ind. Hyg. Assoc. J.,* 47, 189, 1986.
333. **Mikatavage, M., Que Hee, S. S., and Ayer, H. E.,** Permeation of chlorinated aromatic compounds through Viton® and nitrile glove materials, *Am. Ind. Hyg. Assoc. J.,* 45, 617, 1984.
334. **Spence, M. W.,** Chemical permeation through protective glove material: an evaluation of several critical variables, Am. Ind. Hyg. Conf., Portland, OR, Abstr. 395, 1981.
335. **Sumitomo, H. and Kazuhiko, H.,** Polyamides as barrier materials, *Adv. Polym. Sci.,* 64(*Polymer Membranes*), 63, 1985.
336. **Thiel, S. W., Lloyd, D. R., and Dickson, J. M.,** Physicochemical interpretation of the behavior of a pressure-driven membrane separation process, *Am. Chem. Soc. Symp. Ser.,* 281, 125, 1985.
337. **Makhmutov, F. A., Nazarov, V. G., Belyakev, V. K., and Manin, V. N.,** Asymmetry of permeability of multilayer films, *Vysokomol. Soedin. Ser. B,* 25, 387, 1983.
338. **Néel, J., Aptol, P., and Clément, R.,** Basic aspects of pervaporation, *Desalination,* 53, 297, 1985.
339. **Zhu, C.-L., Yuang, C.-W., and Greenburg, D. W.,** Pervaporation membranes — a novel separation technique for trace organics, *Environ. Progr.,* 2, 132, 1983.
340. **Zhu, C.-L., Liu, M., and Xu, W.,** Separation of ethanol-water mixtures by pervaporation membrane separation process, *Desalination,* 62, 299, 1987.
341. **Yoshikawa, M. and Ogata, N.,** Polymer membrane as a reaction field. III. Effect of membrane polarity on selective separation of a water-ethanol binary mixture through synthetic polymer membranes, *J. Membr. Sci.,* 26, 107, 1986; **Yoshikawa, M., Ohsawa, T., Tanigaki, M., and Eguchi, W.,** Application of synthetic polymer membranes. II. Separation of aqueous ethanol solution through poly(*tert*-butyl methacrylate-*co*-styrene) membranes, *J. App. Polym Sci.,* 37, 299, 1989.
342. **Kurokawa, Y., Mukaigawara, H., and Saito, S.,** Reverse osmosis separation of several alcohols from aqueous solution, *Chem. Eng. Commun.,* 36, 333, 1985.
343. **Fiedler, U.,** Optimization of a sodium ion-selective electrode for use in serum measurements, *Analyt. Chim. Acta,* 89, 101, 1977.
344. **Fieldler-Linnersund, U. and Bhatti, K. M.,** Development of polymeric membranes for zinc ion-selective electrodes, *Analyt. Chim. Acta,* 111, 57, 1979.
345. **Nielsen, H. J. and Hansen, E. H.,** New nitrate ion-selective electrodes based on quaternary ammonium compounds in nonporous polymer membranes, *Analyt. Chim. Acta,* 85, 1, 1976.
346. **Igawa, M., Nakamura, Y., Takahashi, K., Tanaka, M., and Senô, M.,** Reverse osmosis separation of alkaline metal ions through hydrophobic membranes, *J. Polym. Sci. Polym. Lett. Ed.,* 20, 165, 1982.
347. **Igawa, M., Torii, K., Tanaka, M., and Senô, M.,** Selective ion permeation through fluorocarbon membrane, *J. Appl. Polym. Sci.,* 29, 117, 1984.
348. **Daughton, C. G., Jones, B. M., and Sakaji, R. H.,** Organic nitrogen determination in oil shale retort waters, *Analyt. Chem.,* 57, 2326, 1985.
349. **Spencer, H. G. and Gaddis, J. L.,** Hyperfiltration of nonelectrolytes: dependence of rejection on solubility parameters, *Desalination,* 28, 117, 1979; *U.S. Environ. Prot. Agency, Off. Res. Dev.,* [*Rep.*] *EPA,* 1978, EPA-600/2-79-104, *Symp. Proc. Text. Ind. Technol.,* PB 299132, 107.
350. **Dickson, J. M., Babal-Pirouz, M., and Lloyd, D. R.,** Aromatic hydrocarbon-water separations by membranes, *Ind. Eng. Chem. Proc. Des. Dev.,* 22, 625, 1983.
351. **Nomura, H., Seno, M., Takahashi, H., and Yamabe, T.,** Permselectivities of various polymer membranes toward some aromatic compounds in organic media in the reverse osmosis process, *J. Membrane Sci.,* 5, 189, 1979.
352. **Nomura, H., Seno, M., Takahashi, H., and Yamabe, T.,** Permeability of phenol through cellulose acetate membranes by reverse osmosis in various alcoholic systems, *J. Appl. Polym. Sci.,* 24, 1191, 1979.
353. **Rånby, B. G.,** Two-component polymer systems: physical properties as related to compatibility and interaction, *J. Polym. Sci. Polym. Symp.,* 51, 89, 1975.
354. **Yampolskii, Yu. P., Krentsel, L. B., Azizov, A. A., Durgaryan, S. G., and Litmanovich, A. D.,** Gas-permeability of chlorinated polyvinyltrimethylsilane, *Vysokomol. Soedin. Ser. A,* 22, 455, 1980; *Polym. Sci. U.S.S.R.,* 22, 506, 1980.
355. **Li, N. N. and Henley, E. J.,** Permeation of gases through polyethylene films at elevated pressures, *Am. Inst. Chem. Eng. J.,* 10, 666, 1964.

356. **Hirose, T., Mizoguchi, K., and Kamiya, Y.,** Gas transport in poly(vinylcyclohexanecarboxylate), *J. Appl. Polym. Sci.*, 35, 517, 1988.
357. **Chern, R. T., Koros, W. J., Hopfenberg, H. B., and Stannett, V. T.,** Material selection for membrane-based gas separations, in *Materials Science of Synthetic Membranes* (*Am. Chem. Soc. Symp. Ser.* 269), Lloyd, D. R., Ed., American Chemical Society, Washington, D.C., 1982, 25.
358. **Rödicker, H. and Kroll, U.,** Contribution to gas phase permeation. Part III: On the solubility behavior of lower hydrocarbons to nonpolar membranes, *Chem. Tech. (Leipzig)*, 27, 724, 1975.
359. **Rollin, J., Rödicker, H., and Wolf, F.,** Comparative study of the diffusion of benzene, cyclohexane and n-hexane vapors in polyethylene, *Plaste Kautsch.*, 33, 335, 1986.
360. **Allen, S. M., Stannett, V., and Hopfenberg, H. B.,** Noble gas transport in poly(methyl vinyl ketone) and poly(methyl vinyl ether), *Polymer,* 22, 912, 1981.
361. **Barnes, J. D.,** Solubility parameters and diffusion coefficients for O_2 and N_2 in polyethylene terephthalate, *Bull. Am. Phys. Soc.*, 21(Abstr.), 414, 1976.
362. **Higashide, F., Omata, K., Nozowa, Y., and Yoshioka, H.,** Permeability to oxygen of poly(methyl methacrylate) films containing silane-treated glass flakes, *J. Polym. Sci. Polym. Chem.*, 15, 2019, 1977.
363. **Hara, H., Okazaki, S., and Fujinaga, T.,** On the enhancement of the selectivity of organic sulfonate ion electrode by alkylphenols, *Bull. Chem. Soc. Jpn.*, 53, 3610, 1980.
364. **Voyutski, S. S., Yagnyatinskaya, S. M., Kaplunova, L. Ya., and Garetovskaya, N. L.,** Elastomer adhesion, *Rubber Age* (NY), 105(2), 37, 1973.
365. **Sushkov, V. I., Gusev, S. S., Perepelkin, A. N., Minyailo, S. A., and Bablyuk, Ye. B.,** Study of the active influence of some solutions on the surface of polyethylene terephthalate films, *Vysokomol. Soedin. Ser. A,* 26, 2291, 1984; *Polym. Sci. U.S.S.R.*, 26, 2560, 1984.
366. **Gerrits, P. O., Horobin, R. W., Wright, D. J., van Goor, H., and Grond, J.,** Retention, localization and staining of lipids in tissues embedded in water-miscible methacrylates: using the solubility parameter to analyse the influences of resins, plasticizers, and other reagents, *Acta Morphol. Neerl.-Scand.*, 25, 29, 1987.
367. **Kinoshita, T., Sasaki, M., Takizawa, A., and Tsujita, Y.,** Solubility and diffusion of diluents in cellulose acetate membrane — a search for separation mechanism by polymer membrane, *Sen-i Gakkaishi,* 37(1), T32, 1981 (English).
368. **Sliwka, W.,** Microencapsulation, *Angew. Chem. Int. Ed.*, 14, 538, 1975.
369. **Michaels, A. S., Vieth, W. R., Hoffman, A. S., and Alcalay, H. A.,** Structure-property relations for liquid transport in modified polypropylene membranes, *J. Appl. Polym. Sci.*, 13, 577, 1969.
370. **Bataille, P., Boissé, S., and Schreiber, H. P.,** Mechanical properties and permeability of polypropylene and poly(ethylene terephthalate) mixtures, *Polym. Eng. Sci.*, 27, 622, 1987.
371. **Ng, H. C., Leung, W. P., and Choy, C. L.,** Sorption and diffusion of toluene in isotropic and oriented linear polyethylene, *J. Polym. Sci. Polym. Phys. Ed.*, 23, 973, 1985.
372. **Salame, M.,** A correlating parameter for liquid permeability to polyethylene, *Org. Coat. Appl. Polym. Sci. Proc.*, 46, 224, 1981.
373. **Frensdorff, H. K.,** Diffusion and sorption of vapors in ethylene-propylene copolymers. I. Equilibrium sorption; II. Diffusion, *J. Polym. Sci. A,* 2, 333 and 241, 1964.
374. **Schuler, M. T. and Stampa, G. B.,** Permeability of nitrile synthetics, *Kunstst.-Plast. (Solothurn),* 21(3), 19, 1974.
375. **Suda, Y. and Lee, K.-T.,** Diffusion of disperse dyes within nylon-6 from normal alcohols, *Nippon Kagaku Kaishi,* 1398, 1975.
376. **Razumovskii, L. P., Artsis, M. I., and Zaikov, G. E.,** Diffusion of organic compounds in polyamide PA-12, *Vysokomol. Soedin Ser. B,* 27, 893, 1985.
377. **Razumovskii, L. P., Markin, V. S., and Zaikov, G. E.,** Study of the solubility of water in polyamides according to the Flory-Huggins theory; Solubility and diffusion coefficient of water in aliphatic polyamides; Diffusion of alcohols into polyamides, *Vysokomol. Soedin. Ser. A,* 26, 1740, 1984; 27, 308 and 863, 1985; *Polym. Sci. U.S.S.R.*, 26, 1948, 1984; 27, 341 and 972, 1985.
378. **Razumovskii, L. P. and Zaikov, G. E.,** Determination of the solubility and diffusion coefficients of water in aliphatic polyamides as a step to studying their hydrolytic stability, *Polym. Degradat. Stability,* 12, 1, 1985.
379. **Razumovskii, L. P., Markin, V. S., and Zaikov, G. E.,** Sorption of water by aliphatic polyamides, *Vysokomol. Soedin. Ser. A,* 27, 675, 1985; *Polym. Sci. U.S.S.R.*, 27, 751, 1985.
380. **Veksli, Z., Miller, W. G., and Thomas, E. L.,** Penetration of nonsolvents into glassy, amorphous polymers, *J. Polym. Sci. Polym. Symp.*, 54, 299, 1976.
381. **Kinzig, B. J. and Ravner, H.,** Factors contributing to the properties of fluoropolymer barrier films, *Am. Soc. Lub. Eng. Trans.*, 21, 291, 1978.
382. **Yeo, R. S.,** Dual cohesive energy densities of perfluorosulfonic acid (Nafion) membrane, *Polymer,* 21, 432, 1980.

383. **Yeo, R. S., Chan, S. F., and Lee, J.,** Swelling behavior of Nafion and radiation-grafted cation-exchange membranes, *J. Membr. Sci.,* 9, 273, 1981.
384. **Shinozaki, T.,** Fiber permeation in solution, II, *Sen'i,* 27, 268, 1975.
385. **Grishin, B. S., Tutorskii, I. A., and Yurovskaya, I. S.,** Application of Gee's equation for assessment of solubility of solid low-molecular substances in polymers, *Vysokomol. Soedin. Ser. B,* 19, 387, 1977.
386. **Anagnostopoulos, C. E., Coran, A. Y., and Gamrath, H. R.,** Polymer-diluent interactions, I. A new micromethod for determining poly(vinyl chloride)-diluent interactions, *J. Appl. Polym. Sci.,* 4, 181, 1960.
387. **Anagnostopoulos, C. E. and Coran, A. Y.,** Polymer-diluent interactions. II. Polyvinyl chloride-diluent interactions, *J. Polym. Sci.,* 57, 1, 1962.
388. **Anagnostopoulos, C. E., Coran, A. Y., and Gamrath, H. R.,** Microtest for PVC-plasticizer compability, *Mod. Plast.,* 43(2), 141, 1965.
389. **Bigg, D. C. H.,** A new micromethod for the determination of plasticizer solvent power for poly(vinyl chloride); A comparison of methods for assessing the degree of interaction between plasticizers and poly(vinyl chloride), *J. Appl. Polym. Sci.,* 19, 1487 and 3119, 1975.
390. **Bigg, D. C. H. and Hill, R. J.,** The effect of plasticizer solvent power on the aging of poly(vinyl chloride) pastes, *J. Appl. Polym. Sci.,* 20, 565, 1976.
391. **Chen, S.-A.,** Polymer miscibility in organic solvents and in plasticizers — a two-dimensional approach, *J. Appl. Polym. Sci.,* 15, 1247, 1971.
392. **Gidanian, K. and Howard, G. J.,** Production of polymers from the seed oil of *Crambe Abyssinica.* II. PVC plasticizers based on tridecanedioic (brassylic) acid, *J. Macromol. Sci. Chem. A,* 10, 1399, 1976.
393. **Aklonis, J. J. and Rele, V. B.,** Relaxation of damped Debye lattices in the primary transition region, *J. Polym. Sci. Polym. Symp.,* 46, 127, 1974.
394. **Blow, C. M.,** Polymer-particulate filler interaction. Bound rubber phenomena, *Polymer,* 14, 309, 1970.
395. **Poimanov, A. M., Trostyanskaya, E. B., Nosov, E. F., and Belnik, A. R.,** Effect of fillers on the structure and properties of cured resins. 2. Effect of a filler on the network interaction and strength of cured epoxy resin, *Mekh. Polim.,* 6, 54, 1970; *Polym. Mech.,* 6, 46, 1970.
396. **Marmo, M. J., Mostafa, M. A., Jinnai, H., Fowkes, F. M., and Manson, J. A.,** Acid-base interaction in filler-matrix systems, *Ind. Eng. Chem. Prod. Res. Dev.,* 15, 206, 1976.
397. **Hansen, C. M.,** The three dimensional solubility parameter — key to paint component affinities. II. Dyes, emulsifiers, mutual solubility and compatibility, and pigments, *J. Paint Technol.,* 39, 505, 1967.
398. **Skaarup, K.,** The three dimensional solubility parameter and its use. II. Pigmented systems, *Skand. Tidskr. Färg Lack,* 14(2), 28; 14(3), 45, 1968.
399. **Rickert, S. E., Balik, C. M., and Hopfinger, A. J.,** Organization and assembly of macromolecules on inorganic substrates, *Adv. Colloid Interface Sci.,* 11, 149, 1979.
400. **Billingham, N. C., Calvert, P. D., and Manke, A. S.,** Solubility of phenolic antioxidants in polyolefins, *J. Appl. Polym. Sci.,* 26, 3543, 1981.
401. **Billingham, N. C.,** personal communication, 1983.
402. **Bohn, L.,** Incompatibility and phase formation in polymer mixtures and graft and block copolymers in the solid state, *Kolloid-Z. Z. Polym.,* 213, 55, 1966.
403. **Bohn, L.,** Incompatibility and phase formation in solid polymer mixtures and graft and block copolymers, *Rubber Chem. Technol.,* 41, 495, 1968.
404. **Bohn, L.,** Compatible polymers, in *Polymer Handbook,* 2nd ed., Brandrup, J. and Immergut, E. H., Eds., John Wiley & Sons, New York, 1974, III-211.
405. **Casper, R. and Morbitzer, L.,** Structure and properties of multiphase plastics. I. Compatibility of polymers in the solid state, *Angew. Makromol. Chem.,* 58/59, 1, 1977.
406. **Cooper, S. L. and Estes, G. M., Eds.,** *Multiphase Polymers (Adv. Chem. Ser.* 176), American Chemical Society, Washington, D.C., 1979.
407. **Lilaonitkul, A. and Cooper, S. L.,** Properties of polyether-polyester thermoplastic elastomers, *Rubber Chem. Technol.,* 50, 1, 1977.
408. **Corish, P. J.,** Fundamental studies of rubber blends, *Rubber Chem. Technol.,* 40, 324, 1967.
409. **Corish, P. J.,** Elastomer blends, in *Science and Technology of Rubber,* Eirich, F. R., Ed., Academic Press, New York, 1978, chap. 12.
410. **Gesner, B. J.,** Polyblends, in *Encyclopedia of Polymer Science and Technology,* Vol. 10, John Wiley & Sons, New York, 1969, 694.
411. **Heidel, K. and Sturzenhofecker, F.,** Development of polyester resins on the basis of fractional factorial design, *Angew. Makromol. Chem.,* 24, 187, 1972.
412. **Krause, S.,** Polymer compatibility, *J. Macromol. Sci. Rev. Macromol. Chem.,* C7, 251, 1972.
413. **Krause, S.,** Polymer-polymer compatibility, in *Polymer Blends,* Vol. 1, Paul, D. R. and Newman, S., Eds., Academic Press, New York, 1978, chap. 2, 19.
414. **Olabisi, O.,** Polyblends, in *Kirk-Othmer Encyclopedia of Chemical Technology,* 3rd ed., Vol. 18, 1982, 443.

415. **Paul, D. R.,** Background and perspective, in *Polymer Blends,* Vol. 1, Paul, D. R. and Newman, S., Eds., Academic Press, New York, 1968, chap. 1.
416. **Paul, D. R. and Barlow, J. W.,** A brief review of polymer blend technology, in *Multiphase Polymers* (*Adv. Chem. Ser.* 176), Cooper S. L. and Estes, G. M., Eds., American Chemical Society, Washington, D.C., 1979, chap. 17.
417. **Paul, D. R. and Barlow, J. W.,** Polymer blends (or alloys), *J. Macromol. Sci. Rev. Macromol. Chem.,* C18, 109, 1980.
418. **Paul, D. R.,** Thermodynamic theory and experimental techniques for polymer blends, in *Polymer Blends and Mixtures,* Walsh, D. J., Higgins, J. S., and Maconnachie, A., Eds., NATO Advanced Science Series E: Applied Sciences No. 89, Martinus Nijhoff, Dordrecht, 1985, 1.
419. **Platzer, N. A. J., Ed.,** *Multicomponent Polymer Systems* (*Adv. Chem. Ser.* 99), American Chemical Society, Washington, D.C., 1971.
420. **Platzer, N. A. J., Ed.,** *Copolymers, Polyblends and Composites* (*Adv. Chem. Ser.* 142), American Chemical Society, Washington, D.C., 1975.
421. **Roovers, J.,** Polymeric alloys, in *Methods of Experimental Physics,* Vol. 16, *Polymers,* Part C, *Physical Properties,* Fava, R. A., Ed., Academic Press, New York, 1980, 275.
422. **Sanchez, I. C.,** Statistical thermodynamics of polymer blends, in *Polymer Blends,* Vol. 1, Paul, D. R. and Newman, S., Eds., Academic Press, New York, 1978, chap 3.
423. **Schmitt, B. J. and Kirste, R. G.,** Molecular dispersion and thermodynamics of amorphous polymer blends, International Union of Pure and Applied Chemistry Meeting, Tashkent, 1978.
424. **Schmitt, B. J.,** Polymer alloys — their structure, morphology and properties, *Angew. Chem.,* 91, 286, 1979; *Angew. Chem. Int. Ed.,* 18, 273, 1979.
425. **Shvarts, A. G.,** The compatibility of high polymers, *Kolloid Zhur.,* 18, 755, 1956; *Colloid J. U.S.S.R.,* 18, 753, 1956.
426. **Slepakova, S. I., Geller, B. E., Koralnik, N. G., Geller, A. A., and Pshedetskaya, V. K.,** Correlation of compatibility and solubility parameters of polymers, *Deposited Doc. VINITI,* 519-78, 1978.
427. **Tlusta, H. and Zelinger, J.,** Miscibility of polymers, *Chem. Listy,* 65, 1143, 1971.
428. **Battaerd, H. A. J.,** The significance of incompatible polymer systems, *J. Polym. Sci. Polym. Symp.,* 49, 149, 1975.
429. **Battaerd, H. A. J. and Tregear, G. W.,** *Graft Copolymers* (*Polymer Reviews* 16), John Wiley & Sons — Interscience, New York, 1967.
430. **Helfand, E. and Tagami, Y.,** Theory of the interface between immiscible polymers. II, *J. Chem. Phys.,* 56, 3592, 1972.
431. **Helfand, E.,** Block copolymers, polymer-polymer interfaces, and the theory of inhomogeneous polymers, *Acc. Chem. Res.,* 8, 295, 1975.
432. **Kuleznev, V. N., Shvarts, A. G., Klykova, V. D., and Dogadkin, B. A.,** Physical and mechanical properties of microheterogeneous mixtures of incompatible polymers polybutadiene and butadiene-nitrile rubber, *Kolloid. Zh.,* 27, 211, 1965; *Colloid J. U.S.S.R.,* 27, 171, 1965.
433. **Rosen, S. L.,** Two-phase polymer systems, *Trans, N.Y. Acad. Sci.,* 35, 480, 1973.
434. **Bernstein, R. E., Cruz, C. A., Paul, D. R., and Barlow, J. W.,** LCST behavior in polymer blends, *Macromolecules,* 10, 681, 1977.
435. **Helfand, E.,** Proposed experiment to estimate the mixing free energy of polymers, *Macromolecules,* 11, 682, 1978.
436. **Koningsveld, R. and Staverman, A. J.,** Determination of critical points in multicomponent polymer solutions, *J. Polym. Sci. C,* 16, 1775, 1967.
437. **Koningsveld, R., Kleintjens, L. A., and Schoffeleers, H. M.,** Thermodynamic aspects of polymer compatibility, *Pure Appl. Chem.,* 39, 1, 1974.
438. **Koningsveld, R. and Kleintjens, L. A.,** Thermodynamics of polymer mixtures, *J. Polym. Sci. Polym. Symp.,* 61, 221, 1977.
439. **Olabisi, O., Robeson, L. M., and Shaw, M. T.,** *Polymer-Polymer Miscibility,* Academic Press, New York, 1979.
440. **Patterson, D. and Robard, A.,** Thermodynamics of polymer compatibility, *Macromolecules,* 11, 690, 1978.
441. **Gaylord, N. G.,** *Compatibilization of Polymers by Adding Graft Copolymers,* U.S. Pat. 3485777, 1969.
442. **Gaylord, N. G.,** Compatibilization concepts in polymer applications, in *Copolymers, Polyblends and Composites* (*Adv. Chem. Ser.* 142), Platzer, N. A., Ed., American Chemical Society, Washington, D.C., 1975, chap. 7.
443. **Gaylord, N. G.,** Role of compatibilization in polymer utilization, *Chemtech,* 6, 392, 1976.
444. **Olabisi, O. and Farnham, A. G.,** Polymer compatibilization: blends of polyarylethers with styrenic interpolymers, in *Multiphase Polyers* (*Adv. Chem. Ser.* 176), Cooper, S. L. and Estes, C. M., Eds., American Chemical Society, Washington, D.C., 1979, 559.

445. **Woods, M. E. and Mass, T. R.,** Fundamental considerations for the convulcanization of elastomer blends, in *Copolymers, Polyblends and Composites* (*Adv. Chem. Ser.* 142), Platzer, N. A. J., Ed., American Chemical Society, Washington, D.C., 1975, chap. 33.
446. **Zapp, R. L.,** Interfacial bonding between different elastomer phases, in *Multicomponent Polymer Systems* (*Adv. Chem. Ser.* 99), Platzer, N. A. J., Ed., American Chemical Society, Washington, D.C., 1971, chap. 6.
447. **Dror, M., Elsabee, M. Z., and Berry, G. C.,** Gradient interpenetrating polymer networks. I. Poly(ether urethane) and polyacrylamide IPN, *J. Appl. Polym. Sci.*, 26, 1741, 1981.
448. **Jehl, D., Widmaier, J. M., and Meyer, G. C.,** The transparency of polyurethane-poly(methyl methacrylate) interpenetrating and semi-interpenetrating polymer networks, *Eur. Polym. J.*, 19, 597, 1983.
449. **Ebdon, J. R., Hourston, D. J., and Klein, P. G.,** Polyurethane-polysiloxane interpenetrating networks: 1. A polyether urethane-poly(dimethylsiloxane) system, *Polymer*, 25, 1633, 1984.
450. **Hourston, D. J. and Satgurunathan, R.,** Latex interpenetrating networks based on acrylic polymers. 1. Predicted and observed compatibilities, *J. Appl. Polym. Sci.*, 29, 2969, 1984.
451. **Krause, S.,** Partial solubility parameter characterization of interpenetrating microphase membranes, in *Materials Science of Synthetic Membranes (Am. Chem. Soc. Symp. Ser.* 269), Lloyd, D. R., Ed., American Chemical Society, Washngton, D.C., 1985, 351; *Polym. Mater. Sci. Eng.*, 50, 102, 1984.
452. **Takemura, A., Tomita, B.-I., and Mizumachi, H.,** Synthesis of IPN-related materials from solution of polyacrylates, *J. Appl. Polym. Sci.*, 32, 2489, 1986.
453. **Patterson, D., Delmas, G., and Somcynsky, T.,** A comparison of lower critical solution temperatures of some polymer solutions, *Polymer*, 8, 503, 1967.
454. **Aharoni, S. M.,** Instataneous shape and segmental density of flexible molecules. II. Observation of individual and aggregate poly-cis-isoprene molecules in polyisobutylene, *J. Macromol. Sci. Phys.*, B15, 635, 1978.
455. **Shaw, M. T.,** Studies of polymer-polymer solubility suing a two-dimensional parameter approach, *J. Appl. Polym. Sci.*, 18, 449, 1974.
456. **Cabasso, I., Jagur-Grodzinski, J., and Vofsi, D.,** Advanced polymer alloys containing polyphosphonate components, in *Polymer Science and Technology*, Vol. 10, *Polymer Alloys: Blends, Blocks, Grafts, and Interpenetrating Networks*, Klempner, D. and Frisch, K. C., Eds., Plenum Press, New York, 1977, 1.
457. **Roe, R.-J.,** Evaluation of the polymer-polymer interaction parameter from the equation of state thermodynamic theory, in *Multiphase Polymers (Adv. Chem. Ser.* 176), Cooper, S. L. and Estes, G. M., Eds., American Chemical Society, Washington, D.C., 1979, chap 31; Roe, R.-J. and Zin, W. C., Determination of the polymer-polymer interaction parameter for the polystyrene-polybutadiene pair, *Macromolecules*, 19, 1221, 1980.
458. **Lin, J.-L. and Roe, R.-J.,** Relationship among polymer-polymer interaction energy densities and the deuterium isotope effect, *Macromolecules*, 20, 2168, 1987.
459. **Cabasso, I.,** A study of compatible polymer systems employing solubility parameter diagrams, *Abstr. Pap. Am. Chem. Soc.*, 173, ORPL-21, 1977.
460. **Barton, A. F. M.,** *Resource Recovery and Recycling*, John Wiley & Sons — Interscience, New York, 1979.
461. **Kuhn, R.,** Fractionation of polymers with demixing solvents, *Makromol. Chem.*, 177, 1525, 1976.
462. **Seymour, R. B. and Sosa, J. M.,** Estimation of solubility parameters of nonelectrolytes, *Nature (London)*, 248, 759, 1974.
463. **Seymour, R. B. and Stahl, G. A.,** Separation of waste plastics. An experiment in solvent fractionation, *J. Chem. Educ.*, 53, 653, 1976.
464. **Robard, A. and Patterson, D.,** Temperature dependence of polystyrene-poly(vinyl methyl ether) compatibility in trichloroethene, *Macromolecules*, 10, 1021, 1977.
465. **Rostami, S. and Walsh, D. J.,** Miscibility of ethylene-vinyl acetate copolymers with chlorinated polyethylenes. 3. Simulation of the spinodal using the equation of state theory, *Macromolecules*, 17, 315, 1984.
466. **Rostami, S. and Walsh, D. J.,** Simulation of upper and lower critical phase diagrams for polymer molecules at various pressures, *Macromolecules*, 18, 1228, 1985.
467. **Waywood, W. J. and Durning, C. J.,** Solvent-induced crystallization of a compatible polymer blend, *Polym. Eng. Sci.*, 27, 1265, 1987.
468. **Olabisi, O.,** Polymer compatibility by gas-liquid chromatography, *Macromolecules*, 8, 316, 1975.
469. **Choe, S. and Aklonis, J. J.,** An investigation of the blend of partially brominated poly(2,6-dimethyl-1,4-phenylene oxide) and polystyrene in the primary (glass-to-rubber) transition region, *Polym. Eng. Sci.*, 27, 1284, 1987.
470. **Sabliovschi, M. and Grigoriu, G.,** Compability of binary systems of poly(methyl methacrylate), poly(vinyl chloride) and poly(vinyl acetate). I. Thermogravimetric, infrared spectroscopic and microscopic studies, *J. Therm. Anal.*, 26, 23, 1983.
471. **Stein, D. J., Jung, R. H., Illers, K.-H., and Hendus, H.,** Phenomenological investigation of the miscibility of polymers, *Angew. Makromol. Chem.*, 36, 89, 1974.

472. **Lyngaae-Jorgensen, J. and Sondergaard, K.,** Phase transitions during shear flow of two phase polymer blends. II. Styrene-acrylonitrile/poly(methylmethacrylate) transition to homogeneous melt state, *Polym. Eng. Sci.*, 27, 351, 1987.
473. **Jelenic, J., Kirste, R. G., Schmitt, B. J., and Schmitt-Strecker, S.,** Neutron scattering experiments on polymer mixtures, International Union of Pure and Applied Chemistry Symposium, Mainz, 1979; Structure and thermodynamics of polymer blends from neutron scattering experiments, *Makromol. Chem.*, 180, 2057, 1979.
474. **Kruse, W. A., Kirste, R. G., Haas, J., Schmitt, B. J., and Stein, D. J.,** Experimental evidence for the molecularly disperse character of the mixture of two polymers and the determination of the chemical potential in these mixtures, *Makromol. Chem.*, 177, 1145, 1976.
475. **Schmitt, B. J., Kirste, R. G., and Jelenic, J.,** Investigation of the thermodynamics and the conformation of macromolecules in polymer blends in the vicinity of phase separation by neutron scattering, *Makromol. Chem.*, 181, 1655, 1980.
476. **Frank, C. W. and Gashgari, M. A.,** Excimer fluorescence as a molecular probe of polymer blend compatibility 1. Blends of poly(2-vinylnaphthalene) with poly(alkyl methacrylates); 4. Effect of temperature in solvent casting, *Macromolecules,* 12, 163, 1979; 14, 1558, 1981.
477. **Frank, C. W.,** Excimer fluorescence: a new tool for studying polymer alloys, *Plast. Compounds,* No. 1/2, 67, 1981.
478. **Frank, C. W. and Gashgari, M. A.,** Excimer fluorescence as a molecular probe of polymer-polymer interactions in the amorphous solid state, *Ann. N.Y. Acad. Sci.*, 366, 387, 1981.
479. **Thomas, J. W., Jr. and Frank, C. W.,** Fluorescence studies of polymer blends: effect of host matrix on copolymers of 1-naphthyl alkyl methacrylates and 9-anthryl methyl methacrylate, *J. Polym. Sci. Polym. Phys. Ed.*, 20, 1749, 1982.
480. **Semerak, S. M. and Frank, C. W.,** Excimer fluorescence as a molecular probe of polymer blend mixcibility. VI. Effect of molecular weight in blends of poly(2-vinylnaphthalene) with poly(methyl methacrylate), *Macromolecules,* 17, 1148, 1984.
481. **Frank, C. W., Gashgari, M. A., and Semerak, S. N.,** Polymer blend thermodynamics: Flory-Huggins theory and its application to excimer fluorescence studies, *NATO Adv. Sci. Ser. Ser. C,* 182 (Photophys. Photochem. Tools Polym. Sci.: Conform., Dyn, Morphol.), 523, 1986.
482. **Ito, J., Mitani, K., and Mizutani, Y.,** Annealing of polypropylene/poly(ethylene-co-propylene) blends. I. Thermal and physical properties of blend. II. Influence of the structure and properties of poly(ethylene-co-propylene) and blended polyethylene on impact strength, *J. Appl. Polym. Sci.*, 29, 75, 1984; 30, 497, 1985.
483. **Stehling, F. C., Huff, T., Speed, C. S., and Wissler, G.,** Structure and properties of rubber-modified polypropylene impact blends, *J. Appl. Polym. Sci.*, 26, 2693, 1988.
484. **Martuscelli, E., Silvestre, C., and Bianchi, L.,** Properties of thin films of isotactic polypropylene blended with polyisobutylene and ethylene-propylene-diene terpolymer rubbers, *Polymer,* 24, 1458, 1983.
485. **Warburton, C. E., Jr.,** Crockfastness of polyacrylate textile pigment printing binders — effect of binder mechanical properties and adhesion to fabric, *J. Adhesion,* 7, 109, 1975.
486. **Martuscelli, E., Silvestre, C., Addonizio, M. L., and Amelino, L.,** Phase structure and compatibility studies in poly(ethylene oxide)/poly(methyl methacrylate) blends, *Makromol. Chem.*, 187, 1557, 1986.
487. **Davis, D. D., Taylor, G. N., and Kwei, T. K.,** Compatibility in a blend of poly(2,3-dichloro-1-propyl acrylate) and poly(glycidyl methacrylate-co-ethyl acrylate), *J. Appl. Polym. Sci.*, 26, 2001, 1981.
488. **Kirshenbaum, S. L., Gazit, S., and Bell, J. P.,** An alternative liquid rubber for epoxy resin toughening. Improving poly(n-butyl acrylate) rubber-epoxy compatibility by the use of acrylonitrile and acrylic acid copolymers and terpolymers, *Adv. Chem. Ser.*, 208, 163, 1984.
489. **Piglowski, J. and Skowronski, T.,** Compatibility of poly(vinyl chloride) with butyl acrylate-acrylonitrile copolymers, *Angew. Makromol. Chem.*, 88, 165, 1980.
490. **Walsh, D. J. and Cheng, G. L.,** The miscibility of polyacrylates and polymethacrylates with PVC. An interpretation of the heats of mixing of oligomers, *Polymer,* 25, 499, 1984.
491. **Katime, I. A. and Quintana, R. J.,** Relationship between physical parameters of the solvent and its influence in stereocomplex formation, *Makromol. Chem.*, 187, 1441, 1986.
492. **Dunn, D. J. and Krause, S.,** Phase separation in diblock copolymers of styrene and α-methylstyrene and in mixtures of the corresponding homopolymers, *J. Polym. Sci. Polym. Lett.*, 12, 591, 1974.
493. **Cowie, J. M. G. and McEwen, I. J.,** Predicted miscibility of polystyrene and poly(α-methylstyrene) and comparison with experiment, *Polymer,* 26, 1662, 1985.
494. **Bowmer, T. N. and Tonelli, A. E.,** Blends of homo- and copolymers of ethylene and vinyl chloride: a compatibility study, *Macromolecules,* 19, 498, 1986.
495. **Bucknall, C. B. and Partridge, I. K.,** Phase-separation in epoxy resins containing polyethersulfone, *Polymer,* 24, 639, 1983.
496. **Bucknall, C. B. and Partridge, I. K.,** Phase separation in crosslinked resins containing polymeric modifiers, *Polym. Eng. Sci.*, 26, 54, 1986.

497. **Partridge, I. K.,** *Effects of Phase Separation in Crosslinked Resins Containing Polymeric Modifiers,* Ph.D. Thesis, Cranfield Institute of Technology, 1984.

498. **Piglowski, J. and Laskawski, W.,** Compatibility of poly(vinyl chloride)/polyurethane blends, *Angew. Makromol. Chem.,* 82, 157, 1979.

499. **Yang, H., Hadziioannou, G., and Stein, R. J.,** The effect of deuteration on the phase equilibrium of the polystyrene/poly(vinyl methyl ether) blend system, *J. Polym. Sci. Polym. Phys. Ed.,* 21, 159, 1983.

500. **Massa, D. J.,** Physical properties of blends of polystyrene with poly(methyl methacrylate) and styrene/(methyl methacrylate) copolymers, in *Multiphase Polymers* (*Adv. Chem. Ser.* 176), Cooper, S. L. and Estes, G. M., Eds., American Chemical Society, Washington, D.C., 1979, chap. 22.

501. **Massa, D. J. and Descamp, M. G.,** Temperature dependence of compatibility of blends of polystyrene with poly(methyl methacrylate) and styrene/(methyl methacrylate) copolymers, Am. Phys. Soc. Meet., New York, 1980.

502. **Rudin, A., Loucks, D. A., and Goldwasser, J. M.,** Oriented monofilaments from blends of poly(ethylene terephthalate) and polypropylene, *Polym. Eng. Sci.,* 20, 741, 1980.

503. **Schreiber, H. P. and Ross, I. C.,** Property modifications in fibers of blends based on poly(ethylene terephthalate), *Text. Res. J.,* 46, 701, 1976.

504. **Chevassus, F.,** Modification of butadiene-acrylonitrile elastomers by certain thermoplastic resins, more particularly vinyl resins, *Chim. Ind. (Paris),* 82, 690, 1959.

505. **Hubbell, D. S. and Cooper, S. L.,** The physical properties and morphology of polyϵ-caprolactone blends, *J. Appl. Polym. Sci.,* 21, 3035, 1977.

506. **Hubbell, D. S. and Cooper, S. L.,** Segmental orientation, physical properties, and morphology of polyϵ-caprolactone blends, in *Multiphase Polymers* (*Adv. Chem. Ser.* 176), Cooper, S. L. and Estes, G. M., Eds., American Chemical Society, Washington, D.C., 1979, chap. 26.

507. **Chirkova, N. V., Epshtein, V. G., and Zakharov, N. D.,** Compatibility of rubbers studied by evaluating their mutual adhesion, *Koll. Zh.,* 32, 912, 1970; *Colloid J. U.S.S.R.,* 32, 767, 1970.

508. **Verkholantsev, V. V.,** Nonhomogeneous-in-layer coatings, *Progr. Org. Coatings,* 13, 71, 1985.

509. **Kundu, A. K., Ray, S. S., Adhikari, B., and Maiti, S.,** Polymer blends. II. Compatibility and thermal behavior of blends of Novolac and polyamideimide from rosin, *Eur. Polym. J.,* 22, 369, 1986.

510. **Kundu, A. K., Ray, S. S., and Maiti, S.,** Polymer blends. IV. Blends of resole with polyamideimide synthesized from rosin, *Eur. Polym. J.,* 22, 821, 1986.

511. **Luce, J. E. and Robertson, A. A.,** The sorption of polymers on cellulose, *J. Polym. Sci.,* 51, 317, 1961.

512. **Rowe, R. C.,** The prediction of compatibility/incompatibility in blends of ethyl cellulose with hydroxpropyl methylcellulose or hydroxypropyl cellulose using 2-dimensional solubility parameter maps, *J. Pharm. Pharmacol.,* 38, 214, 1986.

513. **Burkhanova, N. D., Yugai, S. M., Sinitsyna, G., Tikhonovetskaya, A. D., Peremkulova, Kh., Slepakova, S. I., Nokonovich, G. V., and Usmanov, Kh. U.,** Relation between thermodynamic compatibility and structure formation in polymer mixtures, *Cellul. Chem. Technol.,* 16, 631, 1982.

514. **Lee, L.-H.,** Adhesion of high polymers. III. Mechanisms of adhesion at the rubber-resin interface in heterophase systems, in *Interaction of Liquids at Solid Substrates* (*Adv. Chem. Ser.* 87), Alexander, A. L., Ed., American Chemical Society, Washington, D.C., 1968, chap 7.

515. **Solomon, T. S.,** Systems for tire cord-rubber adhesion, *Rubber Chem. Technol.,* 58, 561, 1985.

516. **Hoffman, D. M. and Caley, L. E.,** Polymer blends as high explosive binders, *Polym. Eng. Sci.,* 26, 1489, 1986.

517. **Burke, J. J. and Weiss, V., Eds.,** *Block and Graft Copolymers,* Syracuse University Press, Syracuse, NY, 1973.

518. **Danon, J., Jobard, M., Lautout, M., Magat, M., Michel, M., Riou, M., and Wippler, C.,** Certain properties of graft polymers in solution, *J. Polym. Sci.,* 34, 517, 1959.

519. **Riess, G. and Callot, P.,** Fractionation of copolymers, in *Fractionation of Synthetic Polymers,* Tung, L. H., Ed., Marcel Dekker, New York, 1977, chap 5.

520. **Shibuya, C., Ogino, K., and Kakagawa, T.,** Chemical composition dependence of polymer-solvent interaction parameter in copolymer solution, *Kobunshi Kagaku,* 25, 570, 1968; Thermodynamic interaction parameters of copolymers in solution, *Gunma Daigaku Kyoyobu Kiyo,* 2, 1, 1968.

521. **Helfand, E. and Wasserman, Z.,** Statistical thermodynamics of microdomain structures in block copolymer systems, *Polym. Eng. Sci.,* 17, 582, 1977.

522. **Klein, D. H.,** Factors involved in designing acrylic copolymers in solution, *J. Paint Technol.,* 42, 335, 1970.

523. **Liepins, R.,** Some observations on internal plasticization of polyvinyl chloride, *Polym. Eng. Sci.,* 9, 86, 1969.

524. **Lautout, M. and Magat, M.,** Thermodynamics of copolymers in solution. I. Solubility of copolymers, *Z. Phys. Chem. (Frankfurt),* 16, 292, 1958.

525. **Imoto, M.,** Senriyama evening talks. 125. Calculation of solubility parameters for random copolymers, *Secchaka,* 16, 436, 1972.

526. **Schneier, B.,** An equation for calculation of the solubility parameter of random copolymers, *J. Polym. Sci. B Polym. Lett.*, 10, 245, 1972.
527. **Schneier, B.,** Solubility parameters of two chlorinated polyethylene polymers, *J. Appl. Polym. Sci.*, 16, 1515, 1972.
528. **Schneier, B.,** Polymer compatibility, *J. Appl. Polym. Sci.*, 17, 3173, 1973.
529. **Pritykin, L. M.,** Calculation of the surface energy of polymers by their refractometric and cohesion characteristics, *Vysokomol. Soedin. Ser. A.*, 23, 757, 1981; *Sow. Beitr. Faserforsch. Textiltech.*, 18, 510, 1981.
530. **Pritykin, L. M., Zyuz, V. T., and Onushko, A. I.,** Possible non-additive calculation evaluation of the solubility parameter of random copolymers, *Vysokomol. Soedin. Ser. B*, 25, 271, 1983.
531. **Pritykin, L. M.,** Calculation of the surface energy of homo- and copolymers from the cohesion parameters and refractometric characteristics of the respective monomers, *J. Colloid Interface Sci.*, 112, 539, 1986.
532. **Mizerovskii, L. N. and Belova, O. V.,** Calculation of the solubility parameter of binary mixtures of liquids and copolymers, *Vysokomol. Soedin. Ser. A*, 29, 1236, 1987.
533. **Beerbower, A. and Dickey, J. R.,** Advanced methods for predicting elastomer/fluids interactions, *Am. Soc. Lubric. Eng. Trans.*, 12, 1, 1969.
534. **Matsuura, T., Blais, P., and Sourirajan, S.,** Polar and nonpolar parameters for polymeric reverse osmosis membrane materials from liquid chromatographic data, *J. Appl. Polym. Sci.*, 20, 1515, 1976.
535. **Beevers, R. B.,** The physical properties of polyacrylonitrile and its copolymers, *Macromol. Rev.*, 3, 113, 1968.
536. **Haaf, F., Breuer, H., and Stabenow, J.,** Investigations of rubber modified PVC, *Angew. Makromol. Chem.*, 58/59, 95, 1977.
537. **Bowen, R. L. Z.** Adhesive bonding of various materials to hard tooth tissues. 18. Synthesis of a polyfunctional surface-active comonomer, *J. Dent. Res.*, 58, 1101, 1979.
538. **Ceausescu, E., Lebada, P., and Petcu, L.,** Determining the solubility parameters of ethylene-propylene co- and ter-polymers, *Mater. Plast. (Bucharest)*, 9, 411, 1972; *Lucr. Conf. Nat. Chim. Anal.*, 3rd, Vol. 4, Inst. Cent. Cerçet. Chim., Bucharest, Romania, 1971, 7.
539. **Kirkham, M. C.,** Properties and microstructures of ethylene-propylene terpolymers, *J. Appl. Polym. Sci.*, 17, 1101, 1973.
540. **Merk, W., Lichtenthaler, R. N., and Prausnitz, J. M.,** Solubilities of fifteen solvents in copolymers of poly(vinyl acetate) and poly(vinyl chloride) from gas-liquid chromatography. Estimation of polymer solubility parameter, *J. Phys. Chem.*, 84, 1694, 1980.
541. **Mailtu, I., Cincu, C., and Uglea, C.,** Characterization of the vinyl chloride (80%)-vinyl acetate (20%) copolymer, *Mater. Plast. (Bucharest)*, 12, 149, 1975.
542. **Vial, J.,** Tridimensional representation of the solubility parameters: study of the solubilization of a vinyl chloride-vinyl acetate copolymer, *Compt. Rend. Acad. Sci. Ser. C*, 270, 683, 1970; *Predicting the Solubility of Polymers: Application to some Vinyl Chloride-Vinyl Acetate Copolymers*, Thesis, Faculty of Science, Paris, 1970.
543. **Dincer, S. and Bonner, D. C.,** Thermodynamic analysis of a ethylene and vinyl acetate copolymer with various solvents by gas chromatography, *Macromolecules*, 11, 107, 1978.
544. **Lath, D., Lathova, E., and Cowie, J. M. G.,** Solubility parameters of ethylene-vinyl acetate copolymers, *Prepr. Short Contrib., Bratislava IUPAC Int. Conf. Modif. Polym. 5th*, 2, 225, 1979.
545. **Rätzsch, M., Freydank, H., and Göbel, G.,** Solubility parameters of polyethylene, poly(vinyl acetate) and ethylene-vinyl acetate copolymers, *Faserforsch. Textiltech.*, 27, 425, 1976.
546. **Miller, R. F.,** Evaluation of ethylene copolymers as pour depressants, in *Macromolecular Solutions: Solvent-Property Relationships in Polymers* (Symp. Proc., Am. Chem. Soc., New York, 1981), Seymour, R. B. and Stahl, G. A., Eds., Pergamon, New York, 1982, 84.
547. **Tachi, Y., Takahashi, A., and Kagawa, I.,** Solubility of quasi-polyelectrolytes. I. Solubility of ethylene-acrylic acid copolymer in organic solvents, *Kogyo Kagaku Zasshi*, 73, 398, 1970.
548. **Dreyfuss, P., Eckstein, Y., Lee, D. P., and Lehmann, J.,** Comparison of Neoprene W graft copolymers having flexible and rigid polyether branches, *Polym. Prep.*, American Chemical Society, Washington, D.C., 1979.
549. **Paul, D. R., St. Lawrence, J. E., and Troell, J. H.,** Flow behavior of concentrated solutions of an SBS (styrene-butadiene-styrene) block copolymer, *Polym. Eng. Sci.*, 10, 70, 1970.
550. **Lyngaae-Jorgensen, Alle, N., and Marten, F. L.,** Domain flow in two-phase polymer systems. A study of the flow properties of a styrene-methylmethacrylate diblock copolymer and SBS triblock copolymers, in *Multiphase Polymers* (*Adv. Chem. Ser.* 176), Cooper, S. L. and Estes, G. M., Eds., American Chemical Society, Washington, D.C., 1979, chap. 28.
551. **Wilkes, G. L., Bagrodia, S., Ophir, Z., and Emerson, J. A.,** A systematic study of the effect of casting solvent on the bulk properties of linear SBS block copolymers, *J. Appl. Phys.*, 49, 5060, 1978.
552. **Fedors, R. F.,** A simple model for the virtually crosslinked block copolymers, *J. Polym. Sci. C.*, 26, 189, 1969.

553. **Morton, M.,** Structure-property relations in amorphous and crystallizable ABA triblock copolymers, *Rubber Chem. Technol.*, 56, 1096, 1983.

554. **Zuchowska, D. and Meissner, W.,** Viscous properties of butadiene/styrene triblock copolymer solutions, *Makromol. Chem.*, 186, 2355, 1985.

555. **Krause, S., Dunn, D. J., Seyed-Mozzaffari, A., and Biswas, A. M.,** Microphase separation in diblock and triblock copolymers of styrene and α-methylstyrene, *Macromolecules,* 10, 786, 1977.

556. **Zurawski, D. E. and Sperling, L. H.,** Characterization of electron beam crosslinked poly(styrene-block-ethylene-co-butylene-block-styrene), *Polym. Eng. Sci.*, 23, 510, 1983.

557. **Lin, O. C. C.,** The change of structure and viscosity of styrene-acrylate copolymers in aqueous media, in *Structure-Solubility Relationships in Polymers,* Harris, F. W. and Seymour, R. B., Eds., Academic Press, New York, 1977, 165.

558. **Vasilova, O. I., Zaitseva, V. V., and Kucher, R. V.,** Solubility parameter of styrene-acrylonitrile-methyl methacrylate terpolymers, *Vysokomol. Soedin. Ser. B,* 29, 912, 1987.

559. **Blanks, R. F. and Shah, B. N.,** Solubility of styrene-acrylonitrile copolymers, in *Structure-Solubility Relationships in Polymers* (Proc. Symp., 1976), Harris, F. W. and Seymour, R. B., Eds., Academic Press, New York, 1977, 111.

560. **Seymour, R. B., Owen, D. R., McGee, M. L., Losado, C. J., and Clark, M.,** Acrylonitrile-styrene copolymer emulsions, *Ind. Eng. Chem. Prod. Res. Dev.*, 12, 198, 1973.

561. **Boyer, R. F. and Keskkula, H.,** Styrene polymers: characterization, in *Encyclopedia of Polymer Science and Technology,* Vol. 13, Bikales, N. M., Ed., John Wiley & Sons, New York, 1970, 206.

562. **Omichi, H. and Araki, K.,** Solvent effects on the radiation-induced graft polymerization of styrene onto poly(isobutyl oxide); Kinetics of radiation-induced graft polymerization of styrene onto poly(ethylene oxide); Kinetics of radiation-induced graft copolymerization of styrene with ethyl acrylate, *J. Polym. Sci. Polym. Chem.*, 14, 2773, 1976; 17, 1401, and 1701, 1979.

563. **Okay, O.,** Styrene-divinylbenzene copolymers. VI. Porosity formation in the presence of toluene-cyclohexanol mixture as inert diluents, *Angew. Makromol. Chem.*, 157, 1, 1988.

564. **Hasegawa, H., Tanaka, H., Yamasaki, K., and Hashimoto, T.,** Bicontinuous microdomain morphology of block copolymers. 1. Tetrapod-network structure of polysyrene-polyisoprene diblock polymers, *Macromolecules,* 20, 1651, 1987.

565. **Veličkovič, J. and Filipovič,** Dilute solution properties of styrene/di-n-alkyl itacontate copolymers, *Polym. Bull.*, 5, 569, 1981.

566. **Gancarz, I. and Laskawskii, W.,** Peroxide-initiated crosslinking of maleic anhydride-modified low molecular weight polybutadiene. II. Crosslinking degree and thermal degradation of cured polymers, *J. Polym. Sci. Polym. Chem. Ed.*, 17, 1523, 1979.

567. **Fritzsche, A. K. and Price, F. P.,** Solvent interactions of poly(diphenylsiloxane-dimethylsiloxane-diphenylsiloxane) block copolymers, in *Colloidal and Morphological Behavior of Block and Graft Copolymers,* Molau, G. E., Ed., Plenum Press, New York, 1971, 85.

568. **Pascault, J. P. and Camberein, Y.,** Quantitative determination of the phase segregation degree in a poly(dimethyl siloxane)-based urethane, *Polym. Commun.*, 27, 230, 1986.

569. **Yerrick, K. B. and Beck, H. N.,** Solvent resistance of silicone elastomers: solvent-polymer interactions, *Rubber Chem. Technol.*, 37, 261, 1964.

570. **Rossi, C., Bianchi, U., and Bianchi, E.,** Internal pressure of block copolymers, *J. Polym. Sci. C.*, 4, 699, 1963.

571. **Ward, T. C., Wnuk, A. J., Shchori, E., Viswanathan, R., and McGrath, J. E.,** Poly(arylene ether sulfone)-poly(aryl carbonate) block copolymers, in *Multiphase Polymers* (*Adv. Chem. Ser.* 176), Cooper, S. L. and Estes, G. M., Eds., American Chemical Society, Washington, D.C., 1979, 293.

572. **Hudson, S. M. and Cuculo, J. A.,** The solubility of cellulose in liquid ammonia/salt solutions, *J. Polym. Sci. Polym. Chem. Ed.*, 18, 3469, 1980.

573. **Hudson, S. M. and Cuculo, J. A.,** The solubility of unmodified cellulose. A critique of the literature, *J. Macromol. Sci. Rev. Macromol. Chem.*, C18, 1, 1980.

574. **Roberston, A. A.,** Cellulose-liquid interactions, *Pulp Pap. Mag. Can,* 65, T171, 1964.

575. **Robertson, A. A.,** Interactions of liquids with cellulose, *Tappi,* 53, 1331, 1970.

576. **Nguyen Huu-Phuoc, Hŏ Nam-Tran, Buchmann, M., and Kesselring, U. W.,** Experimentally optimized determination of the partial and total cohesion parameters of an insoluble polymer (microcrystalline cellulose) by gas-solid chromatography, *Int. J. Pharm.*, 34, 217, 1987.

577. **Yoshino, I.,** Wet-strength paper. XV. Influence of various polar media on the fiber bonding of paper, *Sen-i Gakkaishi,* 14, 11, 1958.

578. **Minhas, P. S. and Robertson, A. A.,** Accessibility of cellulose by the thallous ethylate method — application to the measurement of cellulose liquid interactions, *Text. Res. J.*, 37, 400, 1967.

579. **Philipp, B., Schleicher, H., and Wagenknecht, W.,** Non-aqueous solvents of cellulose, *Chemtech,* 7, 702, 1977.

580. **Seymour, R. B. and Johnston, E. J.,** Solutions of cellulose in organic solvents, in *Structure-Solubility Relationships in Polymers* (Proc. Symp., 1976), Harris, F. W. and Seymour, R. B., Eds., Academic Press, New York, 1977, 241.
581. **Turbak, A. F., Hammer, R. B., Davies, R. E., and Portney, N. A.,** A critical review of cellulose solvent systems, in *Solvent Spun Rayon, Modified Cellulose Fibers and Derivatives* (*Am. Chem. Soc. Symp. Ser.* 58), Turbak, A. F., Ed., American Chemical Society, Washington, D.C., 1977, chap. 2, 12.
582. **Turbak, A. F., Hammer, R. B., Davies, R. E., and Hergert, H. L.,** Cellulose solvents, *Chemtech,* 10, 51, 1980.
583. **Hudson, S. M., Cuculo, J. A., and Wadsworth, L. C.,** The solubility of cellulose in liquid ammonia/ammonium thiocyanate solution: The effect of composition and temperature on dissolution and solution properties, *J. Polym. Sci. Polym. Chem. Ed.,* 21, 651, 1983.
584. **Seymour, R. B., Johnson, E. L., and Stahl, G. A.,** Preparation and properties of cellulose blends in dimethylsulfoxide solution, in *Macromolecular Solutions: Solvent-Property Relationships in Polymers* (Symp. Proc., Am. Chem. Soc., New York, 1981), Seymour, R. B. and Stahl, G. A., Eds., Pergamon, New York, 1982, 90.
585. **Oki, R., Shigematsu, M., Hirabayashi, K., and Arai, M.,** Studies on the mechanochemical reactivity of silk fibroin. II. Mechanical properties of silk fibers in organic solvents, *Nippon Sanshigaku Zasshi,* 49(1), 39, 1980.
586. **Betrabet, S. M., Daruwalla, E. H., and Lokhande, H. T.,** Intercrystalline swelling in morpholine-treated cotton cellulose, *Text. Res. J.,* 36, 684, 1966; *Cotton Technological Research Laboratory, Indian Council of Agricultural Research, Technological Leaflet,* No. 81, Bombay, India, 1966.
587. **Lokhande, H. T.,** Swelling behavior of cotton fibers in morpholine and piperidine, *J. Appl. Polym. Sci.,* 22, 1243, 1978.
588. **Philipp, B., Schleicher, H., and Wagenknecht, W.,** The time-dependence of the swelling of regenerated cellulose fibre and films in organic liquids, *Faserforsch. Textiltech.,* 24, 106, 1973.
589. **Philipp, B., Schleicher, H., and Wagenknecht, W.,** The influence of cellulose structure of the swelling of cellulose in organic liquids, *J. Polym. Sci. Polym. Symp.,* 42, 1531, 1973.
590. **Thode, E. F. and Guide, R. G.,** A thermodynamic interpretation of the swelling of cellulose in organic liquids: the relations among solubility parameter, swelling, and internal surface, *Tappi,* 42, 35, 1959.
591. **Brown, P. F. and Swanson, J.,** Wetting properties of cellulose treated in a corona discharge, *Tappi,* 54, 2012, 1971.
592. **Kawai, T.,** Sorption of water vapor by cellulose and polymers at high humidities, *J. Polym. Sci.,* 37, 181, 1959.
593. **Tanaka, Y. and Shiozaki, H.,** Salt-catalyzed reaction of phenyl glycidyl ether with silk fibroin, *Makromol. Chem.,* 129, 12, 1969.
594. **Shiozaki, H. and Tanaka, Y.,** Reactivity of monoepoxides toward silk fibroin, *Makromol. Chem.,* 143, 25, 1971.
595. **Shiozaki, H. and Tanaka, Y.,** Salt-catalyzed addition reaction of epoxy compounds with various silk fibroins, *Angew. Makromol. Chem.,* 64, 1, 1977.
596. **Tanaka, Y., Shimura, Y., and Shiozaki, H.,** Effect of salts, solvents and substituents on the reaction of cellulose with epoxy compounds, *Makromol. Chem.,* 177, 1301, 1976.
597. **Sakurada, I. and Sakaguchi, Y.,** Liquid-phase fibrous acetylation of cellulose. I. Acetylation of viscose rayon with sulfuric acid catalyst, *Sen-i Gakkaishi,* 13, 194, 1957.
598. **Barton, A. F. M.,** *Handbook of Polymer-Liquid Interaction Parameters and Solubility Parameters,* CRC Press, Boca Raton, FL, 1990.
599. **Moore, W. R., Epstein, J. A., Brown, A. M., and Tidswell, B. M.,** Cellulose derivatives — solvent interaction, *J. Polym. Sci.,* 23 (107), 23, 1957.
600. **Cowie, J. M. G. and Ranson, R. J.,** The dissolution and stability of cellulose triacetate in acetone, *Makromol. Chem.,* 143, 105, 1971.
601. **Howard, P. and Parikh, R. S.,** Solution properties of cellulose triacetate. II. Solubility and viscosity studies, *J. Polym. Sci.* A-1, 6, 537, 1968.
602. **Geller, B. E., Slepakova, S. I., Pshedetskaya, V. K., and Geller, A. A.,** Study of some physicochemical properties of mixtures of cellulose and amylose acetates in solution and in solid state, *Vysokomol. Soedin. Ser. B,* 16, 731, 1974.
603. **Kent, D. J. and Rowe, R. C.,** Solubility studies on ethyl cellulose used in film coating, *J. Pharm. Pharmacol.,* 30, 808, 1978.
604. **Chawla, A. S. and Chang, T. M. S.,** Use of solubility parameters for the preparation of hemodialysis membranes, *J. Appl. Polym. Sci.,* 19, 1723, 1975.
605. **Deinzer, M., Melton, R., and Mitchell, D.,** Trace organic contaminants in drinking water; their concentration by reverse osmosis, *Water Res.,* 9, 799, 1975.

606. **Klein, E. and Smith, J. K.,** The use of solubility parameters for solvent selection in asymmetric membrane formation, in *Reverse Osmosis Membrane Research,* Lonsdale, H. K. and Podall, H. E., Eds., Plenum Press, New York, 1972, 61.
607. **Klein, E. and Smith, J. K.,** Asymmetric membrane formation. Solubility parameters for solvent selection, *Ind. Eng. Chem. Prod. Res. Dev.,* 11, 207, 1972.
608. **Klein, E., Eichelberger, J., Eyer, C., and Smith, J.,** Evaluation of semi permeable membranes for determination of organic contaminants in drinking water, *Water Res.,* 9, 807, 1975.
609. **Kopecek, J. and Sourirajan, S.,** Performance of porous cellulose acetate membranes for the reverse osmosis separation of mixtures of organic liquids, *Ind. Eng. Chem. Proc. Des. Dev.,* 9, 5, 1970.
610. **Kurokawa, Y., Tsuchiya, T., and Yui, N.,** Application of regular solution theory to organic solute rejection by reverse osmosis, *Desalination,* 29, 233, 1979.
611. **Ueno, K., Kurokawa, Y., and Yui, K.,** Reverse osmosis separations of organic solutes in aqueous solutions by complex (cellulose acetate phthalate-polyethylene oxide) membrane, *Membrane,* 4, 103, 1979.
612. **Kurokawa, Y., Ueno, K., and Yui, N.,** Reverse osmosis rejection of organic solute from aqueous solution by using cellulose acetate phthalate-polyvinyl pyrrolidone membrane, *J. Colloid Interface Sci.,* 74, 561, 1980.
613. **Yui, N., Kurokawa, Y., and Shindo, S.,** Application of regular solution theory to the treatment of aqueous organic solutions by reverse osmosis, *Denki Kagaku,* 44, 290, 197.
614. **Yui, N., Kurokawa, Y., and Shindo, S.,** Reverse osmosis rejection of organic solute from aqueous solution (cellulose acetate membrane), *J. Soc. Mat. Sci. Jpn.,* 25, 1175, 1976.
615. **Matsuura, T.,** Reverse osmosis separation of solute from aqueous solution, *Yukui Gosei Kagaku Kyokai (J. Syn. Org. Chem. Jpn.),* 31, 717, 1973.
616. **Matsuura, T.,** Transport processes in membranes for reverse osmosis, *Chem.-Ing. Tech.,* 50, 565, 1978.
617. **Matsuura, T. and Sourirajan, S.,** Reverse osmosis separation of hydrocarbons in aqueous solutions using porous cellulose acetate membranes, *J. Appl. Polym. Sci.,* 17, 3683, 1973.
618. **Matsuura, T. and Sourirajan, S.,** Characterization of reverse osmosis membrane materials by liquid chromatography data, *Proc. 6th Int. Symp., Fresh Water from the Sea,* 3, 227, 1978.
619. **Crowley, J. D., Teague, G. S., Jr., and Lowe, J. W.,** A three-dimensional approach to solubility, *J. Paint Technol.,* 38, 269, 1966; 39, 19, 1967.
620. **Eastman Kodak,** *Three Dimensional Approach to Solubility,* Rochester, NY, (undated).
621. **Hashimoto, I.,** Studies of solvent dyeing. 10. Relation between the swelling and solution behaviors and the three dimensional solubility parameters for various polymers in organic solvents, *Sen'i Gakkaishi,* 34, T469, 1978.
622. **Rooney, M. L.,** Interesterification of starch with methyl palmitate, *Polymer,* 17, 555, 1976.
623. **Guthrie, J. T., Huglin, M. B., Richards, R. W., Shah, V. I., and Simpson, A. H.,** Behavior of cellulose tricarbanilate in dilute solution, *Eur. Polym. J.,* 11, 527, 1975.
624. **Tager, A. A. and Bessonov, Yu. S.,** Thermodynamic study of solutions of poly(vinyl acetate) and cellulose tricarbanilate in the precritical region, *Vysokomol. Soedin. Ser. A,* 17, 2377, 1975; *Polym. Sci. U.S.S.R.,* 17, 2734, 1975.
625. **Roberts, G. A. F. and Thomas, I. M.,** Determination of the solubility parameter of o-hydroxypropyl cellulose, *Polymer,* 19, 459, 1978.
626. **Austin, P. R.,** Chitin solvents and solubility parameters, *Chitin, Chitosan, and Related Enzymes* (Proceedings, Joint US-Japan Seminar), Academic Press, New York, 1984, p. 227; Chitin solutions and purification of chitin, *Methods in Enzymology,* 161, Biomass Part B, 403, 1988; **Austin, P. R., Castle, J. E., and Albisetti, C. J.,** Beta-chitin from liquid: new solvents and plasticizers, in Skjak-Braek, G., Anthonsen, T., and Sanford, P., Eds., *Chitins and Chitosans* (Proc 4th Int. Conf., Trondheim, Norway, 1988), Elsever, NY, 1989, 749.
627. **Moore, G. K. and Roberts, G. A. F.,** Reactions of chitosan: 2. Preparation and reactivity of N-acyl derivatives of chitosan, *Int. J. Biol. Macromol.,* 3, 292, 1981.
628. **Tanaka, Y. and Shiozaki, H.,** Reaction of wool keratin with epoxides, *Proc. Intl. Wool Text. Res. Conf.,* Aachen, Vol. III, Schriftenreihe Deutsches Wollforschungsinstitut an der Technischen Hochschule Aachen e.V., 1975, 192; *Sen'i Kobunshi Zairyo Kenkynsho Kenkyu Hokoku,* 109, 19, 1975; **Shiozaki, H. and Tanaka, Y.,** Solvent effect on the epoxide-wool reactions, *Sen'i Gakkaishi,* 28, 174, 1972.
629. **Mullerschule, D.,** Gamma radiation induced grafting onto wool. 1. Fundamental studies on loose wool; 2. Studies on fabrics and practical implication, *Radiat. Phys. Chem.,* 16, 149 and 159, 1980.
630. **Khanna, B. B.,** Solubility parameter of shellac, *Indian J. Technol.,* 8, 274, 1970.
631. **Banerjee, P. K., Srivastava, B. C., and Kumar, S.,** Cohesive-energy density of shellac, *Polymer,* 23, 417, 1982; Shellac-solvent interaction parameter, *Polymer,* 23, 1244, 1982.
632. **Mishra, M. K.,** Solution properties of shellac. IV. Solubility parameter of shellac, *Angew. Makromol. Chem.,* 147, 107, 1987.
633. **Brauns, F. E. and Brauns, D. A.,** *Chemistry of Lignin,* Supplemental Volume for 1949-1958, Academic Press, New York, 1960, 173.

634. **Dellicolli, H. T.,** Controlled release of pesticides from kraft lignin, in *Controlled Release Pesticides (Am. Chem. Soc. Symp. Ser.* 53), Scher, H. B., Ed., American Chemical Society, Washington, D.C., 1977, chap. 8.

635. **Ekman, K. H. and Lindberg, J. J.,** Notes on the solubility of lignins in binary organic solvent mixtures, *Suom. Kem. B,* 39, 89, 1966.

636. **Lindberg, J. J.,** Studies on thermodynamics of liquids and related polymers. II. Thermodynamics of solubility, *Suom. Kem. B,* 40, 225, 1967; On the association, solubility and molecular weight of lignins, *Finska Kemists. Medd.,* 77, 130, 1968.

637. **Hansen, C. M.,** The universality of the solubility parameter, *Ind. Eng. Chem. Prod. Res. Dev.,* 8, 2, 1969.

638. **Schuerch, G.,** The solvent properties of liquids and their relation to the solubiity, swelling, isolation and fractionation of lignin, *J. Am. Chem. Soc.,* 74, 5061, 1952.

639. **Schuerch, C.,** Physical restrictions to homogeneous reactions on wood, *Ind. Eng. Chem. Prod. Res. Dev.,* 4, 61, 1965.

640. **Brown, W.,** Solution properties of lignin. Thermodynamic properties and molecular weight determination, *J. Appl. Polym. Sci.,* 11, 2381, 1967.

641. **Shumilin, V. A.,** Differences in the values of solubility parameters of lignin precursors and their glycerides as one reason for the compositional heterogeneity of lignin in situ, *Khim. Drev.,* No. 1, 105, 1987.

642. **Teas, J. P.,** Graphic analysis of resin solubilities, *J. Paint Technol.,* 40(516), 19, 1968.

643. **Porter, L. K.,** Factors affecting the solubility and possible fractionation of organic colloids extracted from soil and leonardite with an acetone-H_2O-HCl solvent, *J. Agric. Food Chem.,* 15, 807, 1967.

644. **Chiou, C. T., Porter, P. E., and Schmedding, D. W.,** Partition equilibria of nonionic organic compounds between soil organic matter and water, *Environ. Sci. Technol.,* 17, 227, 1983.

645. **Freeman, D. H. and Cheung, L. S.,** A gel partition model for organic desorption from a pond sediment, *Science,* 214(4522), 790, 1981.

646. **Nayer, A. N. and Hossfeld, R. L.,** Hydrogen bonding and the swelling of wood in various solvents, *J. Am. Chem. Soc.,* 71, 2852, 1949.

647. **Vanasse, C., Chornet, E., and Overend, R. P.,** Liquefaction of lignocellulosics in model solvents: creosote oil and ethylene glycol, *Can. J. Chem. Eng.,* 66, 112, 1988.

648. **van Krevelen, D. W.,** Chemical structure and properties of coal. XXVIII — Coal constitution and solvent extraction, *Fuel,* 44, 229, 1965.

649. **Hombach, H. P.,** General aspects of coal solubility, *Fuel,* 59, 465, 1980.

650. **Kirov, N. Y., O'Shea, J. M., and Sergeant, G. D.,** The determination of solubility parameters for coal, *Fuel,* 46, 415, 1967.

651. **Sanada, Y. and Honda, H.,** Solvent extraction of coal -- relation between the yield of the extract and the solubility parameter of the solvent, *Bull. Chem. Soc. Jpn.,* 35, 1358, 1962.

652. **Sanada, Y. and Honda, H.,** Swelling equilibrium of coal by pyridine at 25°C; Equlibrium swelling of coal by various solvents, *Fuel,* 45, 295 and 451, 1966.

653. **Nelson, J. R., Mahajan, O. P., and Walker, P. L., Jr.,** Measurement of swelling of coals in organic liquids: a new approach, *Fuel,* 59, 831, 1980.

654. **Nelson, J. R.,** Determination of molecular weight between crosslinks of coals from solvent-swelling studies, *Fuel,* 62, 112, 1983.

655. **Hombach, H. P.,** General aspects of coal solubility. II. Indirect determination of the heat of vaporization of coal; Coal solubility. III. Effect of macromolecular structure on the solubility of coal, *Erdöl Kohle Erdgas Petrochem.,* 33, 530, 1980; 34, 65, 1981.

656. **Hombach, H. P. and Weimann, B.,** Solvent polarity and the solubility hypersurface of coal, *Proc. Int. Kohlenwiss. Tag.,* 428, 1981.

657. **Whitehurst, D. D., Mitchell, T. O., and Farcasiu, M.,** *Coal Liquefaction: The Chemistry and Technology of Thermal Processes,* Academic Press, New York, 1980, 287.

658. **Peppas, N. A. and Lucht, L. M.,** Macromolecular structure of coals. X. Thermodynamic interaction parameter for solvents and coal networks, *Polym. Bull. (Berlin),* 16, 375, 1986.

659. **Kiebler, M. W.,** Extraction of a bituminous coal. Influence of the nature of solvents, *Ind. Eng. Chem.,* 32, 1389, 1940.

660. **Roy, J. and Lahiri, A.,** Solvent extraction of coals — a perspective, *Proc. Symp. Chem. Oil Coal 1969,* Central Fuel Research Institute, Dhantad, India, 1972, 447.

661. **Amestica, L. and Wolf, E. E.,** Supercritical liquefaction of an Illinois No. 5 coal, in *Proc. Int. Conf. Coal Sci.,* Cent. Conf. Manage., Pittsburgh, PA, 1983, 3.

662. **Vasilakos, N. P. and Lawson, D. D.,** Solubility parameter spectroscopy in coal extraction, *Liq. Fuels Technol.,* 3, 305, 1985.

663. **van Bodegom, B., van Veen, J. A. R., van Kessel, G. M. M., Sinnige-Nijssen, M. W. A., and Stuiver, H. C. M.,** Action of solvents on coal at low temperature. 1. Low-rank coals, *Fuel,* 63, 346, 1984.

664. **Brower, K. R.,** Evidence for the involvement of quinone rings in reactions of some coals with tetralin, *Fuel,* 56, 245, 1977.
665. **Weinberg, V. L. and Yen, T. F.,** 'Solubility parameters' in coal and coal liquefaction products, *Fuel,* 59, 287, 1980.
666. **Chiba, K., Tagaya, H., and Saito, N.,** Liquefaction of Yallourn coal by binary solvent system, *J. Energy Fuels,* 1, 338, 1987.
667. **Cagniant, D. and Dumay, D.,** The role of solvents in the conversion of coal. Solvolysis reactions. *Biochim. J. Chim.,* 3, 435, 1985.
668. **Cudmore, J. F.,** Solvent extraction of some Australian coals, *Mech. Chem. Eng. Trans.,* 4, 173, 1968.
669. **Angelovich, J. M., Pastor, G. R., and Silver, H. F.,** Solvents used in the conversion of coal, *Ind. Eng. Chem. Proc. Des. Dev.,* 9, 106, 1970.
670. **Altgelt, K. H.,** at Fossil Fuel Chemistry meeting, Laramie, WY, 1976, and personal communication, 1979.
671. **Reucroft, P. J. and Patel, K. B.,** Surface area and swellability of coal; Gas-induced swelling in coal, *Fuel,* 62, 279, 1983; 65, 816, 1986.
672. **Reucroft, P. J., Patel, K. B., and Chiou, C. T.,** Pore volume determination in activated carbon and coal, *Carbon,* 22, 100, 1984.
673. **Reucroft, P. J. and Sethuraman, A. R.,** Effect of pressure on carbon dioxide induced coal swelling, *J. Energy Fuels,* 1, 72, 1987.
674. **Mitchell, D. L. and Speight, J. G.,** The solubility of asphaltenes in hydrocarbon solvents, *Fuel,* 52, 149, 1973.
675. **Hirschberg, A., DeJong, L. N. J., Schipper, B. A., and Meijer, J. G.,** Influence of temperature and pressure on asphaltene flocculation, *Soc. Pet. Eng. J.,* 24, 283, 1984.
676. **Riggs, D. M. and Diefendorf, R. J.,** The solubility of aromatic compounds, *Ext. Abstr. Prog. — Bienn. Conf. Carbon,* 14, 407, 1979.
677. **Riggs, D. M., Chen, S. H., and Diefendorf, R. J.,** Solubility and solubility parameters of polynuclear aromatics systems, *Proc. Gov. Conf. Expanding Use Coal. N.Y. State: Probl. Issues,* Tress, M. H. and Dawson, J. C., Eds., Res. Found. State Univ. New York, Albany, NY, 1981, 539.
678. **Riggs, D. M.** Carbon fiber from solvent extracted pitch, *Prepr. Am. Chem. Soc. Div. Pet. Chem.,* 29, 480, 1984.
679. **Rogacheva, O. V., Gimaev, R. N., Gubaidullin, V. Z., and Danil'yan, T. D.,** Study of asphaltene associate formation in petroleum residues, *Khim. Tekhnol. Pererab. Nefti Gaza,* 43, 1981.
680. **Baltisberger, R. J., Woolsey, N. F., Schwan, J. F., Bolton, G., and Knudson, C. L.,** Solubility parameter relationships between lignite derived asphaltenes and preasphaltenes, *Prep. Pap. Am. Chem. Soc. Div. Fuel Chem.,* 29(5), 43, 1984.
681. **Funk, E. W.,** Study of heavy hydrocarbons by inverse-phase chromatography, *Ind. Eng. Chem. Prod. Res. Dev.,* 16, 115, 1977.
682. **Peck, L. B.,** Recovery of a carbonaceous liquid with a low fines content, *U.S. Pat.* 4572777, 1986.
683. **Kwan, C. L., Chow, E., and Yen, T. F.,** Study of interaction of organic solvents with shale and kerogen using the method of solubility parameter spectrum, *Abstr. Pap. Am. Chem. Soc.,* 177, GEOC 19, 1979.
684. **Imamura, T. and Suzuki, T.,** On the modification of paving binders. I. The solubility of polymer in various binders, *Sekiyu Gakkai Shi (J. Jpn. Pet. Inst.),* 18, 1063, 1975.
685. **Kasahara, Y. and Uemura, T.,** Effect of latex composition on physical properties of rubber asphalt. Relation between compatibility and solubility parameter of malthenes, *Sekiyu Gakkai Shi (J. Jpn. Pet. Inst.),* 17, 672, 1974.
686. **Piskarev, V. A. and Radushkevich, R. L.,** Method for determining the compatibility of bitumens with elastomers, *Polim. Stroit. Mater.,* 42, 163, 1975.
687. **Fu, B. C. H. and Phillips, C. R.,** Plasticizing effectiveness of hydrocarbons in Athabasca bitumen, *Fuel,* 58, 554, 1979.
688. **But'ko, T. A., Pritytskaya, T. S., Lamotkin, A. I., Kaverinskii, V. S., Manerov, V. B., and Tarakanova, A. N.,** Compatibility of components of varnish compositions with high oligostyrene content, *Lakokras. Mater. Ikh. Primen.,* No. 2, 5, 1988.
689. **Painter, P. C., Park, Y., and Coleman, M. M.,** Thermodynamics of coal solutions, *Energy Fuels,* 2, 693, 1988.
690. **Dambis, M., Jakobsone, M., Gravitis, J., and Erins, P.,** Use of an additivity principle for evaluating the solubility and compatibility of biopolymers of wood cell walls, *Khim. Drev.,* No. 1, 14, 1981.
691. **Ham, G. E.,** Role of triad concentration in glass transition temperatures of copolymers. II., *J. Macromol. Sci. Chem.,* A9, 1281, 1975.
692. **Bodily, D. M., Wann, J.-P., and Kopp, V.,** Solvent swelling of coals and coal macerals, *Prepr. Pap. Am. Chem. Soc., Div. Fuel Chem.,* 31(3), 300, 1986.
693. **Aoki, Y.,** Miscibility of poly(acrylonitrile-*co*-styrene) with poly[styrene-*co*-(maleic anhydride)] and poly[styrene-*co*-(*N*-phenylmaleimide)], *Macromolecules,* 21, 1277, 1988.

694. **Ahagon, A., Kobayashi, T., and Misawa, M.,** Friction on ice, *Rubber Chem. Technol.,* 61(1), 14, 1988.
695. **Ruis, M. A., Contreras, M. D., Gallardo, V., and Parera, A.,** Solubility of surfactants: determination of solubility parameters of Methocel HG 90, type K-100, *Comun. Jorn. Com. Esp. Deterg.,* 19, 315, 1988.
696. **Pila, S.,** Optimum solvent mixtures for thick vinyl paints, *Pitture Vernici,* 49, 393, 1973.
697. **Mizerovskii, L. N., Lytkina, N. I., and Vansyatskaya, L. N.,** Prediction of vulcanizate swelling in organic liquids, *Izv. Vyssh. Uchebn. Zaved. Khim. Khim. Tekhnol.,* 30(10), 76, 1987.
698. **Sugihara, A.,** Selection of solvents for paint resins, *Kagaku To Kogyo (Osaka),* 45, 501, 1971.
699. **Krasheninnikov, A. I. and Chervin, V. G.,** Effect of plasticizer nature on the rheological properties of acrylonitrile-methyl acrylate, *Kolloidn. Zh.,* 40, 139, 1978.
700. **Corcoran, W. H., Vasilakos, N. P., and Lawson, D. D.,** Coal desulfurization, *U.S. Pat.* 4334888, 1982.
701. **Nauman, E. B., Ariyapadi, M. V., Balsava, N. P., Grocela, T. A., Furno, J. S., Liu, S. H., and Mallikarjun, R.,** Compositional quenching: a process for forming polymer-in-polymer microdispersions and co-continuous networks, *Chem. Eng. Commun.,* 66, 29, 1988.
702. **Ohm, R. F.,** Practical aspects of curative diffusion, *Rubber World,* 192(3), 34, 1985.
703. **Orofino, T. A. and McNeely, R. L.,** Observations on compatibility of nylon-66/polyethylene mixtures, *J. Macromol. Sci. Phys.,* B27, 31, 1988.
704. **Selinger, E. and Brine, C. J.,** Use of thermal analysis in the optimization of polymeric diffusion barriers in controlled release delivery systems, *Thermochim. Act.,* 134, 275, 1988.
705. **Tanaka, N.,** Derivation of quasi-crysalline enthalpy on melting of copolymers from solubility parameter, *Sen'i Gakkaishi,* 44, 541, 1988.
706. **Balayan, L. I. and Gevorkyan, S. S.,** Multicomponent solvent for chloroprene rubber, *Kauch. Rezina,* No. 12, 48, 1981.
707. **Alfrey, T., Jr. and Lloyd, W. G.,** Microporous polymers, *U.S. Pat. Reissue* 27026, 1971 (Reissue of *U.S. Pat.* 3322695).
708. **Bel'kevich, P. I., Zuev, T. T., and Aksinovich, S. F.,** Solubility of components of release agents, *Vestsi Akad. Navuk BSSR, Ser. Khim. Navuk.,* No. 4, 52, 1982.
709. **Bentz, A. P. and Billing, C. B., Jr.,** Determination of solubility parameters of new suit materials, *ASTM Spec. Tech. Publ.* 989 (Perform. Prot. Clothing, Second Symp.), 209, 1988.
710. **Bentz, A. P. and Billing, C. B., Jr.,** U.S. Coast Guard protective clothing research since Solna (NOKOBETEF II), presented at Third Int. Scand. Symp. (NOKOBETEF III), Norway, 1989.
711. **Hansen, C. M. and Hansen, K. M.,** Solubility parameter prediction of the barrier properties of chemical protective clothing, *ASTM Spec. Tech. Publ.* 989 (Perform. Prot. Clothing, Second Symp.), 197, 1988.
712. **Hansen, C. M. and Anderson, B. H.,** The affinities of organic solvents for biological systems, *Am. Ind. Hyg. Assoc.,* 49, 301, 1988.
713. **Grove, G. L. III, Gardella, J. A., Jr., Chin, R. L., and Salvati, L., Jr.,** Characterization of solution-cast extracts from Biomer® by FT-IR and ESCA, *Appl. Spectros.,* 42, 989, 1988.
714. **Grossman, R. F.,** Internal and external lubricants enhance processing of elastomers, *Elastomerics,* 121(4), 18, 1989.
715. **Iino, M., Takanohashi, T., Ohsuga, H., and Toda, K.,** Extraction of coals with CS_2 — *N*-methyl-2-pyrrolidinone mixed solvent at room temperature, *Fuel,* 67, 1639, 1988.
716. **Ingamells, W. and Yanumet, N.,** Solvent-polymer interactions in the plasticization of poly(ethylene terephthalate), *Bri. Polym. J.,* 12(1), 12, 1980; **Ingamells, W. and Thomas, M. N.,** The mechanism of carrier dyeing, *Text. Chem. Color.,* 16(3), 55, 1984.
717. **Ishimaru, Y. and Adachi, A.,** Swelling anisotropy of wood in organic liquids. I. External swelling and its anisotropy, *Mokuzai Gak,* 34, 200, 1988.
718. **Kalfoglou, N. K., Sotiropoulu, D. D., and Margaritis, A. G.,** Thermal and morphology characterization of blends of poly(ethylene oxide) with poly(vinyl acetate), *Eur. Polym. J.,* 24, 389, 1988.
719. **Kostylev, V. A., Grigor'ev, V. Yu., and Kunin, A. M.,** Use of the three-dimensional solubility parameter of the starting components to calculate the phase diagram of solvent-thinner-polymer systems, *Lakokras. Mater. Ikh. Primen.,* No. 4, 12, 1988.
720. **Larsson, A. and Johns, W. E.,** Acid-base interactions between cellulose/lignocellulose and organic molecules, *J. Adhesion,* 25, 121, 1988.
721. **Iisako, K., Hayama, M., and Fukami, A.,** Dependence of solvent crazing on bisphenol-A polycarbonate on molecular weight of organic liquids, *J. Macromol. Sci. Phys.,* B27, 385, 1988.
722. **Kadir, R., Stempler, D., Liron, Z., and Cohen, S.,** Delivery of theophylline into excised human skin from alkanoic acid solutions. A push-pull mechanism; Penetration of adenosine into excised human skin from binary vehicles: the enhancement factor, *J. Pharm. Sci.,* 76, 774, 1987; 77, 409, 1988.
723. **Aubrey, D. W.,** The nature and action of tackifier resins, *Rubber Chem.,* 61, 448, 1988.
724. **Bodmeier, R. and McGinity, J. W.,** Solvent selection in the preparation of poly(DL-lactide) microspheres prepared by the solvent evaporation method, *Int. J. Pharm.,* 43, 179, 1988.
725. **Bush, S. F., Methven, J. M., and Blackburn, D. R.,** Networks as the basis of pre-thickening SMC, *Biol. Synth. Polym. Networks,* Kramer, O., Ed., Elsevier, London, 1988, 321.

726. **Hoover, J. M., Ward, T. C., and McGrath, J. E.,** Hydrogenated star block copolymers of butylstyrene -isoprene and divinylbenzene, *Adv. Polyolefins* (Proc. Am. Chem. Soc. Int. Symp., 1985), Seymour, R. B. and Cheng, T., Eds., Plenum, New York, 1987, 89.
727. **Dante, M. F., Bittar, A. D., and Caillault, J. J.,** Program calculates solvent properties and solubility parameters, *Mod. Paint Coat.,* 79(9), 46 and 51, 1989.
728. **Sferrazza, R. A. and Gooding, C. H.,** Prediction of sorption selectivity in pervaporation membranes, *Proc. Int. Conf. Pervaporation Processes Chem. Ind.,* 3rd, Bakish, R. A., Ed., Bakish Mater. Corp., Englewood, NJ, 1988, 54.
729. **Niwa, M., Ohya, H., Hara, T., and Negishi, Y.,** Experimental studies of solute distribution coefficient and transfer coefficient in reverse osmosis membrane, *Kagaku Kogaku Ronbunshu,* 15, 726, 1989.
730. **Murata, K.,** Separation of binary mixtures of benzene, cyclohexane, and n-hexane by pervaporation through poly(vinyl chloride) and its plasma-treated membranes, *Osaka Kogyo Gijutsu Shikensho Kiho,* 40(2), 87, 1989.
731. **Kobayashi, T., Hasegawa, H., and Hashimoto, T.,** Thermoreversible sol-gel transition of PDTS (1,3;2,4-p,p′-ditoluylidene sorbitol)/organic solvent and PDTS/oligomer systems, *Nihon Reoroji Gakkaiski,* 17(2), 86, 1989.
732. **Hedrick, J. L., Russell, T. P., Haidar, B., and Yang, A. C. M.,** Structural modifications in hydroxy ether-dimethyldiphenylsiloxane copolymers, *Macromolecules,* 22, 5570, 1989.
733. **Kim, J. H., Barlow, J. W., and Paul, D. R.,** Interactions in SMA - SAN blends, *J. Polym. Sci. Polym. Phys.,* 27, 223, 1989.
734. **Barbari, T. A., Koros, W. J., and Paul, D. R.,** Gas sorption in polymers based on bisphenol A., *J. Polym. Sci. Polym. Phys.,* 26, 729, 1988.
735. **Min, K. E. and Paul, D. R.,** Blends of poly(vinyl methyl ether) with styrenic copolymers, *J. Polym. Sci. Polym. Phys.,* 26, 2257, 1988.
736. **Kelley, S. S., Glasser, W. G., and Ward, T. C.,** Engineering plastics from lignin. XV. Polyurethane films from chain-extended hydroxypropyl lignin, *J. Appl. Polym. Sci.,* 36, 759, 1988.
737. **Singh, D. P., Kalachandra, S., and Turner, D. T.,** Glass transition of polymers made from slurries of poly(ethyl methacrylate) and (methyl methacrylate), *J. Polym. Sci. Polym. Phys.,* 26, 1795, 1988.
738. **Guo, X.,** Study on the properties of ABS resin modified by α-methylstyrene-acrylonitrile copolymer, Shiyou Huagong, 18, 220, 1989.
739. **Hu, J. and Guo, Y.,** The effects of organic solvent in casting solution — computer aided design, *Natl. Res. Counc. Can.* NRCC 29895 (Adv. Reverse Osmosis Ultrafiltr.), 101, 1989.
740. **Novembre, A. E., Jurek, M. J., Kornblit, A., and Reichmanis, E.,** A sub-0.5μm bilevel lithographic process using the deep-UV electron-beam resist: poly(trimethylsilylmethyl methacrylate-chlorostyrene, *Polym. Eng. Sci.,* 29, 920, 1989.
741. **Green, T. K., Chamberlin, J. M., and Lopez-Froedge, L.,** Swelling of coal extracts, *Prepr. Pap. Am. Chem. Soc. Div. Fuel Chem.,* 34, 759, 1989.
742. **Kerber, M. L., Mkhitaryan, M. A., Klubkova, O. M., Voskanyan, P. S., and Barashkov, O. K.,** Study of interaction thermodynamics of plasticizers with poly(vinyl butyral) by inverse gas chromatography, *Plast. Massy,* No. 5, 71, 1989.
743. **Borisova, E. Yu., Erykalova, T. A., Vladychina, S. V., Kulezhev, V. N., Parfeev, V. M., and Peresypkina, S. V.,** Properties of binary blends of thermoplastic polyurethane with butadiene-cyanoethyl methacrylate copolymer, *Plast. Massy,* No. 4, 18, 1989.
744. **Poinescu, I. C. and Beldie, C.,** Styrene divinylbenzene copolymers — influence of diluent mixture on matrix structure, *Angew. Mak.,* 164, 45, 1988; **Poinescu, I. C. and Carpov, A.,** Relationships between crosslinked copolymer synthesis and their features, *Rev. Roum. Chim.,* 34, 1061, 1989.
745. **Margaritis, A. G. and Kalfoglou, N. K.,** Compatibility of poly(vinyl chloride) with polyalkylene oxides. II. Poly(propylene oxide) and poly(tetramethylene oxide), *J. Polym. Sci. Polym. Phys.,* 27, 1767, 1989.
746. **Askadskii, A. A., Gal'pern, E. G., Surov, G. V., Bulatov, V. V., Slonimskii, G. L., and Korshak, V. V.,** Possibility of combination of different physical properties of polymers, *Vysokomol. Soedin. Ser. B,* 31, 308, 1989.
747. **Gaylord, N. G.,** Use surfactants to blend polymers, *Chemtech,* 19, 435, 1989; Compatibilitizing agents: structure and function in polyblends, *J. Macromol. Sci. Chem.,* A26, 1211, 1989.
748. **El-Hibri, M. J., Cheng, W., Hattam, P., and Munk, P.,** Inverse gas chromatography of polymer blends. Theory and practice, *Am. Chem. Soc. Symp. Ser.,* 391 (Inverse Gas Chromatogr.: Charact. Polym. Other Mater.), 121, 1989.
749. **Giuliani, J. F.,** An investigation of the solubility of organic vapors in polymer films using an optical waveguide interfactial probe, *J. Polym. Sci. Polym. Phys.,* 26, 2197, 1988.
750. **Kennedy, V. O.,** Polymeric motions detected by wetting measurements at the film/water interface influenced by conformational changes at the film/substrate interface, *J. Coatings Technol.,* 60(759), 37, 1988.
751. **Momose, T., Tomiie, K., Ishigaki, I., and Okamoto, J.,** Radiation grafting of α,β,β-trifluorostyrene onto various polymer films by preirradiation method, *J. Appl. Polym. Sci.,* 37, 2165, 1989.

752. **Tao, W. C., Thomas, J. W., Jr., and Frank, C. W.,** Excimer fluorescence as a molecular probe of polymer blend miscibility: 8. Polymeric and glassy solvent host matrices, *Polymer,* 29, 1625, 1988.
753. **Uragami, T., Morikawa, T., and Okuno, H.,** Characteristics of permeation and separation of aqueous alcohol solutions through hydrophobic polymer membranes, *Polymer,* 30, 1117, 1989.
754. **Goydan, R., Reed, R. C., and Tseng, H. S.,** Estimation of the solubilities of organic compounds in polymers by group contribution methods, *Ind. Eng. Chem. Res.,* 28, 445, 1989.
755. **Lee, W., Nobile, M. R., DiBenedetto, A. J., and Nicolais, L.,** Growth, kinetics of environmental crazing in polycarbonate, *Int. J. Polym. Mater.,* 12, 185, 1989.
756. **Mizerovskii, L. N., Vansyatskaya, L. N., and Smurova, G. I.,** Investigation of the reaction between a ternary ethylene-propylene copolymer and organic liquids, *Vysokomol. Soedin. Ser. A,* 29, 1512, 1987; *Polym. Sci. U.S.S.R.,* 29, 1663, 1987.
757. **Koningsveld, R., Kleintjens, L. A., and Nies, E.,** Polymers and thermodynamics, *Croat. Chem. Acta,* 60(1), 53, 1987.
758. **Kirikov, A. A., Bablyuk, E. B., and Suvorin, V. V.,** Solubility parameter and its relation to solubility of a poly(ethylene terephthalate) base of photographic film, *Probl. Tekhnol. Fotogr. Mater. i Khimikaty dlya ikh Pr-va M.,* 129, 1987.
759. **Adamchuk, M. R. and Heckman, R. A.,** Propylene-based glycol ethers and acetates as substitutes for methylene chloride in solvent applications, *J. Radiat. Curing,* 15(2), 3, 1988.
760. **Kambour, R. P., Gundlach, P. E., Wong, I. C. W., White, D. M., and Yeager, G. W.,** Miscibility of poly(2,6-dimethyl-1,4-phenylene oxide) with several styrenic homopolymers — dependence of interaction parameters from critical molecular weights on cohesive energy densitites, *Polym. Commun.,* 29(6), 170, 1988.
761. **Plucinski, P. and Szust, J.,** An approach to the mathematical description of the permeation of hydrocarbon through liquid membranes, *J. Membr. Sci.,* 38, 261, 1988.
762. **Chin, Y.-P. and Weber, W. J., Jr.,** Estimating the effects of dispersed organic polymer on the sorption of contaminants by natural solids. 1. A predictive thermodynamic humic substance - organic solute interaction model, *Environ. Sci. Technol.,* 23, 978, 1989.
763. **Coleman, M. M., Lee, J. Y., Serman, C. J., Wang, Z.-Q., and Painter, P. C.,** Poly(ethylene-co-methacrylic) acid - polyether blends, *Polymer,* 30, 1298, 1989; **Serman, C. J., Xu, Y., Painter, P. C., and Coleman, M. M.,** On the phase behavior and the limits of miscibility of styrene-co-vinylphenyl blends with poly(alkyl methacrylates), *Macromolecules,* 22, 2015, 1989.
764. **Karickoff, S. W.,** Organic pollutant sorption in aquatic systems, *J. Hydraul. Eng.,* 10, 707, 1984.
765. **Shah, K., White, J. L., and Min, K.,** A study of the miscibility of blends of polybutadiene and styrene-butadiene copolymers, *Polym. Eng. Sci.,* 29, 586, 1989.
766. **Takahashi, T., Nakajima, N., and Saito, N.,** Morphology and mechanical properties of rubber-modified epoxy systems. Amino-terminated polysiloxane, *Adv. Chem. Ser.,* 222, 243, 1989.
767. **Lee, Y. M., Bourgeois, D., and Belfort, G.,** Sorption, diffusion, and pervaporation of organics in polymer membranes, *J. Membrane Sci.,* 44, 161, 1989.
768. **Bhattacharya, C., Maiti, N., Mandal, B. M., and Bhattacharya, S. N.,** Thermodynamic characterization of miscible blends from very similar polymers by inverse gas chromatography. The poly(ethyl acrylate) poly(vinyl propionate) system, *Macromolecules,* 22, 4062, 1989.
769. **Chirkova, Y. A. and Kreitus, A. E.,** Sorption of a solvent by cellulose and cellulose materials from the liquid phase, *Vysomol. Soedin. Ser. A,* 31, 2079, 1989.
770. **Gauthier, M. and Eisenberg, A.,** Matrix polarity effects on ionic aggregation in a nitrated styrene ionomer, *Macromolecules,* 22, 3756, 1989.
771. **Neumann, H. J.,** Effect of asphaltene solubility on residue processing, *Erdöl Kohle Erdgas Petrochem.,* 42, 278, 1989.
772. **Kotnis, M. A. and Spruiell, J. E.,** Biaxially oriented films prepared from poly(vinyl chloride)/poly(methyl methacrylate co imide) blends, *Polym. Eng. Sci.,* 29, 1528, 1989.
773. **Mikos, A. G. and Peppas, N. A.,** Flory interaction parameter χ for hydrophilic copolymers with water, *Biomaterials,* 9, 419, 1988.
774. **Alessi, P., Orlandini-Visalberghi, M., and Torriano, G.,** New applications of gas chromatography in the field of paints and varnishes, *J. Oil Col. Chem. Assoc.,* 72, 15, 1989.
775. **Curtis, G. P., Reinhard, M., and Roberts, P. V.,** Sorption of hydrophobic organic compounds by sediments, in *Geochemical Processes at Mineral Surfaces* (Am. Chem. Soc. Symp. Ser. 323), Davis, J. A. and Hayes, K. F., Eds., American Chemical Society, Washington, D.C., 1986, chap. 10.
776. **Pussemier, L., De Borger, R., Cloos, P., and Van Bladel, R.,** Relation between the molecular structure and the adsorption of arylcarbamate, phenylurea and anilide pesticides in soil and model organic adsorbents, *Chemosphere,* 18, 1871, 1989.
777. **Ciemniecki, S. L. and Glasser, W. G.,** Polymer blends with hydroxypropyl lignin, in *Lignin: Properties and Materials* (Am. Chem. Soc. Symp. Ser. 397), Glasser. W. G. and Sarkanen, S., Eds., American Chemical Society, Washington, D.C., 1989, 452.

Chapter 17

SURFACES AND INTERFACES

As a result of the close relationship between solubility or phase separation on the one hand and surface activity on the other, cohesion parameters are useful for characterization of heterogeneous as well as homogeneous systems. Interfacial free energy or interfacial tension is the excess free energy due to the existence of an interface arising from the unbalanced molecular forces. An example will illustrate the dependence of interfacial phenomena on bulk solubility. A dissolved component in a mixture may demonstrate "surface activity" by reducing the interfacial free energy of the solvent and by creating an excess concentration at the interface. This can lead to the stabilization of a foam film at a liquid-air interface, as well as to other effects. It is also possible for an *insoluble* component to reduce the interfacial free energy by spontaneously spreading as an insoluble film at the interface, and in the case of a liquid-air interface this has a foam-inhibiting effect. Thus, two very similar compounds with equal surface activity, both reducing interfacial free energy, can differ greatly in their behavior because of a different degree of bulk solubility.[1]

17.1 INTERFACIAL FREE ENERGY

The specific interfacial free energy of a material is the excess free energy per unit area of surface, and in the case of liquid-vapor or solid-vapor interfaces it is known as the specific surface free energy or surface tension, manifested as an internal force which tends to reduce the surface area to a minimum.

The SI unit of surface free energy or interfacial free energy is J m^{-2} or N m^{-1}, but frequently more convenient is the submultiple mJ m^{-2} or mN m^{-1} which is numerically equivalent to the commonly used c.g.s. unit, the erg cm^{-2} or dyn cm^{-1}.

The symbol γ is normally used for specific interfacial free energy, and for a system involving a solid (s), immiscible liquids (i and j) and a vapor phase (v), the following notation is convenient.

- ${}^{i}\gamma$ is the surface free energy or surface tension of liquid i in contact with air or its own vapor, the superscript being omitted if there is no chance of confusion.
- ${}^{s}\gamma$ is the surface free energy of the solid s in contact with air or its own vapor.
- ${}^{is}\gamma$ is the interfacial free energy of liquid i/solid s.
- ${}^{s,i}\gamma$ is the interfacial free energy of solid s/saturated vapor i.
- ${}^{ij}\gamma$ is the interfacial free energy between liquid phases i and j.

This notation is similar to that suggested by Melrose,[2] and is able to accommodate the cases of two immiscible liquid phases and excess concentrations at interfaces. It can be seen that a property referring to the interface between a pure phase and its vapor is denoted by a single superscript, while a double superscript is used for the interface between two phases. A liquid phase component present in the vapor can adsorb onto the surface of the other liquid or the solid and modify its properties: this is denoted by a superscript comma, followed by i or j denoting the component adsorbed. Similarly, adsorption of j, which saturates phase i, can modify the solid-liquid i interface, and this is indicated in the same way.

Solids do not deform reversibly, so their ${}^{s}\gamma$ values cannot be determined directly. However, solids can be wetted reversibly with liquids of known ${}^{i}\gamma$, so "wettability" is an important surface property.[3] The quantity that can be be measured experimentally is the contact angle, ${}^{is}\theta$, of a liquid with a solid. This may be zero for wetting, or ranged from 0° to 90° for spreading systems, and be greater than 90° when the liquid tends to shrink away from the solid. Young's equation relates θ to the interfacial free energy.

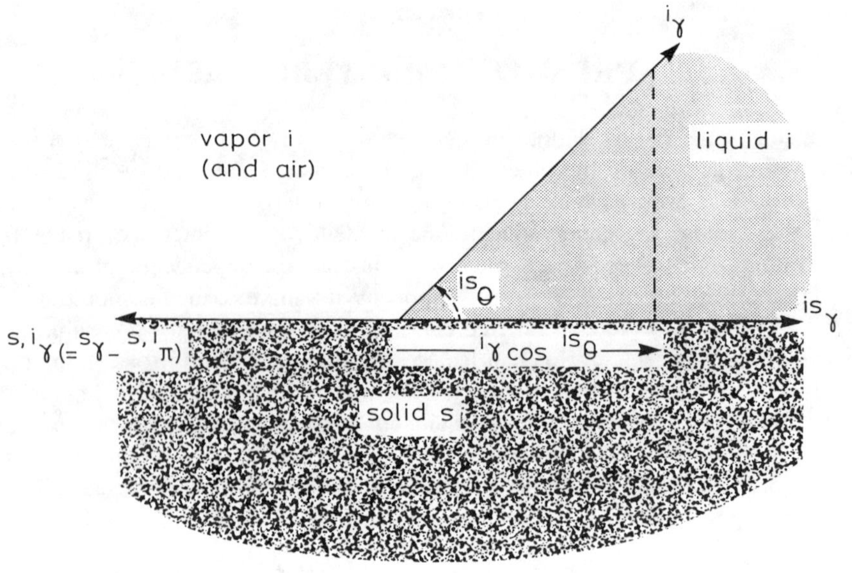

FIGURE 1. Interfacial free energy nomenclature for liquid *i* and solid *s*. (Adapted from Melrose, J. C., *Contact Angle Wettability and Adhesion (Adv. Chem. Ser.* 43), American Chemical Society, Washington, D.C., 1964, chap. 10.)

Consider the case of one liquid, *i*, on a solid surface, *s*, making an equilibrium contact angle ${}^{is}\theta$ (Figure 1). Young's equation is

$$ {}^{s,i}\gamma - {}^{is}\gamma = {}^{i}\gamma \cos {}^{is}\theta \tag{1} $$

where ${}^{s,i}\gamma$ is the free energy of the interface solid *s* with vapor i, ${}^{is}\gamma$ is the free energy of the interface of solid *s* with liquid *i*, and ${}^{i}\gamma$ is the surface free energy of liquid *i* in contact with its vapor. The term ${}^{i}\gamma \cos {}^{is}\theta$ is variously called the adhesion free energy, wetting free energy, or wetting tension, and this difference between ${}^{i,s}\gamma$ and ${}^{is}\gamma$ is the quantity which is experimentally determined. The film pressure or equilibrium spreading pressure is

$$ {}^{s,i}\pi = {}^{s}\gamma - {}^{s,i}\gamma \tag{2} $$

and the surface free energy of solid *s* in contact with its vapor is then

$$ {}^{s}\gamma = {}^{i}\gamma \cos {}^{is}\theta + {}^{is}\gamma + {}^{s,i}\pi \tag{3} $$

For high surface energy liquids on smooth, homogeneous, low energy solids such as polymers, where ${}^{is}\theta > 0$, it is often assumed that ${}^{s,i}\pi = 0$,

$$ {}^{s}\gamma = {}^{i}\gamma \cos {}^{is}\theta + {}^{is}\gamma \tag{4} $$

When ${}^{is}\theta$ decreases and approaches zero, however, ${}^{s,i}\pi$ is often significant and Equation 4 is then invalid. In these conditions ${}^{s,i}\pi = {}^{s}\gamma - {}^{i}\gamma - {}^{is}\gamma$. Liquids spread when the spreading pressure (${}^{s}\gamma - {}^{i}\gamma - {}^{is}\gamma$) is zero or positive, so wetting is favored by large ${}^{s}\gamma$ (high solid surface free energy), small ${}^{i}\gamma$ (low liquid surface free energy), and small ${}^{is}\gamma$ (low $i-s$ interfacial free energy).

Zisman introduced the concept of the critical surface free energy (or tension) of wetting, γ_c, of a solid.[4-7] For a homologous series of some liquids on a given solid the plot of $\cos {}^{is}\theta$

against ${}^{i}\gamma$ is a straight line, and γ_c is the value of ${}^{i}\gamma$ at the intercept of the plotted line with the horizontal line defined by cos ${}^{is}\theta = 1$:

$$^{s,i}\gamma - {}^{is}\gamma = \gamma_c \qquad (5)$$

Thus the critical surface free energy of the solid substrate is the surface free energy of the hypothetical liquid which has cos ${}^{is}\theta = 1$ on this solid and zero solid/liquid surface free energy (${}^{is}\gamma = 0$). Examples of γ_c values are listed in Table 1. As shown in the next section, in many systems γ_c can be identified with ${}^{s}\gamma$. The fundamental importance of γ_c is that liquids having a ${}^{i}\gamma$ value lower than γ_c will spread on the surface, obeying the general rule that liquids of lower surface free energy spread over materials (liquids or solids) of higher surface free energy, so reducing the total system surface free energy.

Not all series of liquids exhibit linear cos θ vs. ${}^{i}\gamma$ plots,[8] but it is possible to use "component" surface free energy methods, separating hydrogen bonding and polar interactions from dispersion effects[9,10] to improve the situation (Section 17.3).

Another function widely used is the molar parachor, an additive quantity defined as

$$P_s = M^{-1}\,\rho^{-1}\,\gamma^{1/4} = V\gamma^{1/4} \qquad (6)$$

P_s is almost temperature-independent. This and other group constant methods can be used for estimating surface free energies in the manner described in Section 6.3.[11-13,357]

17.2 INTERFACIAL FREE ENERGIES AND COHESION PARAMETERS

Since interfacial free energy and adhesion free energy and adhesion properties result from intermolecular forces, they are closely related to cohesion parameters, and this idea has been well explored.*

The work of adhesion, ${}^{is}W$, of liquid i and solid s is the sum of the surface free energies of i and of s, less the ${}^{i-s}$ interfacial free energy:

$$^{is}W = {}^{i}\gamma - {}^{is}\gamma + {}^{s}\gamma(\text{or } {}^{s,i}\gamma) \qquad (7)$$

Each of these surface free energy terms can be estimated in various ways,[26,85-88] including methods based on subdivision into components corresponding to the component cohesion parameters (Section 17.3). Polymer-liquid interaction parameters may be used where appropriate.

Grunberg and Nissan[89] formulated a relationship between the work of cohesion, W, and the surface free energy γ as follows. If the volume per molecule is V/N_A (V = molar volume, N_A = Avogadro constant), the number of molecules per unit surface area of the liquid is $(N_A/V)^{2/3}$. The surface energy per molecule of liquid is $\gamma(N_A/V)^{-2/3}$, so the molar work of cohesion is

* For example by: Anastasiadis;[14] Bagda;[15] Barton;[16] Becher;[17] Beerbower;[18] Bernier and Kambour;[19] Bindschaedler and Peppas;[20] Bonn and van Aartsen;[21] Bulou, Raynal, and Vial;[22] Chatterjee and Ghosh;[23] Cherry;[24] Colin, Guiochon, and Jandera;[25] Crocker;[4] Cross;[26] Dack;[27] Eckert and Prausnitz;[28] Fowkes;[29-32] Fu, Li, and Wang;[33] Gardon and Teas;[34-37] Girifalco and Good;[38,39] Glass;[40,41] Gracia Fadrique;[375] Guiseppi-Elie et al.;[42] Ha, Han, and Lee;[43,44] Hansen;[45-47] Helfand and Tagami;[48] Hobin;[49] Kalz;[373] Kendall;[50] Koenhen and Smolders;[51] Lavygin;[52] Lee;[53-58] Lipatov;[59] Massirio;[60] Michaels;[61] Mitsui et al.;[62] Murgulescu and Terzi;[63] Olabisi et al.;[64] Patterson and Siow;[72] Pritykin;[65,66] Reeves et al.;[67] Roberts and Thomas;[68] Rosseinsky;[69] Rowe;[181-183] Schonhorn;[70] Sears and Darby;[362] Sewell;[71] Spelt;[73,74] Sprow and Prausnitz;[75] van Oss et al.;[351] Vavruch;[76-80] Visser;[81] Wu;[82,83] Yagnyatinskaya et al.;[84] and Zisman.[3]

TABLE 1
Critical Surface Free Energies of Some Polymers and Adsorbed Monolayers

Surface	γ_c/mJ m^{-2}
Perfluorolauric acid monolayer	6
Poly(1,1-dihydroperfluorooctyl methacrylate)	11
Poly(hexafluoropropylene)	16—17
Poly(tetrafluoroethylene)	19
Poly(trifluoroethylene)	22
Poly(dimethylsiloxane) monolayer	24
Stearic acid monolayer	24
Poly(vinylidene fluoride)	25
Poly(1,2-butadiene)	25
Poly(4-methyl-1-pentene), atactic	25
Polyisobutylene	27
Poly(vinyl fluoride)	28
Poly(vinyl methyl ether)	29
Polypropylene	29
Poly(*t*-butylstyrene)	29
trans-Polyisoprene	30
Poly(chlorotrifluoroethylene)	31
cis-Polyisoprene	31
trans-Polybutadiene	31
Polyethylene	31—32
cis-Polybutadiene	32
Poly(propylene oxide)	32
Poly(decamethylene sebacamide)	32
Poly(ethyl methacrylate)	33
Polystyrene	30—36
Poly(11-aminoundecanoic acid)	33
Poly(octamethylene suberamide)	34
Cyclized polyisoprene	34
Poly(ethyl acrylate)	35
Amylopectin	35
Polyepichlorohydrin	35
Polyformaldehyde	36
Poly(nonamethylene azelamide)	36
Rubber hydrochloride	36
Poly(α-methylstyrene)	36
Amylose	37
Poly(vinyl acetate)	37
Poly(vinyl alcohol)	37
Chlorosulfonated polyethylene	37
Chlorinated polyisoprene	37
Polyacrylamide	35—40
Poly(methyl methacrylate)	33—44
Polychloroprene	38
Poly(vinyl chloride)	39
Starch	39
Poly(vinylidene chloride)	40
Poly(methyl acrylate)	41
Poly(6-aminocaproic acid)	42
Poly(2-chlorostyrene)	42
Polycarbonate	42,29
Poly(ethylene terephthalate)	43
Poly(heptamethylene pimelamide)	43
Casein	43
Poly(ethylene oxide)	43
Polyacrylonitrile	44

TABLE 1 (continued)
Critical Surface Free Energies of Some Polymers and Adsorbed Monolayers

Surface	γ_c/mJ m^{-2}
Cellulose, regenerated	44
Wool	45
Poly(hexamethylene adipamide)	46
Polyamide-epichlorohydrin resin	52
Urea-formaldehyde resin	61

Selected from Crocker, G. J., *Rubber Chem. Technol.*, 42(1), 30, 1969. Sources are cited in the original publication.

TABLE 2
Correlation Between Hildebrand Parameter, Surface Free Energy, and Molar Volume of Various Liquids[35-37]

Liquid	$\delta^2 V^{1/3}\gamma^{-1}/10^8mol^{-1/3}$	$\delta^2(V/N_A)^{1/3}\gamma^{-1}$/molecule$^{-1/3}$[a]
Monobasic alcohols, organic acids	8.8—11.6	10.4—13.7
Water, glycerol, ethylene glycol, cyclohexanol	7.8—8.2	9.2—9.7
Other organic liquids	4.9—6.5	5.8—7.7
Liquid metals: Na, K, Bi, Pb, Tl, Sn, Ag, Ga, Al, Au, Cu, Fe	5.4—6.4	6.4—7.6
Liquid metals: Hg, Cd, Mg, Zn	2.1—3.5	2.5—4.1

[a] Calculated by author.

$$W = 2\gamma(N_A/V)^{-2/3} N_A = 2\gamma N_A^{1/3} V^{2/3} \tag{8}$$

The ratio of work of cohesion W evaluated in this way to the cohesion energy $-U$ is a dimensionless constant characteristic of the particular liquid and ranging from about 3.5 for nonpolar liquids to between four and eight for hydrogren-bonded liquids.

In terms of Hildebrand parameter and cohesive energy density or cohesive pressure, an alternative relationship is

$$\delta^2 = c = AV^{2/3}\gamma/V = AV^{-1/3}\gamma \tag{9}$$

where

$$A = \gamma^2 V/(\gamma V^{2/3}) = \delta^2 V^{1/3}\gamma^{-1} \tag{10}$$

A is a constant which varies only between certain limits according to the type of molecule[35-37] (Table 2). It is noteworthy that molten metals and organic liquids have comparable A values, so the properties of these materials are not as divergent as they might at first appear. A can be expressed in molecular terms by introducing the factor

$$1 \text{ mol}^{-1/3} = (6.023 \times 10^{23})^{-1/3} \text{ molecule}^{-1/3}$$
$$= 1.184 \times 10^{-8} \text{ molecule}^{-1/3}$$

and the resulting values of this ratio of cohesive energy to surface energy range from about 5 for spherically symmetric molecules up to about 15 for liquids demonstrating hydrogen bonding which is stronger in three dimensions than in two. Equation 10 is reasonable as long as the area per molecule is proportional to $V^{2/3}$ (as for spherical molecules), but for polymer molecules it is better to interpret δ and V as the Hildebrand parameter and molar volume of polymer segments or repeat units, that is as ${}^u\delta$ and ${}^uV$ (Section 14.8).

Other empirical equations of a similar nature have been used, for example[59,90]

$$\gamma = k\delta^2 V^{1/3} \tag{11}$$

where[72] $k = 1.8 \times 10^{-9}$ mol.$^{1/3}$ Schonhorn,[70] Bonn and van Aartsen,[21] Beerbower,[18] and Becher[17] used the same equation with different constants. Gordon[91] used "cohesion" ($\gamma/V^{1/3}$) to estimate the cohesive nature of molten inorganic salts in relation to polar liquids. Hildebrand and Scott[90] proposed the relationship

$$\delta = k(\gamma/V^{1/3})^{0.43} \tag{12}$$

which may be expressed as

$$\gamma = k'\delta^{2/3}\, V^{1/3} \tag{13}$$

Equation 12 has been widely used. Lee[57] found that 65% of 129 nonpolar and polar liquids obeyed it reasonably well, and it was applied to polymers using for V the polymer repeat unit volume V.[55-57] It follows that group contribution methods (Chapter 6) can be applied in the calculation of surface free energies in the same way as cohesion pressures.[12,71]

Another interesting contribution to the development of the link between cohesion and adhesion, bulk and surface properties, was made by Prausnitz et al.[28,75] The energy of the interphase relative to the ideal gas state was calculated by subtracting the change in energy on forming the surface phase from pure liquid, from the energy of the bulk liquid relative to the ideal gas. Properties of the interfaces could then be determined by assuming that the interphase was a regular solution and defining a *surface cohesion parameter* as the square root of the surface cohesive energy.

Another aspect of the analogy between cohesion energy in liquids and free energy of adhesion at interfaces concerns the nature of molecular interactions and the geometric mean approximation (Section 4.1). The contributions to the general attractive van der Waals interactions between *molecular* species are outlined in Chapter 3: the randomly orienting dipole-dipole Keesom orientation interactions and dipole-induced dipole Debye induction interactions and the fluctuating dipole-induced dipole London dispersion interactions. The extension of the cohesive molecular ideas to adhesive interactions between *macroscopic* bodies resulted in the Lifshitz-van der Waals interaction concept, reviewed recently by van Oss, Chaudhury and Good.[351] In addition, there are the electron acceptor-electron donor or Lewis acid-base interactions, with energies up to two decimal orders of magnitude higher, which occur in both cohesive and adhesive situations.

When only Lifshitz-van der Waals interactions occur, the Good-Girifalco-Fowkes combining rule,[39,92-95] corresponding to the cohesion parameter geometric mean relation, can be used to relate the interfacial free energy for two immiscible phases to the surface free energies of the individual phases:

$${}^{is}\gamma = ({}^{i}\gamma^{1/2} - {}^{s}\gamma^{1/2}) = {}^{i}\gamma + {}^{s}\gamma - 2({}^{i}\gamma\, {}^{s}\gamma)^{1/2} \tag{14a}$$

More generally,

$$^{is}\gamma = {}^{i}\gamma + {}^{s}\gamma - 2\ {}^{is}\phi({}^{i}\gamma\ {}^{s}\gamma)^{1/2} \tag{14b}$$

The value of the interaction parameter $^{is}\phi$ can be computed from molecular properties and is close to unity for organic systems such as the wetting of organic polymers by organic liquids (Section 17.3). It can be shown that if $^{s,i}\pi = 0$, Zisman's critical surface free energy γ_c is related to $^{s}\gamma$ by

$$^{s}\gamma = \gamma_c / {}^{is}\phi^2 \tag{15}$$

so that $^{s}\gamma$ and γ_c are equal when $\theta = 1$. Combination of Equations 1 and 14b to eliminate $^{is}\gamma$ gives

$$^{s}\gamma = \frac{[{}^{i}\gamma(1 + \cos {}^{is}\theta) + {}^{is}\pi]^2}{4\ {}^{is}\theta^2\ {}^{i}\gamma} \approx \frac{{}^{i}\gamma(1 + \cos {}^{is}\theta)^2}{4\ {}^{is}\phi^2} \tag{16}$$

Graphs of cos $^{is}\theta$ against $^{i}\gamma^{-1/2}$ are found to be linear for a wide range of liquids on solid surfaces such as poly(tetrafluoroethylene).

The disadvantage of relationships based on the geometric mean such as Equation 14 is that calculations of $^{is}\theta$ are complex, and a harmonic mean equation (for low surface energy systems) or a geometric-harmonic mean equation (for high surface energy systems) is sometimes preferred.[96] These involve the division of surface free energy into dispersion and polar components, as discussed in the next section.

The close relationship between $^{s}\gamma$ and γ_c implied by Equation 15 explains the success in the correlation of γ_c and the Hildebrand parameter, although in some situations it is claimed that γ_c is more useful in correlation than δ.[97,98]

17.3 COMPONENT SURFACE FREE ENERGIES AND COHESION PARAMETERS

Equation 11 is expected to be valid only for nonpolar, nonassociated liquids, and component cohesion parameters are more likely to be generally useful. Beerbower[18] used Hansen parameters and found different relationships for various classes of liquids:

for nonalcohols $$\delta_d^2 + 0.63\delta_p^2 + 0.63\delta_h^2 = A\gamma V^{-1/3} \tag{17}$$

for most alcohols $$\delta_d^2 + \delta_p^2 + 0.06\delta_h^2 = A\gamma V^{-1/3} \tag{18}$$

for acids, phenols, amines, $$\delta_d^2 + 2\delta_p^2 + 0.48\delta_h^2 = A\gamma V^{-1/3} \tag{19}$$

In each case, $A = 5.9 \times 10^8\ \text{mol}^{-1/3} = 7.0\ \text{molecule}^{-1/3}$ which can be compared with the values in Table 2. It should be noted that for the equation in terms of the commonly used units δ^2/MPa, V/(cm^3 mol^{-1}) and γ/(mJ m^{-2}), the numerical value of A is 59.

Koenhen and Smolders,[51] for a wide range of liquid including monofunctional hydrogen-bonding systems but excluding some cyclic compounds, acetonitrile, carboxylic acids and polyfunctional alcohols, found

$$\delta_d^2 + \delta_p^2 = A\gamma V^{-1/3}$$

where (20)

$$A = 5.8 \times 10^8\ \text{mol}^{-1/3} = 6.9\ \text{molecule}^{-1/3}$$

They attributed the absence of a dependence on δ_h in this relationship to the liquid-vapor interfacial interaction not involving hydrogen bonding. This is also the case with internal pressure and the division of cohesion parameters into "volume dependent" and "residual" components (Section 7.5), although hydrogen bonding is still involved in cohesion properties since during evaporation of liquids hydrogen bonds are being broken. It follows that equations such as 11 are not expected to hold unless $\delta_h = 0$, and that Equation 20 is not valid for polyfunctional alcohols which can form three-dimensional hydrogen bond networks in bulk but not in the surface region. Considerations such as these lead to the suggestion that it may be possible to divide surface free energies up in the same way as cohesion parameters. One set of component surface free energies could be defined by

$$\delta_d^2 = k_d\gamma_d V^{-1/3},\ \delta_p^2 = k_p\gamma_p V^{-1/3},\ \delta_h^2 = \gamma_h = 0 \tag{21}$$

There is considerable theoretical and experimental justification for subdividing the surface free energy into additive components analogous to the component cohesion parameter components described in Section 5.12 and for relating the two sets of properties.*

The surface free energies are additive in the same way as cohesive energies, and various combinations have been used, including

1. That of Koenhen and Smolders,[51] which has already been described:

$$\gamma = \gamma_d + \gamma_p \tag{22}$$

2. Addition of a hydrogen bonding term:

$$\gamma = \gamma_d + \gamma_p + \gamma_h \tag{23}$$

3. Separation of dispersion, dipole-dipole (orientation), induction, and hydrogen bonding:

$$\gamma = \gamma_d + \gamma_o + \gamma_i + \gamma_h \tag{24}$$

4. Incorporation of Lewis acid-base terms such as those discussed in Section 5.4

Because of the empirical and rather arbitrary way in which the component cohesion parameters are evaluated, precise correlations with the component surface free energy terms are not to be expected, but agreement is reasonably good.[10,96]

Hansen[45,47,104,119,120] considered the characterization of surfaces in terms of the liquids which spread spontaneously on them, in contrast to those liquids which yield contact angles, the interactions being described by Hansen parameters (Section 5.9). The results were expressed in the same way as those of solubility studies, with values of ${}^s\delta_d$, ${}^s\delta_p$, ${}^s\delta_h$, ${}^sR$ reported for each surface. These can be used in conjunction with the liquid data of Table 11, Chapter 5 to evaluate ${}^{sl}R$:

$${}^{sl}R^2 = 4({}^s\delta_d - {}^l\delta_d)^2 + ({}^s\delta_p - {}^l\delta_p)^2 + ({}^s\delta_h - {}^l\delta_h)^2 \tag{25}$$

Liquids for which ${}^{sl}R$ is less than or equal to ${}^sR$ are expected to wet the surface, and liquids with coordinates on the spherical boundary surface, ${}^{sl}R = {}^sR$, correspond to critical surface

* For example: Bagda;[15] Beerbower;[18] Chan;[99] Cherry;[24] Crocker;[4] Cuthrel;[100] Dann;[101] Fowkes;[29,31,32,86,102,103] Gardon;[36] Good;[107] Jańczuk et al.;[360] Jensen;[108] Kaelble;[109-112] Kim;[113] Kloubek;[114,115] Köhler;[11] Koenhen and Smolders;[51] Lee;[57] Lipatov;[59] Owens and Wendt;[9] Padday and Uffindell;[86-89] Panzer;[10] Schmidt and Bell;[355] Schröder;[117] Spelt;[73] van Krevelen and Hoftyzer;[13] van Oss et al.;[351] and Wu.[82,96,118]

TABLE 3
Values of Hansen Parameters and Critical Wetting Sphere Radius for Surfaces

Surface	${}^s\delta_d$/MPa$^{1/2}$	${}^s\delta_p$/MPa$^{1/2}$	${}^s\delta_h$/MPa$^{1/2}$	${}^sR$/MPa$^{1/2}$	Ref.
Polyester/melamine	10.2	1.2	10.6	17.8	104
Polyester/melamine/polyvinylidene fluoride	16.6	1.8	3.5	5.5	104
Polyvinylidene fluoride	13.7	10.6	8.2	14.7	104
Polypropylene (high M.W.)	16.6	−2.3	1.0	11.7	104
'Tin plate' (chrome-chrome oxide),					
Sample A	10.2	−0.6	10.6	19.8	104
Sample B	14.1	−0.4	12.9	16.8	104
Bonderite® 1000 (iron phosphate, Parker Division, Hooker Chem.)	14.7	3.3	5.5	19.0	120
Epanol® resin 55-B-40 Shell Chemicals	18.0	0.4	9.2	18.8	120
Polymethylmethacrylate (Cast Optics, Hackensack)	16.4	6.6	9.2		10
Versamid® 930 polyamide (General Mills)	16.8	5.1	8.0		10
Lithium stearate	15.65	7.2			10
	15.71	9.4	17.7, 3.9	(bimodal)	10
Octacosane	16.2	1.8	3.3		10

free energies. The results for some surfaces appear in Table 3. As in the "solubility sphere" studies, the spherical boundaries correspond to situations where there is a predicted zero free energy change for the wetting process. Although cohesion parameters ignore entropy factors and do not provide this kind of information directly in the surface characterization, the limiting boundary conditions correspond to $\Delta H = T\Delta S$ for the adsorption process.

It is also possible to use only the polar and hydrogen-bonding Hansen components by drawing a circle with wetting radius ${}^sR$ on a plot of δ_p against δ_h points for various liquids, or of $(\delta_p - \delta_h)$ lines for liquid mixtures, as illustrated in Figure 2. This example uses the liquid series: ethylbenzene, tetrahydronaphthalene, chlorobenzene, *o*-dichlorobenzene, cyclohexanone, acetone, *N*-methyl-2-pyrrolidone, *N,N*-dimethylformamide, dimethylsulfoxide, 2-pyrrolidone, ethylene cyanohydrin, formamide and water (line A), and ethanol-water mixtures (line B). The critical surface free energy as measured by the series of individual liquids is 44 mJ m^{-2}, and 32 mJ m^{-2} from ethanol-water mixtures. For Epanol® resin 55-B-40 (Figure 3) most of the liquids attacked the substrate, as shown by the fact that the solubility circle overlaps the wetting circle, and data (A) are not useful. The ethanol-water mixture (B) did yield a critical surface free energy value (27 mJ m^{-2}), as this region is outside the solubility circle. One difficulty in this type of approach is that the particular Hansen parameter values to be used for water are not clear[102] (Table 14, Chapter 5).

A related method of surface characterization is used in the Meseran surface analyzer technique,[121-123] which is based on the degree of surface retention of carbon-14 tagged molecules of three bonding types: tridecane, tetrabromethane, and diethyl succinate. A two-dimensional chart is used, with one axis providing Meseran data obtained by subtracting tridecane from diethyl succinate radioactivity counts and other recording tetrabromoethane minus tridecane data. The result is that points corresponding to surfaces of particular organic contaminants on metal surfaces, for example, are distributed on the chart according to their polarity characteristics. This empirical method does not actually use cohesion parameters, but the principle is very similar. Other variations have been used, for example:

- Correlation of γ_c with ${}^s\delta_d$ for polymers[12,82]
- Use of ${}^s\gamma_d$ and δ_d for nonpolar polymers[51]

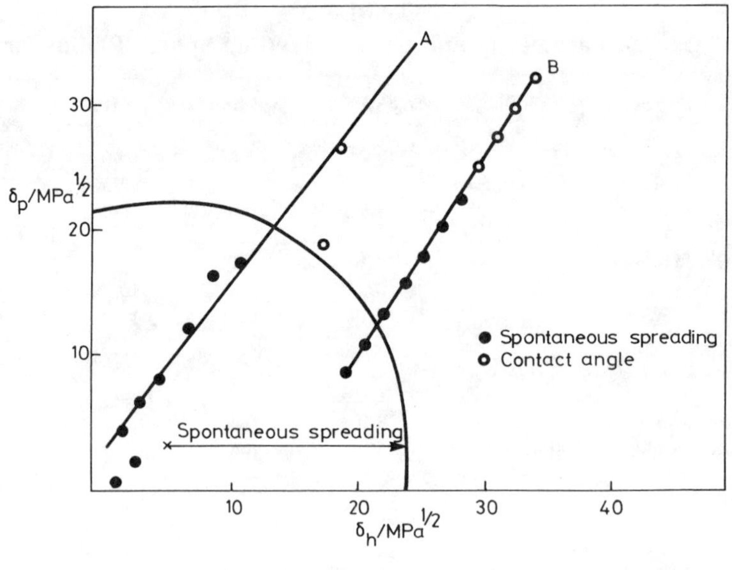

FIGURE 2. Surface evaluation of Bonderite® 1000 (iron phosphate, Parker Division, Hooker Chemical Corporation) (with wetting sphere parameters ${}^s\delta_d$ = 14.7 MPa$^{1/2}$, ${}^s\delta_p$ = 3.3 MPa$^{1/2}$, ${}^s\delta_h$ = 5.5 MPa$^{1/2}$, ${}^sR$ = 19.0 MPa$^{1/2}$) using a series of polar liquids (A) and ethanol-water mixtures (B). Filled circles indicate that spontaneous spreading occurred. (Adapted from Hansen, C. M. and Pierce, P. E., *Ind. Eng. Chem. Prod. Res. Dev.*, 13, 218, 1974.)

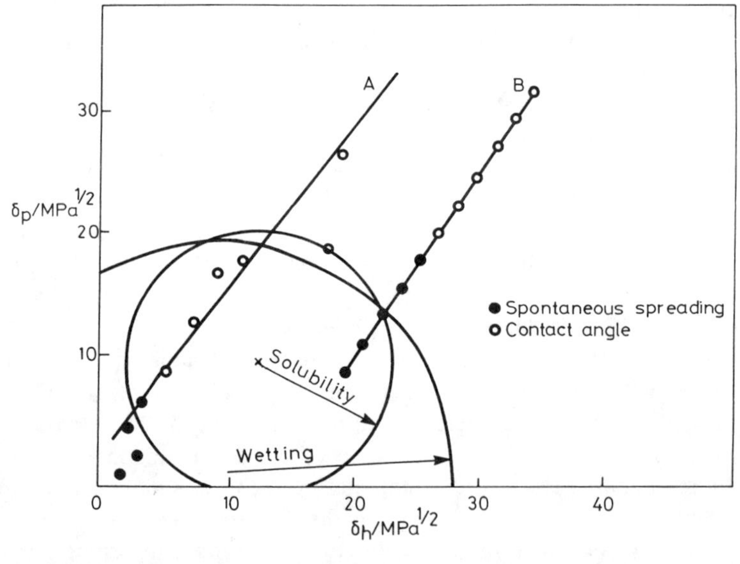

FIGURE 3. Surface evaluation of Epanol® resin 55-B-40 (Shell Chemical Co.) (with wetting sphere parameters ${}^s\delta_d$ = 18.0 MPa$^{1/2}$, ${}^s\delta_p$ = 0.4 MPa$^{1/2}$, ${}^s\delta_h$ = 9.2 MPa$^{1/2}$, ${}^sR$ = 18.8 MPa$^{1/2}$) using a series of polar liquids (A) and ethanol-water mixtures (B). The polar-hydrogen-bonding Hansen parameter solubility circle for the resin is also shown (${}^j\delta_d$ = 20.7 MPa$^{1/2}$, ${}^j\delta_p$ = 9.4 MPa$^{1/2}$, ${}^j\delta_h$ = 12.5 MPa$^{1/2}$, ${}^jR$ = 10.8 MPa$^{1/2}$.) Filled circles indicate that spontaneous spreading occurred. (Adapted from Hansen, C. M. and Pierce, P. E., *Ind. Eng. Chem. Prod. Res. Dev.*, 13, 218, 1974.)

- Description of surface free energy properties of polysiloxanes and fluorocarbons[99] in terms of polar and nonpolar components

Just as the geometric mean equation for cohesion pressure holds only for nonpolar noninteraction systems, Equation 14a is theoretically justified only for the dispersion component of surface free energy. This problem has been approached in various ways. The dimensionless interaction parameter ${}^{is}\phi$ can be expressed in terms of fractional contributions to the cohesive energies of the components

$$ {}^{is}\phi = \left(\frac{{}^{s}C_{d}\,{}^{i}C_{d}}{{}^{s}C\,{}^{i}C}\right)^{1/2} + \left(\frac{{}^{s}C_{o}\,{}^{i}C_{o}}{{}^{s}C\,{}^{i}C}\right)^{1/2} + \left(\frac{{}^{s}C_{i}\,{}^{i}C_{i}}{{}^{s}C\,{}^{i}C}\right)^{1/2} \tag{26} $$

where

$$ C = C_{d} + C_{o} + C_{i} \tag{27} $$

denoting cohesive energy terms that are clearly related to the familiar cohesive pressure c. The interaction parameter ${}^{is}\phi$ is analogous to Walker's specific interchange coefficient (Section 5.7), and it can be seen from Equation 26 that if there is a good match between the C values of the components, ${}^{is}\phi$ is close to unity. But this parameter is less than unity if the contributions to dispersion, orientation and induction are mismatched.

Another possible modification to Equation 14 is

$$ {}^{is}\gamma = {}^{i}\gamma + {}^{s}\gamma - 2({}^{i}\gamma_{d}\,{}^{s}\gamma_{d})^{1/2} - {}^{is}E \tag{28} $$

where ${}^{is}E$ is a ''correction'' term that has been given various forms by different authors, notably Fowkes,[102] Kloubek,[114,115] and Owens and Wendt,[9] for example:

$$ {}^{is}\gamma = {}^{i}\gamma + {}^{s}\gamma - 2({}^{i}\gamma_{d}\,{}^{s}\gamma_{d})^{1/2} - 2({}^{i}\gamma_{p}\,{}^{s}\gamma_{p})^{1/2} \tag{29} $$

An alternative to the geometric mean equation is the harmonic mean expression,

$$ {}^{is}\gamma = {}^{i}\gamma + {}^{s}\gamma - \frac{4\,{}^{i}\gamma_{d}\,{}^{s}\gamma_{d}}{{}^{i}\gamma_{d} + {}^{s}\gamma_{d}} - \frac{4\,{}^{i}\gamma_{p}\,{}^{s}\gamma_{p}}{{}^{i}\gamma_{p}\;{}^{s}\gamma_{p}} \tag{30} $$

which has been applied to water, organic liquids, solid and molten polymers, and organic pigments.[96] The geometric-harmonic mean equation is

$$ {}^{is}\gamma = {}^{i}\gamma + {}^{s}\gamma - 2({}^{i}\gamma_{d}\,{}^{s}\gamma_{d})^{1/2} - \frac{4\,{}^{i}\gamma_{p}\,{}^{s}\gamma_{p}}{{}^{i}\gamma_{p} + {}^{s}\gamma_{p}} \tag{31} $$

Similar equations with γ_h replacing γ_p have also been used.[9,124] Quantities defined as the square root of γ_d, γ_p, etc. (directly analogous to δ_d, δ_p, etc.) may be employed.[111,125]

17.4 MULTICOMPONENT SYSTEMS

When there is more than one liquid and one solid present, complexities arise in both the description and in the understanding of the phenomena, primarily because the composition of the interface is not well defined. It is necessary to assume an interfacial region which differs from the bulk phases, the differences resulting in free energy minimization.

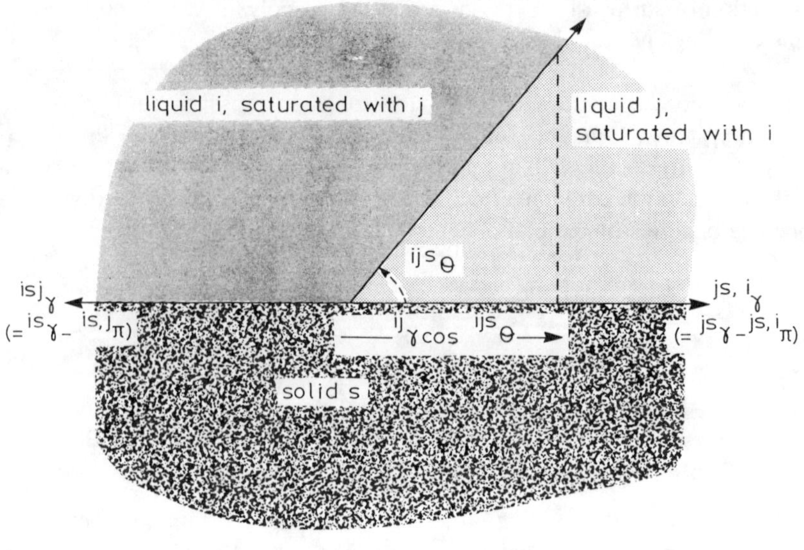

FIGURE 4. Interfacial free energy nomenclature for immiscible liquids *i* and *j* on solid *s*. (Adapted from Melrose, J. C., *Contact Angle Wettability and Adhesion,* American Chemical Society, Washington, D.C., 1964, chap. 10.)

For two liquids *i* and *j*, which form immiscible phases, the notation outlined in Section 17.1 can be extended as illustrated in Figure 4.[2] Young's equation is

$$^{is,j}\gamma = {}^{js,i}\gamma + {}^{ij}\gamma \cos {}^{ijs}\phi \tag{32}$$

where

- $^{is,j}\gamma$ is the interfacial free energy of the interface between solid *s* and liquid *i*, modified by adsorption of *j*
- $^{js,i}\gamma$ is the interfacial free energy of the interface between solid *s* and liquid *j*, modified by adsorption of *i*
- $^{ij}\gamma$ is the liquid-liquid interfacial free energy
- $^{is}\gamma$ and $^{js}\gamma$ are the interfacial free energies of the interfaces between the solid *s* and the pure liquids

The film pressure or equilibrium spreading pressures are

$$^{is,j}\pi + {}^{is}\gamma - {}^{is,j}\gamma \tag{33}$$

and

$$^{js,i}\pi = {}^{js}\gamma - {}^{js,i}\gamma \tag{34}$$

The liquid-liquid interfacial free energy $^{ij}\gamma$ in principle can be deduced from the individual surface free energies $^{i}\gamma$ and $^{j}\gamma$ by an equation analogous to Equation 14:

$$^{ij}\gamma = {}^{i}\gamma + {}^{j}\gamma - 2\,{}^{ij}\phi({}^{i}\gamma\,{}^{j}\gamma)^{1/2} \tag{35}$$

As discussed in the previous section, $^{ij}\phi$ can be evaluated from molecular properties with the aid of relationships deduced theoretically, for example, those presented by Girifalco and Good,[38] who have tabulated $^{ij}\phi$ values for water-organic liquids (Table 4) and for mercury-nonmetallic liquids (Table 5). Alternatively, component surface free energies may be used.[30,102]

The cohesion parameter of a single-phase liquid mixture may be estimated by combining the individual component values on a volume fraction basis (Equation 31, Chapter 4). However, this procedure does not work well for the surface free energies of blends due to enrichment of one component at the solid-liquid interface, although presumably it does give the instantaneous value for a freshly formed surface, which may be useful in rapid processes such as fractional distillation.[93,105] Some success may be achieved by defining a surface phase which is considered a regular solution with properties calculated by means of cohesion parameters.[28,75,126-128] Interfacial activities of 2-hydroxy-5-alkylbenzophenone oximes in octane - water and toluene - water systems have been correlated with Hildebrand parameters.[366]

The same type of procedure has been extended to polymer-polymer interfaces,[48,129-132] making use of the polymer-liquid χ interaction parameter which is closely related to cohesion parameters (Chapter 14). An important aspect of polymer-polymer interfaces that has been discussed in terms of component surface free energies is adhesive bonding and fracture.[111,133-138]

17.5 WETTING, ADHESION, LUBRICATION, POWDERS

An important group of practical surface properties involve "wetting" (spontaneous spreading of an applied liquid droplet on a surface) and "dewetting" (the breaking up or "crawling" of an applied film). In the former case the concern is with advancing contact angles, and in the latter with receding contact angles. The absence of contact angle ($\theta = 0°$) is desirable in the coatings industry in both cases. The Zisman critical surface free energy (Equation 5) has been particularly useful, but cohesion parameter methods have also been applied to various situations, including the wetting properties of

- Coatings[58,104,119]
- Elastomers and copolymers[54,56]
- Powder surfaces[62]
- Lubricants[18,139,140]
- Molten polymers[96]

Separation of dispersion components from "polar" components of the surface free energy or cohesion parameters is important here, as emphasized by Wu.[96] The optimum wettability condition is when the polar surface free energy (nondispersive) components are identical, and this is particularly important when the total surface energies of the two materials are comparable. (When the surface free energy of the liquid is much lower than that of the solid, complete wetting may still occur even though the polarities differ widely.) For the selection of adhesives and dispersion and wetting agents, liquid-solid wetting and solid-solid adhesion are important factors, and both can be discussed in terms of cohesion parameters.*

As shown in Section 17.1, wetting is favored by small $^{l}\gamma$ and small $^{ls}\gamma$. However, polymers which when liquid wet the substrate are not necessarily good adhesives: they should not crystallize, and there should be a matching of the cohesion parameters of the solids. Adhesion performance can be assessed by simple swelling tests, as illustrated in Section

* For example: Andrews;[141] Barton;[16] Bindschaedler;[20] Bleier;[142] Brydson;[143] Chalykh et al.;[144] Crocker;[4] Dyckerhoff and Sell;[145] Fowkes;[32] Gardon and Teas;[36,37] Harrison;[146,147] Huntsberger;[148-150] Iyengar and Erickson;[151] Jackson;[152] Kim;[113] Kitazaki and Hata;[153] Koldunovich et al.;[154] Kuptsov et al.;[155] Lyusov et al.;[377] Runge and Dreyfuss;[157] Sarkar;[158] Schröder;[117] Spelt;[74] Sushkov et al.;[159] Voyutskii et al.;[160] Visser;[81] Wingrave;[161] Wu;[96] and Zisman.[3]

TABLE 4
Interaction Parameter for Water-Organic Liquid Interfaces at 20°C

Liquid	$^{ij}\phi$
Alcohols	
Isobutanol	1.15
1-Butanol	1.13
2-Octanol	1.13
Isoamyl alcohol	1.11
1-Pentanol, *n*-amyl alcohol	1.09
1-Hexanol	1.06
1-Heptanol	1.06
Cyclohexanol	1.04
1-Octanol	1.03
Carboxylic Acids	
Isovaleric acid	1.11
Heptanoic acid	1.04
Octanoic acid	1.03
Undecylenic acid (25°C)	0.98
Oleic acid	0.92
Aldehydes and Ketones	
Methyl hexyl ketone	1.08
Methyl propyl ketone	1.07
Methyl butyl ketone	1.03
Monochloroacetone	1.00
Methyl amyl ketone	0.99
Ethyl propyl ketone	0.98
1-Heptaldehyde	0.97
1,1-Dichloroacetone	0.96
Benzaldehyde	0.90
Esters	
Ethyl acetate	1.08
Ethyl carbonate	0.99
Butyl acetate	0.97
Ethyl butyrate	0.97
Ethyl isovalerate	0.94
Isoamyl butyrate	0.88
Ethyl nonylate	0.85
Ethyl octanoate	0.84
Ethers	
Diethyl ether	1.12
Diisopropyl ether	1.01
Amines	
Dipropylamine	1.17
Dibutylamine	1.06
Aniline	0.98

TABLE 4 (continued)
Interaction Parameter for Water-Organic Liquid Interfaces at 20°C

Liquid	$^{ij}\phi$
Nitriles	
Butyronitrile	1.00
Isovaleronitrile	0.97
Nitro Compounds	
Nitromethane	0.97
Nitrobenzene	0.81
o-Nitrotoluene	0.79
m-Nitrotoluene	0.79
Alkanes	
2,7-Dimethyloctane, diisoamyl	0.60
2-Methylbutane, isopentane	0.59
Pentane	0.58
Hexane	0.55
Heptane	0.55
Octane	0.55
Decane	0.55
Cyclohexane	0.55
Decahydronaphthalene, decalin	0.55
Tetradecane	0.53
Stanolax	0.51
Liquid petroleum, Squibb	0.51
Aromatic Hydrocarbons	
p-Cymene, *p*-isopropyltoluene	0.73
Benzene	0.72
Toluene, methylbenzene	0.71
o-Xylene, 1,2-dimethylbenzene	0.71
m-Xylene, 1,3-dimethylbenzene	0.69
p-Xylene, 1,4-dimethylbenzene	0.69
Ethylbenzene	0.69
Mesitylene, 1,3,5-trimethylbenzene	0.67
Aliphatic Halogenated Compounds	
Isoamyl chloride	0.98
tert-Butyl chloride	0.91
Isobutyl chloride	0.88
Dichloromethane, methylene chloride	0.80
β,β′-Dichlorodiethyl ether	0.78
Trichloromethane, chloroform	0.76
1,2-Dibromoethane, ethylene dibromide	0.71
1,1,2,2-Tetrabromoethane	0.70
Tetrabromomethane, carbon tetrabromide	0.69
1,2,3-Tribromopropane	0.69
Tribromomethane, bromoform	0.67
Tetrachloromethane, carbon tetrachloride	0.61
Diiodomethane, methylene iodide	0.61
Tetrachloroethylene	0.59

TABLE 4 (continued)
Interaction Parameter for Water-Organic Liquid Interfaces at 20°C

Liquid	$^{ij}\phi$
Aromatic Halogenated Compounds	
Chlorobenzene	0.70
Bromobenzene	0.69
o-Chloronaphthalene	0.67
Iodobenzene	0.66
o-Bromotoluene	0.66
o-Bromonaphthalene	0.66
Fluorocarbon Derivatives	
Perfluorotributylamine	0.95
Perfluorodibutyl ether	0.55
Sulfur Compounds	
Ethyl mercaptan	0.85
Phenyl isothiocyanate	0.68
Carbon disulfide	0.58

Adapted from Girifalco, L. A. and Good, R. J., *J. Phys. Chem.*, 61, 904, 1957.

TABLE 5
Interaction Parameter for Mercury-Nonmetallic Liquid Interfaces at 20°C

Liquid	$^{ij}\phi$
Alcohols	
Isobutanol	0.75
1-Octanol	0.70
2-Octanol	0.69
1-Hexanol	0.59
Isoamyl alcohol	0.58
1-Butanol	0.58
1-Propanol	0.57
Methanol	0.56
Ethanol	0.53
Aliphatic Acids	
Acetic acid, ethanoic acid	0.76
Valeric acid	0.76
Oleic acid	0.75
Undecylenic acid	0.64
Ester	
Butyl acetate	0.58
Ether	
Diethyl ether	0.63

TABLE 5 (continued)
Interaction Parameter for Mercury-Nonmetallic Liquid Interfaces at 20°C

Liquid	$^{ij}\phi$
Amines	
Aniline	0.66
Diisoamylamine	0.60
Nitro Compounds	
Nitroethane	0.82
Nitrobenzene	0.59
Alkanes	
Isooctane, 2,2,4-trimethylpentane	0.63
Nonane	0.61
Hexane	0.62
Heptane (25°C)	0.61
Octane	0.60
Aromatic Hydrocarbons	
m-Xylene, 1,3-dimethylbenzene	0.63
p-Xylene, 1,4-dimethylbenzene	0.62
o-Xylene, 1,2-dimethylbenzene	0.61
Toluene, methylbenzene	0.61
Propylbenzene	0.60
Butylbenzene	0.60
Benzene (25°C)	0.59
Aliphatic Halogenated Compounds	
Iodomethane, methyl iodide	0.84
Iodoethane, ethyl iodide	0.79
1,2-Dichloroethane, ethylene dichloride	0.76
1,1-Dichloroethane, ethylidene chloride	0.76
1,1,2,2-Tetrabromoethane	0.72
Dichloromethane	0.70
Trichloromethane, chloroform	0.64
Tetrachloromethane, carbon tetrachloride	0.64
1,2-Dibromoethane, ethylene dibromide	0.59
Aromatic Halogenated Compounds	
Bromobenzene	0.62
Chlorobenzene	0.59
Miscellaneous	
Ethyl mercaptan	0.88
Water (25°C)	0.32

Adapted from Girifalco, L. A. and Good, R. J., *J. Phys. Chem.*, 61, 904, 1957.

14.3. The ideal bonding adhesive is one that exhibits a broad curve when the extent of swelling is plotted against Hildebrand parameter, wide enough to overlap the Hildebrand parameters characterizing both substrates. This can be achieved by using a copolymer formed from both high and low Hildebrand parameter monomers so that it is swollen by a wide range of polymers and is able to bond both poly(vinyl chloride) and rubber, for example.[146,147] It is also useful to evaluate the work of adhesion in terms of component surface free energies.

By analogy with Equation 7, the work of adhesion of two solids *s* and *t* is

$$^{st}W = {}^{s}\gamma + {}^{t}\gamma - {}^{st}\gamma \tag{36}$$

Combining this with the solid-solid form of Equation 14a, for example, there results:

$$^{st}W \approx 2[({}^{s}\gamma_{d}\,{}^{t}\gamma_{d})^{1/2} + ({}^{s}\gamma_{p}\,{}^{t}\gamma_{p})^{1/2}] \tag{37}$$

Similar expressions may be used for predicting if an adhesive *st* joint is stable in the presence of liquid *i*. For spontaneous separation,

$$^{st}\gamma > {}^{si}\gamma + {}^{ti}\gamma \tag{38}$$

The contribution of various factors to the work necessary to separate *s* from *t* has been considered by Schonhorn.[162] The dimensions of tensile strength are the same as those of cohesive pressure, and the ultimate strength of an ideal, flow-free material is predicted to be about one-quarter of δ^2.[35-37] This provides a theoretical upper limit to the performance of polymeric adhesives:

$$\sigma = {}^{st}\phi\,{}^{s}\delta\,{}^{t}\delta/4 \tag{39}$$

where $^{st}\phi$ is the interaction parameter or geometric mean correction factor.

Adhesive failure may be discussed in terms of surface energies;[24,163] see also Section 16.4 for environmentally induced adhesive failure.

From the discussion above it follows that an adhesive must have low $^{i}\gamma$ and low $^{is}\gamma$ for good wetting, but high $^{i}\gamma$ and low $^{is}\gamma$ for good adhesion. In practice, therefore, the surface free energy $^{i}\gamma$ of an adhesive should be moderate, on the low side if good spreading is of critical importance, but high if great joint strength is desired. The surface free energies of polymeric adhesives are often modified by the incorporation of solvents, but again the ease of wetting may be accompanied by strain and discontinuities as solvents evaporate.

The adhesive strengths of surface coatings, as determined by a peel strength method, show a correlation with the cohesive pressure of the solvent from which the polymer film was cast. For a 10% lauryl methacrylate - 90% methyl methacrylate copolymer on tin, the peel strength showed a fairly smooth curve as a function of cohesive pressure and a maximum at 400 MPa (δ = 20 MPa$^{1/2}$).[164] The effect of the nature of coating solvent on film adhesion has applications in such areas as pharmaceutical tablet manufacture.[165] This correlation may be attributed to the state of the polymer in solution, best surface contact being achieved in a good solvent. A similar type of relationship was observed for the peeling of a poly(vinyl chloride) coating formed from an organosol on steel cleaned with different solvents.[166] This was interpreted as being due either to different cleaning actions of the solvents, or to solvent molecules remaining adsorbed on the metal surface and assisting the subsequent polymer adhesion.

In another typical application of Hildebrand parameter, a patent specification[167] states that a metal wire is plated with brass, copper, or zinc, and is then coated with a synthetic resin having a δ value differing by less than 2 MPa$^{1/2}$ from that of the rubber in which it is to be incorporated as reinforcing. Other applications of cohesion parameters to adhesion include:

- Adhesion of lithographic materials[111,137,168,169]
- A surface free energy analysis of bioadhesion[111,138] and artificial blood substitutes[125]
- An optical waveguide device to probe changes in dielectric properties of a polymer-vapor interface[372]
- Adhesive durability and fracture[111,133-138]
- Adhesion of oxypropylene polyol urethane[170] and other polyurethanes[377]
- High-density polyethylene gels as adhesives for polyethylene[369]
- Effect of solvent on the freezing resistance of polychloroprene rubber adhesives[171]
- Adhesive properties of methyl methacrylate-2,3-bis-(diorganophosphone)buta-1,3-diene copolymers[172]
- Adhesion of organics to metal,[164,166,173,174] with addition of acrylic acid copolymers improving resin adhesion to steel, attributed to Hildebrand parameter being similar to that of oil on the steel[368]
- Bonding of cold-curing denture base acrylic resin[175]
- Compatability of adhesives with Mylar® (poly(ethylene terephthalate)) and polyamides[151]
- Adhesion of Kevlar® aramid yarns[176] and other fabrics[352]
- Adhesive properties of polyepichlorohydrin[177] and hydantoin vinyl ether polymer[163]
- Wettability and other adhesion properties of polysiloxanes[58,178,179]
- Wettability of polyacetylene[42]
- Wetting and dispersion of silicon powder[142]
- Organosilicons to reduce magnesia cement viscosity[180]
- Pharmaceutical tablet formulation[165,181-183,376]
- Tackifier resins[364,381]
- Hansen parameters applied to adhesion of photocurable acrylates to solid polymer substrates[382]

Hansen's three-component parameter method (Section 17.3) can be applied to *lubricant* ability, with the spherical surface denoting an arbitrarily defined friction coefficient value, as demonstrated by Beerbower and Panzer[184] with the data of Senior and West.[140] The lubrication abilites of greases based on oils combined with metal soaps such as zinc, lead, and lithium stearates can be correlated with Hildebrand parameters in a rather different way.[185-189] The dissolution of a crystalline soap in a nonpolar liquid is complete at the critical solution temperature, where the combined effects of swelling and thermal loosening of the lattice cause disintegration. Both grease dropping points and critical solution temperatures show marked changes at particular values of the oil Hildebrand parameter. Lubricants for poly(vinyl chloride)[190,191] and for other polymers[358] and ureate greases[192] have also been considered in terms of Hildebrand parameters, and lubricants for polyethylene have been correlated with the components of surface free energies.[140]

17.6 ADSORPTION

The expanded interaction cohesion parameter concept described in Chapter 5 can be extended to adsorption processes — in fact, this approach was first developed by Keller, Karger, Snyder et al.[193-195] for adsorption and chromatography. Equation 22, Chapter 5, is the expression for the change in cohesive pressure during the *i-j* mixing process,

$$
{}^{ij}A = {}^{i}\delta_t^2 + {}^{j}\delta_t^2 - 2\,{}^{i}\delta_d\,{}^{j}\delta_d - 2\,{}^{i}\delta_o\,{}^{j}\delta_o - 2\,{}^{i}\delta_j\,{}^{j}\delta_d
$$

$$
- 2\,{}^{i}\delta_i\,{}^{j}\delta_d - 2\,{}^{i}\delta_a\,{}^{j}\delta_b - 2\,{}^{i}\delta_a\,{}^{j}\delta_b \qquad (40)
$$

where

$$^{i}\delta_t^2 = {^i}\delta_d^2 + {^i}\delta_o^2 + 2\ {^i}\delta_i\ {^j}\delta_d + 2\ {^i}\delta_a\ {^j}\delta_b \tag{41}$$

and the subscripts denote total (t), dispersion (d), induction (i), orientation (o), Lewis acid (proton donor, a), and Lewis base (proton acceptor, b). The corresponding expression for the adsorption of component i from the gas phase onto solid absorbent s is

$$^{is}A = {^i}\delta_d\ {^s}\delta_d + {^i}\delta_o\ {^s}\delta_o + {^i}\delta_i\ {^s}\delta_d + {^s}\delta_i\ {^i}\delta_d + {^i}\delta_a\ {^s}\delta_b + {^s}\delta_a\ {^i}\delta_b \tag{42}$$

Details of the application of this adsorption formalism and other aspects of adsorption are included in Chapter 18 for gas-solid chromatography and liquid-solid or adsorption chromatography. But it is clear from a superficial inspection of Equation 42 that although it is of value if reliable component cohesion parameter values are available, it is not simple to use in this form.

Fowkes and Mostafa[103] in discussing the adsorption of poly(methyl methacrylate) onto a silica filler, considered dispersion, polar, and acid-base components of the adsorption energy, but instead of Equation 42 used a hybrid expression which combined a term of the Drago type (Section 7.12) with a dispersion cohesion pressure term. Balik and Hopfinger[196] used polar-nonpolar cohesion parameter components in describing the role of solvent in the adsorption of poly-γ-benzyl-L-glutamate onto alkali halide crystals. The nonpolar component was correlated with the polymer solubility, and the polar component was related to the state of aggregation of the polymer in solution and with the adsorption process.

McGuire and Suffet[197-200] developed the calculated net adsorption energy concept for discussing the adsorption of trace organics from water onto activated carbon. The net adsorption energy (E_t) is the affinity (^{js}E) of the solute j for the solid surface s, less the affinity of the solvent i for the surface and the affinity (^{ij}E) of the solute j for the solvent i,

$$E_t = {^{js}}E - {^{is}}E - {^{ij}}E \tag{43}$$

For the adsorption of solute and solvent onto nonpolar surfaces such as carbon, dispersion interactions predominate, enabling significant simplification in these terms,

$$^{js}E = {^j}V\ {^j}\delta_d\ {^s}\delta_d,\ {^{is}}E = {^i}V\ {^i}\delta_d\ {^s}\delta_d \tag{44}$$

(The value of δ_d for the carbon surface has been estimated at about 25 MPa$^{1/2}$ for this purpose.) The solute-solvent term is important in the case of aqueous systems, and here δ_a and δ_b terms predominate,

$$^{ij}E = {^j}V\ ({^j}\delta_a\ {^i}\delta_b + {^i}\delta_a\ {^j}\delta_b) \tag{45}$$

Table 6 compares the calculated net adsorption energies for several organic molecules (j) from water (i) onto activated carbon (s). The net adsorption energy method was compared subsequently with alternative approaches.[201,202]

More often, rather than the component cohesion parameters, the simple single-component Hildebrand parameter correlation is used. Kronberg and Stenius[203] used a model based on polymer-liquid interaction parameters and Hildebrand parameters for the adsorption of nonionic surfactants on latices, predicting that adsorption decreases as the latex polarity increases. A parameter χ^s was used to represent the average surfactant-water interaction in the surface "phase".

Sometimes it is found that resin surface areas vary with the nature of the liquid used to wash the resin. This has been explained[379] as follows: if a "poor" solvent replaces a good one, the glass transition temperature is raised, the resin being transformed from the rubbery state into a glassy state, so shrinkage of the resin during evaporation of the washing liquid is suppressed by its lower mobility within the glass and the surface area increases due to the resulting pores.

Other examples of the application of cohesion parameter concepts are:

- Adsorption of iron pentacarbonyl ($\delta \approx 16$ MPa$^{1/2}$) by poly(tetrafluoroethylene)[204,205]
- Adsorption onto carbon[25,206-209]
- Adsorption of polymers on cellulose and on iron powder[210]
- Adsorption of organic species by styrene-divinylbenzene copolymer resins, and their regenerative efficiency[211,212]
- Adsorption at oil-water interfaces[213,214]
- Adsorption of poly(methyl methacrylate) from binary solvent systems onto aluminum and calcium silicates (no simple correlation with Hildebrand parameter)[215]
- Effect of preadsorbed water on sorption by cellulose[216]
- Sorption of dyes[217] (See also Section 17.8)
- Effect of wall adsorption on polybutadiene solution flow properties[218]
- Adsorption on the polymer filler microbarite[219]
- Adsorption selectivities of dilute organics from water on pervaporation membranes[365]
- Gas sorption on bisphenol A polymers[367]
- Adsorption of organic compounds on soil and sediments[376,380,383]

The concept of regular interfacial solutions, already referred to in Section 17.4, is also useful in some adsorption situations.[220,221]

17.7 MICELLES, COLLOIDS, EMULSIONS, FOAMS, SURFACTANTS

There is a continuous range of liquid phase mixture types, from the true solutions that are fully dispersed at a molecular level, through clusters or aggregates of relatively small numbers of solute molecules in micelles, translucent colloidal dispersions (10^{-9} to 10^{-6} m) up to opaque solid-liquid suspensions, liquid-liquid emulsions, and liquid-vapor foams. All can be discussed in terms of cohesion parameters.*

The dispersal efficiency of surfactant or emulsifier molecules is a function of the relative interactions of their polar, hydrophilic "heads" with the aqueous phase and of their nonpolar, lipophilic "tails" with the hydrocarbon phase. (Surfactants may be either nonionic or ionic, but high polarity is achieved only in the ionic surfactants.) Griffin[246] devised a quantitative but empirical hydrophile - lipophile balance (HLB) scale, on which oleic acid has HLB = 1 and potassium oleate has HLB = 20, to describe this behavior. The stabilities of emulsions of hydrocarbons in water may be correlated with the cohesion parameters of the hydrocarbons, while the solubility and aggregation characteristics of surfactants may be related to the cohesion parameters of the surfactants and of the solvents in which they are dispersed. This localized segregation or structuring is not confined to aqueous systems or to liquids: microdomains occur in block and graft copolymers.[130,247]

* For example: Davis and Smith;[222] Deguchi et al.;[223] Fowkes;[85] Funasaki et al.;[224] Hansen;[46,225] Hayashi;[226] Healy et al.;[227] Holtzscherer and Candau;[228] Kertes;[229] Konovalova et al.;[361] Little and co-workers;[185,187-189,230-232] Marszall and van Valkenburg;[233-235] McDonald and co-workers;[236-238] McNicoll et al.;[239] Meusberger;[240,241] Panich et al.;[242] Samaha and Naggar;[243] Schott;[244] Szymanowski et al.;[363] Vijayendran and Bursh;[245] Wellinghoff et al.;[245] and Wingrave.[161]

TABLE 6
A Comparison of Calculated Net Adsorption Energies from Water onto Activated Carbon

Organic compound, j	E/kJ mol^{-1}			
	^{js}E	^{ws}E	^{ij}E	E_t
Decane	75.7	5.9	0.0	69.8
Tetrachloromethane	41.4	5.9	0.0	35.5
p-Nitrophenol	49.8	5.9	4.6	39.3
Nitromethane	19.3	5.9	5.4	8.0
Urea	20.1	5.9	8.8	5.4
Methyl ethyl ketone	32.6	5.9	8.8	17.6
1-Butanol	34.7	5.9	9.2	20.0
1,4-Dioxane	33.9	5.9	16.7	11.3

Adapted from McGuire, M. J. and Suffet, I. H., *Abstr. Pap. ACS,* 176, ENVR 6, 1978

TABLE 7
HLB and Hildebrand Parameter Values for Surfactants

Surfactant	δ/MPa$^{1/2}$	HLB
Oleic acid, 9-octadecenoic acid	16.8	1.0
Glycerol monostearate	17.0	3.8
Sorbitan monolaurate	17.6	8.6
Polyoxyethylene (10) oleyl ether	18.2	12.4
Polyoxyethylene (20) cetyl ether	18.6	15.7
Sodium octadecanoate	19.0	18.0
Sodium hexadecanoate	19.2	19.0
Sodium dodecanoate	19.6	20.9
Sodium octanoate	20.5	22.8
Sodium dinonylnaphthalene sulfonate	21.5	28.5
Dioctyl sodium sulfosuccinate	25.0	32.0
Sodium heptadecyl-1-sulfate	26.0	37.6
Sodium dodecyl-1-sulfate	28.8	40.0
Sodium decyl-1-sulfate	30.3	40.0
Sodium octyl-1-sulfate	32.3	41.9

Adapted from Little, R. C., *J. Colloid Interface Sci.,* 65, 587, 1978.

Little and Singleterry[189,230,232] used Hildebrand parameters, pointing out that the aggregation number (the number of surfactant molecules per micelle) of dinonylnaphthalene sulfonate surfactants is an approximately linear function of the Hildebrand parameter of the oil. Also, there is a clear correlation between HLB data for surfactants and their δ values (Table 7). These data conform to the relationships

$$\text{HLB} = 54\left[\frac{\delta/\text{MPa}^{1/2} - 16.8}{\delta/\text{MPa}^{1/2} - 12.3}\right] \tag{46}$$

$$\delta/\text{MPa}^{1/2} = \frac{243}{54 - \text{HLB}} + 12.3 \tag{47}$$

Other relationships between cohesion parameters and HLB also have been reported.[224,226] A related hydrophile-lipophile function (Equation 98, Chapter 7),

$$D(\delta) = (\delta - {}^{aq}\delta)^2 (\delta - {}^{o}\delta)^2$$

has been used in assessing the efficiency of phase transfer catalysts in the benzene-water system in terms of Hildebrand parameters.[248] Such expressions obviously suffer from using only single-parameter cohesion parameters,[248] and some progress has been made with multi-component parameters.

Beerbower and co-workers[105,249,250] were among those who discussed the dispersing tendencies on the oil and water interfaces of the surfactant or emulsifier region in terms of the ratio of the cohesive energies of the mixtures of oil with the lipophilic portion of the surfactant and of water with the hydrophilic portion. They used Hansen parameters, and for situations where δ_d, δ_p, and δ_h match for oil/lipophile and for water/hydrophile, their expression takes a simplified form. This can be related to the hydrophile-lipophile balance (HLB), for which Beerbower's[105,249] definition is

$$\text{HLB} = \frac{20\ {}^{H}M}{{}^{L}M + {}^{H}M} = \frac{20\ {}^{H}V\ {}^{H}\rho}{{}^{L}V\ {}^{L}\rho + {}^{H}V\ {}^{H}\rho} \tag{48}$$

where ${}^{H}M$ is the molecular weight of the hydrophilic portion of the emulsifier (including the CO group if an acid is involved) and ${}^{L}M$ is the molecular weight of the lipophile starting with the first carbon not attached to oxygen. It is possible to relate this to the *cohesive energy ratio* (originally introduced by Winsor[251]) in the form

$$R_o = {}^{L}V\ {}^{L}\delta_*^2 V\ {}^{H}\delta_*^2 \tag{49}$$

where

$$\delta_*^2 = \delta_d^2 + 0.25\delta_p^2 + 0.25\delta_h^2 \tag{50}$$

(The inclusion of the factor 0.25 is based on the "4" incorporated in Equation 55, Chapter 5; other numerical factors were reported in Section 17.3 for surfaces.) The resulting expression is

$$R_o = ({}^{L}\delta_*^2/{}^{H}\delta_*^2)(20/\text{HLB} - 1)({}^{H}\rho/{}^{L}\rho) \tag{51}$$

Correlation of literature data for the optimum value of R_o for oil (o) in water emulsions yields the optimum HLB value for the emulsifier:

$$\text{HLB}_o = \frac{20\ {}^{o}\delta_*^2/\text{MPa}}{{}^{o}\delta_*^2/\text{MPa} + 161} \tag{52}$$

Other criteria for the selection of the emulsifier are

$${}^{L}\delta_* \approx {}^{o}\delta_*,\ {}^{L}V \approx {}^{o}V \tag{52a}$$

A plot of log R_o against HLB was found to be approximately linear, and it proved convenient to define an extended hydrophile-lipophile balance scale, HLB_E,

$$\log R_o = 0.925 - 0.0963\ \text{HLB}_E \tag{53}$$

This allows the estimation of R_o and subsequent correlation with the cohesion parameters of emulsifiers with HLB_E values covering a wider range than the original definition. The

TABLE 8
Hansen Parameters and Molar Volumes of the Lipophilic Portions of Surfactants

Lipophile	Formula	^{L}V/cm³ mol⁻¹	$^{L}\delta$/MPa$^{1/2}$ $^{L}\delta_d$	$^{L}\delta_p$	$^{L}\delta_h$
Ethyl	CH_3CH_2-	50.3	14.3	0.0	0.0
Isopropyl	$(CH_3)_2CH-$	68.4	13.6	0.0	0.0
Butyl	$CH_3(CH_2)_3-$	82.5	15.5	0.0	0.0
Cyclohexyl	$(CH_2)_5CH-$	94.0	17.6	0.0	0.0
Lauryl	$(CH_3)(CH_2)_{11}-$	211.3	16.7	0.0	0.0
Isooctylphenyl	$(CH_3)_3(CH_2)_3(CH)_2C_6H_4-$	210.9	17.1	0.0	0.0
Isononylphenyl	$(CH_3)_5CH_2(CH)_2C\ C_6H_4-$	232.9	16.1	0.0	0.0
Isotridecyl	$(CH_3)_5(CH_2)_4(CH)_4-$	235.4	15.1	0.0	0.0
Hexadecyl	$CH_3(CH_2)_{15}-$	275.7	16.8	0.0	0.0
Oleyl	$CH_3(CH_2)_7CH{=}CH(CH_2)_8-$	307.9	15.2	0.0	0.0
Laurate	$CH_3(CH_2)_{10}-$	195.7	16.6	0.0	0.0
Palmitate	$CH_3(CH_2)_{14}-$	260.2	16.8	0.0	0.0
Oleate	$CH_3(CH_2)_7CH{=}CH(CH_2)_7-$	283.8	15.6	0.0	0.0
Stearate	$CH_3(CH_2)_{16}-$	293.2	16.8	0.0	0.0
Hydroxystearate	$CH_3(CH_2)_5CHOH(CH_2)_{10}-$	289.0	17.3	1.8	8.2
Alkyl-aryl[a]	$(CH_3)_6(CH_2)_{14}C_6H_3-$	423.0	16.5	0.0	0.0

[a] Approximate formula.

Adapted from Beerbower, A. and Hill, M. W., *Am. Cosmet. Perfum.*, 87(6), 85, 1972.

TABLE 9
Molar Volumes and Cohesive Energies of Ionic Hydrophiles

Hydrophile	^{H}V/cm³ mol⁻¹	$^{H}V^{H}\delta_*^2$/kJ mol⁻¹	$^{H}\delta_*^2$/MPa[a]
KCO_2-	30.9	680	22,000
$NaCO_2-$	22.0	460	20,700
KSO_3-	43.8	310[b]	7,000[b]
$NaSO_3-$	33.8	240	7,000
KSO_4-	46.3	6150[b]	132,600[b]
$NaSO_4-$	36.8	4870	132,600
$Ca(SO_3)_2^{2-}$	63.6	165	2,600
Cl^-	14.6	—	—

[a] Calculated on unhydrated basis.
[b] Estimated from sodium value.

Adapted from Beerbower, A. and Hill, M. W., *Am. Cosmet. Perfum.*, 87(6), 85, 1972.

method of Section 6.5 is used to determine ^{L}V and ^{H}V. Some values resulting from these calculations appear in Tables 8 and 9. Holtzscherer and Candau[228] extended the CER concept to polymerizable microemulsions. Szymanowski et al.[363] considered relations between HLB, cohesion parameters, and chromatographically-determined polarity index for interactions between alcohols and polyoxyethylene glycol dialkyl ethers.

Marszall[252] has proposed an alternative to the cohesive energy ratio, called the cohesive energy balance,

$$^{H}V(^{aq}\delta - {}^{H}\delta)^2/{}^{L}V(^{aq}\delta - {}^{L}\delta)^2 \tag{54}$$

Relative advantages and disadvantages of these approaches are not clear.

Cohesion parameters have been applied to various other commercial surfactants, emulsions, dispersions and suspensions:

- High-molecular weight (PEO-PPO) nonionogenic surface active agents[361]
- Water-based paints[253-256]
- Aqueous emulsions of cellulose nitrate/acrylic polyblends[257]
- Suspensions of drug compounds[258]
- Hansen parameter maps of commercial pastes[259]
- Extraction of organic colloids from soil[260]
- Correlation with δ_h of the sedimentation of the inorganic pigment ultramarine blue from organic liquids[261]
- Pigmented polymer compositions[262]
- Degree of aggregation of Tween® (polyoxyethylene sorbitan aliphatic ester), Span® (sorbitan aliphatic acid ester) and polyoxyethylated ether and alcohol nonionic surfactants and commercial detergents[236,243]
- Properties of alkyl poly(glycol ethers)[237,238,363]
- State of aggregation of poly-γ-benzyl-L-glutamate[196]
- Linear sodium alkylbenzenesulfonate homologs[161]
- Cosmetic oily powder and foundation compositions[263,264]
- Pesticide emulsifiable concentrates[240]
- Solubility of potassium acid soaps of lauric, stearic, and oleic acids[265]
- Hildebrand parameter of Methocel® cellulose[353]

Foaming behavior may also be correlated by means of cohesion parameters. At the beginning of this chapter it was pointed out that there could be an abrupt transition from defoaming (foam inhibiting behavior) to profoaming (foam stabilizing behavior) in a series of surfactants, associated with an insolubility-solubility transition. The same effect occurs with a single surfactant if its solubility is strongly temperature dependent, illustrated by sodium stearate which is much more soluble in hot water than cold, and which foams only at higher temperatures. Block copolymers containing polyethylene oxide act in a similar way.[1] It follows that systems represented by the spherical Hansen solubility surfaces (Sections 5.9 and 15.1) for polymers and other materials represent polymer-liquid combinations which are likely to exhibit foaming. Liquids that readily dissolve polymers do not form foaming solutions, liquids that dissolve no polymer do not foam, but liquids that exhibit limited solvent action exhibit a maximum in foaming ability. This has been illustrated by plotting foam stability as a function of the distance between the solvent and solute points in Hansen three-dimensional solubility space for polymethylmethacrylate, the chlorinated polypropylene Parlon® P10(Hercules, Inc.) and the epoxy-resin Epikote® 1001 (Shell Chemicals).[1] Foaming behavior appears to be a more sensitive indicator of the limits of polymer-solvent interaction that solubility or swelling.

17.8 DISPERSION OF PIGMENTS AND DYES

Particles of pigments in paints range in size from near the upper limit of colloids (about 0.2 μm), to suspensions of about 20 μm, but all exhibit the same general kinds of properties as found in colloids. The dispersions are not thermodynamically stable, their free energies in the absence of favorable interactions with liquids corresponding to spontaneous change in the direction of flocculation or sedimentation, but in the right conditions they can persist

TABLE 10
Pigments Examined by Wet Point Titration

Pigment	Brand and Supplier	Density, g cm^{-3}
Carbon black	Mogul A, SRS 372, Cabot	1.8
Iron blue	50-4066, American Cyanamid	1.8
Titanium dioxide	Titanox RA47, Titanium Pigment	3.5
Phthalo blue	B-4804, Harmon	1.6
Benzidine yellow	Sumatro Yellow X-1940, Hercules	1.4
Talc	Nytal 1300, R. T. Vanderbilt	2.9
Watchung Red	BRT 761D, du Pont	1.4

Adapted from Burrell, H., *Am. Chem. Soc. Div. Org. Coatings Plast. Chem. Pap.*, 35(2), 19, 1975.

indefinitely. The problems inherent in keeping solids in the required physical state in suspension in a liquid phase are considerable, and have attracted wide-ranging theoretical and experimental investigations, including studies of such properties as interfacial free energies, contact angles, and energies of adhesion and cohesion. The inadvertent or intentional poor characterization of the nature of the surfaces of commercial pigments introduces additional difficulties; for example, inorganic pigments may have organic coatings and organic dyes and contain up to 40% resin, factors which are not included in the technical specifications provided by suppliers. The behaviors of pigment dispersions depend on the total coating composition, on manufacturing techniques, and on variations in pigment/resin/liquid combinations.

In the dispersion of pigments in a liquid phase, the wetting process can be distinguished from the subsequent separation of the particles and stabilization of the dispersion. Burrell[350] used a wet-point titration (with the end-point taken as the liquid content which first caused the stirred, dry pigment to cake or ball) and found no correlation with liquid Hildebrand parameter, surface free energy, or water solubility. The result was that for every pigment in Table 10 there is within at least a 95% confidence limit a minimum in the titration value at $\delta = 14.1$ MPa$^{1/2}$. In other words, as far as wetting is concerned, all the pigments acted in the same way. This initially surprising observation was attributed to moisture adsorbed on the pigment, so it is the interaction of liquid with adsorbed water, dominated by hydrogen bonds, which is the same for all the pigments.

However, the overall dispersion process can be correlated in many ways, including cohesion parameter and hydrogen-bonding parameter values.[117,266,267] Hansen and Skaarup[225,268,269] extended the Hansen parameter method (Sections 5.9, 14.5, 17.3) to suspension or deflocculation situations. Pigment surfaces were characterized by shaking pigment with each of 53 liquids, noting the settling tendencies, and defining Hansen parameters ${}^s\delta_d, {}^s\delta_p, {}^s\delta_h$ and interaction radius ${}^sR$ for each surface (Table 11). These data are for use in conjunction with the liquid data of Table 11, Chapter 5, liquids for which ${}^{sl}R \leq {}^sR$ being expected to suspend the pigment:

$$ {}^{sl}R^2 = 4({}^s\delta_d - {}^l\delta_d)^2 + ({}^s\delta_p - {}^l\delta_p)^2 + ({}^s\delta_h - {}^l\delta_h)^2 \qquad (55) $$

Other methods can be used to characterize and correlate pigment surfaces,[270] but the accuracy of this Hansen method was equivalent to the results for polymer and resin solubilities.

Saarnak[271] reported further pigment Hansen parameters (Table 12) when considering polymer adsorption on pigments. It was considered[271-274] that a measure of the strength of interaction in a liquid-polymer-pigment system might be the size of the mutual "volume" of polymer solubility and pigment wetting, but the results did not permit any simple analysis

TABLE 11
Hansen Parameters and Interaction Radius Values of Pigments Based on Beerbower and Hansen's 1971 Liquid Data (Table 11, Chapter 5.)

Pigment, brand, supplier	C.I.	$^s\delta/MPa^{1/2}$				$^sR/MPa^{1/2}$
		$^s\delta_d$	$^s\delta_p$	$^s\delta_h$	$^s\delta_t$	
Titanium dioxide, Kronos RN 57, Titan	–	24.1	14.9	19.4	34.4	17.2
Phthalocyanine blue, BG, du Pont	–	19.0	6.3	7.6	21.5	4.7
Isolbonared 7522, C.I. pigment red 48, Koge (Mn salt)	15865	17.8	7.2	7.2	20.5	5.1
Peerless carbon black	–	21.1	12.3	13.5	27.8	12.3
Isol fast yellow IO GX 2505, C.I. pigment yellow 3, Koge	11710	20.9	9.8	7.8	24.3	9.0
Reflex blau TBK Ext., Hoechst	–	22.1	7.8	13.5	27.0	14.3
Isol ruby BKS 7520, C.I. pigment red 57, (Ca salt) Koge	15850	19.6	6.1	6.6	21.5	8.0
Hansagelb 10G, C.I. pigment yellow 3, Hoechst	11710	18.6	8.2	6.8	21.5	6.8
Fanalrosa G supra pulver, C.I. pigment red 81, BASF	45160	20.0	14.3	10.2	26.6	10.6
Heliogenblau B pulver, C.I. pigment blue 15, BASF	74160	22.1	7.2	8.2	24.6	10.6
Heliogengrun GN, C.I. pigment green 7, BASF	74260	20.5	9.8	9.2	24.6	9.8
Permanentgelb H 10 G, C.I. pigment yellow 81, Hoechst	–	17.2	3.1	4.7	18.0	4.5
Permanent bordeaux FRR, C.I. pigment red 12, Hoechst	12385	21.9	9.8	12.5	27.0	10.6
Permanent violet RL supra, C.I. pigment violet 23, Hoechst	12505	19.6	10.6	7.4	23.5	9.0
Isol benzidine yellow G 2537, C.I. pigment yellow 12, Koge	21090	19.0	6.1	5.9	20.9	8.0
Brilliant sky blue 3862, C.I. pigment blue 3, J. W. & T. A. Smith	42140	19.4	14.7	12.3	27.2	10.4
Permanent orange G, C.I. pigment orange 13, Hoechst	21110	19.8	8.0	9.6	23.5	9.2

TABLE 11 (continued)
Hansen Parameters and Interaction Radius Values of Pigments Based on Beerbower and Hansen's 1971 Liquid Data (Table 11, Chapter 5.)

		$^s\delta/MPa^{1/2}$				
Pigment, brand, supplier	**C.I.**	$^s\delta_d$	$^s\delta_p$	$^s\delta_h$	$^s\delta_t$	$^sR/MPa^{1/2}$
Permanent red, FGR extra pulver, C.I. pigment red 112, Hoechst	12370	20.5	7.2	7.2	22.9	10.2
Isolfast red 2G 2516, C.I. pigment orange 5, Koge	12075	22.3	11.5	14.5	29.1	14.3
Monolite fast blue 3RS, powder C.I. vat blue 4, ICI	69801	22.1	13.3	17.4	31.1	14.3
Heliogenblau LG, pulvcr, C.I. pigment blue 16, BASF	74100	21.9	10.2	13.3	27.6	12.3
Red iron oxide	–	20.7	12.3	14.3	28.0	11.5
Carbon black, Printex V(5519-1), Degussa	–	21.1	12.3	11.3	26.8	11.3
Aluminum pulver black lack 80, Eckart	–	19.0	6.1	7.2	21.3	4.9
Isol benzidine yellow GA-PR, 9500, C.I. pigment yellow, 12, Koge	–	18.4	5.5	4.7	18.6	5.1

Adapted from Hansen, C. M., *J. Paint Technol.*, 39, 505, 1967.

TABLE 12
Hansen Parameters and Interaction Radius Values of Pigments, Based on Hansen's 1967 Liquid Data (Table 12, Chapter 5)

Pigment, brand, supplier	C.I.	$^s\delta$/MPa$^{1/2}$				^sR/MPa$^{1/2}$
		$^s\delta_d$	$^s\delta_p$	$^s\delta_h$	$^s\delta_t$	
Novoperm gelb HR, benzidine yellow, Hoechst, 69 m^2 g^{-1}	PY 83-21108	21.8	5.8	10.8	25.0	12.5
Permanentcarmin FBB, Naftol AS-Red, Hoechst, 38 m^2 g^{-1}	PR 146-12485	9.2	3.5	10.7	14.5	15.2
Paliotolgelb 1820L, Isoimidoline yellow, BASF, 55 m^2 g^{-1}	PY 139	13.0	9.6	12.2	20.3	9.7
Kronos RN 61, 93.5% TiO_2, 14.8 m^2 g^{-1}	PW 6-77891	17.4	6.1	12.5	22.3	7.7

Adapted from Saarnak, A., *Polym. Mater. Sci. Eng.*, 51, 698, 1984.

TABLE 13
Hansen Parameters and Interaction Radius Values of Pigments, Based on Hansen's 1967 Liquid Data (Table 12, Chapter 5)

Pigment	$^s\delta$/MPa$^{1/2}$				^sR/MPa$^{1/2}$
	$^s\delta_d$	$^s\delta_p$	$^s\delta_h$	$^s\delta_t$	
Synthetic red iron oxide, α-Fe_2O_3	16.1	8.6	15.0	23.7	11.3
Synthetic yellow iron oxide, α-$Fe_2O_3.H_2O$[a]	17.3	6.0	14.5	23.3	12.5
Synthetic yellow iron oxide, α-$Fe_2O_3.H_2O$	16.1	8.6	15.0	23.7	11.3
Zinc oxide, ZnO[a]	16.9	7.8	10.6	21.4	13.2
Zinc oxide, ZnO	16.2	10.8	12.7	23.3	9.8
Middle chrome yellow, $PbCrO_4$	16.1	14.5	16.1	26.9	10.5

[a] Including swelling liquids.

Adapted from Shareef, K. M. A., Yaseen, M., Ali, M. M., and Reddy, P. J., *J. Coatings Technol.* 58(733), 35, 1986.

in terms of Hansen parameters for adsorption of polymers on pigments: acid-base interactions were of considerable importance. Shareef, Yaseen, et al.[275] extended similar studies to other pigments, with results in 33 liquids (δ = 16 to 48 MPa$^{1/2}$) on suspension and swelling computer-analyzed to provide the Hansen parameter and interaction radius values in Table 13.

The coordinates of the Hansen parameter sphere can be expressed also in terms of fractional cohesion parameters (Section 5.10); typical values appear in Table 14. This type of data may be presented alternatively in triangular diagrams (Figures 5 and 6), the curved lines defining the areas of most stable dispersion, based on specific sedimentation volumes. These diagrams also illustrate the effect of surface treatment on pigment properties; after treatment with an organic phosphate the zinc oxide powder requires appreciably different dispersing liquids. Information on this comparison has been presented in another way without the tacit assumption made in the triangular formalism that the total cohesion parameter is the same for all materials. Each of a series of curves resulting from a multiple regression analysis[276] shows the effect of varying one parameter while the others are held constant. For untreated zinc oxide the minimum sedimentation volume (most complete dispersion) occurs at a δ_h of 20.1 MPa$^{1/2}$ while δ_d and δ_p volumes are unimportant; for the organic phospate-treated surface the effect of δ_h is unimportant and the parameters characterizing

TABLE 14
Fractional Cohesion Parameters of Some Pigments for Use with the Liquid Data in Table 16, Chapter 5

Pigment	δ_t/MPa$^{1/2}$	100f_d	100f_p	100f_h
Titanium dioxide (Kronos RN57)	34.4	41	26	33
Red iron oxide	28.0	44	26	30
Carbon black (Printex V)	26.8	47	28	25
Pigment green (Heliogengrun GN)	24.6	52	25	23
Permanent violet (LR supra, Hoechst)	23.5	52	28	20
Permanent orange (Hoechst pigment orange 13)	23.5	53	21	26
Permanent red (Hoechst pigment red 112)	22.9	60	20	20
Pigment orange (Isolfast red 2G)	29.1	46	24	30
Fast blue (Monolite 3RS)	31.1	42	25	33
Benzidine yellow (Isol GA-PR)	19.8	65	19	16
Pigment yellow (Hoechst H10G)	18.0	69	12	19

Adapted from Gardon, J. L. and Teas., J. P., in *Treatise on Coatings, Vol. 2, Characterization of Coatings: Physical Techniques,* Part II, Myers, R. R. and Long, J. S., Eds., Marcel Dekker, New York, 1976, chap. 8, p. 413. More complete pigment descriptions appear in Table 11.

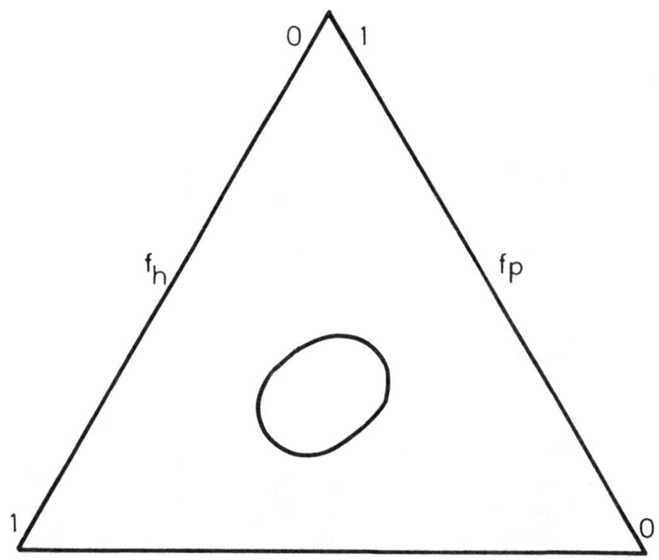

FIGURE 5. Triangular fractional Hansen parameter diagram for the dispersion of untreated zinc oxide powder, for use with the liquids data of Table 16, Chapter 5. (Adapted from Gardon, J. L. and Teas, J. P., *Treatise on Coatings, Vol. 2, Characterization of Coatings: Physical Techniques,* Part II, Myers, R. R. and Long, J. S., Eds., Marcel Dekker, New York, 1976, chap. 8.)

the surface are δ_d = 16.4 MPa$^{1/2}$ and δ_p = 0. (Negative values of δ_p, which appear in the surface wetting characterization, Table 3, are also possible in this case.) For ultramarine blue it was found[261] that the sedimentation behavior is very similar to that of untreated zinc oxide: it was almost exclusively determined by hydrogen bonding, minimum sedimentation volume at δ_h = 21 MPa$^{1/2}$.

In other studies, various alternative functions have been used: δ_p vs. δ_h plots,[277,278] δ

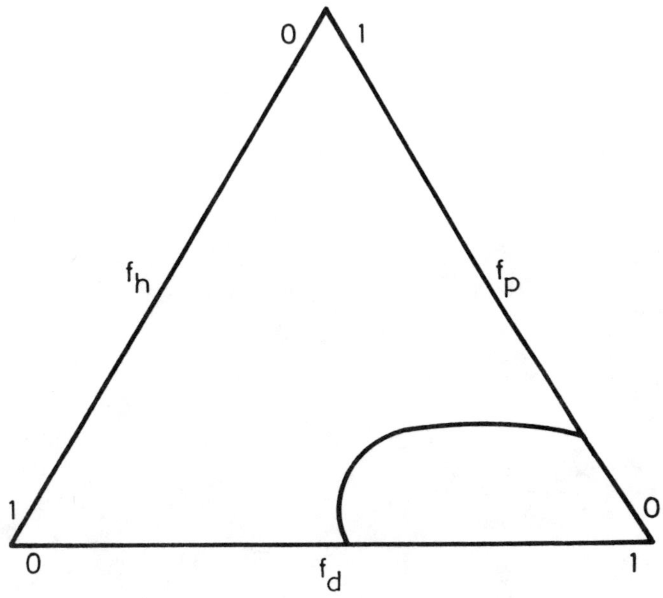

FIGURE 6. Triangular fractional Hansen parameter diagram for the dispersion of zinc oxide powder treated with an organic phosphate, for use with the liquids data of Table 16, Chapter 5. (Adapted from Gardon, J. L. and Teas, J. P., *Treatise on Coatings, Vol. 2, Characterization of Coatings: Physical Techniques,* Part II, Myers, R. R. and Long, J. S., Eds., Marcel Dekker, New York, 1976, chap. 8.)

TABLE 15
Estimated Hansen Parameters at 25°C for Simple Disperse Dyes

Dye	${}^s\delta/\text{MPa}^{1/2}$			
	${}^s\delta_d$	${}^s\delta_{hp}$	${}^s\delta_t$	$V/\text{cm}^3\ \text{mol}^{-1}$
Azobenzene	21.1	2.9	21.3	151
p-Aminobenzene	20.3	8.8	22.1	168
Anthraquinone	22.5	12.9	26.0	145
1-Methylamino-4-β-hydroxyethylanthraquinone	22.7	10.4	25.0	225

Adapted from Urbanik, A., *Text. Chem. Color.*, 15(3), 36, 1983.

vs. γ plots,[279-285] and $(\delta_p^2 + \delta_h^2)^{1/2}$ graphs.[286-288] Table 15 provides estimates of δ, δ_d, and $\delta_{hp} = (\delta_p{}^2 + \delta_h{}^2)^{1/2}$ for simple disperse dyes calculated by Urbanik.[289] Satoh[290,291] found a correlation between δ_h and the adhesion tension of pigments with liquids. Chavan and co-workers[292] calculated δ, δ_d, and δ_{hp} for anionic dyes using for δ the method of van Krevelen (Table 8, Chapter 6) and for δ_d refractive index data[51] (Section 14.8). (The atomic attraction constants of van Krevelen, originally developed for coal, were found preferable to the group constants of Small[293] and Hoy,[294,295] which did not include information on all groups present in anionic dyes.) Results indicated that the dye adsorption to cotton was governed mainly by dispersion interactions. Subsequently, they[296] evaluated the Hildebrand and its δ_d and δ_{hp} components for Disperse Violet 1 and Disperse Red 60. Similar calculations were carried out by Ni and Jen for a large number of azo disperse dyes.[297] Results of Hildebrand parameter calculations are summarized in Table 16.

As pointed out in Chapter 5 and emphasized subsequently in various other contexts, it

TABLE 16
Hildebrand Parameters of Disperse Dyes Calculated by Chavan from the Group Molar Attraction Constants of van Krevelen

	$^{z}F/J^{1/2}\ cm^{3/2}\ mol^{-1}$				
Atom	CI Acid Red 88 (Naphthalene Red JS)	CI Acid Red 13 (Naphthalene Red EA)	CI Acid Red 18 (Naphthalene Scarlet)	CI Disperse Violet 1	CI Disperse Red 60
C	20 × 0	20 × 10	20 × 0	14 × 10	20 × 0
H	13 × 140	12 × 140	11 × 140	10 × 140	13 × 140
N	2 × 235	2 × 235	2 × 235	—	—
N aromatic primary amine	—	—	—	2 × 133	1 × 133
O ester	3 × 256	6 × 256	9 × 256	—	—
O ketone	—	—	—	2 × 685	2 × 685
O phenolic	1 × 511	1 × 511	1 × 511	—	1 × 511
O ether	—	—	—	—	1 × 256
C=C aromatic	11 × 272	11 × 272	11 × 272	6 × 272	9 × 272
Conjugation	—	—	—	1 × 51	2 × 51
Na[a]	1 × 600	2 × 1600	3 × 1600	—	—
S	1 × 460	2 × 460	3 × 460	—	—
$\Sigma_z\ ^{z}F$	8623	11,310	13,997	4721	6643
M	400	502	604	238	331
$\rho/g\ cm^{-3}$	1.53	1.55	1.66	1.37	1.29
$V/cm^3\ mol^{-1}$	261	324	364	174	257
$\delta/MPa^{1/2}$	32.9	35.0	38.5	27.1	25.9

[a] Calculated by Chavan et al.

Adapted from Chavan, R. B. and Jain, A. K., *Dyestuff Rep.*, 76(12), 48, 1987; 77(2), 29, 1988.

is unreasonable to expect either Hildebrand parameters or sets of Hansen parameters to give an adequate description of phenomena where there are significant electron donor-acceptor (Lewis acid-base) interactions. This is just as true for pigment surface interactions as for resin solubilities, as illustrated by Vinther and Sörensen[277,278,280-285] who have used both pigment charges and acid-base properties in the prediction of the behavior of pigment/resin/liquid interactions. In some cases the hydrogen-bonding parameter γ (Section 8.1) can be used to express donor-acceptor capacity, but even a simple classification as basic, acidic, amphoteric, or neutral is useful.[284]

It will be appreciated that the interactions of pigments with liquid-resin or liquid-liquid mixtures make the prediction of surface behavior even more difficult because of selective adsorption. Observations of performance of pigments in pure liquids form an inadequate basis for prediction of performance in complex mixtures. GLC investigations (Section 18.2) provide a greatly extended opportunity for direct characterization of these systems.

Examples of pigment-resin-liquid and dye systems investigated in terms of cohesion parameters and related concepts are:

- Viscosities of liquid ink and lacquer formulations including the phthalocyanine blue pigments in varying liquid compositions[283,284] and the polyamide Versamid® 930 and cellulose nitrate lacquers[280,281]
- Solubility of a polyamide ink resin used in flexographic printing[309,310] (Figure 1, Chapter 15)
- Disperse dyes[286,289,292,296,311-321]
- Solvent and dye selection for nonaqueous dyeing[287,322-325,354]
- Use of cohesion parameters in studying the dyeing of polyamide 6, poly(ethylene terephthalate), polypropylene, polyacrylonitrile, diacetate, and triacetate fibers[298-308,313,326-328,359] (For PET, plasticization is referred to in Section 16.3 and cohesion parameter maps are included in Section 15.1)
- Solubilities of dyes in organic liquids (Figures 11 to 13, Chapter 15)[288] and in liquid crystals.[329]
- Effects of pigments on properties of linseed oil films[330]
- Oligoester chain effects on titanium dioxide dispersions[331]
- Correlation of dispersion-flocculation behavior of pigments and solvent-resin-pigment affinities with Hildebrand parameters and hydrogen bonding parameters[267,332]
- Ink image formation in lithography[137]
- Sublimation-transfer printing on polyester[370]
- Solvents for colorants in art objects[333]
- Adsorption of water-soluble polymers at the water-air interface[334]
- Dispersions of octadecyl-coated silica particles[335]
- Crystallization of soluble vanadyl phthalocyanine dyes in a polymer matrix[336] (See also Section 16.3)
- High solids coatings[337] (Section 16.6)
- Dispersion of antimicrobial agents[338]
- Rheology of red iron oxide (α-Fe_2O_3)-alkyd (66% linseed oil pentaerythritol-phthalate) paints (in which solvents with $\delta \approx 19$ MPa$^{1/2}$ are expected to provide maximum compatibility)[339]
- Hildebrand parameters of Carbopol® 934 and 940 dispersions (33—38 MPa$^{1/2}$) in water-alcohol media[340]
- Interactions between pigments, hydroxypropyl methylcellulose and other pharmaceutical tablet components[341]
- Solvent selection for high solids coatings polymerization and dispersion polymerization (Section 10.4)

- Thick film paste rheology[374]

Polymer-liquid interactions in coating films are reviewed in Section 16.6.

17.9 SURFACE CONTAMINATION AND CLEANING

The contamination and subsequent cleaning of surfaces is of major importance in most areas of endeavor.[342] Both the characterization of surface contaminants and the choice of suitable solvents can be assisted by a consideration of cohesion parameters. Hildebrand parameters[122,342-345,371] and Hansen parameters[346,371] have been used together with the Meseran surface analyzer technique described in Section 17.3. Much of the work on permeability (Section 16.7) is relevant to contamination and cleaning of surfaces, and Table 12a, Chapter 5 provides Hansen parameters of common biological soiling materials. Particular applications have been:

- Removal of photoresists in the microelectronics industry[347] (Section 16.5)
- Etching of polymer surfaces[348]
- Cleaning paint processing equipment[349]
- Sorption of humic-like organic model compounds[380]
- Soiling properties of alpha iron oxide on cellulose fabrics[356]
- Combinations of nonsolvents or poor solvents to replace halogenated liquids for cleaning[371,378]

REFERENCES

1. **Ross, S. and Nishioka, G.,** The relation of foam behavior to phase separations in polymer solutions, *Colloid Polym. Sci.*, 255, 560, 1977.
2. **Melrose, J. C.,** Evidence for solid-fuel interfacial tensions from contact angles, in *Contact Angle Wettability and Adhesion (Adv. Chem. Ser.* 43), American Chemical Society, Washington, D.C., 1964, chap. 10.
3. **Zisman, W. A.,** Relation of the equilibrium contact angle to liquid and solid constitution, in *Contact Angle, Wettability and Adhesion (Adv. Chem. Ser.*, 43), American Chemical Society, Washington, D.C., 1964, chap. 1.
4. **Crocker, G. J.,** Elastomers and their adhesion, *Rubber Chem. Technol.*, 42, 30, 1969.
5. **Zisman, W. A.,** Constitutional effects on adhesion and cohesion, in *Adhesion and Cohesion,* Weiss, P., Ed., Elsevier, Amsterdam, 1962, 176.
6. **Zisman, W. A.,** Influence of constitution on adhesion, *Ind. Eng. Chem.*, 55(10), 18, 1963.
7. **Zorll, U.,** Significance and problem of the critical surface tension, *Adhesion,* 18, 262, 1974.
8. **Dann, J. R.,** Forces involved in the adhesive process. I. Critical surface tensions of polymeric solids as determined with polar liquids, *J. Colloid Interface Sci.*, 32, 302, 1970.
9. **Owens, D. K. and Wendt, R. C.,** Estimation of the surface free energy of polymers, *J. Appl. Polym. Sci.*, 13, 1741, 1969.
10. **Panzer, J.,** Components of solid surface free energy from wetting measurements, *J. Colloid Interface Sci.*, 44, 142, 1973.
11. **Exner, O.,** Additive physical properties. III. Re-examination of the additive character of parachor, *Collect. Czech. Chem. Commun.*, 32, 24, 1967.
12. **Sewell, J. H.,** *Calculation of Critical Surface Tensions of Polymers and Surface Tensions of Liquids from Chemical Structure Only,* Royal Aircraft Establishment Technical Report 69211, Farnborough, Hampshire, 1969, AD 701471.
13. **van Krevelen, D. W. and Hoftyzer, P. J.,** Practical evaluation of the $[\eta]$-M relation, *J. Appl. Polym. Sci.*, 11, 2189, 1967.
14. **Anastasiadis, S. H., Chen, J. K., Koberstein, J. T., Sohn, J. E., and Emerson, J. A.,** The determination of polymer interfacial tension by drop image processing: comparison of theory and experiment for the pair, poly(dimethylsiloxane)/polybutadiene, *Polym. Eng. Sci.*, 26, 1410, 1986.

15. **Bagda, E.,** The relation between surface tension and solubility parameter in liquids, *Farbe Lack,* 84, 212, 1978.
16. **Barton, A. F. M.,** The application of cohesion parameters to wetting and adhesion — a review, *J. Adhes.*, 14, 33, 1982.
17. **Becher, P.,** The calculation of cohesive energy density from the surface tension of liquids, *J. Colloid Interface Sci.*, 38, 291, 1972.
18. **Beerbower, A.,** Surface free energy: a new relationship to bulk energies, *J. Colloid Interface Sci.*, 35, 126, 1971.
19. **Bernier, G. A. and Kambour, R. P.,** The role of organic agents in the stress crazing and cracking of poly(2,6-dimethyl-1,4-phenylene oxide), *Macromolecules,* 1, 393, 1986; General Electric Development Center Reprint 6125, Schenectady, NY, 1986.
20. **Bindschaedler, C. and Peppas, N. A.,** Detachment of small axisymmetric particles from fluid-liquid interfaces, *Chem. Eng. Sci.*, 42, 2531, 1987.
21. **Bonn, R. and van Aartsen, J. J.,** Solubility of polymers in relation to surface tension and index of refraction, *Eur. Polym. J.*, 8, 1055, 1972.
22. **Bulou, J., Raynal, R. M., and Vial, J.,** An interfacial tension treatment with an improved Good-Girifalco equation using solubility parameters, *J. Colloid Interface Sci.*, 98, 168, 1984.
23. **Chatterjee, B. C. and Ghosh, K.,** Cohesive force of a copolymer at the oil-water interface, *J. Indian Chem. Soc.*, 49, 751, 1972.
24. **Cherry, B. W. and Evely, P. B.,** The interaction parameter and the strength of adhesive joints, *J. Adhes.*, 22, 171, 1987.
25. **Colin, H., Guiochon, G., and Jandera, P.,** The solvent eluotropic strength on carbon adsorbents, *Chromatographia,* 15, 133, 1982.
26. **Croll, S. G.,** Effect of solvent on residual strain in clear epoxy coatings, *J. Oil Col. Chem. Assoc.*, 63, 230, 1980.
27. **Dack, M. R. J.,** The importance of solvent internal pressure and cohesion to solution phenomena, *Chem. Soc. Rev.*, 4, 211, 1975.
28. **Eckert, C. A. and Prausnitz, J. M.,** Statistical surface thermodynamics of simple liquid mixtures, *Am. Inst. Chem. Eng. J.*, 10, 677, 1964.
29. **Fowkes, F. M.,** Determination of interfacial tensions, contact angles, and dispersion forces by assuming additivity of intermolecular interactions at surfaces, *J. Phys. Chem.*, 66, 382, 1962.
30. **Fowkes, F. M.,** Attractive forces at interfaces, *Ind. Eng. Chem.*, 56(12), 40, 1964.
31. **Fowkes, F. M.,** Donor-acceptor interactions at interfaces, in *Adhesion and Adsorption of Polymers Part A (Polym. Sci. Technol.*, 12A), Lee, L. H., Ed., Plenum Press, New York, 1980, 43.
32. **Fowkes, F. M.,** Acid-base interactions in polymer adhesion, in *Physicochemical Aspects of Polymer Surfaces,* Vol. 2, Mittal, K. L., Ed., Plenum, New York, 583.
33. **Fu, J., Li, B., and Wang, Z.,** Estimation of fluid-fluid interfacial tensions of multicomponent mixtures, *Chem. Eng. Sci.*, 41, 2673, 1986; **Li, B. and Fu, J.,** Correlation of interfacial tension with mutual solubility for two liquid-phase ternary mixtures, Second International Symposium on Solubility Phenomena, Newark, NJ, 1986.
34. **Gardon, J. L.,** Relationship between cohesive energy densities of polymers and Zisman's critical surface tensions, *J. Phys. Chem.*, 67, 1935, 1963.
35. **Gardon, J. L.,** Cohesive-energy density, in *Encyclopedia of Polymer Science and Technology,* Vol. 3, John Wiley & Sons, New York, 1964, 833.
36. **Gardon, J. L.,** Critical review of concepts common to cohesive energy density, surface tension, tensile strength, heat of mixing, interfacial tension and butt joint strength, *J. Colloid Interface Sci.*, 59, 582, 1977; *Prog. Org. Coatings*, 5, 1, 1977.
37. **Gardon, J. L. and Teas, J. P.,** Solubility parameters, in *Treatise on Coatings: Physical Techniques,* Part II, Myers, R. R. and Long, J. S., Eds., Marcel Dekker, New York, 1976, chap. 8, 413.
38. **Girifalco, L. A. and Good, R. J.,** A theory for the estimation of surface and interfacial energies, I. Derivation and application to interfacial tension, *J. Phys. Chem.*, 61, 904, 1957.
39. **Good, R. J.,** Surface chemistry, *Chemtech,* 10, 100, 1980.
40. **Glass, J. E.,** Adsorption characteristics of water-soluble polymers. I. Poly(vinyl chloride) and poly(vinylpyrrolidine) at the aqueous-air interface; II. Poly(ethylene oxide) at the aqueous-air interface, *J. Phys. Chem.*, 72, 4450 and 4459, 1968.
41. **Glass, J. E.,** The role played by water-soluble polymers in paint performance. Part II. Chemical modeling studies, *J. Oil Col. Chem. Assoc.*, 59(3), 86, 1976.
42. **Guiseppi-Elie, A., Wnek, G. E., and Wesson, S. P.,** Wettability of polyacetylene: surface energetics and determination of material properties, *Langmuir,* 2, 508, 1986.
43. **Lee, K. J. and Ha, W. S.,** Surface characteristics of solvent-cast cellulose acetate, *J. Korean Soc. Text Eng. Chem.*, 19, 310, 1982.

44. **Ha, W. S. and Han, D. S.,** Relationship between surface tension and solubility parameter of polymers, *J. Korean Soc. Text. Eng. Chem.*, 22, 124, 1985.
45. **Hansen, C. M.,** Some answered and unanswered questions about the solubility parameter, in *Macromolecular Solutions: Solvent-Property Relationships in Polymers* (Proc. Symp., Am. Chem. Soc., New York, 1981), Seymour, R. B. and Stahl, G. A., Eds., Pergamon, New York, 1982, 1; *Org. Coat. Plast. Chem.*, 45, 227, 1981.
46. **Hansen, C. M.,** Organic solvents in high solids and water-reducible coatings, *Progr. Org. Coatings*, 10, 331, 1982.
47. **Hansen, C. M. and Wallström, E.,** On the use of cohesion parameters to characterize surfaces, *J. Adhesion*, 15, 275, 1983.
48. **Helfand, E. and Tagami, Y.,** Theory of the interface between immiscible polymers. II. *J. Chem. Phys.*, 56, 3592, 1972.
49. **Hobin, T. P.,** Surface tension in relation to cohesive energy with particular reference to hydrocarbon polymers, *J. Adhesion*, 3, 327, 1972.
50. **Kendall, K.,** Sticky solids, *Contemp. Phys.*, 21, 277, 1980.
51. **Koehnen, D. M. and Smolders, C. A.,** The determination of solubility parameters of solvents and polymers by means of correlation with other physical quantities, *J. Appl. Polym. Sci.*, 19, 1163, 1975.
52. **Lavygin, I. A.,** Evaluation of the polymeric chain flexibility according to the cohesion energy density and the surface tension of the polymer, *Vysokomol. Soedin. Ser. B.*, 18, 426, 1976.
53. **Lee, L.-H.,** Adhesion of high polymers. II. Wettability of elastomers, *J. Polym. Sci. A-2*, 5, 1103, 1967.
54. **Lee, L.-H.,** Adhesion of high polymers. III. Mechanisms of adhesion at the rubber-resin interface in heterophase systems, in *Interaction of Liquids at Solid Substrates (Adv. Chem. Ser.* 87), Alexander, A. L., Ed., American Chemical Society, Washington, D.C., 1968, chap. 7.
55. **Lee, L.-H.,** Relationship between surface wettability and glass transition temperature of high polymers, *J. Appl. Polym. Sci.*, 12, 719, 1968.
56. **Lee, L.-H.,** Adhesion of high polymers. IV. Relationships between surface wettability and bulk properties of high polymers, in *Interaction of Liquids at Solid Substrates (Adv. Chem. Ser.* 87), Alexander, A. L., Ed., American Chemical Society, Washington, D.C., 1968, chap. 8, 106.
57. **Lee, L.-H.,** Relationships between solubility and surface tension of liquids, *J. Paint Technol.*, 42, 365, 1970.
58. **Lee, L.-H.,** Wettability of functional polysiloxanes, *Polym. Sci. Technol.*, 9B, 647, 1975.
59. **Lipatov, Yu and Feinerman, A.,** Surface tension and surface free energy of polymers, *Adv. Colloid Interface Sci.*, 11, 195, 1979.
60. **Massirio, S., Parodi, F., and Turturro, A.,** Mechanical and thermodynamic properties of interfaces between polyamide-6 and aliphatic polyurethanes, *Conv. Ital. Sci., Macromol.*, [Atti] 6th., 237, 1983.
61. **Michaels, A. S.,** Fundamentals of surface chemistry and surface physics, *Am. Soc. Testing Mater. Spec. Tech. Pub.*, 340, 3, 1962 (pub. 1963).
62. **Mitsui, T., Fukushima, S., and Takada, S.,** Evaluation of powder surface properties by an adsorption method, *J. Soc. Cosmet. Chem.*, 23, 525, 1972.
63. **Murgulescu, I. G. and Terzi, M.,** Some physico-chemical properties of simple molten salts. IV. Surface tension, *Rev. Roum. Chim.*, 24, 113, 1979.
64. **Olabisi, O., Robeson, L. M., and Shaw, M. T.,** *Polymer-Polymer Miscibility*, Academic Press, New York, 1979.
65. **Pritykin, L. M.,** Calculation of surface energy of polymers by their refractometric and cohesion characteristics, *Vysokomol. Soedin Ser. A.*, 23, 757, 1981; *Beitr. Faserforsch. Textiltech.*, 18, 510, 1981.
66. **Pritykin, L. M.,** Calculation of the surface energy of homo- and copolymers from the cohesion parameters and refractometric characteristics of the respective monomers, *J. Colloid Interface Sci.*, 112, 539, 1986.
67. **Reeves, R. L., Maggio, M. S., and Costa, L. F.,** Importance of solvent cohesion and structure in solvent effect on binding site probes, *J. Am. Chem. Soc.*, 96, 5917, 1974.
68. **Roberts, G. A. F. and Thomas, I. M.,** Determination of the solubility parameter of o-hydroxypropyl cellulose, *Polymer*, 19, 459, 1978.
69. **Rosseinsky, D. R.,** Surface tension and internal pressure. A simple model, *J. Phys. Chem.*, 81, 1578, 1977.
70. **Schonhorn, H.,** Theoretical relationship between surface tension and cohesive energy density, *J. Chem. Phys.*, 43, 2041, 1965.
71. **Sewell, J. H.,** Polymer critical surface tensions, *Mod. Plast.*, 48(6), 66, 1971.
72. **Siow, K. S. and Patterson, D.,** The prediction of surface tensions of liquid polymers, *Macromolecules*, 4, 26, 1971.
73. **Spelt, J. K., Absolom, D. R., and Neumann, A. W.,** Solid surface tension: the interpretation of contact angles by the equation of state approach and the theory of surface tension components, *Langmuir*, 2, 620, 1986.

74. **Spelt, J. K., Smith, R. P., and Neumann, A. W.,** Attraction and repulsion of solid particles of solidification fronts: evaluation of predictions of the Fowkes equations, *Colloids Surf.*, 28, 85, 1987.
75. **Sprow, F. B. and Prausnitz, J. M.,** Surface tension of simple liquids; Surface tensions of simple liquid mixtures, *Trans. Faraday Soc.*, 62, 1097 and 1105, 1965.
76. **Vavruch, I.,** On the determination of the factor between cohesive energy density and surface tension, *J. Colloid Interface Sci.*, 63, 600, 1978.
77. **Vavruch, I.,** On the predictions of total surface energy of pure liquids, *J. Colloid Interface Sci.*, 73, 267, 1980.
78. **Vavruch, I.,** Stefan's rule as a consequence of cohesive forces, *Colloids Surf.*, 15, 57, 1985.
79. **Vavruch, I.,** On the evaluation of the volume of surface formation for pure liquids, *J. Colloid Interface Sci.*, 114, 290, 1986.
80. **Vavruch, I.,** On the relation between surface energy, internal pressure and molar volume in pure fluids, *Colloids Surf.*, 30, 405, 1988.
81. **Visser, J.,** Colloid and other forces in particle adhesion and particle removal (a review), in *Deposition and Filtration of Particles from Gases and Liquids* (Proc. Symp., Loughborough, England, 1978), Society of Chemical Industry, London, 1978, 121.
82. **Wu, S.,** Estimation of the critical surface tension for polymers from molecular constitution by a modified Hildebrand-Scott equation, *J. Phys. Chem.*, 72, 3332, 1968.
83. **Wu, S.,** Interfacial and surface tensions of polymers, *J. Macromolec. Sci. Rev. Macromolec. Chem.*, C10, 1, 1974.
84. **Yagnyatinskaya, S. M., Voytuskii, S. S., and Kaplunova, L. Ya.,** Surface-tension and solubility parameter of amorphous polymers, *Kolloid Zh.*, 34, 132, 1972; *Colloid J. U.S.S.R.*, 34, 112, 1972.
85. **Fowkes, F. M.,** The interactions of polar molecules, micelles and polymers in nonaqueous media, in *Solvent Properties of Surfactant Solutions,* Shinoda, K., Ed., Marcel Dekker, New York, 1967; Edward Arnold, London, 1968, chap. 3.
86. **Fowkes, F. M.,** Comments on "The calculation of cohesive and adhesive energies", by Padday, J. F. and Uffindell, N. D., *J. Phys. Chem.*, 72, 3700, 1968.
87. **Padday, J. F. and Uffindell, N. D.,** The calculation of cohesive and adhesive energies from intermolecular forces at a surface, *J. Phys. Chem.*, 72, 1407, 1968.
88. **Padday, J. F. and Uffindell, N. D.,** Reply to comments of F. M. Fowkes on "The calculation of cohesive and adhesive energies," *J. Phys. Chem.*, 72, 3700, 1968.
89. **Grunberg, L. and Nissan, A. H.,** The energies of vaporisation, viscosity and cohesion and the structure of liquids, *Trans. Faraday Soc.*, 45, 125, 1949.
90. **Hildebrand, J. H. and Scott, R. L.,** *Solubility of Non-Electrolytes,* 3rd ed., Reinhold, New York, 1950; Dover, New York, 1964.
91. **Gordon, J. E.,** Fused organic salts. IV. Characterization of low melting point quaternary ammonium salts. Phase equilibria for salt-salt and salt-nonelectrolyte systems. Properties of the liquid salt medium, *J. Am. Chem. Soc.*, 87, 4347, 1965; Transition energies for a merocyanine dye in aqueous electrolyte solutions. Solvent polarity indicator transition energy-internal pressure relations, *J. Phys. Chem.*, 70, 2413, 1966.
92. **Good, R. J.,** Theory for the estimation of surface and interfacial energies. VI. Surface energies of some fluorocarbon surfaces from contact angle measurements, in *Contact Angle, Wettability and Adhesion (Adv. Chem. Ser.* 43), American Chemical Society, Washington, D.C., 1964, chap. 4.
93. **Good, R. J.,** Contact angles and the surface free energy of solids, in *Surface and Colloid Science,* Vol. 11, Good, R. J. and Stromberg, R. R., Eds., Plenum Press, New York, 1979, chap. 1.
94. **Good, R. J. and Girifalco, L. A.,** A theory for estimation of surface and interfacial energies. III. Estimation of surface energies of solids from contact angle data, *J. Phys. Chem.*, 64, 561, 1960.
95. **Good, R. J., Girifalco, L. A., and Kraus, G.,** A theory for estimation of interfacial energies. II. Application to surface thermodynamics of Teflon and graphite, *J. Phys. Chem.*, 62, 1418, 1958.
96. **Wu, S.,** Polar and nonpolar interactions in adhesion, *J. Adhesion,* 5, 39, 1973.
97. **Ewane-Ebele, F. and Schreiber, H. P.,** Measurement and use of surface tension data in film-forming polymers, *J. Oil Col. Chem. Assoc.*, 60, 249, 1977.
98. **Schreiber, H. P. and Ewane-Ebele, F.,** On the surface tension and its temperature variation in film-forming polymers, *J. Adhesion,* 9, 175, 1978.
99. **Chan, R. K. S.,** Surface tension of fluoropolymers. I. London dispersion term. II. The polar attraction term, *J. Colloid Interface Sci.*, 32, 492 and 499, 1970.
100. **Cuthrel, R. E.,** Intermolecular forces in polymers and liquids, *Polym. Prepr. Am. Chem. Soc. Div. Polym. Chem.*, 11, 488, 1970; in *Polym. Networks Struct. Mech. Props. (Proc. Symp. Am. Chem. Soc.*, 1970), Chompff, A. J., Ed., Plenum Press, New York, 1971, 121.
101. **Dann, J. R.,** Forces involved in the adhesive process. II. Nondispersion forces at solid-liquid interfaces, *J. Colloid Interface Sci.*, 31, 321, 1970.

102. **Fowkes, F. M.,** Dispersion force contributions to surface and interfacial tensions, contact angles, and heats of immersion, in *Contact Angle, Wettability and Adhesion (Adv. Chem. Ser.* 43), American Chemical Society, Washington, D.C., 1964, 99.
103. **Fowkes, F. M. and Mostafa, M. A.,** Acid-base interactions in polymer adsorption, *Ind. Eng. Chem. Prod. Res. Dev.*, 17, 3, 1978.
104. **Hansen, C. M.,** Characterization of surfaces by spreading liquids, *J. Paint Technol.*, 42, 660, 1970.
105. **Hansen, C. M. and Beerbower, A.,** Solubility parameters, in *Kirk-Othmer Encyclopedia of Chemical Technology*, 2nd ed., Standen, A., Ed., Interscience, New York, Suppl. Vol., 1971, 889.
106. **Hayes, L. J.,** Surface energy of fluorinated surfaces, *J. Fluor. Chem.*, 8, 69, 1976.
107. **Henry, L. F.,** Prediction and evaluation of the susceptibilities of glassy thermoplastics to environmental stress cracking, in *Predicting Plastics Performance (Div. Tech. Conf.)*, Society of Plastics Engineers, Greenwich, CT, 1973, 10; *Polym. Eng. Sci.*, 14, 167, 1974.
108. **Jensen, W. B.,** Lewis acid-base interactions and adhesion theory, *Rubber Chem. Technol.*, 55, 881, 1982; Acids, bases, and adhesion: a synergy, *Chemtech*, 12, 755, 1982.
109. **Kaelble, D. H.,** Dispersion-polar surface tension properties of organic solids, *J. Adhesion*, 2, 66, 1970.
110. **Kaelble, D. H.,** *Physical Chemistry of Adhesion*, John Wiley & Sons-Interscience, New York, 1971.
111. **Kaelble, D. H.,** Interface degradation processes and durability, *Polym. Eng. Sci.*, 17, 474, 1977.
112. **Kaelble, D. H., and Cirlin, E. H.,** Dispersions and polar contributions to surface tension of poly(methylene oxide) and Na-treated polytetrafluoroethylene, *J. Polym. Sci. A-2*, 9, 363, 1971.
113. **Kim, K. U., Park, T. S., and Lee, K. C.,** Solubility parameter and solvent selection of adhesives, *Polymer (Korea)*, 7, 155, 1983.
114. **Kloubek, J.,** Calculation of surface free energy components of ice according to its wettability by water, chlorobenzene and carbon disulfide, *J. Colloid Interface Sci.*, 46, 185, 1974.
115. **Kloubek, J.,** Interaction of polar forces and their contribution to the work of adhesion, *J. Adhesion*, 6, 293, 1974.
116. **Köhler, A. E.,** The orientation effect of substrate surfaces on liquid crystal molecules. I. Model observations, *Z. Phys. Chem. (Leipzig)*, 263, 273, 1982.
117. **Schröder, J.,** Considerations on the energetics of pigment dispersions containing solvents and binders, *Progr. Org. Coatings*, 12, 339, 1984.
118. **Wu, S.,** Interfacial energy, structure and adhesion between polymers, in *Polymer Blends*, Vol. 1, Paul, D. R. and Newman, S., Eds., Academic Press, New York, 1978, chap. 6.
119. **Hansen, C. M.,** Surface dewetting and coatings performance, *J. Paint Technol.*, 44(570), 57, 1972.
120. **Hansen, C. M. and Pierce, P. E.,** Surface effects in coating processes, *Ind. Eng. Chem. Prod. Res. Dev.*, 13, 218, 1974.
121. **Jackson, L. C.,** How to select a substrate cleaning solvent — contaminant removal using solubility parameter technology, *Adhes. Age*, 17(12), 22, 1974.
122. **Jackson, L. C.,** Surface characterization based on solubility parameters, *Adhes. Age*, 19(10), 17, 1976.
123. **Jackson, L. C.,** Contaminant detection, characterization and removal based on solubility parameters, in *Proc. Intl. Symp. Contam. Control 4th*, Inst. Environ. Sci., Mt. Prospect IL, 1978, 226; in *Surface Contamination: Its Genesis, Detection and Control*, Vol. 2, Mittal, K. L., Ed., Plenum Press, New York, 1979, 527.
124. **van Krevelen, D. W.,** *Properties of Polymers: Their Estimation and Correlation with Chemical Structure*, 2nd ed., Elsevier, Amsterdam, 1976.
125. **Kaelble, D. H. and Moacanin, J.,** Surface energetics analysis of artificial blood substitutes, *Med. Biol. Eng. Comput.*, 17, 593, 1979.
126. **Shain, S. A. and Prausnitz, J. M.,** Thermodynamics and interfacial tension of multicomponent liquid-liquid interfaces, *Am. Inst. Chem. Eng. J.*, 10, 766, 1964.
127. **Suri, S. K. and Ramakrishna, V.,** Adsorption from solution on a free liquid surface and on an inert solid, *J. Phys. Chem.*, 72, 1555, 1968.
128. **Suri, S. K. and Ramakrishna, V.,** Surface tension of some binary liquid mixtures, *J. Phys. Chem.*, 72, 3073, 1968.
129. **Aharoni, S. M.,** Instantaneous shape and segmental density of flexible molecules. II. Observation of individual and aggregate poly-cis-isoprene molecules in polyisobutylene, *J. Macromol. Sci. Phys.*, B15, 635, 1978.
130. **Helfand, E.,** Block copolymers, polymer-polymer interfaces, and the theory of inhomogeneous polymers, *Acc. Chem. Res.*, 8, 295, 1975.
131. **Helfand, E.,** Proposed experiment to estimate the mixing free energy of polymers, *Macromolecules*, 11, 682, 1978.
132. **Helfand, E. and Sapse, A. M.,** Theory of unsymmetric polymer-polymer interfaces, *J. Chem. Phys.*, 56, 3592, 1972.
133. **Kaelble, D. H.,** A relationship between the fracture mechanics and surface energetics failure criteria, *J. Appl. Polym. Sci.*, 18, 1869, 1974.

134. **Kaelble, D. H., Dynes, P. J., and Cirlin, E. H.,** Interfacial bonding and environmental stability of polymer matrix composites, *J. Adhesion,* 6, 23, 1974.
135. **Kaelble, D. H., Dynes, P. J., Crane, L. W., and Maus, L.,** Kinetics of environmental degradation in graphite-epoxy laminates, in *Composite Reliability,* ASTM Spec. Tech. Publ. 580, American Society for Testing and Materials, Philadelphia, PA, 1975, 247.
136. **Kaelble, D. H., Dynes, P. J., and Maus, L.,** Surface energy analysis of treated graphite fibers, *J. Adhesion,* 6, 239, 1974.
137. **Kaelble, D. H., Dynes, P. J., and Pav, D.,** Surface energetics analysis of lithography, in *Polymer Science and Technology,* Vol. 9B, Lee, L.-H., Ed., Plenum Press, New York, 1975, 735.
138. **Kaelble, D. H. and Moacanin, J.,** A surface energy analysis of bioadhesion, *Polymer,* 18, 475, 1977.
139. **Beerbower, A.,** *Boundary Lubrication — Scientific and Technical Applications Forecast,* AD 747336, Office of the Chief of Research and Development, Department of the Army, Washington, D.C., 1972.
140. **Senior, J. M. and West, G. H.,** Interaction between lubricants and plastic bearing surfaces, *Wear,* 18, 311, 1971.
141. **Andrews, E. H. and Kinloch, A. J.,** Mechanics of adhesive failure, I and II, *Proc. Roy. Soc. (London) A.,* 332, 385 and 401, 1973.
142. **Bleier, A.,** The roles of van der Waals forces in determining the wetting and dispersion properties of silicon powder, *J. Phys. Chem.,* 87, 3493, 1983.
143. **Brydson, J. A.,** Selection of compounding ingredients, *Plastics (London),* 26, 107, 1961.
144. **Chalykh, T. I., Grivtsova, L. A., and Chuikova, L. F.,** Methods for calculating the hydrophilicity of polymers, *Kozh.-Obuvn. Prom-st.,* 21(9), 50, 1979.
145. **Dyckerhoff, G. A. and Sell, P.-J.,** Influence of interfacial tension on adhesion, *Angew. Makromol. Chem.,* 21, 169, 1972.
146. **Harrison, K. W.,** Solvents in polymer-based adhesives, in *Adhesion,* Vol. 3, Allen, K. W., Ed., Applied Science Publishers, Barking, England, 1979, chap. 9.
147. **Harrison, K. W.,** Solvents in polymer-based adhesives, *Polym. Paint. Col. J.,* 168, 602 and 885, 1978.
148. **Huntsberger, J. R.,** Interfacial energies, contact angles, and adhesion, in *Treatise on Adhesion and Adhesives,* Vol. 5, Patrick, R. L., Ed., Marcel Dekker, New York, 1981, 1.
149. **Huntsberger, J. R.,** Surface energy, wetting, and adhesion, *J. Adhesion,* 12, 3, 1981.
150. **Huntsberger, J. R.,** Adhesion of plasticized poly(vinyl butyral) to glass, *J. Adhesion,* 13, 107, 1981.
151. **Iyengar, Y. and Erickson, D. E.,** Role of adhesive-substrate compatibility in adhesion, *J. Appl. Polym. Sci.,* 11, 2311, 1967.
152. **Jackson, L. C.,** A scientific technique for adhesive development, *Adhes. Age,* 18(3), 30, 1975.
153. **Kitazaki, Y. and Hata, T.,** Surface-chemical criteria for optimum adhesion. II. Variability of critical surface tension and its choice, *J. Adhesion,* 4, 123, 1972.
154. **Koldunovich, G. E., Epshtein, V. G., Chekanova, A. A., and Zakharov, N. D.,** Mutual adhesion of various rubbers, *Nauch. Konf. Yaroslav. Tekhnol. Inst.,* 22nd, 156, 1971.
155. **Kuptsov, Yu. D., Maloshuk, Yu. S., and Voyutskii, S. S.,** Study of influence of ratio of solubility parameters of polymers on adhesion of elastomers to polymeric substrates, *Vysokomol. Soedin. Ser. B,* 13, 908, 1971.
156. **Kuptsov, Yu. D., Maloshuk, Yu. S., and Voyutskii, S. S.,** Correlation between solubility parameters of polymers and their adhesive properties, *Khim. Khim. Teknol., Tr. Yubileinoi Konf., Posvyashch. 70-Letiyu Inst.,* (Mosk. Inst. Tonkoi Khim. Tekhnol.), 1970, Bashkirov, A. N., Ed., 1972, 203.
157. **Runge, M. L. and Dreyfuss, P.,** Adhesion of elastomers to rigid substrates — three substrates with unexpectedly high adhesion, *J. Appl. Polym. Sci.,* 23, 1863, 1979.
158. **Sarkar, A.,** Cohesion and adhesion in high polymers, *J. Indian Chem.,* 56, 1157, 1979.
159. **Sushkov, V. I., Gusev, S. S., Perepelkin, A. N., Minyailo, S. A., and Bablyuk, Ye. B.,** Study of the active influence of some solutions on the surface of polyethylene terephthalate films, *Vysokomol. Soedin. Ser. A,* 26, 2291, 1984; *Polym. Sci. U.S.S.R.,* 26, 2560 1984.
160. **Voyutskii, S. S., Yagnyatinskaya, S. M., Kaplunova, L. Ya., and Geretovskaya, N. L.,** Elastomer adhesion, *Rubber Age,* (NY), 105(2), 37, 1973.
161. **Wingrave, J. A.,** Linear sodium alkylbenzene sulfonate homologs. Comparison of detergency performance with experimental and thermodynamic wetting theories, *Am. Chem. Soc. Symp. Ser.* 253 (Struct. Perform. Relat. Surfactants), 241, 1984.
162. **Schonhorn, H.,** Environmental influence on the fracture strength of coated glass fibers, *J. Appl. Polym. Sci.,* 23, 687, 1979.
163. **Eldin, S. H.,** The adhesive properties of hydantoin vinyl ether polymers, *J. Appl. Polym. Sci.,* 32, 3971, 1986.
164. **Engel, J. H., Jr. and Fitzwater, R. N.,** Adhesion of surface coatings as determined by the peel method, in *Adhesion and Cohesion,* Weiss, P. N. Y., Ed., Elsevier, Amsterdam, 1962, 89.
165. **Nadkarni, P. D., Kildsig, D. O., Kramer, P. A., and Banker, G. S.,** Effect of surface roughness and coating solvent on film adhesion to tablets, *J. Pharm. Sci.,* 64, 1554, 1975.

166. **Sherlock, J. C.,** The effect of surface treatment on the adhesion of an organosol to steel, in *Aspects of Adhesion,* Vol. 6, Alner, D. J., Ed., University of London Press, London, 1969.
167. **Nakagawa, M.,** Adhesions between rubber and wire, *Jpn. Pat.,* 74/60382, 1974.
168. **Davidson, E. B.,** Photoresist adhesion, *Soc. Plast. Eng. Tech. Pap., Reg. Tech Conf., Mid-Hudson Sect.,* 1970, 141.
169. **Mittal, K. L.,** Factors affecting adhesion of lithographic materials, in *Durability of Macromolecular Materials (Am. Chem. Soc. Symp. Ser.* 95), Eby, R. K., Ed., American Chemical Society, Washington, D.C., 1979, chap. 26.
170. **Reegan, S. L.,** Adhesion of urethanes for oxypropylene polyols, *Adv. Urethane Sci. Technol.,* 2, 56, 1973.
171. **Penczek, P., Plocharski, S., and Warchol, W.,** Effect of the composition of solvent mixture on the freezing resistance of polychloroprene rubber based adhesives on storage, *Polimery (Warsaw),* 20, 186, 1975.
172. **Gubanov, E. G., Shulyndin, S. V., Gurskaya, V. Sh., Palikhov, N. A., Zuyer, B. M., and Ivanov, B. Ye.,** Effect of molecular interaction on properties of copolymers containing phosphorous, *Vysokomol. Soedin. Ser. A,* 18, 653, 1976; *Polym. Sci. U.S.S.R.,* 18, 744, 1976.
173. **Wake, W. C.,** *Adhesion and the Formulation of Adhesives,* Applied Science, London, 1964, 163.
174. **Wake, W. C.,** Adhesion of polymers to each other and to fibers, *J. Adhesion,* 3, 315, 1972.
175. **Rupp, N. W., Bowen, R. L., and Paffenberger, G. C.,** Bonding cold-curing denture base acrylic resin to acrylic resin teeth, *J. Am. Dent. Assoc.,* 83, 601, 1971.
176. **Iyengar, Y.,** Adhesion of Kevlar aramid cords to rubber, *J. Appl. Polym. Sci.,* 22, 801, 1978.
177. **Kozakiewicz, J. and Penczek, P.,** Adhesives from polyepichlorohydrin elastomers. I. Studies on the properties of adhesive solutions of polyepichlorohydrin, *Angew. Makromol. Chem.,* 50, 67, 1976.
178. **Plueddemann, E. P.,** Silane coupling agents for thermoplastic resins, in *Soc. Plast. Ind., Proc. Ann. Tech. Conf., Reinforced Plast. Div.,* 20th, Society of Plastics Industry, New York, 1965, 19A, 1.
179. **Plueddemann, E. P.,** Promoting adhesion of coatings through reactive silanes; Mechanism of adhesion of coatings through reactive silanes, *J. Paint. Technol.,* 40(516), 1, 1968; 42, 603, 1970.
180. **Boberski, W. G., Seiner, J. A., and Petracca, V. G.,** Viscosity of magnesia cements with organosilicon compounds, *Ind. Eng. Chem. Prod. Res. Dev.,* 21, 531, 1982.
181. **Rowe, R. C.,** Binder-substrate interactions in tablets: a theoretical approach based on solubility parameters, *Acta Pharm. Technol.,* 34, 144, 1988; Binder-substrate interactions in granulation: a theoretical approach based on surface free energy and polarity; Surface free energy and polarity effects in the granulation of a model system; Polar-nonpolar interactions in the granulation of organic substrates with polymer binding agents, *Int. J. Pharm.,* 52, 149, 1989; 53, 75, 1989; 56, 117, 1989.
182. **Rowe, R. C.,** Adhesion of film coatings to tablet surfaces — a theoretical approach based on solubility parameters; Interaction of lubricants with microcrystalline cellulose and anhydrous lactose — a solubility parameter approach; Interactions in the ternary powder system microcrystalline cellulose, magnesium stearate and colloidal silica — a solubility parameter approach, *Int. J. Pharm.,* 41, 219 and 223, 1988; 45, 259, 1988.
183. **Rowe, R. C.,** Interactions in coloured powders and tablet formulations — a theoretical approach based on solubility parameters, *Int. J. Pharm.,* 53, 47, 1989.
184. **Beerbower, A. and Panzer, J.,** Comments on "Interaction between lubricants and plastic bearing surfaces," *Wear,* 18, 481, 1971.
185. **Bolster, R. N. and Little, R. C.,** Dependence of lithium grease dropping points on solubility parameter of oil component, *Ind. Eng. Chem. Prod. Res. Dev.,* 5, 198, 1966.
186. **Bright, G. S.,** The relationship of the Hildebrand solubility parameter of various oils to the properties of lithium soap greases made therefrom, *Nat. Lubric. Grease Inst. Spokesman,* 34, 156, 1970; *Tex. J. Sci.,* 22(Abstr.), 304, 1971.
187. **Little, R. C.,** Some physical properties of soap/solvent systems and their relation to the solubility parameter of the solvent, *J. Colloid Interface Sci.,* 21, 266, 1966.
188. **Little, R. C.,** Oil solubility parameter as a factor in boundary lubrication effects, *Wear,* 30, 385, 1974.
189. **Little, R. C.,** Some observations on the zinc stearate — toluene system, *J. Colloid Interface Sci.,* 51, 200, 1975.
190. **Ling, L. F. and Noel, F.,** Characterization of lubricants for polyvinyl chloride, *Polym. Eng. Sci.,* 12, 112, 1972.
191. **Overhoff, J., Schuur, G., and Verschoore, V.,** A new lubricant for rigid poly(vinyl chloride): glycol, *Pharm. Acta Helv.,* 39, 305, 1964.
192. **Danilov, A. M. and Sergeeva, A. V.,** Evaluation of compatibility of thickening agents and dispersion media for ureate lubricating greases, *Khim. Tekhnol. Topl. Masel,* (7), 17, 1985; *Chem. Technol. Fuels Oils,* 353, 1986.
193. **Keller, R. A., Karger, B. L., and Snyder, L. R.,** Use of the solubility parameter in predicting chromatographic retention and eluotropic strength, *Gas Chromatography 1970* (Proc. Int. Symp., 8, Dublin, 1970), Stock, R., Ed., The Institute of Petroleum, London, 1971, 125.

194. **Karger, B. L., Snyder, L. R., and Eon, C.,** An expanded solubility parameter treatment for classification and use of chromatographic solvents and adsorbents. Parameters for dispersion, dipole and hydrogen bonding interactions, *J. Chromatogr.*, 125, 71, 1976.
195. **Karger, B. L., Snyder, L. R., and Eon, C.,** Expanded solubility parameter treatment for classification and use of chromatographic solvents and adsorbents, *Analyt. Chem.*, 50, 2126, 1978.
196. **Balik, C. M. and Hopfinger, A. J.,** Quantization of the solvent effect on the adsorption of poly γ-benzyl-L-glutamate, *J. Colloid Interface Sci.*, 67, 118, 1978.
197. **McGuire, M. J.,** *The Optimization of Water Treatment Unit Processes for the Removal of Trace Organic Compounds with an Emphasis on the Adsorption Mechanism,* Ph.D. Thesis, Drexel University, Philadelphia, 1977; *Diss. Abstr. Int. B.*, 38, 1849, 1977.
198. **McGuire, M. J. and Suffet, I. H.,** The calculated net adsorption energy concept, *Abstr. Pap. Am. Chem. Soc.*, 176, ENVR 6, 1978.
199. **McGuire, M. J. and Suffet, I. H.,** Adsorption of organics from domestic water supplies, *J. Am. Waterworks Assoc.*, 70, 621, 1978.
200. **McGuire, M. J., Suffet, I. H., and Radziul, J. V.,** The development of optimum unit processes for the removal of trace organic compounds from drinking water, *Proc. Am. Waterworks Assoc.*, 97th Ann. Conf., Denver, Col., Parts 1 and 2, Paper 16-8, 1977; Assessment of unit processes for the removal of trace organic compounds for drinking water, *J. Am. Waterworks Assoc.*, 70, 565, 1978.
201. **Arbuckle, W. B.,** Estimating equilibrium adsorption of organic compounds on activated carbon from aqueous solution; Comment on 'Estimating equilibrium adsorption of organic compounds on activated carbon from aqueous solution,' *Environ. Sci. Technol.*, 15, 812, 1981; 16, 374, 1982.
202. **Altshuler, G. and Belfort, G.,** Comment on 'Estimating equilibrium adsorption of organic compounds on activated carbon from aqueous solution,' *Environ. Sci. Technol.*, 16, 374, 1982.
203. **Kronberg, B. and Stenius, P.,** The effect of surface polarity on the adsorption of nonionic surfactants. I. Thermodynamic considerations, *J. Colloid Interface Sci.*, 102, 410, 1984.
204. **Galembeck, F.,** Sorption of iron pentacarbonyl in Teflon, *J. Polym. Sci., Polym. Chem.*, 16, 3015, 1978.
205. **Galembeck, F., Galembeck, S. E., and Vicente, E. J.,** Investigation of the interaction of certain low cohesive energy liquids with Teflon and polyethylene and its implications in the contamination of polymeric surfaces, in *Proc. Int. Symp. Contam. Control,* 4th, Institute of Environmental Science, Mt. Prospect, IL, 1978, 329.
206. **Blow, C. M.,** Polymer-particulate filler interaction. Bound rubber phenomena, *Polymer,* 14, 309, 1973.
207. **Abe, I., Hayashi, K., Kitagawa, M., and Urahata, T.,** Adsorptive mechanism on activated carbon in the liquid phase. 3. The relationship between the physical constants of organic compounds and their adsorbabilities on activated carbon from an aqueous solution, *Bull. Chem. Soc. Jpn,* 53, 1199, 1980.
208. **Radeke, K. H.,** On the 'solubility coefficient' of activated carbons, *Carbon,* 24, 514, 1986.
209. **Rudling, J.,** Simple model based on solubility parameters for liquid desorption of organic solvents adsorbed on activated carbon, *J. Chromatogr.*, 362, 175, 1986.
210. **Luce, J. E. and Robertson, A. A.,** The sorption of polymers on cellulose, *J. Polym. Sci.*, 51, 317, 1961.
211. **Gustafson, R. L., Albright, R. L., Heisler, J., Lirio, J. A., and Reid, O. T., Jr.,** Adsorption of organic species by high-surface-area styrene-divinylbenzene copolymers, *Ind. Eng. Chem. Proc. Res. Dev.*, 7, 107, 1968.
212. **Gustafson, R. L. and Lirio, J. A.,** Adsorption of organic ions by anion exchange resins, *Ind. Eng. Chem. Proc. Res. Dev.*, 7, 116, 1968.
213. **Sakai, I., Nakamae, K., Okawa, T., Inubishi, H., and Matsumoto, T.,** Studies on suspension and emulsion. 33. Adsorption behavior of polybutadiene having carboxyl end groups at water-oil interface, *Kobunshi Ronbunshu,* 34, 661, 1977.
214. **Vijayendran, B. R. and Bursh, T. P.,** Effect of oil phase polarity on the saturation adsorption of sodium lauryl sulfate at oil-water interfaces, *J. Colloid Interface Sci.*, 68, 383, 1979.
215. **Patel, V. M., Patel, K. C., and Patel, R. D.,** Adsorption of polymethylmethacrylate from binary solvent systems, *Angew. Makromol. Chem.*, 62, 177, 1977.
216. **Friedman, H. L., Miller, B., and Meiser, C. H., Jr.,** Influence of preadsorbed water on chromatographic detention of organic volatiles by cellulosic substrates, *Text. Res. J.*, 55, 726, 1985.
217. **Belkin, A. I. and Karpov, V. V.,** Solubility parameter and sorption of disperse dyes by man-made fibers in perchloroethylene, *Zh. Prikl. Khim.*, 53, 2107, 1980; *J. Appl. Chem. U.S.S.R.*, 53, 1568, 1980.
218. **Lee, K. W. and Boger, D. V.,** The importance of solvent composition and adsorption effects on the flow properties of polybutadiene solutions, *Chem. Eng. Commun.*, 53, 97, 1987.
219. **Stepin, S. N., Gusev, A. V., Bogatov, F. R., and Svetlakov, N. V.,** Adsorption activity on microbarite surface, *Zh. Prikl. Khim.*, 60, 1410, 1987; *J. Appl. Chem. U.S.S.R.*, 60, 1335, 1987.
220. **Bennes, R.,** Adsorption isotherms of neutral molecules for regular interfacial solutions, *J. Electroanalyt. Chem.*, 105, 85, 1979.
221. **Bennes, R. and Hamdi, M.,** Adsorption isotherms of neutral molecules for regular interfacial solutions, *Inorg. Chim. Acta Articles,* 40, X145, 1980.

222. **Davis, S. S. and Smith, A.,** The stability of hydrocarbon oil droplets at the surfactant/oil interface, *Colloid Polym. Sci.*, 254, 82, 1976.
223. **Denguchi, K., Mizuno, T., and Meguro, K.,** Solvent effects on the micellar destruction of nonionic surfactant, *J. Colloid Interface Sci.*, 48, 474, 1974.
224. **Funasaki, N., Hada, S., and Neya, S.,** Volumetric study of solubilization of hydrophobic liquids in nonionic micelles; Volumetric study of intermolecular interactions in mixed micelles, *J. Phys. Chem.*, 88, 1243, 1984; 90, 5469, 1986.
225. **Hansen, C. M.,** The universality of the solubility parameter, *Ind. Eng. Chem. Prod. Res. Dev.*, 8, 2, 1969.
226. **Hayashi, S.,** Emulsification. III. Required hydrophile-lipophile balance value for interfacial viscosity and solubility parameter of oils, *Yukagaku,* 16, 554, 1967.
227. **Healy, R. N., Reed, R. L., and Stenmark, D. G.,** Multiphase microemulsion systems, *Soc. Pet. Eng. J.*, 16, 147, 1976.
228. **Holtzscherer, C. and Candau, F.,** Application of the cohesive energy ratio concept (CER) to the formation of polymerizable microemulsions, *Colloids Surfaces,* 29, 411, 1988.
229. **Kertes, A. S.,** Solubility parameters of long-chain alkylamines and their activities in organic solvents; Solubility and activity of high-molecular amine hydrochlorides in organic solvents, *J. Inorg. Nucl. Chem.*, 26, 1764, 1964; 27, 209, 1965.
230. **Little, R. C.,** Micelle size of barium dinonylnapthalenesulfonate in low polarity solvents by vapor phase osmometry, *J. Phys. Chem.*, 74, 1817, 1970.
231. **Little, R. C.,** Correlation of surfactant hydrophile-lipophile balance (HLB) with solubility parameter, *J. Colloid Interface Sci.*, 65, 587, 1978.
232. **Little, R. C. and Singleterry, C. R.,** The solubility of alkali dinonylnaphthalenesulfonates in different solvents and a theory for the solubility of oil-soluble soaps, *J. Phys. Chem.*, 68, 3453, 1964.
233. **Marszall, L. and van Valkenburg, J. W.,** The effect of glycols on the hydrophile-lipophile balance and the micelle formation of nonionic surfactants, *J. Am. Oil. Chem. Soc.*, 59, 84, 1982.
234. **Marszall, M.,** Emulsion inversion point and properties of the oil phase, *J. Colloid Interface Sci.*, 107, 572, 1985.
235. **Marszall, M.,** The effect of the kind of hydrocarbon on the emulsion inversion, *Tluszcze Srodki Piorace Kosmet.*, 29(3-4), 41, 1985.
236. **Attwood, D., Jagielski, C. E., McDonald, C., and Wilkinson, A. E.,** The solution properties of some nonionic surface-active agents in non-aqueous solvents, *Colloid Polym. Sci.*, 252, 991, 1974.
237. **McDonald, C.,** Hydrophilic-lipophilic balance values and solubility parameters, *Can. J. Pharm. Sci.*, 5, 81, 1970.
238. **McDonald, C.,** The effect of change of solvent on the critical micelle concentration of a non-ionic surfactant, *J. Pharm. Pharmac.*, 22, 148, 1970.
239. **McNicoll, B., Sangster, J., and Schreiber, H. P.,** Some observations on the micellar behavior of surfactants in water and aqueous solvents, in *Solution Chemistry of Surfactants,* Vol. 1, Mittal, K. L., Ed., Plenum Press, New York, 1979, 367.
240. **Meusberger, K. E.,** Computerized optimization of emulsifiers for pesticide emulsifiable concentrates, in *Advances in Pesticide Formulation Technology,* Scher, H., Ed., American Chemical Society, Washington, D.C., 1984, 121.
241. **Meusberger, K. E.,** Determination of cohesive energy density parameters for pesticide formulation development, *Am. Chem. Soc. Symp. Ser.*, 371 *(Pesticide Formulations),* 151, 1988.
242. **Panich, R. M., Konovalova, N. V., Snezhko, A. G., and Gul, V. E.,** Stabilization of dispersed systems by non-ionic surfactants in *Khim. Khim. Tekhnol., Tr. Yubileinoi Konf., Posryoshch. 70-Letiyu Inst.*, (Mosk. Inst. Tonkoi Khim. Tekhnol.) 1970, Baskkirov., A. N., Ed., 1972, 200.
243. **Samaha, M. W. and Naggar, V. F.,** Micellar properties of non-ionic surfactants in relation to their solubility parameters, *Int. J. Pharm.*, 42, 1, 1988.
244. **Schott, H.,** Solubility parameter and hydrophilic-lipophilic balance of nonionic surfactants, *J. Pharm. Sci.*, 73, 790, 1984.
245. **Wellinghoff, S., Shaw, J., and Baer, E.,** Polymeric materials from the gel state — development of fringed micelle structure in a glass, *Macromolecules,* 12, 932, 1979.
246. **Griffin, W. C.,** Classification of surface-active agents by 'HLB', *J. Soc. Cosmet. Chem.*, 1, 311, 1949.
247. **Helfand, E. and Wasserman, Z.,** Statistical thermodynamics of microdomain structures in block copolymer systems, *Polym. Eng. Sci.*, 17, 582, 1977.
248. **Fukunaga, K., Shirai, H., Ide, S., and Kimura, M.,** Solubility parameters of phase-transfer catalysts and their applications, *Nippon Kagaku Kaishi,* 1148, 1980.
249. **Beerbower, A. and Hill, M. W.,** Application of the cohesive energy ratio (CER) concept to anionic emulsifiers, *Am. Cosmet. Perfum.*, 87(6), 85, 1972.
250. **Beerbower, A. and Nixon, J.,** Properties of high-internal-phase emulsions. II. Theory of emulsifier selection, *Am. Chem. Soc. Div. Pet. Chem. Prepr.*, 14(1), 62, 1969.

251. **Winsor, P. A.,** *Solvent Properties of Amphiphilic Compounds,* Butterworths, London, 1954.
252. **Marszall, L.,** Further approach to the energy balance of nonionic surfactants, *Colloid Polym. Sci.,* 254, 440, 1976.
253. **Brendley, W. H.,** Fundamentals of acrylic polymers, *Paint Varnish Prod.,* 63(7), 19, 1973.
254. **Brendley, W. H. and Haag, T. H.,** Physical characterization of water dispersed and soluble acrylic polymers, in *Nonpolluting Coatings and Coating Processes,* Gardon, J. L. and Prane, J. W., Eds., Plenum Press, New York, 1973, 5.
255. **Paul, S.,** Water-borne acrylic emulsion paints, *Prog. Org. Coatings,* 5, 79, 1977.
256. **Seymour, R. B. and Owens, D. R.,** Recent advances in water-based polymer systems, *Am. Paint J.,* 58(8), 14, 1973.
257. **Zunker, D. W.,** Unique aqueous nitrocellulose/acrylic air-dry systems for industrial finishing, *J. Coatings Technol.,* 48(616), 37, 1976.
258. **Lewis, G. A. and Enever, R. P.,** Solution thermodynamics of some potentially long-acting norethindrone derivatives. II. Solutions in 2,2,4-trimethylpentane (iso-octane), *Int. J. Phar.,* 3, 275, 1979.
259. **Meyer zu Bexten, J. H.,** Solubility parameters and their practical application, *Farbe Lack,* 78, 813, 1972.
260. **Porter, L. K.,** Factors affecting the solubility and possible fractionation of organic colloids extracted from soil and leonardite with an acetone-H_2O-HCl solvent, *J. Agric. Food Chem.,* 15, 807, 1967.
261. **Schmitz, R.,** Information on the sedimentation behavior of pigments. I. Effect of the sedimentation liquid, *Colloid Polym. Sci.,* 256, 578, 1978.
262. **Schreiber, H. P.,** Solvent balance, dispersion and rheological properties of pigmented polymer compositions, *J. Paint Technol.,* 46(598), 35, 1974.
263. **Nasuno, T. and Sano, I.,** Oily powder make up compositions, *Jpn. Pat.,* 55/51011, 1980.
264. Kanebo, Ltd., Improved oil foundation compositions, *Jpn. Pat.,* 55/57510, 1980.
265. **Schott, H.,** The solubility of potassium acid soaps in organic solvents, *J. Colloid Interface Sci.,* 117, 94, 1987.
266. **Takehara, B., Asada, T., Tani, T., Suezawa, M., Yamamoto, A., and Tawara, T.,** Evaluation of pigment dispersion by solubility parameters. Determination of dispersion parameters, *Shikizai Kyokaishi,* 47, 411, 1974.
267. **Trudgian, L.** The pattern of solvent-resin-pigment affinities, *Off. Dig. Fed. Soc. Paint Technol.,* 35, 1211, 1963.
268. **Hansen, C. M.,** The three dimensional solubility parameter — key to paint component affinities. II. Dyes, emulsifiers, mutual solubility and compatibility, and pigments, *J. Paint Technol.,* 39, 505, 1967.
269. **Skaarup, K.,** The three dimensional solubility parameter and its use. II. Pigmented systems, *Skand. Tidskr. Farg Lack,* 14(2), 28; (3), 45, 1968.
270. **Weisberg, H. E.,** Dispersion-flocculation phenomena and the dielectric constant, *Off. Dig. Fed. Soc. Paint Technol.,* 34, 1154, 1962.
271. **Saarnak, A.,** Adsorption on pigment surfaces, *Polym. Mater. Sci. Eng.,* 51, 698, 1984; *Org. Coat.,* 7, 357, 1984.
272. **Saarnak, A.,** Pigment-binder-solvent interaction Part 1. Effect of the solvent on the state of dispersion, *Skand. Tidskr. Farg Lack,* 23, 317, 1977.
273. **Saarnak, A. and Hansen, C. M.,** *Solubility Parameter Characterisation of Pigment-Polymer Interactions,* Research Report T10-82, M + T, Scandinavian Paint and Printing Ink Research Institute, 1982.
274. **Saarnak, A. and Hansen, C. M.,** Solubility parameter and acid-base studies of pigment dispersion, *Abstr. Pap. Am. Chem. Soc.,* 188, 140, 1984.
275. **Shareef, K. M. A., Yaseen, M., Ali, M. M., and Reddy, P. J.,** Suspension interaction of pigments in solvents: characterization of pigment surfaces in terms of three-dimensional solubility parameters of solvents, *J. Coatings Technol.,* 58(733), 35, 1986.
276. **Eissler, R. L., Zgol, R., and Stolp, J. A.,** Solubility parameters applied to sedimentation volumes of pigments, *Abstr. Pap. Am. Chem. Soc.,* 157, ORPL 13, 1969; *J. Paint Technol.,* 42, 483, 1970.
277. **Vinther, A.,** Application of the concepts solubility parameter and pigment charge, *Chim. Peintures (Engl.),* 34, 363, 1971.
278. **Vinther, A. and Sörensen, P.,** Use of the concepts of solubility parameter and pigment charge in the formulation of pigmented systems, *Farbe Lack,* 77, 317, 1971.
279. **Sausaman, D. K.,** A practical approach to solvent applications in coatings and inks, in *Solvents Theory and Practice (Adv. Chem. Ser. 124),* Tess, R. W., Ed., American Chemical Society, Washington, D.C., 1973, chap. 8.
280. **Sörensen, P.,** The solubility parameter concept in the formulation of liquid inks, *J. Oil. Col. Chem. Assoc.,* 50, 226, 1967.
281. **Sörensen, P.,** The solubility parameter concept in pigment-resin-solvent systems, *Deut. Farben-Z.,* 24, 473, 1970; *PIRA Translation,* No. 876 (PR), Print Abstr. 963/1971, Research Association for the Paper Board, Printing and Packaging Industries, Leatherhead, Surrey, U.K., 1971.
282. **Sörensen, P.,** Pigment charges and the solubility parameter concept, *Am. Ink Maker,* 49(November), 30, 1971.

283. **Sörensen, P.,** Electrical charge and solubility parameter concept used for pigmented system; New aspects on the application of the solubility parameters, *Skand. Tidskr. Farg. Lack.,* 17, 200, 1971; 20(2), 9, 1974.
284. **Sörensen, P.,** Application of the acid/base concept describing the interaction between pigments, binders and solvents, *J. Paint Technol.,* 47(602), 31, 1975.
285. **Sörensen, P.,** New formulation technology can reduce high costs, *Pigm. Resin Technol.,* 4(7), 5, 1975.
286. **Belkin, A. I., Korobanova, I. A., and Karpov, V. V.,** Utilization of solubility parameters in selection of dispersed dyes for coloring of hydrophobic fibers from a medium of organic solvents, *Zh. Vses. Khim. O-va,* 21, 221, 1976.
287. **Skelly, J. K.,** Alternatives to water in textile processing, *J. Soc. Dye Col.,* 91, 177, 1975.
288. **Riedel, G.,** Solubility of dyes in organic solvents, *Farbe Lack,* 82, 281, 1976.
289. **Urbanik, A.,** Effects of solubility parameters on disperse dyeing of polyester, *Text. Chem. Color.,* 15(3), 36, 1983.
290. **Satoh, T.,** Correlation between color stability of paint systems and pigment-solvent wettability, *Org. Coat.,* 5, 147, 1983.
291. **Fujitani, T. and Satoh, T.,** The stability of pigment dispersions. II. Correlations between adhesion tension of solvents to pigments and solubility parameters, *Shikizai Kyokaishi,* 55, 459, 1982.
292. **Chavna, R. B., Venkata Rao, J., and Ahmad, A.,** Solubility parameters for dyeing water swollen, solvent exchanged cotton with anionic dyes from mixed solvents, *Text. Res. J.,* 53, 347, 1983.
293. **Small, P. A.,** Some factors affecting the solubility of polymers, *J. Appl. Chem.,* 3, 71, 1953.
294. **Hoy, K. L.,** *Tables of Solubility Parameters,* Union Carbide Corporation, Chemicals and Plastics Research and Development Dept., South Charleston, WV, 1969, 1975, and 1985.
295. **Hoy, K. L.,** New values of the solubility parameters from vapor pressure data, *J. Paint Technol.,* 42(541), 76 and 115, 1970.
296. **Chavan, R. B. and Jain, A. K.,** Solubility parameters for transfer printing of cotton pretreated with ethylene glycols, *Am. Dyestuff Rep.,* 76(12), 48, 1987; 77(2), 29, 1988.
297. **Ni, J.-B. and Jen, S.-W.,** Molecular design and structure selection for monoazo disperse dyes through their solubility parameters, *Hua-tung Hua Kung Hsueh Yuan Hsueh Pao (J. East China Inst. Chem. Technol.),* No. 1, 55, 1981.
298. **Dodd, G. S. and Ingamells, W.,** The significance of theoretical results in the dyeing of Acrilan 16 with basic dyes from an anhydrous perchlorethylene dyebath. I. The selection of carriers. II. The dyeing of Acrilan 16 with basic dyes from perchloroethylene, *J. Soc. Dyers Colour.,* 93, 74 and 79, 1977.
299. **Gur-Arieh, W. and Ingamells, W.,** The plasticization and swelling of Acrilan under dyebath conditions, *J. Soc. Dyers Colour.,* 90, 12, 1974.
300. **Gur-Arieh, Z., Ingamells, W., and Peters, R. H.,** The effect of plasticizing compounds on acrylic fibers, *J. Appl. Polym. Sci.,* 20, 31, 1976.
301. **Ingamells, W. C.,** The response of fibres to dyeing processes, in *Theory of Coloration of Textiles,* Bird, C. L. and Boston, W. S., Eds., Dyers Trust, Bradford, England, 1975, chap. 6.
302. **Ingamells, W. C.,** Controlled diffusion of cationic dyes into poly(acrylonitrile) from a nonpolar organic solvent containing plasticizing agents, in *Solvents — Neglected Parameters, 2nd Solvents Symp.,* Kakabadse, G., Ed., University of Manchester Institute of Science and Technology, Manchester, U.K., 1977, 56.
303. **Ingamells, W.,** Important concepts in the dyeing and finishing of man-made fibers, *Adv. Prep., Color. Finish, Pap. Symp.,* 1979, 1, Paper No. 6, Soc. Dyers Colour, Bradford, England; *J. Soc. Dyers Colour,* 96, 466, 1980.
304. **Ingamells, W. and Peters, R. H.,** The carrier dyeing of synthetic fibers, *Polym. Eng. Sci.,* 20, 176, 1980.
305. **Ingamells, W. and Yabani, A.,** The swelling and plasticization of poly(ethylene terephthalate) during carrier dyeing, *J. Soc. Dyers Colour.,* 93, 417, 1977.
306. **Ingamells, W. and Yabani, A. M.,** The effect of carriers on drawn poly(ethylene terephthalate) filaments, *J. Appl. Polym. Sci.,* 22, 1583, 1978.
307. **Lemons, J. H., Katar, S. K., and Cates, D. M.,** The effect of various organic compounds on poly(ethylene terephthalate) fibers. II. Plasticization and acceleration of dyeing, *Am. Dyestuff Rep.,* 55(3), 76, 1966.
308. **Lim, Y. J., Choi, H. J., Yoon, N. S., and Lee, H. K.,** Dyeability of poly(ethylene terephthalate) pretreated with organic solvent, *Han'guk Somyu Konghahoechi,* 21, 281, 1984; **Lim, Y. J., Yoon, N. S., and Chang, J. G.,** A study on the carrier dyeing for PET fiber(I), *Han' guk Somyu Konhakhoechi,* 24, 438, 1987.
309. **Hinden, E. R. and Whyzmusis, P. D.,** Polyamide resin solubility parameters, *Abstr. Pap. Am. Chem. Soc.,* 162, ORPL 65, 1971; *Am. Chem. Soc. Div. Org. Coatings Plast. Chem. Pap.,* 31, 496, 1971.
310. **Hinden, E. R. and Whyzmusis, P. D.,** Polyamide resin solubility, in *Solvents Theory and Practice (Adv. Chem. Ser.* 124), Tess, R. W., Ed., American Chemical Society, Washington, D.C., 1973, chap. 11.
311. **Gerber, H.,** Relations between the structure and properties of azo disperse dyes, *J. Soc. Dyers Colour.,* 94, 298, 1978.
312. **Ibe, E. C.,** The solubility parameter concept applied to disperse dyeing of polymers, *J. Appl. Polym. Sci.,* 14, 837, 1970.
313. **Suda, Y. and Lee, K.-T.,** Diffusion of disperse dyes within nylon-6 from normal alcohols, *Nippon Kagaku Kaishi,* 1398, 1975.

314. **Kuwabara, S.,** Effects of solubility parameter on disperse dyeing of poly(ethylene terephthalate) films, *Sen-i Gakkaishi,* 42, T-240, 1986.
315. **Siddiqui, S. A.,** Studies of solvent dyeing, Part 1: Preparation of disperse dyes and determination of the solubility parameters, *Text. Res. J.,* 51, 527, 1981.
316. **Siddiqui, S. A. and Needles, H. L.** Solubility parameters, *Text. Res. J.,* 52, 570, 1982.
317. **Siddiqui, S. A.,** Determining the solubility parameter of dyes; Dye solubility by solubility method, *Am. Dyest. Rep.,* 72(3), 30; 72(4), 36, 1983.
318. **Seu, G. and Mura, L.,** Effect of substituents on colour and dying properties of some pyridine disperse aminoazo dyes, *Annal. Chem.,* 76, 473, 1986.
319. **Koch, U., Hartmann, H. P., and Wolf, F.,** Interaction of disperse dyes with water, *Zh. Prikl. Khim.,* 57, 2551, 1984; *J. Appl. Chem. U.S.S.R.,* 57, 2356, 1984.
320. **Hou, Y.-F. and Li, Z.-S.,** Study of methine-type disperse dyes of benzothiophenone derivatives. II. Fixation-structure relations, *Huagong Xuebao (Chin. Ed.)* (3), 312, 1987.
321. **Qian, X-H., Lin, Y., and Ren, S.-W.,** Calculation and application of the solubility parameter of disperse dyes, *Huadong Huagong Xueyuan Xuebao,* 13, 573, 1987.
322. **Guion, T. H. and Harris, F. O.,** Solvent and dye selection for non-aqueous dyeing, *Colourage* (Nov. 15), 1973; Guide to solvent, dye selection for nonaqueous dying processes, *Mod. Text.,* 55, 19, 1974.
323. **Harris, F. O.,** *Experimental Study of the Application of the Solubility Parameter Concept to Solvent Dyeing,* Ph.D. Thesis, North Carolina State University, Raleigh, 1974; *Diss. Abstr. Int.* B, 36, 256, 1975.
324. **Harris, F. O. and Guion, T. H.,** A new approach to solvent dyeing with non ionic dyes, *Text. Res. J.,* 42, 626, 1972.
325. **Harris, F. O. and Guion, T. H.,** Solvent selection for non-aqueous dyeing, in *Textile Solvent Technology — Update '73,* American Association of Textile Chemists and Colorists, Research Triangle Park, NC, 1973, 163.
326. **Hashimoto, I.,** Studies of solvent dyeing. 1. Behavior of disperse dyes when dyeing synthetic fibers in organic solvents, *J. Text. Mach. Soc. Jpn.* 21, 78, 1975.
327. **Hashimoto, I.,** Studies of solvent dyeing. 10. Relation between the swelling and solution behaviors and the three dimensional solubility parameters for various polymers in organic solvents, *Sen'i Gakkaishi,* 34, T469, 1978.
328. **Alberti, G., Cerniani, A., and de Giorgi, R.,** Solvent effects on short time dyeings of acrylics, *Text. Res. J.,* 55, 635, 1985.
329. **Jones, F. and Kirby, F. A.,** Naphthoquinone dyes in liquid crystalline media, *Mol. Cryst. Liq. Cryst.,* 108, 165, 1984.
330. **Eissler, R. L. and Princen, L. H.,** Effect of some pigments on tensile and swelling properties of linseed oil films, *J. Paint Technol.,* 40(518), 105, 1968.
331. **Doroszkowski, A. and Lambourne, R.,** Effect of molecular architecture of long chain fatty acids on the dispersion properties of titanium dioxide in non-aqueous liquids, *Faraday Disc. Chem. Soc.,* 65, 252, 1978.
332. **Stephen, H. G.,** Hydrogen bonding — key to dispersion?; Parameters controlling color acceptance in latex paints, *J. Oil Col. Chem. Assoc.,* 65, 191, 1982; 69, 53, 1986.
333. **Billmeyer, F. W., Jr., Kumar, R., and Saltzman, M.,** Identification of organic colorants in art objects by solution spectrophotometry, *J. Chem. Educ.,* 58, 307, 1981.
334. **Glass, J. E., Ahmed, H., and Glass, M. E.,** Adsorption of nonionic water-soluble polymers (W-SPs) at aqueous-air interfaces, *Am. Chem. Soc. Symp. Ser.,* 240 (Polym. Adsorpt. Dispersion Stab.), 113, 1984.
335. **Jansen, J. W., de Kruif, C. G., and Vrij, A.,** Attractions in sterically stabilized silica dispersions. II. Experiments on phase separation induced by temperature variation, *J. Colloid Interface Sci.,* 114, 481, 1986.
336. **Law, K.-Y.,** An investigation of solvent-vapor-induced crystallization of soluble vanadyl phthalocyanine dyes in a polymer matrix, *J. Phys. Chem.,* 89, 2652, 1985.
337. **But'ko, T. A., Pritytskaya, T. S., Lamotkin, A., Kaverinskii, V. S., and Manerov, V. B.,** Compatability of components of varnish compositions with high oligostyrene content, *Lakokras. Mater. Ikh. Primen,* No. 2, 5, 1988.
338. **Malysheva, I. I., Belkin, A. I., and Korenev, A. D.,** Production of commercial forms of antimicrobial agents for treatment of textile materials in aqueous media, *Zh. Prikl. Khim.,* 56, 1623, 1983; *J. Appl. Chem. U.S.S.R.,* 56, 1522, 1983.
339. **Shareef, K. M. A. and Yaseen, M.,** Rheological parameters of red iron oxide-alkyd paints: their dependence on pigment content and on solubility parameters of solvents, *J. Oil Col. Chem. Assoc.,* 68, 138, 1985; **Thangavel K., Subbi Reddy, M., and Yaseen, M.,** Application of rheological parameters in the selection of solvent blends for paint formulations, *Paint Resin,* 57(5), 15 and 40, 1987.
340. **Ruiz, M. A., Contreras, M. D., Parera, A., and Cerezo, A.,** Determination of the solubility parameter of Carbopol 940 in ethanol-water, methanol-water mixtures, using apparent viscosity measurements; Determination of the solubility parameter of Carbopol 934 in water-alcohol medium by measuring the apparent viscosity, *Farm. Clin.,* 3, 257 and 344, 1986.

341. **Gibson, S. H. M., Rowe, R. C., and White, E. F. T.,** Quantitative assessment of additive-polymer interaction in pigmented hydroxypropyl methylcellulose formulation using immersion calorimetry, *Int. J. Pharm.*, 48, 113, 1988.

342. **Mittal, K. L.,** Surface contamination: an overview, in *Surface Contamination: its Genesis, Detection and Control,* Vol. 1, Mittal, K. L., Ed., Plenum Press, New York, 1979, 3.

343. **Jackson, L. C.,** *Solubility Parameter Technology and Evaporative Rate Analysis Used to Evaluate Contaminants,* Report BDX-613-859(Rev), Bendix Corporation, Kansas City, MO, 1973.

344. **Jackson, L. C.,** *Solubility Parameters and Evaporation Rate Analysis in Organic Residue Characterization,* Report BDX-613-1099(Rev.), Bendix Corporation, Kansas City, MO, 1974.

345. **Jackson, L. C.,** Solvent cleaning process efficiency — contaminant removal in the electronics industry; Removal of silicone grease and oil contaminants, *Adhes. Age,* 19(7), 31, 1976; 20(4), 29, 1977.

346. **Zorll, U.,** Aspects of the interchangeability of solvent in the cleaning of surfaces, *Oberflachentechnik,* 53(1-2), 11, 1976.

347. **Ritchie, K. and Hughes, H. G.,** Comparative studies of photoresist processing of microelectronics regarding chemical pollution, shortage, and cost, *Sol. State Technol.*, 19, 45, 1976.

348. **Mijovic, J. S. and Koutsky, J. A.,** Etching of polymeric surfaces, a review, *Polym.-Plast. Technol. Eng.*, 9, 139, 1977.

349. **Saarnak, A.,** Selection of solvents for cleaning production equipment for paint processing — a solubility parameter consideration, *Skand. Tidskr. Farg Lack,* 24, 195, 1978.

350. **Burrell, H.,** Pigment wetting parameter; Pigment dispersion parameter, *Am. Chem. Soc. Div. Org. Coat. Plast. Chem.* Pap., 35(2), 18 and 25, 1975.

351. **van Oss, C. J., Chaudhury, M. K., and Good, R. J.,** Interfacial Lifshitz-van der Waals and polar interactions in macroscopic systems, *Chem. Rev.*, 88,927, 1988; **van Oss, C. J., Good, R. J., and Chaudhury, M. K.,** Additive and nonadditive surface tension components and the interpretation of contact angles, *Langmuir,* 4, 884, 1988; **van Oss, C. J. and Good, R. J.,** On the mechanism of 'hydrophobic' interactions, *J. Dispersion Sci. Technol.*, 9, 355, 1988; **van Oss, C. J., Ju, L., Chaudhury, M. K., and Good, R. J.,** Estimation of the polar parameters of the surface tension of liquids by contact angle measurements on gels, *J. Colloid Interface Sci.*, 128, 313, 1989; **van Oss, C. J.,** Coacervation, complex-coacervation and flocculation, *J. Dispersion Sci. Technol.*, 9, 561, 1988/89; Energetics of cell-cell and cell-biopolymer interactions, *Cell Biophys.*, 14, 1, 1989.

352. **Sakamoto, Y.,** Saving of sewing labor with a low-melting-point polyamide copolymer, *Kobunshi Kako,* 20, 674, 1971.

353. **Ruiz, M. A., Contreras, M. D., Gallardo, V., and Parera, A.,** Solubility of surfactants: determination of solubility parameters of Methocel HG 90, type K-100, *Comun. Jorn. Com. Esp. Deterg.*, 19, 315, 1988.

354. **Seckin, M. L.,** *Low Temperature Solvent Dyeing of Synthetic Polymers: a Solubility Parameter Approach,* Thesis, University of California, Davis, CA, 1988: *Diss. Abstr. Int. B,* 49, 2364, 1988.

355. **Schmidt, R. G. and Bell, J. P.,** Consideration of thermodynamic factors in the analysis of steel/polymer coupling agent/epoxy adhesion systems, *J. Adhesion Sci. Technol.*, 3, 515, 1989.

356. **Doi, C. and Matsumoto, T.,** Coagulation of α-iron oxide (Fe_2O_3) particles and their soiling properties on cellulose fabrics, *Kaseigaku Zasshi,* 29, 162, 1978.

357. **Carré, A. and Vial, J.,** Determination of surface free energy of polymers by the group contribution method, *J. Chim. Phys.*, 85, 621, 1988; Surface free energy calculation for polymers by a group contribution method, in *Adhesion,* 13(26th Ann. Conf., Adhesion and Adhesives, London), Allen, K. W., Ed., Elsevier, 1989, 23.

358. **Grossman, R. F.,** Internal and external lubricants enhance processing of elastomers, *Elastomerics,* 121(4), 18, 1989.

359. **Ingamells, W. and Thomas, M. N.,** The mechanism of carrier dyeing, *Text. Chem. Color.*, 16(3), 55, 1984.

360. **Jańczuk B., Bialopiotrowicz, T., and Wójcik, W.,** The components of surface tension of liquids and their usefulness in determinations of surface free energy of solids; **Jańczuk, B. and Bialopiotrowicz, T.,** Surface free-energy components of liquids and low energy solids and contact angles, *J. Colloid Interface Sci.*, 127, 59 and 189, 1989.

361. **Konovalova, N. V., Panich, R. M., Snezhko, A. G., Rogova, N. I., and Gul', V. E.,** Reasons for selecting high-molecular emulsifiers — block copolymers of propylene oxide and ethylene oxide — for emulsifying some organic liquids in water, *Dokl. Akad. Nauk SSSR,* 204, 1385, 1972; *Proc. Acad. Sci. USSR, Phys. Chem.*, 204, 525, 1972.

362. **Sears, J. K. and Darby, J. R.,** *The Technology of Plasticizers,* Wiley-Interscience, New York, 1982.

363. **Szymanowski, J., Sobzynska, A., and Voelkel, A.,** Solubility and polarity parameters of polyoxyethylene glycol dialkyl ethers, *J. Pharm. Sci.*, 77, 893, 1988.

364. **Aubrey, D. W.,** The nature and action of tackifier resins, *Rubber Chem.*, 61, 448, 1988.

365. **Sferrazza, R. A. and Gooding, C. H.,** Prediction of sorption selectivity in pervaporation membranes, *Proc. Int. Conf. Pervaporation Processes Chem. Ind.*, 3rd, Bakish, R. A., Ed., Bakish Mater. Corp., Englewood, NJ, 1988, 54.

366. **Szymanowski, J. and Prochaska, K.,** Interfacial activity of model 2-hydroxy-5-alkylbenzophenone oximes and their intermediates, *J. Colloid Interface Sci.*, 123, 456, 1988.

367. **Barbaric, T. A., Koros, W. J., and Paul, D. R.,** Gas sorption in polymers based on bisphenol A, *J. Polym. Sci. Polym. Phys.*, 26, 729, 1988.

368. **Arai, T., Ohkita, M., and Arai, K.,** Corrosion-preventive treatment for steel pipes by UV-curable coatings (I). Effect of formation on coating properties of UV-curable resins for rust prevention, *Shikizai Kyokaishi*, 62, 208, 1989.

369. **Fujimatsu, H., Ogasawara, S., Satoh, N., Komori, K., and Kuroiwa, S.,** Adhesive effect of high-density polyethylene gels on polyethylene moldings, *Colloid Polym. Sci.*, 267, 500, 1989.

370. **Chavan, R. B. and Jain, A. K.,** The importance of film thickener, dye vapor pressure and solubility parameter in sublimation transfer printing, *J. Soc. Dyers Colour.*, 105(2), 73, 1989.

371. **Durbut, P. and Broze, G.,** Application of the solubility parameter theory to fatty soil removal, *Comun. Jorn. Com. Esp. Deterg.*, 20, 383, 1989.

372. **Giuliani, J. F.,** An investigation of the solubility of organic vapors in polymer films using an optical wave guide interfacial probe, *J. Polym. Sci. Polym. Phys.*, 26, 2197, 1988.

373. **Kalz, G.,** Solubility and interfacial parameters of chlorinated olefinic and vinyl polymers; Intra- and intermolecular energies of moist dispersed solids, *Plaste Kautsch.*, 35, 154, 1988; 36, 23, 1989.

374. **Stanton, R. M.,** Rheological aspects of thick film technology: an investigation of the flow properties of ethyl cellulose vehicle systems, *Int. J. Hybrid Microelectronics*, 6, 419, 1983.

375. **Gracia Fadrique, J.,** Prediction of the standard energy of adsorption, by means of hydrophilic-lipophilic contributions to the solubility parameter: the case of the *n*-alcohols, *Comun. Jorn. Com. Esp. Deterg.*, 20, 221, 1989.

376. **Pussemier, L., De Borger, R., Cloos, P., and Van Bladel, R.,** Relation between the molecular structure and the adsorption of arylcarbamate, phenylurea and anilide pesticides in soil and model organic adsorbents, *Chemosphere*, 18, 1871, 1989.

377. **Lyusov, Yu. N., Krokhina, L. S., Lyusova, L. R., and Kuleznev, V. N.,** Bond strength between polyurethane coatings and vulcanizates, *Kauch. Rezina*, 44(10), 37, 1985; *Int. Polym. Sci. Technol.*, 13(4), 22, 1986.

378. **Adamchuk, M. R. and Heckman, R. A.,** Propylene-based glycol ethers and acetates as substitutes for methylene chloride in solvent applications, *J. Radiat. Curing*, 15(2), 3, 1988.

379. **Toshima, N., Asanuma, H., and Hirai, H.,** The removal of nitrogen monoxide by the dry iron(II) complex immobilized on chelate resin. The preparation and characterization of the complex, *Bull. Chem. Soc. Jpn.*, 62, 893, 1989.

380. **Chin, Y.-P. and Weber, W. J., Jr.,** Estimating the effects of dispersed organic polymers on the sorption of contaminants by natural solids. 1. A predictive thermodynamic humic substance-organic solute interaction model, *Environ. Sci. Technol.*, 23, 978, 1989.

381. **Han, C. D., Kim, J., and Baek, D. M.,** Viscoelastic behavior, thermodynamic compatibility, and phase equilibria in block copolymer-based pressure-sensitive adhesives, *J. Adhesion*, 28, 201, 1989.

382. **Jönsson, S., Gölander, C. G., Biverstedt, A., Göthe, S., and Stenius, P.,** Adhesion of photocurable acrylates to solid polymer substrates, *J. Appl. Polym. Sci.*, 38, 2037, 1989.

383. **Curtis, G. P., Reinhard, M., and Roberts, P. V.,** Sorption of hydrophobic organic compounds by sediments, in *Geochemical Processes at Mineral Surfaces* (Am. Chem. Soc. Symp. Ser. 323), Davis, J. A. and Hayes, K. F., Eds., American Chemical Society, Washington, D.C., 1986, chap. 10.

Chapter 18

CHROMATOGRAPHY

The general term "chromatography" is used for separation processes relying on the relative movement of two phases, one *stationary* (solid in adsorption chromatography or liquid in partition chromatography) and the other *mobile* (liquid or gas). From what has been said so far in this book about the interaction of solids, liquids, and gases it can be seen that cohesion parameters are of particular importance in the description and correlation of chromatographic separations. Hildebrand parameters have been applied to the evaluation of chromatographic phases for at least 20 years.[1,2]

In general, cohesion parameters have proven useful in the *characterization* of chromatographic materials and in the description or correlation of retention, but they have had limited success in the *prediction* of retention.[3]

18.1 COMPONENT COHESION PARAMETER EXPRESSIONS

Among the various measures of chromatographic solvent selectivity is the modification by Snyder[4-6] for liquid chromatography of Rohrschneider's gas chromatography scheme[7-11] (Section 8.5). The resulting system has much in common with the subdivision of cohesion parameters described in Section 5.6, and a general treatment of chromatographic phenomena in terms of component cohesion parameters was presented by Snyder, Keller, Karger, and co-workers.[12-15] Subsequently, the topic has been reviewed on several occasions.[16-22]

What follows is a simplified outline based on this treatment. The reviews of Shoenmakers[20-22] are recommended for further detail; Laffort and co-workers[23] have compared the "solubility parameter" approach with "solubility factor" method.[24] An alternative approach is the use of dispersion cohesion parameters combined with other measures of polarity, basicity, and acidity.[25]

An important aim in planning any type of chromatographic separation is the estimation of solvent *strength* and solvent *selectivity*. The former term refers to the ability to dissolve what may be described in general terms as "more polar" compounds (those with higher Hildebrand parameters), and the latter is the ability to interact selectively with one compound rather than another of similar polarity (distinguished in terms of different component cohesion parameters, particularly acid-base or hydrogen-bonding parameters). Various types of polarity index (section 8.5) derived from chromatographic measurements can be related to cohesion parameters.[190,196]

The partition coefficient, K, of a solute between two phases is related to the excess partial molar thermodynamic functions by

$$-RT \ln K = \Delta^E G \tag{1}$$

and for ideal entropy of mixing and zero volume change (that is, regular solution behavior),

$$-RT \ln K = \Delta^E U = \Delta U \tag{2}$$

where ΔU is the molar internal energy change for the transfer of j from one phase to another. ΔU is related to the exchange cohesive pressure A by expressions such as Equation 4 Chapter 14 (which relates the internal energy change on mixing to ${}^{ij}A$). As seen in Equation 22, Chapter 5, for *mixing*

$$^{ij}A = {}^{i}\delta_t^2 + {}^{j}\delta_t^2 - 2\,{}^{i}\delta_d\,{}^{j}\delta_d - 2\,{}^{i}\delta_o\,{}^{j}\delta_o - 2\,{}^{i}\delta_i\,{}^{j}\delta_d - 2\,{}^{j}\delta_i\,{}^{i}\delta_d - 2\,{}^{i}\delta_a\,{}^{j}\delta_b - 2\,{}^{j}\delta_a\,{}^{i}\delta_b \qquad (3)$$

Another transfer process involved in chromatography is *vaporization*, for which the cohesive pressure expression can be shown to be

$$^{j}A^{v} = {}^{j}\delta_d^2 + {}^{j}\delta_o^2 + 2\,{}^{j}\delta_i\,{}^{j}\delta_d + 2\,{}^{j}\delta_a\,{}^{j}\delta_b \qquad (4)$$

The *adsorption* of component j on adsorbent s is accompanied by a cohesive pressure term (Equation 42, Chapter 17),

$$^{ja}A = {}^{j}\delta_d\,{}^{a}\delta_d + {}^{j}\delta_o\,{}^{a}\delta_o + {}^{j}\delta_i\,{}^{a}\delta_d + {}^{a}\delta_i\,{}^{j}\delta_d + {}^{j}\delta_a\,{}^{a}\delta_b + {}^{a}\delta_a\,{}^{j}\delta_b \qquad (5)$$

and for *dissolution* of gaseous j in liquid i, ${}^{j}\delta_t = 0$, so

$$^{ij}A = {}^{i}\delta_t^2 - 2\,{}^{i}\delta_d\,{}^{j}\delta_d - 2\,{}^{i}\delta_i\,{}^{j}\delta_d - 2\,{}^{j}\delta_i\,{}^{i}\delta_d - 2\,{}^{i}\delta_a\,{}^{j}\delta_b - 2\,{}^{j}\delta_a\,{}^{i}\delta_b \qquad (6)$$

Gas-*liquid* (partition) chromatography (GLC) involves the process

$$^{j}g \rightleftharpoons {}^{j}i \qquad (7)$$

with solute j being partitioned between the stationary liquid phase i and the gas phase g, so the energy change is simply that corresponding to dissolution of j in liquid i, Equation 6. *Gas-solid* chromatography (GSC) is associated with the adsorption of solute j onto adsorbent a, and the energy change is given by Equation 5, the adsorption expression. *Liquid-liquid* chromatography (LLC), or liquid partition chromatography, involves partition of j molecules between a mobile phase m and a stationary phase s,

$$^{j}m \rightleftharpoons {}^{j}s \qquad (8)$$

$$^{j}K = {}^{j}x_s/{}^{j}x_m \qquad (9)$$

This ratio is determined by the interaction or mixing energy of the sample molecule j with the stationary phase, minus the interaction energy of j with the mobile phase, i.e., it is a function of $({}^{js}A - {}^{jm}A)$. Inspection of Equations 3 and 6 shows that this can also be expressed as a function of the differences in the dissolution expressions (Equation 6). This has the form

$$[{}^{s}\delta^2 - {}^{m}\delta^2 - 2\,{}^{j}\delta_d\,({}^{s}\delta_d - {}^{m}\delta_d)] - [2\,{}^{j}\delta_i\,({}^{s}\delta_d - {}^{m}\delta_d) + 2\,{}^{j}\delta_d\,({}^{s}\delta_i - {}^{m}\delta_i)] - [2\,{}^{j}\delta_o\,({}^{s}\delta_o - {}^{m}\delta_o)] - [2\,{}^{j}\delta_a\,({}^{s}\delta_b - {}^{m}\delta_b) + 2\,{}^{j}\delta_b\,({}^{s}\delta_a - {}^{m}\delta_a)] \qquad (10)$$

The square brackets separate the dispersion, induction, orientation, and Lewis acid-base terms, respectively.

Liquid-solid chromatography (LSC), or adsorption chromatography, results in eluotropic series of liquids which are much less sensitive to sample type than in liquid-liquid chromatography. The liquid-solid solvent strength parameter, ϵ^o, varies in absolute value between adsorbents (e.g., alumina or silica: Table 9) but the *order* of solvent strengths does not

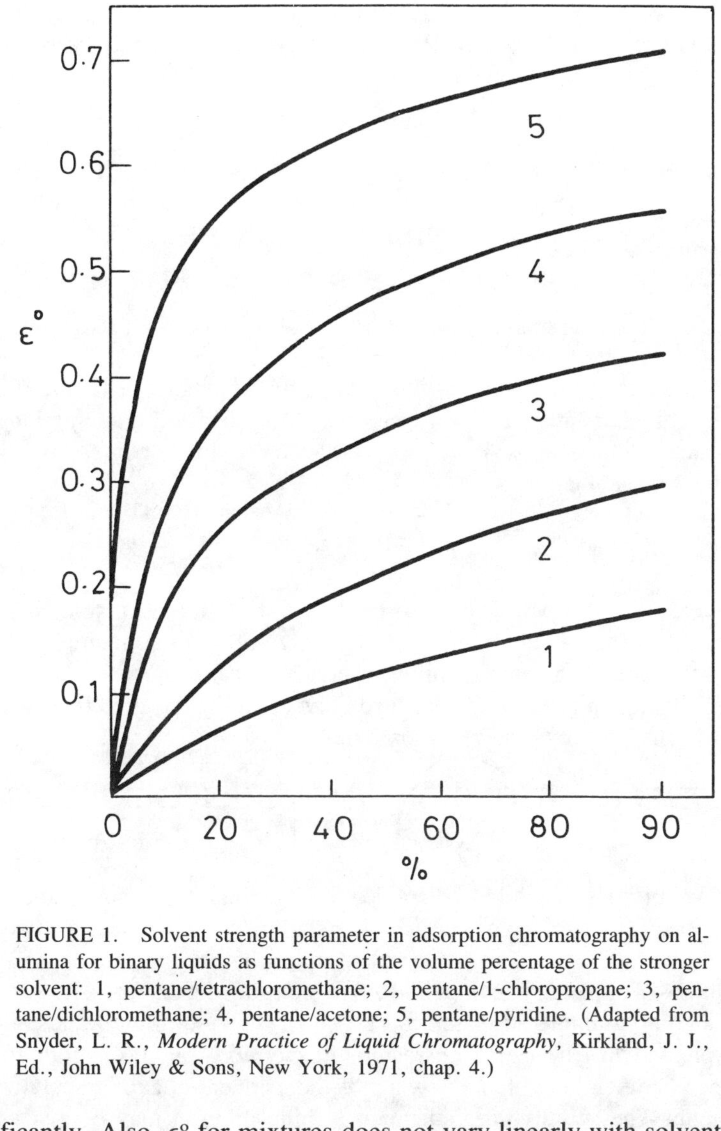

FIGURE 1. Solvent strength parameter in adsorption chromatography on alumina for binary liquids as functions of the volume percentage of the stronger solvent: 1, pentane/tetrachloromethane; 2, pentane/1-chloropropane; 3, pentane/dichloromethane; 4, pentane/acetone; 5, pentane/pyridine. (Adapted from Snyder, L. R., *Modern Practice of Liquid Chromatography,* Kirkland, J. J., Ed., John Wiley & Sons, New York, 1971, chap. 4.)

change significantly. Also, ϵ^o for mixtures does not vary linearly with solvent composition (although cohesion parameters usually do) but the stronger (higher ϵ^o) solvent concentrates into the adsorbed phase. As the concentration of the stronger solvent component increases from zero, ϵ^o rises rather steeply and then levels off (Figure 1).

In liquid-solid chromatography there is a competitive process between solvent i and sample solute j molecules for adsorption on the surface. Adsorption of j (initially in the solvent phase) requires the displacement of solvent molecules i (initially adsorbed) from the surface. The LSC eluotropic series is less sensitive to the nature of the solute j than in the case of LLC, because the interactions involving the adsorbent, a, are usually much more important than the solute/solvent (i-j) interactions. It can be shown that the interaction energy is approximately represented by the function

$$({}^{ja}A - n\ {}^{ia}A) \tag{11}$$

where n is the ratio of the adsorption areas of j and i in the adsorbed state. ${}^{ja}A$ is independent of solute i, and ${}^{ia}A$ is independent of j, so the strength of solvent i depends only on ${}^{ia}A$. The

TABLE 1
Component Cohesion Parameters of Silica Bonded-Phase Chromatographic Materials

	δ/MPa$^{1/2}$				
Column material	δ_d	δ_o	δ_b	δ_a	δ_t
Cyanopropyl	3.1	3.7	0.4	0.6	4.9
Aminopropyl	11.9	7.6	8.8	1.8	15.1
1,2-Dihydroxypropyl propyl ether	6.3	5.7	4.1	1.0	9.0

Adapted from Smith, P. L. and Cooper, W. T., *Chromatographia,* 25, 1988.

empirical *solvent strength parameter,* ϵ^o, is based on a scale such that $\epsilon^o = 0$ for retention of the reference solvent pentane on alumina, where only dispersion interactions are present. Neglecting induction interactions, it follows that the approximate expression relating ϵ^o, and the partial cohesion parameters is

$$\epsilon^o = C[({}^i\delta_d - {}^o\delta_d)\ {}^a\delta_d + {}^i\delta_o\ {}^a\delta_o + {}^i\delta_a\ {}^a\delta_b + {}^a\delta_a\ {}^i\delta_b] \tag{12}$$

where C is a constant for a particular adsorbent and ${}^o\delta_d$ is the dispersion parameter for the reference solvent. Consequently, ϵ^o depends strongly only on the natures of i and a.

Smith and Cooper[26] evaluated component cohesion parameters for HPLC column materials (Table 1) using empirically determined solvent strengths, a method which appeared to be superior to the use of solute retention data.

18.2 GLC INTERACTIONS OF POLYMERS AND PLASTICIZERS WITH LIQUIDS

Polymers and plasticizers and their interactions with liquids may be studied conveniently by using polymer, polymer blend or plasticizer as stationary phase and the volatile compound as mobile phase on gas-liquid chromatographic columns.*

The retention of a volatile compound on a column depends on its saturation vapor pressure at the column temperature and the extent of interaction with the stationary phase. It may be expressed in terms of a characteristic thermodynamic quantity for the system called the ''specific retention volume'', which is independent of experimental variables such as column length. The activity coefficients at infinite dilution may be evaluated from the specific retention volume, the saturation vapor pressure and the molecular weight of the stationary phase. A second virial coefficient correction term should be included, but this is sometimes omitted and ''apparent'' activity coefficients based on pressure rather than fugacity may be evaluated. The activity coefficients correspond to the situation of polymers or plasticizers acting as solvents and low molecular weight solutes at low concentration. Although concentration dependent, these activity coefficients are constant at infinite dilution, with values less than unity for interactions favorable for mixing and greater than one for unfavorable interactions. (If a polymer-liquid system shows very limited miscibility, separating into two phases, each of the phases contains one of the components in a state fairly close to its standard state, essentially at ''infinite dilution''.) Typical activity coefficient values are listed in Tables 2 and 3. From such activity coefficients, other thermodynamic

* For example: Alessi et al.;[27-32] Barcelo et al.,[33-35] Dincer and Bonner;[36-37] DiPaolo-Baranyi, Guillet, et al.,[38,39] El-Hibri et al.;[195] Fernandez-Sanchez et al.;[189] Figgins et al.;[40,41] Funk;[42] Kerber et al.;[193] Lafosse et al.;[43-45] Leca et al.;[46,47] Letcher;[48] Leung;[49] Newman and Prausnitz;[50] Risby et al.;[51] Schreiber and Lloyd;[197] Senich and Sanchez;[52,53] Szymanowski et al.;[190] Takamiya et al.;[54] Varsano and Gilbert,[55] and Vilcu and Leca.[188]

TABLE 2
Apparent Activity Coefficients Determined by Gas-Liquid Chromatography for Organic Liquids in Polymers at 25°C

Polyethylene (Marlex EMN TR-885, Phillips Petroleum) as solvent		Poly(vinyl chloride) (Allied Chemical Co.) as solvent	
Low M. W. solute	**Activity coefficient**	**Low M. W. solute**	**Activity coefficient**
1-Butanol	3.800	Propylparaben	1.335
Methylparaben	1.070	Octane	1.000
2-Nitropropane	0.813	*tert*-Butanol	0.902
Propylparaben	0.776	Methylparaben	0.891
tert-Butanol	0.432	Hexane	0.393
Chlorobutanol	0.367	1-Butanol	0.237
Butyl acetate	0.097	Chlorobutanol	0.182
Benzyl alcohol	0.089	Methyl isobutyl ketone	0.062
Methyl isobutyl ketone	0.040	Phenol	0.048
Phenol	0.038	2-Nitropropane	0.042
Toluene	0.017	Benzyl alcohol	0.039
Hexane	0.009	Butyl acetate	0.030
Octane	0.009	Toluene	0.026

Adapted from Varsano, J. L. and Gilbert, S. G., *J. Pharm. Sci.*, 62, 87, 1973.

functions may be determined, and polymer-liquid behavior patterns may be established. It should be noted that retention characteristics and, therefore, the derived quantities also, do not become independent of molecular weight until fairly high values are reached, probably 20,000 in the case of poly(ethylene glycol).[58]

This type of investigation has been described as "inverse chromatography": using mobile phases to investigate the properties of stationary phases, rather than the reverse. From these activity coefficient values and the relationships described in Section 4.2 and Chapter 13, cohesion parameters may be estimated for polymers and plasticizers. These may be in the form of a set of component cohesion parameters (δ_d, δ_o, δ_i, δ_a, δ_b), a set of Hansen parameters (δ_d, δ_p, δ_h), a pair of polar-nonpolar parameters (δ_λ δ_τ), or a single Hildebrand parameter. Examples appear in Tables 4 to 6.

There have been numerous studies of polymers by GLC using cohesion parameter concepts including:

- Classification and selection of stationary phases[23,57,189]
- Properties of Chromosorb® supports[33-35]
- Organic liquid interactions with copolymers of vinyl acetate and vinyl chloride,[58] vinyl acetate and vinyl alcohol,[59] and vinyl acetate and ethylene[36]
- Organic liquid interactions (including Hildebrand parameter determination) with polyethylene and other polyolefins[60-62,185] polystyrene,[46,63] polysulfone,[64] poly(ethylene glycol),[65,66] poly(vinyl acetate),[59,67] poly(ethylene terephthalate)[68] poly(ethyl methacrylate),[69,70] and poly(butyl methacrylate)[62]
- Interaction of low molecular weight molecules with polymer packaging[55]
- Polymer-polymer compatibility[19,71,203]
- Gas chromatographic study of the interactions between binder and solvent in paint films[72]
- Determination of molecular weights and Hildebrand parameters of high polymers [silicone oil, poly(ethylene glycol)][73]
- Effect of preadsorbed water on chromatographic behavior of cellulosic materials[74]

TABLE 3
Weight Fraction Activity Coefficients at Infinite Dilution and 150°C for Organic Solutes in Polymer, Determined by Gas-Liquid Chromatography[50]

Description of Polymers

	Polymer	Source
1.	Acryloid® A-11 high M.W. poly(methyl acrylate), T_g = 105°C	Rohm & Haas
2.	Beckosol®-23 alkyd resin (phthalic anhydride, glycerin, lauric acid), M.W. 2000-3000, T_g below room temperature	Reichhold Chemicals
3.	Lucite® 2044 high M.W. poly(butyl methacrylate), T_g = 20°C	du Pont
4.	Eponol® 55 thermoplastic epoxy resin, M.W. > 2000	Shell Chemical
5.	Epotuf® hardener 37-612, liquid polyamide for curing of epoxy resins, T_g below room temperature	Reichhold Chemicals
6.	Poly(vinyl acetate), $\overline{M}_w$ = 331,400, $\overline{M}_n$ = 83,350	Cellomer Corp.
7.	Polyurethane, T_g below room temperature	Reichhold Chemicals
8.	Polystyrene, $\overline{M}_n$ = 97,000	Pressure Chemical Co.

Activity Coefficients

Solute	Weight fraction infinite dilution activity coefficient in polymer							
	1	2	3	4	5	6	7	8
Hexane	—	22.0	10.4	—	—	—	—	—
Cyclohexane	—	22.0	10.4	—	—	—	—	12.2
Benzene	8.5	6.6	4.6	7.8	6.1	5.9	7.4	5.4
Toluene, methylbenzene	9.0	6.7	4.5	8.1	6.3	6.4	7.6	5.2
p-Xylene	10.2	—	4.7	—	—	—	—	—
Trichloromethane	2.9	3.0	1.9	4.0	2.1	2.2	—	—
1,2-Dichloroethane	4.2	4.1	3.3	4.1	3.0	—	—	4.6
2-Propanol	—	9.0	9.7	13.5	5.9	7.4	—	—
1-Butanol	—	7.5	7.7	9.9	4.4	6.6	9.6	—
Methyl ethyl ketone	—	6.5	7.4	8.6	8.3	6.3	—	9.4
Methyl isobutyl ketone	—	7.8	7.5	10.3	9.2	8.4	11.5	—
Cyclohexanone	—	3.7	5.1	4.0	—	4.6	4.8	5.8
Methyl Cellosolve®	—	6.8	10.7	8.4	4.8	6.2	8.2	—
Cellosolve	—	6.5	8.7	7.9	4.7	6.0	—	—
Tetrahydrofuran	—	—	—	—	—	—	5.5	—
1,4-Dioxane	—	—	4.6	—	4.7	—	4.7	5.2
Vinyl acetate	—	—	—	—	—	5.7	—	—
Ethyl acetate	—	6.7	6.7	8.0	8.4	5.9	—	—
Butyl acetate	—	6.7	5.9	8.2	8.5	6.9	9.0	—
N,N-Dimethylformamide	—	3.1	—	3.9	—	—	4.2	10.6
Acetonitrile	—	11.7	17.3	14.8	13.7	—	—	19.8

- Infinite dilution activity coefficients[186], including studies of final stages of solvent loss from drying polymer coatings.[204]

18.3 GLC STUDIES OF NONPOLYMER SYSTEMS

Gas-liquid chromatography has been used widely for the determination of partition coefficients and the characterization of solvents[7-11,54,75-78] and of low volatility monomeric compounds.[63,79]

In particular, the study of activity coefficients (discussed in Chapter 7) by gas chromatography is becoming increasingly important. The significance of limiting infinite dilution activity coefficients, accessible by gas chromatography, is emphasized in Section 7.8.[80]

Applications of cohesion parameters in these GLC studies include:

TABLE 4
Hansen Parameters for Resins and Plasticizers at 125°C by Gas-Liquid Chromatography[27,32]

	$\delta/MPa^{1/2}$			
Resin	δ_d	δ_p	δ_t	δ_t
Araldit® 488 (CIBA)	13.1	5.5	8.4	16.5
D. E. R.® 684 (Dow)	10.4	5.0	8.1	14.1
Bakelite® PKHH (Union Carbide)	11.3	4.1	8.0	14.4
Plasticizer				
Butyl benzyl phthalate	16.2	4.3	7.9	18.6
Diethyl phthalate	16.4	4.4	8.4	18.9
Diisodecyl phthalate	16.5	4.2	7.3	18.5
Tricresyl phosphate	17.0	4.6	8.0	19.4
Trioctyl phosphate	15.6	3.4	8.7	18.2
Triphenyl phosphate	17.1	4.8	8.1	19.5
Trixilenyl phosphate	16.6	4.2	8.0	18.9

TABLE 5
Polar and Nonpolar Cohesion Parameters for Plasticizers at 25°C by Gas-Liquid Chromatography[30,31]

		$\delta/MPa^{1/2}$		
Plasticizer	$V/cm^3\ mol^{-1}$	δ_λ	δ_τ	δ_t
Diethyl phthalate	200	17.8	11.0	21.4
Dibutyl phthalate	274	16.3	8.4	18.3
Diisobutyl phthalate	268	17.3	10.1	20.0
Dicyclohexyl phthalate	287	17.5	9.5	19.9
Di-2-ethylhexyl phthalate	396	16.8	8.4	18.8
Diisodecyl phthalate	461	16.1	8.6	18.3
Butyl benzyl phthalate	282	18.5	10.4	21.2
Butyl ethylhexyl phthalate	335	15.6	7.0	17.1

TABLE 6
Estimated Hildebrand Parameters for Polymers by Gas-Liquid Chromatography[38,39,60]

Polymer	t/°C	$\delta/MPa^{1/2}$
Anionic polystyrene, $\overline{M}_w$ = 120,000, Polymer Corp.	193 25	15.6 20.1
Poly(methyl acrylate), $\overline{M}_w$ = 200,000, $\overline{M}_n$ = 63,200, Aldrich	100 25	17.8 20.7
Poly(vinyl acetate) B1000, $\overline{M}_n$ = 86,000, Wacker Chemicals	135 25	17.4 20.9
Ethylene (40%)-propylene (60%) copolymer, Exxon Vistalon® 404	73 30	14.8 15.8
cis-Polyisoprene, $\overline{M}_n$ = 126,000, Polysar	30	16.3
Amorphous polypropylene, Eastman Epolene® D10	30	15.7

- Determination of activity coefficients of dilute solutions of hydrocarbons in glycols,[29,81] phthalates[30,31] octadecane,[82] eicosane and squalane,[83] tetracosane, triacontane and hexatriacontane,[84] esters and propionitriles[85]
- Activity coefficients in binary nonelectrolyte solution by gas chromatographic analysis of the vapor phase[86-88]
- Thermodynamics of liquids in propylene glycol[89]
- Hildebrand parameters of lipophilic liquids and other liquids[19,90]
- Thermodynamics of benzene-diphenyl and terphenyl systems[91]
- Use of functional group partial molal thermodynamic quantities in the characterization of stationary phases[40,41,51,92]
- Relative Lewis acidities of polar organic solvents from chromatographic measurements[93]
- A four-parameter modification of the Rohrschneider characterization of gas chromatographic liquid phases[75,94]
- Hildebrand parameters of surfactants (Table 1, Chapter 12)[95]
- Biological studies such as olfactory properties (Chapter 19)[24,96,97]
- Gas-liquid interactions between solutes and free nitrile groups in 1,3,5-tricyanopentane, 1,2,3-tricyanoethoxypropane, stearonitrile, and squalane liquid stationary phases[43-45]
- Homomorphs and Hansen parameters of polar substances[46,47]
- Hansen parameters of alkyl derivatives of phenol and resorcinol (Table 19, Chapter 5)[98]
- Relation between GLC retention volume and boiling point[99] and its application to halogens[100] and Group II, IV, and V halides[101,102]
- Partition coefficients and activity coefficients of pinenes in polypropylene sebacate and of hydrocarbons in silicone oil[103]
- Solution thermodynamic studies of metal β-diketonates[104]
- Classification of liquid stationary phases including silicones and squalane[105]
- Isolation of volatile trace organics by air:water partitioning and gas chromatograph head space analysis[86,106]
- Charge-transfer interactions between tetrabromomethane and aromatic donors[107]
- Reversed flow gas chromatography (RFGC) to determine liquid Hildebrand parameters[202]

18.4 LIQUID CHROMATOGRAPHY

Examination of Equation 10 shows the reason for the observation that any single eluotropic series for liquid-liquid chromatography is only approximately constant for different samples. Although any one of the parameters δ_d, δ_o and δ_b may be a suitable measure of solvent strength for a particular class of samples, in general it is necessary to consider the variation of all the parameters in predicting an eluotropic series. If the sequence of solvent strengths experienced by all samples is to be the same, each parameter must change regularly for each liquid. This is almost impossible to achieve with pure liquids, but blends of liquids do show this behavior (Table 7). Series of liquid pairs are listed in Table 8. Occasionally, component cohesion parameters, Hansen parameters, or polar-nonpolar parameters have been used, but most studies have used only Hildebrand parameters.

In terms of Hildebrand parameters, Equation 88, Chapter 7, leads to the following expression for the LLC distribution constant, ${}^{j}K$:[16,21,108]

$$RT \ln {}^{j}K = {}^{j}V\,[({}^{j}\delta - {}^{m}\delta)^2 - ({}^{j}\delta - {}^{s}\delta)^2] \qquad (13a)$$

or

$$RT \ln {}^{j}K = {}^{j}V\,[({}^{m}\delta + {}^{s}\delta - 2\,{}^{j}\delta)({}^{m}\delta - {}^{s}\delta)] \qquad (13b)$$

TABLE 7
Effective Component Cohesion Parameters of Blends of Acetone and Pentane

	$\delta/MPa^{1/2}$			
	$\bar{\delta}_d$	$\bar{\delta}_o$	$\bar{\delta}_a$	$\bar{\delta}_b$
Pentane	14.5	0.0	0.0	0.0
20/80 (v/v) pentane/acetone	14.3	2.0	1.0	0.0
40/60 (v/v) pentane/acetone	14.3	4.1	2.0	0.0
60/40 (v/v) pentane/acetone	14.3	6.1	3.1	0.0
80/20 (v/v) pentane/acetone	14.1	8.2	4.1	0.0
Acetone, 2-propanone	13.9	10.2	5.1	0.0

Adapted from Snyder, L. R., *Modern Practice of Liquid Chromatography,* Kirkland, J. J., Ed., John Wiley & Sons, New York, 1971, chap. 4.

TABLE 8
Series of Liquid Pairs Covering a Range of Solvent Strengths, with Effective Dispersion Cohesion Parameters

Components	Range of $\bar{\delta}_d/MPa^{1/2}$
Series A	
Pentane/methyl acetate	14.5—20.1
Methyl acetate/methanol	20.1—26.4
Methanol/water	26.4—43.0
Series B	
Pentane/1,4-dioxane	14.5—20.1
1,4-Dioxane/dimethylsulfoxide	20.1—26.2
Dimethylsulfoxide/water	26.2—43.0
Series C	
50/50 (v/v) pentane/1,2-dichloroethane	14.5—19.6
50/50 (v/v)1,2-dichloroethane/nitromethane	19.6—22.5
Nitromethane/1,3-dicyanopropane	22.5—26.6

Adapted from Snyder, L. R., *Modern Practice of Liquid Chromatography,* Kirkland, J. J., Ed., John Wiley & Sons, New York, 1971, chap. 4.

Thus, the value of ${}^{j}K$ varies exponentially with ${}^{j}\delta$, and in order for ${}^{j}K$ to be of the order of unity (neither too high nor too low for practical purposes) the quantity in the square brackets must be of the order of zero. This can be achieved either by making ${}^{m}\delta \approx {}^{s}\delta$ (which makes the system nonselective) or by making ${}^{j}\delta \approx ({}^{m}\delta + {}^{s}\delta)/2$, a very convenient first approximation in the choice of a liquid chromatography *m-s* system for solute *j*.

For ${}^{j}\delta = ({}^{m}\delta + {}^{s}\delta)/2$, the Hildebrand parameter of *j* is midway in value between that of stationary phase *s* and that of mobile phase *m*, and as *j* and *s* become more similar, ${}^{j}K$ becomes larger. The nature of Equation 13 is illustrated in Figure 2. (Although *V* has been held constant in this figure to illustrate the relationship between ${}^{j}K$ and the Hildebrand parameters, it is a significant variable and cannot be neglected.) Usually ${}^{s}\delta > {}^{m}\delta$, but for the separation of meterials of low ${}^{j}\delta$ the phases are reversed (''reversed phase liquid chromatography''), so that ${}^{s}\delta > {}^{m}\delta$, and all ${}^{j}K$ values in Figure 2 are replaced by their reciprocals. If account is taken of molar volume dissimilarities, another term is required in Equation 13, for example $\ln({}^{s}V/{}^{m}V)$.[109] Other forms of the equation are given in Section 7.11. These simple equations in terms of Hildebrand parameters are sometimes useful,[110] particularly if combined with qualitative consideration of the polarity and acid-base terms. The alternative approach is to use the component cohesion parameter values of Table 1, Chapter 5, for solvents,

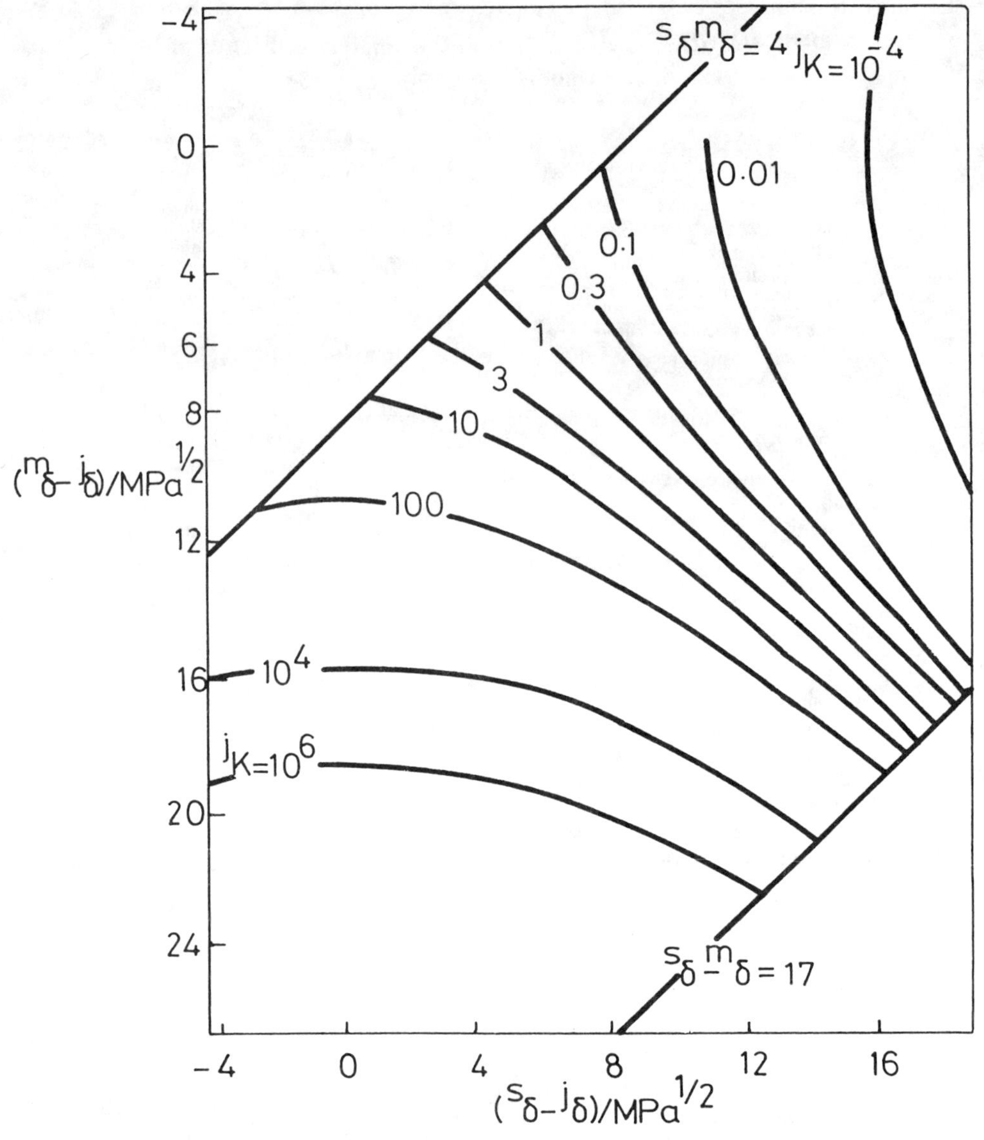

FIGURE 2. Solute (j) retention as a function of the composition of mobile (m) and stationary (s) phases in normal liquid-liquid chromatography. ${}^jK$ is plotted as a function of (${}^m\delta - {}^j\delta$) and (${}^s\delta - {}^j\delta$) for ${}^jV$ = 100 cm^3 mol^{-1} and t = 25°C. ${}^jK$ values are indicated on the curves. (Adapted from Karger, B. L., Snyder, L. R., and Horvath, C., *An Introduction to Separation Science,* John Wiley & Sons, New York, 1973.)

together with estimates of these quantities for the solute, inserting them in the relations based on Equation 10. Procedures of intermediate precision are the use of Hansen parameters δ_d, δ_p, δ_h,[27,28,32,184] or polar-nonpolar parameters δ_τ, δ_λ.[29,31,32,81]

Liquid-liquid chromatography from some points of view is an extension of gas-liquid chromatography, the techniques differing in the density of the components and the extent of interactions. As indicated by Equation 13, only the *ratio* of the activity coefficients at infinite dilution is available from liquid chromatography, but if the activity coefficient of the solute in one phase can be determined by gas-liquid chromatography, the other activity coefficient can be evaluated. LLC and GLC are therefore complementary in this respect.

Liquid chromatography is becoming increasingly important for studying larger molecules. Some examples where cohesion parameters have been involved are:

- Solvent selection[184] and the role of gas solubility[111] in high-performance liquid chromatography
- High pressure liquid chromatography of free and conjugated 17-ketosteroids[110] and in the determination of therapeutic agents[16]
- Study of heavy hydrocarbons, including asphaltenes and tar sand and bitumen[42]
- Theory on selectivity for pairs of similar solutes in liquid-liquid chromatography[112]
- HPLC separation of metal (II) chelates of thiothenoyltrifluoroacetone on polystyrene beads[113]
- Prediction of selectivity and retention in liquid chromatography, including gradient selection[3,20-22,26,114-127]
- Characterization of liquid chromatography stationary phases by inverse size exclusion chromatography[128]
- Swollen polysiloxanes as stationary phases[201]
- Separation of chloramphenicol intermediates by HPLC[129]
- Separation of complex hydroxyl aromatic mixtures by HPLC[124]
- Reversed phase liquid chromatography[130,131,191,194,201]
- Liquid organic salts as mobile phases in reversed phase liquid chromatography[132]
- Retention of phenols in reversed phase liquid chromatography[183]
- Elastic inner wall coating of capillary liquid chromatography columns[200]

18.5 SUPERCRITICAL FLUID CHROMATOGRAPHY

The application of high pressure gases and supercritical fluids (Section 11.3) to chromatography is becoming increasingly important.[133-145,192,198]

Most of the reports cited here make some reference to the use of cohesion parameters, but it is not yet clear if they will be used extensively in this application.

18.6 LIQUID-SOLID CHROMATOGRAPHY AND GEL CHROMATOGRAPHY

The Hildebrand parameter is a measure of the ability of a liquid molecule to interact with surrounding molecules of any type, including adsorbent surfaces, so it is not surprising that there is an approximate correlation between the liquid-solid solvent strength parameter ϵ^o and the Hildebrand parameter,[15,135,146-149], such as,

$$\text{for alumina:} \quad \epsilon^o = -0.786 + 0.0622\,(\delta/\text{MPa}^{1/2}) \tag{14}$$

$$\text{for silica:} \quad \epsilon^o = -0.519 + 0.0430\,(\delta/\text{MPa}^{1/2}) \tag{15}$$

However, there are significant discrepancies, and at high ϵ^o values variability of ϵ^o with sample type begins to show up, so the component cohesion parameters δ_o, δ_a, and δ_b become more important (Equation 12). Table 9 gives the eluotropic series for alumina and silica, together with the 1971 estimates of component cohesion parameters.[146] (A more complete and more recent compilation of component cohesion parameters appears in Table 1, Chapter 5.)

Closely related to LSC are gas-solid chromatography, and thin layer chromatography and gel chromatography.

Cohesion parameters have been used for:

TABLE 9
Solvent Strength Parameters for LSC on Alumina, Silica, and Component Cohesion Parameters for Chromatographic Solvents in Order of Increasing Hildebrand Parameter[15,146]

			δ/MPa$^{1/2}$				
Solvent	ε° (alumina)	ε° (silica)	δ_t[a]	δ_d	δ_o[b]	δ_a[b]	δ_b[b]
Perfluoroalkanes[c]	−0.25	—	12.3	12.3	0.0	2.0	0.0
Isooctane, 2,2,4-trimethylpentane	0.01	—	14.3	14.3	0.0	0.0	0.0
Diisopropyl ether	0.28	0.21	14.3	14.1	1.0	1.0	0.0
Pentane (reference)	0.00	0.00	14.5	14.5	0.0	0.0	0.0
Hexane	0.01	—	14.9	14.9	0.0	0.0	0.0
Heptane	0.01	0.01	15.1	15.1	0.0	0.0	0.0
Diethyl ether	0.38	0.38	15.1	13.7	4.1	4.1	0.0
Cyclopentane	0.05	—	16.6	16.6	0.0	0.0	0.0
Cyclohexane	0.04	0.03	16.8	16.8	0.0	0.0	0.0
Chloropropane	0.30	0.22	17.0	14.9	6.1	0.0	0.0
Tetrachloromethane	0.18	0.11	17.6	17.6	0.0	1.0	0.0
Diethyl sulfide	0.38	—	17.6	16.8	4.1	1.0	0.0
Ethyl acetate	0.58	0.38	17.6	14.3	6.1	4.1	0.0
Ethyl bromide, bromoethane	0.37	—	18.0	16.0	6.1	0.0	0.0
m-Xylene, 1,3-dimethylbenzene	0.26	—	18.0	18.0	0.0	1.0	1.0
Toluene, methylbenzene	0.29	0.23	18.2	18.2	0.0	1.0	0.0
Chloroform, trichloromethane	0.40	0.26	18.6	16.6	6.1	1.0	0.0[d]
Tetrahydrofuran	0.45	—	18.6	15.6	8.2	6.1	0.0
Methyl acetate	0.60	—	18.8	13.9	9.2	4.1	0.0
Benzene	0.32	0.25	18.8	18.8	0.0	1.0	0.0
Acetone, 2-propanone	0.56	0.47	19.2	13.9	10.2	5.1	0.0
1,2-Dichloroethylene	0.42	—	19.6	13.1	11.3	1.0	0.0
Chlorobenzene	0.30	0.23	19.6	18.8	4.1	1.0	0.0
1,2-Dichloroethane	0.49	—	19.8	16.8	8.2	0.0	0.0
1,4-Dioxane	0.56	0.49	20.1	16.0	8.2	6.1	0.0
Carbon disulfide	0.15	0.20	20.5	20.5	0.0	1.0	0.0
1-Propanol	0.82	—	20.9	14.7	5.1	8.2	8.2
Pyridine	0.71	—	21.3	18.4	8.2	10.2	0.0
Nitromethane	0.64	—	22.5	14.9	16.4	2.1	0.0
Ethanol	0.88	—	22.9	13.9	8.2	10.2	10.2
Acetonitrile	0.65	0.05	24.1	13.3	16.4	5.1	0.0
Acetic acid, ethanoic acid	1.0+	—	25.4	14.3	—	—	—
Dimethylsulfoxide	0.6	—	26.2	17.2	15.3	10.2	0.0
Methanol	0.95	0.73	26.4	12.7	10.2	15.3	15.3

[a] Calculated from the boiling point.
[b] Approximate values.
[c] Average values of different compounds.
[d] Other work suggests a larger value.

- Polymer-liquid interaction parameter χ for the high T_g (405°C) polyquinoxaline with 40 volatile liquids by GSC[37]
- Enantiomeric separation of tertiary phosphine oxides[150]
- Component cohesion parameters by GSC[199] of cellulose,[151] caffeine, theophylline, and methyl *p*-hydroxybenzoate[152] and lactose[153]
- Solvent strength in adsorption thin-layer
- chromatography[182]
- Polymer separation and characterization by TLC[154-156]
- Metal complexes of porphine and porphyrin derivatives[157-160]

- Chlorophylls and their derivatives[127]
- Characterization of reverse osmosis membrane materials in terms of δ_d and δ_h (Table 20, Chapter 14)[161,162]
- Gel chromatography with mixed eluents[163]
- Gel permeation chromatography[109,164-167]
- Elution behavior of solutes in gel chromatography with polystyrene,[168,169] porous polystyrene-divinylbenzene copolymer,[170,171] cross-linked poly(vinyl acetate) gels,[172,173] polypropylene,[174] and poly(dimethylsiloxane) and poly(methyl methacrylate)[175]
- Partition of acetylacetone (2,4-pentanedione) and its metal chelates on gel[176-178]
- Chromatography of heavy petroleum fractions[179]
- Size exclusion chromatographic analysis of refuse-derived fuel for mycotoxins,[180] polymer-substrate interactions in silica size-exclusion chromatography[181] and correlation of exclusion limit of polyacrylamide terpolymer[187]

REFERENCES

1. **Waksmundzki, A., Suprynowics, Z., and Niedzielska, K.,** Tentative representation of gas chromatographic spectra in terms of the solubility parameter of stationary phases, *Chem. Anal.,* 14, 611, 1969.
2. **Mironenko, V. F. and Garber, Yu. N.,** Evaluation of chromatographic phases from solubility parameters, *Zh. Prikl. Khim.,* 46, 1605, 1973; *J. Appl. Chem.,* 46, 1710, 1973.
3. **Schoenmakers, P. J., Billiet, H. A. H., and de Galan, L.,** Solubility parameter as a tool in understanding liquid chromatography, *Chromatographia,* 15, 205, 1982 (English); 15, 387, 1982 (German).
4. **Snyder, L. R.,** Classification of the solvent properties of common liquids, *J. Chromatogr.,* 92, 223, 1974.
5. **Snyder, L. R.,** Solvent selection for separation processes, in *Separation and Purification,* 3rd ed., Perry, E. S. and Weissberger, A., Eds., John Wiley & Sons-Interscience, New York, 1978, 25.
6. **Snyder, L. R.,** Classification of the solvent properties of common liquids, *J. Chromatogr. Sci.,* 16, 223, 1978.
7. **Rohrschneider, L.,** A method of characterization for gas chromatographic separation of liquids, *J. Chromatogr.,* 22, 6, 1966.
8. **Rohrschneider, L.,** The polarity of stationary liquid phase in gas chromatography, *Adv. Chromatogr.,* 4, 333, 1967.
9. **Rohrschneider, L.,** Gas chromatographic separation in regular solutions, *J. Gas Chromatogr.,* 6, 5, 1968.
10. **Rohrschneider, L.,** Solvent characterization by gas-liquid partition coefficients of selected solutes, *Analyt. Chem.,* 45, 1241, 1973.
11. **Vernon, F.,** The characterization of solute-solvent interactions in gas-liquid chromatography, *Dev. Chromatogr.,* 1, 1, 1978.
12. **Karger, B. L., Snyder, L. R., and Eon, C.,** An expanded solubility parameter treatment for classification and use of chromatographic solvents and adsorbents. Parameters for dispersion, dipole and hydrogen bonding interactions, *J. Chromatogr.,* 125, 71, 1976.
13. **Karger, B. L., Snyder, L. R., and Eon, C.,** Expanded solubility parameter treatment for classification and use of chromatographic solvents and adsorbents, *Analyt. Chem.,* 50, 2126, 1978.
14. **Keller, R. A., Karger, B. L., and Snyder, L. R.,** Use of the solubility parameter in predicting chromatographic retention and eluotropic strength, *Gas Chromatography 1970* (Proc. Int. Symp., *8*, Dublin, 1970), Stock, R., Ed., The Institute of Petroleum, London, 1971, 125.
15. **Keller, R. A. and Snyder, L. R.,** Relation between the solubility parameter and the liquid-solid solvent strength parameter, *J. Chromatogr. Sci.,* 9, 346, 1971.
16. **Bombaugh, K. J.,** Column selection in HPLC, in *GLC and HPLC Determination of Therapeutic Agents,* Part 1 (Chromatographic Science: A Series of Monographs, Vol. 9), Tsuji, K. and Morozowich, W., Eds., Marcel Dekker, New York, 1978.
17. **Feng, Y.-Q., Zhu, P.-L., and Hu, Z.,** Applications of solubility parameter in chromatography, *Fenxi Huazue,* 15, 469, 1987.
18. **Korol, A. N.,** The possibility of predicting the retention parameters in gas-liquid chromatography, *Usp. Khim.,* 51, 1225, 1982; *Russ. Chem. Chem.,* 51, 698, 1982.
19. **Munafo, A., Buchmann, M., Hô Nam-Tran, and Kesselring, U. W.,** Determination of the total and partial cohesion parameters of lipophilic liquids by gas-liquid chromatography and from molecular properties, *J. Pharm. Sci.,* 77, 169, 1988.

20. **Shoenmakers, P. J., Billiet, H. A. H., Tijssen, R., and de Galan, L.,** Gradient selection in reversed-phase liquid chromatography, *J. Chromatogr.*, 149, 519, 1978.
21. **Schoenmakers, P. J.,** *A Systematic Approach to Mobile Phase Effects in Reversed Phase Liquid Chromatography,* Thesis, Delft Technical College, Netherlands, 1981.
22. **Schoenmakers, P. J.,** *Optimization of Chromatographic Selectivity: A Guide to Method Development* (Journal of Chromatography Library, Vol. 35), Elsevier, Amsterdam, 1986.
23. **Patte, F., Etcheto, M., and Laffort, P.,** Solubility factors for 240 solutes and 207 stationary phases in gas-liquid chromatography, *Analyt. Chem.*, 54, 2239, 1982.
24. **Laffort, P. and Patte, F.,** Solubility factors in gas-liquid chromatography. Comparison between two approaches and application to some biological studies, *J. Chromatogr.*, 126, 625, 1976.
25. **Burns, W. and Hawkes, S. J.,** Choice of stationary phase in gas chromatography, *J. Chromatogr. Sci.*, 15, 185, 1977.
26. **Smith, P. L. and Cooper, W. T.,** Extended solubility parameters and stationary phase selectivities of amino-, cyano- and diol-silica normal bonded-phase liquid chromatography columns, *Chromatographia,* 25, 55, 1988.
27. **Alessi, P., Kikic, I., Papo, A., and Torriano, G.,** Gas chromatographic study of solvent-plasticizer-resin interactions, *Industria Vernica,* 32(11), 24, 1978; *Boll. Assoc. Ital. Ind. Vernici Affini,* 66, 24, 1978.
28. **Alessi, P., Kikic, I., Papo, A., and Torriano, G.,** Epoxy resin characterization by G.L.C., *FATIPEC (Fed. Assoc. Tech. Ind. Paint, Vernis, Emaux Encres Impr. Eur. Cont.) Congr.*, 14, 525, 1978.
29. **Alessi, P., Kikic, I., and Stele, L.,** Gas chromatographic investigation of solvents with partial miscibility, *Croat. Chem. Acta,* 47, 1, 1975.
30. **Alessi, P., Kikic, I., and Torriano, G.,** Activity coefficients of hydrocarbons in phthalates, *J. Chromatogr.*, 105, 257, 1975.
31. **Alessi, P., Kikic, I., and Torriano, G.,** Correlation of hydrocarbon activity coefficients with solubility parameters for phthalate esters, *J. Chromatogr.*, 106, 17, 1975.
32. **Alessi, P., Kikic, I., Torriano, G., and Papo, A.,** Solvent-plasticizer-resin interactions by gas-liquid chromatography, *J. Coatings Technol.*, 51(650), 62, 1979.
33. **Barceló, D., Galcerán, M. T., and Eek, L.,** Gas chromatographic determination of adsorption terms on Chromosorb 102 coated with squalane and Fractonitril VI. Study of adsorption phenomena using porous polymers coated with liquid phases, *Chromatographia,* 21, 169, 1986.
34. **Barceló, D., Galcerán, M. T., and Eek, L.,** Comparison of the porous Chromosorb 101 and Chromosorb 102 as supports coated with Ethofat 60/25 in gas chromatography, in *Chromatography, the State of the Art,* Kalász, H. and Ettre, L. S., Eds., Akádemiai Kiadó, Budapest, 1986, 697.
35. **Barceló, D., Galcerán, M. T., and Eek, L.,** Prediction of the net retention volume in gas chromatography using porous polymers coated with different stationary phases, *Chromatographia,* 23, 481, 1987.
36. **Dincer, S. and Bonner, D. C.,** Thermodynamic analysis of an ethylene and vinyl acetate copolymer with various solvents by gas chromatography, *Macromolecules,* 11, 107, 1978.
37. **Dincer, S. and Bonner, D. C.,** Solubility criteria for a thermally stable polymer determined by gas-solid chromatography, *J. Appl. Polym. Sci.*, 22, 3234, 1978.
38. **DiPaola-Baranyi, G. and Guillet, J. E.,** Estimation of polymer solubility parameters by gas chromatography, *Macromolecules,* 11, 228, 1978.
39. **DiPaola-Baranyi, G., Guillet, J. E., Klein, J., and Jeberien, H. E.,** Estimation of solubility parameters for poly(vinyl acetate) by inverse gas chromatography, *J. Chromatogr.*, 166, 349, 1978.
40. **Figgins, C. E., Risby, T. H., and Jurs, P. C.,** The use of partial molal entropies and free energies in the characterization of gas chromatographic stationary phases, *J. Chromatogr. Sci.*, 14, 453, 1976.
41. **Figgins, C. E., Reinhold, B. L., and Risby, T. H.,** Partial molal and mole fraction free energies as characterizers for stationary liquid phases, *J. Chromatogr. Sci.*, 15, 208, 1977.
42. **Funk, E. W.,** Study of heavy hydrocarbons by inverse-phase chromatography, *Ind. Eng. Chem. Prod. Res. Dev.*, 16, 115, 1977.
43. **Lafosse, M. and Durand, M.,** Squalane-1,3,5-tricyanopentane mixtures in gas-liquid chromatography. Partition of several solutes in nitrile phases, *J. Chromatogr.*, 171, 183, 1979.
44. **Lafosse, M. and Durand, M.,** Gas-liquid adsorption and solubility phenomena as studied by gas chromatography. Comparison of both concurrent effects on two similar nitrile liquid phases, *J. Colloid Interface Sci.*, 77, 363, 1980.
45. **Lafosse, M. and Dreux, M.,** Gas-liquid chromatographic determination of equilibrium constants of interactions between solutes and free nitrile groups, *J. Chromatogr.*, 193, 9, 1980.
46. **Leca, M., Groswald-Cremer, L., and Surmeian, M.,** Determination of solubility parameters of polymers using gas-liquid chromatography, *Mater. Plast.*, 16, 114, 1979.
47. **Leca, M., Surmeian, M., and Groswald-Cremer, L.,** Determination of homomorphs of polar substances by gas-liquid chromatography, *Rev. Chim.*, 30, 172, 1979.
48. **Letcher, T. M.,** Activity coefficient at infinite dilution from gas-liquid chromatography, *Chem. Soc. Spec. Per. Rep. Chem. Thermodyn.*, 2, 46, 1978.

49. **Leung, Y.-K.,** Thermodynamics of α,ω-methoxypoly(ethylene oxide)/*n*-alkane systems by gas-liquid chromatography, *Polymer,* 17, 374, 1976.
50. **Newman, R. D. and Prausnitz, J. M.,** Polymer-solvent interactions from gas-liquid partition chromatography, *J. Phys. Chem.,* 76, 1492, 1972; Polymer-solvent interactions from gas-liquid chromatography, *J. Paint Technol.,* 45(585), 33, 1973.
51. **Risby, T. H., Jurs, P. C., and Reinbold, B. L.,** Use of thermodynamic solubility parameters for the characterization of liquid phases, *J. Chromatogr.,* 99, 173, 1974.
52. **Senich, G. A. and Sanchez, I. C.,** Inverse gas chromatography, a method for studying polymer-migrant interactions in polyolefin packaging materials, *Org. Coat. Plast. Chem.,* 41, 345, 1979.
53. **Senich, G. A.,** Migration to and from plastics, *Chemtech,* 11, 360, 1981.
54. **Takamiya, N., Kojima, J., and Tomari, Y.,** A method of evaluating solubility parameters from gas chromatographic data, *Kogyo Kagaku Zasshi,* 62, 1371, 1959.
55. **Varsano, J. L. and Gilbert, S. G.,** Evaluation of interactions between polymers and low molecular weight compounds by GLC. I. Methodology and interaction evaluation; II. Thermodynamics and elucidation, *J. Pharm. Sci.,* 62, 87, and 92, 1973.
56. **Hawkes, E. C. and Hawkes, S. J.,** Variation of retention volume with molecular weight of polyethylene glycol and with carbon number of n-alkanols, *J. Chromatogr. Sci.,* 17, 285, 1979.
57. **McCloskey, D. H. and Hawkes, S. J.,** Choosing standard stationary phases for gas chromatography, *J. Chromatogr. Sci.,* 13, 1, 1975.
58. **Merk, W., Lichtenthaler, R. N., and Prausnitz, J. M.,** Solubilities of fifteen solvents in copolymers of poly(vinyl acetate) and poly(vinyl chloride) from gas-liquid chromatography. Estimation of polymer solubility parameter, *J. Phys. Chem.,* 84, 1694, 1980.
59. **Castells, R. C. and Mazza, G. D.,** Study of polymer-solvent interactions by gas chromatography. Copolymers of vinyl acetate and vinyl alcohol with hydrocarbons and alcohols, *J. Appl. Polym. Sci.,* 32, 5917, 1986; *An.-CIDEPINT,* 123, 1987.
60. **Ito, K. and Guillet, J. E.,** Estimation of solubility parameter for some olefin polymers and copolymers by inverse gas chromatography, *Macromolecules,* 12, 1163, 1979.
61. **Ghosh, S. K.,** Solubility parameter and hydrocarbon sorption of low density polyethylene, *Makromol. Chem.,* 143, 181, 1971.
62. **Grozdov, A. G. and Stepanov, B. N.,** Determination of solubility parameter of polymers by gas chromatography, *Vysokomol. Soedin, Ser.* B., 17, 907, 1975.
63. **Leca, M. and Enache, M.,** Determination of polymer partial solubility parameters by gas-liquid chromatography, *Rev. Roum. Chim.,* 28, 287, 1983.
64. **Dangayach, K. C. and Bonner, D. C.,** Solvent interactions with polysulfone, *Polym. Eng. Sci.,* 20, 59, 1980; *Org. Coat. Plast. Chem.,* 39, 309, 1978.
65. **Kawakami, M., Egashira, M., and Kagawa, S.,** Measurements of the interactions between polyethylene glycol and organic compounds by gas chromatographic technique, *Bull. Chem. Soc. Jpn.,* 49, 3449, 1976.
66. **Shoji, K. and Takeda, M.,** Study on interaction of organic vapor and polyethyleneglycol by gas-chromatography, *Kobunshi Ronbunshi,* 31, 420, 1974; (English Ed.), 3, 1606, 1974.
67. **Castells, R. C., Mazza, G. D., and Arancibia, G. L.,** Study of polymer-solvent interaction by gas chromatography systems formed by hydrocarbons and alcohols with poly(vinyl acetate), *An. Assoc. Quim. Bras.,* 73, 519, 1985.
68. **Vial, J.,** Studies of solvents — polyethylene terephthalate interactions by inverse gas chromatography: application to adhesion phenomena, *FATIPEC (Fed. Assoc. Tech. Ind. Paint Vernis, Emaux Encres Impr. Eur. Cont.) Congr.,* 16th, Vol. 1, 321, 1982.
69. **Karim, K. A. and Bonner, D. C.,** Thermodynamic interpretation of solute-polymer interactions at infinite dilution, *J. Appl. Polym. Sci.,* 22, 1277, 1978; erratum *J. Appl. Polym. Sci.,* 23, 2825, 1979.
70. **Karim, K. A. and Bonner, D. C.,** An improved concept in solubility parameter theory applied to amorphous polymers. I. Poly(ethyl methacrylate), *Polym. Eng. Sci.,* 19, 1174, 1979.
71. **Olabisi, O.,** Polymer compatibility by gas-liquid chromatography, *Macromolecules,* 8, 316, 1975.
72. **Reichert, K.-H.,** Gas chromatographic study of the interaction forces between binder and solvent in paint films, *J. Oil Col. Chem. Assoc.,* 54, 887, 1971.
73. **Otozai, K. and Tohyama, I.,** Determination of molecular weights and solubility parameters of high-polymers by gas chromatography, *Z. Anal. Chem.,* 281, 131, 1976.
74. **Friedman, H. L., Miller, B., and Meiser, C. H., Jr.,** Influence of preadsorbed water on chromatographic detention of organic volatiles by cellulosic substrates, *Text. Res. J.,* 55, 726, 1985.
75. **Hartkopf, A., Grubfeld, S., and Delumyea, R.,** A reconsideration of the Rohrschneider approach for characterizing gas chromatographic liquid phases. II. Use of four functional probes, *J. Chromatogr. Sci.,* 12, 119, 1974.
76. **Pierotti, G. J., Deal, C. H., Derr, E. L., and Porter, P. E.,** Solvent effects in gas-liquid partition chromatography, *J. Am. Chem. Soc.,* 78, 2989, 1956.

77. **Porter, P. E., Deal, C. H., and Stross, F. H.,** The determination of partition coefficients from gas-liquid partition chromatography, *J. Am. Chem. Soc.*, 78, 2999, 1956.
78. **Souter, P.,** Calculation of Rohrschneider constants, *J. Chromatogr.*, 92, 231, 1974.
79. **Bermejo, J. and Guillén, M. D.,** The estimation of the solubility parameter of low volatile compounds from gas chromatographic data; Some remarks about the estimation of the solubility parameter of low volatile compounds from GC data, *J. Chem. Tech. Biotechnol.*, 37, 101, 1987; 41, 41, 1988.
80. **Park, J. H. and Carr, P. W.,** Predictive ability of the MOSCED and UNIFAC activity coefficient estimation methods, *Analyt. Chem.*, 59, 2596, 1987.
81. **Alessi, P., Kikic, I., and Tlustos, G.,** Prediction of activity coefficients of hydrocarbons in glycols using solubility parameters, *Chim. Ind.*, 53, 1148, 1971.
82. **Cruickshank, A. J. B., Gainey, B. W., and Young, C. L.,** Activity coefficients of hydrocarbons C_4 to C_8 in *n*-octadecane at 35°C. Gas chromatographic investigation, *Trans. Faraday Soc.*, 64, 337, 1968.
83. **Martire, D. E.,** Thermodynamics of dilute solutions: aliphatic solutes in aromatic solvents and various solutes in paraffinic solvents, in *Gas Chromatography 1966*, Littlewood, A. B., Ed., Institute of Petroleum, London, 1967, 21.
84. **Tewari, Y. B., Martire, D. E., and Sheridan, J. P.,** Gas-liquid partition chromatographic determination and theoretical interpretation of activity coefficients for hydrocarbon solutes in alkane solvents, *J. Phys. Chem.*, 74, 2345, 1970.
85. **Meen, D. L., Morris, F., Purnell, J. H., and Srivastava, O. P.,** Gas chromatography measured activity coefficients in non-electrolyte solutions, *J. Chem. Soc. Faraday I*, 69, 2080, 1973.
86. **Li, Y.-G., Teng, T., Lu, J.-F., Chen, G., and Li, J. D.,** A study on Scatchard-Hildebrand solution theory for metal solvent extraction systems, *Fluid Phase Equil.*, 30, 297, 1986.
87. **Milanová, E. and Cave, G. C. B.,** Limiting activity coefficients in dilute solutions of nonelectrolytes. I. Determination for polar-nonpolar binary mixtures by a novel apparatus, and a solubility-parameter treatment of results, *Can. J. Chem.*, 60, 2697, 1982.
88. **Afrashtehfar, S. and Cave, G. C. B.,** Limiting activity coefficients in dilute solutions of nonelectrolytes. II. Determination for polar-nonpolar and polar-polar binary mixtures, and the application of some solubility parameter treatments, *Can. J. Chem.*, 64, 198, 1986.
89. **Castells, R. C., Nardillo, A. M., Arancibia, E. L., and Delfino, M. R.,** Solution and adsorption thermodynamics in propylene glycol by gas chromatography. A comparative study with other polyhydroxylated solvents. *J. Chromatogr.*, 259, 413, 1983.
90. **Leca, M.,** A new route for determining the nonpolar solubility parameter of organic solvents by gas liquid chromatography, *Rev. Roum. Chim.*, 27, 987, 1982.
91. **Clark, R. K. and Schmidt, H. H.,** Gas-liquid partition chromatography in benzene-polyphenol systems and polymer statistical thermodynamics, *J. Phys. Chem.*, 69, 3682, 1965.
92. **Reinbold, B. L. and Risby, T. H.,** Use of partial molar enthalpies for the characterization of stationary liquid phases, *J. Chromatogr. Sci.*, 13, 372, 1975.
93. **Harris, H. G. and Prausnitz, J. M.,** Relative Lewis acidities of polar organic solvents from chromatographic measurements, *J. Chromatogr. Sci.*, 7, 685, 1969.
94. **Hartkopf, A.,** A reconsideration of the Rohrschneider approach for characterising gas chromatographic liquid phases. I. Review and description of basic approach, *J. Chromatogr. Sci.*, 12, 113, 1974.
95. **Kon-no, K., Sakiyama, M., Suzuki, T., and Kitahara, A.,** The determination of solubility parameter of surfaces by gas chromatography, *Nippon Kagaku Kaishi*, 2211, 1974.
96. **Laffort, P., Patte, F., and Etcheto, M.,** Olfactory coding on the basis of physicochemical properties, *Ann. N.Y. Acad. Sci.*, 237, 193, 1974.
97. **Patte, F.,** Actual status of olfactory properties predictability from chromatographic measurements; Molecular Aspects of chemoreception, *First* and *Third Congresses of European Chemoreception Research Organization*, European Chemoreception Research Organization, London, 1974 and 1978.
98. **Lille, Ü., Kundel, H., and Eisen, O.,** Solubility parameters of alkyl derivatives of phenol and resorcinol, *J. Chromatogr.*, 116, 1, 1976.
99. **Otozai, K.,** Semi-empirical relation between retention volume and boiling point in gas-chromatography, *Z. Anal. Chem.*, 268, 257, 1974.
100. **Otozai, K. and Tohyama, I.,** Gas chromatography of halogens, *Z. Anal. Chem.*, 279, 195, 1976.
101. **Otozai, K. and Tohyama, I.,** Gas chromatography of chlorides and bromides of III-, IV- and V-group elements in the periodic table, *Z. Anal. Chem.*, 274, 353, 1975.
102. **Tohyama, I. and Otozai, K.,** Gas chromatography of $SiCl_4$, $GeCl_4$, $SnCl_4$, and $TiCl_4$, *Z. Anal. Chem.*, 271, 117, 1974.
103. **Sunal, A. B. and Barker, P. E.,** Partition coefficient and activity coefficient determinations of pinenes in polypropylene sebacate and of halocarbons in silicone oil using gas/liquid chromatographic techniques, *J. Chromatogr. Sci.*, 13, 541, 1975.

104. **Wolf, W. R., Sievers, R. E., and Brown, G. H.,** Vapor pressure measurements and gas chromatographic studies of the solution thermodynamics of metal β-diketonates, *Inorg. Chem.*, 11, 1995, 1972; Aerospace Research Laboratories Reprint 72-0164, Wright-Patterson Air Force Base, OH, 1972.
105. **Miedziak, I. and Waksumundzki, A.,** Some problems of classification of liquid stationary phases in gas chromatography, *Rocz. Chem.*, 51, 313, 1977.
106. **Friant, S. L. and Suffet, I. H.,** Interactive effects of temperature, salt concentration and pH on head space analysis for isolating trace organics in aqueous environmental samples, *Analyt. Chem.*, 51, 2167, 1979.
107. **King, J. W. and Quinney, P. R.,** Investigations of charge-transfer interactions between carbon tetrabromide and aromatic donors by gas chromatography, *J. Chromatogr.*, 49, 161, 1970.
108. **Karger, B. L., Snyder, L. R., and Horvath, C.,** *An Introduction to Separation Science*, John Wiley & Sons, New York, 1973.
109. **Freeman, D. H. and Killion, D.,** Gel partition/permeation chromatography using n-alkane solutes, *J. Polym. Sci. Polym. Phys.*, 15, 2047, 1977.
110. **Lafosse, M., Kéravis, G., and Durand, M. H.,** High pressure liquid chromatography of free and conjugated 17-kerosteroids, *J. Chromatogr.*, 118, 283, 1976.
111. **Bakalyar, S. R., Bradley, M. P. T., and Honganen, R.,** The role of dissolved gases in high-performance liquid chromatography, *J. Chromatogr.*, 158, 277, 1978.
112. **Martire, D. E. and Locke, D. C.,** Selectivity in liquid-liquid chromatography, *Analyt. Chem.*, 43, 68, 1971.
113. **Suzuki, N., Suzuki, J., and Saitoh, K.,** Liquid chromatographic separation of metal (II) chelates of thiothenoyltrifluoroacetone on polystyrene beads, *J. Chromatogr.*, 177, 166, 1979.
114. **Bakalyar, S. R., McIlwrick, R., and Roggendorf, E.,** Solvent selectivity in reversed-phase high-pressure liquid chromatography, *J. Chromatog.*, 142, 353, 1977.
115. **Tijssen, R., Billiet, H. A. H., and Schoenmakers, P. J.,** Use of the solubility parameter for predicting selectivity and retention in chromatography, *J. Chromatogr.*, 122, 185, 1976.
116. **Rabel, F. M., Butts, E. T., Hellman, M. L., and Popovich, D. J.,** Solvent effects in reversed phase liquid chromatography for high and low loadings of octadecylsilane on microparticle silica. I. *J. Liq. Chromatogr.*, 1, 631, 1978.
117. **Lewis, J. J. and Rogers, L.B.,** Solvent selectivity in reversed-phase high-performance liquid chromatographic separation of isomers of polystyrene oligomers, *J. Chromatogr.*, 264, 339, 1983.
118. **Billiet, H. A. H., Schoenmakers, P. J., and de Galan, L.,** Retention and selectivity characteristics of a non-polar perfluorinated stationary phase for liquid chromatography, *J. Chromatogr.*, 218, 443, 1981.
119. **Schoenmakers, P. J., Billiet, H. A. H., and de Galan, L.,** Systematic study of ternary solvent behaviour in reversed-phase liquid chromatography; Description of solute retention over the full range of mobile phase compositions in reversed-phase liquid chromatography, *J. Chromatogr.*, 218, 261, 1981; 282, 107, 1983.
120. **Schoenmakers, P. J., Drouen, A. C. J. H., Billiet, H. A. H., and de Galan, L.,** A simple procedure for the rapid optimization of reversed-phase separations with ternary mobile-phase mixtures, *Chromatographia*, 15, 688, 1982.
121. **Drouen, A. C. J. H., Billiet, H. A. H., Schoenmakers, P. J., and de Galan, L.,** An improved optimization procedure for the selection of mixed mobile phases in reversed phase liquid chromatography, *Chromatographia*, 16, 48, 1982.
122. **Weber, W. J., Jr., Chin, Y.-P., and Rice, C. P.,** Determination of partition coefficients and aqueous solubilities by reverse phase chromatography — I. Theory and background, *Wat. Res.*, 20, 1433, 1986.
123. **Chin, Y.-P., Weber, W. J., Jr., and Voice, T. C.,** Determination of partition coefficients and aqueous solubilities by reverse phase chromatography — II. Evaluation of partitioning and solubility models, *Wat. Res.*, 20, 1443, 1986.
124. **Cooper, H. A. and Hurtubise, R. J.,** Mobile phase optimization for the reversed-phase liquid chromatography of complex hydroxyl aromatic mixtures; Solubility parameter and window diagram optimization methods for the reversed-phase chromatographic optimization of complex hydroxyl aromatic mixtures; Prediction of retention of hydroxyl aromatics in normal-phase liquid chromatography with slope-intercept relations, *J. Chromatgr.*, 324, 1, 1985; 328, 81, 1985; 360, 327, 1986.
125. **Hafkenscheid, T. L. and Tomlinson, E.,** Observations on capacity factor determination for reversed-phase liquid chromatography with aqueous methanol eluents using the solubility parameter concept model and its derivatives, *J. Chromatogr.*, 264, 47, 1984.
126. **Szántó, J. I. and Veress, T.,** Comparative study on the thermodynamic retention models in HPLC, *Chromatographia*, 20, 596, 1985.
127. **Suzuki, N., Saitoh, K., and Adachi, K.,** Reversed-phase high-performance thin-layer chromatography and column liquid chromatography of chlorophylls and their derivatives, *J. Chromatogr.*, 408, 181, 1987.
128. **Schram, S. B. and Freeman, D. H.,** Characterization of LC stationary phases by inverse size exclusion chromatography, *J. Liq. Chromatogr.*, 3, 403, 1980.

129. **Vigh, G. and Inczédy, J.,** Separation of chloramphenicol intermediates by high-performance liquid chromatography of micropak-NH_2 columns, *J. Chromatogr.,* 129, 81, 1976.
130. **Nikolov, R. N.,** Mixed retention mechanisms in reversed-phase high-performance liquid chromatography with alkyl-bonded silica, *J. Chromatogr.,* 286, 147, 1984.
131. **Nikolov, R. N. and Darzhalieva, M. M.,** Hydrophobic-dispersive partition coefficient in alkyl-bonded reverse-phase systems with water-methanol mobile phases, *J. Chromatogr.,* 370, 377, 1986.
132. **Shetty, P. H., Youngberg, P. J., Kersten, B. R., and Poole, C. F.,** Solvent properties of liquid organic salts used as mobile phases in microcolumn reversed-phase liquid chromatography, *J. Chromatogr.,* 411, 61, 1987.
133. **Czubryt, J. J., Myers, M. N., and Giddings, J. C.,** Solubility phenomena in dense carbon dioxide gas in the range 270-1900 atmospheres, *J. Phys. Chem.,* 74, 4260, 1970.
134. **Giddings, J. C., Myers, M. N., and King, J. W.,** Dense gas chromatography at pressures up to 2000 atmospheres, *J. Chromatogr. Sci.,* 7, 276, 1969.
135. **Giddings, J. C., Myers, M. N., McLaren, L., and Keller, R. A.,** High pressure gas chromatography of nonvolatile species, *Science,* 162, 67, 1968.
136. **Myers, M. N. and Giddings, J. C.,** High-pressure gas chromatography, in *Progress in Separation and Purification,* Vol. 3, Perry, E. S. and Van Oss, C. J., Eds., John Wiley & Sons-Interscience, New York, 1970.
137. **Novotny, M., Springston, S. R., Peaden, P. A., Fjeldsted, J. C., and Lee, M. L.,** Capillary supercritical fluid chromatography, *Analyt. Chem.,* 53, 407A, 1981.
138. **Smith, R. M., Ed.,** *Supercritical Fluid Chromatography* (Royal Society of Chemistry Chromatography Monograph Series), Royal Society of Chemistry, London, 1988.
139. **Bartle, K. D.,** Theory and principles of supercritical fluid chromatography, in *Supercritical Fluid Chromatography,* Smith, R. M., Ed., Royal Society of Chemistry, London, 1988, 1.
140. **Randall, L. G.,** The present status of dense (supercritical) gas extraction and dense gas chromatography. Impetus for DGC.MS development: review, *Sep. Sci. Technol.,* 17, 1, 1982.
141. **Schoenmakers, P. J.,** Thermodynamic model for supercritical fluid chromatography, *J. Chromatogr.,* 315, 1, 1984.
142. **Mourier, P., Sassiat, P., Caude, M., and Rosset, R.,** Supercritical fluid chromatography, *Analusis,* 12, 229, 1984.
143. **Mourier, P., Caude, M., and Rosset, R.,** Retention, selectivity and efficiency in supercritical fluid chromatography, *Analusis,* 13, 299, 1985.
144. **Yonker, C. R. and Smith, R. D.,** Study of retention processes in capillary supercritical fluid chromatography with binary fluid mobile phases, *J. Chromatogr.,* 361, 25, 1986.
145. **Leyendecker, D.,** Selection of conditions for an SFC separation, in *Supercritical Fluid Chromatography,* R. M. Smith, Ed., Royal Society of Chemistry, London, 1988, 53.
146. **Snyder, L. R.,** The role of the mobile phase in liquid chromatography, in *Modern Practice of Liquid Chromatography,* Kirkland, J. J., Ed., John Wiley & Sons, New York, 1971, chap. 4.
147. **Snyder, L. R. and Kirkland, J. J.,** *Introduction to Modern Liquid Chromatography,* John Wiley & Sons, New York, 1974.
148. **Colin, H., Guiochin, G., and Jandera, P.,** The solvent eluotropic strength on carbon adsorbents, *Chromatographia,* 15, 133, 1982.
149. **Buchmann, M. L. and Kesselring, U. W.,** A practical eluotropic series of solvents based on the expanded solubility parameter, *Pharm. Acta Helv.,* 56, 166, 1981.
150. **Pescher, P., Caude, M., Rosset, R., and Tambute, A.,** Enantiomeric separation of tertiary phosphine oxides on Pirkle chiral stationary phase - mobile phase and temperature optimization, *J. Chromatogr.,* 371, 159, 1986.
151. **Nguyen, Huu-Phouc., Hô, Nam-Tran., Buchmann, M., and Kesselring, U. W.,** Experimentally optimized determination of the partial and total cohesion parameters of an insoluble polymer (microcrystalline cellulose) b gas-solid chromatography, *Int. J. Pharm.,* 34, 217, 1987.
152. **Nguyne, Huu-Phouc., Hô, Nam-Tran., Buchmann, M., and Kesselring, U. W.,** Determination of partial and total cohesion parameters of caffeine, theophylline and methyl p-hydroxybenzoate by gas-solid chromatography, *J. Pharm. Sci.,* 76, 406, 1987.
153. **Nguyen, Huu-Phouc., Luu, R. P. T., Munafo, A., Ruelle, P., Hô, N. T., Buchmann, M., and Kesselring, U. W.,** Determination of partial solubility parameters of lactose by gas solid chromatography, *J. Pharm. Sci.,* 75, 68, 1986.
154. **Inagaki, H.,** Polymer separation and characterization by thin-layer chromatography, *Adv. Polym. Sci.,* 24 (Molecular Properties), 189, 1977.
155. **Song, X. and Xu, C.,** Determination of solubility parameter of polystyrene by TLC method, *Gaodeng Xuexiao Huaxue Xuebao,* 3, 271, 1982.
156. **Belkin, A. I. and Karpov, V. V.,** Determination of solubility parameter of dispersed (nonionic) dyes by thin-layer chromatography, *Zh. Prikl. Khim.,* 55, 948, 1982; *J. Appl. Chem. U.S.S.R.,* 55, 877, 1982.

157. **Wakui, Y., Saitoh, K., and Suzuki, N.,** Non-aqueous, reversed-phase, high-performance, thin-layer chromatography and column liquid chromatography of metal complexes of porphine, *Chromatographia,* 22, 160, 1986.
158. **Kobayashi, M., Saitoh, K., and Suzuki, N.,** Reversed-phase, high-performance, thin-layer chromatography of metal tetrakis(p-tolyl)porphine complexes on octadecyl bonded stationary phase, *Chromatographia,* 20, 49, 1985.
159. **Kobayashi, M., Saitoh, K., and Suzuki, N.,** High-performance thin-layer chromatography of metal complexes on cellulose and silica gel, *Chromatographia,* 18, 441, 1984.
160. **Miyake, S., Saitoh, K., and Suzuki, N.,** High-performance thin layer chromatography of etioporphyrin and its vanadium (IV), nickel (II) and copper (II), *Chromatographia,* 20, 417, 1985.
161. **Matsuura, T., Blais, P., and Sourirajan, S.,** Polar and nonpolar parameters for polymeric reverse osmosis membrane materials from liqid chromatographic data, *J. Appl. Polym. Sci.,* 20, 1515, 1976.
162. **Matsuura, T. and Sourirajan, S.,** Characterization of reverse osmosis membrane materials by liquid chromatography data, *Proc. 6th Int. Symp., Fresh Water from the Sea,* 3, 227, 1978.
163. **Bleha, T. and Berek, D.,** Gel chromatography with mixed eluents. Partition of solute due to preferential solvation of gel, *Chromatographia,* 14, 163, 1981.
164. **Figueruelo, J. E., Sorlo, V., and Campos, A.,** Polymer retention mechanism in GPC on active gels. 2. Low molecular weight polystyrene, *J. Liq. Chromatogr.,* 3, 367, 1980.
165. **Stejskal, J., Janca, J., and Kratochvil, P.,** Solution properties of poly(2-methoxyethyl methacrylate), *Polym. J.,* 8, 549, 1976.
166. **Spychaj, T. and Berek, D.,** Cyclohexanone/1,2-dichloroethane as eluents in GPC of polymers, *Acta Polym.,* 33, 477, 1982.
167. **Biagini, E., Costa, G., Gattiglia, E., Imperato, A. and Russo, S.,** Gel permeation chromatographic analysis of *N*-trifluoroacetylated poly(ϵ-caprolactam): calibration procedures, *Polymer,* 28, 114, 1987.
168. **Ozaki, E., Saitoh, K., and Suzuki, N.,** Gel chromatography of *n*-alkanes on a polystyrene gel with various organic solvents, *J. Chromatogr.,* 177, 122, 1979.
169. **Bicking, M. K. L. and Serwon, S. J.,** Molecular interactions on polystyrene stationary phases, *J. Liq. Chromatogr.,* 10, 1369, 1987.
170. **Mori, S.,** Elution behavior of some solutes on a porous polystyrene gel in high performance liquid chromatography, *Analyt. Chem.,* 50, 745, 1978.
171. **Saitoh, K., Ozaki, E., and Suzuki, N.,** Solvent dependence of gel chromatographic retention of low-molecular-weight compounds on polystyrene-divinylbenzene gel, *J. Chromatogr.,* 239, 661, 1982.
172. **Saitoh, K., Ozawa, T., and Suzuki, N.,** Solvent dependence of swelling of Merckogel OR-2000 and various organic solvents, *J. Chomatogr.,* 124, 231, 1976.
173. **Saitoh, K. and Suzuki, M.,** Solvent effects in the gel chromatography of n-alkanes on Merckogel OR-2000, *J. Chromatogr.,* 111, 29, 1975.
174. **Ying, Q., Xie, P., and Ye, M.,** A new eluting solvent for gel permeation chromatography of isotactic polypropylene, *Makromol. Chem. Rapid Commun.,* 6, 105, 1985.
175. **Figueruelo, J. E., Soria, V., and Campos, A.,** Polymer retention mechanisms in GPC on active gels, 3. Poly(dimethylsiloxane) and poly(methyl methacrylate), *Makromol. Chem.,* 182, 1525, 1981.
176. **Saitoh, K. and Suzuki, N.,** Gel chromatography of β-diketones and their metal complexes. IV. The partition concept in gel chromatography, *Bull. Chem. Soc. Jpn.,* 51, 116, 1978.
177. **Yamamoto, M. and Yamamoto, Y.,** Solvent effects in gel chromatography: relationship between distribution coefficients of metal acetylacetonates and solubility parameters of solvents, *Analyt. Chim. Acta,* 87, 375, 1976.
178. **Yamamoto, Y., Yamamoto, M., Ebisui, S., Takagi, T., Hashimoto, T., and Izuhara, M.,** Gel chromatography of metal acetylacetonates in organic solvents, *Analyt. Lett.,* 6, 451, 1973.
179. **Altgelt, K. H. and Gouw, T. H.,** Chromatography of heavy petroleum fractions, *Adv. Chromatogr.,* 13, 71, 1975.
180. **Bicking, M. K. and Kniseley, R. N.,** Size exclusion chromatographic analysis of refuse-derived fuel for mycotoxins, *Analyt. Chem.,* 52, 2164, 1980.
181. **Chiantore, O.,** Polymer-substrate interactions in size exclusion chromatography with silica gels and pure solvents, *J. Liq. Chromatogr.,* 7, 1, 1984.
182. **Rozylo, J. K., Chojnacka, G., and Kolodziejczyk, H.,** Influence of solvent strength in possible optimization of adsorption thin-layer chromatography, *J. High Resolut. Chromatogr. Chromatogr. Commun.,* 5, 139, 1982.
183. **Oh, D. S., Lee, S. H., Kim, S. H., and Kim, S. T.,** A new solubility parameter for retention behavior in liquid chromatography, *Taehan Hwahakhoe Chi,* 32, 458, 1988.
184. **Shuster, J. E.,** *The Application of Solubility Parameters to the Selection of Solvents for High-Performance Liquid Chromatography,* Thesis, Temple University, Philadelphia, PA, 1988; Diss. Abstr. Int. B., 49, 1662, 1988.

185. **Galin, M.,** Gas-liquid chromatography study of the thermodynamics of interactions between linear polyethylene and non-polar and polar solutes, *Polymer,* 30, 2074, 1989.
186. **Holten-Andersen, J. and Eng, K.,** Activity coefficients in polymer solutions, *Progr. Org. Coatings,* 16, 77, 1988.
187. **Nakazato, K. and Suzuki, K.,** Correlation between solubility parameter and exclusion limit on polyacrylamide terpolymer, *Macromolecules,* 22, 1497, 1989.
188. **Vilcu, R. and Leca, M.,** Thermodynamic characterization of polymers by chromatography with gas mobile phase, *Rev. Roum. Chim.,* 34, 387, 1989; *Polymer Thermodynamics by Gas Chromatography,* Elsevier, Amsterdam, 1990.
189. **Fernández-Sánchez, E., Fernández-Torres, A., Garcia-Domínguez, J. A., and Santiuste, J. M.,** Solubility parameters of gas chromatographic mixed stationary phases, *J. Chromatogr.,* 457, 55, 1988.
190. **Szymanowski, J., Sobzyńska, A., and Voelkel, A.,** Solubility and polarity parameters of polyoxyethylene glycol dialkyl ethers, *J. Pharm. Sci.,* 77, 893, 1988.
191. **Sun, Z.-L., Liu, M.-C., and Hu, Z.,** Study on retention behavior of solutes in reversed phase liquid chromatography — discussion on retention mechanism and a simple method for calculating expanded solubility parameter, *Fenxi Ceshi Tongbao (Bull. Anal. Testing),* 7(5), 15, 1988.
192. **Ikushima, Y., Saito, N., Hatakeda, K., Ito, S., and Goto, T.,** Development of extraction and separation monitoring in situ with an on-line SFE-SFC/FTIR system, *Chem. Lett.,* No. 10, 1707, 1989.
193. **Kerber, M. L., Mkhitaryan, M. A., Klubkova, O. M., Voskanyan, P. S., and Barashkov, O. K.,** Study of interaction thermodynamics of plasticizers with poly(vinyl butyral) by inverse gas chromatography, *Plast. Massy,* No. 5, 71, 1989.
194. **Lee, D. W., Yook, K. S., Kim, H. S., and Lee, W.,** A study of the parameters of the retention of monosubstituted benzenes in reversed-phase liquid chromatography, *Bull. Korean Chem. Soc.,* 10(1), 34, 1989.
195. **El-Hibri, M. J., Cheng, W., Hattam, P., and Munk, P.,** Inverse gas chromatography of polymer blends. Theory and practice, *Am. Chem. Soc. Symp. Ser.,* 391 (Inverse Gas Chromatogr.: Charact. Polym. Other Mater.), 121, 1989.
196. **Rutan, S. C., Carr, P. W., Cheong, W. J., Park, J. H., and Snyder, L. R.,** Re-evaluation of the solvent triangle and comparison to solvatochromic based scales of solvent strength and selectivity, *J. Chromatogr.,* 463, 21, 1989.
197. **Schreiber, H. P. and Lloyd, D. R.,** Overview of inverse gas chromatography, *Am. Chem. Soc. Symp. Ser.,* 391 (Inverse Gas Chromatogr.: Charact. Polym. Other Mater.), 1, 1989.
198. **King, J. W.,** Fundamentals and application of supercritical fluid extraction in chromatographic science, *J. Chromatogr. Sci.,* 27, 355, 1989.
199. **Nguyen, Huu-Phuoc.,** *Determination of Partial Cohesion Parameters by Gas-Solid Chromatography of Solid Substances of Pharmaceutical Interest,* Ph.D. Thesis, University of Lausanne, Switzerland, 1984.
200. **Verzele, M., Dewaele, C., De Weerdt, M., and Abbott, S.,** Inner wall coating of micro-LC columns. Wall effects in LC, *J. High Resolut. Chromatogr.,* 12, 164, 1989.
201. **Folestad, S. and Larsson, M.,** Swollen polysiloxanes as stationary phases in reversed-phase open-tubular column liquid chromatography, *J. Chromatogr.,* 394, 455, 1987.
202. **Koliadima, A., Agathonos, P., and Karaiskakis, G.,** Estimation of solubility and interaction parameters in binary liquid mixtures by reversed-flow gas chromatography, *Chromatographia,* 26, 29, 1988.
203. **Bhattacharya, C., Maiti, N., Mandal, B. M., and Bhattacharya, S. N.,** Thermodynamic characterization of miscible blends from very similar polymers by inverse gas chromatography. The poly(ethyl acrylate) poly(vinyl propionate) system, *Macromolecules,* 22, 4062, 1989.
204. **Alessi, P., Orlandini-Visalberghi, M., and Torriano, G.,** New applications of gas chromatography in the field of paints and varnishes, *J. Oil Col. Chem. Assoc.,* 72, 15, 1989.

Chapter 19

BIOLOGICAL SYSTEMS

Solubility is of major importance in biological processes, and correlations with cohesion properties have been explored for such purposes as:

- Transport of molecules through biological tissues
- Rationalization of physicochemical influences on biological responses (structure-activity relations)
- Formulation of drugs in liquid and tablet dosage form
- Design of insecticides
- Compatibility of prostheses with biological materials

Several reviews have been compiled, including those of Beezer,[3] Cammarata et al.,[2,3] Dikstein,[4] Hansch et al.,[5,6] Mullins,[7] Nizhnii and Epshtein,[8] Osman et al.,[9] Smith and Lonsdale,[10] and Wolkoff.[11]

There is a wide variety of approaches to quantitative structure-activity correlations, including those based on partition coefficients, polarity, solubility or activity in lipids and stereochemical "fit". Many are of the linear free energy type[12] described in Section 7.12, which separate the influence of substituents from the effect of the medium on the action of these substituents, and it is this latter medium effect which may be described in terms of cohesion parameters. An example is the linear solvation energy relationship of Abraham, Carr, Doherty, Kamlet, Taft and co-workers described in Sections 8.2 and 8.4, which has been applied to octanol/water partition coefficients of polychlorinated biphenyls and polycyclic aromatic hydrocarbons.[142]

Biological systems inevitably involve water, and therefore hydrogen bonding. As emphasized elsewhere (particularly Chapter 5), cohesion parameters (like most other simple approaches to the interaction of materials) are least successful when applied to systems with specific interactions like hydrogen bonding. Because of large contributions to cohesive properties from hydrogen bonding, Hildebrand parameter values tend to be high and of limited value for discriminating purposes, and different Hansen hydrogen-bonding parameters must be used for dilute water-in-organic and dilute organic-in-water systems (Table 14, Chapter 5). Even with these limitations, cohesion parameters are useful for many biological systems, even those involving alcohols as well as water.

Hildebrand parameters may be determined by the various methods outlined in previous Chapters, particularly Chapter 12 for solids, including the measurement of dilute solution densities;[13] and some values appear in Table 1 of Chapter 12. Hansen parameters[141] for materials of biological interest appear in Table 12a, Chapter 5, and Rey-Mermet, Kesselring and co-workers[138,139] have conducted an extensive study of Hansen parameters of pharmaceutical solids.

19.1 HANSCH LIPOPHILIC PARAMETER

It was shown at the beginning of this century that the relative effectiveness of some drugs correlates with their partition coefficients for distribution between water and oils or lipids, and the term "lipophilicity" is commonly used to describe the tendency for a chemical agent to partition itself between aqueous and organic phases. This includes the penetration of membranes by drugs. Subsequently, it was suggested that it was the *concentration* of the drug in the organic phase which was critical, while Ferguson[14] provided expressions in terms of the thermodynamic *chemical potential*.

One of the parameters developed during the study of structure-activity relationships was the Hansch or lipophilic parameter π[6,15-17]. This was derived from partition studies and defined as the difference between the logarithms of the partition coefficients or distribution coefficients of a parent compound RH and its substituent derivative RX between aqueous and organic phases:

$$^{x}\pi = \log {}^{RX}D_{x} - \log {}^{RH}D_{x} \tag{1}$$

Water and 1-octanol were used as the reference solvents. This parameter may be expressed in terms of the Hildebrand parameters of the species involved, as follows.[2,3,18]

From equation 88, Chapter 7,

$$\ln {}^{RX}D_{x} = ({}^{RX}V/RT)[({}^{RX}\delta - {}^{aq}\delta)^2 - ({}^{RX}\delta - {}^{o}\delta)^2] \tag{2a}$$

and

$$\ln {}^{RH}D_{x} = ({}^{RH}V/RT[({}^{RH}\delta - {}^{aq}\delta)^2 - ({}^{RH}\delta - {}^{o}\delta)^2] \tag{2b}$$

so

$$\begin{aligned} 2.303RT\,{}^{x}\pi = {} & {}^{o}\delta^2 ({}^{RH}V - {}^{RX}V) + {}^{aq}\delta^2 ({}^{RX}V - {}^{RH}V) \\ & + 2\,{}^{o}\delta ({}^{RX}V\,{}^{RX}\delta - {}^{RH}V\,{}^{RH}\delta) \\ & + 2\,{}^{aq}\delta ({}^{RH}V\,{}^{RH}\delta - {}^{RX}V\,{}^{RX}\delta) \end{aligned} \tag{3}$$

This may be simplified in terms of group molar attraction constants, defined in Equation 17, Chapter 6:

$$\delta = \Sigma_z\,{}^{z}F/\Sigma_z\,{}^{z}V$$

$$2.303RT\,{}^{x}\pi = {}^{x}V({}^{aq}\delta - {}^{o}\delta)^2 + 2\,{}^{x}V({}^{o}\delta - {}^{aq}\delta) \tag{4}$$

${}^{x}\pi$ and ${}^{x}F$ are therefore related by an expression which includes the molar volume ${}^{x}V$ of the functional group X, and the Hildebrand parameters of the aqueous and organic phases. If the (often unjustified) assumption is made that molar volumes are approximately constant in a series of different functional groups, F alone may be used to correlate biological data as demonstrated by Ostrenga,[19,20] although there are examples where F gives a good correlation but π does not. The polar functional molar attraction constant F_p/F and other additive parameters[8-23] have also been used (see Section 6.9). Although 1-octanol/water distribution has been used widely, a recent study[136] on barbiturates suggests trichloromethane/water provides a better correlation of properties. Octanol-water partitioning models also have been applied in the interpretation of the behavior of natural polymer systems (Section 16.11).

Partial molar volume correlations are also possible,[13,24,25] and the UNIQUAC method (Sections 6.10 and 7.13)[26,27] has been used in the examination of distribution coefficient expressions such as Equation 1.

The Hansch parameter π and similar measures of solubility and activity have been related to other energy expressions, including those based on the electrostatic properties of permittivity and polarizability introduced in Chapter 3.[28-33] Such expressions, of course, can be correlated in turn with cohesion parameters, resulting in a variety of other relationships. This aspect of the cohesion parameter approach is not well developed.

19.2 EFFECTS AND TRANSMISSION RATES OF BIOLOGICALLY ACTIVE MATERIALS

The Hildebrand parameters and related properties have been used in various biochemical systems, and it is believed that this is a fairly complete list:

- Drug cosolvency[34] (see also Section 7.7)
- Binding of drugs by plasma proteins[35,36]
- Absorption of barbiturates by gambusia fish[37] and other measures and correlations of barbiturate activity[19,37,38,136] (Table 1, Chapter 12)
- Transport of carboxylsalicyclic acid[39] and sulfonamides[40]
- Transmission rates of drugs and liquids through polymers,[41-43] model membranes,[39,44,148] and human and other animal membranes[45-50,146,148]
- Comparative permeabilities of narcotic analgesics[146,149]
- Prolan® (1,1-bis(*p*-chlorophenyl)-2-nitropropane) activity as an insecticide[51]
- Action of anesthetics[52-54]
- Interaction of fluorinated ether anesthetics and fluorinated artificial membranes with solvents by ^{19}F NMR[55,56]
- Solution thermodynamics of norethindrone derivatives in 2,2,4-trimethylpentane (isooctane)[57]
- Prediction of the in vitro diffusion rate of active agents dissolved in a gel with a Carbopol® base[58]
- Drug release rates from suppositories[59]
- Poly(vinyl acetate phthalate) as an enteric coating material[60]
- Vehicle effects in the delivery of 6-mercaptopurine[61,62]
- Permeation of 5-fluorouracil[63]
- Solution thermodynamics of theophylline in polar binary liquids[64]
- Frog muscle and tadpole narcosis[6,65,66]
- Surface energetics analysis of artificial blood substitutes[67] and bioadhesion[68,69]
- Erythrocyte membrane hemolysis[3]
- Binding to bovine serum albumin[6]
- Chick embryo hatching[6]
- Transport of steroids[70] and benzocaine[41] through polymers
- Solubilities and related properties of parabens,[71-74] sulfonamides,[75,76] caffeine,[77] barbiturates,[39,78-80] testosterone derivatives,[81-83] quinidine,[144] narcotic analgesics,[149] and other pharmaceuticals[84,85,140] and environmentally hazardous compounds such as chlordane, DDT, dieldrin, and lindane[86] and polychlorinated biphenyls and polycyclic aromatic hydrocarbons[142]
- Drug analyses by high pressure liquid chromatography (Chapter 18)
- Solubilities in binary liquids of phenobarbitals[80]
- Biologically active modification of poly(vinyl alcohol)[87]
- Olfactory properties, solubility factors[88-91] and interaction of polymer packaging with low molecular weight molecules[73] from GLC (Chapter 8)
- Lipid extraction from biological tissues[92]
- Partition into lipid bilayer membranes[93]
- Permeability to alkanoic acids[46,47]
- Role of specific interactions in solubility of polar organic solutes in nonaqueous systems[94]
- Drug microspheres[144]
- Design of DDT analogs[95,96]
- Pesticide solubility[143] and adsorption[150]

One particularly interesting example is the action of fluorinated ethers as either inhalation convulsants or anesthetics.[97-101] Those such as hexafluorodiethyl ether with $\delta < 15$ MPa$^{1/2}$ are powerful convulsants, whereas those with $\delta > 15$ MPa$^{1/2}$, such as methoxyflurane, are anesthetics. Although all the ethers dissolve equally well in bulk lipids, and have similar octanol-water partition values, they dissolve differentially into specific local microenvironments or subregions which can be considered to have δ values different from the bulk δ value. This shows that the cohesion parameter concept is valuable both in models concerned with general partition, concentration, or activity and in those which assume more specific mechanisms.

A hydrogen bonding parameter has been investigated[102] to extend the cohesion parameter approach, but partial parameters have not been employed widely in structure-activity studies, and there is considerable potential for further investigation.

19.3 PROSTHETIC MATERIALS

The long-term compatibility and performance of polymeric materials in physiological environments is of increasing importance. As well as adsorption of biological materials, polymer swelling resulting from absorption is also possible. Both adsorption and absorption are now receiving considerable attention, with occasional discussion in terms of Hildebrand parameters.[103] Bioadhesion may be discussed in terms of dispersion and polar components of surface free energy.[69] Generally, however, studies in the field of biocompatibility are not making use of advances in cohesion parameter techniques which have proved so useful in the polymer coatings industry. Even as recently as the early 1960s, some silicone ball-shaped occluders for aortic valve replacements were being tested only in water.

Consideration of materials for prosthetic devices and similar applications in terms of cohesion parameters include:

- Effect of food and drink components on artificial teeth resins[104-107]
- Plasticizer choice in denture lining materials[108]
- Dental adhesives[109]
- Absorption of blood lipid by silicone rubber heart valves[110,111]
- Calcification of porcine xenograft heart valves and other elastomers
- Polymer-liquid interaction parameters for hydrogels with water[151]

19.4 OTHER BIOLOGICAL SYSTEMS

Cohesion parameters also have been applied to

- Pesticide formulation[95,96,113,114]
- Controlled release pesticides[115]
- Pharmaceutical liquid formulations[116]
- Pharmaceutical tablet formulations[117,122,147] (see also Section 17.5)
- Antimicrobial agents for textiles[123]
- Antimicrobial activity correlated with Hildebrand parameter[124]
- Solvent extraction of freezing curd from alfalfa (*Medicago sativa*) leaf juice[125]
- Lipid extractions from biological tissues[92] and flour[126-131]
- Solvent selection in biocatalysis (see also Section 10.3)[132]
- Fluorochemicals as oxygen and carbon dioxide transporters[133,145]
- Behavior of lipids in histology tissue samples embedded in methacrylate[134]
- Hydrogels for biomedial applications[135,151]
- Acetylcholinostearase inhibitors:[137] isopropyl methylphosphorofluoridate (sarin), 18.6 MPa$^{1/2}$; diisopropylphosphorofluoridate, 17.2 MPa$^{1/2}$; methanesulfonyl fluoride, 23.1 MPa$^{1/2}$.

REFERENCES

1. **Beezer, A. E.,** Solubility and pharmacological importance, *Pure Appl. Chem.,* 61, 153, 1989.
2. **Cammarata, A., Yau, S. J., and Rogers, K. S.,** Structure-activity relationships having a basis in regular solution theory, *J. Med. Chem.,* 14, 211, 1971.
3. **Cammarata, A., Yau, S. J., and Rogers, K. S.,** Some biological applications of regular solution theory, *Pure Appl. Chem.,* 35, 495, 1973.
4. **Dikstein, S.,** Physicochemical models of drug action, in *Biophysics of Physiological and Pharmacological Actions,* Science, Washington, D.C., Publ. No. 69, 1961, 49.
5. **Hansch, C.,** Quantitative approach to pharmacological structure-activity relationships, in *Structure-Activity Relationships,* Cavallito, C. J., Ed., Pergamon Press, Oxford, 1973, chap. 3.
6. **Leo, A., Hansch, C., and Church, C.,** Comparison of parameters currently used in the study of structure-activity relationships, *J. Med. Chem.,* 12, 766, 1969.
7. **Mullins, L. J.,** Some physical mechanisms in narcosis, *Chem. Rev.,* 54, 289, 1954.
8. **Nizhnii, S. V. and Epshtein, N. A.,** Quantitative ''chemical structure — biological activity'' relations, *Usp. Khim.,* 47, 739, 1978; *Russ. Chem. Rev.,* 47, 383, 1978.
9. **Osman, R., Weinstein, H., and Green, J. P.,** Parameters and methods in quantitative structure-activity relationships, in *Computer-Assisted Drug Design (Am. Chem. Soc. Symp. Ser.* 112), Olson, E. C. and Christofferson, R. E., Eds., American Chemical Society, Washington, D.C., 1979, 21.
10. **Smith, K. L. and Lonsdale, H. K.,** Membrane systems: theoretical aspects, in *Methods in Enzymology,* Vol. 112, Academic Press, 1985, chap. 37, 495.
11. **Wolkoff, H. N.,** Liquids, in *The Theory and Practice of Industrial Pharmacy,* Leachman, L., Lieberman, H. A., and Kanig, J. L., Eds., Lea & Febiger, Philadelphia, PA., 1970, chap. 15.
12. **Cammarata, A. and Rogers, K. S.,** The interpretation of drug action through linear free energy relationships, in *Advances in Linear Free Energy Relationships,* Chapman, N. B. and Shorter, J. Eds., Plenum Press, London, 1972, 401.
13. **Liron, Z. and Cohen, S.,** Densitometric determination of the solubility parameter and molal volume of compounds of medicinal relevance, *J. Pharm. Sc.,* 72, 499, 1983.
14. **Ferguson, J.,** The use of chemical potentials as indices of toxicity, *Proc. Roy. Soc. B.,* 127, 387, 1939.
15. **Fujita, T., Iwasa, J., and Hansch, C.,** A new substituent constant, π, derived from partition coefficients, *J. Am. Chem. Soc.,* 86, 5175, 1964.
16. **Hansch, C. and Fujita, T.,** ρ-γ-π analysis. A method for the correlation of biological activity and chemical structure, *J. Am. Chem. Soc.,* 86, 1616, 1964.
17. **Iwasa, J., Fujita, T., and Hansch, C.,** Substituent constants for aliphatic functions obtained from partition coefficients, *J. Med. Chem.,* 8, 150, 1965.
18. **Davis, S. S.,** Molar attraction constants applied to structure activity relationships, *Experientia,* 26, 671, 1970.
19. **Ostrenga, J. A.,** Correlation of biological activity with chemical structure. Use of molar attraction constants, *J. Med. Chem.,* 12, 349, 1969.
20. **Ostrenga, J. A.,** On the additivity of molar attraction constants, *J. Pharm. Sci.,* 58, 1281, 1969.
21. **Davis, S. S.,** Determination of the thermodynamics of the methyl group in solutions of drug molecules, *J. Pharm. Pharmacol.,* 25, 1, 1973.
22. **Davis, S. S.,** The effect of mutual solubility on the distribution of the methylene group between different organic solvents and water, *Sep. Sci.,* 10, 1, 1975.
23. **Davis, S. S., Higuchi, T., and Rytting, J. H.,** Determination of thermodynamics of the methylene group in solutions of drug molecules, *J. Pharm. Pharmacol.,* 24 (Suppl.), 30P, 1972; Determination of thermodynamics of functional groups in solutions of drug molecules, *Adv. Pharm. Sci.,* 4, 73, 1974.
24. **Cohen, S. and Haberman, F.,** The molal volumes of atropine and hyoscine in relation to their respective potencies, *Br. J. Pharmacol.,* 83, 807, 1984.
25. **Cohen, S. and Haberman, F.,** Enthalpy-entropy relationship in drug-cholinceptor interaction: a new approach, *Br. J. Pharmacol.,* 85, 889, 1985.
26. **Grünbauer, H. J. M. and Tomlinson, E.,** Local composition models in pharmaceutical chemistry. I. Liquid-liquid distribution coefficients from COMMON UNIQUAC parameters, *Int. J. Pharm.,* 21, 61, 1984.
27. **de Meere, A. L. J., Taylor, P. J., Grünbauer, H. J. M., and Tomlinson, E.,** Local composition models in pharmaceutical chemistry. II. Differentiation of hydrophobic fragmental constants, *Int. J. Pharm.,* 31, 209, 1986.
28. **Cammarata, A. and Rogers, K. S.,** Electronic representation of the lipophilic parameter, π, *J. Med. Chem.,* 14, 269, 1971.
29. **Cammarata, A. and Yau, S. J.,** Molecular estimates for molar attraction constants, *J. Pharm. Sci.,* 61, 723, 1972.

30. **Moore, W. E.,** The use of an approximate dielectric constant to blend solvent systems, *J. Am. Pharm. Assoc. Sci. Ed.*, 47, 855, 1958.
31. **Moore, W. E.,** Use of an approximate dielectric constant in solubility study, *J. Pharm. Sci.*, 51, 391, 1962.
32. **Rogers, K. S. and Cammarata, A.,** A molecular orbital description of the partitioning of aromatic compounds between polar and nonpolar phases, *Biochem. Biophys. Acta,* 193, 22, 1969.
33. **Rogers, K. S. and Cammarata, A.,** Superdelocalizability and charge density. A correlation with partition coefficients, *J. Med. Chem.*, 12, 692, 1969.
34. **Rubino, J. T. and Yalkowsky, S. H.,** Cosolvency and cosolvent polarity, *Pharm. Res.*, 4, 220, 1987.
35. **Bustamante, P. and Sellés, E.,** Relationship between the solubility parameter and the binding of drugs by plasma proteins, *J. Pharm. Sci.*, 75, 639, 1986.
36. **Bustamante, P. and Sellés, E.,** Solubility parameter as an index of drug binding to plasma proteins, in *Second European Colloquium of Biopharmaceutics and Pharmacokinetics,* Vol. 2, 1984, 385.
37. **Khalil, S. A., Abdallah, O. Y., and Moustafa, M. A.,** Absorption of some barbiturates by gambusia fish and its correlation to solubility parameter, *Can. J. Pharm. Sci.*, 11, 26, 1976.
38. **Khalil, S. A., Abdallah, O. Y., and Moustafa, M. A.,** The in vitro availability of some barbituric acid derivatives and its correlation to solubility parameter, *Pharm. Acta Helv.*, 58, 307, 1983.
39. **Khalil, S. A. and Martin, A. N.,** Drug transport through model membranes and its correlation with solubility parameters, *J. Pharm. Sci.*, 56, 1225, 1967.
40. **Schumacker, G. E. and Nogwekar, J. B.,** Kinetic and thermodynamic aspects of *in vitro* interphase transfer of sulfonamides. II. Influence of interface composition on transfer of unionized sulfonamides, *J. Pharm. Sci.*, 63, 244, 1974.
41. **Most, C. F.,** Co-permeant enhancement of drug transmission rates through silicone rubber, *J. Biomed. Mater. Res.*, 6, 3, 1972.
42. **Pitt, C. G., Andrady, L. A., Bao, Y. T., and Samuel, N. K. P.,** Estimation of rates of drug diffusion in polymers, *Am. Chem. Soc. Symp. Ser.*, 348, 49, 1987.
43. **Bao, Y. T., Samuel, N. K. P., and Pitt, C. G.,** The prediction of drug solubilities in polymers, *J. Polym. Sci. Polym. Lett.*, 26, 41, 1988.
44. **Shroff, A. P. and Shaw, C. J.,** A model for estimating availability for specific drugs, *Can. J. Pharm. Sci.*, 6(1), 24, 1971.
45. **Adjei, A., Newburger, J., Stavchansky, S., and Martin, A.,** Membrane solubility parameter and in situ release of theophylline, *J. Pharm. Sci.*, 73, 742, 1984.
46. **Liron, Z. and Cohen, S.,** Percutaneous absorption of alkanoic acids. I. A study of operational conditions, *J. Pharm. Sci.*, 73, 534, 1984.
47. **Liron, Z. and Cohen, S.,** Percutaneous absorption of alkanoic acids. II. Application of regular solution theory, *J. Pharm Sci.*, 73, 538, 1984.
48. **Sloan, K. B., Koch, S. A. M., Siver, K. G., and Flowers, F. P.,** Use of solubility parameters of drug and vehicle to predict flux through skin, *J. Invest. Dermatol.*, 87, 244, 1986.
49. **Sloan, K. B., Siver, K. G., and Koch, S. A. M.,** The effect of vehicle on the diffusion of salicyclic acid through hairless mouse skin, *J. Pharm. Sci.*, 75, 744, 1986.
50. **Kadir, R., Stempler, D., Liron, Z., and Cohen, S.,** Delivery of theopohylline into excised human skin from alkanoic acid solutions. A push-pull mechanism; Penetration of adenosine into excised human skin from binary vehicles: the enhancement factor; Penetration of theophylline and adenosine into excised human skin from binary and ternary vehicles: effect of a nonionic surfactant, *J. Pharm. Sci.*, 76, 774, 1987; 77, 409, 1988; 78, 149, 1989.
51. **Lee, A. H., Metcalf, R. L., Williams, J. W., Hirwe, A. S., Sanhorn, J. R., Coats, J. R., and Fukuto, T. R.,** Structure-activity correlations of the insecticide Prolan and its analogs, *Pesticide Biochem. Physiol.*, 7, 426, 1977.
52. **Lever, M. J., Miller, K. W., Patton, W. D. M., and Smith, E. B.,** Pressure reversal of anaesthesia, *Nature (London),* 231, 368, 1971.
53. **Miller, K. W., Paton, W. D. M., and Smith, E. B.,** Site of action of general anaesthetics, *Nature (London),* 206, 574, 1965.
54. **Dumas, J. M. and Kribii, A.,** Anaesthesia and molecular interactions — the case of halogenated anaesthetics, *Nouv. J. Chim.*, 8, 349, 1984.
55. **Koehler, L. S., Curley, W., and Koehler, K. A.,** solvent effects on halothane: ^{19}F nuclear magnetic resonance in solvents and artificial membranes, *Mol. Pharmacol.*, 13, 113, 1977.
56. **Koehler, K. A., Stone, E. E., Shelton, R. A., Jarnagin, F., Koehler, L. S., and Fossel, E. T.,** Interaction of fluorinated ether anesthetics with solvents, *J. Mag. Res.*, 30, 75, 1978.
57. **Lewis, G. A. and Enever, R. P.,** Solution thermodynamics of some potentially long-acting norethindrone derivatives. II. Solutions in 2,2,4-trimethylpentane (iso-octane); III. Measurement of aqueous solubilities and the use of group free-energy contributions in predicting partition coefficients, *Int. J. Pharm.*, 3, 275, and 319, 1989.

58. **Lhoest, G.,** Use of certain physicochemical parameters to predict the relative *in vitro* diffusion rate of an agent dissolved in a gel with a Carbopol® base, *Pharm. Acta Helv.*, 48, 549, 1973.
59. **Liversidge, G. G. and Grant, D. J. W.,** Influence of physicochemical interactions on the properties of suppositories. II. Interactions between the constituents of fatty suppository bases and ketoprofen or metronidazole, *Drug. Dev. Ind. Pharm.*, 9, 223, 1983.
60. **Nesbitt, R. U., Goodhart, F. W., and Gordon, R. H.,** Evluation of polyvinyl acetate phthalate as an enteric coating material, *Int. J. Pharm.*, 26, 215, 1985.
61. **Waranis, R. P., Siver, K. G., and Sloan, K. B.,** The solubility parameter of vehicles as a predictor of relative vehicle effects on the diffusion of 6-mercaptopurine, *Int. J. Pharm.*, 36, 211, 1987.
62. **Waranis,R. P. and Sloan, K. B.,** Effects of vehicles and prodrug properties and their interactions on the delivery of 6-mercaptopurine through skin: bisacyloxymethyl-6-mercaptopurine drugs; S^6-Acyloxymethyl-6-mercaptopurine prodrugs, *J. Pharm. Sci.*, 76, 587, 1987; 77, 210, 1988.
63. **Sherertz, E. F., Sloan, K. B., and McTiernan, R. G.,** Use of theoretical partition coefficient determined from solubility parameters to predict permeability coefficients for 5-fluorouracil, *J. Invest. Dermatol.*, 86, 506, 1986 (abstr.); 89, 147, 1987.
64. **Martin, A., Newburger, J., and Adjei, A.,** Extended Hildebrand solubility approach. II. Solubility of theophylline in polar binary solvents, *J. Pharm. Sci.*, 69, 487, 1980.
65. **Moriguchi, I.,** Quantitative structure-activity studies. I. Parameters relating to hydrophobicity, *Chem. Pharm. Bull.*, 23, 247, 1975.
66. **Moriguchi, I., Kanada, Y., and Komatsu, K.,** Quantitative structure-activity studies. II. van der Waals volume and the related parameters for hydrophobicity in structure-activity studies. *Chem. Pharm. Bull.*, 24, 1799, 1976.
67. **Kaelble, D. H. and Moacanin, J.,** Surface energetics analysis of artificial blood substitutes, *Med. Biol. Eng. Comput.*, 17, 593, 1979.
68. **Kaelble, D. H.,** Interface degradation processes and durability, *Polym. Eng. Sci.*, 17, 474, 1977.
69. **Kaeble, D. H. and Moacanin, J.,** A surface energy analysis of bioadhesion, *Polymer,* 18, 475, 1977.
70. **Michaels, A. S., Wong, P. S. L., Prather, R., and Gale, R. M.,** A thermodynamic method of predicting the transport of steroids in polymer matrices, *Am. Inst. Chem. Eng. J.*, 21, 1073, 1975.
71. **Paruta, A. N.,** Solubility of parabens in alcohols, *J. Pharm. Sci.*, 58, 216, 1969.
72. **Restaino, F. A. and Martin, A. N.,** Solubility of benzoic acid and related compounds in a series of *n*-alkanols, *J. Pharm. Sci.*, 53, 636, 1964.
73. **Varsano, J. L. and Gilbert, S. C.,** Evaluation of interactions between polymers and low molecular weight compounds by GLC. I. Methodology and interaction evaluation; II. Thermodynamics and elucidation, *J. Pharm. Sci.*, 62, 87 and 92, 1973.
74. **Yalkowsky, S. H., Flynn, G. L., and Slunick, T. G.,** Importance of chain length on physicochemical and crystalline properties of organic homologs, *J. Pharm. Sci.*, 61, 852, 1972.
75. **Mauger, J. W., Paruta, A. N., and Gerraughty, R. J.,** Solubilities of sulfadiazine, sulfisomidine, sulfadimethoxine in several normal alcohols, *J. Pharm Sci.*, 61, 94, 1972.
76. **Sunwoo, C. and Eisen, H.,** Solubility parameter of selected sulfonamides, *J. Pharm. Sci.*, 60, 238, 1971.
77. **Adjei, A., Newburger, J., and Martin, A.,** Extended Hildebrand approach. III. Solubility of caffeine in dioxane-water mixtures, *J. Pharm. Sci.*, 69, 659, 1980.
78. **Ghosal, S. K. and Gupta, B. K.,** Determination of the solubility parameters of several barbiturates: their significance in drug activity, *J. Inst. Chem. (India),* 51, 117, 1979.
79. **Khalil, S. A., Moustafa, M. A., and Abdallah, O. Y.,** The use of the solubility parameter as an index of drug activity, *Can. J. Pharm. Sci.*, 11, 121, 1976.
80. **Maher, N. and Sirois, G.,** Correlation between the solubility of phenobarbitol and the polarity of various solvent systems, *Can. J. Pharm. Sci.*, 6(3), 55, 1971.
81. **Bowen, D. B. and James, K. C.,** Effect of temperature on the solubilities of testosterone propionate in low polarity solvents, *J. Pharm. Pharmacol.*, 22, 104S, 1970.
82. **James, K. C., Ng, C. T., and Noyce, P. R.,** Solubilities of testosterone propionate and related esters in organic solvents, *J. Pharm. Sci.*, 65, 656, 1976.
83. **James, K. C. and Roberts, M.,** The solubilities of lower testoesterone esters, *J. Pharm. Pharmacol.*, 20, 709, 1968.
84. **Sellés, E. and Bustamante, P.,** Solubility of Ketoprofen in binary solvents, *27th National Congress, Asoc. Espanola de Farmaceuticos,* Hospital Tomo I, Granada, Spain, 1982, 95.
85. **Parera, A., Bustamante, P., and Sellés, E.,** Application of the regular solution theory to the solubilization of indomethacin, *An. Real. Acad. Farm.*, 48, 37, 1982.
86. **Thomas, A. M.,** Petroleum fractions as DDT solvents, *J. Agric. Food Chem.*, 12, 442, 1948.
87. **Gebelein, C. G. and Burnfield, K. E.,** Biologically active modification of polyvinyl alcohol: the reaction of phenyl isocyanate with polyvinyl alcohol, in *Modification of Polymers (Am. Chem. Soc. Symp. Ser.* 121), Carraher, C. E. and Tsuda, M., Eds., American Chemical Society, Washington, D.C., 1980, 83.

88. **Laffort, P., Patte, F., and Etcheto, M.,** Olfactory coding on the basis of physicochemical properties, *Ann. N.Y. Acad. Sci.*, 237, 193, 1974.
89. **Laffort, P. and Patte, F.,** Solubility factors in gas-liquid chromatography. Comparison between two approaches and application to some biological studies, *J. Chromatogr.*, 126, 625, 1976.
90. **Patte, F.,** Actual status of olfactory properties predictability from chromatographic measurements, *First Congress of European Chemoreception Research Organization*, European Chemoreception Research Organization, London, 1974.
91. **Patte, F.,** Molecular aspects of chemoreception, *Third Congress European Chemoreception Research Organization*, European Chemoreception Research Organization, London, 1978.
92. **Schmid, P.,** Extraction and purification of lipids. I. Solubility of lipids in biologically important solvents; II. Why is chloroform-methanol such a good lipid solvent? *Physiol. Chem. Phys.*, 3, 98, 1971; 5, 141, 1973.
93. **Simon, S. A., Stone, W. L., and Bennett, P. B.,** Can regular solution theory be applied to lipid bilayer membranes?, *Biochim. Biophys. Acta*, 550, 38, 1979.
94. **Anderson, B. D., Rytting, J. H., and Higuchi, T.,** Solubility of polar organic solutes in non-aqueous systems — role of specific interactions, *J. Pharm. Sci.*, 69, 676, 1980.
95. **Coats, J. P., Metcalf, R. L., and Kapoor, I. P.,** Effective DDT analogues with altered aliphatic moieties. Isobutane and chloropropanes, *J. Agric. Food. Chem.*, 25, 859, 1977.
96. **Kapoor, I. P., Metcalf, R. L., Hirwe, A. S., Coats, J. R., and Khalsa, M. S.,** Structure activity correlations of biodegradability of DDT analogs, *J. Agric. Food. Chem.*, 21, 310, 1973.
97. **Cohen, S., Goldschmid, A., Shtacher, G., Srebrenik, S., and Gitter, S.,** The inhalation convulsants: a pharmacodynamic approach, *Molec. Pharmacol.*, 11, 379, 1975.
98. **Landau, E. M.,** Function and structure of the ACh receptor at the muscle end-plate, *Prog. Neurobiol.*, 10, 253, 1978.
99. **Landau, E. M., Richter, J., and Cohen, S.,** Differential solubilities in subregions of the membrane: a nonsteric mechanism of drug specificity, *J. Med. Chem.*, 22, 325, 1979.
100. **Richter, J., Landau, E. M., and Cohen, S.,** The action of volatile anesthetics and convulsants on synaptic transmission; a unified concept, *Molec. Pharmacol.*, 13, 548, 1977.
101. **Richter, J., Landau, E. M., and Cohen, S.,** Anaesthetic and convulsant ethers act on different sites at the crab neuromuscular junctions *in vitro, Nature (London),* 266, 70, 1977.
102. **Fujita, T., Nishioka, T., and Nakajima, M.,** Hydrogen bonding parameter and its significance in quantitative studies, *J. Med. Chem.*, 20, 1071, 1977.
103. **Bruck, S. D.,** Effect of the physiological environment on the mechanical properties of biomaterials in cardiovascular applications, *Biomat. Med. Devices Artif. Organs,* 6, 341, 1978.
104. **Asmussen, E.,** Softening of BISGMA-based polymers by ethanol and by organic acids of plaque, *Scand. J. Dent. Res.*, 92, 257, 1984.
105. **Wu, W. and McKinney, J. E.,** Influence of chemicals on wear of dental composites, *J. Dent. Res.*, 61, 1180, 1982.
106. **McKinney, J. E. and Wu, W.,** Chemical softening and wear of dental composites, *J. Dent. Res.*, 64, 1326, 1985.
107. **McKinney, J. E. and Wu, W.,** Relationships between subsurface damage and wear of dental restorative composites, *J. Dent. Res.*, 61, 1083, 1982.
108. **Jones, D. W.,** Influence of plasticizer on soft polymer gelation, *J. Dent. Res.*, 65, 634, 1986.
109. **Rupp, N. W., Bowen, R. L., and Paffenberger, G. C.,** Bonding cold-curing denture base acrylic resin to acrylic resin teeth, *J. Am. Dent. Assoc,* 83, 601, 1971.
110. **Fedors, R. F.,** Absorption of liquids by rubber vulcanizates, *Polymer,* 20, 1087, 1979.
111. **Moacanin, J., Lawson, D. D., Chin, H. P., Harrison, E. C., and Blankenhorn, D. H.,** Prediction of lipid uptake by prosthetic heart valve poppets from solubility parameters, *Biomat. Med. Devices Artif. Organs,* 1, 183, 1973.
112. **Bruck, S. D.,** Possible causes for the calcification of glutaraldehyde-treated tissue heart valves and blood contacting elastomers during prolonged use in medical devices: a physico-chemical view, *Biomaterials,* 2, 14, 1981.
113. **Meusberger, K. E.,** Determination of cohesive energy density parameters for pesticide formulation development, *Am. Chem. Soc. Symp. Ser.* 371 (Pesticide Formulations), 151, 1988.
114. **Meusberger, K. E.,** Computerized optimization of emulsifiers for pesticide emulsifiable concentrates, Personal communication, 1988.
115. **Dellicolli, H. T.,** Controlled release of pesticides from kraft lignin, in *Controlled Release Pesticides, (Am. Chem. Soc. Symp. Ser.* 53), Scher, H. B., Ed., American Chemical Society, Washington, D.C., 1977, chap. 8.
116. **Bustamante, P., Parra, A., and Sellés, E.,** Applications of the multiple solubility parameter to the formulation of liquid drug forms, *An. Real. Acad. Farm.*, 49, 221, 1983.

117. **Rowe, R. C.,** The prediction of compatibility/incompatibility in blends of ethyl cellulose with hydroxypropyl methylcellulose or hydroxypropyl cellulose using two-dimensional solubility parameter maps, *J. Pharm. Sci.*, 38, 214, 1986.
118. **Rowe, R. C.,** Adhesion of film coatings to tablet surfaces — a theoretical approach based on solubility parameters; Interaction of lubricants with microcrystalline cellulose and anhydrous lactose — a solubility parameter approach; Interactions in the ternary power system microcrystalline cellulose, magnesium stearate and colloidal silica — a solubility parameter approach, *Int. J. Pharm.*, 41, 219 and 223, 1988; 45, 259, 1988.
119. **Rowe, R. C.,** Binding-substrate interactions in tablets: a theoretical approach based on solubility parameters, *Acta Pharm. Technol.*, 34, 144, 1988.
120. **Rowe, R. C.,** Interactions in coloured powders and tablet formulations — a theoretical approach based on solubility parameters, *Int. J. Pharm.*, 53, 47, 1989.
121. **Sakellariou, P., Rowe, R. C., and White, E. F. T.,** The solubility parameters of some cellulose derivatives and polyetheylene glycols used in tablet film coating, *Int. J. Pharm.*, 31, 175, 1986.
122. **Gibson, S. H. M., Rowe, R. C., and White, E. F. T.,** Quantitative assessment of additive-polymer interactions in pigmented hydroxypropyl methylcellulose formulations using immersion calorimetry, *Int. J. Pharm.*, 48, 113, 1988.
123. **Malysheva, I. I., Belkin, A. I., and Korenev, A. D.,** Production of commercial forms of antimicrobial agents for treatment of textile materials in aqueous media, *Zh. Prikl. Khim.*, 56, 1623, 1983; *J. Appl. Chem., U.S.S.R.*, 56, 1522, 1983.
124. **Vaughan, C. D. and Wright, F. J.,** Solubility parameters and antimicrobial activity, *Pharm. Acta Helv*, 61(3), 95, 1986.
125. **Hernandez, A., Martinez, C., and Gonzalez, G.,** Freezing of alfalfa leaf juice — formation and solvent extraction of freezing curd, *J. Sci. Food Agric.*, 42, 173, 1988.
126. **Chung, O. K., Pomeranz, Y., Finney, K. F., and Shogren, M. D.,** Defatted and reconstituted wheat flours. I. Effects of solvents and Soxhelt types on functional (breadmaking) properties, *Cereal Chem.*, 54, 454, 1977.
127. **Chung, O. K., Pomeranz, Y., Hwang, E. C., and Dikeman, E.,** Defatted and reconstituted wheat flours. IV. Effects of flour lipids on protein extractibility from flours that vary in bread-making quality, *Cereal Chem.*, 56, 220, 1979.
128. **Chung, O. K., Pomeranz, Y., and Jacobs, R. M.,** Extractibiltiy of wheat-flour lipids by solvents that vary in solubility parameters, *J. Am. Oil. Chem. Soc.*, 56 (Abstr.), A194, 1979.
129. **Chung, O. K., Pomeranz, Y., Finney, K. F., Shogren, M. D., and Carville, D.,** Defatted and reconstituted wheat flours, V. Breadmaking response to shortening of flour differentially defatted by varying solvent and temperature, *Cereal Chem.*, 57, 106, 1980.
130. **Chung, O. K., Pomeranz, Y., Jacobs, R. M., and Howard, B. G.,** Extraction conditions to differentiate among hard red winter wheats that vary in breadmaking, *J. Food Sci.*, 45, 1168, 1980.
131. **Chung, O. K., Pomeranz, Y., and Jacobs, R. M.,** Solvent solubility parameter and flour mositure effects on lipid extractibility, *J. Am. Oil Chem. Soc.*, 61, 793, 1984.
132. **Brink, L. E. S. and Tramper, J.,** Optimization of organic solvent in multiphase biocatalysis, *Biotech. Bioeng.*, 27, 1258, 1985.
133. **Yamanouchi, K., Tanaka, M., Tsuda, Y., Yokoyama, K., Awazu, S., and Kobayashi, Y.,** Quantitative structure — *in vivo* half-life relationships of perfluorochemicals for use as oxygen transporters, *Chem. Pharm. Bull.*, 33, 1221, 1985.
134. **Gerrits, P. O., Horobin, R. W., Wright, D. J., van Goor, H., and Grond, J.,** Retention, localization and staining of lipids in tissues embedded in water-miscible methacrylates using the solubility parameter to analyse the influences of resins, pasticizers and other reagents, *Acta Morphol. Neerl.-Scand.*, 25, 29, 1987.
135. **Bruck, S. D.,** Aspects of three types of hydrogels for biomedical applications, *J. Biomed. Mater. Res.*, 7, 387, 1973.
136. **Guillén sans, R. and Guzmán chozas M.,** Determination of dielectric requirement and solubility parameter of 5-(anisylidene)-2-thiobarbituric acid, *Pharm. Acta Helv.*, 64(2), 51, 1989.
137. **Snow, A. W. and Barger, W. R.,** A chemical comparison of methanesulfonyl fluoride with organofluorophosphorus ester anticholinesterase compounds, *Chem. Res. Toxicol.*, 1, 379, 1988.
138. **Rey-Mermet, C.,** *Determination of Partial Cohesion Parameters from Binary Thermodynamic Quantities*, Doctoral Thesis, School of Pharmacy, University of Lausanne, Switzerland, 1989.
139. **Rey-Mermet, C., Ruelle, P., Hô Nam-Tran, Negre, J. C., Buchmann, M., and Kesselring, U. W.,** Determination of cohesion parameters of three xanthines by heat-flow calorimetry, to be published.
140. **Grant, D. J. W. and Abougela, I. K. A.,** Prediction of the solubility of griseofulvin in glycerides and other solvents of relatively low polarity from simple regular solution theory, *Int. J. Pharm.*, 17, 77, 1983; Solubility behavior of griseofulvin in solvents of relatively low polarity, *Expo.-Congr. Int. Technol. Pharm., 3rd*, Vol. 4, Assoc. Pharm. Galenique Ind., Chatenay-Malabry, France, 1983, 118.

141. **Hansen, C. M. and Anderson, B. H.,** The affinities of organic solvents for biological systems, *Am. Ind. Hyg. Assoc.*, 49, 301, 1988.
142. **Kamlet, M. J., Doherty, R. M., Carr, P. W., Mackay, D., Abraham, M. H., and Taft, R. W.,** Linear solvation energy relationships. 44. Parameter estimation rules that allow accurate prediction of octanol/water partition coefficients and other solubility and toxicity properties of polychlorinated biphenyls and polycyclic aromatic hydrocarbons, *Environ. Sci. Technol.*, 22, 503, 1988.
143. **Bozzay, J., Dávid, Á., Ékes, G., and Rusznák, I.,** On the solubility of pesticides and compounds of pesticide type, *Period. Polytech. Chem. Eng.*, 23(2), 87, 1979.
144. **Bodmeier, R. and McGinity, J. W.,** Solvent selection in the preparation of poly(DL-lactide) microspheres prepared by the solvent evaporation method, *Int. J. Pharm.*, 43, 179, 1988.
145. **Devallez, B., Gauffreteau, Y., LeBlanc, M., and Riess, J. G.,** Solubility of respiratory gases in 1,2-bis(fluoroalkyl) ethenes, *J. Chim. Phys. Phys-Chim. Biol.*, 85, 947, 1988.
146. **Roy, S. D. and Flynn, G. L.,** Transdermal delivery of narcotic analgesis: comparative permeabilities of narcotic analgesis through human cadaver skin, *Pharm. Res.*, 6, 825, 1989.
147. **Rowe, R. C.,** Binder substrate interactions in granulation — a theoretical approach based on surface free energy and polarity, *Int. J. Pharm.*, 52, 149, 1989.
148. **Berner, B., Juang, R.-H., and Mazzenga, G. C.,** Ethanol and water sorption into stratum corneum and model systems, *J. Pharm. Sci.*, 78, 472, 1989.
149. **Roy, S. D. and Flynn, G. L.,** Solubility and related physicochemical properties of narcotic analgesis, *Pharm. Res.*, 5, 580, 1988.
150. **Pussemier, L., De Borger, R., Cloos, P., and Van Bladel, R.,** Relation between the molecular structure and the adsorption of arylcarbamate, phenylurea and anilide pesticides in soil and model organic adsorbents, *Chemospheres*, 18, 1871, 1989.
151. **Mikos, A. G. and Peppas, N. A.,** Flory interaction parameter χ for hydrophilic copolymers with water, *Biomaterials*, 9, 419, 1988.

GLOSSARY OF SYMBOLS

Symbol	Definition	Section in which introduced
A	Empirical parameter in various relationships	2.5, 6.6, 8.2, 14.7, 16.4
A	Parameter linking surface energy and cohesion parameter	17.2
A	First Antoine parameter	7.2
A	Polymer contribution to dissolving power	8.4
A	Rate equation preexponential parameter	10.2
${}^{ij}A$	Exchange energy density, interchange cohesive pressure	4.2
${}^{j}A^{v}$	Vaporization cohesive pressure	18.1
${}^{ja}A$	Adsorption cohesive pressure	18.1
A	(Superscript) acceptor	7.9
AB	Lewis acid-base term	7.10
B	Empirical parameter in various relationships	6.6, 8.2, 14.7, 14.8, 16.4
B	Second Antoinc parameter	7.2
B	Lewis basicity scale parameter	8.3
B	Interaction energy density	13.3, 14.5
B	Nonideality correction coefficient	7.2
B_{λ}	Correlation parameter for δ_{λ}	7.8
C	Empirical parameter in various relationships	13.7, 16.4, 18.1, 19.1
C	Third Antoine equation	7.2
C	Drago parameter	7.12
C	Covalent heat capacity	12.1
C	Segment-segment hydrogen-bonding term in dissolving power	8.4
C	(Superscript) relating to carbon	16.11
C	(Subscript) chemical	5.5
C	(Unit) coulomb	2.5
°C	(Unit) degree celsius	2.5
D	Distribution ratio	7.11, 19.1
D	Empirical parameter	16.4
D	Dispersion term	7.10
D	Cohesion parameter term in dissolving power	8.4
D	(Superscript or subscript) donor	7.9
DA	(Superscript) donor-acceptor	7.9
D(δ)	Correlation function	7.11
D	(Unit) debye	2.5
D	(Subscript) sodium "D" line for refractive index	3.1
E	Electric field strength	3.1
E	Drago parameter	7.12
E	Electrostatic bonding energy	5.7
E	Empirical parameter	16.4
E	Adsorption energy	17.6
E	Entropy term	10.2
E	Activation energy	10.2

GLOSSARY OF SYMBOLS (continued)

Symbol	Definition	Section in which introduced
E_T	Reichardt parameter	8.4
E_t	Calculated net adsorption energy	17.6
E_η	Activation energy for viscous flow	10.5
E	(Superscript) excess	2.6
EPA	Electron-pair acceptor	5.4
EPD	Electron-pair donor	5.4
EF	Electrostatic factor	8.2
F	Molar attraction	6.2
${}^zF$	Group molar attraction constant	6.2, 14.8
G	Molar Gibbs free energy function	2.6
H	Molar enthalpy	2.1
H	Liquid-polymer hydrogen-bonding term in dissolving power	8.4
H, ${}^{ij}H$	Henry's constant	11.1
H	(Superscript) hydrophile	17.7
H	(Subscript) relating to enthalpy	13.3
HLB	Hydrophile-lipophile balance	17.7
$\mathrm{HLB_E}$	Extended hydrophile-lipophile balance	17.7
I	Ionization potential	3.4, 4.1
I	Induction term	7.10
I	(Superscript) ideal	
J	(Unit) joule	
K	Equilibrium constant	7.9
K	Empirical parameter in various relationships	14.7, 16.4
K	UOP characterization factor	8.5
K	Vaporization ratio	11.5
K	Partition coefficient, distribution constant	18.1
K	Walden's constant	12.2
K	Mark-Houwink parameter	14.7
${}^{ij}K$	Empirical geometric mean deviation parameter	5.7
K	(Unit) kelvin	
KB	Kauri-butanol number	8.5
L	(Superscript) lipophile	17.7
LCST	Lower critical solution temperature	13.4
LSER	Linear solvation energy relation	8.4
[L]	Normal lyoparachor	6.3
M	Molar mass, molecular weight	3.1, 6.3
${}^CM$	Molar mass per carbon atom	16.11
$\underline{M}$	Miscibility number	8.5
$\overline{M}$	Polymer viscosity-average molecular weight	14.7
$\overline{M}_n$	Polymer number-average molecular weight, $\Sigma_i n_i M_i / \Sigma_i n_i$	
M	(Unit prefix) mega, 10^6	
MPa	(Unit multiple) megapascal	
N_A	Avogadro constant	6.5, 13.1, 17.2
N	(Unit) newton	2.5

GLOSSARY OF SYMBOLS (continued)

Symbol	Definition	Section in which introduced
O	Orientation term	7.10
P	Parameter in rate equation	10.2
P'	Polarity index (chromatographic strength parameter)	8.5
P	(Subscript) physical	5.5
$\mathrm{P_S}$	Molar parachor	17.1
Pa	(Unit) pascal, N m^{-2}	2.5
Q	Parameter in rate equation	10.2
Q	Swelling coefficient	14.3
R	Gas constant	2.1
R	Molar refraction	3.1
${}^{ij}R$ ${}^{sl}R$	Distance in Hansen space	5.9, 14.5, 17.3
${}^{j}R$, ${}^{s}R$	Radius of interaction circle in Hansen space	5.9, 14.5, 17.3
R_o	Cohesive energy ratio	17.7
RH	(Superscript) parent compound of series RX	19.1
RX	(Superscript) substituted derivative of RH	19.1
R	(Superscript or subscript) residual	5.12, 7.5, 13.2
S	Molar entropy	2.6
S	Softness value	5.12
S	(Subscript) relating to entropy	13.3
T	Absolute temperature	2.1
T_L	Correlation parameter for δ_λ	7.8
T_b	(Normal) boiling point	7.3
T_c	Critical temperature	5.7
T_{cs}	Critical solution temperature	4.2
T_g	Glass transition temperature	13.1
T_m	Melting point	12.1
T_r	Reduced temperature	9.1
T_t	Triple point	
T_*	Temperature of parameter determination	5.8
T	(Subscript) isothermal, constant temperature	
U	Molar internal energy	2.1
U	Multivariate analysis parameter	8.4
$-U$	Molar cohesive energy	2.1
$-U^*$	Effective molar cohesive energy	6.6
U^∞	Molar internal energy at infinite expansion	11.3
UCST	Upper critical solution temperature	13.4
V	Molar volume	2.1
V	Multivariate analysis parameter	8.4
V	(Subscript) isochoric, constant volume	4.3
V_c	Critical molar volume	7.4
${}^{C}V$	Molar volume per carbon atom	16.11
V_o	Van der Waals molar volume	9.1
${}^{z}V$	Group molar volume constant	6.2
$\overline{V}$	Partial molar volume	7.5
${}^{z}V_g$	Group molar volume at T_g for group z	5.9

GLOSSARY OF SYMBOLS (continued)

Symbol	Definition	Section in which introduced
${}^{n}V_{r}$	Corresponding states functions	7.4
${}^{\ddagger}V$	Molar volume of activated complex	10.2
V	(Subscript) relating to volume-dependent part	5.12, 7.5, 13.2
W	Work of adhesion	17.5
W	Multivariate analysis parameter	8.4
${}^{ij}W$	Binary interaction parameter	5.7
X	Electronegativity	5.7
X	Complex formation parameter	7.9
X	(Superscript) substituent in RX	19.1
Y	Molar additive property	6.3
Z	Compressibility factor	7.2, 11.3
Z	Kosower parameter	8.4
Z_{c}	Critical compressibility factor	7.4
a	Activity	2.6
a	Empirical parameters in various relationships	6.7, 7.4, 7.8, 7.9, 8.4, 11.2, 13.3
a	van der Waals constant	2.4
a	Exponent in Mark-Houwink expression	14.7
a_0,a_1,a_2	Empirical parameters in H-T-δ relation	11.6
a	(Subscript) Lewis acid	5.4
a	(Superscript or subscript) adsorbed, adsorbent	18.1
ab	(Subscript) Lewis acid-Lewis base	5.4
ad	(Subscript) adhesion	
add	(Subscript) additional	5.12
am	(Subscript) amorphous	13.1
ap	(Subscript) apparent	7.2, 14.4
aq	(Subscript or superscript) aqueous	4.2, 7.11, 18.1
atm	(Unit) atmosphere	2.5
b	van der Waals constant	2.4
b	Empirical parameter in various relationships	6.6, 6.7, 7.4, 7.8, 7.9, 8.4, 11.2, 13.3
b	Liquid structure parameter in dissolving power	4.4, 8.4
b	Coefficient in homomorph equation	6.7
${}^{i}b$	Hansen weighting factor	5.8
${}^{ij}b$	Induction cohesive pressure correction factor	5.8
b	(Subscript) Lewis base	5.4
b	(Subscript) boiling point	7.3
c	Cohesive pressure, cohesive energy density	2.2
c	Empirical parameter in various relationships	7.4
c	(Superscript or subscript) crystalline solid	
c	(Unit prefix) centi, 10^{-2}	
c	(Subscript) critical	5.7, 13.3
c	(Subscript) molar concentration scale	7.9
cal	(Unit) calorie	
ca	(Subscript) calculated	7.2
co	(Subscript) combinatorial	13.2

GLOSSARY OF SYMBOLS (continued)

Symbol	Definition	Section in which introduced
cr	(Subscript) crystalline	13.1
cs	(Subscript) critical solution	4.2
d	Empirical parameter in various relationships	7.4, 8.4
d	Correlating parameter in induction	7.8
d	U_d/U	8.2
d	(Superscript) diluent	7.11
^{ij}d	Flory-Huggins combinatorial size effect term	4.2, 5.13
d	(Subscript) dispersion	3.4
e	Empirical parameter in various relationships	7.4, 7.8, 8.4
e	(Superscript) end group	6.2, 6.5, 14.8
e	(Superscript) extractant	7.11
exp	Exponent	
f	Empirical parameter in various relationships	7.4, 7.8
f	Function	
f	Fugacity	11.1
f	Fractional cohesion parameter	5.10
f	Alternative symbol for ^{ij}K	5.5
f_a	Coal aromaticity	16.11
f_x	Activity coefficient (mole fraction scale)	2.6
f_x^∞	Infinite dilution activity coefficient (mole fraction scale)	4.2
f_1,f_2,f_3	Corresponding states functions	7.4
fv	(Subscript) free volume	3.1, 13.4
g	Material parameter	
g	Energy of interaction between molecules or polymer segments	4.6, 13.8
g	Coefficient in linear solvation energy relationship	8.4
g,g_0,g_1	Empirical free energy interaction parameters	13.3
g	(Superscript or subscript) gas	2.1
gal	(Unit) gallon	2.5
gl	(Subscript) glassy	13.1
h	Coefficient in linear solvation energy relationship	8.4
h	(Subscript) hydrogen-bonding	5.12, 7.8
i	U_i/U	8.2
i	(Superscript) component i, usually solvent	2.6
ij	(Superscript) i-j mixture	4.1
i	(Subscript) induction	3.3
id	(Subscript) ideal	11.2, 12.1, 13.2
j	(Superscript) component j, usually solute	3.2
k	Boltzmann constant	3.2
k	Distribution coefficient	
k	Hydrogen-bonding weighting factor	8.3
k	Rate or equilibrium constant	10.1
k	Molecular packing coefficient	6.5, 13.1
k,k'	Empirical parameter in various relationships	5.7, 6.6, 7.4
^{ij}k	Gas phase geometric mean deviation parameter	5.7

GLOSSARY OF SYMBOLS (continued)

Symbol	Definition	Section in which introduced
k	(Superscript) component k, usually solute in i-j solvent mixture	4.4
k	(Unit prefix) kilo, 10^3	
^{ij}l	Geometric mean correction parameter	5.7
l	(Superscript) lower	14.3
l	(Superscript or subscript) liquid	2.1
lb	(Unit) pound	2.5
ln	Natural logarithm	
log	Logarithm, base 10	
m	Molar volume ratio	4.2
m	Empirical parameter in various relationships	6.6, 7.4, 9.1
m	Number of C-H bonds	6.2
m	(Superscript) mixture	14.5
m	(Superscript or subscript) mobile phase	18.1, 18.4
m	(Unit) meter	
m	(Subscript) mixing	2.6
m	(Subscript) melting point	12.1
max	(Subscript) maximum	
mol	(Unit) mole	
n	Amount of substance (number of moles)	4.4
n	Number of carbon atoms	6.6
n	Number of polymer repeat units	14.8
n	Empirical parameter, π/c	2.3
n	Ratio of adsorption areas	18.1
n	Refractive index	3.1
n_D	Refractive index at sodium 'D' line	3.1, 16.1
n	(Superscript) term n	
np	(Subscript) nonparticipating	6.9
o	(Superscript or subscript) organic phase, oil	4.2, 7.11, 18.1
o	(Subscript) orientation	3.2
o	(Subscript) optimum	17.7
ob	(Subscript) observed	7.2
p	Pressure	2.1
p	Fractional polarity	8.2
^ip,^jp	Partial pressure	11.1
$^ip^o$	Vapor pressure of pure component i	7.2
p_r	Reduced vapor pressure	7.4
p_s	Saturation vapor pressure	2.1
p_x	Characteristic pressure	2.4
p	(Subscript) isobaric, constant pressure	2.6
p	(Subscript) polar	5.12
q	Electric charge	3.1
q	Degree of unsaturation parameter	5.13, 7.8
q	(Subscript) quadrupole	5.7
r	Separation of particles or charges	3.1
r	Number of reactants	10.2

GLOSSARY OF SYMBOLS (continued)

Symbol	Definition	Section in which introduced
r	Atom ratio	16.11
r	Number of segments in polymer	13.3
${}^{ij}r$	Distance in Hansen space, ${}^{ij}A^{1/2}$	5.9, 14.5
${}^{j}r$	Radius of cohesion pressure interaction circle	5.9, 14.5
r	(Subscript) reduced	7.4
rel	(Subscript) relative	
ru	(Subscript) rubbery	13.1
s	Selectivity	7.10, 11.5
s	Coefficient in linear solvation energy relationship	8.4
s	(Superscript or subscript) first solid surface, stationary phase	4.4, 18.1
s	(Superscript or subscript) solvent phase	
s	(Subscript) saturation	2.1
sc	(Subscript) semicrystalline	13.1
sp	(Subscript) specific	6.9
t	Temperature (°C)	2.1
t	(Superscript) (second) solid surface	4.4
t	(Superscript or subscript) total	2.2, 4.2
t	(Subscript) triple point	12.1
u	(Superscript) repeat unit	6.2, 6.5, 13.1, 14.8
v	Specific volume	6.5, 13.1
v	(Subscript) relating to van der Waals model	6.5, 13.1
v	(Superscript) vaporization process	
w	(Subscript) weight fraction scale	14.7
x	Mole fraction in liquid phase	2.6
x	Representative physical property	6.3
x_c	Degree of polymer crystallinity	13.1
x_d	Proton donor selectivity parameter	8.5
x_e	Proton acceptor selectivity parameter	8.5
x_{id}	Ideal mole fraction solubility	11.2
x_n	Dipole selectivity parameter	8.5
x_o	Average reference solubility	11.2
x_s	Mole fraction solubility	4.2
x	(Subscript) mole fraction scale	14.7
x	(Subscript) indicating Hansen d, p, or h	14.5
y	Mole fraction in gas phase	11.1
z	(Superscript) contributing group or atom z	6.2, 16.11
Γ	Chameleonic cohesion parameter	5.5
Δ	Characteristic parameter for polar, nonassociated liquids	11.2
Δ	Polarizability correction term in linear solvation energy relationship	8.4
${}^{E}\Delta$	Excess change	
Δ_m	Molar change on mixing	2.6
${}_{l}\Delta_{g}, \Delta$	Molar change on vaporization	2.1
${}_{s}\Delta_{l}, {}_{c}\Delta_{l}$	Molar change on fusion	12.1

GLOSSARY OF SYMBOLS (continued)

Symbol	Definition	Section in which introduced
${}_{g}\Delta_{\infty}$	Molar change on expansion of gas to infinite volume	2.1
${}^{z}\Delta_{T}$	Lyderson group constant for critical temperature	5.11
${}^{z}\Delta_{T}^{P}$	Polymer aggregation constant	14.8
Δf	Change in effective number of degrees of freedom	7.5
$\Delta V^{\ddagger}$	Volume of activation	10.2
$\Delta\nu$	Spectroscopic absorption shift	8.3
$\delta\nu_{A}$	Electron-accepting power	8.3
$\Delta\nu_{D}$	Electron-donating power	8.3
Θ	Theta temperature	13.7, 14.9
Θ	(Subscript) theta condition	14.7
$\Theta(\epsilon)$	Solvent polarity function	8.2
Λ	Virtual cohesion parameter	5.5
Λ	Polymer-polymer interaction energy density	13.3
$[\Lambda]$	True lyoparachor	6.3
Π_{r}	Product of all *r*	10.2
Σ_{i}	Sum over all *i*	6.1
Φ	Surface interaction parameter	17.2, 17.3
Ψ	Induction cohesive pressure parameter	5.8
Ω	Weight fraction activity coefficient	14.2
α	Isobaric thermal expansion coefficient	2.3
α	Electric polarizability	3.1
α	Complex free energy transfer parameter	7.9
α	Hydrogen-bond donor acidity in linear solvation energy relationship	8.4
α	Parameter in relation between δ and $\ln E_{\eta}$	10.5
α	Mark-Houwink parameter	14.7
α	Aggregation number (Hoy)	5.11
α^{*}	Chain aggregation number (Hoy)	14.8
${}^{ij}\alpha$	Separation factor, relative volatility	7.2, 7.9, 11.5
β	Isochoric thermal pressure coefficient	2.3
β	Entropy of mixing term in dissolving power	8.4
β	Hydrogen-bond acceptor basicity in linear solvation energy relationship	8.4
β	Parameter in relation between δ and $\ln E_{\eta}$	10.5
β	Spectroscopic proton-accepting parameter	5.12
${}^{ij}\beta$	Binary interaction parameter	5.8
${}^{i}\gamma$	Surface free energy of liquid *i* in air or its own vapor	17.1
${}^{s}\gamma$	Surface free energy of solid *s* in contact with air or its own vapor	17.1
${}^{ij}\gamma$	Interfacial free energy of immiscible liquid phases *i* and *j*	17.1
${}^{is}\gamma$	Interfacial free energy of liquid *i* and solid *s*	17.1
${}^{s,i}\gamma$	Interfacial free energy of solid *s* and saturated vapor *i*	17.1

GLOSSARY OF SYMBOLS (continued)

Symbol	Definition	Section in which introduced
γ_c	Critical surface free energy	17.2
γ_C	Crowley hydrogen-bonding parameter	8.3
γ_D	du Pont hydrogen-bonding parameter	8.3
γ_G	Gordy hydrogen-bonding parameter	8.3
γ_L	Lieberman hydrogen-bonding parameter	8.3
γ_O	Orr hydrogen-bonding parameter	8.3
γ_S	Small iodine-bonding number	8.3
γ_w	Sound velocity hydrogen-bonding parameter	8.1
δ, δ_t	Hildebrand parameter, total cohesion parameter	2.2, 5.9, 7.5
$\overline{\delta}, {}^{ij}\delta$	Volume fraction average δ, effective Hildebrand parameter	4.2, 4.4, 5.9, 7.8
δ_a	Lewis acid cohesion parameter	5.4
δ_b	Lewis base cohesion parameter	5.4
δ_c	Hildebrand parameter at critical temperature	7.4
δ_c	Complexing cohesion parameter	5.11
δ_d	Dispersion cohesion parameter	5.1
δ_{dp}	$(\delta_d^2 + \delta_p^2)^{1/2}$	16.4
δ_f	Orientation-independent cohesion parameter	5.11
δ_h	Hydrogen-bonding cohesion parameter	5.12
δ_{hp}	$(\delta_h^2 + \delta_p^2)^{1/2}$	5.12
δ_i	Induction cohesion parameter	5.3
δ_K	Kumar cohesion parameter	5.13
δ_o	Orientation cohesion parameter	5.2
δ_{np}	Non-participating cohesion parameter	6.9
δ_p	Polar cohesion parameter	5.12
δ_R	Residual cohesion parameter	5.12, 7.5
δ_{sp}	Participating cohesion parameter	6.9
δ_V	Volume-dependent cohesion parameter	5.12, 7.5
δ_α	Acidity cohesion parameter	5.13
δ_β	Basicity cohesion parameter	5.13
δ_λ	Nonpolar, polarizability cohesion parameter	5.8, 5.11, 5.13
δ_ξ	Association cohesion parameter	5.11
δ_σ	Proton-donor cohesion parameter	5.12
δ_τ	Polar, polarity cohesion parameter	5.8, 5.11, 5.13, 13.5
δ_ψ	Polar-nonpolar (inductive) cohesion parameter	5.13
δ'	Modified non-specific cohesion parameter	8.4
${}^{ij}\delta'$	Binary interaction cohesion parameter	5.7
δ_O	Limiting low temperature Hildebrand parameter	9.1
δ^*	Cohesion parameter in cohesive energy ratio	17.7
${}^{\ddagger}\delta$	Cohesion parameter of activated complex	10.2
ε	Permittivity	3.1
ε	Strain	16.4
ε_c	Critical strain	16.4
ε_o	Permittivity of vacuum	3.1
ε_r	Relative permittivity, dielectric constant	3.1
ε^o	Solvent strength parameter	18.1

GLOSSARY OF SYMBOLS (continued)

Symbol	Definition	Section in which introduced
${}^{z}\varepsilon$	Group molar cohesive energy constant	6.2
η	Viscosity	14.7
$[\eta]$	Intrinsic viscosity	14.7
$[\eta]_w$	Weight intrinsic viscosity	14.7
η_{rel}	Relative viscosity	14.7
η_{sp}	Specific viscosity	14.7
${}^{ij}\eta$	Correlation parameter for induction corrections	7.8
θ	Net hydrogen-bond accepting ability	8.3
θ	Exponent in homomorph equation	6.7
θ	Contact angle	17.1
κ	Isothermal compressibility	2.3
κ	Enthalpy polymer-liquid interaction parameter	13.7
λ	(Subscript) nonpolar	5.8, 5.11, 5.13
μ	Dipole moment	3.1
μ_i	Induced dipole moment	3.1
μ	Chemical potential	2.6
μ^o	Chemical potential of pure compound	2.6
${}^{z}\mu$	Chain-length correction factor	6.2
v	Specific volume	13.1
v	Fugacity coefficient of hypothetical liquid	11.1
ξ	Local volume fraction	4.6, 13.8
ξ	Empirical coordinate covalency parameter in linear solvation energy relation	8.4
ξ	Hydrogen-bonding asymmetry factor	5.13
ξ	(Subscript) association	5.11
π	Internal pressure	2.3, 14.9
π	Film pressure	17.1
π	Lipophilic parameter	19.1
π^*	Solvatochromic factor in linear solvation energy relation	8.2, 8.4
ρ	Density	3.1
ρ	Hammett constant	7.12
σ	Tensile strength	16.2
σ	Hammett constant	7.12
σ	(Subscript) proton donor	5.12
τ	(Subscript) polar	5.8, 5.11, 5.13
ϐ	Fugacity coefficient in gas phase	11.1
ϕ	Volume fraction	4.2
ϕ_c	Critical volume fraction	13.3
ϕ^*	Effective volume fraction	4.4
ϕ	(Subscript) volume fraction scale	
χ	Polymer-liquid interaction parameter	13.3
χ_c	Critical polymer-liquid interaction parameter	13.3
χ_H	Enthalpy component of χ	13.3
χ_S	Entropy component of χ	13.3
χ^s	Interaction parameter for surface phase	17.5

GLOSSARY OF SYMBOLS (continued)

Symbol	Definition	Section in which introduced
χ_0, χ_1, χ_2	Polymer-liquid interaction concentration-dependence coefficients	13.3
χ', χ'_S, χ'_H	Chen polymer-liquid interaction parameters	13.5
ψ	Induction cohesive pressure parameter	5.8
ψ	Entropy polymer-liquid interaction parameter	13.7
$^i\psi$	Polar asymmetry factor	5.13
ω	Pitzer acentric factor	7.4
‡	(Superscript) activated complex	10.2
∞	(Subscript or superscript) infinite dilution, infinite expansion	4.2
$	Dissolving power	8.4
$\circ$	(Subscript or superscript) reference	2.6, 7.2, 7.9, 9.1

CONCORDANCE BETWEEN EDITIONS

Sections

First Edition	Second Edition	First Edition	Second Edition
2.1	2.1	8.2	8.1
2.2	2.2	8.3	5.9
2.3	2.3	8.4	5.10
2.4	2.4	8.5	8.2
2.5	2.5	8.6	8.3
2.6	2.6	8.7	8.4
2.7	2.7	8.8	8.5
3.1	3.1	9.1	9.1
3.2	3.2	9.2	9.2
3.3	3.3	9.3	9.3
3.4	3.4	9.4	12.3
3.5	3.5	10.1	10.1
3.6	3.6	10.2	10.2
4.1	4.1	10.3	10.3
4.2	4.2	—	10.4
4.3	4.3	10.4	10.5
4.4	4.4	10.5	10.6
8.1	4.5	11.1	11.1
—	4.6	11.2	11.2
5.1	5.1	11.3	11.3
5.2	5.2	—	11.4
5.3	5.3	11.4	11.5
5.4	5.4	11.5	11.6
5.5	5.5	12.1	12.1
5.6	5.6	12.2	12.2
5.7	5.7	9.4, 12.3	12.3
5.8	5.8	13.1	13.1
8.3	5.9	13.2	13.2
8.4	5.10	13.3	13.3
5.9	5.11	13.4	13.4
5.9	5.12	13.5	13.5
—	5.13	—	13.6
—	5.14	13.6	13.7
6.1	6.1	13.7	13.8
6.2	6.2	—	14.1
6.3	6.3	14.1	14.2
6.4	6.4	14.2	14.3
6.5	6.5	—	14.4
6.6	6.6	14.3	14.5
6.7	6.7	—	14.6
6.8	6.8	14.4	14.7
6.9	6.9	14.5	14.8
6.10	6.10	14.6	14.9
—	7.1	15.1	15.1
7.1	7.2	15.2	15.2
7.2	7.3	16.1	16.1
7.3	7.4	16.2	16.2
7.4	7.5	16.4	16.3
—	7.6	16.3	16.4
—	7.7	—	16.5
7.5	7.8	16.5	16.6
7.6	7.9	—	16.7
7.7	7.10	16.6	16.8
7.8	7.11	16.7	16.9
7.9	7.12	16.8	16.10
8.1	4.5	16.9	16.11

CONCORDANCE BETWEEN EDITIONS (continued)

Sections

First Edition	Second Edition	First Edition	Second Edition
17.1	17.1	18.2	18.2
17.2	17.2	18.3	18.3
17.3	17.3	18.4	18.4
17.4	17.4	—	18.5
17.5	17.5	18.5	18.6
17.6	17.6	18.6	18.7
17.7	17.7	19.1	19.1
17.8	17.8	—	19.2
17.9	17.9	19.3	19.3
18.1	18.1	19.2	19.4

Tables

First Edition		Second Edition		First Edition		Second Edition	
Chapter	Table	Chapter	Table	Chapter	Table	Chapter	Table
1	1	1	1	6	1	6	1
1	2	1	2	6	2	6	2
1	3	—	—	6	3	6	3
—	—	1	3	6	4	6	4
2	1	2	1	6	5	6	5
—	—	2	2	6	6	6	6
2	2	2	3	6	7	6	7
—	—	2	3a	6	8	6	8
12	1	2	4	6	9	6	9
2	3	2	5	6	10	6	10
2	4	2	6	6	11	6	11
8	1	4	1	6	12	6	12
—	—	4	2	6	13	6	13
5	1	5	1	6	14	6	14
—	—	5	2	—	—	6	15
—	—	5	3	6	15	6	16
12	2	5	4	6	16	6	17
5	2	5	5a	13	9	13	8
5	3	5	5b	13	10	13	9
5	4	5	6	13	11	13	10
5	5	5	7	13	12	13	11
5	6	5	8	13	13	13	12
5	7	5	9	13	14	13	13
5	8	5	10	—	—	14	1
8	5	5	11	14	1	14	2
8	6	5	12	14	2	14	3
—	—	5	12a	14	3	13	7
—	—	5	13	14	4	14	4
8	7	5	14	14	5	14	5
8	8	5	15	14	6	14	6
8	9	5	16	14	7	14	7
—	—	5	16a	14	8	14	8
—	—	5	17	14	9	14	9
7	2	5	18	—	—	14	10
18	6	5	19	14	10	14	11
—	—	5	20	14	11	14	12
—	—	5	21	14	12	14	13
—	—	5	22	14	13	14	14
—	—	5	23	—	—	14	15
—	—	5	24	—	—	14	15a

Tables

First Edition		Second Edition	
Chapter	Table	Chapter	Table
14	14	14	16
14	15	14	17
14	16	14	18
14	17	14	19
14	18	14	20
14	19	14	21
15	1	15	1
15	2	15	2
16	1	16	1
16	2	16	2
16	3	16	3
17	1	17	1
17	2	17	2
17	3	17	3
17	4	17	4
17	5	17	5
17	6	17	6
17	7	17	7
17	8	17	8
17	9	17	9
17	10	17	10
17	11	17	11
—	—	17	12
—	—	17	13
17	12	17	14
—	—	17	15
—	—	17	16
—	—	18	1
18	1	18	2
9	2	6	18
7	1	7	1
—	—	7	2
—	—	7	3
7	2	5	18
7	3	7	4
—	—	7	5
7	4	7	6
7	5	—	—
7	6	7	7
7	7	—	—
7	8	7	8
7	9	7	9
7	10	7	10
—	—	7	11
—	—	7	12
7	11	7	13
7	12	7	14
—	—	7	15
7	13	7	16
8	1	4	1
8	2	8	1
8	3	8	2
8	4	8	3
8	5	5	11
8	6	5	12

First Edition		Second Edition	
Chapter	Table	Chapter	Table
8	7	5	14
8	8	5	15
8	9	5	16
8	10	8	4
8	11	8	5
8	12	8	6
8	13	8	7
8	14	8	8
8	15	8	9
8	16	8	10
8	17	8	11
—	—	8	12
8	18	8	13
8	19	8	14
8	20	8	15
8	21	8	16
—	—	9	1
9	1	9	2
9	2	6	18
9	3	9	3
—	—	10	1
11	1	11	1
11	2	11	2
11	3	11	3
—	—	11	4
—	—	11	5
11	4	11	6
12	1	2	4
12	2	5	4
12	3	12	1
—	—	12	1a
12	4	12	2
12	5	12	3
12	6	12	4
9	4	12	5
—	—	12	6
—	—	12	7
13	1	13	1
13	2	13	2
13	3	13	3
13	4	13	4
13	5	—	—
13	6	13	5
13	7	13	6
14	3	13	7
13	8	—	—
18	2	18	3
18	3	18	4
18	4	18	5
18	5	18	6
18	6	5	19
18	7	18	7
18	8	18	8
18	9	18	9

CONCORDANCE BETWEEN EDITIONS (continued)

Figures

First Edition		Second Edition	
Chapter	Figure	Chapter	Figure
5	1	5	1
5	2	5	2
5	3	5	3
5	4	5	4
8	1	5	5
8	2	5	6
8	3	8	1
8	4	8	2
8	5	8	3
8	6	8	4
8	7	8	5
8	8	8	6
8	9	8	7
11	1	11	1
11	2	11	2
11	3	11	3
—	—	11	4
11	4	11	5

Figure numbers are unchanged, with the exception of those in Chapters 5, 8, and 11.

INDEX

A

B

C

D

E

F

G

H

I

J

K

L

M

N

O

P

Q

R

S

T

U

V

W

X

Y

Z